GENES V

GENES V

Benjamin Lewin

OXFORD UNIVERSITY PRESS
Oxford   New York   Tokyo

Oxford University Press, Walton Street, Oxford OX2 6DP

Oxford New York
Athens Auckland Bangkok Bombay
Calcutta Cape Town Dar es Salaam Delhi
Florence Hong Kong Istanbul Karachi
Kuala Lumpur Madras Madrid Melbourne
Mexico City Nairobi Paris Singapore
Taipei Tokyo Toronto

Oxford is a trade mark of Oxford University Press

Published in the United States
by Oxford University Press Inc., New York

A catalogue record for this book is available from the British Library

Library of Congress Cataloging in Publication Data

Printed in U.S.A.

PREFACE

Comparison between this edition and the first edition of *GENES* shows remarkable advances in the past decade. The initial edition claimed that the eponomyous title described a new field, whose extraordinary progress had all but overwhelmed the traditional discipline of genetics. In the second edition, I suggested that studying genetics essentially meant dealing with DNA, and in many respects *GENES* could be viewed as a textbook of modern genetics. By the time of the fourth edition, the claim that the purpose of *GENES* was to explain heredity in terms of molecular structures began to seem a reality that impacted equally on prokaryotes and eukaryotes. The idea that the operations of both types of genomes should be considered in a unified approach seems unarguable now, and the appearance of five editions of *GENES* in only ten years is a tribute to the speed with which the subject has advanced.

This present edition marks more changes than any of its predecessors. Major portions of the book have been rethought. We start, as always, with the view of DNA and how its sequence is translated into protein, but the resolution of the underlying events has increased steadily over ten years. There is new emphasis on the importance of post-translational events, and the discussion of how proteins are appropriately localized makes a new section of the book, together with a molecular account of the cell cycle; this marks a movement of this text towards cell biology. We understand the control of bacterial gene expression in more depth; and now this is discussed in terms of protein–DNA interactions rather than the original genetic characterization. Replication and maintenance of DNA are also more extensively understood, and now can be linked to the cell cycle. An enormous change is evident in the characterization of the eukaryotic genome, where we move from an anectodal account of individual genes and gene families to the first systematic account of the constitution of the genome, in terms of both DNA and chromosome structure. There have also been great advances in characterizing eukaryotic gene expression, where transcription now is understood at a level comparable to bacterial systems, and RNA splicing becomes approachable in terms of individual components and reactions. Advances on the catalytic role of RNA, and the process of RNA editing, are simply startling. Major increases in understanding are also evident with regards to gene regulation in development and in the formation of tumors, where in both cases the broad outlines of the relevant events, and their connection into regulatory circuits, are becoming evident.

An implicit theme in this edition is that the advances of the past few years suggest unifying views in many areas; although this book rests on a broader data base and covers more ground than any previous edition, I believe it is in fact easier to follow, because simplifying principles have in many cases replaced individual descriptions. Consistent with the view that the subject of the book is central to life sciences, and that it should be accessible to students at introductory levels, the book has been entirely freshly illustrated; and indeed the illustrations extend now to overviews and summaries that depict virtually every major point of the book (conceptual as well as factual) in diagrammatic form.

It is as always a pleasure to thank colleagues who have read chapters or discussed individual topics (in many cases stimulating new directions of thought and much improving the book), including particularly Tania Baker, Michael Chamberlin, Albert Dahlberg, Douglas Engel, Martin Gellert, Tony Hunter, Michael Levine, Richard Losick, Ira Mellman, Paul Nurse, Paul Schimmel, Matthew Scott, Philip Sharp, Robert Thach, and Andrew Travers.

GENES retains the aim of providing a *tour d'horizon* of the current state of the art, and also identifying the questions that need now to be answered. By keeping the current version of the text up to date with the latest facts and thoughts in the field, I hope to make it possible for both teachers and students to keep abreast of modern genetics.

Cambridge, Massachusetts   B.L.
May 1993

Outline

Outline

Contents

Cells as macromolecular assemblies

We shall assume the structure of the gene to be that of a huge molecule, capable only of discontinuous change, which consists in a rearrangement of the atoms and leads to an isomeric molecule. The rearrangement [mutation] may affect only a small region of the gene, and a vast number of different rearrangements may be possible.

Erwin Schrödinger, 1945

CHAPTER 1

Cells obey the laws of physics and chemistry

Ever since it was realized that an organism does not pass on a simulacrum of itself to the next generation, but instead provides it with **genetic material** containing the **information** needed to construct a progeny organism, we have wanted to define the nature of this material and the manner in which its information is utilized. The genetic material consists of a particular type of molecule (nucleic acid), whose structure is distinct from the other types of molecules that comprise the organism (proteins, lipids, saccharides, etc.). Now that we know the physical structure of the genetic material, we may state the aim of molecular biology as defining the complexity of living organisms in terms of the properties of their constituent molecules.

The **gene** is the unit of genetic information. The crucial feature of Mendel's work, a century ago, was the realization that the gene is a distinct entity. The era of the molecular biology of the gene began in 1945 when Schrödinger developed the view that the laws of physics might be inadequate to account for the properties of the genetic material, in particular its stability during innumerable generations of inheritance. The gene was expected to obey the laws of physics so far established, but it was thought that characterizing the genetic material might lead to the discovery of new laws of physics, a prospect that brought many physicists into biology.

Now, of course, we know that a gene is a huge molecule, in fact part of a vast length of genetic material containing many genes. A gene does not function autonomously, but relies upon other cellular components for its perpetuation and function. All of these activities obey the known laws of physics and chemistry; and it has not, in the end, been necessary to invoke new laws.

There are two central questions to ask in characterizing the genetic material:

◆ How is it faithfully reproduced so that it is inherited generation after generation?

◆ How is information transferred from genetic material to specify construction of the many other types of structures that constitute a living organism?

Although we have posed these questions in terms of molecular structure, the individual structural characteristics are not the sole determinant of the functions of a molecule. Location is also important. A cell does not consist merely of protoplasm contained in a membranous bag: each structural component occupies a specific location. This is true not just to the extent that (for example) membranes provide the circumference while the nucleus contains the genetic material; but the perpetuation of the cell from one generation to the next depends on maintaining a highly ordered structure, in which particular functions can be exercised only at particular places.

A profound influence during the development of a multicellular organism from the fertilized egg is the need for components to be located at the right place. We know that the contents of the first cells of the embryos of certain species cannot be distributed homogeneously, because different parts of the early embryo give rise to descendants with different

Figure 1.1

Macromolecules can be viewed on a scale from 10-100Å (1-10 nm).

Macromolecular scale

$100Å = 10\ nm = 10^{-2}\mu m$

DNA
20Å diameter

Globular protein
50,000 daltons
Diameter = 50Å

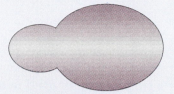

Bacterial RNA polymerase
500,000 daltons
90 x 90 x 160Å

Nucleosome
300,000 daltons
60 x 110 x 110Å

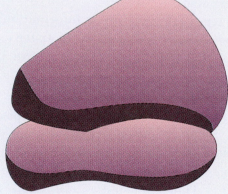

Ribosome
2.5×10^6 daltons
200 x 200 x 230Å

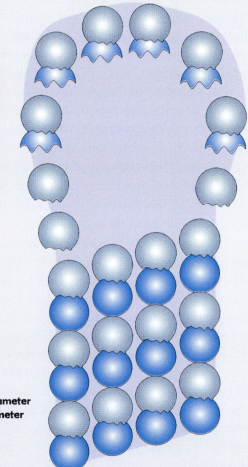

Microtubule
24 nm outside diameter
15 nm inside diameter

Figure 1.2

Organelles can be viewed on a scale of 0.1 μm, bacteria at scale 1 μm, and eukaryotic cells at a scale of 10 μm. The structure from each level is shown at the next level at the appropriate scale for comparison.

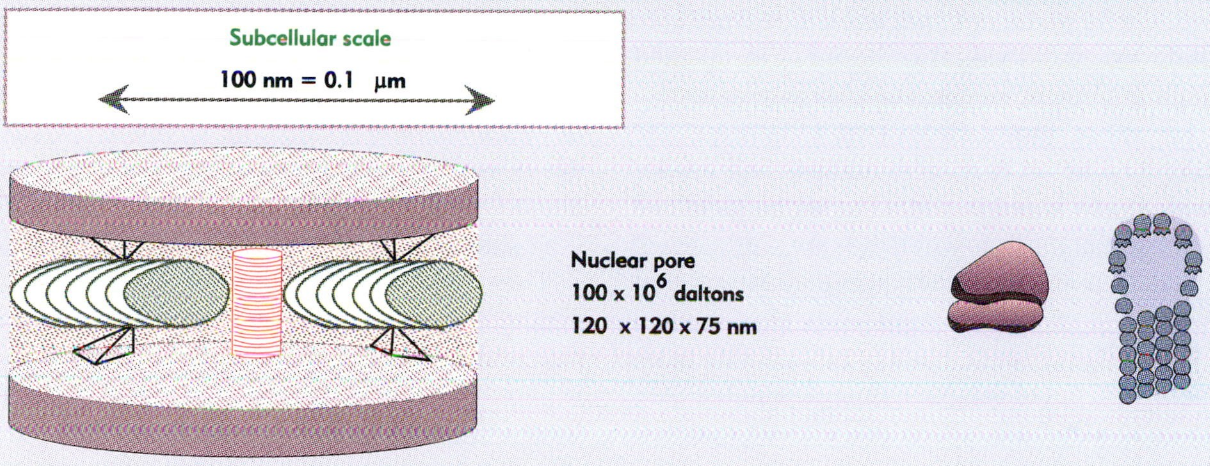

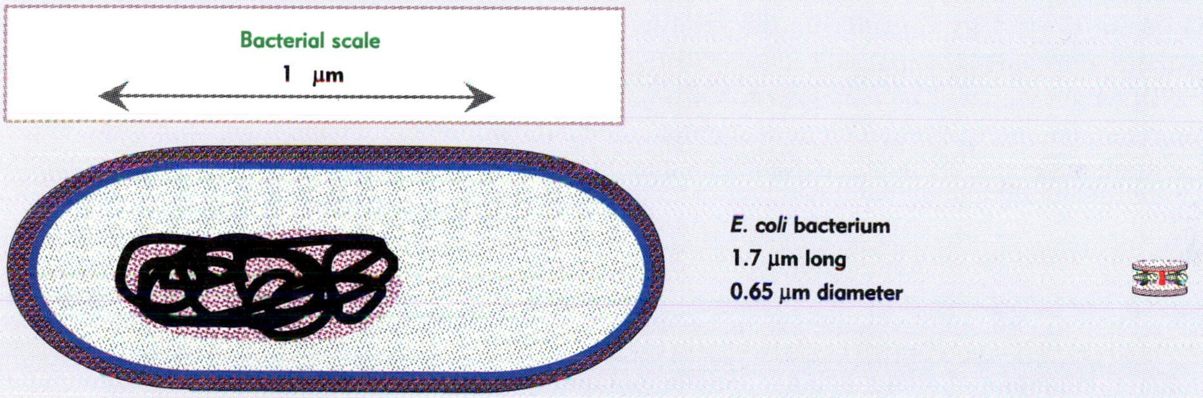

Subcellular scale
100 nm = 0.1 μm

Nuclear pore
100×10^6 daltons
$120 \times 120 \times 75$ nm

Bacterial scale
1 μm

E. coli bacterium
1.7 μm long
0.65 μm diameter

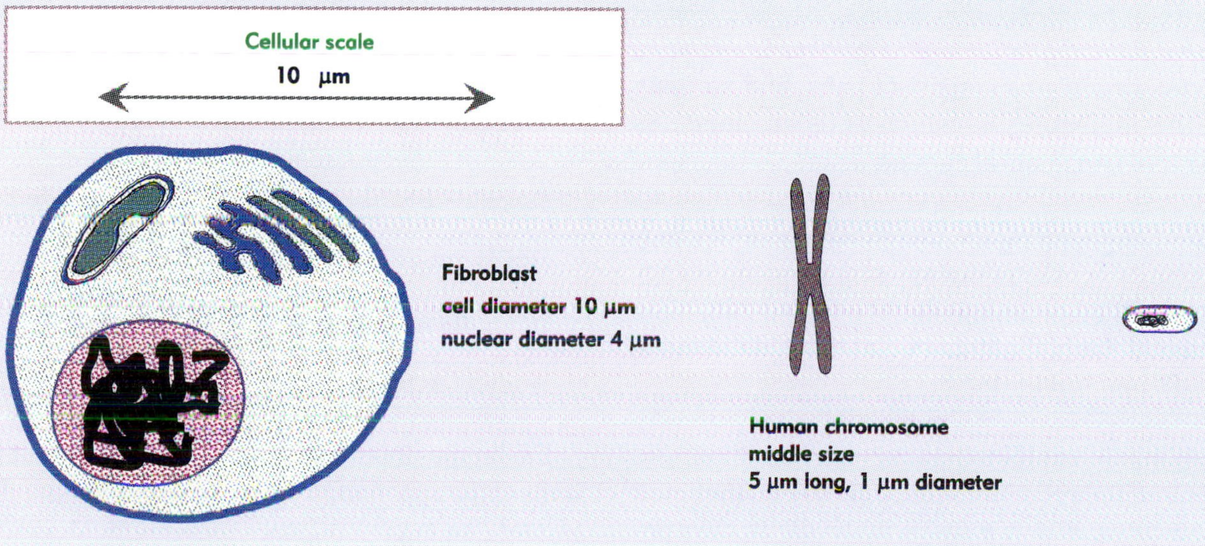

Cellular scale
10 μm

Fibroblast
cell diameter 10 μm
nuclear diameter 4 μm

Human chromosome
middle size
5 μm long, 1 μm diameter

properties. The relationship between location and function is sometimes called **positional information,** a phrase that implies the importance of location as well as structure *per se.*

Are the properties of individual molecular structures sufficient to account for all the properties of the cell? Could we reconstruct a cell if we were able to obtain every individual component in purified form? Or is some other form of information necessary for these molecules to find their proper locations? Some components can find their proper location at any time, but others can be localized properly only at the time when they are synthesized.

In this chapter, we review the parameters that influence the structural properties of biological molecules. The purpose is not to recapitulate basic biochemistry, but to establish the relevant features of the context within which the genetic material functions.

It is useful to start by considering the relative scales of the common molecular and cellular structures with which we shall be concerned. Starting with macromolecules in **Figure 1.1**, and anticipating the components that we discuss in detail later, a

DNA molecule has a diameter of only 20Å, but may be (almost) indefinitely long. A protein of 50,000 daltons organized as a sphere would have a diameter of ~50Å, so a protein large enough to surround a DNA duplex might have a mass as low as 25,000 daltons. Proteins that function in gene expression by synthesizing RNA or replicating DNA may be much larger; bacterial RNA polymerase is 500,000 daltons, and DNA polymerase is 900,000 daltons; they have dimensions of the order of 100–200Å. DNA in the cells of higher organisms is organized into a 300,000 dalton body called the nucleosome, which is a cylinder of diameter 110Å and depth 60Å.

The diameter of the ribosome (the machine that synthesizes proteins) is comparable to that of a microtubule (a common filamentous structure that is part of the skeleton of the cell). The ribosome is large compared to an individual nucleic acid or protein. But moving to the cellular scale in **Figure 1.2**, the ribosome seems small relative to a nuclear pore (the means of communication between nucleus and cytoplasm). The nucleus of an animal cell is ~50× the size of a bacterium; and a single human chromosome may be larger than a bacterium.

Macromolecules are assembled by polymerizing small molecules

All living organisms consist of cells, and we may view the molecules of which cells are made as falling into two general classes:

◆ **Small molecules** are the substrates and products of metabolic pathways, providing the energy needed for cell survival. They fall into four general classes: **sugars, fatty acids, amino acids,** and **nucleotides.**

◆ **Polymeric molecules**—the structural components of the cell—are synthesized from the small molecules. When a small molecule is incorporated into a polymer, it is sometimes described as

a **subunit** of that structure (see Table 1.2). The four types of these assembly reactions are:

polysaccharides are assembled from sugars;

lipids are assembled from fatty acids;

proteins are assembled from amino acids;

nucleic acids are assembled from nucleotides.

Each type of polymer consists of a series of subunits of the appropriate type, usually connected end to end. As a general rule, a biological polymer is assembled by a process in which individual subunits are added one by one to the chain so far assembled.

Addition of each subunit to the polymeric chain involves the formation of a **covalent bond**. Covalent bonds are intrinsically stable under physiological conditions. A biological polymer can survive indefinitely in a living cell in the absence of any specific attack on the bond.

Each polymerization reaction is accompanied by the loss of a molecule of water for every subunit added, giving rise to the name **condensation reaction**. To reverse the polymerization reaction by introducing a break in a chain requires the addition of a water molecule, and this therefore falls into the general class of a **hydrolytic** reaction. Reactions in which protein chains are cleaved are called **proteolytic**; those in which nucleic acids are cleaved are called **nucleolytic**.

Table 1.1 views the bacterial cell in terms of its molecular components. The major part of the mass, of course, consists of water (actually a dilute salt solution containing many inorganic ions). Proteins provide the major component of the dry mass

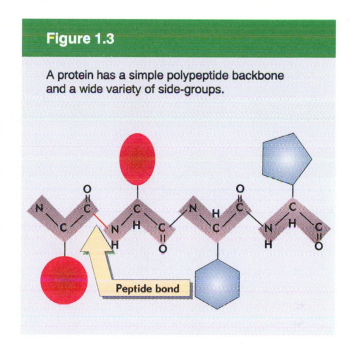

Figure 1.3

A protein has a simple polypeptide backbone and a wide variety of side-groups.

Peptide bond

(~15%), and the nucleic acids (DNA and RNA) altogether comprise another appreciable component (~5%). The small molecules, comprising metabolic intermediates and the precursors to the macromolecules, together make up <5% of the mass.

Proteins and nucleic acids are *very large molecules,* known as **macromolecules**. They are responsible for conveying genetic information. Nucleic acids carry the information, while proteins provide the means of executing it. In each case, *the sequence in which the individual building blocks are joined together is the critical feature that determines the property of the resulting macromolecule.*

Polysaccharides also may be large enough to qualify as macromolecules, although they do not have the complexity (in the informational sense) of proteins and nucleic acids. Lipids are not usually large enough to classify formally as macromolecules.

Biological polymers can be described in terms of a general type of organization. A polymer has:

◆ A **backbone** consisting of a regularly repeating series of bonds.

◆ **Side-groups** of characteristic diversity that stick out from the backbone.

Table 1.1

Considering a bacterium in terms of its molecular components. (The overall mass is ~ 7×10^{-13} g).

Component	Proportion of Cell Mass
Inorganic	
Water	70%
Inorganic ions	1%
Organic	
Carbohydrate	3%
Amino acids	0.5%
Nucleotides	0.5%
Large molecules	
Proteins	15%
DNA	0.5%
RNA	6%
Polysaccharides	2%
Lipids	2%

Figure 1.4

A nucleic acid has a sugar-phosphate backbone, and only four types of side-group.

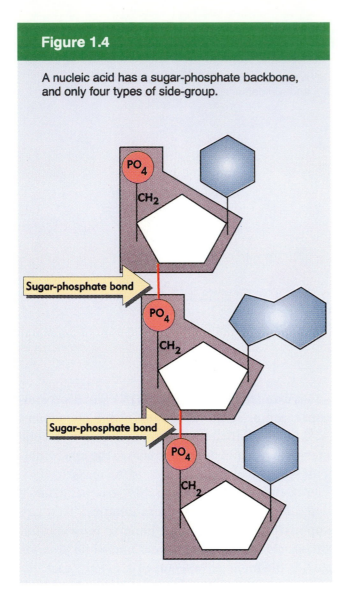

Sugar-phosphate bond

Sugar-phosphate bond

Figures 1.3–1.5 illustrate the general structure of each type of biological polymer diagrammatically. The backbone is constructed by joining together the repeating units (shaded in purple); the side-groups (shown in blue or red) are joined to the backbone. Proteins and nucleic acids are strictly linear; polysaccharides can have branches. **Table 1.2** summarizes the types of links in each backbone and the extent of diversity that can be conferred by the side-groups. Proteins are highly diverse, nucleic acids have some variety, but polysaccharides have little diversity in the types of side-groups. We now consider the structures of proteins in more detail; and in Chapter 4 we consider the properties of nucleic acids.

Table 1.2

Structural diversity in biological polymers is determined by side-groups.

Subunit	Nature of Backbone	Nature of Side-Group	Number of Side-Groups
Amino acid	Peptide bond	Amino acid	20
Nucleotide	Sugar-phosphate	Nitrogenous base	4
Sugar	Glycosidic bond	Hydroxyl or substituent	-

Figure 1.5

A polysaccharide has a backbone of linked sugars, characterized by the positions of their hydroxyl groups. There is little diversity in any given polysaccharide, although the chain may be branched.

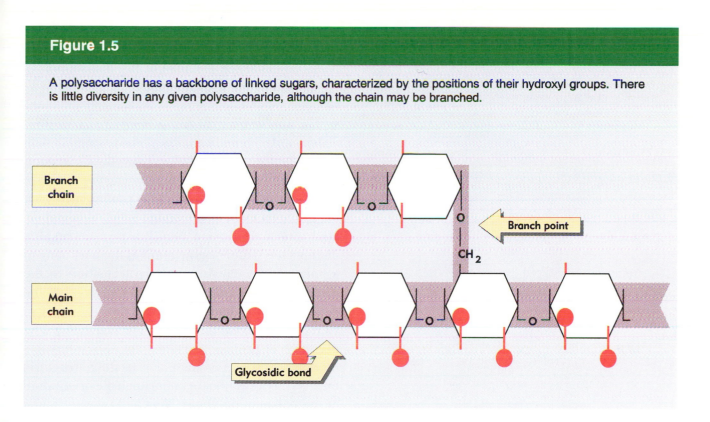

Branch chain

Branch point

Main chain

Glycosidic bond

Proteins consist of chains of amino acids

Each protein consists of a unique sequence of amino acids. A free amino acid has the general structure

$$
\begin{array}{c}
NH_2 \\
| \\
R{-}CH \\
| \\
HO{-}C{=}O
\end{array}
$$

where the side group is denoted R.

Amino acids are joined together to form a chain by **peptide bonds,** which are created by the condensation of the carboxyl (COOH) group of one amino acid with the amino (NH_2) group of the next, as illustrated in **Figure 1.6**. Since peptide bonds are covalent, they are relatively stable, and in the living cell are broken only rarely, usually as the result of a specific action.

A **peptide** consists of a small number of amino acids connected by peptide bonds. A longer chain of amino acids joined in this manner is called a **polypeptide**. The term **protein** usually is used to describe the functional unit, which may consist of one or more polypeptide chains.

We can define the direction of a polypeptide chain according to the orientation of the peptide bonds. The amino acid at one end of the chain has a free NH_2 group, and thus defines the **amino-** or **N-terminal** end. The amino acid at the other end has a free COOH group, and thus defines the **carboxy-** or **C-terminal** end. Protein sequences are conventionally written from N-terminus (at the left) to C-terminus (at the right).

Figure 1.6

A peptide bond is formed by a condensation reaction in which a water molecule is lost.

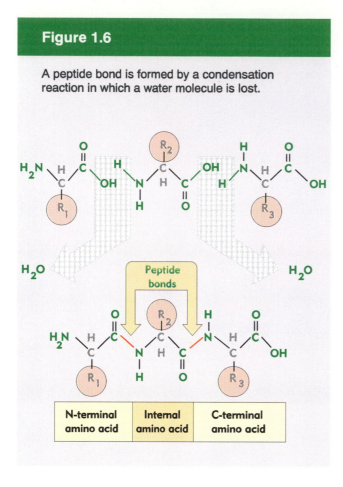

| N-terminal amino acid | Internal amino acid | C-terminal amino acid |

As illustrated in two dimensions, the peptide bonds form a zig-zag backbone, from which the side-groups protrude (in Figure 1.3). The side-group is different for each amino acid and determines the nature of its contribution to the overall protein structure. Twenty amino acids are used to synthesize proteins, and **Figure 1.7** shows their structures.

Classified by their ionic charges, the amino acids fall into four groups:

◆ Lysine, arginine and histidine are **basic**. Addition of a hydrogen ion converts the free (second) amino group of lysine or arginine to the positively charged form NH_3^+. A proton similarly can be added to the histidine ring.

◆ Aspartic acid and glutamic acid are **acidic**, because the carboxyl group can lose a hydrogen ion to exist in the negatively charged form $O{=}C{-}O^-$.

◆ Amino acids that have no net charge are **neutral**. Some of the neutral amino acids are **polar** (electrically charged because of the distribution of charges within the molecule).

◆ The **apolar neutral** amino acids are **hydrophobic** (water-repelling). They tend to interact with one another and with other hydrophobic groups.

An exceptional amino acid is proline, in which the nitrogen atom of the amino group is incorporated into a ring. As a result, a proline residue disrupts the usual organization of the backbone of a polypeptide, causing a sharp transition in the direction of the chain. **Figure 1.8** shows that the peptidyl–prolyl bond can exist in either of two stereochemical configurations (*cis* and *trans*). The presence of proline therefore not only interrupts the formation of any regular repeating structure, but also has a significant effect upon the orientation of the polypeptide chain. This is important when proteins acquire their correct structure (see later in this chapter).

In addition to the 'standard' 20 amino acids, certain others are occasionally found in proteins. They are created by **modifying** one of the standard amino acids *after it has been incorporated into protein*. These modifications change the properties of the side-group and often play important roles in protein function. Some examples are shown in **Figure 1.9**.

Some common modifications consist of the addition of a small group that changes the ionic charge of an amino acid:

◆ **Phosphorylation** results from addition of a phosphate group, usually to the hydroxyl group of serine or tyrosine, occasionally threonine. A protein with phosphate groups is sometimes called a **phosphoprotein**. The phosphate group introduces negative charges, and therefore has a significant effect upon the electrostatic properties of the protein.

◆ **Acetylation** and **methylation** result from addition of acetyl or methyl groups, usually to the

Figure 1.7

Amino acids are classified according to the nature of their side-groups. Each amino acid may be described by either a three letter abbreviation or a one letter code.

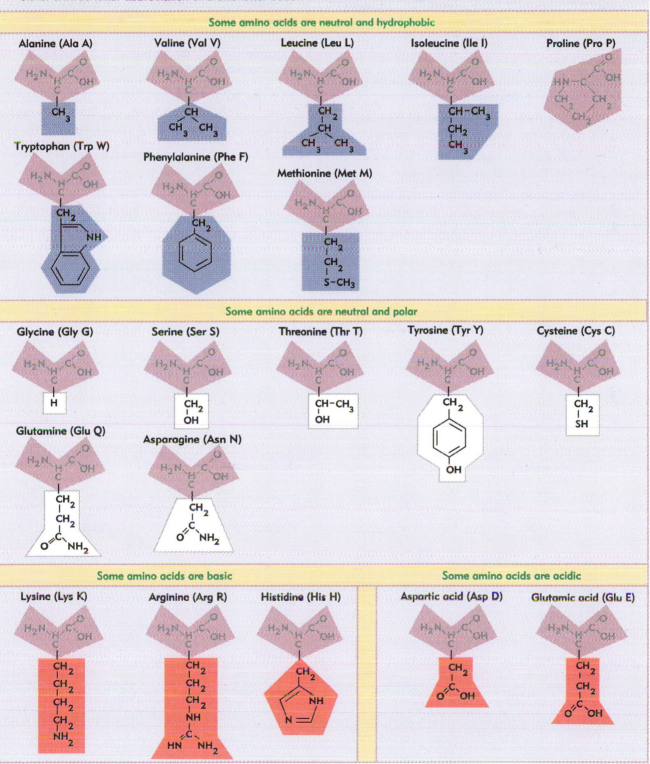

basic amino acid lysine. The addition prevents positive charges from forming on the amino group.

Certain modified amino acids are found in particular proteins or types of cells. For example, hydroxyproline and hydroxylysine are variants of proline and lysine that have an additional –OH group; they are found in the collagens, which are proteins of connective tissue.

Modifications that involve covalent addition of groups to proteins are catalyzed by specific enzymes. The groups can be removed by other enzymes. The types of enzymes that add and remove these groups are summarized in **Table 1.3**. A modifying enzyme may act specifically on a particular target protein or group of target proteins.

Table 1.3

Enzymes are needed to add covalent groups to proteins or to remove them.

Target Amino Acid	Modifying Enzyme	Unmodifying Enzyme
Tyr Ser Thr	Kinase	Phosphatase
Lys	Methylase	Demethylase
Lys	Acetylase	Deacetylase
Pro Lys	Hydroxylase	-
Glu	γ-Carboxylase	-

Figure 1.8

Proline introduces a bend in a polypeptide chain, because the nitrogen atom is restrained by the ring structure. The existence (and interconversion) of two stereochemical forms of the peptidyl-proline link is an important feature of protein structure.

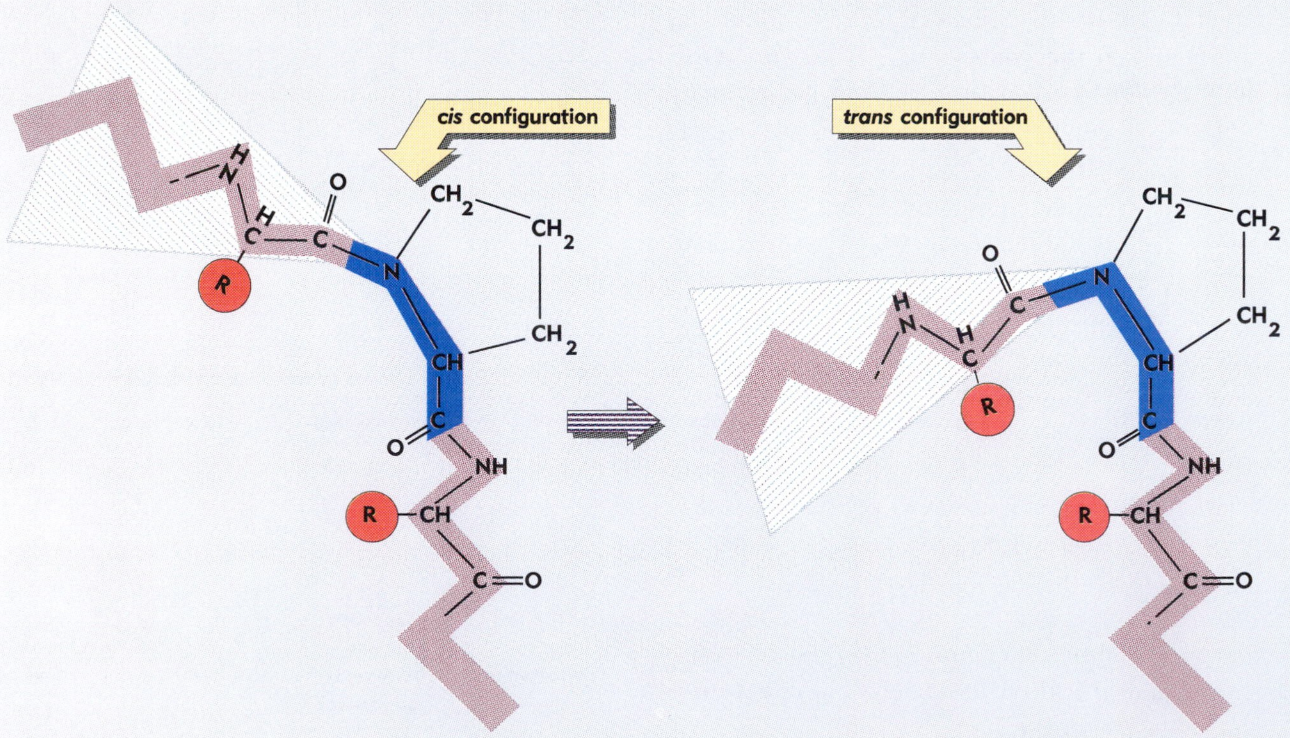

Figure 1.9

Modified amino acids are created by adding additional groups to reactive moieties of the side-groups of amino acids that have already been incorporated into proteins.

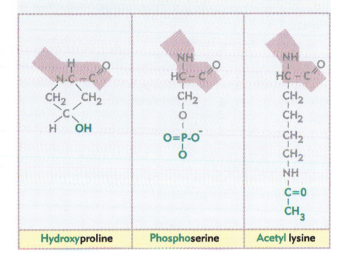

| Hydroxy**proline** | Phospho**serine** | Acetyl **lysine** |

The carbohydrate 'side chains' of glycoproteins can be extremely large, consisting of complex oligosaccharides, not just simple sugar residues. They may comprise an appreciable proportion of the protein by mass, and have a considerable effect on its properties. An example of a junction between asparagine and N-acetyl-glucosamine (an amino derivative of glucose) in an N-linked chain is shown diagrammatically in **Figure 1.10**. Glycosylation is important not just for its structural consequences, but also because it occurs as part of a complex apparatus for transporting proteins through the cell, which provides the means by which they find their proper location (which is the subject of Chapter 11).

These modifications are by no means the only ones that occur to proteins, although they are the most prominent. The important point about modification is that its existence makes it possible to extend yet further the repertoire of structures and functions that proteins can display. And in addition, of course, proteins can be associated noncovalently with a variety of groups that contribute to their structure or function.

The types of these reactions with which we shall be most concerned are the addition and removal of phosphate groups by kinases and phosphatases, respectively, in the context of regulating the cell cycle (Chapter 13), transmitting messages from the exterior (Chapter 12), and generating tumorigenic cells (Chapter 39).

A more extensive change in structure occurs in proteins that are **glycosylated**, when a carbohydrate side chain is attached to an amino acid. Such proteins are called **glycoproteins**; they include many proteins that function in an extracellular capacity. The carbohydrate can be attached to the protein in either of two ways:

◆ The most common linkages are the **N-linked oligosaccharides**, in which the sugar is linked to the amino group of asparagine.

◆ In the less common **O-linked oligosaccharides**, the sugar is linked to the hydroxyl group of serine or threonine.

Figure 1.10

An N-linked glycoprotein has an oligosaccharide chain linked to the amino group of an asparagine residue.

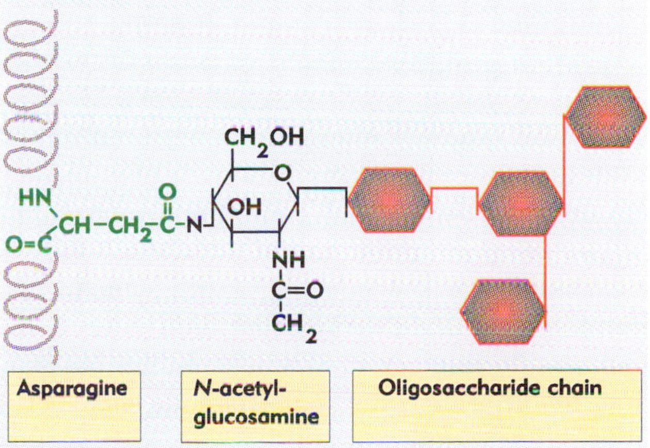

| **Asparagine** | *N*-acetyl-glucosamine | Oligosaccharide chain |

Protein conformation depends upon the aqueous environment

A critical feature of a protein is its ability to fold into a three dimensional **conformation**. Each protein exists in a unique conformation (or sometimes a series of alternative conformations). The conformation can be described in terms of several levels of structure.

◆ The series of amino acids linked into the polypeptide chain comprises its **primary structure**. This linear sequence is the essential information that is coded by the genetic material.

◆ The primary structure folds into the **secondary structure**, which describes the path that the polypeptide backbone of the protein follows in space. This reaction may be an intrinsic property of the primary structure, or require assistance from other proteins.

◆ The **tertiary structure** describes the organization in three dimensions of all the atoms in the polypeptide chain, including the side (R) groups as well as the polypeptide backbone. In a protein consisting of a single polypeptide chain, this level describes the complete structure.

◆ The highest level of organization is recognized in **multimeric proteins**, which consist of aggregates of more than one polypeptide chain. The individual polypeptide chains that make up a multimeric protein are often described as the **protein subunits**.

Several types of interactions between amino acids determine the conformation of a protein. Both covalent and noncovalent bonds are involved.

The most common use of covalent bonds in establishing the secondary structure is the formation of S–S disulfide 'bridges' between two cysteine residues (forming cystine). Each cysteine is separately incorporated into the polypeptide chain at the appropriate location when the protein is synthesized; the condensation of the two –SH groups into

the S–S bridge occurs later, when they are brought into apposition as the chain begins to fold into the correct conformation. An example is presented in **Figure 1.11**.

A major force underlying the acquisition of conformation in all proteins is the formation of **noncovalent bonds**. Four types of noncovalent interactions occur in protein structures:

◆ ionic bonds

◆ hydrogen bonds

◆ hydrophobic interactions

◆ Van der Waals attractions.

Figure 1.11

Formation of a disulfide bridge between the sulfhydryl groups of two cysteine residues may connect two different parts of a polypeptide chain.

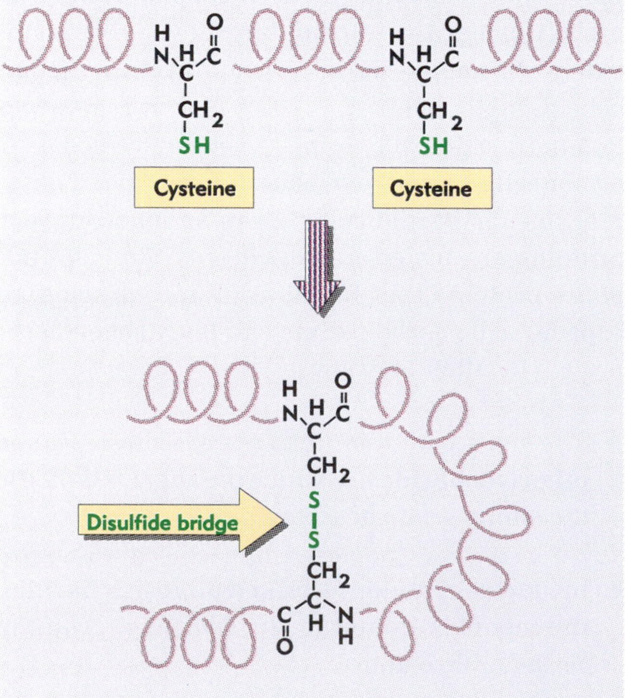

Table 1.4

Covalent and noncovalent bonds differ greatly in strength.

Type of Bond	Strength (kcal/mol)		
Covalent	-50	to	-100
Ionic	-80	or	-1
Hydrogen	-3	to	-6
Van der Waals	-0.5	to	-1
Hydrophobic	-0.5	to	-3

Noncovalent bonds are much weaker than covalent bonds, by a factor of more than 10. **Table 1.4** summarizes some typical values for the strengths of different types of bonds. Strength is assessed by the free energy (given as a negative value) that is released by the formation of the bond; the greater the release of energy, the more stable the bond (because a corresponding amount of energy would then be required to break the bond).

The formation of noncovalent bonds is strongly influenced by the aqueous environment. A major factor that influences these reactions is the structure of water itself, which forms a transient network of connections between individual molecules. The major force in this network is the hydrogen bond.

Hydrogen bonds (H-bonds) are weak electrostatic bonds that form between a partially negatively charged oxygen atom and a partially positively charged hydrogen atom, as in the examples:

$$\overset{\delta^+}{-C}\overset{\delta^-}{=O} \cdots\cdots\cdots\cdots \overset{\delta^+}{H}\overset{\delta^-}{-N}-$$

$$\overset{\delta^+}{-C}\overset{\delta^-}{=O} \cdots\cdots\cdots\cdots \overset{\delta^+}{H}\overset{\delta^-}{-O}-$$

(The δ indicates the partial nature of the electric charge.) The polarization of the C=O and the N–H or O–H bonds in effect allows the hydrogen atom to be shared between the reacting groups (accounting for the name of the bond).

Hydrogen bonds commonly form between the NH and CO groups of the peptide backbone (see Figures 1.16 and 1.17 later). **Figure 1.12** presents an example of hydrogen bonding involving side-groups in a protein chain.

Hydrogen bonding is an important feature of the structure of liquid water. Most of the molecules in water are hydrogen bonded at any given moment, as illustrated in **Figure 1.13**. But an individual bond exists for a period that is short indeed, with an average half life of $<10^{-9}$ sec.

When groups that can form hydrogen bonds find themselves in water, they interact with the water instead of or as well as with one another directly. As a result, the strength of the interaction between them is less in water than it would be *in vacuo*. The aqueous environment therefore has an

Figure 1.12

Hydrogen bonds may form between the side-groups of polar amino acids.

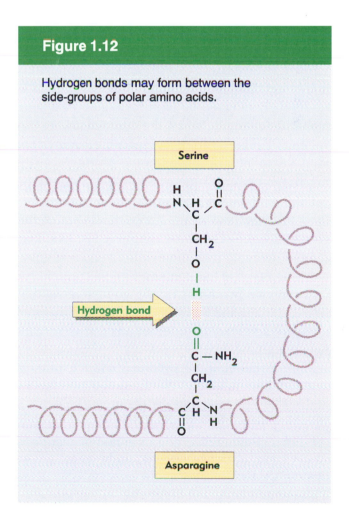

Figure 1.13

Water contains a network of transient hydrogen bonds, in which the O of one H_2O is linked to the H of another water molecule.

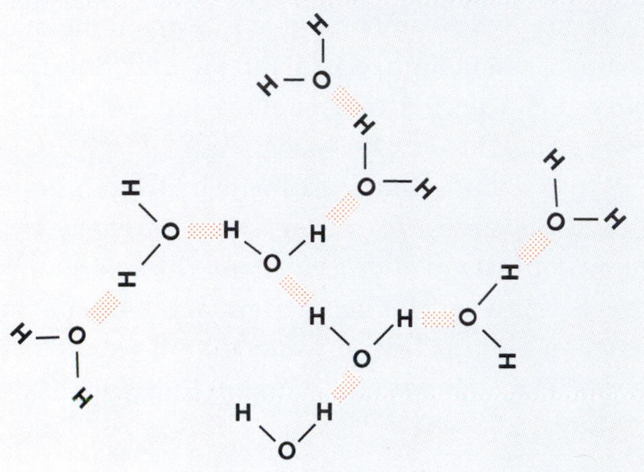

important influence on the structure of a protein.

Ionic interactions occur between oppositely charged groups. The strength of the ionic bond is enormously affected by circumstances. In a solid crystal, for example, in salt, the interaction between Na^+ and Cl^- has a strength comparable to a covalent bond. But in solution, water molecules interact with the charged groups—the water is said to **shield** the charge—and the strength of an ionic bond is weakened to about the same strength as that of the hydrogen bond. The basic amino acids, Lys, Arg, and His, interact with the acidic amino acids, Asp and Glu. A typical ionic interaction in a protein (or between two proteins) might involve attraction between acidic and basic amino acids, as illustrated in **Figure 1.14**. Similar interactions can occur with the polar amino acids.

Ionic forces may be involved when a protein binds another molecule. Thus a protein may have a binding site that includes a positive group (or a series of positive groups) designed to interact with a negatively charged group or series of groups in some target molecule. For example, basic amino acids in a protein can interact with the negatively charged phosphate groups in a nucleic acid.

Because many hydrogen bonds can be formed in a macromolecule, their overall contribution to the stability of the conformation can be substantial. But the weakness of the individual hydrogen bond (and of other noncovalent bonds) allows them to be broken relatively easily under physiological conditions. This is an important aspect of the function of both proteins and nucleic acids.

Hydrophobic interactions occur between amino acids with apolar side-chains, which aggregate together to exclude water. Viewed from the perspective of the water, the hydrophobic groups have disruptive effects on the interactions between water molecules; so water tends to force the hydrophobic groups into juxtaposition (just as an oil drop will form in aqueous surroundings). The hydrophobic effect is illustrated in **Figure 1.15**.

Figure 1.14

Basic and acidic amino acids may be attracted via ionic bonds, shielded by water in the aqueous environment of a protein.

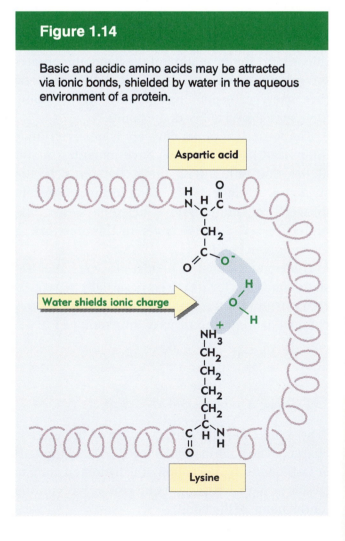

Figure 1.15

Water may force hydrophobic groups into juxtaposition to prevent them from disrupting its hydrogen-bonded network.

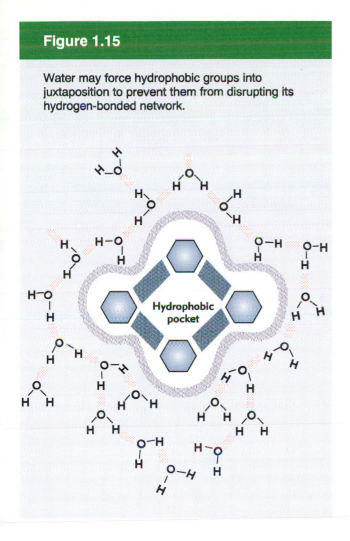

Hydrophobic pocket

The stability of noncovalent bonds can be judged by comparison with the average energy of kinetic motion at 25°C, which is −0.6 kcal/mol. For a bond to be stable, its formation must release more free energy than could be gained by kinetic motion. By comparison with the bond strengths given in Table 1.4, we see that at least some noncovalent bonds are barely stable.

The amino acids in any region of a protein can be viewed as forming a surface whose properties reflect the composition of the side-groups. Two types of surfaces can be formed in a protein structure that react in opposite ways to water:

◆ A **hydrophilic surface** contains charged or polarized groups that react with water; it therefore functions in an aqueous environment. This describes the majority of protein sequences (or at least one might say that they are not sufficiently hydrophobic to exclude water).

◆ A **hydrophobic surface** contains groups that repel water. Several hydrophobic amino acids are needed to generate such a surface. Usually a hydrophobic surface is a small part of a protein that forms a pocket in the interior or finds a non-aqueous environment. The hydrophobicities of individual amino acids can be summed to estimate the probability that a region of a protein will form a hydrophobic region, which has important consequences for protein behavior, as we shall see in Chapter 2.

When amino acids with hydrophobic side-groups are driven away from water, they form a molecular interior in the protein, where they are inaccessible to the solvent (water). Amino acids can be classified on a hydrophobicity scale, on which the hydrophobicity of the individual amino acid is correlated with the area that the hydrophobic residue contributes to the interior of the protein. On this scale, Trp, Phe, and Tyr, the amino acids with aromatic rings, are the most hydrophobic; Leu, Val, and Met are less strongly hydrophobic.

Van der Waals forces come into play at very close distances, when any two atoms experience an attraction. The individual forces are very weak, but can become relevant in closely packed regions of a macromolecule, or when two macromolecular surfaces are closely aligned.

To summarize, protein and nucleic acid structures exist largely in an aqueous environment, in which water shields the hydrogen and ionic bonds. The biological polymers have many loose connections to the hydrogen-bonded network of water. All noncovalent bonds are rather weak in this environment, providing at most only a few percent of the strength of the covalent bond. Ionic and hydrogen bonds are weakened directly by the shielding effect of the water. Hydrophobic bonds are created indirectly by the effect of water in forcing hydrophobic groups together to prevent them from disrupting the network of hydrogen bonds in the water itself.

Protein structures are extremely versatile

Proteins show enormous diversity of form as a result of their ability to generate a huge range of conformations, but certain types of secondary structure are relatively common. They result from hydrogen bonding between the NH and CO groups of the polypeptide backbone. Often several regions organized in a particular type of secondary structure are packed close together.

Hydrogen bonding between groups on the same polypeptide chain causes the backbone to twist into a helix, most often the form known as the α-**helix**, illustrated in **Figure 1.16**. We may think of the polypeptide backbone as winding in a coil on the surface of a cylinder, with the side-groups protruding from the cylindrical surface. The α-helix is stabilized by hydrogen bonds formed between the C=O group of one peptide bond and the NH group of the peptide bond four residues farther along the polypeptide chain.

The α-helix is a common component of protein secondary structures. Individual segments of α-helix often are quite short; for example, in some proteins that bind nucleic acids, a critical structural feature is the presence of particular stretches of α-helix that are less than 10 amino acid residues long.

A polypeptide chain can be extended into a sheet-like structure by hydrogen bonding with another chain that runs in the opposite direction. Once again, the hydrogen bonding involves the C=O group of one peptide bond and the NH group of another. The β-**sheet** structure is illustrated in **Figure 1.17**; it is favored by the presence of glycine and alanine residues. As well as occurring between different polypeptide chains, this type of structure can be formed between two sections of a single polypeptide chain arranged so that the adjacent regions are in reverse orientation.

Proteins can be divided into two general classes (which represent extremes of structure) on the basis of their tertiary structures:

◆ **Fibrous proteins** have elongated structures, with the polypeptide chains arranged in long strands. The tertiary structure may be based on

Figure 1.16

A polypeptide α-helix is stabilized by hydrogen bonding between amino acids four residues apart in the same polypeptide. (H-bonds are indicated by dots).

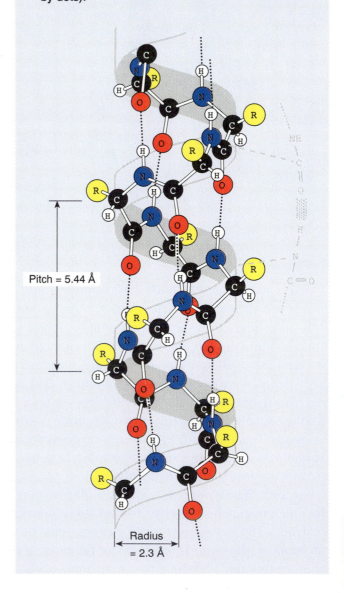

Pitch = 5.44 Å

Radius = 2.3 Å

Figure 1.17

A β-sheet is stabilized by hydrogen bonding between amino acids in different polypeptide chains (or in parts of the same chain that run in opposite directions).

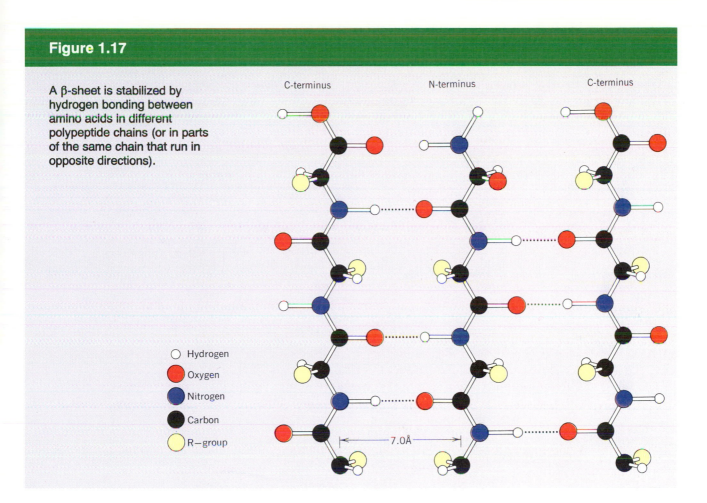

○ Hydrogen
🔴 Oxygen
🔵 Nitrogen
⚫ Carbon
🟡 R−group

an α-helix or β-sheet. The particular form of secondary structure stretches through a large part of the protein. The fibrous proteins are major structural components of the cell or tissue; for example, they are prominent in connective tissues. Thus their role tends to be static in providing a structural framework.

◆ **Globular proteins** have more compact structures. The tertiary structure is generally rather irregular, often containing many different types of secondary structure. For example, although a part of the structure often takes the form of an α-helix, it is rare for it to comprise the entire structure. Individual stretches of α-helix are likely to be just a few amino acids in length. Globular proteins include most enzymes and most of the proteins involved in gene expression and regulation with which we shall

be involved. Their roles tend to be dynamic, involving the ability to catalyze reactions or change conformation.

The versatility of protein structures can be illustrated by considering the alternative forms depicted in **Figure 1.18** for a polypeptide chain of (say) 300 amino acids:

◆ As a fully extended array of amino acids, it would stretch for 100 nm.

◆ Coiled in an α-helix, it would extend for 45 nm.

◆ But organized as a β-sheet it could become a flat box of 7 × 7 nm only 0.8 nm deep.

◆ And as a sphere utilizing various forms of secondary structure, it could have a diameter of only 2 nm.

Figure 1.18

Protein structures may be highly extended or very compact, as seen in various possible forms for a single polypeptide chain containing the same number of residues.

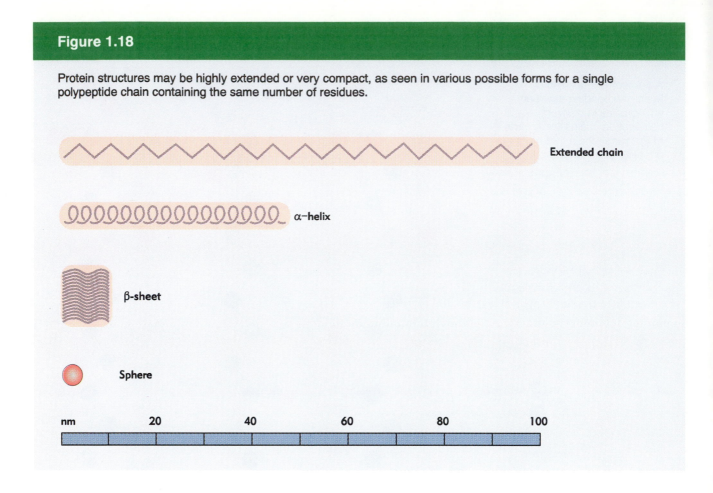

Alternative conformations are made possible by the ability of atoms to rotate freely about single covalent bonds. Protein conformations *are interconvertible by rotation about such bonds to change the positions of individual atoms. A change from one conformation to another therefore can be accomplished by making and breaking noncovalent bonds only.*

Because individual noncovalent bonds are relatively weak, the amount of energy involved in exchanging some noncovalent contacts for others can be relatively small, allowing changes in protein conformation to occur with some facility under physiological conditions.

Each protein in principle has available to it an almost indefinite number of conformations, but in practice, one or a few conformations are favored. In saying that a protein exists in a particular conformation, we do not imply that the position of every atom is static: there is a small amount of motion about the 'average' position that characterizes the conformation.

Noncovalent bonds are also important in the interactions between proteins and other molecules. A small molecule that binds to a protein is called a **ligand**. The specificity of protein–ligand binding depends on the protein's possession of a surface containing a number of groups each positioned so that it can make contact with a corresponding group in the ligand.

The relationship between structure and function depends on the type of protein. Some proteins, especially enzymes that catalyze metabolic reactions, have an **active site** at which the catalytic reaction occurs. The structure of this site, created by the juxtaposition of a handful of amino acids, may be absolutely crucial, and admitting of no variation. But it may be possible to change other regions of the protein without abolishing its catalytic function.

An active site includes amino acids whose side-groups are involved in a particular chemical reaction or that make specific contacts with another macromolecule. In the example of chymotrypsin, illustrated in **Figure 1.19**, three amino acids, located at distant parts of the primary chain, come together in the tertiary structure to form the active site. The OH group of serine is involved in the catalytic reaction, the cleavage of a peptide bond in a target protein. The role of the other two amino acids is uncertain, but a substitution of any one of the three by another amino acid abolishes catalytic activity.

In proteins that bind to nucleic acids, the active site is likely to be more extensive, because the target to which the protein binds is larger, and side-groups of particular amino acids are involved in making certain contacts with bases in the nucleic acid. Changes in these amino acids alter the ability of the protein to recognize its target.

The main characteristic of an active site, therefore, is that, *within the conformation established by the protein as a whole (which involves regions outside the site itself), the active site contains amino acids whose side-groups play an essential role in making specific contacts with groups in the target molecule or macromolecule.*

In contrast with the localization of a discrete active site, the functions of certain proteins, in particular some structural proteins, involve most or all of the conformation; and any change at all is therefore likely to prevent function. Proteins therefore differ in the extent to which their structure can vary without preventing their function.

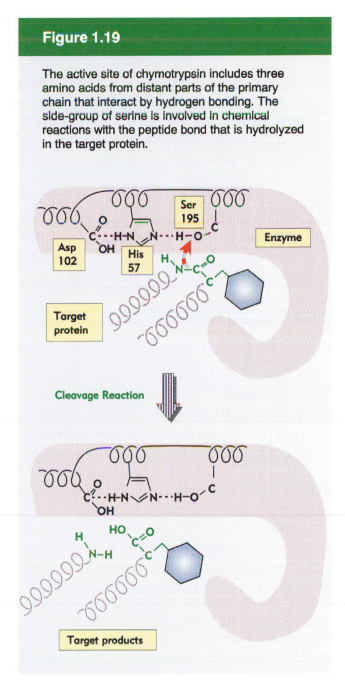

Figure 1.19

The active site of chymotrypsin includes three amino acids from distant parts of the primary chain that interact by hydrogen bonding. The side-group of serine is involved in chemical reactions with the peptide bond that is hydrolyzed in the target protein.

How do proteins fold into the correct conformation?

A fundamental principle is that *higher-order structures are determined by lower-order structures.* This means that the primary sequence of amino acids carries the information required to fold into the correct conformation. The folding reaction may involve the formation of both noncovalent and covalent bonds.

What conditions are necessary for the primary

sequence to fold into the correct higher-order structure?

◆ Is the folding an intrinsic feature of the primary sequence? In this case, the final structure must always be the most stable thermodynamically and can be generated at any time after synthesis of the polypeptide chain is complete.

◆ Or can the correct structure be generated only during the synthesis of the polypeptide? Then it becomes possible that an intrinsically less stable structure could prevail because the protein becomes 'trapped' in it during synthesis.

◆ Does folding occur spontaneously or is it assisted by other proteins?

We still lack detailed knowledge about the processes involved in protein folding. The reaction is usually rapid, occurring within seconds or less. It begins even before a protein has been completely synthesized. Probably it involves a **sequential folding** mechanism, in which the reaction passes through discrete (although highly transient) intermediates. The process appears to be cooperative, so that formation of one region of secondary structure enhances formation of the next region, and so on.

The relationship between higher-order structures and the primary structure may be revealed when a protein is **denatured** by heating or by chemical treatments that disrupt its conformation. Most denaturing events involve the breakage of hydrogen and other noncovalent bonds. An exception is the disruption of S–S bridges that results from treatment with reducing agents. However, all of these changes affect the *conformation*; the primary sequence of amino acids in the polypeptide chain remains unaltered.

In some cases, the higher-order structure follows ineluctably from the primary sequence. The enzyme ribonuclease is the classic example. After the protein has been denatured, its active conformation can be regained by reversing the denaturing procedure. *All the information necessary to form the secondary structure resides in the primary sequence.* Thus the production of active ribonuclease is an inevitable event whenever the intact primary chain is placed in the appropriate conditions. Renaturation *in vitro* starts with the collapse of hydrophobic regions into the interior of the protein, after which secondary structures form successively until the whole protein is folded; formation of disulfide bonds occurs at a late stage.

In other cases, proteins can be irreversibly denatured. Thus under certain (nonphysiological) conditions, a protein may have alternative stable conformations. It is possible (although not proven) that in some cases the correct conformation can be attained *only* during synthesis of the protein; the conformation could depend on specific interactions between regions of the protein that can occur only in the absence of other regions (that is, those that have not yet been synthesized).

In some instances, a **cofactor** that is part of the active protein (such as the iron-binding heme group of the cytochromes) must be present in order for the polypeptide chain to take up its proper conformation. In the case of multimeric proteins, it may be necessary for one subunit to be present in order for another to acquire the proper conformation.

The acquisition of structure when a protein is synthesized is not a spontaneous process, but may require assistance. More precisely, we should say that spontaneous folding is a slow reaction, which under normal cellular conditions is a rate-limiting step. The rate is significantly increased by several types of additional functions. These are summarized in **Table 1.5**. They fall into two groups: enzymes that catalyze specific isomerization steps; and factors that act stoichiometrically to influence folding directly.

Formation of disulfide bridges is influenced by both environment and specific accessory proteins. Disulfide bonds are rare in cytoplasmic proteins, but common in exported proteins. This may be related to a difference in the thiol/disulfide redox state between internal and external conditions. For example, it may help to prevent bond formation in a bacterium, but helps to drive it in the periplasm (the layer surrounding the bacterial cell).

Table 1.5

Both catalytic and structural functions may be required to assist protein folding.

Protein Activity	Families	Function	Effect
Catalytic			
Protein disulfide isomerase	Thioredoxin	Formation of S-S bonds	Increases rate of folding
Peptidyl prolyl isomerase	Cyclophilin PPI FKBP PPI	*cis/trans* bond conversion	Increases rate of folding
Stoichiometric			
Chaperonins	Hsp-70 (DnaK) Hsp-60 (GroEL) Hsp-90	Binds improperly folded protein	Inhibits wrong folding pathway

Disulfide bond formation can occur spontaneously *in vitro,* but the rate is slow. It has a $t_{\frac{1}{2}} > 15$ min, compared with the ability to form disulfide bonds correctly within a few seconds *in vivo.* (The reaction was shown previously in Figure 1.11.) The process is catalyzed *in vivo* by an enzyme, protein disulfide isomerase (PDI), although direct evidence for its participation has been difficult to obtain. PDI is a curious protein, which participates in a variety of functions concerned with protein modification, in addition to its sponsorship of disulfide bridge formation. It is not entirely clear whether it simply helps the initial formation of disulfide bonds or whether it also catalyzes rearrangement of disulfide bonds that have formed incorrectly.

Proline has a major effect upon protein structure because of the restrictions imposed by its ring structure. Proteins containing proline fold slowly because the peptidyl–proline link does not necessarily form in correct stereochemical conformation. The enzyme peptidyl-prolyl isomerase (PPI) catalyzes the *cis–trans* conversion illustrated previously in Figure 1.8, and by this means significantly accelerates the folding reaction. Enzymes with PPI activity fall into two major groups, named for their abilities to bind certain drugs: cyclophilin PPI binds the drug cyclosporin A, and FKBP PPI

binds the drug FK506. Members of the cyclophilin class are better characterized, and they vary in their specificity of action from those that appear to be generic (able to act on any protein) to those that appear to work only with specific proteins. This makes the point that, although control of proline isomerization is a general feature of many proteins, it can also be used to control specifically the maturation of an individual protein.

Proteins that act stoichiometrically on the folding of other proteins are called **molecular chaperones**, or **chaperonins**. A chaperone forms a complex with a protein during folding, *but is required only during assembly, and is not part of the mature structure.* The major role of a chaperone is to prevent the formation of incorrectly folded structures, in which the substrate protein might otherwise become trapped during folding, as illustrated in **Figure 1.20**.

Chaperones act directly upon higher-order structure. They may recognize features of a protein structure that become accessible only during its synthesis, or when for some other reason it enters an unfolded state; by sequestering such regions, the chaperone prevents their participation in incorrect folding reactions. A chaperone does not itself impose a structure upon the folding protein, but plays a preventative role. Chaperoning is an energy-dependent

Figure 1.20

A chaperone may recognize a target protein by its unfolded state during synthesis and may sequester regions that would otherwise interfere with proper folding. This drives the target protein to fold in the correct pathway. Chaperones do not bind (or remain bound) to properly folded proteins.

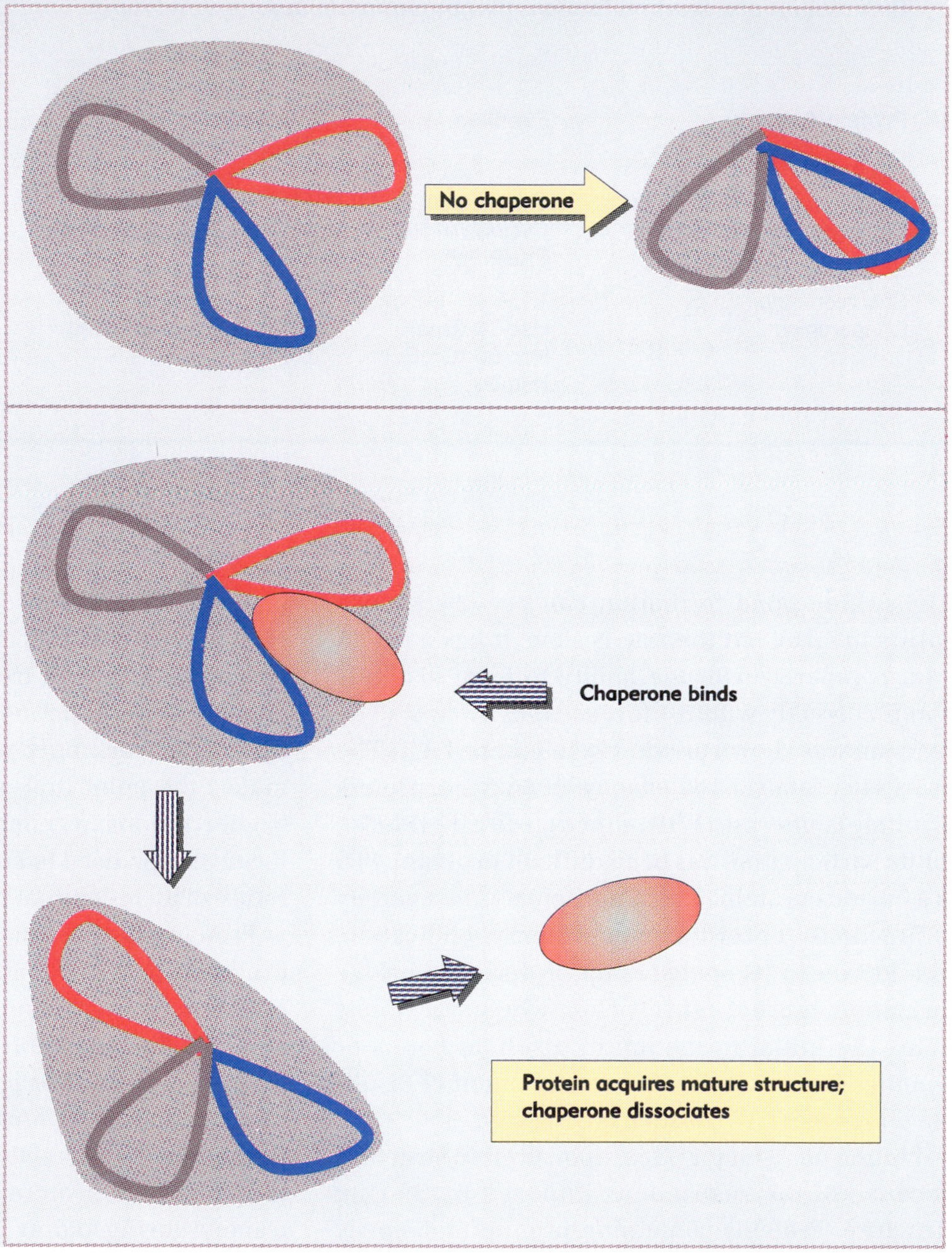

No chaperone

Chaperone binds

Protein acquires mature structure; chaperone dissociates

process that requires the hydrolysis of significant amounts of ATP.

There are three common families of chaperones, defined by their relationship (or identity) with stress proteins—proteins that act to minimize the effects of temperature or other stresses. The families are called **Hsp**, which stands for heat shock protein, because they were originally discovered as proteins whose synthesis is induced by a heat shock. Each family is identified by a number, which corresponds to the mass of the first heat shock protein to have been discovered in its group. The Hsp families have members in probably all organisms. The best characterized is the Hsp70 family, which includes the eponymous Hsp70 in animals, a common member called Bip, and the DnaK protein in *E. coli.* The Hsp60 family includes GroEL of *E. coli.*

When stress generates denatured proteins, chaperones recognize the unfolded state and bind to the denatured proteins. In the same way that the reaction prevents acquisition of an incorrectly folded state during synthesis, it prevents a denatured protein from entering an irreversibly denatured conformation. Thus the function during stress is related or identical to the function of a chaperone.

Drastic changes in protein conformation occur when a protein is transported from one region of the cell to another. It is necessary for a protein to be largely unfolded when, for example, it is transported through a membrane. This makes it impossible for its final conformation to be acquired during synthesis. Chaperones may be needed both before and after transport, because the protein must be maintained in an unfolded form prior to transport, and then subsequently requires assistance from chaperones to acquire its final conformation at its destination (see Chapter 11).

Protein folding therefore turns out to be a more intricate process than had been thought originally. We are left with the knowledge that the primary sequence of a protein is a crucial determinant of its higher-order structures; sometimes it is in fact the sole determinant, but in other cases additional interactions are involved in acquiring the final conformation. In each case, however, if the primary sequence is synthesized within the appropriate environment, it will acquire the proper higher-order structures.

Although higher-order structure follows from primary sequence, the same general tertiary structure can be determined by different primary sequences. For example, the globin (red blood cell) proteins of different species vary substantially in sequence, but have the same general tertiary structure.

An important concept is that a protein may consist of **domains**. A domain is a (relatively) independent region of the protein; in some cases, its conformation can be acquired independently by the relevant fragment of the polypeptide chain. Some globular proteins consist of discrete domains connected by 'clefts'.

A domain may represent a functional unit which is identified with a particular activity of the protein, for example, its ability to bind a certain ligand. A domain may represent an evolutionary unit; it may have arisen as a functional polypeptide or region of a polypeptide and later have associated with other domains to generate a new protein with more widely ranging abilities. Sometimes a substrate binds to the cleft between domains.

Allosteric proteins exist in alternative confor-

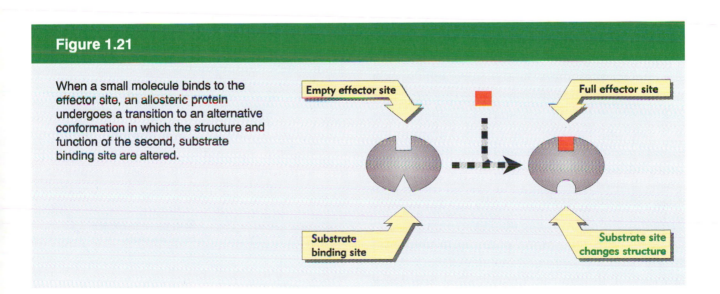

Figure 1.21

When a small molecule binds to the effector site, an allosteric protein undergoes a transition to an alternative conformation in which the structure and function of the second, substrate binding site are altered.

Empty effector site

Full effector site

Substrate binding site

Substrate site changes structure

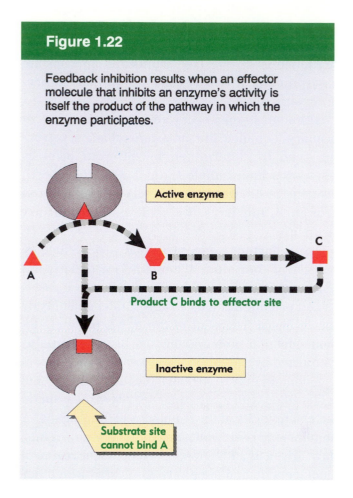

Figure 1.22

Feedback inhibition results when an effector molecule that inhibits an enzyme's activity is itself the product of the pathway in which the enzyme participates.

Active enzyme

A   B   C

Product C binds to effector site

Inactive enzyme

Substrate site cannot bind A

inhibition. It is illustrated in **Figure 1.22**. The crucial feature is the ability of a small molecule to bind to a specific site on the protein, thereby changing the conformation in such a way that the activity of the protein is altered. The same effect is used in genetic regulation, to enable the activity of a protein that controls gene expression to respond to a small molecule.

Allosteric proteins are often multimeric. In such cases, the conformation of one subunit can influence the conformation of the other subunit(s).

One common form of interaction occurs when the protein consists of two different types of subunit, one bearing the regulatory (effector) site, and the other bearing the catalytic (substrate-binding) site. When we redraw the interaction of Figure 1.21 in terms of a dimer, we see the result illustrated in **Figure 1.23**; binding of the effector molecule to the regulatory subunit causes a change in the structure of the substrate site on the catalytic subunit. Such a change may in principle be responsible either for activating or for inhibiting the catalytic activity.

Another interesting interaction occurs when an

mations and have different biological properties in each conformation. The transition between conformations is influenced by the interaction of the protein with a cofactor or with another protein.

Figure 1.21 illustrates the consequences of an allosteric transition. The change in conformation brought about by an interaction at one binding site (the effector site) alters the structure and thus the function of another binding site (the substrate binding, or active site). Allosteric transitions affect *only* the conformation—they do not change the primary sequence—and they rely heavily on making and breaking hydrogen bonds.

Allosteric proteins play an important role in both metabolic and genetic regulation. Often the product of a metabolic pathway can bind to an enzyme catalyzing an early step in the pathway to prevent it from sending further small molecules through the pathway. This interaction is called **feedback**

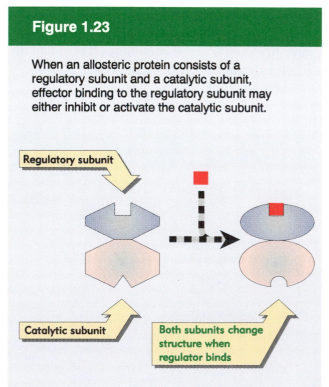

Figure 1.23

When an allosteric protein consists of a regulatory subunit and a catalytic subunit, effector binding to the regulatory subunit may either inhibit or activate the catalytic subunit.

Regulatory subunit

Catalytic subunit   Both subunits change structure when regulator binds

allosteric protein consists of identical subunits. Binding of an effector molecule to one subunit makes it much easier for the other subunits to bind the effector. As illustrated in **Figure 1.24**, this effect amplifies the effect of binding the first small molecule in such a way that the protein characteristically flips very rapidly from one state to another. This is an important ability in a regulatory protein.

Figure 1.24

When an allosteric protein is a multimer of identical subunits, binding to one subunit may cause a structural change that enhances binding at other subunits.

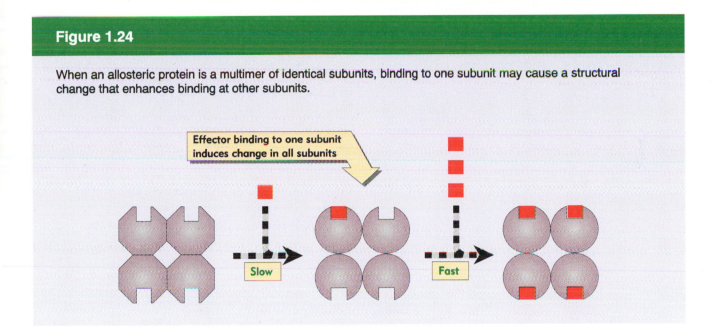

Effector binding to one subunit induces change in all subunits

Slow

Fast

Summary

The structural, catalytic, and regulatory activities of the proteins of a cell are responsible, directly or indirectly, for creating its ultrastructure and interconverting the small molecules. The primary structure of a protein consists of a covalently linked linear series of amino acids. The higher-order structures are determined by the primary structure, sometimes independently, sometimes with the participation of ligands or other proteins. Protein conformation depends on a large number of individually weak noncovalent interactions. Proteins generally function in an aqueous environment, but may have local regions that are hydrophobic.

Proteins are synthesized according to the directions of the nucleic acids that constitute the genetic apparatus. Obeying the laws of physics and chemistry, together these macromolecules are responsible for the survival of living cells. We have no need to postulate additional forces, although there remain questions about how the properties of macromolecules are related to the nature of positional information.

Further reading

Historical

Schrödinger's provocative view *What is Life?* was published by Cambridge University Press in 1944.

Reviews

Protein folding has been reviewed by **Gething and Sambrook** (*Nature* **355**, 33–45, 1992).
Chaperones have been reviewed by **Ellis and van der Vies** (*Ann. Rev. Biochem.* **60**, 321–347, 1991).

Discoveries

Monod, Changeaux, and Jacob (*J. Mol. Biol.* **6**, 306–329, 1963) introduced the concept of regulation via allosteric changes in protein structure.

The principle that the amino acid sequence dictates protein conformation was enunciated by **Anfinsen** (*Science* **181**, 223–230, 1973).

CHAPTER 2

Cells are organized into compartments

The cell theory established in the middle of the nineteenth century proposed that all living organisms are composed of cells. Cells can arise only from pre-existing cells. We may distinguish two general types of cells:

◆ Bacteria are the **prokaryotes**, organisms in which nominally there is only a single cell compartment, bounded by a membrane or membranes that give security against the outside world.

◆ **Eukaryotes** are defined by the division of each cell into a **nucleus** that contains the genetic material, surrounded by a **cytoplasm**, which in turn is bounded by the **plasma membrane** that marks the periphery of the cell. The cytoplasm contains other discrete compartments, also bounded by membranes.

Unicellular organisms (either prokaryotic or eukaryotic) consist of individual cells, able to survive, reproduce, and (in some cases) mate. **Multicellular** organisms (usually eukaryotic) exist by virtue of cooperation between many cells, which have different specialties to contribute to the survival of the individual.

All organisms, prokaryotic or eukaryotic, unicellular or multicellular, are descended by evolution from an original population of primitive 'cells'. The events involved in early evolution are unknowable in any detail, but we may surmise that the development of the very first 'cells' involved two general types of event:

◆ Some form of self-replicating material, probably a nucleic acid, must have allowed the 'cell' to make copies of itself. From this material evolved the genetic information.

◆ This material, and any other contents of the 'cell', must have been kept together, or at least prevented from diffusing into the primeval soup, by a boundary material that generated some sort of compartment. From this boundary evolved the membranes that surround a cell and also create compartments within it.

The nature of present day genetic information and the general means (although not every detail) of its expression are the same in prokaryotes and eukaryotes. In describing how the inheritance of genetic information accounts for the perpetuation of both prokaryotic and eukaryotic cells, we may regard their genetic instructions as being of the same kind, although the process of reading the instructions involves extra steps in eukaryotes.

To account for the execution of genetic instructions in the fullest sense—the maintenance of the structure of a living cell—we must define the cellular environment. We must therefore consider the nature of the various compartments that are found in prokaryotic and eukaryotic cells, and ask how their components are assembled and how they function to create regions with different properties. A major question to bear in mind is *whether each particular structure can self-assemble from its components or whether it requires an example of a pre-existing structure to use as a template.*

Figure 2.1 illustrates the structure of a typical

Figure 2.1

A bacterial cell may consist of a single compartment, bounded by a membrane, although its genetic material is organized into a compact nucleoid.

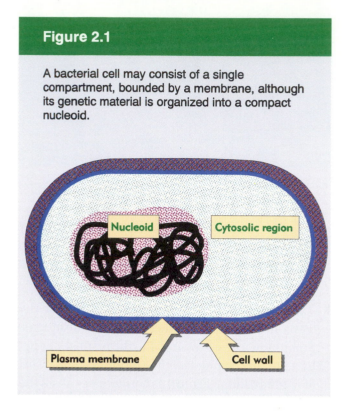

Nucleoid

Cytosolic region

Plasma membrane

Cell wall

In a bacterial cell, the superficial membrane provides the sole membranous surface. But a eukaryotic cell contains many internal membrane surfaces in addition to its plasma membrane, which typically comprises <5% of the total membranous material of the cell.

The internal membranes divide a eukaryotic cell into the compartments drawn in **Figure 2.2**. The general area of the cytoplasm excluding the compartments is called the **cytosol.** Genetic material is contained in the nucleus. Other closed compartments within the cell are called **organelles. Table 2.1** lists the major types of compartments. Some organelles are delineated by a single, closed membrane; others—the nucleus, mitochondrion and (in plants) chloroplasts—are surrounded by an **envelope** that consists of two membrane layers. The closed region confined by the organelle membrane (such as the single membrane of the endoplasmic reticulum or the inner membrane of the mitochondrion) is known as the **lumen.**

The environment within and at the surface of each individual cellular compartment is controlled to suit local needs. This affects the structure and function of the proteins and other molecules located there. We may distinguish two general types of environment that influence the nature of biochemical reactions:

bacterium, in which there is only a single formal compartment. The membrane(s) defining this compartment may be connected to, or may have inserted within them, a coat of more rigid material that makes a cell wall.

Table 2.1

A eukaryotic cell can be divided into organelles and cytosol. The proportion of volume is only approximate, and refers to a fibroblast, a small animal cell that can be grown readily as an individual cell in culture conditions.

Compartment	Type of Boundary	Proportion of Volume	Functions
Nucleus	Envelope	5%	Gene expression
Cytosol	Plasma membrane	55%	Protein synthesis & metabolism
Mitochondrion (>1000/cell)	Envelope	25%	Energy production
Chloroplast	Envelope		Photosynthesis (in plants)
Endoplasmic reticulum	Folded membrane	10%	Protein modification
Golgi apparatus	Membrane stacks	5%	Protein sorting
Lysosome (˜400/cell)	Closed membrane	<1%	Protein degradation
Peroxisome (˜300/cell)	Closed membrane	<1%	Oxidation reactions

Figure 2.2

A eukaryotic cell may have several compartments, each bounded by a membrane.

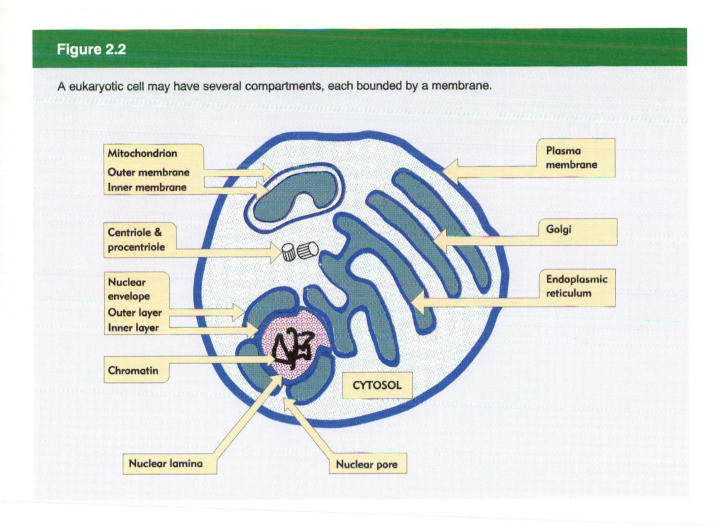

- The cytosol and the nucleus provide the general **aqueous environment** described in Chapter 1. The lumen of organelles such as the endoplasmic reticulum also forms an aqueous environment. Of course, there are differences in the aqueous environments of the cytosol, endoplasmic reticulum, mitochondrion, etc., such as in concentrations of particular ions.

- The membranes that surround these and other compartments provide an environment that is hydrophobic, but with a hydrophilic surface that interacts with the aqueous environment on either side. Some important biochemical reactions—for example, oxidative phosphorylation and photosynthesis—occur in the internal **hydrophobic environment** of the membranes.

Cellular compartments are bounded by membranes

The characteristic properties of membranes result from their high contents of lipids. A crucial feature of lipids creates the membranous environment: they are **amphipathic.** One end of the molecule consists of a polar 'head', while the other end consists of a hydrophobic 'tail'.

The major bulk of a lipid is provided by its hydrophobic tails, which differ in overall length and

Figure 2.3

A lipid has one of three types of polar head and a hydrophobic tail(s).

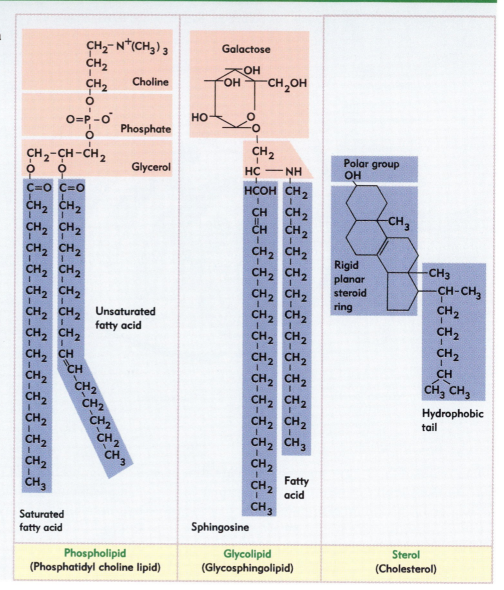

| Phospholipid (Phosphatidyl choline lipid) | Glycolipid (Glycosphingolipid) | Sterol (Cholesterol) |

in the nature of the carbon–carbon bonds. One type of fatty acid tail is *saturated:* all the carbon–carbon links are single bonds. The other type of tail is *unsaturated:* one or more carbon–carbon links consist of double bonds. Because rotation is restricted around the double bond, the unsaturated tail has a bend, while the saturated tail can extend freely. Fatty acid tails are usually ~20 residues long. Distinguished by their polar heads, membranes contain the three principal types of lipids illustrated in **Figure 2.3**:

◆ In **phospholipids** the head has a positively charged group linked via a negatively charged phosphate group to the rest of the molecule. The example of Figure 2.3 has a head consisting of choline–phosphate–glycerol, attached to two hydrophobic tails. Lipids based on glycerol have one saturated and one unsaturated fatty acid tail.

◆ **Glycolipids** are characterized by the presence of oligosaccharide. The chain of sugars typically

Figure 2.4

In an aqueous environment, the formation of a lipid bilayer satisfies the amphipathic nature of lipids by allowing the hydrophobic tails to segregate away from water while the polar heads are immersed in water.

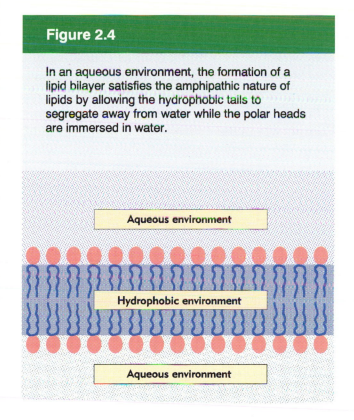

Aqueous environment

Hydrophobic environment

Aqueous environment

consists of 1–15 residues. In animal cells, the connection between the saccharide head and the fatty acid tail is sphingosine (a long amino alcohol). Lipids based on sphingosine have a fatty acid chain in addition to the long hydrocarbon chain of sphingosine itself. In plants and bacteria, glycerol connects the head and tail.

◆ **Sterols** contain a planar steroid ring. They lend rigidity to a membrane because the steroid ring is planar. Cholesterol, a prominent component of animal cell membranes, has a polar hydroxyl group at the terminus.

In an aqueous environment, a lipid is happy to have its polar head exposed, but endeavors to bury its hydrophobic tail away from the water. **Figure 2.4** illustrates how this is accomplished in the cell. Two lipid **monolayers** are juxtaposed to form a **lipid bilayer**, a sheet in which the polar heads of the lipids face out toward the aqueous environment on either side, while the hydrophobic tails face in to create a hydrophobic environment.

Although a membrane consists of a specific type of structure, the lipid bilayer, there is variety in the constitution of different membranes. Overall lipid compositions of membranes vary considerably, as summarized in outline in **Table 2.2**. The differences mean that different membranes have different biophysical properties.

One of the important properties of a membrane is the ability of the constituent lipids to move within it. Lipid molecules rarely move from one

Table 2.2

Membranes from different sources have different proportions of protein and amounts of lipids.

Molecular Class	Plasma Membrane	Endoplasmic Reticulum	Mitochondrial	Bacterial
Protein/membrane	50%	50%	75%	50%
Lipids				
Cholesterol	20%	5%	<5%	70%
Phospholipid	55%	65%	75%	30%
Glycolipid	5%	30%	20%	-
Others	20%	-	-	-

monolayer to the other in a bilayer, but frequently they move laterally to exchange places with their neighbors within the monolayer. The property of movement is called **fluidity**; and a membrane is often regarded as a 'two dimensional fluid'.

The more readily the tails of adjacent lipids can pack together, the more crystalline the membrane structure can become, and the less fluid. The major determinants of membrane fluidity are therefore the types and lengths of the lipid tails. The proportion of saturated versus unsaturated residues in the tails has a major effect on fluidity; unsaturated chains are more difficult to pack, and therefore give a more fluid structure.

The plasma membrane circumscribes a cell. It marks the boundary between the cellular milieu inside and the environment outside. In the case of a unicellular organism, the surroundings constitute the environment in which the organism lives. In a multicellular organism, the environment for any one cell is created by other cells.

The role of the plasma membrane extends beyond providing a mere barrier to the outside. It controls ingress and egress for molecules both small and large. Within the membrane reside specific transport systems that pump ions in or out, that allow proteins to be secreted from the cell into the environment (see Chapter 11), and that recognize molecules outside and as a result transmit messages to the interior (see Chapter 12).

Plasma membranes of animal cells contain relatively large amounts of cholesterol, which increases mechanical stability because of the steroid rings near its polar head. (Plant cells lack cholesterol, but have other sterols instead.) Plasma membranes also are the only membranes to contain significant amounts of glycolipids.

Phospholipids are always present in membranes of eukaryotic cells, but the amount and type varies.

A plasma membrane contains about equal masses of lipid and protein. Internal membranes, such as those surrounding mitochondria, have a greater proportion of protein. The mass of an individual protein molecule is much larger than any

lipid, so there are 10–100 times more lipid molecules than protein molecules. We might view the basic structure of a membrane as consisting of a lipid bilayer that provides a residence for relatively large protein molecules.

Protein components of the membrane can move laterally within the lipid bilayer, although they diffuse much more slowly than the lipid components. As the result of a stimulus, proteins can be 'internalized', when they are removed to the interior of the cell. Other proteins are secreted from the interior of the cell to the exterior by passing through the membrane. The lipid bilayer itself, and the proteins associated with it, therefore comprises a dynamic structure.

A typical membrane protein has only part of its

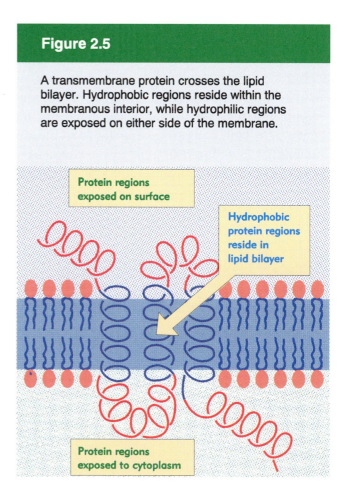

Figure 2.5

A transmembrane protein crosses the lipid bilayer. Hydrophobic regions reside within the membranous interior, while hydrophilic regions are exposed on either side of the membrane.

Protein regions exposed on surface

Hydrophobic protein regions reside in lipid bilayer

Protein regions exposed to cytoplasm

Figure 2.6

A hydropathy plot identifies potential membrane-spanning regions as the most hydrophobic sequences of the protein. Each potential membrane segment is identified by shading in color.

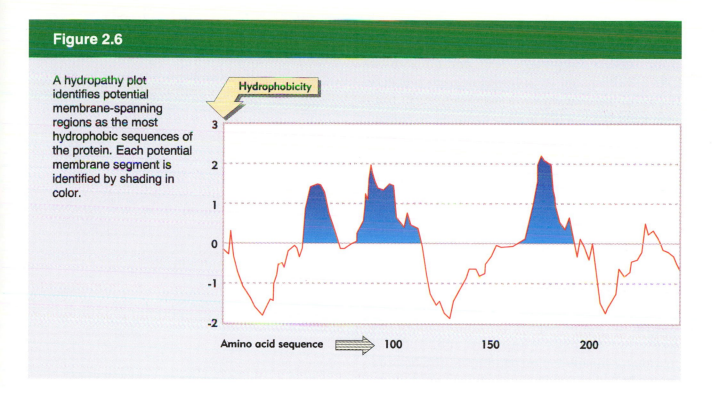

structure incorporated into the membrane. It extends across the membrane, with some regions of the protein exposed on one or both sides, as indicated in **Figure 2.5**. A protein that has parts of its chain available in this manner to react with either the exterior or interior of the cell is called a **transmembrane protein**.

Proteins or parts of proteins that reside in the cytosol are generally hydrophilic. By contrast, a transmembrane protein requires a hydrophobic region that is comfortable in the environment of the lipid bilayer. Such a region is called a **transmembrane domain**. The transmembrane domain is flanked by hydrophilic regions that protrude into the interior of the cell or out into the surrounding environment. A protein may have several transmembrane domains, in which case the parts of the polypeptide chain connecting them can be viewed as 'looping out' into the cytoplasm or the exterior.

The hydrophobicity of a sequence of amino acids can be used to predict (rather imperfectly) whether it is likely to reside in a membrane. A **hydropathy plot** shows the sequence of a protein in terms of the hydrophobicity of overlapping segments. There are various means of measuring hydrophobicity, but whichever scale is used, it is now conventional to assess hydrophobicity as a positive score and hydrophilicity as a negative score. Most of the scales utilize the energy required in kcal/mol to transfer from a hydrophobic to a hydrophilic phase.

In the example of **Figure 2.6**, the hydrophobicity is calculated for each position in the protein by summing the scores of the individual amino acids in the next 21 positions. (A length of 21 amino acids in α-helical form would just span a lipid bilayer.) A region with a positive score is therefore a candidate to provide a transmembrane domain that resides in a membrane.

Within the lipid bilayer, water is effectively excluded. As a result, the regions of the proteins located within the bilayer are not subjected to the aqueous environment described in Chapter 1. The lipid 'solvent' does not form hydrogen bonds with the protein, and therefore solvates neither the groups of the peptide backbone nor polar side chains. Hydrogen bonding occurs solely between groups within the protein itself, and functions to form α-helices and (to a lesser degree) β-sheets. The resulting conformation may be different from what would be reached in an aqueous environment; indeed, it may well be that membrane proteins can attain their natural conformation *only* in the hydrophobic environment.

An important class of transmembrane proteins consists of **receptors.** A receptor has an active site that recognizes some ligand on the exterior side of the membrane. Binding of the ligand triggers a change in the protein, which is transmitted to the cytoplasmic face by a conformational change in the protein, or even reflected as movement of the whole protein into the interior. This change in turn triggers other changes within the cell, and thus provides a means for responding to the environment. This type of relationship is called **signal transduction**, and is a major topic of Chapter 12.

The membranes bounding different cellular compartments are different not only in their overall composition, but also in the particular proteins that reside within them. The proteins in each type of membrane serve, for example, to control transport into or out of the particular compartment, and are therefore designed to recognize the particular molecules or macromolecules that will be travelling this route.

Each membrane has two 'faces', as indicated in **Figure 2.7**. In all membranes, the **cytoplasmic face** is defined as the surface that contacts the general cytosol. The noncytoplasmic face is given various names, depending on the membrane. On a plasma membrane, it provides the outside surface of the cell. In a membrane within the cell, it provides a limit for an interior compartment, comprising the surface that separates the lumen of the compartment from the cytosol.

A major feature of the lipid bilayer is an asymmetry created by biochemical differences between the two faces of the membrane. Three components of the plasma membrane are unevenly distributed:

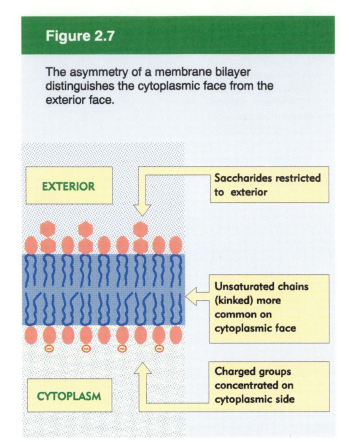

Figure 2.7

The asymmetry of a membrane bilayer distinguishes the cytoplasmic face from the exterior face.

EXTERIOR

Saccharides restricted to exterior

Unsaturated chains (kinked) more common on cytoplasmic face

Charged groups concentrated on cytoplasmic side

CYTOPLASM

◆ Different lipids are concentrated in the cytoplasmic and extracellular monolayers. This affects both the polar heads and the hydrophobic tails. Lipids on the cytoplasmic face are more highly charged and tend to be unsaturated. How is the difference between the monolayers established? Lipids are synthesized within the cell, and initially inserted into the cytoplasmic surface of the membrane. A specific protein, a 'flippase', may be responsible for transporting a lipid from one monolayer to the other. The existence of specific flippases for different lipids could be responsible for creating some of the asymmetries in lipid distribution between the bilayers.

◆ Proteins are oriented so that different sequences (or even entire proteins) are present on each face. The location of a protein is determined by its sequence, which contains signals that cause it to be inserted in the membrane in a particular orientation. This is discussed in detail in Chapter 11.

◆ Carbohydrate groups (on glycolipids or glyco-proteins) are found exclusively on the extra-cellular face. The consequence of this organiza-tion for the plasma membrane is that the exterior of the cell has a surface rich in oligosaccharides.

In some cases, the plasma membrane is extended by the presence of additional glyco-proteins, connected to those actually included in the membrane. This type of arrangement may form a cell coat. In some cases, the cell coat is extended into an **extracellular matrix,** rich in glycoproteins, and providing a thicker layer at the cell surface.

The cytoplasm contains networks of membranes

Membranes occupy a major part of the eukaryotic cell. In addition to surrounding individual organ-elles, large sheets of membranes are a prominent feature of many cells (especially those involved in secreting proteins). The electron micrograph in **Figure 2.8** shows an extensive sheet of membranes that extends from the nucleus. This is the **endo-plasmic reticulum (ER),** a highly convoluted sheet of membranes representing 30–60% of total mem-brane. Its overall structure is hard to define, but probably comprises a single surface enclosing the considerable volume between the membrane folds. This interior space comprises the **lumen.**

Visualized in the electron micrograph, the endo-plasmic reticulum can be divided into two types: **rough ER** and **smooth ER.** They are part of the same membrane sheet. The characteristic appear-ance of the rough ER results from the presence of **ribosomes** on its cytoplasmic surface. The ribo-somes are small particles concerned with the synthesis of proteins; their presence is an indication that proteins are being synthesized at the cytoplas-mic surface of the ER, which then processes them for assignment to various cell compartments. This is discussed in more detail in Chapter 11.

Between the ER and the plasma membrane lies

Figure 2.8

The endoplasmic reticulum consists of a highly folded sheet of membranes that extends from the nucleus. The small objects attached to the outer surface of the membranes are ribosomes. Photograph kindly provided by Lelio Orcl.

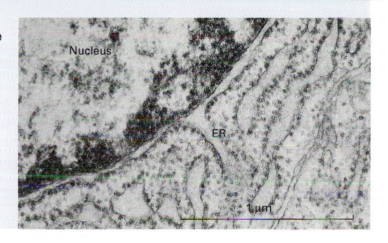

Figure 2.9

The Golgi apparatus consists of a series of individual membrane stacks. Photograph kindly provided by Alain Rambourg.

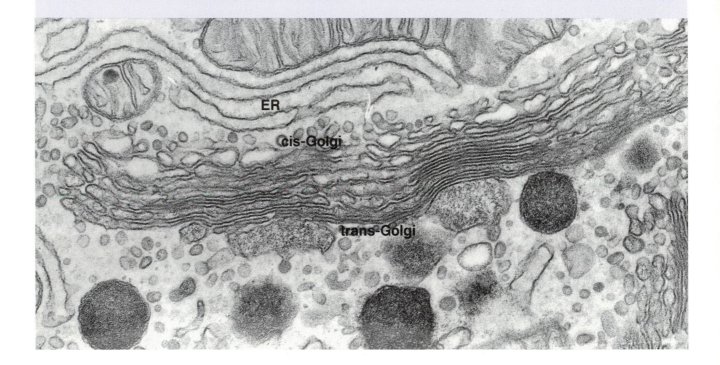

ER

cis-Golgi

trans-Golgi

the **Golgi apparatus**. The electron micrograph in **Figure 2.9** shows the Golgi as a stack of individual membrane sheets, quite tightly packed together.

The relationship between the ER and Golgi is depicted diagrammatically in **Figure 2.10.** (Note that the orientation of this figure is reversed from the preceding electron micrographs, and shows the nucleus at the bottom with the plasma membrane at the top.) The ER is a sheet made from the folding of a single lipid bilayer that extends from the outer membrane surrounding the nucleus. The Golgi consists of a 'stack' of flat **cisternae**, like a pile of discs. Each cisterna consists of a closed structure bounded by a single continuous membrane. A stack usually consists of <10 cisternae, but the number of

stacks varies considerably among different types of cell. The cisternae appear to be separate structures, each contained by its own membrane.

The Golgi apparatus is polarized. In secretory cells, the **cis face** is associated with the ER. The **trans face** lies toward the plasma membrane. Proteins that have been modified in the ER enter the Golgi at the *cis* face, pass through further modifications while they are transported to the *trans* face, and then exit.

A major function of the ER and Golgi apparatus is to sort proteins according to destination, using signals inherent in the protein sequence. Some of the modifications that occur in the ER and Golgi provide further signals for sorting. The process of directing proteins to their final destination is called **protein sorting** or **trafficking** and is the subject of Chapter 11.

Most lipids are synthesized in the ER. Other membranes presumably gain their lipid components by transport from the ER, although little is known about this process of **lipid trafficking**. The lipid transport system must be responsible for the differences in lipid composition between different cellular membranes.

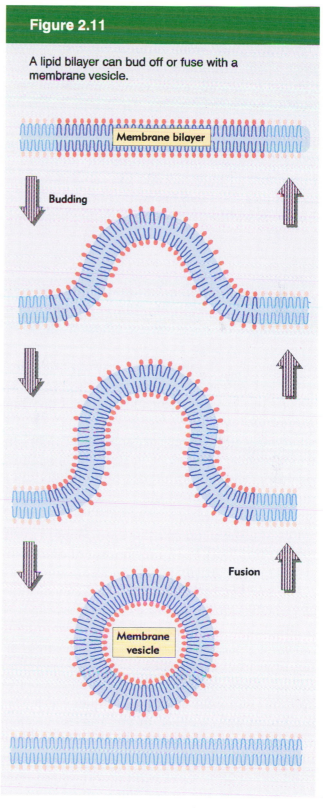

Figure 2.11

A lipid bilayer can bud off or fuse with a membrane vesicle.

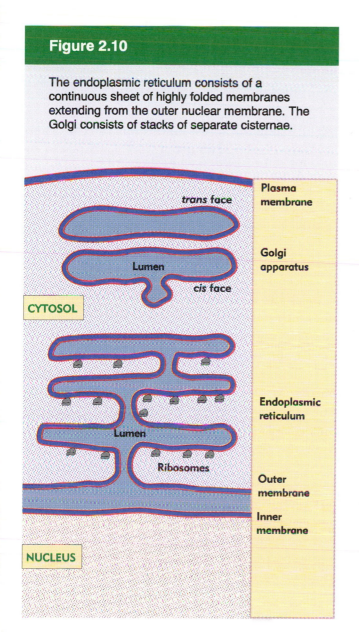

Figure 2.10

The endoplasmic reticulum consists of a continuous sheet of highly folded membranes extending from the outer nuclear membrane. The Golgi consists of stacks of separate cisternae.

The fluid structure of membranes is connected with an important property. As illustrated in **Figure 2.11,** membrane surfaces can fuse together into a single membrane; or a single membrane may give rise to multiple surfaces by a process of 'budding' to generate **vesicles.** Originating as small membrane-enclosed bodies that bud off from a membrane, vesicles are involved in transporting proteins from one location to another by virtue of their ability to move around the cell and through membranes. They may fuse with a target membrane in a reversal of the process by which they originated.

Lysosomes are rather small, spherical membrane-enclosed bodies that contain hydrolytic enzymes. They are formed by budding these bags of enzymes from the Golgi apparatus. Lysosomes are heterogeneous; different vesicles contain different enzymes. Their properties depend on the particular hydrolytic activities of the particular enzymes.

An organelle of rather similar size is the **peroxisome,** another membrane-enclosed body, which contains the enzyme catalase. Together with other enzymes, it is responsible for a series of oxidizing reactions.

Cell shape is determined by the cytoskeleton

The 'typical' cell, eukaryotic or prokaryotic, is a myth. Certain features are characteristic of all cells, such as the existence of the plasma membrane and (almost always) the presence of organelles in eukaryotic cells, but cells are specialized to fit particular niches, whether ecological for unicellular organisms, or comprising particular locations in a multicellular organism. As a result, they have a variety of structures and functions. We can therefore catalog the components that are found in a cell, but the overall structure must be described individually for each particular type of cell.

The set of features that together account for the shape of the cell, internal and external, is called the **cytoskeleton.** A striking feature of cell structure is that many components of the cytoskeleton are common to a variety of cell types, but they are organized into quite different patterns to comprise distinctive cell structures in the different cell phenotypes. This of course brings us back to the crucial question: how are the interactions of these components determined so that upon division a cell generates daughters of the same structure; or (perhaps more difficult) how do these interactions change when a cell generates a daughter or daughters that are different from itself?

The cytoskeleton contains networks of protein fibers, extending across and around the cell. There are three classes of fibers:

◆ **actin filaments**

◆ **microtubules**

◆ **intermediate filaments**.

In addition to the proteins that comprise the fibers themselves, other 'accessory' proteins connect fibers to one another, to the plasma membrane, or to other cellular structures.

Actin filaments are related to the thin filaments of muscle, which contain actin and other proteins. The form of actin that polymerizes into filaments in the cytoskeleton is closely related to, but different from, the actin protein contained in muscle. The actin filament itself consists of two chains, each containing a series of actin subunits twisted around one another in helical form. Accessory proteins are associated with these chains in a regular manner.

Actin filaments form bundles. Two forms are of particular interest. The **stress fibers** illustrated in **Figure 2.12** are evident in cultured cells, and are connected with the ability to spread out on a surface; this property is lost by cells converted to a cancerous state (see Chapter 39). And in all dividing cells, a **contractile ring** forms at the end of cell division to separate the daughter cells (see later).

A major function of actin filaments is to exert force, either to move a cell on a surface, or to change shape internally. The ability of actin to generate force resides in its interaction with myosin. The contractile reaction has been characterized in detail for the example of muscle; we assume that it rests on the same general principles for cytoskeletal actin filaments.

Generation of force requires the filament to be

Figure 2.12

Actin bundles consist of intertwined actin filaments, which may be linked at their ends to the plasma membrane. Photograph kindly provided by Elias Lazarides.

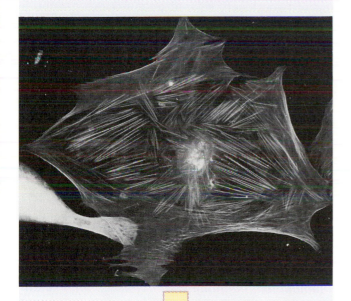

An actin filament consists of a helix of two strands

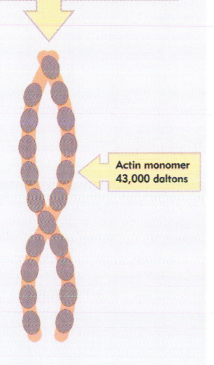

Actin monomer 43,000 daltons

Figure 2.13

Microtubules form an extensive network through the cytoplasm, stretching out to the edges of the cell. Photograph kindly provided by Frank Solomon.

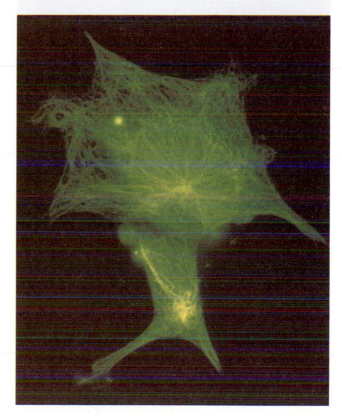

Figure 2.14

A microtubule consists of a hollow cylinder made from 13 protofilaments, each formed from tubulin dimers.

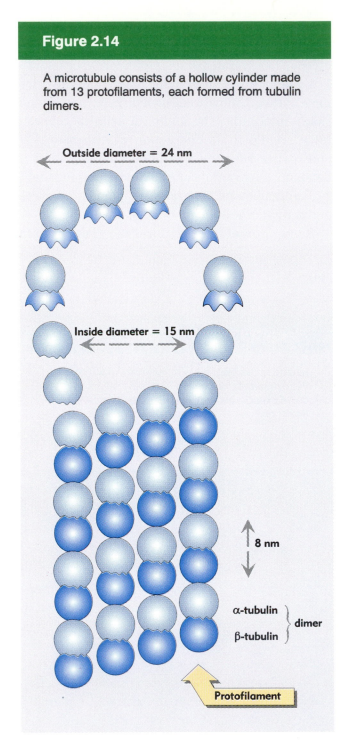

Outside diameter = 24 nm

Inside diameter = 15 nm

8 nm

α-tubulin
β-tubulin
} dimer

Protofilament

through another protein or proteins, to the end of an actin filament.

Microtubules are a ubiquitous component of eukaryotic cells. Seen as a network of fibers, they are organized as shown in **Figure 2.13**, criss-crossing the cytoplasm.

The subunit of microtubules is a dimer consisting of two closely related polypeptides, α-tubulin and β-tubulin. The dimers form protofilaments, and 13 protofilaments are organized to form a hollow cylinder, as illustrated in **Figure 2.14**.

Microtubules grow by the addition of tubulin dimers to their ends. Sites from which microtubules extend are called **microtubule organizing centers (MTOCs)**. Microtubules are dynamic structures: the assembled tubule is in equilibrium with a pool of tubulin subunits. There is continuous flux of tubulin into and out of the assembled form. Accessory proteins called **microtubule associated proteins (MAPs)** influence the state of the equilibrium. The dynamic equilibrium is an important aspect of microtubule function: it allows the tubules to disassemble and reassemble in response to cellular conditions.

Intermediate filaments (IFs) include five different types of filament. Each is found in a particular type of cell. Usually a particular type of cell contains only one type of IF. **Table 2.3** classifies the IFs.

Although the IF subunits are diverse, their structure is united by a common feature: each has a central region >300 amino acids long that forms a rod based on α-helical organization. The N-terminal and C-terminal domains on either side of this rod vary with the particular type of filament.

Many of the intermediate filaments have a structural role, for example, they provide rigidity for the cell shape. With the exception of the keratins, we know relatively little about their organization and function.

We assume that the overall organization of the three types of network is coordinated in any given cell, indeed, that they are connected, although the details of their relationship are obscure. The filaments may also be connected to organelles such as mitochondria.

anchored to a surface; usually this is the plasma membrane. How is an actin filament connected to the membrane? We assume that some integral protein of the membrane is connected, probably

Table 2.3

Intermediate filaments fall into five general classes.

Cell Type	Filament Type/Protein	Subunit Types per Filament	Variety
Epithelial (skin)	Keratin	2	Many types each
Myogenic (muscle)	Desmin	1	Single protein
Mesenchymal (cultured)	Vimentin	1	Single protein
Astroglial (brain)	GFAP	1	Single protein
Neuronal	Neurofilament	3	Many types each

The organization of these structures brings to a molecular level the issue of how macromolecular assemblies are put together. Would it be possible to assemble networks with a physiological pattern of connections from purified components *in vitro?* Or may they be assembled only by using the pre-existing network as a template that is extended by addition of new subunits?

Some organelles are surrounded by an envelope

The plasma membrane surrounding the cell and the internal membranes of the ER and Golgi apparatus each consist of a lipid bilayer, the basic construct of all membranes. A more complex organization surrounds the nucleus, mitochondrion and chloroplast (in plant cells). These organelles are bounded by an **envelope**, which consists of two, concentric membranes, as illustrated in **Figure 2.15.** Bacteria also are circumscribed by envelopes, connected with a more rigid structure; in the example of *E. coli*, rigidity is conferred by a proteoglycan wall between the two membranes.

Each membrane of an organelle envelope consists of the usual lipid bilayer, but the protein contents of the inner and outer layers are different. They are described as the **inner membrane** and **outer membrane**; the space between them is the **intermembrane space.** The two membranes are connected at **adhesion sites.** The compartment within the inner membrane comprises the lumen.

Both small molecules and large molecules are transported across envelopes. An important feature of an envelope is that each membrane layer contains proteins or receptors that transport particular molecules across it in the appropriate direction. The envelope is therefore asymmetric with regard to both structure and function.

Proteins are synthesized in the cytosol, but they may reside specifically in the outer membrane, the intermembrane space, the inner membrane, or the lumen of an organelle. Different sequences within a protein will determine whether it proceeds all the way through the series of membranes or stops at an appropriate site. The adhesion site could be

Figure 2.15

An envelope consists of adjacent membrane layers. Each membrane layer consists of a lipid bilayer with resident proteins. The layers may be connected across the intermembrane space at regions that are called adhesion sites, but whose structure is not well defined.

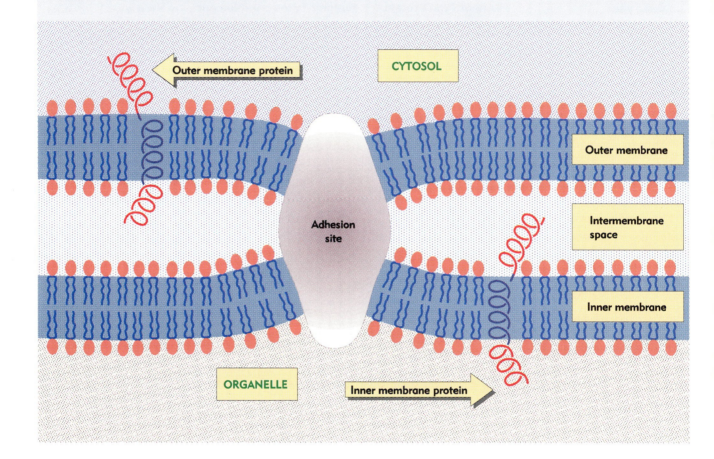

connected with the transfer of proteins from the outer to the inner membrane layer.

Both inner and outer membranes contain specific **pumps** or **channels** that are responsible for transport through the membrane. A small molecule may be transported via one pump into the intermembrane space, and then by another pump into the lumen. Some small molecules are concentrated in the intermembrane space. The aqueous constitutions of the cytosol, inter-

membrane space, and lumen are therefore distinct.

The construction of an envelope from two membrane bilayers allows one of the bilayers to extend as a continuous sheet within or beyond the organelle. Such extensions occur within the mitochondrion and outside the nucleus.

The mitochondrial outer membrane surrounds the organelle, but the inner membrane has a highly convoluted structure, folding within the compartment to generate a series of **cristae**, as illustrated in **Figure 2.16**. The large surface area of the inner membrane is an important feature of mitochondrial function.

The (eukaryotic) nucleus is surrounded by an envelope consisting of a double membrane. The space between the inner and outer membranes is called the **perinuclear space**; the interior of the nucleus is called the **nucleoplasm**. The general features of the nuclear envelope are illustrated in **Figure 2.17**.

The outer membrane faces the cytoplasm, and is connected to the ER, as illustrated previously in Figure 2.10. The inner membrane contains the contents of the nucleus.

Nuclear pores connect the two membranes, and provide sites where proteins are transported into the nucleus. The inner and outer membranes fuse at a pore. The pore itself is a protein complex of $\sim 10^8$ daltons that is connected to the membrane. The pore is a symmetrical structure consisting of two annuli, one facing the cytosol and the other the nucleoplasm. We discuss its structure in Chapter 12.

Immediately within the nucleus, in contact with the inner membrane of the envelope, is the **nuclear lamina**. It is a dense network of proteinaceous fibers, connected to proteins that reside in the lipid bilayer of the inner membrane. The lamina has a major role in organizing the structure of the nucleus. The outer layer, in contact with the inner membrane of the envelope, helps to hold nuclear pores in place. The inner layer provides a surface to which the genetic material is attached. The continuity of the lamina is interrupted at nuclear pores, where the nucleoplasmic annulus is connected to the nuclear lamina.

Figure 2.16

The mitochondrial envelope has a semi-permeable outer layer; the inner layer is highly folded within the mitochondrion to create a multitude of cristae. Photograph kindly provided by Lan Bo Chen.

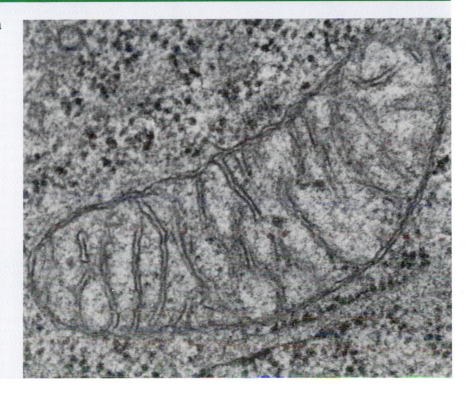

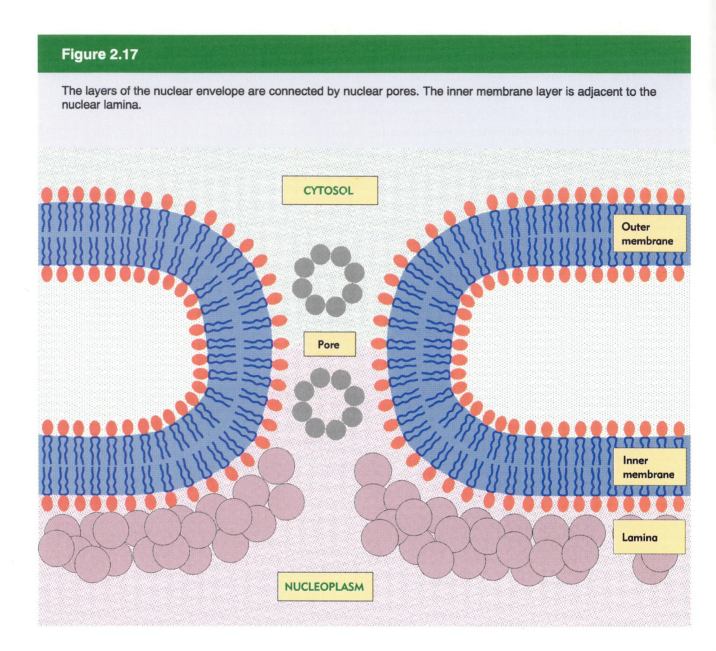

Figure 2.17

The layers of the nuclear envelope are connected by nuclear pores. The inner membrane layer is adjacent to the nuclear lamina.

CYTOSOL

Outer membrane

Pore

Inner membrane

Lamina

NUCLEOPLASM

The environment of the nucleus and its reorganization

The most important feature of the nucleus is the genetic material. It was identified in early cytology as a granular region, called **chromatin**, that can be recognized by its reaction with certain stains. Between cell divisions, chromatin forms a single dense mass. But when a cell divides, its chromatin can be seen to consist of a discrete number of thread-like particles, the **chromosomes**.

The nucleus exists as a discrete organelle between cell divisions. The nuclear environment is cramped. The mass of chromatin occupies a major part of the nuclear volume, connected at (presumably) discrete points to the nuclear lamina. We assume that these connections are important for genetic structure and function, but we know little about them.

A feature of all cells except those that have reached a final, specialized state of development, is their ability to divide. Many structural changes in the cell occur when it divides by **mitosis**. There is extensive reorganization of membranous components and the cytoskeleton. The former organization of the cell is replaced by a new structure, the **spindle**, whose function is to allow chromosomes to be distributed to the daughter cells produced by the division. The result of these changes is to bring many of the former activities of the cell—including gene expression, protein synthesis and secretion, and cell motility—to a halt. All its attention now is devoted to the act of division.

Proceeding away from the nucleus, the types of change are:

◆ Chromatin becomes much more densely packed and individual chromosomes become visible; they are attached to microtubules of the spindle.

◆ The nuclear envelope breaks down. The membrane becomes converted into small vesicles, while the lamina dissociates into its protein subunits. (This is true of higher eukaryotes; in lower eukaryotes, membranes may remain intact throughout division, which occurs by a 'closed' mitosis.)

◆ Membranes of the ER and Golgi apparatus break down into small vesicles, causing protein movement to halt.

◆ The cytoskeleton dissociates. Microtubules dissociate into tubulin dimers, which are reassembled into the spindle. Actin filaments are reorganized and a contractile ring forms at the end of mitosis.

The purpose of these changes is to divide the components of the parent cell equally between the two daughter cells. The mechanism used for distribution depends on the nature of each organelle. Organelles that are present in multiple

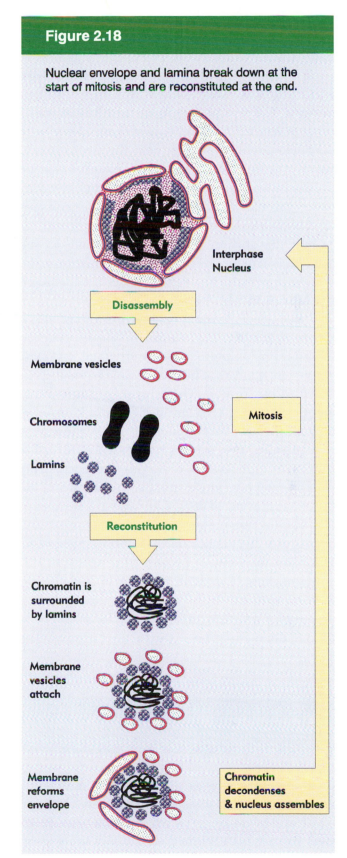

Figure 2.18

Nuclear envelope and lamina break down at the start of mitosis and are reconstituted at the end.

Interphase Nucleus

Disassembly

Membrane vesicles

Chromosomes

Lamins

Mitosis

Reconstitution

Chromatin is surrounded by lamins

Membrane vesicles attach

Membrane reforms envelope

Chromatin decondenses & nucleus assembles

copies, such as mitochondria, are segregated by dint of their distribution in the cytoplasm: approximately half of the cytoplasm is gained by each daughter cell. The single nucleus dissolves, and two nuclei are reconstituted from its components. The vesicularization of the ER allows parts of what had been a single membrane sheet to be distributed to the daughter cells for reconstitution into a coherent sheet. The genetic material must be precisely duplicated and then segregated, one copy of the parental genome to each daughter cell (see next section).

The dissolution and reconstruction of the nucleus is a major feature of the perpetuation of the genetic material. The overall events are illustrated in **Figure 2.18**.

The nuclear lamina consists of three proteins, lamins A, B, and C. Each lamin is phosphorylated at mitosis; this solubilizes the individual proteins. This event appears to be responsible for the breakdown of the lamina. The trigger for lamina dissolution is the activation of a protein kinase, an enzyme that specifically adds the phosphate groups to the lamins. The process is reversed at the end of mitosis by a phosphatase enzyme that specifically removes the phosphate groups. Breakdown of the lamina can occur *in vitro* while the membranes of the nuclear envelope remain intact, so lamina breakdown and membrane dissolution appear to be independent events.

Membrane dissolution appears to rely on production of some factor needed in stoichiometric amounts. A possible role for the factor would be to bind to the membrane to cause it to form vesicles.

The process could be reversed by removing or inactivating the factor at the end of mitosis.

Chromosome condensation occurs independently of lamina breakdown and membrane dissolution. It involves a variety of events, including both the production of factors needed in stoichiometric amounts and enzymes that modify chromosomal proteins. Although temporally coordinated with the other events, it does not appear to be mechanistically connected to them.

These various events are triggered by a common factor, M phase kinase, an enzyme that phosphorylates target proteins at mitosis, including the laminins and chromosomal proteins. The regulation of cell division is the subject of Chapter 13.

Reconstitution of the nuclear envelope is a gradual process. First, fragments of the nuclear membrane associate with individual chromosomes, partially enclosing each chromosome. Then they fuse to give a membranous shell surrounding the entire set of chromosomes. The lamins are dephosphorylated, attach to chromatin following the condensation stage, and reassociate to form the intact lamina.

The envelope must reconstitute not merely the integrity of the boundary around the nucleus, but also the ancillary structures of nuclear pores and the connections to the ER. The manner in which the nuclear pores disaggregate, and how they are reconstructed, is unknown. Since the pre-existing structure can be dissolved and reconstituted, we assume that the information for overall organization resides within the structures of the individual components.

The role of chromosomes in heredity

Chromosomes can be visualized in most cells only during the process of cell division. The two types of division in sexually reproducing organisms explain both the perpetuation of the genetic material within an organism and the process of inheritance as predicted by Mendel's laws.

Starting as a fertilized egg, an organism develops through many cell divisions, each division generating two cells from one. The **cell cycle** comprises the period between the release of a cell as one of the progeny of a division and its own subsequent division into two daughter cells.

The cell cycle falls into two parts:

◆ A relatively long **interphase** represents the time during which there is no visible change, but the (single) cell engages in its synthetic activities and reproduces its components. During interphase, the cell has a discrete nuclear compartment, containing a compact mass of chromatin.

◆ The relatively short period of **mitosis** provides an interlude during which the actual process of visible division into two daughter cells is accomplished. During mitosis, the internal organization of the cell is replaced by the spindle. The nuclear compartment has been lost and individual chromosomes are apparent.

The products of the series of mitotic divisions that generate the entire organism are called the **somatic cells**. During embryonic development, many or most of the somatic cells will be proceeding through the cell cycle. In the adult organism, however, many cells are **terminally differentiated** and no longer divide; they remain in a stationary phase in which chromatin remains permanently in an interphase state.

At the end of a mitosis, each daughter cell can be seen to start its life with two copies of each chromosome. These copies are called **homologues**. The *total number* of chromosomes is called the **diploid set** and has **2n** members. The typical somatic cell exists in the diploid ($2n$) state (except when it is preparing for or is actually engaged in mitosis).

The chromosomal complement of any cell, as visualized at mitosis, is called its **karyotype**. The term is used also to describe the chromosomal constitution of a species: for example, the human (diploid) karyotype has 46 chromosomes.

During interphase, a growing cell duplicates its chromosomal material. This action is not evident at the time and becomes apparent only at the beginning of the subsequent mitosis, when each chromosome appears to have split longitudinally to generate two copies. These copies are called **sister chromatids**. Since each of the original homologues has been duplicated to form sister chromatids, the cell now contains $4n$ chromosomes, organized as $2n$ pairs of sister chromatids.

The use of the term 'chromosome' is ambiguous. In traditional cytological nomenclature, 'chromosome' refers to the entity that possesses the **centromere**, a region that is responsible for its movement at mitosis. The centromere is often visible as a constricted region in the chromosome. Its structure and function is discussed in Chapter 27.

When the chromosomal material doubles to form sister chromatids, each pair of sister chromatids has only a single centromere. During mitosis, the centromere is functionally duplicated, and at this point each sister chromatid is regarded as a chromosome in its own right. Yet we should not allow this traditional focus on the centromere to cloud the main point, which is that *in terms of genetic information the cell has four copies of each chromosome when it enters mitosis*. We shall regard division from this perspective.

The spindle is the major morphological feature of a mitotic cell. It contains a series of fibers running from the poles of the cell. The spindle fibers consist of microtubules, reorganized from the tubulin comprising the microtubules of the interphase cell. The fibers emanate from regions at each pole called the **centrosomes**, which contain the microtubule organizing centers (MTOCs). 'Centrosome' is an operational definition, corresponding to the structures equated with MTOCs; it does not necessarily take the same structural form in every type of cell.

Some fibers extend between the poles of the cell; others connect the polar regions to the chromosomes. The point of connection on each chromosome is a dense granular body called the **kinetochore**, which lies in the constricted region defined by the centromere. The microtubules comprising the fibers running across the spindle or terminating on the chromosomes can change their length by the addition or removal of tubulin subunits from their ends. Elongation is involved in the establishment of the spindle; shortening is the mechanism by which chromosomes are moved during mitosis.

Figure 2.19

The centriole consists of nine microtubule triplets, apparent in cross-section as the wall of a hollow cylinder. Photograph kindly provided by A. Ross.

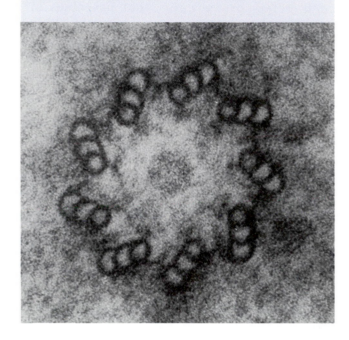

The structure of centrosomes is not well defined, but in animal cells a centrosome contains a pair of **centrioles**, surrounded by a dense amorphous region. The centriole is a small hollow cylinder whose wall is comprised of a series of triplet fused microtubules. A centriole is shown in cross-section in **Figure 2.19**.

The function of the centriole in mitosis is not clear. Originally it was thought that it might provide the structure to which microtubules are anchored at the pole, but the fibers seem instead to terminate in the amorphous region around the centrioles. It is possible that the centriole is concerned with orienting the spindle; it may also have a role in establishing directionality for cell movement. However, there are cell types in which centrosomes do not appear to contain centrioles.

Centrioles have their own cycle of duplication. When born at mitosis, a cell inherits two centrioles. During interphase they reproduce, so that at the start of mitosis there are four centrioles, two at each pole. Probably only the parental centriole is functional.

The centriole cycle is illustrated in **Figure 2.20**. Soon after mitosis, a **procentriole** is elaborated perpendicular to the parental centriole. It has the same structure as the mature parental centriole, but is only about half its length. Later during interphase, it is extended to full length. It plays no role in the next mitosis, but becomes a parental centriole when it is distributed to one of the daughter cells. The orientation of the parental centriole at the mitotic pole is responsible for establishing the direction of the spindle.

How are centrioles reproduced? The precise elaboration of the procentriole adjacent to the parental centriole suggests that some sort of template function is involved. The parental centriole cannot itself be seen to reproduce or divide, but it could provide some nucleating structure onto which tubulin dimers assemble to extend the procentriole. Could a centriole be assembled in the absence of a pre-existing centriole?

Figure 2.21 illustrates the series of events by which mitotic division is accomplished. It is usually divided into several periods:

◆ During **prophase** the individual pairs of chromosomes become apparent.

◆ Chromosomes move toward the equator during **prometaphase.**

◆ The chromosomes are at their most compact state during **metaphase,** when they become aligned at the equator of the cell.

◆ During **anaphase** chromosomes are pulled to

Figure 2.20

A centriole reproduces by forming a procentriole on a perpendicular axis; the procentriole is subsequently extended into a mature centriole. Photograph kindly provided by J. B. Rattner and S. G. Phillips.

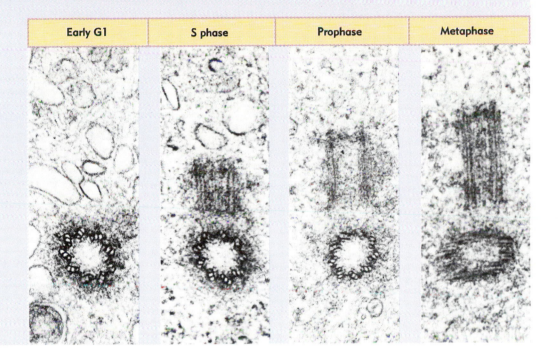

Early G1	S phase	Prophase	Metaphase

the poles, a process called anaphase movement. The cause of the movement is a shortening of the fibers connecting them to the poles.

◆ The chromosomes reach the poles at **telophase**, and then decondense into chromatin while the nucleus is reformed.

◆ In the final phase of **cytokinesis**, the two daughter cells separate and are pinched apart by the formation of a contractile ring consisting of actin filaments.

The essence of mitosis therefore is that the sister chromatids are pulled toward opposite poles of the cell, so that each daughter cell receives one member of each sister chromatid pair (now an individual chromosome). The $4n$ chromosomes present at the start of division have been divided into two sets of $2n$ chromosomes. This process is repeated in the next cell cycle. Thus mitotic division ensures the constancy of the chromosomal complement in the somatic cells.

A different purpose is served by the production of germ cells (eggs or spermatozoa), which is accomplished by another type of division. **Meiosis** generates cells that contain the **haploid** chromosome number, n. Once again, the chromosomes have been duplicated before the process of division begins. The cell enters meiosis with a complement of $4n$ chromosomes, which are then divided by two successive divisions into four sets.

Figure 2.22 illustrates the two divisions of meiosis. When the first division begins, the homologous

Figure 2.21

Mitosis perpetuates the chomosome constitution of the cell. A single pair of homologous chromosomes is shown, but a eukaryotic cell has many such pairs.

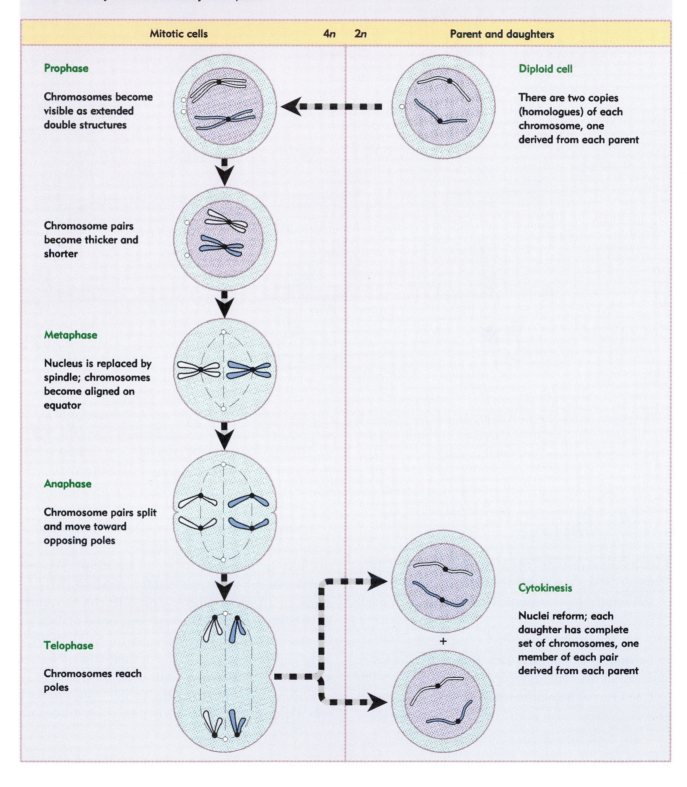

Mitotic cells		4n	2n	Parent and daughters

Prophase

Chromosomes become visible as extended double structures

Diploid cell

There are two copies (homologues) of each chromosome, one derived from each parent

Chromosome pairs become thicker and shorter

Metaphase

Nucleus is replaced by spindle; chromosomes become aligned on equator

Anaphase

Chromosome pairs split and move toward opposing poles

Cytokinesis

Nuclei reform; each daughter has complete set of chromosomes, one member of each pair derived from each parent

Telophase

Chromosomes reach poles

Figure 2.22

Meiosis halves the chromosome number.

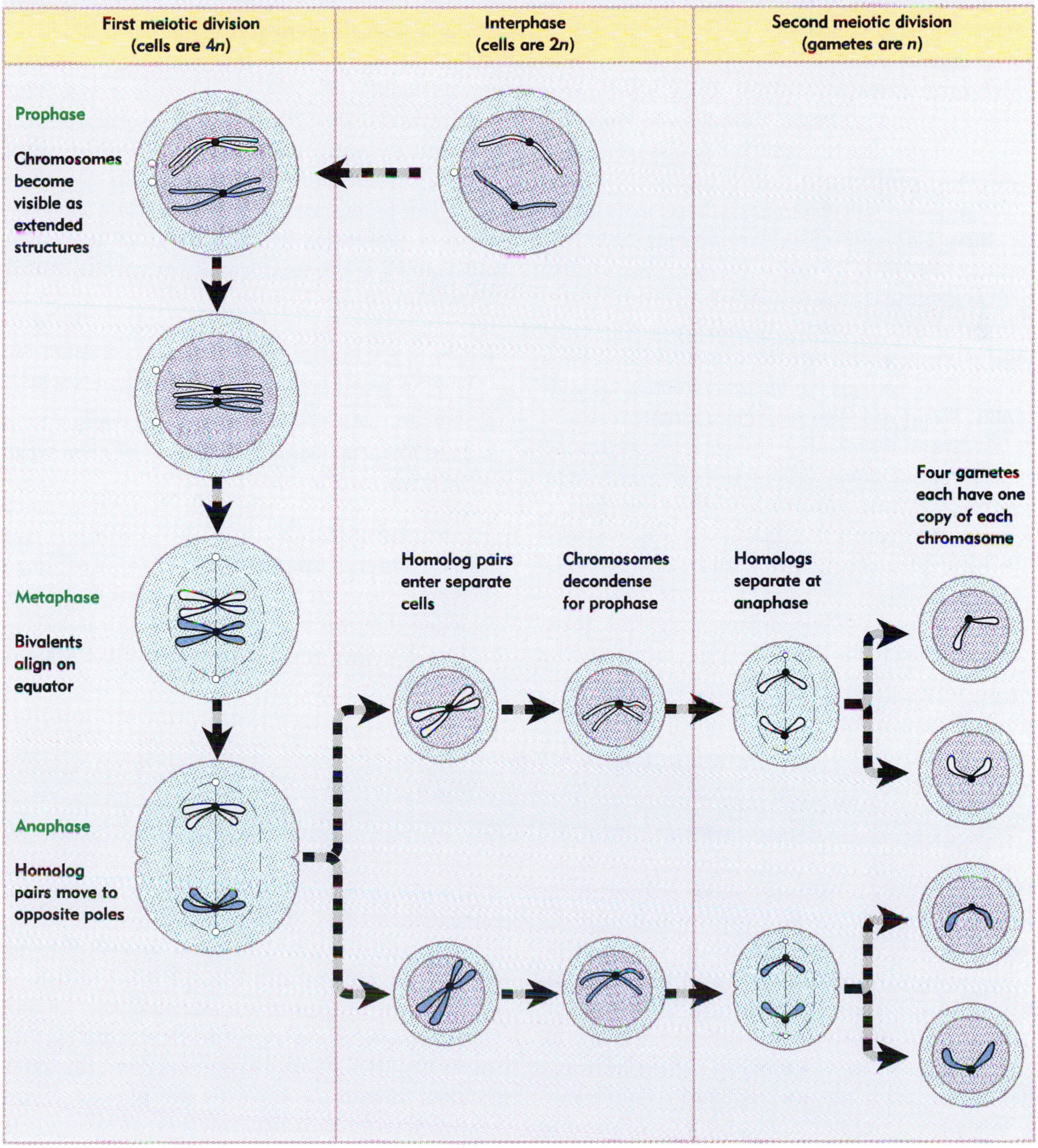

First meiotic division (cells are 4n)	Interphase (cells are 2n)	Second meiotic division (gametes are n)

Prophase

Chromosomes become visible as extended structures

Metaphase

Bivalents align on equator

Anaphase

Homolog pairs move to opposite poles

Homolog pairs enter separate cells

Chromosomes decondense for prophase

Homologs separate at anaphase

Four gametes each have one copy of each chromosome

pairs of sister chromatids **synapse** or **pair** to form **bivalents.** *Each bivalent contains all four of the cell's copies of one chromosome.* The first division causes each bivalent to segregate into its constituent sister chromatid pairs. This generates two sets of $2n$ chromosomes, each set consisting of n sister chromatid pairs.

Now the second meiotic division follows, in which both of the sets of $2n$ divide again. This division is formally like a mitotic division, since one member of each sister chromatid pair segregates to a different daughter cell.

The overall result of meiosis is therefore to divide the starting number of $4n$ chromosomes into four haploid (n) cells. These cells then give rise to mature **gametes** (eggs or sperm).

In forming the gametes, homologues of paternal and maternal origin are separated, so that each gamete gains only *one of the two homologues* of its parent. A critical feature in relating this process to the predictions of Mendel's laws was the realization that *nonhomologous chromosomes segregate independently*, so that either member of one pair of homologues enters the gamete at random with either member of a different pair of homologues (see Chapter 3).

In many sexually reproducing organisms, there is an exception to the rule that chromosomes occur in homologous pairs whose separation at meiosis produces identical haploid sets. Male and female sets differ visibly in chromosome constitution; the most common form of difference is that one particular pair of homologues takes different forms in each sex. In one sex it is present as a pair of identical chromosomes, like any other, but in the other sex one member of this pair is replaced by a different chromosome.

This pair is called the **sex chromosomes,** and the remaining homologous pairs are called **autosomes.** The chromosome complements of the two sexes can be described as 2A + XX and 2A + XY, where A represents the haploid set of autosomes and X and Y are the individual sex chromosomes. When gametes are formed:

◆ The sex with the homogametic complement 2A + XX forms gametes only of the type A + X.

◆ The sex with the heterogametic complement 2A + XY forms equal proportions of gametes of the types A + X and A + Y.

The union of gametes from one sex with gametes from the other sex perpetuates the equal sex ratio at zygote formation. This maintains the basic principle that the diploid organism gives rise to haploid gametes, which unite to reform a diploid.

Summary

A prokaryotic cell consists of a single compartment, whereas a eukaryotic cell has several discrete compartments, or organelles, all bounded by membranes. Membranes consist of lipid bilayers that contain proteins. The interior of the membrane provides a hydrophobic environment that contrasts with the aqueous environment on either side. Some membrane proteins are receptors that transduce signals from one side of the membrane to the other, most notably from the exterior to the interior of the cell.

The plasma membrane circumscribing the cell consists of a single lipid bilayer. In a eukaryotic cell, the nucleus, mitochondrion, and chloroplast each is bounded by an envelope that consists of two concentric membrane bilayers. The ER and Golgi apparatus consist respectively of folds and stacks of a lipid bilayer extending from the outer nuclear membrane towards the plasma membrane. The overall architecture of the cell is determined by its cytoskeleton, which consists of networks of protein fibers, including actin filaments, microtubules, and IFs.

Eukaryotic genetic material is organized into discrete chromosomes, which are visible only during cell division. During the interphase of the cell cycle they are confined to the nucleus in the form of a tangled mass of chromatin. The karyotype is the chromosomal complement of a cell; a diploid organism has two copies of each chromosome in its karyotype. Somatic (diploid) cells produce copies of themselves by mitotic division; chromosomes are reproduced prior to the division, so that one copy of each duplicate can segregate to each daughter cell. Germ cells are produced by meiosis, when two successive divisions segregate and reduce the chromosome number so that there is only one copy of each (diploid) pair in the oocyte or sperm.

Further reading

Reviews

An extensive review of membrane structure and biosynthesis is to be found in **Gennis**, *Biomembranes* (Springer, New York, 1989). The problems of lipid trafficking have been considered by **van Meer** (*Ann. Rev. Cell Biol.* **5**, 247–275, 1989).

Organization of the nuclear envelope has been reviewed by **Gerace and Burke** (*Ann. Rev. Cell Biol.* **4**, 336–376, 1988).

Reorganization of microtubules and their role in mitosis has been reviewed by **Mitchison** (*Ann. Rev. Cell Biol.* **4**, 527–550, 1988).

Chromosome segregation has been reviewed by **Murray** (*Ann. Rev. Cell Biol.* **1**, 289–316, 1985).

The mechanics of mitosis have been considered by **MacIntosh and Koonce** (*Science* **246**, 622–628, 1989).

Discoveries

The notion of the membrane as a two dimensional fluid was introduced by **Singer and Nicolson** (*Science* **175**, 720–731, 1972).

PART 1

DNA as a store of information

When alcohol reaches a concentration of about 9/10 volume there separates out a fibrous substance which on stirring the mixture wraps itself about the glass rod like thread on a spool and the other impurities stay behind as a granular precipitate. The fibrous material is redissolved and the process repeated several times. In short, this substance is highly reactive and on elementary analysis conforms *very* closely to the theoretical values of pure DNA (who could have guessed it).

Oswald Avery, 1944

CHAPTER 3

Genes are mutable units

Inherited traits are defined by their ability to be passed from one generation to the next in a predictable manner. Before we consider the nature of the unit of inheritance, it is important to realize the distinction between the appearance of the organism (what we observe) and the underlying genetic constitution (which we must infer). Visible or otherwise measurable properties are called the **phenotype**, while the genetic factors responsible for creating the phenotype are called the **genotype**.

As seen by their effects on the phenotype, inherited traits vary widely in complexity. Some appear in principle to be relatively limited—for example, humans may have either brown or blue eyes. Some are apparently more complex; for example, the inheritance of the shape of a nose. Through the concept of the gene, genetics finds common ground for the inheritance of traits ranging from the ability to perform a simple metabolic reaction to the construction of a complex shape.

The gene is the unit of inheritance. *Each gene is a nucleic acid sequence that carries the information representing a particular polypeptide.* A gene is a stable entity, but can suffer a change in sequence. Such a change is called a **mutation**. When a mutation occurs, the new form of the gene is inherited in a stable manner, just like the previous form.

The organism carrying the altered gene is called a **mutant**; an organism carrying the normal (unaltered) gene is called **wild type**. 'Wild type' is used to describe either the genotype or phenotype. We distinguish between genotype and phenotype by *italicizing* terms that refer to the genotype; the same term in roman characters refers to the phenotype caused by the gene.

The phenotypic results of mutations vary from the undetectable (no defect appears to result from the absence of a protein) to lethal (the organism cannot survive without the protein). However, since *most detectable mutations damage the function of a protein*, the study of mutations is biased toward situations in which a gene fails to function properly.

As revealed by the effects of mutations, some phenotypic traits are determined by single genes, while others are determined by several genes. So some features are altered only when a specific gene is mutated, while others can be affected by mutation in any one of several genes.

Complex changes in the phenotype can result from single mutations. A classic example of a deleterious mutation that exerts its effect by interfering with a metabolic pathway is phenylketonuria, an inherited disease of man. The disease results from the absence of the enzyme that converts the amino acid phenylalanine into another amino acid, tyrosine. The failure results in an accumulation of phenylalanine, which is then converted into the toxic compound phenylpyruvic acid. Among the resulting defects is mental retardation.

The basis for many complex phenotypic traits is not yet known (for example, how the shape of a nose is determined), but we assume that the interactions of proteins represented by as yet unidentified genes will explain the phenotypic differences between individuals.

We usually regard genetics from the perspective of the organism, but the alternation of generations illustrated in **Figure 3.1** recognizes the sense in

Figure 3.1

When eukaryotes perpetuate their genes through an alternation of diploid and haploid states, one state provides the predominant type while the other is represented in the gametes. The mammals and mosses are extreme examples in which the conspicuous generations are diploid and haploid, respectively.

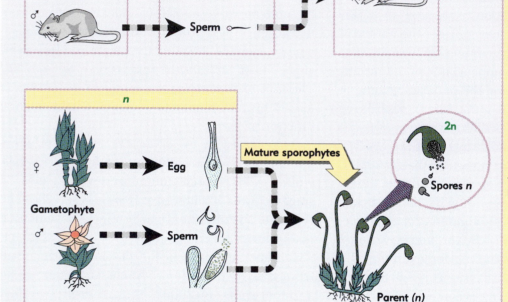

In mammals the haploid stage is only a transient intermediate between diploid parents and progeny

In mosses the diploid stage is only a transient intermediate concerned with passage between haploid generations

which the organism is the gene's way of expressing and perpetuating itself.

Consider the life cycle of organisms that pass through a **diploid** stage, during which there are two copies of each gene. One of the two copies is passed from the parent to a **gamete** (germ cell: egg or sperm). The gamete therefore contains one copy of every gene of the organism; this is called the **haploid** set. The alternative types of gamete produced by parents of different sexes unite to form a **zygote** (fertilized egg). The zygote gains a haploid set of genes from each parent, restoring the situation in which every diploid organism has one set of genes of paternal origin and one of maternal origin.

Note, however, that lower eukaryotes may exist predominantly in the haploid stage, using the diploid stage only as a transient intermediate. In mosses, for example, the plant is essentially haploid, and it generates gametes whose fusion gives a diploid organ (the sporophyte); this in turn can be used to generate haploid spores that give rise to new plants.

Bacteria and some lower eukaryotes exist exclusively in haploid form, perpetuating themselves by simple cell division (mitosis), without producing gametes (by meiosis) for the gymnastics of sexual mating.

The concept that genotype determines phenotype applies to viruses as well as to living organisms. Viruses take the physical form of

Figure 3.2

Viral inheritance proceeds through a cycle in which a virus infects a host cell to produce progeny virus like itself. Photograph kindly provided by Lee Simon.

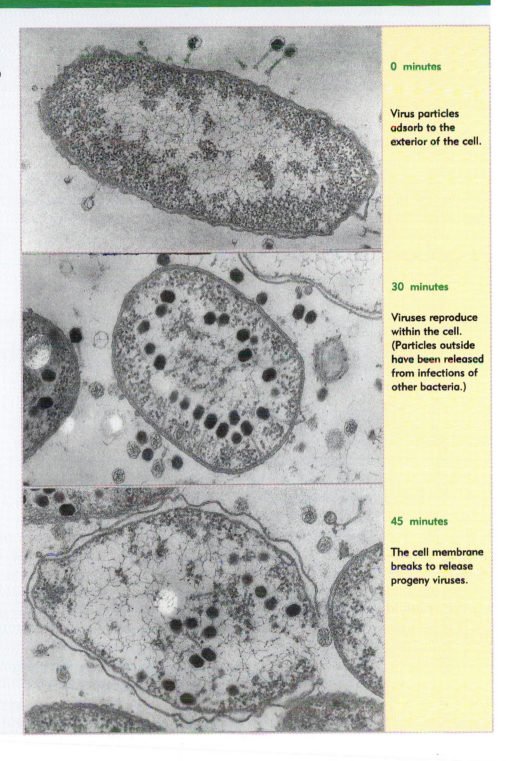

0 minutes

Virus particles adsorb to the exterior of the cell.

30 minutes

Viruses reproduce within the cell. (Particles outside have been released from infections of other bacteria.)

45 minutes

The cell membrane breaks to release progeny viruses.

exceedingly small particles. They share with organisms the property that one generation gives rise to the next; they differ in lacking a cellular structure of their own, instead needing to infect a **host cell**.

Both prokaryotic and eukaryotic cells are subject to viral infections; viruses that infect bacteria are usually called **bacteriophages**, abbreviated to **phages**. An example of a phage infection is shown in **Figure 3.2**. Viruses vary in complexity and in their effects upon an infected cell, but have the general property that the result of infection is the production of progeny virus particles like those that originally infected the host cell.

The phenotype of a virus is represented not just by the structure of the virus particle itself, but also by its effect upon the infected cell. Changes in the genotype of the virus may alter either the particle itself or the phenotype developed by the infected cell. In the same way as cells or organisms, virus strains are characterized as wild type or mutant, their type is inherited by their progeny, and one type may be stably converted into the other by mutation. The inheritance of viruses therefore shows the same hallmarks that characterize heredity in living organisms.

Viral inheritance is carried by genes; and the nature and behavior of viral genes is the same as that of cellular genes. The cell propagates a rather complex set of genes in an autonomous manner; a virus propagates a smaller set of genes nonautonomously.

Discovery of the gene

The essential attributes of the gene were defined by Mendel more than a century ago. As he concluded in his analysis of pea genetics in 1865: 'The law of combination of different characters, which governs the development of the hybrids, finds therefore its explanation in the principle enunciated, that the hybrids produce egg cells and pollen cells, which in equal numbers, represent all constant forms which result from combinations of the characters brought together in fertilization.' Summarized in his two laws, the gene was recognized as a 'particulate factor' that passes unchanged from parent to progeny.

A gene may exist in alternative forms that determine the expression of some particular characteristic. For example, the color of a flower may be red or white. The forms of the gene are called **alleles**. Mendel's first law describes the **segregation of alleles**: *alleles have no permanent effect on one another when present in the same plant, but segregate unchanged by passing into different gametes.*

When an organism has two identical alleles of a gene, it is said to be **homozygous** (or true-breeding) for the trait conveyed by that gene. If the alleles are different, the organism is **heterozygous** (or hybrid). The phenotype of a homozygote directly reflects the genotype of the allele, but the phenotype of a heterozygote depends on the relationship between the types of alleles that are present.

Mendel's first law recognizes that the genotype of a heterozygote includes both alleles, irrespective of the phenotype that is displayed. When a homozygote for one allele is crossed with a homozygote for another allele, all the progeny in the first (F1) generation are heterozygotes with the same phenotype. But when the heterozygotes are crossed with one another to generate a second (F2) generation, the genotypes of the original parents reappear. The critical point is that the alleles must consist of discrete physical entities that contribute independently (or fail to contribute) to the phenotype.

Figure 3.3 illustrates the principle of the experiment that revealed this situation, and shows how the observed results differ according to the type of relationship between alleles.

We may distinguish three classes of relationship between alleles in a heterozygote:

Figure 3.3

Alleles segregate each generation.

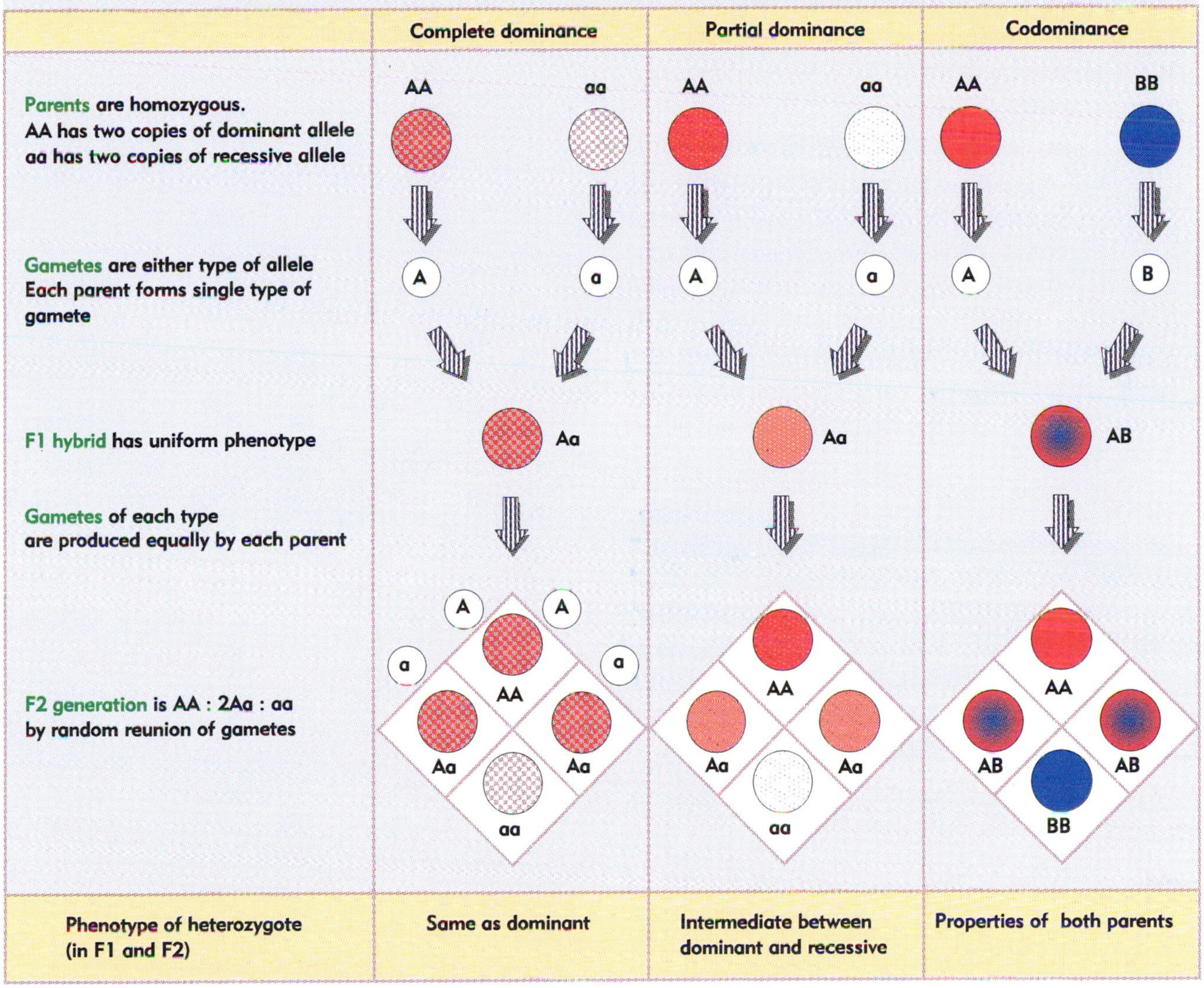

	Complete dominance	Partial dominance	Codominance
Parents are homozygous. AA has two copies of dominant allele aa has two copies of recessive allele	AA aa	AA aa	AA BB
Gametes are either type of allele Each parent forms single type of gamete	A a	A a	A B
F1 hybrid has uniform phenotype	Aa	Aa	AB
Gametes of each type are produced equally by each parent			
F2 generation is AA : 2Aa : aa by random reunion of gametes	A A a a AA Aa Aa aa	AA Aa Aa aa	AA AB AB BB
Phenotype of heterozygote (in F1 and F2)	Same as dominant	Intermediate between dominant and recessive	Properties of both parents

◆ The case analyzed by Mendel corresponds to **complete dominance,** and is shown in the first column of results. When one allele is **dominant** and the other is **recessive,** the phenotype of a heterozygote is determined by the dominant allele. The recessive allele makes no contribution. The single dominant allele produces the same phenotype that is seen in a wild-type homozygote. The appearance of the heterozygote is indistinguishable from that of the homozygous dominant parent. The F1 resembles the dominant parent. Complete dominance generates the classic 3:1 ratio of dominant to recessive phenotypes in the F2.

◆ Some alleles exhibit **incomplete (or partial) dominance,** as shown in the middle column. The phenotype of the heterozygote is intermediate between that of the two homozygotes. In the snapdragon, for example, a cross between red and white generates heterozygotes with pink flowers. However, the same rule is observed that the first hybrid (F1) generation is uniform in phenotype; and the same ratios are generated in the second (F2) generation, except that three phenotypes can be distinguished instead of two (*AA* is red, 2 *Aa* are pink, *aa* is white). This type of situation arises through quantitative effects; the single red allele in the heterozygote produces half as much pigment as the two red alleles in a homozygote.

◆ Alleles are said to be **codominant** when they contribute equally to the phenotype, as shown in the final column. In human blood groups, the *AA* and *BB* combinations are homozygous, and *AB* is a codominant heterozygote in which the *A* and *B* groups are equally expressed (see Figure 3.14). The result is that each genotypic class produces a different phenotype. The F1 has the properties of both parents, and the F2 has the phenotypes A, 2 AB, B.

Mendel's second law summarizes the **independent assortment of different genes**. When a homozygote that is dominant for *two different characters* is crossed with a homozygote that is recessive for both characters, as before the F1 consists of plants whose phenotype is the same as the dominant parent. But in the next (F2) generation, two general classes of progeny are found:

◆ One class consists of the two **parental types**.

◆ The other class consists of *new* phenotypes, representing plants with the dominant feature of one parent and the recessive feature of the other. These are called **recombinant types**; and they occur in both possible (**reciprocal**) combinations.

Figure 3.4 shows that the ratios of the four phenotypes comprising the F2 can be explained by supposing that gamete formation involves an entirely random association between one of the two alleles for the first character and one of the two alleles for the second character. All four possible types of gamete are formed in equal proportion; and then they associate at random to form the zygotes of the next generation. Once again, the phenotypes conceal a greater variety of genotypes.

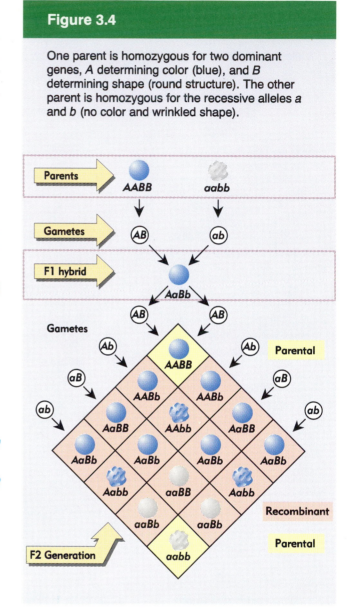

Figure 3.4

One parent is homozygous for two dominant genes, *A* determining color (blue), and *B* determining shape (round structure). The other parent is homozygous for the recessive alleles *a* and *b* (no color and wrinkled shape).

The law of independent assortment establishes the principle that the behavior of any pair (or greater number) of genes can be predicted overall by the rules of mathematical combination. *The assortment of one gene does not influence the assortment of another*. Implicit in this concept is the view that assortment is a matter of *statistical probability* and not an exact result. The ratio of progeny types will approximate increasingly closely to the predicted proportions as the number of crosses is increased.

Appreciation of Mendel's discoveries was

inhibited by the lack of any known physical basis for the postulated factors (genes). When the chromosomal theory of inheritance was subsequently proposed, however, it was realized that the behavior of chromosomes at meiosis and fertilization correspond precisely with the properties of Mendel's particulate units of inheritance.

To summarize the parallels between chromosomes and Mendel's units of inheritance:

◆ Genes occur in allelic pairs. One member of each pair is contributed by each parent; so the diploid

set of chromosomes results from the contribution of a haploid set by each parent.

◆ The assortment of nonallelic genes into gametes is independent of (parental) origin; correspondingly, nonhomologous chromosomes undergo independent segregation (see Figure 2.22).

The critical proviso is that each gamete obtains a complete haploid set, and this condition is fulfilled whether viewed in terms of Mendel's factors or chromosomes.

Genes lie in a linear array on chromosomes

The correlation between the behavior of genes and chromosomes poses the question: can particular genes be identified with particular chromosomes? Proof that a specific gene always lies on a certain chromosome was provided by a mutant of the fruit fly *Drosophila melanogaster* obtained by Morgan in 1910. This white-eyed male appeared spontaneously in a line of flies with the usual (wild-type) red eye color.

The *white* mutation could be located on a particular chromosome because of its association with sexual type. We noted in Chapter 2 that the sex chromosomes provide the sole exception to the rule that a diploid organism has 2 copies of every chromosome. An animal has 2 copies of every autosome, and 2 sex chromosomes, which are the same (XX) in one sex, and different (XY) in the other sex.

In *Drosophila,* males are XY and females are XX. As a result, a recessive allele located on the X chromosome shows a characteristic pattern of behavior called **sex linkage**. Because the Y chromosome does not contain any allele, the phenotype of a male will be determined by the single allele present on the X chromosome, which was derived from the mother. As a result, a recessive mutation is transmitted from a mother to all of her sons and to none of her daughters. The coincidence of the inher-

itance patterns of the *white* mutation and the X chromosome thus provided the first instance in which an individual gene was localized on a particular chromosome.

The independent assortment of chromosomes at meiosis explains the independent assortment of genes that are carried on different chromosomes. But the number of genetic factors is much greater than the number of chromosomes. If all genes lie on chromosomes, many genes must be present on each chromosome. What is the relationship between these genes?

The behavior of two genes carried on the same chromosome may deviate from the predictions of Mendel's second law. Instead of generating the proportions depicted in Figure 3.4, the proportion of parental genotypes in the F2 is greater than expected, *because there is a reduction in the formation of recombinant genotypes.* The propensity of some characters to remain associated instead of assorting independently is called **linkage**.

Figure 3.5 shows how a **backcross** to a recessive homozygote is used to measure linkage. The alleles of the recessive parent make no contribution to the phenotype of the progeny. As a result, the backcross essentially makes it possible to examine directly the genotype of the organism being investigated.

66 ▪ CHAPTER THREE

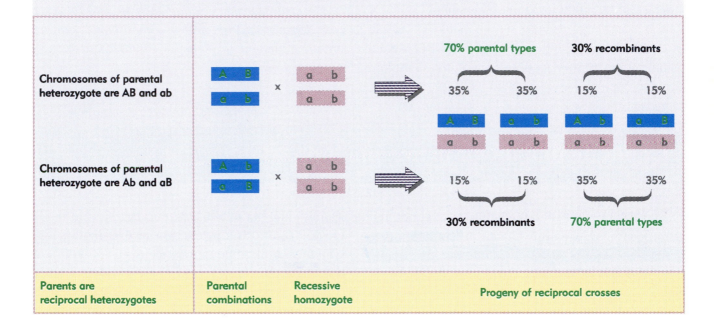

Figure 3.5

Linkage can be measured by a backcross with a double homozygote. The genotype of an unknown gamete is revealed directly by the result of its combination with a gamete that brings in only the recessive alleles. In each cross the progeny show an increase in the proportion of parental types (70%) and a decrease in the proportion of recombinant types (30%), compared with the 50% of each type that is expected from independent assortment. The linkage between *A* and *B* is measured as 30%.

The smaller the proportion of recombinants in the progeny, the tighter the linkage. A crucial characteristic is that the *same proportion of recombinants is obtained with each of the reciprocal crosses;* and in each case, both of the recombinant types are present in the same amount.

Morgan proposed that genetic linkage is the 'simple mechanical result of the location of the (genes) in the chromosomes'. He suggested that the production of recombinant classes can be equated with the process of **crossing-over** that is visible during meiosis (see Figure 2.22). Early in meiosis, at the stage when all four copies of each chromosome are organized in a bivalent, pairwise exchanges of material occur between the closely associated (synapsed) chromatids.

The visible result of a crossing-over event is called a **chiasma**, and is illustrated diagrammatically in **Figure 3.6**. A chiasma represents a site at which two of the chromatids in a bivalent have been broken at corresponding points. The broken ends have been rejoined crosswise, generating new chromatids. Each new chromatid consists of material derived from one chromatid on one side of the junction point, with material from the other chromatid on the opposite side. The two recombinant chromatids have reciprocal structures. The event is described as a **breakage and reunion**.

Proof that crossing-over is responsible for recombination requires a correlation between chromosome structure and genotype. This is made possible by the existence of **translocations, in which part of one chromosome has broken off and become attached to another.** The translocation chromosome can therefore be distinguished by its appearance from the normal chromosome. In suitable crosses, as indicated in **Figure 3.7**, it is possible to show that the formation of genetic recombinants occurs only when there has been a physical crossing-over between the appropriate chromosomal

Figure 3.6

Chiasma formation is responsible for generating recombinants.

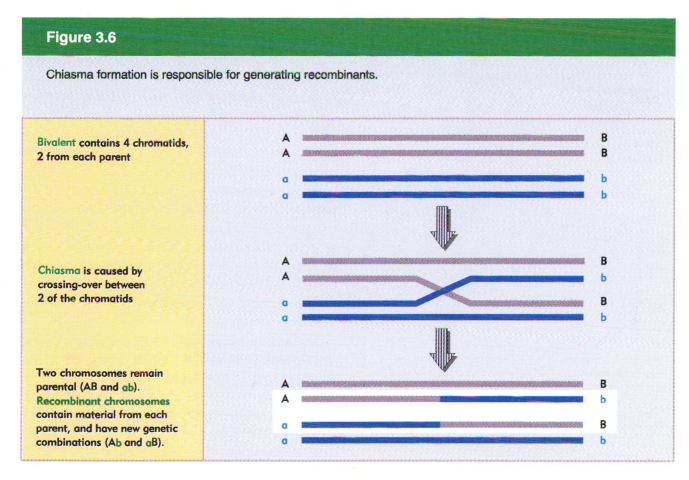

Bivalent contains 4 chromatids, 2 from each parent

Chiasma is caused by crossing-over between 2 of the chromatids

Two chromosomes remain parental (AB and ab). **Recombinant chromosomes** contain material from each parent, and have new genetic combinations (Ab and aB).

regions. We investigate the mechanisms responsible for genetic recombination in Chapter 33.

If the likelihood that a chiasma will form between two points on a chromosome depends on their distance apart, genes located near each other will tend to remain together. As the distance decreases, the probability of crossing-over between them will decrease. If crossing-over is responsible for recombination, the closer genes lie to one another, the more tightly they will be linked. Reversing the argument, genetic linkage can be taken to be a measure of physical distance.

The extent of recombination between two genes on the same chromosome can be used as a **map distance** to measure their relative locations. The formula to measure genetic distance is:

$$\text{Map distance} = \frac{\text{Number of recombinants} \times 100}{\text{Total number of progeny}}$$

Map units are defined as 1 unit (or centi-Morgan) equals 1% crossover. For short distances (<10%), map units are given directly by the percent recombinants. However, when two crossovers occur near one another, they restore the parental arrangement of the loci on either side. This reduces the number of recombinants, so that recombination frequency underestimates the map distance.

A critical feature is observed when multiple characters are followed together. For genes carried on the same chromosome, *individual map distances are (approximately) additive.* If two genes *A* and *B* are 10 units apart, and gene *C* lies a further 5 units beyond *B*, the direct measure of distance between *A* and *C* will be close to 15 units. The genes can therefore be placed in a **linear order.**

A crucial concept in the construction of a genetic map is that the distance between genes does not depend on the particular *alleles* that are used, but only on the genetic **loci**. *The locus defines the*

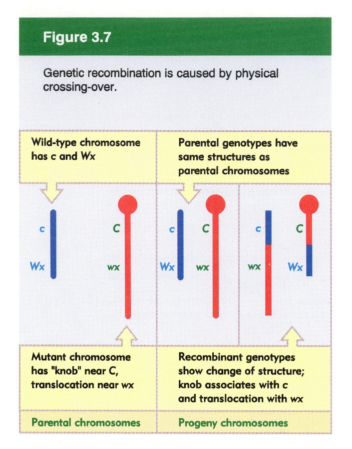

Figure 3.7

Genetic recombination is caused by physical crossing-over.

Wild-type chromosome has c and Wx	Parental genotypes have same structures as parental chromosomes	
Mutant chromosome has "knob" near C, translocation near wx	Recombinant genotypes show change of structure; knob associates with c and translocation with wx	
Parental chromosomes	**Progeny chromosomes**	

position occupied on the chromosome by the gene representing a particular trait. The various alternative forms of a gene—that is, the alleles used in mapping—all reside at the same location on its particular chromosome.

So genetic mapping is concerned with identifying the positions of genetic loci, which are fixed and lie in a linear order. In a mapping experiment, the same result is obtained irrespective of the particular combination of alleles (see Figure 3.5). We take up the principles of genetic mapping in Chapter 6; and we consider some of the aspects of the overall genetic map of an organism in Chapter 27. For the moment, we conclude that the genetic map represents (approximately) the physical length of the genetic material.

Linkage is not displayed between all pairs of genes located on a single chromosome. The maximum recombination between two loci is the 50% corresponding to the independent segregation predicted by Mendel's second law. (Although there is a high probability that recombination will occur between two genes lying far apart on a chromosome, each individual recombination event involves only two of the four associated chromatids, so there is a limit of 50% recombination between the genes.)

In spite of their presence on the same chromosome, genes that are far apart therefore assort independently. But although they show no direct linkage, each can be linked to genes that lie between them, and so a genetic map can be extended beyond the limit of 50% recombination directly measurable between any pair of genes. A genetic map is usually based on measurements involving genes that are fairly close together, and is subject to corrections from the simple percent recombination.

A **linkage group** includes all those genes that can be connected either directly or indirectly by linkage relationships. Genes lying close together show direct linkage; those more than 50 map units apart assort independently. As linkage relationships are extended, the genes of any organism fall into a discrete number of linkage groups. Each gene identified in the organism can be placed into one of the linkage groups. Genes in one linkage group always show independent assortment with regard to genes located in other linkage groups.

The number of linkage groups is the same as the (haploid) number of chromosomes. The relative lengths of the linkage groups are similar to the actual relative sizes of the chromosomes. The example of *D. melanogaster* (where it happens to be particularly easy to measure the relative chromosome lengths) is illustrated in **Figure 3.8**. The basis for the measurement of length is the existence of 'giant' chromosomes in some tissues, where a pattern of bands can be seen, and the number of bands can be counted. We consider the structures of these chromosomes in Chapter 27.

Mendel's concept of the gene as a discrete particulate factor can therefore be extended into the concept that *the chromosome constitutes a linkage group, divided into many genes, whose physical arrangement underlies their genetic behavior.*

Figure 3.8

The linkage map can be related to the physical size of the chromosomes. Each chromosome in *D. melanogaster* can be visualized as a number of "bands". The relative number of bands reflects the physical length of each chromosome. The relationship between map units and bands is similar for each chromosome.

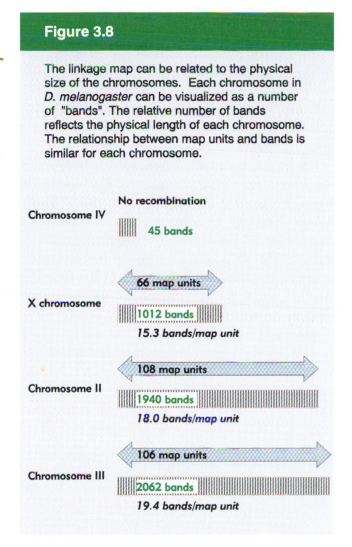

The resolution of the recombination map of a higher eukaryote is restricted by the small number of progeny that can be obtained from each mating. Recombination occurs so infrequently between nearby points that it is rarely observed between different mutations in the same gene. Does recombination occur within a gene; and can its frequency at these close quarters be used to arrange sites of mutation in a linear order?

To answer these questions by conventional genetic means requires a microbial system in which a very large number of progeny can be obtained from each genetic cross. A suitable system is provided by **phage T4**, a virus that infects the bacterium *E. coli*. Infection of a single bacterium leads to the production of ~100 progeny phages in <30 minutes (see the example of Figure 3.2).

The constitution of an individual locus was investigated by Benzer in a series of intensive studies of the *rII* genes of the phage, which are responsible for a change in the pattern of bacterial killing known as **rapid lysis**. When two different *rII* mutant phages are used to infect a bacterium simultaneously, the conditions can be arranged so that progeny phages will be produced *only* if recombination has occurred between the two mutations to generate a wild-type recombinant. The frequency of recombination depends on the distance between sites, just as in the eukaryotic chromosome.

The selective power of this technique in distinguishing recombinants of the desired type allows even the rarest recombination events to be quantitated, so that the map distance between *any* pair of mutations can be measured. About 2400 mutations fall into 304 different mutant sites. (When two mutations fail to recombine, they are assumed to represent independent and spontaneous occurrences at the *same* genetic site.)

The mutations can be arranged into a linear order, showing that *the gene itself has the same linear construction as the array of genes on a chromosome*. So the genetic map is linear within as well as between loci: it consists of an unbroken sequence within which the genes reside. This conclusion has of course now been extended in molecular terms to all known genetic systems.

We sometimes use the term **genetic marker** to describe a gene of interest, for example, one being used in a mapping experiment or identifying a particular region. Thus a chromosome may be said to carry a particular set of markers, that is, alleles.

On the genetic maps of higher organisms established during the first half of this century, the genes are arranged like beads on a string. They occur in a fixed order, and genetic recombination involves transfer of corresponding portions of the string between homologous chromosomes. The gene is to all intents and purposes a mysterious object (the bead), whose relationship to its surroundings (the string) is unclear.

One gene–one protein: the basic paradigm

Until 1945, the gene was considered to be the fundamental unit of inheritance, but there was no unifying explanation for its function. Genes could be identified only by mutations that produced some aberration in the phenotype. The main difficulty was the need to rely on those mutants that happened to be available, which were not necessarily suitable for biochemical studies. Many examples accumulated in which defects in particular biochemical reactions could be associated with specific mutations; but the physical nature of the gene and its relationship to biochemical defects remained unknown.

A systematic attempt to associate genes with enzymes was started by Beadle and Ephrussi in the 1930s, when they were able to conclude that the development of the normal red eye color in the fruit fly *Drosophila* passes through a discrete series of stages. Blockage at different stages results in the production of different mutant colors. But it was not until much later that the complete pathway could be worked out. Beadle and Tatum thus attempted to approach the problem from the other direction. As Beadle recollected, 'it suddenly occurred to me that it ought to be possible to reverse the procedure we had been following and instead of attempting to work out the chemistry of known genetic differences we should be able to select mutants in which known chemical reactions were blocked'.

Using the fungus *Neurospora*, mutants were generated (by irradiation with X-rays) and selected for their inability to grow on a medium that could support wild-type cells. The mutants fail to grow because they have lost the ability to produce some compound that wild-type cells can produce. The biochemical nature of the defect in each mutant strain could be identified by finding an additional compound whose addition to the medium allowed that mutant strain to grow. Each mutant proved to be blocked in a particular metabolic step, undertaken in the wild-type strain by a single

enzyme. Blockage at each step leads to accumulation of the metabolic intermediate immediately prior to the step.

Figure 3.9 illustrates the pathway subsequently elucidated for production of brown eye pigment in *D. melanogaster*. Tryptophan is converted into brown pigment by a series of reactions. The absence of brown pigment changes the color of the eye. Mutation at each stage of the metabolic pathway results in the absence of a particular enzyme, causing accumulation of the intermediate on which it acts. Eye color is influenced by the effect (or lack thereof) of this intermediate, so a variety of eye color phenotypes results from the various blockages.

Figure 3.9

Genes control metabolic steps, as shown by mutations affecting eye color of *D. melanogaster*, which act by blocking different stages in the pathway for converting tryptophan to brown pigment.

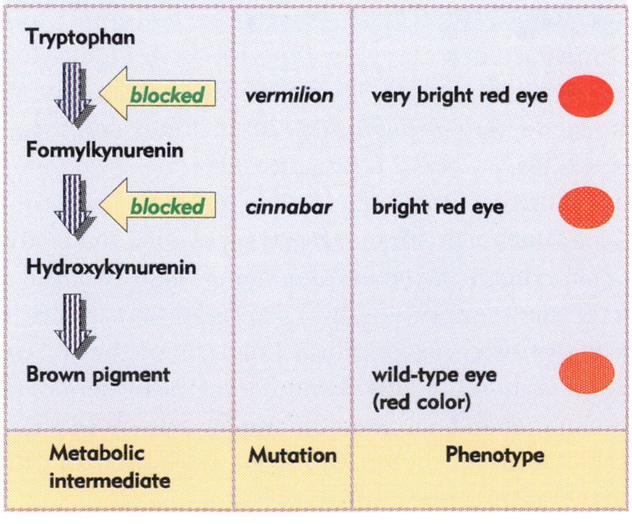

By 1945 the results of the analysis had become known in common parlance as the **one gene–one enzyme hypothesis.** This proposed that each metabolic step is catalyzed by a particular enzyme, whose production is the responsibility of a single gene. A mutation in the gene alters the activity of the protein for which it is responsible.

Identifying which protein represents a particular gene can be a protracted task. The mutation responsible for creating Mendel's wrinkled-pea mutant was identified only in 1990 as an alteration that inactivates the gene for a starch branching enzyme!

Since a mutation is a random event with regard to the structure of the gene, the greatest probability is that it will damage or even abolish gene function. This explains the nature of recessive mutations: they represent an absence of function, because the mutant gene has been prevented from producing its usual enzyme. As illustrated in **Figure 3.10**, however, in a heterozygote containing one wild-type and one mutant allele, the wild-type allele is able to direct production of the enzyme. The wild-type allele is therefore dominant. (In terms of Mendel's analysis, alleles pass through generations without affecting one another's hereditary properties, and also function (or not) independently of one another in each generation.)

(This assumes that an adequate *amount* of protein is made by the single wild-type allele. When this is not true, the smaller amount made by one allele as compared to two results in the intermediate phenotype of a partially dominant allele in a heterozygote.)

Direct proof that a gene actually is responsible for controlling the structure of a protein had to wait until 1957, when Ingram showed that the single-gene trait of sickle-cell anemia can be accounted for by a change in the amino acid composition of the protein hemoglobin.

A modification in the hypothesis is needed to accommodate proteins that consist of more than one subunit. If the subunits are all the same, the protein is a **homomultimer,** represented by a single gene. If the subunits are different, the protein is a **heteromultimer.**

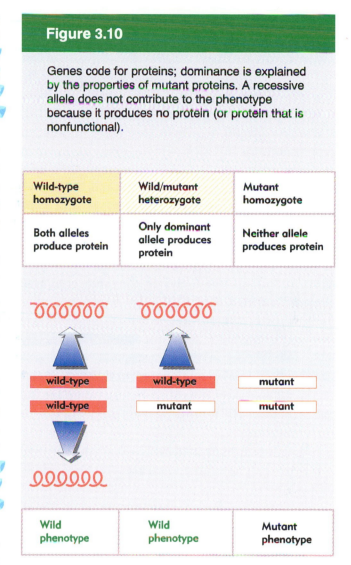

Figure 3.10

Genes code for proteins; dominance is explained by the properties of mutant proteins. A recessive allele does not contribute to the phenotype because it produces no protein (or protein that is nonfunctional).

Wild-type homozygote	Wild/mutant heterozygote	Mutant homozygote
Both alleles produce protein	Only dominant allele produces protein	Neither allele produces protein
wild-type	wild-type	mutant
wild-type	mutant	mutant
Wild phenotype	Wild phenotype	Mutant phenotype

Hemoglobin provides an example of a protein that consists of more than one type of polypeptide chain; a heme (iron-binding) group is associated with two α subunits and two β subunits. Each type of subunit comprises a different polypeptide chain and is represented by its own gene. The function of hemoglobin is inhibited by mutation in the genes coding for either the α or β polypeptides.

Stated as a more general rule applicable to any heteromultimeric protein, the one gene–one enzyme hypothesis becomes more precisely expressed as **one gene–one polypeptide chain.**

A modern definition: the cistron

If a recessive mutation is produced by every change in a gene that prevents the production of an active protein, there should be a large number of such mutations in any one gene; many amino acid replacements may change the structure of the protein sufficiently to impede its function. Different variants of the same gene are called **multiple alleles**, and their existence makes it possible to create a heterozygote between mutant alleles.

When two mutations have the same phenotypic effect and map close together, they may comprise alleles. However, they could also represent mutations in two *different* genes whose proteins are involved in the same function. The **complementation test** is used to determine whether two mutations lie in the same gene or in different genes. The test consists of making a heterozygote for the two mutations (by mating parents homozygous for each mutation).

If the mutations lie in the same gene, the parental genotypes can be represented as:

$$\frac{m_1}{m_1} \quad \text{and} \quad \frac{m_2}{m_2}$$

The first parent provides an m_1 mutant allele and the second parent provides an m_2 allele, so that the heterozygote has the constitution:

$$\frac{m_1}{m_2}$$

in which *no wild-type gene is present*, so the heterozygote has mutant phenotype.

If the mutations lie in different genes, the parental genotypes can be represented as:

$$\frac{m_1 +}{m_1 +} \quad \text{and} \quad \frac{+ m_2}{+ m_2}$$

where each chromosome has a wild-type copy of one gene (represented by the plus sign) and a mutant copy of the other. Then the heterozygote has the constitution:

$$\frac{m_1 +}{+ m_2}$$

in which the two parents between them have provided a wild-type copy of each gene. The heterozygote has wild phenotype; the two genes are said to **complement**.

Figure 3.11 provides a more elaborate description of the complementation test. If we consider just the individual sites of mutation (without regard to whether they lie in the same or in different genes), the double heterozygote may have either of two configurations. In the *cis* configuration, both mutations are present on the *same* chromosome. In the *trans* configuration, they are present on opposite chromosomes. The relative effects of these configurations are determined by whether the mutations lie in the same or in different genes.

First consider the situation shown in the upper part of the figure in which the mutations lie in the same gene:

◆ The *trans* configuration corresponds to the test we have just described. Both copies of the gene are mutant.

◆ In the *cis* configuration, however, one genome elaborates a protein that has two mutations, while the other has none and is therefore wild type.

Thus when two mutations lie in the same gene, the phenotype of a heterozygote is determined by the configuration. It is mutant when the mutations lie in *trans*, and must be wild type when they lie in *cis*. This comparison provides the basis for the *cis/trans*

Figure 3.11

The cistron is defined by the complementation test. Genes are represented by bars; red circles identify sites of mutation.

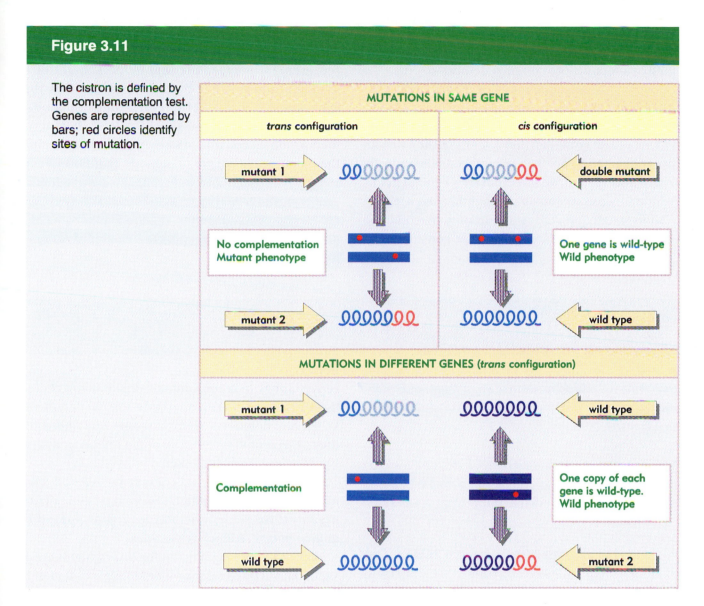

complementation test. The complementation of two mutations is tested in *trans*; and if they fail to complement, the *cis* configuration is used as a control to measure the presence of wild-type function.

By contrast, when the mutations lie in different genes, as shown in the lower part of the figure, the configuration is irrelevant. In either case there is one copy of each mutant gene and one copy of each wild-type gene.

Complementation is tested in practice by determining whether the *trans* heterozygote shows wild phenotype (the mutations lie in different genes) or mutant phenotype (the mutations lie in the same gene). For diploid organisms, the experiment is performed by constructing the appropriate double heterozygote. For viruses a host cell is simultaneously infected with the two mutant types.

When two mutations *fail* to complement in *trans*, the inference is that both affect the same function. They are therefore assigned to the same **complementation group.** We expect that a complementation group will correspond to a discrete genetic unit; this unit is formally called the **cistron.** Two mutations in the same cistron *cannot* complement in *trans*; the occurrence of complementation shows that the mutations lie in *different* cistrons. A 'cistron' is essentially the same as a 'gene'. We take up this question again in

Chapter 6, and in more detail in Chapter 31, when it becomes clear that in some situations we need to adopt a more complex definition.

If two genes complement in *trans*, each must be responsible for producing a protein that is able to function independently of the protein made by the other gene, so that together they create the wild phenotype. The products are said to be **trans-acting**; and we infer that they represent diffusible molecules able to act together irrespective of their origins in different genomes. The genetic consequences of this situation are considered in Chapter 15.

An exception to the rule that only different genes can complement is sometimes found when a gene represents a polypeptide that is the subunit of a homomultimeric protein. In the wild-type cell, the active protein consists of several *identical* subunits. In a cell containing two mutant alleles, however, their products can mix to form multimeric proteins that contain *both types* of subunit. Sometimes the two mutations compensate, so that the mixed-subunit protein is active, even though the proteins consisting solely of either type of mutant subunit are inactive. This effect is called **interallelic** complementation.

Mapping mutations at the molecular level

A mutation is any change in the sequence of DNA in a genome. We can divide mutations into two general classes:

◆ A **point mutation** is a change affecting a single position in a gene. This causes a corresponding change in the protein that the gene produces (see Chapter 4).

◆ A **rearrangement** affects a large region. The simplest types of rearrangements are **insertions** of additional material or **deletions** of a stretch of the gene.

Deletions have a critical use in genetic mapping. The genetic extent of a deletion is defined by its inability to recombine with a series of adjacent mutations (because it lacks the entire corresponding stretch of wild-type DNA). The deletion can recombine with point mutations on either side. **Figure 3.12** shows that a deletion is visualized genetically by its failure to recombine with a series of point mutations.

By obtaining a series of partially overlapping deletions, we can map any point mutation by testing its ability to recombine with them. **Figure 3.13** illustrates the protocol. When two deletions both fail to recombine with a point mutation, the site of mutation must lie in the region *common* to the deletions. When one deletion recombines and one does not, the site of mutation must lie in the region in which the deletions do *not overlap*.

In characterizing a genome, we should like ideally to be able to mutate every single gene of the organism. But the extent to which we can apply this approach is restricted in two ways.

First, we need a criterion for distinguishing the mutant from the wild type. The simplest criterion is visible change, for example, in eye color. Other criteria can be established by devising suitable selective procedures; for example, by adjusting the conditions of growth so that the absence or presence of some enzyme is required for survival.

The problem with this approach is that there may be functions of whose existence we are ignorant, and for which we therefore fail to devise suitable tests. Taken to an extreme, there is the possibility that genes exist whose products are unnecessary and whose absence therefore has no effect (at least under the conditions that we use). How are they to be detected? We take up this question in Chapter 24.

Our second difficulty concerns functions that are *essential* for viability. A mutation in such a gene is likely to kill the organism. To isolate mutants in

Figure 3.12

A deletion can recombine with a series of point mutations on either side of the deleted region, but cannot recombine with point mutations that lie within the deleted region. The boundaries of the deletion are indicated by the switch between ability and inability to recombine.

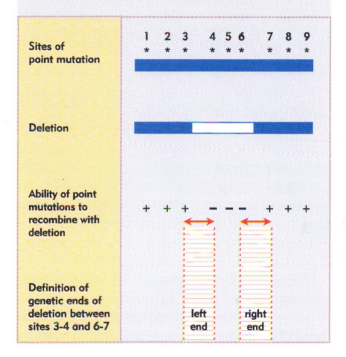

these genes, they must be obtained in the form of **conditional lethal** mutations. These mutations are lethal under one set of conditions, but exhibit no deficiency, or a reduced deficiency, under alternative conditions that allow the cell to be perpetuated.

Thus the same (mutant) organism can be studied under two conditions. In **permissive** conditions, it does not display the mutant phenotype and may therefore be perpetuated. In **nonpermissive** conditions, it dies or becomes severely ill, but it can be studied during the transition from permissive to nonpermissive conditions. (Of course, conditional mutations are not restricted to essential functions, but in principle can be found in any gene.) **Table 3.1** summarizes some type of conditional mutations and the systems in which they can be used.

A 'nonsense mutation' in a gene prevents the protein from being synthesized. Other mutations,

called suppressors, allow the nonsense mutation to be overcome so that the protein can be synthesized (see Chapter 8). This situation provides a conditional system that can be used for viruses. A virus carrying a nonsense mutation can be grown on a permissive host cell that has the suppressor mutation, but does not grow on a wild-type host cell.

Temperature is a common parameter used to provide permissive versus nonpermissive conditions. Usually a gene fails to function at high temperature, but functions normally at low temperature. More properly such mutations are called

Figure 3.13

Deletion mapping can be used to locate point mutations between the ends of overlapping deletions. If a point mutation cannot recombine with a deletion, the site of mutation must lie within the region that has been deleted. If a point mutation recombines with a deletion, the site of mutation must lie outside the deleted region. By comparing the ability of a series of deletions to recombine with a point mutation, the site of mutation can be identified.

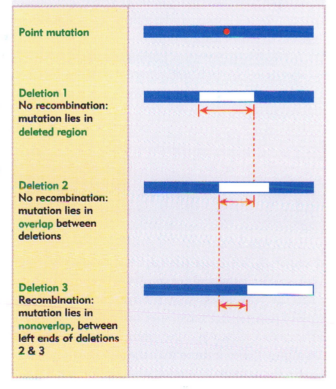

Table 3.1

Conditional mutations may affect protein synthesis or function.

Type of Mutation	Activity Affected	Permissive Condition	Nonpermissive Condition	System
Nonsense	Protein synthesis	Suppressor	No suppressor	Viral/host
Temperature-sensitive	Protein function	Normal temp.	High temp.	Virus or cell
Cold-sensitive	Protein function	Normal temp.	Low temp.	Virus or cell
D_2O-sensitive	Protein function	Growth on H_2O	Growth on D_2O	Virus or cell

heat-sensitive. This effect rests upon the susceptibility of the protein conformation to change to an inactive form as a function of temperature. Such mutations can occur in genes of any organism and so have general utility.

Temperature also can work in the opposite direction. Cold-sensitive mutants function normally at higher temperatures, but fail to function at a reduced temperature. This type of effect sometimes occurs in proteins that are incorporated into macromolecular structures in a temperature-dependent process.

When a gene has been identified, insight into its function in principle can be gained by generating a mutant organism that entirely lacks the gene. A mutation that completely eliminates gene function, usually because the gene has been deleted, is called a **null mutation**. If a gene is essential, a null mutation is of course lethal.

To determine whether and what effect a gene has upon the phenotype, it is essential to characterize a null mutant. When a mutation fails to affect the phenotype, it is always possible that this is because it is leaky—enough active product is made to fulfill its function, even though the activity is quantitatively reduced or qualitatively different from the wild type. But if a null mutant fails to affect a phenotype, we may safely conclude that the gene function is not necessary.

Null mutations, or other mutations that impede gene function (but do not necessarily abolish it entirely) are also called **loss-of-function** mutations. A loss-of-function mutation is recessive (as in the example of Figure 3.10). Sometimes a mutation has the opposite effect and causes a protein to acquire a new function; such a change is called a **gain-of-function** mutation. A gain-of-function mutation is dominant.

The nature of multiple alleles

We have so far discussed the relationship between alleles largely in terms of combinations of wild-type with mutant alleles. Although any individual diploid animal can have only two alleles, the population as a whole contains many different alleles of the gene. The relationship between these multiple alleles takes various forms.

In the simplest case, a wild-type gene codes for a protein product that is functional, and mutant allele(s) code for proteins that are nonfunctional.

But there are often cases in which a series of mutant alleles have different phenotypes. For example, wild-type function of the *white* locus of *D. melanogaster* is required for development of the normal red color of the eye. The locus is named for the effect of extreme (null) mutations, which cause the fly to have a white eye in mutant homozygotes.

To describe wild-type and mutant alleles, wild genotype is indicated by a plus superscript after the name of the locus (w^+ is the wild-type allele for [red] eye color in *D. melanogaster*). Sometimes + is used by itself to describe the wild-type allele, and only the mutant alleles are indicated by the name of the locus.

An entirely defective form of the gene (or absence of phenotype) may be indicated by a minus superscript. To distinguish among a variety of mutant alleles with different effects, other super-scripts may be introduced, such as w^i or w^a.

The w^+ allele is dominant over any other allele in heterozygotes. There are many different mutant alleles. **Table 3.2** shows a (small) sample. Although some alleles (including the very first mutant to be isolated) have no eye color, many alleles produce some color. Each of these mutant alleles must therefore represent a different mutation of the gene, which does not eliminate its function entirely, but leaves a residual activity that produces a char-acteristic phenotype. These allele are named for the color of the eye in a homozygote. (Most w alleles affect the quantity of pigment in the eye, and the examples in the table are arranged in (roughly) declining amount of color, but others, such as w^{sp}, affect the pattern in which it is deposited.)

When multiple alleles exist, an animal may be a heterozygote that carries two different mutant alleles. The phenotype of such a heterozygote depends on the nature of the residual activity of each allele. The relationship between two mutant alleles is in principle no different from that between wild-type and mutant alleles: one allele may be dominant, there may be partial dominance, or there may be codominance.

There is not necessarily a unique wild-type allele at any particular locus. Control of the human blood group system provides an example. Lack of func-tion is represented by the null type, *O* group. But the functional alleles *A* and *B* provide activities that are codominant with one another and dominant over *O* group. The basis for this relationship is illustrated in **Figure 3.14**.

Table 3.2	
The *w* locus has an extensive series of alleles, whose phenotypes extend from wild-type (red) color to complete lack of pigment.	
Allele	**Phenotype of Homozygote**
w^+	Red eye (wild-type)
w^{bl}	Blood
w^{ch}	Cherry
w^{bf}	Buff
w^h	Honey
w^a	Apricot
w^e	Eosin
w^i	Ivory
w^z	Zeste (lemon-yellow)
w^{sp}	Mottled, color varies
w^1	White (no color)

The **O** (or **H**) antigen is generated in all indi-viduals, and consists of a particular carbohydrate group that is added to proteins. The *ABO* locus codes for a galactosyltransferase enzyme that adds a further sugar group to the O antigen. The specificity of this enzyme determines the blood group. The *A* allele produces an enzyme that uses the cofactor UDP-*N*-acetylgalactose, creating the A antigen; while the *B* allele produces an enzyme that uses the cofactor UDP-galactose, creating the B antigen. The A and B versions of the transferase protein differ in 4 amino acids that presumably affect its recognition of the type of cofactor. The *O* allele has suffered a mutation (a small deletion) that eliminates activity, so no modification of the O antigen occurs.

This explains why *A* and *B* alleles are dominant in the *AO* and *BO* heterozygotes: the corresponding transferase activity creates the A or B antigen. The *A* and *B* alleles are codominant in *AB* heterozygotes, because both transferase activities are expressed. The *OO* homozygote is a null that has neither activity, and therefore lacks both antigens.

Neither *A* nor *B* can be regarded as uniquely wild type, since they represent alternative activities rather than loss or gain of function. A situation such as this, in which there are multiple alleles in a

Figure 3.14

The ABO blood group locus codes for a galactosyl transferase whose specificity determines the blood group.

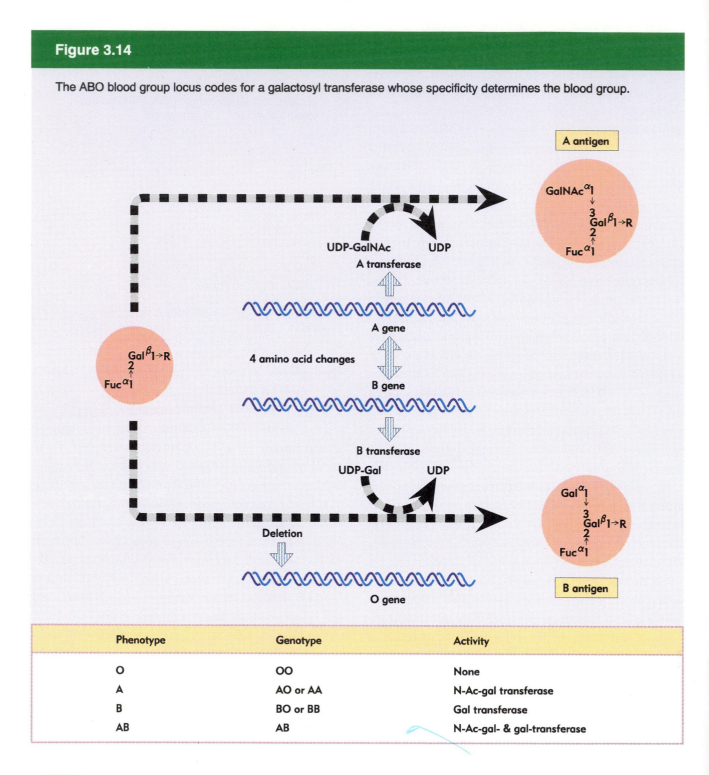

Phenotype	Genotype	Activity
O	OO	None
A	AO or AA	N-Ac-gal transferase
B	BO or BB	Gal transferase
AB	AB	N-Ac-gal- & gal-transferase

population, is described as a **polymorphism.** This may refer either to the distribution of different functional alleles (as in the case of *AB* blood groups) or to the presence of various nonfunctional alleles (as in the case of *white*).

Genes are usually named by abbreviations. In bacteria, the standard nomenclature is to use a three-letter, lowercase abbreviation for all genes that affect a particular characteristic. Each individual gene is given an additional capital letter. Thus there is a series of *lac* genes, known as *lacZ, lacY,* and *lacA*. The wild-type allele of *lacZ* would be *lacZ*+; a defective mutant would be *lacZ*−. The same abbreviations are used to describe the phenotype, but the first letter is capitalized. Thus wild-type or mutant *lac* alleles would give rise to the respective bacterial phenotypes of Lac+ or Lac−. The product of a gene is a feature of the phenotype and is referred to accordingly; for example, the LacZ protein is the enzyme β-galactosidase.

Summary

The genotype consists of the complete set of genetic information inherited by an organism; its expression is responsible for generating the phenotype, the physical form of the organism. The genotype includes many genes, organized into chromosomes. Mendel's laws treat genes as discrete factors: alleles segregate and genes assort independently. The process of meiosis provides a physical basis for the behavior that is responsible for Mendel's laws.

Genes that are on different chromosomes (or that are far apart on the same chromosome) recombine independently. Genes on a chromosome form a linear linkage group, in which those genes near one another tend to be inherited together. A linkage map provides a linear representation of the locations of the genes on a chromosome. The genetic material of a chromosome forms a continuous structure in which the gene itself has the same linear construction in miniature as the chromosome at large.

Wild-type describes the usual genotype or phenotype. A wild-type gene produces a functional protein. Mutations are heritable alterations in genetic information. A mutant gene may produce an altered protein or fail to produce any functional protein. A null mutant produces no protein. If the function of a single allele is sufficient in the diploid cell, a wild-type allele is (fully) dominant over a recessive mutant. Partial dominance occurs when gene functions are quantitative (two alleles produce twice the activity of one allele). Codominance results when different alleles have distinct specificities, so that a heterozygote possesses the properties of both parents. Mutations can be conditional, showing mutant phenotype under non-permissive conditions, but appearing wild type in permissive conditions.

The relationship between two mutations that map near one another and have a similar phenotype can be determined by the complementation test; if the mutations fail to complement in *trans* configuration, they are assigned to the same cistron, or gene. A cistron codes for a single polypeptide chain.

Further reading

Historical

The concept of the gene can be traced through a series of classic papers: **Morgan** (*Science* **32**, 120–122, 1910); **Sturtevant** (*J. Exp. Zool.* **14**, 39–45, 1913); **Muller** (*Science* **46**, 84–87, 1927); **McClintock and Creighton** (*Proc. Nat. Acad. Sci. USA* **17**, 492–497, 1931).

Analysis of gene function began with **Beadle and Tatum** (*Proc. Nat. Acad. Sci. USA* **27**, 499–506, 1941); the modern view of the gene/cistron started with **Benzer's** paper (*Proc. Nat. Acad. Sci. USA* **41**, 344, 1955). The first demonstration that a mutation in a gene changes an amino acid in the corresponding proteins was provided by **Ingram** (*Nature* **180**, 326–328, 1957).

Reviews

Work with bacteriophages is reviewed in the volume edited by **Cairns, Stent, and Watson**, *Phage and the Origins of Molecular Biology* (Cold Spring Harbor Laboratory Press, New York, 1966).

Discoveries

The basis for determination of human blood groups was reported by **Yamamoto** *et al.* (*Nature* **345**, 229–233, 1990).

CHAPTER 4

DNA is the genetic material

Information is passed in two forms from one generation to the next. A fertilized egg (in sexually reproducing species) or a daughter cell (in asexually reproducing species) receives a set of preexisting organized structures—its very existence reflects the features of cellular structure. And it receives a set of genes, needed for the manufacture of further structures during development of the organism.

The genetic material functions by virtue of its ability to specify a large variety of proteins. Early thoughts about the nature of the genetic material were biased by an erroneous assumption: that the structure of genetic material must be as complex as the proteins whose production it specifies. It was thought for a long time that *only* proteins could have sufficient diversity to specify other proteins. This assumption was jettisoned when it was realized that the genetic material carries the information needed to specify the protein in an enciphered form, a **code**.

Each gene functions by representing a particular polypeptide chain. The concept that each protein consists of a particular series of amino acids dates from Sanger's characterization of insulin in the 1950s. The discovery that a gene consists of DNA faces us with the issue of how a sequence of nucleotides in DNA represents a sequence of amino acids in protein.

A crucial feature of the general structure of DNA is that *it is independent of the particular sequence of its component nucleotides*. The sequence of nucleotides in DNA is important not because of its

structure *per se*, but because it *codes* for the sequence of amino acids that constitutes the corresponding polypeptide. The relationship between a sequence of DNA and the sequence of the corresponding protein is called the **genetic code**.

The structure and/or enzymatic activity of each protein follows from its primary sequence of amino acids. By determining the sequence of amino acids in each protein, the gene is able to carry all the information needed to specify an active polypeptide chain. In this way, a single type of structure—the gene—is able to represent itself in innumerable polypeptide forms.

Together the various protein products of a cell undertake the catalytic and structural activities that are responsible for establishing its phenotype. Of course, in addition to the sequences of genes that code for proteins, DNA also contains certain sequences whose function is to be recognized by regulator molecules, usually proteins. Here the function of the DNA is determined by its sequence directly, not via any intermediary code. Both types of region, genes expressed as proteins and sequences recognized as such, constitute genetic information.

Mutations are heritable alterations that change genetic information. From the discovery that DNA is the genetic material, the concept that a mutation is a change in the sequence of nucleotides follows naturally. Through the genetic code, a change in the nucleotide sequence leads to a change in the amino acid sequence, thus altering or abolishing the activity of the protein.

The existence of mutations allows us to compare the properties of a wild-type (normal) gene with a defective gene. By identifying the protein that is altered by a mutation, we may characterize the product of a gene. And by analyzing the changes that occur in the phenotype of the organism, we may identify the function of the gene.

The structural design of DNA enables it to accomplish the purpose of maintaining and perpetuating its sequence. Consisting of two strands, each of whose sequence corresponds to the other in a predictable manner, an individual molecule of DNA in effect contains redundant information. This enables the sequence of each strand to be checked against the other, so that inconsistencies can be corrected. Because the two strands are held together only by noncovalent forces, their separation can be accomplished with a moderate input of energy, thus allowing replication under physiological conditions.

The discovery of DNA

The idea that genetic material is nucleic acid had its roots in the discovery of **transformation** by Griffith in 1928. The bacterium *Pneumococcus* kills mice by causing pneumonia. The virulence of the bacterium is determined by the **capsular polysaccharide**, a component of the surface, which allows the bacterium to escape destruction by the host. Several **types** (I, II, III) of *Pneumococcus* have different capsular polysaccharides, which have a **smooth** (S) appearance.

Each of the smooth *Pneumococcal* types can give rise to variants that fail to produce the capsular polysaccharide. These bacteria have a **rough** (R) surface (consisting of the material that was beneath the capsular polysaccharide). They are **avirulent**; they do not kill the mice, because the absence of the polysaccharide allows the animal to destroy the bacteria.

When smooth bacteria are killed by heat treatment, they lose their ability to harm the animal. But inactive heat-killed S bacteria and the ineffectual variant R bacteria together have a quite different effect from either bacterium by itself. **Figure 4.1** shows that when they are injected together into an animal, the mouse dies as the result of a *Pneumococcal* infection. Virulent S bacteria can be recovered from the mouse postmortem.

In this experiment, the dead S bacteria were of type III. The live R bacteria had been derived from type II. The virulent bacteria recovered from the mixed infection had the smooth coat of type III. Thus some property of the dead type III S bacteria **transform** the live R bacteria so that they make the type III capsular polysaccharide, and as a result become virulent.

The component of the dead bacteria responsible for transformation was called the **transforming principle**. It was purified by developing a cell-free system, in which extracts of the dead S bacteria could be added to the live R bacteria before injection into the animal. The classic studies of Avery and his colleagues showed chemically in 1944 that the isolated transforming principle is **deoxyribonucleic acid (DNA)**.

The surprise of this result is indicated by the fact that, at this time, DNA was not even known to be a component of *Pneumococcus*, although of course it had been recognized for many decades as a major component of eukaryotic chromosomes. In showing

Figure 4.1

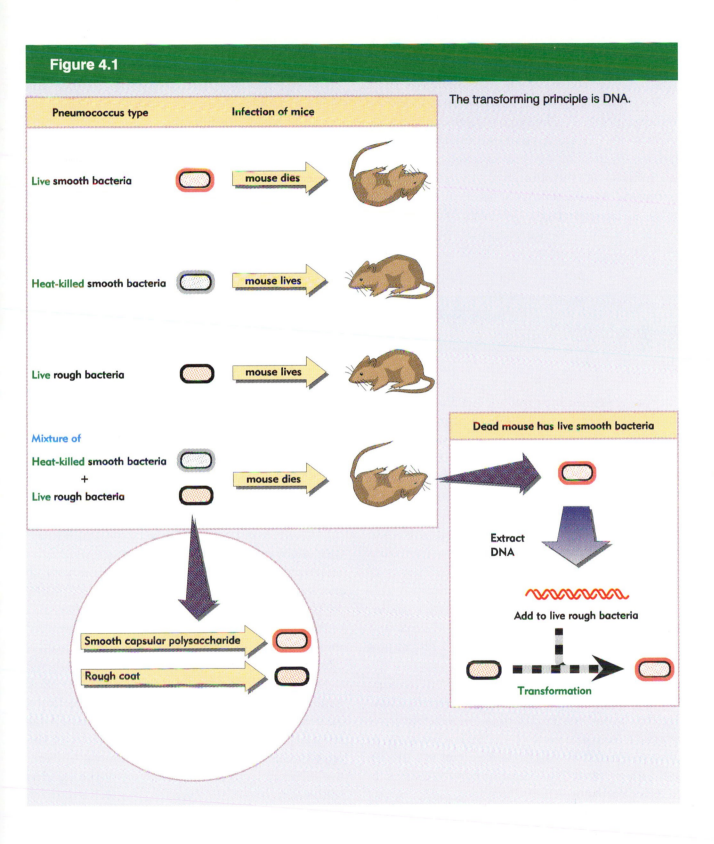

The transforming principle is DNA.

that the genetic material of a prokaryote is DNA, this result therefore offered a unifying view for the basis of heredity in bacteria and higher organisms.

The implications of the result were captured by the original paper. 'The inducing substance, on the basis of its chemical and physical properties, appears to be a highly polymerized and viscous form of DNA. On the other hand, the type III capsular polysaccharide, the synthesis of which is evoked by this transforming agent, consists chiefly of a non-nitrogenous polysaccharide.... Thus it is evident that the inducing substance and the substance produced in turn are chemically distinct and biologically specific in their action and that both are requisite in determining the type specificity of the cells of which they form a part.'

This discussion marked the introduction of a distinction between the genetic material and the products of its expression, a view that became an implicit basis for subsequent studies.

DNA is the (almost) universal genetic material

After the transforming principle had been shown to consist of DNA, the next step was to demonstrate that DNA provides the genetic material in a quite different system. Phage T2 is a virus that infects the bacterium *E. coli*. When phage particles are added to bacteria, they adsorb to the outside surface, some material enters the bacterium, and then ~20 minutes later each bacterium bursts open (lyses) to release a large number of progeny phage particles (see Figure 3.2).

In 1952, Hershey and Chase infected bacteria with T2 phages that had been radioactively labeled *either* in their DNA component (with ^{32}P) *or* in their protein component (with ^{35}S). **Figure 4.2** illustrates the results of this experiment.

The infected bacteria were agitated in a blender, and two fractions were separated by centrifugation. One contained the empty phage coats that were released from the surface of the bacteria; these consist of protein and therefore carried the ^{35}S radioactive label. The other fraction consisted of the infected bacteria themselves.

Most of the ^{32}P label was present in the infected bacteria. The progeny phage particles produced by the infection contained ~30% of the original ^{32}P label. The progeny received very little—less than 1%—of the protein contained in the original phage population. This experiment therefore showed directly that the DNA of parent phages enters the bacteria and then becomes part of the progeny phages, exactly the pattern of inheritance expected of genetic material.

A phage (virus) reproduces by commandeering the machinery of an infected host cell to manufacture more copies of itself. The phage possesses genetic material whose behavior is analogous to that of cellular genomes: its traits are faithfully reproduced, and they are subject to the same rules that govern inheritance. The case of T2 reinforces the general conclusion that the genetic material is DNA, whether part of the genome of a cell or virus.

Bacteria and phages clearly have DNA as their genetic material. But what about eukaryotes? For a long time the evidence was only inferential. DNA is present in the right location and behaves in the appropriate manner. Direct evidence became available only long after the matter was regarded as settled.

In discussing his results on transformation, Avery made a comment with wider-ranging implications than he could have known. 'If we are right', he wrote, 'it means that nucleic acids are not merely

Figure 4.2

The genetic material of phage T2 is DNA.

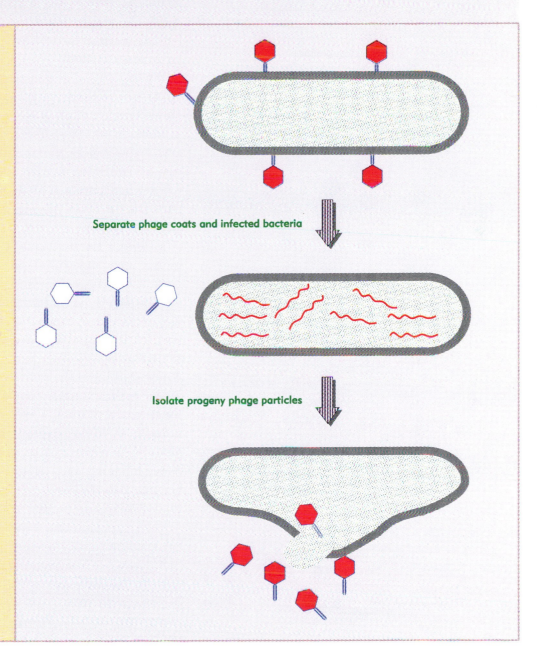

Bacteria are infected with phages labeled with P^{32} (●) in DNA or with S^{35} (●) in protein

Separate phage coats and infected bacteria

Phage coats contain 80% of S^{35} label

Infected bacteria contain 70% of P^{32} label

Isolate progeny phage particles

Progeny phages have 30% of P^{32} label and <1% of S^{35} label

Figure 4.3

Eukaryotic cells can acquire a new phenotype as the result of transfection by added DNA.

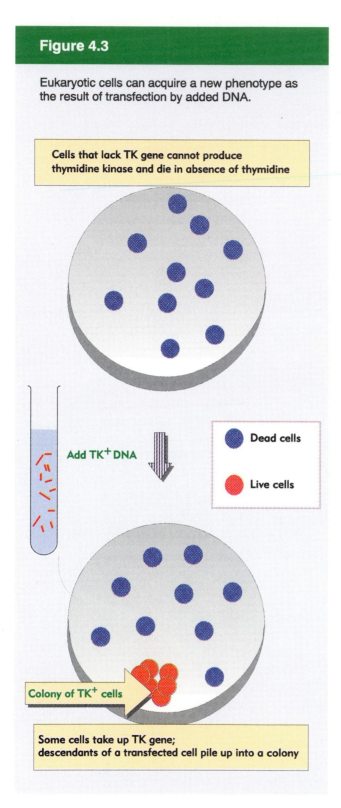

Cells that lack TK gene cannot produce thymidine kinase and die in absence of thymidine

Add TK⁺ DNA

Dead cells

Live cells

Colony of TK⁺ cells

Some cells take up TK gene; descendants of a transfected cell pile up into a colony

structurally important but functionally active substances in determining the biochemical activities and specific characteristics of cells and that by means of a known chemical substance it is possible to induce predictable and hereditary changes in cells'. He had in mind the bacterial system, but similar results now have been obtained with eukaryotes as well.

When DNA is added to populations of single eukaryotic cells growing in culture, the nucleic acid enters the cells, and in some of them results in the production of new proteins. At first performed with DNA extracted *en masse*, these experiments now can be routinely performed with purified DNA whose incorporation leads to the production of a particular protein. **Figure 4.3** depicts one of the standard systems.

Although for historical reasons these experiments are described as **transfection** when performed with eukaryotic cells, they are a direct counterpart to bacterial transformation. The DNA that is introduced into the recipient cell becomes part of its genetic material, inherited in the same way as any other part. Its expression confers a new trait upon the cells (synthesis of thymidine kinase in the example shown in the figure). At first, these experiments were successful only with individual cells adapted to grow in a culture medium. Since then, however, DNA has been introduced in mouse eggs by microinjection; and it may become a stable part of the genetic material of the mouse (see Chapter 36).

Such experiments show directly not only that DNA is the genetic material in eukaryotes, but also that *it can be transferred between different species and yet remain functional.*

The genetic material of all known organisms and many viruses is DNA. However, some viruses use an alternative nucleic acid, **ribonucleic acid (RNA)**, as the genetic material. Although its chemical formula is slightly different from that of DNA, in these circumstances RNA exercises the same role. The general principle of the nature of the genetic material, then, is that it is always nucleic acid; in fact, it is DNA except in the RNA viruses.

The components of DNA

A nucleic acid consists of a chemically linked sequence of subunits (with the general structure shown previously in Figure 1.4). Each subunit contains a **nitrogenous base** (a heterocyclic ring of carbon and nitrogen atoms), a **pentose** sugar (a five-carbon sugar in ring form), and a **phosphate** group.

The nitrogenous bases fall into the two types shown in **Figure 4.4: pyrimidines** and **purines**. Pyrimidines have a six-member ring; purines have fused five- and six-member rings.

Each nucleic acid contains 4 types of base. The same two purines, adenine and guanine, are present in both DNA and RNA. The two pyrimidines in DNA are cytosine and thymine; in RNA uracil is found instead of thymine. The only difference between uracil and thymine is the presence of a methyl substituent at position C_5. The bases are usually referred to by their initial letters; so DNA contains A, G, C, T, while RNA contains A, G, C, U.

Two types of pentose are found in nucleic acids. They distinguish DNA and RNA and give rise to the general names for the two types of nucleic acid. **Figure 4.5** shows their structures. In DNA the pentose is **2-deoxyribose**; whereas in RNA it is **ribose**. The difference lies in the absence/presence of the hydroxyl group at position 2 of the sugar ring.

The nitrogenous base is linked to position 1 on the pentose ring by a glycosidic bond from N_1 of pyrimidines or N_9 of purines. To avoid ambiguity between the numbering systems of the heterocyclic rings and the sugar, positions on the pentose are given a prime ('`).

A base linked to a sugar is called a **nucleoside**; when a phosphate group is added, the base–sugar–phosphate is called a **nucleotide**. The

Figure 4.4

Purines and pyrimidines provide the nitrogenous bases in nucleic acids. The "purine" and "pyrimidine" rings show the general structures of each type of base; the numbers identify the positions on the ring.

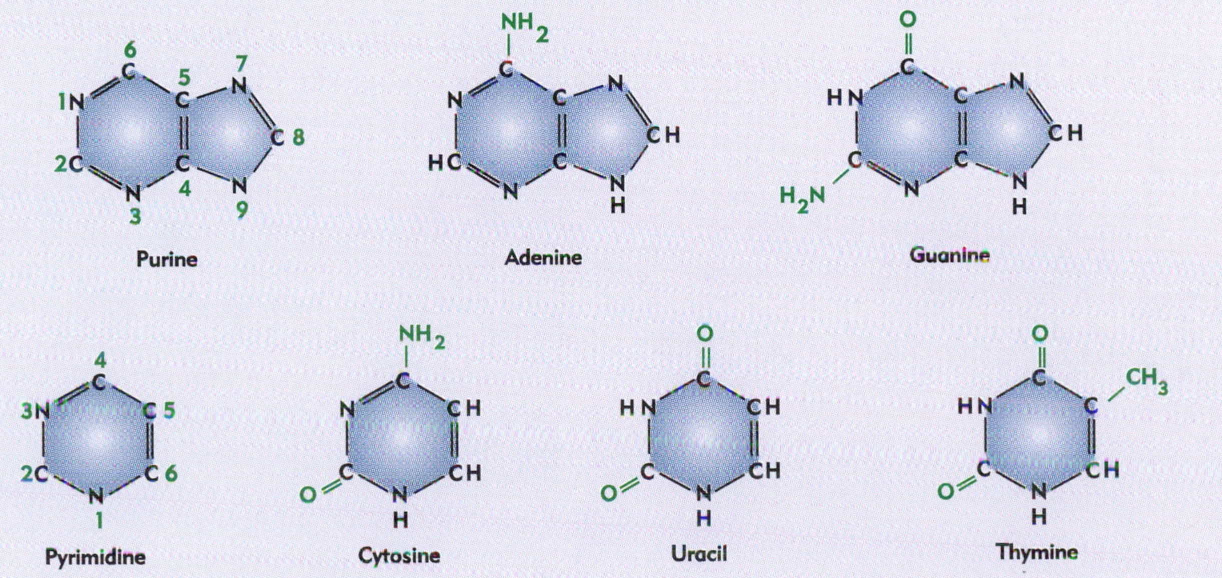

Figure 4.5

2-Deoxyribose is the sugar in DNA and ribose is the sugar in RNA. The carbon atoms are numbered as indicated for deoxyribose. The sugar is connected to the nitrogenous base via position 1.

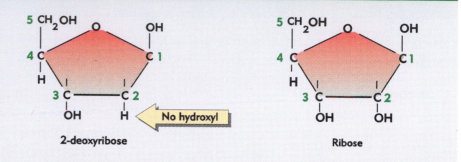

2-deoxyribose

Ribose

No hydroxyl

Figure 4.6

A polynucleotide chain consists of a series of 5'–3' sugar–phosphate links that form a backbone from which the bases protrude.

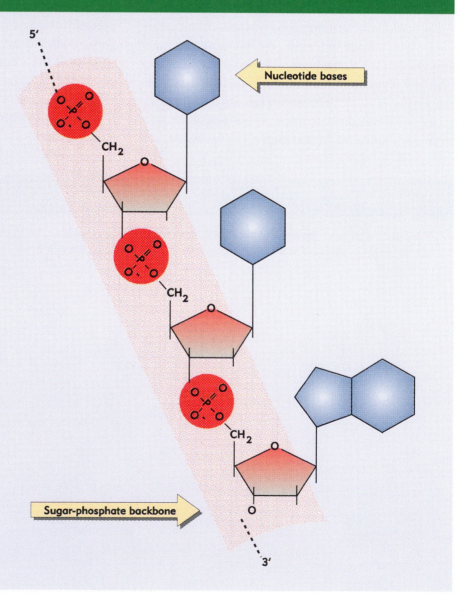

Nucleotide bases

Sugar-phosphate backbone

Table 4.1

Bases, nucleosomes, and nucleotides have related names. Abbreviations of the form NMP stand for nucleoside monophosphate; "d" indicates the 2'-deoxy form.

Base	Nucleoside	Nucleotide	Abbreviation RNA	DNA
Adenine	Adenosine	Adenylic acid	AMP	dAMP
Guanine	Guanosine	Guanylic acid	GMP	dGMP
Cytosine	Cytidine	Cytidylic acid	CMP	dCMP
Thymine	Thymidine	Thymidylic acid		dTMP
Uracil	Uridine	Uridylic acid	UMP	

nomenclature of the individual units is described in **Table 4.1**.

Nucleotides provide the building blocks from which nucleic acids are constructed. The nucleotides are linked together into a **polynucleotide chain** by a backbone consisting of an alternating series of sugar and phosphate residues. The 5' position of one pentose ring is connected to the 3' position of the next pentose ring via a phosphate group, as shown in **Figure 4.6**. Thus the sugar–phosphate backbone is said to consist of 5'–3' phosphodiester linkages. The nitrogenous bases 'stick out' from the backbone.

The terminal nucleotide at one end of the chain has a free 5' group; the terminal nucleotide at the other end has a free 3' group. It is conventional to write nucleic acid sequences in the 5'–3' direction—that is, from the 5' terminus at the left to the 3' terminus at the right.

Figure 4.7

Nucleotides may carry phosphate in the 5' or 3' position.

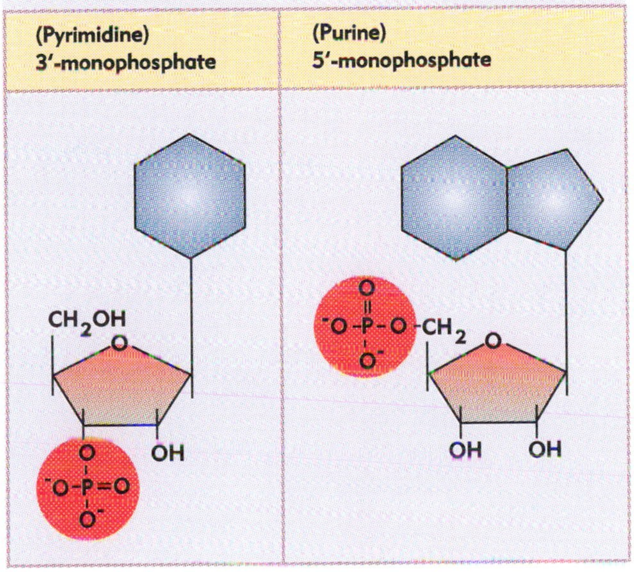

(Pyrimidine) 3'-monophosphate	(Purine) 5'-monophosphate

Figure 4.8

A nucleoside-5'-triphosphate has energy rich phosphate bonds.

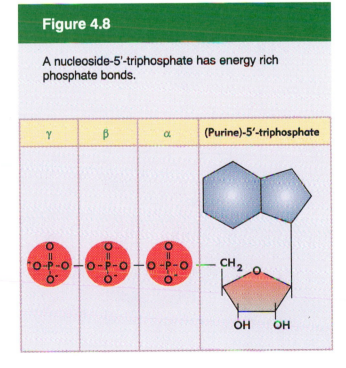

γ	β	α	(Purine)-5'-triphosphate

When DNA or RNA is broken into its constituent nucleotides, the cleavage may take place on either side of the phosphodiester bonds. Depending on the circumstances, nucleotides have their phosphate group attached to either the 5' or the 3' position of the pentose, as shown in **Figure 4.7**. The two types of nucleotides released from nucleic acids are therefore the nucleoside-3'-monophosphates and nucleoside-5'-monophosphates.

All the nucleotides can exist in a form in which there is more than one phosphate group linked to the 5' position. An example is shown in **Figure 4.8**. The bonds between the first (α) and second (β), and between the second (β) and third (γ), phosphate groups are **energy-rich** and are used to provide an energy source for various cellular activities. The abbreviation for a nucleoside triphosphate takes the form NTP; the abbreviation for a nucleoside diphosphate is NDP.

The 5' triphosphates are the precursors for nucleic acid synthesis. **Figure 4.9** shows the reaction, in which the 5' end of the triphosphate reacts with a 3'-OH group at the end of the polynucleotide chain. A bond is formed from the α phosphate to the 3'-OH of the sugar at the end of the polynucleotide chain, and the two terminal phosphate groups (γ and β) of the triphosphate are released (in the form of a single molecule, called pyrophosphate).

Figure 4.9

Nucleic acid synthesis occurs by adding the nucleoside-5'-monophosphate moiety of a nucleoside triphosphate to the 3'-OH end of the polynucleotide chain.

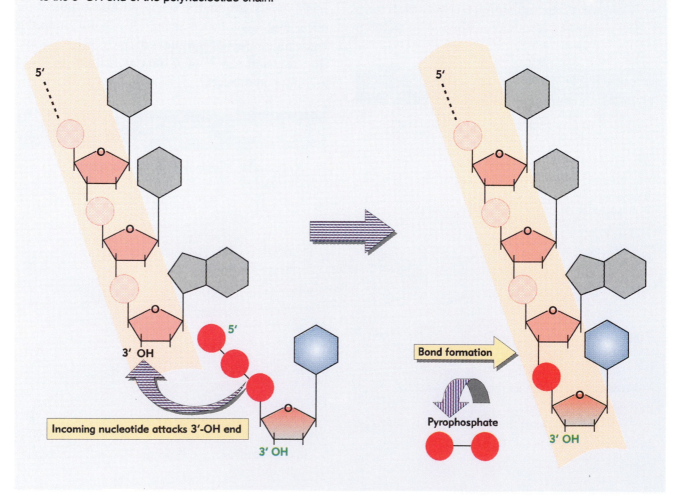

DNA is a double helix

The observation that the bases are present in different amounts in the DNAs of different species led to the concept that the *sequence of bases is the form in which genetic information is carried*. By the 1950s, the concept of genetic information was common: the twin problems it posed were working out the structure of the nucleic acid, and explaining how a sequence of bases in DNA could represent the sequence of amino acids in a protein.

Three notions converged in the construction of the double helix model for DNA by Watson and Crick in 1953:

◆ X-ray diffraction data showed that DNA has the form of a regular helix, making a complete turn every 34Å (3.4 nm), with a diameter of ~20Å (2 nm). Since the distance between adjacent nucleotides is 3.4Å, there must be 10 nucleotides per turn.

◆ The density of DNA suggests that the helix must contain two polynucleotide chains. The constant diameter of the helix can be explained if the bases in each chain face inward and are restricted so that a purine is always opposite a pyrimidine, avoiding purine–purine (too thick) or pyrimidine–pyrimidine (too thin) partnerships.

◆ Irrespective of the actual amounts of each base, the proportion of G is always the same as the proportion of C in DNA, and the proportion of A is always the same as that of T. Thus the composition of any DNA can be described by the proportion of its bases that is G + C, which ranges from 26% to 74% for different species.

Watson and Crick proposed that the two polynucleotide chains in the double helix associate by *hydrogen bonding between the nitrogenous bases*. **Figure 4.10** demonstrates that, in their usual forms, G can hydrogen bond specifically only with

Figure 4.10

Complementary base pairing involves the formation of two hydrogen bonds between A and T, and of three hydrogen bonds between G and C. No other pairs form in DNA.

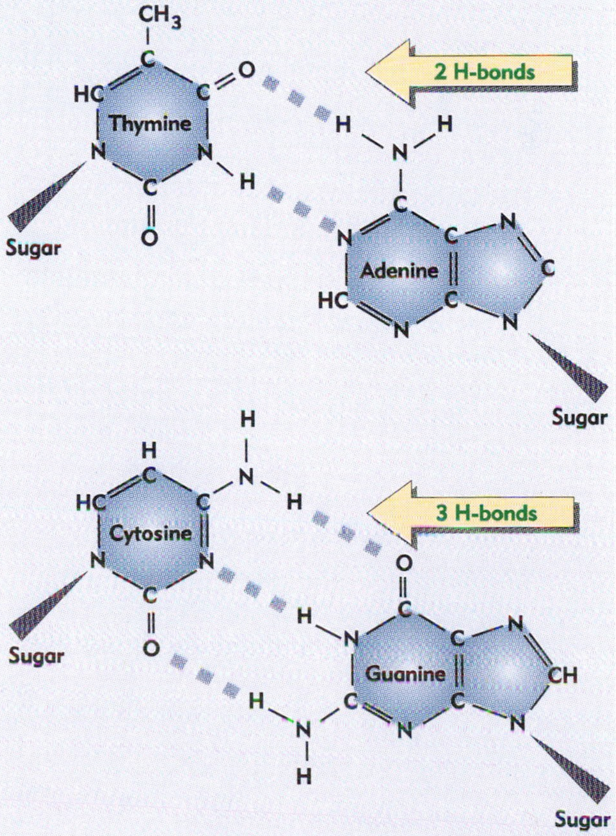

Figure 4.11

The double helix maintains a constant width because purines always face pyrimidines in the complementary A-T and G-C base pairs.

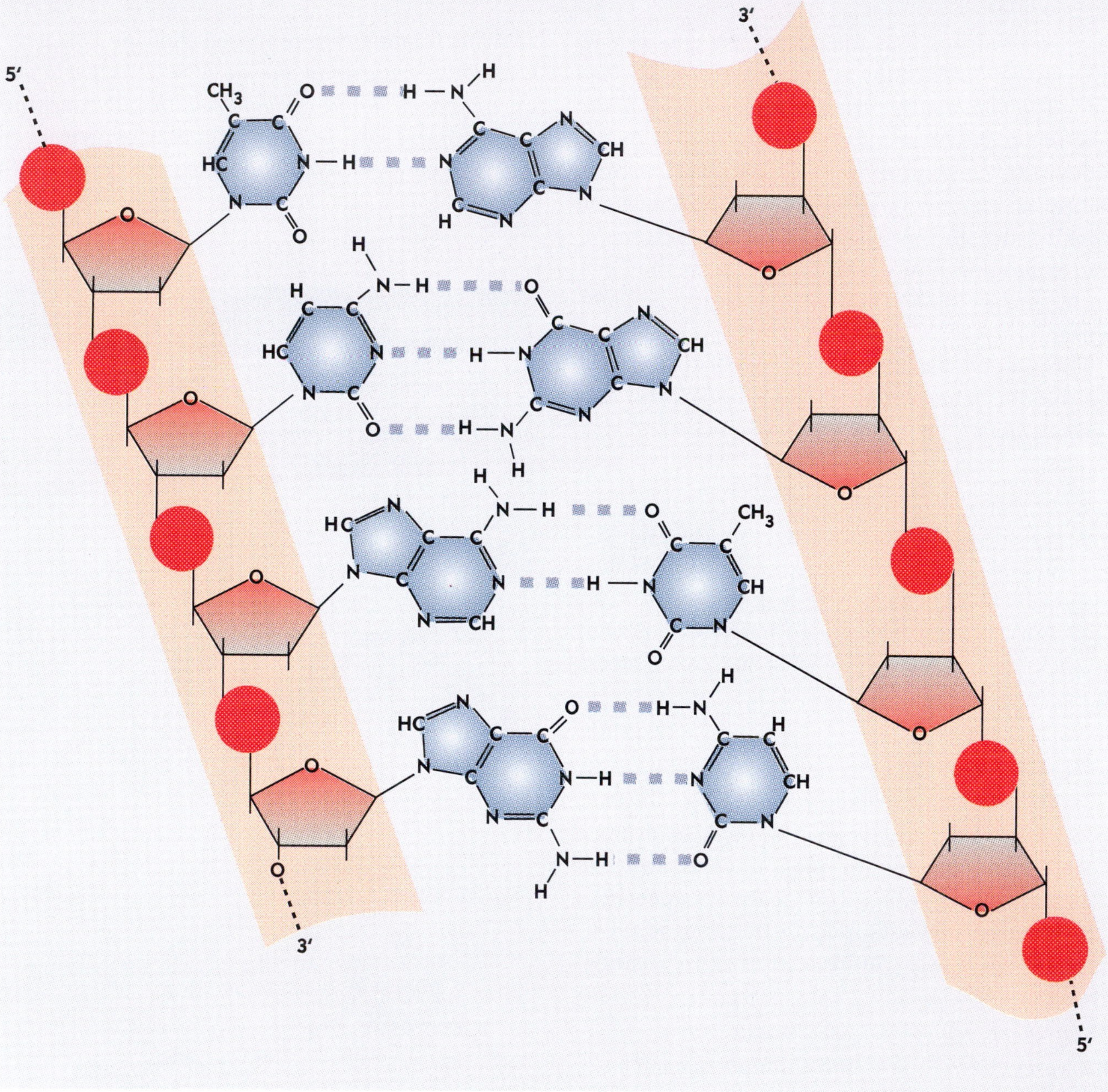

Figure 4.12

Flat base pairs lie perpendicular to the sugar-phosphate backbone.

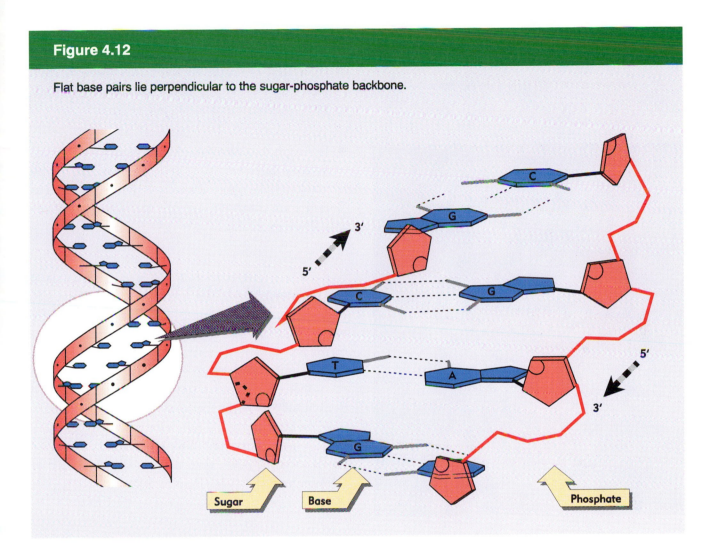

C, while A can bond specifically only with T. These reactions are described as **base pairing**, and the paired bases (G with C, or A with T) are said to be **complementary.**

The model requires the two polynucleotide chains to run in opposite directions (**antiparallel**), as illustrated in **Figure 4.11**. Looking along the helix, therefore, one strand runs in the 5'–3' direction, while its partner runs 3'–5'.

The sugar–phosphate backbone is on the outside and carries negative charges on the phosphate groups. When DNA is in solution *in vitro*, the charges are neutralized by the binding of metal ions; usually Na$^+$ is provided. In the natural state *in vivo*, positively charged proteins provide some of the neutralizing force. These proteins play an important role in determining the organization of DNA in the cell.

The bases lie on the inside. They are flat structures, lying in pairs perpendicular to the axis of the helix. Consider the double helix in terms of a spiral staircase: the base pairs form the treads, as illustrated schematically in **Figure 4.12**. Proceeding along the helix, bases are stacked above one another, in a sense like a pile of plates.

The base pairs contribute to the thermodynamic stability of the double helix in two ways. Energy is released by the formation of base pairs

Figure 4.13

The two strands of DNA form a double helix.

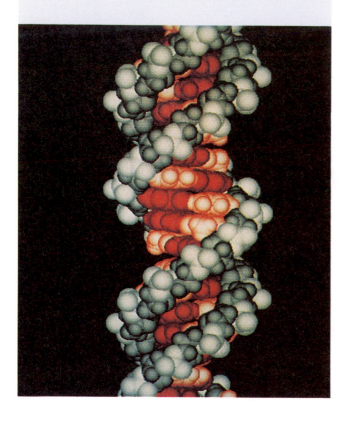

and by interactions between adjacent base pairs:

◆ Hydrogen bonding between the bases in each pair releases energy corresponding to 3 bonds per G•C pair and 2 bonds per A•T pair.

◆ Hydrophobic **base-stacking** results from interactions between the electron systems of the stacked base pairs. Each base pair is involved in two such interactions, one with the other base pair on each side.

Each base pair is rotated ~36° around the axis of the helix relative to the next base pair. Thus ~10 base pairs make a complete turn of 360°. The twisting of the two strands around one another forms a double helix with a **narrow groove** (~12Å across) and a **wide groove** (~22Å across), as can be seen from the scale model of **Figure 4.13**. The double helix is **right-handed**; the turns run clockwise looking along the helical axis. These features represent the accepted model for what is known as the **B-form** of DNA. (We discuss the structure of DNA in more detail in Chapter 5.)

DNA replication is semiconservative

It is crucial that the genetic material is reproduced accurately. Because the two polynucleotide strands are joined only by hydrogen bonds, they are able to separate without requiring breakage of covalent bonds. The specificity of base pairing suggests that each of the separated **parental strands** could act as a **template** for the synthesis of a complementary **daughter strand**, as depicted in **Figure 4.14**. The principle is that a new daughter strand is assembled on each parental strand. The sequence of the daughter strand is dictated by the parental strand; an A in the parental strand causes a T to be placed in the daughter strand, a parental G directs incorporation of a daughter C, and so on.

Figure 4.14

Base pairing provides the mechanism for faithfully replicating DNA.

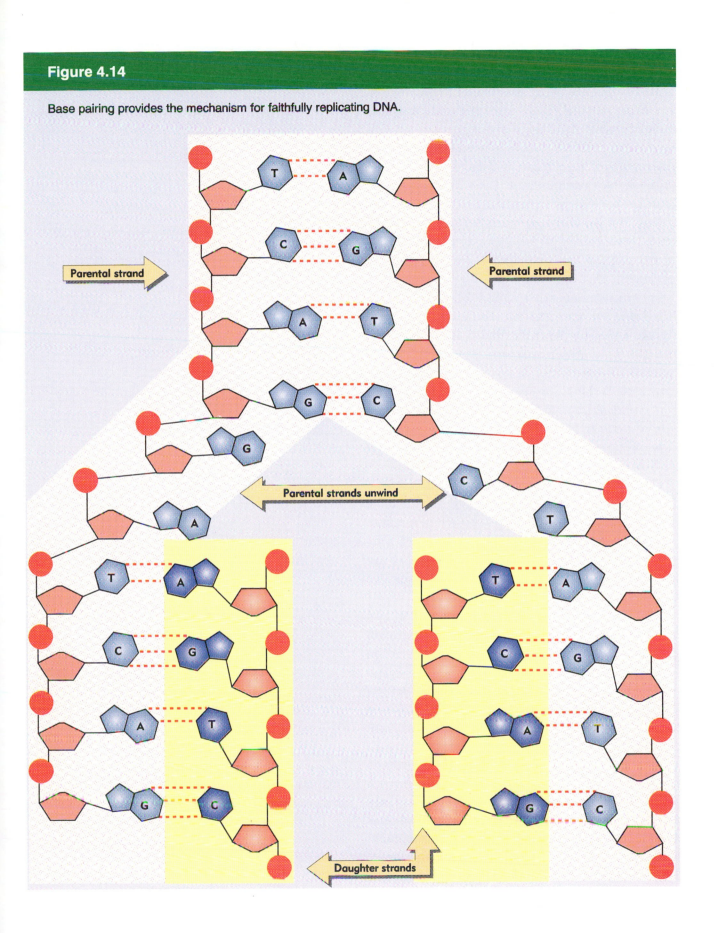

The top part of the figure shows a **parental duplex** and the lower part shows the two **daughter duplexes** that are being produced by complementary base pairing. The (unreplicated) parental duplex consists of the original two parental strands. Replication requires the two parental strands to be separated so that each can be used as a template for synthesis of a complement. Each of the daughter duplexes is identical in sequence with the original parent, and contains one parental strand and one newly synthesized strand. *The structure of DNA carries the information needed to perpetuate its sequence.*

The consequences of this mode of replication are illustrated for the DNA molecule as a whole in **Figure 4.15**. The parental duplex is replicated to form two daughter duplexes, each of which consists of one parental strand and one (newly synthesized) daughter strand. *The unit conserved from one generation to the next is one of the two individual strands comprising the parental duplex.* This behavior is called **semi-conservative replication**.

The figure illustrates a prediction of this model. If the parental DNA carries a 'heavy' density label because the organism has been grown in medium containing a suitable isotope (such as ^{15}N), its strands can be distinguished from those that are synthesized when the organism is transferred to a medium containing normal 'light' isotopes.

The parental DNA consists of a duplex of two heavy strands (red). After one generation of growth in light medium, the duplex DNA is 'hybrid' in density—it consists of one heavy parental strand (red) and one light daughter strand (blue). After a second generation, the two strands of each hybrid duplex have separated; each gains a light partner, so that now half of the duplex DNA remains hybrid while half is entirely light (both strands are blue).

The individual strands of these duplexes all are entirely heavy or entirely light. This pattern was confirmed experimentally in the Meselson–Stahl experiment of 1958, which followed the semi-conservative replication of DNA through three

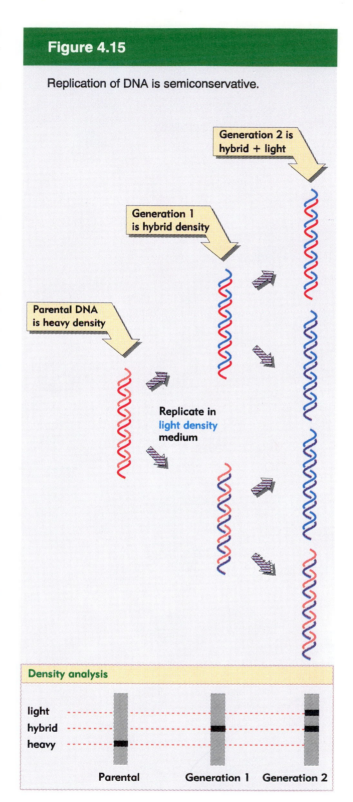

Figure 4.15

Replication of DNA is semiconservative.

Generation 2 is hybrid + light

Generation 1 is hybrid density

Parental DNA is heavy density

Replicate in light density medium

Density analysis

light
hybrid
heavy

Parental   Generation 1   Generation 2

generations of growth of *E. coli*. When DNA was extracted from bacteria and its density measured by centrifugation, the DNA formed bands corresponding to its density—heavy for parental, hybrid for the first generation, and half hybrid and half light in the second generation.

Replication involves a major disruption of the structure of DNA. However, although the two strands of the parental duplex must separate, they do not exist as single strands. The disruption of structure is only transient and is reversed as the daughter duplex is formed. So only a small part of the DNA loses the duplex structure at any moment.

Consider a molecule of DNA engaged in replication. Its helical structure is illustrated in **Figure 4.16**. The nonreplicated region consists of the parental duplex, opening into the replicated region where the two daughter duplexes have formed. The double helical structure is disrupted at the junction between the two regions, called the **replication fork**. Replication involves movement of the replication fork along the parental DNA, so there is a continuous unwinding of the

Figure 4.16

The replication fork is the region of DNA in which there is a transition from the unwound parental duplex to the newly replicated daughter duplexes.

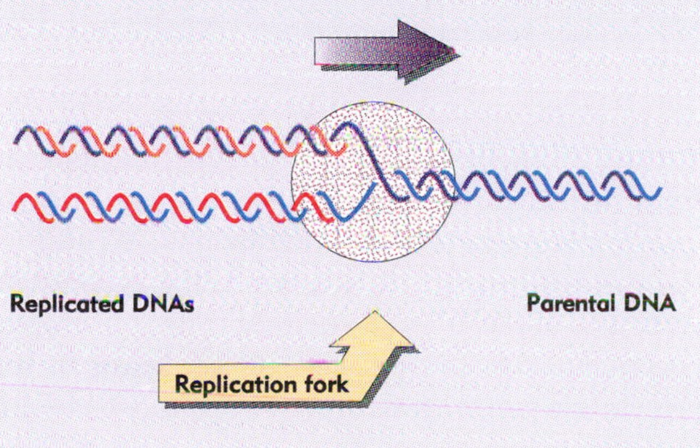

Replicated DNAs   Parental DNA

Replication fork

parental strands and rewinding into daughter duplexes.

The synthesis of nucleic acids is catalyzed by specific enzymes, which recognize the template and undertake the task of catalyzing the addition of subunits to the polynucleotide chain that is being synthesized. The enzymes are named according to the type of chain that is synthesized: **DNA polymerases** synthesize DNA, and **RNA polymerases** synthesize RNA.

Degradation of nucleic acids also is undertaken by specific enzymes: **deoxyribonucleases (DNAases)** degrade DNA, and **ribonucleases (RNAases)** degrade RNA. The nucleases fall into the general classes of **exonucleases** and **endonucleases**.

Endonucleases cut individual bonds *within* RNA molecules, generating discrete fragments. They are involved in cutting reactions, when the mature sequence is separated from the flanking sequence.

Exonucleases remove residues one at a time from the end of the molecule, generating mononucleotides. They are involved in trimming reactions. From the perspective of the RNA substrate, the extra residues are whittled away, base by base. All known exonucleases proceed along the nucleic acid chain from the 3′ end. Their attitude toward the substrate may be **random** or **processive**. A randomly acting enzyme removes a base from one RNA molecule and then dissociates; for its next catalytic event, the enzyme attacks a different RNA substrate molecule. Processive action means that the enzyme stays with one substrate molecule, removing further bases until its mission is accomplished.

When the phosphodiester bond linking two nucleotides is cleaved, it can in principle be cut on either side of the phosphate group. The consequences are illustrated in **Figure 4.17**. Cleavage on one side generates 3′-hydroxyl and 5′-phosphate termini. Cleavage on the other side generates 3′-phosphate and 5′-hydroxyl termini.

Figure 4.17

The nature of the new termini is determined by which side of the phosphodiester bond is cleaved.

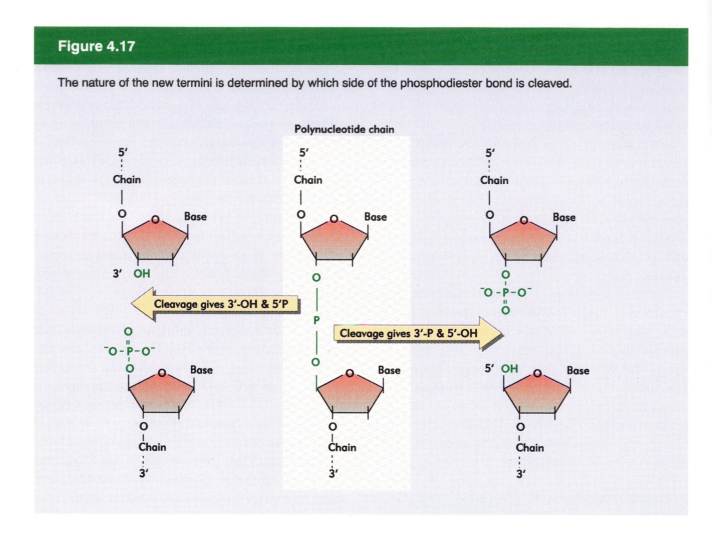

The genetic code is read in triplets

The genetic code is deciphered by a complex apparatus that stands between nucleic acid and protein. This apparatus is essential if the information carried in DNA is to have meaning. In any given region, only one of the two strands of DNA codes for protein, so we write the genetic code as a sequence of bases (rather than base pairs).

The genetic code is read in groups of three nucleotides, each group representing one amino acid. Each trinucleotide sequence is called a **codon**.

A gene includes a series of codons that is read in series from a starting point at one end to a termination point at the other end. Written in the conventional 5′–3′ direction, the nucleotide sequence of the DNA strand that codes for protein corresponds to the amino acid sequence of the protein written in the direction from N-terminus to C-terminus.

The general basis of the code was discovered by genetic analysis of mutants of the *rII* region of the bacterial virus, phage T6. In 1961, Crick and his colleagues showed that the code must be read in *nonoverlapping triplets from a fixed starting point*:

◆ *Nonoverlapping* implies that each codon consists of three nucleotides and that successive codons are represented by successive trinucleotides.

◆ The use of a *fixed starting point* means that assembly of a protein must start at one end and work to the other, so that different parts of the coding sequence cannot be read independently.

If the genetic code is read in nonoverlapping triplets, there are three possible ways of translating a nucleotide sequence into protein, depending on the starting point. These are called **reading frames**. For the sequence

A C G A C G A C G A C G A C G A C G A C G

the three possible frames of reading are

ACG ACG ACG ACG ACG ACG
CGA CGA CGA CGA CGA CGA
GAC GAC GAC GAC GAC GAC

A mutation that inserts or deletes a single base will change the reading frame for the entire subsequent sequence. A change of this sort is called a **frameshift**. Because the sequence of the new reading frame is completely different from the old one, the entire amino acid sequence of the protein is altered beyond the site of mutation. Thus the function of the protein is likely to be lost completely.

Frameshift mutations are induced by the **acridines**, compounds that bind to DNA and distort the structure of the double helix, causing additional bases to be incorporated or omitted during replication. Each mutagenic event sponsored by an acridine results in the addition or removal of a single base pair.

If an acridine mutant is produced by, say, addition of a nucleotide, it should revert to wild type by deletion of the nucleotide. But reversion can also be caused by deletion of a different base, at a site close to the first. Combinations of such mutations provided revealing evidence about the nature of the genetic code.

Figure 4.18 illustrates the properties of frameshift mutations. An insertion or a deletion changes the entire protein sequence following the site of mutation. But the combination of an insertion and a deletion causes the code to be read in the incorrect frame only between the two sites of mutation; correct reading resumes after the second site.

The original analysis was performed by genetic means, in which all acridine mutations could be classified into one of two sets, described as (+) and (−). Either type of mutation by itself causes a frameshift, the (+) type by virtue of a base addition, the (−) type by virtue of a base deletion. Double mutant combinations of the types (+ +) and (− −) continue to show mutant behavior. But combinations of the types (+ −) or (− +) suppress one another, giving rise to a description in which one mutation is described as a **suppressor** of the other. (In the context of this work, 'suppressor' is used in an unusual sense; see later).

These results show that the genetic code must be read as a sequence in a reading frame that is fixed by the starting point, so additions or deletions compensate for each other, whereas double additions or double deletions remain mutant. But this does not reveal how many nucleotides make up each codon.

When triple mutants are constructed, only (+ + +) and (− − −) combinations show the wild phenotype, while other combinations remain mutant. If we take three additions or three deletions to correspond respectively to the addition or omission overall of a single amino acid, this implies that the code is read in triplets. An incorrect amino acid sequence is found between the two outside sites of mutation, and the sequence on either side remains wild type, as indicated in Figure 4.18.

Figure 4.18

Frameshift mutations show that the genetic code is read in triplets from a fixed starting point.

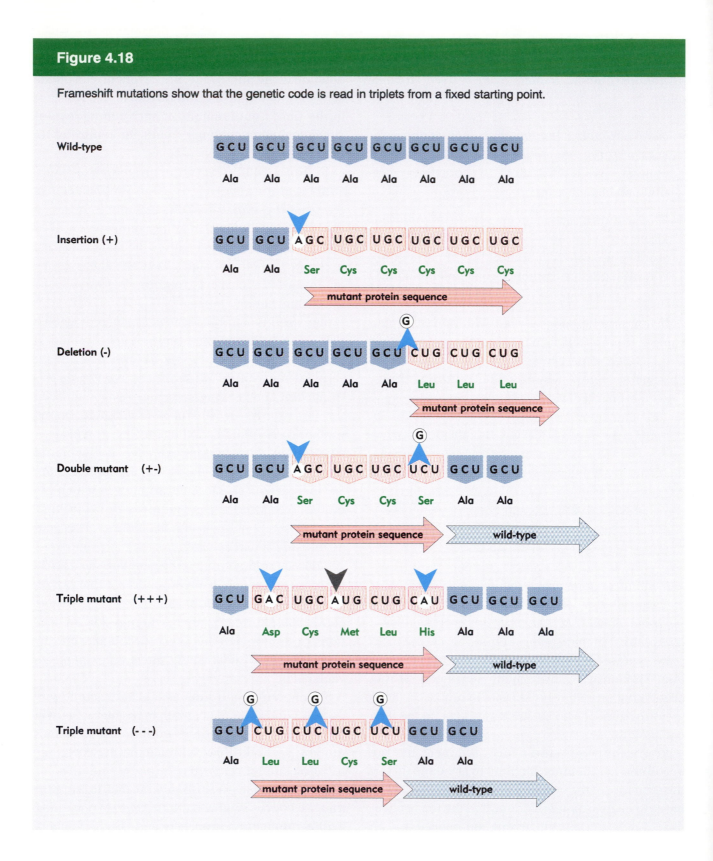

Mutations change the sequence of DNA

Mutations provide decisive evidence that DNA is the genetic material. When a change in the sequence of DNA causes an alteration in the sequence of a protein, we may conclude that the DNA codes for that protein. Furthermore, from the nature of the consequent effects on the organism, we may be able to identify the function of the protein. The existence of many mutations in a gene may allow many variant forms of a protein to be compared, and a detailed analysis can be used to identify regions of the protein responsible for individual enzymatic or other functions.

All organisms suffer a certain number of mutations as the result of normal cellular operations or random interactions with the environment. Such mutations are called **spontaneous**; the rate at which they occur is characteristic for any particular organism and sometimes is called the **background level**. Mutations are rare events, and of course those that damage a gene are selected against during evolution. It is therefore difficult to obtain large numbers of spontaneous mutants to study from natural populations.

The occurrence of mutations can be increased by treatment with certain compounds. These are called **mutagens**, and the changes they cause are referred to as **induced mutations**. Most mutagens act directly by virtue of an ability either to act on a particular base of DNA or to become incorporated into the nucleic acid. The effectiveness of a mutagen is judged by the degree to which it increases the rate of mutation above background. By using mutagens, it becomes possible to induce many changes in any gene.

Any base pair of DNA can be mutated. A **point mutation** changes only a single base pair, and can be caused by either of two types of event:

◆ Chemical modification of DNA directly changes one base into a different base.

◆ A malfunction during the replication of DNA causes the wrong base to be inserted into a polynucleotide chain during DNA synthesis.

Point mutations can be divided into two types, depending on the nature of the change when one base is substituted for another:

◆ The more common class is the **transition**, comprising the substitution of one pyrimidine by the other, or of one purine by the other; thus a G•C pair is exchanged with an A•T pair or vice versa.

◆ The less common class is the **transversion**, in which a purine is replaced by a pyrimidine or vice versa, so that an A•T pair becomes a T•A or C•G pair.

A classic example of a transition caused by the chemical conversion of one base into another is provided by the effects of nitrous acid. **Figure 4.19** shows that nitrous acid performs an oxidative deamination that converts cytosine into uracil. In the replication cycle following the transition, the U pairs with an A, instead of with the G with which the original C would have paired. So the C•G pair is replaced by a T•A pair when the A pairs with the T in the next replication cycle. Nitrous acid also deaminates adenine, causing the reverse transition from A•T to G•C.

Transitions are also caused by **base mispairing**, when unusual partners pair in defiance of the usual restriction to Watson–Crick pairs. Base mispairing usually occurs as an aberration resulting from the introduction of an abnormal base.

Some mutagens are analogs of the usual bases that have ambiguous pairing properties; their mutagenic action results from their incorporation into DNA in place of one of the regular bases. **Figure 4.20** shows the example of bromouracil (BrdU), which is an analog of thymine that contains a bromine atom in place of the methyl group of thymine. BrdU is incorporated into DNA in place of thymine. But it has ambiguous pairing properties, because the presence of the bromine atom allows a shift to occur in which the base changes structure from a keto (=O) form to an enol (–OH) form. The enol form can base pair with guanine, which leads to substitution of the original A•T pair by a G•C pair.

The mistaken pairing can occur either during the original incorporation of the base or in a subsequent replication cycle. The transition is induced with a certain probability in each replication cycle, so the incorporation of BrdU has continuing effects on the sequence of DNA.

Mutations induced by base substitution often are **leaky**: the mutant has some residual function. This situation arises when the sequence change in the corresponding protein does not entirely abolish its activity. The nature of the genetic code explains how this occurs. A point mutation that alters only a single base will change only the one codon in which that base is located. So only one amino acid is affected in the protein. While this substitution may reduce the activity of the protein, it does not necessarily abolish it entirely. This contrasts with the mutations induced by acridines, where a long series of codons may be altered by a shift of reading frame, completely abolishing protein function.

Point mutations were thought for a long time to be the principal means of change in individual genes. However, we now know that **insertions** of stretches of additional material are quite frequent. The source of the inserted material lies with **transposable elements**, sequences of DNA with the ability to move from one site to another. (We discuss these elements in detail in Chapters 34 and 35.) An insertion usually abolishes the activity of a gene. Where such insertions have occurred, **deletions** of part or all of the inserted material, and sometimes of the adjacent regions, may subsequently occur.

A significant difference between point mutations and the insertions/deletions is that the frequency of point mutation can be increased by mutagens, whereas the occurrence of changes caused by transposable elements is indifferent to these reagents. However, insertions and deletions can also occur by other mechanisms—for example, involving mistakes made during replication or recombination—although probably these are less common. The changes induced by the acridines also constitute (very small) insertions and deletions.

The isolation of **revertants** is an important characteristic that distinguishes point mutations and insertions from deletions:

◆ A point mutation can revert by restoring the original sequence or by gaining a compensatory mutation elsewhere in the gene.

◆ An **insertion** of additional material can revert by deletion of the inserted material.

◆ A **deletion** of part of a gene cannot revert.

Mutations can also occur in other genes to circumvent the effects of mutation in the original gene. This effect is called **suppression**, or, more formally, intercistronic suppression. This is the normal use of the term 'suppression'.

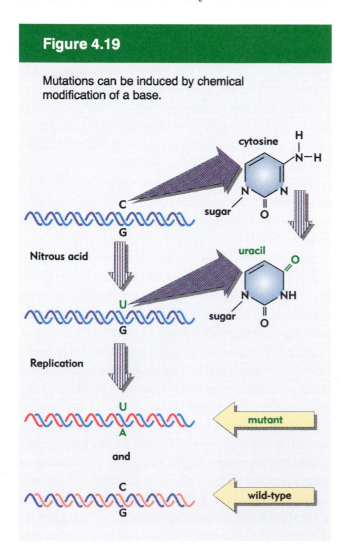

Figure 4.19

Mutations can be induced by chemical modification of a base.

Figure 4.20

Mutations can be induced by the incorporation of base analogs into DNA.

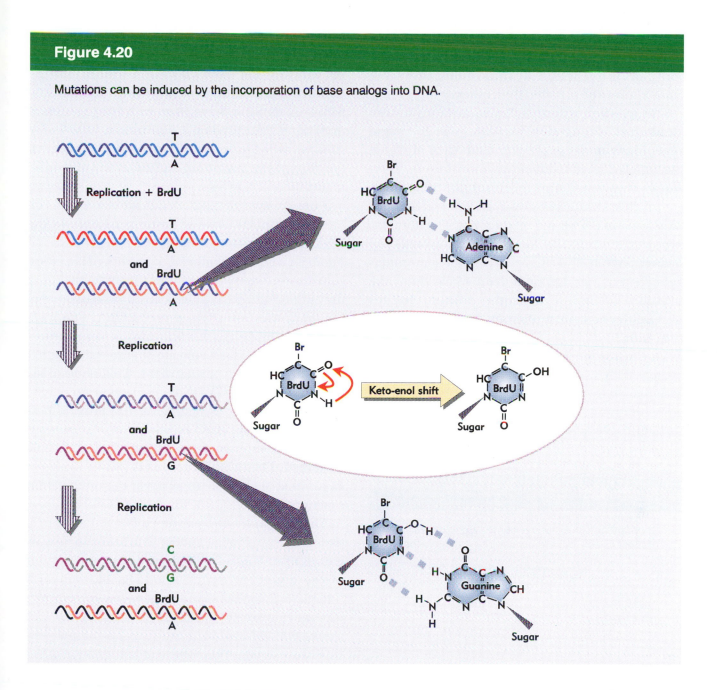

Mutations are concentrated at hotspots

So far we have dealt with mutations in terms of individual changes in the sequence of DNA that influence the activity of the genetic unit in which they occur. When we consider mutations in terms of the inactivation of the gene, most genes within a species show more or less similar rates of mutation relative to their size. This suggests that the gene can be regarded as a target for mutation, and that damage to any part of it can abolish its function. As a result, susceptibility to mutation is roughly proportional to the size of the gene. But consider the sites of mutation within the sequence of DNA; are all base pairs in a gene equally susceptible or are some more likely to be mutated than others?

This question is approached by isolating a large number of independent mutations in the same gene. Many mutants are obtained, each of which has suffered an individual mutational event. Then the site of each mutation is determined. Most mutations will lie at different sites, but some will lie at the same position. Two independently isolated mutations at the same site may constitute exactly the same change in DNA (in which case the same mutational event has happened on more than one occasion), or they may constitute different changes (three different point mutations are possible at each base pair).

The histogram of **Figure 4.21** shows the frequency with which mutations are found at each base pair in the *lacI* gene of *E. coli*. The statistical probability that more than one mutation occurs at a particular site is given by random-hit kinetics (as seen in the Poisson distribution). So some sites will gain one, two, or three mutations, while others will not gain any. But some sites gain far more than the number of mutations expected from a random distribution; they may have 10 or even 100 times more mutations than predicted by random hits. These sites are called **hotspots**. Hotspots are not universal for all types of mutation; and different mutagens have different hotspots.

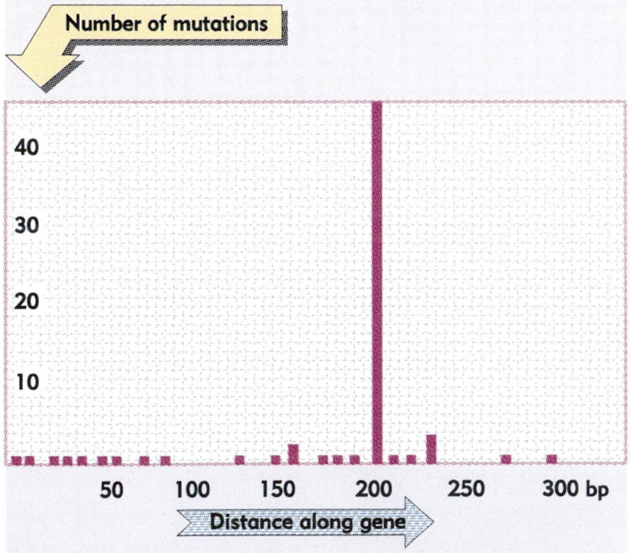

A major cause of spontaneous mutation in *E. coli* has been pinned down to an unusual base in the DNA. In addition to the four bases that are inserted into DNA when it is synthesized, **modified bases** are sometimes found. The name reflects their origin; they are produced by chemically modifying one of the four bases already present in DNA. The most common modified base is 5-methylcytosine, generated by a methylase enzyme that adds a methyl group to a small proportion of the cytosine residues (at specific sites in the DNA).

Sites containing 5-methylcytosine provide hotspots for spontaneous point mutation. In each case, the mutation takes the form of a G•C to A•T transition. The hotspots are not found in strains of *E. coli* that are unable to perform the methylation reaction.

The reason for the existence of the hotspots is that 5-methylcytosine suffers spontaneous deamination at an appreciable frequency; replacement of the amino group by a keto group converts 5-methylcytosine to thymine (the structures of the pyrimidines are shown in Figure 4.4). **Figure 4.22** shows why deaminating the (rare) 5-methylcytosine causes a mutation, whereas deamination of the more common cytosine does not have this effect.

The deamination of cytosine generates uracil. However, *E. coli* contains an enzyme, uracil-DNA-glycosidase, that removes uracil residues from DNA. This action leaves an unpaired G residue, and a repair system then inserts a C base to partner it. The net result of these reactions is to restore the original sequence of the DNA. Presumably this system serves to protect DNA against the consequences of spontaneous deamination of cytosine (although it is not active enough to prevent the effects of nitrous acid; see Figure 4.19).

Figure 4.21

Spontaneous mutations occur throughout the *lacI* gene of *E. coli*, but are concentrated at a hotspot.

Figure 4.22

The deamination of 5-methylcytosine produces thymine (causing C-G to T-A transitions), while the deamination of cytosine produces uracil (which usually is removed and then replaced by cytosine).

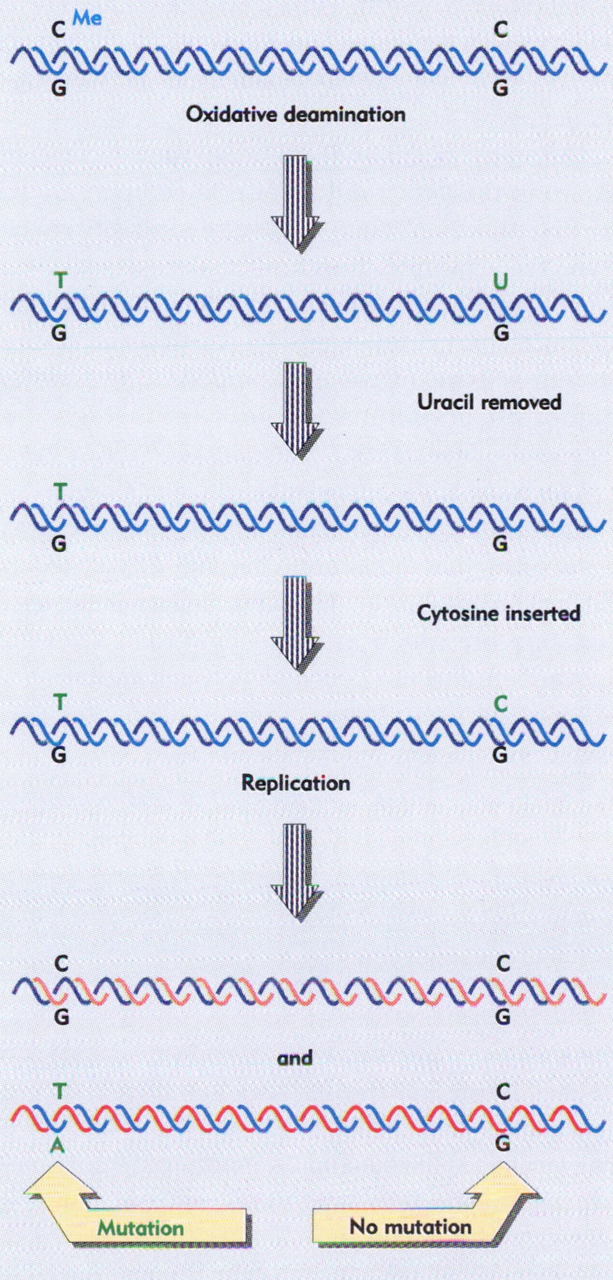

C^Me — C

G — G

Oxidative deamination

T — U

G — G

Uracil removed

T —

G — G

Cytosine inserted

T — C

G — G

Replication

C — C

G — G

and

T — C

A — G

Mutation   **No mutation**

But the deamination of 5-methylcytosine leaves thymine; because this base is a respectable constituent of DNA in its own right, the system does not operate in these circumstances, and a mutation results. The conversion creates a mispaired G•T partnership, whose separation at the subsequent replication produces one wild-type G•C pair and one mutant A•T pair.

The operation of this system casts an interesting light on the use of T in DNA compared with U in RNA. Perhaps it relates to the need of DNA for stability of sequence; the use of T means that any deaminations of C are immediately recognized, because they generate a base (U) not usually present in the DNA.

A mutation consists of a change in the sequence of DNA that takes the form of adding, deleting, or substituting A•T or G•C base pairs. Such a change inevitably becomes a permanent part of the genetic information.

But when a stable change is observed in the properties of a gene, it is not necessarily true that it has been caused by a mutation. A change in the properties of a cell that is inherited but that does not represent a change in genetic information is called an **epigenetic** change. (Formally an epigenetic change is defined as an alteration of phenotype without change in the genotype.) Modification of the sequence of DNA after it has been synthesized provides one means of epigenetic change. The introduction of a methyl group at a particular cytosine is an example of a signal that may change the properties of a gene.

The presence of methylcytosine in DNA is not maintained by the replication system itself. The change can be perpetuated: this requires the methyl group to be added each generation after the DNA has been replicated. But the change can be reversed by removing the methyl group. Since this form of information requires the intervention of an additional system to add or remove the methyl group, and is not intrinsic to the sequence of DNA, it is distinct from a genetic mutation, which constitutes an alteration in the sequence of DNA that is inherited by virtue of the restrictions of base pairing.

The rate of mutation

Spontaneous mutations that inactivate gene function occur in bacteria at a rate of $\sim 10^{-5}$–10^{-6} events per locus per generation. We have no really accurate measurement of the rate of mutation in eukaryotes, although usually it is thought to be somewhat similar to that of bacteria on a per-locus per-generation basis. We do not know what proportion of the spontaneous events results from point mutations.

Suppose that a bacterial gene consists of 1200 base pairs, coding for a protein of 400 amino acids (~45,000 daltons of mass). The average mutation rate corresponds to changes at individual nucleotides of 10^{-9}–10^{-10} per generation. Even if all the changes were due to point mutations, this calculation is an over-simplification, because not all mutations in DNA actually lead to a detectable change in the phenotype.

Mutations without apparent effect are called **silent mutations**. They fall into two types. Some involve base changes in DNA that do not cause any change in the amino acid present in the corresponding protein. Others change the amino acid, but the replacement in the protein does not affect its activity; these are called **neutral substitutions**.

Mutations that inactivate a gene are called **forward mutations**. Their effects are reversed by **back mutations**, which are of two types.

An exact reversal of the original mutation is called **true reversion**. Thus if an A•T pair has been replaced by a G•C pair, another mutation to restore the A•T pair will exactly regenerate the wild-type sequence.

Alternatively, another mutation may occur elsewhere in the gene, and its effects compensate for the first mutation. This is called **second-site reversion**. For example, one amino acid change in a protein may abolish gene function, but a second alteration may compensate for the first and restore protein activity. (Thus the compensating acridine frameshift mutations fall into the category of second site revertants.)

A forward mutation results from any change that inactivates a gene, whereas a back mutation must restore function to a protein damaged by a particular forward mutation. Thus the demands for back mutation are much more specific than those for forward mutation. The rate of back mutation is correspondingly lower than that of forward mutation, typically by a factor of ~10.

Summary

Two classic experiments proved that DNA is the genetic material. DNA isolated from one strain of *Pneumococcus* bacteria can confer properties of that strain upon another strain. And only the DNA is physically inherited by progeny phages from the parental phages. More recently, DNA has been used to transfect new properties into eukaryotic cells.

DNA is a double helix consisting of antiparallel strands in which the nucleotide units are linked by 5′–3′ phosphodiester bonds. The backbone provides the exterior; purine and pyrimidine bases are stacked in the interior in pairs in which A is complementary to T while G is complementary to C. The strands separate and use complementary base pairing to assemble daughter strands in semiconservative replication.

A stretch of DNA may code for protein. The genetic code describes the relationship between the sequence of DNA and the sequence of the protein. Only one of the two strands of DNA codes for protein. A coding sequence of DNA consists of a series of codons, read as nonoverlapping triplets from a fixed starting point.

A mutation consists of a change in the sequence of A•T and G•C base pairs comprising DNA. A mutation in a coding sequence may change the sequence of amino acids in the corresponding protein. A frameshift mutation alters the subsequent reading frame by inserting or deleting a base. A point mutation changes only the amino acid represented by the codon in which the mutation occurs. Point mutations may be reverted by back mutation of the original mutation, insertions may revert by loss of the inserted material, but deletions cannot revert. Mutations may also be suppressed when another mutation elsewhere counters the original defect.

The natural incidence of mutations is increased by mutagens. Mutations may be concentrated at hotspots. A type of hotspot responsible for some point mutations is caused by deamination of the modified base 5-methylcytosine.

Forward mutations occur at a rate of ~10^{-6} per locus per generation; back mutations are rarer. Not all mutations have an effect on the phenotype.

Epigenetic changes consist of events that change the state or structure of a gene without altering the sequence of base pairs. One example of such a change is the addition of a methyl group to cytosine to generate 5-methylcytosine.

Further reading

Historical

The classic model for the structure of DNA is **Watson and Crick** (*Nature* 171, 737–738, 1953), accompanied by the data of **Wilkins** *et al.* (*Nature* 171, 738–740, 1953), and later followed by the further views of **Watson and Crick** (*Nature* 171, 964–967, 1953). Semiconservative replication of DNA was proved by the analysis of **Meselson and Stahl** (*Proc. Nat. Acad. Sci. USA* 44, 671–682, 1958).

The use of frameshifts to define the nature of the genetic code was reported in the classic paper by **Crick** *et al.* (*Nature* 192, 1227–1232, 1961).

The discovery of hotspots was reported by **Benzer** (*Proc. Nat. Acad. Sci. USA* 47, 403–416, 1961) and (nearly twenty years later) they were equated with modified bases by **Coulondre** *et al.* (*Nature* 274, 775–780, 1978).

Reviews

Recollections of the development of molecular biology were edited by **Cairns, Stent, and Watson**, in *Phage and the Origins of Molecular Biology* (Cold Spring Harbor Laboratory Press, New York, 1966).

Historical accounts of this period have been written by **Olby**, *The Path to the Double Helix* (MacMillan, London 1974) and **Judson**, *The Eighth Day of Creation* (Knopf, New York, 1978).

The drive to characterize mutations was reviewed by **Drake and Balz** (*Ann. Rev. Biochem.* 45, 11–37, 1976). The basis of frameshift mutations has been reviewed by **Roth** (*Ann. Rev. Genet.* 8, 319–346, 1974).

CHAPTER 5

The topology of nucleic acids

There is more to DNA than a sequence of base pairs organized into an invariant double-helical structure. The great majority of the genome is organized into the duplex B-form of DNA described in Chapter 4. But the duplex structure is flexible, and the ability to change its organization is a crucial aspect of the function of DNA.

The range of changes extends from minor alterations in the parameters describing the B-form duplex to separation into single strands. Features of physiological importance are:

◆ The number of base pairs per turn in B-DNA is not fixed, but can be adjusted slightly, depending on the circumstances.

◆ The double helix does not exist as a long straight rod, but is coiled in space to fit into the dimensions of the cell or virus whose genome it provides. This additional level of organization places the duplex under stress in such a way that its own structure is affected.

◆ Discontinuities in the structure can take the form of particular sequences at which 'bends' occur.

◆ For DNA to be replicated or expressed, the strands of the double helix must separate. Strand separation is never more than temporary, and only rather short regions of the genome are single-stranded at any one moment.

From the perspective of its functions in replicating itself and being expressed as protein, the central property of the double helix is the ability of the two strands to separate without needing to disrupt covalent bonds. This makes it possible for the strands to separate and reform under physiological conditions at the (very rapid) rates needed to sustain genetic functions. The specificity of the process is determined by complementary base pairing.

Although DNA usually is found in the form of a double helix, the genomes of some viruses consist of single-stranded DNA. Within the cell there are several forms of the other nucleic acid, RNA, usually present as single strands. However, a single-stranded nucleic acid (either DNA or RNA) may generate double-helical regions. An intramolecular duplex region can form within a single-stranded molecule that contains two complementary sequences if these sequences base pair with one another. Or a single-stranded molecule may base pair with an independent, complementary single-stranded molecule to form an intermolecular duplex. Base pairing between independent complementary single strands is not restricted to DNA–DNA or RNA–RNA interactions, but can also occur between a DNA molecule and an RNA molecule.

The concept of base pairing is central to all processes involving nucleic acids. Disruption of the base pairs is a crucial aspect of the function of a double-stranded molecule, while the ability to form base pairs is essential for the activity of a single-stranded nucleic acid.

The winding of the two strands of DNA around one another in the double helix has some crucial consequences for its function. An important effect is seen whenever a double helix of DNA is itself twisted in space. If it is twisted in the opposite sense from the winding of the two strands, a torsional force is created that can be relieved by partially unwinding the strands. This means that the higher order structure of DNA (as created by its path in space) can affect the structure of the double helix itself. Such effects can be important in initiating the unwinding of DNA that is needed for its replication or expression.

DNA can be denatured and renatured

The same features that allow DNA to fulfill its biological role make it possible to manipulate the nucleic acid *in vitro*, and ultimately to isolate the segment of DNA that represents a particular protein.

The noncovalent forces that stabilize the double helix are disrupted by heating or by exposure to low salt concentration. The two strands of a double helix separate entirely when all the hydrogen bonds between them are broken.

The process of strand separation is called **denaturation** or (more colloquially) **melting**. ('Denaturation' is also used to describe loss of authentic protein structure; it is a general term implying that the natural conformation of a macromolecule has been converted to some other form.)

Denaturation of DNA occurs over a narrow temperature range and results in striking changes in many of its physical properties. A particularly useful change occurs in the optical density. The heterocyclic rings of nucleotides adsorb light strongly in the ultraviolet range (with a maximum close to 260 nm that is characteristic for each base). But the adsorption of DNA itself is some 40% less than would be displayed by a mixture of free nucleotides of the same composition.

This is called the **hypochromic** effect; it results from interactions between the electron systems of the bases, made possible by their stacking in the parallel array of the double helix. Any departure from the duplex state is immediately reflected by a decline in this effect—that is, by an increase in optical density toward the value characteristic of free bases. The denaturation of DNA can therefore be followed by this hyperchromicity.

The midpoint of the temperature range over which the strands of DNA separate is called the **melting temperature**, denoted T_m. An example of a melting curve determined by change in optical adsorbance is shown in **Figure 5.1**. The curve always takes the same form, but its absolute position on the temperature scale (that is, its T_m) is influenced by both the base composition of the DNA and the conditions employed for denaturation.

T_m depends on the proportion of G•C base pairs. Because each G•C base pair has three hydrogen

Figure 5.1

Denaturation of DNA can be followed by the increase in optical density and is described by the T_m.

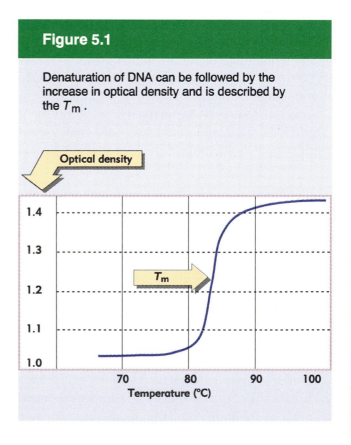

bonds, it is more stable than an A•T base pair, which has only two hydrogen bonds. The more G•C base-pairs are contained in a DNA, the greater the energy that is needed to separate the two strands; the T_m increases ~0.4°C for every 1% increase in G•C content. When DNA is in solution under approximately physiological conditions, the T_m usually lies in a range of 85–95°C. (A DNA that is 40% G•C—a value typical of mammalian genomes—denatures with a T_m of about 87°C under approximately physiological conditions; a DNA that is 60% G•C has a T_m of ~95°C under the same conditions.) Thus without intervention from cellular systems, duplex DNA is stable at the temperature prevailing in the cell.

Nucleic acids hybridize by base pairing

Nucleic acid sequences can be assessed in terms of either similarity or complementarity.

◆ *Similarity* between two sequences is given in principle by the proportion of bases (for single-stranded sequences) or base pairs (for double-stranded sequences) that is identical. Without determining the actual sequences, however, there is no direct way to measure similarity.

◆ *Complementarity* is determined by the rules for base pairing between A•T and G•C. In a perfect duplex of DNA, the strands are precisely complementary. If we compare two different but related double-stranded molecules, therefore, each strand of the first molecule will be similar to one strand of the second molecule and will be (partly) complementary to the other strand of the second molecule. Complementarity can be measured directly by the ability of two single-stranded nucleic acids to base pair with each other. If double-stranded molecules are denatured into single strands, the complementarity between the single strands can be used to indicate the similarity between the original duplex molecules.

Figure 5.2

Denatured single strands of DNA can renature to give the duplex form.

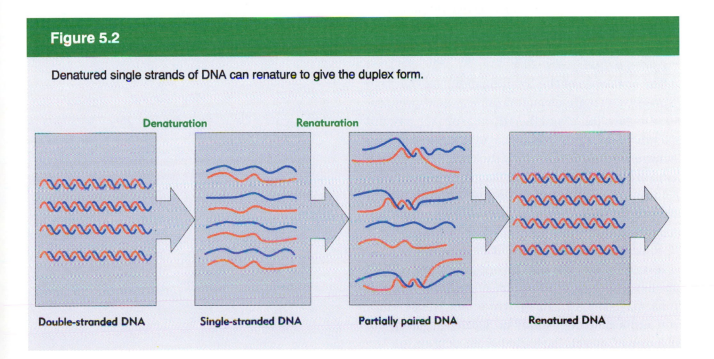

Denaturation   Renaturation

Double-stranded DNA   Single-stranded DNA   Partially paired DNA   Renatured DNA

It is possible to measure complementarity because the denaturation of DNA is reversible under appropriate conditions. The ability of the two separated complementary strands to reform into a double helix is called **renaturation**. It is illustrated in **Figure** 5.2.

Renaturation depends on specific base pairing between the complementary strands. The reaction takes place in two stages. First, single strands of DNA in the solution encounter one another by chance; if their sequences are complementary, the two strands base pair to generate a short double-helical region. Then the region of base pairing extends along the molecule by a zipper-like effect to form a lengthy duplex molecule. Renaturation of the double helix restores the original properties that were lost when the DNA was denatured. Renaturation describes the reaction between two complementary sequences that were separated by denaturation. However, the technique can be extended to allow any two complementary nucleic acid sequences to **anneal** with each other to form a duplex structure.

The reaction is generally described as **hybridization** when nucleic acids from different sources are involved, as in the case when one preparation consists of DNA and the other consists of RNA. *The ability of two nucleic acid preparations to hybridize constitutes a precise test for their complementarity since only complementary sequences can form a duplex structure.*

The principle of the hybridization reaction is to expose two single-stranded nucleic acid preparations to each other and then to measure the amount of double-stranded material that forms. There are two common ways of performing the reaction: **solution (liquid) hybridization** and **filter hybridization**.

Liquid hybridization is described by its name: the two preparations of single-stranded DNA are mixed together in solution. When large amounts of material are involved, the reaction can be followed by the change in optical density. With smaller amounts of material, one of the preparations may carry a radioactive label, whose entry into duplex form is followed by determining the amount of double-stranded DNA containing the

label. Double-stranded DNA can be assayed either by using chromatography to separate duplex DNA from single strands or by degrading all the single strands that have not reacted and then

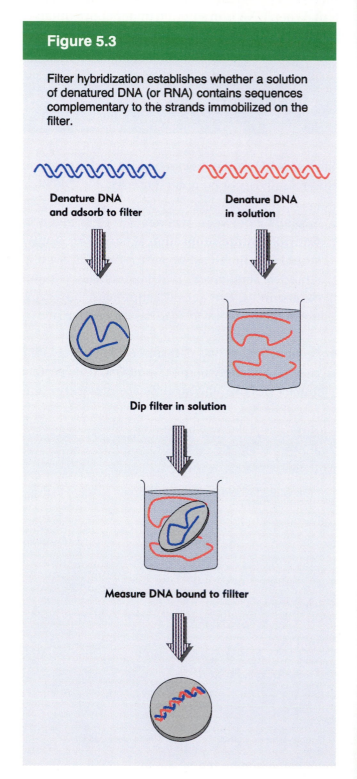

Figure 5.3

Filter hybridization establishes whether a solution of denatured DNA (or RNA) contains sequences complementary to the strands immobilized on the filter.

Denature DNA and adsorb to filter

Denature DNA in solution

Dip filter in solution

Measure DNA bound to filter

measuring the amount of material that remains.

Solution hybridization is not an appropriate technique for investigating the relationship of two preparations if one or both consist of duplex DNA. The problem is that if two duplex DNA preparations are denatured and then the single strands are mixed, two types of reaction occur. The *original* complementary single strands can renature. Or each single strand can hybridize with a complementary sequence in the *other* DNA. The competition between the two reactions makes it difficult to assess the extent of hybridization.

This difficulty can be overcome by immobilizing one of the DNA preparations so that it cannot renature. Nitrocellulose filters have the useful property of adsorbing single strands of DNA but not RNA; and once a filter has been used to adsorb DNA, it can be treated to prevent any further adsorption of single strands.

Figure 5.3 illustrates the resulting procedure in which a DNA preparation is denatured and the single strands are adsorbed to the filter. Then a second denatured DNA (or RNA) preparation is added. This material adsorbs to the filter only if it is able to base pair with the DNA that was originally adsorbed. The usual form of the experimental procedure is to add a radioactively labeled RNA or DNA preparation to the filter, allowing the extent of reaction to be measured as the amount of radioactive label retained by the filter.

The extent of hybridization between two single-stranded nucleic acids can be taken in principle to represent their degree of complementarity. Two sequences need not be *perfectly* complementary to hybridize; if they are closely related but not identical, an imperfect duplex is formed in which base pairing is interrupted at positions where the two single strands do not correspond.

Single-stranded nucleic acids may have secondary structure

The stability of the double helix results from the hydrogen bonding between the complementary A•T and G•C pairs and also from interactions between the bases as they are 'stacked' above each other along the axis of the helix. These forces can be used to predict the stability of a double helix between two complementary sequences. Because RNA is the predominant single-stranded nucleic acid, the formation of double-stranded regions from a single strand is usually analyzed in terms of RNA, but the technique is equally valid for single-stranded DNA.

The primary structure of RNA is the same as that of DNA: a polynucleotide chain with 5'–3' sugar–phosphate links. Considered as a single strand, the molecule follows a random path in space, but base pairing within it can fix the location of one region relative to another.

When a sequence of bases is followed by a complementary sequence nearby in the same molecule, the chain may fold back on itself to generate an antiparallel duplex structure, called a **hairpin**. It consists of a base-paired, double-helical region, the **stem**, with a **loop** of unpaired bases at one end. **Figure 5.4** shows an example. When the complementary sequences are relatively distant in the molecule, their juxtaposition to form a double-stranded region essentially creates a stem with a very long single-stranded loop.

Our ability to measure secondary structure is rather crude. The *overall* extent of base pairing is reflected in the biophysical properties of a molecule. However, this does not reveal which individual regions are involved. Single-stranded and double-stranded regions have different susceptibilities to some **nucleases** (enzymes that degrade nucleic acids), and this provides a test for analyzing the involvement of particular regions in base

Figure 5.4

A single-stranded nucleic acid may fold back on itself to form a duplex hairpin by base pairing between complementary sequences.

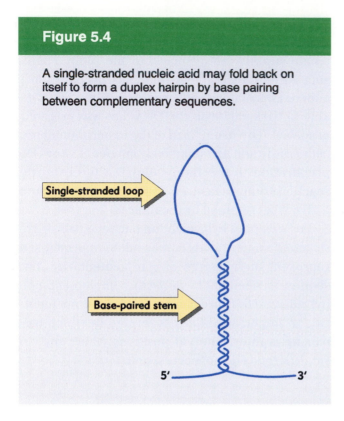

Single-stranded loop

Base-paired stem

5′   3′

pairing. In general, however, such data are effective only with relatively short molecules.

To what extent can we predict the occurrence of secondary structure in single-stranded nucleic acid sequences? The situation is more complicated than merely deciding whether secondary structure is likely to exist, because a single-stranded molecule may have several regions that are able potentially to base pair with one another in alternative arrangements; so it is necessary to resolve which (if any) actually occur.

The *plausibility* of a particular base-paired structure can be predicted by **rules** that describe the interactions of the base pairs. When alternative structures exist, their relative stabilities can be assessed by these rules. Of course, this approach treats the RNA as an isolated and stable structure, ignoring any other factors that may intervene to influence the structure (for example, binding by proteins). However, the application of these

rules provides a first step in considering the probability that a particular structure will be formed.

The basis of these rules is to calculate the **free energy** for the formation of each structure. (The free energy is a thermodynamic constant that gives the amount of energy required for or released by a reaction. It is measured in kcal/mol as the parameter ΔG. Reactions that *require* energy have a *positive* value. Reactions that *release* free energy have *negative* value.) *Energy must be released overall to form a base-paired structure; the stability of the structure is determined by the amount of energy released.* Thus if alternate structures have free energies of formation of $\Delta G = -21$ kcal/mol and $\Delta G = -35$ kcal/mol, the latter is intrinsically more likely to be formed.

The overall free energy of a double-stranded structure is given by the general equation:

$$\Delta G_{total} = \Delta G_i + \Sigma \Delta G_x + \Sigma \Delta G_u$$

The individual terms in the equation are calculated as:

◆ ΔG_i is the free energy for initiation of a double helix. It takes a *positive* value of +3.4 kcal/mol, representing the energy *required* to form the first base pair. This applies to *inter*molecular duplex formation, and does not apply to *intra*molecular duplexes (such as hairpins).

◆ $\Sigma \Delta G_x$ is the sum of the individual reactions involved in propagating the double helix as each base pair is added. The formation of each base pair *releases* energy, so this term is *negative*.

◆ $\Sigma \Delta G_u$ is the sum of individual instances encountered as the double helix is propagated in which the opposing bases are not complementary. It represents the energy *required* to hold these bases in an unpaired state, so it is *positive*.

The overall free energy of reaction, ΔG_{total} must be negative for a double helix to form. In effect, the energy released by the individual base pairing reactions ($\Sigma \Delta G_x$) must significantly exceed the energy required to initiate the double helix if applicable ($\Sigma\Delta G_i$), and to overcome any unpaired regions ($\Sigma\Delta G_u$).

In calculating the energy released by formation of a base pair, we must take account both of effects that are intrinsic to the individual base pair and of those that are influenced by the relationship between adjacent base pairs:

◆ The formation of each hydrogen bond releases rather a small amount of energy. Since G•C base pairs have three hydrogen bonds, they are more stable than A•U base pairs, which have only two hydrogen bonds. This explains why the stability of a double helix increases with the proportion of G•C base pairs.

◆ Hydrophobic interactions occur between the bases as they are stacked on top of one another within the helical axis (water is excluded from the interior of the helix, which thus forms a hydrophobic environment). The free energy of base stacking depends on the particular combination of adjacent nitrogenous bases, so the overall free energy differs for each possible doublet combination of base pairs.

The exact amount of energy released in each reaction therefore depends on the combinations of individual base pairs. This means that the free energy calculated for each possible doublet of base pairs is influenced by both composition and sequence, as can be seen from the summary in **Table 5.1.** Doublets containing only A•U pairs have ΔG values between -0.9 and -1.1, doublets containing one A•U pair and one G•C pair vary between -1.7 and -2.3, and doublets containing only G•C pairs vary from -2.0 to -3.4 kcal/mol.

Another base pair that can form in RNA is the irregular partnership G•U (and G•T can form in double-stranded regions of single-stranded DNA, although it is excluded from a regular double helix of DNA). The ΔG for a G•U pair varies from -0.5 to -1.5, depending on the neighboring base pair.

Table 5.1

Free energy of base pairing is determined by doublet sequences. The doublet sequences are written so that the top row represents a strand running 5'–3' from left to right; the lower strand is the complement running 3'–5' from left to right.

Doublet Sequence	ΔG
A-U doublets	
AA	
UU	-0.9
AU	
UA	-0.9
UA	
AU	-1.1
Mixed doublets	
CA	
GU	-1.8
CU	
GA	-1.7
GA	
CU	-2.3
GU	
CA	-2.1
G-C doublets	
CG	
GC	-2.0
GC	
CG	-3.4
GG	
CC	-2.9

The free energy released in propagating a duplex region is calculated by summing the free energies for each doublet of adjacent base pairs. Each base pair within the duplex region is involved in the calculation for two doublets, one for the base pair on each side. (A base pair that lies at

Figure 5.5

The free energy of formation for a potential base-paired region can be calculated by summing the ΔG values released (by base pair formation) or demanded (to maintain single-stranded regions in a restricted conformation).

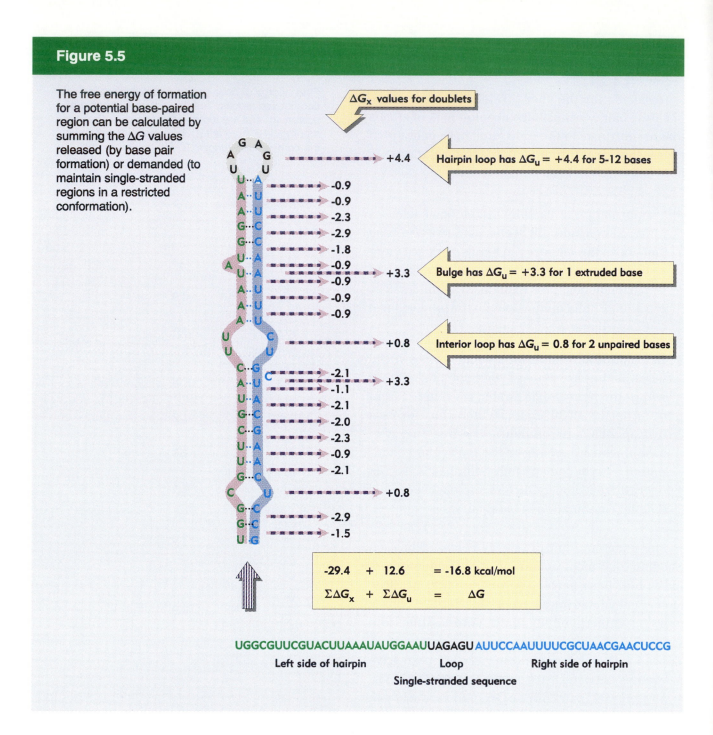

ΔG_x values for doublets

+4.4 — Hairpin loop has $\Delta G_u = +4.4$ for 5-12 bases

-0.9
-0.9
-2.3
-2.9
-1.8
-0.9
+3.3 — Bulge has $\Delta G_u = +3.3$ for 1 extruded base
-0.9
-0.9
-0.9

+0.8 — Interior loop has $\Delta G_u = 0.8$ for 2 unpaired bases

-2.1
-1.1 +3.3
-2.1
-2.0
-2.3
-0.9
-2.1

+0.8

-2.9
-1.5

-29.4	+	12.6	= -16.8 kcal/mol
$\Sigma\Delta G_x$	+	$\Sigma\Delta G_u$	= ΔG

UGGCGUUCGUACUUAAAUAUGGAAU UAGAGU AUUCCAAUUUUCGCUAACGAACUCCG

Left side of hairpin   Loop   Right side of hairpin

Single-stranded sequence

one end of the duplex participates in the formation of only one doublet, with its single neighbor.)

Unlike double-stranded DNA, which is maintained as a perfect duplex structure, double helical regions of RNA usually form between single strands that are not perfectly complementary. Interruptions in the integrity of a duplex region take several forms:

◆ **Hairpin loops** occur at the end of the duplex region when the double helix is formed by looping back to pair with an immediately adjacent complementary sequence.

◆ **Interior loops** form when there are corresponding regions in each potential complement that do not base pair.

◆ **Bulge loops** form when one of the potential complements contains an additional sequence that does not pair.

All of these unpaired regions *hinder* the formation of the double helix. Their effect is taken into account in the free energy calculation by assigning each type of interruption a positive free energy value that is included in the parameter $\Sigma \Delta G_u$. In other words, holding these regions in a particular structure requires the *input* of energy, which must be offset against the free energy released by the base pairing and stacking.

Table 5.2 summarizes the free energy demands of these structures. For interior loops and bulge loops, the energy increases with the size of the loop that must be supported. Hairpin loops require more energy at very small and large sizes; the minimum requirement is shown by a loop size of ~5 bases. An example of a calculation for the free energy of formation from the sequence of the

Table 5.2

Free energy is required to support loops and mismatches in a duplex structure.

Structure	ΔG (kcal/mol)	
	Small Loop (1–3 bases)	30 Base Loop
Hairpin loop	+7.4	+8.9
Interior loop	+0.8	+8.4
Bulge loop	+3.3	+15.8

duplex is illustrated in the example of **Figure 5.5**.

The stability of a potential double helix is determined by the free energy calculation, which must produce a sufficiently negative value overall, or the secondary structure will be unable to form.

Inverted repeats and secondary structure

When a single-stranded nucleic acid contains two sequences that are complementary, they can base pair to form a hairpin as described in Figure 5.4. The sequence in a double-stranded DNA that corresponds to such a structure consists of two copies of an identical sequence present in the reverse orientation. They are called **inverted repeats**.

The sequence of the inverted repeats together is called a **palindrome**. It is defined as a sequence of duplex DNA that is the same when either of its strands is read in a defined direction. The sequence

      5′   GGTACC   3′
      3′   CCATGG   5′

is palindromic because when either strand is read in the direction 5′ to 3′, it generates the sequence GGTACC (that is, reading left to right on the upper strand, and reading right to left on the lower strand).

A palindromic sequence is formally described as a **region of dyad symmetry**, in which the **axis of symmetry** separates the inverted repeats. By drawing a line through the axis of symmetry,

we see that the same sequence $^{GGT}_{CCA}$ is present on either side of the axis of symmetry, in inverse orientation, as indicated by the arrows. Each arrow identifies one of the inverted repeats.

The two copies of an inverted repeat need not necessarily be contiguous. Consider the same inverted repeat in a different situation

      GGTNN | NNACC
      CCANN | NNTGG

The same triplet sequence still provides an inverted

Figure 5.6

Inverted repeats could generate cruciforms in DNA or hairpins in RNA.

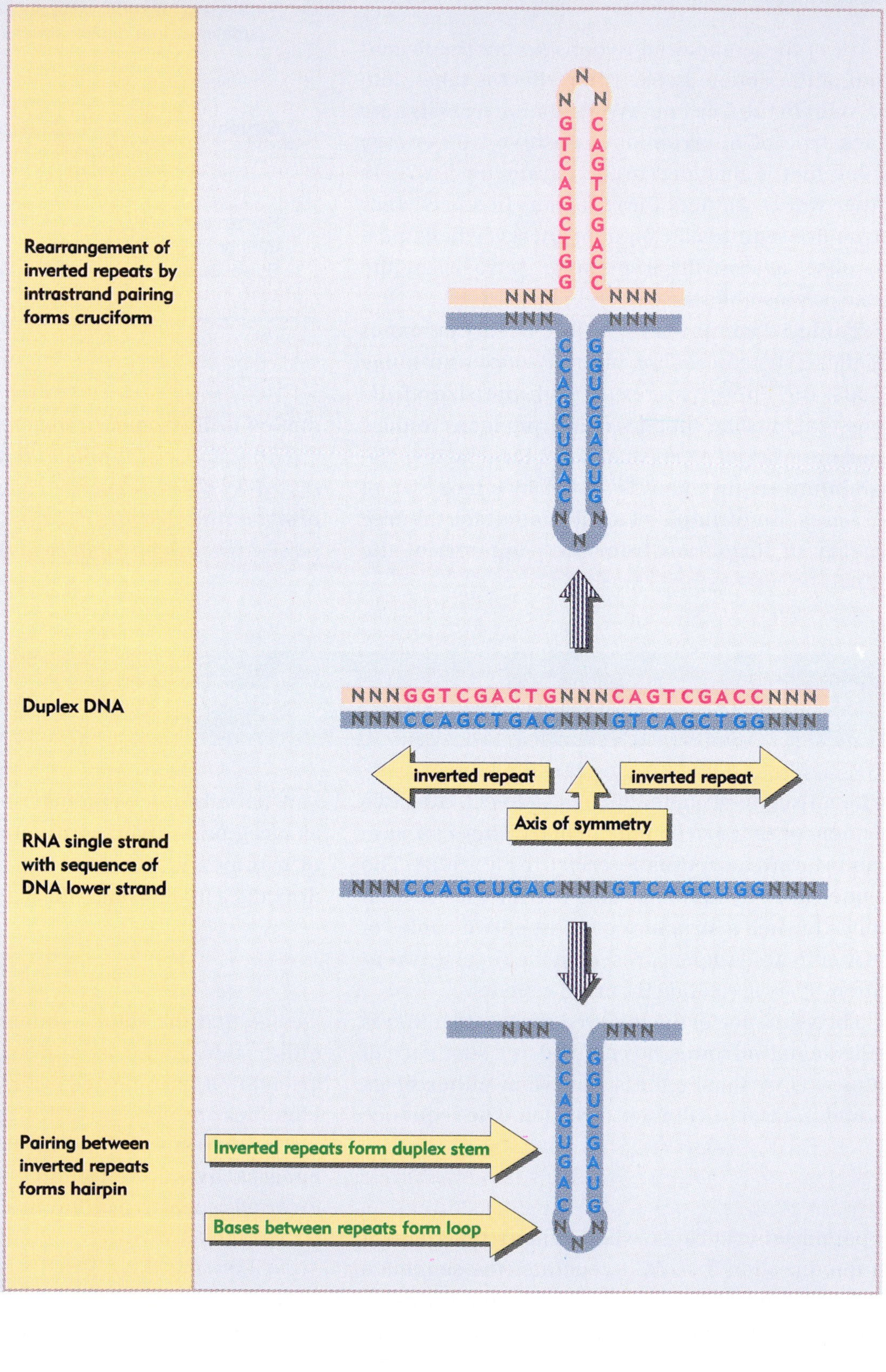
Rearrangement of inverted repeats by intrastrand pairing forms cruciform
Duplex DNA
NNNGGTCGACTGNNNCAGTCGACCNNN
NNNCCAGCTGACNNNGTCAGCTGGNNN
inverted repeat
Axis of symmetry
inverted repeat
RNA single strand with sequence of DNA lower strand
NNNCCAGCUGACNNNGUCAGCUGGNNN
Pairing between inverted repeats forms hairpin
Inverted repeats form duplex stem
Bases between repeats form loop

repeat, but the axis of symmetry has been shifted into the center of the additional four base pairs that separate the two copies of the repeat.

An inverted repeat can be reflected in the secondary structure of either single-stranded or double-stranded nucleic acid, as illustrated in **Figure 5.6**. The copies of the repeat are complementary when read in opposite directions on a single strand. (In the short example above, reading away from the axis of symmetry, we obtain TGG going left and ACC going right.) As shown in the figure for a longer palindrome, the complementary sequences can base pair to form a hairpin.

In a double-stranded DNA, the complementary sequences on one strand have the opportunity to base pair only if the strand separates from its partner. The situation is the same on each strand: a hairpin could be formed. The formation of the two apposed hairpins creates a **cruciform**, so called because it represents the junction of four duplex regions. The original duplex extends on either side, and the intrastrand hairpins protrude from it.

The conditions under which cruciforms might form *in vivo* are controversial. A double helix is unlikely to abandon its continuous duplex structure to form the protruding hairpin loops spontaneously. A considerable amount of energy would be required to separate the two strands ($\Delta G \approx +50$ kcal/mol), and only some of it would be regained by the formation of the cruciform, which is energetically less stable than a continuous duplex (by about $\Delta G = +18$ kcal/mol). Yet cruciforms can be

Figure 5.7

An electron micrograph of a cruciform generated in DNA *in vitro*. Photograph kindly provided by Martin Gellert.

generated by providing appropriate conditions *in vitro*, as visualized in the example of **Figure 5.7**.

We do not know whether cruciforms occur *in vivo*. Current opinion favors the view that probably cruciforms do not form naturally in cells; and if they are induced by constructing DNA *in vitro* with appropriate sequences, they are likely to be unstable when introduced into a cell.

Duplex DNA has alternative double-helical structures

It has been known for a long time that DNA can exist in more than one type of double-helical structure, although under physiological conditions it appeared always to take the form of the now classic B-type duplex. We still think that duplex DNA is almost entirely in the B-form within the cell, although some of the parameters that describe the precise structure of the B-type duplex have been adjusted, in particular the number of base pairs per turn. But under certain conditions, duplex DNA can make a transition from the B-form to another form. Probably this affects rather a small proportion of the DNA.

Measurements of the number of base pairs per

turn in double-stranded DNA give an answer of 10.4 instead of the 10.0 predicted by the classic B-model. The change requires a slight adjustment in the angle of rotation between adjacent base pairs along the helix, to 34.6°, so that it takes slightly longer to accomplish the full 360° turn.

The B-duplex provides a model to fit *average* data; variations could occur in particular regions, either as the result of a particular base sequence or because of conditions imposed by the environment. The tight coiling and compaction that is necessary to fit DNA into the cell influences its structure *in vivo*. Thus we take the value of 10.4 base pairs per turn to represent DNA as a whole under physiological conditions, and we assume that the value for any particular sequence will be close to, but not necessarily identical with, this estimate.

The idea that the DNA double helix has a single structure has been superseded by the view that there are families of structures. Each family represents a characteristic type of double helix, as described by the parameters n (the number of nucleotides per turn) and h (the distance between adjacent repeating units). The variation is achieved by changes in the rotation of groups about bonds with rotational freedom. Within each family, the parameters can vary slightly; for example, for B-DNA n could be 10.0–10.6.

Table 5.3 summarizes the average properties of four families of structure. The existence of the A-, B- and C-forms has been known for a considerable time, and transitions between these right-handed structures occur when appropriate changes are made in the conditions of solution. However, the Z-form represents a rather different structure that was discovered more recently.

Figure 5.8 compares the common forms of double-stranded nucleic acids. The right-handed structures of physiological significance are the B-form and A-form, which present different superficial features:

◆ The B-form represents the general structure of DNA in the conditions of the living cell. B-DNA has a major (wide) groove and a minor (narrow) groove. Differences between the bases are most easily recognized in the wide groove, which is a major point of contact for proteins that bind specific sequences of DNA.

◆ The A-form is probably very close to the conformation of double-stranded regions of RNA, where the presence of the 2′ hydroxyl group prevents adoption of the B-form. Hybrid duplexes with one strand of DNA and one strand of RNA also probably lie in the A-form.

In the relatively compact structure of A-form RNA, instead of lying flat, the bases are tilted with regard to the helical axis; and there are more base pairs per turn. The major groove is less accessible, because it is much deeper and phosphate groups overhang it. The result is that the features of individual bases are buried deep in the groove. The minor groove, however, is superficial.

Table 5.3

DNA can exist in several types of structure families. Rotation per base pair is indicated as (+) for a right-handed duplex and (-) for a left-handed duplex.

Helix Type	Base Pairs/Turn	Rotation/Base Pair	Vertical Rise/Base Pair	Helical Diameter
A	11	+34.7°	2.56Å	23Å
B	10	+34.0°	3.38Å	19Å
C	9.33	+38.6°	3.32Å	19Å
Z	12	- 30.0°	5.71Å	18Å

Figure 5.8

Nucleic acids can form several types of double helix.

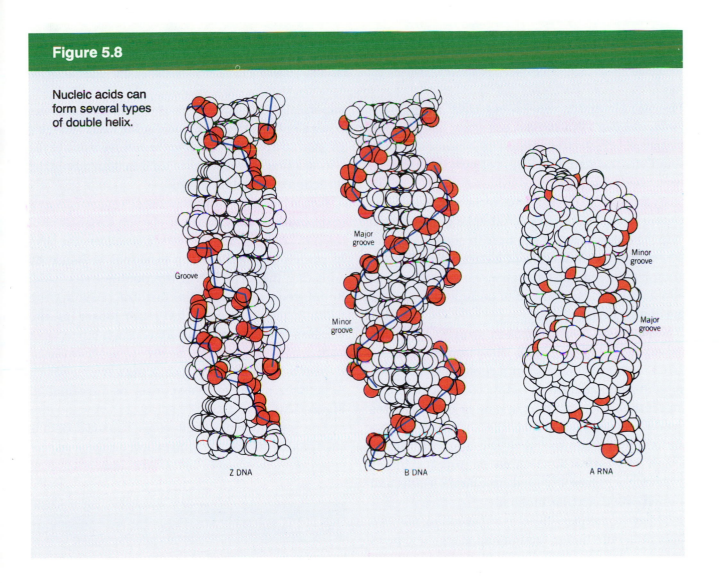

The **Z-form** provides the most striking contrast with the classic structure families. It is the only **left-handed** helix. Its structure is illustrated above in Figure 5.8. Z-DNA has the most base pairs per turn of any duplex form, and so has the least twisted structure; it is skinny, and its name is taken from the zigzag path that the sugar–phosphate backbone follows along the helix, quite different from the smoothly curving path of the backbone of B-form DNA. Z-DNA has only a single groove, with a greater density of negative charges than either of the two grooves in B-DNA.

The Z-form double helix occurs in polymers that have a sequence of alternating purines and pyrimidines. Two that have been examined have a simple repeating dinucleotide sequence: poly $d_{\{CG\}}^{\{GC\}}$ and poly $d_{\{TG\}}^{\{AC\}}$. (The d indicates that these are the deoxy forms, that the sequence is DNA and not RNA.)

Z-form DNA was discovered *in vitro* under somewhat unusual conditions (using a high salt concentration to counter the increased electrostatic repulsion between the nucleotides compressed into the more slender double helix of Z-DNA). It is not clear whether Z-DNA forms *in vivo*.

Closed DNA can be supercoiled

The double helix represents DNA as a linear molecule. But DNA *in vivo* generally has a **closed** structure: it lacks free ends. The genomes of some small viruses actually consist of **circular DNA, in** which both strands of the double helix run continuously around the circle. In bacterial and eukaryotic genomes, the DNA exists as rather large loops; each loop is held together at the base in such a way that it becomes a simulacrum of a circle. This form of organization is important because it allows an additional structural constraint to be superimposed on the double helix.

Supercoils are introduced into DNA when a duplex is twisted in space around its own axis. The usual analogy is to consider a rubber band twisted about itself to generate a tightly coiled structure in which the rubber band (the double helix) crosses over itself in space. Possible forms of a DNA molecule are compared in the electron micrograph of **Figure 5.9**, which captures a linear molecule, a nonsupercoiled circular molecule, and a supercoiled circular molecule.

The twisting introduced by supercoiling places a DNA molecule under torsion and gives the type of structure depicted in **Figure 5.10**. Supercoiling can occur *only* in closed structures, because an **open** molecule can release the torsion simply by untwisting. A closed molecule must have no breaks on *either* strand of DNA; a break even in one strand of a circular molecule allows untwisting. A molecule that lacks supercoiling, whether closed or open, is said to be **relaxed**.

The principle behind supercoiling is that *the structure cannot unravel when a duplex is wound around itself in space and the ends are fixed.* One supercoil (or more formally, supercoiled turn) is introduced every time that the duplex thread is twisted about its axis; and the greater the number of supercoils, the greater the torsion in a closed molecule.

For a rubber band, it does not matter in which *direction* we apply the twist that generates the

supercoils. (The two edges of the rubber band are equivalent.) But because the double helix is itself a twisted structure (as seen in the intertwining of the two strands), its response to torsion depends on the direction of the supercoiling.

Negative supercoils twist the DNA about its axis in the *opposite* direction from the clockwise turns of the right-handed double helix. This allows the DNA in principle to relieve the torsional pressure by adjusting the structure of the double helix itself. Relief generally takes the form of reducing the rotation per base pair, that is, of loosening the winding of the two strands about each other. DNA with negative supercoils is said to be **underwound**. If the torsion is great enough, it may even lead to a limited disruption of base pairing, as illustrated in Figure 5.10.

The opposite types of effect are caused if DNA is supercoiled in the *same* direction as the intrinsic winding of the double helix. The introduction of

Figure 5.9

Supercoiled DNA has a compact twisted structure compared with a non-supercoiled circle or linear molecule. Photograph kindly provided by Svend O. Freytag.

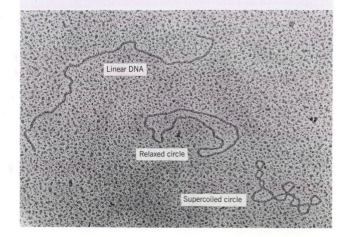

Linear DNA

Relaxed circle

Supercoiled circle

Figure 5.10

Supercoiling causes a DNA duplex to be twisted about itself in space; negative supercoiling can be relieved by disrupting base pairs.

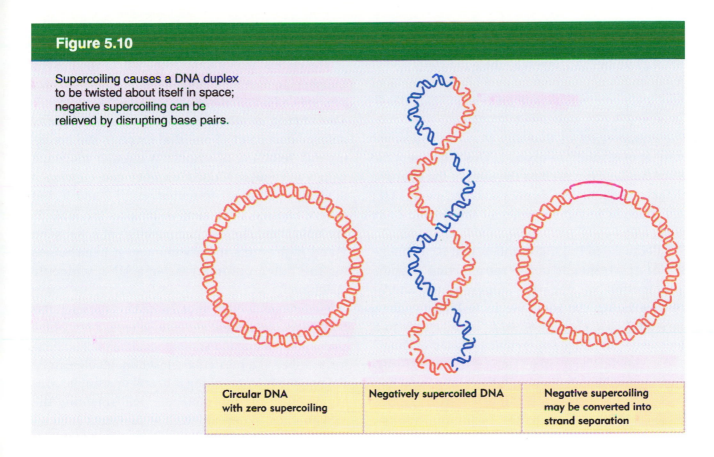

| Circular DNA with zero supercoiling | Negatively supercoiled DNA | Negative supercoiling may be converted into strand separation |

positive supercoils tightens the structure, applying torsional pressure to wind the double helix even more tightly. Positively supercoiled DNA is said to be **overwound**. This state can be created by treatment *in vitro*.

The supercoiling of DNA is often described in terms of the **superhelical density**, the number of superhelical turns per turn of the double helix. It was originally described as the parameter $\sigma = \tau/\beta$, where τ is the number of superhelical turns and β is the number of turns of the duplex, approximately the number of base pairs divided by 10. Thus σ corresponds to the number of superhelical turns per ~10 bp. It is negative for negative supercoiling and positive for positive supercoiling.

Current nomenclature to describe supercoiling is based on the concept of the **linking number**, which specifies the number of times that the two strands of the double helix of a closed molecule cross each other *in toto*. The linking number is the number of revolutions that one strand makes around the other when the DNA is considered (hypothetically) to lie flat on a plane surface. The convention is to count the linking number so that it is positive for each crossover in a right-handed double helix. It is necessarily an integer.

The linking number has two components, the twist and the writhe, defined by the equation

$$L = W + T$$

The **twisting number**, T, is a property of the double helical structure itself, representing the rotation of one strand about the other. It represents the *total number of turns of the duplex*. It is determined by the number of base pairs per turn. For a relaxed closed circular DNA lying flat in a plane, the twist is the total number of base pairs divided by the number of base pairs per turn.

The **writhing number**, W, represents the *turning of the axis of the duplex in space*. It corresponds to the intuitive concept of supercoiling, but does not have exactly the same quantitative definition or measurement. For a relaxed molecule, $W = 0$.

In this case, the linking number equals the twist.

We are often concerned with the *change* in linking number, ΔL, given by the equation

$$\Delta L = \Delta W + \Delta T$$

The equation states that any change in the total number of revolutions of one DNA strand about the other can be expressed as the sum of the changes of the coiling of the duplex axis in space (ΔW) and changes in the screwing of the double helix itself (ΔT).

A decrease in linking number, that is, a change of $-\Delta L$, corresponds to the introduction of some combination of negative supercoiling and/or underwinding. An increase in linking number, measured as a change of $+\Delta L$, corresponds to a decrease in negative supercoiling/underwinding. We can describe the change in state of any DNA by the specific linking difference, $\sigma = \Delta L/L^0$, where L^0 is the linking number when the DNA is relaxed. If all of the change in linking number were due to change in W (that is, $\Delta T = 0$), the specific linking difference equals the supercoiling density. In effect, σ as defined in terms of $\Delta L/L^0$ can be assumed to correspond to superhelix density so long as the structure of the double helix itself remains constant.

The critical feature about the use of the linking number is that *this parameter is an invariant property of any individual closed DNA molecule.* The linking number cannot be changed by any deformation short of one that involves the breaking and rejoining of strands. A circular molecule with a particular linking number can express it in terms of different combinations of T and W, but cannot change their sum so long as the strands are unbroken. (In fact, the partition of L between T and W prevents the assignment of fixed values for the latter parameters for a DNA molecule in solution.)

We now see the utility of the linking number. *It is related to the actual enzymatic events by which changes are made in the topology of DNA.* The linking number of a particular closed molecule can be changed only by breaking a strand or strands, coiling or uncoiling the molecule, and rejoining the broken ends. When an enzyme performs such an action, it must change the linking number by an integer; this value can be determined as a characteristic of the reaction. Then we can consider the effects of this change in terms of ΔW and ΔT.

Supercoiling influences the structure of the double helix

Changes in the structure of DNA do not occur spontaneously *in vivo*—they happen in an environment in which DNA is under a variety of constraints. Some of these changes are assisted by negative supercoiling.

The introduction of negative supercoiling requires energy; the supercoiled molecule has more energy through its possession of the supercoils. We can regard them as a store of energy. So negative supercoiling influences the equilibrium of a structural change; and this enables supercoiled DNA to undergo structural transitions in conditions in which relaxed DNA would be unable to do so.

Can we quantitate the effect of negative supercoiling? All genomes that have been examined exhibit some negative supercoiling. A typical level *in vivo* is ~1 negative turn for every 200 bp, which is described as a superhelical density of -0.05. This confers an energy on the molecule of about -9 kcal/mol.

Supercoiling *in vivo* is controlled by a balance between the activities of enzymes that introduce supercoils into DNA and enzymes that remove them (see Chapter 33). Mutations that affect the activities of these enzymes have been found in bacteria. Strains with altered levels of supercoiling usually grow more slowly; the original growth rate is resumed by strains that have gained compensating mutations tending to restore the original level.

The average level of supercoiling is therefore important *in vivo*. However, the level is not necessarily even throughout the genome, but fluctuates. It is possible that increased supercoiling in particular regions could assist structural transitions in DNA.

One such transition is strand separation, which is assisted by the effect of negative supercoiling in underwinding the molecule. *The excess energy possessed by a negatively supercoiled molecule can be used to help provide the energy needed to separate the strands of DNA.*

The energy needed to unwind DNA depends on the sequence of base pairs; it is in effect the reverse of the energy released when a double helix forms as described in Figure 5.5. Some 12–50 kcal/mol are needed to separate 10 bp. So the actual level of supercoiling probably corresponds to enough energy to assist the unwinding of just a very few base pairs. But such effects could be important, for example, in initiating strand separation in order to replicate DNA.

The ability of circular molecules of DNA to take up different forms shows that long chains of DNA are quite flexible. Deformation from the monotonous linear duplex can occur either gradually (by ΔT or ΔW that is distributed along the molecule) or abruptly, in the form of a bend. Some sequences of DNA bend intrinsically, because of interactions between the adjacent base pairs; in other cases, bends are introduced or caused by binding of proteins that deform the double helix. We take up the characterization of bends later (see Figure 15.23).

Summary

The ability of DNA strands of a double helix to separate can be mimicked *in vitro* by denaturation. The T_m is one measure of the stability of a double helix. Complementary single strands can renature or hybridize (when from different sources).

For a double helix to form between complementary sequences, base pairing reactions must release sufficient free energy to meet the requirement for initiation of pairing and to overcome restraints imposed by noncomplementary pairs in the duplex region. Free energy released by base pairing depends on doublet sequences. Inverted repeats can in principle generate a cruciform by intrastrand pairing instead of by interstrand pairing, but this requires significant energy and is unlikely to occur *in vivo*.

Each family of double-helical nucleic acid structures allows local variation in base pairs per turn and rotation per base pair. The A, B, and C forms are right-handed. DNA usually occurs in B-form; RNA and DNA–RNA hybrids usually occur in A-form. Z-DNA takes a left-handed conformation and is formed *in vitro* by certain simple repeating sequences, such as alternating pyrimidines and purines.

Supercoiling is caused when a DNA double helix is itself coiled in space, and may be described either by the older terminology of supercoiling density or the more recent terminology of linking number. Negative supercoiling twists the DNA about its axis in the opposite direction from the clockwise turns of the right-handed double helix. The torsion is relieved by reducing the rotation per base pair in the double helix or ultimately by unwinding the two strands. A typical value for negative supercoiling *in vivo* is −1 turn/200 bp, or a superhelical density of −0.05. Supercoiling may be important locally in assisting unwinding of DNA and generally in reactions that require strand separation.

In contrast with the view of DNA as a monotonous, linear double helix, DNA in the cell is compacted into the nucleus by deformations of structure that include supercoiling and bending. This may involve significant variations from the average parameters of B-form DNA.

Further reading

Reviews

Forms of DNA characterized in crystals have been summarized by **Drew, McCall, and Calladine** (*Ann. Rev. Cell Biol.* 4, 1–20, 1988).

Structural features of DNA have been discussed thoroughly by **Kornberg and Baker** (*DNA Replication*, W. H. Freeman & Co., New York, 1992).

Research Articles

Conditions for cruciform generation have been analyzed by **Wang** (*Cell* 33, 817–829, 1983).

CHAPTER 6

Isolating the gene

A chromosome contains a long, uninterrupted thread of DNA along which lie many genes. Identifiable sites in the genetic material are provided by mutations that create changes in the sequence of base pairs. On the linkage map of the chromosome, distances are represented by the frequencies of recombination. A eukaryotic chromosome contains thousands of genes, and at the molecular level, a map unit is a very large measure indeed.

Placing a gene on a linkage group requires that at least one previously identified marker lies within ~40 map units of it. To have markers available at intervals of 40 map units, ~100 evenly spaced genes would be needed for the human genome. Of course, obtaining 100 evenly spaced markers requires mapping of a much larger number of loci.

Equating genes with linkage groups is only the start of genetic mapping. More detailed localization requires a high concentration of markers, a situation achieved in higher eukaryotes with the nematode worm (*C. elegans*) and the fruit fly (*D. melanogaster*). In the most intensively mapped mammal, the mouse, markers are located every map unit or so in some parts of the genome, but there are other regions where the nearest markers are >30 map units apart.

In mapping genes we are not restricted solely to genetic analysis. We may also use techniques for directly visualizing the location of particular sequences on the chromosome. When the sequence corresponds to a known protein, these methods allow us to see directly where its gene is located. This approach reverses the old order of events: instead of using genetic mapping to isolate

the gene, we use the isolated gene for mapping.

By comparing the total map length of all the linkage groups in a genome with the total amount of DNA, we can obtain an idea of how frequently genetic exchange occurs on the average relative to length of DNA. We take 50% recombination to correspond to the average distance between independent recombination events. Values for a range of genomes are summarized in **Table 6.1**.

Phages engage in *exceedingly* frequent genetic exchanges (although any quantitation of the frequency is only *very* approximate). Also, it is possible to generate enormous numbers of progeny, so that even the rarest recombination events (between mutations in adjacent base pairs) can be measured. These features allow the detailed mapping within a gene described in Chapter 3. The rate of genetic exchange in bacteria is also high, and again it is possible to identify recombinants from very large numbers of progeny.

The high rate of recombination in prokaryotes makes it difficult to apply the concept of the map unit consistently, but the table shows that there are ~2000 bp per map unit. The prokaryotic genetic map is therefore constructed more or less at the level of the gene itself.

In lower eukaryotes, at least as typified by yeast and other fungi, the frequency of recombination remains high, but in higher eukaryotes it is roughly a hundred times lower. Mapping of genes to linkage groups is feasible, but the difficulties of mapping at the molecular level are compounded by the large size of the map unit, 10^5–10^6 bp, and the small number of progeny from each mating.

Scrutinized from the perspective of the genetic

Table 6.1

The frequency of genetic recombination is much lower in eukaryotes than in prokaryotes.

Species	Size of Haploid Genome	Units in Genetic Map	Size of Map Unit	Average Distance Between Crossovers
Phage T4	1.6×10^5 bp	800	200 bp	1.0×10^4 bp
E. coli	4.2×10^6 bp	1750	2,400 bp	1.2×10^5 bp
Yeast	2.0×10^7 bp	4200	5,000 bp	2.5×10^5 bp
Fungus	2.7×10^7 bp	1000	27,000 bp	1.3×10^6 bp
Nematode	8.0×10^7 bp	320	250,000 bp	1.2×10^7 bp
Fruit fly	1.4×10^8 bp	280	500,000 bp	2.5×10^7 bp
Mouse	3.0×10^9 bp	1700	1,800,000 bp	9.0×10^7 bp
Human (male)	3.3×10^9 bp	2809	1,200,000 bp	6.0×10^7 bp
Human (female)	3.3×10^9 bp	4782	700,000 bp	3.5×10^7 bp

map, the outlook for characterizing individual genes is rather poor. Suppose a gene in the fruit fly consists of 5000 bp. Its length corresponds to 0.01% recombination. To detect recombination between mutations located at opposite ends of the gene, it would be necessary to isolate 1 fly in 10,000 progeny. The result is that conventional genetic mapping within genes of the fruit fly is just possible. It is not at all practical with mammals, in which the frequency of recombination is lower and the number of progeny from each mating is less. In mouse or man, for example, crossovers are separated by ~10^7 bp, so their distance apart is much greater than the length of the average gene (~10,000 bp).

How accurately does the genetic map represent the actual length of the genetic material? At the gross level of the chromosome, the genetic map usually provides a reasonable representation of the apparent organization (as illustrated in Figure 3.8), although the recombination distance between any pair of markers does not accurately reflect their physical distance apart (see Figure 6.11). And, of course, the map unit depends on the relative frequency of recombination; since this is strikingly different in each species, the genetic maps of different organisms cannot be compared.

Even in prokaryotes, the information to be gained from fine structure mapping is limited by the nature of the recombination event. At the level of mapping within a gene, recombination frequencies are not independent of the particular mutations used in each cross, but may be influenced by the local sequence of DNA. In other words, they do not have the ideal property of **allele-independence** described in Chapter 3, but actually show allele-specific effects. Our view of the gene in terms of the genetic map is therefore distorted by the characteristics of the recombination systems.

So at the molecular level, the map distances between mutations do not necessarily correspond with the distance that actually separates them on DNA. The resolution of the genetic map also is limited by the availability of mutations; a gap in the map could represent a region in which no mutations have been found, or could be caused by a local increase in recombination frequency.

We have no reason to suppose that the situation would be any different in eukaryotes if we were able to perform fine structure mapping. So for both theoretical and practical reasons, we need another source of information. How are we to establish the molecular structure of the gene and relate it to the structure of the protein product? How far apart are adjacent genes and how are we to recognize the regions between them?

The ultimate aim in characterizing a gene is to

determine its nucleotide sequence and to extend the sequence to the neighboring genes on either side. Genes can be isolated by working back from a protein, using nucleic acid hybridization techniques to isolate the corresponding DNA. It has become almost routine to isolate the cellular DNA for which a protein product is available (although this can be a lengthy process if the product is scarce).

Given an isolated segment of DNA, we can determine its sequence between points that are within ~300 bp. By choosing suitable points, short regions can be connected into a sequence of an entire gene and its environs. By comparing the sequence of the DNA with the sequence of the protein that the gene represents, we can delineate the regions that code for the polypeptide; and by extending the sequence in either direction, the distance to the next gene can be determined.

By comparing the sequence of a wild-type DNA with that of a mutant allele, we can determine the nature of the mutation and its exact site of occurrence. This defines the relationship between the genetic map (based entirely on sites of mutation) and the physical map (based on or even comprising the sequence of DNA). Ultimately, the map of a region of the genome can be expressed in base pairs of DNA rather than in the relative map units of formal genetics. A map in terms of DNA sequence gives the genotype at the highest degree of resolution.

Since genes now can be identified by virtue of their protein products, or even sometimes by DNA sequences alone, we are no longer entirely dependent on mutations to provide the raw material for constructing a map of the genome. Of course, mutants remain essential for identifying the functions of gene products. One means of compensating for the absence of natural mutations in any particular gene is provided by techniques that allow genes to be disrupted in animal genomes, allowing the consequences of null mutations to be visualized. Together with techniques that allow the introduction of new genes, mutant genes can be constructed *in vitro* and then tested *in vivo*; this extends the ability to characterize genes well beyond what is provided by Nature. We discuss the use of these techniques to extend and substitute for conventional genetics in Chapter 36.

A restriction map can be constructed by cleaving DNA into specific fragments

Once a segment of DNA has been isolated, a crucial step *en route* to obtaining its sequence is to map the nucleic acid at the molecular level. A physical map of any DNA molecule can be obtained by breaking it at defined points whose distance apart can be accurately determined. Specific breakages are made possible by the ability of certain **restriction enzymes** to recognize rather short sequences of double-stranded DNA as targets for cleavage.

Each restriction enzyme has a particular target in duplex DNA, usually a specific sequence of 4–6 bp. The enzyme cuts the DNA at every point at which its target sequence occurs. Different restriction enzymes have different target sequences, and a large range of these activities (obtained from a wide variety of bacteria) is now available. (They are discussed in the context of their natural habitat in Chapter 20.)

A map of DNA obtained by identifying these points of breakage is known as a **restriction map**. The map represents a linear sequence of the sites at which particular restriction enzymes find their targets. Distance along such maps is measured directly in base pairs (abbreviated **bp**) for short distances; longer distances are given in **kb**, corresponding to kilobase pairs in DNA or to kilobases in RNA. At the level of the chromosome, a map is described in megabase pairs ($1\ \text{Mb} = 10^6\ \text{bp}$).

When a DNA molecule is cut with a suitable restriction enzyme, it is cleaved into distinct fragments. These fragments can be separated on the basis of their size by **gel electrophoresis**. In this technique, the cleaved DNA is placed on top of a gel made of agarose or polyacrylamide. When an electric current is passed through the gel, each fragment moves down it at a rate that is inversely related to the log of its molecular weight.

This movement produces a series of **bands**. Each band corresponds to a fragment of particular size, decreasing down the gel. The length of any particular fragment can be determined by **calibrating** the gel. This is done by running a control in parallel in another slot of the same gel. The control has a mixture of standard fragments all of known size (often called **markers**). The migration of the size markers defines the relationship between fragment length and distance moved for the particular gel.

Figure 6.1 shows an example of this technique. A DNA molecule of length 5000 bp is incubated separately with two restriction enzymes, A and B. After cleavage the DNA is electrophoresed. The sizes of the individual fragments generated by enzyme A (left) or enzyme B (right) are determined by comparison with the positions of fragments of known size, such as the control shown in the center. This demonstrates that enzyme A has cut the substrate DNA into four fragments (of lengths 2100, 1400, 1000, and 500 bp), while enzyme B has generated three fragments (of lengths 2500, 1300, and

---

Figure 6.1

DNA can be cleaved by restriction enzymes into fragments that can be separated by gel electrophoresis.

Cleave with enzyme A

Cleave with enzyme B

Electrophorese A-fragments

Electrophorese control DNA

Electrophorese B-fragments

2100	2500	2500
	2000	
1400	1500	1300
1000	1000	1200
500	500	

| Fragment sizes compared with control | Control consists of fragments of known size | Fragment sizes compared with control |

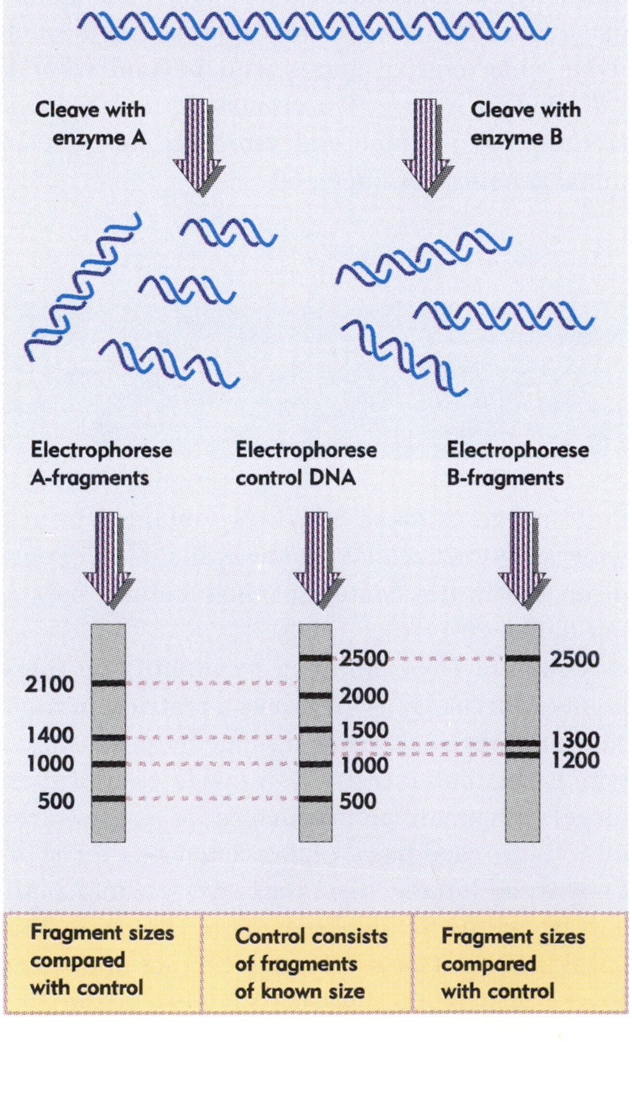

Figure 6.2

Double digests define the cleavage positions of one enzyme with regard to the other.

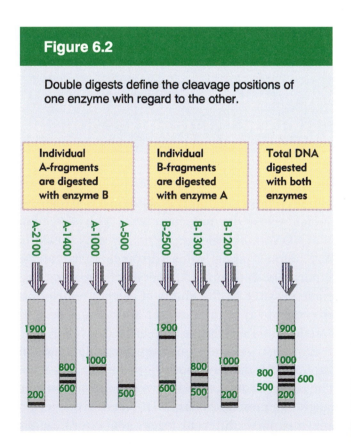

| Individual A-fragments are digested with enzyme B | Individual B-fragments are digested with enzyme A | Total DNA digested with both enzymes |

Figure 6.3

A restriction map can be constructed by relating the A-fragments and B-fragments through the overlaps seen with double digest fragments.

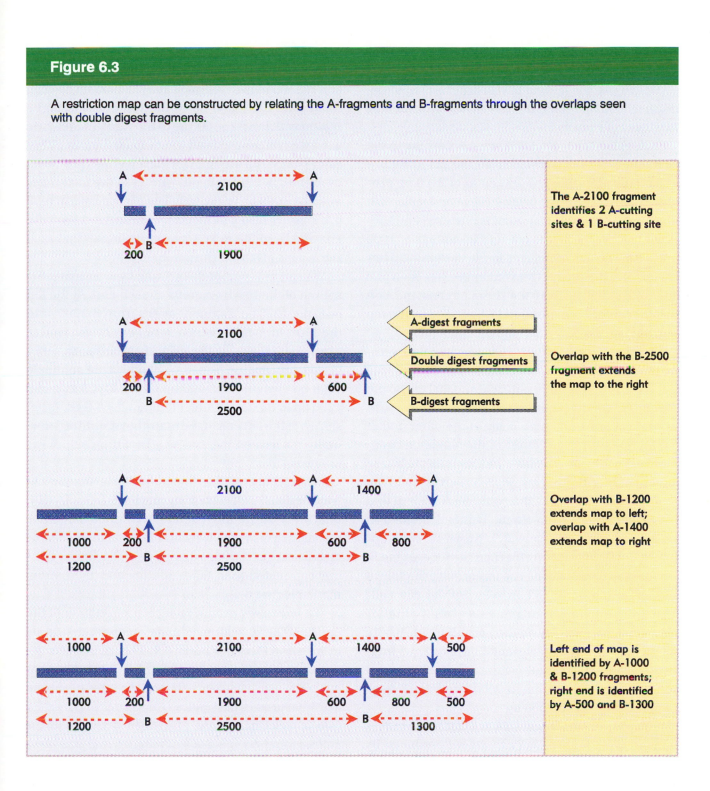

	The A-2100 fragment identifies 2 A-cutting sites & 1 B-cutting site
A-digest fragments	Overlap with the B-2500 fragment extends the map to the right
Double digest fragments	
B-digest fragments	
	Overlap with B-1200 extends map to left; overlap with A-1400 extends map to right
	Left end of map is identified by A-1000 & B-1200 fragments; right end is identified by A-500 and B-1300

1200 bp). Can we proceed further from these data to generate a map that places the sites of breakage at defined positions on the DNA?

The patterns of cutting by the two enzymes can be related by several means. **Figure 6.2** illustrates the principle of analysis by **double digestion**. In this technique, the DNA is cleaved simultaneously with two enzymes as well as with either one by itself. The most decisive way to use this technique is to extract each fragment produced in the individual digests with either enzyme A or enzyme B and then to cleave it with the other enzyme. The products of cleavage are analyzed again by electrophoresis.

We can use these data to construct a map of the original 5000 bp molecule of DNA, as illustrated by the stages of **Figure 6.3**.

Each gel in Figure 6.2 is labeled according to the fragment that was isolated from the gel in Figure 6.1. A-2100 identifies the fragment of 2100 bp produced by degrading the original DNA molecule with enzyme A. When this fragment is retrieved and subjected to enzyme B, it is cut into fragments of 1900 and 200 bp. So one of the cuts made by enzyme B lies 200 bp from the nearest site cut by enzyme A on one side, and is 1900 bp from the site cut by enzyme A on the other side. This situation is described by the top map in Figure 6.3.

A related pattern of cuts is seen when we examine the susceptibility of fragment B-2500 to enzyme A. It is cut into fragments of 1900 and 600 bp. So the 1900 bp fragment is generated by double cuts, with an A site at one end and a B site at the other end. It can be released from either of the single-cut fragments (A-2100 or B-2500) that contain it. These single-cut fragments must therefore **overlap** in the region of the 1900 bp of the common fragment that can be generated from them. This is described in the second map of Figure 6.3, which extends our map to the right to add a cleavage site for enzyme B.

This map demonstrates an important principle of restriction mapping. When we consider the construction of larger fragments from smaller fragments, we can rely on the **complete additivity** of lengths (within experimental limits). Thus fragment A-2100 consists of fragments of 200 bp and 1900 bp, while fragment B-2500 consists of fragments of 1900 bp and 600 bp.

When all of the fragments are analyzed in this manner, we see that every fragment produced by cutting an original A fragment with the B enzyme *also* is found in one of the double digests in which an original B fragment was cut with the A enzyme. The entire pattern can be seen in the complete double-digest (the gel at the right of Figure 6.2), in which every double-cut fragment occurs once. These data allow the sites of cutting to be placed into an unequivocal map.

The key to restriction mapping is the use of overlapping fragments. Because of the overlap of A-2100 and B-2500 in the central region of 1900 bp, we can relate the A site 200 bp to the left of the 1900 bp region with the B site 600 bp to the right. In the same way, we can now extend the map farther on either side. The 200 bp fragment at the left is also produced by cutting B-1200 with enzyme A, so the next B site must lie 1000 bp to the left. The 600 bp fragment at the right is also produced by cutting A-1400 with enzyme B, so the next A site must lie 800 bp to the right. This gives the third map in Figure 6.3.

We can now complete the map by identifying the source of the two fragments at each end. At the left end, the 1000 bp fragment arises from B-1200 or in the form of A-1000, which is not cut by enzyme B. Thus A-1000 lies at the end of the map; in other words, proceeding from the left end of the complete 5000 bp region, it is 1000 bp to the first A site and 1200 bp to the first B site. (This is why a B cut is not shown at the left end of the map above, although formally we treated the end as a B-cutting site in the analysis.)

At the right end of the map, the 800 bp double-cut fragment is generated by cutting B-1300 with enzyme A, so we must add a fragment of 500 bp to the right. This is the terminal fragment, as seen by its presence as A-500 in the single-cut A digest. Thus our completed map takes the form of the bottom map in Figure 6.3.

We have now produced a restriction map of the entire 5000 bp region. This is recapitulated in its more usual form in **Figure 6.4**. The map shows the positions at which particular restriction enzymes cut DNA; the distances between the sites of cutting are measured in base pairs. *Thus the DNA is divided into a series of regions of defined lengths that lie between sites recognized by the restriction enzymes.*

The actual construction of a restriction map usually requires recourse to several enzymes, so it becomes necessary to resolve quite a complex pattern of the overlapping fragments generated by the various enzymes. Several further techniques are used to assist the mapping.

One approach is to make **partial digests**, by using conditions in which an enzyme does not

Figure 6.4

A restriction map is a linear sequence of sites separated by defined distances on DNA. The map identifies the sites cleaved by enzymes A and B, as defined by the individual fragments produced by the single and double digests.

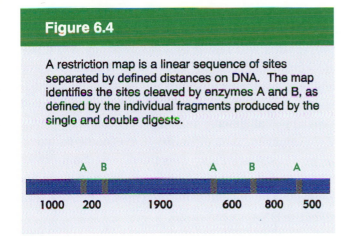

recognize every target in every DNA molecule, but instead sometimes fails to cleave some of its target sites. The conditions can be set so that the enzyme has a specified probability, for example, 50%, of cutting at any particular one of its recognition sites.

With this technique, enzyme A might generate partial cleavage fragments of 3500, 3100 and 1400 bp with the DNA segment mapped in Figure 6.2. By comparing the partial products with the complete digest pattern, the 1000 and 2100 bp fragments can be assigned as adjacent components of the 3100 bp partial fragment. Similarly, the 2100 and 1400 bp fragments could be placed as adjacent components of the 3500 bp partial fragment. Thus this technique allows the sites cut by a single enzyme to be ordered.

Another useful technique is **end-labeling**, in which the ends of the DNA molecule are labeled with a radioactive phosphate (certain enzymes can add phosphate moieties specifically to 5′ or to 3′ ends). This allows the fragments containing the ends to be identified directly by their radioactive label. Thus in the fragment A preparation, A-1000 and A-500 would be placed immediately at opposite ends of the map; similarly, fragments B-1200 and B-1300 would be identified as ends.

By combining partial digestion with end-labeling, a series of sites cut by one enzyme can be mapped directly relative to the end. **Figure 6.5** shows that if fragments are identified *only by their radioactively labeled ends*, those cut from within the molecule are ignored, while a series of fragments

Figure 6.5

When restriction fragments are identified by their possession of a labeled end, each fragment directly shows the distance of a cutting site from the end. Successive fragments increase in length by the distance between adjacent restriction sites.

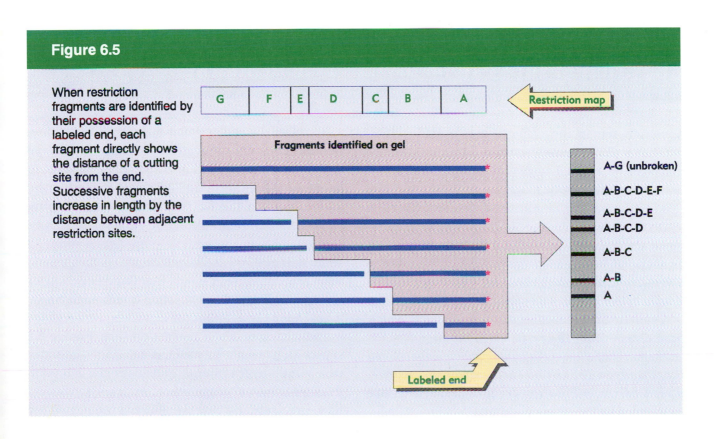

identifies the distance of each cutting site from the labeled end. If the left end of the 5000 bp fragment of Figure 6.4 were labeled, partial cleavage with enzyme A would immediately identify cutting sites at 1000, 3100, and 4500 bp from the end.

In analyzing the products of cleavage by multiple enzymes, fragments are often compared by the technique of nucleic acid hybridization. This greatly strengthens the analysis. For example, in constructing the second map of Figure 6.3, we inferred that the A-2100 and B-2500 fragments must overlap because they both generate a fragment of 1900 bp in double digests. However, in an analysis of a longer DNA, there could be several fragments of the same size. Hybridization allows us to test directly whether two fragments overlap; thus A-2100 and B-2500 would hybridize.

All of these techniques require that we have at our disposal a complete set of unique restriction fragments that together account for the entire DNA region being mapped. In some circumstances this expectation may not be fulfilled.

If several sites for one enzyme lie *very close* *together* (say, within 50 bp), very small fragments are generated that are lost from the agarose gel. Of course, this will result in a discrepancy when the molecular weights of the other fragments are totalled, but this would not necessarily be considered sinister by itself, since there are always modest experimental discrepancies when fragment sizes are compared. This difficulty is overcome by using several restriction enzymes to construct a map, so there are sufficient overlaps between the fragments generated by the different enzymes to be sure that all of the DNA is contained.

A complication is provided by the assumption that every restriction fragment is seen as an individual band on the gel. This will not be true when more than one fragment has the same molecular weight (or, more practically, whenever fragments exist with sizes so similar that they cannot be resolved.) To check whether each band is unique, its intensity can be quantitated. Bands that contain more than one fragment should be correspondingly more intense. Multiple fragments of the same size can be resolved by using further restriction enzymes.

Restriction sites can be used as genetic markers

A restriction map and a genetic map provide different views of the same material. The two types of map can be related to reveal the physical basis underlying the genetic map.

A restriction map identifies a linear series of sites in DNA, separated from one another by actual distance along the nucleic acid. *A restriction map can be obtained for any sequence of DNA, irrespective of whether mutations have been identified in it, or, indeed, whether we have any knowledge of its function.*

A genetic map identifies a series of sites at which mutations occur. *The existence of the sites depends on the fact that changes in base sequence have altered the phenotype. The distance between the sites is determined by recombination frequencies*, which are influenced by the local conditions.

To relate the restriction map to the genetic map, we must compare the restriction maps of wild-type and corresponding mutant DNAs. The relationship between the genetic and physical maps can be determined on the basis of restriction mapping alone when there are mutations that make significant changes in the genetic map. For example, a small deletion or insertion that affects the genetic map should cause a corresponding reduction or increase in size of the restriction fragment in which it lies. A longer deletion (or insertion) removes (or introduces) a series of restriction sites. **Figure 6.6** shows an example in which a deletion eliminates two complete fragments and fuses the adjacent fragments into a single piece. An insertion would have precisely the reverse effect; a mutant would

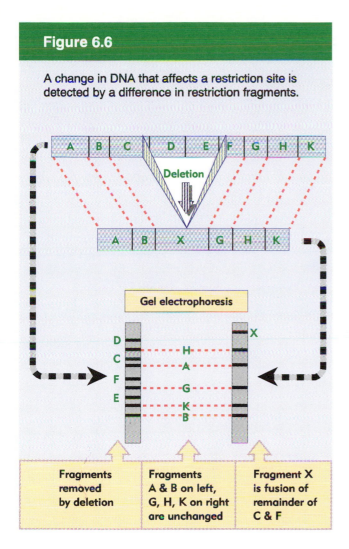

Figure 6.6

A change in DNA that affects a restriction site is detected by a difference in restriction fragments.

Deletion

Gel electrophoresis

| Fragments removed by deletion | Fragments A & B on left, G, H, K on right are unchanged | Fragment X is fusion of remainder of C & F |

existence of variations in the genetic constitution of the population. The coexistence of more than one variant is called **genetic polymorphism**. Any site at which multiple alleles exist as stable components of the population is by definition polymorphic. For example, we have seen that the fruit fly *D. melanogaster* is polymorphic for the series of alleles at the *white* locus, w^+, w^i, w^a, etc. (see Table 3.2).

Considered in terms of the phenotype, the polymorphism involves the existence in a fly population of a wild-type allele and a series of mutant alleles. Think about the basis for the polymorphism among the mutant alleles. They possess mutations that alter the gene product in such a way that the protein function is altered, thus producing a change in phenotype. If we compare the restriction maps or the DNA sequences of the relevant alleles, they too will be polymorphic in the sense that each map or sequence will be different from the others.

Although not evident from the phenotype, the wild type may itself be polymorphic. Multiple versions of the wild-type allele may be distinguished by differences in sequence that do not affect their function, and which therefore are not detected in the form of phenotypic variants. Considered in terms of genotype, a fly population may have extensive polymorphism; many different sequence variants exist at the w locus, some of them evident because they affect the phenotype, others hidden because they have no visible effect.

Some polymorphisms in the genome can be detected by comparing the restriction maps of different individuals. The criterion is a change in the pattern of fragments produced by cleavage with a restriction enzyme. **Figure 6.7** shows that when a target site is present in the genome of one individual and absent from another, the extra cleavage in the first genome will generate two fragments corresponding to the single fragment in the second genome.

Because the restriction map is independent of gene function, a polymorphism at this level can be detected *irrespective of whether the sequence change affects the phenotype.* Probably only a minority of the restriction site polymorphisms in a genome actually affect the phenotype. The majority involve

contain additional fragments that are not present in the wild type, or existing fragments would become larger.

Locating point mutations on the restriction map is more difficult. Occasionally they change target sites for restriction enzymes, but otherwise they remain undetectable, since the sizes of the restriction fragments remain the same in wild-type and mutant DNAs. Sometimes it is possible to detect changes in sequence by their effects on the migration of short fragments of single-stranded DNA, in a technique called single strand conformation polymorphism (SSCP). But to locate base substitutions within larger regions of DNA, it may be necessary to determine the sequence of the DNA.

Construction of a genetic map is based on the

Figure 6.7

A change in DNA that affects a restriction site is detected by a difference in restriction fragments.

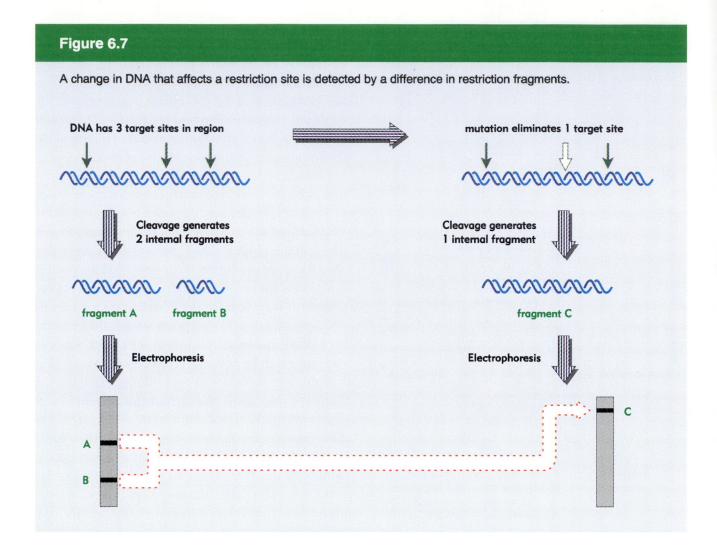

DNA has 3 target sites in region

mutation eliminates 1 target site

Cleavage generates 2 internal fragments

Cleavage generates 1 internal fragment

fragment A   fragment B

fragment C

Electrophoresis

Electrophoresis

A

B

C

sequence changes that have no effect on the production of proteins (for example, because they lie between genes).

A difference in restriction maps between two individuals is called a **restriction fragment length polymorphism (RFLP)**. It can be used as a genetic marker in exactly the same way as any other marker. Instead of examining some feature of the phenotype, we directly assess the genotype, as revealed by the restriction map. **Figure 6.8** shows a pedigree of a restriction polymorphism followed through three generations. It displays Mendelian segregation at the level of DNA marker fragments.

Recombination frequency can be measured between a restriction marker and a visible phenotypic marker as illustrated in **Figure 6.9**. Thus a genetic map can include both genotypic and phenotypic markers.

Because restriction markers are not restricted to those genome changes that affect the phenotype, they provide the basis for an extremely powerful technique for identifying genetic loci at the molecular level. A typical problem concerns a mutation with known effects on the phenotype, where the relevant genetic locus can be placed on a genetic map, but for which we have no knowledge about the corresponding gene or protein. Many damaging or fatal human diseases fall into this category. For example cystic fibrosis shows Mendelian inheritance, but the molecular nature of the mutant function was unknown until it could be identified as a result of characterizing the gene.

If restriction polymorphisms occur at random in the genome, some should occur near any particular target gene. We can identify such restriction markers by virtue of their tight linkage to the mutant phenotype. If we compare the restriction map of DNA from patients suffering from a disease with the DNA of normal people, we may find that a particular restriction site is always present (or always absent) from the patients.

A hypothetical example is shown in **Figure 6.10**. This situation corresponds to finding 100% linkage between the restriction marker and the phenotype. It would imply that the restriction marker lies so close to the mutant gene that it is never separated from it by recombination.

The identification of such a marker has two important consequences:

◆ It may offer a diagnostic procedure for detecting the disease. Some of the human diseases that are genetically well characterized but ill-defined in molecular terms cannot be easily diagnosed. If a restriction marker is reliably linked to the phenotype, then its presence can be used to diagnose the disease, either at a prenatal stage or subsequently.

◆ It may lead to isolation of the gene. The restriction marker must lie relatively near the gene on the genetic map if the two loci rarely or never recombine. Although 'relatively near' in genetic terms can be a substantial distance in terms of base pairs of DNA (see Table 6.1), nonetheless it provides a starting point from which we can proceed along the DNA to the gene itself.

Figure 6.8

Restriction site polymorphisms are inherited according to Mendelian rules. Four alleles for a restriction marker are found in all possible pairwise combinations, and segregate independently at each generation. Photograph kindly provided by Ray White.

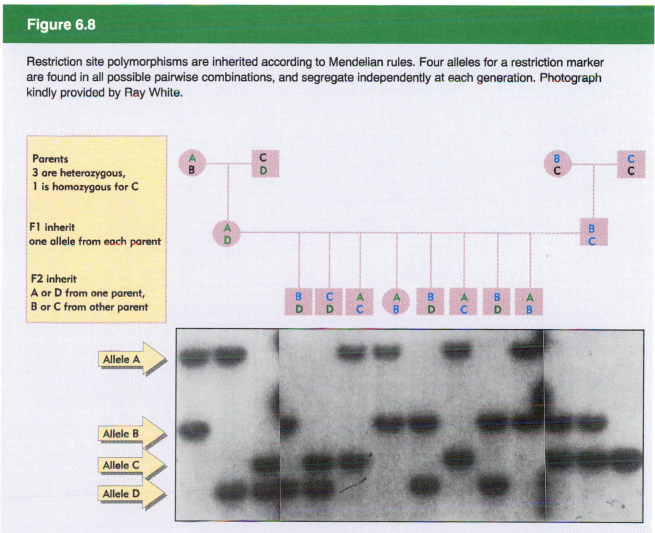

When we seek to identify the gene responsible for a disease, unless gross deletions or other obvious changes identify the damaged gene in patients, it may be difficult to identify the locus. Any gene that is not separated by recombination from the genetic marker for the disease is a *candidate* for the locus.

This means that an RFLP at this gene must occur in all cases of the disease. There could be several such genes in a region of DNA known to be closely linked to the disease. However, although genetic mapping cannot prove that any particular gene is responsible for the disease, it can exclude target genes. The

Figure 6.9

A restriction polymorphism can be used as a genetic marker to measure recombination distance from a phenotypic marker (such as eye color). Note that the figure simplifies the real situation by showing only the DNA bands corresponding to a haploid genome; actually, there would also be bands corresponding to the allele of the other genome in a diploid.

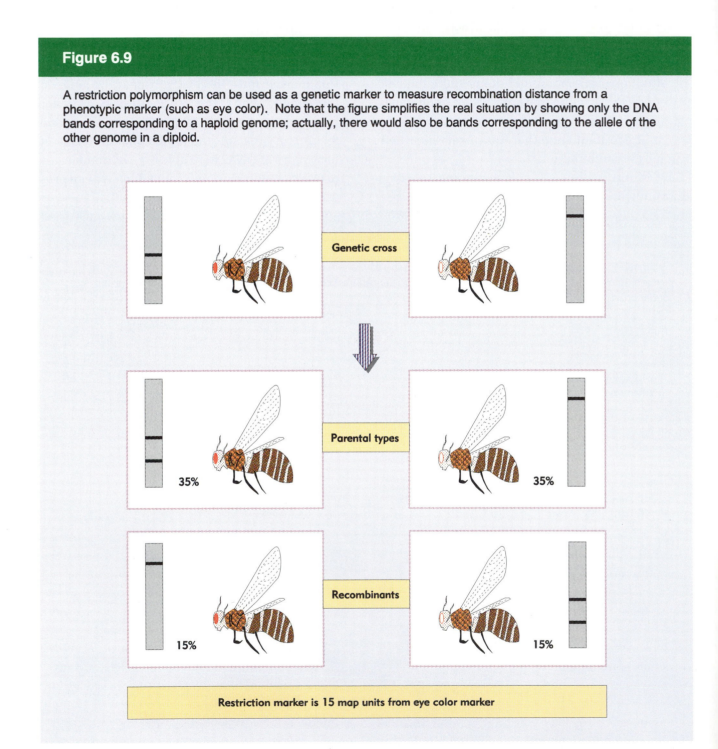

Genetic cross

Parental types

35%   35%

Recombinants

15%   15%

Restriction marker is 15 map units from eye color marker

Figure 6.10

If a restriction marker is associated with a phenotypic characteristic, the restriction site must be located near the gene responsible for the phenotype.

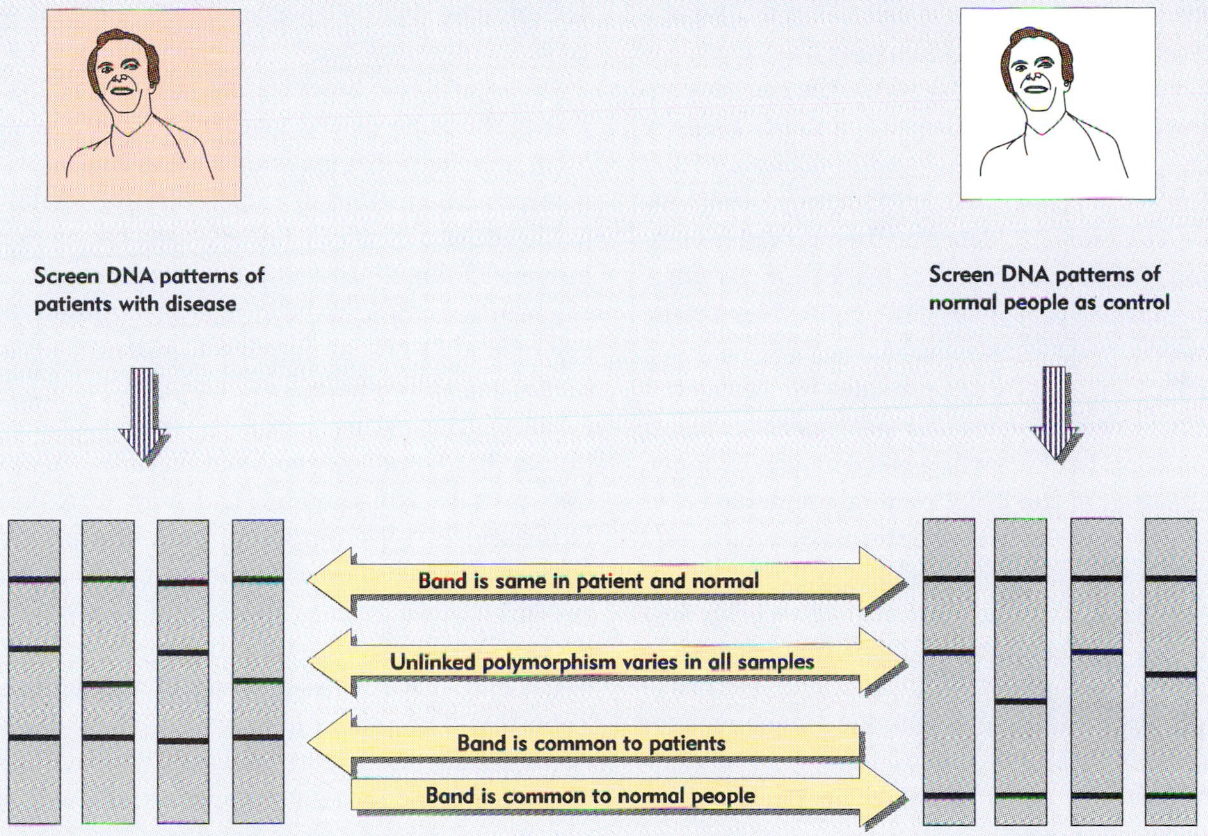

Screen DNA patterns of patients with disease

Screen DNA patterns of normal people as control

Band is same in patient and normal

Unlinked polymorphism varies in all samples

Band is common to patients

Band is common to normal people

existence of only one patient with a disease who shows recombination with an RFLP at a target locus is sufficient to rule out that locus.

The converse is the difficulty of proving that a defect in a particular gene is responsible for a disease. It is necessary to show that every patient with the disease has a mutation in that gene. In cases where the mutations associated with a disease are rather large deletions that affect more than one gene, it may not be obvious which gene is associated with the disease. The only satisfactory evidence that implicates a gene is to demonstrate that it is affected in all cases and that in some cases it is the *only* gene affected. This requires the identification of mutations that are *internal*, that is, taking the form of point mutations or small deletions.

The large size of the human genome makes it a far from trivial task to detect a particular restriction polymorphism. There are practical difficulties just in identifying the relative part of the genome that should be examined. As with conventional genetic markers, we need a battery of restriction markers that cover the entire genome. With such a battery at hand, however, it becomes possible to scan a new marker (phenotypic or genotypic) for linkage with known markers.

RFLPs occur frequently enough in the human genome to be useful for genetic mapping. If allelic sequences are compared between any two individual chromosomes, differences in individual base pairs occur at a frequency of >1 per 1000 bp. Those base changes that affect restriction sites can be detected as RFLPs.

The frequency of polymorphism means that every individual has a unique constellation of restriction sites. The particular combination of sites found in a specific region is called a **haplotype**, a genotype in miniature. Haplotype was originally introduced as a concept to describe the genetic constitution of the major histocompatibility locus, a region specifying proteins of importance in the immune system (see Chapter 37). The concept now has been extended to describe the particular combination of alleles or restriction sites (or any other genetic marker) present in some defined area of the genome.

The existence of RFLPs provides the basis for a technique to establish unequivocal parent-progeny relationships. In cases where parentage is in doubt, a comparison of the RFLP map in a suitable chromosome region between potential parents and child allows absolute assignment of the relationship. The use of DNA restriction analysis to identify individuals has been called **DNA fingerprinting**. We discuss in more detail in Chapter 26 the use of particularly variable 'minisatellite' sequences for mapping in the human genome.

Difficulties in collecting data on human pedigrees result from the small number of progeny per mating and the low number of generations available for analysis. As a result, genetic linkage cannot be measured by a simple Mendelian analysis as described in Chapter 3. We are forced to seek **informative meioses**, in which a mating has generated crossovers between RFLP sites that are useful for determining recombination frequencies. The data then are analyzed by a multilocus analysis performed by means of computer programs.

The first procedure in this statistical analysis is to assign a new RFLP to a linkage group. The criterion used to assess linkage is the **LOD score**. This is defined as the $\log_{10}$ of the ratio of the probability that the data would have arisen if the loci are linked, to the probability that the data would have arisen from unlinked loci. The conventional threshold for linkage is a LOD score >3.0, which corresponds to a

ratio of 1000:1 in favor of linkage. However, any arbitrary pair of loci is 50× more likely to be unlinked than linked, so a LOD score of 3.0 actually provides odds of 20:1 in favor of linkage. This means that for every 20 linkages detected at a LOD score of 3.0, 1 will turn out to be false when further data accumulate. By a LOD score of 4.0, the chance of error is only 1 in 200.

Once an RFLP has been assigned to a linkage group, it can be placed in position on the genetic map, and map distances to its flanking markers determined. An effort to map RFLPs in man has led to the construction of a linkage map for the entire human genome. Most human chromosomes now can be represented in the form of a continuous linkage map of RFLPs, although some significant gaps in the map still remain.

Some interesting features emerge from the human RFLP map. Recombination rates are different in male and female. The map of a typical autosome is 1.9× longer in female than male, indicating that there are almost twice as many recombination events in oocytes compared to sperm. Thus there are ~1.2×10^7 bp per map unit in males, but ~7×10^6 bp per map unit in females (see Table 6.1). The genetic length of each chromosome (in map units) is proportional to its physical length (demonstrating for another species and a greater number of chromosomes the overall relationship that we discussed previously for *D. melanogaster* in Figure 3.8). The relationship is not completely uniform, however, and local differences in recombination frequency occur in each sex. There is a tendency toward increased polymorphism and increased recombination at the ends of chromosomes.

Substantial progress is being made with the mapping technology toward identifying genes for many human diseases. For example, **Figure 6.11** shows a linkage map of the human X chromosome, which consists of ~184 Mb and has a linkage map (in female) of ~220 cM. The map is based on ~20 restriction polymorphisms that allow any X-linked disease locus to be mapped within 10 map units (~10,000 kb) of at least two markers.

The ideogram in the center shows the physical structure of the chromosome in terms of bands that can be visualized under certain conditions (the banding technique is discussed in Chapter 27). The map on the left shows the limits for the localization of known genes. (The vertical bar at the base of the triangle connecting the gene to the chromosome shows the limits of the gene's location.) The map on the right shows the relative locations of the restriction polymorphisms. The distance apart of the polymorphic markers relative to the scale on the right reflects linkage as measured in centiMorgans;

the distortion relative to the physical map can be seen by comparing the actual locations of the markers (which, if known, are shown by the triangles connecting them to the chromosome).

Note that this map is a *linkage map*; the difference from a conventional genetic map is that it is based on RFLPs measured by the sizes of DNA restriction fragments rather than on the phenotypic effects of mutations. The distance between the RFLPs is measured in genetic map units. The next step in characterizing the genome will be to proceed to a *restriction map*, that is, a physical map of the actual distances between restriction sites.

We should like really to relate the genetic map to the sequence of chromosomal DNA, but it is difficult to isolate extensive lengths of DNA. Even the smallest human chromosome contains ~50,000 kb of DNA. However, the DNA of relatively small chromosomes of lower eukaryotes can now be isolated intact, allowing direct characterization of the karyotype.

The example of **Figure 6.12** visualizes the yeast karyotype by directly separating the chromosomal DNAs through a technique called pulse field gel electrophoresis (**PFGE**). *S. cerevisiae* has 16 linkage groups, varying in length from ~400 map units to one chromosome identified by only a single mutation. We expect the chromosomal DNAs to range in length from ~2000 kb to as little as ~300 kb, which seems to be borne out by the results (although the sizes of the larger bands are uncertain).

A powerful application made possible by this technique is that a gene can be mapped directly to any chromosome by identifying the DNA band with which it hybridizes. By using yeast strains with known genetic translocations, mapping within chromosomal regions becomes possible. With this technique, we see for the first time the possibility of directly relating a linkage map to the chromosomal DNA.

In terms of mapping the human genome, PFGE is important because it allows relatively large fragments of DNA to be manipulated and mapped relative to one another. This significantly reduces the difficulties to be expected in constructing a restriction map.

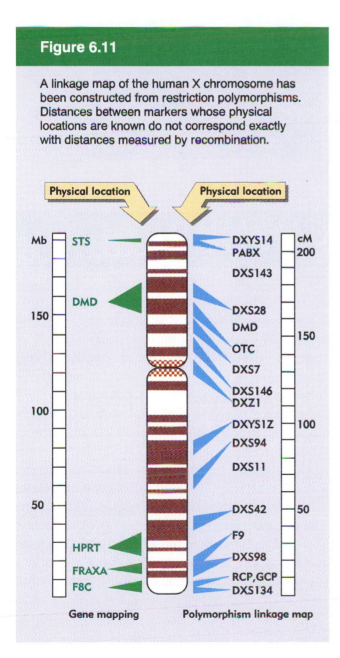

Figure 6.11

A linkage map of the human X chromosome has been constructed from restriction polymorphisms. Distances between markers whose physical locations are known do not correspond exactly with distances measured by recombination.

Figure 6.12

Chromosomal DNAs of the yeast *S. cerevisiae* can be separated into 12 bands, 9 of which represent individual chromosomes, and 3 of which each represent two chromosomes. The components of the doublets can be separated in other gels. Photograph kindly provided by Maynard Olson.

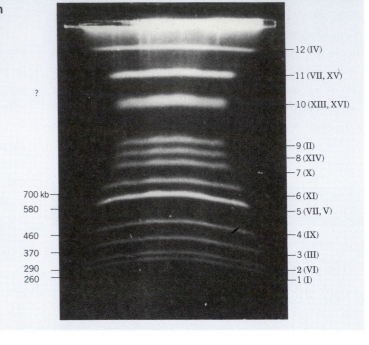

Obtaining the sequence of DNA

Through the specificity of base pairing, a double helix of DNA maintains a constant structure irrespective of its particular sequence. The difference between individual molecules of DNA lies in their particular sequences of base pairs, not in gross changes of structure. Very small differences in DNA sequence can be highly significant; mutations consist of changes in DNA sequence as small as one base pair.

How can the exact nucleotide pair sequence of DNA be determined? The first approach to nucleic acid sequencing followed that of protein sequencing: break the molecule into small fragments, determine their base composition, and by obtaining overlapping fragments deduce the exact sequence. With proteins this is practical because of the variety

offered by the 20 amino acids; with nucleic acids, it poses a problem because there are only four bases. The old techniques were therefore restricted to determining the sequences of very short nucleic acids.

But now it is possible to determine DNA sequences *directly*. In fact, DNA sequences can be obtained much more rapidly than protein sequences.

All sequencing protocols share a general principle. Each protocol generates a series of single-stranded DNA molecules, each molecule one base longer than the last. DNA molecules of the *same sequence*, but differing in length by as little as one base at one end, can be separated by electrophoresis on acrylamide gels. Bands

Figure 6.13

The locations of a particular base in single-stranded DNA can be determined by electrophoresing end-labeled fragments after treatment with a suitable reagent.

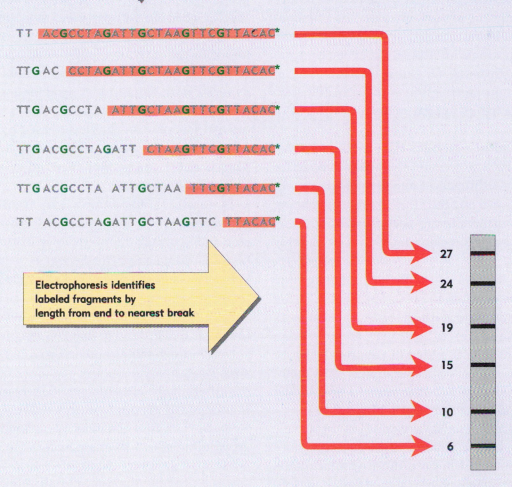

corresponding to molecules of increasing length form a 'ladder' that can be followed for ~300 nucleotides.

Two types of approach have been used to obtain these sets of bands. One is to use chemical reactions that cleave DNA at individual bases. The other is to use an enzymatic reaction in which DNA is synthesized *in vitro* in such a way that the reaction terminates specifically at the position corresponding to a given base.

To determine the sequence of the molecule by either approach, it is subjected to the appropriate protocol in four separate reactions, each reaction specific for one of the bases. The products are electrophoresed in four parallel slots of the same gel. The band corresponding to a particular length appears in only one of the four slots. The slot identifies the nucleotide that is present at the corresponding position in DNA. By proceeding from each band to the band of the next size, the series of nucleotides can be 'read' according to the slots in which the bands appear. This gives the DNA sequence.

In the chemical method, a DNA fragment previously labeled at one (unique) end with radioactive ^{32}P is broken specifically at one of the four types of base. The key to using the cleavage for sequencing is to ensure that there is only partial breakage at any susceptible position, occurring with a probability of, say, 1–2%.

So we start with a preparation that contains a large number of identical molecules, all labeled at the same end. When subjected to the chemical treatment, each molecule is likely to be cleaved at only a small number of the sites at which the appropriate base occurs. But in the entire preparation of DNA molecules, every position will theoretically be attacked on some occasion, as illustrated in **Figure 6.13**.

The chain is broken at G residues in this example. Some chains are broken only once, some more than once, but in each case the fragment that is identified is the one with the labeled end (indicated by the asterisk). Altogether, the labeled molecules extend from the common end to every position at which a G was present in the original DNA. The

treated molecules are electrophoresed, and those carrying the radioactive label are identified (by autoradiography). Each will form a band whose length depends on the distance from the labeled end to the site of cleavage. The principle is precisely the same as in the end-labeling of restriction fragments shown in Figure 6.5.

In the chemical sequencing method, each of the chemical treatments generates a series of bands like this, identifying every position at which the particular base occurs relative to the end of the molecule.

In the original (Maxam–Gilbert) method, specificity for the purines was provided by using dimethylsulfate. This reagent methylates the N7 position of guanine ~5 times more effectively than the N3 position of adenine. Upon heating, the methylated base is lost from the DNA, generating a break in the polynucleotide chain. Because the reaction is more effective with guanine, G residues generate dark bands and A residues generate light bands (that is, the cleavage of A has been less frequent). The pattern can be reversed by using acid instead of heat to release the methylated bases. Thus G and A can be identified individually in the appropriate reactions.

Pyrimidines are analyzed by treatment with hydrazine, which acts equally effectively on cytosine and thymine; but they can be distinguished in high salt, when only cytosine reacts. So two series of bands are generated, one representing C only and one representing the combination of C + T.

In the enzymatic sequencing method, four reaction mixtures are set up that contain all four of the substrate nucleotides and an enzyme that synthesizes DNA. A small amount of a dideoxy derivative of a different one of the nucleotide precursors (lacking the 3'-OH group) is included in each mixture. The incorporation of a dideoxy nucleotide into the growing nucleic acid chain causes the reaction to cease with that base. So in one reaction mixture a series of bands is generated all ending in A, in another the bands all end in T, and so on.

Four reactions are performed and the products are electrophoresed in parallel, as indicated in

Figure 6.14

DNA can be sequenced by a set of reactions that together are specific for all four bases.

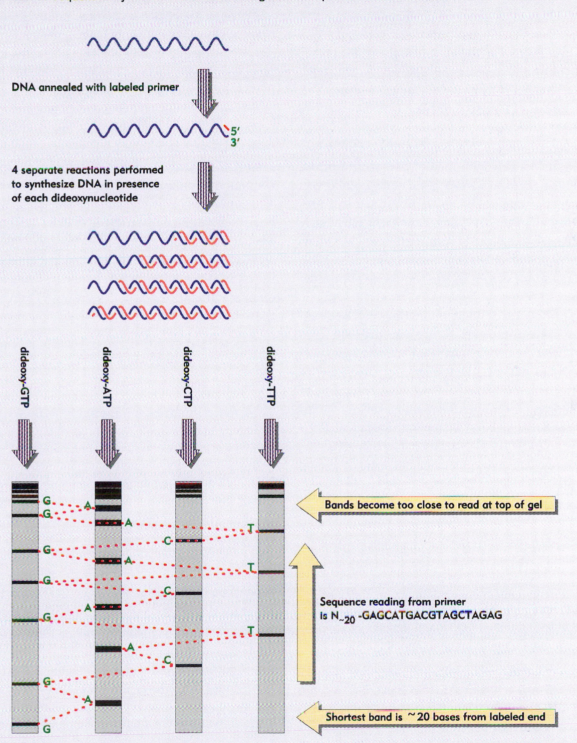

DNA annealed with labeled primer

4 separate reactions performed to synthesize DNA in presence of each dideoxynucleotide

dideoxy-GTP dideoxy-ATP dideoxy-CTP dideoxy-TTP

Bands become too close to read at top of gel

Sequence reading from primer is N$_{\sim20}$-GAGCATGACGTAGCTAGAG

Shortest band is ~20 bases from labeled end

Figure 6.14. Each of the four reactions generates a series of bands identifying molecules in which DNA synthesis terminated at a position where the dideoxynucleotide was inserted. The sequence of DNA is obtained by proceeding from each band to the band that is one base different in length. The radioactive label end that identifies the fragments is carried by the 5′ end of the primer, so the DNA sequence is read up from the bottom of the gel, following the bands successively across as indicated by the zigzag.

How accurate is DNA sequencing? The set of reactions gives the sequence of one strand of the DNA. There is always a risk of an occasional error. However, the accuracy can be improved by independently sequencing *both* strands of DNA. Any sites at which they are not complementary are identified as possible errors that can be corrected. With this precaution, errors in the sequence should be very rare.

DNA molecules of <300 nucleotides can be separated on acrylamide gels. How do we proceed from determining the sequences of these relatively short fragments to that of a longer region? Just as with restriction mapping itself, the key concept is the use of overlapping fragments. Consider a series of adjacent fragments, ordered on a restriction map as

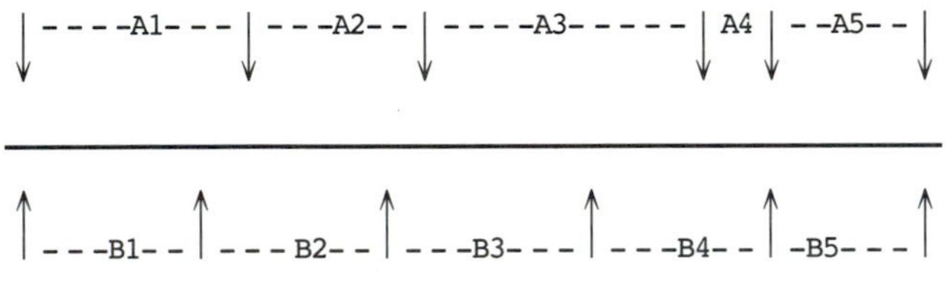

```
 | - - - -A1- - - | - - -A2- - | - - - -A3- - - - - | A4 | - -A5- - |
 ↓             ↓          ↓               ↓  ↓         ↓
 _____

 ↑ - - -B1- - ↑ - - -B2- - ↑ - - -B3- - - ↑ - - -B4- - ↑ -B5- - - |
```

Suppose that we determine the sequence of each member of the A series of fragments. We know that they are adjacent from the restriction map, so their individual sequences should in theory abut directly and give us the entire sequence. But to be sure that there are no small missing fragments generated by closely adjacent cleavages by the same enzyme (suppose there were two cleavage sites, 10 bp apart, between A2 and A3!), it is essential to sequence *across each junction.* Thus we require another set of fragments, one in which all the junctions between the A fragments are intact. This is provided by the B series of fragments, each of which contains the sequence on either side of the junction between two A fragments.

Just as we can construct the restriction map from overlapping fragments, we can construct the overall DNA sequence by overlapping the ends of each individually sequenced fragment. Suppose, for example, that the ends of fragments A1 and A2 have the sequences

```
A1.......CGTAGGGTCAAGTCAT AATGCTGCCCAAA..........A2
```

Then in fragment B2 we should find the sequence

```
......CGTAGGGTCAAGTCATAATGCTGCCCAAA..........
```

in which the two ends run continuously from one to the other.

Prokaryotic genes and proteins are colinear

By comparing the nucleotide sequence of a gene with the amino acid sequence of a protein, we can determine directly whether the gene and the protein are **colinear**: whether the sequence of nucleotides in the gene corresponds exactly with the sequence of amino acids in the protein. In bacteria and their viruses, there is an exact equivalence. Each gene contains a continuous stretch of DNA whose length is directly related to the number of amino acids in the protein that it represents. A gene of $3N$ base pairs is required to code for a protein of N amino acids, according to the genetic code.

The equivalence of the prokaryotic gene and its

product means that a restriction map of DNA will exactly match an amino acid map of the protein. How well do these maps fit with the recombination map?

The colinearity of gene and protein was originally investigated in the tryptophan synthetase gene of *E. coli*. Genetic distance was measured by the percent recombination between mutations; protein distance was measured by the number of amino acids separating sites of replacement. **Figure 6.15** compares the two maps. The order of seven sites of mutation is the same as the order of the corresponding sites of amino acid replacement. And the recombination distances are relatively

Figure 6.15

The recombination map of the tryptophan synthetase gene corresponds with the amino acid sequence of the protein.

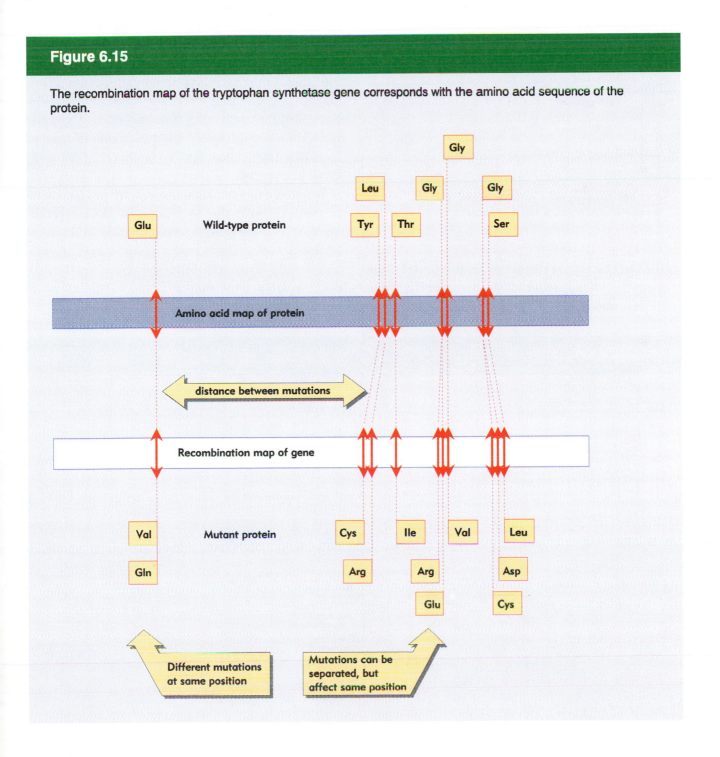

Figure 6.16

RNA is synthesized by using one strand of DNA as a template for complementary base pairing.

DNA consists of two base paired strands

5′ ATGCCGTTAGACCGTTAGCGGACCTGAC top strand
3′ TACGGCAATCTGGCAATCGCCTGGACTG bottom strand

⬇ RNA synthesis

5′ AUGCCGUUAGACCGUUAGCGGACCUGAC   3′

RNA has same sequence as DNA top strand;
is complementary to DNA bottom strand

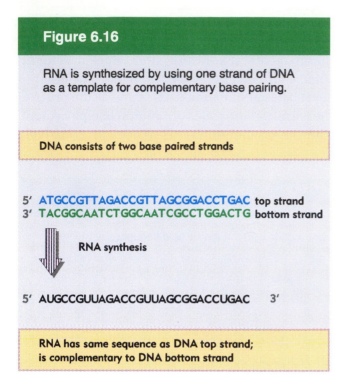

tortion of the recombination map relative to the physical map.

The recombination map makes two further general points about the organization of the gene. Different mutations may cause a wild-type amino acid to be replaced with different substituents. If two such mutations cannot recombine, they must involve different point mutations at the same position in DNA. If the mutations can be separated on the genetic map, but affect the same amino acid on the upper map (the connecting lines converge in the figure), they must involve point mutations at different positions that affect the same amino acid. Formally this indicates that the unit of genetic recombination (actually one base pair) is smaller than the unit coding for the amino acid (actually three base pairs).

In comparing gene and protein, we are restricted to dealing with the sequence of DNA stretching between the points corresponding to the ends of the protein. However, a gene is not directly translated into protein, but is expressed via the production of a **messenger RNA**, a nucleic acid intermediate actually used to synthesize a protein

similar to the actual distances in the protein; the recombination map expands the distances between some mutations, but otherwise there is little dis-

Figure 6.17

The gene may be longer than the sequence coding for protein.

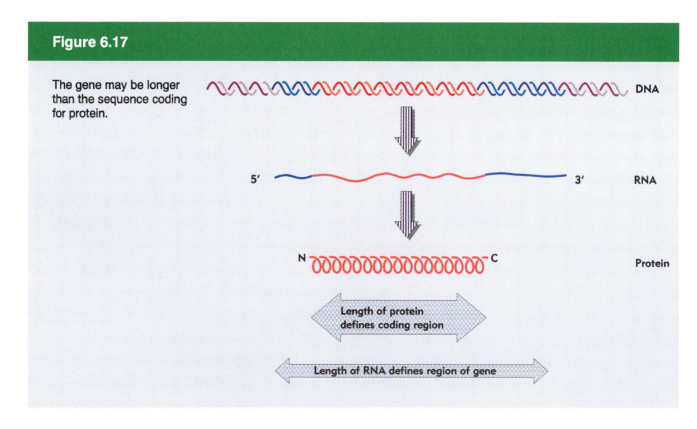

DNA

5′   3′   RNA

N 𝄆𝄆𝄆𝄆𝄆𝄆𝄆𝄆𝄆𝄆𝄆𝄆 C   Protein

Length of protein defines coding region

Length of RNA defines region of gene

(as we see in detail in Part 2). Messenger RNA is synthesized by the same process of complementary base pairing used to replicate DNA, with the important difference that it corresponds to only one strand of the DNA double helix. **Figure 6.16** shows that the sequence of messenger RNA is complementary with the sequence of one strand of DNA and is identical (apart from the replacement of T with U) with the other strand of DNA. The convention for writing DNA sequences is that the top strand runs 5'–3', with the sequence that is the same as RNA.

A messenger RNA includes a sequence of nucleotides that corresponds with the sequence of amino acids in the protein. This part of the nucleic acid is called the **coding region**. But the messenger RNA includes additional sequences on either end; these sequences do not directly represent protein. The gene is considered to include the entire sequence represented in messenger RNA. Sometimes mutations impeding gene function are found in the additional, noncoding regions, confirming the view that these comprise a legitimate part of the genetic unit.

Figure 6.17 illustrates this situation, in which the gene is considered to comprise a continuous stretch of DNA, needed to produce a particular protein. It includes the sequence coding for that protein, but also includes sequences on either side of the coding region.

Eukaryotic genes can be interrupted

The simple view of the gene was upset in 1977 by the discovery of **interrupted genes**. The primary evidence for the existence of these interruptions was a comparison between the structure of DNA and the corresponding messenger RNA. The messenger RNA always includes a nucleotide sequence that corresponds exactly with the protein product according to the rules of the genetic code. *But the gene includes additional sequences that lie within the coding region, interrupting the sequence that represents the protein.*

This discrepancy between the structure of the DNA and messenger RNA is common in eukaryotes, and has been found in an archebacterium (representing another evolutionary branch of the bacteria). It has not been found in the eubacteria (which provide the typical prokaryotes), although it is present in a bacteriophage of *E.coli.*

The sequences of DNA comprising an interrupted gene are divided into the two categories depicted in **Figure 6.18**:

◆ **Exons** are the regions that are represented in the messenger RNA.

◆ **Introns** are regions that are missing from the messenger RNA.

The process of gene expression requires a new step, one that does not occur in eubacteria. The DNA gives rise to an RNA copy that exactly represents the genome sequence. But this RNA is only a precursor; it cannot be used for producing protein. First the introns must be removed from the RNA to give a messenger RNA that consists only of the series of exons. This process is called RNA **splicing**.

How does this change our view of the gene? Following splicing, the exons are always joined together in the same order in which they lie in DNA. Thus the colinearity of gene and protein is maintained between the individual exons and the corresponding parts of the protein chain. The *order* of mutations in the gene remains the same as the order of amino acid replacements in the protein. But the *distances* in the gene do not correspond at all with the distances in the protein. The length of the gene is defined by the length of the initial (precursor) RNA instead of by the length of the messenger RNA.

Figure 6.18

Interrupted genes are expressed via a precursor RNA, from which the introns are removed when the exons are spliced together to give the messenger RNA.

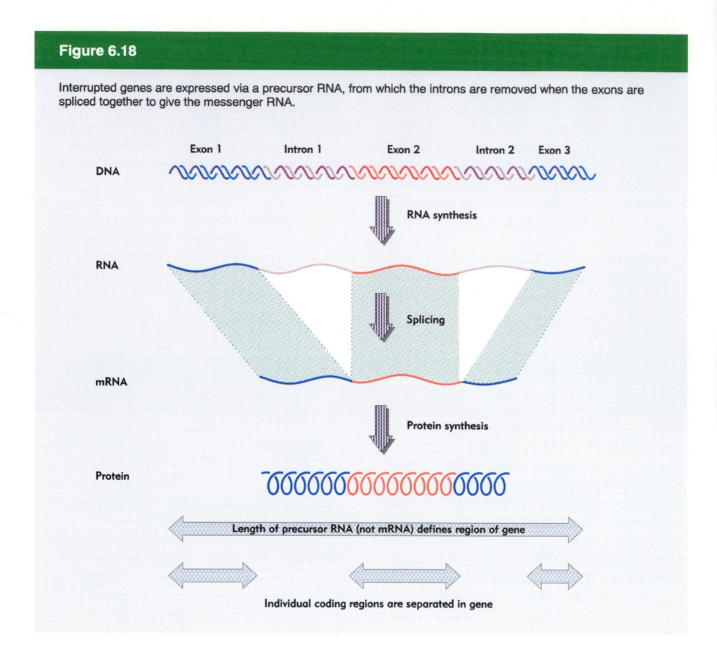

All the exons are represented on the same molecule of RNA, and their splicing together occurs only as an *intra*molecular reaction. There is usually no joining of exons carried by *different* RNA molecules, so that the mechanism excludes any splicing together of sequences representing different alleles. So mutations located in different exons of a gene cannot complement one another; thus they continue to be defined as members of the same complementation group.

If mutations that affect the sequence of a higher eukaryotic protein could be mapped with sufficient resolution, they should be found in clusters. Each cluster should correspond to an exon, and should be separated from the next cluster on the genetic map by a distance corresponding to the relative length of the intron. Thus recombination frequencies cannot be taken as a guide to relative distance within an interrupted gene.

What are the effects of mutations in the introns? Since the introns are not part of the messenger RNA, mutations in them cannot directly affect protein structure. However, they can prevent the production of the messenger RNA—for example, by inhibiting the splicing together of exons. A mutation of this sort acts only on the allele that carries it, therefore fails to complement any other mutation in that allele, and so constitutes part of the same complementation group as the exons.

In the interrupted genes of eukaryotes, most introns appear to serve no function other than to be removed during gene expression. However, there are some exceptions, most notably in the yeast mitochondrion, in which an intron itself codes for the production of a protein that functions independently from the protein coded by the exons. In this case, mutations in the intron fall into a different complementation group from that represented by mutations in the exons. Thus one complementation group may lie between clusters of mutations that comprise another group.

Eukaryotic genes are not necessarily interrupted. Some correspond directly with the protein product in the same manner as eubacterial genes. In yeast, the majority of genes in fact is uninterrupted. In higher eukaryotes, most genes are interrupted; and in fact the introns are usually much longer than exons, creating genes that are very much larger than their coding regions. The structure of such genes is the subject of Chapter 23.

Some DNA sequences code for more than one protein

Most genes consist of a sequence of DNA that is devoted solely to the purpose of coding for one protein (although the gene may include noncoding regions at either end and introns within the coding region). However, there are some cases in which a sequence of DNA does not have a unique function in

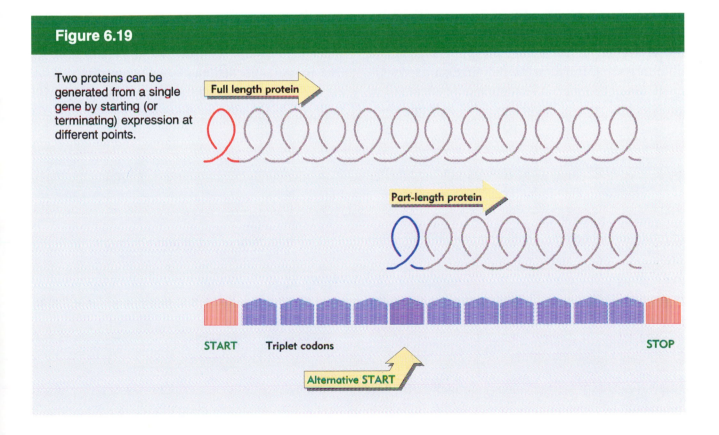

Figure 6.19

Two proteins can be generated from a single gene by starting (or terminating) expression at different points.

Full length protein

Part-length protein

START Triplet codons STOP

Alternative START

representing protein, because a single sequence of DNA codes for more than one protein.

Overlapping genes occur in the relatively simple situation in which one gene is part of the other. The first half (or second half) of a gene is used independently to specify a protein that represents the first (or second) half of the protein specified by the full gene. This relationship is illustrated in **Figure 6.19**. It does not present any particular *genetic* problem (although it does require adjustments in the act of synthesizing the protein). The end result is much the same as though a partial cleavage took place in the protein product to generate part-length as well as full-length forms.

Two genes overlap in a more subtle manner when the same sequence of DNA is shared between two *nonhomologous* proteins. This situation arises when the same sequence of DNA is translated in more than one reading frame. In cellular genes, a DNA sequence usually is read in only one of the three potential reading frames, but in some viral and mitochondrial genes, there is an overlap between two adjacent genes that are read in different reading frames. This situation is illustrated in **Figure 6.20**. The distance of overlap is usually relatively short, so that most of the sequence representing the protein retains a unique coding function. Depending on its type, a mutation in the shared sequence may affect either gene or both; thus it could be part of only one com-

plementation group or might belong to two groups.

In the usual form of the interrupted gene, each exon codes for a single sequence of amino acids, representing an appropriate part of the protein, and each intron plays no part in the final production of protein. Their roles are distinct. But in some genes a sequence of DNA is used in more than one way, so it cannot simply be characterized as an exon or intron.

In these genes, *alternative* patterns of gene expression create switches in the pathway for connecting the exons. Thus a particular exon may be connected to any one of several alternative exons to form a messenger RNA. The alternative forms produce proteins in which one part is common while the other part is different. An example is illustrated in **Figure 6.21**, which demonstrates that some regions behave as exons when expressed via one pathway, but are introns when expressed via another.

In one pathway, *exon 1* is joined to *exon 2* by removing from RNA the regions marked *segment* and *intron*. In the other pathway, the region of *exon 1–segment* is joined directly to *exon 2*, removing only the intron. Thus in the first pathway the *segment* region is an intron, but in the second pathway it is an exon. The pathways produce two proteins that are the same at their ends, but one of which has an additional sequence in the middle. Thus the region of DNA codes for more than one protein. We

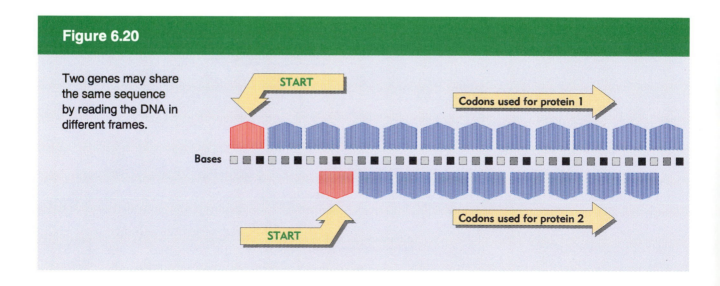

Figure 6.20

Two genes may share the same sequence by reading the DNA in different frames.

START

Codons used for protein 1

Bases

START

Codons used for protein 2

Figure 6.21

Genes may be difficult to define when there are alternative pathways for expression.

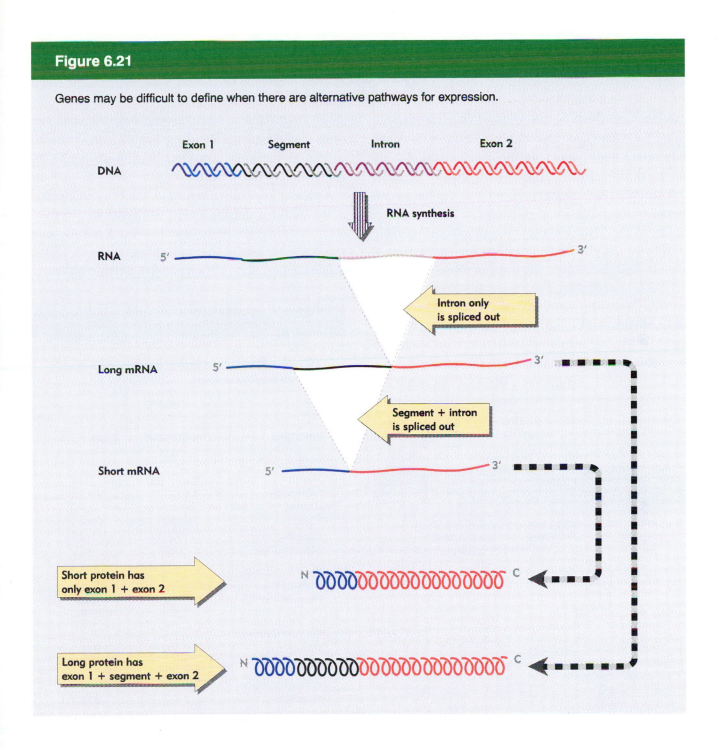

discuss such situations in more detail in Chapter 31.

Sometimes two pathways operate simultaneously, a certain proportion of the RNA being spliced in each way; sometimes the pathways are alternatives that are expressed under different conditions, one in one cell type and one in another cell type. The complementation properties of such mutations can in principle be complex, and it is not necessarily possible to assign all mutations to one or another independent complementation group.

Genetic information can be provided by DNA or RNA

Nucleic acid fulfills the mandate for the genetic material with a unique combination of stability and flexibility. Replication via the mechanism of complementary base pairing allows genetic information to be reliably inherited. Yet the occurrence of the occasional mutation allows heritable changes to occur, providing the substrate for evolution. The genetic code provides for expression of genetic information in the form of proteins.

These mechanisms are equally effective for the cellular genetic information of prokaryotes or eukaryotes, and for the information carried by viruses. The genomes of all living organisms consist of duplex DNA. Viruses have genomes that consist of DNA or RNA; and there are examples of each type that are double-stranded (ds) or single-stranded (ss). Details of the mechanism used to replicate the nucleic acid vary among the viral systems, but the principle of replication via synthesis of complementary strands remains the same, as illustrated in **Figure 6.22**.

Cellular genomes reproduce DNA by the mechanism of semi-conservative replication. Double-stranded virus genomes, whether DNA or RNA, also replicate by using the individual strands of the duplex as templates to synthesize partner strands.

Figure 6.22

Double-stranded and single-stranded nucleic acids both replicate by synthesis of complementary strands governed by the rules of base pairing.

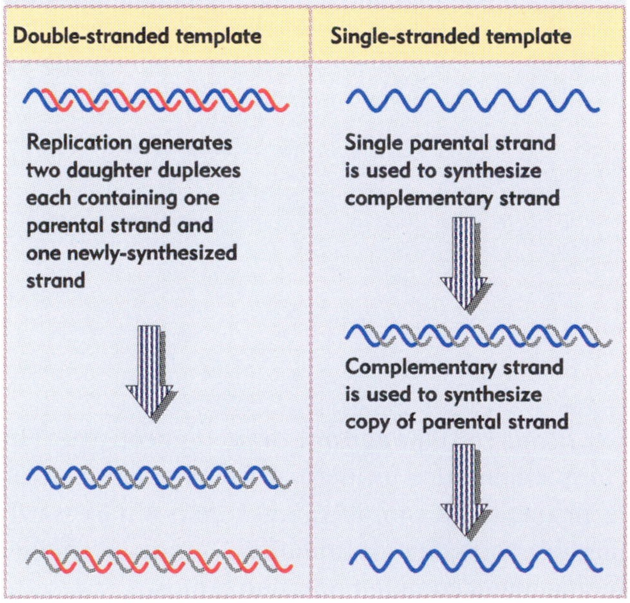

Double-stranded template

Replication generates two daughter duplexes each containing one parental strand and one newly-synthesized strand

Single-stranded template

Single parental strand is used to synthesize complementary strand

Complementary strand is used to synthesize copy of parental strand

Table 6.2

The amount of nucleic acid in the genome varies over an enormous range.

Genome	Gene Number	Base Pairs
Organisms		
Plants	<50,000?	$<10^{11}$
Mammals	<25,000?	$~3 \times 10^8$
Worms	~5,000?	$~10^8$
Fungi	~4,000?	$~4 \times 10^7$
Bacteria	~2,000?	$<10^7$
Mycoplasma	~750?	$<10^6$
dsDNA Viruses		
Vaccinia	<300	187,000
Papova (SV40)	~6	5,226
Phage T4	~200	165,000
ssDNA Viruses		
Parvovirus	5	5,000
Phage fX174	11	5,387
dsRNA Viruses		
Reovirus	22	23,000
ssRNA Viruses		
Coronavirus	7	20,000
Influenza	12	13,500
TMV	4	6,400
Phage MS2	4	3,569
STNV	1	1,300
Viroids		
PSTV RNA	0	359
Scrapie		
Prion	?	?

Viruses with single-stranded genomes use the single strand as template to synthesize a complementary strand; and this complementary strand in turn is used to synthesize its complement, which is, of course, identical with the original starting strand. Replication may involve the formation of stable double-stranded intermediates or use double-stranded nucleic acid only as a transient stage.

Thus the same principles are followed to perpetuate genetic information from the massive genomes of plants or amphibians to the tiny genomes of mycoplasma and the yet smaller genetic information of DNA or RNA viruses. **Table 6.2** summarizes some examples that illustrate the range of genome types and sizes.

The DNA genomes of all living cells consist of very long molecules; all but the bacteria and mycoplasma have more than one such molecule. Some of the larger viruses (such as influenza) have segmented genomes, consisting of more than one nucleic acid molecule. Other viral genomes consist of a single nucleic acid molecule. For the smaller viruses (SV40, øX, MS2), the number of functions includes overlapping genes.

Throughout the range of organisms, with genomes varying in total content over a 100,000 fold range, a common principle prevails. *The DNA codes for all the proteins that the cell(s) of the organism must synthesize; and the proteins in turn (directly or indirectly) provide the functions needed for survival.* The total number of genes is difficult to estimate (an issue taken up in more detail in Chapter 22). The very smallest living organism, the mycoplasma, has a genome size only about twice that of the very largest virus.

A similar principle describes the function of the genetic information of viruses, be it DNA or RNA. *The nucleic acid codes for the protein(s) needed to package the genome and also for any functions additional to those provided by the host cell that are needed to reproduce the virus during its infective cycle.* (The very smallest virus, the satellite tobacco necrosis virus [STNV], cannot replicate independently, but requires the simultaneous presence of a 'helper' virus [tobacco necrosis virus, TNV], which is itself a normally infectious virus.)

Viroids are infectious agents that cause diseases in higher plants. They are very small circular molecules of RNA. Unlike viruses, where the infectious agent consists of a **virion**, a genome encapsulated

Figure 6.23

PSTV RNA is a circular molecule that forms an extensive double-stranded structure, interrupted by many interior loops. The severe and mild forms differ at three sites.

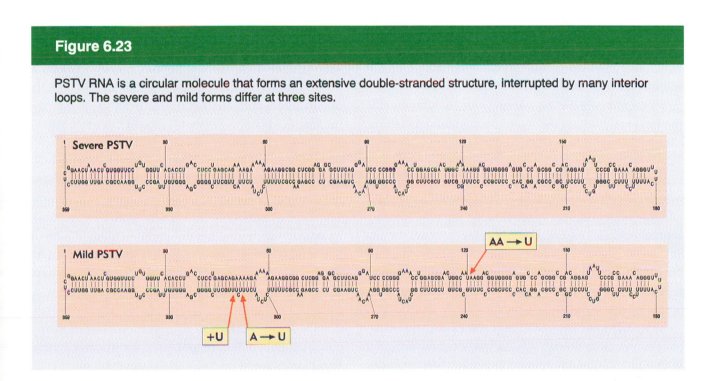

in a protein coat, *the viroid RNA is itself the infectious agent*. The viroid consists solely of the RNA, which is extensively but imperfectly base paired, forming a characteristic rod like the example shown in **Figure 6.23**. Mutations that interfere with the structure of the rod reduce infectivity.

A viroid RNA consists of a single molecular species that is replicated autonomously in infected cells. Its sequence is faithfully perpetuated in its descendants. Viroids fall into several groups. A given viroid is identified with a group by its similarity of sequence with other members of the group. For example, four viroids related to PSTV (potato spindle tuber viroid) have 70–83% similarity of sequence with it. Different isolates of a particular viroid strain vary from one another, and the change may affect the phenotype of infected cells. For example, the *mild* and *severe* strains of PSTV differ by three nucleotide substitutions.

Viroids resemble viruses in having nucleic acid genomes whose sequences are heritable, subject to mutation, and determine the phenotype. They fulfill the criteria for genetic information. Yet viroids differ from viruses in both structure and function. They are sometimes called **subviral pathogens**.

Viroid RNA does not appear to be translated into protein, in which case it cannot itself code for the functions needed for its survival. This situation poses two questions. How does viroid RNA replicate? And how does it affect the phenotype of the infected plant cell?

Replication must be carried out by enzymes of the host cell, subverted from their normal function. The heritability of the viroid sequence indicates that viroid RNA provides the template.

Viroids are presumably pathogenic because they interfere with normal cellular processes. They might do this in a relatively random way, for example, by sequestering an essential enzyme for their own replication or by interfering with the production of necessary cellular RNAs. Alternatively, they might represent abnormal regulatory molecules, with particular effects upon the expression of individual genes.

An even more unusual agent is **scrapie**, the cause of a degenerative neurological disease of sheep and goats. The disease is related to the human diseases of kuru and Creutzfeldt–Jacob syndrome, which affect brain function.

The infectious agent of scrapie appears to contain no nucleic acid. Persistent attempts to identify even a small RNA or DNA have failed. The only component so far identified is a protein that is extremely resistant to attack by proteases.

This extraordinary agent is sometimes described as a **prion**. Its predominant component is a 28,000 dalton hydrophobic glycoprotein, **PrP**. PrP is coded by a cellular gene (conserved among the mammals) that is expressed in normal brain. The cellular product differs from the prion: it is entirely degraded by proteinase, in contrast with the resistance of PrP. We assume that some modification conferring proteinase resistance is involved in the formation of PrP.

If PrP is the infectious agent of scrapie, it must in some way modify the synthesis of its normal cellular counterpart so that it becomes infectious instead of harmless. Such an arrangement would comply with the formal restrictions of the central dogma, but certainly this idea would be at odds with the spirit of the paradigm. Mice that lack a PrP gene cannot be infected to develop scrapie, which demonstrates that PrP is essential for development of the disease.

Either PrP is a novel type of infectious agent, or prion preparations contain extremely small amounts of a rather small and highly infectious nucleic acid (which requires PrP for infectivity). To be undetectable in the conditions of assay, the nucleic acid (virus or viroid) would have to be more infectious than known agents.

How can we resolve the basis for infectivity of scrapie? Synthesis of PrP outside the animal, *in vitro*, in bacteria, or in cultured mammalian cells should allow the isolation of absolutely pure preparations in which there can be no contaminant. If such a preparation were infectious, this would be proof for the role of the protein. Such experiments so far have proved negative, but since (unknown) modifications of PrP could be involved in its infectivity, this does not prove its lack of involvement. If some other component is in fact responsible for infectivity, that component must be isolated to prove its role.

The scope of the paradigm

The exceptional types of agents represented by viroids and (perhaps) scrapie demonstrate that hereditary information can be carried in forms other than a gene coding for protein. Yet in each case we have been compelled to analyze the behavior of the agent in terms of its interaction with, or subversion of, cellular genes.

The concept of the gene itself, however, has recently evolved further. The question of what's in a name is especially appropriate for the gene. Clearly we can no longer say that a gene is a sequence of DNA that continuously and uniquely codes for a particular protein. In situations in which a stretch of DNA is responsible for production of one particular protein, current usage regards the entire sequence of DNA, from the first point represented in the messenger RNA to the last point corresponding to its end, as comprising the 'gene', exons, introns, and all.

When the sequences representing proteins overlap or have alternative forms of expression, we may reverse the usual description of the gene. Instead of saying 'one gene–one polypeptide', we may describe the relationship as 'one polypeptide–one gene'. Thus we regard the sequence actually responsible for production of the polypeptide (including introns as well as exons) as constituting the gene, while recognizing that from the perspective of another protein, part of this same sequence also belongs to *its* gene. This allows the use of descriptions such as 'overlapping' or 'alternative' genes.

We can now see how far we have come from the original hypothesis of Beadle and Tatum. Up to that time, the driving question was the nature of the gene. Once it was discovered that genes represent proteins, the paradigm became fixed in the form of the concept that every genetic unit functions through the synthesis of a particular protein.

This view remains the central paradigm of molecular biology: a sequence of DNA functions either by directly coding for a particular protein or by being necessary for the use of an adjacent segment that actually codes for the protein. How far does this paradigm take us beyond explaining the basic relationship between genes and proteins?

The development of multicellular organisms rests on the use of different genes to generate the different cell phenotypes of each tissue. The expression of genes is determined by a regulatory network that probably takes the form of a cascade. Expression of the first set of genes at the start of embryonic development leads to expression of the genes involved in the next stage of development, which in turn leads to a further stage, and so on until all the tissues of the adult are functioning. The molecular nature of this regulatory network is largely unknown, but we assume that it consists of genes that code for products (probably protein, perhaps sometimes RNA) that act on other genes.

While such a series of interactions is almost certainly the means by which the developmental program is executed, we can ask whether it is entirely sufficient. One specific question concerns the nature and role of **positional information**. We know that all parts of a fertilized egg are not equal; one of the features responsible for development of different tissue parts from different regions of the egg is location of information (presumably specific macromolecules) within the cell.

We do not know how these particular regions are formed. But we may speculate that the existence of positional information in the egg leads to the differential expression of genes in the cells subsequently formed in these regions, which leads to the development of the adult organism, which leads to the development of an egg with the appropriate positional information...

This possibility prompts us to ask whether some information needed for development of the organism is contained in a form that we cannot directly attribute to a sequence of DNA (although the expression of particular sequences may be needed to perpetuate the positional information). Put in a more general way, we might ask: if we could read out the entire sequence of DNA comprising the genome of some organism and interpret it in terms of proteins and regulatory regions, could we then construct an organism (or even a single living cell) by controlled expression of the proper genes?

Summary

Restriction fragments cleave DNA at short, specific sites. The cleavage reaction can be used to generate a restriction map of any DNA, representing the nucleic acid in terms of actual distance between sites of cleavage. The existence of sequence polymorphism in genomes creates polymorphism for the sites of restriction cleavage. Restriction sites can be used like any other genetic marker. An RFLP linkage map is generated by analyzing recombination between RFLPs. The entire human genome has been mapped by this means. At the overall level of the chromosome, the length of the genetic linkage map is proportional to the physical length of the genetic material.

The sequence of DNA can be determined directly by chemical or enzymatic techniques. They share the principle that single-stranded fragments of DNA are produced, identical at one (labeled) end, but differing in length at the other end by single bases. The fragments fall into 4 sets, one for each possible base. The positions at which the chain is broken (in the chemical technique) or at which synthesis is terminated (in the enzymatic technique) in each set identifies the locations of one of the 4 bases.

Using overlapping fragments (for DNA sequencing or restriction mapping) ensures that no small parts of the sequence are omitted. A secure sequence is based upon sequencing both strands of a double helix and comparing them to check for complementarity.

Comparing DNA and protein sequences shows that they are colinear in prokaryotes; but in eukaryotes coding regions may be interrupted. An interrupted eukaryotic gene consists of exons that are spliced together in RNA, removing the introns. Alternative patterns of gene expression allow a single sequence of DNA to represent more than one sequence of protein.

Although all genetic information in cells is carried by DNA, viruses have genomes of double-stranded or single-stranded DNA or RNA. Viroids are subviral pathogens that consist solely of small circular molecules of RNA, with no protective packaging. The RNA does not code for protein and its mode of perpetuation and of pathogenesis is unknown. Scrapie presents a puzzle since no nucleic acid has been detected to provide the infectious agent. It is not known whether all genetic information of an organism is provided by its genome, or whether additional information exists in the form of positional information or self-templating cellular structures.

Further reading

Reviews

The principle of restriction mapping was first adumbrated by **Danna, Sack and Nathans** (*J. Mol. Biol.* **78**, 363–376, 1973) and was reviewed by **Nathans and Smith** (*Ann. Rev. Biochem.* **46**, 273–293, 1975).

The construction of linkage maps from restriction polymorphisms was reviewed by **White** *et al.* (*Nature* **313**, 101–105, 1985). The use of DNA polymorphism to analyze human diseases was reviewed by **Gusella** (*Ann. Rev. Biochem.* **55**, 831–854, 1986).

Methods for sequence analysis of DNA have been reviewed by **Wu** (*Ann. Rev. Biochem.* **47**, 607–734, 1978).

Viroids have been encapsulated by **Diener** (*Adv. Virus Res.* **28**, 241–283, 1983; *Proc. Nat. Acad. Sci. USA* **83**, 58–62, 1986).

Discoveries

Sequencing protocols were introduced by **Maxam and Gilbert** (*Proc. Nat. Acad. Sci. USA* **74**, 560–564, 1977) and **Sanger, Nicklen and Coulson** (*Proc. Nat. Acad. Sci. USA* **74**, 5463–5467, 1977).

An RFLP map of the human genome was reported by **Donis-Keller** *et al.* (*Cell,* **51**, 319–337, 1987).

Colinearity of a bacterial gene and protein was established by **Yanofsky** *et al.* (*Proc. Nat. Acad. Sci. USA* **57**, 296–298, 1967).

The scrapie prion has been equated with protein PrP by **McKinley, Bolton, and Prusiner** (*Cell* **35**, 57–62, 1983), and its gene identified by **Oesch** *et al.* (*Cell* **40**, 735–746, 1985).

TRANSLATION: EXPRESSING GENES AS PROTEINS

PART

2

The central dogma states that once 'information' has passed into protein it cannot get out again. The transfer of information from nucleic acid to nucleic acid, or from nucleic acid to protein, may be possible, but transfer from protein to protein, or from protein to nucleic acid, is impossible. Information means here the precise determination of sequence, either of bases in the nucleic acid or of amino acid residues in the protein.

Francis Crick, 1958

CHAPTER 7

The assembly line for protein synthesis

The **central dogma** defines the paradigm of molecular biology: genes are perpetuated as sequences of nucleic acid, but function by being expressed in the form of proteins.

Three types of processes are responsible for the inheritance of genetic information and for its conversion from one form to another:

◆ Information is *perpetuated* by **replication**; a double-stranded nucleic acid is duplicated to give identical copies.

◆ Information is *expressed* by a two stage process.

◆ **Transcription** generates a single-stranded RNA identical in sequence with one of the strands of the duplex DNA.

◆ **Translation** converts the nucleotide sequence of the RNA into the sequence of amino acids comprising a protein.

The breaking of the genetic code showed that genetic information is stored in the form of nucleotide triplets (codons), but did not reveal *how* each codon specifies its corresponding amino acid. The concept that there must be a code evolved together with the idea that the process of translation must involve a **template** that is separate from the DNA. Because the genetic material in the nucleus is physically separated from the site of protein synthesis in the cytoplasm of a eukaryotic cell, it was clear that the DNA could not *itself* be translated into protein.

The template is generated by transcription, in the form of a **messenger RNA** (abbreviated **mRNA**) that is identical to one strand of the DNA duplex (see Figure 6.16). We might think of the cell as keeping a 'master set' of sequences in the nucleus, while a 'working set' consists of cytoplasmic mRNA copies of the sequences that are to be expressed.

We distinguish the two strands of DNA as follows:

◆ The DNA strand that bears the *same* sequence as the mRNA (except for possessing T instead of U) is called the **coding strand** or **sense strand**.

◆ The other strand of DNA, which directs synthesis of the mRNA via complementary base pairing, is called the **template strand** or **antisense strand**. (We see later that 'antisense' is used as a general term to describe a sequence of DNA or RNA that is complementary to mRNA.)

Since the genetic code is actually *read* on the mRNA, usually it is described in terms of the four bases present in RNA: U, C, A, and G.

The use of the term *messenger* RNA reflects its ability (in eukaryotes) to move from the nucleus where it is synthesized to the cytoplasm where it functions. Translation of mRNA into protein is accomplished by reading the genetic code: each triplet of nucleotides is converted into one amino acid. Thus 'translation' describes the step at which

Figure 7.1

Overview: the central dogma states that information in nucleic acid can be perpetuated or transferred, but the transfer of information into protein is irreversible.

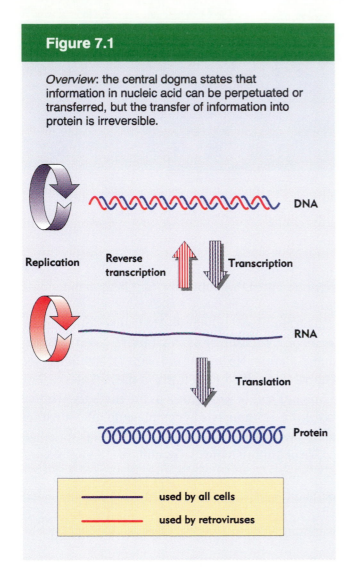

Replication   Reverse transcription   Transcription

DNA

RNA

Translation

Protein

used by all cells
used by retroviruses

Figure 7.1 illustrates the roles of replication, transcription, and translation, viewed from the perspective of the central dogma:

◆ *The perpetuation of nucleic acid may involve either DNA or RNA as the genetic material* (see Figure 6.22). Cells use only DNA. Some viruses use RNA, and replication of viral RNA occurs in the infected cell.

◆ *The expression of cellular genetic information usually is unidirectional.* Transcription of DNA generates RNA molecules that can be used further *only* to generate protein sequences; generally they cannot be retrieved for use as genetic information. Translation of RNA into protein is always irreversible.

The restriction to unidirectional transfer from DNA to RNA is not absolute. It is overcome by the **retroviruses**, whose genomes consist of single-stranded RNA molecules. During the infective cycle, the RNA is converted by the process of **reverse transcription** into a single-stranded DNA, which in turn is converted into a double-stranded DNA. This duplex DNA becomes part of the genome of the cell, and is inherited like any other gene. *Thus reverse transcription allows a sequence of RNA to be retrieved and used as genetic information.*

The existence of RNA replication and reverse transcription establishes the general principle that *information in the form of either type of nucleic acid sequence can be converted into the other type.* In the usual course of events, however, neither of these mechanisms is used by the cell itself, which relies on the processes of replication, transcription, and translation. But on rare occasions (possibly mediated by an RNA virus), information in the form of a cellular RNA is retrieved and inserted into the genome. Although reverse transcription plays no role in the regular operations of the cell, it becomes a mechanism of potential importance when we consider the evolution of the genome. (We discuss this issue in more detail in Chapter 35.)

the nucleotide sequence is deciphered as representing individual amino acids.

The two stages of expression are delineated by systems for synthesizing proteins *in vitro*. Such systems can be prepared in the form of cell-free extracts, by breaking cells and centrifuging the mixture to remove matter such as cell walls, membrane fragments, and nuclei (from eukaryotic cells). When provided with a suitable source of energy and precursors, the 'supernatant' (or components purified from it) can translate most mRNAs that are added to it. Thus we can distinguish the mRNA from the apparatus responsible for its translation.

Transfer RNA is the adaptor

Figure 7.2

The secondary structure of tRNA has four arms. Each tRNA has the dual properties of an adaptor that recognizes both the amino acid and codon.

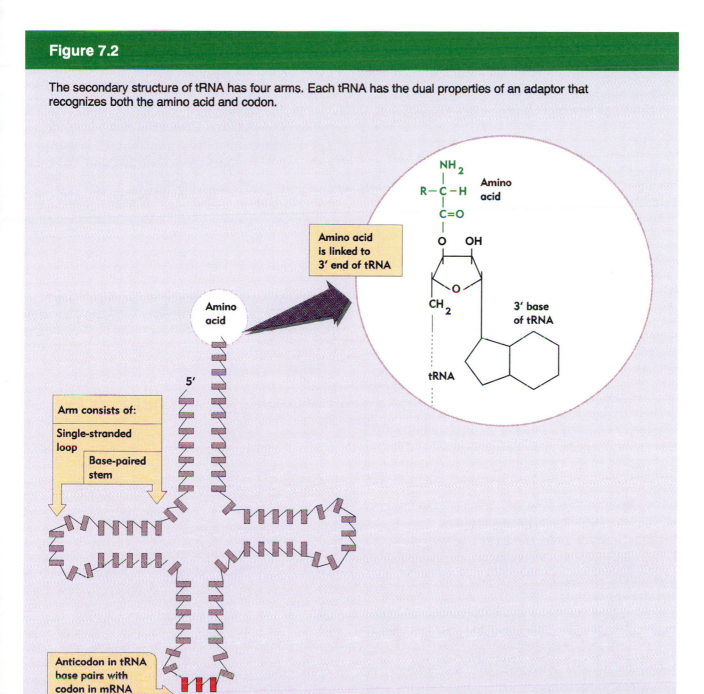

Each amino acid is represented by (at least) one codon that consists of a nucleotide triplet in the mRNA. The incongruity of structure between trinucleotide and amino acid immediately raises the question of how each codon is matched to its particular amino acid. Even before the exact form of the code had been discovered, Crick suggested that translation might be mediated by an 'adaptor' molecule. This adaptor is **transfer RNA** (abbreviated **tRNA**), a small molecule whose polynucleotide chain is only 75–85 bases long.

A tRNA has two crucial properties:

It is able to represent only one amino acid, to which it is *covalently linked.*

It contains a trinucleotide sequence, the **anticodon**, which is *complementary to the codon representing its amino acid.* The anticodon enables the tRNA to recognize the codon via complementary base pairing.

Transfer RNA has a characteristic secondary structure. The nucleotide sequence of every tRNA can be written in the form of a **cloverleaf**, illustrated in **Figure 7.2**, in which complementary base pairing forms **stems** for single-stranded **loops**. The stem–loop structures are called the **arms** of tRNA.

When a tRNA is **charged** with the amino acid corresponding to its anticodon, it becomes **aminoacyl-tRNA**. The amino acid is linked by an ester bond from its carboxyl group to the 2′ or 3′ hydroxyl group of the ribose of the 3′ terminal base of the tRNA (which is always adenine).

There is at least one tRNA (but usually more) for each amino acid. A tRNA is named by using the three letter abbreviation for the amino acid as a superscript. For example, tRNA^Ala is a tRNA for alanine. If there is more than one tRNA for the same amino acid, subscript numerals are used to distinguish them. Thus two tRNAs for tyrosine would be described as $tRNA_1^{Tyr}$ and $tRNA_2^{Tyr}$. A tRNA carrying an amino acid—that is, an aminoacyl-tRNA—is indicated by a prefix that identifies the amino acid. Ala-tRNA describes tRNA^Ala carrying its amino acid.

The secondary structure of each tRNA folds into a compact L-shaped tertiary structure in which the 3′ end that binds amino acid is distant from the anticodon that binds mRNA. This form is depicted diagrammatically in **Figure 7.3**, and we discuss the structure in more detail in Chapter 8.

The process of charging a tRNA is catalyzed by a specific enzyme, **aminoacyl-tRNA synthetase**. There are (at least) 20 aminoacyl-tRNA synthetases. Each recognizes a single amino acid and all the tRNAs onto which it can legitimately be placed.

Does the anticodon sequence alone allow aminoacyl–tRNA to recognize the correct codon? An experiment to test this question is illustrated in **Figure 7.4**. Reductive desulfuration converts the amino acid of cysteinyl-tRNA into alanine, generating alanyl-tRNA^Cys. The tRNA has an anticodon that responds to the codon UGU. Modification of the amino acid does not influence the

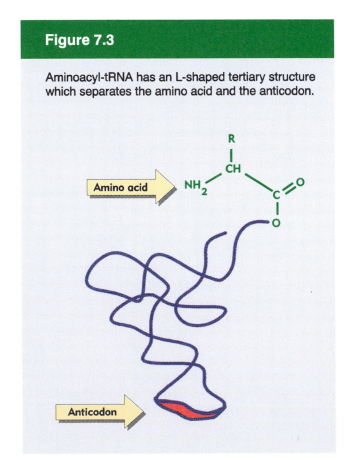

Figure 7.3

Aminoacyl-tRNA has an L-shaped tertiary structure which separates the amino acid and the anticodon.

Figure 7.4

The meaning of tRNA is determined by its anticodon and not by its amino acid.

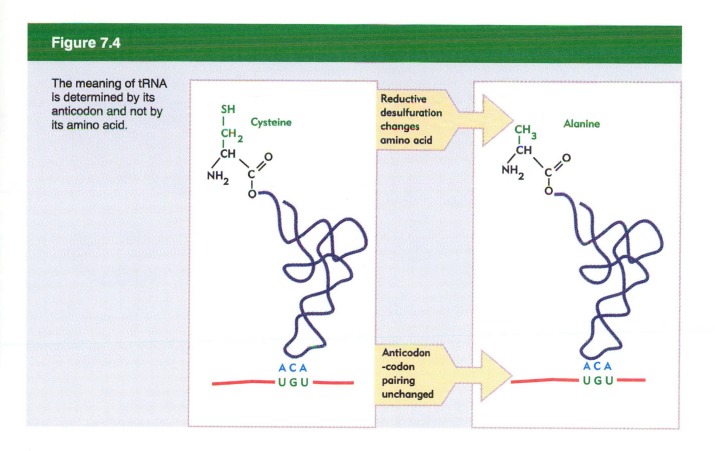

specificity of the anticodon–codon interaction, so the alanine residue is incorporated into protein in place of cysteine. *Once a tRNA has been charged, the amino acid plays no further role in its specificity, which is determined exclusively by the anticodon.*

Messenger RNA is translated by ribosomes

Reading the genetic code as a series of adjacent triplets, protein synthesis proceeds from the start of a coding region to the end. *Proteins are assembled in the direction from the N-terminus to the C-terminus.*

Amino acids are assembled into proteins by the **ribosome**, a compact **ribonucleoprotein particle** consisting of two **subunits**. Each ribosome subunit consists of several proteins associated with a long RNA molecule; the RNAs are known as **ribosomal RNA** (abbreviated **rRNA**).

Ribosomes are traditionally described in terms of their (approximate) rate of sedimentation (measured in Svedbergs, in which a higher S value indicates a greater rate of sedimentation and a larger mass). Bacterial ribosomes generally sediment at ~70S. The ribosomes of the cytoplasm of higher eukaryotic cells are larger, usually sedimenting at ~80S.

The relationship between a ribosome and its subunits is depicted in **Figure 7.5**. The two subunits dissociate *in vitro* when the concentration of Mg^{2+} ions is reduced. Bacterial (70S) ribosomes have subunits that sediment at roughly 50S and 30S. The

Figure 7.5

A ribosome consists of two subunits.

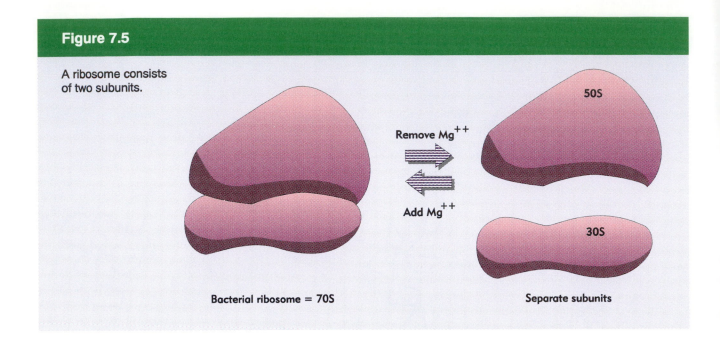

Bacterial ribosome = 70S

Remove Mg^{++}

Add Mg^{++}

50S

30S

Separate subunits

subunits of eukaryotic cytoplasmic (80S) ribosomes sediment at ~60S and ~40S. In each case, the larger subunit is about twice the mass of the smaller subunit. The two subunits work together as part of the complete ribosome, but each undertakes distinct reactions in protein synthesis.

All the ribosomes of a given cell compartment are identical. *They undertake the synthesis of different proteins by associating with the different mRNAs that provide the actual coding sequences.*

The ribosome provides an environment that controls the recognition between a codon of mRNA and the anticodon of tRNA. To accomplish the sequential synthesis of a protein, *the ribosome moves along the mRNA,* one codon at a time.

A ribosome attaches to mRNA at or near the 5′ end of a coding region; moving along the RNA toward the 3′ end, it translates each triplet codon into an amino acid *en route.* As the ribosome proceeds, the appropriate aminoacyl-tRNAs associate with it, donating their amino acids to the polypeptide chain. At any given moment, the ribosome can accommodate the two aminoacyl-tRNAs corresponding to successive codons, making it possible for a peptide bond to form between the two corresponding amino acids. At each step, the growing polypeptide

chain becomes longer by one amino acid.

When active ribosomes are isolated in the form of the fraction associated with newly synthesized proteins, they are found in the form of a unit consisting of an mRNA associated with several ribosomes. This is the **polyribosome** or **polysome**.

Each ribosome in the polysome independently synthesizes a single polypeptide during its traverse of the messenger sequence. Thus the mRNA has a series of ribosomes that carry increasing lengths of the protein product, moving from the 5′ to the 3′ end, as illustrated in **Figure 7.6**. A polypeptide chain in the process of synthesis is sometimes called a **nascent protein**.

Roughly the last 30–35 amino acids added to a growing polypeptide chain are protected from the environment by the structure of the ribosome. Probably all of the preceding part of the polypeptide protrudes and is free to start folding into its proper conformation. Thus proteins can display parts of the mature conformation even before synthesis has been completed.

A classic characterization of polysomes is shown in the electron micrograph of **Figure 7.7**. Globin protein is synthesized by a set of 5 ribosomes attached to each mRNA (pentasomes). The

Figure 7.6

A polyribosome consists of an mRNA being translated simultaneously by several ribosomes moving in the direction from 5' to 3'. Each ribosome has two tRNA molecules, the first carrying the last amino acid added to the chain (connected to the nascent protein, i.e., the polypeptide chain synthesized so far), the second carrying the next amino acid to be added.

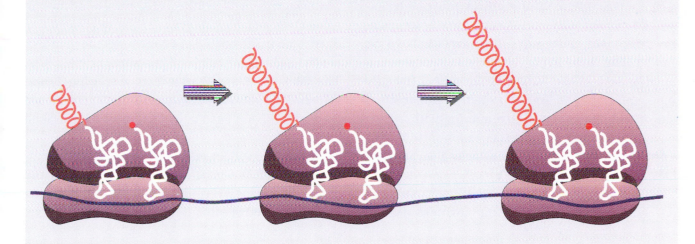

ribosomes appear as (roughly) squashed spherical objects of ~7 nm (70 Å) in diameter, connected by a thread of mRNA. The ribosomes are located at various positions along the messenger. Those at one end have just started protein synthesis; those at the other end are about to complete production of a polypeptide chain.

The size of the polysome depends on several variables. In bacteria, it is very large, with tens of ribosomes simultaneously engaged in translation.

Figure 7.7

Protein synthesis occurs on polysomes. Photograph kindly provided by Alex Rich.

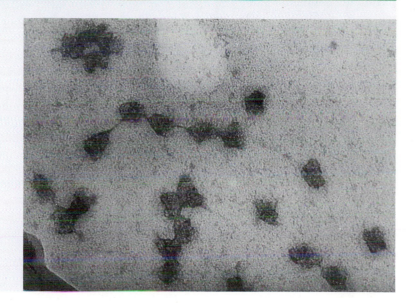

Partly the size is due to the length of the mRNA (which usually codes for several proteins); partly it is due to the high efficiency with which the ribosomes attach to the mRNA. Since ribosomes attach to bacterial mRNA even before its transcription has been completed, the polysome is likely still to be attached to DNA.

In a eukaryotic cell, the mRNA must be transported from the nucleus to reach the ribosomes in the cytoplasm. The polysomes are likely to be smaller than those in bacteria; again, their size is a function both of the length of the mRNA (usually representing only a single protein in eukaryotes) and of the characteristic frequency with which ribosomes attach. An average eukaryotic mRNA probably has ~8 ribosomes attached at any one time.

The number of ribosomes on each mRNA molecule synthesizing a particular protein is not precisely determined, in either bacteria or eukaryotes, but is a matter of statistical fluctuation, determined by the variables of mRNA size and efficiency.

Bacterial mRNA originally proved elusive, because it is unstable and represents only a very small proportion of the mass of the bacterial RNA. In bacteria, mRNA is synthesized, translated by the ribosomes, and degraded, all in rapid succession, as illustrated in **Figure 7.8**. A given molecule of mRNA survives for only a matter of minutes or even less. The ribosomes are drawn from a free pool, and return to the pool to be used again after their release from an mRNA. (The state of the pool is described later.)

Eukaryotic mRNA also constitutes only a small proportion of the total cellular RNA (~3% of the mass). However, it is relatively stable, often surviving for a period of some hours in the cell.

Figure 7.8

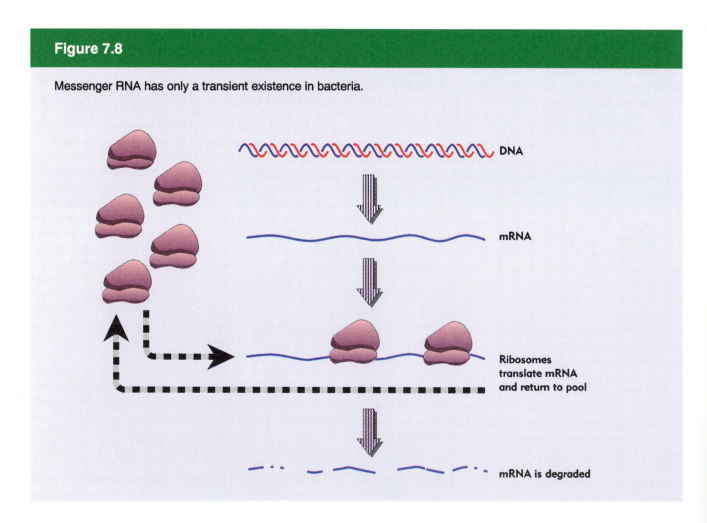

Messenger RNA has only a transient existence in bacteria.

DNA

mRNA

Ribosomes translate mRNA and return to pool

mRNA is degraded

Table 7.1

Considering *E. coli* in terms of its macromolecular components.

Component	Proportion of Dry Cell Mass	Molecules/Cell	Number of different kinds	Copies of each kind
Wall	10%	1	1	1
Membrane	10%	2	2	1
DNA	1.5%	1	1	1
mRNA	1%	1,500	600	2-3
tRNA	3%	200,000	60	>3000
rRNA	16%	38,000	2	19,000
Ribosomal protein	9%	10^6	52	19,000
Soluble protein	46%	2.0×10^6	1850	>1000
Small molecules	3%	7.5×10^6	800	

An overall view of the attention devoted to protein synthesis in the intact bacterium is given in **Table 7.1**. The 20,000 or so ribosomes account for a quarter of the cell mass. There are >3000 copies of each tRNA, and altogether, the tRNA molecules outnumber the ribosomes by almost tenfold; most of them are present as aminoacyl-tRNAs, that is, ready to be used at once in protein synthesis. Because of their instability, it is difficult to calculate the number of mRNA molecules, but a reasonable guess would be 1500, in varying states of synthesis and decomposition. There are ~600 different types of mRNA in a bacterium. This suggests that there are usually only 2–3 copies of each mRNA per bacterium. On average, each probably codes for ~3 proteins. If there are 1850 different soluble proteins, there must on average be >1000 copies of each protein in a bacterium.

The meaning of the genetic code

Any one of four possible nucleotides can occupy each of the three positions of the codon, so that there are $4^3 = 64$ possible combinations of trinucleotide sequences. Thus there are more codons than the 20 amino acids from which proteins are synthesized.

Soon after the discovery of the triplet nature of the code, two methods were developed to allow codons to be assigned systematically to amino acids. Both used *in vitro* systems consisting of components of the apparatus that synthesizes proteins:

◆ A protein-synthesizing system from *E. coli* was used to translate *synthetic polynucleotides*. The first report of success with such a system was Nirenberg's demonstration in 1961 that polyuridylic acid [poly(U)] can act as an mRNA to direct the assembly of phenylalanine into polyphenylalanine. This result means that UUU must be a codon for phenylalanine. Subsequently, many other synthetic polynucleotides, consisting of known sequences of different bases, were used by Khorana; they allowed the meaning of about half of the 64 codons to be assigned.

◆ The **ribosome-binding assay** for making codon assignments was developed by Nirenberg and Leder in 1964. A trinucleotide can be used to mimic a codon, by causing the corresponding aminoacyl-tRNA to bind to a ribosome. A triple complex of trinucleotide•aminoacyl-tRNA•ribosome can be isolated by taking advantage of the ability of ribosomes to bind to nitrocellulose filters. The aminoacyl-tRNA itself does not bind to such filters, but is retained as part of the triple complex. Its retention is detected by means of a radioactive label in the amino acid component. Then the meaning of each trinucleotide

can be determined by testing which one of 20 labeled aminoacyl-tRNA preparations is retained on the filter.

The two techniques together assigned meaning to all of the codons that represent amino acids. Since then, the sequencing of DNA has made it possible to compare corresponding nucleotide and amino acid sequences directly. *The sequence of the coding strand of DNA, read in the direction from 5' to 3', consists of triplets corresponding to the amino acid sequence of the protein read from N-terminus to C-terminus.* The entire genetic code has been confirmed in overwhelming detail from such analysis.

The code is summarized in **Figure 7.9.** A striking feature is its **degeneracy**: almost every amino acid

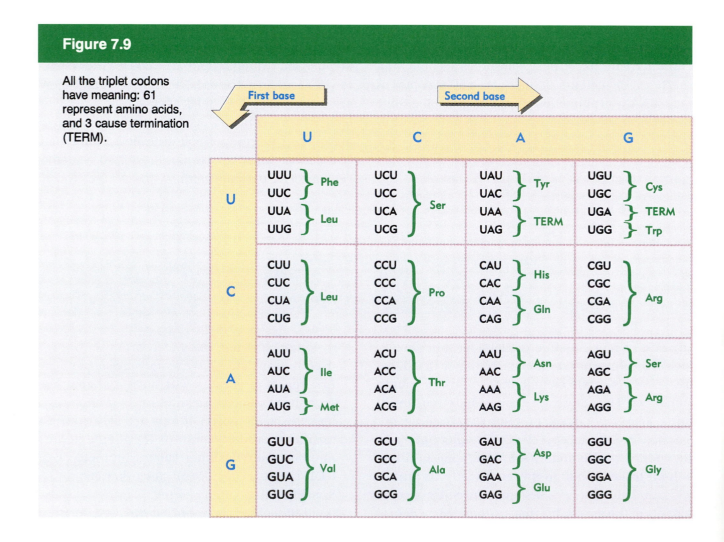

Figure 7.9

All the triplet codons have meaning: 61 represent amino acids, and 3 cause termination (TERM).

Figure 7.10

The number of codons for each amino acid does not correlate with its frequency of use in proteins.

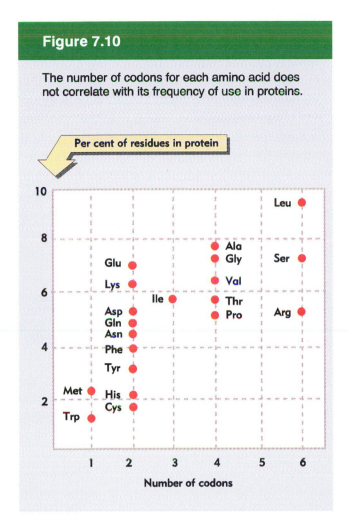

is represented by several codons. The only exceptions are methionine and tryptophan. Codons that have the same meaning are called **synonyms**. **Figure 7.10** plots the number of codons representing each amino acid against the frequency with which the amino acid is used in proteins (in *E. coli*). Although there is a tendency for amino acids that are more common to be represented by more codons, this is slight, and therefore it does not seem that the genetic code has been optimized with regard to the utilization of amino acids.

Codons representing the same or related amino acids tend to be similar in sequence. Often the base in the third position of a codon is not significant, because the four codons differing only in the third base represent the same amino acid. Sometimes a distinction is made only between a purine versus a

pyrimidine in this position. The reduced specificity at the last position is known as **third-base degeneracy**.

The tendency for similar amino acids to be represented by related codons minimizes the effects of mutations. It increases the probability that a single random base change will result in no amino acid substitution or in one involving amino acids of similar character. For example, a mutation of CUC to CUG has no effect, since both codons represent leucine; a mutation of CUU to AUU results in replacement of leucine with isoleucine, a closely related amino acid.

Three codons (UAA, UAG and UGA) do not represent amino acids. They are used specifically to terminate protein synthesis; one of these **stop codons** marks the end of every gene.

Is the genetic code the same in all living organisms?

Comparisons of DNA sequences with the corresponding protein sequences reveal that the *identical set of codon assignments is used in bacteria and in eukaryotic cytoplasm.* As a result, mRNA from one species usually can be translated correctly *in vitro* or *in vivo* by the protein synthetic apparatus of another species. Thus the codons used in the mRNA of one species have the same meaning for the ribosomes and tRNAs of other species.

The universality of the code argues that it must have been established very early in evolution. Originally there may have been a stereochemical relationship between amino acids and the codons representing them. Then the system now used for protein synthesis evolved by selection for features such as greater efficiency or accuracy.

Perhaps the code started in a primitive form in which a small number of codons were used to represent comparatively few amino acids, possibly even with one codon corresponding to any member of a group of amino acids. More precise codon meanings and additional amino acids could have been introduced later. One possibility is that at first only two of the three bases in each codon were used; discrimination at the third position could have evolved later.

Evolution of the code could have become 'frozen'

at a point at which the system had become so complex that any changes in codon meaning would disrupt existing proteins by substituting amino acids. Its universality implies that this must have happened at such an early stage that all living organisms are descended from a single pool of primitive cells in which this occurred.

Exceptions to the universal genetic code are rare. Changes in meaning in the principal genome of a species usually concern the termination codons. For example, in a mycoplasma, UGA codes for tryptophan; and in certain species of the ciliates *Tetrahymena* and *Paramecium*, UAA and UAG code for glutamine.

Systematic alterations of the code have occurred only in mitochondrial DNA. Such changes may have been possible because the mitochondrion is a relatively closed system, concerned with the synthesis of a few specific proteins. The majority of these changes affect initiation and termination, but other substitutions of meaning are also found. We discuss changes in the genetic code in Chapter 8, in the context of the interaction between aminoacyl-tRNA and messenger RNA.

The ribosomal sites of action

Synthesis of proteins involves an assembly line in which the ribosomes proceed inexorably along the messenger, bringing in the aminoacyl-tRNAs that provide the actual building blocks of the protein product. The ribosome itself constitutes a small mobile factory, in which a compact package of proteins and rRNAs forms several active centers able to undertake various catalytic activities. Different sets of accessory factors assist the ribosome in each of the three stages of protein synthesis: initiation, elongation, and termination. Energy for ribosome movement is provided by hydrolysis of GTP.

◆ **Initiation** involves the reactions that precede formation of the peptide bond between the first two amino acids of the protein. It requires the ribosome to bind to the mRNA, forming an initiation complex that contains the first aminoacyl-tRNA. This is a relatively slow step in protein synthesis, and usually determines the rate at which an mRNA is translated.

◆ **Elongation** includes all the reactions from synthesis of the first peptide bond to addition of the last amino acid. Amino acids are added to the chain one at a time; the addition of an amino acid is the most rapid step in protein synthesis.

◆ **Termination** encompasses the steps that are needed to release the completed polypeptide chain; at the same time, the ribosome dissociates from the mRNA.

Protein synthesis overall is a rapid process, although the rate depends strongly on temperature. In bacteria at 37°C, ~15 amino acids are added to a growing polypeptide chain every second. So it takes only ~20 seconds to synthesize an average protein of 300 amino acids. In eukaryotes, the rate of protein synthesis is slower; in red blood cells at 37°C, elongation typically sees ~2 amino acids added to the chain per second.

Most of the experiments to define the stages of protein synthesis have been performed with *in vitro* systems, consisting of ribosomes, aminoacyl-tRNAs, other enzymatic factors, and an energy source. In these systems, the rate of protein synthesis is slower by an order of magnitude than the rate *in vivo*.

Messenger RNA is associated with the small subunit, ~30 bases of the mRNA being bound at any time. But only two molecules of tRNA are involved in peptide bond synthesis at any moment. So polypeptide elongation involves reactions taking place at just two of the (roughly) ten codons covered by the ribosome.

Figure 7.11 shows that each tRNA lies in a distinct site. The two sites have different features:

◆ The only site that can be entered by an incoming aminoacyl-tRNA is the **A site** (or **entry site**). Prior to the entry of aminoacyl-tRNA, the site exposes the codon representing the next amino acid due to be added to the chain.

Figure 7.11

The ribosome has two sites for binding charged tRNA.

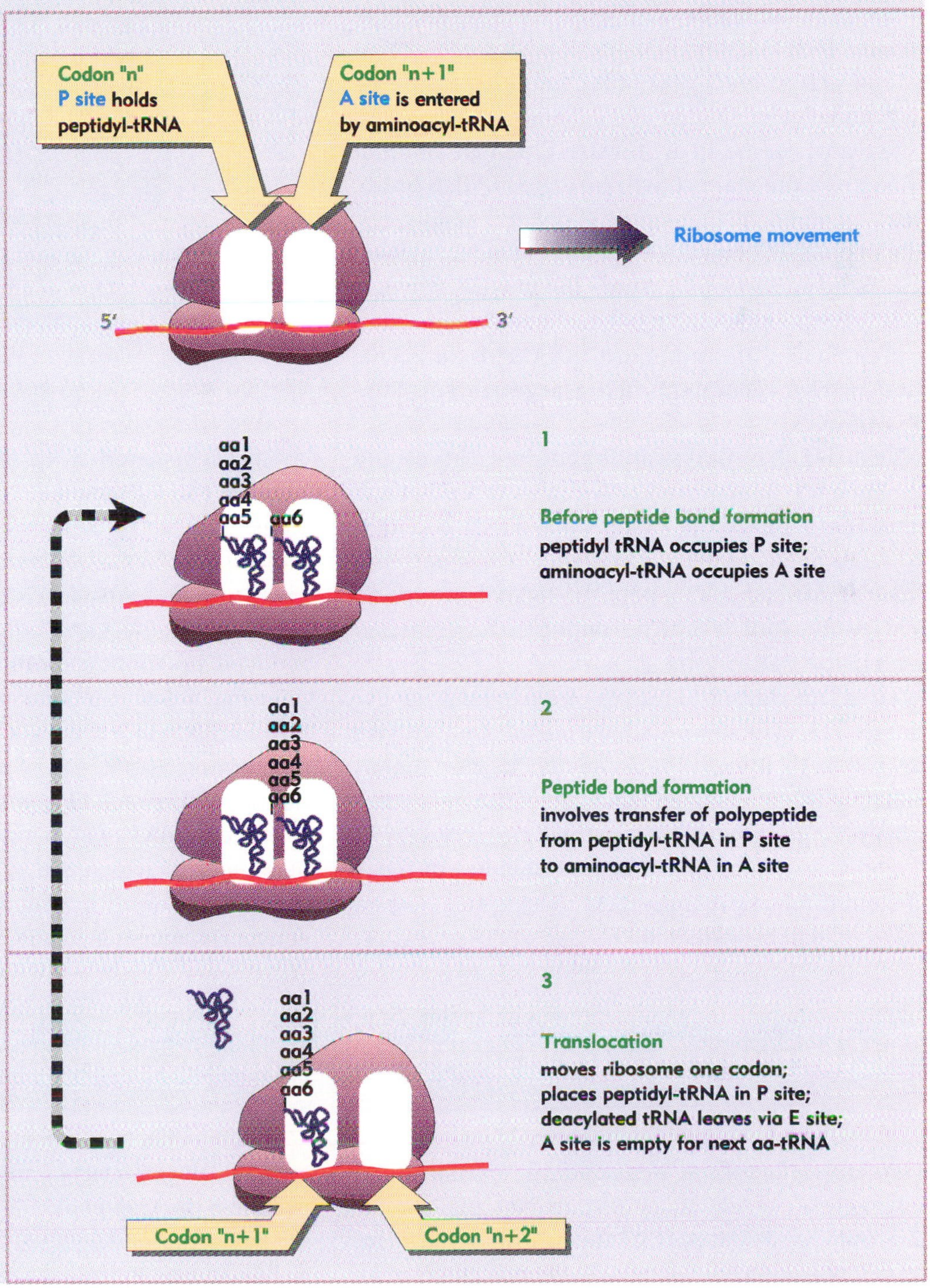
Codon "n"
P site holds
peptidyl-tRNA
Codon "n+1"
A site is entered
by aminoacyl-tRNA
Ribosome movement
5′
3′
1
aa1
aa2
aa3
aa4
aa5
aa6
Before peptide bond formation
peptidyl tRNA occupies P site;
aminoacyl-tRNA occupies A site
2
aa1
aa2
aa3
aa4
aa5
aa6
Peptide bond formation
involves transfer of polypeptide
from peptidyl-tRNA in P site
to aminoacyl-tRNA in A site
3
aa1
aa2
aa3
aa4
aa5
aa6
Translocation
moves ribosome one codon;
places peptidyl-tRNA in P site;
deacylated tRNA leaves via E site;
A site is empty for next aa-tRNA
Codon "n+1"
Codon "n+2"

◆ The codon representing the most recent amino acid to have been added to the nascent polypeptide chain lies in the **P site** (or **donor site**). This site is occupied by **peptidyl-tRNA**, a tRNA to which the nascent polypeptide chain is attached.

The end of the tRNA that carries an amino acid is located on the large subunit, while the anticodon at the other end interacts with the mRNA bound by the small subunit. So the P and A sites must extend across both ribosomal subunits, as drawn in Figure 7.11.

For a ribosome to synthesize a peptide bond, it must be in the state shown in *step 1* in the Figure, when peptidyl-tRNA is in the P site and aminoacyl-tRNA is in the A site. Then peptide bond formation occurs by a reaction in which the polypeptide carried by the peptidyl-tRNA is transferred to the amino acid carried by the aminoacyl-tRNA. This transfer is catalyzed by constituents of the large subunit of the ribosome.

Transfer of the polypeptide generates a ribosome in which the **deacylated tRNA**, now devoid of any amino acid, lies in the P site, while a new peptidyl-tRNA has been created in the A site. This situation is illustrated in *step 2*. This peptidyl-tRNA is one amino acid residue longer than the peptidyl-tRNA that had been in the P site in *step 1*.

Then the ribosome moves one triplet along the messenger. This step is called translocation, and we discuss it in more detail shortly. Its movement transfers the deacylated tRNA out of the P site, and moves the peptidyl-tRNA into the P site (see *step 3*). The next codon to be translated now lies in the A site, ready for a new aminoacyl-tRNA to enter, when the cycle will be repeated.

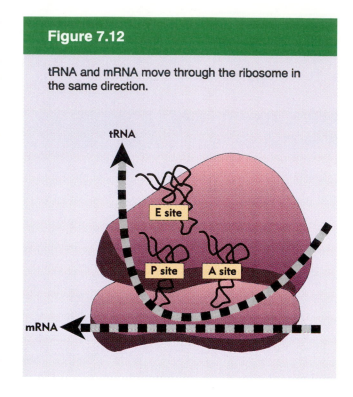

Figure 7.12

tRNA and mRNA move through the ribosome in the same direction.

The deacylated tRNA leaves the ribosome via another tRNA-binding site, the E site. This site is transiently occupied by the tRNA *en route* between leaving the P site and being released from the ribosome into the cytosol. While the E site is occupied, the affinity of the A site for aminoacyl-tRNA is much reduced, preventing another aminoacyl-tRNA from entering before the ribosome is ready. Thus the flow of tRNA, into the A site, through the P site, and out through the E site, is controlled in an orderly manner. **Figure 7.12** compares the movement of tRNA and mRNA, which may be thought of as a sort of ratchet in which the reaction is driven by the codon–anticodon interaction. (We consider this concept in more detail in Chapter 8.)

Initiation in bacteria needs 30S subunits and accessory factors

Bacterial ribosomes engaged in elongating a polypeptide chain exist as 70S particles. At termination, they are released from the mRNA in this form; then they enter a pool of free ribosomes. In growing bacteria, the majority of ribosomes are synthesizing proteins; the free pool is likely to contain ~20% of the ribosomes.

Ribosomes in the free pool can dissociate into

separate subunits; thus free 70S ribosomes are in dynamic equilibrium with 30S and 50S subunits. *Initiation of protein synthesis is not a function of intact ribosomes, but is undertaken by the separate subunits, which reassociate during the initiation reaction.* **Figure 7.13** summarizes the ribosomal subunit cycle during protein synthesis in bacteria. (The details differ in eukaryotic organelles and cytoplasm, as discussed later.)

The sequence on mRNA at which initiation occurs is called the **ribosome-binding site**: it is a short sequence of bases that precedes the actual

coding region (see Chapter 10). The reaction occurs in two steps:

◆ Recognition of mRNA occurs when a *small subunit* binds to form an **initiation complex** at the ribosome-binding site.

◆ Then a *large subunit* joins the complex to generate a complete ribosome.

Thus the ribosome-binding site is a sequence at which the small and large subunits associate on mRNA to form an intact ribosome, rather than a sequence to which the ribosome binds as such.

Although the 30S subunit is involved in initiation, it is not by itself competent to undertake the reactions of binding mRNA and tRNA. It requires additional proteins called **initiation factors (IF)**.

The factors were originally discovered by the effects of 'washing' 30S subunits with ammonium chloride. This treatment leaves them unable to sponsor initiation of new chains. (Because the initiation factors are bound loosely, they are released from the subunits by the ammonium chloride wash, which does not have any effect on the 'permanent' ribosomal proteins.)

The initiation factors are found only on 'free' 30S subunits, which are derived from the pool of ribosomes not currently engaged in protein synthesis. The number of copies of each factor is much less than the number of free ribosomes.

Initiation factors are released when 50S subunits associate with the 30S•mRNA complexes to generate 70S ribosomes. Loss of the factors therefore marks the transition between the two stages of initiation. This behavior distinguishes initiation factors from the structural proteins of the ribosome. *The initiation factors are concerned solely with formation of the initiation complex, they are absent from 70S ribosomes, and they play no part in the stages of elongation.*

Bacteria use three initiation factors, numbered **IF-1, IF-2,** and **IF-3.** They are needed for both mRNA and tRNA to enter the initiation complex:

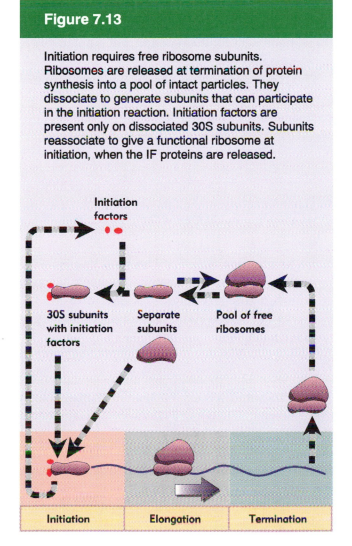

Figure 7.13

Initiation requires free ribosome subunits. Ribosomes are released at termination of protein synthesis into a pool of intact particles. They dissociate to generate subunits that can participate in the initiation reaction. Initiation factors are present only on dissociated 30S subunits. Subunits reassociate to give a functional ribosome at initiation, when the IF proteins are released.

Figure 7.14

Initiation requires 30S subunits that carry IF3.

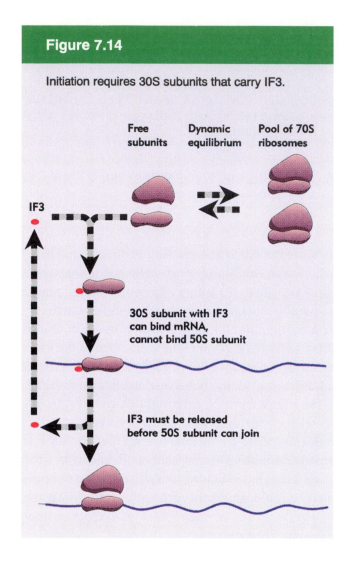

Free subunits   Dynamic equilibrium   Pool of 70S ribosomes

IF3

30S subunit with IF3 can bind mRNA, cannot bind 50S subunit

IF3 must be released before 50S subunit can join

◆ IF-3 is needed for 30S subunits to bind specifically to initiation sites in mRNA.

◆ IF-2 binds a special initiator tRNA and brings it to the initiation complex.

◆ The role of IF-1 has yet to be pinned down; it binds to 30S subunits only as a part of the complete initiation complex, and could be involved in stabilizing it, rather than in recognizing any specific component.

The role of IF-3 is illustrated in **Figure 7.14.** The factor has dual functions: it is needed first to stabilize (free) 30S subunits; and then it enables them to bind to mRNA. IF-3 essentially controls the freedom of 30S subunits, which lasts from their dissociation from the pool of ribosomes to their reassociation with a 50S subunit at initiation.

The first function of IF-3 controls the equilibrium between ribosomal states. IF-3 binds to free 30S subunits, but not to 70S ribosomes. When IF-3 binds to a 30S subunit, it prevents the subunit from reassociating with a 50S subunit. In this capacity, its role is to act as an *antiassociation factor*. The reaction between IF-3 and the 30S subunit is stoichiometric: one molecule of IF-3 binds per subunit. Since there is a relatively small amount of IF-3, its availability determines the number of free 30S subunits.

The second function of IF-3 controls the ability of 30S subunits to bind to mRNA. Small subunits that lack the factor cannot form initiation complexes with mRNA. However, the *specificity* with which initiation sites are selected on mRNA appears to be entirely a function of the ribosome subunit. Thus IF-3 is necessary for the reaction, but is not involved in selecting the site.

IF-3 is released before the 30S subunit is joined by the 50S subunit. A 30S subunit cannot simultaneously bind IF-3 and a 50S subunit. The options open to the 30S subunit are therefore either:

◆ When carrying IF-3, it exists as a 'free' subunit that can bind to mRNA.

◆ In the absence of IF-3, it can bind to a 50S subunit, forming a 70S ribosome engaged in translation (if the 30S subunit is bound to mRNA) or a free ribosome (if the 30S subunit is not bound to mRNA).

The 30S subunit is driven around its life cycle by these alternatives. When a ribosome is a member of the free pool, its dynamic equilibrium with the subunits allows IF-3 to replace the 50S subunit. When an initiation complex has been formed, the IF-3 is released, and the 30S subunit is joined by a 50S subunit. On its release, IF-3 immediately recycles by finding another 30S subunit.

A special initiator tRNA starts the polypeptide chain

What initiates protein synthesis: how is the first codon of the gene recognized as providing the starting point for translation? The reaction takes place in two stages:

◆ The sequence of the ribosome-binding site on a bacterial mRNA is recognized by a 30S subunit carrying IF-3. The part of the site recognized directly by the ribosome is called the **Shine–Dalgarno** sequence; we discuss the interaction between this sequence and the ribosome in Chapter 9.

◆ The ribosome-binding site also contains a signal that marks the start of the reading frame, a special **initiation codon**. Usually the initiation codon is the triplet AUG, but in bacteria, GUG or UUG are also used.

The AUG codon represents methionine, and two types of tRNA can carry this amino acid. One is used for initiation, the other for recognizing AUG codons during elongation.

Figure 7.15

The initiator *N*-formyl-methionyl-tRNA (fMet-tRNA$_f$) is generated by formylation of methionyl-tRNA, using formyl-tetrahydrofolate as cofactor.

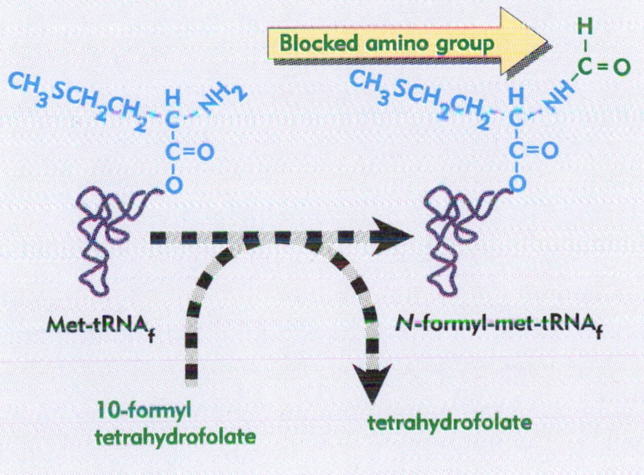

In bacteria and in eukaryotic organelles, the initiator tRNA carries a methionine residue that has been formylated on its amino group, forming a molecule of *N*-formyl-methionyl-tRNA. The tRNA is known as tRNA$_f^{Met}$. The name of the aminoacyl–tRNA is usually abbreviated to fMet-tRNA$_f$.

The initiator tRNA gains its modified amino acid in a two stage reaction. First, it is charged with the amino acid to generate Met-tRNA$_f$; then the formylation reaction shown in **Figure 7.15** blocks the free NH_2 group. Although the blocked amino acid group would prevent the initiator from participating in chain elongation, it does not interfere with the ability to initiate a protein.

This tRNA is used only for initiation. It recognizes the codons AUG or GUG (occasionally UUG). The codons are not recognized equally well: the extent of initiation declines about half when AUG is replaced by GUG, and declines by about half yet again when UUG is employed.

The species responsible for recognizing AUG codons in internal locations is tRNA$_m^{Met}$. *This tRNA responds only to internal AUG codons.* Its methionine cannot be formylated.

Thus there are two differences between the initiating and elongating Met-tRNAs: the tRNA moieties themselves are different; and the amino acids differ in the state of the amino group.

The meaning of the AUG and GUG codons depends on their **context**. When the AUG codon is used for initiation, it is read as formyl-methionine; when used within the coding region, it represents methionine. The meaning of the GUG codon is even more dependent on its location. When present as the *first* codon, it is read via the initiation reaction as formyl-methionine. Yet when present *within* a gene, it is read by Val-tRNA, one of the regular members of the tRNA set, to provide valine as required by the genetic code (see Figure 7.9).

How is the context of AUG and GUG codons interpreted? **Figure 7.16** illustrates the decisive role of the ribosome.

Figure 7.16

Only fMet-tRNA_f can be used for initiation by 30S subunits; only other aminoacyl-tRNAs can be used for elongation by 70S ribosomes.

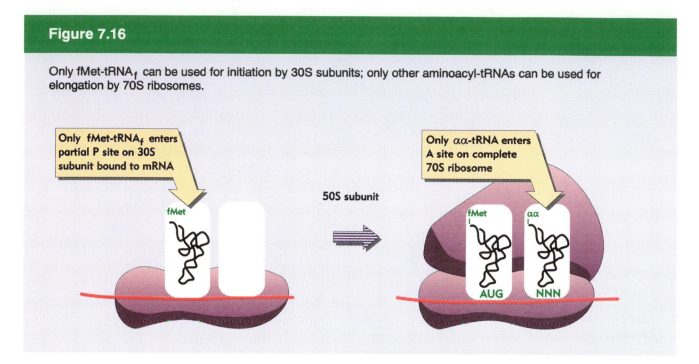

Only fMet-tRNA_f enters partial P site on 30S subunit bound to mRNA

50S subunit

Only αα-tRNA enters A site on complete 70S ribosome

fMet

fMet   αα

AUG   NNN

In an initiation complex, the small subunit sits on the mRNA in such a way that the initiation codon lies within the part of the P site carried by the subunit. The *only aminoacyl-tRNA (αα-tRNA) that can become part of the initiation complex is the initiator, which has the unique property of being able to enter directly into the partial P site to recognize its codon.*

When the large subunit joins the complex, the initiator fMet-tRNA_f lies in the now-intact P site; and the A site is available for entry of the aminoacyl-tRNA complementary to the second codon of the gene. The first peptide bond forms between the initiator and the next aminoacyl-tRNA. The initiator tRNA behaves as an analog of peptidyl-tRNA (an analog in the sense that it donates a peptidyl chain consisting of only one amino acid).

So initiation prevails when an AUG (or GUG) codon lies within a ribosome-binding site, because only the initiator tRNA can enter the partial P site generated when the 30S subunit binds *de novo* to the mRNA. The internal reading prevails subsequently, when the codons are encountered by a ribosome that is *continuing* to translate an mRNA, because only the regular aminoacyl-tRNAs can enter the (complete) A site.

What features distinguish the fMet-tRNA_f initiator from Met-tRNA_m and other tRNAs used during

Figure 7.17

fMet-tRNA_f has unique features that distinguish it as the initiator tRNA.

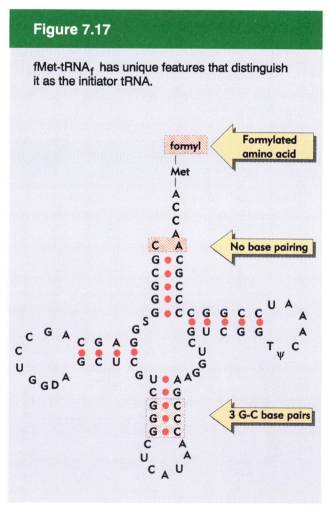

formyl — Formylated amino acid

Met

No base pairing

3 G-C base pairs

Figure 7.18

IF2 binds fmet-tRNA$_f$ to the 30S-mRNA complex. After after 50S binding, IF2 and other factors are released and GTP is cleaved.

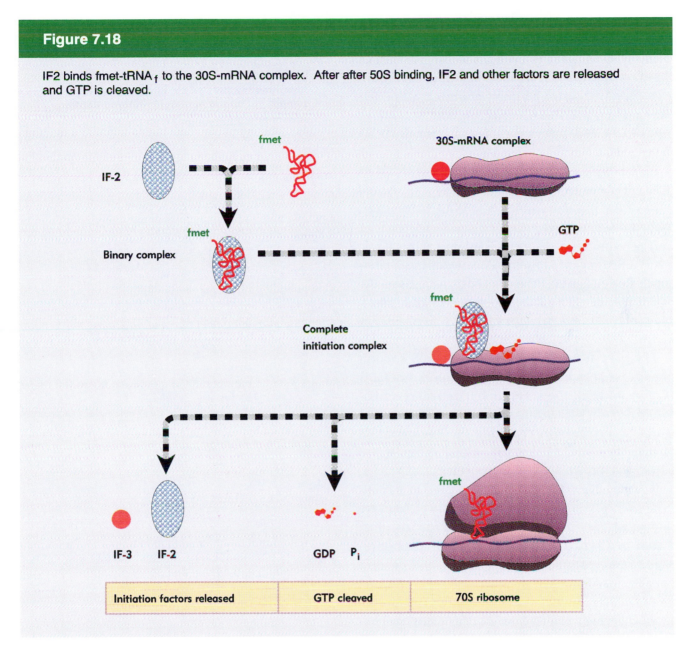

Initiation factors released	GTP cleaved	70S ribosome

elongation? The modification of the methionine would prevent its use in elongation, but is not strictly necessary, since nonformylated Met-tRNA$_f$ can function as an initiator. Some characteristic features of the tRNA sequence are important, as summarized in **Figure 7.17**. (We discuss the structure of tRNA in more detail in Chapter 8, and just for now concentrate on the features that are unique to the initiator tRNA.) Some of these features are needed to *prevent* the initiator from being used in elongation, others are *necessary* for it to function in initiation:

◆ The last two bases in the stem to which the amino acid is connected are paired in all tRNAs except tRNA$_f^{Met}$. Mutations that create a base pair in this position of tRNA$_f^{Met}$ allow it to function in elongation. The absence of this pair is therefore important in preventing tRNA$_f^{Met}$ from being used in elongation.

◆ A series of 3 G–C pairs in the stem that precedes the loop containing the anticodon is exclusive to tRNA$_f^{Met}$. Mutations in these base pairs prevent the fMet-tRNA$_f$ from being inserted into the P site.

The ability of aminoacyl-tRNAs to enter the ribosome is controlled by accessory factors. The fMet-tRNA_f initiator is brought to the 30S•mRNA complex by IF-2 in the manner illustrated in **Figure 7.18**. First IF-2 forms a **binary complex** with fMet-tRNA_f. Then the IF-2•fMet-tRNA_f binary complex binds to the 30S•mRNA complex, placing the tRNA in the partial P site.

By forming a complex specifically with fMet-tRNA_f, IF-2 ensures that only the initiator tRNA, and none of the regular aminoacyl-tRNAs, participates in the initiation reaction.

IF-2 remains part of the 30S subunit at this stage; it has a further role to play. This factor has a **ribosome-dependent GTPase activity**: it sponsors the hydrolysis of GTP in the presence of ribosomes, releasing the energy stored in the high-energy bond.

The GTP is hydrolyzed when the 50S subunit joins to generate a complete ribosome. Probably IF-2 is not itself the GTPase, but activates a ribosomal protein with this function. The GTP cleavage could be involved in changing the conformation of the ribosome, so that the joined subunits are converted into an active 70S ribosome.

Because the same amino acid is used to start all protein chains, it is sometimes necessary to remove unwanted residues from the N-terminus, as illustrated in **Figure 7.19**. The removal reaction(s) occur rather rapidly, probably when the nascent polypeptide chain has reached a length of 15–30 amino acids.

In bacteria and mitochondria, the formyl residue is removed by a specific deformylase enzyme to generate a normal NH_2 terminus. If methionine is to be the N-terminal amino acid of the protein, this is the only necessary step. In about half the proteins, the methionine at the terminus is removed by an aminopeptidase, creating a new terminus of R_2 (originally the second amino acid incorporated into the chain). When both steps are necessary, they occur sequentially.

Figure 7.19

Newly synthesized proteins in bacteria start with formyl-methionine, but the formyl group, and sometimes the methionine, is removed during protein synthesis.

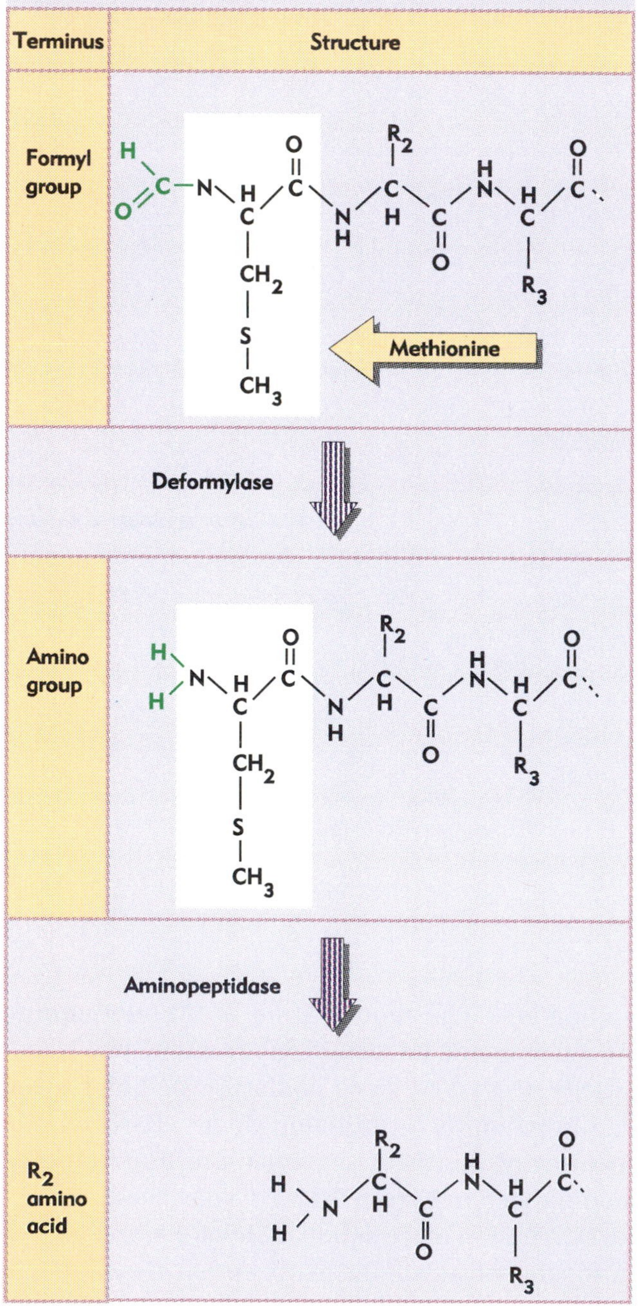

Eukaryotic initiation involves many factors

Similar components are involved in initiation in eukaryotic cytoplasm, but the codon AUG is almost always used as the initiator; GUG is found only very rarely. The initiator tRNA is a distinct species, but its methionine does *not* become formylated. It is called $tRNA_i^{Met}$. Thus the difference between the initiating and elongating Met-tRNAs lies solely in the tRNA moiety, with Met-tRNA$_i$ used

Figure 7.20

In eukaryotic initiation, eIF2 forms a ternary complex with Met-tRNA$_i$. The ternary complex binds to free 40S subunits, which attach to the 5' end of mRNA.

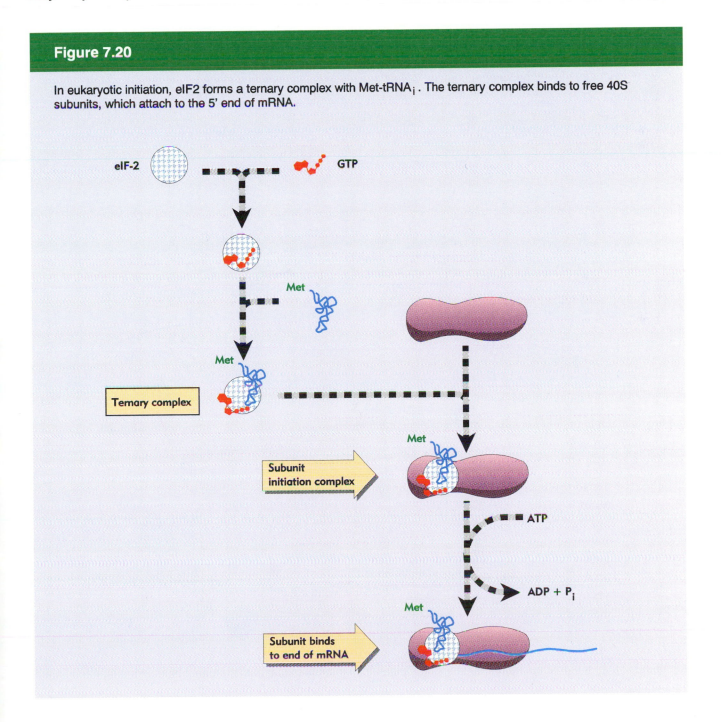

Figure 7.21

Several eukaryotic initiation factors are required to unwind mRNA, bind the subunit initiation complex, and support joining with the large subunit. The functions of the principal factors are known, but others are also involved.

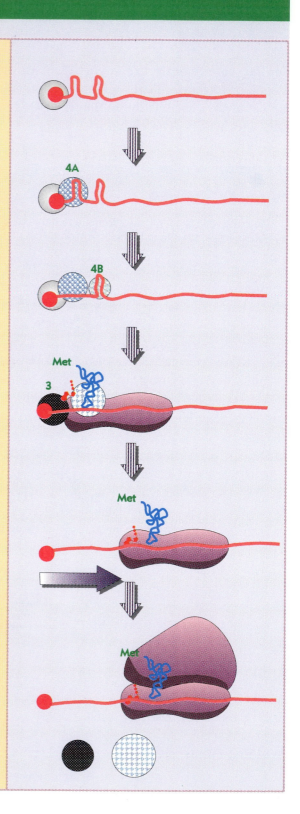

CBP
CAP binding protein binds to 5′ end

eIF4A
unwinds structure at 5′ end

eIF4B
assists further unwinding

eIF3
required for 40S subunit with ternary complex to bind to 5′ end

40S subunit migrates along mRNA to AUG codon

eIF5
required for 60S joining, and release of eIF-2 & eIF-3

for initiation and Met-tRNA$_m$ used for elongation.

There is a difference in the way that bacterial 30S and eukaryotic 40S subunits find their binding sites for initiating protein synthesis on mRNA. In bacteria, the initiation complex forms directly at a sequence surrounding the AUG initiation codon. In eukaryotes, small subunits first recognize the 5′ end of the mRNA, and then move to the initiation site, where they are joined by large subunits. We discuss this process in more detail in Chapter 10.

Aside from this difference, the process of initiation in eukaryotes appears to be generally analogous to that in *E. coli*. There are more initiation factors—at least nine already have been found in reticulocytes (immature red blood cells), in which the most work has been done. The factors are named similarly to those in bacteria, but with a prefix 'e' to indicate their eukaryotic origin.

In eukaryotes initiation proceeds through the formation of a **ternary complex** containing Met-tRNA$_i$, eIF-2, and GTP. The complex is formed in two stages. GTP binds to eIF-2; and this increases the factor's affinity for Met-tRNA$_i$, which then is bound. **Figure 7.20** shows that the ternary complex associates directly with free 40S subunits. The reaction is independent of the presence of mRNA. In fact, *the Met-tRNA$_i$ initiator must be present in order for the 40S subunit to bind to mRNA.* So eIF-2 is the eukaryotic counterpart to IF-2, but the interaction with the small ribosome subunit occurs at an earlier stage of initiation.

The recognition of mRNA requires several additional factors; an important part of their function is to remove any secondary structure in the mRNA. The roles of the initiation factors involved in binding the subunit initiation complex to mRNA are expanded in **Figure 7.21**.

The eIF-4F factor is a multimer that includes CPB (the cap binding protein) and eIF-4A. CPB recognizes the 'cap' at the 5′ end of mRNA (see Chapter 10), and eIF-4A unwinds any secondary structure that exists in the first 15 bases of the mRNA. Energy for the unwinding is provided by hydrolysis of ATP. Unwinding of structure farther along the mRNA is accomplished by eIF-4A together with another factor, eIF-4B.

Binding of the 40S ternary complex to mRNA depends on eIF-3. Unlike its bacterial counterpart (IF-3), the main function of eIF-3 is concerned with mRNA binding. Other factors (including eIF-4A and eIF-4B) are also involved.

When the small subunit has bound mRNA, it migrates to (usually) the first AUG codon. Little is known about this process; we assume it requires expenditure of energy in the form of ATP. When the small subunit reaches the initiation site, it stops, and can be joined by a large subunit.

Another factor, eIF-6, is required to maintain subunits in their dissociated state; it may act on the large ribosomal subunit. It is released when the large subunit joins the initiation complex.

Junction of the 60S subunits with the initiation complex cannot occur until eIF-2 and eIF-3 have been released from the initiation complex, a function mediated by eIF-5. The 40S-60S joining reaction also depends directly on eIF-4C. These two factors therefore fulfill a role that is not necessary in bacteria. Probably all of the remaining factors are released when the complete 80S ribosome is formed.

Elongation factor T brings aminoacyl-tRNA into the A site

Once the complete ribosome is formed at the initiation codon, the stage is set for a cycle in which aminoacyl-tRNA enters the A site of a ribosome whose P site is occupied by peptidyl-tRNA. *Any aminoacyl-tRNA except the initiator can enter the A site.* Its entry is mediated by an **elongation factor (EF-Tu in bacteria).** (The process is similar in eukaryotes.)

Just like its counterpart in initiation (IF-2), this factor, EF-Tu, is associated with the ribosome only during its sponsorship of aminoacyl-tRNA entry. Once the aminoacyl-tRNA is in place, EF-Tu leaves the ribosome, to work again with another aminoacyl-tRNA. Thus it displays the cyclic association with, and dissociation from, the ribosome that is the hallmark of the accessory factors.

The pathway for aminoacyl-tRNA entry to the A site is illustrated in **Figure 7.22**. EF-Tu carries a guanine nucleotide. The factor provides another example of a protein whose activity is controlled by the state of the guanine nucleotide:

Figure 7.22

EF-Tu-GTP places aminoacyl-tRNA on the ribosome and then is released as EF-Tu-GDP, which requires EF-Ts to mediate the replacement of GDP by GTP. The reaction consumes GTP and releases GDP. The only aminoacyl-tRNA that cannot be recognized by EF-Tu-GTP is fMet-tRNA$_f$, whose failure to bind prevents it from responding to internal AUG or GUG codons.

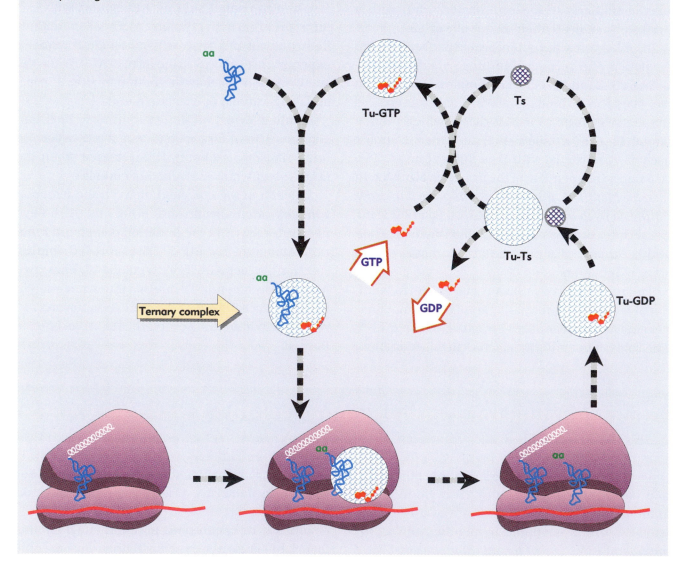

◆ When GTP is present, the factor is in its active state.

◆ When the GTP is hydrolyzed to GDP, the factor becomes inactive.

◆ Activity is restored when the GDP is replaced by GTP.

The **binary complex** of EF-Tu•GTP binds aminoacyl-tRNA to form a ternary complex of aminoacyl-tRNA•EF-Tu•GTP. *The ternary complex binds only to the A site of ribosomes whose P site is already occupied by peptidyl-tRNA.* This is the critical reaction in ensuring that the aminoacyl-tRNA and peptidyl-tRNA are correctly positioned for peptide bond formation.

After aminoacyl-tRNA has been placed in the A site, the GTP is cleaved; and then the binary complex EF-Tu•GDP is released. This form of EF-Tu is inactive and does not bind aminoacyl-tRNA effectively.

Another factor, EF-Ts, mediates the regeneration of the used form, EF-Tu•GDP, into the active form, EF-Tu•GTP. First, EF-Ts displaces the GDP from EF-Tu, forming the combined factor EF-Tu•EF-Ts. Then the EF-Ts is in turn displaced by GTP, reforming EF-Tu•GTP. The active binary complex binds aminoacyl-tRNA; and the released EF-Ts can recycle.

From the perspective of the elongation factors, EF-Ts cycles between freedom and binding EF-Tu, while EF-Tu oscillates between binding GDP or GTP in the binary and ternary complexes or being associated with EF-Ts (the EF-Tu•EF-Ts complex is called the T factor).

There are ~70,000 molecules of EF-Tu per bacterium, which approaches the number of aminoacyl-tRNA molecules. This implies that most aminoacyl-tRNAs are likely to be present in ternary complexes. Coded by two genes, EF-Tu provides ~5% of the total bacterial protein. Most of the EF-Tu is in the form of binary or ternary complexes, because there are only ~10,000 molecules of EF-Ts per cell (an amount approaching the number of ribosomes).

The role of GTP in the ternary complex has been studied by substituting an analog that cannot be hydrolyzed. The compound **GMP-PCP** has a methylene bridge in place of the oxygen that links the β and γ phosphates in GTP. In the presence of GMP-PCP, a ternary complex can be formed that binds aminoacyl-tRNA to the ribosome. But the peptide bond cannot be formed. Thus the *presence of GTP* is needed for aminoacyl-tRNA to be bound at the A site; but the *hydrolysis* is not required until later.

Kirromycin is an antibiotic that inhibits the function of EF-Tu. When EF-Tu is bound by kirromycin, it remains able to bind aminoacyl-tRNA to the A site. But the EF-Tu•GDP complex cannot be released from the ribosome. Its continued presence prevents formation of the peptide bond between the peptidyl-tRNA and the aminoacyl-tRNA. As a result, the ribosome becomes 'stalled' on mRNA, bringing protein synthesis to a halt.

This effect of kirromycin demonstrates that inhibiting one step in protein synthesis blocks the next step. In this case, the release of EF-Tu•GDP is needed for the ribosome to acquire the right conformation to sponsor peptide bond formation. The same principle is seen at other stages of protein synthesis: one reaction must be completed properly before the next can occur.

In eukaryotes, the factor **eEF-1** is responsible for bringing aminoacyl-tRNA to the ribosome, again in a reaction that involves cleavage of a high-energy bond in GTP. The active factor consists of aggregates of polypeptide chains of various sizes. The details of how GTP is regenerated after cleavage are not known, but the factor may have components analogous to EF-Tu and EF-Ts. Quantitatively, the situation is similar in eukaryotes and prokaryotes, with eEF-1 (like EF-Tu) constituting a major protein of the cell.

An analogous set of reactions also occurs in eukaryotic initiation, where GDP bound to eIF-2 must be replaced by GTP before the factor can be used. This 'guanylate exchange' reaction is an important target for a variety of regulatory reactions that link the overall translation rate in a cell to other metabolic pathways.

Translocation moves the ribosome

The ribosome remains in place while the polypeptide chain is elongated by transferring the polypeptide attached to the tRNA in the P site to the aminoacyl-tRNA present in the A site. The reaction is shown in **Figure 7.23**. The activity responsible for

synthesis of the peptide bond is called **peptidyl transferase**.

The nature of the transfer reaction is revealed by the ability of the antibiotic **puromycin** to inhibit protein synthesis. Puromycin closely resembles an amino acid attached to the terminal adenosine of tRNA. **Figure 7.24** shows that puromycin has an N instead of the O that joins an amino acid to tRNA. The antibiotic is treated by the ribosome as though it were an incoming aminoacyl-tRNA; the polypeptide attached to peptidyl-tRNA is transferred to the NH_2 group of the puromycin.

Because the puromycin moiety is not anchored to the A site of the ribosome, the polypeptidyl-puromycin adduct is released from the ribosome in the form of polypeptidyl-puromycin. This premature termination of protein synthesis is responsible for the lethal action of the antibiotic.

Peptidyl transferase is a function of the large (50S or 60S) ribosomal subunit. The transferase is part of a ribosomal site at which the ends of the peptidyl-tRNA and aminoacyl-tRNA are brought close together. From the effects of erythromycin and chloramphenicol, summarized later in Table 9.1, we know that both rRNA and 50S subunit proteins are necessary for this activity. We see in Chapter 9 that the catalytic activity actually is a property of the ribosomal RNA of the 50S subunit, rather than of any individual protein.

The cycle of addition of amino acids to the growing polypeptide chain is completed by the **translocation** illustrated in **Figure 7.25**, in which the ribosome advances three nucleotides along the mRNA. The result of translocation is to expel the uncharged tRNA from the P site, so that the new peptidyl-tRNA can enter. The ribosome then has an empty A site ready for entry of the aminoacyl-tRNA corresponding to the next codon.

In bacteria the discharged tRNA leaves the ribosome via another site, the E site. (In eukaryotes it is expelled directly into the cytosol.) The process of relocating the discharged tRNA and peptidyl-tRNA

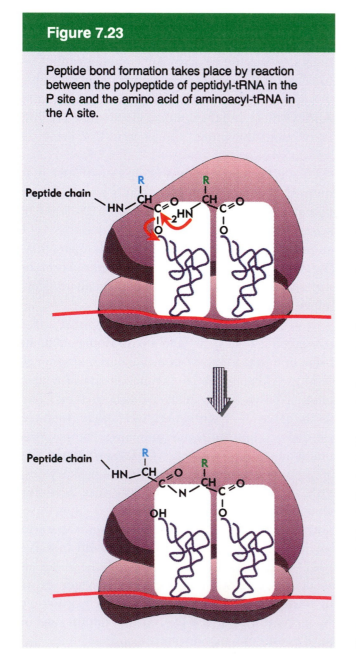

Figure 7.23

Peptide bond formation takes place by reaction between the polypeptide of peptidyl-tRNA in the P site and the amino acid of aminoacyl-tRNA in the A site.

Figure 7.24

Puromycin mimics aminoacyl-tRNA because it resembles an aromatic amino acid (shaded part) linked to a sugar-base moiety.

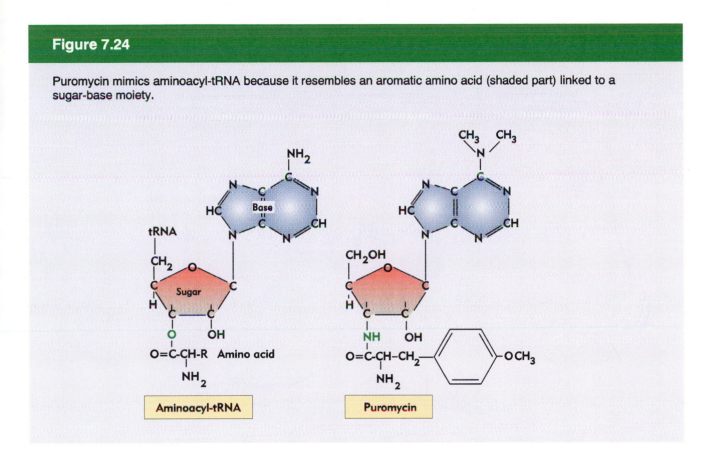

takes place in two stages. Essentially the regions of the tRNAs located in the 50S subunit move into the new sites first. At this stage, the tRNAs are effectively bound in hybrid sites, consisting of the 50S E/30S P and the 50S P/30S A sites. Then movement is extended to the 30S subunits, so that the anticodon-codon pairing region finds itself in the right site.

The figure shows two possible models for translocation. In one it is imagined to occur by a movement of tRNA relative to the ribosome, so that the aminoacyl end of tRNA moves within the 50S subunit; the anticodon end moves later when translocation occurs. In the second model, the entire 50S subunit first moves relative to the 30S subunit, so that translocation in effect involves two stages, the normal structure of the ribosome being restored by the second stage.

Translocation requires GTP and another elongation factor, **EF-G**. This factor is a major constituent of the cell; it is present at a level of ~1 copy per ribosome (20,000 molecules per cell).

EF-G binds to the ribosome to sponsor trans-

location; and then is released when GTP hydrolysis occurs. Binding can still occur when GMP-PCP is substituted for GTP; thus the presence of a guanine nucleotide is needed for binding, but its hydrolysis is not needed until the actual translocation reaction occurs. The hydrolysis of GTP is not catalyzed by the factor, but is a ribosomal function.

Another factor involved in protein synthesis is 4.5S RNA. This is a 114 base, largely double-stranded species that is essential in *E.coli*. There are ~3000 molecules per bacterium, but only a small minority are associated with ribosomes. Mutants in the 4.5S RNA grow poorly; they can be suppressed by mutations in the gene for EF-G, which suggests that 4.5S RNA interacts with EF-G in some way. The 4.5S RNA is needed for secretion of some, but not all, secreted proteins (see Chapter 11). We do not understand the nature of its interaction with EF-G.

Ribosomes cannot bind EF-Tu and EF-G simultaneously, so protein synthesis follows the cycle illustrated in **Figure 7.26** in which the factors are alternately bound to, and released from, the

Figure 7.25

Two models for translocation both involve two stages. The first stage occurs at peptide bond formation, when the aminoacyl end of the tRNA in the A site becomes located in the P site, either because the tRNA itself moves or because the 50S subunit moves relative to the 30S subunit. In the second stage, the anticodon end of the tRNA becomes located in the P site.

tRNA moves within 50S subunit

50S subunit moves relative to 30S

Discharged tRNA leaves via E site

Incoming aa-tRNA

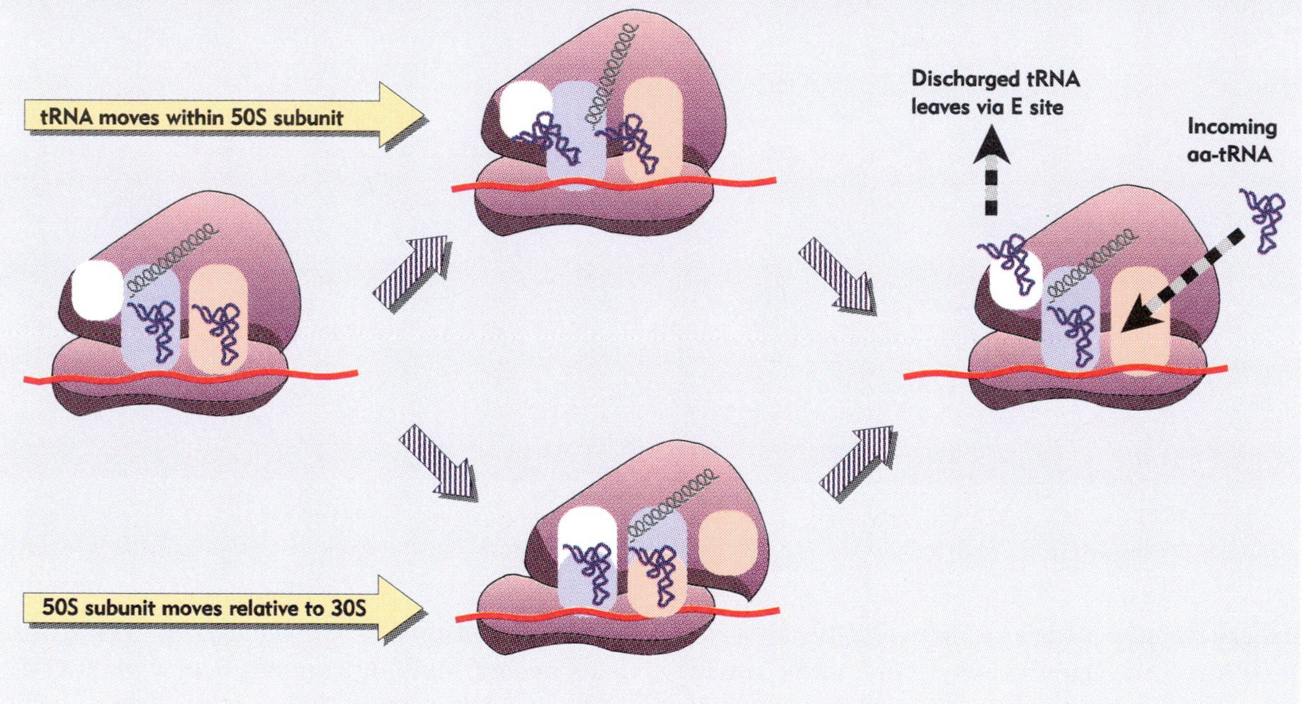

ribosome. Thus EF-Tu•GDP must be released before EF-G can bind; and then EF-G must be released before aminoacyl-tRNA•EF-Tu•GTP can bind.

The need for EF-G release was discovered by the effects of the steroid antibiotic fusidic acid, which 'jams' the ribosome in its post-translocation state. In the presence of fusidic acid, one round of translocation occurs: EF-G binds to the ribosome, GTP is hydrolyzed, and the ribosome moves three nucleotides. But fusidic acid stabilizes the ribosome•EF-G•GDP complex, so that EF-G and GDP remain on the ribosome instead of being released. Because the ribosome then cannot bind aminoacyl-tRNA, no further amino acids can be added to the chain.

We do not know whether the ability of each elongation factor to exclude the other is mediated via an effect on the overall conformation of the ribosome

or by direct competition for overlapping binding sites. The need for each factor to be released before the other can bind ensures that the events of protein synthesis proceed in an orderly manner.

Both factors require the presence of GTP for binding to the ribosome, but need to hydrolyze it only later. Since neither factor can use GDP to support its binding to the ribosome, the triphosphate form is probably needed for the factor to acquire an active conformation. This mechanism ensures that factors obtain access to the ribosome only in the company of the GTP that they will need later to fulfill their function. Hydrolysis of the GTP provides energy to change the conformation of the ribosome; with EF-Tu, it is needed for the aminoacyl-tRNA in the A site to be reactive; with EF-G it is needed for ribosome movement.

The eukaryotic counterpart to EF-G is the protein

Figure 7.26

Binding of factors EF-Tu and EF-G alternates as ribosomes accept new aminoacyl-tRNA, form peptide bonds, and translocate. Fusidic acid diverts the ribosome into a post-translocation state that is jammed with EF-G-GDP.

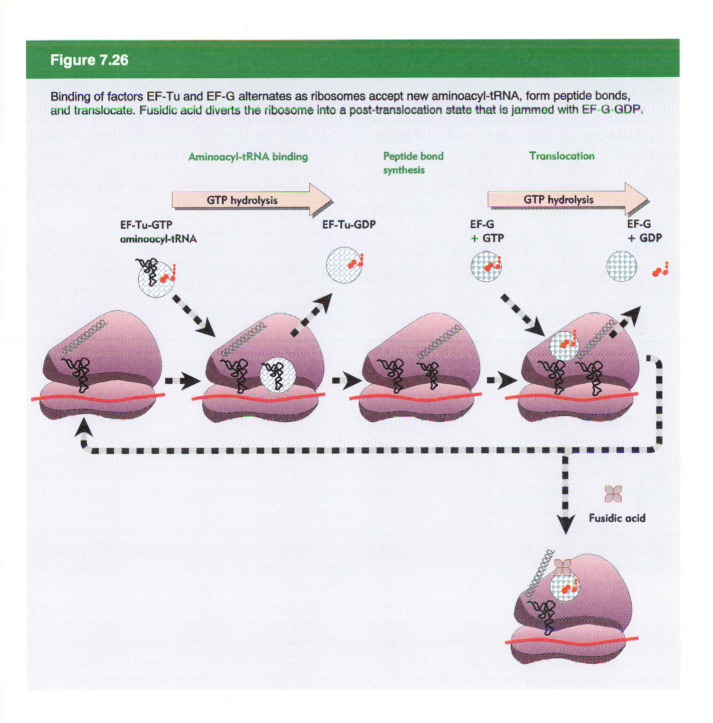

eEF2, which functions in a similar manner, as a translocase dependent on GTP hydrolysis. Its action also is inhibited by fusidic acid. A stable complex of eEF2 with GTP can be isolated; and the complex can bind to ribosomes with consequent hydrolysis of its GTP.

A unique reaction of eEF2 is its susceptibility to diphtheria toxin. The toxin uses NAD (nicotinamide adenine dinucleotide) as a cofactor to transfer an ADPR moiety (adenosine diphosphate ribosyl) onto the eEF2. The ADPR–eEF2 conjugate is inactive in protein synthesis. The substrate for the attachment is an unusual amino acid, produced by modifying a histidine; it is common to the eEF2 of many species.

The ADP-ribosylation is responsible for the lethal effects of diphtheria toxin. The reaction is extraordinarily effective: a single molecule of toxin can modify sufficient eEF2 molecules to kill a cell.

Finishing off: three codons terminate protein synthesis

Only 61 triplets are assigned to amino acids. The other three triplets are **termination codons** (or stop codons) that end protein synthesis. They have casual names from the history of their discovery. The UAG triplet is called the **amber** codon; UAA is the **ochre** codon; and UGA is sometimes called the **opal** codon.

The nature of these triplets was originally shown by a genetic test that distinguished two types of point mutation:

◆ A point mutation that changes a codon to represent a different amino acid is called a **missense** mutation. One amino acid replaces the other in the protein; the effect on protein function depends on the site of mutation and the nature of the amino acid replacement.

◆ When a point mutation creates one of the three termination codons, it causes **premature termination** of protein synthesis at the mutant codon. This is likely to abolish protein function, since only the first part of the protein is made in the mutant cell. A change of this sort is called a **nonsense** mutation.

(Sometimes the term *nonsense codon* is used to describe the termination triplets. 'Nonsense' is really a misnomer, since the codons do have meaning, albeit a disruptive one in a mutant gene.)

In every gene that has been sequenced, one of the termination codons lies immediately after the codon representing the C-terminal amino acid of the wild-type sequence. Nonsense mutations show that any one of the three codons is sufficient to terminate protein synthesis within a gene. The UAG, UAA, and UGA triplet sequences are therefore necessary and sufficient to end protein synthesis, whether occurring naturally at the end of a gene or created by mutation within a coding sequence.

In bacterial genes, UAA is the most commonly used termination codon. UGA is used more heavily than UAG, although there appear to be more errors reading UGA. (An error in reading a termination codon, when an aminoacyl-tRNA improperly responds to it, results in the continuation of protein synthesis until another termination codon is encountered.)

None of the termination codons is represented by a tRNA. They function in an entirely different manner from other codons, and are recognized directly by protein factors. (Since the reaction does not depend on codon–anticodon recognition, there seems to be no particular reason why it should require a triplet sequence. Presumably this reflects the evolution of the genetic code.)

In *E. coli* two related proteins catalyze termination. They are called **release factors** (RF), and are specific for different sequences. **RF1** recognizes UAA and UAG; **RF2** recognizes UGA and UAA. The factors act at the ribosomal A site and require polypeptidyl-tRNA in the P site. The release factors are present at much lower levels than initiation or elongation factors; there are ~600 molecules of each per cell, equivalent to 1 RF per 50 ribosomes. Probably at one time there was only a single release factor, recognizing all termination codons, and later it evolved into two factors with specificities for particular codons.

Mutations in the RF genes reduce the efficiency of termination, as seen by an increased ability to continue protein synthesis past the termination codon. Over-expression of either gene increases the efficiency of termination at the codons on which it acts. This suggests that codon recognition by either RF competes with aminoacyl-tRNAs that erroneously recognize the termination codons.

In eukaryotic systems, there is only a single release factor, **eRF**. GTP is needed for eRF to bind to ribosomes (it is not involved in bacteria). Probably the GTP is cleaved after the termination step has occurred; the hydrolysis is needed to allow eRF to dissociate from the ribosome.

The termination reaction involves release of the

completed polypeptide from the last tRNA, expulsion of the tRNA from the ribosome, and dissociation of the ribosome from mRNA. Cleavage of polypeptide from tRNA could take place by a reaction analogous to the usual transfer from peptidyl-tRNA to aminoacyl-tRNA during elongation; perhaps the release factor diverts the reaction. We do not yet understand how the ribosome dissociates from mRNA. The dissociation may result from a conformational change triggered by the RF factor, but it is possible that another protein factor(s) could be involved.

A sequence of DNA can be read in three possible reading frames. A frame that consists exclusively of triplets that represent amino acids is called an **open reading frame** or **ORF**. A reading frame that cannot be read into protein because termination codons occur frequently is said to be **blocked**. If a sequence is blocked in all three reading frames, it cannot have the function of coding for protein. A sequence that is translated into protein has a reading frame that starts with an AUG initiation codon and that extends through a series of triplets representing amino acids until it ends at a termination codon. Usually the alternative reading frames are blocked by frequent termination codons, as illustrated in **Figure 7.27**.

When the sequence of a DNA region of unknown function is obtained, each possible reading frame is analyzed to determine whether it is open or blocked. Usually no more than one of the three possible frames of reading is open in any single stretch of DNA. An extensive ORF is unlikely to exist by chance; if it were not translated into protein, there would have been no selective pressure to prevent the accumulation of nonsense codons. Thus the identification of a lengthy ORF is taken to be *prima facie* evidence that the sequence is translated into protein in that frame. An ORF for which no protein product has been identified is sometimes called an unidentified reading frame (URF).

Figure 7.27

An open reading frame starts with AUG and continues in triplets to a termination codon. Blocked reading frames may be interrupted frequently by termination codons.

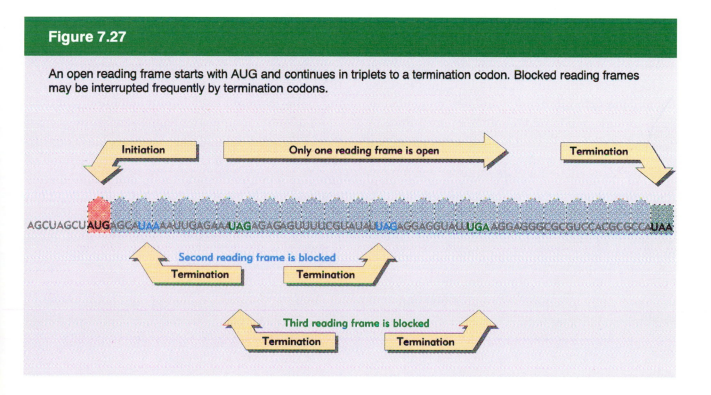

Summary

Genetic information carried by DNA is expressed in two stages: transcription of one DNA strand into mRNA; and translation of the mRNA into protein. The sequence of mRNA, in triplets 5′–3′, is related to the amino acid sequence of protein, N- to C-terminal. Of the 64 triplets, 61 code for amino acids and 3 provide termination signals. Synonym codons that represent the same amino acid are related, often by a change in the third base of the codon. This third base degeneracy, coupled with the arrangement in which related amino acids tend to be coded by related codons, minimizes the effects of mutations. The genetic code is universal.

A codon in mRNA is recognized by an aminoacyl-tRNA, which has an anticodon complementary to the codon and carries the amino acid corresponding to the codon. A special initiator tRNA (fMet-tRNA$_f$ in prokaryotes or Met-tRNA$_i$ in eukaryotes) recognizes the AUG codon, which is used to start all coding sequences. In prokaryotes, GUG is also used. Only the termination (nonsense) codons UAA, UAG and UGA are not recognized by aminoacyl-tRNAs.

Ribosomes catalyze the translation of mRNA. Prokaryotic (70S) ribosomes consist of small (30S) and large (50S) subunits. Eukaryotic ribosomes are 80S, and the subunits are 40S and 60S, respectively. Small subunits bind to mRNA and then are joined by large subunits to generate an intact ribosome that undertakes protein synthesis. A single mRNA molecule is translated simultaneously by several ribosomes, each carrying a nascent polypeptide chain at a different stage of completion.

A ribosome can carry two aminoacyl-tRNAs simultaneously: its P site is occupied by a polypeptidyl-tRNA, which carries the polypeptide chain synthesized so far, while the A site is used for entry by an aminoacyl-tRNA carrying the next amino acid to be added to the chain. The polypeptide chain in the P site is transferred to the aminoacyl-tRNA in the A site and then the ribosome translocates one codon along the mRNA. Translocation and several other stages of ribosome function require hydrolysis of GTP, which is probably used to drive conformational changes in the ribosome.

Protein synthesis is an expensive process. ATP is used to provide energy at several stages, including the charging of tRNA with its amino acid, and the unwinding of mRNA. It has been estimated that upto 90% of all the ATP molecules synthesized in a rapidly growing bacterium are consumed in assembling amino acids into protein!

Additional factors are required at each stage of protein synthesis. They are defined by their cyclic association with, and dissociation from, the ribosome. IF factors involved in prokaryotic initiation include IF-3 (needed for 30S subunits to bind to mRNA) and IF-2 (needed for fMet-tRNA$_f$ to bind to the 30S subunit). Prokaryotic EF factors involved in elongation are EF-Tu, which binds aminoacyl-tRNA to the 70S ribosome, EF-Ts which regenerates the active form of EF-Tu after it has been used, and EF-G which is required for translocation. Binding of the EF-Tu and EF-G factors to ribosomes is mutually exclusive, which ensures that each step must be completed before the next can be started. RF factors are required for termination. Protein synthesis in eukaryotes is generally similar to the process in prokaryotes, but involves a more complex set of accessory factors.

Further reading

Reviews

The breaking of the genetic code was described in **Lewin's** *Gene Expression,* **1,** *Bacterial Genomes* (Wiley, New York, 1974).

Since the basic pathway for protein synthesis was worked out, most reviews have concentrated on the roles of the factors that control the reactions. A general description of factors was provided by **Maitra** *et al.* (*Ann. Rev Biochem.* **51,** 869–900, 1982). Eukaryotic factors were described by **Merrick** (*Microbiol. Rev.* **56,** 291–315, 1992), and their role in regulation was analyzed by **Hershey** (*Ann. Rev. Biochem.* **60,** 717-755, 1991).

The processes involved in prokaryotic, eukaryotic and organelle initiation have been compared by **Kozak** (*Microbiol. Rev.* **47,** 1–45, 1983).

CHAPTER 8

Transfer RNA is the translational adaptor

Transfer RNA occupies a pivotal position in protein synthesis, providing the adaptor molecule that accomplishes the translation of each nucleotide triplet into an amino acid. In Crick's phrase, tRNA represents Nature's attempt to make a nucleic acid fulfill the sort of role more usually performed by a protein. For tRNAs are involved in a multiplicity of reactions in which it is necessary for them to have certain characteristics in common, yet be distinguished by others. The crucial feature that confers this capacity is the ability of tRNA to fold into a specific tertiary structure.

Working backward from the final action of tRNA in protein synthesis, all tRNAs are able to sit in the P and A sites of the ribosome, where at one end they are associated with mRNA via codon–anticodon pairing, while at the other end the polypeptide is being transferred. For the P and A sites to welcome all tRNAs, the entire set must conform to the same general dictates for size and shape.

Similarly, all tRNAs (except the initiator) share the ability to be recognized by the translation factors (EF-Tu or eEF1) for binding to the ribosome. The initiator tRNA is recognized instead by IF2 or eIF2. So the tRNA set must possess common features for interaction with elongation factors, but the initi-ator tRNA must lack this feature, or possess some contradictory feature. We have some information about the distinctive behavior of the initiator tRNA, but we have yet to define the features required by the other tRNAs for interaction with EF-Tu and the ribosome.

And there must be critical differences between small groups of tRNAs. Usually, each amino acid is represented by more than one tRNA. Multiple tRNAs representing the same amino acid are called iso-accepting tRNAs. *A group of isoaccepting tRNAs must be charged only by the single aminoacyl-tRNA synthetase specific for their amino acid.* Thus isoaccepting tRNAs must share some common feature(s) enabling the enzyme to pick them out from the other tRNAs. The entire complement of tRNAs is divided into 20 isoaccepting groups; each group is able to identify itself to its particular synthetase.

We lack a systematic view of what features identify tRNAs to the proteins with which they interact. Early attempts to deduce common features focused on seeking out short nucleotide stretches that are common to all tRNAs or to a particular group of tRNAs. However, although some features in the primary sequence of tRNA are highly conserved, and are presumably essential for one or more of its functions, attempts to correlate individual regions or sequences with particular protein recognition reactions have proved difficult. Some particular regions of tRNA usually are involved in binding to the synthetase, but the exact contact points differ between different groups of tRNAs.

A factor that contributes to the diversity of tRNAs is the modification of certain bases to unusual forms following synthesis of the molecule. Modification around the anticodon directly influences the ability of the tRNA to interact with its intended codon(s), and modifications elsewhere presumably influence the other functions of tRNA. We do not know whether and what role the modifications play in tRNA tertiary structure.

The common (or distinctive) features of tRNAs transcend the primary and secondary structures and are conveyed at least in part by the tertiary structure. To analyze the function of tertiary structure in any detail requires comparison of several (perhaps all) of the complement of tRNAs. This will be a long job, since at present only a few individual tRNA molecules have been analyzed at the level of tertiary structure. From such analyses, however, it seems that all tRNAs will conform to the same general tertiary structure, *a rather compact L-shape which has the amino acid at one end and the anticodon at the other end.* Changes in the details of this structure, such as the angle of the two arms of the 'L' or the protrusion of individual bases, may distinguish the individual tRNAs.

The universal cloverleaf

The sequences of several hundred tRNAs from a wide variety of bacteria and eukaryotes have been determined. All conform to the same general secondary structure. Each tRNA sequence can be written in the form of a **cloverleaf**, maintained by base pairing between short complementary regions.

A general form of the cloverleaf is illustrated in **Figure 8.1**. The four major **arms** are named for their structure or function:

◆ The **acceptor arm** consists of a base-paired stem that ends in an unpaired sequence whose free 2'- or 3'-OH group is aminoacylated.

   The other arms consist of base-paired **stems** and unpaired **loops**.

◆ The **TψC arm** is named for the presence of this triplet sequence. (ψ stands for pseudouridine, one of the 'unusual' bases in tRNA that are discussed later).

◆ The **anticodon arm** always contains the anticodon triplet in the center of the loop.

◆ The **D arm** is named for its content of the base dihydrouridine (another of the modified bases in tRNA).

The numbering system for tRNA illustrates the constancy of the structure. Positions are numbered from 5' to 3' according to the most common tRNA structure, which has 76 residues. The overall range of tRNA lengths is from 74 to 95 bases. The variation in length is caused by differences in the structure of two of the arms.

In the D loop, there is variation of up to four residues. The extra nucleotides relative to the most common structure are denoted 17:1 (lying between 17 and 18) and 20:1 and 20:2 (lying between 20 and 21). However, in the smallest D loops, residue 17 as well as these three is absent.

The most variable feature of tRNA is the so-called **extra arm**, which lies between the TψC and anticodon arms. Depending on the nature of the extra arm, tRNAs can be divided into two classes:

◆ **Class 1 tRNAs** have a small extra arm, consisting of only 3-5 bases. They represent ~75% of all tRNAs.

◆ **Class 2 tRNAs** have a large extra arm—it may even be the longest in the tRNA—with 13–21 bases, and ~5 base pairs in the stem. The additional bases are numbered from 47:1 through 47:18. The functional significance of the extra arm is unknown.

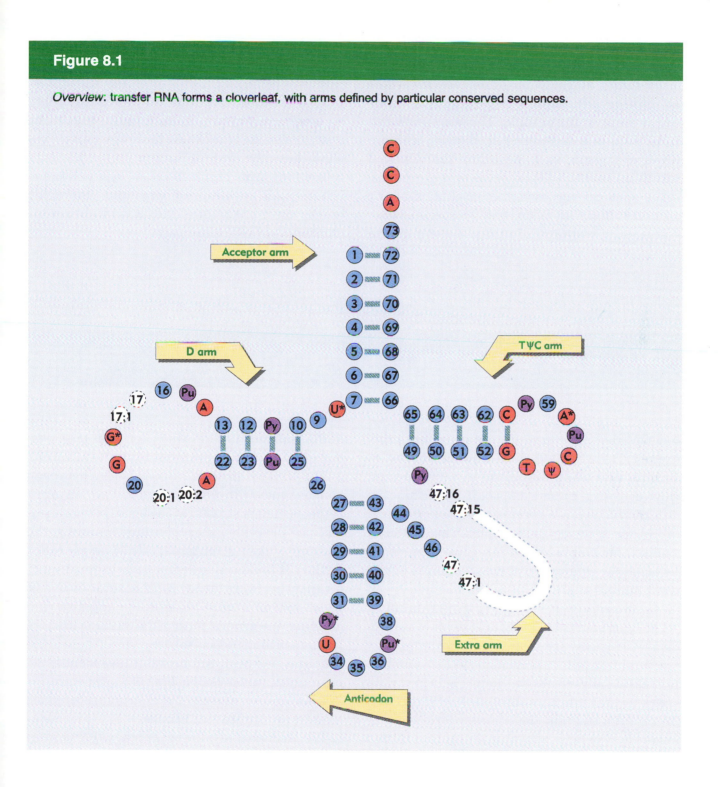

Figure 8.1

Overview: transfer RNA forms a cloverleaf, with arms defined by particular conserved sequences.

The base pairing that maintains the secondary structure is virtually invariant. Going clockwise around the cloverleaf, there are always 7 base pairs in the acceptor stem, 5 in the TψC arm, 5 in the anticodon arm, and usually 3 (sometimes 4) in the D arm. Within a given tRNA, most of the base pairings will be conventional partnerships of A•U and G•C, but occasional G•U, G•ψ, or A•ψ pairs are found. The additional types of base pairs are less stable than the regular pairs, but still allow a double-helical structure to form in RNA.

When the sequences of tRNAs are compared, the bases found at some positions are **invariant** (or **conserved**); almost always a particular base is found at the position. These are indicated in the sequence of Figure 8.1. Actually, as more tRNAs are sequenced, positions that seemed entirely invariant do display occasional exceptions. So for practical purposes, the description of any position as invariant means that the specified base is present in >90–95% of tRNAs. Sometimes the exceptions are individual; sometimes they fall into groups representing some peculiarity of a particular cell.

Some positions are described as **semiinvariant** (or **semiconserved**) because they seem to be restricted to one type of base (purine versus pyrimidine), but either base of that type may be present.

The four arms of the cloverleaf typify tRNAs in virtually all species and situations, including prokaryotes, eukaryotic cytosol, and eukaryotic mitochondria and chloroplasts.

The tertiary structure is L-shaped

The cloverleaf form in which the secondary structure of tRNA is written should not be allowed to convey a misleading view of the tertiary structure, which actually is rather compact. To determine the tertiary structure, it is necessary to grow crystals of a tRNA for X-ray crystallography. This has been achieved for only a few tRNAs, most from yeast. However, the similarities of the structures suggest that a common tertiary theme is honored by all tRNAs, although each will present its own variation.

The base-paired double-helical stems of the secondary structure are maintained in the tertiary structure, but their arrangement in three dimensions essentially creates two double helices at right angles to each other, as illustrated in **Figure 8.2**. The acceptor stem and the TψC stem form one continuous double helix with a single gap; the D stem and anticodon stem form another continuous double helix, also with a gap. The region between the double helices, where the turn in the L-shape is made, contains the TψC loop and the D loop. Thus the amino acid resides at the extremity of one arm

of the L-shape, and the anticodon loop forms the other end.

The tertiary structure is created by hydrogen bonding, mostly involving bases that are unpaired in the secondary structure. The bonds of the cloverleaf are described as **secondary H bonds**; the additional bonds of the tertiary structure are called **tertiary H bonds**. Many of the invariant and semiinvariant bases are involved in the tertiary H bonds, which explains their conservation and also suggests that the general form of the tertiary structure is common to all tRNAs. Not every one of these interactions is universal, but probably they identify the *general* pattern for establishing tRNA structure.

A molecular model of the structure of yeast tRNA^Phe is shown in **Figure 8.3**. Differences in the structure are found in other tRNAs, thus accommodating the dilemma that all tRNAs must have a similar shape, yet it must be possible to recognize differences between them. For example, in tRNA^Asp, the angle between the two axes is slightly greater, so the molecule

Figure 8.2

Transfer RNA folds into a compact L-shaped tertiary structure.

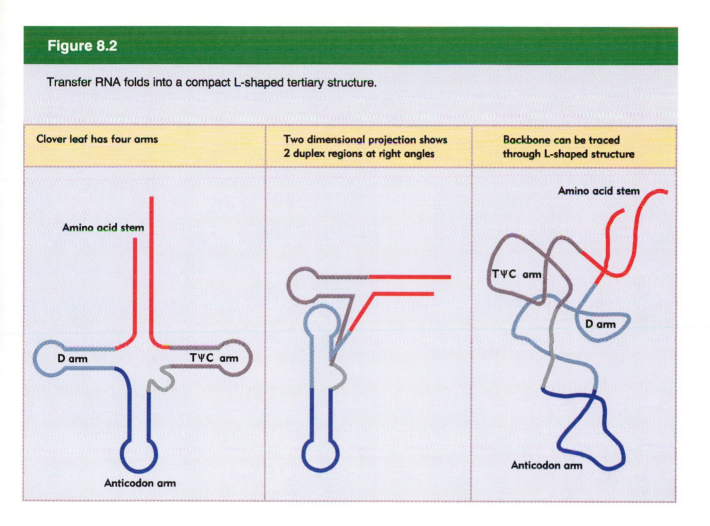

Clover leaf has four arms	Two dimensional projection shows 2 duplex regions at right angles	Backbone can be traced through L-shaped structure

has a slightly more open conformation.

The structure visualized by X-ray crystallography represents a stable organization of the molecule. In solution, or in association with proteins, some flexibility is shown. The conformation of tRNA can change in both circumstances, but we do not know yet how these changes are related to the crystal structure.

The structure suggests a general conclusion about the function of tRNA. *Its sites for exercising particular functions are maximally separated.* The amino acid is as far distant from the anticodon as possible, which is consistent with the need for the aminoacyl group to be near the peptidyl transferase site on the large subunit of the ribosome, while the anticodon pairs with mRNA on the small subunit. The TψC sequence lies at the junction of the arms of the L-shape, possibly a critical location for controlling changes in the tertiary structure. The structure accommodates the various demands on tRNA by providing different sites, analogous to the active sites of a protein.

Figure 8.3

A space-filling model shows that tRNA tertiary structure is compact. The two views of tRNA*Phe* are rotated by 90°. The view on the right corresponds with the right panel in Figure 8.2. Photograph kindly provided by S. H. Kim.

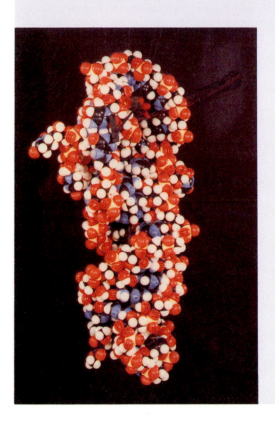
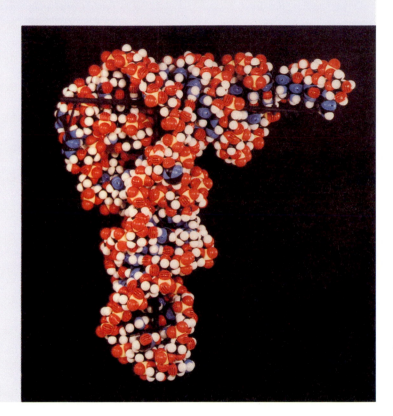

Synthetases fall into two classes that recognize similar features in tRNA

Amino acids enter the protein synthesis pathway through the aminoacyl-tRNA synthetases, which provide the interface for connection with nucleic acid. All synthetases function by the two-step mechanism depicted in Figure 8.4:

◆ First, the amino acid reacts with ATP to form aminoacyl~adenylate, releasing pyrophosphate.

Energy for the reaction is provided by cleaving the high energy bond of the ATP.

◆ Then the activated amino acid is transferred to the tRNA, releasing AMP.

The synthetases sort the tRNAs and amino acids into corresponding sets, each synthetase

Figure 8.4

An aminoacyl-tRNA synthetase charges tRNA with an amino acid.

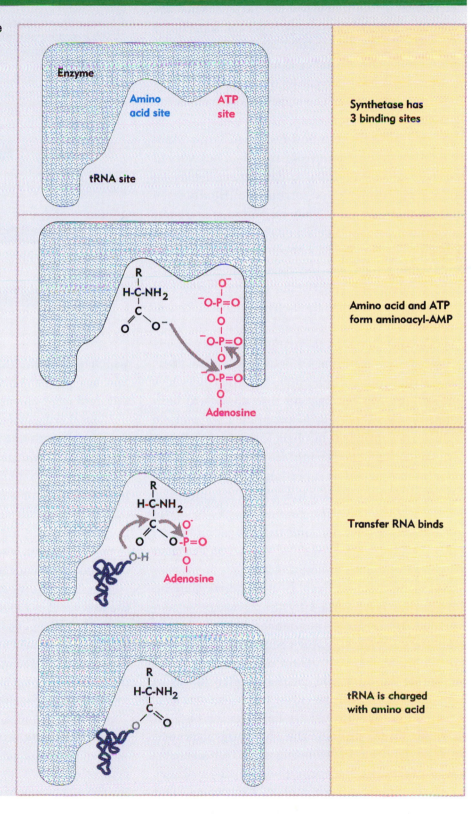

Synthetase has 3 binding sites

Amino acid and ATP form aminoacyl-AMP

Transfer RNA binds

tRNA is charged with amino acid

recognizing a single amino acid and all the tRNAs that should be charged with it. (There may be many such tRNAs. In addition to the several tRNAs that are needed to respond to synonym codons, sometimes there are multiple species of tRNA reacting with the same codon.) The tRNAs recognized by a synthetase are described as its **cognate** tRNAs. Usually they differ in other parts of the sequence as well as the anticodon.

In spite of their common function, synthetases are a rather diverse group of proteins. They vary in size from 40,000 to 100,000 daltons, and may be monomeric, dimeric, or tetrameric. Homologies between them are rare. They may have arisen early in evolution, and have retained the interaction with tRNA and amino acid as their sole function. Of course, the active site that recognizes tRNA comprises a rather small part of the molecule; it will be interesting to compare the active sites of different synthetases.

Synthetases have been divided into two general groups, each containing 10 enzymes, on the basis of the structures of their tRNA-binding regions. A general type of organization that applies to both groups is represented in **Figure 8.5**. The catalytic domain includes the binding sites for ATP and amino acid. It can be recognized as an large region that is interrupted by an insertion of the domain that binds the acceptor helix of the tRNA. This places the terminus of the tRNA in proximity to the catalytic site. A separate domain binds the anticodon region of tRNA. Those synthetases that are multimeric also possess an oligomerization domain.

Class I synthetases have an N-terminal catalytic domain that is identified by the presence of two short, partly conserved sequences of amino acids, sometimes called 'signature sequences'. The catalytic domain takes the form of a motif called a nucleotide-binding fold (which is also found in other classes of enzymes that bind nucleotides). The nucleotide fold consists of alternating parallel β-strands and α-helices; the signature sequence forms part of the ATP-binding site. The insertion that contacts the acceptor helix of tRNA differs widely between different class I enzymes. The C-terminal domains of the class I synthetases,

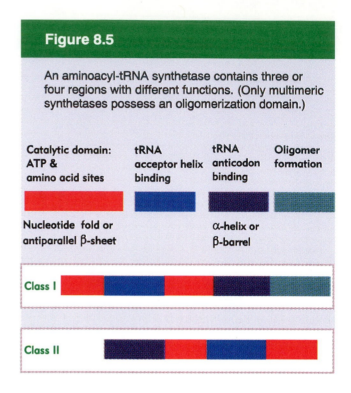

Figure 8.5

An aminoacyl-tRNA synthetase contains three or four regions with different functions. (Only multimeric synthetases possess an oligomerization domain.)

which include the tRNA anticodon-binding domain and any oligomerization domain, also are quite different from one another.

Class II enzymes share three rather general similarities of sequence in their catalytic domains. The active site contains a large antiparallel β-sheet surrounded by α-helices. Again, the acceptor helix-binding domain which interrupts the catalytic domain has a structure that differs with the individual enzyme. The anticodon-binding domain tends to be N-terminal. The location of any oligomerization domain is widely variable.

The lack of any apparent relationship between the groups of synthetases is a puzzle. It may be the case that they evolved independently of one another; this makes it seem possible even that an early form of life could have existed with proteins that were made up of just the 10 amino acids coded by one type or the other.

A general model for synthetase•tRNA binding suggests that the protein binds the tRNA along the 'side' of the L-shaped molecule. The same general principle applies for all synthetase•tRNA binding: the tRNA is bound principally at its two extremities, and most of the tRNA sequence is not involved in recognition by a synthetase. However, the detailed nature of the interaction is different between class I and class II enzymes, as can be seen from the models of **Figure 8.6**, which are based on crystal structures. The two types of enzyme approach the tRNA from opposite sides, with the result that the tRNA–protein models look almost like mirror images of one another.

A class I enzyme (Gln-tRNA synthetase) approaches the D-loop side of the tRNA, and recognizes the minor groove of the acceptor stem at one end of the binding site, and the anticodon loop at the other end. **Figure 8.7** is a diagrammatic representation of the crystal structure of the tRNA^Gln •synthetase complex. A revealing feature of the structure is that contacts with the enzyme change the structure of the tRNA at two important points. These can be seen by comparing the dotted and solid lines in the anticodon loop and acceptor stem:

◆ Bases U35 and U36 in the anticodon loop are pulled farther out of the tRNA into the protein.

◆ The end of the acceptor stem is seriously distorted, with the result that base pairing between U1 and A72 is disrupted. The single-stranded end of the stem pokes into a deep pocket in the synthetase protein, which also contains the binding site for ATP.

This structure confirms the results of earlier studies that showed that changes in U35, G73, or the U1-A72 base pair affected the recognition of the tRNA by its synthetase. At all of these positions, hydrogen bonding occurs between the protein and tRNA.

A class II enzyme (Asp-tRNA synthetase) approaches the tRNA from the other side, and recognizes the variable loop, and the major groove of the acceptor stem. **Figure 8.8** shows a diagram-

Figure 8.6

Crystal structures show that class I and class II aminoacyl-tRNA synthetases bind the opposite faces of their tRNA substrates. The tRNA is shown in red, and the protein in blue. Photograph kindly provided by Dino Moras.

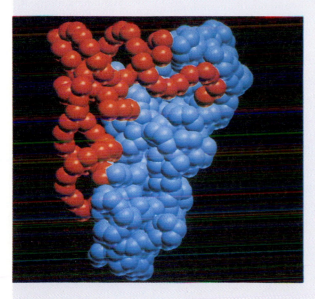

Class I

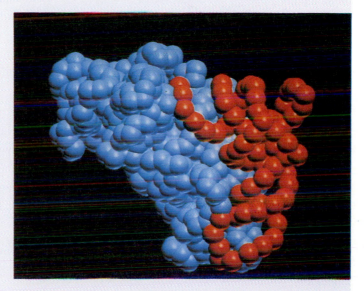

Class II

Figure 8.7

A class I tRNA synthetase contacts tRNA at the minor groove of the acceptor stem and at the anticodon.

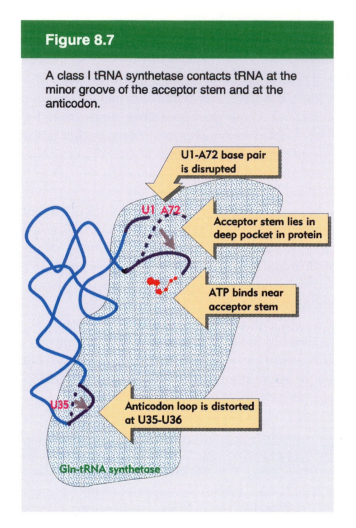

U1-A72 base pair is disrupted

U1 A72

Acceptor stem lies in deep pocket in protein

ATP binds near acceptor stem

U35

Anticodon loop is distorted at U35-U36

Gln-tRNA synthetase

alter recognition of a tRNA are found in these two regions. (The anticodon itself is not necessarily recognized; for example, the 'suppressor' mutations discussed later in this chapter change a base in the anticodon, and therefore the codons to which a tRNA responds, without altering its charging with amino acids.)

Table 8.1 summarizes the positions that identify some tRNAs to their synthetases. A common feature is that the number of critical positions is small, varying from 1 to 5. Often, but not always, the anticodon provides at least some of the crucial contacts. It is exclusively involved for valine and methionine tRNAs. An extreme case is represented by alanine tRNA, which is identified by a single

Figure 8.8

A class II aminoacyl-tRNA synthetase contacts tRNA at the major groove of the acceptor helix and at the anticodon loop.

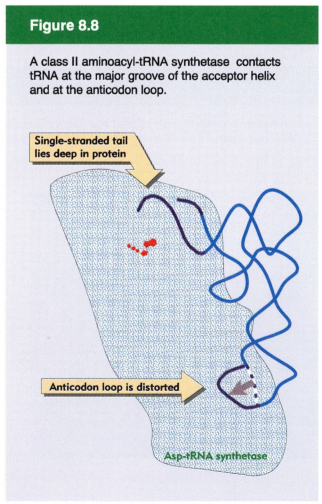

Single-stranded tail lies deep in protein

Anticodon loop is distorted

Asp-tRNA synthetase

matic representation. The acceptor stem remains in its regular helical conformation. ATP is probably bound near to the terminal adenine. At the other end of the binding site, there is a tight contact with the anticodon loop, which has a conformation change that allows the anticodon to be in close contact with the protein.

Many attempts to deduce similarities in sequence between cognate tRNAs, or to induce chemical alterations that affect their charging, have shown that the basis for recognition is different for different tRNAs, and does not necessarily lie in some feature of primary or secondary structure alone. We know from the crystal structure that the acceptor stem and the anticodon stem make tight contacts with the synthetase, and mutations that

Table 8.1

Each synthetase recognizes its cognate tRNAs by means of a few specific bases

tRNA	Mutations that Alter Recognition by Synthetase
Class I enzymes	
Valine	3 bases of anticodon
Methionine	3 bases of anticodon
Isoleucine	modification at C34 of anticodon
Glutamine	U35 in anticodon; U1-A72 and G73 in acceptor stem
Class II enzymes	
Phenylalanine	3 bases of anticodon; G20 in D loop; A73 at terminus
Serine	G1-C72; G2-C71; A3-U70 bp in acceptor stem; C11-G24 bp in D stem
Alanine	G3-U70 bp in acceptor stem

unique base pair in the acceptor stem. Since one base pair can scarcely constitute a general means of distinguishing 20 sets of tRNAs, such effects fortify the conclusion that recognition of tRNAs appears to be idiosyncratic, each following its own rules!

Several synthetases can specifically charge a 'minihelix' consisting only of the acceptor and TψC arms (equivalent to one arm of the L-shaped molecule) with the correct amino acid. The efficiency of aminoacylation of these substrates is much higher for the class II enzymes; for example, seryl- and alanyl-tRNA synthetases are especially effective, which fits with the concentration of mutations in their acceptor stems. For these tRNAs, specificity depends exclusively on the acceptor stem. However, it is clear that there are significant variations between tRNAs, since mutations in the anticodon of the class II phe-tRNA synthetase are important; and minihelices from the tRNA^{Val} and tRNA^{Met} (where we know that the anticodon is important *in vivo*) can react specifically with their class I synthetases.

Thus recognition depends on an interaction between a few points of contact in the tRNA, concentrated at the extremities, and a few amino acids constituting the active site in the protein. The relative importance of the roles played by the acceptor stem and anticodon are different for each tRNA•synthetase interaction.

Discrimination in the charging step

The nature of discriminatory events is a general issue raised by several steps in gene expression. How do synthetases recognize just the corresponding tRNAs and amino acids? How does a ribosome recognize only the tRNA corresponding to the codon in the A site? How do the enzymes that synthesize DNA or RNA recognize only the base complementary to the template? Each case poses a similar

Figure 8.9

Recognition of the correct tRNA by synthetase is controlled at two steps. *First,* the enzyme has a greater affinity for its cognate tRNA. *Second,* the aminoacylation of the incorrect tRNA is very slow.

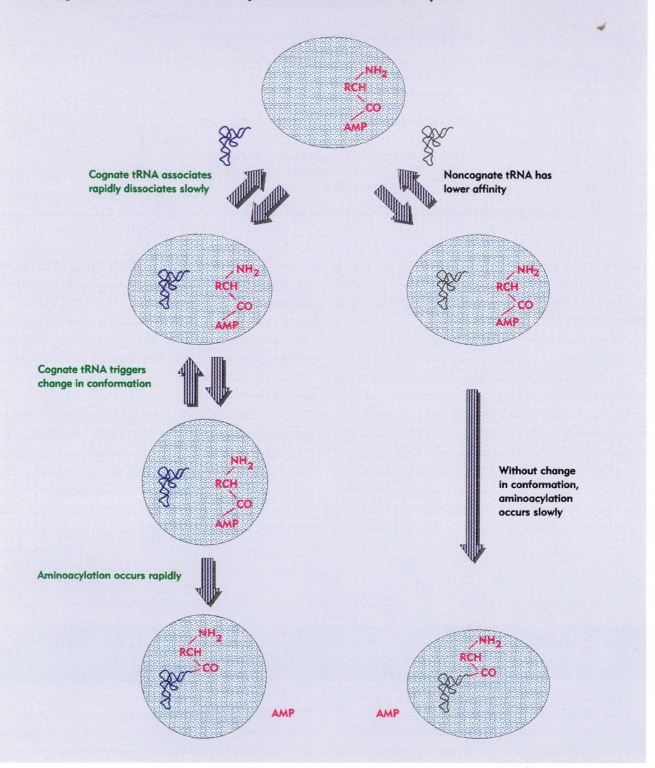

Cognate tRNA associates rapidly dissociates slowly

Noncognate tRNA has lower affinity

Cognate tRNA triggers change in conformation

Without change in conformation, aminoacylation occurs slowly

Aminoacylation occurs rapidly

problem: *how to distinguish one particular member from the entire set, all of which share the same general features.*

Probably any member initially can contact the active center by a random-hit process, but then the wrong members are rejected and only the appropriate one is accepted. The appropriate member is always in a minority (1 of 20 amino acids, 1 of ~40 tRNAs, 1 of 4 bases), so the criteria for discrimination must be strict. We can imagine two general ways in which the decision whether to reject or accept might be taken:

◆ The cycle of admittance, scrutiny, rejection/acceptance could represent a single binding step that *precedes all other stages* of whatever reaction is involved. This is tantamount to saying that the affinity of the binding site is controlled in such a way that only the appropriate species is comfortable there. In the case of synthetases, this would mean that only the cognate tRNAs could form a stable attachment at the site.

◆ Alternatively, *the reaction proceeds through some of its stages*, after which a decision is reached on whether the correct species is present. If it is not present, the reaction is reversed, or a bypass route is taken, and the wrong member is expelled. This sort of postbinding scrutiny is generally described as **proofreading**. In the example of synthetases, it would require that the charging reaction proceeds through certain stages even if the wrong tRNA or amino acid is present.

Synthetases use proofreading mechanisms to control the recognition of both tRNA and amino acid.

Transfer RNA binds to synthetase by the two stage reaction depicted in **Figure 8.9**. Cognate tRNAs have a greater intrinsic affinity for the binding site, so they are bound more rapidly and dissociate more slowly.

Following binding, the enzyme scrutinizes the tRNA that has been bound. If the correct tRNA is present, binding is stabilized by a conformational change in the enzyme. This allows aminoacylation to occur rapidly.

If the wrong tRNA is present, the conformational change does not occur. As a result, the reaction proceeds much more slowly; this increases the chance that the tRNA will dissociate from the enzyme before it is charged.

Specificity for amino acids varies among the synthetases. Some are highly specific for initially binding a single amino acid, but others can also activate amino acids closely related to the proper substrate. Although the analog amino acid can sometimes be converted to the adenylate form, in none of these cases is an incorrectly activated amino acid actually used to form a stable aminoacyl-tRNA.

Editing by means of a reaction that proceeds part way and is then reversed is called **chemical proofreading**. There are two stages at which proofreading of an incorrect aminoacyl-adenylate may occur during formation of aminoacyl-tRNA. **Figure 8.10** shows that both require the presence of the cognate tRNA. The extent to which one pathway or the other predominates varies with the individual synthetase:

◆ The noncognate aminoacyl-adenylate may be hydrolyzed when the cognate tRNA binds. This mechanism is used predominantly by several synthetases, including those for methionine, isoleucine, and valine. (Usually, the reaction cannot be seen *in vivo*, but it can be followed when the incorrectly activated amino acid is homocysteine (which lacks the methyl group of methionine). Proofreading releases the amino acid in an altered form, as homocysteine thiolactone. In fact, homocysteine thiolactone is produced in *E. coli* as a byproduct of the charging reaction of Met-tRNA synthetase. This shows that continuous proofreading is part of the process of charging a tRNA with its amino acid.)

◆ Some synthetases predominantly use chemical proofreading at a later stage. The wrong amino acid is actually transferred to tRNA, is then recognized as incorrect by its structure in the

Figure 8.10

When a synthetase binds the incorrect amino acid, proofreading requires binding of the cognate tRNA. It may take place either by a conformation change that causes hydrolysis of the incorrect aminoacyl- adenylate, or by transfer of the amino acid to tRNA, followed by hydrolysis.

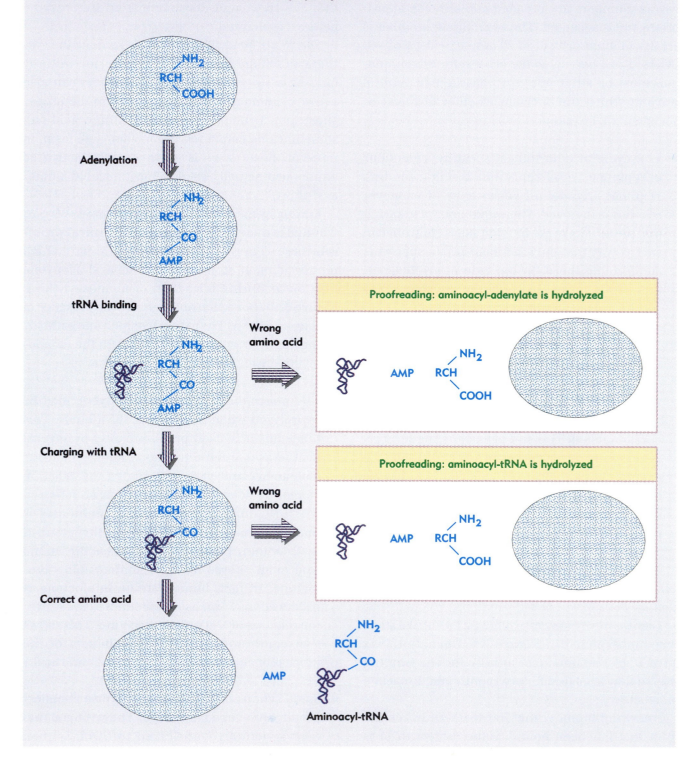

tRNA binding site, and so is hydrolyzed and released. The process requires a continual cycle of linkage and hydrolysis until the correct amino acid is transferred to the tRNA.

The presence of the cognate tRNA is needed to trigger proofreading, even if the reaction occurs at the stage before formation of aminoacyl-adenylate. A classic example in which discrimination between amino acids depends on the presence of tRNA is provided by the Ile-tRNA synthetase of *E. coli*. The enzyme can charge valine with AMP, but hydrolyzes the valyl-adenylate when tRNA^{Ile} is added. The overall error rate depends on the specificities of the individual steps, as summarized in **Table 8.2**. The overall error rate of 1.5×10^{-5} is less than the measured rate at which valine is substituted for isoleucine (in rabbit globin), which is $2–5 \times 10^{-4}$.

Table 8.2

The accuracy of charging tRNA^{Ile} by its synthetase depends on error control at two stages.

Step	Frequency of Error
Activation of Valine to Val-AMP	1/225
Release of Val-tRNA^{Ile}	1/270
Overall rate of error	1/225 × 1/270 = 1/60,000

Thus mischarging probably provides only a small fraction of the errors that actually occur in protein synthesis.

Codon–anticodon recognition involves wobbling

The function of tRNA in protein synthesis is fulfilled when it recognizes the codon in the ribosomal A site. The interaction between anticodon and codon takes place by base pairing, but under rules that extend pairing beyond the usual G•C and A•U partnerships.

We can deduce the rules governing the interaction from the sequences of the anticodons that correspond to particular codons. The ability of any tRNA to respond to a given codon can be measured by the trinucleotide binding assay or by its use in an *in vitro* protein synthetic system (the procedures used originally to define the genetic code, discussed in Chapter 7).

The genetic code itself yields some important clues about the process of codon recognition. The pattern of third-base degeneracy is drawn in

Figure 8.11, which shows that in almost all cases either the third base is irrelevant or a distinction is made only between purines and pyrimidines.

There are eight **codon families** in which all four codons sharing the same first two bases have the same meaning, so that the third base has no role at all in specifying the amino acid. There are seven **codon pairs** in which the meaning is the same whichever pyrimidine is present at the third position; and there are five codon pairs in which either purine may be present without changing the amino acid that is coded.

It may be more significant to look at the code from the reverse perspective. *There are only three cases in which a unique meaning is conferred by the presence of a particular base at the third position*: AUG (for methionine), UGG (for tryptophan), and UGA

Figure 8.11

Third bases have the least influence on codon meanings. Boxes indicate groups of codons within which third-base degeneracy ensures that the meaning is the same.

UUU	UCU	UAU	UGU
UUC	UCC	UAC	UGC
UUA	UCA	UAA	UGA
UUG	UCG	UAG	UGG
CUU	CCU	CAU	CGU
CUC	CCC	CAC	CGC
CUA	CCA	CAA	CGA
CUG	CCG	CAG	CGG
AUU	ACU	AAU	AGU
AUC	ACC	AAC	AGC
AUA	ACA	AAA	AGA
AUG	ACG	AAG	AGG
GUU	GCU	GAU	GGU
GUC	GCC	GAC	GGC
GUA	GCA	GAA	GGA
GUG	GCG	GAG	GGG

Third base relationship	Third bases with same meaning	Number of codons
third base irrelevant	U, C, A, G	32
purines differ from pyrimidines	U or C	14
	A or G	12
unique definitions	U, C, A	3
	G only	2

(nonsense). Thus C and U never have a unique meaning in the third position, and A never signifies a unique amino acid.

Because the anticodon is complementary to the codon, it is the *first* base in the anticodon sequence written conventionally in the direction from 5′ to 3′ that pairs with the *third* base in the codon sequence written by the same convention. Thus the combination

Codon	5:	A C G	3:
Anticodon	3:	U G C	5:

is usually written as codon ACG/anticodon CGU, where the anticodon sequence must be read backward for complementarity with the codon.

To avoid confusion, we shall retain the usual convention in which all sequences are written 5′–3′, but indicate anticodon sequences with a backward arrow as a reminder of the relationship with the codon. Thus the codon–anticodon pair shown above will be written as ACG and $\overleftarrow{\text{CGU}}$, respectively.

Does each triplet codon demand its own tRNA with a complementary anticodon? Or can a single tRNA respond to both members of a codon pair and to all (or at least some) of the four members of a codon family?

Often one tRNA can recognize more than one codon. *This means that the base in the first position of the anticodon must be able to partner alternative bases in the corresponding third position of the codon.* Base pairing at this position cannot be limited to the usual G•C and A•U partnerships.

The rules governing the recognition patterns are summarized in the **wobble hypothesis**, which states that the pairing between codon and anticodon at the first two codon positions always follows the usual rules, but that exceptional 'wobbles' occur at the third position. Wobbling occurs because the conformation of the tRNA anticodon loop

Table 8.3

Codon-anticodon pairing involves wobbling at the third position.

Base in First Position of Anticodon	Base(s) Recognized in Third Position of Codon
U	A or G
C	G only
A	U only
G	C or U

permits flexibility at the first base of the anticodon.

The rules for recognition of the third base of the codon admit pairing between G and U in addition to the usual pairs (see Figure 8.13). This single change creates a pattern of base pairing in which A can no longer have a unique meaning in the codon (because the U that recognizes it must also recognize G). Similarly, C also no longer has a unique meaning (because the G that recognizes it also must recognize U). **Table 8.3** summarizes the pattern of recognition.

It is therefore possible to recognize unique codons only when the third bases are G or U; this option is not used often, since UGG and AUG are the only examples of the first type, and there is none of the second type.

(G•U pairs are common in RNA duplex structures, where they neither assist nor impede formation of the double helix, as described in Figure 5.5. But the formation of stable contacts between codon and anticodon, when only 3 base pairs can be formed, is much more constrained, and thus G•U pairs can contribute only in the last position of the codon.)

tRNA contains many modified bases

Transfer RNA is unique among nucleic acids in its content of 'unusual' bases. An unusual base is any purine or pyrimidine ring except the usual A, G, C, and U from which all RNAs are synthesized. All other bases are produced by **modification** of one of the four bases after it has been incorporated into the polyribonucleotide chain.

All classes of RNA display some degree of modification, but in all cases except tRNA this is confined to rather simple events, such as the addition of methyl groups. In tRNA, there is a vast range of modifications, ranging from simple methylation to wholesale restructuring of the purine ring. These modifications increase the structural versatility of tRNA, which is important in its various functions. Modifications occur in all parts of the tRNA molecule; the list of modified nucleosides extends to ~50 bases.

Figure 8.12 shows some of the more common modified bases. Modifications of pyrimidines (C and U) are less complex than those of purines (A and G). In addition to the modifications of the bases themselves, methylation at the 2'-O position of the ribose ring also occurs.

The most common modifications of uridine are straightforward. Methylation at position 5 creates ribothymidine (T). The base is the same commonly found in DNA; but here it is attached to ribose, not deoxyribose. In RNA, thymine constitutes an unusual base, originating by modification of U.

Dihydrouridine (D) is generated by the saturation of a double bond, changing the ring structure. Pseudouridine (ψ) interchanges the positions of N and C atoms. And 4-thiouridine has sulfur substituted for oxygen.

The nucleoside inosine (I) is found normally in the cell as an intermediate in the purine biosynthetic pathway. However, it is not incorporated directly into RNA, where instead its existence depends on modification of A to create I. Other modifications of A include the addition of complex groups.

Two complex series of nucleotides depend on modification of G. The Q bases, such as queuosine, have an additional pentenyl ring added via an NH linkage to the methyl group of 7-methylguanosine. The pentenyl ring may carry various further groups. The Y bases, such as wyosine, have an additional ring fused with the purine ring itself; the extra ring carries a long carbon chain, again to which further groups are added in different cases.

The modification reaction usually involves the alteration of, or addition to, existing bases in the tRNA. An exception is the synthesis of Q bases, where a special enzyme exchanges free queuosine

Figure 8.12

Some of the modified nucleosides found in tRNA (modifications are indicated by red shading).

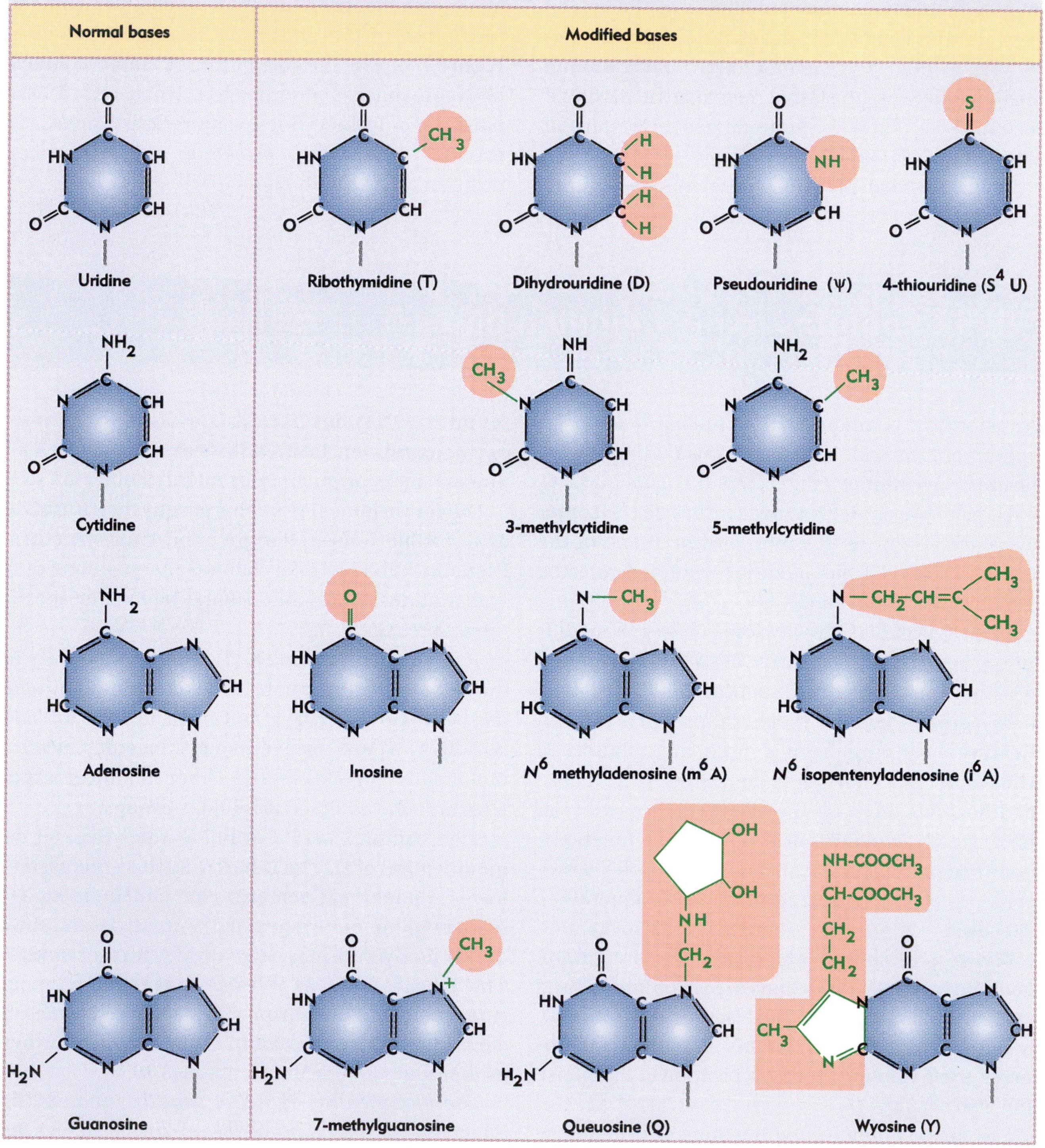

with a guanosine residue in the tRNA. The reaction involves breaking and remaking bonds on either side of the nucleoside.

The modified nucleosides are synthesized by specific tRNA-modifying enzymes. The original nucleoside present at each position can be determined either by comparing the sequence of tRNA with that of its gene or (less efficiently) by isolating precursor molecules that lack some or all of the modifications. The sequences of precursors show that different modifications are introduced at different stages during the maturation of tRNA.

Some modifications are constant features of all tRNA molecules—for example, the D residues that give rise to the name of the D arm, and the ψ found in the TψC sequence. On the 3′ side of the anticodon there is always a modified purine, although the modification varies widely.

Other modifications are specific for particular tRNAs or groups of tRNAs. For example, wyosine bases are characteristic of tRNA^{Phe} in bacteria, yeast, and mammals. There are also some species-specific patterns.

The features recognized by the tRNA-modifying enzymes are unknown. When a particular modification is found at more than one position in a tRNA, the same enzyme does not necessarily make all the changes; for example, a different enzyme may be needed to synthesize the pseudouridine at each location. We do not know what controls the specificity of the modifying enzymes, but it is clear that there are many enzymes, with varying specificities. Some enzymes undertake single reactions with individual tRNAs; others have a range of substrate molecules. Some modifications require the successive actions of more than one enzyme.

Base modification may control codon recognition

The most direct effect of modification is seen in the anticodon, where change of sequence influences the ability to pair with the codon, thus determining the meaning of the tRNA. Modifications elsewhere in the vicinity of the anticodon also influence its pairing.

When bases in the anticodon are modified, further pairing patterns become possible in addition to those predicted by the regular and wobble pairing involving A, C, U, and G. **Figure 8.13** illustrates some of the variations in codon–anticodon pairing that are allowed by the combination of wobbling and modification.

Actually, some of the predicted regular combinations do not occur, because some bases are *always* modified; in particular, U and A are not employed at the first position of the anticodon. Usually, U at this position is converted to a modified form that has altered pairing properties. There seems to be an absolute ban on the employment of A; usually it is converted to I.

Inosine is often present at the first position of the anticodon, where it is able to pair with any one of three bases, U, C, and A. This ability is especially important in the isoleucine codons, where AUA codes for isoleucine, while AUG codes for methionine. Because with the usual bases it is not possible to recognize A alone in the third position, any tRNA with U starting its anticodon would have to recognize AUG as well as AUA. So AUA must be read together with AUU and AUC, a problem that is solved by the existence of tRNA with I in the anticodon.

Some modifications create preferential readings of some codons with respect to others. Anticodons with uridine-5-oxyacetic acid and 5-methoxyuridine in the first position recognize A and G efficiently as third bases of the codon, but recognize U less efficiently. Another case in which multiple pairings can occur, but with some preferred to others is provided by the series of queuosine and its derivatives. These modified G bases continue

Figure 8.13

Wobble in base pairing allows some bases at the first position of the anticodon to recognize more than one base in the third position of the codon. Pairing between standard bases is extended from the G-C and A-U pairs to the G-U wobble pair .Base modifications may restrict or extend the pattern. Modification to 2-thiouridine restricts pairing to A alone because only one H-bond could form with G. Modification to inosine allows pairing with U, C, and A.

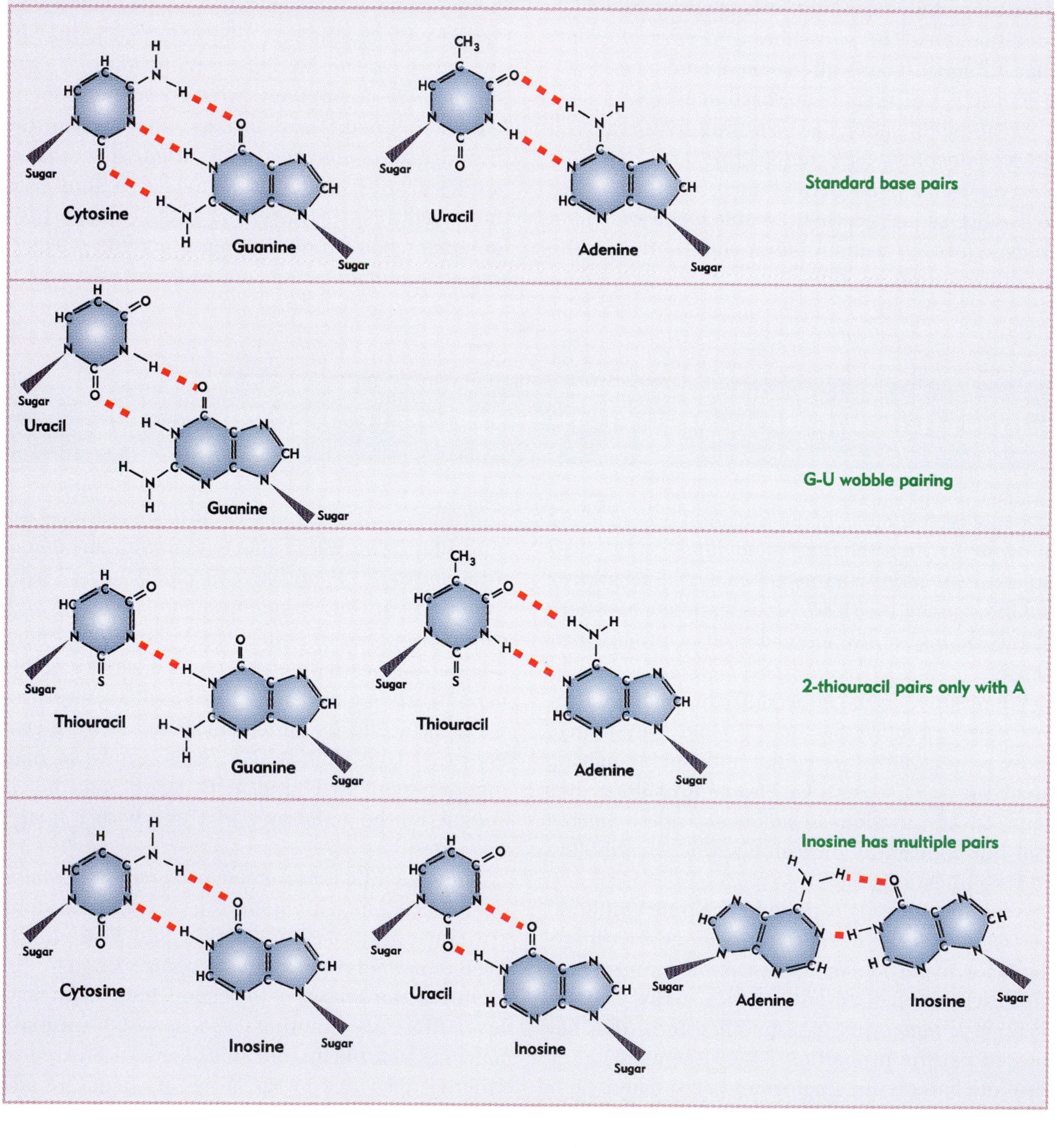

to recognize both C and U, but pair with U more readily.

A restriction not allowed by the usual rules can be achieved by the employment of 2-thiouridine in the anticodon. This modification allows the base to continue to pair with A, but prevents it from indulging in wobble pairing with G.

These and other pairing relationships make the general point that *there are multiple ways to construct a set of tRNAs able to recognize all the 61 codons representing amino acids*. No particular pattern predominates in any given organism, although the absence of a certain pathway for modification can prevent the use of some recognition patterns. Thus a particular codon family is read by tRNAs with different anticodons in different organisms.

Often the tRNAs will have overlapping responses, so that a particular codon is read by more than one tRNA. In such cases there may be differences in the efficiencies of the alternative recognitions. (As a general rule, codons that are commonly used tend to be more efficiently read.) And in addition to the construction of a set of tRNAs able to recognize all the codons, there may be multiple tRNAs that respond to the same codons.

The predictions of wobble pairing accord very well with the observed abilities of almost all tRNAs. But there are exceptions in which the codons recognized by a tRNA differ from those predicted by the wobble rules. Such effects probably result from the influence of neighboring bases and/or the conformation of the anticodon loop in the overall tertiary structure of the tRNA. Indeed, the importance of the structure of the anticodon loop is inherent in the idea of the wobble hypothesis itself. Further support for the influence of the surrounding structure is provided by the isolation of occasional mutants in which a change in a base in some other region of the molecule alters the ability of the anticodon to recognize codons (see later).

Another unexpected pairing reaction is presented by the ability of the bacterial initiator, fMet-tRNA$_f$, to recognize both AUG and GUG. This misbehavior involves the third base of the anticodon.

The genetic code is altered in ciliates and mitochondria

The universality of the genetic code is striking, but some exceptions exist. They tend to affect the codons involved in initiation or termination and result from the production (or absence) of tRNAs representing certain codons. Three examples of changes in the genetic code in bacterial or nuclear genomes illustrate the nature of such situations.

In the prokaryote *Mycoplasma capricolum*, UGA is not used for termination, but instead codes for tryptophan. In fact, it is the predominant Trp codon, and UGG is used only rarely. Two Trp-tRNA species exist, with the anticodons UCA (reads UGA and UGG) and CCA (reads only UGG).

Some ciliates (unicellular protozoa) read UAA and UAG as glutamine instead of termination signals. *Tetrahymena thermophila*, one of the ciliates, contains three tRNA^{Glu} species. One recognizes the usual codons CAA and CAG for glutamine, one recognizes both UAA and UAG (in accordance with the wobble hypothesis), and the last recognizes only UAG. We assume that a further change is that the release factor eRF has a restricted specificity, compared with that of other eukaryotes.

In another ciliate (*Euplotes octacarinatus*), UGA codes for cysteine. Only UAA is used as a termination codon, and UAG is not found. The change in meaning of UGA might be accomplished by a modification in the anticodon of tRNA^{Cys} to allow it to read UGA with the usual codons UGU and UGC.

Ciliates branched from other eukaryotes at an early stage in evolution, and we can offer two explanations for their idiosyncratic genetic codes. One is that at this time the code was still quite error-prone, with termination in particular being an erratic process. Perhaps in fact termination was so leaky that many proteins ended (if not at random sites) at any one of many possible sites. In the major branch of evolution, termination became effective at all three termination codons as the precision of protein synthesis increased, but in *Mycoplasma* and ciliates only some of the termination codons were retained. A more likely explanation, however, is that the divergent uses of the termination codons represents their 'capture' for normal coding purposes; if some termination codons were used only rarely, they could be recruited to coding purposes by changes that allowed tRNAs to recognize them.

Exceptions to the universal genetic code also occur in the mitochondria from several species, as summarized in **Table 8.4**.

A common change is that UGA has the same meaning as UGG, and therefore represents tryptophan instead of termination. This change is found in yeasts, invertebrates and vertebrates, but not in plants. UGA is certainly the most flexible codon in evolution.

Other changes are characteristic for particular species, including some changes in internal codons. The use of AUA to represent methionine instead of isoleucine occurs in several instances; in some the isoleucine can be used as an initiator, in others only for elongation.

Some of these changes make the code simpler, by replacing two codons that had different meanings with a pair that has a single meaning. Pairs treated like this include UGG and UGA (both Trp instead of one Trp and one termination) and AUG and AUA (both Met instead of one Met and the other Ile).

Why have changes been able to evolve in the mitochondrial code? Because the mitochondrion synthesizes only a small number of proteins (~10), the problem of disruption by changes in meaning is much less severe. Probably the codons that are altered were not used extensively in locations where amino acid substitutions would have been deleterious. The variety of changes found in mitochondria of different species suggests that they have evolved separately, and not by common descent from an ancestral mitochondrial code.

According to the wobble hypothesis, a minimum of 31 tRNAs (excluding the initiator) are required to recognize all 61 codons (at least 2 tRNAs are required for each codon family and 1 tRNA is needed per codon pair or single codon). But an unusual situation exists in (at least) mammalian mitochondria in which there are only 22 different tRNAs. How does this limited set of tRNAs accommodate all the codons?

The critical feature lies in a simplification of codon–anticodon pairing, in which one tRNA recognizes all four members of a codon family. This reduces to 23 the minimum number of tRNAs required to respond to all usual codons. The use of

Table 8.4

Changes occur in the mitochondrial genetic code.

Organism	Codon	Meaning in Mitochondrion	Usual Meaning
Common	UGA	Tryptophan	Termination
Mammal	AG$_G^A$	Termination	Arginine
Mammal	AUA	Met (initiation)	Isoleucine
Fruit fly	AUA	Met (initiation)	Isoleucine
Yeast	AUA	Met (elongation)	Isoleucine
Yeast	CUA	Threonine	Leucine
Fruit fly	AGA	Serine	Arginine

AG^A_G for termination reduces the requirement by one further tRNA, to 22.

In all eight codon families, the sequence of the tRNA contains an unmodified U at the first position of the anticodon. The remaining codons are grouped into pairs in which all the codons ending in pyrimidines are read by G in the anticodon, and all the codons ending in purines are read by a modified U in the anticodon, as predicted by the wobble hypothesis. The complication of the single UGG codon is avoided by the change in the code to read UGA with UGG as tryptophan; and in mammals, AUA ceases to represent isoleucine and instead is read with AUG as methionine. This allows all the nonfamily codons to be read as 14 pairs.

The 22 identified tRNA genes therefore code for 14 tRNAs representing pairs, and 8 tRNAs representing families. This leaves the two usual termination codons UAG and UAA unrecognized by tRNA, together with the codon pair AG^A_G. Similar rules are followed in the mitochondria of fungi.

As well as these general changes in the code, specific changes in reading occur in individual genes. The specificity of such changes implies that the reading of the particular codon must be influenced by the surrounding bases (also see later).

A striking example is the incorporation of the modified amino acid seleno-cysteine at certain UGA codons within the genes that code for seleno-proteins in both prokaryotes and eukaryotes. Usually these proteins catalyze oxidation–reduction reactions, and contain a single seleno-cysteine residue, which forms part of the active site. The most is known about the use of the UGA codons in three *E. coli* genes coding for formate dehydrogenase isozymes. The internal UGA codon is read by a seleno-cys-tRNA. This unusual reaction is determined by the local secondary structure of mRNA, in particular by the presence of a hairpin loop downstream of the UGA.

Mutations in four *sel* genes create a deficiency in selenoprotein synthesis. *selC* codes for tRNA (with the anticodon ACU) that is charged with serine. *selA* and *selD* are required to modify the serine to seleno-cysteine. *selB* codes for a guanine nucleotide-binding protein that acts as a specific translation factor for entry of seleno-cys-tRNA into the A site; it thus provides (for this single tRNA) a replacement for factor EF-Tu. The sequence of SelB is related to both EF-Tu and IF-2.

Suppressor tRNAs have mutated anticodons that read new codons

Isolation of mutant tRNAs has been one of the most potent tools for analyzing the ability of a tRNA to respond to its codon(s) in mRNA, and for determining the effects that different parts of the tRNA molecule have on codon–anticodon recognition.

Mutant tRNAs are isolated by virtue of their ability to overcome the effects of mutations in genes coding for proteins. We have already described the terminology in which a mutation that is able to overcome the effects of another is called a **suppressor** (see Chapter 3).

In tRNA suppressor systems, the primary mutation changes a codon in an mRNA so that the protein product is no longer functional; the secondary, suppressor mutation changes the anticodon of a tRNA, so that it recognizes the mutant codon instead of (or as well as) its original target codon. The amino acid that is now inserted restores protein function. The suppressors are named as **nonsense** or **missense**, depending on the nature of the original mutation.

In a wild-type cell, a nonsense mutation is recognized only by a release factor, terminating protein synthesis. The suppressor mutation creates an aminoacyl-tRNA that can recognize the termination codon; by inserting an amino acid, it allows protein

Figure 8.14

Nonsense mutations can be suppressed by a tRNA with a mutant anticodon. The wild-type gene contains a UUG codon specifying leucine, which is recognized by a tRNA with the anticodon CAA. A mutation changes the codon to UAG, so it is now recognized by release factor. A second mutation changes the anticodon of tyrosine tRNA from GUA to CUA, so that it recognizes the amber codon. Suppression results in synthesis of a full length protein in which the original Leu residue has been replaced by Tyr.

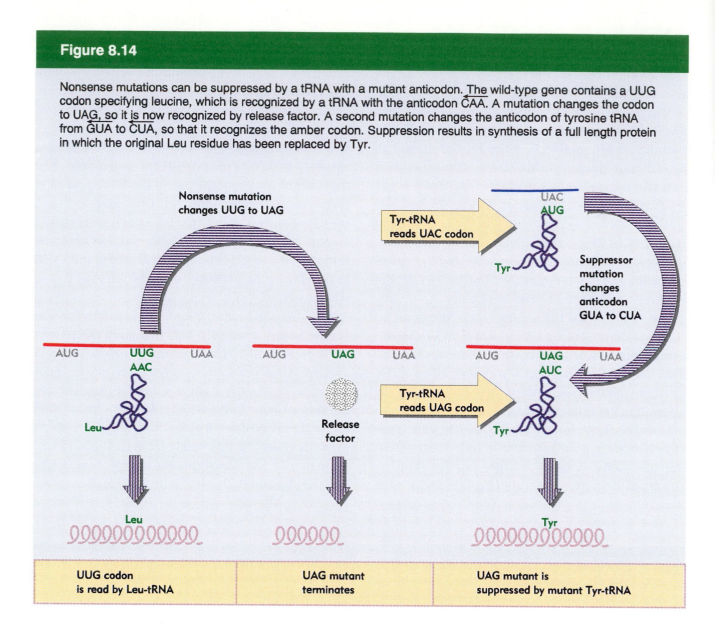

| UUG codon is read by Leu-tRNA | UAG mutant terminates | UAG mutant is suppressed by mutant Tyr-tRNA |

synthesis to continue beyond the site of nonsense mutation.

This new capacity of the translation system allows a full-length protein to be synthesized, as illustrated in **Figure 8.14**. If the amino acid inserted by suppression is different from the amino acid that was originally present at this site in the wild-type protein, the activity of the protein may be reduced.

Nonsense suppressors fall into three classes, one for each type of termination codon. **Table 8.5** describes the properties of some of the best characterized suppressors.

The easiest to characterize have been amber suppressors. In *E. coli*, at least 6 tRNAs have been mutated to recognize UAG codons. All of the amber suppressor tRNAs have the anticodon CUA, in each case derived from wild-type by a single base change. The site of mutation can be any one of the three bases of the anticodon, as seen from *supD*, *supE*, and *supF*. Each suppressor tRNA recognizes *only* the UAG codon, instead of its former codon(s). The amino acids inserted are serine, glutamine, or tyrosine, the same as those carried by the corresponding wild-type tRNAs.

Ochre suppressors also arise by mutations in the anticodon. The best known are *supC* and *supG*, which insert tyrosine or lysine in response to *both* ochre (UAA) and amber (UAG) codons. This conforms with the prediction of the wobble hypothesis that UAA cannot be recognized alone.

The suppressors of UGA are interesting. One type (*supU*) is a mutant of tryptophan tRNA. The anticodon of the wild-type tRNA^Trp is CCA, which recognizes only the UGG codon. The suppressor tRNA has the mutant anticodon UCA, which recognizes *both* its original codon and the termination codon UGA.

Another UGA suppressor has an unexpected property. It is derived from the same tRNA^Trp, but its only mutation is the substitution of A in place of G *at position 24*. This change replaces a G•U pair in the D stem with an A•U pair, increasing the stability of the helix. The sequence of the anticodon remains the same as the wild-type, CCA. So the mutation in the D stem must in some way alter the conformation of the anticodon loop, allowing CCA to pair with UGA in an unusual wobble pairing of C with A. The suppressor tRNA continues to recognize its usual codon, UGG.

A related response is seen with a eukaryotic tRNA. Bovine liver contains a tRNA^Ser with the anticodon ^mCCA. The wobble rules predict that this tRNA should respond to the tryptophan codon UGG; but in fact it responds to the termination codon UGA. Thus it is possible that UGA is suppressed naturally in this situation.

The general importance of these observations lies in the demonstration that *codon-anticodon recognition of either wild-type or mutant tRNA cannot be predicted entirely from the relevant triplet sequences, but is influenced by other features of the molecule.*

All of these nonsense suppressors are created by point mutation in a structural gene coding for the tRNA. In principle, however, suppressor tRNAs also could be caused by changes in the modification of bases in the anticodon. Such changes could result from mutation of the gene coding for a tRNA-modifying enzyme.

Missense mutations alter a codon representing one amino acid into a codon representing another amino acid, one that cannot function in the protein in place of the original residue. (Formally, any substitution of amino acids constitutes a missense mutation, but in practice it is detected only if it changes the activity of the protein.) The mutation can be suppressed by the insertion either of the original amino acid or of some other amino acid that is acceptable to the protein.

Table 8.5

Nonsense suppressor tRNAs are generated by mutations in the anticodon.

Locus	tRNA	Wild Type		Suppressor	
		Codons recognized	Anticodon	Anticodon	Codons recognized
supD (su1)	Ser	UCG	CGA	CUA	UAG
supE (su2)	Gln	CAG	CUG	CUA	UAG
supF (su3)	Tyr	UAU/C	GUA	CUA	UAG
supC (su4)	Tyr	UAU/C	GUA	UUA	UAG/A
supG (su5)	Lys	AAA/G	UUU	UUA	UAG/A
supU (su7)	Trp	UGG	CCA	UCA	UGG/A

Nomenclature:

The loci were originally known as *su1, su2,* etc., as indicated in parentheses, with su^+ indicating the suppressor mutant form of the gene, and su^- indicating the wild-type form. The loci are now named *sup* and lettered from *A* to *V*. All the loci described here are structural genes coding for tRNAs.

Figure 8.15

Missense suppression occurs when the anticodon of tRNA is mutated so that it responds to the "wrong" codon. Mutation of a codon from GGA to AGA causes the wild-type Gly to be replaced by Arg in the mutant protein. If the anticodon in a tRNA^{Gly} is mutated from UCC to UCU, it inserts Gly in response to the codon AGA that should represent Arg. Note that both the wild-type tRNA^{Arg} and the suppressor tRNA^{Gly} will respond to AGA, so that suppression is only partial.

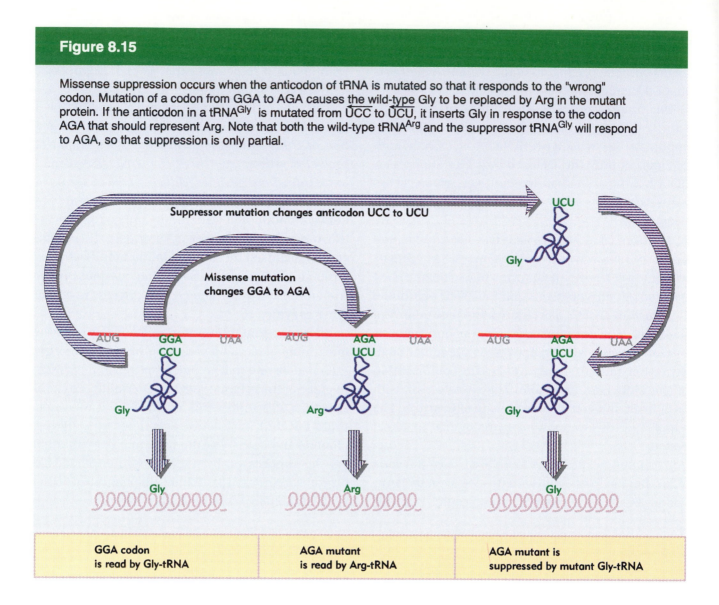

| GGA codon is read by Gly-tRNA | AGA mutant is read by Arg-tRNA | AGA mutant is suppressed by mutant Gly-tRNA |

Figure 8.15 demonstrates that missense suppression can be accomplished in the same way as nonsense suppression, by mutating the anticodon of a tRNA carrying an acceptable amino acid so that it responds to the mutant codon. Thus missense suppression involves a change in the meaning of the codon from one amino acid to another.

An alternative way to create missense suppressors is for a mutation in the tRNA (or synthetase) to change the amino acid with which a tRNA is charged. Although the tRNA would continue to respond to the same codon(s), the insertion of a different amino acid might cause suppression. This type of suppression is rather unusual, probably because it demands a mutation that *creates* a new recognition site.

Most suppressors have been characterized in *E. coli*, although some comparable mutants have been obtained in other bacteria (notably *S. typhimurium*). Much less is known about the occurrence and possible suppression of nonsense and missense mutants in eukaryotes, but generally the picture is probably similar for ochre and amber codons, although there are hints that UGA sometimes is partially suppressed during natural translation.

There is an interesting difference between the usual recognition of a codon by its proper aminoacyl-tRNA and the situation in which mutation allows a suppressor tRNA to recognize a new codon. In the wild-type cell, *only one meaning can be attributed to a given codon*, which represents either a particular amino acid or a signal for termination. But in a cell carrying a suppressor mutation, the mutant codon has the *alternatives* of being recognized by the suppressor tRNA or of being read with its usual meaning.

A nonsense suppressor tRNA must compete with the release factors that recognize the termination codon(s). A missense suppressor tRNA must compete with the tRNAs that respond properly to its new codon. The extent of competition will influence the efficiency of suppression; so the effectiveness of a particular suppressor depends not only on the affinity between its anticodon and the target codon, but also on its concentration in the cell, and on the parameters governing the competing termination or insertion reactions.

The efficiency with which any particular codon is read is influenced by its location. Thus the extent of nonsense suppression by a given tRNA can vary quite widely, depending on the **context** of the codon. We do not understand the effect that neighboring bases in mRNA have on codon–anticodon recognition, but the context can change the frequency with which a codon is recognized by a particular tRNA by more than an order of magnitude. The base on the 3′ side of a codon appears to have a particularly strong effect.

A nonsense suppressor is isolated by its ability to respond to a mutant nonsense codon. But the same triplet sequence constitutes one of the normal termination signals of the cell! The mutant tRNA that suppresses the nonsense mutation must in principle be able to suppress natural termination at the end of any gene that uses this codon. **Figure 8.16** shows that this **readthrough** results in the synthesis of a longer protein, with additional C-terminal material. The extended protein will end at the next termination triplet sequence found in the phase of the reading frame. Any extensive suppression of

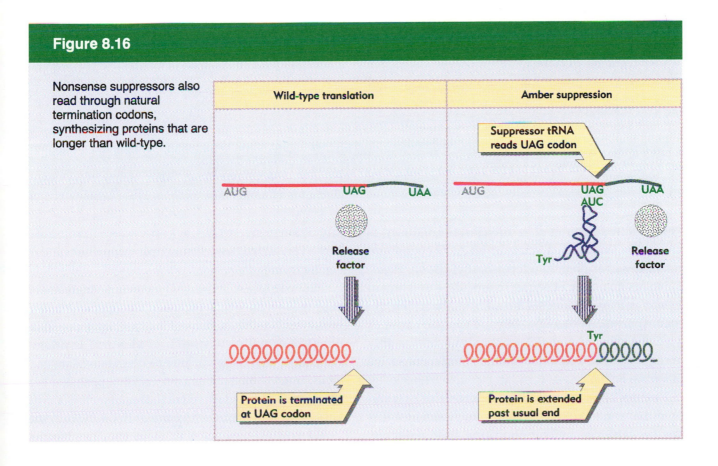

Figure 8.16

Nonsense suppressors also read through natural termination codons, synthesizing proteins that are longer than wild-type.

Wild-type translation | Amber suppression

Suppressor tRNA reads UAG codon

AUG   UAG   UAA

Release factor

Protein is terminated at UAG codon

AUG   UAG   UAA
       AUC

Tyr

Release factor

Tyr

Protein is extended past usual end

termination is likely to be deleterious to the cell by producing extended proteins whose functions are thereby altered.

Amber suppressors tend to be relatively efficient, usually in the range of 10–50%, depending on the system. If we suppose that the cell could not tolerate such a high level of readthrough at the ends of natural proteins, this indicates that amber codons are used infrequently to terminate protein synthesis in *E. coli*. Indeed, they turn out to be the least frequently used of all three termination codons.

Ochre suppressors are difficult to isolate. They are always much less efficient, usually with activities below 10%. All ochre suppressors grow rather poorly, which indicates that suppression of both UAA and UAG is damaging to *E. coli*, probably because the ochre codon is used most frequently as a natural termination signal.

UGA is the least efficient of the termination codons in its natural function; it is misread by Trp-tRNA as frequently as 1–3% in wild-type situations. In spite of this deficiency, however, it is used more commonly than the amber triplet to terminate bacterial genes.

One gene's missense suppressor is likely to be another gene's mutator. A suppressor corrects a mutation by substituting one amino acid for another at the mutant site. But in other locations, the same substitution will replace the wild-type amino acid with a new amino acid. The change may inhibit normal protein function.

This poses a dilemma for the cell: it must suppress what is a mutant codon at one location, while failing to change too extensively its normal meaning at other locations. The absence of any strong missense suppressors is therefore explained by the damaging effects that would be caused by a general and efficient substitution of amino acids.

A mutation that creates a suppressor tRNA can have two consequences. First, it allows the tRNA to recognize a new codon. Second, sometimes it *prevents* the tRNA from recognizing the codons to which it previously responded. It is significant that all the high-efficiency amber suppressors are derived by mutation of one copy of a redundant tRNA set. In these cases, the cell has several tRNAs able to respond to the codon originally recognized by the wild-type tRNA. Thus the mutation does not abolish recognition of the old codons, which continue to be served adequately by the tRNAs of the set. (In the unusual situation in which there is only a single tRNA that responds to a particular codon, any mutation that prevents the response is lethal.)

tRNA may influence the reading frame

The reading frame of a messenger usually is invariant. Translation starts at an AUG codon and continues in triplets to a termination codon. Reading takes no notice of sense: insertion or deletion of a base causes a frameshift mutation, in which the reading frame is changed beyond the site of mutation. Ribosomes and tRNAs continue ineluctably in triplets, synthesizing an entirely different series of amino acids.

Frameshift suppression restores the original reading frame. We have already discussed how this can be achieved by compensating base deletions and insertions within a gene (see Chapter 4). However, *extragenic* frameshift suppressors also can be found.

The simplest type of external frameshift suppressor corrects the reading frame when a mutation has been caused by inserting an additional base within a stretch of identical residues. For example, a G may be inserted in a run of several contiguous G bases. The frameshift suppressor is a tRNA^{Gly} that has an extra base inserted in its anticodon loop, converting the anticodon from the usual triplet sequence CCC to the quadruplet sequence CCCC. The suppressor tRNA recognizes a 4-base 'codon.'

Some frameshift suppressors can recognize more than one 4-base 'codon'. For example, a bacterial tRNA^{Lys} suppressor can respond to either AAAA or AAAU, instead of the usual codon AAA. Another suppressor can read any 4-base 'codon' with ACC in the first three positions; the next base is irrelevant. In these cases, the alternative bases that are acceptable in the fourth position of the longer 'codon' are not related by the usual wobble rules. The suppressor tRNA probably recognizes a 3-base codon, but for some other reason—most likely steric hindrance—the adjacent base is blocked. This forces one base to be skipped before the next tRNA can find a codon.

A related type of frameshift suppression has been found in the yeast mitochondrion, in which mutations that insert or delete a single T residue in a run of five T residues all are leaky. The reason seems to be that the tRNA^{Phe} responding to the UUU codon sometimes allows the ribosome to 'slip' a base in either direction, thus suppressing the mutation. This slippage seems to be a normal, if infrequent, occurrence (<5%).

The existence of frameshifting suggests a uniform model for control of ribosome translocation. When the ribosome moves, it shifts the tRNA that was in the A site so that now it properly occupies the P site. When a tRNA is bound to a 4-base 'codon', or if it extends sterically beyond the usual 3-base codon, the next aminoacyl-tRNA binds to a triplet that is out of phase in the A site. But when translocation occurs, the ribosome places this species properly into the P site, so that subsequent aminoacyl-tRNAs encounter a triplet codon in the A site in the usual way. Thus the geometry of the tRNA/mRNA interaction is used to measure distance for the ribosome.

Situations in which an occasional frameshift is a normal event are presented by phages and viruses. Such events may affect the continuation or termination of protein synthesis.

In phage MS2, a frameshift causes the ribosome to recognize a termination codon at an early position in its new reading frame. The terminating ribosome then can recognize the initiation codon of the lysis gene, which lies just a few bases farther along. When the ribosome does not terminate, it reads right over the lysis gene initiation codon. *So the frameshift-dependent termination event is a prerequisite for initiation of lysis gene expression.*

In retroviruses, translation of the first gene is terminated by a nonsense codon in phase with the reading frame. The second gene lies in a different reading frame, and (in some viruses) is translated by a frameshift that changes into the second reading frame and therefore bypasses the termination codon (see Chapter 35). **Figure 8.17** illustrates the similar situation of the yeast Ty element, in which the nonsense codon of *tya* must be bypassed by a frameshift in order to read the subsequent *tyb* gene.

Such situations makes the important point that *the rare (but predictable) occurrence of 'misreading' events can be relied on as a necessary step in natural translation.*

These frameshifting events represent a response of the normal translation apparatus to a particular sequence in the mRNA. A related phenomenon that casts some light on the basis of the effect is the ability to cause a shift in phase by manipulating *in vitro* systems that contain only normal components. A vast excess or a deficiency of a particular aminoacyl-tRNA can produce frameshifts. Probably the long delay in responding to the 3-base codon in the A site allows a tRNA that recognizes an overlapping (out-of-phase) codon to gain access.

Context-dependent effects, such as the use of unusual codons or the formation of secondary structure in the mRNA, are sometimes necessary to produce frameshifting *in vivo*. The common feature is that the ribosome is made to pause, allowing time for the tRNA frameshift to occur. In the case of retroviruses, a −1 frameshift requires secondary structure in the mRNA (which takes the form of a 'pseudoknot', a particular conformation of RNA). In the case of the frameshift illustrated in Figure 8.17, a particular pair of codons is required, one recognized by a scarce Arg-tRNA (which causes a delay), the other recognized by the Leu-tRNA that shifts +1 in reading frame.

Figure 8.17

A +1 frameshift is required for expression of the *tyb* gene of the yeast Ty element. The shift occurs at a 7 base sequence at which two Leu codon(s) are followed by a scarce Arg codon.

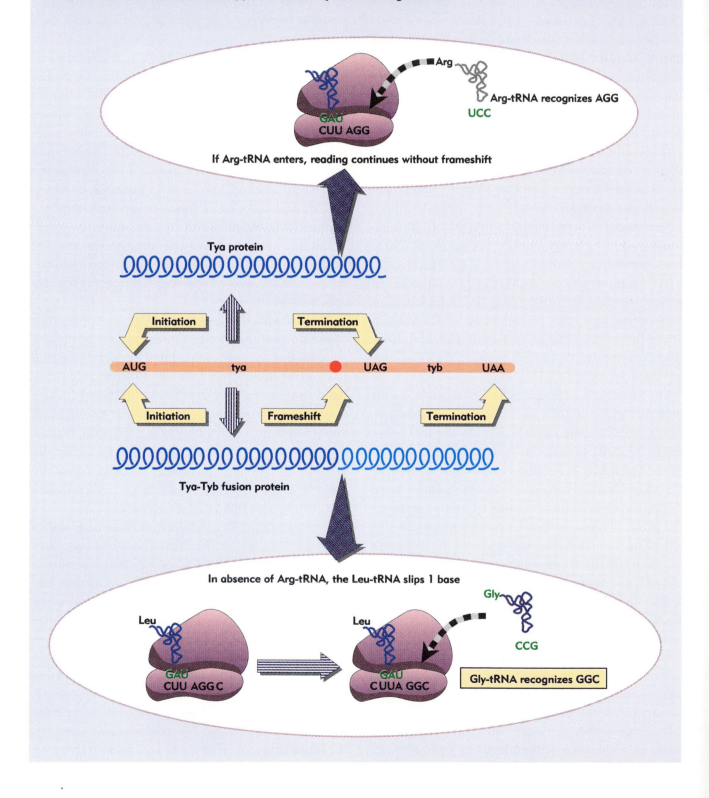

tRNA transcripts are cut and trimmed from clusters by several enzymes

We have only an incomplete picture of the organization of tRNA genes in either bacteria or eukaryotes, but there seems to be a tendency for tRNA genes to lie in clusters.

In eukaryotes, the scanty information available is provided by cases where genomic sequences corresponding to a particular tRNA probe have been cloned. Often there turn out to be sequences close by that code for other tRNAs. In contrast to tandem gene clusters, the tRNAs coded in the same region all may be different. We do not know whether the clustered tRNA genes are transcribed independently or form a common unit of gene expression.

A similar situation is seen in *E. coli*, in which tRNA genes form clusters, each of which is concerned with several amino acid acceptors. At least some of them form single units of expression, being transcribed into long precursor RNAs.

In every known case, a common principle applies. *Each tRNA is formed as part of a longer precursor RNA from which it is released by cleavage and other processing reactions.*

Two of the tRNA gene clusters in *E. coli* have been characterized in detail, and both prove to contain other sequences as well as the tRNAs. These two clusters are named for their possession of tyrosine tRNA genes. The *tyrU* cluster includes the genes for four tRNAs, $tRNA^{Thr}_4$, $tRNA^{Tyr}_2$, $tRNA^{Gly}_2$, and $tRNA^{Thr}_3$. The *tyrT* cluster contains two identical genes coding for $tRNA^{Tyr}_1$.

In each cluster, a single startpoint for transcription is located before the first tRNA gene, as shown in the example of **Figure 8.18**. Beyond the last tRNA gene, there is a sequence coding for protein. In the

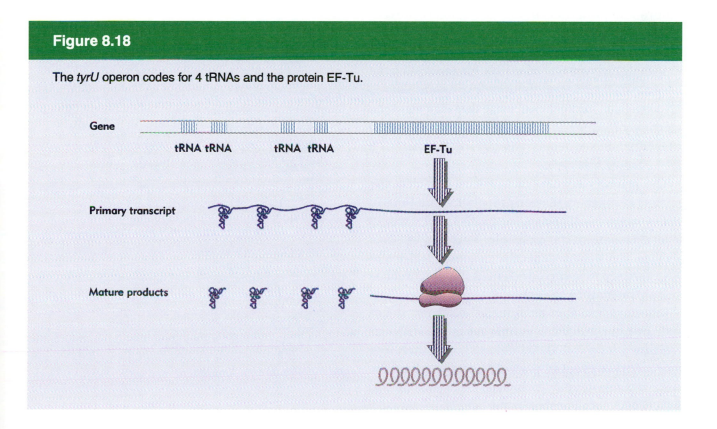

Figure 8.18

The *tyrU* operon codes for 4 tRNAs and the protein EF-Tu.

Gene

tRNA tRNA   tRNA tRNA   EF-Tu

Primary transcript

Mature products

tyrU cluster this is the gene *tufB* (one of two coding for the protein synthesis factor EF-Tu). In the *tyrT* cluster, the coding sequence represents protein P, a protamine-like polypeptide (protamines are very basic proteins associated with DNA in some spermatozoa).

Presumably the mRNA for each of these proteins is generated by cleavage of the primary transcript between the last tRNA sequence and the start of the structural gene. There is a short common sequence purpose.

Processing is a necessary step in the production of all known tRNAs. In bacteria, tRNAs are synthesized in the form of precursors, containing one or more tRNAs, with additional sequences at both the 5′ and 3′ ends. In eukaryotic cytoplasm, newly synthesized tRNAs sediment at ~4.5S, corresponding to molecules of roughly 100 nucleotides or so, compared with the mature size of 4S tRNA at 70–80 nucleotides.

So far as we know, the tRNA part of the precursor usually takes up the same secondary structure that is displayed by mature tRNA; this is an important feature in its recognition by ribonucleases during maturation. Mutations in the tRNA moiety prevent processing, with the result that defective mature tRNAs do not accumulate.

A single enzyme is responsible for generating the 5′ end of all tRNA molecules in *E. coli*. This is **RNAase P**, an endoribonuclease that cleaves at the junction of either the 5′ leader sequence or the intercistronic sequence (when a tRNA is part of a polycistronic transcript).

The primary sequences in which the cleavage occurs are different in each precursor. Mutations in tRNAs that prevent processing by RNAase P occur in several regions of the molecule. This suggests that RNAase P must recognize a common feature of tRNA tertiary structure, present in the precursor as well as the mature molecule.

Ribonuclease P is an unusual enzyme. It has both protein and RNA components. The RNA is 375 bases long (~130,000 daltons), and therefore represents a greater part of the mass than the protein component, which is only ~20,000 daltons. Both the RNA and protein components of RNAase P

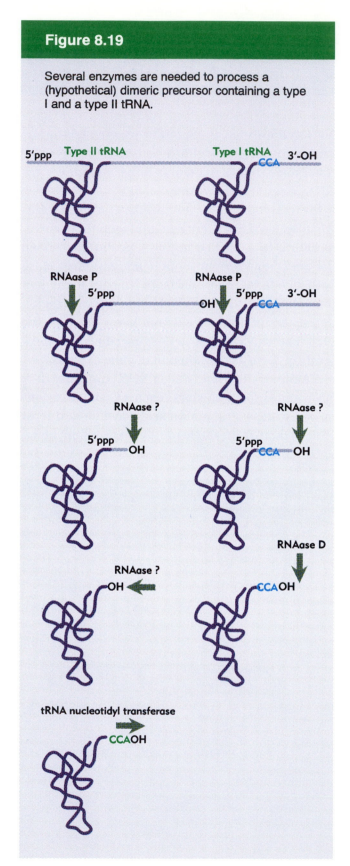

Figure 8.19

Several enzymes are needed to process a (hypothetical) dimeric precursor containing a type I and a type II tRNA.

are needed for its catalytic activity, although the basic responsibility for the cleavage reaction rests with the RNA (see Chapter 32). Potential counterparts to RNAase P have been found in eukaryotes.

In *E. coli* mutants that lack RNAase P, many precursor tRNA molecules accumulate. Some of the polycistronic precursors are cleaved into individual monocistronic precursors, although none has an authentic 5′ terminus. Such cleavage indicates the existence of another enzyme able to cleave in the intercistronic regions. The enzyme responsible may be RNAase III.

Why are some but not other precursors cleaved to monocistronic precursors in the absence of RNAase P? Even if the same set of enzymes are involved in cleavage of all tRNA precursors, they may act in a different order in different precursors. This idea is borne out by studies of individual precursors, in which analogous events occur at different stages. The compulsion for a particular order of events may be that only one cleavage site is accessible in the original transcript, but its cleavage then changes the structure to expose the site for the next enzyme.

The other steps in tRNA maturation are not so well-established. A summary of the necessary events is given in **Figure 8.19**.

Endonucleolytic cleavage probably occurs also on the 3′ side of tRNA sequences. We do not yet know which enzyme(s) is responsible.

The actual 3′ terminus is generated by an exonucleolytic activity. An enzyme called **RNAase D** removes bases one at a time from the 3′ end (possibly of the original precursor, possibly generated by cleavage) until it arrives at the -CCA terminus common to all mature tRNAs. The enzyme seems to recognize the overall tRNA structure rather than the CCA sequence itself as a signal to stop, because it is able to remove an additional CCA sequence added to the authentic 3′-terminal sequence. RNAase D is a randomly acting endonuclease.

Another enzyme with exonucleolytic activity is **RNAase II**. Once thought to be involved in tRNA processing, it is able to degrade tRNA sequences completely. So probably it is concerned with general degradation rather than with specific maturation. Its action is processive.

In bacteria, two types of tRNA precursor are distinguished by their 3′ sequences. Type I molecules have a CCA triplet that is the demarcation between the mature tRNA sequence and the additional 3′ material. But type II molecules (coded by certain phages) have no CCA sequence. After the additional 3′ nucleotides have been removed from the precursor, the CCA must be added. We do not know how the 3′ terminus for CCA addition is generated in the type II tRNAs, whether it is a function of RNAase D like the type I tRNAs, or whether a different enzyme is involved. In eukaryotes, probably all tRNAs are of type II.

The addition of CCA is the function of the enzyme **tRNA nucleotidyl transferase**, which adds the triplet (or part of it) to any tRNA sequence lacking it. Mutants of *E. coli* with reduced levels of this enzyme (called *cca*−) grow slowly, which suggests that it is essential for tRNA biosynthesis. The terminal A residue of all bacterial tRNAs turns over constantly, so this enzyme is needed also in a repair function as well as for the initial production of type II tRNAs.

Summary

All tRNAs form the secondary structure of a cloverleaf with four (sometimes five) arms. The secondary structure folds into a compact L-shaped tertiary structure, with the anticodon at one extremity and the amino acid at the other.

Multiple tRNAs may respond to a particular codon. The set of tRNAs responding to the various codons for each amino acid is distinctive for each organism. Each amino acid is recognized by a particular aminoacyl-tRNA synthetase, which also recognizes all of the tRNAs coding for that amino acid. Aminoacyl-tRNA synthetases have a proofreading function that scrutinizes the aminoacyl-tRNA products, and hydrolyzes incorrectly joined aminoacyl-tRNAs.

Aminoacyl-tRNA synthetases vary widely, but fall into two general groups according to the structure of the catalytic domain. Synthetases of each group bind the tRNA from the side, making contacts principally with the extremities of the acceptor stem and the anticodon stem–loop; the two types of synthetase bind tRNA from opposite sides. The relative importance attached to the acceptor stem and the anticodon region for specific recognition varies with the individual tRNA.

Codon–anticodon recognition involves wobbling at the first position of the anticodon (third position of the codon), which allows some tRNAs to recognize multiple codons. All tRNAs have modified bases, introduced by enzymes that recognize target bases in the tRNA structure. Codon–anticodon pairing is influenced by modifications of the anticodon itself and also by the context of adjacent bases, especially on the 3′ side of the anticodon. Taking advantage of codon–anticodon wobble allows vertebrate mitochondria to use only 22 tRNAs to recognize all codons, compared with the usual minimum of 31 tRNAs; this is assisted by changes in the genetic code in mitochondria. Changes in the code are found also in certain other situations.

Mutations may allow a tRNA to read different codons; the most common form of such mutations occurs in the anticodon itself. Alteration of its specificity may allow a tRNA to suppress a mutation in a gene coding for protein. A tRNA that recognizes a termination codon provides a nonsense suppressor; one that changes the amino acid responding to a codon is a missense suppressor. Suppressors of UAG and UGA codons are more efficient than those of UAA codons, which is explained by the fact that UAA is the most commonly used natural termination codons. But the efficiency of all suppressors depends on the context of the individual target codon.

Some tRNAs appear to read 'codons' of four bases, thus generating a frameshift. Such effects suggest that codon–anticodon recognition is involved in setting the distance that the ribosome moves in a translocation event. Frameshifts determined by the mRNA sequence may be required for expression of natural genes.

Further reading

Historical
The first sequence of a tRNA was reported by **Holley** *et al.* (*Science* **147**, 1462–1465, 1965).

Reviews
The literature on tRNA has been well served by some review books. In **Altman's** (Ed.) *Transfer RNA* (MIT Press, Cambridge, 1978), primary and secondary structure were discussed by **Clark** (pp. 14–47) and tertiary structure by **Kim** (pp. 248–293). Modified bases were discussed by **Nishimura** (pp. 168–195).

A two volume set of reviews and research papers also gives a broad view of tRNA: **Schimmel, Soll and Abelson**, *Transfer RNA: Structure, Properties and Recognition*, and **Soll, Abelson and Schimmel**, *Transfer RNA: Biological Aspects* (Cold Spring Harbor Lab., New York, 1979).

Modification of bases in tRNA has been reviewed by **Bjork** (*Ann. Rev. Biochem.* **56**, 263–287, 1987).

The variety of aminoacyl-tRNA synthetases has been discussed by **Schimmel** (*Ann. Rev. Biochem.* **56**, 125–158, 1987), who has also analyzed their recognition of tRNA (*Biochemistry* **28**, 2747–2759, 1988). Features responsible for tRNA identity have been summarized by **Normanly and Abelson** (*Ann. Rev. Biochem.* **58**, 1029–1049, 1989).

Variations in the genetic code have been summarized by **Fox** (*Ann. Rev. Genet.* **21**, 67–91, 1987). The use of UGA to represent seleno-cysteine has been reviewed by **Bock** *et al.* (*Trends. Biochem. Sci.* **16**, 463–467, 1991).

Suppression has been reviewed by **Murgola** (*Ann. Rev. Genet.* **19**, 57–80, 1985) and by **Eggertsson and Soll** (*Microbiol. Rev.* **52**, 354–374, 1988). Options for frameshifting have been reviewed by **Atkins** *et al.* (*Ann. Rev. Genet.* (**25**, 201–228, 1991).

Discoveries
The first look at a class I tRNA•synthetase crystal structure was provided by **Rould** *et al.* (*Science* **246**, 1135–1142, 1989). A comparison with class II structure was provided by **Ruff** *et al.* (*Science* **252**, 1682–1689, 1991).

Proofreading *in vivo* was followed by **Jakubowski** (*Proc. Nat. Acad. Sci* **87**, 4504–4508, 1990).

CHAPTER 9

Ribosomes provide a translation factory

The ribosome behaves like a small migrating factory that travels along the template engaging in rapid cycles of peptide bond synthesis. Aminoacyl-tRNAs shoot in and out of the particle at a fearsome rate, depositing amino acids; and elongation factors cyclically associate and dissociate. Together with its accessory factors, the ribosome provides the full range of activities required for all the steps of protein synthesis.

Ribosomes are a major cellular component. In an actively growing bacterium, there are roughly 20,000 ribosomes per genome (see Table 7.1). They contain ~10% of the total bacterial protein and account for ~80% of the total mass of cellular RNA. In eukaryotic cells, their proportion of total protein is less, but their absolute number is greater, and still they contain most of the mass of RNA of the cell. The number of ribosomes (in prokaryotes or eukaryotes) is directly related to the protein-synthesizing activity of the cell.

Bacterial ribosomes are attached to mRNAs that are themselves still connected with the DNA (because translation starts at the 5′ end of the mRNA before transcription has completed synthesis of the 3′ end). In eukaryotic cytosol, the ribosomes commonly are associated with the cytoskeleton or with membranes of the endoplasmic reticulum. The common feature is that ribosomes engaged in translation are not free in the cell, but are associated, directly or indirectly, with cellular structures.

All ribosomes can be dissociated into two subunits, one roughly twice the size of the other. Each subunit contains a major rRNA component and many different protein molecules, almost all present in only one copy per subunit. The larger subunit contains smaller RNA molecule(s) in addition to the major rRNA. A pointer to the importance of rRNA is that its sequence remains almost invariant within a cell, although it is coded by many genes. This conservation argues that there is selection against changes in its sequence.

A major function of the rRNAs is structural. Proteins bind to each major rRNA at particular sites, in the specific order required to assemble each subunit. Indeed, ribosomes are interesting not only for their function, but also for the process by which they self-assemble from the constituent RNA and protein molecules.

The ribosome possesses several active centers, each of which is constructed from a particular group of proteins associated with a region of rRNA. The active centers require the direct participation of rRNA in a structural or even catalytic role. Some catalytic functions require individual proteins, but none of the activities can be reproduced by isolated proteins or groups of proteins; in fact, there is virtually no evidence that ribosomal proteins are *directly* involved in any of the catalytic functions.

The ribosome represents a collection of many

proteins, each active only in the context of the proper overall structure, whose coordinated activities together accomplish the act of translation. Many of the proteins (and the rRNA) are concerned principally with establishing the overall structure that brings the various active sites into the right relationship; they need not necessarily participate directly in the synthetic reactions. We should think of the ribosome as an *interactive* structure, in which changes in conformation transmit information from one site to another.

Two rather different types of information are important in analyzing the ribosome. Mutations implicate particular ribosomal proteins or bases in rRNA in participating in particular reactions. Structural analysis, including both direct modification of components of the ribosome and comparisons to identify conserved features in rRNA, identifies the physical locations of components involved in particular functions.

Ribosomes are compact particles in which most proteins interact with rRNA

Ribosomes can be constructed in several ways. There are appreciable variations in the overall size and proportions of RNA and protein in the ribosomes of bacteria, eukaryotic cytoplasm, and organelles. **Figure 9.1** compares the components of bacterial and mammalian ribosomes. Both are ribonucleoprotein particles that contain more RNA than protein. The ribosomal proteins are known as **r-proteins**.

All ribosomes in a bacterium are identical. In *E. coli,* the ribosomal RNAs have been sequenced, and the amino acid sequences of the proteins have been determined. The small (30S) subunit consists of the 16S rRNA and 21 proteins. The large (50S) subunit contains 23S rRNA, the small 5S RNA, and 31 proteins. With the exception of one protein present at four copies per ribosome, there is one copy of each protein.

The ribosomes of higher eukaryotic cytoplasm are larger than those of bacteria. The total content of both RNA and protein is greater; the major RNA molecules are longer, and there are more proteins. Probably most or all of the proteins are present in stoichiometric amounts. RNA is still the pre-

dominant component by mass.

Organelle ribosomes are distinct from the ribosomes of the cytosol, and take varied forms. The largest are almost the size of bacterial ribosomes and have 70% RNA; the smallest are only 60S and have <30% RNA.

The shapes of bacterial, chloroplast, and eukaryotic cytoplasmic ribosomes are generally similar (mitochondrial ribosomes have not been well characterized). **Figure 9.2** is a diagrammatic representation of the subunits of the bacterial ribosome. The small subunit has a somewhat flat shape, and the large subunit has a more spherical structure.

The complete 70S ribosome has an asymmetrical construction. The partition between the head and body of the small subunit is aligned with the notch of the large subunit, so that the platform of the small subunit fits into the large subunit. There could be a space or 'tunnel' between the subunits.

Some electron micrographs of subunits and complete bacterial ribosomes are shown in **Figure 9.3**, together with models in the corresponding orientation.

Figure 9.1

Ribosomes are large ribonucleoprotein particles that contain more RNA than protein and dissociate into large and small subunits.

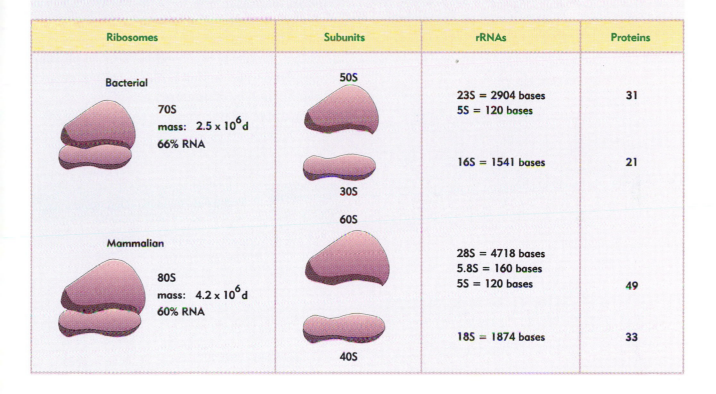

Ribosomes	Subunits	rRNAs	Proteins
Bacterial 70S mass: 2.5×10^6 d 66% RNA	50S	23S = 2904 bases 5S = 120 bases	31
	30S	16S = 1541 bases	21
Mammalian 80S mass: 4.2×10^6 d 60% RNA	60S	28S = 4718 bases 5.8S = 160 bases 5S = 120 bases	49
	40S	18S = 1874 bases	33

The apparent structure of mammalian cytoplasmic ribosomes is similar, although they are larger. There is a suggestive similarity between the structures of bacterial and eukaryotic small subunits; the large subunits seem more divergent. Although a detailed model has not yet been constructed, the relationship between the mammalian subunits seems similar to that of prokaryotic ribosomes.

Much attention has been paid to the location of mRNA in the ribosome. One popular idea is that it fits between or close to the junction of the subunits. Some models depict the mRNA as 'threaded' through the ribosome, but its location is just as likely to be superficial. As evident from **Figure 9.4**, the two tRNAs are quite large relative to the ribosome; they seem more likely to be inserted into large 'clefts' opening from the surface than to be inserted completely into holes in the interior.

The RNAs constitute the major part of the mass of the ribosome. Their presence is pervasive, and probably most or all of the ribosomal proteins actually contact rRNA. Thus the major rRNAs form what is sometimes thought of as the backbone of each subunit, a continuous thread whose presence dominates the structure, and which determines the positions of the ribosomal proteins.

Both major rRNAs have considerable secondary structure. Although the sequence of an RNA can be used to predict the formation of base-paired regions, in molecules as large as the rRNAs there are many alternative conformations. It is not possible to predict which would be chosen even if the molecule were simply free in solution. Different

Figure 9.2

The 30S subunit has an elongated and asymmetrical shape, the 50S subunit has a fairly compact body, from which a "central protuberance" and a "stalk" stick out, and the 70S ribosome may be held together by associations between discrete areas of the two subunits, as indicated in these two views (which display the subunit rotated by 90°).

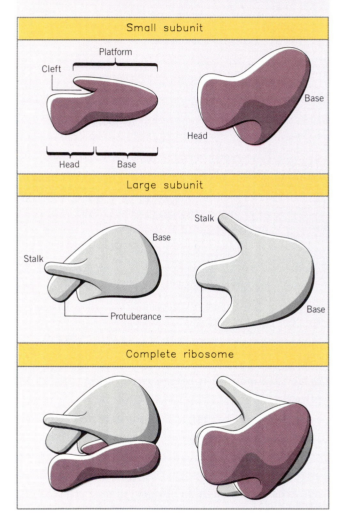

Small subunit

Large subunit

Complete ribosome

ture are conserved; so if a base pair is required, it can form at the same relative position in each rRNA. From such comparisons, models have been constructed for both 16S and 23S rRNA.

The current model for *E. coli* 16S rRNA is illustrated in diagrammatic form in **Figure 9.5**. The molecule forms four general domains, in which just under half of the sequence is base-paired. The individual double-helical regions tend to be short (<8 bp). Often the duplex regions are not perfect, but contain bulges of unpaired bases.

Comparable models have been drawn for mito-

Figure 9.3

Electron microscopic images of bacterial ribosomes and subunits reveal their shapes. Photographs kindly provided by James Lake.

models can be distinguished by their predictions of the relative stabilities of particular base-paired regions; but data of this type are somewhat limited in scope.

The most penetrating approach to analyzing secondary structure is to compare the sequences of corresponding rRNAs in related organisms. Those regions that are important in the secondary struc-

Figure 9.4

Size comparisons show that the ribosome is large enough to bind two tRNAs (as well as 40 bases of mRNA).

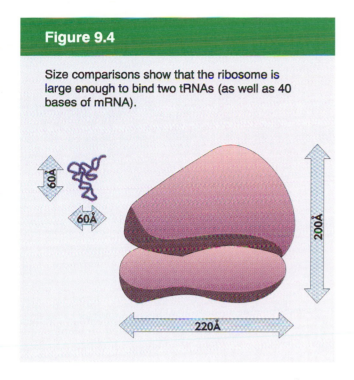

chondrial rRNAs (which are shorter and have fewer domains) and for eukaryotic cytosolic rRNAs (which are longer and have more domains). The increase in length in eukaryotic rRNAs is due largely to the acquisition of sequences representing additional domains.

How well does this structure correspond with the organization of rRNA in the ribosomal subunit? Methods that have been used to investigate structure in the subunit include determining which bases can react with kethoxal (a reagent that attacks unpaired guanines), which regions become cross-linked to each other on treatment with the reagent psoralen, where preferential sites of cleavage by nucleases are located, and also which regions of the rRNA can be linked to specific proteins. The results are consistent with the model for free rRNA structure, but there is less base pairing in the subunit than in the model.

Is the structure of rRNA in the subunit invariant? Some differences in the reactivity of 16S rRNA are found when 30S subunits are compared with 70S ribosomes; also there are some differences between free ribosomes and those engaged in protein

synthesis. This suggests that changes in the conformation of the rRNA occur when mRNA is bound, when the subunits associate, or when tRNA is bound. We do not know whether such changes reflect a direct interaction of the rRNA with mRNA or tRNA, or are caused indirectly by other changes in ribosome structure. The main point is that ribosome conformation is flexible during protein synthesis.

A feature of the primary structure of rRNA is the presence of methylated residues, sometimes in regions that are well-conserved. There are ~10 methyl groups in 16S rRNA (located mostly toward the 3′ end of the molecule) and probably ~20 in 23S rRNA. In mammalian cells, the 18S and 28S rRNAs carry 43 and 74 methyl groups, respectively, so ~2% of the nucleotides are methylated (about three times the proportion methylated in bacteria).

The large ribosomal subunit also contains a molecule of a 120 base **5S RNA** (in all ribosomes except those of mitochondria). Prokaryotic 5S RNAs show some conservation of sequence, especially in the regions that bind to ribosomal proteins. Similarly, there is conservation of eukaryotic 5S RNAs, with one sequence (for example) predominating in the mammals.

All 5S RNA molecules display a highly base-paired structure, although the exact organization could not be settled by direct analysis of the RNA (more than 20 models were proposed). This emphasizes the difficulty of distinguishing between alternative pairing possibilities even in quite small molecules. As with the large rRNAs, a model was resolved by comparisons between the structures that can be formed in 5S RNAs of different species.

In eukaryotic cytosolic ribosomes, another small RNA is present in the large subunit. This is the **5.8S RNA**. Its sequence corresponds to the 5′ end of the prokaryotic 23S rRNA.

The major and smaller rRNA molecules both bind proteins at specific sites. The positions on the rRNA at which proteins bind can be defined by characterizing the parts of the nucleic acid that are protected from cleavage by nucleases. Such experiments give a linear map of sites on the rRNA. A common feature of the binding sites is the presence

Figure 9.5

Four domains of 16S rRNA each contain many short double-stranded regions.

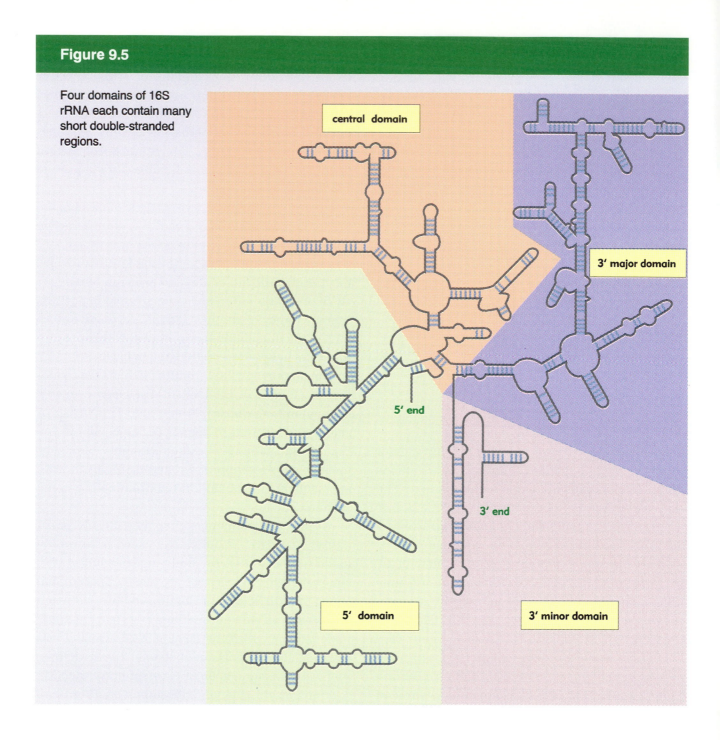

central domain

3' major domain

5' end

3' end

5' domain

3' minor domain

of secondary structure, often a hairpin whose duplex stem contains unpaired bulges.

Some ribosomal proteins bind strongly to isolated rRNA. Some do not bind to free rRNA, but can bind after other proteins have bound. This suggests that the conformation of the rRNA is important in determining whether binding sites exist for some proteins. As each protein binds, it induces conformational changes in the rRNA that make it possible for other proteins to bind. In *E. coli*, virtually all the ribosomal proteins can bind (albeit with varying affinities) to one of the rRNAs.

Techniques used to investigate ribosome structure have included identifying the locations of individual ribosomal proteins by immune electron microscopy, determining the relationships of

Figure 9.6

A diagrammatic representation of the 30S subunit shows that each domain of rRNA occupies a discrete location, and the positions of all ribosomal proteins have been mapped. Note that this is a two dimensional projection of a three dimensional reconstruction and is not to scale.

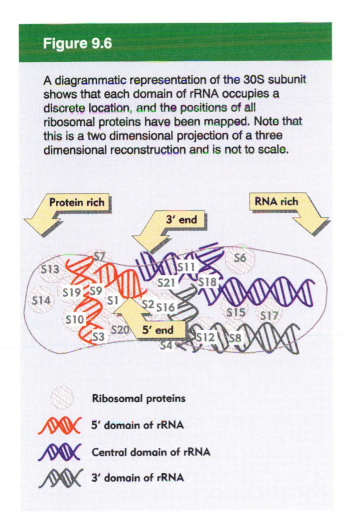

Ribosomal proteins

5' domain of rRNA

Central domain of rRNA

3' domain of rRNA

adjacent proteins by cross-linking them, analyzing the interactions between particular ribosomal proteins and rRNA, and using more biophysical approaches such as neutron scattering of subunits containing specifically deuterated proteins. A somewhat similar picture emerges from all the techniques. Each ribosomal protein can be located at a particular position in the structure; some proteins can be equated with individual features of ribosomal appearance.

Progress in mapping the locations of the ribosomal proteins and rRNA is summarized in **Figure 9.6**. We do not yet have a good impression of the shapes of the individual proteins, which are therefore represented by spheres. The configuration of rRNA in the particle has been mapped by using immune electron microscopy to identify the regions that hybridize with DNA probes complementary to short sequences of the rRNA.

The domains of rRNA identified by the secondary structure of Figure 9.5 occupy relatively discrete regions of the small subunit. One interesting feature is that the centers of mass of the protein and RNA components are displaced by ~25Å. This generates a concentration of rRNA at one end of the subunit; the subunit is relatively free of protein at the right end in Figure 9.6.

Subunit assembly is linked to topology

Ribosome assembly takes place by a series of reactions in which groups of proteins associate with rRNA, and the structure then folds so that the next group of proteins can join. Each ribosomal subunit of *E. coli* can be assembled *in vitro* from its component rRNAs and proteins. The group reactions that occur during assembly can be reversed *in vitro*.

When bacterial ribosomal subunits are centrifuged in CsCl, a discrete group of proteins (the **split proteins**) is lost from each subunit. The disso-

ciation generates 23S and 42S particle **cores** from the 30S and 50S subunits, respectively. In fact, when ribosomes are exposed to increasing concentrations of either CsCl or LiCl, groups of proteins are lost in successive disruptions. This suggests that the ribosome contains groups of cooperatively organized proteins, so that disruption of a group effectively causes loss of all its members together.

The stepwise dissociation can be reversed by

Figure 9.7

30S subunits can be dissociated into subparticles that are related to assembly intermediates. RI particles and 21S precursors have similar, but not identical, protein contents.

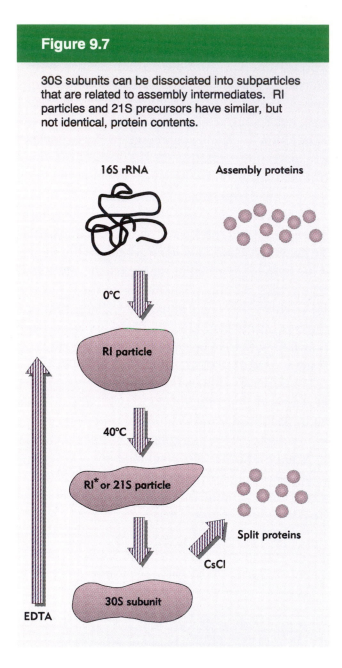

16S rRNA

Assembly proteins

0°C

RI particle

40°C

RI* or 21S particle

Split proteins

CsCl

30S subunit

EDTA

removing the CsCl and providing Mg²⁺ ions. The reconstituted subunits are active in protein synthesis. **Figure 9.7** summarizes the reconstitution procedure.

The first stage in reconstituting a 30S subunit is to assemble the 16S rRNA with a group of ~15 proteins. They react in the cold to form **RI particles**. Then these particles must be heated to allow a unimolecular rearrangement to take place,

generating **RI* particles**. Finally, the remaining 6 proteins can join.

Particles similar to RI particles can be generated by treating 30S subunits with EDTA. The chelating agent removes magnesium ions and causes the subunit to unfold and lose some proteins. The resulting particles require heating before they can reassociate with the proteins that have been lost. *So the limiting step in assembly is a temperature-dependent conversion in which the assembling particle folds into a more compact structure.*

This temperature dependence suggests that mutations blocking assembly *in vivo* might be detected as **cold-sensitive** mutants. Such *sad* (subunit assembly defective) mutants can be recovered as bacteria in which protein synthesis is prevented at low temperature, because ribosomes cannot be assembled. Also, some mutants isolated originally for other properties (such as resistance to antibiotics that act on the ribosome) have similar deficiencies.

The *sad* mutants accumulate smaller particles that are precursors in the assembly of the affected ribosome subunit. The precursor particles contain rRNA, associated with some, but not all, of the subunit proteins. A single precursor has been identified for the 30S subunit, sedimenting at ~21S; two precursors to 50S subunits sediment at 32S and 43S. Thus ribosome assembly passes through discrete stages in which groups of proteins are added to the rRNA.

The small subunit precursor found *in vivo* and the RI particles reconstituted *in vitro* have similar protein components. The RNA molecules are slightly different, because the *in vivo* precursor has a precursor form of the rRNA, while the *in vitro* particle has mature 16S rRNA. It is interesting (and fortunate) that mature rRNA can be used for assembly.

The precursor RNA is a little longer than the mature rRNA and is only slightly methylated. Probably the conformational change in the precursor is associated with removal of the surplus sequences and methylation of the rRNA; then the remaining proteins are added. Free 16S rRNA cannot be methylated *in vitro* at several positions, but the core particle is a substrate for methylase activity, which is consistent with this scheme.

The proteins present in the precursors and RI particles also are very similar to the proteins that are the last to be removed by treatment with LiCl or CsCl. This suggests that the topology of the 30S subunit reflects the assembly process. Those proteins that are part of the precursor tend to be the more secure components of the 30S subunit, and are harder to remove by abusive treatment. A naïve view would be that proteins are added to the surface of the assembling subunit, so those added last are easiest to remove.

The roles of particular components in assembly can be analyzed by attempting to reconstitute subunits in the absence of individual proteins. This type of partial reconstitution can be used to examine whether the omission of one protein affects the ability of other proteins to assemble into the particle. Essentially the protocol is to leave out one protein and then to see which other proteins fail to bind.

The *in vitro* assembly map of **Figure 9.8** was constructed by determining which proteins must be bound to 16S rRNA for another particular protein to bind. The results show that ribosomal proteins must bind to the assembling particle in a certain order. The map is read down the page, from the rRNA toward the split proteins. An arrow indicates an interaction between a protein and rRNA, or between proteins. Interactions closer to the rRNA must take place before those farther away can occur. The individual proteins in the group that initially recognizes rRNA do not bind independently, but show cooperative effects; at low concentrations of rRNA, all the proteins bind to the same rRNA molecule rather than to different molecules.

Some of the proteins that show dependent relationships in the assembly map in fact lie physically close to one another. This is consistent with a scheme in which dependence largely involves physical interactions; but clearly this is not the whole story, since (for example) the binding of one protein opens a site elsewhere on rRNA for another to bind.

Assembly of the 30S subunit probably starts while the rRNA is being transcribed. In this context, it may be significant that most of the strongest rRNA binding sites are clustered in the 5' region of the molecule, which is transcribed first. Probably by the time the rRNA is processed to the mature form, it already has several ribosomal proteins attached, as illustrated in **Figure 9.9**.

Figure 9.8

Ribosome assembly involves a series of sequential reactions. Arrows to rRNA identify the most strongly RNA-binding proteins, but others also bind. Arrows to proteins indicate that the target must be present for the second protein to bind. Proteins closest to the rRNA are those found most persistently in precursor particles; those farther away are assembled into the particle later and are removed more easily by salt.

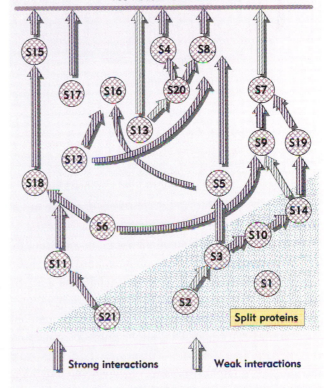

Figure 9.9

Ribosome subunit assembly is an ordered process in which changes in conformation occur as each protein is added.

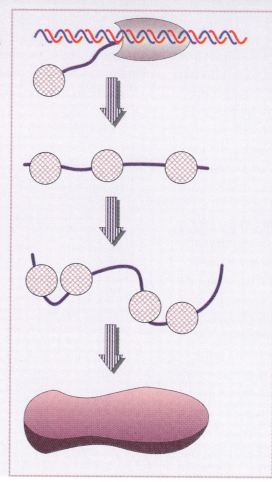

Transcription

Ribosomal proteins associate with 16S rRNA during transcription

Release of p16 rRNA

rRNA is released as p16 precursor, bound to ~50% of the r-proteins

21S precursor particle

rRNA changes conformation. ~10-20% of methyl groups are bound, more proteins bind

30S ribosome subunit

rRNA is cleaved, remaining methyl groups added, final r-proteins bind,

The role of ribosomal RNA in protein synthesis

The ribosome was originally viewed as a collection of proteins with various catalytic activities, held together by protein–protein interactions and by binding to rRNA. But the discovery of RNA molecules with catalytic activities (see Chapter 32) immediately suggests that rRNA might play a more active role in ribosome function. There is now evidence that rRNA interacts with mRNA or tRNA at each stage of translation, and that in fact the proteins are necessary to maintain the rRNA in a structure in which it can perform the catalytic functions. Several interactions involve specific regions of rRNA:

◆ The 3′ terminus of the rRNA interacts directly with mRNA at initiation.

◆ Specific regions of 16S rRNA interact directly with the anticodon regions of tRNAs in both the A site and the P site; 23S rRNA interacts with the CCA terminus of peptidyl-tRNA.

◆ Subunit interaction involves both 16S and 23S rRNAs, although it is not yet clear whether association is mediated directly by RNA–RNA interactions or by RNA–protein interactions.

Much information about the individual steps of bacterial protein synthesis has been obtained by using antibiotics that inhibit the process at particular stages. The properties of some of the more widely used inhibitors are summarized in **Table 9.1**. It is striking that many of the antibiotics act on rRNA, which suggests that it is involved with many or even all of the functions of the ribosome.

The functions of rRNA have been investigated by two types of approach. Structural studies show that particular regions of rRNA are located in important sites of the ribosome, and that chemical modifications of these bases impede particular ribosomal functions. And mutations identify bases in rRNA that are required for particular ribosomal functions. **Figure 9.10** summarizes sites in 16S rRNA that have been identified by these means.

An indication of the importance of the 3′ end of 16S rRNA is given by its susceptibility to the lethal

agent colicin E3. Produced by some bacteria, the colicin cleaves ~50 nucleotides from the 3′ end of the 16S rRNA of *E. coli*. The cleavage entirely abolishes initiation of protein synthesis. Several important functions require the region that is cleaved: binding the factor IF3; recognition of mRNA; and binding of tRNA.

The 3′ end of the 16S rRNA is directly involved in binding mRNA in the initiation reaction. It possesses a sequence of 6 bases that is (imperfectly) complementary to a sequence found just prior to each AUG initiation codon in bacterial mRNA. As a general rule, mutations that decrease complementarity between the rRNA and mRNA sequences prevent translation, and mutations that increase complementarity improve the efficiency of translation. This suggests that complementary base pairing between these sequences is involved in the initial binding of ribosomes to initiation sites on mRNA. We discuss this in more detail in Chapter 10.

Another direct role for the 3′ end of 16S rRNA in protein synthesis is shown by the properties of kasugamycin-resistant mutants, which lack certain

Table 9.1

Inhibitors of protein synthesis act at a variety of stages.

Target	Inhibitor	Effective on	Action or Nature of Resistant Mutations
Initiation	Kasugamycin	Bacteria	Mutants lack methyl groups in 16S rRNA
	Streptomycin	Bacteria	Mutations in S12 30S protein
Elongation	Kirromycin	Bacteria	EF-Tu-GDP cannot be released
	Puromycin	Bacteria	Accepts peptide chain from peptidyl-tRNA mutations occur in 23S rRNA modification
Peptidyl transferase	Erythromycin	Bacteria & mitochondria	Mutations occur in 23S rRNA sequence
	Chloramphenicol	Bacteria & mitochondria	Mutations occur in 50S proteins Mutations occur in 23S rRNA sequence
Translocation	Cycloheximide	Eukaryotic Cytoplasm	Inhibits peptidyl transferase on 60S
	Fusidic acid	Bacteria	EF-G-GDP cannot be released
	Thiostrepton	Bacteria	Binds 23S rRNA & inhibits GTPase activity

Figure 9.10

Some sites in 16S rRNA are protected from chemical probes when 50S subunits join 30S subunits or when aminoacyl-tRNA binds to the A site. Others are the sites of mutations that affect protein synthesis. TERM suppression sites may affect termination at some or several termination codons.

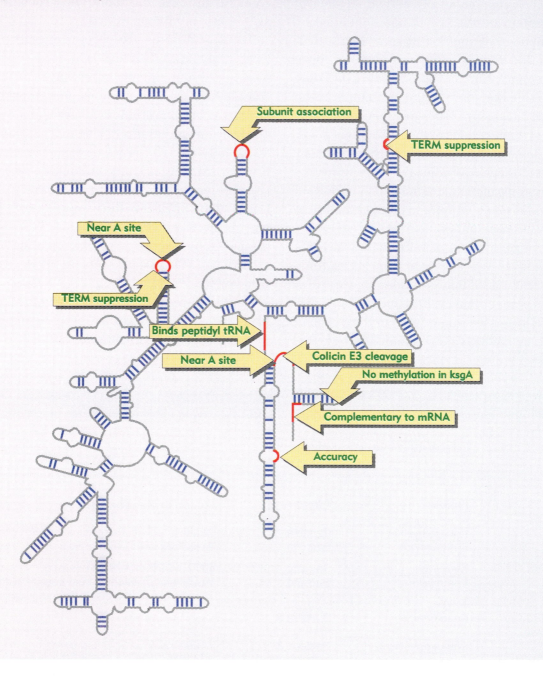

modifications in 16S rRNA. Kasugamycin blocks initiation of protein synthesis. Resistant mutants of the type *ksgA* lack a methylase enzyme that introduces four methyl groups into two adjacent adenines at a site near the 3′ terminus of the 16S rRNA. The methylation generates the highly conserved sequence G–m$_2^6$A–m$_2^6$A, found in both prokaryotic and eukaryotic small rRNA. The methylated sequence is involved in the joining of the 30S and 50S subunits, which in turn is connected also with the retention of initiator tRNA in the complete ribosome. Kasugamycin causes fmet-tRNA$_f$ to be released from the sensitive (methylated) ribsomes, but the resistant ribosomes are able to retain the initiator.

Mutations in rRNA also can influence the specificity of protein synthesis. One example is that a mutation in the 3′ major domain of 16S rRNA suppresses UGA codons. It is not clear whether this effect is specific for UGA (perhaps caused by pairing between rRNA and the UGA codon), or whether it represents a general interference with termination at all three termination codons. In another case, a recognition reaction between rRNA and mRNA appears to be involved in a frameshift in which one base is skipped in reading the messenger. These results raise the possibility that rRNA is continually exposed to the sequence of mRNA as the ribosome moves along the message.

Changes in the structure of 16S rRNA occur when ribosomes are engaged in protein synthesis, as seen by protection of particular bases against chemical attack. The individual sites fall into a few groups, concentrated in the 3′ minor and central domains. Although the locations are dispersed in the linear sequence of 16S rRNA, it seems likely that base positions involved in the same function are actually close together in the tertiary structure.

Some of the changes in 16S rRNA are triggered by joining with 50S subunits, binding of mRNA, or binding of tRNA. They indicate that these events are associated with changes in ribosome conformation that affect the exposure of rRNA. They do not necessarily indicate direct participation of rRNA in these functions.

The 16S rRNA is involved in both A site and P site function, since significant changes in its structure occur when these sites are occupied. Two distinct regions are protected by tRNA bound in the A site. One is in the 3′ terminal region; the other is called the 530 loop (and also is the site of a mutation that prevents termination at the UAA, UAG, and UGA codons).

A variety of bases in different positions are pro-tected by tRNA in the P site; probably the bases lie near one another in the tertiary structure. One prominent stretch of 16S rRNA (known as the 1400 region) can be directly cross-linked to peptidyl-tRNA, which suggests that this region is a structural component of the P site. All of the effects that tRNA binding has on 16S rRNA can be produced by the isolated oligonucleotide of the anticodon stem–loop, so that tRNA–30S subunit binding must involve this region.

The sites involved in the functions of 23S rRNA are less well-identified than those of 16S rRNA, but the same general pattern is observed: bases at certain positions affect specific functions. Positions in 23S rRNA can be identified that are affected by the conformation of the A site or P site. Each of these sites therefore interacts with the 23S rRNA as well as with the 16S rRNA. In particular, oligonucleotides derived from the 3′ CCA terminus of tRNA protect a set of bases in 23S rRNA which essentially are the same as those protected by peptidyl-tRNA. This suggests that the major interaction of 23S rRNA with peptidyl-tRNA in the P site involves the 3′ end of the tRNA.

Another site that binds tRNA is the E site, which provides the region by which deacylated tRNA leaves the ribosome after its amino acid has been used. The E site is localized almost exclusively on the 50S subunit, and correspondingly, bases affected by its conformation can be identified in 23S rRNA.

What is the nature of the site on the 50S subunit that provides peptidyl transferase function? It seems likely that the rRNA at the least provides an important part of the peptidyl transferase site, because a particular region of the rRNA is the site of mutations that confer resistance to antibiotics that are known to inhibit peptidyl transferase.

A long search for ribosomal proteins that might possess the catalytic activity has been unsuccessful. More recent results suggest that the ribosomal RNA of the large subunit has the catalytic activity. Extraction of almost all the protein content of 50S subunits leaves the 23S rRNA associated largely with fragments of proteins, amounting to <5% of the mass of the ribosomal proteins. This preparation retains peptidyl transferase activity. Treatments that destroy secondary structure in RNA abolish the catalytic activity. It is not possible yet to prove directly that 23S rRNA provides the enzymatic function, because of the difficulty of preparing the RNA in a completely purified form that retains its secondary structure in the absence of proteins; indeed, it may well be the case that certain proteins of the 50S subunit are required to enable the rRNA to form the proper structure *in vivo*. But the idea that rRNA is the catalytic component is consistent with the results discussed in Chapter 32 that identify catalytic properties in RNA that are involved with several RNA processing reactions. It fits with the notion that the ribosome evolved from functions originally possessed by RNA.

Less is known about eukaryotic ribosome function, but it seems reasonable to assume that the role of rRNA is analogous to that played in bacterial ribosomes. Eukaryotic rRNA is a target for agents that inactivate eukaryotic ribosomes; for example, the lectin ricin inhibits protein synthesis by depurinating a single adenine from mammalian 28S rRNA.

Ribosomes have several active centers

We can distinguish several ribosomal activities on the basis of function or location. The 30S subunits bind mRNA and the initiator-tRNA•initiation factor complex; then they bind 50S subunits. The 70S ribosomes possess the functionally distinct P and A sites at which tRNA is bound; the peptidyl transferase center is carried on the 50S subunit. The EF-G binding site, and hence responsibility for translocation, is carried on the 50S subunit.

The binding sites are large, each occupying a relatively substantial part of the ribosomal structure. They are not small, discrete regions like the active centers of enzymes. A simplified view of the ribosomal sites is drawn in **Figure 9.11**. They comprise about two thirds of the ribosomal structure and are known as the **translational domain**. The other part of the ribosome comprises the **exit domain**; ribosomes that are attached to membranes (see Chapter 11) are connected through this domain. A polypeptide chain emerges from one domain into the other through an exit site located at some distance from the peptidyl transferase.

We have seen in Chapter 7 that tRNA enters the A site, is transferred by translocation into the P site, and then leaves the (bacterial) ribosome by the E site. The A and P sites must extend across both ribosome subunits. Since tRNA is paired with mRNA in the 30S subunit, but peptide transfer takes place in the 50S subunit. The A and P sites are usually visualized as being adjacent, since translocation moves the tRNA from one site into the other. The E site is presumed to be located near the P site; it has an allosteric interaction with the A site, since the presence of deacylated tRNA in the E site reduces the affinity of the A site for aminoacyl-tRNA.

How are these sites related to the actual topology of the ribosome? Which ribosomal components are involved in its various functions? Several approaches have been used to analyze the relationship between structure and function.

One way to identify particular sites is affinity labeling. This technique uses analogs of components that bind to the ribosome; the analogs are either themselves chemically reactive or can be activated *in situ*. For example, by using tRNA with appropriate modifications, it is possible to identify the ribosomal components to which a label is transferred.

Modification of complete subunits has been used extensively. One approach is to treat subunits with

Figure 9.11

The ribosome has several active centers in the translational domain, and may be associated with a membrane at the exit domain. The locations of the various sites are strictly diagrammatic and cannot yet be related to the precise three dimensional structure of the ribosome.

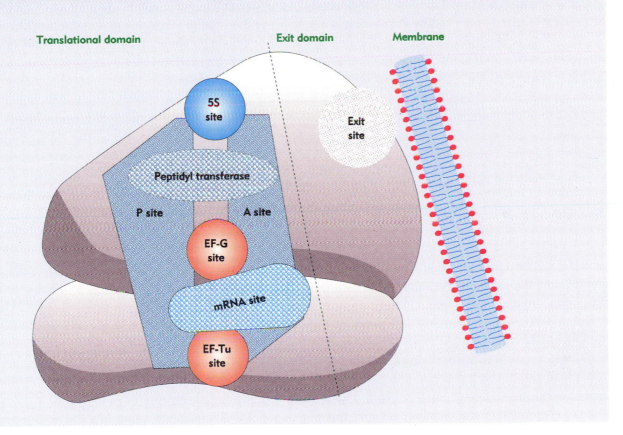

reagents that damage protein or RNA and then to correlate particular damage with specific loss of function. For example, kethoxal has been used to modify the guanines of rRNA, and tetranitromethane to nitrate the proteins. A problem is to limit the damage so that a particular event can be closely correlated with a given loss of function.

Initial binding of 30S subunits to mRNA requires protein S1, which has a strong affinity for single-stranded nucleic acid. It is responsible for maintaining mRNA in the single-stranded state upon binding to the 30S subunit. This action is necessary to prevent the mRNA from taking up a base-paired conformation that would be unsuitable for translation.

Analyzing the location of S1 in the ribosome is complicated by its extremely elongated structure. Cross-linking studies show that it is closely related to S18 and S21, which are among the proteins that react in affinity labeling experiments when initiator tRNA is bound to an AUG codon. The three proteins constitute a domain that is involved in both the initial binding of mRNA and binding initiator tRNA. This would locate the mRNA-binding site in the vicinity of the cleft of the small subunit. The 3′ end of rRNA, which pairs with the mRNA initiation site, is located in this region.

The initiation factors bind in the same region of the ribosome. IF3 can be cross-linked to the 3′ end of the rRNA, as well as to several ribosomal proteins, including those probably involved in binding mRNA. If this area of the small subunit is the same region involved in binding the large subunit, the role of IF3 could be to bind there to stabilize mRNA•30S subunit binding, then to be displaced when the 50S subunit joins.

Not very much is known about the mechanism of subunit joining. It is affected by a mutation in a loop of 16S rRNA (at position 791) that is located at the subunit interface. Also, there is some complementarity between a part of 23S rRNA and (again) the 3′ region of 16S rRNA. One possibility is that base pairing is involved in bringing the subunits together. In this case, the ability of IF3 to bind to 16S rRNA might be incompatible with subunit association.

The 16S rRNA is part of the A site. The rRNA has a conserved sequence in the 3′ minor domain, residues 1392–1407, called the 1400 region, that may interact with tRNA. Cross-linking studies show that peptidyl-tRNA can be linked to a point in this sequence. When aminoacyl-tRNA binds to the ribosome, some bases in this region and elsewhere are shielded from attack by chemical probes.

Cross-linking studies involving EF-Tu show that the A site lies in the vicinity of a group of several proteins, but we do not at present know much about how these (and other) proteins interact to provide the tRNA-binding site. Similarly, a group of proteins has been identified around the 3′ end of the peptidyl-tRNA in the P site.

The incorporation of 5S RNA into 50S subunits that are assembled *in vitro* depends on the ability of three proteins, L5, L8, and L25, to form a stoichiometric complex with it. The complex can bind to 23S rRNA, although none of the isolated components can do so. It lies in the vicinity of the P and A sites.

The peptidyl transferase site has been localized on the central protuberance by the binding of puromycin. A group of several proteins and the 23S rRNA are involved in creating the site.

The growing polypeptide chain appears to be extruded from the 50S subunit ~150Å away from the peptidyl transferase site. It probably extends through the ribosome as an unfolded polypeptide chain until it leaves the exit domain, when it is free to start folding.

The only exception to the unimolarity of ribosomal proteins is presented by L7/L12. L7 differs from L12 only in the presence of an acetyl group on the N-terminus; there are two copies of the dimer per ribosome. The L7/L12 aggregate forms the stalk of the large subunit. When it is removed, the particles become unable to perform GTP hydrolysis at the behest of any accessory factor. This does not necessarily mean that L7/L12 is the GTPase; it could instead be necessary for the activity of another protein.

Although translocation involves movement of the mRNA through the 30S subunit, the translocation factor EF-G binds to the 50S subunit. The binding site for EF-G is close to S12, one of the proteins of the mRNA-binding site on the 30S subunit. This places EF-G at the interface between subunits, in the vicinity of the L7/L12 dimers.

An indication that the EF-G and EF-Tu sites are connected is provided by changes in 23S rRNA. Binding of EF-G to the ribosome protects bases at two locations in 23S rRNA. Bases at one of these locations, the 2660 loop, are also protected by binding of EF-Tu. This is a counterpart to the location in eukaryotic 28S rRNA that is the target for lectins that inhibit elongation (ricin depurinates rRNA and α-sarcin cleaves it). The structure of this loop is therefore important for elongation.

The P and A sites must lie close together, since their tRNAs are bound to adjacent triplets on mRNA. The P site influences the activity of the A site, since peptidyl-tRNA must be present for aminoacyl-tRNA to bind. It remains a problem to see how two tRNA molecules might fit into the ribosome next to each other. The distance between the anticodons cannot be greater than ~10Å, yet the diameter of the tRNA is ~20Å.

The conformation of the tRNA seems to remain the same in both the P site and A site. One solution to the stereochemical problem would be to have a twist or kink in the mRNA between the codons, so that the two tRNAs fit onto the tRNA from different directions. Then ribosome movement might be a matter of angling the tRNA from the A site into the P site. This would allow mRNA advancement to be a function of tRNA movement, which would let anticodon–codon pairing measure the distance for translocation, as discussed in Chapter 8.

The functional relationships between the various ribosomal sites have yet to be defined. The ribosome probably has a highly interactive structure, in which a change at one point greatly affects the activity of another site elsewhere.

The accuracy of translation

The lack of detectable variation when the sequence of a protein is analyzed demonstrates that protein synthesis must be extremely accurate: very few mistakes are apparent in the form of substitutions of one amino acid for another. There are two stages in protein synthesis at which errors might be made:

◆ Charging a tRNA only with its correct amino acid clearly is critical. We have seen in Chapter 6 that this is a function of the aminoacyl-tRNA synthetase enzyme. Probably the error rate varies with the particular enzyme, but current estimates are that mistakes occur in less than 1 in 10^5 aminoacylations.

◆ The specificity of codon–anticodon recognition is crucial, but puzzling. Although binding constants vary with the individual codon–anticodon reaction, the specificity is always much too low to provide an error rate of $<10^{-5}$. When free in solution, tRNAs bind to their trinucleotide codon sequences only relatively weakly; and related, but erroneous triplets (with two correct bases out of three) are recognized 10^{-1}–10^{-2} times as efficiently as the correct triplets.

Codon–anticodon base pairing therefore seems to be a weak point in the accuracy of translation. This suggests that the ribosome has some function that directly or indirectly acts as a 'proofreader', distinguishing correct and incorrect codon–anticodon pairs, and thus amplifying the rather modest intrinsic difference. Suppose that there is no specificity in the initial collision between the aminoacyl-tRNA•EF-Tu•GTP complex and the ribosome. If any complex, irrespective of its tRNA, can enter the A site, the number of incorrect entries must far exceed the number of correct entries.

So there must be some mechanism for stabilizing the correct aminoacyl-tRNA, allowing its amino acid to be accepted as a substrate for receipt of the polypeptide chain; contacts with an incorrect aminoacyl-tRNA must be rapidly broken, so that the complex leaves without reacting. How does a ribosome assess the codon–anticodon reaction in the A site to determine whether a proper fit has been achieved?

The ability of the ribosome to influence the accuracy of translation was first shown by the effects of mutations that confer resistance to streptomycin. One effect of streptomycin is to increase the level of misreading of the pyrimidines U and C (usually one is mistaken for the other, occasionally for A). The site at which streptomycin acts is influenced by the S12 protein; the sequence of this protein is altered in resistant mutants. Ribosomes with an S12 protein derived from resistant bacteria show a reduction in the level of misreading compared with wild-type ribosomes. This compensates for the effect of streptomycin on misreading. Streptomycin probably binds the rRNA, and S12 affects this reaction by influencing the accessibility of the rRNA.

Mutations at two other loci, coding for proteins S4 and S5, influence misreading, since revertants showing the usual level of misreading can be isolated at either locus. So the accuracy of translation is controlled by the interactions of these three proteins. The level of misreading and the nature of the response to streptomycin depend on the versions of these proteins that are present. (Some combinations even make the ribosome **dependent** on the presence of streptomycin for correct translation.) We can interpret the role of these proteins in two ways:

◆ A direct mechanism for controlling accuracy would be for the stereochemistry of the A site to determine the latitude of codon–anticodon recognition. The geometry of the ribosome could be designed so that codon–anticodon binding is scrutinized in such a way that the criteria for accepting aminoacyl-tRNA could be made more or less precise by the structures of proteins S12, S4, and S5 (or by their effects on the structure of rRNA).

◆ The effect on accuracy could be an indirect result of the speed of ribosome movement. Ribosome velocity might determine availability for tRNA recognition, and thus the efficiency of the process. This model explains the effect of streptomycin by adjusting the kinetics of chain elongation. The relevant parameter is the speed of ribosome action relative to the time required to make and break contacts. If the velocity of peptide bond formation is increased, incorrect aminoacyl-tRNAs are more likely to be trapped by bond formation before the aminoacyl-tRNA escapes. Slowing the rate of protein synthesis gives more time to correct errors.

One idea is that the making of a correct contact between codon and anticodon could be signaled to the other end of the tRNA by a change in conformation. The change could be needed to place the amino acid in the appropriate location to accept the polypeptide from the peptidyl-tRNA.

Irrespective of any mechanism that is used to indicate the presence of aminoacyl-tRNA in the A site, the fact is that mismatched aminoacyl-tRNA dissociates more rapidly than correctly matched aminoacyl-tRNA, probably by a factor of ~500×. Increasing time spent in the A site before peptide bond formation occurs therefore increases the probability that the correct aminoacyl-tRNA will be utilized. This type of mechanism is called **kinetic proofreading.** The overall error rate in protein synthesis is ~5×10^{-4} per codon, and the majority of errors probably occur by recognition of mistaken aminoacyl-tRNAs in the A site.

An important question in calculating the cost of protein synthesis is the stage at which the decision is taken on whether to accept a tRNA. If a decision occurs immediately to release an aminoacyl-tRNA•EF-Tu•GTP complex, there is little extra cost for rejecting the large number of incorrect tRNAs that are likely (statistically) to enter the A site before the correct tRNA is recognized. But if the GTP is hydrolyzed when the complex binds, an additional high-energy bond will be cleaved for every incorrectly associating tRNA. This would increase the cost of protein synthesis well above the three high-energy bonds that are used in adding every (correct) amino acid to the chain. There is some evidence that the use of GTP *in vivo* is greater than had been expected, possibly involving an extra 3–4 GTP cleavages per amino acid.

The specificity of decoding has been assumed to reside with the ribosome itself, but some recent results suggest that translation factors influence the process at both the P site and A site.

A striking case concerns initiation. Mutation of the AUG initiation codon to UUG in the yeast gene *HIS4* prevents initiation. Extragenic suppressor mutations can be found that allow protein synthesis to be initiated at the mutant UUG codon. Two of these suppressors prove to be in genes coding for the α and β subunits of eIF2, the factor that binds Met-tRNA$_i$ to the P site. The mutation in eIFβ2 resides in a part of the protein that is almost certainly involved in binding nucleic acid. It seems likely that its target is either the initiation sequence of mRNA as such or the base-paired association between the mRNA codon and tRNA$_i^{Met}$ anticodon. This suggests that eIF2 participates in the discrimination of initiation codons as well as bringing the initiator tRNA to the P site.

An indication that EF-Tu is involved in maintaining the reading frame is provided by mutants of the factor that suppress frameshifting. This probably means that EF-Tu does not merely bring aminoacyl-tRNA to the A site, but also is involved in positioning the incoming aminoacyl-tRNA relative to the peptidyl-tRNA in the P site.

Summary

Ribosomes are ribonucleoprotein particles in which a majority of the mass is provided by rRNA. The shapes of all ribosomes are generally similar, but only those of bacteria have been characterized in detail. The small subunit has a squashed shape, with a 'body' containing about two-thirds of the mass divided from the 'head' by a cleft. The large subunit is more spherical, with a prominent 'stalk' on the right and a 'central protuberance'. Locations of all proteins are known approximately in the small subunit.

Each subunit contains a single major rRNA, 16S and 23S in prokaryotes, 18S and 28S in eukaryotic cytosol. There are also minor rRNAs, most notably 5S rRNA in the large subunit. Both major rRNAs have extensive base pairing, mostly in the form of short, imperfectly paired duplex stems with single-stranded loops. Conserved features in the rRNA can be identified by comparing sequences and the secondary structures that can be drawn for rRNA of a variety of organisms. The 16S rRNA has four distinct domains; the three major domains have been mapped into regions of the small subunit. Eukaryotic 18S rRNA has additional domains. One end of the 30S subunit may consist largely or entirely of rRNA.

The pathway of ribosome subunit assembly is related to subunit structure: proteins are added in a defined order, which is the reverse of the ease with which they are dissociated from the subunit. The order of addition defines an assembly map. Assembly *in vivo* utilizes precursor rRNA, which is incorporated into a precursor particle with some of the ribosomal proteins. The rRNA is cleaved and methyl groups are added at this stage.

Each subunit has several active centers, concentrated in the translational domain of the ribosome where proteins are synthesized. Proteins leave the ribosome through the exit domain, which may be associated with a membrane. The major active sites are the P and A sites, the EF-Tu and EF-G binding sites, peptidyl transferase, and mRNA-binding site. Ribosomal proteins required for the function of some of these sites have been identified, but the sites have yet to be mapped in terms of three-dimensional ribosome structure. Ribosome conformation may change at stages during protein synthesis; differences in the accessibility of particular regions of the major rRNAs have been detected.

The major rRNAs contain regions that are localized at some of these sites, most notably the mRNA-binding site and P site on the 30S subunit. The 3′ terminal region of the rRNA seems to be of particular importance. Functional involvement of the rRNA in ribosomal sites is best established for the mRNA-binding site, where mutations in 16S rRNA affect the initiation reaction. Ribosomal RNA is also the target for some antibiotics or other agents that inhibit protein synthesis. It appears to possess the essential catalytic activity of peptidyl transferase.

The ribosome has an important role in controlling the specificity of protein synthesis via the codon–anticodon pairing reaction. The cost of protein synthesis in terms of high-energy bonds could be increased by proofreading processes. In addition to the role of the ribosome itself, the factors that place initiator- and aminoacyl-tRNAs in the ribosome also may influence the pairing reaction.

Further reading

Reviews

Although now outdated, the reference work on ribosomes edited by **Nomura, Tissieres, and Lengyel** *Ribosomes* (Cold Spring Harbor Lab., New York, 1974) provides a useful general historical view of both structural and functional aspects. A more recent compendium was edited by **Hill** *et al*. *The Ribosome* (American Society for Microbiology, Washington, 1990).

Locations of components in the ribosome were reviewed by **Wittman** (*Ann. Rev. Biochem.* **52**, 35–65, 1983). Ribosome structure has been unified by **Lake** (*Ann. Rev. Biochem.* **54**, 507–530, 1985). Reviews of the current structure have been prepared by **Noller and Nomura** (in *E. coli and S. typhimurium,* Ed. Neidhardt, American Society for Microbiology, Washington DC, 104–125, 1987) and by **Moore** (*Nature* **331**, 223–227, 1988).

Ribosomal RNA has been analyzed in minute detail by **Woese** *et al.* (*Microbiol. Rev.* **47**, 621–669, 1983) and **Noller** (*Ann. Rev. Biochem.* **53**, 119–162, 1984). Its role in ribosomal function has been analyzed by **Noller** (*Ann. Rev. Biochem.* **60**, 191–227, 1991).

Proofreading was reviewed by **Kurland** (*Ann. Rev. Genet.* **26**, 29–50, 1992).

Discoveries

The concept of proofreading was introduced by **Hopfield** (*Proc. Nat. Acad. Sci. USA* **71**, 4135–4139, 1974).

The probable responsibility of 23S rRNA for peptidyl transferase activity was identified by **Noller, Hoffarth, and Zimniak** (*Science* **256**, 1416–1419, 1992).

CHAPTER 10

Messenger RNA is the template

The existence of mRNA was first suspected because of the need for an intermediary in eukaryotic cells to carry genetic information from the nucleus (where DNA resides) into the cytoplasm (where proteins are synthesized). Conceived as providing a template on which amino acids would be assembled into polypeptides, the messenger was first sought in bacteria. The ribosomes, established as the site of protein synthesis, were excluded from the role of providing the template; and then mRNA was found in the form of a transient species that associates with the ribosomes to be translated into proteins (see Chapter 7).

Although its transient existence at first prevented the isolation of bacterial mRNA, the properties of the messenger were deduced in detail from the features of its translation into protein. It was some time before mRNA could be isolated from eukaryotes, but then it proved to be a relatively more stable component of the cytoplasm. Because of its (relative) stability, eukaryotic mRNA can be isolated with some facility; now it is possible in principle to isolate the mRNA for any particular protein.

Messenger RNA is synthesized on a DNA template by the process of transcription. Only one DNA strand is transcribed. Base pairing is used to synthesize an RNA complementary to the template strand of DNA. The RNA is therefore identical in sequence with the other (non-template) strand of DNA (see Figure 6.16).

Messenger RNA is fixed of purpose in all living cells: to be translated via the genetic code into protein. Yet there are important differences in the details of the synthesis and structure of prokaryotic and eukaryotic mRNA.

A major difference in the production of mRNA depends on the locations where transcription and translocation occur:

◆ In bacteria mRNA is transcribed and translated in the single cellular compartment; and the two processes are so closely linked that they occur simultaneously.

◆ In a eukaryotic cell, synthesis and maturation of mRNA occurs exclusively in the nucleus. Only after these events are completed is the mRNA exported to the cytoplasm, where it is translated by ribosomes.

Significant differences in the translation of bacterial and eukaryotic mRNAs stem from their characteristic structures and stabilities:

◆ A structural difference is that a bacterial mRNA codes for several proteins, whereas a eukaryotic mRNA invariably codes for only one polypeptide chain.

◆ A functional difference is that bacterial mRNA usually is unstable, and is therefore translated into proteins for only a short period of time (typically a few minutes). Eukaryotic mRNA is more stable and continues to be translated for several hours.

In this chapter, we discuss the general properties of mRNA in prokaryotes and eukaryotes. In later chapters, we discuss the transcription of bacterial mRNA and its regulation (Part 4), procedures for identifying mRNAs for particular eukaryotic genes (Chapter 21), and the intricate processes by which eukaryotic mRNA is synthesized and processed (Part 7).

The life cycle of messenger RNA

An overview of the processes of transcription and translation in bacteria and eukaryotic cells is illustrated in Figures 10.1 and 10.3. **Table 10.1** compares the nature of the events that are involved at each stage of production of mRNAs in these two types of cells.

Figure 10.1 shows that transcription and translation are intimately related in bacteria. Transcription begins when the enzyme RNA polymerase binds to DNA and then moves along making a copy of one strand. As soon as transcription begins, ribosomes attach to the 5' end of the mRNA and start translation, even before the rest of the message has been synthesized. A group of ribosomes moves along the mRNA while it is being synthesized. The 3' end of the mRNA is generated when transcription terminates. Ribosomes continue to translate the mRNA while it survives, but it is degraded in the 5'–3' direction quite rapidly.

Bacterial transcription and translation take place at similar rates:

◆ At 37°C, transcription of mRNA occurs at ~40 nucleotides/second. This is very close to the rate of protein synthesis, roughly 15 amino acids/second. It therefore takes ~2 minutes to transcribe and translate an mRNA of 5000 bp, corresponding to 180,000 daltons of protein.

◆ When expression of a new gene is initiated, its mRNA typically will appear in the cell within ~2.5 minutes. The corresponding protein will appear within perhaps another 0.5 minute.

Table 10.1

Prokaryotic and eukaryotic mRNAs are organized differently.

	Bacterial mRNA	Animal mRNA
Transcription unit	Polycistronic	Monocistronic
Initiation (5' end)	RNA polymerase binds DNA	RNA polymerase binds DNA
Elongation	˜40 nucs/sec	˜40 nucs/sec
5' end	Triphosphate	Methylated cap
3' end generation	Termination releases mRNA	Cleavage releases mRNA
3' end	Last base of mRNA	Poly(A)
Transport	Not necessary	Necessary
Translation	˜45 nucs/sec	˜20 nucs/sec
Half life	< 5 min	> 4 hours
Ribosomes/mRNA	> 20	> 10

Figure 10.1

Overview: transcription, translation, and degradation of mRNA all proceed simultaneously in bacteria. The times apply to a unit of 5000 bp, whose expression starts at time 0.

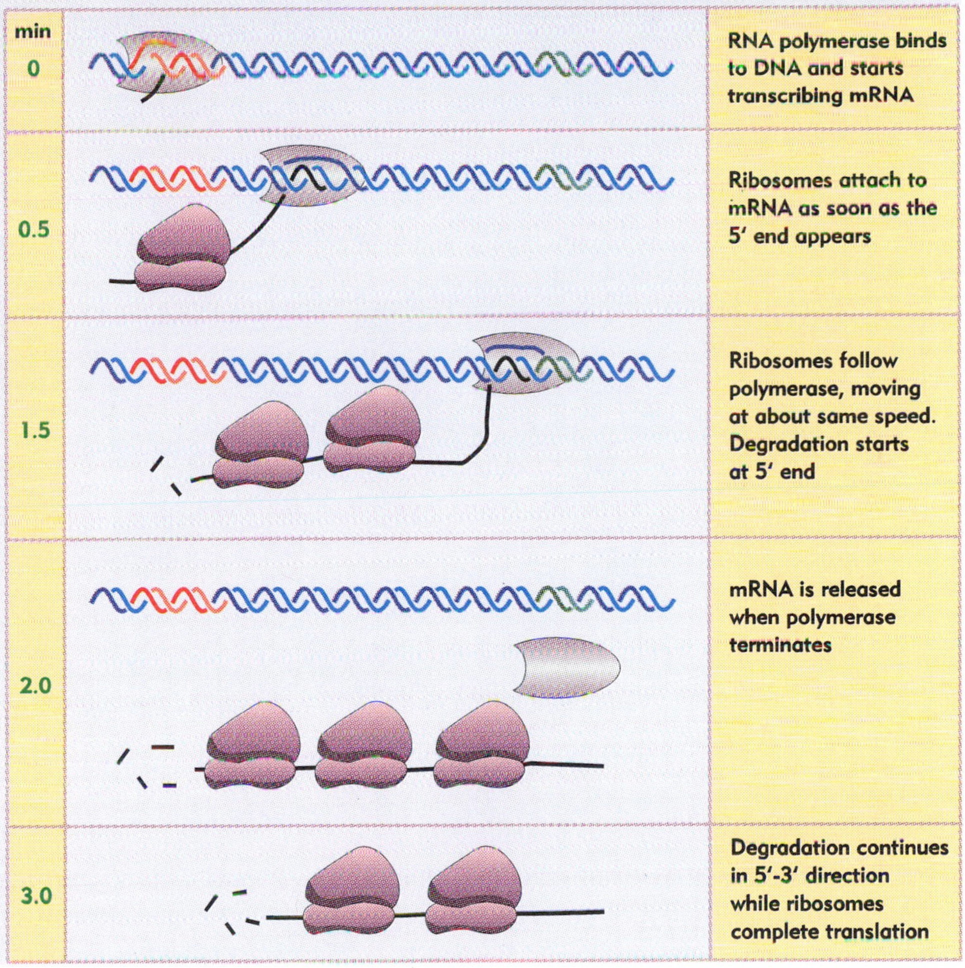

min		
0		RNA polymerase binds to DNA and starts transcribing mRNA
0.5		Ribosomes attach to mRNA as soon as the 5′ end appears
1.5		Ribosomes follow polymerase, moving at about same speed. Degradation starts at 5′ end
2.0		mRNA is released when polymerase terminates
3.0		Degradation continues in 5′-3′ direction while ribosomes complete translation

Bacterial translation is very efficient, and most mRNAs are translated by a large number of tightly packed ribosomes. In the example of the *tryptophan* operon, about 15 initiations of transcription occur every minute, and each of the 15 mRNAs probably is translated by ~30 ribosomes in the interval between its transcription and degradation.

The instability of most bacterial mRNAs is striking. Degradation of mRNA closely follows its translation. Probably it begins within 1 minute of the start of transcription. The 5′ end of the mRNA starts to decay before the 3′ end has been synthesized or translated. Degradation seems to follow the last ribosome of the convoy along the mRNA. But degradation proceeds more slowly, probably at about half the speed of transcription or translation.

This series of events is only possible, of course, because transcription, translation, and degradation all occur in the same direction. The dynamics of gene expression have been caught *in flagrante delicto* in the electron micrograph of **Figure 10.2**. In these (unknown) transcription units, several mRNAs are under synthesis simultaneously; and each carries many ribosomes engaged in translation. (This corresponds to the stage shown in the third panel in Figure 10.1.) An RNA whose synthesis has not yet been completed is often called a **nascent RNA**.

Figure 10.2

Transcription units can be visualized in bacteria. Photograph kindly provided by Oscar Miller.

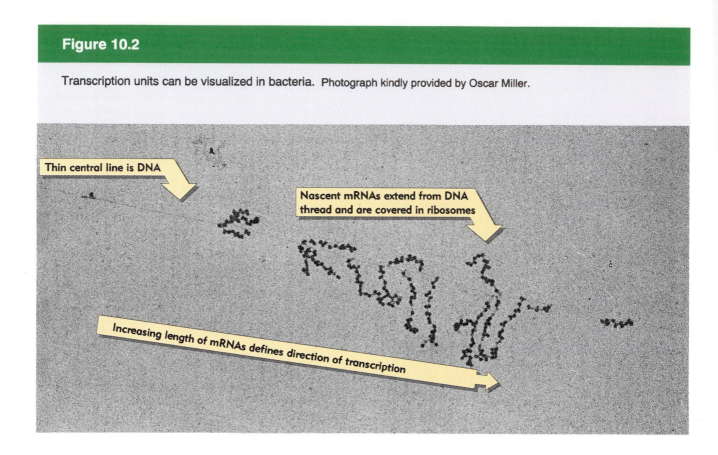

Thin central line is DNA

Nascent mRNAs extend from DNA thread and are covered in ribosomes

Increasing length of mRNAs defines direction of transcription

Figure 10.3 shows that the life cycle of eukaryotic mRNA is more protracted. Transcription occurs at about the same speed as in bacteria, ~40 nucleotides per second in animal cells (the speed is lower in insects, ~20 nucleotides/second). Many eukaryotic genes are large; a gene of 10,000 bp would take ~5 minutes to transcribe. Transcription of mRNA is not terminated by the release of enzyme from the DNA; instead the enzyme continues past the end of the gene, and the 3′ end of the RNA is generated by a cleavage event.

The 5′ end of the RNA is modified by addition of some extra groups virtually as soon as it appears. The 3′ end is modified by addition of further nucleotides (all adenylic acid) immediately after its cleavage. Those RNAs that are derived from interrupted genes require splicing to remove the introns, generating a smaller mRNA that contains an intact coding sequence to be translated. The order in which these events occur is not entirely clear: some splicing events could occur during transcription, but others may happen after the transcript has been completed. Only after the completion of all modification and processing events can the mRNA be exported from the nucleus to the cytoplasm. The average delay in leaving for the cytoplasm is 20 minutes. Once the mRNA has entered the cytoplasm, it is recognized by ribosomes and translated.

The stability of mRNA has a major influence on the amount of protein that is produced. It is usually expressed in terms of the half-life. This is usually ~2 minutes in bacteria. In other words, the amount of new protein that an individual mRNA can synthesize is halved about every 2 minutes. However, within the context that bacterial mRNAs are unstable, there are variations between mRNAs representing different genes. A range of half-lives is seen also in eukaryotes, but they are much longer: animal cells typically have mRNAs with half-lives ranging from 4 to 24 hours.

Figure 10.3

Overview: expression of mRNA in animal cells requires transcription, modification and processing, nucleocytoplasmic transport, and translation. The times apply to a unit of 10,000 bp.

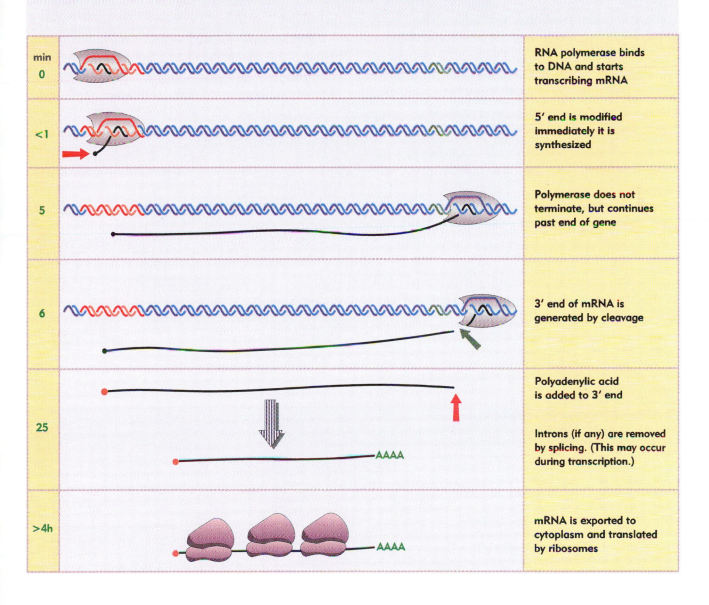

min 0	RNA polymerase binds to DNA and starts transcribing mRNA
<1	5' end is modified immediately it is synthesized
5	Polymerase does not terminate, but continues past end of gene
6	3' end of mRNA is generated by cleavage
	Polyadenylic acid is added to 3' end
25	Introns (if any) are removed by splicing. (This may occur during transcription.)
>4h	mRNA is exported to cytoplasm and translated by ribosomes

Most bacterial genes are expressed via polycistronic messengers

Because of their instability *in vivo,* bacterial mRNAs can rarely be isolated intact. However, the region of DNA represented in the mRNA can be defined by hybridization assays, which measure the ability of radioactively labeled mRNA to hybridize with DNA fragments representing specific parts of the gene. In this way, a detailed picture of the structure and function of bacterial mRNA can be constructed without characterizing the mRNA molecules directly.

An important approach for working directly with bacterial mRNA is the use of cell-free systems for transcription and translation. An mRNA can be transcribed *in vitro* from the appropriate template, usually a 'cloned' copy of the gene. It can be translated by *E. coli* ribosomes, aminoacyl-tRNAs, etc., into the protein. The structure as well as the function of

the mRNA is amenable to analysis in such systems.

Bacterial mRNAs vary greatly in the number of proteins for which they code. Some mRNAs represent only a single gene: they are **monocistronic**. Others (the majority) carry sequences coding for several proteins: they are **polycistronic**. In these cases, a single mRNA is transcribed from a group of adjacent genes. (Such a cluster of genes constitutes an *operon* that is controlled as a single genetic unit; see Chapter 15.)

All mRNAs contain two types of region. The **coding region** consists of a series of codons representing the amino acid sequence of the protein, starting (usually) with AUG and ending with a termination codon. But the mRNA is always longer than the coding region. In a monocistronic mRNA, extra regions are present at both ends. An additional sequence at the 5' end, preceding the

Figure 10.4

Bacterial mRNA includes nontranslated as well as translated regions. Each coding region possesses its own initiation and termination signals. A typical mRNA may have several coding regions.

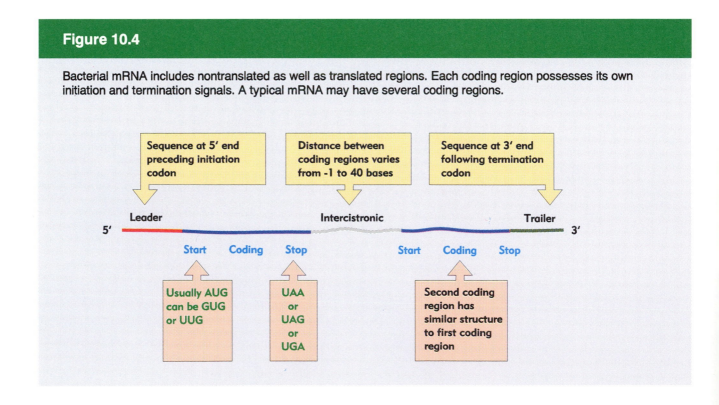

Figure 10.5

Initiation may occur independently at each cistron in a polycistronic mRNA. When the intercistronic region is longer than the span of the ribosome, dissociation at the termination site is followed by independent reinitiation at the next cistron.

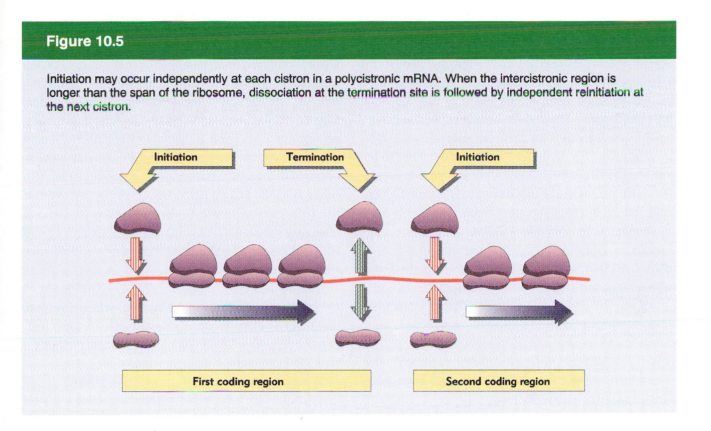

start of the coding region, is described as a **leader.** An additional sequence following the termination signal, forming the 3′ end, is called a **trailer.** Although part of the transcription unit, these sequences are not used to code for protein.

The structure of a polycistronic mRNA is illustrated in **Figure 10.4.** The **intercistronic regions** that lie between the various coding regions vary greatly in size. They may be as long as 30 nucleotides in bacterial mRNAs (and even longer in phage RNAs), but they can also be very short, with as little as one or two nucleotides separating the termination codon for one protein from the initiation codon for the next. In an extreme case, two genes actually overlap, so that the last base of the UGA termination codon at the end of one coding region is also the first base of the AUG initiation codon at the start of the next gene.

The number of ribosomes engaged in translating a particular cistron depends on the efficiency of its initiation site. The initiation site for the first cistron becomes available as soon as the 5′ end of the mRNA is synthesized. How are subsequent cistrons translated? Are the several coding regions in a polycistronic mRNA translated independently or is their expression connected? Is the mechanism of initiation the same for all cistrons, or is it different for the first cistron and the internal cistrons?

Translation of a bacterial mRNA proceeds sequentially through its cistrons. At the time when ribosomes attach to the first coding region, the subsequent coding regions have not yet even been transcribed. By the time the second ribosome site is available, translation is well under way through the first cistron.

What happens between the coding regions depends on the individual mRNA. Probably in most cases the ribosomes bind independently at the beginning of each cistron. The most common series of events is illustrated in **Figure 10.5.** When synthesis of the first protein terminates, the ribosomes dissociate into subunits and leave the mRNA. Then a new 30S subunit must attach at the next initiation codon, be joined by a 50S subunit, and set out to translate the next cistron.

In some bacterial mRNAs, translation between

adjacent cistrons is directly linked, because ribosomes gain access to the initiation codon of the second cistron *as they complete translation of the first cistron*. This effect requires the space between the two coding regions to be small. It may depend on the high local density of ribosomes; or the juxtaposition of termination and initiation sites could allow some of the usual intercistronic events to be bypassed. A ribosome physically spans ~35 bases of mRNA, so that it may simultaneously contact a termination codon and the next initiation site if they are separated by only a few bases.

In an extreme case, the 30S subunit of a terminating ribosome might fail to dissociate from mRNA, because it is instantly attracted to the initiation site of which it is already virtually in possession. As illustrated in **Figure 10.6**, this would mean that, when the 50S subunits and the completed polypeptide chain are released, the 30S subunit would remain *in situ* to reinitiate translation of the next cistron.

Translation of one cistron may require changes in secondary structure that depend on translation of a preceding cistron. An effect of this nature is seen normally in the translation of the RNA phages, whose cistrons always are expressed in a set order. **Figure 10.7** shows that the phage RNA takes up a secondary structure in which only one initiation sequence is accessible; the second cannot be recognized by ribosomes because it is base-paired with other regions of the RNA. However, translation of the first cistron disrupts the secondary structure, allowing ribosomes to bind to the initiation site of the next cistron. In this mRNA, secondary structure controls translatability.

Usually this type of relationship does not apply in bacterial mRNAs, because the ribosomes translate the mRNA as soon as it is synthesized, precluding the option of base pairing into stable double-stranded regions.

Whether the mechanism of dependence relies on changes in secondary structure or on direct

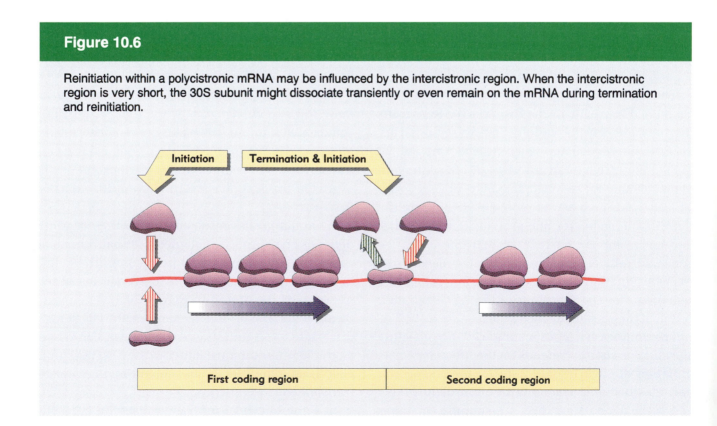

Figure 10.6

Reinitiation within a polycistronic mRNA may be influenced by the intercistronic region. When the intercistronic region is very short, the 30S subunit might dissociate transiently or even remain on the mRNA during termination and reinitiation.

Initiation

Termination & Initiation

First coding region

Second coding region

Figure 10.7

Secondary structure can control initiation. Only one initiation site is available in the RNA phage, but translation of the first cistron changes the conformation of the RNA so that other initiation site(s) become available.

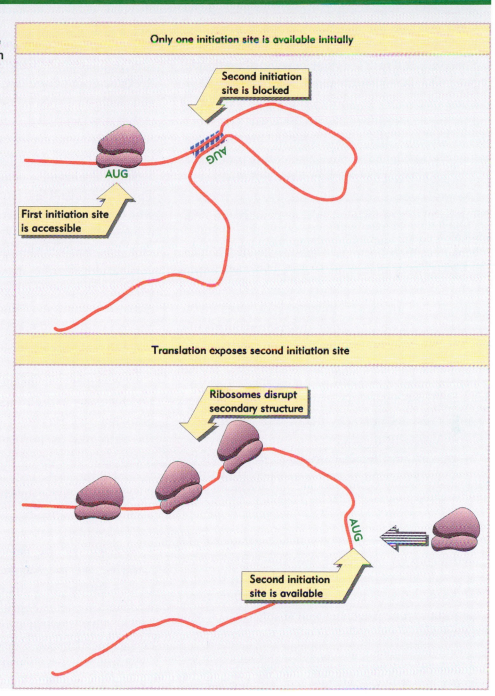

Only one initiation site is available initially

Second initiation site is blocked

First initiation site is accessible

AUG

Translation exposes second initiation site

Ribosomes disrupt secondary structure

Second initiation site is available

AUG

ribosome transfer from termination to initiation sites, a consequence of such a relationship is that the translation of two (or more) cistrons on a polycistronic mRNA may be controlled coordinately. We know of cases in bacteria in which preventing translation of the first cistron in an mRNA also prevents all the subsequent cistrons from being translated. This effect is called **polarity** and results from interference with either the production or translation of mRNA for the downstream cistrons.

The translation of eukaryotic mRNA

Eukaryotic polyribosomes are reasonably stable. But because the mRNA is only a minor component of the total mass of the RNA in the isolated polysome fraction (the overwhelming proportion is rRNA), it cannot be isolated directly by the usual fractionation techniques. Attempts to label the mRNA preferentially by using radioactive nucleotide precursors (the approach used in bacteria) proved unsuccessful, because some contaminating cytoplasmic RNA fractions also are labeled just as rapidly and efficiently.

The first technique to distinguish mRNA from other RNAs took advantage of the results of treating polysomes with the chelating agent EDTA (which removes Mg^{2+} ions from ribosomes). As a result, the ribosomes dissociate into individual subunits, releasing the mRNA as a **ribonucleoprotein (RNP)** fraction sedimenting at ~18S. This fraction consists of mRNA associated with proteins. The nontranslated material that was contaminating the polysomes is not disrupted by EDTA and continues to sediment rapidly at ~200S.

In spite of considerable advances since this technique was developed, the release of mRNA by EDTA remains important for two reasons:

◆ It is a *functional* assay: it identifies mRNA actually in the process of translation by the ribosomes. The presence of an mRNA in this fraction is taken as proof that it is indeed used to direct protein synthesis in the cell from which it was obtained.

◆ The assay allows the mRNA to be isolated in what is presumably its natural form, the ribonucleoprotein particle.

The ribonucleoprotein fraction that contaminates the polysomes is 'contaminating' only in the sense that it is an unwanted component when mRNA is being isolated. The RNP is presumably a legitimate component of the cell in its own right.

The fraction includes RNA molecules that resemble mRNA. But based on the EDTA-release assay, these RNAs are not active in translation. Their relationship with the mRNA is not clear, but one possibility is that this fraction includes legitimate mRNAs that (although currently inactive) will be used for translation later or under different circumstances.

Some embryonic cells contain 'stored' mRNAs. They are found as ribonucleoprotein particles whose mRNAs are not being translated, but that are used to direct protein synthesis at a later stage of embryogenesis. Although we do not know whether comparable storage mechanisms are used in adult cells, it is now well-established (especially in marine organisms) that an appreciable part of newly synthesized mRNA is not immediately translated in the egg or early embryo, but is stored for later use.

The ability to translate mRNA *in vitro* provides a crucial assay in defining the process of translation. Showing that the product of translation *in vitro* has the authentic properties of a protein found *in vivo* provides unequivocal evidence for the coding function of an mRNA. This can be confirmed in more detail by showing that the sequence of the mRNA includes a coding region corresponding with the sequence of the protein.

In vitro systems provide the only approach for defining the *mechanism* of either translation or transcription. In both cases, the critical step in expression occurs at initiation. The *in vitro* systems allow definition of the sites on mRNA that are recognized by ribosomes.

Although *in vitro* translation systems generally should mimic the *in vivo* process, there are two important situations in which *differences* between the products of translation *in vitro* and *in vivo* provide interesting information.

Sometimes mRNAs are found that direct protein synthesis *in vitro*, although the corresponding proteins are not synthesized in the cells from which the mRNA was taken. The ability of the mRNA to be translated into authentic proteins *in vitro* demonstrates that it is capable of functioning as a template. So its failure to be translated *in vivo* can be taken as evidence for **translational control**. Some mechanism must act *in vivo* to prevent translation. The 'stored' RNP particles of marine embryos are a good example of this phenomenon.

Sometimes the product of translation *in vitro* is related to the authentic *in vivo* protein, but possesses additional amino acids, usually a short length at the N-terminal end. This discrepancy usually means that the protein found *in vivo* is not the initial product of translation, but is derived by cleaving a precursor that contains the extra residues. Processing of such precursors *in vivo* generally is rapid, and this makes them difficult to detect, unless inhibitors are used to block the cleavage reaction.

Isolated mRNAs have been characterized by translation in the two types of system depicted in **Figure 10.8**:

◆ Reconstituted **cell-free systems** include ribosomes, protein synthesis factors, and tRNAs. Many such systems have been developed; the original systems were derived from *E. coli*, wheat germ, and rabbit reticulocyte. The systems are somewhat inefficient, since each mRNA is translated a relatively low number of times, at a rate much below that found *in vivo*. At best the systems function for 90–120 minutes before stopping. In all cases, there is some background due to translation of endogenous mRNAs that were not removed; its level varies with the system.

◆ An alternative translation system is presented by the intact *Xenopus* oocyte. Injected mRNAs are translated by the natural protein synthetic apparatus. The system is efficient and treats the injected mRNAs as though they were endogenous mRNAs, so that they are used for repeated rounds of translation. The only limit is that too

Figure 10.8

Exogenous mRNAs can be translated by cell-free systems or by injection into *Xenopus* oocytes.

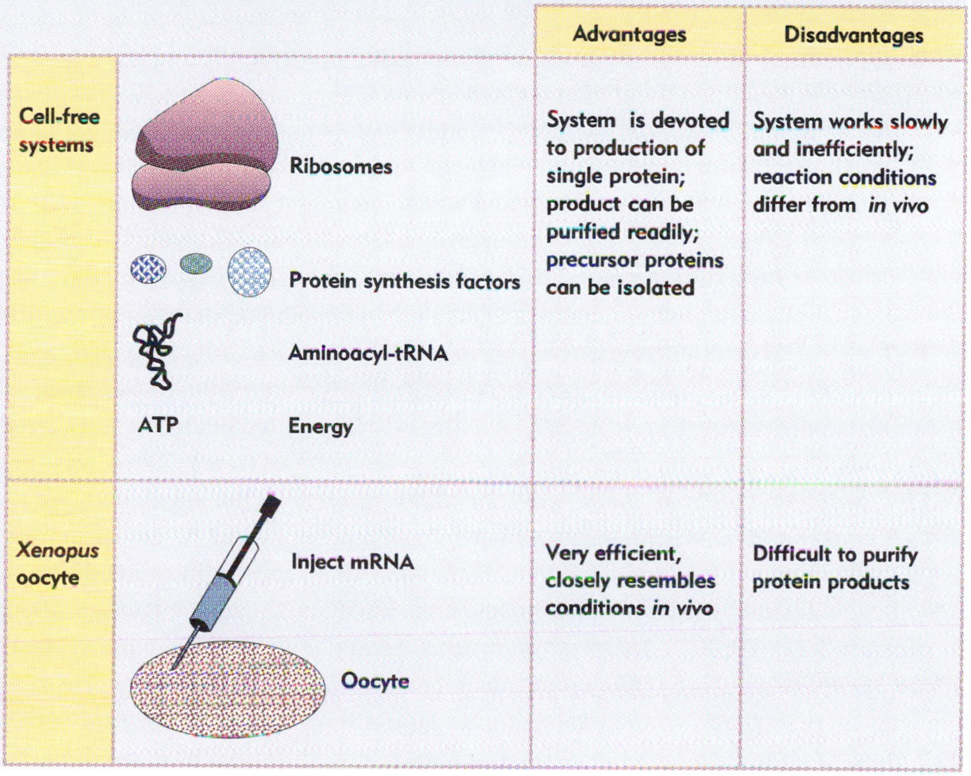

		Advantages	Disadvantages
Cell-free systems	Ribosomes Protein synthesis factors Aminoacyl-tRNA ATP Energy	System is devoted to production of single protein; product can be purified readily; precursor proteins can be isolated	System works slowly and inefficiently; reaction conditions differ from *in vivo*
***Xenopus* oocyte**	Inject mRNA Oocyte	Very efficient, closely resembles conditions *in vivo*	Difficult to purify protein products

much mRNA may saturate the translation system (which is present in great excess relative to the demands placed on it by endogenous mRNAs). Generally the system continues to be active for 24–48 hours.

Neither type of system displays any tissue or species specificity. This indicates that mRNAs and the protein-synthesizing apparatus of (perhaps) any eukaryotic cytoplasm are interchangeable. Control is not exercised by ribosomes or translation factors that function with one set of mRNAs but not with another. The translational machinery is not pre-programmed, but will accept any mRNA as template.

Translational control must take the form of preventing mRNAs from reaching the protein-synthesizing apparatus. This may be achieved either by sequestering them in a form in which they are physically unavailable, or by creating competitive conditions in which more efficient mRNAs are translated at the expense of less efficient messengers.

The oocyte system is able to process at least some proteins that usually are cleaved during or soon after synthesis. In some cases, it can even direct protein products to enter facsimiles of the appropriate cell compartment. This ability implies that signals for processing maybe common in different cell types and species. As a practical consequence, it means that the isolation of precursor proteins must be undertaken in cell-free translation systems.

Most eukaryotic mRNAs are polyadenylated at the 3′ end

Most eukaryotic mRNAs have a sequence of poly-adenylic acid at the 3′ end. This terminal stretch of A residues is often described as the **poly(A) tail**; and mRNA with this feature is denoted **poly(A)⁺**.

The poly(A) sequence is not coded in the DNA, but is added to the RNA in the nucleus after transcription. The addition of poly(A) is catalyzed by the enzyme **poly(A) polymerase**, which adds ~200 A residues to the free 3′-OH end of the mRNA. We do not know how the length of the added stretch is controlled. It is shortened after the mRNA has entered the cytoplasm.

The poly(A) tract of both nuclear RNA and mRNA is associated with a particular protein, the poly(A)-binding protein (ABP). Related forms of this protein are found in many eukaryotes. One PABP monomer of ~70,000 daltons is bound every 10–20 bases of the poly(A) tail. Thus a common feature in many or most eukaryotes is that the structure of the 3′ end of the mRNA consists of a stretch of poly(A) bound to a large mass of protein.

What is the role of poly(A)? In spite of many suggestions that it confers stability upon mRNA, it has not been possible to demonstrate a systematic correlation between the presence or length of poly(A) and the survival of mRNA. However, the removal of the poly(A) tail does seem to precede degradation of certain mRNAs; stability of mRNA is likely to be connected with poly(A), although the relationship remains to be defined. The ability of the poly(A) to protect mRNA against degradation requires binding of the PABP.

Removal of poly(A) inhibits the initiation of translation *in vitro*, and depletion of PABP has the same effect in yeast *in vivo*, but it is not clear whether these effects are due to a direct influence of poly(A)-PABP on the initiation reaction or have some indirect cause. However, there are many examples in early embryonic development where polyadenylation of a particular mRNA is correlated with its translation. In some cases, mRNAs are stored in a nonpolyadenylated form, and poly(A) is added when their translation is required; in other cases, poly(A)⁺ mRNAs are de-adenylated, and their translation is reduced. But we still do not understand how the poly- or de-adenylation is related to the control of utilization of the mRNA.

The presence of poly(A) has an important practical consequence. The poly(A) region of mRNA can bind by base pairing to oligo(U) or oligo(dT); and this reaction can be used to isolate poly(A)⁺ mRNA. The most convenient technique is to immobilize the oligo(U or dT) on a solid support material. Then when an RNA population is applied to the column, as illustrated in **Figure 10.9**, only the poly(A)⁺ RNA is retained. It can be retrieved by treating the column with a solution that breaks the bonding to release the RNA.

The only drawback to this procedure is that it isolates *all* the RNA that contains poly(A). If RNA of the whole cell is used, for example, both nuclear and cytoplasmic poly(A)⁺ RNA will be retained. If preparations of polysomes are used (a common procedure), most of the isolated poly(A)⁺ RNA will be active mRNA; but some of the RNA in the contaminating RNP fraction also carries poly(A) and therefore will be included. This makes it necessary to use the EDTA-release assay as a functional test when it is important to isolate precisely the active mRNA population. (People often forget this caveat and treat cytoplasmic poly(A)⁺ RNA as though it were equivalent with mRNA.)

The 'cloning' approach for purifying mRNA uses a procedure in which the mRNA is copied to make a complementary DNA strand (known as **cDNA**). Then the cDNA can be used as a template to synthesize a DNA strand that is identical with the original mRNA sequence. The product of these reactions is a double-stranded DNA corresponding to the sequence of the mRNA. This DNA can be reproduced in large amounts by the techniques described in Chapter 21.

The availability of a cloned DNA makes it easy to isolate the corresponding mRNA by hybridization techniques. Even mRNAs that are present in only a very few copies per cell can be isolated by this

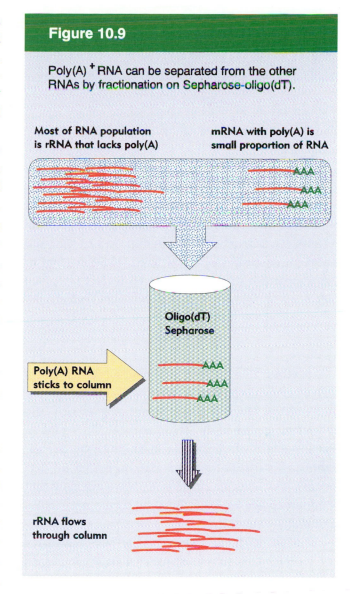

Figure 10.9

Poly(A)⁺ RNA can be separated from the other RNAs by fractionation on Sepharose-oligo(dT).

Most of RNA population is rRNA that lacks poly(A)

mRNA with poly(A) is small proportion of RNA

Oligo(dT) Sepharose

Poly(A) RNA sticks to column

rRNA flows through column

approach. Indeed, only mRNAs that are present in relatively large amounts can be isolated directly without using a cloning step.

Almost all cellular mRNAs possess poly(A). A significant exception is provided by the mRNAs that code for the histone proteins (a major structural component of chromosomal material). These mRNA comprise most or all of the **poly(A)⁻** fraction. The significance of the absence of poly(A) from histone mRNAs is not clear, and there is no particular aspect of their function for which this appears to be necessary.

All eukaryotic mRNAs have a methylated cap at the 5′ end

The 5′ end of mRNA is modified in the eukaryotic cytoplasm (but not in the mitochondrion or chloroplast). The modification reactions are probably common to all eukaryotes.

Transcription starts with a nucleoside triphosphate (usually a purine, A or G). The first nucleotide retains its 5′ triphosphate group and makes the usual phosphodiester bond from its 3′ position to the 5′ position of the next nucleotide. The initial sequence of the transcript can be represented as:

$$5' \quad ppp^A_GpNpNpNp\ldots$$

But when the mature mRNA is treated *in vitro* with enzymes that should degrade it into individual nucleotides, the 5′ end does not give rise to the expected nucleoside triphosphate. Instead it contains *two* nucleotides, connected by a *5′–5′ triphosphate linkage* and also bearing methyl groups. The terminal base is always a guanine that is added to the original RNA molecule *after transcription.*

Addition of the 5′ terminal G is catalyzed by a nuclear enzyme, guanylyl transferase. The reaction occurs so soon after transcription has started that it is not possible to detect more than trace amounts of the original 5′ triphosphate end in the nuclear RNA. The overall reaction can be represented as a condensation between GTP and the original 5′ triphosphate terminus of the RNA. Thus

$$\begin{array}{cc} 5' & 5' \\ Gppp & + \quad pppApNpNp\ldots \end{array}$$

$$\downarrow$$

$$\begin{array}{cc} 5' & 5' \\ GpppApNpNp\ldots & + \quad pp \quad + \quad p \end{array}$$

The new G residue added to the end of the RNA is in the reverse orientation from all the other nucleotides.

This structure is called a **cap.** It is a substrate for several methylations. **Figure 10.10** shows the full structure of a cap after all possible methyl groups have been added. Types of caps are distinguished by how many of these methylations have occurred:

◆ The first methylation occurs in all eukaryotes and consists of the addition of a methyl group to the 7 position of the terminal guanine. A cap that possesses this single methyl group is known as a **cap 0.** This is as far as the reaction proceeds in unicellular eukaryotes. The enzyme responsible for this modification is called guanine-7-methyltransferase.

◆ The next step is to add another methyl group, to the 2′-O position of the penultimate base (which was actually the original first base of the transcript before any modifications were made). This reaction is catalyzed by another enzyme (2′-O-methyl-transferase). A cap with the two methyl groups is called **cap 1.** This is the predominant type of cap in all eukaryotes except unicellular organisms.

◆ In a small minority of cases in higher eukaryotes, another methyl group is added to the second base. This happens only when the position is occupied by adenine; the reaction involves addition of a methyl group at the N^6 position. The enzyme responsible acts only on an adenosine substrate that already has the methyl group in the 2′-O position.

◆ In some species, a methyl group is added to the third base of the capped mRNA. The substrate for this reaction is the cap 1 mRNA that already possesses two methyl groups. The third base modification is always a 2'-O ribose methylation. This creates the **cap 2** type. This cap usually represents less than 10–15% of the total capped population.

In a population of eukaryotic mRNAs, every molecule is capped. The proportions of the different types of cap are characteristic for a particular organism. We do not know whether the structure of a particular mRNA is invariant or can have more than one type of cap.

In addition to the methylation involved in capping, a low frequency of *internal methylation* occurs in the mRNA only of higher eukaryotes. This is accomplished by the generation of N^6 methyladenine residues at a frequency of about one modification per 1000 bases. There are 1–2 methyladenines in a typical higher eukaryotic mRNA, although their presence is not obligatory, since some mRNAs do not have any.

Figure 10.10

The cap blocks the 5' end of mRNA and may be methylated at several positions.

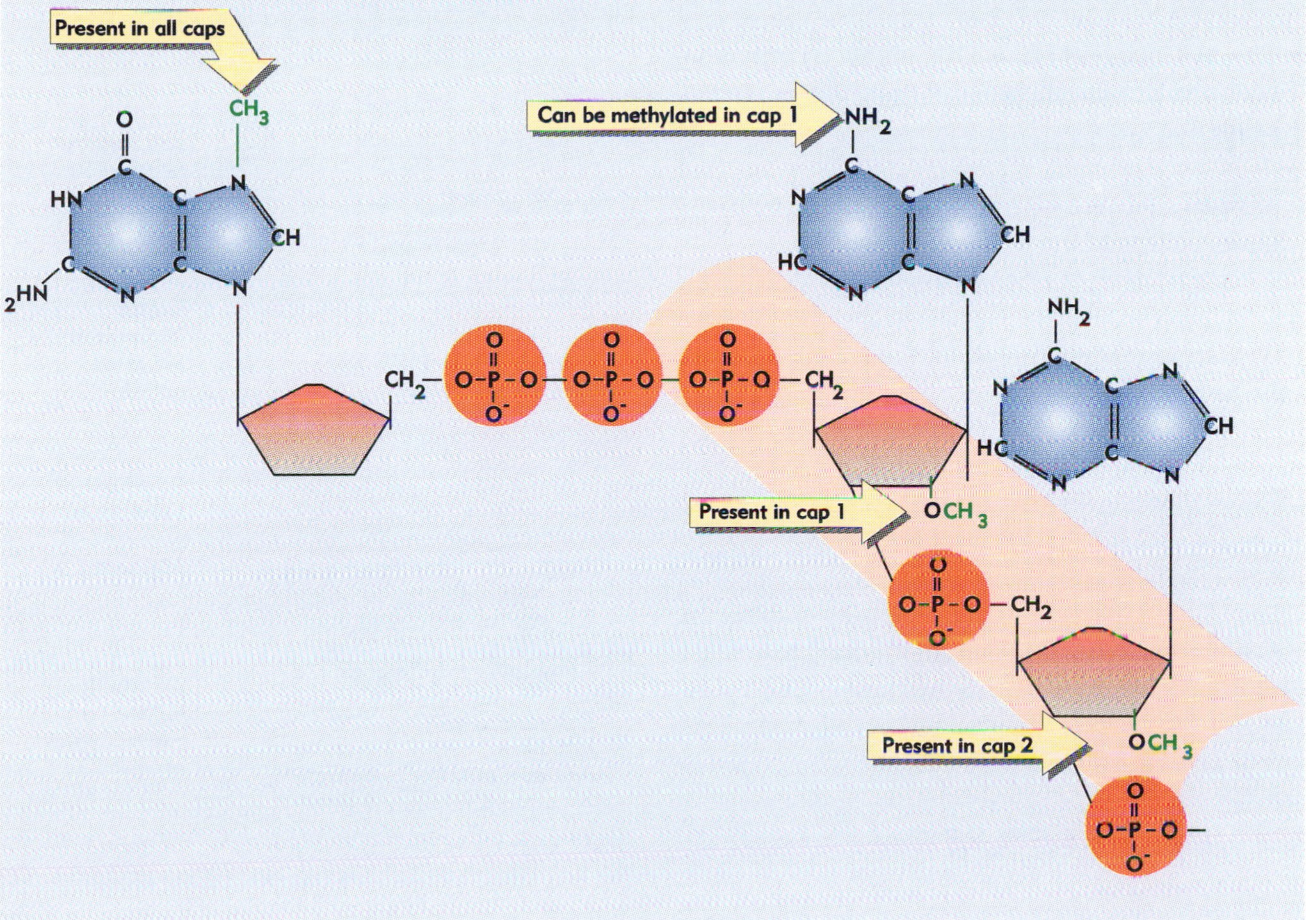

Initiation involves base pairing between mRNA and rRNA

The sites on mRNA where protein synthesis is initiated can be identified by binding the ribosome to mRNA under conditions that block elongation. Then the ribosome remains at the initiation site. When ribonuclease is added to the blocked initiation complex, all the regions of mRNA outside the ribosome are degraded, but those actually bound to it are protected, as illustrated in **Figure 10.11**. The protected fragments can be recovered and characterized.

The initiation sequences protected by bacterial ribosomes are 35–40 bases long. Very little homology exists between the ribosome-binding sites of different bacterial mRNAs. They display only two common features:

◆ The AUG (or less often, GUG or UUG) initiation codon is always included within the protected sequence.

◆ A short sequence is complementary to a sequence close to the 3′ end of 16S rRNA.

Written in reverse direction, the rRNA sequence is the hexamer:

$$3' \ldots U\ C\ C\ U\ C\ C \ldots 5'$$

Virtually all of the known *E. coli* mRNA initiation sites include a sequence complementary to at least a trinucleotide part of this hexamer, and more usually to 4–5 bases. Thus bacterial mRNA contains part or all of the oligonucleotide

$$5' \ldots A\ G\ G\ A\ G\ G \ldots 3'$$

This polypurine stretch is known as the **Shine–Dalgarno** sequence. It lies ~7 bases before the AUG codon.

Does the Shine–Dalgarno sequence pair with its complement in rRNA during mRNA-ribosome bind-

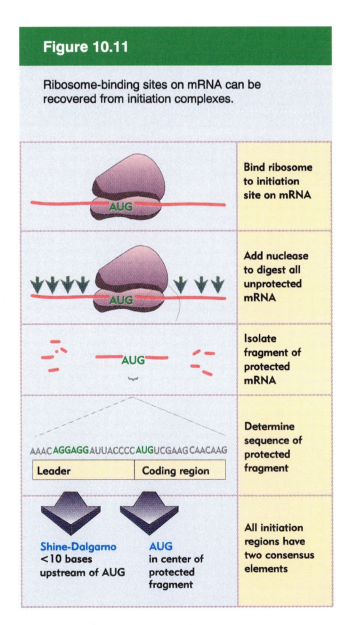

Figure 10.11

Ribosome-binding sites on mRNA can be recovered from initiation complexes.

Bind ribosome to initiation site on mRNA

Add nuclease to digest all unprotected mRNA

Isolate fragment of protected mRNA

Determine sequence of protected fragment

AAAC**AGGAGG**AUUACCCC**AUG**UCGAAGCAACAAG

| Leader | Coding region |

Shine-Dalgarno <10 bases upstream of AUG

AUG in center of protected fragment

All initiation regions have two consensus elements

ing? Mutations of both partners in this reaction demonstrate its importance in initiation. Point mutations in the Shine–Dalgarno sequence can prevent an mRNA from being translated. And the introduction of mutations into the complement in rRNA is deleterious to the cell and changes the pattern of protein synthesis (see Chapter 9). As further confirmation of the base pairing reaction, compensating changes in the Shine–Dalgarno sequence of an mRNA and in an rRNA may create populations of mRNAs whose defect in initiation can be suppressed specifically by ribosomes carrying the mutant rRNA.

The 3′ end of 16S rRNA is highly conserved among bacteria. It is self-complementary and could form the base-paired hairpin drawn in **Figure 10.12**. The sequence complementary to mRNA is part of this potential hairpin, as highlighted in the figure.

The complementary sequence in the rRNA cannot simultaneously be part of the intramolecular hairpin and bind to mRNA. These partnerships could be alternatives, in which case the initiation reaction may involve disrupting the terminal hairpin to allow base pairing between mRNA and rRNA, as shown in the figure. After initiation, the mRNA–rRNA duplex might be broken by reconstitution of the rRNA base-paired hairpin. This mechanism could reconcile the need to form a stable initiation complex at a particular site with the need later to move off along the mRNA.

The sequence at the 3′ end of rRNA is well-conserved between prokaryotes and eukaryotes. In the 20 nucleotides between the adjacent double-methylated adenines and the 3′ end, there are only two significant changes. In bacteria, there are two U residues, whereas in higher eukaryotes there are two A residues (in lower eukaryotes the sequence is an intermediate AU). And in all eukaryotes there is a deletion of the 5-base sequence CCUCC that is the principal complement to the Shine–Dalgarno sequence.

There have been many suggestions that some

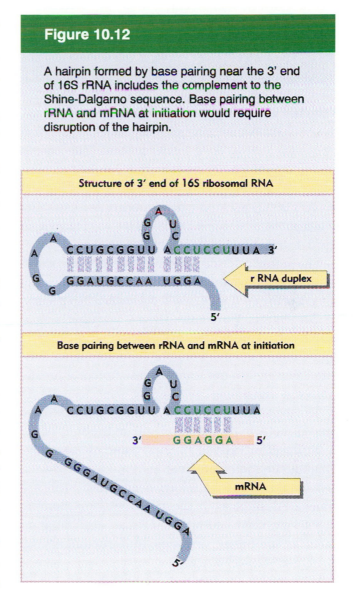

Figure 10.12

A hairpin formed by base pairing near the 3′ end of 16S rRNA includes the complement to the Shine-Dalgarno sequence. Base pairing between rRNA and mRNA at initiation would require disruption of the hairpin.

other sequence in eukaryotic rRNA might provide a counterpart to the Shine–Dalgarno prokaryotic consensus. However, we have no evidence to support the occurrence of base pairing between eukaryotic mRNA and 18S rRNA. If indeed there is no such mechanism, this constitutes a significant difference in the mechanism of initiation.

Small subunits migrate to initiation sites on eukaryotic mRNA

Virtually all eukaryotic mRNAs are monocistronic, but each mRNA usually is substantially longer than necessary just to code for its protein. The coding region therefore occupies only a part of the messenger. The average mRNA in eukaryotic cytoplasm is 1000–2000 bases long, has a methylated cap at the 5′ terminus, and carries 100–200 bases of poly(A) at the 3′ terminus.

The nontranslated 5′ leader is relatively short, usually (but not always) less than 100 bases. The length of the coding region is determined by the size of the protein. The nontranslated 3′ trailer is often rather long, sometimes ~1000 bases. By virtue of its location, the leader cannot be ignored during initiation, but we do not know of any function for the trailer in translation.

The ribosomes of eukaryotic cytoplasm do not bind directly to the initiation site at the start of the coding region. Instead, the first feature to be recognized is the methylated cap that marks the 5′ end. Messengers whose caps have been removed are not translated efficiently in the *in vitro* systems. Binding of 40S subunits to mRNA requires several initiation factors, including proteins that recognize the structure of the cap.

We have dealt with the process of initiation as though the ribosome-binding site is always freely available. However, its availability may be impeded by secondary structure, reducing its ability to be recognized for initiation. One function of cap-binding proteins is to unwind the leader region, and thus to help ensure that ribosome-binding sites are available in the single-stranded state. The summary of eukaryotic initiation factors discussed previously in Figure 7.21 shows that the factors eIF4A and eIF4B play a role in unwinding the structure of the leader of mRNA.

Modification at the 5′ end occurs to almost all cellular or viral mRNAs, and is essential for their translation in eukaryotic cytoplasm (although it is not needed in organelles). The sole exception to this rule is provided by a few viral mRNAs (such as poliovirus) that are not capped; *only* these exceptional viral mRNAs can be translated *in vitro*

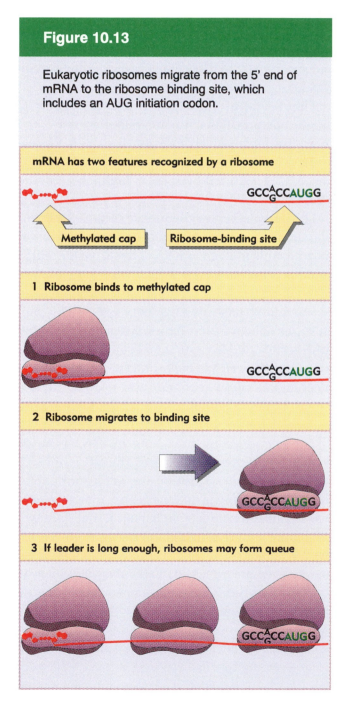

Figure 10.13

Eukaryotic ribosomes migrate from the 5′ end of mRNA to the ribosome binding site, which includes an AUG initiation codon.

mRNA has two features recognized by a ribosome

Methylated cap   Ribosome-binding site

1 Ribosome binds to methylated cap

2 Ribosome migrates to binding site

3 If leader is long enough, ribosomes may form queue

without caps, so they must have some alternative feature that renders capping unnecessary.

Some viruses take advantage of this difference. Poliovirus infection inhibits the translation of host mRNAs. This is accomplished by interfering with the cap binding proteins that are needed for initiation of cellular mRNAs, but that are superfluous for the noncapped poliovirus mRNA.

Sometimes the AUG initiation codon lies within 40 bases of the 5′ terminus of the mRNA, so that both the cap and AUG lie within the span of ribosome binding. But in many mRNAs the cap and AUG are farther apart, in extreme cases ~1000 bases distant. Yet the presence of the cap still is necessary for a stable complex to be formed at the initiation codon. How can the ribosome rely on two sites so far apart?

Figure 10.13 illustrates the 'scanning' model, which supposes that the 40S subunit initially recognizes the 5′ cap and then 'migrates' along the mRNA. Scanning from the 5′ end is a linear process. When 40S subunits scan the leader region, they melt secondary structure hairpins with stability of <-30 kcal, but hairpins of greater stability impede or prevent migration.

Migration stops when the 40S subunit encounters the AUG initiation codon. Usually, although not always, the first AUG triplet sequence will be the initiation codon. The AUG triplet by itself does not seem sufficient to halt migration; probably it is recognized efficiently as an initiation codon only when it is in the right context (which allows an inappropriate AUG triplet to be bypassed). The optimal context consists of the sequence $GCC_G^ACCAUGG$. The purine (A or G) 3 bases before the AUG codon, and the G immediately following it, are the most important, and influence efficiency of translation by 10×; the other bases have much smaller effects. When the leader sequence is long, further 40S subunits can recognize the 5′ end before the first has left the initiation site, creating a queue of subunits proceeding along the leader to the initiation site.

Binding is stabilized at the initiation site. When the 40S subunit is joined by a 60S subunit, the intact ribosome is located at the site identified by the protection assay. A 40S subunit protects a region of up to 60 bases; when the 60S subunits join the complex, the protected region contracts to about the same length of 30–40 bases seen in prokaryotes.

The idea that ribosomes must start at the 5′ end is consistent with the monocistronic nature of eukaryotic mRNAs. Some exceptional viral mRNAs contain more than one coding region. But in these cases, only the coding region nearest the 5′ end is translated from the intact mRNA. The other(s) can be read only after the mRNA has been cleaved to generate a new 5′ end in the vicinity of the internal initiation codon. This behavior supports the idea that internal initiation sites cannot be recognized directly.

Processing is necessary to produce some RNAs

Processing of a mature molecule from its precursor is a crucial stage in the production of many forms of RNA. Prokaryotic and eukaryotic rRNAs and tRNAs are synthesized in the form of more complex primary transcripts from which the mature RNA molecule must be released. By contrast, most prokaryotic mRNAs *are* primary transcripts, so only in the exceptional case is processing necessary for translation. In eukaryotes, however, producing an mRNA from an interrupted gene is the most labor-intensive of all RNA processing, and involves a series of splicing reactions in which the exons are connected via removal of the introns (see Chapter 31).

Processing reactions are highly specific, producing mature RNA molecules with unique 5′ and 3′ ends. The reactions are the responsibility of particular ribonucleases, and both endonucleases and exonucleases are involved. Endonucleases are involved in releasing a mature molecule from a longer primary transcript. Exonucleases are involved in trimming reactions in which the extra residues are whittled away, base by base. All known exonucleases proceed along the nucleic acid chain from the 3′ end. *Thus additional material at the 3′ end can be trimmed, but additional material at the 5′ end can be released only by cutting.*

As well as participating in specific maturation reactions, ribonucleases accomplish the degradation of superfluous RNA. 'Superfluous' molecules include mature species that turn over, as well as the extraneous material discarded during the cutting of precursors. In particular, mRNA whose time has come must be degraded (its time comes rapidly in bacteria, more slowly in eukaryotes, but in very few cases is mRNA stable enough to survive for a protracted period).

We might expect that the ribonucleases engaged in RNA turnover should be less specific than those involved in particular maturation pathways, since their role is to degrade the RNA completely rather than to generate specific termini. Both endoribonucleases and exoribonucleases are found in this class. The degradation of an RNA to mononucleotides is its final maturation, returning its components for reutilization via the appropriate metabolic pathways.

Enzymes with greater specificity may recognize the conformation of the RNA substrate rather than a particular sequence of bases. Substrate recognition takes the form of seeking a specific feature of secondary structure (such as a hairpin of certain size), or examining the overall secondary/tertiary structure (as seems to apply in tRNA processing).

To understand the flow of nucleotides into and out of RNA, we should like to define the entire set of ribonucleases for some cell type. The only case in which this seems possible is presented by *E. coli*. Here not only can the enzymes be defined biochemically, but their effects *in vivo* can be determined by selecting bacteria in which a particular enzyme has been mutated. Indeed, both lines of evidence often are needed to distinguish enzymes or to show that two apparently different

Table 10.2

E.coli contains a relatively small number of ribonucleases. Some are concerned exclusively with processing tRNA molecules from longer precursors. Others can degrade RNAs of more than one type.

Enzyme	Locus	RNA Substrate Class	Type
Ribonuclease P	rnpA,B	tRNA 5′ end	Endo
Ribonuclease BN	?	tRNA 3′ end	Exo
Ribonuclease D	rnd	tRNA 3′ end	Exo
Ribonuclease T	?	tRNA 3′ CCA	Exo
Ribonuclease III	rnc	rRNA & mRNA	Endo
Ribonuclease R	?	rRNA & mRNA	Exo
Ribonuclease E	rne	5S rRNA	Endo
Ribonuclease I	rna	Most RNA	Endo
Ribonuclease II	rnb	RNA unstructured 3′	Exo
Polynucleotide phosphorylase	pnp	RNA unstructured 3′	Exo
Ribonuclease H	rnhA,B	RNA-DNA hybrids	Endo

activities reside in the same enzyme. (We may expect that yeast also will become approachable in these terms.)

We do not yet know precisely how many ribonucleases there are in *E. coli,* but the current count is ~12. The enzymes that have been purified are summarized in **Table 10.2**. It seems likely that combinations of a small number of ribonuclease activities provide the necessary range of reactions to process rRNA and tRNA. Although mRNA is actively degraded, the process has been difficult to analyze. Only RNAase E has been implicated so far (see below).

Do these ribonucleases play essential roles in the cell? Mutants in the endoribonucleases (except ribonuclease I, which is without effect) accumulate unprocessed precursors, but are viable. Mutants in the exonucleases often have apparently unaltered phenotypes, which suggests that one enzyme can substitute for the absence of another. Mutants lacking multiple enzymes sometimes are inviable.

Several of the enzymes can be implicated in tRNA processing, because mutants accumulate uncleaved precursors to tRNA. RNAase III can be implicated in the processing of rRNAs and some phage mRNAs. Less is known about mRNA degradation than about the processing of any class of RNA.

Stability of mRNA is determined by particular sequences

The stability of mRNA can be measured in two ways. Both rely on halting the transcription of mRNA and then following the fate of the existing mRNA molecules in the cell:

◆ The **functional half-life** measures the ability of the mRNA to serve as template for synthesis of its protein product.

◆ The **chemical half-life** is determined by measuring the decline in the amount of mRNA able to hybridize with the DNA of its gene. Generally the chemical decline of bacterial mRNA lags slightly behind the functional decline.

The discrepancy between functional and chemical half-lives suggests that the degradation of mRNA involves an initial step that prevents its use as a template, but which is not detectable as a structural alteration that impedes its hybridization. For example, a single cleavage in the mRNA would not affect its ability to hybridize with DNA, but could prevent its translation.

We should not allow our need to visualize the 'average' case obscure the harsh reality that every mRNA is in statistical jeopardy at all times, with a constant probability that its decay will begin. Thus 'young' mRNAs are as likely to be attacked as 'old' mRNAs. Some copies of an mRNA are translated many times, while others function hardly at all. This random life expectancy is a feature of individual mRNA molecules of both prokaryotes and eukaryotes. But the overall translational yield of any messenger sequence is predictable.

We understand the process of degrading mRNA only in outline. Both endonucleases and exonucleases are involved in bacteria. Susceptibility to degradation is conferred by sequences that are targets for endonucleolytic attack. Resistance to degradation is conferred by regions that impede the progress of exonucleases. But we have yet to identify the endonucleases and exonucleases involved in these processes. And we do not yet have enough data to form a systematic view of how particular sequences determine the stability of an RNA.

A rather general model for the degradation of

Figure 10.14

Degradation of mRNA is a two stage process, in which endonucleolytic cleavages occur generally moving 5'-3' behind the ribosomes, after which the released fragments are degraded by exonucleases that move 3'-5'.

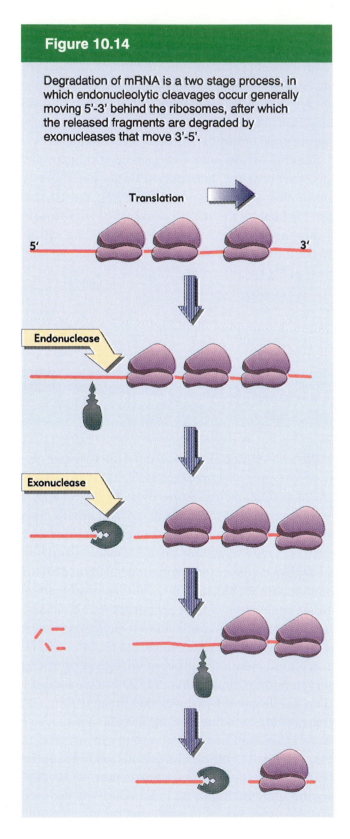

mRNA is shown in **Figure 10.14**. The *overall* direction of degradation (as measured by loss of ability to synthesize proteins) is 5'–3', which probably results from a succession of endonucleolytic cleavages following the last ribosome. Degradation of the released fragments of mRNA into nucleotides proceeds by exonucleolytic attack from the free 3'-OH end toward the 5' terminus (that is, in the opposite direction from transcription). Several 3' ends may be generated by endonucleolytic cleavage within the mRNA, and we now know of several cases in which a bacterial mRNA has a specific site that is a target for attack by an endonuclease. Once the endonuclease has cleaved the mRNA, exonucleolytic degradation starts at the new 3' end and degrades the 5' fragment of the mRNA.

Bacterial mutants that have a defective ribonuclease E have increased stability (2- to 3-fold) of mRNA as well as of the small 5S ribosomal RNA. RNAase E is the enzyme that is responsible for processing 5S rRNA from the primary transcript, which is a specific processing event. It is also involved in the general decay of mRNA, which requires much less specificity. It may be the case that in addition it undertakes specific degradation of certain highly unstable bacterial mRNAs.

An endonucleolytic attack (by an enzyme such as RNAase E) releases fragments that may have different susceptibilities to exonucleases. A region of secondary structure within the mRNA may provide an obstacle to the exonuclease, thus allowing the regions on its 5' side to survive longer and be translated more often.

In one case, the increased stability of a phage mRNA is an intrinsic property of the messenger. The 5' leader of a stable mRNA coded by phage T4 confers stability on a bacterial mRNA to which it is attached. This is interesting because it implies that the mere absence of sites subject to endonucleolytic cleavage is not the sole determinant of stability in this case.

Measurements of the stability of eukaryotic

mRNA often identify more than one distinct component. Typically about half of the mRNA of mammalian cells in tissue culture has a half-life of ~6 hours or so, while the other half has a stability roughly equivalent to the length of the cell cycle, ~24 hours. In differentiated cells devoted to the synthesis of specific products, some mRNAs are even more stable. The stability of an mRNA may be regulated by external agents, for example, hormones, providing another point at which translatability can be controlled.

Some eukaryotic mRNAs are less stable. A common feature is that decreased stability is conferred upon eukaryotic mRNA by an AU-rich sequence of ~50 bases that is found naturally in the 3' trailer region of several unstable mRNAs.

Current information about the stability of mRNA rests upon a few individual cases, and is therefore anecdotal. However, it demonstrates that stability is influenced by particular nucleotide sequences that are targets for endonucleases or that impart a secondary structure that affects overall message stability. These sequences may be located in different regions of the mRNA molecule.

Summary

Messenger RNA is transcribed from one strand of DNA and is therefore complementary to this (noncoding) strand and is identical with the other (coding) strand. A typical mRNA contains both a nontranslated 5' leader and 3' trailer as well as coding region(s). Bacterial mRNA is usually polycistronic, with nontranslated regions between the cistrons. It possesses a hexamer upstream of each AUG initiation codon, which is complementary to rRNA and pairs with it during initiation. Bacterial mRNA has an extremely short half-life, only a few minutes. The 5' end starts translation even while the downstream sequences are being transcribed. Little is known about the enzymes responsible for degrading mRNA or about the specificities with which different mRNAs are attacked.

Eukaryotic mRNA is stable for several hours, and must be processed in the nucleus before it is transported to the cytoplasm for translation. A methylated cap is added to the 5' end. Most eukaryotic mRNA has an ~200 base sequence of poly(A) added to its 3' terminus, but poly(A)⁻ mRNAs appear to be translated and degraded with the same kinetics as poly(A)+ mRNAs. Eukaryotic mRNA exists as a ribonucleoprotein particle; in some cases mRNPs are stored that fail to be translated. The mRNA of eukaryotic cytosol is monocistronic. Small ribosomal subunits bind to the cap at the 5' end and migrate downstream, scanning the message for an AUG codon. Usually initiation occurs at the first AUG triplet.

Further reading

Reviews

Bacterial mRNAs have been reviewed mostly in the context of the individual genes that they represent; citations will be found in Chapters 14–17. The production of eukaryotic mRNA is usually treated in terms of the processing reactions; citations will be found in Chapter 29.

Interation of bacterial mRNAs with ribosomes has been reviewed by **Gold** (*Ann. Rev. Biochem.* **57**, 199–233, 1988).

Capping was reviewed by **Bannerjee** (*Microbiol. Rev.* **44**, 175–205, 1980).

The relationship between polyadenylation and function of mRNA has been considered by **Jackson and Standard** (*Cell* **62**, 15–24, 1990)

The scanning hypothesis has been propagated by **Kozak** (*Cell* **15** 1109–1123, 1978).

CONSTRUCTING THE CELL

The code-script must itself be the operative factor bringing about the development [of the organism]. But ... with the molecular picture of the gene it is no longer inconceivable that the miniature code should precisely correspond with a highly complicated and specified plan of development and should somehow contain the means to put it into operation.

Erwin Schrödinger, 1945

CHAPTER 11

The apparatus for protein localization

The eukaryotic cell is a highly ordered structure. It is obvious that proteins that are components of macromolecular structures are incorporated into specific locations, but it is equally true that all proteins except cytosolic enzymes are directed towards specific destinations. Examples of protein localization extend from the egg to the terminally differentiated cell. In oocytes of some species, an initial asymmetry in the distribution of proteins may be required for subsequent development of the organism; similarly, the functions of specialized cells are often determined by macromolecular assemblies of particular proteins. Localization is the *sine qua non* of the internal structure of every eukaryotic cell, and a major determinant of structure is the ability of individual proteins to be localized in the appropriate compartments. We can classify proteins by their sites of synthesis and ultimate locations:

◆ 'Soluble' proteins are not localized in any particular organelle. They are synthesized in the cytosol, and remain there, where they function as individual catalytic centers, acting on metabolites that are in solution in the cytosol.

◆ Macromolecular structures constructed from cytosolic proteins (and sometimes incorporating other components) may be located at particular sites in the cytoplasm; for example, centrioles are associated with the regions that become the poles of the mitotic spindle.

◆ Nuclear proteins must be transported from their site of synthesis in the cytosol through the nuclear membrane before they can take their place. Many nuclear proteins are components of chromatin itself, but others are part of the nuclear lamina, matrix, or envelope.

◆ Cytoplasmic organelles contain proteins synthesized in the cytosol and transported specifically to (and through) the organelle membrane, for example, to the mitochondrion or (in plant cells) to the chloroplast. Mitochondria and chloroplasts also possess the ability to synthesize proteins, and some of their proteins are synthesized within the organelle, where they remain. (We discuss protein synthesis in organelles in Chapter 25.)

◆ The cytoplasm contains a series of membranous bodies, including endoplasmic reticulum, Golgi apparatus, endosomes, and lysosomes. This is sometimes referred to as the 'reticuloendothelial system' or the 'vacuolar apparatus'. Proteins destined to reside within these compartments are inserted at the start of their synthesis into the membranes of endoplasmic reticulum, and then are directed to their particular locations by the transport system of the Golgi apparatus.

◆ Proteins that are secreted from the cell must pass through the plasma membrane to the exterior. They start their synthesis in the same way as proteins associated with the reticuloendothelial system, but pass entirely through the system instead of halting at some particular point within it.

Figure 11.1 maps the cell in terms of these systems and the possible ultimate destinations for a newly

Figure 11.1

Overview: proteins may be synthesized on "free" or "membrane-bound" ribosomes. Proteins synthesized on free ribosomes are released into the cytosol. Some have signals for targeting to organelles such as the nucleus or mitochondria. Proteins synthesized on "membrane-bound" ribosomes pass into the endoplasmic reticulum, along to the Golgi, and then through the plasma membrane, unless they possess signals that cause retention at one of the steps on the pathway. They may also be directed to other organelles, such as lysosomes.

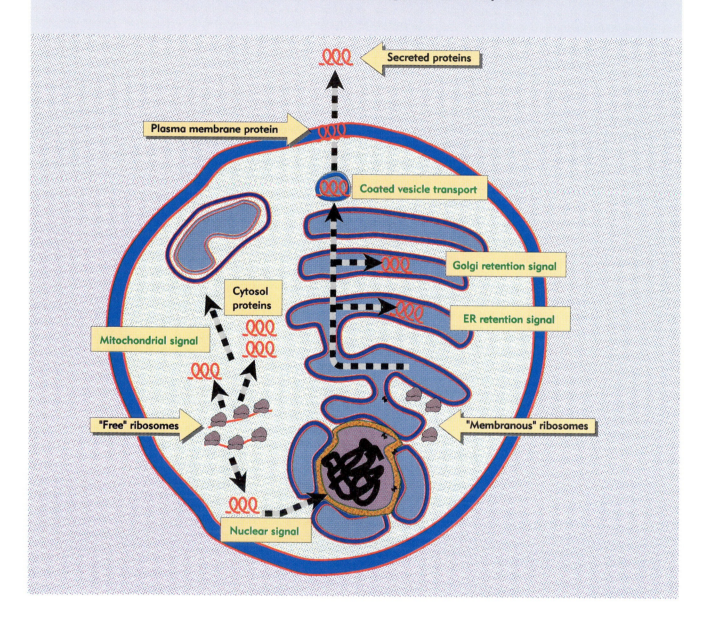

synthesized protein. Except for soluble cytosolic proteins, protein synthesis is only the beginning of the pathway by which a protein finds its appropriate location in the cell. The process of **protein sorting** or **protein trafficking** depends on the ability of the protein to interact with the membrane systems of the cell. By virtue either of its sequence or of covalent groups added to it, a protein is recognized by receptors in a particular membrane. The process of passing through a membrane into an organelle is called **protein translocation**. Depending on its destination, a traveling protein may require a single translocation into its organelle, or may pass through several membranes.

The starting point for protein synthesis is the polyribosome. Polyribosomes can be divided into two classes: 'free' and 'membrane-bound'. They are engaged in the synthesis of different groups of proteins:

◆ 'Free' polysomes are responsible for synthesizing proteins that are released directly into the cytosol. The destination of the protein depends upon whether it possesses specific localization signals. In the absence of any specific signal, the 'default' is for a protein to remain in the cytosol in quasi-soluble form. Various types of signals are used to route proteins to specific destinations; they all take the form of short sequence motifs. These signals function after synthesis of the protein has been completed, so the process is called **post-translational translocation**. **Figure 11.2** summarizes some signals used by proteins released from cytosolic ribosomes. The usual route to mitochondria and chloroplasts rests on N-terminal sequences of ~25 amino acids in length. The sequences are recognized by receptors on the organelle envelope, and usually are cleaved during translocation. Import into the nucleus results from the presence of a variety of rather short sequences within proteins; these 'nuclear localization signals' enable the proteins to pass through nuclear pores (which we discuss in Chapter 12). Transport to the peroxisome is determined by a very short C-terminal sequence.

◆ Membrane-bound polysomes synthesize proteins which start their sorting route *by entering the endoplasmic reticulum while they are being synthesized*. This process is called **co-translational translocation**. Often it is initiated by an N-terminal sequence, but in some cases a sequence within the protein is used. The ultimate locations of these proteins within the cell depend on how they are directed as they transit the endoplasmic reticulum and Golgi apparatus. Some of the destinations and signals are summarized in **Figure 11.3**. The principle is that

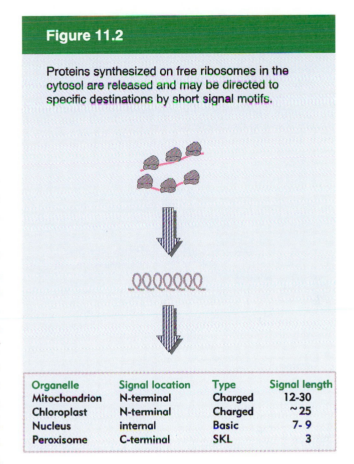

Figure 11.2

Proteins synthesized on free ribosomes in the cytosol are released and may be directed to specific destinations by short signal motifs.

Organelle	Signal location	Type	Signal length
Mitochondrion	N-terminal	Charged	12-30
Chloroplast	N-terminal	Charged	~25
Nucleus	internal	Basic	7- 9
Peroxisome	C-terminal	SKL	3

some sort of permit is required for residence in any particular part of the membrane system; the permit most often takes the form of a short amino acid sequence, but there are also other types of information. The 'default pathway' takes a protein through the ER, into the Golgi, and on to the plasma membrane. Proteins that reside in the ER possess a C-terminal tetrapeptide (which actually provides a signal for them to return to the ER from the Golgi). The signal that diverts a protein to the lysosome is a covalent modification: the addition of a particular sugar residue.

A common feature is found in proteins that use N-terminal sequences to be transported co-translationally to the ER or post-translationally to mitochondria or chloroplasts. *The N-terminal sequence is cleaved from the protein during the act of protein translocation*. The N-terminal sequence comprises a **leader** that is not part of the mature

Figure 11.3

Membrane-bound ribosomes have proteins with N-terminal sequences that enter the ER during synthesis. The proteins may flow through to the plasma membrane or may be diverted to other destinations by specific signals.

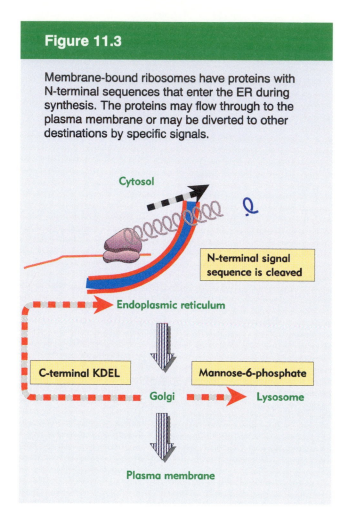

sequence that describes the additional regions present on proteins that exist as *stable* precursors. Some proteins have both. For example, insulin is initially synthesized as **preproinsulin**; the **pre** sequence is cleaved early in secretion, generating **proinsulin**, which is the substrate for subsequent processing to mature insulin.)

Proteins that associate with membranes via N-terminal leaders use a hierarchy of signals to find their final destination. The leader sequence itself introduces the protein to the membrane; the intrinsic consequence of the interaction is for the protein to pass through the membrane into the compartment on the other side. However, other sequences within the protein may cause it to remain within the membrane, and also determine the particular conformation that it adopts. A protein that resides within a membrane is called an **integral membrane protein**. (We discussed the ability of proteins to reside in membranes in Chapter 2.)

The ability to enter a membrane seems to be a function solely of the leader sequence. However, a persistent theme in membrane passage is that control (or delay) of protein folding is an important feature. It may be necessary to maintain a protein in an unfolded state because of the geometry of passage: the mature protein could simply be too large to fit into the available channel. Another reason could be the need to minimize the exposure of hydrophilic groups as the protein passes through the hydrophobic lipid bilayer. Once through the membrane, the protein refolds to its mature conformation.

protein. The protein carrying this leader is called a **preprotein**, and is a transient precursor to the mature protein.

(The **pre** leader sequence is distinct from the **pro**

Post-translational membrane insertion depends on leader sequences

Mitochondria and chloroplasts both are able to synthesize all of their nucleic acids and some of their proteins. Mitochondria synthesize only a few (~10) organelle proteins; chloroplasts synthesize ~20% of total organelle protein. Organelle proteins that are synthesized in the cytosol are produced by the same pool of free ribosomes that synthesizes cytosolic proteins. They must then be imported into the organelle.

Proteins that enter mitochondria or chloroplasts by a post-translational process usually are synthesized in the form of a precursor 12–70 amino acids longer than the mature protein. The leader sequence is responsible for primary recognition of the outer membrane of the organelle. As shown in the simplified diagram of **Figure 11.4**, the leader sequence initiates the interaction between the precursor and the organelle membrane. The leader sequence passes through the membrane, and is cleaved by a protease on the organelle side. Translocation continues until the entire protein has passed through the membrane, or until some internal sequence causes translocation to terminate.

Leader sequences of proteins imported into mitochondria and chloroplasts have little homology. The leaders are usually hydrophilic, consisting of stretches of uncharged amino acids

interrupted by basic amino acids, and they lack acidic amino acids. An example is given in **Figure 11.5**. The lack of homology among leader sequences implies that features of the secondary or tertiary structure, or the general nature of the region, must be involved in recognition. One possibility is that the leader forms an amphiphilic helix (in which the charged groups are on one face and the uncharged groups are on the other side).

The leader sequence contains all the information needed to localize an organelle protein. The ability of a leader sequence can be tested by constructing an artificial protein in which a leader from an organelle protein has been joined to some other protein, one that usually is located in the cytosol.

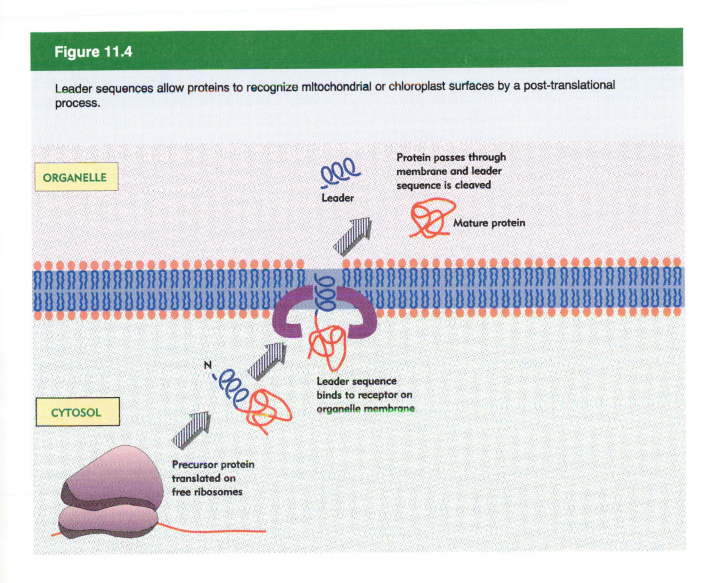

Figure 11.4

Leader sequences allow proteins to recognize mitochondrial or chloroplast surfaces by a post-translational process.

ORGANELLE

Protein passes through membrane and leader sequence is cleaved

Leader

Mature protein

Leader sequence binds to receptor on organelle membrane

N

CYTOSOL

Precursor protein translated on free ribosomes

Figure 11.5

The leader sequence of yeast cytochrome c oxidase subunit IV consists of 25 neutral and basic amino acids. The first 12 amino acids are sufficient to transport any attached polypeptide into the mitochondrial matrix.

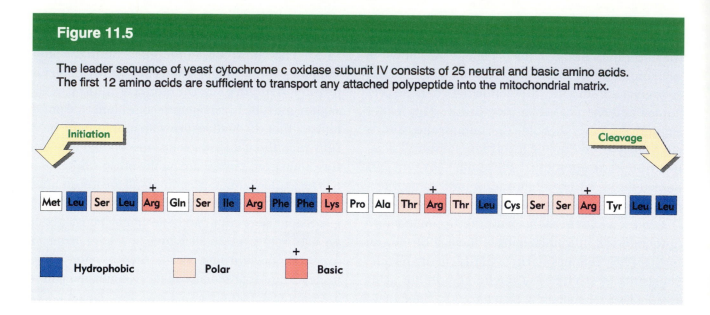

The experiment is usually performed by constructing a hybrid gene, which is then translated into the hybrid protein.

Several leader sequences have been shown by such experiments to function independently to target any attached sequence to the mitochondrion or chloroplast. For example, if the leader sequence given in Figure 11.5 is attached to the murine cytosolic protein DHFR (dihydrofolate reductase), the DHFR becomes localized in the mitochondrion.

The leader sequence and the transported protein appear to represent domains that fold independently. Irrespective of the sequence to which it is attached, the leader must be able to fold into an appropriate structure to be recognized by receptors on the organelle envelope. The attached polypeptide sequence plays no part in recognition of the envelope.

Hydrolysis of ATP is required for translocation across the membrane. The ATP could be required directly to phosphorylate participants in protein translocation, to provide an energy donor for a component of the transport apparatus, or to unfold the polypeptide that is being transported. ATP hydrolysis does not seem to be needed for the early stages of association with the membrane, and may therefore be involved with entry into the membrane or with the translocation event itself.

What provides the motive force for membrane passage? Mitochondrial import (and bacterial export) require an electrochemical potential across the inner membrane to transfer the N-terminal part of the leader. The membrane potential is not needed for the subsequent transfer of the rest of the protein, which implies that passage of the leader sequence is sufficient to overcome the barrier posed by the structure of the lipid bilayer.

What restrictions are there upon transporting a hydrophilic protein through the hydrophobic membrane? An insight into this question is given by the observation that methotrexate, a ligand for the enzyme DHFR, blocks transport into mitochondria of DHFR fused to a mitochondrial leader. The tight binding of methotrexate prevents the enzyme from unfolding when it is translocated through the membrane. So although the sequence of the transported protein is irrelevant for targeting purposes, in order to follow its leader through the membrane, it requires the flexibility to assume an unfolded conformation.

Transport through the mitochondrial membrane may thus require a series of folding and unfolding steps. When a protein is released into the cytosol upon completion of synthesis, it becomes folded. To pass through the mitochondrial membrane, it must be unfolded. After a protein has passed through the membrane in an unfolded state, it must fold into its mature conformation. This final folding appears to require the participation of an accessory protein, a 'chaperonin'.

Many of the components of the mitochondrial import apparatus have been identified. The first stage in import is binding to a receptor located in the outer membrane that passes the protein into the import channel. There are a variety of receptors with varying specificities for imported proteins: some proteins have unique receptors, while others pass through a generalized apparatus.

When the imported protein reaches the matrix, it is bound by the matrix protein hsp70, which is a chaperone related to other hsp70 'stress proteins'. (Stress—or heat shock—proteins are chaperones that are induced by stress, and which have the role of binding to improperly folded proteins, as described previously in Chapter 2.) It is possible that the high affinity of hsp70 for the unfolded conformation of the protein as it emerges from the inner membrane helps to 'pull' the protein through the channel.

A major chaperone activity in the mitochondrial matrix is provided by the stress protein hsp60, which exists as a large oligomer (14 subunits of 60,000 daltons). It is homologous to the bacterial chaperone GroEL. When hsp60 binds to imported proteins, it maintains them in a loosely folded form. It releases them in a process requiring ATP cleavage. The hsp60 oligomer could provide a framework on which the matrix proteins assemble into their mature form. Association with hsp60 is necessary for joining of the subunits of imported proteins that form oligomeric complexes. An imported protein may be 'passed on' from hsp70 to hsp60 in the process of acquiring its proper conformation.

A mitochondrial protein therefore folds under two different conditions. Ionic conditions and the chaperones that are present are different in the cytosol and in the mitochondrial matrix. Folding in the mitochondrion is specifically assisted by hsp60. It is therefore possible that different conformations are achieved, that is, that a mitochondrial protein can attain its mature conformation *only* in the mitochondrion.

Leader sequences determine protein location within mitochondria and chloroplasts

The mitochondrion is surrounded by an envelope consisting of two membranes. Proteins imported into mitochondria may be located in the outer membrane, the intermembrane space, the inner membrane, or the matrix. A protein that is a component of one of the membranes may be oriented so that it faces one side or the other.

What is responsible for directing a mitochondrial protein to the appropriate compartment? The 'default' pathway for a protein imported into a mitochondrion is to move through both membranes into the matrix. This property is conferred by the N-terminal part of the leader sequence. A protein that is localized within the intermembrane space or in

the inner membrane itself requires an additional signal, which specifies its destination within the organelle. A multipart leader contains signals that function in a hierarchical manner, as summarized in **Figure 11.6**. The first part of the leader targets the protein to the organelle, and the second part is required if its destination is prior to the matrix. The two parts of the leader are removed by successive cleavages.

Cytochrome c1 is an example. It is bound to the inner membrane and faces the intermembrane space. Its leader sequence consists of 61 amino acids, and can be divided into regions with different functions. The intact leader transports an attached

Figure 11.6

When a leader sequence is recognized by the outer membrane, the protein enters the membrane. In the absence of further signals, it is transferred into the matrix, where the leader is cleaved. If it has a membrane-targeting signal, it may become located between or within the membranes.

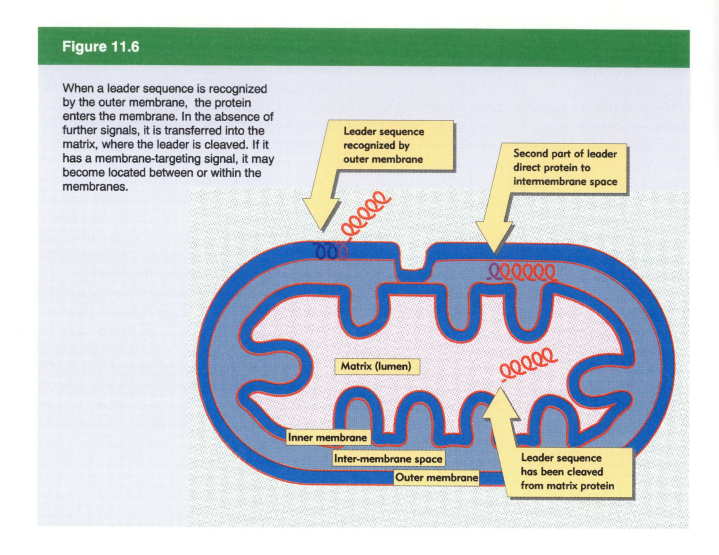

Leader sequence recognized by outer membrane

Second part of leader direct protein to intermembrane space

Matrix (lumen)

Inner membrane

Inter-membrane space

Outer membrane

Leader sequence has been cleaved from matrix protein

sequence—such as murine DHFR—into the inter-membrane space. But the sequence of the first 32 amino acids alone, or even the N-terminal half of this region, can transport DHFR all the way into the matrix. *So the first part of the leader sequence (32 N-terminal amino acids) comprises a matrix-targeting signal.*

Why does the intact leader direct a protein into the intermembrane space? *The region following the matrix-targeting signal (comprising 19 amino acids of the leader) provides another signal that localizes the protein at the inner membrane or within the inter-membrane space.* For working purposes, we call this the *membrane-targeting signal.*

Cleavage of the matrix-targeting signal is the sole processing event required for proteins that reside in the matrix. This signal must also be cleaved from proteins that reside in the intermembrane space; but following this cleavage, the membrane-targeting signal directs the protein to its destination in the outer membrane, intermembrane space, or inner membrane. Then it in turn is cleaved.

A single protease is involved in cleaving the matrix-targeting signal, irrespective of the final destination of the protein. This protease is a water soluble, Mg^{2+}-dependent enzyme that is located in the matrix. *So the N-terminal sequence must reach the matrix, even if the protein ultimately will reside in the intermembrane space.*

The N-terminal matrix-targeting signal probably functions in the same manner for all mitochondrial proteins. Its recognition by a receptor on the outer membrane leads to transport through the two membranes as described in the previous section. So the initial transfer event is the same for all proteins entering the mitochondrion, irrespective of their final destination.

The nature of the leader determines the subsequent events. Residence in the matrix occurs in the absence of any other signal. If there is a membrane-targeting signal, however, it is rendered functional by cleavage of the matrix-targeting signal. Then the remaining part of the leader (which is now N-terminal) causes the protein to take up its final destination.

How does a protein make its way across the intermembrane space from one membrane to the next? Transfer probably occurs at locations where the outer and inner membranes are in contact or even

fused. By slowing the transport process at low temperature, it is possible to trap intermediates in which the leader is cleaved by the matrix protease, while a major part of the precursor remains exposed on the cytosolic surface of the envelope. This suggests that a protein spans the two membranes during passage. The hydrophilic nature of the leader suggests that the imported protein passes through an environment such as a hydrophilic pore that shields it from the lipid bilayer.

The nature of the membrane-targeting signal is controversial. One model holds that the entire protein enters the matrix, after which the membrane-targeting signal causes it to be re-exported into or through the inner membrane. An

Figure 11.7

The leader of yeast cytochrome c1 contains an N-terminal region that targets the protein to the mitochondrion, followed by a region that halts transfer through the inner membrane. The leader is removed by two cleavage events, the first at an unidentified site between the two regions, the second at the start of the mature protein.

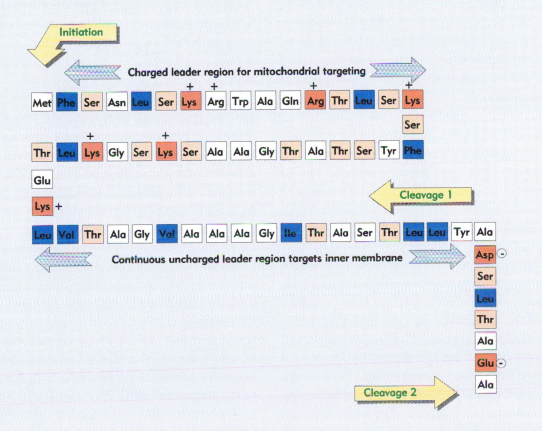

alternative model proposes that the membrane-targeting sequence simply prevents the rest of the protein from following the leader through the inner membrane into the matrix. Whichever model applies, another protease (located within the intermembrane space) completes the removal of leader sequences.

The two parts of a leader that contains both types of signal have different compositions. As indicated in **Figure 11.7**, the 35 N-terminal amino acids resemble other organelle leader sequences in the high content of uncharged amino acids, punctuated by basic amino acids. The next 19 amino acids, however, comprise an uninterrupted stretch of uncharged amino acids, long enough to span a lipid bilayer. This sequence resembles the sequences that are involved in protein translocation into membranes of the endoplasmic reticulum (see later).

A cleavable leader is not the only acceptable form of information that causes a protein to be located in an organelle. Some mitochondrial proteins are recognized as such in their mature form, and have a sequence, located at the N-terminus or internally, that can sponsor membrane passage without cleavage. In any case, cleavage does not appear to be mechanically linked to organelle recognition, since mutation of the cleavage site does not prevent import of a protein.

The need to regulate passage through two membranes also applies to chloroplast proteins. **Figure 11.8** illustrates the variety of locations for chloroplast proteins. They pass the outer and inner membranes of the envelope into the stroma, a process involving the same types of passage as into the mitochondrial matrix. But some proteins are transported yet further, across the stacks of the thylakoid membrane into the lumen. Proteins destined for the thylakoid membrane or lumen must cross the stroma *en route.*

Chloroplast targeting signals resemble mitochondrial targeting signals. The leader consists of ~50 amino acids, and the N-terminal half is needed to recognize the chloroplast envelope. A cleavage between positions 20–25 occurs during or following passage across the envelope, and proteins destined for the thylakoid membrane or lumen have the

other half of the original leader as a (now) N-terminal leader that guides recognition of the thylakoid membrane.

The general principle governing protein transport into mitochondria and chloroplasts therefore is that *the N-terminal part of the leader targets a protein to the organelle matrix, and an additional sequence (within the leader) is needed to localize the protein at the outer membrane, intermembrane space, or inner membrane.*

The leader sequences of proteins that are transported to either mitochondrion or chloroplast are very similar in nature. In fact, they are so closely related that a hybrid protein can be constructed in which a chloroplast leader from a green alga can be used in yeast to direct transport of an attached DHFR sequence to the mitochondrion. However, it functions less efficiently than an authentic mitochondrial transfer sequence.

What distinguishes the mitochondrion and chloroplast as targets for proteins synthesized in the cytosol of a plant cell? The similarities of the leader sequences suggest that the mechanism of transport into each organelle is similar. Differences in the relative affinities of the receptors of each organelle for the appropriate leader sequences must be based on features that we do not yet understand.

Another example of targeting by means of specific amino acid sequences utilizes C-terminal regions. Peroxisomes are small bodies enclosed by a single membrane. They contain enzymes concerned with oxygen utilization. They convert oxygen to hydrogen peroxide by removing hydrogen atoms from substrates. Catalase then uses the hydrogen peroxide to oxidize a variety of other substrates. All of the enzymes in the peroxisome are imported from the cytosol. Like transport into the nucleus, transport into peroxisomes occurs post-translationally by means of a short sequence. Several peroxisomal enzymes have the C-terminal sequence SKL (Ser-Lys-Leu); and the addition of this tripeptide to the C-terminus of cytosolic proteins is sufficient to ensure their import into the organelle. This very short sequence therefore constitutes one means of entry; there may be others.

Figure 11.8

The chloroplast is surrounded by an envelope consisting of outer and inner membranes. The stroma lies within the outer membrane. Within the stroma are the stacks of the thylakoid membrane. A protein approaches from the cytosol with a ˜50 residue leader. The N-terminal half of the leader sponsors passage into or through the envelope. After the cleavage associated with envelope transit, the remaining part of the leader functions to direct proteins that must cross the thylakoid membrane.

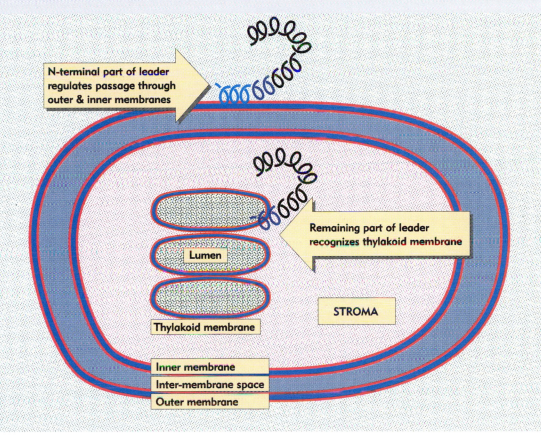

N-terminal part of leader regulates passage through outer & inner membranes

Remaining part of leader recognizes thylakoid membrane

Lumen

Thylakoid membrane

STROMA

Inner membrane

Inter-membrane space

Outer membrane

Signal sequences link protein synthesis to membranes during co-translational transfer

The use of a leader sequence that conveys information about destination is a feature of almost all proteins that associate with or pass through membranes. The role of the leader is probably similar in all cases: *to initiate entry through the barrier of the lipid bilayer.* The way in which the leader initially associates with the membrane, however, is different in post-translational recognition of an organelle envelope, and co-translational entry into the secretory apparatus.

Proteins that reside within the endoplasmic reticulum, Golgi apparatus, or plasma membrane, or that are secreted from the cell, have a common starting point for their association with membranes. The ribosomes synthesizing these proteins become associated with the endoplasmic reticulum so that the nascent protein can be co-translationally transferred to the membrane.

The endoplasmic reticulum can be divided into two types of region:

◆ Membrane-bound polysomes are associated with the sheets of the endoplasmic reticulum, giving the **rough ER** its characteristic appearance.

◆ The regions of **smooth ER** lack associated polysomes, and have a tubular, rather than sheet-like, appearance.

The endoplasmic reticulum is particularly prominent in cells that synthesize large amounts of secreted proteins.

The proteins synthesized at the rough endoplasmic reticulum pass from the ribosome directly to the membrane. From the endoplasmic reticulum membranes, the proteins are transferred to the Golgi apparatus, and then are directed to their ultimate destination, such as the lysosome or secretory vesicle or plasma membrane (this is the route by which proteins such as immunoglobulins and many polypeptide hormones are secreted from the cell).

A model for the mechanism of membrane insertion originated from work with mammalian microsomal systems (which contain ribosomes and endoplasmic reticulum membranes). These systems can package *nascent* proteins into membranes; but they do not work when isolated preproteins are added post-translationally. This shows that *the protein must associate with the membrane while it is being synthesized.*

The N-terminus of a secreted protein usually consists of a cleavable leader of 15–30 amino acids; this is called the **signal sequence**. At or close to the N-terminus are 2–3 polar residues, and within the leader is a hydrophobic core consisting exclusively or very largely of hydrophobic amino acids. There is no other conservation of sequence. There are no acidic groups in the leader, and usually its net charge is ~+1.7. **Figure 11.9** gives an example.

Like the leader sequence of proteins targeted to organelles, the signal sequence is both necessary and sufficient to sponsor transfer of any attached polypeptide into the target membrane. A signal

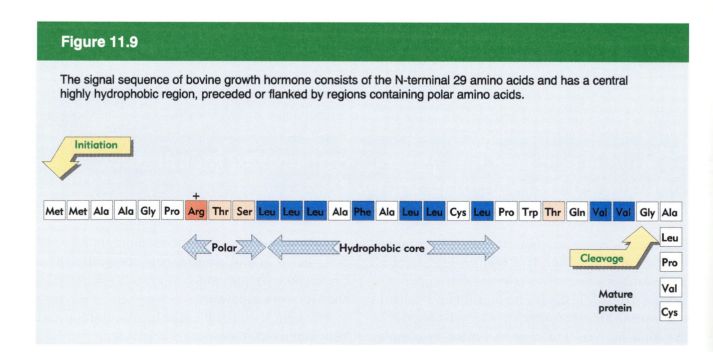

Figure 11.9

The signal sequence of bovine growth hormone consists of the N-terminal 29 amino acids and has a central highly hydrophobic region, preceded or flanked by regions containing polar amino acids.

sequence added to the N-terminus of a globin protein, for example, causes it to be secreted through cellular membranes instead of remaining in the cytosol.

The signal sequence provides the means by which ribosomes translating the mRNA can attach to the membrane. Responsibility for the initial membrane attachment rests solely with the signal sequence; the ribosome attaches by virtue of its synthesis of the secreted protein. *Thus there is no intrinsic difference between ribosomes in the free fraction and ribosomes in the membrane-bound fraction.*

The signal sequence is recognized as a signal for attachment to the ER membrane, possibly by virtue of hydrophobicity, as soon as the first few N-terminal amino acids have been synthesized. After the protein chain is inserted into the membrane, the signal sequence is cleaved by a protease embedded within the membrane. By the time the ribosome has completed translation, the protein is already well on its way through the membrane.

Salt-washed membranes cannot sponsor ribosomal attachment; but this ability can be recovered by adding back the salt wash. The active component of the salt wash is called the **signal recognition particle (SRP)**. It has the structure of a rod 5–6 nm wide and 23–24 nm long and can be isolated as an 11S ribonucleoprotein complex, containing six proteins (total mass 240,000 daltons) and a small (305 base, 100,000 dalton) 7S RNA. The 7S RNA provides the structural backbone of the particle; the individual proteins do not assemble in its absence.

The SRP has two important abilities:

◆ It can bind to the signal sequence of nascent secretory proteins.

◆ And it can bind to a receptor protein located in the membrane.

SRP activity can be reconstituted *in vitro* from the individual components. In fact, a functional SRP can be assembled from 7S RNA of one species and proteins of another. Like the rest of the apparatus that transports and processes membrane proteins, the SRP is well conserved throughout the eukaryotic kingdom.

The SRP and SRP receptor function catalytically to transfer a ribosome carrying a nascent protein to the membrane. The first step is the recognition of the signal sequence by the SRP. Then the SRP binds to the SRP receptor and the ribosome binds to the membrane. The stages of translation of membrane proteins are summarized in **Figure 11.10**.

The role of the SRP receptor in protein translocation is transient. When the SRP binds to the signal sequence, it impedes translation. Protein synthesis pauses or is even arrested. This happens when the ~70 amino acids have been incorporated into the polypeptide chain (so that the 25–30 residue leader has become exposed, with the next ~40 amino acids still buried in the ribosome).

When the SRP binds to the SRP receptor, the SRP releases the signal sequence; the ribosome then is bound by some other (unidentified) component of the membrane. At this point, translation can resume. When the ribosome has been passed on to the membrane, the role of SRP and SRP receptor has been played; they now recycle, and are free to sponsor the association of another nascent polysome with the membrane.

The 7S RNA of the SRP particle is divided into two parts. The 100 bases at the 5′ end and 45 bases at the 3′ end are closely related to the sequence of Alu RNA, a common mammalian small RNA. They therefore define the **Alu domain**. The remaining part of the RNA comprises the **S domain**.

The SRP structure depicted in **Figure 11.11** partitions three functions in protein targeting to different parts of the ribonucleoprotein. Using SRP reconstituted *in vitro* allows the functions of each component to be delineated. The 54 kilodalton protein is the only one that does not bind directly to the RNA; it binds via the 19 kilodalton protein, which binds to two extremities of the RNA. The 54 kilodalton protein is needed for recognition of the signal sequence. The 68–72 kilodalton dimer binds to the central region of the RNA; it is needed for recognizing the SRP receptor and for protein translocation through the membrane. The 9–14 kilodalton dimer binds in the vicinity of the other end of the molecule; it is responsible for elongation arrest.

Figure 11.10

The signal hypothesis proposes that ribosomes synthesizing secretory proteins are attached to the membrane via the leader sequence on the nascent polypeptide.

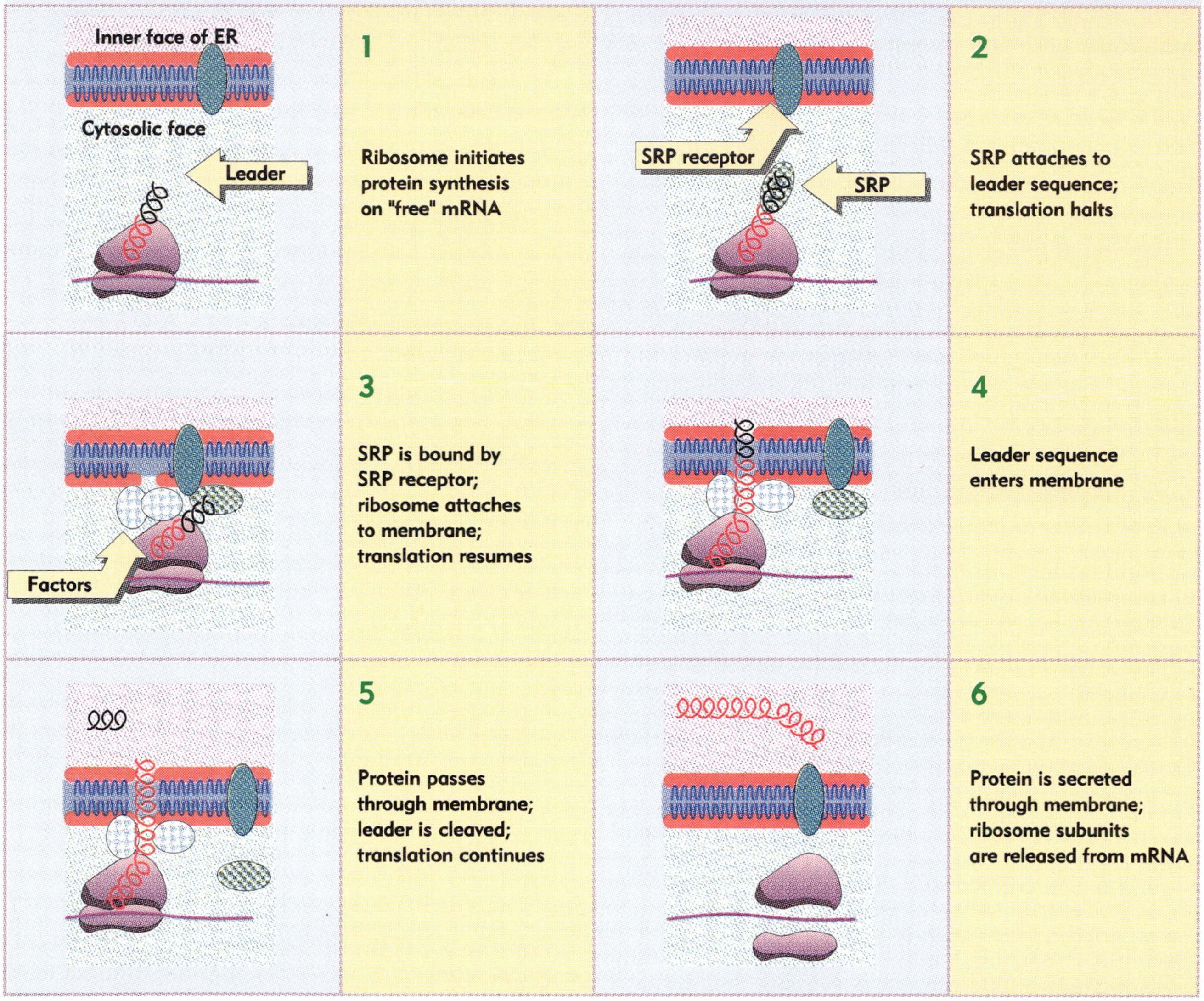

1 Ribosome initiates protein synthesis on "free" mRNA

2 SRP attaches to leader sequence; translation halts

3 SRP is bound by SRP receptor; ribosome attaches to membrane; translation resumes

4 Leader sequence enters membrane

5 Protein passes through membrane; leader is cleaved; translation continues

6 Protein is secreted through membrane; ribosome subunits are released from mRNA

The SRP receptor is a dimer containing subunits of 72,000 and 30,000 daltons. The N-terminal end of the large subunit is anchored in the endoplasmic reticulum. The bulk of the protein protrudes into the cytosol. A large part of the sequence of the cytoplasmic region of the protein resembles a nucleic acid-binding protein, with many positive residues, which suggests the possibility that the SRP receptor recognizes the 7S RNA in the SRP. The structure and function of the smaller subunit of the SRP receptor are unknown.

Why are proteins that enter the reticuloendothelial system inserted co-translationally, while proteins that enter mitochondria or chloroplasts can be targeted post-translationally? No clear answer is evident, and the problem is exacerbated by the fact that in yeast, protein translocation into the ER occurs post-translationally. The energy requirements for the two types of process are different: translocation into mitochondria or chloroplasts requires an electrochemical potential, while entry into the ER requires ATP. It does not seem that the act of protein synthesis is involved in providing energy. The presence of ribosomes is necessary to maintain the proper geometry for protein entry into the membrane in co-translational translocation.

One aspect of this process may be involved with controlling the conformation of the protein. If enough of the protein sequence was released into the cytoplasm, it could take up a conformation determined by the aqueous environment; in this conformation, it might be unable to traverse the membrane. The ability of the SRP to inhibit translation while the ribosome is being handed over to the membrane is therefore important in preventing the protein from being released into the aqueous environment. However, conformation can be controlled in this way only when the signal sequence is N-terminal, and not in other cases, so transport through membranes could also involve factors that bind to the transported proteins to influence their structure directly.

The signal peptidase (identified *in vitro*) consists of a complex of 6 proteins. The actual peptidase activity is probably carried by one of the proteins. The others could have other functions concerned with protein modification or might have a structural role—for example, they might be concerned with location in the membrane or with forming a channel for protein transfer. The complex is several times more abundant than the SRP and SRP receptor. Its amount is equivalent roughly to the amount of bound ribosomes, suggesting that it functions in a structural capacity. It is probably located on the lumenal face of the ER membrane, which implies that the entire signal sequence must cross the membrane before the cleavage event occurs.

Figure 11.11

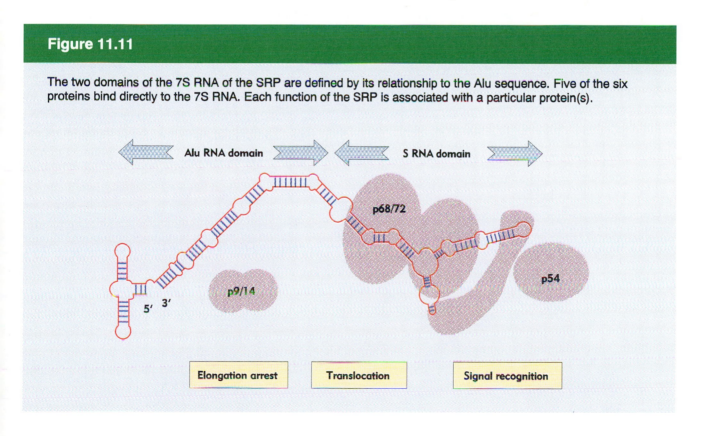

The two domains of the 7S RNA of the SRP are defined by its relationship to the Alu sequence. Five of the six proteins bind directly to the 7S RNA. Each function of the SRP is associated with a particular protein(s).

Anchor sequences cause proteins to be retained in membranes

What processes are involved in allowing a (largely) hydrophilic protein to pass through a hydrophobic membrane? The role of an N-terminal signal sequence is essentially to lead the way, as illustrated previously in Figure 11.10. *We assume that once the barrier of the lipid bilayer has been broken by entry of the signal sequence, transfer continues through the membrane until the entire polypeptide has been translocated to the other side. Translocation is initiated by the signal sequence, and is independent of the attached sequences.*

A protein in the process of translocation across the ER membrane can be extracted by denaturants that are effective in an aqueous environment. The same denaturants do not extract proteins that are resident components of the membrane. This suggests the model for translocation illustrated in **Figure 11.12.** Either the signal sequence inserts into the lipid bilayer at a specific site, or its insertion provides a signal for the creation of such a site, at which proteins of the ER membrane form an aqueous channel through the bilayer. A translocating protein moves through this channel, interacting with the resident proteins rather than with the lipid bilayer.

Such channels can be detected by their ability to allow the passage of ions (measured as a localized change in electrical conductance). The channel

Figure 11.12

Only the signal sequence need interact directly with the hydrophobic environment of the lipid bilayer. The remaining sequences of a protein translocating through the membrane may move through an aqueous tunnel created by resident ER membrane proteins.

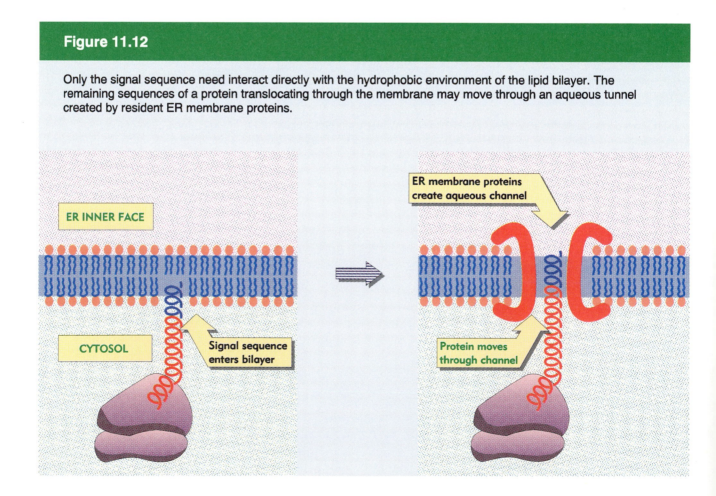

Figure 11.13

Proteins that reside in membranes enter by the same route as secreted proteins, but transfer is halted when an anchor sequence passes into the membrane. If the anchor is at the C-terminus, the bulk of the protein passes through the membrane and is exposed on the far surface.

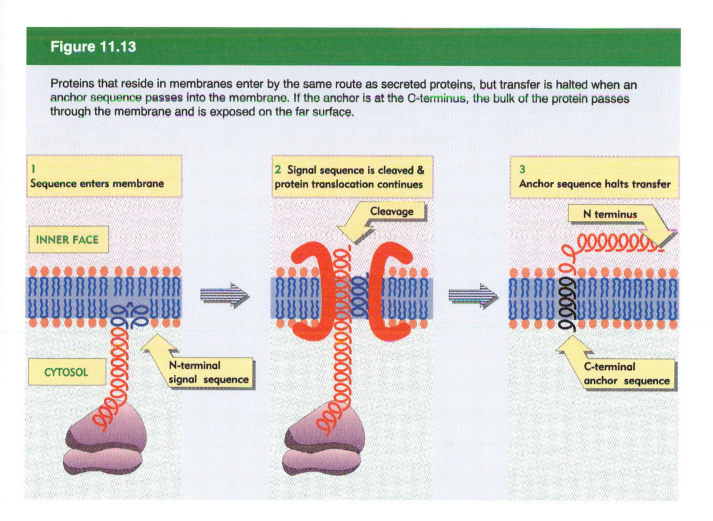

1 Sequence enters membrane

INNER FACE

CYTOSOL

N-terminal signal sequence

2 Signal sequence is cleaved & protein translocation continues

Cleavage

3 Anchor sequence halts transfer

N terminus

C-terminal anchor sequence

appears to open when a nascent polypeptide is transferred from a ribosome to the ER membrane. The translocating protein fills the channel completely, because ions cannot pass through during translocation. But if the protein is released by treatment with puromycin, then the channel becomes freely permeable. If the ribosomes are removed from the membrane, the channel closes, suggesting that the open state requires the presence of the ribosome.

How are proteins that are secreted *through* the ER membrane distinguished from those that reside *within* it? Integral membrane proteins have hydrophobic regions that actually are located within the lipid bilayer, while external domains are located on one or the other side of the membrane. A hydrophobic region provides a **transmembrane domain** that enables the polypeptide to float in the membrane. It is organized as an α-helix, 21–26 residues long, which forms a coil that can span the lipid bilayer (see Figure 2.6).

The signals that are responsible for insertion of proteins in membranes determine the orientation of the protein. Proteins that have the N-terminus on the far side of the membrane, while the C-terminus remains exposed to the cytosol where it was synthesized, are called **type I** membrane proteins. They comprise the majority of membrane proteins. A smaller number of **type II** integral membrane proteins have the opposite orientation; they display an N-terminal domain on the cytosolic side and expose the C-terminal domain on the extracellular side.

The pathway by which proteins of either type I or type II are inserted into the membrane follows the same initial route as that of secretory proteins, relying on a signal sequence that functions co-translationally. But proteins that are to remain within the membrane possess a second,

Figure 11.14

A combined signal-anchor sequence causes a protein to reverse its orientation, so that the N-terminus remains on the inner face and the C-terminus is exposed on the outer face of the membrane.

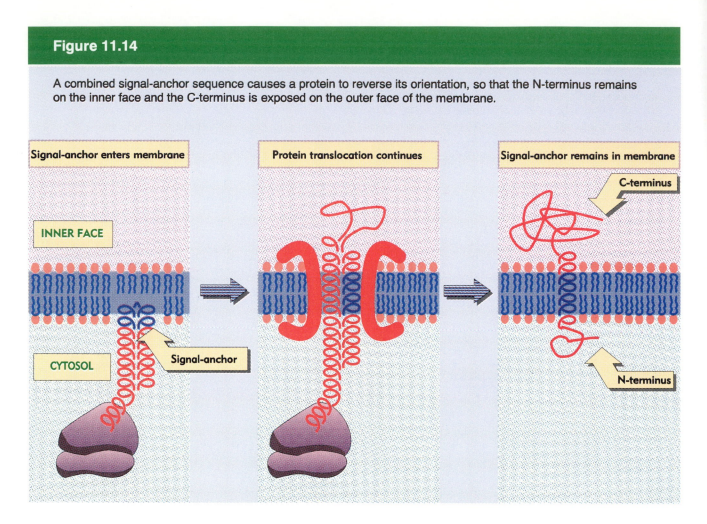

Signal-anchor enters membrane

INNER FACE

CYTOSOL

Signal-anchor

Protein translocation continues

Signal-anchor remains in membrane

C-terminus

N-terminus

stop–transfer signal. This takes the form of a cluster of hydrophobic amino acids adjacent to some ionic residues. The cluster serves as an **anchor** that latches onto the membrane and stops the protein from passing right through. We do not know how it functions.

Membrane insertion starts for both types of protein by the insertion of a signal sequence in the form of a hairpin loop, in which the N-terminus remains on the cytoplasmic side. Two features determine the position and orientation of a protein in the membrane: whether the signal sequence is cleaved; and the location of the anchor sequence.

The insertion of type I proteins is illustrated in **Figure 11.13**. The signal sequence is always N-terminal. When it is attacked by the peptidase on the lumenal side of the membrane, the cleavage releases a new N-terminus that can continue into the lumen. The location of the anchor signal

determines at what point transfer of the protein is halted. When the anchor sequence takes root in the membrane, domains on the N-terminal side will be located in the lumen, while domains on the C-terminal side are located facing the cytosol.

A common location for a stop–transfer sequence of this type is at the C-terminus. As shown in the figure, transfer is halted only as the last sequences of the protein enter the membrane. This type of arrangement is responsible for the location in the membrane of many proteins, including cell surface proteins. Most of the protein sequence is exposed on the lumenal side of the membrane, and it is a sequence at or close to the C-terminus that anchors the protein in the membrane.

Type II proteins do not have a cleavable leader sequence at the N-terminus. Instead the signal sequence is combined with an anchor sequence. We imagine that the general pathway for the integration of type I proteins into the membrane involves the steps illustrated in **Figure 11.14**. The signal sequence enters the membrane, but the joint signal–anchor sequence does not pass through. Instead it sticks in the lipid phase of the membrane, while the rest of the growing polypeptide continues to loop into the endoplasmic reticulum.

The signal–anchor sequence is usually internal, and its location determines which parts of the protein remain in the cytosol and which are extracellular. Essentially all the N-terminal sequences that precede the signal–anchor are exposed to the cytosol. Usually this cytosolic tail is short, ~6–30 amino acids. In effect the N-terminus remains constrained while the rest of the protein passes through the membrane. This reverses the orientation of the protein with regard to the membrane. (We discuss various types of orientation for membrane proteins in more detail in Chapter 12.)

Anchor sequences are less well defined than signal sequences, but fit the general criteria for transmembrane domains. A surprising property of anchor sequences is that they can function as signal sequences when engineered into a different location. When placed into a protein lacking other signals, such a sequence may sponsor membrane translocation. One possible explanation for these results is that the signal sequence and anchor sequence interact with some common component of the apparatus for translocation. Binding of the signal sequence initiates translocation, but the appearance of the anchor sequence displaces the signal sequence and halts transfer. In the absence of any prior signal sequence, an anchor sequence therefore might be able to initiate translocation.

The combined signal–anchor sequence of type II proteins resembles a cleavable signal sequence. **Figure 11.15** gives an example. Like cleavable leader sequences, the amino acid composition is more important than the actual sequence. The regions at the extremities of the signal–anchor carry positive charges; the central region is uncharged and resembles a hydrophobic core of a cleavable leader. Mutations to introduce charged amino acids

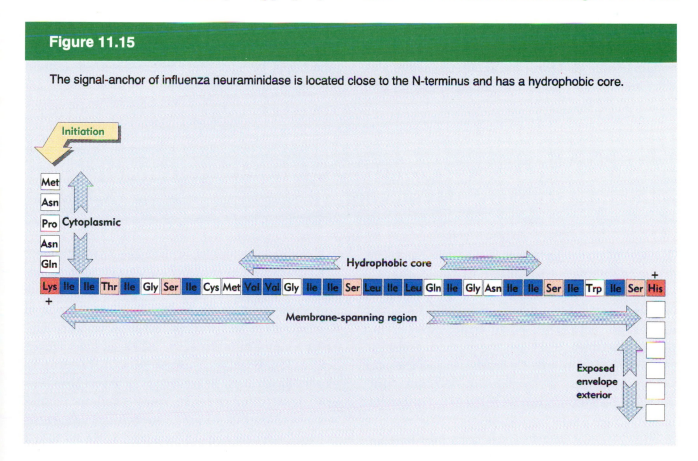

Figure 11.15

The signal-anchor of influenza neuraminidase is located close to the N-terminus and has a hydrophobic core.

in the core region prevent membrane insertion; mutations on either side prevent the anchor from working, so the protein is secreted or located in an incorrect compartment.

The distribution of charges around the anchor sequence has an important effect on the orientation of the protein. More positive charges are usually found on the cytoplasmic side (N-terminal side in type II proteins). If the positive charges are removed by mutation, the orientation of the protein can be reversed. The effect of charges on orientation is summarized by the 'positive inside' rule, which says that the side of the anchor with the most positive charges will be located in the cytoplasm. The positive charges in effect prove a hook that latches on to the cytoplasmic side of the membrane, controlling the direction in which the hydrophobic region is inserted, and thus determining the orientation of the protein.

The process of insertion into a membrane has been characterized for both type I proteins (Figure 11.13) and type II proteins (Figure 11.14), in which there is a single transmembrane domain. But some membrane proteins have a more complex organization in which there are multiple transmembrane domains, separated from one another by sequences that loop out into the extracellular space on one side or the cytosol on the other (see Figure 12.4).

An interesting question is how an integral membrane protein initially moves into the lipid bilayer. As it passes through the channel illustrated in Figure 11.12, it must reach a point at which the channel is occupied by the hydrophobic transmembrane region. Then this region must be transferred laterally from the channel to the lipid bilayer. This could be accomplished by a dissociation of the channel into its components or perhaps by some rearrangement that exposes the hydrophobic region to the surrounding lipids.

How is a protein with multiple membrane-spanning regions inserted into a membrane? Much less is known about this process, but we assume that it relies on sequences that provide signal and/or anchor capabilities. One model is to suppose that there is an alternating series of signal and anchor sequences. Translocation is initiated at the first signal sequence and continues until stopped by the first anchor. Then it is reinitiated by a subsequent signal sequence, until stopped by the next anchor. Another possibility is that, once one transmembrane domain has been inserted, subsequent transmembrane domains become localized in the membrane because they interact with a domain already located there.

Bacterial proteins are transported by both co-translational and post-translational mechanisms

The secretion of proteins from bacteria relies on mechanisms very similar to those characterized for eukaryotic cells. Transport from the bacterial cytoplasm passes through the inner membrane into the periplasmic space and then (sometimes) through the outer membrane into the environment. Co-translational transfer is common in *E. coli,* but is not universal. Some proteins are secreted both co-translationally and post-translationally. The relative kinetics of translation versus secretion through the membrane could determine the balance.

Exported bacterial proteins have N-terminal leader sequences, with a hydrophilic N-terminus and an adjacent hydrophobic core. Mutations in N-terminal leaders prevent secretion; they are suppressed by mutations in other genes, which are thus defined as components of the protein export apparatus. Several genes given the general description *sec* are implicated in coding for components of the secretory apparatus by the occurrence of mutations that block secretion of many or all exported proteins.

As with eukaryotes, secondary signals are necessary for proper location after the protein has entered the membrane. For example, the C-terminal region of the *E. coli* β-lactamase is necessary for the protein to leave the membrane to enter the periplasmic space on the other side.

Control of folding is important in protein transport. Changes in conformation occur during transit of a membrane. For example, β-lactamase exists in a trypsin-sensitive form before and during passage through the bacterial inner membrane, but changes its conformation to a trypsin-resistant form when it is released into the periplasm.

Bacterial protein translocation passes through the stages summarized in **Figure 11.16**. A chaperone binds to the nascent protein to control its folding; the protein associates with the secretion apparatus;

it is translocated through the membrane; and a peptidase cleaves the leader from proteins with a cleavable N-terminus.

Several chaperones can increase the efficiency of bacterial protein export by preventing premature folding; they include trigger factor (characterized as a chaperone that assists export), GroEL (characterized originally for its effects on phage morphogenesis), and SecB (identified as the product of one of the *sec* mutants). Although SecB is the least abundant of these proteins, it has the major role in promoting export, probably because it has two functions. It can behave as a chaperone and bind to a nascent protein; and it has an affinity for the protein SecA.

In its function as a chaperone, SecB binds to a precursor protein to retard folding, but it cannot reverse the change in structure of a folded protein.

Figure 11.16

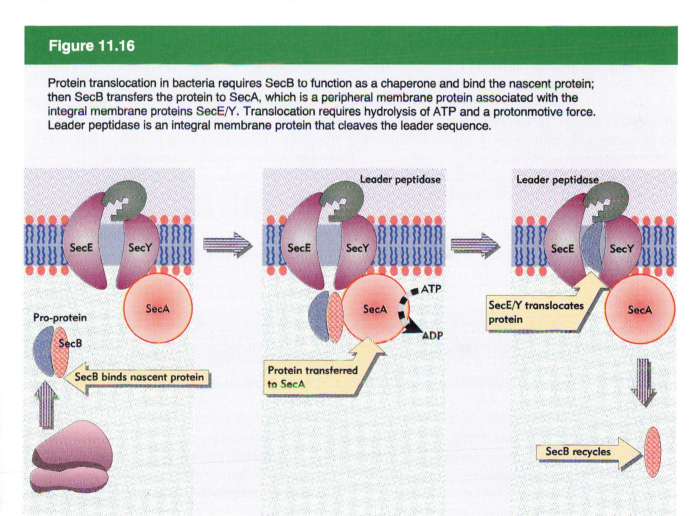

Protein translocation in bacteria requires SecB to function as a chaperone and bind the nascent protein; then SecB transfers the protein to SecA, which is a peripheral membrane protein associated with the integral membrane proteins SecE/Y. Translocation requires hydrolysis of ATP and a protonmotive force. Leader peptidase is an integral membrane protein that cleaves the leader sequence.

So it does not function as an unfolding factor; its role is strictly to inhibit improper folding of the newly synthesized protein. The affinity of SecB for SecA allows it to target a precursor protein to the membrane.

SecA is a large peripheral membrane protein which associates with the membrane by virtue of its affinity for acidic lipids and for the protein SecY. SecY and SecE are integral membrane proteins (SecY has 10 transmembrane segments and SecE has 3 transmembrane segments), which are part of a multisubunit complex that provides the translocase function.

SecA recognizes both SecB and the precursor protein that it chaperones; probably features of the mature protein sequence as well as its leader are required for recognition. SecA has an ATPase activity which depends on binding to lipids, SecY, and a precursor protein. The ATPase functions in a cyclical manner during translocation. After SecA binds a precursor protein, it binds ATP, and ~20 amino acids are translocated through the membrane. Hydrolysis of ATP is required to release the precursor from SecA. Then the cycle may be repeated. Precursor protein is bound again to provide the spur to bind more ATP, translocate another segment of protein, and release the precursor.

Another process is also involved in translocation. When a precursor is released by SecA, it can be driven through the membrane by a protonmotive force (that is, an electrical potential across the membrane). This process cannot initiate transfer through the membrane, but can continue the process initiated by a cycle of SecA ATPase action. Usually only a small number of cycles (or even only one initiating cycle) of the SecA ATP-driven reaction occurs, after which the protonmotive force completes translocation of the precursor.

Many bacterial proteins are synthesized as pro-proteins that have cleavable N-terminal leader sequences. The leader is cleaved by an enzyme called **leader peptidase** that recognizes precursor forms of several exported proteins. The leader peptidase is an integral membrane protein, located in the inner membrane.

A notable feature of eukaryotic systems is the use of the ribonucleoprotein SRP to provide the initial recognition of a signal sequence. There is a counterpart to SRP in bacteria. *E. coli* contains a 4.5S RNA that associates with ribosomes and is homologous to the 7S RNA of the SRP. It associates with a 48 kilodalton protein that is homologous to the 54 kilodalton protein component of the SRP. The complex appears to play a role in the secretion of certain, but not all, secreted proteins. The role of this complex may be to associate with the leader sequences of nascent proteins as they emerge from ribosomes, perhaps to keep the nascent protein in an appropriate conformation until it interacts with other components of the secretory apparatus. Perhaps this is the original connection between protein synthesis and secretion; and in eukaryotes the SRP has acquired the additional roles of causing translational arrest and targeting to the membrane.

Oligosaccharides are added to proteins in the endoplasmic reticulum and Golgi

Proteins enter the secretory apparatus by co-translational transfer into the membranes of the endoplasmic reticulum. They are then transferred to the Golgi apparatus, where they are **sorted** according to their final intended destination. Once a protein has entered a membranous environment, it remains inserted within the membrane until it reaches its final destination. A protein that enters the endoplasmic reticulum is inserted into the membrane with the appropriate orientation; if it is destined to reside in the Golgi, lysosome, or plasma membrane, or to be secreted from the cell, it is transported in membrane vesicles along the secretory pathway to the appropriate stage. At the appropriate destination, some structural feature of the protein is recognized and it is permanently secured (or secreted from the cell).

Two important changes occur to a protein in the endoplasmic reticulum: it becomes folded into its proper conformation; and it is modified by glycosylation.

As with transport through other membranes, a protein is translocated into the ER in unfolded form. Folding occurs as the protein enters the lumen; probably a series of domains each folds independently from N-terminus to C-terminus. Folding of a 50,000 dalton protein is complete in < 3–4 minutes, compared with the ~1 minute required to synthesize the chain.

Folding in the ER is associated with modification and is assisted by accessory proteins. Addition of carbohydrate may be required for correct folding; in fact, this may be an important function of the modification. Reshuffling of disulfide bonds by the enzyme PDI (protein disulfide isomerase) may be involved. And association with specific ER proteins may be necessary. Some or all of these activities could be exercised by a complex of enzymes as a protein enters the ER, that is, the necessary functions all could associate with the protein at the site of translocation through the membrane. Calculations of the spontaneous rates of folding and oligomerization suggest that the accessory enzymes are needed to catalyze the process in order to enable it to occur rapidly enough in the cell.

One protein with a function in folding is BiP, a member of the Hsp70 family of chaperonins (see Table 1.5), which binds incompletely folded heavy chain immunoglobulins and influenza hemagglutinin. BiP facilitates oligomerization and/or folding of these and other proteins in the lumen of the ER. The ER may contain several such accessory proteins whose function is to recognize the partially folded forms of proteins, and to assist them in acquiring a conformation that allows them to be transported to the next destination.

Most membrane glycoproteins are oligomeric, and usually they oligomerize in the ER. In fact,

oligomerization may be necessary for further transport. Oligomers are rapidly transported from the ER to the Golgi, but unassembled subunits or misassembled proteins are held back. Misfolded proteins are often associated with BiP. In due course, they are removed by degradation. Thus a protein is found in the Golgi or subsequent stages of transport only if it has been properly folded.

BiP has two functions: to assist in folding newly translocated proteins; and to remove misfolded proteins. These activities could result from the same basic mode of function. Suppose that BiP recognizes certain peptide sequences that are inaccessible within the conformation of a mature, properly folded protein. These sequences are exposed and attract BiP when the protein enters the ER lumen in an essentially one dimensional form. And if a protein is misfolded or denatured, they may become exposed on its surface instead of being properly buried. We do not yet know how a permanently misfolded protein is targeted for degradation and whether this is connected with the presence of BiP.

Virtually all proteins that pass through the secretory apparatus are glycosylated. Glycoproteins are generated by the addition of oligosaccharide groups either to the NH_2 group of asparagine (N-linked glycosylation; see Figure 1.10) or to the OH group of serine, threonine or hydroxylysine (O-linked glycosylation). N-linked glycosylation is initiated in the endoplasmic reticulum and completed in the Golgi; O-linked glycosylation occurs in the Golgi alone. The stages of N-glycosylation are illustrated in the next three figures.

The addition of all N-linked oligosaccharides starts in the ER by a common route, as illustrated in **Figure 11.17**. An oligosaccharide containing 2 N-acetyl-glucosamine, 9 mannose, and 3 glucose residues is formed on a special lipid, **dolichol**. Dolichol is a highly hydrophobic lipid that resides within the ER membrane, with its active group facing the lumen. The oligosaccharide is constructed by adding sugar residues individually; it is linked to dolichol by a pyrophosphate group, and is transferred as a unit to a target protein by a membrane-bound glycosyl transferase enzyme

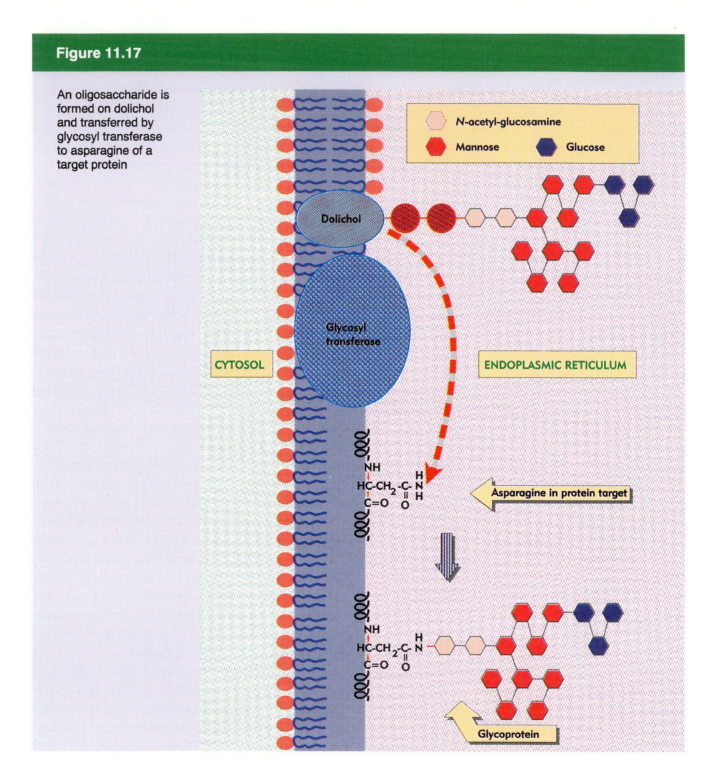

Figure 11.17

An oligosaccharide is formed on dolichol and transferred by glycosyl transferase to asparagine of a target protein

N-acetyl-glucosamine

Mannose   Glucose

Dolichol

Glycosyl transferase

CYTOSOL   ENDOPLASMIC RETICULUM

NH
HC-CH₂-C-   H N H
C=O   O

Asparagine in protein target

NH
HC-CH₂-C- N
C=O   O

Glycoprotein

whose active site is exposed in the lumen of the endoplasmic reticulum.

The acceptor group is an asparagine residue, located within the sequence Asn-X-Ser or Asn-X-Thr (where X is any amino acid except proline). It is recognized as soon as the target sequence is exposed in the lumen, when the nascent protein crosses the ER membrane.

Some trimming of the oligosaccharide occurs in the ER, after which a nascent glycoprotein is handed over to the Golgi. The oligosaccharide structures generated during transport through the ER and Golgi fall into two classes, determined by the fate of the mannose residues. Mannose residues are added only in the ER, although they can be trimmed subsequently. The number that is removed in the ER varies; the ER mannosidase attacks the first mannose quickly, and the next 3 more slowly.

High mannose oligosaccharides contain only the sugar residues added in the ER, as left by the trimming process summarized in **Figure 11.18**. Almost immediately following addition of the oligosaccharide, most of its residues are removed from the protein. The 3 glucose residues are removed by the enzymes glucosidases I and II that reside in the ER. A mannosidase present in the ER removes 2–4 of the mannose residues from proteins that reside in the ER, generating the final structure of the oligosaccharide.

Complex oligosaccharides result from additional trimming and further additions carried out in the Golgi. Golgi modifications occur in the fixed order illustrated in **Figure 11.19**. The first step is further trimming of mannose residues by Golgi mannosidase I. Then a single sugar residue is added by the enzyme *N*-acetyl-glucosamine transferase. This allows further processing to occur, when the Golgi enzyme mannosidase II removes further mannose residues. At this point, the oligosaccharide becomes resistant to degradation by the enzyme endoglycosidase H (Endo H). *Susceptibility to Endo H is therefore used as an operational test to determine when a glycoprotein has left the ER for the Golgi.*

During modification in the Golgi, all but the **inner core**, consisting of the sequence NAc-Glc•NAc-Glc•Man₃, is eventually stripped off. Additions to the inner core generate the **terminal region**. The residues that can be added to a complex oligosaccharide include *N*-acetyl-glucosamine, galactose, and sialic acid (*N*-acetyl-neuraminic acid). The pathway for processing and glycosylation is highly ordered, and the two types of reaction

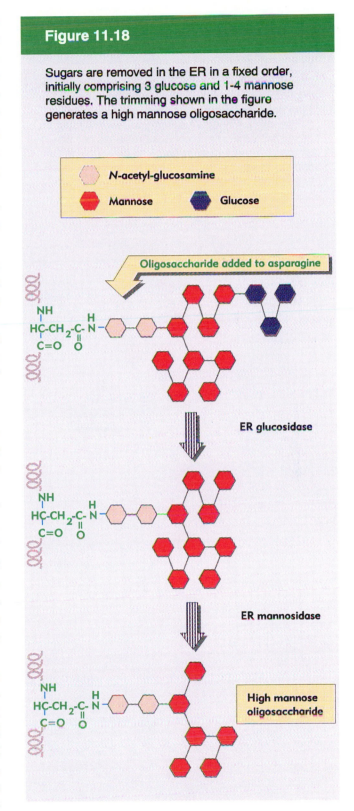

Figure 11.18

Sugars are removed in the ER in a fixed order, initially comprising 3 glucose and 1-4 mannose residues. The trimming shown in the figure generates a high mannose oligosaccharide.

N-acetyl-glucosamine

Mannose

Glucose

Oligosaccharide added to asparagine

ER glucosidase

ER mannosidase

High mannose oligosaccharide

Figure 11.19

Processing for a complex oligosaccharide occurs in the Golgi and trims the original preformed unit to the inner core consisting of 2 N-acetyl-glucosamine and 3 mannose residues. N-acetyl-glucosamine must be added before the final mannose residues can be removed. Then further sugars can be added, in the order in which the transfer enzymes are encountered, to generate a terminal region containing N-acetyl-glucosamine, galactose, and sialic acid.

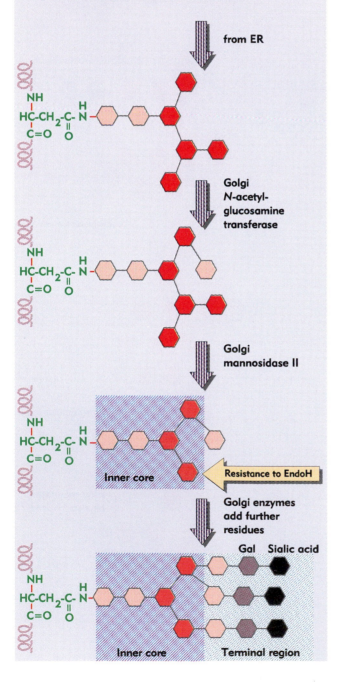

from ER

Golgi N-acetyl-glucosamine transferase

Golgi mannosidase II

Inner core

Resistance to EndoH

Golgi enzymes add further residues

Gal   Sialic acid

Inner core   Terminal region

are interspersed in it. Addition of one sugar residue may be needed for removal of another, as in the example of the addition of N-acetyl-glucosamine before the final mannose residues are removed.

We do not know the basis of the information by which each protein undergoes its particular characteristic pattern of processing and glycosylation. We assume that it resides in the structure of the polypeptide chain; it cannot lie in the oligosaccharide, since all proteins subject to N-linked glycosylation start the pathway by addition of the same (preformed) oligosaccharide.

The individual cisternae of the Golgi are organized into a series of **stacks**, somewhat resembling a pile of plates (see Figure 2.10). A typical stack consists of 4–8 flattened cisternae. A major feature of the Golgi apparatus is its *polarity*. The *cis* side faces the endoplasmic reticulum; the *trans* side in a secretory cell faces the plasma membrane. The Golgi consists of compartments, which are named *cis, medial, trans,* and *TGN (trans-Golgi network)*, proceeding from the *cis* to the *trans* face. Proteins enter a Golgi stack at the *cis* face and are modified during their transport through the cisternae of the various compartments. When they reach the *trans* face, they are directed to their destination.

Membrane structure changes across the Golgi stack. The main difference is an increase in the content of cholesterol proceeding from *cis* to *trans*. As a result, fractionation of Golgi preparations generates a gradient in which the densest fractions represent the *cis* cisternae, and the lightest fractions represent the *trans* cisternae. The positions of enzymes on the gradient, and *in situ* immunochemistry with antibodies against individual enzymes, suggests that some of the enzymes are differentially distributed proceeding from *cis* to *trans*. The difference between the *cis* and *trans* faces of the Golgi is clear, but it is not clear how the concept of compartments relates to individual cisternae; there may in fact be a continuous series of changes proceeding through the cisternae.

Nascent proteins encounter the modifying enzymes as they are transported through the Golgi stack. **Figure 11.20** illustrates the order in which the enzymes function. This may be partly determined

Figure 11.20

A Golgi stack consists of a series of cisternae, organized with *cis* to *trans* polarity. Protein modifications occur in order as a protein moves from the *cis* face to the *trans* face.

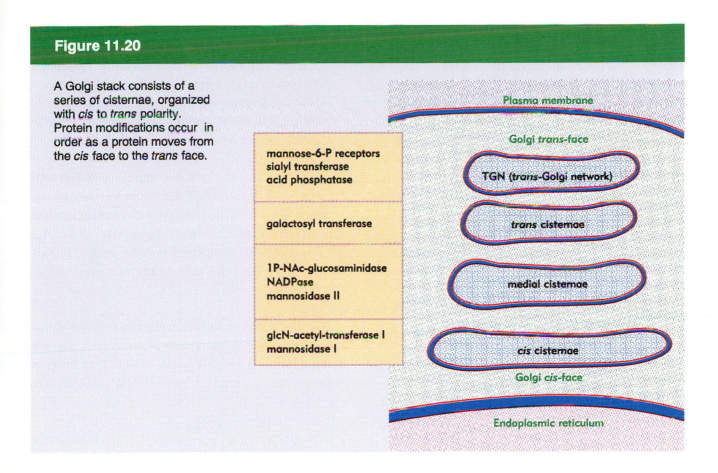

by the fact that the modification introduced by one enzyme is needed to provide the substrate for the next, and partly by the availability of the enzymes proceeding through the cisternae.

The addition of a complex oligosaccharide can make a considerable difference to the mass of a protein. Glycoproteins often have a mass that is much greater than would be expected simply from their primary sequence. What is the significance of these extensive glycosylations? In some cases, the saccharide moieties play a structural role, for example, in the behavior of surface proteins that are involved in cell adhesion. Another possible role could be in promoting folding into a particular conformation. At least one modification—the addition of mannose-6-phosphate—confers a targeting signal. Are other carbohydrate moieties used to direct protein sorting? None is yet known.

Coated vesicles transport both exported and imported proteins

How is a protein transported from the ER, through the Golgi stacks, to the plasma membrane or other destinations? Transported proteins are packaged into membrane-coated vesicles. A vesicle is generated by budding off from a membrane; it fulfills its function by fusing with another membrane to release its contents (see Figure 2.11).

Vesicles are used to transport proteins both out of

the cell and into the cell. (Import is discussed in more detail in Chapter 12.) The movement of coated vesicles is pictured in **Figure 11.21**. The cycle for each type of vesicle is similar, whether they are involved in export or import of proteins: *budding from the donor membrane is succeeded by fusion with the target membrane.* There are two pathways for export and one for import:

◆ Some proteins are **constitutively secreted**, moving from the ER into the Golgi and then to the plasma membrane. This represents a mechanism for bulk flow of proteins and lipids. It takes place by a cycle in which a membrane vesicle containing protein cargo buds from one membrane, and then moves to fuse with the next membrane. The cycle is repeated moving from ER to Golgi, through the Golgi stacks, and to the plasma membrane. The vesicles that undertake this process are called **transport vesicles**.

◆ The export of some proteins is regulated. This process is called **exocytosis**. These proteins are packaged into **exocytic** or **secretory vesicles** at the *trans* Golgi. These vesicles provide storage medium, and release their contents at the plasma membrane only following receipt of a particular signal (triggered, for example, by a hormone).

◆ Proteins enter the cell by **endocytosis**. They are packaged into **endocytic vesicles**, which are released from the plasma membrane, and transport their contents toward the interior of the cell. The protein is released when the endocytic vesicle fuses with the membrane of a target compartment.

The transport of a protein starts when it is incorporated into a transport vesicle at the endoplasmic reticulum for transport to the Golgi. This stage of transport requires ATP. The protein is carried along

Figure 11.21

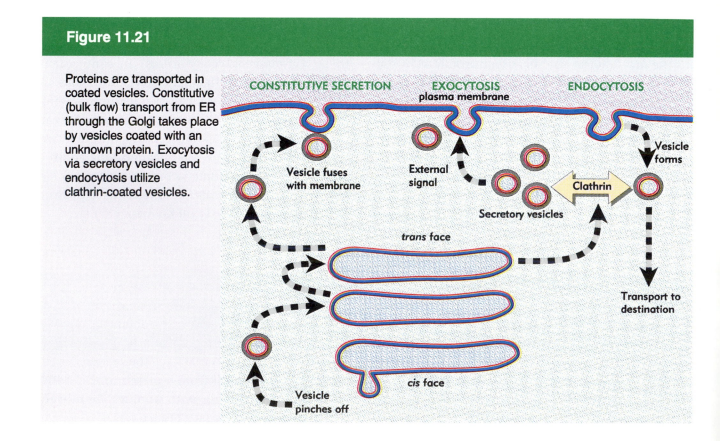

Proteins are transported in coated vesicles. Constitutive (bulk flow) transport from ER through the Golgi takes place by vesicles coated with an unknown protein. Exocytosis via secretory vesicles and endocytosis utilize clathrin-coated vesicles.

Figure 11.22

Vesicles are released from the *trans* face of the Golgi. Photograph kindly provided by Lelio Orci.

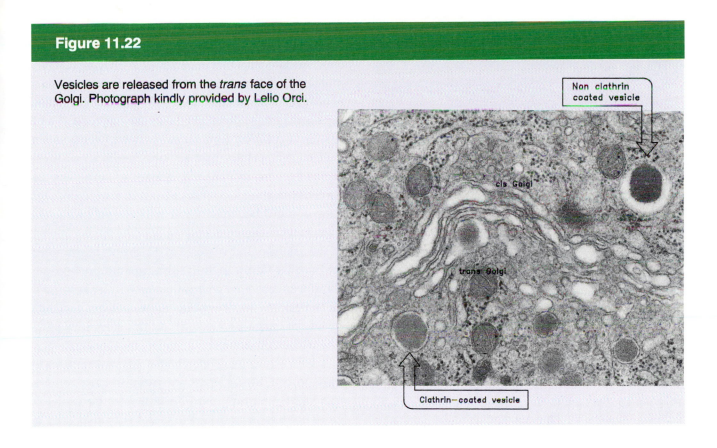

Non clathrin coated vesicle

cis Golgi

trans Golgi

Clathrin—coated vesicle

the Golgi stacks by secretory vesicles, changing its state of glycosylation as it passes from *cis* to *trans* compartments. Finally it is transported to the plasma membrane.

All vesicles involved in transporting proteins have a protein layer surrounding their membranes, and for this reason are called **coated vesicles**. They are shown in the electron micrograph of **Figure 11.22**. We discuss the two types of protein coats below. The coat serves two purposes: it enables the type of vesicle to be identified, so that it is directed to the appropriate target membrane; and it is involved with the processes of budding and fusion. It is possible that it is also concerned with the selection of proteins to be transported.

The process of vesicle generation requires a membrane bilayer to exude a vesicle that eventually buds (as illustrated previously in Figure 2.9.) Such events require deformation of the membrane, as illustrated in **Figure 11.23**. Proteins concerned with this process are required specifically for budding, and could include some that become part of the coat.

In the reverse reaction, fusion is a property of membrane surfaces, and, in order to fuse with a target membrane, a vesicle must be 'uncoated' by removal of the protein layer. The uncoating reaction is catalyzed by an enzyme with ATPase activity. Probably a coated vesicle recognizes its destination by a reaction between a component of the vesicle coat and a receptor in the target membrane.

A vesicle therefore follows a cycle in which it is released from a donor membrane, gains its coat, moves to the next membrane, becomes uncoated, and fuses with the target membrane. When a vesicle is generated, it carries proteins that were resident in (or associated with) the stretch of membrane that was pinched off. The interior of the vesicle has the constitution of the lumen of the organelle from which it was generated. When the vesicle fuses with its target membrane, its components become part of that

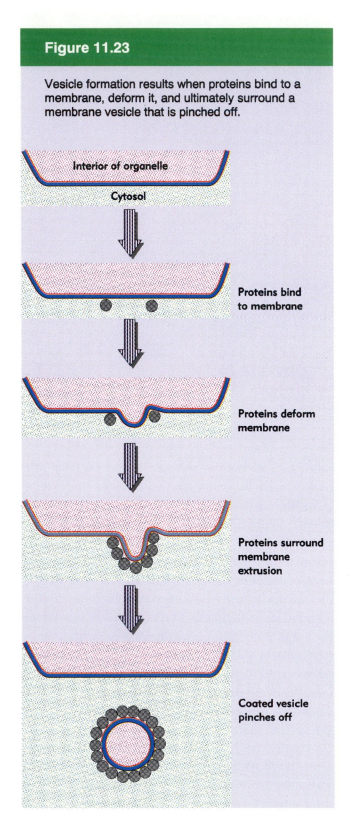

Figure 11.23

Vesicle formation results when proteins bind to a membrane, deform it, and ultimately surround a membrane vesicle that is pinched off.

Interior of organelle

Cytosol

Proteins bind to membrane

Proteins deform membrane

Proteins surround membrane extrusion

Coated vesicle pinches off

membrane or the lumen of the compartment.

Several components of the secretory pathway are concerned with vesicle formation or fusion. Yeast mutations of the *sec* genotype, which are unable to export proteins from the ER–Golgi pathway, identify some genes whose products are concerned with vesicle formation and others whose products are needed for vesicle fusion. The specificity of these proteins is an interesting question: some are involved with the mechanics of budding or fusion, but others could be concerned with establishing the specificity of the vesicles.

We assume that vesicles at different stages along the pathway differ in their target specificity because of differences in their protein coats. At least some of these vesicles may be 'bulk carriers', transporting a variety of cargo proteins along the secretory pathway without specificity for individual targets.

Are proteins that enter the ER automatically transported along this pathway or are special signals needed for incorporation into the vesicles? Probably any protein in the Golgi is transported by vesicles along the exocytic pathway *unless it has a retention signal that prevents its transport*. Thus once a protein has entered the endoplasmic reticulum by co-translational transfer, transport via vesicles functions as a 'default' pathway, which the protein can avoid only by possessing a special signal.

The various components of the membrane have different fates. The dilemma is that secreted proteins must leave for the plasma membrane, but proteins resident in the Golgi or ER membranes must stay there; while the net amount (and types) of lipid in each membrane must remain unperturbed. The transport process illustrated previously in Figure 11.21 suggests that there is a bulk flow of membranes and proteins from the ER, through the Golgi, towards the plasma membrane. In the course of ~20 minutes, a typical protein can pass through the system. A significant proportion of the membrane surface of the ER and Golgi is incorporated into vesicles that move to the plasma membrane. Such a flow of membrane would rapidly denude the Golgi apparatus and enormously enlarge the plasma membrane, yet both are stable in size.

Is there a pathway for returning membrane segments from the Golgi to ER, so that there is no net flow of membrane? It would require the average lipid to spend only a few minutes in the Golgi before recycling to the ER. How would the lipid recycle without perturbing the protein constitutions of the ER and Golgi? We do not yet understand the balance of forward and reverse flow, but one possibility is that different types of structures are used in each direction. Forward transport occurs in vesicles, which are essentially spheres with a low surface to volume ratio. Perhaps reverse flow occurs by structures that have a high surface to volume ratio, such as tubules, which could thus return large amounts of material.

We can divide coated vesicles into two general types, distinguished by their outer coats.

◆ Endocytic and exocytic vesicles have **clathrin** as the most prominent protein in their coats, and are therefore known as **clathrin-coated vesicles**.

◆ Transport vesicles (in the constitutive pathway) and transition vesicles (shuttling from ER to Golgi and between the Golgi stacks) do not have clathrin, and are therefore called **nonclathrin-coated vesicles**.

The structure of clathrin-coated vesicles is known in some detail. The 180,000 dalton chain of clathrin, together with a smaller chain of 35,000 daltons, forms a polyhedral coat on the surface of the coated vesicle. The subunit of the coat consists of a **triskelion**, a three-pronged protein complex consisting of 3 light and 3 heavy chains. The triskelions form a lattice-like network on the surface of the coated vesicle, as revealed in the electron micrograph of **Figure 11.24**. Endocytic vesicles form and are coated at invaginations of the plasma membrane that are called **coated pits**. Similar

Figure 11.24

Coated vesicles have a polyhedral lattice on the surface, created by triskelions of clathrin. Photograph kindly provided by Tom Kirchhausen.

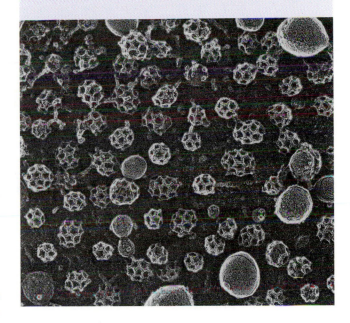

structures can be observed at the *trans* face of the Golgi, where exocytic vesicles originate.

The structure of clathrin-coated vesicles is shown in outline in **Figure 11.25**. The inner shell of the coat is made by proteins called **adaptors**, which bind both to clathrin and to integral membrane proteins of the vesicle. Two types of adaptors have been characterized so far; their distribution suggests that they correspond to coated vesicles with different origins. The HA-2 adaptor is found at plasma membrane–coated pits and characterizes endocytic vesicles; the HA-1 adaptor is found at Golgi–coated pits, and presumably identifies exocytic vesicles. Each adaptor is a heterotetramer, consisting of individual subunits that are called **adaptins**. The β and β' adaptins are the corresponding subunits in the two adaptors that bind clathrin.

Figure 11.25

Clathrin-coated vesicles have a coat consisting of two layers: the outer layer is formed by clathrin, and the inner layer is formed by adaptors, which lie between clathrin and the integral membrane proteins.

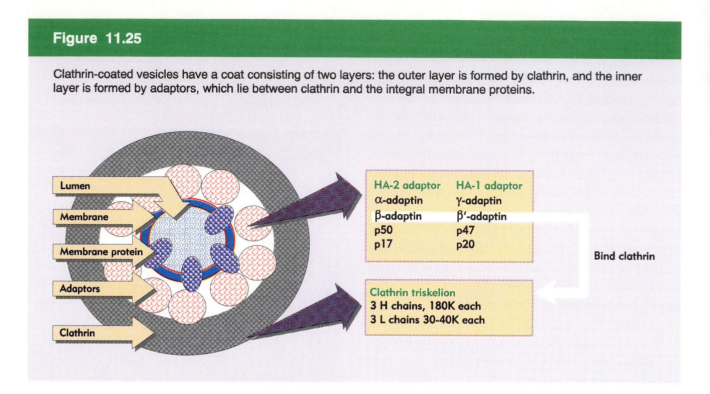

A significant proportion of the clathrin and the adaptors in the cell are found in a pool of free molecules, which suggests that both components are removed when vesicles become uncoated prior to fusion with their membrane targets. Clathrin is common to all endocytic and exocytic coated vesicles, so other proteins in the coat must distinguish different types of clathrin-coated vesicle, ensuring that they transport their protein cargo to appropriate destinations. The adaptors bind to the cytoplasmic tails of membrane proteins that are carried by the vesicles, and thus appear to have responsibility for recognizing the appropriate cargo proteins to load into the vesicle. The adaptors may also select the correct target membranes for fusion and direct the vesicle to dump the proteins.

How many types of cargo protein can be carried by a single endocytic vesicle? It is not yet clear how many types of vesicle exist and what variety they display on the coats and in their cargo. We do know that some endocytic vesicles carry more than one type of cargo protein. Generally they are viewed as fairly specific carriers.

The coats of the nonclathrin-coated vesicles involved in bulk flow through the Golgi have seven major protein components, called COPs. They exist as a high molecular weight complex (~700,000 daltons), called **coatomer**, which is probably the precursor to the nonclathrin coat. The β-COP component has significant homology to β-adaptin, the component of clathrin-coated vesicles that binds clathrin. It seems likely that β-COP plays a similar role in connecting an outer coat protein (an analog of clathrin) to the membrane proteins in the vesicle. Of course, a difference is that clathrin-coated vesicles tend to carry rather specific target proteins, whereas nonclathrin-coated vesicles undertake bulk flow. The mechanism for recognizing the protein cargo could therefore be different.

Budding of both nonclathrin- and clathrin-coated vesicles requires a small GTP-binding protein called ARF (ADP-ribosylation factor), and an energy supply in the form of ATP. Budding is initiated by ARF, which is myristylated at the N-terminus and can insert spontaneously into lipid bilayers. Only ARF that is bound to GTP has this ability; ARF complexed with GDP lacks the function, which suggests that its activity (and ability to recycle) could be controlled by GTP hydrolysis. One possibility is that the type of guanine nucleotide controls the conformation, so that the myristylated N-terminus is exposed only when GTP is bound.

Budding of nonclathrin-coated vesicles requires only ARF and coatomer. As shown in **Figure 11.26**, ARF enables coatomer to bind, allowing the coat complex to surround the membrane and cause vesicle budding. ARF is present in stoichiometric amounts (~1 ARF per 2–3 coatomers), and therefore becomes a component of the vesicle. The reaction is presumably similar for clathrin-coated vesicles.

One route to investigating the specificity of vesicle movement has been provided by the identification of a class of small GTP-binding proteins. Like other GTP-binding proteins, they are active in the form bound to GTP; but hydrolysis of the GTP converts the protein to an inactive form

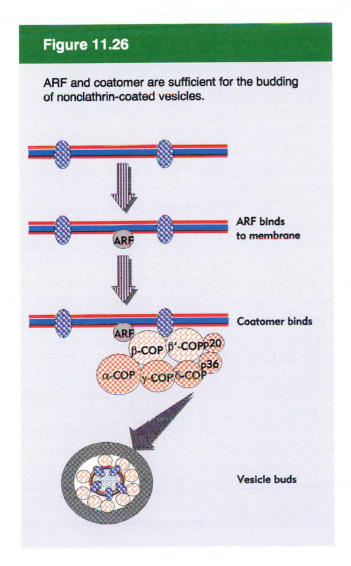

Figure 11.26

ARF and coatomer are sufficient for the budding of nonclathrin-coated vesicles.

ARF binds to membrane

Coatomer binds

Vesicle buds

that requires regeneration of GTP before it can participate in another cycle of activity. Mutations in the yeast genes *YPT1* or *SEC4* that code for two such (related) proteins block transport and cause the accumulation of vesicles in the Golgi stacks or between Golgi and plasma membrane, respectively. This shows that at least some proteins are involved only in specific transport stages. The general involvement of proteins that hydrolyze GTP is shown also by the ability of a nonhydrolyzable analog to block secretion in animal cells.

The counterparts to these proteins in mammalian cells are called **rabs**. They are attached to membranes via the addition of prenyl or palmityl groups at the C-terminus. There are >15 rabs, distributed to different membranous systems in the cell; they are found in the ER and Golgi, on endosomes, and on some vesicles. One speculation is that rabs may provide targeting information that allows a vesicle to dock at an appropriate membrane. The hydrolysis of GTP controls their activities.

The coat of the vesicle is an impediment to fusion with a target membrane, and must be removed when a vesicle reaches its destination. The components involved in fusion were identified via a mammalian protein called NSF, identified by its sensitivity to the sulfhydryl agent NEM (*N*-ethylmaleimide). NSF is the homolog for the product of yeast gene *sec18*, which is required for vesicle fusion during passage of the Golgi stacks, and also subsequently for fusion with the plasma membrane. **Figure 11.27** shows that fusion requires a 20S complex that consists of NSF (a soluble ATPase), 2 SNAPs (soluble proteins that have an affinity for hydrophobic surfaces), and a SNAP-receptor located in the target membrane. ATP hydrolysis is involved in the reaction; probably the NSF–SNAP complex dissociates during fusion, when ATP is hydrolyzed, allowing NSF and SNAPs to recycle.

If we suppose that uncoated vesicles would have an ability to fuse spontaneously with membranes in the cell, we can view the coat as a protective layer that preserves the vesicle until it reaches its destination. One possibility is that the uncoating reaction is triggered by hydrolyzing the GTP carried by ARF. This could cause ARF to be released from

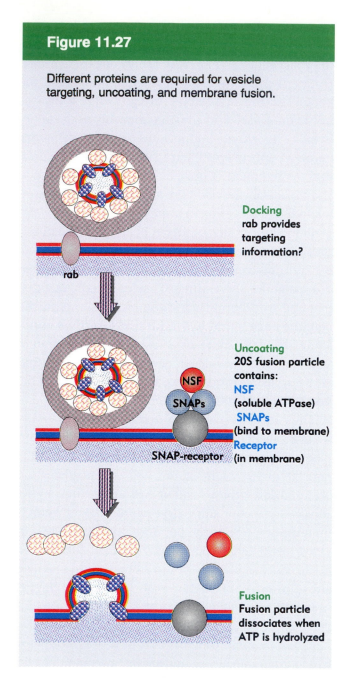

Figure 11.27

Different proteins are required for vesicle targeting, uncoating, and membrane fusion.

Docking
rab provides targeting information?

rab

Uncoating
20S fusion particle contains:
NSF (soluble ATPase)
SNAPs (bind to membrane)
Receptor (in membrane)

NSF
SNAPs
SNAP-receptor

Fusion
Fusion particle dissociates when ATP is hydrolyzed

reached by endocytosed proteins within ~1 minute. **Late endosomes** are closer to the nucleus, and are reached within 5–10 minutes.

The early endosome provides the main location for sorting proteins on the endocytic pathway. Its role is a counterpart to that played by the Golgi for newly synthesized proteins. The interior of the endosome is acidic, with a pH <6. Proteins that are transported to the endosome change their structure in response to the lowering of pH; this change can be important in determining their fate. For example, receptors bound to ligands dissociate from them as a result of the pH change (see Chapter 12).

Receptors that have been endocytosed to the early endosome behave in one of two ways. They may return to the plasma membrane (by vesicular transport), most commonly to the same region from which they were derived, sometimes to a different region. Or they may be transported further to the **lysosome**, where they are degraded.

The lysosome is another small body bounded by membranes. It contains the cellular supply of hydrolytic enzymes, which are responsible for degradation of macromolecules. Like endosomes, the lysosome is an acidic compartment, whose pH is maintained at the low level of 5.

The relationship between the various types of endosomes and lysosomes is not yet clear. Vesicles may be used to transport proteins along the pathway from one pre-existing structure to the next; or early endosomes may 'mature' into late endosomes, which in turn 'mature' into lysosomes. At all events, the pathway is one way, and a protein that has left the early endosome for the late endosome will end up in the lysosome.

There are two routes to the lysosome. Proteins endocytosed from the plasma membrane may be directed via the early endosome to the late endosome. Newly synthesized proteins may be directed from the *trans* Golgi to the late endosome.

Protein transport is a unidirectional process: proteins enter the ER and are transported through the Golgi, unless stopped *en route*. This direction of transport is called **anterograde**. However, transport in the opposite **retrograde** direction occurs

the membrane, in turn releasing the clathrin or coatomer complex.

The immediate destination for endocytic (clathrin-coated) vesicles is the **endosome**, a rather heterogeneous structure consisting of tubules and vesicles bounded by membranes. There are at least two types of endosome, as indicated in **Figure 11.28**. **Early endosomes** lie just beneath the plasma membrane and are

when cells are treated with the drug brefeldin A (BFA). The drug blocks conversion of ARF from the GDP-bound to the GTP-bound form, and as a result prevents the β-COP protein of nonclathrin-coated vesicles from binding to the membranes. This action blocks budding of the coated vesicles, thereby preventing forward transport.

As a result of the block, a network of tubules forms between the cisternae of the Golgi (abolishing their usual independence) and joins them to the endoplasmic reticulum. There is resorption of most of the membranes of the *cis*-medial Golgi into the endoplasmic reticulum, which is accompanied by the redistribution of Golgi proteins into the ER, effecting a retrograde transport.

There has been a long controversy about the interpretation of this effect. One view is that brefeldin treatment reveals a continuing process of retrograde transport, which usually is obscured by anterograde transport, but which is revealed when brefeldin specifically blocks the forward direction of transport. If retrograde transport is a normal function, it could be part of a system for retrieving membrane components to compensate for the anterograde movement of bulk flow. An alternative view is that membranous surfaces, including ER and Golgi membranes, have a spontaneous ability to fuse, but this is usually prevented, perhaps because coatomer binds to the fusogenic sites. Inability to protect these sites leads to unrestrained

Figure 11.28

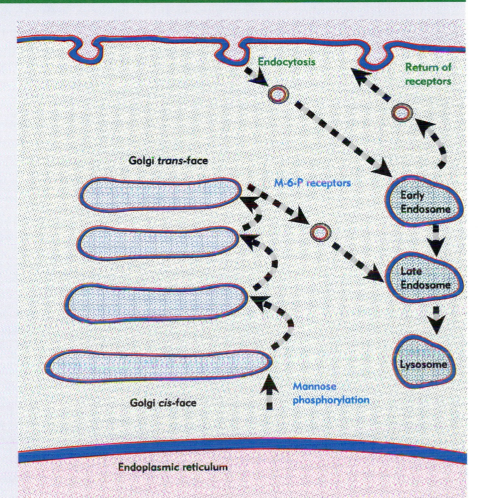

Endosomes sort proteins that have been endocytosed and provide one route to the lysosome. Proteins are transported via clathrin-coated vesicles from the plasma membrane to the early endosome, and may then either return to the plasma membrane or proceed further to late endosomes and lysosomes. Newly synthesized proteins may be directed to late endosomes (and then to lysosomes) from the Golgi stacks. The common signal in lysosomal targeting is the recognition of mannose-6-phosphate by a specific receptor.

fusion, and induces large scale retrograde transport. Of course, it may still be the case that retrograde transport occurs normally, but at a lower level, or for only certain specific proteins (as described in the next section.)

The effects of brefeldin on ER–Golgi transport are universal, but in addition it inhibits other transport processes differently in different cells. In some cell types it inhibits transcytosis (transport from the basolateral surface to the apical surface in polarized cells); in other cells it inhibits transport from the *trans* Golgi network to endosomes. This probably means that there is a variety of adaptin-like proteins that characterize coated vesicles involved in these particular transport processes, and that only some of these proteins can bind brefeldin (so that the functions of their particular types of vesicles are inhibited).

Protein localization depends on further signals

We view the process of transport from the endoplasmic reticulum through the Golgi and into and/or through the plasma membrane as the ineluctable fate of any protein that is not specifically halted *en route*: in the absence of a signal, a protein continues by bulk flow to the plasma membrane. What signals are responsible for recognition at each stage? There are two (perhaps three) types of signal: a specific amino acid sequence (or possibly a specific conformation or exposed residue in the protein); and groups that have been added covalently. We have detailed information about several types of signal: a conformation that is required for proteins to be internalized by endocytosis; an amino acid sequence that targets proteins to the ER; and a modification that targets proteins to lysosomes.

Internalization of receptors via coated pits requires information in their cytoplasmic tails. This is discussed in more detail in Chapter 12, but we note now that the sequence NPXY (Asn-Pro-X-Tyr) is located close to the C-terminus, and adopts the conformation of a β-turn that exposes the tyrosine. Although this is a basic signal for internalization, other sequences in the cytoplasmic tail influence the efficiency.

Enzymes destined to be transported to lysosomes are transferred to the endoplasmic reticulum cotranslationally. They are recognized as targets for high mannose glycosylation, and are trimmed in the ER as described in Figure 11.19. Then mannose-6-phosphate residues are generated by a two stage process. First the moiety N-acetyl-glucosamine-1-phosphate is added to the 6 position of mannose by GlcNAc-phosphotransferase; then a glucosaminidase removes the N-acetyl-glucosamine (GlcNAc).

The enzymes act sequentially in the ER–Golgi pathway (see Figure 11.20). The action of the phosphotransferase provides the critical step in marking a protein for lysosomal transport. It occurs early in ER–Golgi transfer, possibly between the ER and the *cis* Golgi. The basis for the enzyme's specificity is its ability to recognize a structure that is common to lysosomal proteins. The structure consists of two short sequences, which are separated in the primary sequence, but form a common surface in the tertiary structure. Each of these sequences has a crucial lysine residue. The nature of this signal explains how proteins with little identity of sequence may share a common pathway for localization.

Lysosomal proteins continue to be transported along the Golgi stacks until they encounter receptors for mannose-6-phosphate. Two related proteins serve as receptors: a large (215,00 dalton) species, and a smaller (46,000 dalton) species. Recognition of mannose-6-phosphate targets a protein for transport in a coated vesicle to the lysosome. This final stage of sorting for the lysosome occurs in the *trans* Golgi, where the proteins are collected by (presumably) specific transport vesicles that are (probably) coated with clathrin. The vesicles transport the lysosomal proteins to the late endosome, where they join the pathway for movement to the lysosome.

A single pool of mannose-6-phosphate receptors is probably used for directing proteins to the lysosome whether they are newly synthesized or endocytosed. Most of the receptors in fact are located on endosomes, where they could recognize endocytosed proteins.

How far does the receptor continue along the pathway toward the lysosome? Probably the low pH of endosomal compartments allows the receptor to dissociate from the protein it has targeted, whereupon the protein continues to the lysosome, but the receptor recycles to the plasma membrane or to the Golgi.

Proteins that reside in the lumen of the endoplasmic reticulum possess a short sequence at the C-terminus, Lys-Asp-Glu-Leu (KDEL in single letter code). The alternative signals HDEL or DDEL are used in yeast. If this sequence is deleted, or if it is extended by the addition of other amino acids, the protein is secreted from the cell instead of remaining in the lumen. Conversely, if this tetrapeptide sequence is added to the C-terminus of lysozyme, the enzyme is held in the ER lumen instead of being secreted from the cell. This suggests that there is a mechanism to recognize the C-terminal tetrapeptide and cause it to be localized in the lumen.

An interesting point emerges from the behavior of proteins that have the ER-localization signal. Does this signal cause a protein to be held so that it cannot pass beyond the ER or is it the target for a more active localization process? The model shown in **Figure 11.29** suggests that *the KDEL sequence causes a protein to be returned to the ER from an early Golgi stack.*

Because the modification of proteins as they pass through the Golgi is ordered, we can use the types of sugar groups that are present on any particular species as a marker for its progress. When a KDEL sequence is added to a protein that usually is targeted to the lysosome (because its oligosaccharide gains 6-mannose-P residues), it causes the protein

to be held in the ER. But the protein is modified by the addition of GlcNAc-P, which happens only in the Golgi. The GlcNAc is not removed, so the protein cannot have proceeded far enough through the Golgi stacks to encounter the second of the enzymes in the mannose-6-P pathway. This suggests that KDEL is recognized by a receptor located after entering the Golgi, but before the stack containing

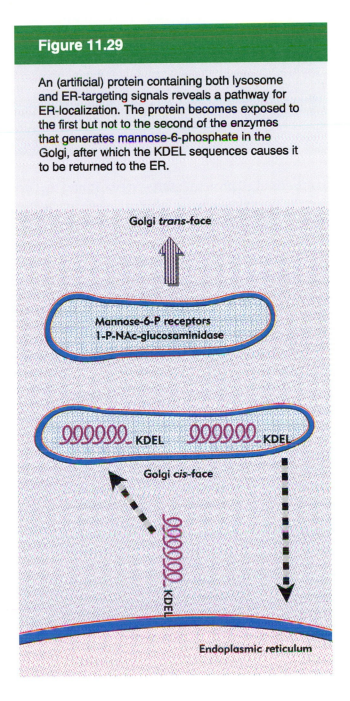

Figure 11.29

An (artificial) protein containing both lysosome and ER-targeting signals reveals a pathway for ER-localization. The protein becomes exposed to the first but not to the second of the enzymes that generates mannose-6-phosphate in the Golgi, after which the KDEL sequences causes it to be returned to the ER.

the second enzyme. This could occur in a salvage compartment located between the ER and *cis* Golgi.

Mutations in the *S. cerevisiae* genes *ERD1* and *ERD2* prevent retention of proteins with the HDEL signal in the ER; instead the proteins are secreted from the cell. The products of both these genes are integral membrane proteins. The *ERD1* mutation causes a general defect in the Golgi; this supports the idea that sorting of the ER proteins occurs by salvage from the Golgi. The *ERD2* mutation identifies the receptor for the HDEL sequence. One model for its role is that it cycles between the Golgi salvage compartment and the ER. This idea is supported by the localization of the corresponding receptor in mammalian cells: it is found largely in the Golgi, but overexpression of a protein with KDEL sequence causes it to concentrate in the ER. Thus binding of its ligand induces the protein to move from Golgi to ER. It may have a high affinity for the HDEL sequence under the conditions prevailing in the Golgi, but a low affinity in the ER. This could enable ERD2 to seize proteins by their HDEL tails in the Golgi, and take them back to the ER, where they are then released.

Summary

Synthesis of all proteins starts on ribosomes that are 'free' in the cytosol. In the absence of any particular signal, a protein is released into the cytosol when its synthesis is completed. Proteins that are imported into mitochondria or chloroplasts possess N-terminal leader sequences that target them to the organelle envelope; then they are transported through the double membrane into the matrix. Translocation requires ATP and a potential across the outer membrane. The N-terminal leader is cleaved by a protease within the organelle. Proteins that reside within the membranes or intermembrane space possess a signal (which becomes N-terminal when the first part of the leader is removed) that causes export from the matrix to the appropriate location. Control of folding, by hsp70 and hsp60 in the mitochondrial matrix, is an important feature of the process. Requirements for export from bacteria also show strong dependence on control of protein conformation.

The N-terminal region of a secreted protein provides a signal sequence that causes the nascent protein and its ribosome to become attached to the membrane of the endoplasmic reticulum. The signal sequence is recognized by the SRP, which arrests translation; then the SRP binds to a receptor in the ER membrane, and transfers the signal sequence to the membrane. Synthesis resumes, and the protein is translocated through the membrane while it is being synthesized, although there is no mechanical connection between the processes. The SRP is a ribonucleoprotein particle.

For type I proteins, the N-terminal signal sequence is cleaved and the protein becomes oriented in the membrane with its N-terminus on the far side and its C-terminus in the cytosol; transfer through the membrane is halted by an anchor sequence. Type II proteins do not have a cleavable N-terminal signal, but instead have a combined signal–anchor sequence, which enters the membrane and becomes embedded in it, causing the C-terminus to be located on the far side, while the N-terminus remains in the cytosol. The orientation of the signal-anchor is determined by the 'positive inside' rule that the side of the anchor with more positive charges will be located in the cytoplasm.

Proteins may have multiple membrane-spanning regions, with loops between them protruding on either side of the membrane. The mechanism of insertion of multiple segments is unknown.

Proteins that enter the endoplasmic reticulum continue with the bulk flow into the Golgi unless they possess specific ER-targeting signals. The C-terminal signal KDEL (HDEL in yeast) causes proteins to reside in the ER, by a mechanism that involves retrieval from the Golgi (or possibly from an intermediate compartment between the ER and Golgi) by a salvage procedure. Bulk flow transports proteins along the Golgi from the *cis* to the *trans* face. Modification of proteins by addition of a preformed oligosaccharide starts in the endoplasmic reticulum. High mannose oligosaccharides are trimmed. Complex oligosaccharides are generated by further modifications that are made during transport through the Golgi, determined by the order in which the protein encounters the enzymes localized in the various Golgi stacks. Proteins are sorted for different destinations in the *trans* Golgi. The signal for sorting to lysosomes is the presence of mannose-6-phosphate residue.

Proteins are transported from the ER and along the Golgi stacks as cargos in membrane-bound vesicles. The vesicles form by budding from donor membranes; they unload their cargos by fusing with target membranes. These nonclathrin-coated vesicles are coated with coatomer. One of the proteins of coatomer, β-COP, is related to the β-adaptin of clathrin-coated vesicles, suggesting the possibility of a common type of structure between nonclathrin-coated and clathrin-coated vesicles. The protein coats are added when the vesicles are formed and must be removed before they can fuse with target membranes. There is no net flow of membrane from the ER to the Golgi and/or plasma membrane, so membrane moving with the bulk flow must be returned to the ER. Retrograde transport can be observed when brefeldin A is used to inhibit forward transport, but its mechanism (and relationship to normal events) is controversial. Brefeldin functions by binding to β-COP and preventing budding of nonclathrin-coated vesicles.

Exocytosis describes the pathway for regulated secretion of proteins. Proteins are sorted into secretory vesicles at the Golgi *trans* face. The predominant protein in the outer coat of the vesicle is clathrin. The inner coat contains adaptins; β-adaptin binds to clathrin. Secretory vesicles are stimulated to unload their cargos at the plasma membrane by extracellular signals. Similar vesicles are used for endocytosis, the pathway by which proteins are internalized from the cell surface. The specificity of clathrin-coated vesicles presumably is conferred by other proteins located in the coats.

Further reading

Reviews

The signal hypothesis has been reviewed by **Walter and Lingappa** (*Ann. Rev. Cell Biol.*, **2**, 499–16, 1986).

The variety of patterns of membrane insertion have been summarized by **Wickner and Lodish** (*Science* **230**, 400–407, 1986).

Transport and sorting in the ER has been analyzed by **Pfeffer and Rothman** (*Ann. Rev. Biochem.* **56**, 829–852, 1987) and **Rose and Doms** (*Ann. Rev. Cell Biol.* **4**, 257–288, 1988). Retrieval of ER proteins has been analyzed by **Pelham** (*Ann. Rev. Cell Biol.* **5**, 1–23, 1989).

Modification and sorting in the Golgi have been reviewed by **Farquhar** (*Ann. Rev. Cell Biol.* **1**, 447–488, 1985), **Griffiths and Simons** (*Science* **234**, 438–443, 1986), and **Rothman and Orci** (*Nature* **355**, 409–415, 1992). Lysosomal targeting has been reviewed by **von Figura and Hasilik** (*Ann. Rev. Biochem.* **55**, 167–193, 1986) and by **Kornfeld and Mellman** (*Ann. Rev. Cell Biol.* **5**, 483–525, 1989). The structure and functions of clathrin-coated vesicles have been reviewed by **Pearse and Robinson** (*Ann. Rev. Cell Biol.* **6**, 151–171, 1990).

Export of bacterial proteins has been reviewed by **Oliver** (*Ann. Rev. Microbiol* **39**, 615–648, 1985) and by **Lee and Beckwith** (*Ann. Rev. Cell Biol.* **2**, 315–336, 1986).

The components of the mitochondrial import apparatus have been reviewed by **Baker and Schatz** (*Nature* **349**, 205–208, 1991).

Transport into the nucleus has been reviewed by **Dingwall and Laskey** (*Ann. Rev. Cell Biol.* **2**, 367–390, 1986).

The issues involved in folding proteins have been reviewed by **Hurtley and Helenius** (*Ann. Rev. Cell Biol.* **5**, 277–307, 1989) and **Rothman** (*Cell* **59**, 591–601, 1989).

Discoveries

Different models for protein transport into mitochondria were developed by **van Loon, Brandi, and Schatz** (*Cell* **44**, 801–812, 1986) and **Hartl** *et al.* (*Cell* **51**, 1027–1037, 1988).

Protein folding during membrane transport has been characterized in several systems by **Eilers and Schatz** (*Nature* **322**, 228–232, 1986), **Collier** *et al.* (*Cell* **53**, 273–283, 1988), **Crooke** *et al.* (*Cell* **54**, 1003–1011, 1988), and **Ostermann** *et al.* (*Nature* **341**, 124–130, 1989).

Functions of the SRP were dissected by **Siegel and Walter** (*Cell* **52**, 39–49, 1988).

A purified system for bacterial translocation was developed by **Brundage** *et al.* (*Cell* **62**, 649–657, 1990).

Bulk flow through the Golgi was measured by **Wieland** *et al.* (*Cell* **50**, 289–300, 1987). Interference with recycling was reported by **Lippincott-Schwartz** *et al.* (*Cell* **56**, 801–813, 1989). The KDEL ER-targeting signal was uncovered by **Munro and Pelham** (*Cell* **48**, 899–907, 1987), and further characterized by **Semenza** *et al.* (*Cell* **61**, 1349–1357, 1990) and **Lewis** *et al.* (*Cell* **61**, 1359–1363, 1990).

The channel for protein translocation through the ER membrane was characterized by **Simon and Blobel** (*Cell* **65**, 371–380, 1991).

CHAPTER 12

Receptors and signal transduction; channels and ion uptake

Ability to respond to the surrounding environment, and control of the entry and exit of molecules through the plasma membrane, are crucial features of any cell. These processes are the responsibilities of proteins that reside within the plasma membrane. The focus of this chapter is to consider these functions of the plasma membrane in terms of the structures of the resident proteins.

The plasma membrane is impermeable to water-soluble material, including ions, small inorganic molecules, and polypeptides or proteins. To enter or influence the cell, any hydrophilic material must react with a protein resident in the plasma membrane. The extracellular material (whether an inorganic molecule or polypeptide) is called a **ligand** (a common term for a component that binds to a protein to alter its function; see Chapter 1). The target protein to which the ligand binds is called a **receptor**. By contrast, such interactions are not necessary for a lipid-soluble molecule, such as a steroid hormone, which can diffuse through the plasma membrane into the cytoplasm, where it binds its target protein.

Two fundamental types of response to an external stimulatory molecule that cannot cross the membrane are reviewed in **Figure 12.1**:

◆ *A signal* is transmitted by means of a change in the conformation of a membrane protein. As a result of this change, there is in turn a change in the activity of the cytosolic domain of the membrane protein or of some other protein with which it associates.

◆ *Material*—molecular or macromolecular—is physically transmitted from the outside of the membrane to the inside by transport through the lipid bilayer.

When a receptor remains in the membrane and responds to ligand-binding by triggering a response pathway in the cytosol, the process is called **signal transduction**, because a signal has in effect been transduced across the membrane.

Signal transduction provides a means for **amplification** of the original signal. When an extracellular ligand interacts with a membrane receptor, it converts it from an inactive to an active form. In its active form, the receptor stimulates a catalytic activity that generates a cytosolic signal whose amplitude is much greater than the original extracellular signal (the ligand). Sometimes the cytosolic signal takes the form of increasing the quantity

Figure 12.1

Overview: information may be transmitted from the exterior to the interior of the cell by signal transduction or by movement of a ligand.

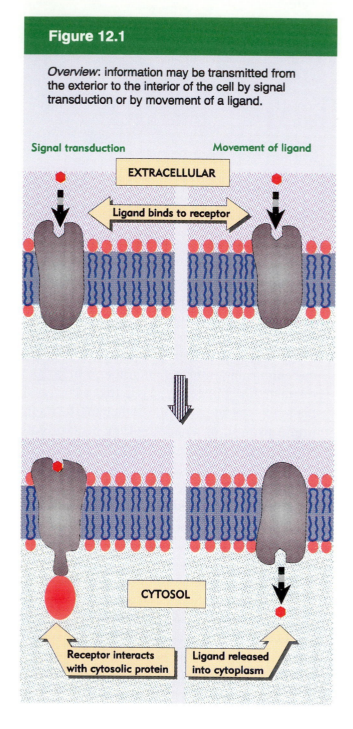

Signal transduction   **Movement of ligand**

EXTRACELLULAR

Ligand binds to receptor

CYTOSOL

Receptor interacts with cytosolic protein   Ligand released into cytoplasm

◆ The receptor may be a transmembrane protein with domains on both the extracellular side and the cytoplasmic side. Ligand binding on the extracellular side influences the activity of the domain on the cytoplasmic side. A typical reaction is for ligand-binding to activate a protein tyrosine kinase activity possessed by a cytoplasmic domain. The kinase phosphorylates its own cytoplasmic domain; the phosphorylation enables the receptor to associate with and activate a target protein, which in turn acts upon new substrates within the cell.

◆ The receptor may interact with a **G-protein** that is associated with the membrane. G-proteins are named for their ability to bind guanine nucleotides. The inactive form of the G-protein is a trimer bound to GDP. Upon binding ligand, a receptor acts upon the G-protein to cause the GDP to be replaced with GTP; as a result, the G protein dissociates into a single subunit carrying GTP and a dimer. Either the monomer or the dimer then acts upon a target protein, often also associated with the membrane, which in turn reacts with a target(s) in the cytoplasm. This chain of events often stimulates the production of second messengers, the classic example being the production of cyclic AMP.

The physical transfer of material extends from ions to small molecules such as sugars, and to macromolecules such as proteins. Three major transport routes are reviewed in **Figure 12.3**:

◆ Proteins resident in the membrane may constitute **channels** that control the passage of ions: different channels exist for potassium, sodium, and calcium ions. By opening and closing in response to appropriate signals, the channels establish ionic levels within the cell, a feature of particular significance for cells of the neural network.

◆ One means to import small molecules is for a receptor itself to transport the molecule from one side of the membrane to the other. Receptors are responsible for the transport of small

of a small molecule inside the cell. A molecule produced in response to transduction of an extracellular signal is called a **second messenger** (by contrast with the first messenger, which was the extracellular ligand).

Two major types of signal transduction are reviewed in **Figure 12.2**:

Figure 12.2

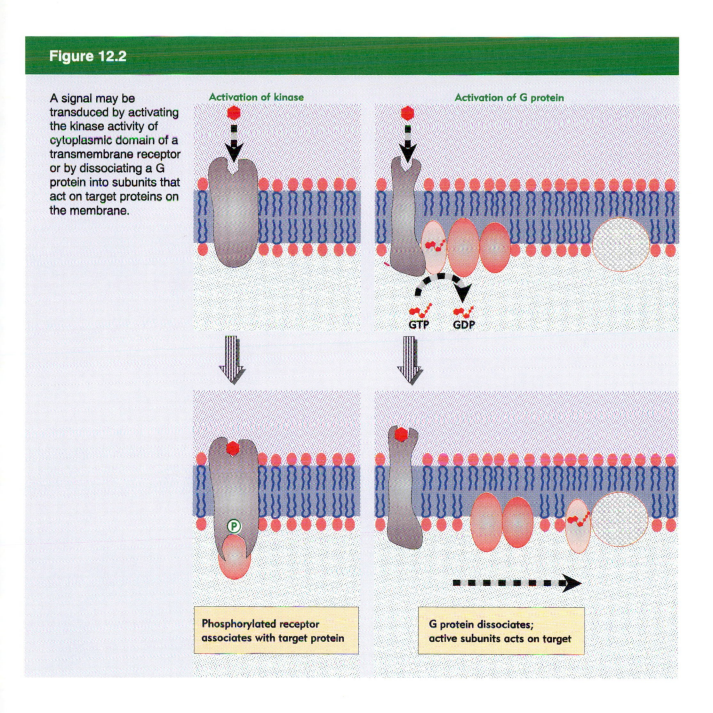

A signal may be transduced by activating the kinase activity of cytoplasmic domain of a transmembrane receptor or by dissociating a G protein into subunits that act on target proteins on the membrane.

Activation of kinase

Activation of G protein

GTP   GDP

Phosphorylated receptor associates with target protein

G protein dissociates; active subunits acts on target

molecules (such as sugars) across the membrane by a process in which the target molecule binds to the receptor on the extracellular side, but then is released on the cytoplasmic side.

◆ Ligand-binding may trigger the process of **internalization**, in which the receptor–ligand combination is brought into the cell by the process of **endocytosis**. In due course, the

receptor and ligand are separated; the receptor may be returned to the surface for another cycle, or may be degraded. Endocytosis provides a means to bring a ligand into the cell. Transport of proteins into and out of the cell is conducted by means of the endocytic and exocytic pathways, respectively. As described in Chapter 11, these pathways involve the passage of membrane proteins from one

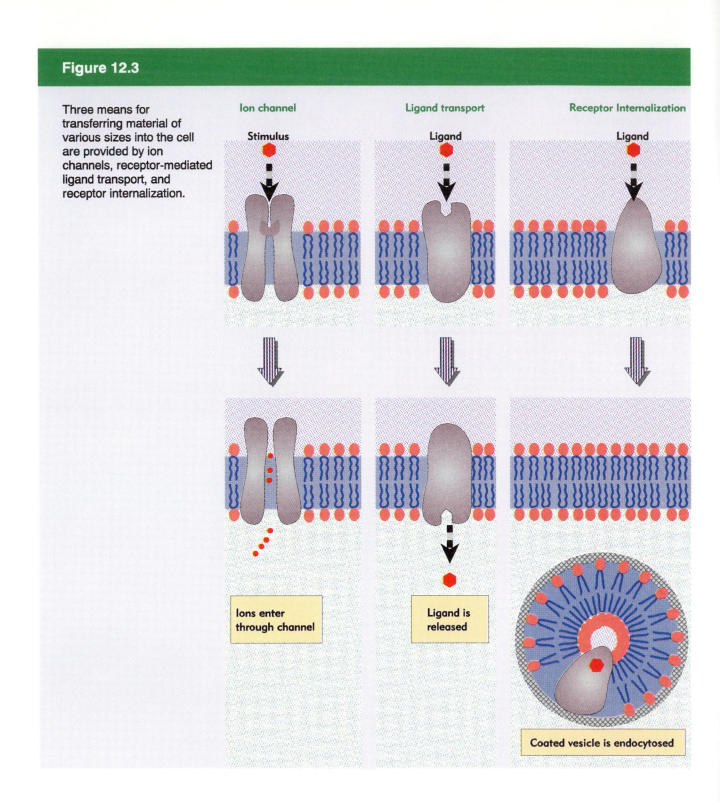

Figure 12.3

Three means for transferring material of various sizes into the cell are provided by ion channels, receptor-mediated ligand transport, and receptor internalization.

Ion channel

Stimulus

Ions enter through channel

Ligand transport

Ligand

Ligand is released

Receptor Internalization

Ligand

Coated vesicle is endocytosed

surface to another via coated vesicles.

An interesting contrast with another situation in which macromolecules must be specifically transported across a membrane in both directions is provided by the relationship between nucleus and cytoplasm. All classes of RNA must be exported from the nucleus; and all nuclear proteins must be imported from the cytosol. These processes occur via large structures called **nuclear pores** (see later).

A more detailed view of the plasma membrane

The operational definition of an **integral membrane protein** is that it requires disruption of the lipid bilayer in order to be released from the membrane. This property can be predicted from the amino acid sequence, since a common feature in all such proteins is the presence of at least one membrane-spanning region, consisting of a stretch of 21–26 hydrophobic amino acids, coiled into an α-helix. A sequence that fits the criteria for membrane insertion can be identified by a hydropathy plot, as described previously in Figure 2.6 (although not every sequence so identified is in fact found in the membrane).

Various forms of organization for membrane proteins are summarized in **Figure 12.4**. An important feature is the number of membrane-spanning regions. When there is a single membrane-spanning region, its position determines how much of the protein is exposed on either side of the membrane. A protein may have extensive domains exposed on both sides of the membrane or may have a site of insertion close to one end, so that little or no material is exposed on one side. The length of the N-terminal or C-terminal tail that protrudes from the membrane near the site of insertion varies from insignificant to quite bulky. A protein that has domains exposed on both sides of the membrane is called a **transmembrane protein**. In considering the activities of various types of membrane proteins we must remember that the membrane fluid provides the environment within which these proteins must function.

Proteins with a single transmembrane domain fall into two classes. Group I proteins in which the

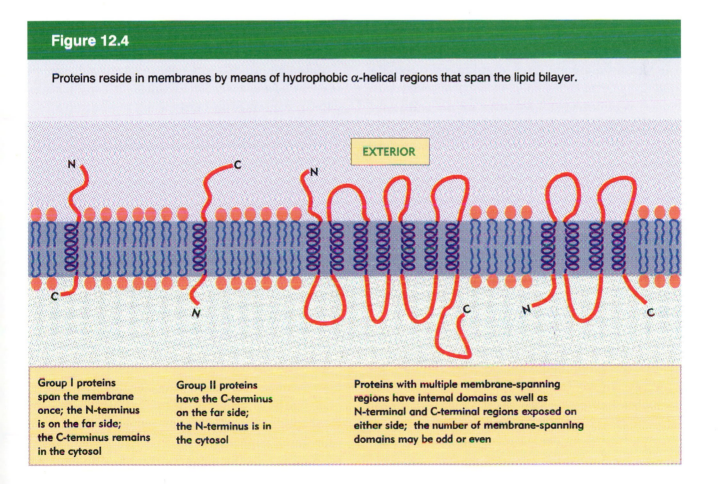

Figure 12.4

Proteins reside in membranes by means of hydrophobic α-helical regions that span the lipid bilayer.

EXTERIOR

| Group I proteins span the membrane once; the N-terminus is on the far side; the C-terminus remains in the cytosol | Group II proteins have the C-terminus on the far side; the N-terminus is in the cytosol | Proteins with multiple membrane-spanning regions have internal domains as well as N-terminal and C-terminal regions exposed on either side; the number of membrane-spanning domains may be odd or even |

N-terminus faces the extracellular space are more common than group II proteins in which the orientation has been reversed so that the N-terminus faces the cytoplasm. Orientation is determined during the insertion of the protein into the endoplasmic reticulum following its synthesis (see Chapter 11). Receptors of the tyrosine kinase type typically are group I, with the kinase activity located in the C-terminal part of the cytoplasmic domain.

When there are multiple membrane-spanning domains, an even number means that both termini of the protein are on the same side of the membrane, whereas an odd number implies that the termini are on opposite faces. The extent of the domains exposed on one or both sides is determined by the locations of the transmembrane domains. Domains at either terminus may be exposed, and internal sequences between the domains may 'loop out' into the extracellular space or cytoplasm. One common type of structure is the seven-membrane pass or 'serpentine' receptor, such as the example shown in the figure, which typically couples to a G protein.

Does a transmembrane domain itself play any role besides allowing the protein to insert into the lipid bilayer? In cases such as the group I tyrosine kinase receptors, it has little or no additional function; often it can be replaced by any other transmembrane domain. However, transmembrane domains play an important role in the function of proteins that make multiple passes through the membrane or that have subunits that oligomerize within the membrane. The transmembrane domains in such cases often contain polar residues, which are absent entirely from the single membrane-spanning domains of group I and group II proteins. If polar regions in the membrane-spanning domains do not interact with the lipid bilayer, but instead interact with one another, they could form a polar pore or channel within the lipid bilayer. Interaction between such transmembrane domains can create a hydrophilic passage through the hydrophobic interior of the membrane. This can allow highly charged ions or molecules to pass through the membrane, and is important for the function of ion channels and transport of ligands.

Another case in which conformation of the transmembrane domains is important is provided by certain receptors that bind lipophilic ligands; in such cases, the transmembrane domains (rather than the extracellular domains) bind the ligand within the plane of the membrane.

Membrane proteins have a restricted ability to diffuse laterally, in contrast with the continuous motion of the lipids in the bilayer. Lateral movement may be important, however, in signal transduction, since it enables membrane proteins to interact with one another. It plays a role in resolving the problem of how information is transmitted from one side of the membrane to the other when signal transduction is caused by a ligand that binds to an extracellular domain. Binding of ligand causes the extracellular domains of group I receptors to dimerize; this causes the transmembrane domains to diffuse laterally, bringing the cytoplasmic domains into juxtaposition. The cytoplasmic domains then interact with one another to trigger a conformational change that activates the appropriate function, such as a kinase activity, or interaction with proteins present on the cytoplasmic surface of a coated pit. An extreme case of lateral diffusion is seen in some cases of receptor internalization, when receptors of a given type aggregate into a 'cap' in response to an extracellular stimulus.

Proteins may also be associated with a membrane by means of covalent linkage to a fatty acid that is incorporated into the lipid bilayer. **Figure 12.5** depicts four forms of such association. In each case, a fatty acid or lipid is attached to an amino acid near to or at one terminus of the protein, with the result that the entire polypeptide chain resides on one side of the membrane, but is attached to it.

Prenylation is used to attach proteins to both the plasma membrane and internal membranes. Two types of prenyl groups have been identified: farnesyl is a 15 carbon isoprenoid (shown in the figure), and geranylgeranyl is a 20 carbon chain. They are added to cysteine residues by a thioester linkage; the cysteine is always located at the fourth position from the C-terminus, as part of the sequence CAAX, where A represents aliphatic amino acids and X is methionine or serine for farnesylation, and leucine for

Figure 12.5

Proteins may be associated with one face of a membrane by acyl linkages to fatty acids.

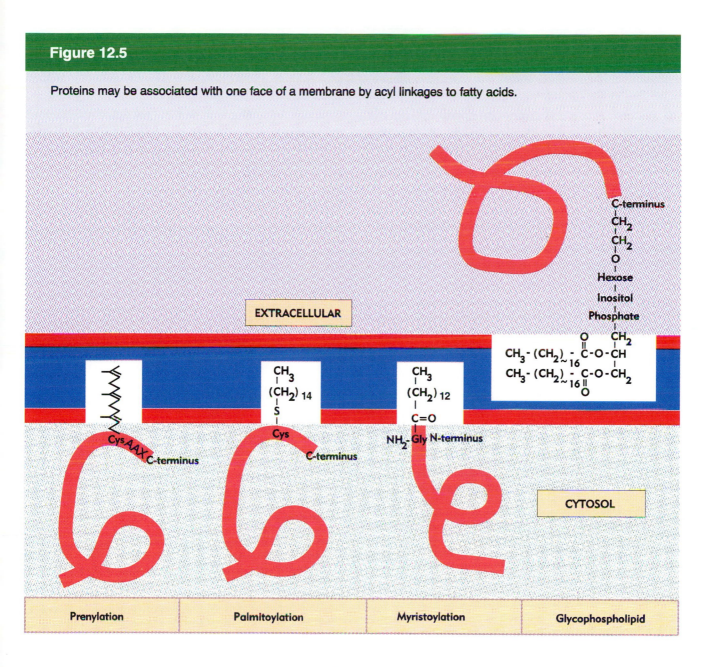

| Prenylation | Palmitoylation | Myristoylation | Glycophospholipid |

geranylgeranylation. The usefulness of the prenyl groups for assisting attachment to the membrane is obvious, but we do not yet know how specificity is conferred with regards to the choice of membrane.

Two fatty acids are used to anchor proteins on the cytoplasmic side of the plasma membrane. Palmitic acid, a 16 carbon-chain saturated fatty acid, is linked through a sulfide bond to a cysteine residue located close to the terminus (usually the C-terminus, but sometimes the N-terminus). Myristic acid, a 14 carbon-chain saturated fatty acid, is linked to the amino group of N-terminal glycine.

Myristoylated proteins are often, but not always, associated with a membrane.

The more complex structure of a glycosyl-phosphatidyl-inositol (GPI) anchor is linked to the carboxyl group of the C-terminal amino acid of protein exposed on the extracellular side of the membrane. Addition of the GPI anchor actually involves cleavage of the original polypeptide chain near the C-terminus, generating a new C-terminus that is linked to the anchor. Enzymes exist that can cleave the GPI anchor from the protein, releasing the protein into the extracellular medium.

Receptors recycle via endocytosis

The systems involved in importing proteins into the cell are closely related to those used for exporting secreted proteins (as reviewed in Chapter 11). A protein that is to be transported is incorporated into a vesicle that buds off from the membrane in which it resides. The vesicle is coated with clathrin and makes its way to a target membrane, where the coat is removed and the membranes fuse, releasing the protein into its new compartment. The process is called **receptor-mediated endocytosis**.

Ingestion of receptors starts by a common route, which leads to several pathways in which receptors have different fates. Some receptors are internalized continuously, but others remain exposed on the surface until a ligand is bound, after which they become susceptible to endocytosis. In either case, the receptors slide laterally into **coated pits,** which are indented regions of the plasma membrane surrounded by clathrin. It is not clear whether simple lateral diffusion can adequately explain the movement into coated pits or whether some additional force is required. The signals involved in internalization are different for ligand-independent and ligand-induced endocytosis.

Coated pits invaginate into the cell and pinch off to form clathrin-coated endocytic vesicles. These vesicles move to early endosomes, are uncoated, fuse with the target membrane, and release their contents as described earlier in the previous chapter in Figure 11.21. At this point, material may either be recycled to the surface in coated vesicles or transported further to the lysosome, where it is degraded. Transport to the lysosome is the default pathway, and applies to any material that does not possess a signal specifically directing it elsewhere.

The fate of a receptor–ligand complex depends upon its response to the acidic environment of the endosome. Exposure to low pH changes the conformation of the external domain of the receptor, causing its ligand to be released, or changing the structure of the ligand. But the receptor must avoid becoming irreversibly denatured by the acid environment; the presence of multiple disulfide bridges in the external domain may play an important role in maintaining this unusual stability.

Four possible fates for a receptor–ligand complex are described in the alternative pathways of Figures 12.6–12.9:

◆ *Receptor recycles to the surface in coated vesicles, while the ligand is degraded.* **Figure 12.6** shows that this pathway is used by receptors that transport ligands into cells at high rates. A receptor recycles every 1–20 minutes, and can undertake >100 cycles during its lifetime of ~20 hours. The classic example of this pathway is the LDL receptor, whose ligands are the plasma low density lipoproteins apolipoprotein E and apolipoprotein B (collectively known as the LDLs). Apo-B is a very large (500,000 dalton) protein that carries cholesterol and cholesterol esters. The LDL is released from its receptor in the endosome. The receptor recycles to the surface to be used again. The LDL and its cholesterol separate in the endosome; the LDL is sent on to the lysosome, where it is degraded, and the cholesterol is released for use by the cell. This constitutes the major route for removing cholesterol from the circulation, and people with mutations in the LDL receptor accumulate large amounts of plasma cholesterol that cause the disease of familial hypercholesterolemia.

◆ *Receptor and ligand both recycle.* The transferrin receptor provides the classic example of this pathway, illustrated in **Figure 12.7**. The ligand for the receptor is the iron-carrying form of transferrin. When this reaches the endosome, the acid environment causes transferrin to release the iron. The iron-free ligand, called apo-transferrin

Figure 12.6

LDL receptor transports apo-B (and apo-E) into endosomes, where receptor and ligand separate. The receptor recycles to the surface, apo-B (or apo-E) continues to the lysosome and is degraded, and cholesterol is released (not shown).

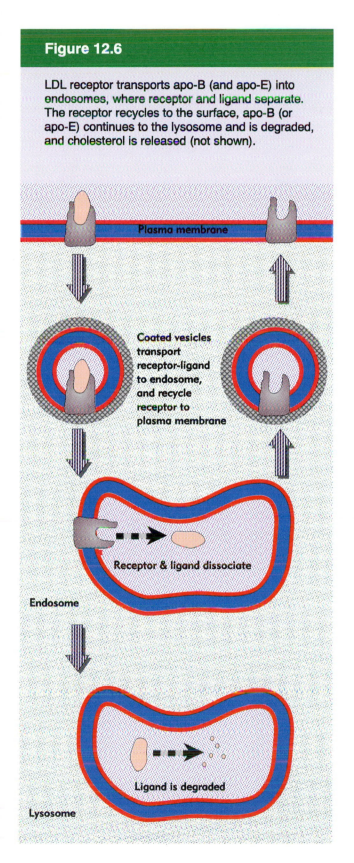

Plasma membrane

Coated vesicles transport receptor-ligand to endosome, and recycle receptor to plasma membrane

Receptor & ligand dissociate

Endosome

Ligand is degraded

Lysosome

remains bound to the transferrin receptor, and recycles to the plasma membrane. In the neutral pH of the plasma membrane, apo-transferrin dissociates from the receptor. This leaves apo-transferrin free to bind another iron, while the transferrin receptor is available to internalize another iron-carrying transferrin.

Figure 12.7

Transferrin receptor bound to transferrin carrying iron releases the iron in the endosome; the receptor now bound to apo-transferrin (lacking iron) recycles to the surface, where receptor and ligand dissociate.

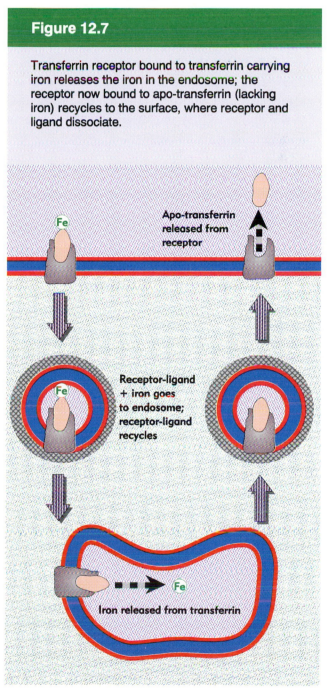

Apo-transferrin released from receptor

Receptor-ligand + iron goes to endosome; receptor-ligand recycles

Iron released from transferrin

Again this cycle is quite intensively used; a transferrin receptor recycles every 15–20 minutes, and has a half-life of >30 hours. It provides the cell with the means of taking up iron.

◆ *Receptor and ligand both are degraded.* The EGF receptor binds its ligand, the small polypeptide epidermal growth factor, as a requirement for internalization. Although EGF and its receptor appear to dissociate at low pH, they are both carried on to the lysosome, where they are degraded, as indicated in **Figure 12.8**. We do not know how and whether these events are related to the ability of EGF to change the phenotype of a target cell via binding to the receptor.

◆ *Receptor and ligand are transported elsewhere.* The route illustrated in **Figure 12.9** is available in certain polarized cells. A receptor–ligand combination is taken up at one cell surface, transported to the endosome, and then released for transport to the far surface of the cell. This is called **transcytosis**. By this means, receptors can transport immunoglobulins across epithelial cells.

Rapid recycling in general occurs for receptors that bring ligands into the cell, not for those that trigger pathways of signal transduction. Receptors involved in signalling changes from the surface are usually degraded if they are endocytosed.

What features of protein structure are required for endocytosis? Mutations that prevent internalization can be used to identify the relevant sequences in the receptors. In fact, the characterization of an internalization defect provided evidence that entry into coated pits is needed for receptor-mediated endocytosis of LDL. In cells from human patients with such defects in the LDL receptor, the receptor gathers in small clusters over the plasma membrane, and cannot enter coated pits in the manner observed for wild-type cells.

The mutations responsible for this type of defect all affect the *cytoplasmic domain* of the receptor, which functions independently in sponsoring endocytosis. A recombinant protein whose extracellular

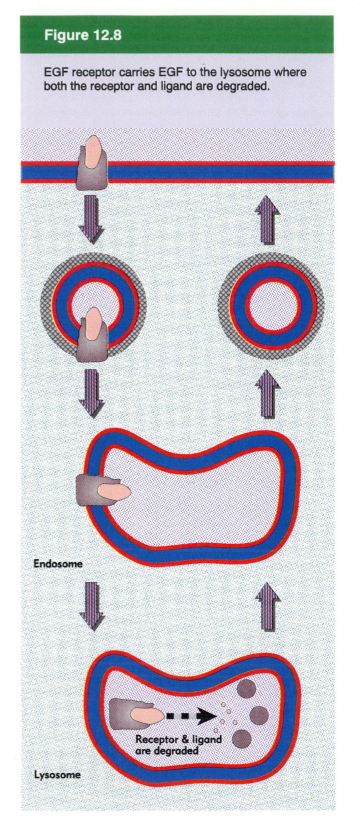

Figure 12.8

EGF receptor carries EGF to the lysosome where both the receptor and ligand are degraded.

Endosome

Receptor & ligand are degraded

Lysosome

domain is derived from influenza virus hemagglutinin, which is not usually endocytosed, can be internalized if it is provided with a cytoplasmic domain from an endocytosed protein.

We do not yet have a clear view of all of the signals that mediate endocytosis, but two features are common. The relevant region of the protein comprises a relatively short part of the cytoplasmic tail close to the plasma membrane. And the presence of a tyrosine residue in this region often is necessary. Removal of the tyrosine prevents internalization; conversely, substituting tyrosine in the relevant region of a protein that is not endocytosed (such as influenza virus hemagglutinin) allows it to be internalized.

Mutational analysis of the LDL receptor shows that the short amino acid sequence NPXY (Asn-Pro-X-Tyr) is required for internalization. The essential tyrosine can be replaced by other aromatic amino acids, but not by other types of amino acids. The NPXY motif is found in several group I proteins that

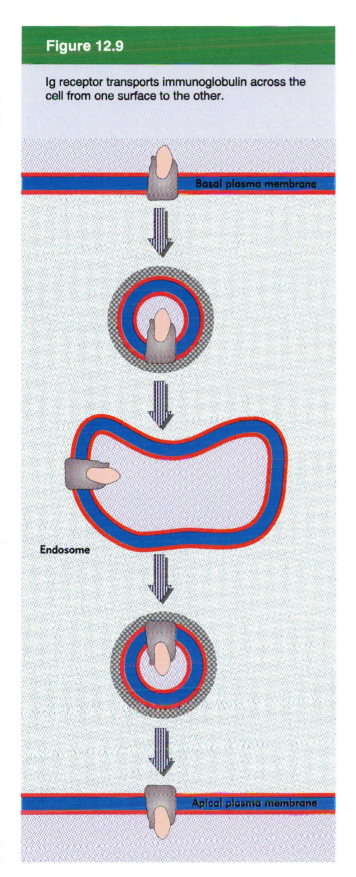

Figure 12.9

Ig receptor transports immunoglobulin across the cell from one surface to the other.

Basal plasma membrane

Endosome

Apical plasma membrane

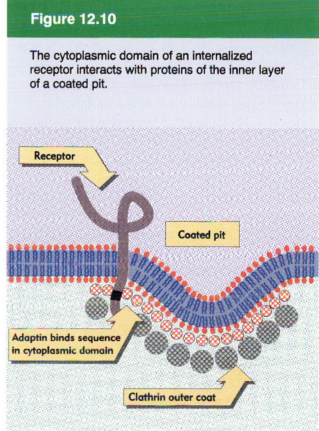

Figure 12.10

The cytoplasmic domain of an internalized receptor interacts with proteins of the inner layer of a coated pit.

Receptor

Coated pit

Adaptin binds sequence in cytoplasmic domain

Clathrin outer coat

are internalized, and is usually located close to the plasma membrane. It could be a general signal for endocytosis. It is not present in all internalized proteins. It forms a tight β-turn in the protein, and may function by presenting the essential tyrosine to the surface of the coated pit. In proteins that are internalized in response to ligand binding, the internalization signal may be generated by a change in conformation as a result of the binding.

Do internalized receptors interact directly with proteins on the vesicles that transport them?

Clathrin forms an outer polyhedral layer on clathrin-coated vesicles, but other proteins form an inner layer. The adaptins recognize the appropriate sequences in the cytoplasmic domains of receptors that are to be internalized, as discussed in Chapter 11 (see Figure 11.25). **Figure 12.10** illustrates a model in which, as a coated pit forms, the adaptins bind to the receptor cytoplasmic domain, immobilizing the receptor in the pit. As a result, the receptor is retained by the coated vesicle when it pinches off from the plasma membrane.

Protein tyrosine kinases induce phosphorylation cascades

Growth factor receptors take their name from the nature of their ligands, which are usually small polypeptides (growth factors) that stimulate the growth of particular classes of cells. The factors have a variety of effects, including changes in the uptake of small molecules, initiation or stimulation of the cell cycle, and ultimately cell division. Examples are EGF, PDGF (platelet-derived growth factor), and insulin. The receptors share a general characteristic structure: there are large protein domains on both the extracellular and cytoplasmic sides of the membrane, and the individual transmembrane polypeptides span the membrane only once. Some receptors, such as those for EGF or PDGF, consist of single polypeptide chains; others, such as the insulin receptor, are dimers.

These receptors have a common mode of function. They are **protein kinases** which transfer phosphate groups to target proteins. The best characterized growth factor receptors phosphorylate the target amino acid tyrosine, and are called **protein tyrosine kinases**. The most common target amino acids for other types of protein kinases are serine or (less often) threonine; these kinases are therefore called **protein serine/threonine kinases**. They are responsible for the vast majority

of phosphorylation events in the cell. Receptor serine kinases are rare. Most protein serine kinases are cytosolic, but some function in the nucleus (for example, with transcription factors as their targets). The catalytic domains of all these kinases are large (~250 amino acids), and they show the presence of conserved amino acid sequences.

We visualize the process initiated by a receptor kinase as constituting a **cascade**. The original idea about such cascades was that the activation of the receptor tyrosine kinase causes other protein tyrosine kinases to be activated, leading farther along the chain to the activation of protein serine kinases. We now know that other components besides kinases are found in the cascades, and indeed that some pathways involving other types of events are triggered, but the basic idea remains that the signal is amplified as it passes from one component of the pathway to the next, and that some components have multiple targets, so that the pathway branches.

When a ligand binds to the extracellular domain of a growth factor receptor, the catalytic activity of the cytoplasmic domain is activated. *Phosphorylation of tyrosine is identified as the key event by which the growth factor receptors function because mutants in the tyrosine kinase domain are biologically inactive, although they continue to be able to bind ligand.* Two types of event follow from the activation of the kinase activity:

◆ The receptor phosphorylates itself in the cytoplasmic region, an event called **auto-phosphorylation**. This is important for two reasons. First, the phosphorylation changes the properties of the cytoplasmic domain of the receptor in a way that allows it to bind the next protein in the cascade. Second, it may further increase the catalytic activity, and therefore comprises a **positive feedback**.

◆ The receptor also phosphorylates target proteins in the cytoplasm, therefore activating (or inactivating) pathways that change the cell phenotype. The phosphorylation of target proteins is a major pathway for passing the signal along the cascade, but it is not the only means of activating target proteins; some proteins that bind to phosphorylated receptors appear to be activated without themselves being phosphorylated.

A key question in the concept of how a signal is transduced across a membrane is how binding of the ligand to the domain on the extracellular side activates the catalytic domain on the cytoplasmic side. The general principle is that a conformational change is induced that affects the overall organization of the receptor. **Figure 12.11** shows the general nature of this change and its consequence. The ligand-binding domain in the extracellular N-terminal region of the protein has a regulatory role. This has been characterized most fully for the EGF receptor. When ligand binds, the structure of this region changes in such a way that the individual monomer dimerizes with another monomer (the second monomer does not necessarily have a ligand). Dimerization of the extracellular domains brings the cytosolic domains into juxtaposition. The stabilization of contacts between the C-terminal

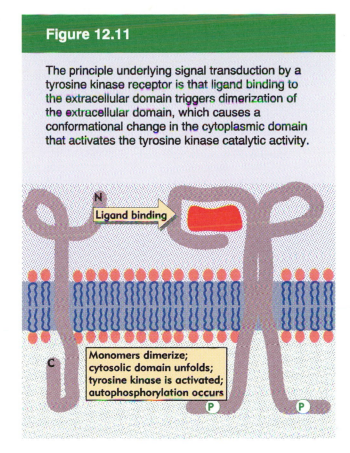

Figure 12.11

The principle underlying signal transduction by a tyrosine kinase receptor is that ligand binding to the extracellular domain triggers dimerization of the extracellular domain, which causes a conformational change in the cytoplasmic domain that activates the tyrosine kinase catalytic activity.

Ligand binding

Monomers dimerize; cytosolic domain unfolds; tyrosine kinase is activated; autophosphorylation occurs

cytosolic domains causes an unfolding of the conformation, which enables the kinase activity of one subunit to phosphorylate the other subunit.

Figure 12.12 shows that the dimerization can take several forms. A ligand binds to one or to both monomers to induce them to dimerize, a dimeric ligand binds to two monomers to bring them together, or a ligand binds to a dimeric receptor (one stabilized by extracellular disulfide bridges) to cause an intramolecular change of conformation. A major consequence of this mechanism is to allow transmission of a conformational change from the extracellular domain to the cytoplasmic domain without requiring a change in the structure of the transmembrane region.

The basic chain of events that is triggered by activation of a receptor of this type is illustrated in **Figure 12.13**. The dimerization of the cytoplasmic domains triggers autophosphorylation, in which

Figure 12.12

Binding of ligand to the extracellular domain induces dimerization of a receptor, causing new contacts to form between the cytoplasmic domains.

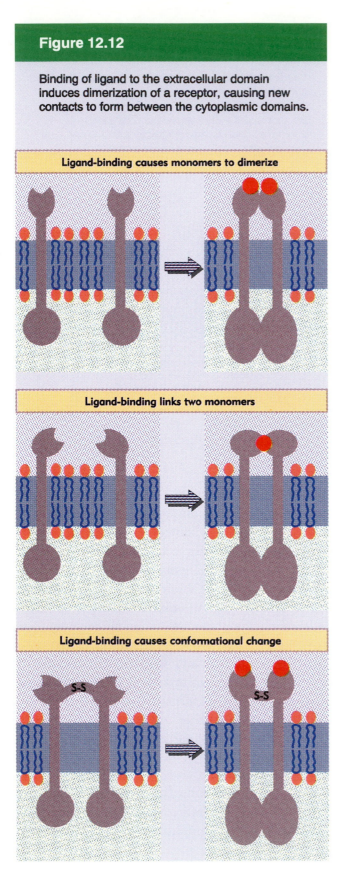

Ligand-binding causes monomers to dimerize

Ligand-binding links two monomers

Ligand-binding causes conformational change

one subunit phosphorylates the other in the dimer. In fact, it is necessary for *both* subunits to have kinase activity for the receptor to be activated; if one subunit is defective in kinase activity, the dimer cannot be activated.

The existence of the phosphorylated tyrosine(s) causes the cytoplasmic domain to associate with its target. The target protein utilizes a region called the SH2 domain, which binds specifically to the phosphotyrosine in the receptor. The figure shows that binding may be followed by phosphorylation of the target protein, which in turn changes its properties. However, as we have noted, there are cases in which a substrate binds to the phosphorylated receptor, and becomes active without being phosphorylated.

Some targets that are activated by receptors are themselves kinases. They in turn may phosphorylate further kinases, creating a cascade in which a series of phosphorylation events activates successive kinases, ultimately leading to the phosphorylation of target proteins that change transcription or cell structure. One of the pathways involves a transition from the initial type of kinase (phosphorylating tyrosine) to a series of kinases that phosphorylate serine or threonine residues. In Chapter 39, we discuss the cascade in more detail (see Figure 39.15), analyze how mutations of the growth factor receptor may activate signal transduction inappropriately (see Figure 39.17), and consider in more detail the origin of the SH2 domain.

The relationship between kinase activity and endocytosis is unclear. Phosphorylation at particular residues may be needed for endocytosis; whether the kinase activity as such is needed may differ for various receptors. It is possible that endocytosis of receptor kinases serves principally to clear receptor (and ligand) from the surface following the response to ligand binding (thus terminating the response). However, in some cases, movement of receptors to coated pits followed by internalization could be necessary for them to act on the target proteins.

Because growth factor receptors generate signals that lead to cell division, their activation in the wrong circumstances is potentially damaging to an

Figure 12.13

Autophosphorylation triggers the kinase activity of the cytoplasmic domain of a receptor. The target protein may be recognized by an SH2 domain, and itself may be a kinase that is activated by phosphorylation. The signal may be passed along a cascade of kinases.

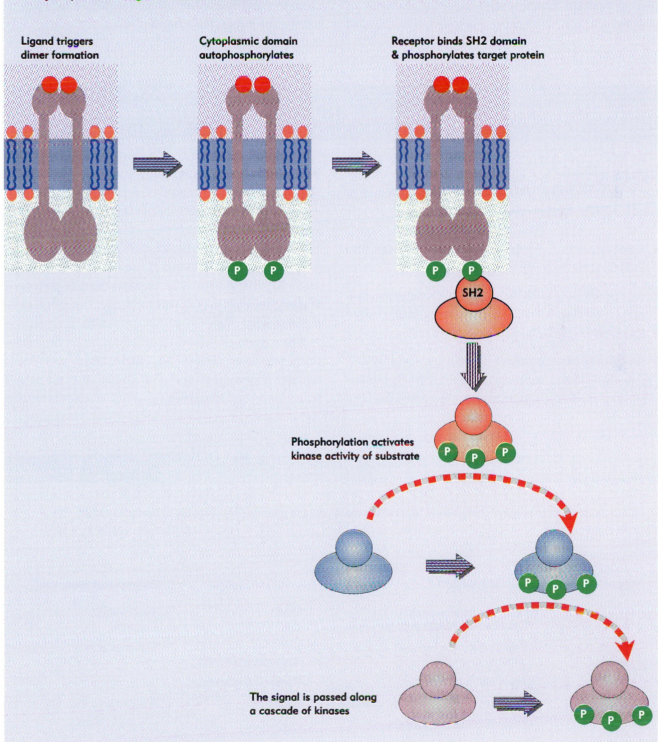

Ligand triggers
dimer formation

Cytoplasmic domain
autophosphorylates

Receptor binds SH2 domain
& phosphorylates target protein

SH2

Phosphorylation activates
kinase activity of substrate

The signal is passed along
a cascade of kinases

organism, and can lead to uncontrolled growth of cells. Many of the growth factor receptor genes are represented in the **oncogenes,** a class of mutant genes active in cancers. The mutant genes are derived by changes in cellular genes; often the mutant protein is truncated in either or both of its N-terminal or C-terminal regions. The mutant protein usually displays two properties: the tyrosine kinase has been activated; and there is no longer any response to the usual ligand. As a result, the tyrosine kinase activity of the receptor is either increased or directed against new targets. The nature of these changes in generating tumorigenic phenotypes in cells is the subject of Chapter 39.

G proteins may activate or inhibit target proteins

G proteins transduce signals from a variety of receptors to a variety of targets. The components of the general pathway can be described as:

◆ a **receptor** is a resident membrane protein that is activated by an extracellular signal;

◆ a **G protein** is converted into active form when an interaction with the activated receptor causes its bound GDP to be replaced with GTP;

◆ an **effector** is the target protein that is activated (or—less often—inhibited) by the G protein; sometimes it is another membrane-associated protein.

◆ a **second messenger** is a small molecule that is released as the result of activation of (certain types) of effectors.

Another terminology that is sometimes used to describe the relationship of the components of the transduction pathway is to say that the receptor is *upstream* of the G protein, while the effector is *downstream.*

The effectors linked to different types of G proteins are summarized in **Table 12.1**. G proteins occupy a classic position in the history of signal transduction. Their role fulfills a requirement for

Table 12.1

Classes of G proteins are distinguished by their effectors and are activated by a variety of transmembrane receptors.

G protein	Effector function	Second messenger	Example of receptor
s	Stimulates adenylate cyclase	↑ cAMP	β-adrenergic
olf	Stimulates adenylate cyclase	↑ cAMP	Odorant
i	Inhibits adenylate cyclase Opens K$^+$ channels	↓ cAMP ↑ Membrane potential	Somatostatin "
o	Closes Ca^{++} channels	↓ Membrane potential	m2 acetylcholine
t (transducin)	Stimulates cGMP phosphodiesterase	↓ cGMP	Rhodopsin
q	Activates phospholipase Cβ	↑ InsP3, DAG	m1 acetylcholine

a regulatory molecule to mediate the ability of hormone receptors to activate the enzyme adenylate cyclase, resulting in release of cyclic AMP (the classic second messenger). A variety of receptors, responding to many types of hormones, rely upon G proteins to influence the activity of adenylate cyclase. The G protein that stimulates adenylate cyclase is called G_s; the G protein that inhibits adenylate cyclase (a less effective reaction) is called G_i.

In addition to the classic G proteins, G_s and G_i, there are other G proteins that trigger different signaling pathways. A protein called G_q is involved in activating the enzyme phospholipase $C\beta$, which hydrolyzes phosphoinositide lipids into inositol phosphates and diacylglycerol. These products provide a second class of second messengers that activate a variety of response pathways.

Another well characterized pathway is found in the retina. Rhodopsin is a serpentine receptor that binds the ligand retinol, and responds to light by activating a G protein (G_t) called transducin. Transducin in turn activates a cyclic GMP phosphodiesterase, reducing the level of cyclic GMP, which is a crucial determinant of the level of visual excitation in retinal rods. Similar components are found in pathways for recognition of odorants.

G proteins are trimers whose function depends on the ability to dissociate into an α monomer and a $\beta\gamma$ dimer. The α subunit binds a guanine nucleotide, and this reaction controls association with the $\beta\gamma$ dimer. An α subunit bound to GDP associates stably with $\beta\gamma$ to form the intact G protein. Displacement of the GDP by GTP generates an α-GTP subunit, which dissociates from the $\beta\gamma$ dimer.

In the trimeric state, a G protein is inactive. **Figure 12.14** illustrates the classic model for G protein action, which proposes that its function is mediated by the dissociated α-GTP subunit. In the example of the adenylate cyclase pathway, the $\beta\gamma$ dimer is bound to α_s in the stimulatory G_s protein or to α_i in the inhibitory G_i protein. The stimulatory activity of G_s on adenylate cyclase or of G_t on cyclic GMP phosphodiesterase is determined directly by the ability of the released α-GTP subunit to stimulate the effector enzyme effectively. How the

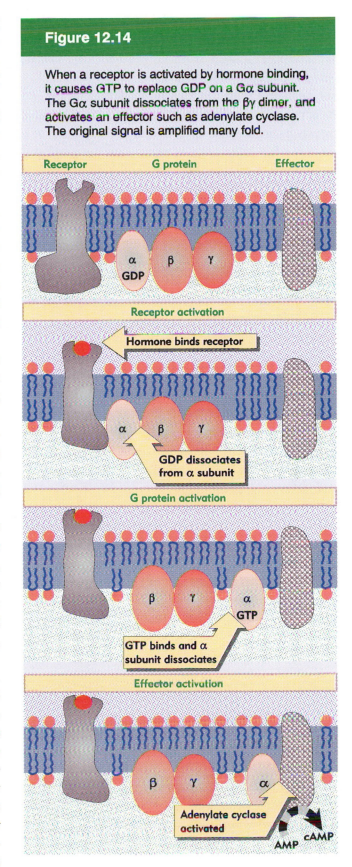

Figure 12.14

When a receptor is activated by hormone binding, it causes GTP to replace GDP on a Gα subunit. The Gα subunit dissociates from the $\beta\gamma$ dimer, and activates an effector such as adenylate cyclase. The original signal is amplified many fold.

Receptor   G protein   Effector

α GDP β γ

Receptor activation

Hormone binds receptor

α β γ

GDP dissociates from α subunit

G protein activation

β γ α GTP

GTP binds and α subunit dissociates

Effector activation

β γ α

Adenylate cyclase activated

AMP   cAMP

activated G_i inhibits adenylate cyclase is less clear; it is possible that the effect is indirect, because the $\beta\gamma$ dimers released from G_i bind to any free α_s subunits released from G_s.

Consistent with the idea that α subunits vary in their reactions with effectors, there is much more variety in the sequences of α subunits than of β or γ subunits. But although it is most often the α subunit that activates effector function, some reactions are mediated by the $\beta\gamma$ dimer. The first to be discovered was a reaction in the mating type pathway of yeast (see Chapter 36). Another is the ability of the $\beta\gamma$ dimer to simulate the enzyme phospholipase A_2. *The common feature in all of these reactions is that a G protein acts upon an effector enzyme that changes the concentration of some small molecule(s) in the cell.*

How does a receptor activate a G protein? An activated hormone–receptor complex binds to an intact G protein. This causes GDP to dissociate from the single guanine nucleotide-binding site on the α subunit. This generates a hormone–receptor–G protein complex, which is stable but transient. In the presence of the relatively high cellular concentrations of GTP, the G_α subunit rapidly binds GTP, and the G protein is released from the hormone–receptor complex. The reaction is catalytic: one hormone–receptor complex can activate many G protein molecules in a short period, thus amplifying the original signal.

In either the intact or dissociated state, G proteins are associated with the cytoplasmic face of the plasma membrane. But the individual subunits are quite hydrophilic, and none of them appears to have a transmembrane domain. The $\beta\gamma$ dimer has an intrinsic affinity for the membrane because the γ subunit is prenylated. The α_i and α_o types of subunit are myristoylated (see Figure 12.5), which explains their ability to remain associated with the membrane after release from the $\beta\gamma$ dimer. The basis for attachment of α_s is unknown.

As proteins associated with the cytoplasmic face of the membrane, G proteins effectively function in the cytosol. It is the receptor that must actually transmit the initial signal across the membrane. Again this poses the question of how a reaction with ligand on the extracellular side of the membrane gives rise to the ability to interact with G protein on the cytoplasmic side.

Since several receptors can activate the same G proteins, and since (at least in some cases) a given G protein has more than one effector, we must ask how specificity is controlled. The most common model is to suppose that receptors, G proteins, and effectors all are free to diffuse in the plane of the membrane. In this case, the concentrations of the components of the pathway, and their relative affinities for one another, are the important parameters that regulate its activity. We might imagine that an activated α-GTP subunit scurries along the cytoplasmic face of the membrane from receptor to effector. But it is also possible that the membrane constrains the locations of the proteins, possibly in a way that restricts interactions to local areas. Such compartmentation could allow localized responses to occur.

Carriers and channels form water-soluble paths through the membrane

A major purpose of a membrane is to allow different aqueous conditions to be maintained on either side. Within the cytoplasm, different organelles offer different ionic environments. The most striking example is the maintenance of an acid pH in endosomes and lysosomes, with immediate implications for the functions of the proteins that enter them (as described earlier). A major exception from the ability of organelles to differ from the cytosol is provided by the nucleus, which essentially is subjected to the

Table 12.2

Concentrations of Ions (mM) differ inside and outside the cell.

Ion	Intracellular	Extracellular
Na^+	˜10	145
K^+	140	5
Mg^{++}	0.5	˜1.5
Ca^{++}	0.0001	˜1.5
H^+	8×10^{-5}	4×10^{-5}
Cl^-	˜10	110

same conditions as the cytosol. This is related to the fact that passage into and out of the nucleus occurs through nuclear pores—relatively large openings in the nuclear envelope, through which ions and other small molecules can diffuse freely.

The plasma membrane is essentially impermeable to water-soluble compounds. The ionic environment of the cytosol is quite different from the extracellular ionic milieu. The general concentrations of ions on either side of the plasma membrane are summarized in **Table 12.2**.

A notable feature of the intracellular environment is that there are more free cations (positively charged) (~150 mM) than anions (~10 mM). The reason is that many cellular constituents are negatively charged—for example, nucleic acids have multiple negative charges for every phosphate group in the phosphodiester backbone. The superfluity of cations therefore establishes electrical neutrality by balancing these fixed charges. (In addition to the free ion concentrations summarized in the table, cells store ions; for example, Ca^{2+} is stored in various organelles.)

Comparison between the intracellular and extracellular environments shows large imbalances in Na^+ ions (with an external concentration an order of magnitude greater than the internal concentration), K^+ ions (with the imbalance reversed), and Cl^- ions (with an extracellular excess). This creates a **concentration gradient** across the membrane for each ion.

The plasma membrane is electrically charged. There is an **electrical gradient** in which the inside is negative compared to the outside. This voltage difference favors the entry of cations and opposes the entry of anions.

Together the concentration gradient and electrical gradient constitute the **electrochemical gradient**, which is characteristic for each solute. A solute whose gradient is favorable can enter the cell when a channel opens; the gradient is sufficient to drive **passive transport** of a solute such as Na^+ or Cl^- into the cell. But a solute that faces an unfavorable gradient requires **active transport** in which energy is used to pump it into the cell against the gradient.

The passage of ions (and other small solutes) through the plasma membrane is mediated by resident transmembrane proteins. A common feature of these proteins is their large size and the presence of multiple membrane-spanning regions, features which together argue that they provide a relatively static feature of the membrane. We may imagine that they function *in situ,* as individual structures, constrained by the membrane. **Figure 12.15** illustrates two general means of transport across the membrane:

◆ A **carrier protein** binds a solute on one side of the membrane and then experiences a conformational change that transports the solute to the other side of the membrane. By binding the solute on one side and releasing it on the other, the carrier in effect directly transports the solute across the membrane. There are several types of carriers, distinguished by the number of solutes that they transport, and the directions in which they transport them. Carriers that transport a single solute across the membrane are called **uniporters**; carriers that simultaneously or sequentially transport two different solutes are called **symporters**; and carriers that transport one solute in one direction while transporting a different solute in the opposite direction are called **antiporters**.

Carrier proteins may be used for passive transport or linked to an energy source to provide

Figure 12.15

A carrier (porter) transports a solute into the cell by a conformational change that brings the solute-binding site from the exterior to the interior, while an ion channel is controlled by the opening of a gate (which might in principle be located on either side of the membrane).

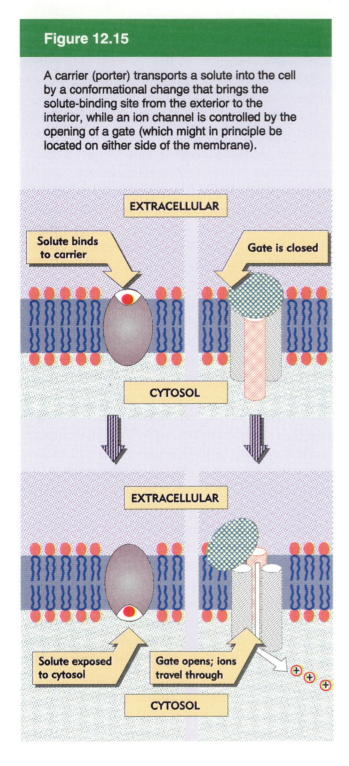

EXTRACELLULAR

Solute binds to carrier

Gate is closed

CYTOSOL

EXTRACELLULAR

Solute exposed to cytosol

Gate opens; ions travel through

CYTOSOL

active transport. Energy for active transport is provided by hydrolysis of ATP, the classic example being the Na^+/K^+ pump that functions as an antiporter, pumping sodium out of the cell and potassium into it. Another source of energy is the electrochemical gradient itself; a symporter brings Na^+ into the cell together with some other solute, using the favorable gradient of sodium to overcome the unfavorable gradient of the other solute.

◆ An **ion channel** comprises a water-soluble pore in the membrane. Its activity is controlled by regulation of the opening and closing of the channel. When it is open, ions can diffuse passively, as driven by the electrochemical gradient. Ion channels allow *only* passive transport. The resting state of an ion channel is closed, and the **gates** that control channel activity usually open only briefly, in response to a specific signal. Ion channels are classified in **Table 12.3** according to the type of signal that controls the gate. **Ligand-gated** channels are receptors that respond to binding of particular molecules, amongst which the neurotransmitters acetylcholine, glycine, GABA (γ-amino-butyric acid), and glutamate are prominent examples. **Voltage-gated** channels respond to electric changes, again a prominent feature of the neural system. **Second-messenger** gated channels provide yet another means for signal transduction, one interesting example comprising channels that respond to activation of G-proteins.

The structures of both carriers and channels present a paradox. They are transmembrane proteins that have multiple membrane-spanning domains, each consisting of a stretch of amino acids of sufficient hydrophobicity to reside in the lipid bilayer. Yet within these hydrophobic regions must be a highly selective, water-filled path that permits ions to travel through the membrane.

One solution to this problem lies in the structure of the transmembrane regions. Instead of comprising unremittingly hydrophobic stretches like those of single membrane-pass proteins, they contain some polar amino acids. They are likely to be organized as illustrated in **Figure 12.16** as amphipathic helices in which the hydrophobic face associates with the lipid bilayer, while the polar faces are aligned with one another to create the channel.

Table 12.3

Ion channels can be classified into several groups.

Class of Channel	Ion selectivity	Channel
Gap junction	Cations & anions	
Ligand-gated	Cations (Na$^+$, K$^+$)	Acetylcholine receptor
	Anion (Cl$^-$)	Glycine receptor
	Anion (Cl$^-$)	GABA receptor
Voltage-gated	Individual cations	Na$^+$ channel
		K$^+$ channel
		Ca^{2+} channel
Second messenger-gated	Cations or anions	

One property that argues for the importance of the interior of the channel is the **ion selectivity**. Different channels permit the passages of different ions or groups of ions. The channels are thought to be extremely narrow, so that ions must be stripped of their associated water molecules in order to pass

Figure 12.16

A channel may be created by amphipathic helices, which present their hydrophobic faces to the lipid bilayer, while juxtaposing their charged faces away from the bilayer. In this example, the channel is lined with positive charges, which would encourage the passage of anions.

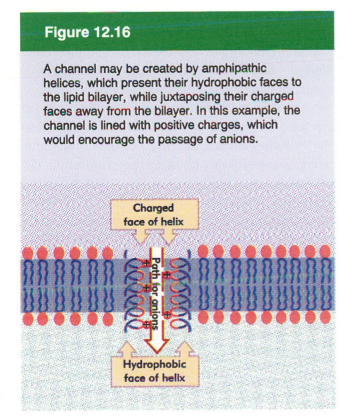

through. One parameter that may control passage is the diameter of the channel, but its role is probably to force the ions into contact with the walls, rather than to filter them directly. For example, potassium channels seem to have the narrowest constrictions, ~3Å, but Na$^+$, K$^+$, and Ca^{2+} ions all are <2Å in diameter, and a channel's permeability for K$^+$ ions may be >10^3 × greater than for the other cations. The channel possesses a 'filter' at the entrance to the pore that has specificity for the desired ion, presumably based upon its geometry and electrostatic charge.

The structures of particular ion channels are beginning to reveal their general features. A common feature is that the constituent proteins are large and have several membrane-spanning regions. A channel probably consists of a 'ring' of 4, 5, or 6 subunits, organized in a symmetrical or quasi-symmetrical manner. The water-filled pore is found at the central axis of symmetry. The size of the pore depends at least partly on the number of subunits in the ring, since 4-member channels are the most selective, while 6-member channels are the least selective. The subunits are always related in structure, and sometimes are identical. They may consist of separate proteins or of related domains in a single large protein.

Voltage-gated sodium channels have a single type of subunit, a protein of 1820 amino acids with a repetitive structure that consists of 4 related domains. Each domain has several membrane-spanning regions. The four domains are probably arranged in the membrane in a pseudo-symmetrical structure. Two smaller subunits are associated with the large protein.

Potassium channels have a smaller subunit, equivalent to one of the domains of the sodium channel; 4 identical subunits associate to create the channel. Six transmembrane domains are identified in the protein subunit by hydrophobicity analysis; they are numbered S1–S6. The S4 domain has an unusual structure for a transmembrane region: it is highly positively charged, with arginine or lysine residues present at every third or fourth position. The S4 motif is found in voltage-gated K+, Na+, and Ca2+ channels, so it seems likely that it is involved with a common property, thought to be channel opening.

Analysis of the *shaker* potassium channel of the fly has revealed some novel features, illustrated in **Figure 12.17**. The region that forms the pore has been identified by mutations that alter the response to toxins that inhibit channel function. It turns out to occupy the region between transmembrane domains S5 and S6, which apparently forms two membrane-spanning stretches that are not organized in the usual hydrophobic α-helical structure. The structure could be a rather extended β-hairpin. The state of the channel (open or closed) is controlled by the N-terminal end, which resembles a ball on a chain. The ball is in effect tethered to the channel by a chain, and plugs it on the cytoplasmic side. The length of the chain controls the rate with which the ball can plug the channel after it has been opened.

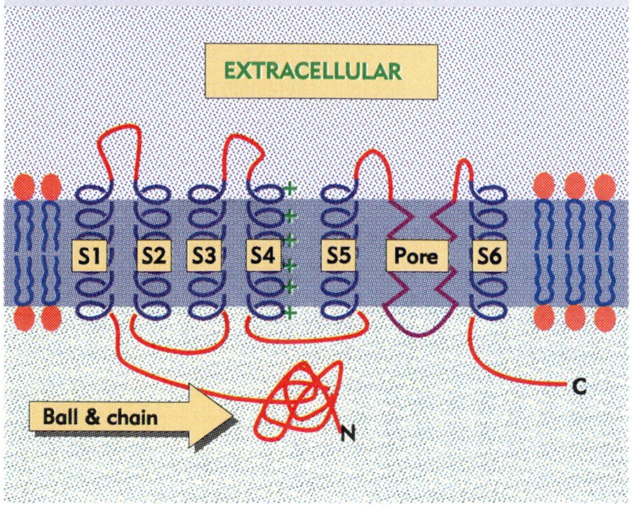

Neurotransmitter-gated receptors form a super-family of related proteins. They appear to comprise a 5-member channel class. The nicotinic acetylcholine receptor has been characterized in the most detail, and is a pentamer with the structure $\alpha_2\beta\delta\gamma$. As illustrated in **Figure 12.18**, the bulk of the 5 subunits projects above the plasma membrane into the extracellular space. The openings to the channel narrow from a diameter of ~25Å until reaching the pore itself. The entrance on the extracellular side is very deep, ~60Å; the distance on the cellular side is shorter, 20Å. The pore extends through the 30Å of the lipid bilayer and is only ~7Å in diameter.

Ligand binding occurs on the α subunits. Both α subunits must bind an acetylcholine for the gate to open. Where is the gate? Since the channel is really narrow only in the region within the lipid bilayer, the gate seems likely to be located well within the receptor. Structural changes that occur upon opening seem greatest just by the cytoplasmic side of the lipid bilayer, so it is possible that the gate is located at the level of the phospholipid heads on the cytoplasmic boundary. Thus the acetylcholine receptor, like many other receptors, must transmit information about ligand binding internally, from the extracellular acetylcholine binding site to the near-cytoplasmic gate.

How does the gate function? It might consist of an electrostatic repulsion, in which positive groups are extruded into the channel to prevent passage of

Figure 12.17

A potassium channel has a pore consisting of unusual transmembrane regions, with a gate whose mechanism of action resembles a ball and chain.

EXTRACELLULAR

S1 | S2 | S3 | S4 | S5 | Pore | S6

Ball & chain

Figure 12.18

The acetylcholine receptor consists of a ring of 5 subunits, protruding into the extracellular space, and narrowing to form an ion channel through the membrane.

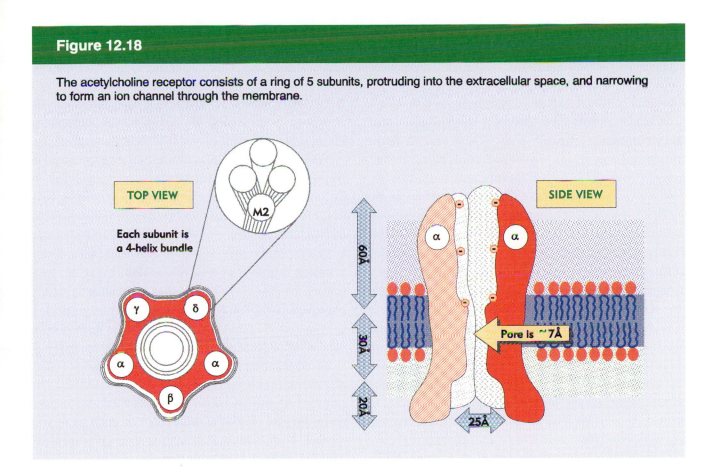

cations. Or it may take the form of a physical impediment to passage, in which a conformational change brings bulky groups to block the pore.

Ion selectivity may be determined by the walls of the wide entry passage. The walls lining the entrances to the pore have negatively charged groups; each subunit carries ~10 negative charges in its extracellular region. These charge clusters could modify the ionic environment at the entrance to the channel, concentrating the desired ions and diluting ions that are selected against. The structure of the acetylcholine receptor allows passage of Na^+, K^+ or Ca^{2+} ions, but because of the prevailing gradients, its main use in practice is to allow the entry of Na^+ into the cell.

The acetylcholine receptor is an example of a superfamily of receptors gated by neurotransmitters. All appear to have the same general organization, consisting of 5 subunits whose structures are related to one another. All the subunits are about the same size (~50,000 daltons), and each is probably organized in the membrane as a bundle of 4 helices (each helix containing a transmembrane domain). In each case, one of the four transmembrane domains (called M2) has an amphipathic structure and seems likely to be involved in lining the walls of the pore itself. The presence of serine and threonine residues, and some paired acid–basic residues, may assist ion passage. The sequences of subunits of the glycine and GABA receptors are related to the acetylcholine receptor subunits. Some changes in the sequences seem likely to reflect the ion selectivity. Thus the glycine and GABA receptors have positively charged groups in the entrance walls, consistent with their transport of anions.

Pores control nuclear ingress and egress

An entirely different type of system is used to transport macromolecules between the nucleus and the cytoplasm. The nucleus is segregated from the cytoplasm by an envelope consisting of two membranes. The inner membrane contacts the nuclear lamina, providing in effect a surface layer for the nucleus. The outer membrane is continuous with the endoplasmic reticulum in the cytosol (see Figure 2.2). The two membranes come into contact at openings called **nuclear pore complexes**, very large structures that are used apparently for both import and export of material.

Nuclear pore complexes are large structures. They are in fact the only structures in the nuclear membrane that appear large enough to support the transit of proteins and RNA (or ribonucleoproteins). At the center of each complex is a pore that appears to provide a water-soluble channel between nucleus and cytoplasm. There are ~3000 pore complexes on the nuclear envelope of an animal cell.

Transport between nucleus and cytoplasm proceeds in both directions. Since all proteins are synthesized in the cytosol, any proteins required in the nucleus must be transported there. Since all RNA is synthesized in the nucleus, the entire cytoplasmic complement of RNA (mRNA, rRNA, tRNA and other small RNAs) must be derived by export from the nucleus. **Table 12.4** summarizes the frequency with which the pores are used for some of the more prominent substrates of passage.

We can form an impression of the magnitude of import by considering the histones, the major protein components of chromatin. In a dividing cell, enough histones must be imported into the nucleus during the period of DNA synthesis to associate with a diploid complement of chromosomes. Since histones form about half the protein mass of chromatin, we may conclude that overall about 200 chromosomal protein molecules must be imported through each pore per minute. This amounts to ~2.7×10^8 protein molecules per cell cycle.

Uncertainties about the processing and stability of mRNA make it more difficult to calculate the number of mRNA molecules exported, but to account for the ~250,000 molecules of mRNA per cell probably requires ~1 event per pore per minute. The major RNA synthetic activity of the nucleus is of course the production of rRNA to produce the ~3×10^6 ribosomes in the cytoplasm. The rRNA is exported in the form of assembled ribosomal subunits. Just to double the number of ribosomes during one cell cycle would require the export of ~5 ribosomal subunits (60S and 40S) through each pore per minute.

Of course, in order for the ribosomal proteins to assemble with the rRNA, they must first be imported into the nucleus, so ribosomal proteins must undergo a rapid shuttling into the nucleus as free proteins and out again as assembled ribosomal subunits. Given ~80 proteins per ribosome, their import must be comparable in magnitude to that of the chromosomal proteins.

How does a nuclear pore accommodate the transit of material of varied sizes and characteristics in either direction? Nuclear pore complexes have a uniform appearance when examined by microscopy. The pores can be released from the nuclear envelope by detergent, and **Figure 12.19** shows that they appear as annular structures, consisting

Table 12.4

Nuclear pores are used for import and export.

Direction	Substrate	Passages/Pore/Min
Import	Histones	100
	Nonhistone proteins	100
	Ribosomal proteins	150
Export	Ribosomal subunits	~5
	mRNA	<1

Figure 12.19

Nuclear pores appear as annular structures by electron microscopy. The bar is 0.5 μm. Photograph kindly provided by Ronald Milligan.

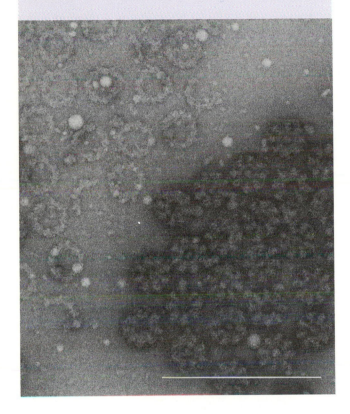

of rosettes made of 8 spokes. **Figure 12.20** shows a model for the pore based on three-dimensional reconstruction of electron microscopic images. It consists of an upper ring and a lower ring, connected by a lattice of 8 structures.

The basis for the 8-fold symmetry is explained in terms of individual components in the schematic view from above shown in **Figure 12.21**. The outside of the pore complex consists of a ring of diameter ~120 nm. The ring itself consists of 8 subunits. The 8 radial arms outside the ring may be responsible for anchoring the pore complex in the nuclear envelope; they penetrate the membrane. The 8 interior spokes project from the ring, closing the opening to a diameter of ~48 nm. Within this region is the transporter, which contains a pore that approximates a cylinder <10 nm

in diameter. The model of Figure 12.20 shows the basic framework of the nuclear core complex of Figure 12.21, that is, it does not have the radial arms or the central transporter.

The pore provides a passage across the outer and inner membranes of the nuclear envelope. As illustrated in **Figure 12.22**, the side view has 2-fold symmetry about a horizontal axis in the plane of the nuclear envelope. There are matching annuli at the outer and inner membranes, comprising the surfaces that project into the cytosol and into the nucleus, and each is connected to the spokes, which form a central ring. (Only 2 of the 8 spokes are seen in this side view.) The spokes are symmetrical about the horizontal axis. The central pore projects for the distance across the envelope. Sometimes material can be seen within the pore, but it has been difficult to equate such sightings with the transport of particular material.

The size of the nuclear pore complex corres-

Figure 12.20

A model for the nuclear pore shows 8-fold symmetry. Two rings form the upper and lower surfaces (shown in yellow); they are connected by the spokes (shown in green on the inside and blue on the outside). Photograph kindly provided by Ronald Milligan.

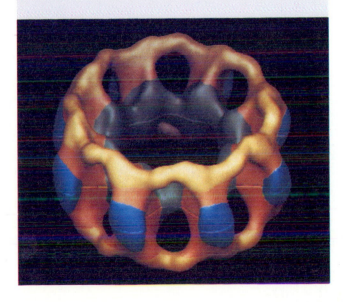

Figure 12.21

The outsides of the nuclear coaxial (cytoplasmic and nucleoplasmic) rings are connected to radial arms. The interior is connected to spokes that project towards the transporter that contains the central pore.

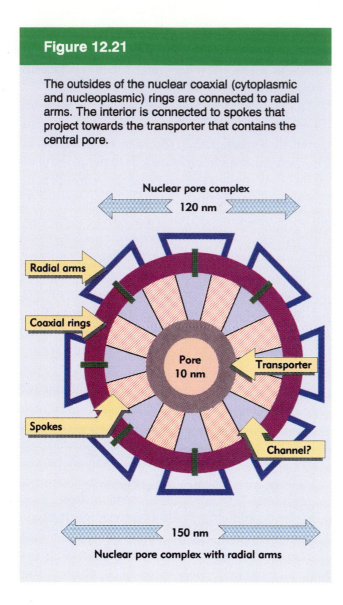

Nuclear pore complex with radial arms

◆ Molecules of <5000 daltons that are injected into the cytoplasm appear virtually instantaneously in the nucleus: we may conclude that the nuclear envelope is freely permeable to ions, nucleotides, and other small molecules. Thus the aqueous milieu of the nucleus is identical with that of the cytoplasm.

◆ Proteins of 5000–50,000 daltons diffuse at a rate that is inversely related to their size, presumably determined by random contacts with, and passage through, the pore. It takes a few hours for the levels of an injected protein to equilibrate between cytoplasm and nucleus. We may conclude that small proteins enter the nucleus by passive diffusion. The nuclear envelope in effect provides a mesh or molecular sieve that permit passage of material <50,000 daltons.

Figure 12.22

The nuclear pore complex spans the nuclear envelope by means of a triple ring structure. The side view shows two-fold symmetry from either horizontal or perpendicular axes.

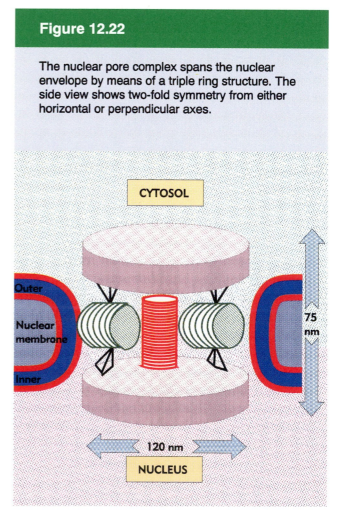

ponds to a total mass >100 × 10⁶ daltons (compare this with the 80S ribosome at 4 ×10⁶ daltons). We can identify the smallest subunit component by using the 8-fold symmetry as seen in cross-section (see Figure 12.21) and the 2-fold symmetry seen from the side (see Figure 12.22). This divides the scaffold into 16 identical units, that is, ~6 × 10⁶ daltons. Each of these units must contain a large number of individual proteins. The central pore constitutes only a small part of the overall complex.

The ability of compounds to diffuse freely through the pores is limited by their size, and it is convenient to consider the material in three size classes:

◆ Proteins >50,000 daltons in size do not enter the nucleus by passive diffusion; a mechanism of active transport must be required for their passage.

The lattice-like structure of the nuclear pore suggests that different features could be responsible for active transport of large proteins versus passive transport of smaller proteins.

Eight channels are created by the open regions between the spokes, that is, around the periphery between the inner and outer annuli. These have a rigid structure. They are oval in cross-section, with a diameter of ~10 nm. This is large enough to provide a mesh that could allow passive diffusion for small proteins; a globular protein of 50,000 daltons in mass would be expected to have a diameter of ~5 nm if it were spherical. Presumably objects need to be somewhat smaller than the mesh size in order to be able to pass through by diffusion.

If these channels allow free passage, material below the size limit will in principle equilibrate in concentration between the nucleus and the cytoplasm (although as the material approaches the size of the mesh, equilibration is slow). Proteins that enter the nucleus in this way would be retained, and therefore accumulate in the nucleus, if they participate in some nuclear function. For example, if a protein becomes incorporated into a large structure, such as a chromosome, this removes it from the equilibrating pool, and thus pulls more protein into the nucleus.

The central pore is used to transport larger material. A protein probably requires a specific signal to pass through the central pore. Smaller proteins may be transported in this way (as well as through the peripheral channels), and larger proteins *must* use an active transport mechanism that overcomes the apparent size restriction of the pores. But how can a protein or ribonucleoprotein with a diameter exceeding that of the pore pass through it?

Evidence about the nature of the transport through the pore has been gained by the use of colloidal gold particles coated with a nuclear protein. When these particles are injected into the cytoplasm, they cluster at the nuclear pores,

and then accumulate in the nucleus. Two important conclusions follow. Transport indeed occurs through nuclear pores. And the pore structure can widen to accommodate objects of the size of the coated gold particles (~20 nm). Similar experiments have shown that gold particles coated with polynucleotides can be exported from the nucleus via pores, and in fact at the present comprise the sole direct evidence for the involvement of the pores in export.

The rigidity of the gold particle excludes the possibility that transport through the pore requires the protein to change into a conformation with a diameter physically smaller than the pore. Thus the function of the pore in opening to accommodate the substrate is more analogous to the (much smaller) channels in the plasma membrane than to the transport through other organelle membranes (such as the mitochondrion) where conformational changes are *de rigeur*. However, very large substrates, such as exported ribonucleoprotein particles, may change their conformation, being restricted to a diameter <20 nm.

We conclude that the nuclear pore has a 'gating' mechanism that allows the interior to expand as material passes through. Pores containing material apparently under transport appear to be opened to a diameter of ~20 nm, possibly by a mechanism akin to the iris of a camera lens. It is possible that two irises, one connected with the cytoplasmic ring and one connected with the nucleoplasmic ring, open in turn as material proceeds through the pore.

Specific signals required for nuclear transport can be identified by the results of proteolytic cleavage to divide (large) proteins into regions that possess the signal and can enter the nucleus, while regions that lack the signal stay in the cytoplasm. Mutation of the signal can prevent the entire protein from entering the nucleus. Addition of the **nuclear localization signal (NLS)** can cause cytosolic proteins to be imported into the nucleus.

The relevant signals reside in different regions of different proteins. Short, rather basic amino acid sequences have been identified as nuclear import signals in some proteins. Often there is a proline residue to break α-helix formation upstream of the

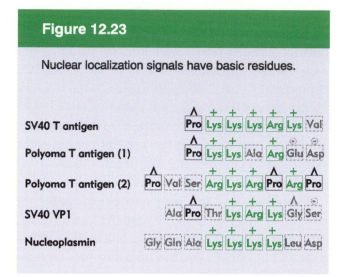

Figure 12.23

Nuclear localization signals have basic residues.

of transport. When ATP is provided, proteins can be translocated through the pore.

Translocation is inhibited by wheat germ agglutinin, a lectin (glycoprotein). Several of the proteins in the core can bind wheat germ agglutinin; this reaction inhibits translocation. The lectin-binding proteins are called **nucleoporins** and appear to be localized at or near the region of the central pore. They can be removed from the *in vitro* system. The pore complexes remain normal in appearance, but can no longer function to transport large material. Material smaller than the pore size continues to be able to move through by diffusion. When the depleted proteins are added back, they restore full activity to the deficient pores. This suggests that the nucleoporins control the gating of the pore necessary to admit and transport material larger than the resting diameter.

The nuclear pore complex provides a structural framework that supports the proteins actually responsible for binding and transporting material into (or out of) the nucleus; it does not include all of the active components that are involved in binding and translocation. These are being identified by the functional assays for binding and translocation. A major question about nuclear pores is whether they are in fact all identical, or whether their uniform appearance disguises functional differences. The accessory proteins could vary; for example, there might be different receptors for different types of NLSs, in which case there could be classes of pores with differing substrate specificities. We should like to know whether the same pores undertake transport into and out of the nucleus or whether different classes of pores exist.

basic residues. Hydrophobic residues are rare. The best characterized of these signals is that of SV40 T antigen, Pro-Lys-Lys-Lys-Arg-Lys-Val. This peptide sequence is sufficient to cause nuclear import if it is linked to a cytoplasmic protein. The summary of nuclear localization signals in **Figure 12.23** shows that there is no evidence yet for any conservation of an NLS; perhaps the shape of the region or its basicity are the important features.

A connection between the NLS and movement through nuclear pores has been established by the use of *in vitro* systems for transport. The transport process can be divided into two stages: binding to the pore; and movement through it. In the absence of ATP, proteins containing a nuclear import signal can bind at the pore, but they remain at the cytoplasmic face. Binding involves recognition of a nuclear localization signal by a receptor; we do not know whether the receptor is also involved in the subsequent stages

Summary

Integral proteins of the plasma membrane offer several means for communication between the extracellular fluid and the cytoplasm. Proteins may reside in the plasma membrane by means of a single

transmembrane domain: group I proteins, with the N-terminus facing the extracellular space and the C-terminus in the cytoplasm are more common than group II proteins with the reverse orientation.

Some proteins have multiple membrane-spanning regions, in which case the individual transmembrane regions may contain polar amino acids, and possibly interact as a result.

Receptors may be internalized either continuously or as the result of binding to an extracellular ligand. Receptor-mediated endocytosis initiates when the receptor moves laterally into a coated pit. The cytoplasmic domain of the receptor has a signal that is recognized by proteins that are presumed to be associated with the coated pit. An exposed tyrosine located near the transmembrane domain is a common signal; it may be part of the sequence NPXY, which forms a tight β-turn. When a receptor has entered a pit, the clathrin coat pinches off a vesicle, which then migrates to the early endosome.

The acid environment of the endosome causes some receptors to release their ligands; the ligands are carried to lysosomes, where they are degraded, and the receptors are recycled back to the plasma membrane by means of coated vesicles. A ligand that does not dissociate may recycle with its receptor. In some cases, the receptor–ligand complex is carried to the lysosome and degraded.

Many receptors for growth factors possess protein tyrosine kinase activity in their cytoplasmic domains. These receptors often have autophosphorylation activity, and this may be important in activating their ability to act on other targets.

Some receptors function by activating G proteins. All G proteins are trimers of a guanine nucleotide-binding α subunit, and a $\beta\gamma$ dimer. In the resting state, the G_α subunit carries GDP, and the G protein exists as the $\alpha\beta\gamma$ trimer. A receptor activates the G protein by displacing the GDP, which allows GTP to bind to G_α. The α-GTP subunit then dissociates from the $\beta\gamma$ dimer. The ability of dissociated α-GTP subunits to activate the next protein in a transduction pathway is well characterized. A more recent development concerns discovery of the ability of the $\beta\gamma$ dimer to stimulate effector functions.

Specific transport mechanisms exist to promote the passage of charged small molecules across the plasma membrane. An electrochemical gradient across the plasma membrane exists for each charged molecule. The membrane itself is negatively charged on the cytoplasmic face. Ions may be transported by carrier proteins, which may utilize passive diffusion or may be connected to energy sources to undertake active diffusion. The detailed mechanism of movement via a carrier is not clear, but is presumed to involve conformational changes in the carrier protein that directly or indirectly allow the ion to move from one side of the membrane to the other. Ion channels can be used for passive diffusion (driven by the gradient). They may be gated by voltage, extracellular ligands, or cytoplasmic second messengers. Ion channels vary in their selectivity, but have a common general structure in which subunits each containing several transmembrane domains aggregate in such a way that one transmembrane domain of each subunit contributes to the face of the aqueous channel.

Nuclear pore complexes are massive structures embedded in the nuclear membrane, and are responsible for all transport of protein into the nucleus and RNA out of the nucleus. Whether nuclear pore complexes are heterogeneous remains to be established. Proteins of <50,000 daltons equilibrate between nucleus and cytoplasm at a rate that depends inversely on the size of the protein. Each nuclear pore complex contains a central pore, which forms a channel of diameter <10 nm. Additional channels are present round the periphery. The central channel can be opened to a diameter of ~20 nm to allow passage of larger material, some of which may need to undergo conformational changes to fit. Proteins that are actively transported into the nucleus require specific transport signals, which are short, but do not seem to share common features except for their basicity. Nuclear entry is a two stage process, involving binding followed by an ATP-dependent translocation.

Further reading

Reviews

The classic pathway of receptor-mediated endocytosis for LDL has been reviewed by **Goldstein** *et al.* (*Ann. Rev. Cell Biol.* **1**, 1–40, 1985).

Factors that influence signal transduction by G proteins have been reviewed by **Neer and Clapham** (*Nature* **333**, 129–134, 1988).

The structure and function of ligand-gated ion channels has been reviewed by **Unwin** (*Neuron* **3**, 665–676, 1989). Voltage gated channels were revisited by **Miller** (*Science* **252**, 1092–1096, 1990).

Signal transduction by tyrosine kinase receptors has been reviewed by **Ullrich and Schlessinger** (*Cell* **61**, 203–212, 1990).

Nuclear pore structure and function were reviewed by **Forbes** (*Ann. Rev. Cell Biol.* **8**, 495–527, 1992).

Discoveries

A possible role for the motif NPXY in internalization was proposed by **Collawn** *et al.* (*Cell* **63**, 1061–1072, 1990).

The ability of nuclear pores to transport rigid structures was demonstrated by **Feldherr, Kallenbach, and Schulz** (*J. Cell Biol.* **99**, 2216–2222, 1984). An updated view of nuclear pore structure was developed by **Akey and Goldfarb** (*J. Cell Biol.* **109**, 955–970 and 971–982, 1989). A model based on 3D analysis was proposed by **Hinshaw, Carragher, and Milligan** (*Cell* **69**, 1133–1141, 1992). Dissection of transport into the nucleus was begun by **Newmeyer and Forbes** (*Cell* **52**, 641–653, 1988) and **Richardson** *et al.* (*Cell* **52**, 655–664, 1988).

CHAPTER 13

Cell cycle and growth regulation

The act of division is the culmination of a series of events that have occurred either continuously or discretely since the last time a cell divided. The period between two mitotic divisions defines the somatic **cell cycle**. The time from the end of one mitosis to the start of the next is called **interphase**. The period of actual division, corresponding to the visible mitosis, is sometimes called **M phase**. A cell cycle may also end in meiosis, producing gametes with a haploid chromosome constitution. There are common features in the regulation of mitosis and meiosis.

In order to divide, a eukaryotic somatic cell must double its mass and then apportion its components equally between the two **daughter cells**. Doubling of size is a continuous process, resulting from transcription and translation of the genes that code for the proteins constituting the particular cell phenotype. By contrast, reproduction of the genome occurs only during a specific period of DNA synthesis. (Variants on this type of cycle occur in embryogenesis; for example, there may be specific cell divisions that are unequal, and of course the egg presents an example where sufficient mass is accumulated to support the early division(s).)

Mitosis of a somatic cell generates two identical daughter cells, each bearing a diploid complement of chromosomes (with a total amount of $2n$ DNA). Interphase is divided into periods that are defined by reference to the timing of DNA synthesis, as summarized in **Figure 13.1**:

◆ Cells are released from mitosis into a condition called **G1 phase**, during which RNAs and proteins are synthesized, but there is no DNA replication.

◆ The initiation of DNA replication marks the transition from G1 phase to the period of **S phase**. S phase is defined as lasting until all of the DNA has been replicated. During S phase, the total content of DNA increases from the diploid value of $2n$ to the fully replicated value of $4n$.

◆ The period from the end of S phase until mitosis is called **G2 phase**; during this period, the cell has two complete diploid sets of chromosomes.

(S phase was so called as the synthetic period when DNA is replicated, G1 and G2 standing for the two 'gaps' in the cell cycle when there is no DNA synthesis.)

The changes in cellular components are summarized in **Figure 13.2**. During interphase, there is little visible change in the appearance of the cell. The more or less continuous increase of RNA and protein contrasts with the discrete doubling of DNA, so that protein/DNA ratios first increase and then are restored during the cycle. The nucleus

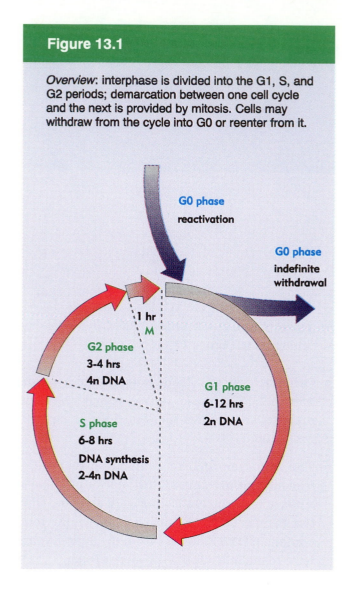

Figure 13.1

Overview: interphase is divided into the G1, S, and G2 periods; demarcation between one cell cycle and the next is provided by mitosis. Cells may withdraw from the cycle into G0 or reenter from it.

G0 phase
reactivation

G0 phase
indefinite
withdrawal

1 hr
M

G2 phase
3-4 hrs
4n DNA

G1 phase
6-12 hrs
2n DNA

S phase
6-8 hrs
DNA synthesis
2-4n DNA

activities come to a halt during mitosis.

In a cycling somatic animal cell, this period of events is repeated every 18–24 hours. Figure 13.1 shows that G1 phase usually occupies the bulk of this period, varying from ~6 hrs in a fairly rapidly growing animal cell to ~12 hours in a more slowly growing cell. It is G1 phase that varies most when cells with different cycle lengths are compared. The duration of S phase is determined by length of time required to replicate all the genome, and a period of 6–8 hrs is typical. G2 phase is usually the shortest part of interphase, possibly comprising the preparations for mitosis. M phase (or mitosis itself) is a brief interlude in the cell cycle, usually <1 hr in duration.

Figure 13.2

Synthesis of RNA and proteins occurs continuously, but DNA synthesis occurs only in the discrete period of S phase. The units of mass are arbitrary.

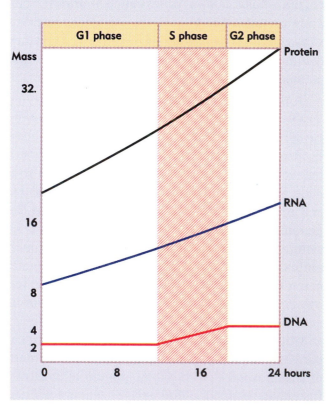

increases in size predominantly during S phase, when proteins accumulate to match the amount of DNA. Chromatin remains a compact mass in which no change of state is visible.

Mitosis segregates one diploid set of chromosomes to each daughter cell by the mechanisms described previously in Chapter 2. Individual chromosomes become visible only during this period, when the nuclear envelope dissolves, and the cell is reorganized on a spindle. The mechanism for specific segregation of material applies only to chromosomes, and other components are divided essentially by the flow of cytoplasm into the two daughter cells. Virtually all synthetic

Replication and mass cycles are coordinated

There are two points in the cycle at which a decision may be taken on whether to proceed:

◆ Commitment to chromosome replication occurs in G1 phase. If conditions to pass the commitment point are satisfied, after a lag a cell will enter S phase. The conditions that are tested are likely to assess the nutritional state of the medium, the mass of the cell, and so on. The commitment point has been defined most clearly in yeast cells, where it is called **START**. The comparable feature in animal cells is called the **restriction point**.

◆ Commitment to mitotic division occurs at the end of G2. This provides another opportunity for the cell to check that its mass has increased to a level adequate for division; and there are also systems to ensure that replication has been completed and that DNA is undamaged.

How do cells use these two control points?

For animal cells growing in culture, G1 control is the major point of decision, and G2/M control is subsidiary. Cells spend the longest part of their cycle in G1, and it is the length of G1 that is adjusted in response to growth conditions. When a cell proceeds past G1, barring accidents, it will complete S phase, proceed through G2, and divide. Cultured cells do not halt in G2. Control at G1 is probably typical of most diploid cells, in culture or *in vivo*.

Some cell phenotypes do not divide at all. These cells are often considered to have withdrawn from the cell cycle into another state, resembling G1 but distinct from it because they are unable to proceed into S phase. This noncycling state is called G0. Certain types of cell can be stimulated to leave G0 and reenter a cell cycle. Withdrawal from, or reentry into, the cell cycle in effect occurs at an early part of G1 phase.

Some cell types do halt in G2. In the diploid world, these are usually cells likely to be called upon to divide again; for example, nuclei at some stages of insect embryogenesis divide and rest in the tetraploid state. In the haploid world, it is more common for cells to rest in G2; this affords some protection against damage to DNA, since there will be two copies of the genome instead of the single copy present in G1. Some yeasts can use either G1 or G2 as the primary control point, depending on the nutritional conditions. Some (haploid) mosses usually use G2 as the control point.

Two types of cycle must be coordinated for a cell to divide.

◆ A cell must replicate every sequence of DNA once and only once. *Having begun replication, it must complete it; and it must not try to divide until replication has been completed.*

◆ The mass of the cell must double, so that there is sufficient material to apportion to the daughter cells. Thus *a cell must not try to start a replication cycle unless its mass will be sufficient to support division.* The effect of mass is coupled to the growth rate, that is, a cell is permitted to divide at a mass that is not absolute but that is determined by a control that itself responds to the growth rate.

We may view the somatic cell cycle in terms of the need to use the two control points (G1/S and G2/M) to measure readiness to proceed in terms of the cell mass and the state of its DNA. Circuits that respond to features such as completion of replication or cell mass are called **checkpoints**; their role is to prevent the cycle from proceeding until a particular condition has been satisfied. They function by acting directly on the factors that control progression through the cell cycle. Some embryonic cycles bypass some of these controls and respond instead to a timer or oscillator. Thus the control

Figure 13.3

Cell mass doubles by continuous growth during the cycle, whereas DNA is replicated during a discrete period. Control points exist during G1, at S phase, and at mitosis.

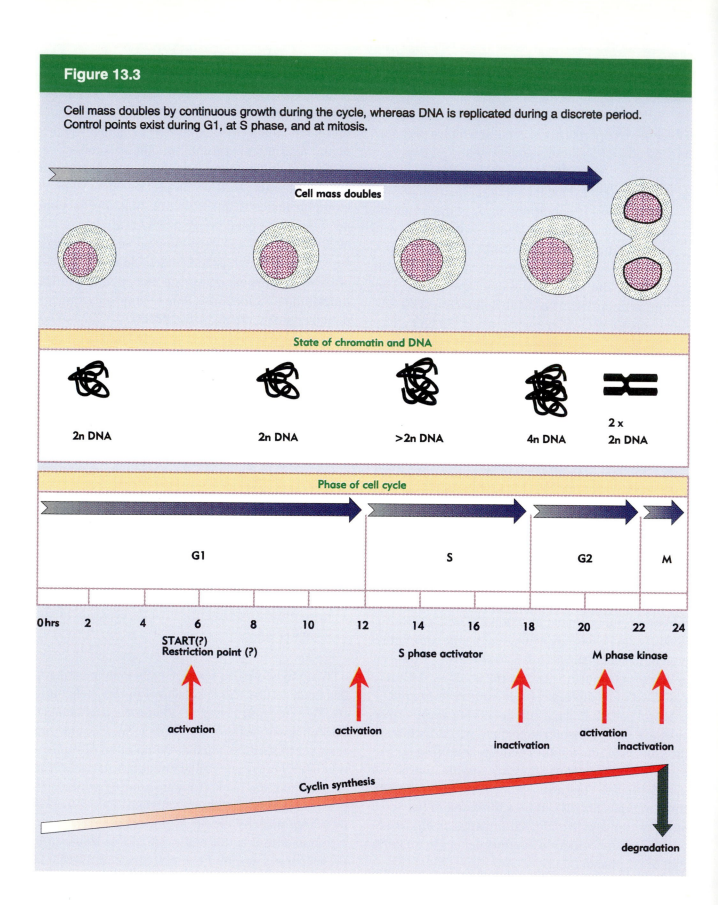

of the cell cycle can be coupled as required to time, growth rate, mass, and the completion of replication.

Figure 13.3 illustrates a view of the cell cycle that shows the molecular basis for some of the regulatory cyclical events by which the replication and mass cycles are coordinated. Some of these events require the synthesis of new proteins or the degradation of existing proteins; other events occur by reversible activation or inactivation of pre-existing components. A minimum of three molecular activities must exist:

◆ During G1, an event whose molecular basis is unknown commits the cell to enter S phase. This involves an activation event that may be related to the activation that occurs at mitosis. We do not understand the nature of the lag period before S phase actually begins.

◆ The period of S phase is marked by the presence of an **S phase activator**. This is not yet defined in molecular terms.

◆ Mitosis itself depends upon the activation of a pre-existing protein, the **M phase kinase**, which has two subunits. One is activated by modification at the start of M phase. The other subunit is a **cyclin,** so named because it accumulates by continuous synthesis during interphase, but is destroyed during mitosis. Its destruction is responsible for inactivating M phase kinase and releasing the daughter cells to leave mitosis.

A striking feature of cell cycle regulation is that similar regulatory elements are employed in a great variety of eukaryotic systems. Some of these sytems have cycles that superficially appear quite different from the normal somatic cycle. Thus very rapid divisions in which S phase alternates directly with mitosis characterize the development of the *Xenopus* egg, where entry into mitosis is controlled by M phase kinase, the very same factor that controls somatic mitosis. Yeast cells exist in a unicellular state, and certain species divide by an asymmetrical budding process; but control of entry into S phase and control of mitosis are determined by procedures and proteins related to those employed in the *Xenopus* egg. As the genes involved in cell cycle regulation are characterized, it is becoming apparent that homologous genes play related roles in organisms as distant as yeasts, insects, and mammals, and that the same regulators are employed in the unusual conditions of early embryogenesis and in normal somatic cycles.

Regulatory activities are found at S phase and at M phase

The existence of different regulators at different stages of the cell cycle was revealed by early experiments that fused together cells in different stages of the cycle. As illustrated in **Figure 13.4**, cell fusion is performed by mixing the cells in the presence of either a chemical or viral agent that causes their plasma membranes to fuse, generating a hybrid cell (called a **heterokaryon** which contains two (or more) nuclei in a common cytoplasm).

When a cell in S phase is fused with a cell in G1, both nuclei in the heterokaryon replicate DNA. This suggests that *the cytoplasm of the S phase cell contains an activator of DNA replication.* The *quantity* of the activator may be important, because in fusions involving multiple cells, an increase in the ratio of S phase to G1 phase nuclei increases the rate

Figure 13.4

Cell fusions generate heterokaryons whose nuclei behave in a manner determined by the states of the cells that were fused.

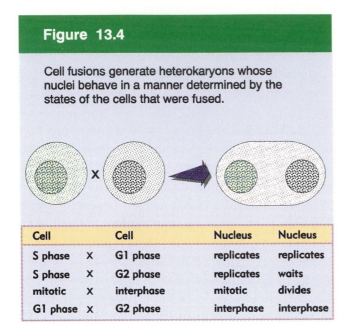

Cell		Cell	Nucleus	Nucleus
S phase	X	G1 phase	replicates	replicates
S phase	X	G2 phase	replicates	waits
mitotic	X	interphase	mitotic	divides
G1 phase	X	G2 phase	interphase	interphase

at which G1 nuclei enter replication. The regulator identified by these fusions is called the **S phase activator**.

When an S phase cell is fused with a G2 cell, the S phase nucleus continues to replicate, but the G2 nucleus does not replicate. This suggests that DNA that has replicated becomes refractory to the effects of the S phase activator, a feature to ensure that each sequence of DNA replicates only once. However, a compromise in the timing of mitosis is seen in such fusions. The S phase nucleus enters M phase sooner than it would have in its previous cyto-plasm, but *the G2 cell does not enter mitosis until after the S phase nucleus has completed replication.* This could mean that some regulator in the S phase cell—perhaps the S phase activator itself—inhibits the start of mitosis.

The nature of the S phase activator is unknown. It could be a regulator whose activation is deter-mined when cells in G1 become committed to a cycle of replication. Or it might comprise some limiting component(s) of the replication apparatus itself.

When a mitotic cell is fused with a cell at any stage of interphase, it causes the interphase nuc-leus to enter a pseudo-mitosis, characterized by

a premature condensation of its chromosomes. This suggests that *an M phase inducer is present in dividing cells.*

Both the S phase and M phase inducers are present only transiently, because fusions between G1 and G2 cells do not induce replication or mitosis in either nucleus of the heterokaryon.

Two general types of regulation for the cell cycle could be imagined, distinguished by the relation-ship between these regulatory activities:

◆ Progress could depend on a *regulatory cascade,* in which completion of one stage is required for the next stage to begin. Such a cascade could be imposed extrinsically by regulators that specif-ically inhibit or activate stages of the cycle. For example, the S phase activator might inhibit mitosis. Alternatively, the cascade could be an intrinsic consequence of the synthesis of par-ticular structures, in which the structure syn-thesized at one stage is a substrate for the next stage. For example, the structure of unrepli-cated DNA could block mitosis. In either case, blocking the cycle at one stage should block all the subsequent stages.

◆ A *master timing* mechanism could directly regulate each stage. Although the mechanism usually would work so that (for example) DNA replication precedes mitosis, a mutation that prevented one event would not necessarily prevent the subsequent events. Thus inhibition of an early stage of the cell could allow the later events to occur anyway.

At least some events in the cell cycle fall into a dependent cascade; for example, we know that blocking DNA replication (with inhibitors of DNA synthesis such as hydroxyurea) prevents somatic cells from progressing through S phase into G2 and M phase. Thus a common feature in the cycle of probably all somatic eukaryotic cells is that completion of DNA replication is a prerequisite for cell division.

M phase kinase is a dimer that regulates entry into mitosis

The early development of eggs of the toad *X. laevis* has provided a particularly powerful system to analyze the features that drive a cell into mitosis. **Figure 13.5** summarizes the early divisions of *Xenopus* embryogenesis.

In its immature form, the *Xenopus* oocyte is arrested in its first meiotic division cycle, just at the initial stage of chromosome condensation. The closest correspondence to a somatic cycle seems to be to G2. Ovulation occurs when hormones trigger progress into the second meiotic division; and when the egg is laid, it is arrested towards the end of the second meiosis in a condition that corresponds to a somatic M phase.

The egg is a large structure (~1 mm diameter in the case of *X. laevis*), which contains a vast store of material needed for early divisions. Fertilization triggers a series of very rapid division cycles. The initial division takes ~90 minutes; then during the **cleavage stage**, another 11 divisions occur synchronously, each lasting ~30 minutes. These divisions in effect represent an alternation of S

Figure 13.5

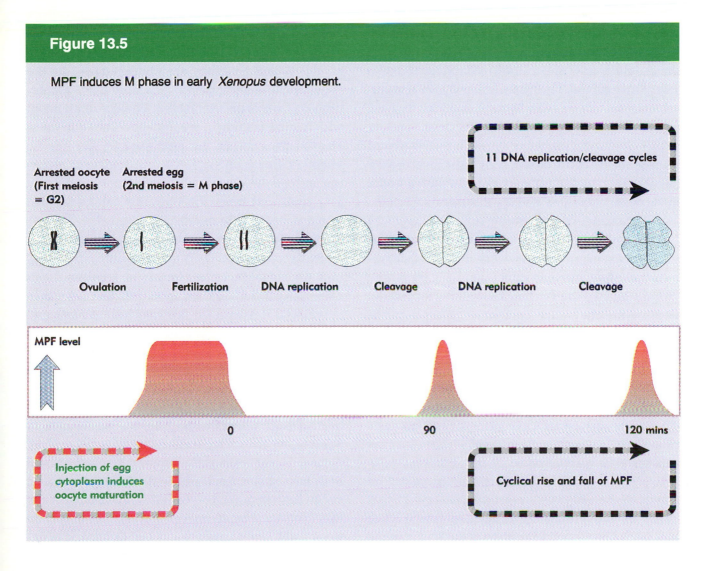

MPF induces M phase in early *Xenopus* development.

phase and mitosis; the major synthetic activity of the cleavage egg is the replication of DNA, all the required proteins having been previously synthesized and stored in the oocyte.

The size of the egg allows material to be purified from it. It is particularly useful that arrested oocytes (equivalent to G2 somatic cells) and arrested eggs (equivalent to M phase somatic cells) can be readily obtained. A factor that stimulates mitosis was discovered by the results of injecting arrested oocytes with cytoplasm extracted from arrested eggs. The injection releases the immature oocytes from the G2 block, and induces them to enter M phase. The active component of the extract was therefore called **maturation promoting factor (MPF)**. (Although MPF is present in arrested eggs, other factors prevent division from initiating.)

Because MPF turns out to have a general responsibility for causing somatic cells to enter M phase, MPF is now understood to stand for **M phase promoting factor**. The importance of MPF in inducing mitosis can be seen from its cyclical increase and decrease at the next stage of development, in the cleaving egg. As a synchronous wave of mitoses occurs in the cleavage egg, the level of MPF activity rises; as the mitoses are completed and S phase occurs, the MPF activity disappears.

MPF causes germinal vesicle (nuclear envelope) breakdown when injected into *Xenopus* oocytes, and induces several mitotic events in a cell-free system, including nuclear envelope disaggregation, chromosome condensation, and spindle formation. By purifying the active component from the cell-free system, MPF has been isolated from *Xenopus* eggs. It consists of two protein subunits, p34 and p45 (the numbers indicating molecular weight; p34 = 34,000 daltons). The protein is a dimer that has kinase activity, and can phosphorylate a variety of protein substrates. This immediately suggests a mode for its action: *by phosphorylating target proteins at a specific point in the cell cycle, MPF controls their ability to function.* In fact, the activity of the enzyme is most often assayed by its ability to phosphorylate target proteins (rather than by the induction of mitosis), and so the name **M phase kinase** provides a better description. Its

two types of subunit have different functions:

◆ The sequence of p34 suggests that it is the *catalytic subunit* which phosphorylates serine and threonine residues in target proteins.

◆ The other subunit is a *regulatory subunit* which is necessary for the kinase to function with appropriate substrates. This subunit is a cyclin: there are multiple cyclins, providing alternative partners for p34, in a given cell.

Cyclins were originally named for their property of accumulating continuously through the cell cycle; then they are destroyed abruptly by proteolysis during mitosis (see Figure 13.3). The mitotic cyclins can be classified into two general types, A and B, which are only weakly related to each other (with 30% overall identity in the case of clam cyclins A and B). They are characterized by the sequence of a stretch of ~150 amino acids (sometimes known as the cyclin box). The B cyclins, about which more is known, are phosphorylated in mitosis; but preventing the phosphorylation has no effect on the cyclin's ability to participate in inducing mitosis or on its subsequent destruction. In mammals and frogs, the B cyclins can be divided into the subtypes B1 and B2.

The timing of cyclin destruction is characteristic; typically A precedes B (by a few minutes in embryonic divisions, by rather longer in a cultured cell cycle), and this difference appears to be common to all cells. Differences in the relative amounts and timing of accumulation suggest that the control of cyclin A and B production is different. For example, in both clam and frog, cyclin B is provided as a maternal product and is an inducer for meiosis I, but cyclin A must be synthesized *de novo*. Cyclin destruction appears to be under a single type of control, because treatments always affect the destruction of both cyclins equally. Cyclins A and B both have a short motif near the N-terminus—the cyclin destruction box—which is required to make the cyclin a target for proteolysis. The system responsible for degradation is one employed to degrade other proteins also, and requires the polypeptide ubiquitin to be added to the cyclin to identify it as a target.

Thus there are (at least) two general forms of the M phase kinase: p34-cyclin A and p34-cyclin B. The common properties of the cyclins suggest that they have the same type of function: to influence the activity of the catalytic subunit of M phase kinase. Yet the two classes of cyclins have only weak similarity and follow a different temporal and spatial pattern of behavior. Although they are involved in the timing and localization of M phase kinase activity, we do not know how they function as regulatory subunits in determining the activity of the catalytic (p34) subunit. We should like to know in particular whether p34-cyclin A and p34-cyclin B recognize different proteins as substrates.

The events that activate M phase kinase at G2/M and inactivate it during M phase identify crucial points in the cell cycle. Activation and inactivation are achieved by different types of action:

◆ *Activation* requires modification of the p34 catalytic subunit. In most cells, the level of p34 remains constant through the cycle, and is in excess compared to the cyclins. Cyclins are necessary to turn on the M phase specific kinase activity of p34, but they do not provide the activating event since they accumulate to a maximum level before the kinase activity appears. The intact p34-cyclin dimer accumulates in an inactive form.

◆ *Inactivation* is achieved by the physical destruction of the cyclin component, that is, by proteolysis of the protein.

What activates the catalytic activity of the M phase kinase? p34 is itself a phosphoprotein, and its state of phosphorylation is a crucial determinant of activity. The events involved in the cycle of activation and inactivation of M phase kinase are summarized in **Figure 13.6**. For M phase kinase to be active, a phosphate group must be absent at some positions, but present at another position.

Two residues located within the ATP-binding site of p34 must be *dephosphorylated* in order to activate the kinase (in mouse cells). Since these residues are different types of amino acids, Thr-14 and Tyr-15, either the reaction requires an unusual type of phosphatase (able to act on both types of phosphorylated amino acid) or two different phosphatase activities are required to provide the activating event (one to remove the phosphate from each amino acid). There is some reason to believe that M phase kinase is autocatalytic, that is, that the activation of a small amount of the kinase is sufficient to trigger activation of the rest. This could be explained if M phase kinase itself activates the phosphatase.

Another phosphorylation occurs on Thr-161 of p34. The timing and maintenance of this phosphorylation are not resolved for the entire cycle, but it appears that the phosphate group is added in G2 and is removed at the end of mitosis. This phosphate is *required* for activity of p34; mutations that introduce an amino acid that cannot be phosphorylated at this site inactivate the kinase activity.

A common supposition had been that cyclin would associate with the inactive form of p34 to generate a quiescent kinase. However, the order of events is surprising. Cyclin B associates specifically with the tyrosine-*dephosphorylated* form of p34 *in vitro*. This generates a potentially active dimer. However, formation of the dimer then causes Thr-14/Tyr-15 to be phosphorylated. Thus *association with cyclin induces the inactivating event*. The dimer is then maintained in its inactive form until the phosphates are removed.

Destruction of the cyclin subunit is responsible for inactivating M phase kinase during mitosis, and indeed is necessary for cells to exit mitosis. A truncated cyclin B that lacks the N-terminal region is resistant to proteolysis. When the truncated protein is synthesized in *Xenopus* eggs, or in a cell-free extract that undertakes some of the typical cycling reactions, it causes metaphase arrest. This suggests that *loss of kinase activity is a prerequisite for completing mitosis*. An amusing speculation is that, if the protease were itself activated (directly or indirectly) by M phase kinase (or if cyclin's susceptibility were determined by a modification dependent on M phase kinase), then the activity of

Figure 13.6

The activity of M phase kinase is regulated by phosphorylation, dephosphorylation, and protein proteolysis. The 3 phosphorylated amino acids in order are: Thr-14, Tyr-15, Thr-167. The first two are in the ATP-binding site. The relative timing of all the events is not certain.

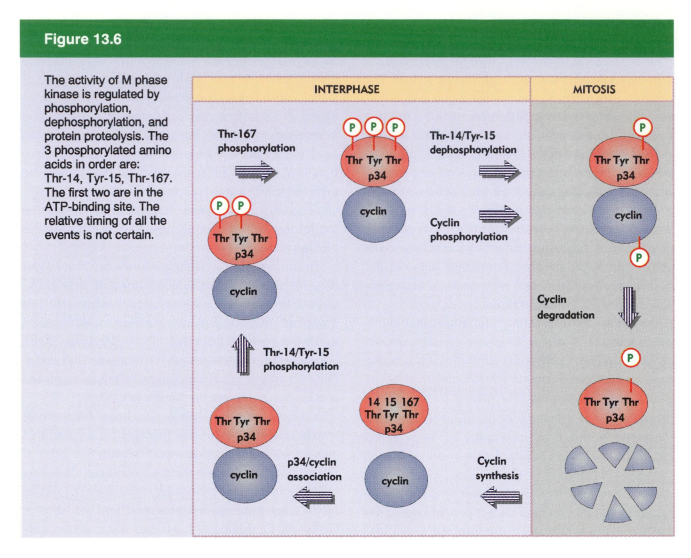

M phase kinase would be self-limiting, thus automatically placing a limit on the duration of mitosis.

An important question is how the roles of cyclin A and cyclin B might differ in mitosis. A hint that their functions are different is provided by the difference in the timing of their synthesis and destruction, but there is as yet no direct evidence to show whether both cyclins are required for passage through mitosis in one cell, or whether different cyclins are required in different cell types.

Cyclin stabilization or destabilization provides an attractive point for control in animal cells, although data so far are lacking. A possible connection is provided by the properties of CSF (cytostatic factor), a component of *Xenopus* that arrests oocytes at meiosis II. (MPF induces cells to enter meiosis or mitosis, but does not hold them there.

CSF, by contrast, causes cells at M phase to remain there.) The relationship between these regulators could be explained if the action of CSF is to stabilize MPF, that is, by preventing the loss of MPF activity, the oocytes are prevented from leaving metaphase.

CSF activity can be provided by the product of c-*mos*, a proto-oncogene which codes for a serine/threonine kinase (see Chapter 39). CSF is itself destroyed in order to release cells from their metaphase arrest. The target upon which CSF acts is unknown, but one possibility is that it is a cyclin—cyclin B2 is a substrate *in vitro* for the c-mos kinase activity. Whether this phosphorylation actually stabilizes the cyclin and thereby maintains active MPF remains to be seen. Another possibility is that the substrate for c-mos is the protease that degrades cyclin, and that it is inactivated by phosphorylation.

Protein phosphorylation and dephosphorylation control the cell cycle

Activation of M phase kinase is the event that triggers onset of M phase. Inactivation is necessary to exit M phase. This suggests that the events regulated by M phase kinase are reversible: *phosphorylation of substrates is required for the reorganization of the cell into a mitotic spindle, and dephosphorylation of the same substrates is required to return to an interphase organization.*

A major reorganization of the cell occurs at mitosis, and the ability of MPF to induce mitosis implies that the M phase kinase, directly or indirectly, triggers these activities. The relevant events include:

◆ Condensation of chromatin to give recognizable chromosomes.

◆ Dissolution of the nuclear lamina and breakdown of the nuclear envelope.

◆ Dissociation and reconstruction of microtubules into a spindle.

◆ Reorganization of actin filaments for cytokinesis.

All of these events are reversible, and modification of appropriate substrate proteins (which could be either the structural subunits themselves or proteins associated with them) provides a plausible means to control the passage of mitosis. (Not all of these events are required in every system; for example, in the yeasts discussed later, the nuclear envelope does not break down.)

Protein phosphorylation and dephosphorylation are major activities that control or assist a variety of events in the interphase cycle and at mitosis. Does M phase kinase act directly or indirectly upon the various potential substrates? Two general models could be proposed for its role:

◆ It may be a 'master regulator' that phosphorylates target proteins that in turn act to regulate other necessary functions—a classic cascade.

◆ Or it may be a 'workhorse' that itself directly phosphorylates the crucial substrates needed to execute the regulatory events or cell reorganization involved in the cycle.

The only common feature in substrates that are phosphorylated by M phase kinase is the presence of the duo Ser-Pro, flanked by basic residues (most often in the form Ser-Pro-X-Lys). The list of potential substrates summarized in **Table 13.1** is based

Table 13.1

Potential substrates for M phase kinase could be involved in reorganization of cell structure at mitosis.

Target for p34	Timing	Potential Consequences
H1 histone	M phase	Chromosome condensation at M phase?
Lamins	early M phase	Nuclear envelope breakdown
Nucleolin	early M phase	Arrest ribosome synthesis?
Myosin light chain	M phase	Inhibit cytokinesis?
p60^{src}	M phase	Phosphorylate further targets?
SWI5	G2 & M phase	Repress nuclear localization

upon the ability of M phase kinase preparations to phosphorylate targets *in vitro*. The strength of the evidence varies as to which of these targets is phosphorylated *in vivo* in a cyclic manner and whether M phase kinase is in fact the active enzyme.

What criteria should be applied to conclude that a potential substrate is an authentic target for p34 in the cell cycle? The same sites should be phosphorylated by p34 *in vitro* that are cyclically phosphorylated *in vivo* at time(s) when p34 is known to be active. Ideally it should be possible to show that a mutation in p34 kinase activity blocks the phosphorylation *in vivo*, but this is at present practical only in yeast. To conclude that the phosphorylation is a significant event in the cycle, some function of the protein must be altered by the presence of phosphate. This can be tested by making mutations at the amino acid that is phosphorylated to determine whether the absence of phosphorylation blocks a mitotic function.

The most established data relate to phosphorylation of H1 histone (one of the 5 histones that are the major protein constituents of chromatin; see Chapter 28). It has been known for a long time that H1 is phosphorylated during the cell cycle, with 2 phosphate groups added during S phase, and 4 further phosphate groups added during mitosis. The major H1 kinase activity of the cell is provided by M phase kinase.

What purpose the phosphorylation of H1 serves in the cell cycle remains a matter for speculation, since no effects upon chromatin structure have been directly demonstrated. It is reasonable to suppose that it might be connected with chromosome condensation at M phase; not enough is known about the timing of modification at S phase to wonder whether it is concerned with preparations for replication (which might require uncoiling) or with the consequences of replication (when preparations for mitosis could begin). But H1 histone is an exceedingly good substrate for kinases based on the p34 engine, with the result that H1 kinase activity has become the usual means by which this enzyme is assayed *in vitro*. An illustration of the appeal of this assay is its application to *S. cerevisiae*, where H1 kinase activity is routinely measured as a cyclic event, although this yeast in fact is unusual in containing no H1 histone!

Nuclear integrity is abandoned when the underlying lamina dissociates into its constituent lamins, and the nuclear envelope breaks down into vesicles (see Figure 2.18). Lamins are phosphorylated during mitosis, and apparently the presence of phosphate groups on only two serine residues per lamin is sufficient to cause dissociation of the lamina. Mutations that change these serines into alanines prevent the lamina from dissociating at mitosis; both sites must be inactivated for a full effect. And if a lamin that has deletions of both serines together with the flanking amino acids is introduced into a cell, it blocks the completion of mitosis. These results suggest that the reversible phosphorylation of these two serine residues induces a structural change in the individual lamin subunits that controls their ability to associate into the lamina.

This phosphorylation event appears to be the direct responsibility of the M phase kinase, which can cause the nuclear lamina to dissociate *in vitro*. Although several kinases can phosphorylate lamins, only treatment with the M phase kinase causes solubilization. M phase kinase phosphorylates lamins at the same two serines that are the targets for modification *in vivo*.

Kinases appear to recognizes specific targets at specific times in the cell cycle. It is not yet clear whether there will be a comparable diversity of phosphatases. There is relatively little variety in known Ser/Thr phosphatases, which can be classified into four groups, 1, 2A, 2B, and 2C. There are at least two isozymes of each type in plants and mammals, which have similar spectra of activities. Generally these phosphatases seem to have broad specificities, at least as characterized *in vitro*.

Although dephosphorylation seems to be an event of comparable frequency and importance to phosphorylation in the cell cycle, much less is known about the actions of specific phosphatases. Some cell cycle mutants in type 1 phosphatases are known that block the cell cycle; most of them act within mitosis itself. The importance of these phosphatases is clear in principle, but we have yet to define their diversity and specificity. They could be responsible for removing the phosphate groups from the substrates of p34.

Will there be a cascade of phosphatases intermingled with the cascade of kinases? It is curious that cell cycle mutants often include kinases (see below), but less often include phosphatases. One possibility is that the phosphatases have broader specificities, so that mutations tend to be lethal. The critical regulatory event could be phosphorylation by a kinase, whose success is measured against a constitutive level of dephosphorylation. Another possibility is that the classification of phosphatases refers to catalytic subunits (of broad specificity), and that the specific cell cycle mutants will be located in regulatory subunits, whose sequences are unknown. In this case, existing cell cycle mutants could identify genes that control phosphatase activities, although this has not yet been realized. Yet another possibility is that the phosphatases are functionally redundant, with more than one enzyme available to catalyze each necessary event (thus ensuring that a cell that has entered mitosis will be able to leave it); in this case, mutations in individual phosphatases would not block the cell cycle.

A major insight into the nature and function of M phase kinase was provided by the observation that p34 is the *Xenopus* homologue to genes involved in the cell cycle in two yeasts, *cdc2* of *S. pombe* and *CDC28* of *S. cerevisiae*. These two yeast genes are homologues, as identified by their sequence relationship and by the ability of one to correct the deficiency in mutants of the other. The protein products are called p34^{cdc2} and p34^{CDC28}, respectively, or sometimes are referred to just as cdc2 and CDC28. In both yeasts, proteins that resemble B-type cyclins associate with p34 at G2/M.

The existence of a p34 catalytic subunit, in organisms as diverse in evolution as yeasts, frogs, and mammals, is a common feature in cell cycle control. The evolutionary conservation of p34 is indicated by the relationship between the yeast and human homologues, which are 63% identical in sequence. And conservation of function is indicated by the ability of the cloned human gene to complement the deficiency in *cdc2* mutants of *S. pombe*.

(Ability to compensate for the deficiency of a specific yeast mutant has been used with great effect to identify higher eukaryotic genes homologous to several cell cycle regulators of yeast. The assay is essentially to introduce the cloned gene into a yeast mutant and to identify cells that resume growth. It is quite remarkable that the control of the cell cycle has been so well conserved as to make this possible. However, it should be remembered that this is sometimes a lax assay that leads to the identification of a gene that is only loosely related to the mutant function.)

The regulatory partner for p34 is always a cyclin. In some cases, several cyclins have been characterized, such as *Xenopus*, which contains A, B1, and B2; in others, a single cyclin is predominantly found. We see later that similar components constitute a kinase that is active earlier in the cell cycle, possibly controlling the G1/S transition.

p34 is the key regulator in yeasts

To define the complexity of cell cycle control, we should like to identify the genes directly involved with regulating progression through the cycle. We define a mutation in the cell cycle as one that *blocks the cell cycle at a specific stage,* a definition that excludes mutations in genes that control continuous processes of growth and metabolism. But the approach of characterizing mutants that are unable to proceed through the cycle has not been straightforward to develop. A mutation that prevents a cell from dividing will be lethal; and the reverse is also true, insofar as any lethal mutation will stop cells from dividing. It has therefore been difficult to devise procedures to isolate cell cycle mutants of animal cells, or, indeed, to demonstrate that potential mutants have specific blocks in the cell cycle.

Because of their nature, cell cycle mutants must be obtained as conditional lethals, so that although they are unable to grow under the conditions of isolation, they can be maintained by growth in other conditions (see Chapter 3). A series of such mutants has been isolated in two yeasts, in which the block to the cell cycle can be seen to affect the visible phenotype of the cell.

The mitotic cycles of fission yeast (*S. pombe*) and baker's yeast (*S. cerevisiae*) are summarized in **Figure 13.7**.

S. pombe has a conventional cell cycle, divided into the usual phases. The cell grows longitudinally and then divides; progress through the cell cycle can be assessed (approximately) by the physical length of the cell (which doubles from 7–8 μm to 14–16 μm).

The cycle of *S. cerevisiae* is unusual. Cells proceed almost directly from S phase into division, so there is effectively very little or no G2 phase. (A short G2 phase is shown in this and subsequent figures for the purpose of localizing the relative timing of events concerned with the transition into M phase.) And instead of an equal division of the cell, the daughter cell grows as a **bud** off the mother cell, eventually obtaining its independence when it is released as a small separate cell. Again the cell cycle can be followed visually in terms of the growth of the bud. Mitosis itself shares some unusual features in both types of yeast; the nuclear membrane does not break down, and segregation of chromosomes therefore is compelled to occur within the nucleus.

Extensive screens for **cell division cycle** or *cdc* mutants have been performed in both yeasts. Initial isolation relies upon the criterion that cells accumulate at a particular stage of the cell cycle at an elevated temperature (36°C), but continue normally through the cycle at 23°C. The mutant phenotype allows cells to continue growing while the cycle is blocked, causing an obvious aberration; in *S. pombe* the cells become highly elongated, and in *S. cerevisiae* they fail to bud. **Figure 13.8** compares cell cycle mutants with wild-type *S. pombe*.

Figure 13.7

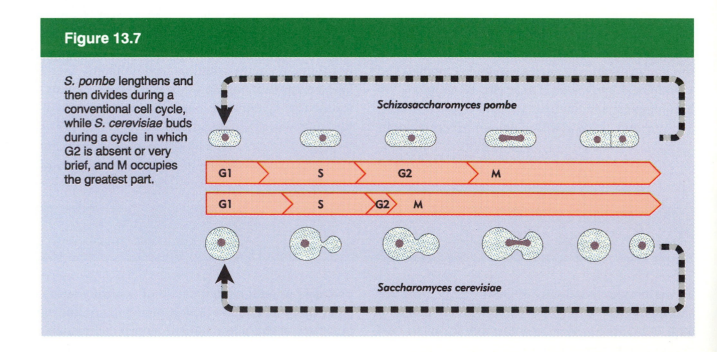

S. pombe lengthens and then divides during a conventional cell cycle, while *S. cerevisiae* buds during a cycle in which G2 is absent or very brief, and M occupies the greatest part.

Figure 13.8

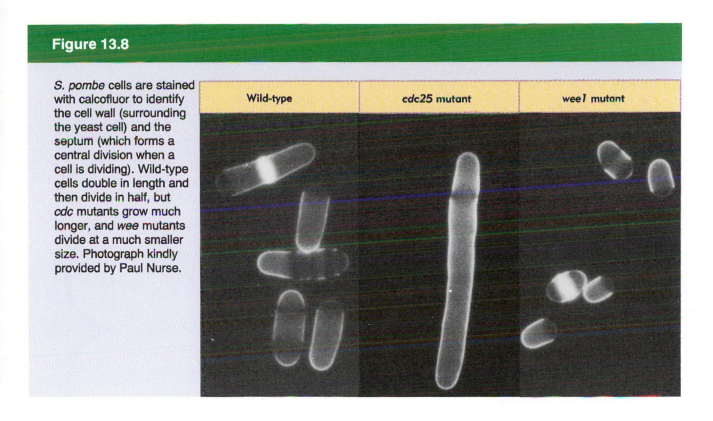

S. pombe cells are stained with calcofluor to identify the cell wall (surrounding the yeast cell) and the septum (which forms a central division when a cell is dividing). Wild-type cells double in length and then divide in half, but *cdc* mutants grow much longer, and *wee* mutants divide at a much smaller size. Photograph kindly provided by Paul Nurse.

Wild-type   *cdc25* mutant   *wee1* mutant

The left panel shows a group of normal cells; the center panel shows a *cdc* mutant that is blocked in the ability to divide, and has therefore become elongated, because it has continued to grow.

Because the mutants are temperature sensitive, the time at which a mutation takes effect can be determined by temperature shift protocols in which cells are shifted up in temperature at a specific point in the cycle. If this point is prior to the point at which the mutation takes effect, the cells halt at the execution point, but if the point is past the time when the mutation acts, the cells continue their cycle.

Of the order of 80 *cdc* genes have been identified in each species, but not all of these loci are concerned with regulating the cell cycle. A significant number represents genes whose products are needed for structural purposes; for example, absence of enzymes that replicate DNA or synthesize the nucleotide precursors can block progression through S phase.

Yeast cells may exist in either haploid or diploid form. They have two forms of life cycle, as illustrated in **Figure 13.9**. Haploid cells double by mitosis. A haploid cell has a **mating type** of either **a** or α. Haploids of opposite types enter the sexual mating pathway, in which they conjugate (fuse) to form diploid cells. The diploid cells in turn sporulate to form haploid cells by meiosis. (The control of mating type is the subject of Chapter 36.)

A crucial point in the cell cycle is defined by the behavior of haploid cells. A haploid cell decides at a point early in G1 whether to proceed through a division cycle or to mate. The decision is influenced by environmental factors; for example, cells of opposite mating type must be present for conjugation to occur. In fact, a mating factor (a polypeptide hormone) secreted by a cell of one type causes a cell of the other type to arrest its cycle (see Chapter 36). But a mating factor can divert a cell into the mating pathway *only early in G1*; later in G1 the cell becomes committed to the division cycle and cannot be stimulated to enter the mating pathway.

Figure 13.9

Haploid yeast cells of either a or α mating type may reproduce by a mitotic cycle; cells of opposite type may mate to form an a/α diploid; the diploid may sporulate to generate haploid spores of both a and α types.

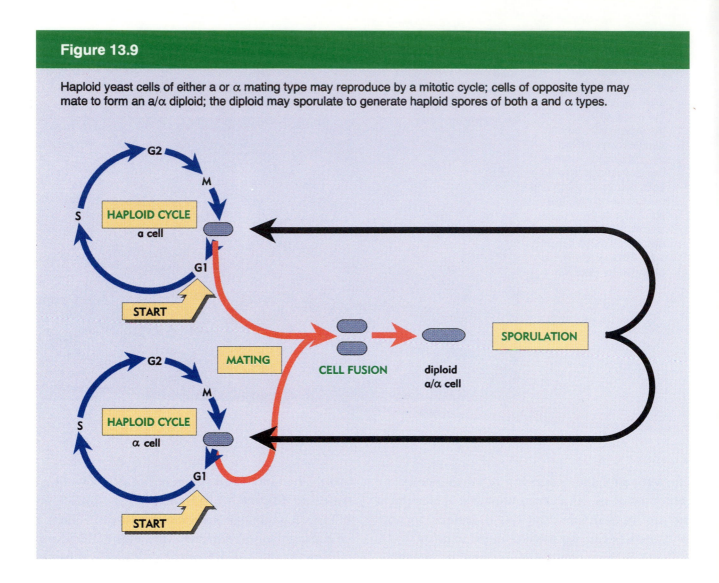

The commitment point in G1 is called **START**.

The events and genes involved in the cycle of *S. pombe* are summarized in **Figure 13.10**. Various *cdc* mutants of *S. pombe* block the cell cycle at one of two stages: at the boundary between G2 and M phase; or in G1 at START.

cdc2 is identified as a crucial regulator by its involvement at *both* stages of cell cycle block: mutants of *cdc2* may be blocked prior to START or prior to M phase (depending on the point a cell had reached in the cycle when the mutation took effect). Recall that *cdc2* is the homologue of the *Xenopus* gene for p34, which identifies *cdc2* as the catalytic component of the key regulator at both control

points. We can divide the other *cdc* genes into two classes, depending on which stage they function at:

◆ After the act of division is accomplished via cytokinesis, a cell enters the mating type pathway if stimulated by mating factor. To continue through the cell cycle, that is, to pass START, it requires the genes *cdc2* and *cdc10* (the latter codes for a subunit of a transcription factor that activates genes required for S phase).

◆ After S phase, *cdc2* is required again, together with other *cdc* genes, including *cdc13* and *cdc25*, in order to enter M phase.

Most is known about function of *cdc2* at the G2/M boundary. The partner for p34cdc2 is the product of *cdc13*, which by sequence resembles a B cyclin. Activation of M phase kinase activity requires events similar to those summarized for mammalian p34 in Figure 13.6. The tyrosine in the ATP-binding site at position 15 of p34cdc2 must be *dephosphorylated*, while the threonine at residue 161 is *phosphorylated*; there is no modification comparable to Thr-14 of mammalian p34.

How is the activity of p34cdc2 controlled? Other cell cycle genes code for kinase and phosphatase activities that respond to environmental signals or to checkpoints and act accordingly on p34cdc2.

The products of two further genes interact directly with the dimer of p34cdc2/cdc13 to form protein complexes.

The product of *suc1*, p13, retains p34cdc2 when bound to Sepharose columns. The use of p13 columns was in fact a crucial technique in the original characterization of p34cdc2, and has since proved to be a powerful approach for isolating proteins related to p34cdc2. There have been conflicting results on the role of p13suc1, and we do not understand its ability to interact with p34cdc2 (why is not found associated with M phase kinase?)

Figure 13.10

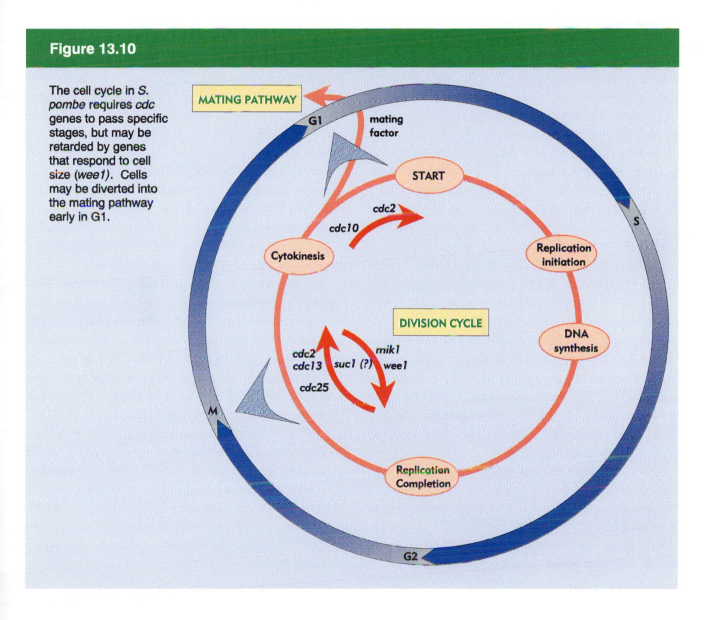

The cell cycle in *S. pombe* requires *cdc* genes to pass specific stages, but may be retarded by genes that respond to cell size (*wee1*). Cells may be diverted into the mating pathway early in G1.

The product of *cdc25* is required for the dephosphorylation that activates the *cdc2* function. The sequence of *cdc25* does not resemble any of the standard phosphatases, so it was originally thought that it might regulate a phosphatase. However, we now know that it has an intrinsic tyrosine phosphatase activity, and it seems likely that Tyr-15 of phosphorylated cdc2 is its natural target. It is probably responsible for the key dephosphorylating event in activating the M phase kinase. It is not a very powerful phosphatase, and the quantities of *cdc25* and *cdc2* proteins are comparable, so the reaction appears almost to be stoichiometric rather than catalytic.

The level of cdc25 increases at mitosis. Its accumulation over a threshold level could be important. In mutants of *cdc2* that do not require *cdc25*, or in strains that overexpress *cdc25*, blocking DNA replication does not impede mitosis. (Wild-type cells arrest in the cell cycle if DNA synthesis is prevented, for example, by treatment with hydroxyurea. But these mutant cells attempt to divide in spite of the deficiency in replication, with lethal consequences.) This suggests that activity of *cdc25* responds to the checkpoint that ensures S phase is completed before M phase can be activated.

Under normal conditions, the cell division cycle is related to the size of the cell. In poor growth conditions, when the cells increase in size more slowly, G1 becomes longer, because START does not occur until the cells attain a critical size. This is a protection against starting a cell division cycle and risking division before the amount of material is adequate to support two daughter cells. *cdc* mutants typically delay the onset of mitosis and lead either to cell cycle arrest or to division at increased size (as shown for *cdc25* in Figure 13.8.)

Genes involved in cell size control are identified by mutants with the opposite property: they advance cells into mitosis and therefore divide at reduced size.

Mutations that eliminate the *wee1* gene divide at reduced size (hence its name). An example is shown in the right panel of Figure 13.8. This behavior suggests that *wee1* usually inhibits cells from initiating mitosis until their size is adequate. The general model suggested by these results is that *wee1* is part of a checkpoint that prevents the activation of p34^{cdc2} until an adequate mass has been attained. *wee1* codes for a kinase of an unusual sort: it can phosphorylate serine/threonine and tyrosine (we discuss other kinases of this type in Chapter 39). It can phosphorylate Tyr-15 of p34^{cdc2}, which inhibits *cdc2* activity. Another gene, *mik1*, has rather similar effects. The deletion of both *wee1* and *mik1* is lethal, suggesting that either gene can fulfill the same function. We shall see that this is a common theme in the yeast cell cycle.

The products of *wee1* and *cdc25* play antagonistic roles. The kinase activity of *wee1* acts on Tyr-15 to inhibit cdc2 function. The phosphatase activity of *cdc25* acts on the same site to activate cdc2. Mutants that over-express *cdc25* have the same phenotype as mutants that lack *wee1*. Some mutants of *cdc2* cause cells to divide at a reduced size; they have an extra active product of *cdc2*. Regulation of cdc2 activity is therefore important for determining *when* the yeast cell is ready to commit itself to a division cycle; inhibition by wee1 and activation by cdc25 allow the cell to respond to environmental or other cues that control these regulators.

It is striking that all of the genes known to affect the G2/M boundary appear to have been widely conserved in evolution. Extending beyond the conservation of the components of the M phase kinase, *cdc25* has a counterpart in the *string* gene of *D. melanogaster*, which has the phenotype of causing cell cycle arrest at the G2/M boundary at the 14th embryonic division cycle. And analogous proteins are found in amphibian and mammalian cells.

The basic principle established by this work is that the activity of the key regulator, p34^{cdc2}, is controlled by kinase and phosphatase activities that themselves respond to other signals. p34^{cdc2} is the means by which all of these various signals are ultimately integrated into a decision on whether to proceed through the cycle.

CDC28 acts at both START and mitosis in *S. cerevisiae*

The analysis of *cdc* mutants in *S. cerevisiae* shows that there is more than one type of cycle, although the cycles are connected at crucial points. (Remember that mitosis in *S. cerevisiae* is unusual morphogically, since chromosome segregation occurs within the intact nucleus; see Figure 13.7.) **Figure 13.11** shows how the three cycles of *S. cerevisiae* relate to the conventional phases of the overall cell cycle:

Figure 13.11

The cell cycle in *S. cerevisiae* consists of three cycles that separate after START and join before cytokinesis. Cells may be diverted into the mating pathway early in G1.

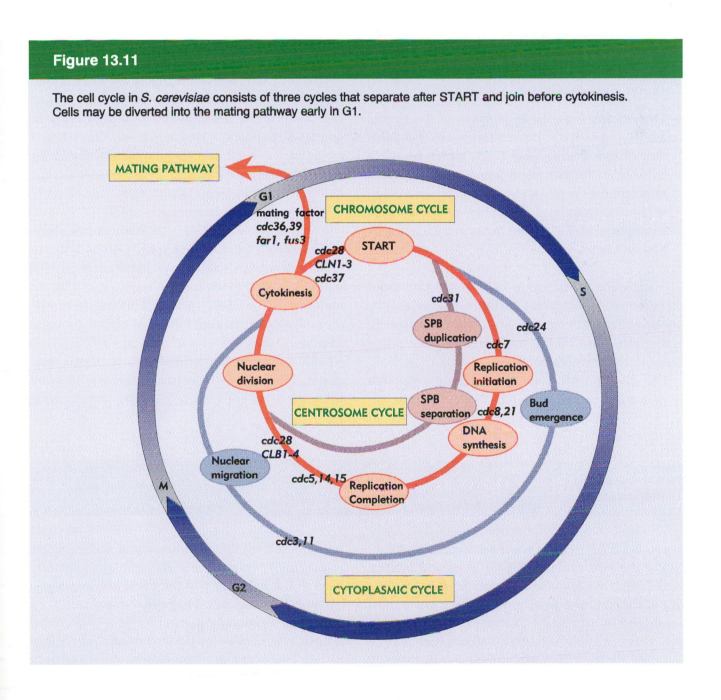

◆ The **chromosome cycle** comprises the events required to duplicate and separate the chromosomes, consisting of the initiation, continuation, and completion of S phase, and nuclear division. Thus a mutation such as *cdc8* stops this cycle in S phase.

◆ Mutations in the chromosome cycle do not stop the **cytoplasmic cycle**, which consists of bud emergence and nuclear migration into the bud (visualized at the start of M phase in Figure 13.7). This cycle can be halted before bud emergence by *cdc24,* but the mutation does not prevent chromosome replication and division.

◆ The **centrosome cycle** consists of the events associated with the duplication and then separation of the spindle pole body, which in effect substitutes for the centrosome and organizes microtubules to allow chromosome segregation within the nucleus. Blocking this cycle, for example with *cdc31,* does not prevent S phase or bud emergence.

Completion of an entire cell cycle requires all three constituent cycles to be functional, since nuclear division requires both the chromosome and centrosome cycles, and cytokinesis requires these and the cytoplasmic cycle.

The decision on whether to initiate a division cycle is made before the point START. The cells can be diverted into the mating type pathway by mating factors; and *cdc36* and *cdc39*, which appear to block the cell cycle before START, really function in the mating type pathway. The mutants probably block the cell cycle by diverting cells into mating even in the absence of mating factor.

The crucial gene in passing START is *CDC28* (the homologue of *cdc2* in *S. pombe*). Mutations in *CDC28* prevent all three cycles from proceeding. The ability to pass START is determined by environmental conditions, since stationary phase populations that are limited by nutrients are arrested at START. Morphologically, the cells of such a population are arrested at point after cell separation and prior to SPB duplication, bud emergence, and DNA replication, which is to say that when *S. cerevisiae* cells exhaust their nutrients, they complete the current cycle and arrest all three cycles before START. In terms of phases of the cell cycle, this corresponds to a point early in G1.

The function of *CDC28* is required at two stages in *S. cerevisiae:* before START; and prior to M phase. Most mutants in *CDC28* are blocked at START, which has therefore been characterized as the major point for *CDC28* action. This contrasts with the more common function of M phase kinase at M phase. However, one mutant in *CDC28* passes START normally, but is inhibited in mitosis, suggesting that p34^{CDC28} is required to act at both points in the cycle.

What controls the G1/S transition?

There is a difference in emphasis concerning control of the cell cycle in fission yeast and bakers' yeast, since in *S. pombe* we know most about control of mitosis, and in *S. cerevisiae* we know most about control of START. However, in both yeasts, the same catalytic subunit (cdc2/CDC28) is required for the G2/M and the G1/S transitions. It has different regulatory partners at each transition, using B-like cyclins at mitosis and different, but related, partners at START. The regulatory subunits used at the two stages of the cycle are sometimes called **G2 cyclins** and **G1 cyclins**, respectively.

A single regulatory partner for p34^{cdc2} is used at mitosis in *S. pombe,* the product of *cdc13.* In *S. cerevisiae,* there are multiple alternative partners. These are coded by the *CLB1-4* genes, which code for products that resemble B cyclins and associate with p34^{CDC28} at mitosis. The genes fall into two pairs, *CLB1-2* and *CLB3-4* on the basis of sequence relationships. Mutation in any one of these genes fails to block division, but loss of the *CLB1-2* pair of genes is lethal. Constitutive expression of *CLB1* prevents cells from exiting mitosis.

G1 cyclins have been characterized in *S. cerevisiae*, where they were not immediately revealed by mutations that block the cell cycle in G1. The absence of such mutants has been explained by the discovery that three independent genes, *CLN1, CLN2, CLN3 all must be inactivated* to block passage through START in *S. cerevisiae*. Mutations in any one or even any two of these genes fail to block the cell cycle; thus the *CLN* genes are **functionally redundant**. The *CLN* genes show a weak relationship to cyclins (resembling neither the A nor B class particularly well), although they are usually described as G1 cyclins.

Accumulation of the CLN proteins is the rate-limiting step for controlling the G1/S transition. Blockage of protein synthesis arrests the cycle by preventing the proteins from accumulating. The half life of CLN2 protein is ~15 minutes; its accumulation to exceed a critical threshold level could be the event that triggers passage. Its intrinsic instability presents a different type of control from that shown by the abrupt destruction of the cyclin A and B types. The instability is crucial to the CLN proteins' functions: dominant mutations that truncate the protein by removing the C-terminal stretch (which contains sequences that target the protein for degradation) advance G1. In these mutants, the protein is probably stable, and G1 phase is basically absent, with cells proceeding directly from M phase into S phase. Similar behavior is shown by the product of *CLN3*.

A prerequisite for a cell to leave the cycle and enter the mating type pathway is that the actions of all three *CLN* gene products must be inhibited. Different regulator genes function on individual *CLN* gene products, ensuring that the cycle is halted only when all of the conditions required for mating have been satisfied. Each of these inhibitory actions involves a different type of mechanism.

Although the *CLN* and *CLB* genes are identified as cyclin-like by their functions and by their sequences, the proteins do not appear to possess the feature of cyclic proteolysis by which the cyclins were originally identified. It seems instead that their periodic expression is controlled at the level of transcription.

The redundancy of the *CLN* genes and the *CLB* genes is a feature found at several other stages of the cell cycle. The inhibition of the G2/M transition by *wee1* and *mik1* identifies a pair of redundant kinases; and redundant phosphatases have been found at mitosis in *S. pombe*. In each case, the hallmark is that deletion of an individual gene produces a cyclic phenotype, but deletion of both or all members of the group is lethal. Members of a group may play overlapping rather than identical functions. This form of organization has the practical consequence of making it difficult to identify mutants in the corresponding function.

The use of the *same* p34 subunit at the stage of both G2/M and START in the yeasts is an interesting contrast with the redundancy of the cyclins. It contrasts also with the probable use of multiple p34-like proteins in higher eukaryotes (see below).

The use of a kinase consisting of a catalytic subunit plus regulatory partner to regulate both START and mitosis in fission yeast and bakers' yeast suggests that a similar scheme could control the cycle in animal cells. One early model proposed that p34^{cdc2} associates with cyclins of the A and/or B type to regulate passage through G2/M, and associates with different partners to regulate passage through G1. The involvement of the p34 kinase engine (and/or another related to it) at two regulatory points is consistent with an increase in H1 kinase activity at S phase as well as at M phase. On the assumption that the partner that p34 requires for G1/S could be generally similar in nature to its partner at G2/M, the proposed G1/S partners were called 'G1 cyclins', and sought as proteins that associate specifically with p34 at the G1 stage of the cycle.

But it is probably the case in higher eukaryotes that the catalytic subunit, p34, is required *only* at mitosis. Higher eukaryotes possess a large number of genes (~12) related to the true p34cdc2 homologue. It is by no means clear how many of these genes code for kinases with other functions and how many might be involved with the cell cycle. However, their products have the potential to associate with certain cyclins, and are therefore called cyclin-dependent kinases (CDKs).

Just as families of cyclins can be defined by ability to interact with p34, so may families of catalytic subunits be defined by the ability to interact with cyclins. The pairwise associations between the members of each class are not exclusive, and a particular cyclin may associate with several potential catalytic subunits, while a catalytic subunit may associate with several potential cyclins. The trick is to determine which of these pairwise combinations form in the cell and are concerned with regulating the cell cycle.

Two of the *CDK* genes, *CDK2* and *CDK4* code for proteins that form pairwise combinations with potential G1 cyclins. The first and best characterized is *CDK2*; it has 66% similarity to the *cdc2* homologue in the same organism. It codes for a protein that was initially called p33, which has a different cycle of activity from p34. Its H1 kinase activity remains relatively constant through the cell cycle. Since histone H1 is probably not its natural target, it is difficult to assess to what degree its activity really fluctuates. The key result is that CDK2 is required for DNA replication in *Xenopus* extracts, but is not required for entry into mitosis. It appears to associate with several alternative partners, including cyclin A, but not including the B cyclins. Thus cyclin A could be involved at both stages of the cycle control. The other proteins associated with CDK2 may be G1-specific cyclins that are needed for its activation at the appropriate point in the cell cycle.

G1 cyclins have been sought in higher eukaryotes by identifying genes that can overcome the deficiency of *CLN* mutants in *S. cerevisiae*. Three new types of cyclin (C, D, and E) have been identified by this means. They are distantly related to one another and to other cyclins. Cyclins C and E

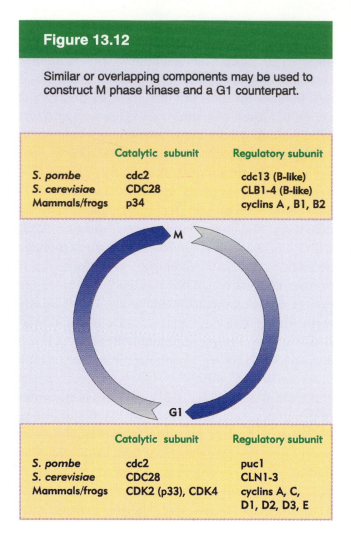

Figure 13.12

Similar or overlapping components may be used to construct M phase kinase and a G1 counterpart.

	Catalytic subunit	Regulatory subunit
S. pombe	cdc2	cdc13 (B-like)
S. cerevisiae	CDC28	CLB1-4 (B-like)
Mammals/frogs	p34	cyclins A , B1, B2

	Catalytic subunit	Regulatory subunit
S. pombe	cdc2	puc1
S. cerevisiae	CDC28	CLN1-3
Mammals/frogs	CDK2 (p33), CDK4	cyclins A, C, D1, D2, D3, E

accumulate in a periodic manner through the cycle, but are regulated by control of transcription: they do not show periodic destruction of protein. Cyclin E can bind to CDK2, constituting a dimer with H1 kinase activity. There are 3 D-type cyclins; they form dimers with CDK2 and/or CDK4. Cyclin D2 is turned on when quiescent cells are reactivated, oscillates in amount during subsequent cell cycles, and forms a complex with CDK2.

Proteins described as 'cyclins' are therefore now significantly more diverse than the A and B classes encompassed by the original definition. For working purposes, we class as cyclins various proteins that have some sequence relationship to the original class, and which can participate in formation of a kinase by pairing with a p34 or CDK2 or related catalytic subunit. It is not yet certain that all of the proteins presently classed as cyclins in this way in fact have this function *in vivo*.

The potential components of the regulatory kinases are summarized in **Figure 15.12**. The unifying theme is that the kinase consists of a catalytic subunit and a regulatory partner. At mitosis, the catalytic subunit is provided by a single gene (p34 is protein of *cdc2* and *CDC28*). The regulatory partner at mitosis usually is not unique, but is provided by a family of B-type, and possibly A-type, cyclins. At G1/S, the catalytic subunit in yeast does not change, but in animal cells it is a related protein, CDK2 or CDK4. The regulatory partner is one of a large group of G1 cyclins, only loosely related in sequence to the mitotic cyclins.

Both the beginning and end of S phase are important points in the cell cycle. Just as a cell must know when it is ready to initiate replication, so it must have some means of recognizing the successful completion of replication. This may be accomplished by examining the state of DNA.

The ability of inhibitors of DNA replication to block the cell cycle shows directly that the presence of nonreplicated DNA can provide an impediment to passage into or through G2. How does the cell assay unreplicated DNA? Treatment of *Xenopus* eggs with aphidicolin (a reagent that blocks DNA synthesis) entirely dissociates the replication and cell division cycles, and allows pseudo-divisions to occur in the absence of any duplication of DNA. This probably reflects the fact that the concentration of unreplicated DNA is too small to trigger the feedback inhibition of the cycle. (The ratio of DNA to cytoplasmic mass in the egg is $<10^{-4}$ of that of a somatic cell.) By adding DNA, this system can be used to assay the level at which the DNA concentration can inhibit cell division.

Extracts of *Xenopus* eggs can reproduce *in vitro* the rapid alternation between S phase and mitosis that characterizes early embryogenesis. When sperm DNA is added to the extracts, the cycle slows, and ultimately is arrested in S phase. The magnitude of the effect depends upon the concentration of added DNA. The addition of aphidicolin reduces the concentration of DNA required for inhibition by ~100-fold, presumably because it causes the accumulation of unreplicated DNA. The magnitude of the effect suggests that unreplicated DNA, at concentrations equivalent to ~1% of the genome, can block progress from S phase to mitosis.

How does unreplicated DNA block cell cycle progress? It does not act upon M phase kinase activity, since addition of active M phase kinase releases the extracts into mitosis. This suggests the existence of some signal that prevents the activation of M phase kinase.

Another interesting connection between DNA and the cell cycle has been revealed by the properties of the *RAD9* mutant of *S. cerevisiae*. Wild-type yeast cells cannot progress from G2 into M if they have damaged DNA, introduced by X-irradiation or by the result of replication in a mutant such as *cdc9* (DNA ligase). Mutation of *RAD9* allows these cells to divide in spite of the damage. *RAD9* therefore exercises a checkpoint that inhibits mitosis in response to the presence of damaged DNA. The reaction may be triggered by the existence of double-strand breaks. *RAD9*-dependent arrest and recovery from arrest can occur in the presence of cycloheximide, suggesting that the *RAD9* pathway functions at the post-translational level. At least 6 other genes are involved in this pathway; its ultimate regulatory target remains to be found.

Other controls that ensure an orderly progression through the cell cycle and that can be visualized in the context of checkpoints include monitoring the completion of DNA replication and checking cell size. It may be necessary to bypass some of these checkpoints in early embryogenesis; for example, early divisions in *Drosophila* in effect are nuclear only (because there are no cellular compartments: see Chapter 38). It may be significant that the product of *string* (the counterpart of *S. pombe cdc25* which exercises a size-dependent control) is present at high levels in these early cycles; after the 14th cycle, division becomes dependent on *string* expression, and it is at this point that cellular compartments develop, and it becomes appropriate for there to be a checkpoint for cell size. *string* may provide this checkpoint.

The concept of checkpoints provides a way to visualize a variety of controls by which a cell ensures that mitosis occurs only when all the conditions have been satisfied for a successful division. We have probably not yet identified all of the types of events that are controlled by checkpoints.

Summary

The cell cycle consists of transitions from one regulatory state to another. The change in regulatory state is separable from the subsequent changes in cell phenotype. The transitions take the form of activating or inactivating a kinase(s), which modifies substrates that determine the physical state of the cell. Checkpoints can retard a transition until some intrinsic or extrinsic condition has been satisfied.

The two key control points in the cell cycle are in G1 and at the end of G2. During G1, a commitment is made to enter a replication cycle; the decision is identified by the restriction point in animal cells, and by START in yeast cells. After this decision has been taken, cells are committed to beginning an S phase, although there is a lag period before DNA replication initiates. The end of G2 is marked by a decision that is executed immediately to enter mitosis.

A unifying feature in the cell cycles of yeasts and animals is the existence of an M phase kinase, consisting of two subunits: p34, with serine/threonine protein kinase catalytic activity; and cyclin of either the A or B class. Homologous subunits can be recognized in (probably) all eukaryotic cells. In fission yeast *S. pombe*, p34 is coded by the cell cycle gene *cdc2*; in budding yeast *S. cerevisiae*, p34 is coded by the homologous gene *CDC28*. Animal cells seem usually to contain multiple cyclins; in *S. pombe*, there appears to be only a single cyclin at M phase, a B class coded by *cdc13*, although *S. pombe* has several *CLB* proteins.

Activity of the M phase kinase controls the transitions into and out of mitosis. The M phase kinase is activated by dephosphorylating one (in yeast) or two (in animal cells) amino acid residues in the ATP-binding site of p34 and by phosphorylating a residue elsewhere in p34. This event causes entry into mitosis. The cyclins are also phosphorylated, but the significance of this modification is not known. The kinase is inactivated by degradation of the cyclin component, which occurs abruptly during mitosis. Cyclins of the A type are typically degraded before cyclins of the B type. Destruction of at least the B cyclins, and probably of both classes of cyclin, is required for cells to exit mitosis.

By phosphorylating appropriate substrates, the kinase provides MPF activity, which stimulates mitosis or meiosis (as originally defined in *Xenopus oocytes*). A prominent substrate is histone H1, and H1 kinase activity is now used as a routine assay for M phase kinase. Phosphorylation of H1 could be concerned with the need to condense chromatin at mitosis. Another class of substrates comprises the lamins, whose phosphorylation causes the dissolution of the nuclear lamina. A general principle governing these (and presumably other) events is that the state of the substrates is controlled reversibly in response to phosphorylation, so that the phosphorylated form of the protein is required for mitotic organization, while the dephosphorylated form is required for interphase organization. Phosphatases are required to reverse the modifications introduced by M phase kinase; mutations in some phosphatases block yeast cells in mitosis.

Initiation of S phase may require a similar kinase. In yeasts, the catalytic subunit is identical with that of the M phase kinase. In mammalian cells, a family of potential catalytic subunits is provided by the *CDK* genes. The best characterized is *CDK2*, which codes for the protein p33, which is well related to p34; another member that appears in complexes at appropriate times is the product of

CDK4. The regulatory subunit may be cyclin A or one of several G1 cyclins (C, D1, D2, D3, or E) in mammalian cells. In *S. cerevisiae,* it is one of the CLN cyclins.

A comprehensive analysis of genes that affect the cell cycle has identified *cdc* mutants in both *S. pombe* and *S. cerevisiae.* The best characterized mutations are those that affect the components or activity of M phase kinase. Mutations of *cdc25* and *wee1* in *S. pombe* regulate M phase kinase in response to cell size. The existence of *cdc25* homologues in *Drosophila* and man suggests that the apparatus for cell cycle control is widely conserved in evolution. Other mutations identify checkpoints that respond to the integrity of DNA or the completion of replication.

Further reading

Historical
The existence of S phase was discovered in plant cells by **Howard and Pelc** (*Heredity suppl.* **6**, 261–273, 1953).

The original screen for *cdc* mutants of *S. cerevisiae* was performed by **Hartwell** *et al*. (*Science* **183**, 46–51, 1974).

Reviews
Mechanisms for cell cycle control in animals have been reviewed by **Murray and Kirschner** (*Science* **246**, 614–622, 1989) and by **Norbury and Nurse** (*Ann. Rev. Biochem.* **61**, 441–470, 1992).

The activation and inactivation of M phase kinase has been reviewed by **Nurse** (*Nature* **344**, 503–508, 1990). The role of related kinases in the G1/S transition has been reviewed by **Reed** (*Ann. Rev. Cell Biol.* **8**, 529–561, 1992).

The concept of checkpoints was discussed by **Hartwell and Weinert** (*Science* **246**, 629–635, 1989).

The classic review of the *S. cerevisiae* cell cycle(s) was written by **Pringle and Hartwell** (pp. 97–142 in *The*

Molecular Biology of the Yeast Saccharomyces, eds Strathern, Jones, and Broach, Cold Spring Harbor, New York, 1981.) A more recent account of yeast cell cycles was provided by **Forsberg and Nurse** (*Ann. Rev. Cell Biol.* **8**, 227–256, 1991).

Discoveries
The crucial insight that MPF is the equivalent of *S. pombe cdc2* was reached by **Dunphy** *et al*. (*Cell* **54**, 423–431, 1988) and **Gautier** *et al*. (*Cell* **54**, 433–439, 1988). MPF was equated with H1 kinase by **Arion** *et al*. (*Cell* **55**, 371–378, 1988).

Cyclins were discovered by **Evans** *et al*. (*Cell* **33**, 389–396, 1983). Their role in M phase kinase was uncovered by **Draetta** *et al*. (*Cell* **56**, 829–838, 1989).

The roles of the components of M phase kinase in activating and inactivating its activity were analyzed by **Gould and Nurse** (*Nature* **342**, 39–45, 1989) and **Murray. Solomon, and Kirschner** (*Nature* **339**, 280–286, 1989).

Control of prokaryotic gene expression

P A R T

4

According to the strictly structural concept, the genome is considered as a mosaic of independent molecular blueprints for the building of individual cellular constituents. In the execution of these plans, however, coordination is evidently of absolute survival value. The discovery of regulator and operator genes, and of repressive regulation of the activity of structural genes, reveals that the genome contains not only a series of blueprints, but a coordinated programme of protein synthesis and the means of controlling its execution.

François Jacob and Jacques Monod,
1961

CHAPTER 14

Control at initiation: RNA polymerase–promoter interactions

Transcription involves synthesis of an RNA chain representing one strand of a DNA duplex. By 'representing' we mean that the RNA is *identical in sequence* with one strand of the DNA, which is called the **coding strand.** It is *complementary* to the other strand, which provides the **template** for its synthesis. This relationship between double-stranded DNA and its single-stranded RNA transcript is recapitulated in **Figure 14.1**.

RNA synthesis is catalyzed by an enzyme, called **RNA polymerase**. Transcription starts when RNA polymerase binds to a special region, the **promoter**, at the start of the gene. The promoter surrounds the first base pair that is actually transcribed into RNA, the **startpoint**. From this point, RNA polymerase moves along the template, synthesizing RNA, until it reaches a **terminator** sequence. This action defines a **transcription unit** that extends from the

Figure 14.1

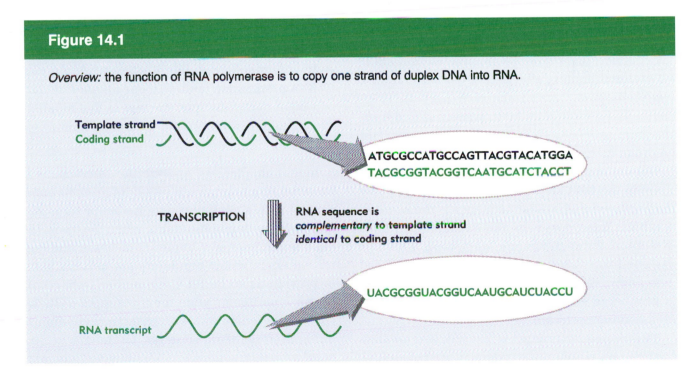

Overview: the function of RNA polymerase is to copy one strand of duplex DNA into RNA.

Template strand
Coding strand

ATGCGCCATGCCAGTTACGTACATGGA
TACGCGGTACGGTCAATGCATCTACCT

TRANSCRIPTION

RNA sequence is
complementary to template strand
identical to coding strand

UACGCGGUACGGUCAAUGCAUCUACCU

RNA transcript

Figure 14.2

Overview: a transcription unit is a sequence of DNA transcribed into a single RNA, starting at the promoter and ending at the terminator.

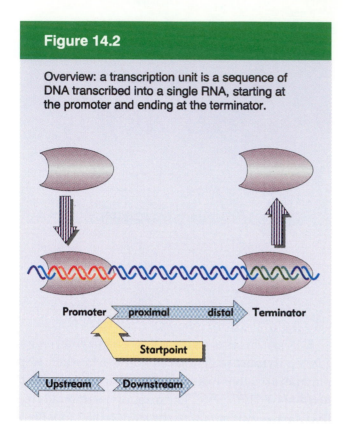

Promoter   proximal   distal   Terminator

Startpoint

Upstream   Downstream

promoter to the terminator. The critical feature of the transcription unit, depicted in **Figure 14.2**, is that it constitutes a stretch of DNA *expressed via the production of a single RNA molecule*. A transcription unit may include only one or more than one gene.

Sequences prior to the startpoint are described as **upstream** of it; those after the startpoint (within the transcribed sequence) are **downstream** of it. Sequences are conventionally written so that transcription proceeds from left (upstream) to right (downstream). This corresponds to writing the mRNA in the usual 5′ to 3′ direction.

Often the DNA sequence is written just to show the coding strand, whose sequence is the same as the RNA. Base positions are numbered in both directions away from the startpoint, which is assigned the value +1; numbers are increased going downstream. The base before the startpoint is numbered −1, and the negative numbers increase going upstream.

The immediate product of transcription is called the **primary transcript**. It would consist of an RNA extending from the promoter to the terminator, possessing its original 5′ and 3′ ends. However, the primary transcript is almost always unstable, and therefore is difficult to characterize *in vivo*. In prokaryotes, it is rapidly degraded (mRNA) or cleaved to give mature products (rRNA and tRNA). In eukaryotes, it is modified at the ends (mRNA) and/or cleaved to give mature products (all RNA).

Transcription is undertaken exclusively by RNA polymerase, but genes are not transcribed indiscriminately by the enzyme; other proteins, whose function is to regulate transcription, determine whether a particular gene is available to be transcribed.

Transcription is the first stage in gene expression, and the principal step at which it is controlled. The initial (and sometimes the only) step in regulation is the decision on whether or not to transcribe a gene. In considering the various stages of transcription, we should therefore keep in mind the opportunities they offer for regulating gene activity.

Within this context, there are two basic questions in gene expression:

◆ How does RNA polymerase find promoters on DNA? This is a particular example of a more general question: how do proteins distinguish their specific binding sites in DNA from other sequences?

◆ How do specific regulatory proteins interact with RNA polymerase (and with one another) to activate or to repress specific steps in the initiation, elongation, or termination of transcription?

In this chapter, we analyze the interactions of bacterial RNA polymerase with DNA, from its initial contact with a gene, through the act of transcription, culminating in its release when the transcript has been completed. Chapter 15 discusses the various means by which regulatory proteins can assist or prevent bacterial RNA polymerase from recognizing a particular gene for transcription. In the succeeding chapters, we consider regulation at different stages of transcription, and how individual regulatory interactions can be connected into more complex networks. In Chapters 29 and 30, we consider the analogous reactions between eukaryotic RNA polymerases and their templates.

(Transcription is not the only means by which RNA can be synthesized. Viruses with RNA genomes specify enzymes able to synthesize RNA on a template itself consisting of RNA. Such reactions produce mRNAs coding for proteins needed in the infective cycle (RNA transcription) and provide genomic RNAs to perpetuate the infective cycle (RNA replication). But the cell synthesizes RNA only by transcription of a DNA template.)

Transcription is catalyzed by RNA polymerase

Transcription takes place by the usual process of complementary base pairing, catalyzed and scrutinized by the enzyme RNA polymerase. **Figure 14.3** illustrates the general nature of transcription. RNA synthesis takes place within a 'transcription bubble', in which DNA is transiently separated into its single strands, and one strand is used as a template for synthesis of the RNA strand. As RNA polymerase moves along the DNA, the bubble moves with it, and the RNA chain grows longer.

The structure of the bubble within RNA polymerase is shown in the expanded view of **Figure 14.4**. As RNA polymerase moves along the DNA template, it unwinds the duplex at the front of the bubble (the unwinding point), and rewinds the DNA at the back (the rewinding point). The length of the transcription bubble is ~18 bp, but the length of the RNA–DNA hybrid region within it is shorter.

The classic view of transcription has been that the RNA–DNA hybrid extends for ~12 bp, but this idea was derived from indirect evidence about the structure of the RNA region within RNA polymerase. In fact the distance of RNA–DNA hybrid has never been measured directly. More recent results show that bases in the RNA as close as 3 positions from the growing point can be cleaved by ribonucleases that recognize single-stranded RNA. (This type of analysis is made possible by stopping an elongating RNA polymerase at a particular position by withholding the nucleotide that it requires in order to add the next base to the RNA chain.)

RNA therefore remains associated with the DNA for only 2–3 bases after the growing point, after

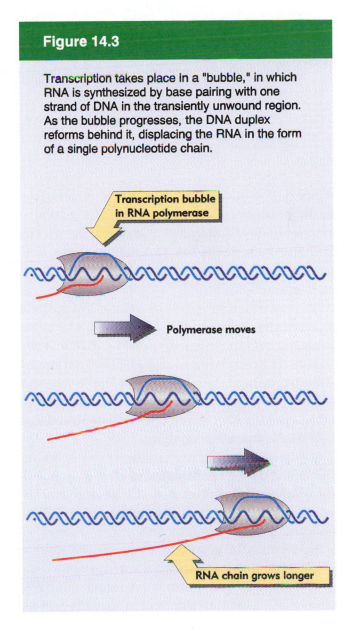

Figure 14.3

Transcription takes place in a "bubble," in which RNA is synthesized by base pairing with one strand of DNA in the transiently unwound region. As the bubble progresses, the DNA duplex reforms behind it, displacing the RNA in the form of a single polynucleotide chain.

Transcription bubble in RNA polymerase

Polymerase moves

RNA chain grows longer

Figure 14.4

During transcription, the bubble is maintained within bacterial RNA polymerase, which unwinds and rewinds DNA, maintains the conditions of the partner and template DNA strands, and synthesizes RNA.

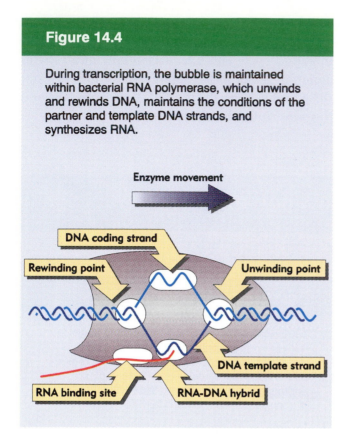

Enzyme movement

DNA coding strand
Rewinding point
Unwinding point
DNA template strand
RNA binding site
RNA-DNA hybrid

Structural analysis shows that they share a common type of structure, in which there is a 'channel' or groove about 25Å wide on the surface that could be the path for DNA. This is illustrated in **Figure 14.5** for the example of yeast RNA polymerase. The length of the groove could hold 16 bp in the bacterial enzyme, and ~25 bp in the yeast enzyme, but this represents only part of the total length of DNA bound during transcription. Aside from this general description, we cannot yet relate the features of the enzyme to the generalized structure shown in Figure 14.4.

Figure 14.5

Yeast RNA polymerase has grooves that could be binding sites for nucleic acids. The pink beads show a possible path for DNA that is ˜25Å wide and 5-10Å deep. The green beads show a narrower channel, 12-15Å wide and ˜20Å deep, that could hold RNA. Photograph kindly provided by Roger Kornberg.

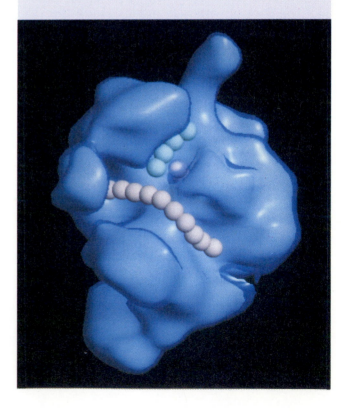

which it probably enters a high affinity binding site in the polymerase. Thus the RNA–DNA hybrid is extremely short and transient, extending in fact no farther than is needed to give stability to the base pairing reaction that determines the specificity of nucleotide addition.

The separated strands of unwound DNA are held apart in the enzyme structure. The DNA coding strand is held in single-stranded condition in the bubble. The template strand is released into single-stranded form at the unwinding point. Except for the very short region where it is paired with the end of the growing RNA chain, the template strand also is in single-stranded condition.

As the enzyme moves on, the DNA duplex reforms, and the RNA is displaced as a free polynucleotide chain. About the last 25 ribonucleotides added to a growing chain are complexed with DNA and/or enzyme at any moment.

Bacterial RNA polymerase has overall dimensions of ~90 × 95 × 160Å as an isolated protein. Yeast RNA polymerase is somewhat larger (but less elongated), with dimensions of ~140 × 136 × 110Å.

All nucleic acids are synthesized from nucleoside 5' triphosphate precursors. **Figure 14.6** shows the condensation reaction between the 5' triphosphate group of the incoming nucleotide and the 3'-OH group of the last nucleotide to have been added to the chain. The incoming nucleotide loses its terminal two phosphate groups (γ and β); its α group is used in the phosphodiester bond linking it to the chain. Thus the RNA chain is synthesized from the 5' end toward the 3' end. The overall reaction rate is ~40 nucleotides/second at 37°C (for the bacterial RNA polymerase); this is about the same as the rate of translation (15 amino acids/sec), but much slower than the rate of DNA replication (800 bp/sec).

The acceptability of an incoming nucleotide is judged by its fit with the template strand of DNA, an action apparently supervised by the enzyme. Probably the enzyme allows phosphodiester bond formation to proceed only when a complementary nucleotide matches the template base. The nucleotide is expelled if its fit is deemed inadequate; then

Figure 14.6

Phosphodiester bond formation involves a hydrophilic attack by the 3'-OH group of the last nucleotide of the chain on the 5' triphosphate of the incoming nucleotide, with release of pyrophosphate.

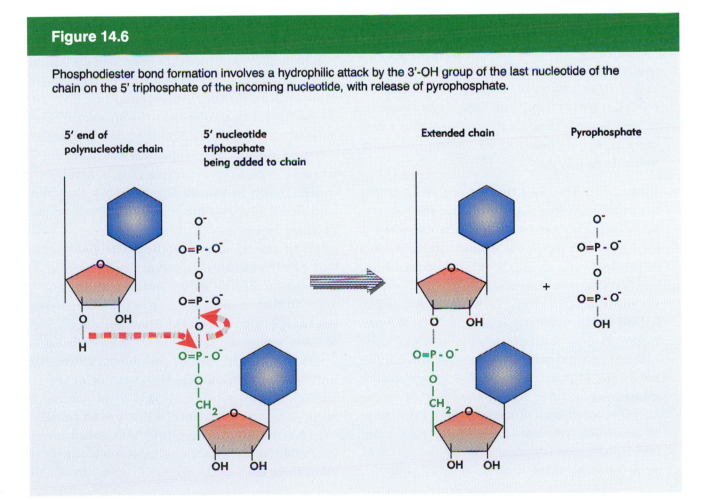

another can enter. The mechanism for discrimination is not known, but probably does not rely upon base pairing directly, because some base analogs that cannot pair are incorporated well.

The transcription reaction can be divided into the stages illustrated in **Figure 14.7**, in which a bubble is created, RNA synthesis begins, the bubble moves along the DNA, and finally is terminated:

◆ **Template recognition** begins with the binding of RNA polymerase to the double-stranded DNA. Then the strands of DNA are separated to make the template strand available for base pairing with ribonucleotides. The transcription bubble is created by a local unwinding that begins at the site bound by RNA polymerase. *The sequence of DNA needed for polymerase binding defines the* **promoter***.*

◆ **Initiation** describes the synthesis of the first nucleotide bonds in RNA. The enzyme remains at the promoter while it synthesizes the first ~9 nucleotide bonds. (The initiation phase is protracted by the occurrence of abortive events, in which the enzyme makes short transcripts (<9 bases), releases them, and then starts synthesis of RNA again.) The initiation phase ends when the enzyme succeeds in extending the chain beyond this length.

◆ **Elongation** describes the phase during which the enzyme moves along the DNA and extends the growing RNA chain. As the enzyme moves, it unwinds the DNA helix to expose a new segment of the template in single-stranded condition. Nucleotides are covalently added to the 3′ end of the growing RNA chain, forming an RNA–DNA hybrid in the unwound region. Behind the unwound region, the DNA template strand pairs with its original partner to reform the double helix. The RNA emerges as a free single strand. *Elongation involves the movement of the transcription bubble by a disruption of DNA structure, in which the template strand of the transiently unwound region is paired with the nascent RNA at the growing point.*

◆ **Termination** involves recognition of the point at which no further bases should be added to the chain. To terminate transcription, the formation of phosphodiester bonds must cease, and the transcription complex must come apart. When the last base is added to the RNA chain, the transcription bubble collapses as the RNA–DNA hybrid is disrupted, the DNA reforms in duplex state, and the enzyme and RNA are both released. *The sequence of DNA required for these reactions is called the* **terminator***.*

Originally defined simply by its ability to incorporate nucleotides into RNA under the direction of a DNA template, the enzyme RNA polymerase now is seen as part of a more complex apparatus involved in transcription. *The ability to catalyze RNA synthesis defines the minimum component that can be described as RNA polymerase.* It supervises the base pairing of the substrate ribonucleotides with DNA and catalyzes the formation of phosphodiester bonds between them.

But ancillary activities are needed to initiate and to terminate the synthesis of RNA, when the enzyme must associate with, or dissociate from, a specific site on DNA. The analogy with the division of labors between the ribosome and the protein synthesis factors is obvious. Sometimes it is difficult to decide whether a particular protein that is involved in transcription at one of these stages should be considered as part of the 'RNA polymerase' or as an ancillary factor.

All of the subunits of the basic polymerase that participate in elongation are necessary for initiation and termination. But transcription units differ in their dependence on additional polypeptides at the initiation and termination stages. Some of these additional polypeptides are needed at all genes, but others may be needed specifically for initiation or termination at particular genes. An additional polypeptide needed to recognize all promoters (or terminators) is likely to be classified as part of the enzyme. A polypeptide needed only for the initiation (or termination) at particular genes is likely to be classified as an ancillary control factor.

Figure 14.7

Transcription has four stages, which involve different types of interaction between RNA polymerase and DNA. The enzyme binds to the promoter and melts DNA, remains stationary during initiation, moves along the template during elongation, and dissociates at termination. (The change in shape of RNA polymerase between initiation and elongation indicates the involvement of different forms of the enzyme, as described in the next section.)

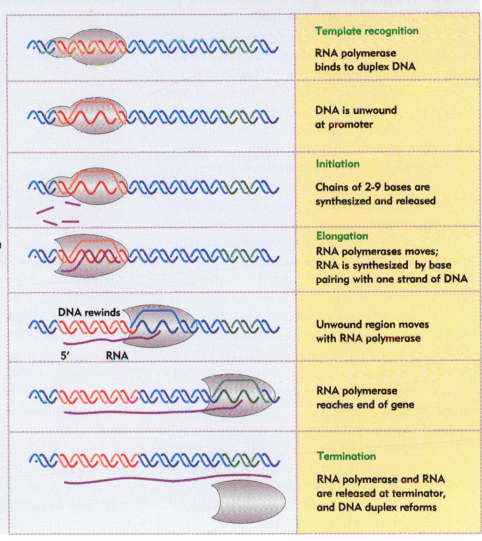

Template recognition

RNA polymerase binds to duplex DNA

DNA is unwound at promoter

Initiation

Chains of 2-9 bases are synthesized and released

Elongation

RNA polymerases moves; RNA is synthesized by base pairing with one strand of DNA

Unwound region moves with RNA polymerase

RNA polymerase reaches end of gene

Termination

RNA polymerase and RNA are released at terminator, and DNA duplex reforms

Bacterial RNA polymerase consists of core enzyme and sigma factor

The best characterized RNA polymerases are those of eubacteria, of which *E. coli* is a typical case. (The enzyme is similar in most eubacteria, but has a different structure in cyanobacteria and archae-bacteria.) *A single type of RNA polymerase appears to be responsible for almost all synthesis of mRNA, and all rRNA and tRNA, in a eubacterium.* About 7000 RNA polymerase molecules are present in an

E. coli cell. Many of them are actually engaged in transcription; probably between 2000 and 5000 enzymes are synthesizing RNA at any one time, the number depending on the growth conditions.

The **complete enzyme** or **holoenzyme** in *E. coli* has a molecular weight of ~480,000 daltons. Its subunit composition is summarized in **Table 14.1**.

The holoenzyme ($\alpha_2\beta\beta'\sigma$) can be separated into two components, the **core enzyme** ($\alpha_2\beta\beta'$) and the **sigma factor** (the σ polypeptide). The names reflect the fact that *only the holoenzyme can initiate transcription.* The sigma 'factor' is released when the RNA chain reaches 8–9 bases, after the stage of abortive initiation, leaving the core enzyme to undertake elongation. *The core enzyme has the ability to synthesize RNA on a DNA template, but cannot initiate transcription at the proper sites.*

The release of sigma is indicated by the change in the shape of the RNA polymerase at the transition from initiation to elongation in Figure 14.7. Once sigma has been released, the core enzyme begins to move along the template extending the RNA chain; the region of local unwinding moves with it (as shown previously in Figure 14.3). The unwound segment comprises only part of the total length of DNA bound by the polymerase.

The elongated dimension of the RNA polymerase protein extends along the DNA, but some interesting changes in shape occur during transcription. Transitions in shape and size identify three forms of the complex, as illustrated in **Figure 14.8**:

♦ When RNA polymerase holoenzyme initially binds to DNA, it covers some 77–80 bp, extending from −55 to +20. It remains in this location during the initiation phase, which involves RNA chain extension in the unwound region extending immediately beyond the startpoint.

♦ As soon as the enzyme loses sigma factor and makes the transition to the elongation phase, it loses its contacts in the −55 to −35 region and covers only ~60 bp of DNA. This corresponds with the concept that the more upstream part of the promoter is involved in initial recognition by RNA polymerase, but is not required for the later stages of initiation (see later).

♦ When the RNA chain extends to 15–20 bases, the enzyme makes a further transition, to form the complex that undertakes elongation; now it covers only 30 bp. It is not known whether this change involve a significant alteration in the shape of the enzyme or whether the DNA is released into a less compact form. (The long dimension of RNA polymerase (160Å) could cover ~50 bp of DNA in extended form, which implies that binding of a longer stretch of DNA must involve some bending of the nucleic acid.)

The traditional view of elongation is that it is a monotonous process, in which the enzyme moves forward 1 bp along DNA for every nucleotide added to the RNA chain. But in fact the process is much less

Table 14.1

Eubacterial RNA polymerases have four types of subunit; α, β and β' have rather constant sizes in different bacterial species, but σ varies more widely.

Subunit	Gene	Mass (daltons)	Number	Location	Possible Functions
α	rpoA	˜40,000 each	2	core enzyme	enzyme assembly
β	rpoB	˜155,000	1	core enzyme	nucleotide binding
β'	rpoC	˜160,000	1	core enzyme	template binding
σ	rpoD	32-92,000	1	sigma factor	promoter binding

Figure 14.8

RNA polymerase initially contacts the region from -55 to +20. When sigma dissociates, the core enzyme contracts beyond -30, and when the enzyme has moved a few base pairs, it becomes more compactly organized into the general elongation complex.

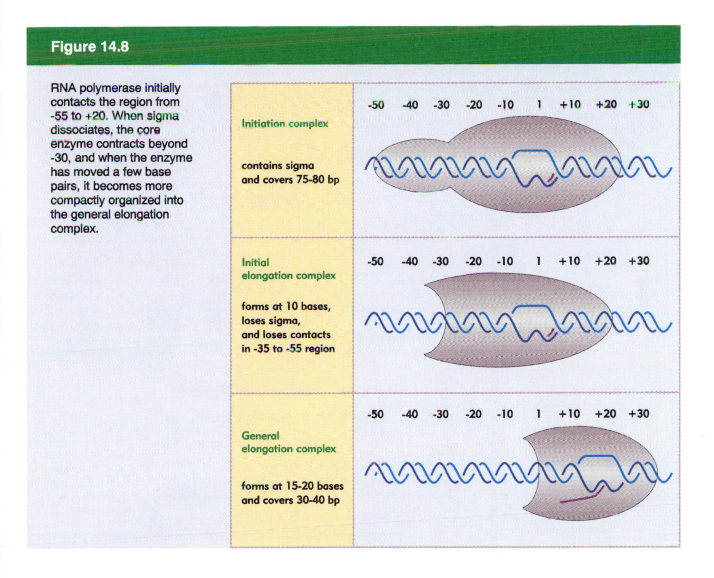

Initiation complex

contains sigma
and covers 75-80 bp

-50 -40 -30 -20 -10 1 +10 +20 +30

Initial elongation complex

forms at 10 bases,
loses sigma,
and loses contacts
in -35 to -55 region

-50 -40 -30 -20 -10 1 +10 +20 +30

General elongation complex

forms at 15-20 bases
and covers 30-40 bp

-50 -40 -30 -20 -10 1 +10 +20 +30

---

regular, and the form of the enzyme illustrated in Figure 14.8 represents only an average. During elongation, the upstream (left) boundary of the enzyme moves steadily as the RNA chain is extended, but the downstream (right) boundary remains stationary while several nucleotides are added; then suddenly it moves 7–8 bp along the DNA. **Figure 14.9** illustrates this mode of elongation, which involves steady compression and sudden expansion of the footprint on DNA; it contracts steadily from ~35 bp to ~28 bp, and then expands back to ~35 bp.

Our knowledge of the topology of the core enzyme is really very primitive, and the best we can do at present is to make a diagrammatic representation of the sites defined by the various enzymatic functions, as illustrated previously in Figure 14.4. None of these sites has yet been physically located on the polypeptide subunits. However, there is some general information about the roles of individual subunits.

Two types of antibiotic act on the β subunit, as defined by the location of mutations conferring resistance. The **rifamycins** (of which rifampicin is the most used) prevent completion of the initiation phase. They prevent addition of the third or fourth nucleotide, probably by blocking transfer of the nascent RNA to the RNA-binding site. **Streptolydigins** inhibit chain elongation. The β subunit is the target for both types of antibiotic; also

Figure 14.9

RNA polymerase moves like a bellows during elongation, when it compresses from the back end and releases from the front end.

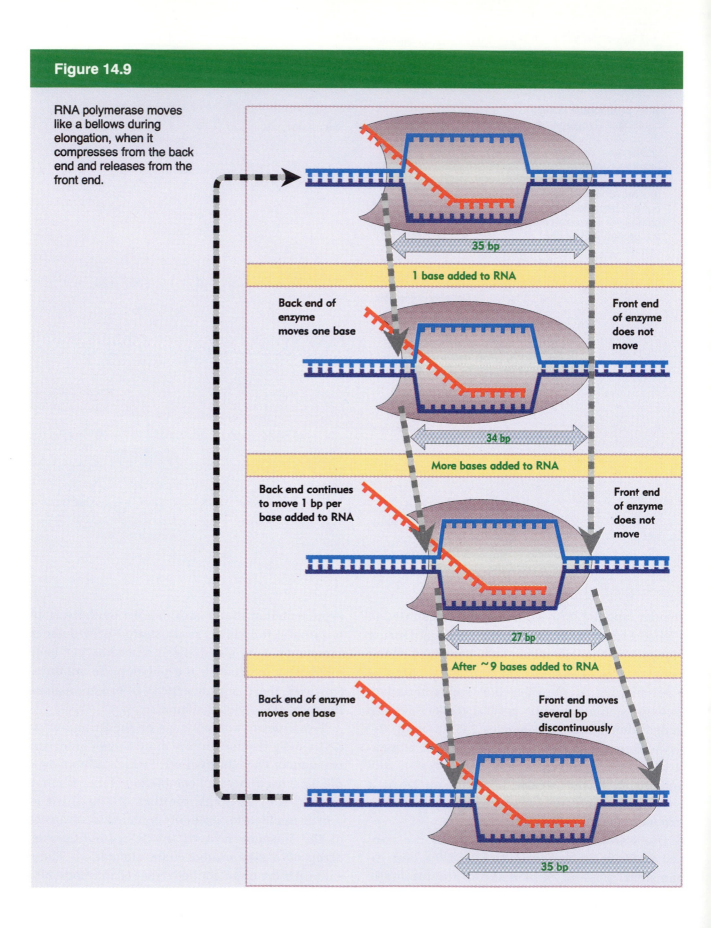

35 bp

1 base added to RNA

Back end of enzyme moves one base

Front end of enzyme does not move

34 bp

More bases added to RNA

Back end continues to move 1 bp per base added to RNA

Front end of enzyme does not move

27 bp

After ~9 bases added to RNA

Back end of enzyme moves one base

Front end moves several bp discontinuously

35 bp

it is labeled by certain affinity analogs of the nucleoside triphosphates. Together these results suggest that the β subunit is involved in binding the ribonucleotide substrates.

Heparin is a polyanion that binds to the β' subunit and inhibits transcription *in vitro*. Heparin competes with DNA for binding to the polymerase, which suggests that the β' subunit is involved in binding the template DNA. The β and β' subunits can both be crosslinked to DNA.

The α subunit is required for assembly of the core enzyme; no direct role for it has been identified in transcription *per se*. However, when phage T4 infects *E. coli*, the α subunit is modified by ADP-ribosylation of an arginine. The modification is associated with a reduced affinity for the promoters formerly recognized by the holoenzyme, so the α subunit plays a role in promoter recognition. The α subunit also plays a role in the interaction of RNA polymerase with some regulatory factors.

These assignments of individual functions are very approximate; probably each subunit contributes to the activity of the core enzyme as a whole, and we cannot compartmentalize its actions. Note that on the most critical question—the detailed nature of the sites that bind to DNA—we have little information.

Why does bacterial RNA polymerase require a large, multimeric structure? The existence of much smaller RNA polymerases, comprising single polypeptide chains coded by certain phages, demonstrates that the apparatus required for RNA synthesis can be much smaller than that of the host enzyme.

These enzymes give some idea of the 'minimum' apparatus necessary for transcription. They recognize a very few promoters on the phage DNA; and they have no ability to change the set of promoters to which they respond. Thus they are limited to the intrinsic ability to recognize a very few specific DNA-binding sequences and to synthesize RNA. How complex are they?

The RNA polymerases coded by the related phages T3 and T7 are single polypeptide chains of <100,000 daltons each. They synthesize RNA at rates of ~200 nucleotides/second at 37°C, more rapidly than bacterial RNA polymerase. The initiation reaction shows very little variation.

By contrast, the enzyme of the host bacterium can transcribe any one of many (>1000) transcription units. Some of these units are transcribed directly, with no further assistance. But many units can be transcribed only in the presence of further protein factors. Some of these factors are specific for a single transcription unit; others are involved in coordinating transcription from many units. Certain phages induce general changes in the affinity of host RNA polymerase, so that it stops recognizing host genes and instead initiates at phage promoters.

So the host enzyme requires the ability to interact with a variety of host and phage functions that modify its intrinsic transcriptional activities. The complexity of the enzyme therefore at least in part reflects its need to interact with a multiplicity of other factors, rather than any demand inherent in its catalytic activity.

Sigma factor controls binding to DNA

The function of the sigma factor is to ensure that bacterial RNA polymerase binds stably to DNA *only at promoters, not at other sites*.

The core enzyme itself has an affinity for DNA, in which electrostatic attraction between the basic protein and the acidic nucleic acid plays a major

role. Probably this general ability to bind to any DNA, irrespective of its particular sequence, is a feature of all proteins that have specific binding sites on DNA (see Chapter 15).

Any (random) sequence of DNA that is bound by RNA polymerase in this general binding reaction is described as a **loose binding site**. The complex at such a site is stable; the half-life for dissociation of the enzyme from DNA is ~60 minutes. Core enzyme does not distinguish between promoters and other sequences of DNA.

Sigma factor introduces a major change in the affinity of RNA polymerase for DNA. *The holoenzyme has a drastically reduced ability to recognize loose binding sites*—that is, to bind to any general sequence of DNA. The association constant for the reaction is reduced by a factor of ~10^4, and the half-life of the complex is <1 second. Thus sigma factor destabilizes the general binding ability very considerably.

But sigma factor also *confers the ability to recognize specific binding sites*. The holoenzyme binds to promoters very tightly, with an association constant increased from that of core enzyme by (on average) 1000 times and a half-life of several hours.

Figure 14.10

RNA polymerase passes through several steps prior to elongation. A closed binary complex is converted to an open form and then into a ternary complex.

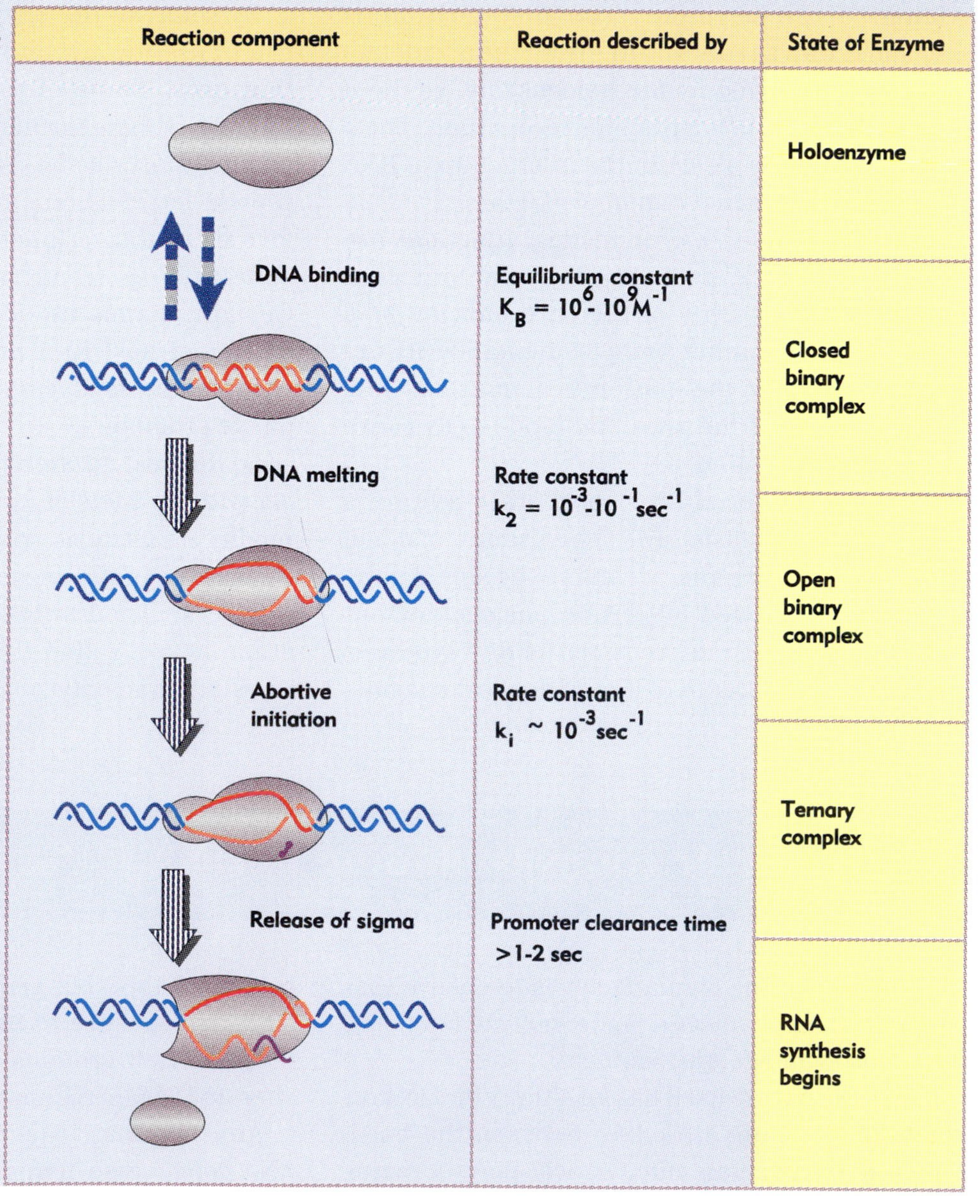

Reaction component	Reaction described by	State of Enzyme
		Holoenzyme
DNA binding	Equilibrium constant $K_B = 10^6 - 10^9 M^{-1}$	Closed binary complex
DNA melting	Rate constant $k_2 = 10^{-3} - 10^{-1} sec^{-1}$	Open binary complex
Abortive initiation	Rate constant $k_i \sim 10^{-3} sec^{-1}$	Ternary complex
Release of sigma	Promoter clearance time >1-2 sec	RNA synthesis begins

The specificity of holoenzyme for promoters compared to other sequences is therefore $\sim 10^7$, but the association constant can be quoted only as an average, because there is ~ 100 fold variation in the rate at which the holoenzyme binds to different promoter sequences; this is an important factor in determining the efficiency of an individual promoter in initiating transcription.

We are now in a position to describe the stages of transcription in terms of the interactions between different forms of RNA polymerase and the DNA template. Expanding the view that we showed previously in Figure 14.7, the initiation reaction can be described by the parameters for which typical values are summarized in **Figure 14.10**:

◆ The holoenzyme•promoter reaction starts in the same way as the loose binding reaction, by forming a closed binary complex. Because the formation of the closed binary complex is reversible, it is usually described by an equilibrium constant (K_B). There is a wide range in values of the equilibrium constant for forming the closed complex.

◆ The closed complex is converted into an **open complex** by 'melting' of a short region of DNA within the sequence bound by the enzyme. The series of events leading to formation of an open complex is called **tight binding**. Conversion into an open binary complex is in practice irreversible, so this reaction is described by a rate constant (k_2). This reaction is fast (k_2 is high).

◆ The next step is to incorporate the first two nucleotides; then a phosphodiester bond forms between them. This generates a **ternary complex** that contains RNA as well as DNA and enzyme. Formation of the ternary complex is described by the rate constant k_i; this is even faster than the rate constant k_2. Further nucleotides can be added without any enzyme movement to generate an RNA chain of up to 9 bases. After each base is added, there is a certain probability that the enzyme will release the chain. This comprises an **abortive initiation**,

after which the enzyme begins again with the first base. A cycle of abortive initiations usually occurs to generate a series of short (2–9 base) oligonucleotides.

◆ When initiation succeeds, the enzyme releases sigma and makes the transition to the elongation ternary complex of core polymerase•DNA• nascent RNA. The critical parameter here is *how long it takes for the polymerase to leave the promoter so another polymerase can initiate*. This parameter is the promoter clearance time; its minimum value of 1-2 seconds establishes the maximum frequency of initiation as <1 event per second. The enzyme then moves along the template, and the RNA chain extends beyond 10 bases.

What do these values mean for the distribution of RNA polymerase? A (somewhat speculative) picture of the enzyme's situation is depicted in **Figure 14.11**:

◆ Excess core enzyme exists largely in the form of closed loose complexes, because the enzyme enters into them rapidly and leaves them slowly. There is very little, if any, free core enzyme.

◆ There is enough sigma factor for about one third of the polymerases to exist as holoenzymes, and they are distributed between loose complexes at nonspecific sites and binary complexes (mostly closed) at promoters.

◆ About half of the RNA polymerase consists of core enzymes engaged in transcription.

◆ How much holoenzyme is free? We do not know, but we suspect that the amount is very small.

RNA polymerase must find promoters within the context of the genome. Suppose that a promoter is a stretch of ~ 60 bp; how is it picked out from the 4×10^6 bp that comprise the *E. coli* genome? **Figure 14.12** illustrates the principles of three models.

The simplest model is to suppose that RNA polymerase moves by random diffusion. Holoenzyme

Figure 14.11

Core enzyme and holoenzyme are distributed on DNA, and very little RNA polymerase is free.

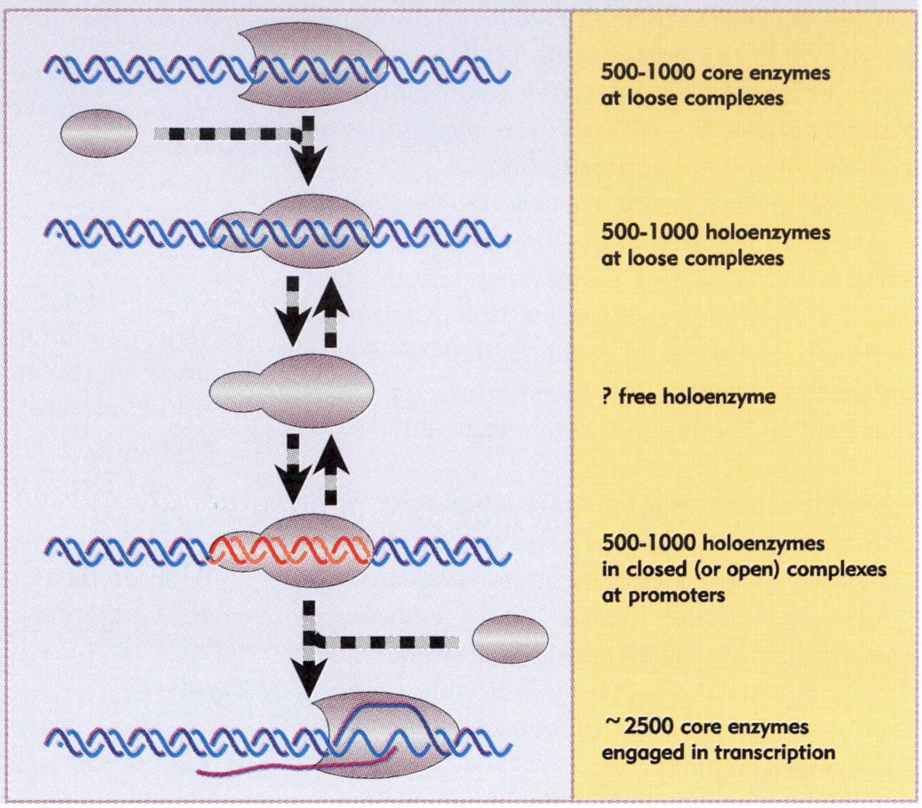

500-1000 core enzymes at loose complexes

500-1000 holoenzymes at loose complexes

? free holoenzyme

500-1000 holoenzymes in closed (or open) complexes at promoters

~2500 core enzymes engaged in transcription

very rapidly associates with, and dissociates from, loose binding sites. So it could continue to make and break a series of closed complexes in an agitated manner until (by chance) it encounters a promoter. Then its recognition of the specific sequence would allow tight binding to occur by formation of an open complex.

For RNA polymerase to move from one binding site on DNA to another, it must dissociate from the first site, find the second site, and then associate with it. Movement from one site to another is limited by the speed of diffusion through the medium. Diffusion sets an upper limit for the rate constant for associating with a 60 bp target of $<10^8$ M^{-1} sec^{-1}. But the actual forward rate constant for some promoters *in vitro* appears to be ~10^8 M^{-1} sec^{-1}, at or above the diffusion limit. *If this value applies* in vivo, *the time required for random cycles of successive associations and dissociations at loose binding sites is too great to account for*

the way RNA polymerase finds its promoter.

RNA polymerase must therefore use some other means to seek its binding sites. The process could be speeded up if the initial target for RNA polymerase is the whole genome, not just a specific promoter sequence. By increasing the target size, the rate constant for diffusion to DNA is correspondingly increased, and is no longer limiting.

If this idea is correct, a free RNA polymerase binds DNA and then remains in contact with it. How does the enzyme move from a random (loose) binding site on DNA to a promoter? The most likely model is to suppose that the bound sequence is directly displaced by another sequence. Having taken hold of DNA, the enzyme exchanges this sequence with another sequence very rapidly, and continues to exchange sequences in this promiscuous manner until a promoter is found. Then the enzyme forms a stable, open complex, after which initiation occurs. The search process becomes

Figure 14.12

How does RNA polymerase find target promoters so rapidly on DNA?

Model	Reaction	Rate constant
Random diffusion to target	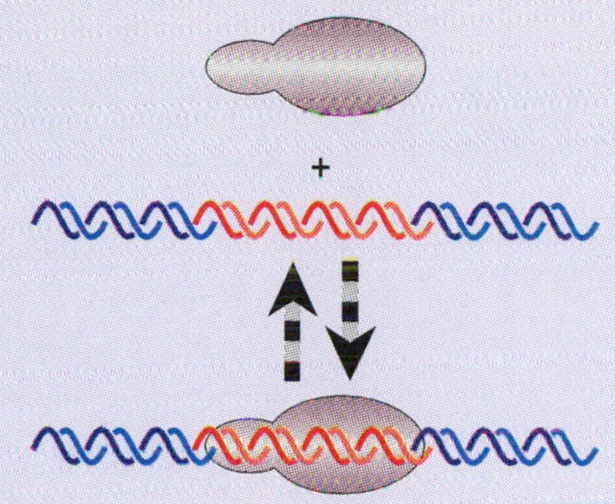	$<10^8 \ M^{-1}sec^{-1}$
Random diffusion to any DNA followed by Random displacement between DNA	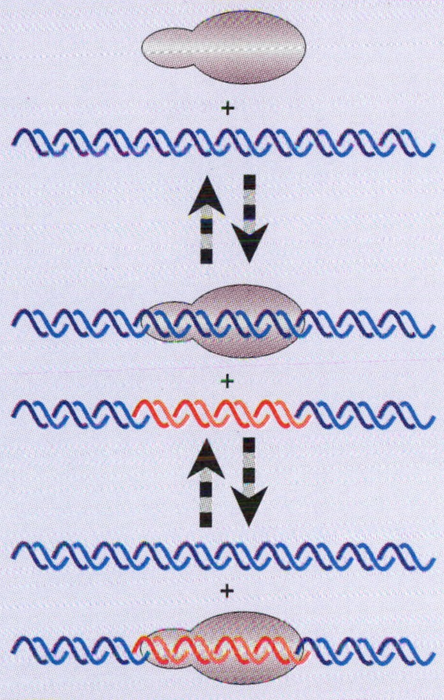	$\sim 10^{14} \ M^{-1}sec^{-1}$?
Sliding along DNA	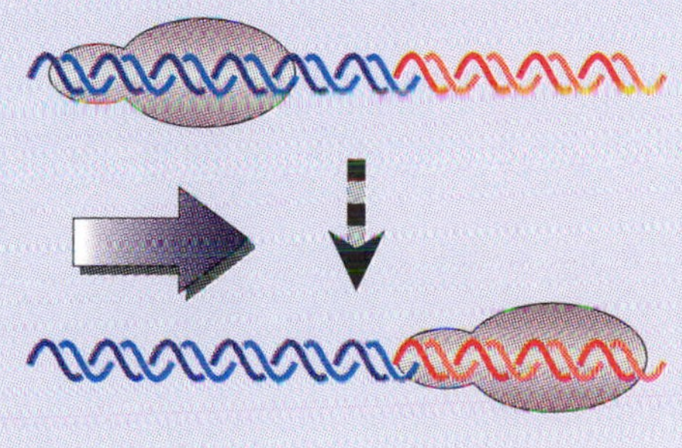	Does not occur

much faster because association and dissociation are virtually simultaneous, and time is not spent commuting between sites. Direct displacement can give a 'directed walk', in which the enzyme moves preferentially from a weak site to a stronger site.

Another idea supposes that the enzyme slides along the DNA by a one dimensional random walk, being halted only when it encounters a promoter. However, there is no evidence that RNA polymerase (or other DNA-binding proteins) can function in this manner.

RNA polymerase encounters a dilemma in re-

Figure 14.13

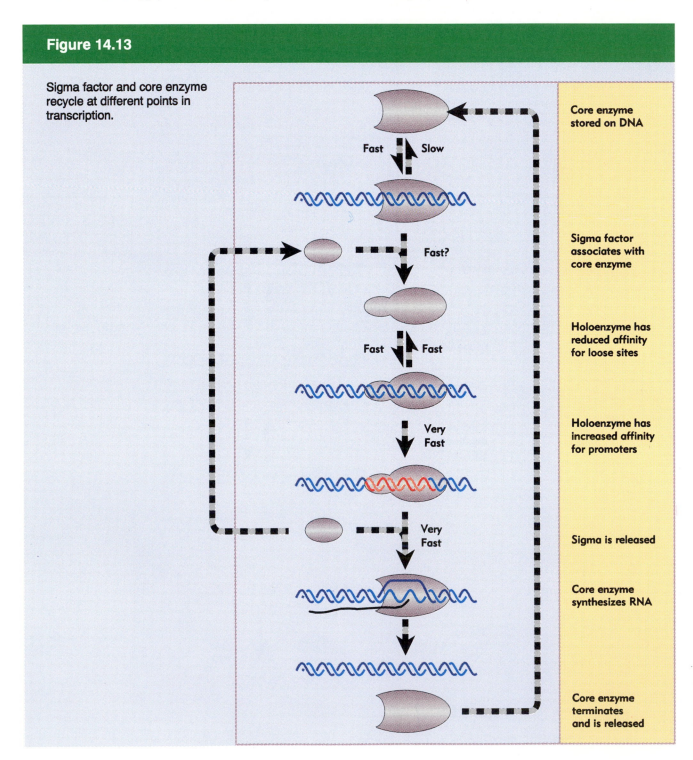

Sigma factor and core enzyme recycle at different points in transcription.

Core enzyme stored on DNA

Sigma factor associates with core enzyme

Holoenzyme has reduced affinity for loose sites

Holoenzyme has increased affinity for promoters

Sigma is released

Core enzyme synthesizes RNA

Core enzyme terminates and is released

conciling its needs for initiation with those for elongation. Initiation requires tight binding *only* to particular sequences (promoters), while elongation requires close association with *all* sequences that the enzyme encounters during transcription. **Figure 14.13** illustrates how the dilemma is solved by the reversible association between sigma factor and core enzyme.

Sigma factor is involved only in initiation. It is released from the core enzyme when abortive initiation is concluded and RNA synthesis has been successfully initiated. Free sigma factor becomes immediately available for use by another core enzyme. We do not know whether sigma is released as a *consequence* of overcoming abortive initiation, or whether instead it is the *release of sigma factor that ends abortive initiation* and allows elongation to commence.

Release of sigma leaves the core enzyme on DNA. The core enzyme in the ternary complex is bound very tightly to DNA. It is essentially 'locked in' until elongation has been completed. When transcription terminates, the core enzyme is released from DNA as a free protein tetramer. It is then 'stored' by binding to a loose site on DNA. It must find another sigma factor in order to undertake a further cycle of transcription.

Core enzyme has a high intrinsic affinity for DNA, which is increased by the presence of nascent RNA. But its affinity for loose binding sites is too high to allow the enzyme to distinguish promoters efficiently from other sequences. By reducing the stability of the loose complexes, sigma allows the process to occur much more rapidly; and by stabilizing the association at tight binding sites, the factor drives the reaction irreversibly into the formation of open complexes. To avoid becoming paralyzed by its specific affinity for the promoter, the enzyme releases sigma, and thus reverts to a general affinity for all DNA, irrespective of sequence, that suits it to continue transcription.

What is responsible for the ability of holoenzyme to bind specifically to promoters? Sigma factor has domains that recognize the promoter DNA. As an independent polypeptide, sigma does not bind to DNA, but when holoenzyme forms a tight binding complex, σ contacts the DNA in the region upstream of the startpoint. This difference is due to a change in the conformation of sigma factor when it binds to core enzyme. The N-terminal region of free sigma factor suppresses the activity of the DNA-binding region; when sigma binds to core, this inhibition is released, and it becomes able to bind specifically to promoter sequences (see also Figure 14.18 later). The inability of free sigma factor to recognize promoter sequences may be important: if σ could freely bind to promoters, it might block holoenzyme from initiating transcription. We do not know what role the core subunits play in promoter recognition.

Promoter recognition depends on consensus sequences

As sequences of DNA whose function is to be *recognized by proteins*, a promoter and any adjacent control sites differ from other sequences whose role is exerted by being transcribed or translated. The information for promoter function is provided directly by the DNA sequence: its structure is the signal. By contrast, expressed regions gain their meaning only after the information is transferred into the form of some other nucleic acid or protein.

A key question in examining the interaction between an RNA polymerase and its promoter is how the protein recognizes a specific promoter sequence. Does the enzyme have an active site that distinguishes the chemical structure of a particular sequence of bases in the DNA double helix? How specific are its requirements? Promoters vary in their affinities for RNA polymerase; this can be an important factor in controlling the frequency of initiation and thus the extent of gene expression.

One way to design a promoter would be for a particular sequence of DNA to be recognized by RNA polymerase. Every promoter would consist of, or at least include, this sequence. In the bacterial genome, the minimum length that could provide an adequate signal is 12 bp. (Any shorter sequence is likely to occur—just by chance—a sufficient number of additional times to provide false signals. The minimum length required for unique recognition increases with the size of genome.) The 12 bp sequence need not be contiguous; and, in fact, if a specific number of base pairs separates two constant shorter sequences, their combined length could be less than 12 bp, since the *distance* of separation itself provides a part of the signal (even if the intermediate *sequence* is itself irrelevant).

Attempts to identify the features in DNA that are necessary for RNA polymerase binding started by comparing the sequences of different promoters. Any essential nucleotide sequence should be present in all the promoters. Such a sequence is said to be **conserved**. However, a conserved sequence need not necessarily be conserved at every single position; some variation is permitted. How do we analyze a sequence of DNA to determine whether it is sufficiently conserved to constitute a recognizable signal?

Putative DNA recognition sites can be defined in terms of an idealized sequence that represents the base most often present at each position. A **consensus sequence** is defined by aligning all known examples so as to maximize their homology. For a sequence to be accepted as a consensus, each particular base must be reasonably predominant at its position, and most of the actual examples must be related to the consensus by rather few substitutions, say, no more than 1 or 2.

More than 100 promoters have been sequenced in *E. coli*, and a striking feature is the *lack of any extensive conservation of sequence* over the 60 bp associated with RNA polymerase. The sequence of much of the binding site is irrelevant. But some short stretches within the promoter are conserved, and they are critical for its function. *Conservation of only very short consensus sequences is a typical feature of regulatory sites (such as promoters) in both prokaryotic and eukaryotic genomes.*

There are four conserved features in a bacterial promoter: the startpoint; the −10 sequence; the −35 sequence; and the distance between the −10 and −35 sequences:

◆ The startpoint is usually (>90% of the time) a purine. It is quite common for the startpoint to be the central base in the sequence CAT, but the conservation of this triplet is not great enough to regard it as an obligatory signal.

◆ Just upstream of the startpoint, a 6 bp region is recognizable in almost all promoters. The center of the hexamer generally is close to 10 bp upstream of the startpoint; the distance varies in known promoters from position −18 to −9. Named for its location, the hexamer is often called the **−10 sequence**. Its consensus is **TATAAT**, and can be summarized in the form

$$T_{80} A_{95} t_{45} A_{60} a_{50} T_{96}$$

where the subscript denotes the percent occurrence of the most frequently found base, varying from 45 to 96%. (Capital letters are used to indicate bases conserved >54%; lower case letters are used to indicate bases not so well conserved, but nonetheless present more often than predicted from a random distribution. A position at which there is no discernible preference for any base would be indicated by N.) If the frequency of occurrence indicates likely importance in binding RNA polymerase, we would expect the initial highly conserved TA and the final almost completely conserved T in the −10 sequence to be the most important bases.

◆ Another conserved hexamer is centered ~35 bp upstream of the startpoint. This is called the **−35 sequence**. The consensus is **TTGACA**; in more detailed form, the conservation is

$$T_{82} T_{84} G_{78} A_{65} C_{54} a_{45}$$

◆ The distance separating the −35 and −10 sites is between 16 and 18 bp in 90% of promoters; in the exceptions, it is as little as 15 or as great as 20 bp. *Although the actual sequence in the intervening region is unimportant, the distance is critical in holding the two sites at the appropriate separation for the geometry of RNA polymerase.*

A major source of information about promoter function is provided by mutations. Mutations in promoters affect the level of expression of the gene(s) they control, without altering the gene products themselves. Most are identified in the form of bacterial mutants that have lost, or have very much reduced, transcription of the adjacent genes. They are known as **down mutations**. Less often, mutants are found in which there is increased transcription from the promoter. They are called **up mutations**.

It is important to remember that 'up' and 'down' mutations are defined relative to the *usual* efficiency with which a particular promoter functions. This varies widely. So a change that is recognized as a down mutation in one promoter might never have been isolated in another (which in its wild-type state could be even less efficient than the mutant form of the first promoter). Thus information gained from studies *in vivo* simply identifies the overall direction of the change caused by mutation.

Almost all of the point mutations that affect promoter function fall within the two consensus sequences. Occasionally a mutation is found just upstream of either consensus sequence (see Figure 14.16). The bases present at other positions in the vicinity are clearly much less important or even irrelevant in most promoters. In addition to these mutations, deletions or insertions between the consensus sequences may alter their separation.

From data collected on many promoters, *we can define the optimal promoter as a sequence consisting of the −35 hexamer, separated by 17 bp from the −10 hexamer, lying 7 bp upstream of the startpoint.* The structure of a promoter, showing the permitted range of variation from this optimum, is illustrated in **Figure 14.14**.

Is the most effective promoter one that has the consensus sequences themselves? This expectation

is borne out by the simple rule that up mutations usually increase homology with one of the consensus sequences or bring the distance between them closer to 17 bp. Down mutations usually decrease the resemblance of either site with the consensus or make the distance between them more distant from 17 bp. Down mutations tend to be concentrated in the most highly conserved positions, which confirms their particular importance as the main determinant of promoter efficiency.

Occasional exceptions to these rules demonstrate that promoter efficiency cannot be predicted entirely from conformity to the consensus. Virtually all actual promoters vary from the consensus, so the neighbors of any particular base differ from promoter to promoter, even if the base itself is conserved. We cannot predict the effects of context; it is possible that a nonconsensus base at some position can function effectively in one promoter but not in another. It is also possible that an important feature is the *exclusion* of some base from a particular position, rather than the presence of one particular base.

To determine the absolute effects of promoter mutations, we must measure the affinity of RNA polymerase for wild-type and mutant promoters *in vitro*. There is ~100 fold variation in the rate at which RNA polymerase binds to different promoters *in vitro,* which correlates well with the frequencies of transcription when their genes are expressed *in vivo*. Taking this analysis further, we can investigate the stage at which a mutation influences the capacity of the promoter. Does it change the affinity of the promoter for binding RNA

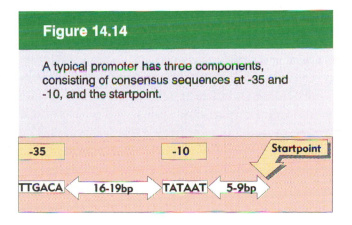

Figure 14.14

A typical promoter has three components, consisting of consensus sequences at -35 and -10, and the startpoint.

polymerase? Does it leave the enzyme able to bind but unable to initiate? Is the influence of an ancillary factor altered?

By measuring the kinetic constants for formation of a closed complex and its conversion to an open complex, as defined in Figure 14.10, we can dissect the two stages of the initiation reaction:

◆ Down mutations in the −35 sequence reduce the rate of closed complex formation (they reduce K_B), but do not inhibit the conversion to an open complex.

◆ Down mutations in the −10 sequence do not slow the initial formation of a closed complex, but they slow its conversion to the open form (they reduce k_2).

These results suggest that *the function of the −35 sequence is to provide the signal for recognition by RNA polymerase, while the −10 sequence allows the complex to convert from closed to open form*. We might view the −35 sequence as comprising a 'recognition domain', while the −10 sequence comprises an 'unwinding domain' of the promoter.

The consensus sequence of the −10 site consists exclusively of A•T base pairs, which assists the initial melting of DNA into single strands. The lower energy needed to disrupt A•T pairs compared with G•C pairs means that a stretch of A•T pairs demands the minimum amount of energy for strand separation.

The sequence immediately around the startpoint influences the initiation event. And the initial transcribed region (from +1 to +30) influences the rate at which RNA polymerase clears the promoter, and therefore has an effect upon promoter strength. So the overall strength of a promoter cannot be predicted entirely from its −35 and −10 consensus sequences.

A 'typical' promoter relies upon its −35 and −10 sequences to be recognized by RNA polymerase, but one or the other of these sequences can be absent from some (exceptional) promoters. In at least some of these cases, the promoter cannot be recognized by RNA polymerase alone, and the reaction requires the intercession of ancillary proteins. Possibly their reaction with adjacent sequences overcomes the deficiency in the promoter. We do not know how many alternatives there are to the usual promoter elements, nor exactly how their interaction with other factors substitutes for the absence of the consensus sequences.

RNA polymerase binds to one face of DNA

The ability of RNA polymerase (or indeed any protein) to recognize DNA can be characterized by **footprinting**. A sequence of DNA bound to the protein is *partially* digested with an endonuclease, an enzyme that attacks individual phosphodiester bonds *within* a nucleic acid. Under appropriate conditions, any particular phosphodiester bond is broken in some, but not in all, DNA molecules. The positions that are cleaved are recognized by using DNA labeled on one strand at one end only. The principle is the same as that involved in DNA sequencing; partial cleavage of an end-labeled molecule at a susceptible site creates a fragment of unique length.

As **Figure 14.15** shows, following the nuclease treatment, the broken DNA fragments are recovered and electrophoresed on a gel that separates them according to length. Each fragment that retains a labeled end produces a radioactive band. The position of the band corresponds to the number of bases in the fragment. The shortest fragments move the fastest, so distance from the labeled end is counted up from the bottom of the gel (see Figure 6.13).

Figure 14.15

Footprinting identifies DNA-binding sites for proteins by their protection against nicking.

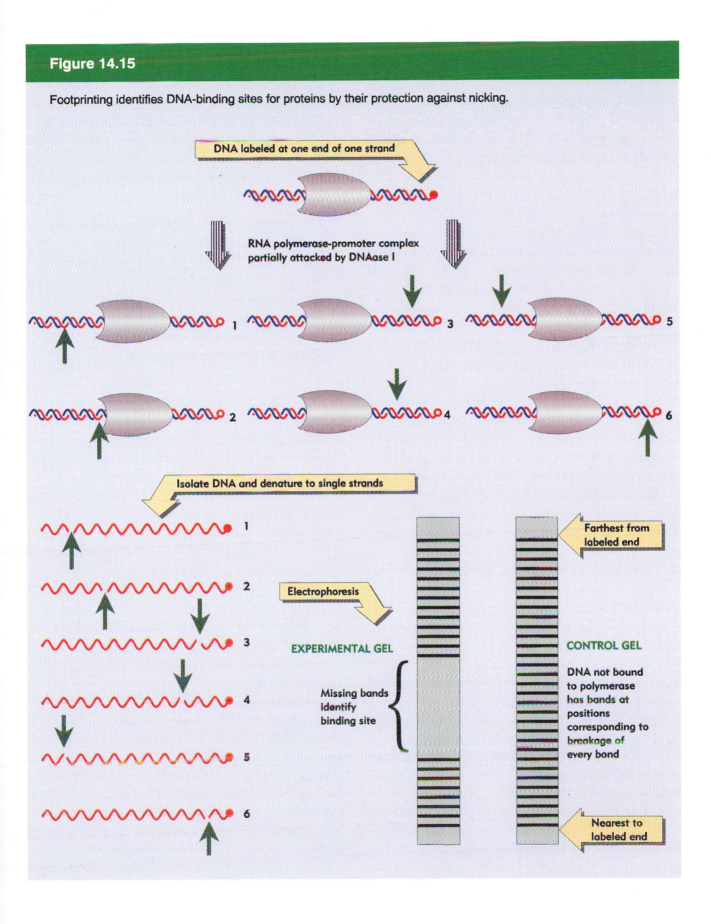

DNA labeled at one end of one strand

RNA polymerase-promoter complex partially attacked by DNAase I

1 3 5

2 4 6

Isolate DNA and denature to single strands

1
2
3
4
5
6

Electrophoresis

EXPERIMENTAL GEL

Missing bands identify binding site

CONTROL GEL

Farthest from labeled end

DNA not bound to polymerase has bands at positions corresponding to breakage of every bond

Nearest to labeled end

In a free DNA, *every* susceptible bond position is broken in one or another molecule. But when the DNA is complexed with a protein, the region covered by the DNA-binding protein is protected in every molecule. So two reactions are run in parallel: a control of pure DNA; and an experimental mixture containing molecules of DNA bound to the protein. *Only if a bound protein blocks access of the nuclease to DNA will a particular bond fail to be broken in the experimental mixture.*

In the control, every bond is broken, generating a series of bands, one representing each base. In the figure, 31 bands can be counted. In the protected fragment, bonds cannot be broken in the region bound by the protein, so bands representing fragments of the corresponding sizes are not generated. The absence of bands 9–18 in the figure identifies a protein-binding site covering the region located 9–18 bases from the labeled end of the DNA. By comparing the control and experimental lanes with a sequencing reaction that is run in parallel (see Figure 6.14) it becomes possible to 'read off' the corresponding sequence directly, thus identifying the nucleotide sequence of the binding site.

Analysis of promoters bound by RNA polymerase shows that the binding site extends farther upstream than downstream, relative to the startpoint, covering the region from about −50 to +20. This stretch includes all the sequences needed for binding and initiation.

The two strands of DNA are not equally well protected in all regions of the promoter, especially at the ends of the binding site. This implies that RNA polymerase has an asymmetric conformation when bound to DNA, which accords with the need to transcribe only one strand of DNA.

The points at which RNA polymerase contacts the promoter can be identified by treating RNA polymerase–promoter complexes with reagents that modify particular bases. The common feature of all the types of modification is that *they allow a breakage to be made at the corresponding bond in the polynucleotide chain.* The site of breakage can be identified by the same approach used in footprinting with endonucleases (see Figure 14.15). We can perform the experiment in two ways:

◆ The direct analogy with footprinting is to treat an RNA polymerase–DNA complex with a modifying agent and to compare its susceptibility with that of free DNA. Some bands disappear, identifying sites at which the enzyme has protected the promoter against modification. Other bands increase in intensity, identifying sites at which the DNA must be held in a conformation in which it is more exposed.

◆ The reverse experiment can be performed by modifying the DNA *first*; then it is bound to RNA polymerase. Those DNA molecules that cannot bind RNA polymerase are recovered and treated in the usual way to generate strand breakages whose positions can be identified. This locates points at which prior modification *prevents* RNA polymerase from binding to DNA.

These changes in sensitivity reveal the geometry of the complex, as summarized in **Figure 14.16** for a typical promoter, with consensus sequences at −35 and −10, and a region for initial unwinding that extends from within the −10 sequence to just past the startpoint. The regions at −35 and −10 contain most of the contact points for the enzyme. Within these regions, the same sets of positions tend both to prevent binding if previously modified, and to show increased or decreased susceptibility to modification after binding. Although the points of contact do not coincide completely with sites of mutation, they occur in the same limited region. At both consensus sites, the region of contact extends for 12–15 bp, somewhat longer than the conserved region.

It is noteworthy that the same *positions* in different promoters provide the contact points, even though a different base is present. This indicates that there is a common mechanism for RNA polymerase binding, although the reaction does not depend on the presence of particular bases at some of the points of contact. This model explains why some of the points of contact are not sites of mutation. Also, not every mutation lies in a point of contact; could some influence the neighborhood without actually being touched by the enzyme?

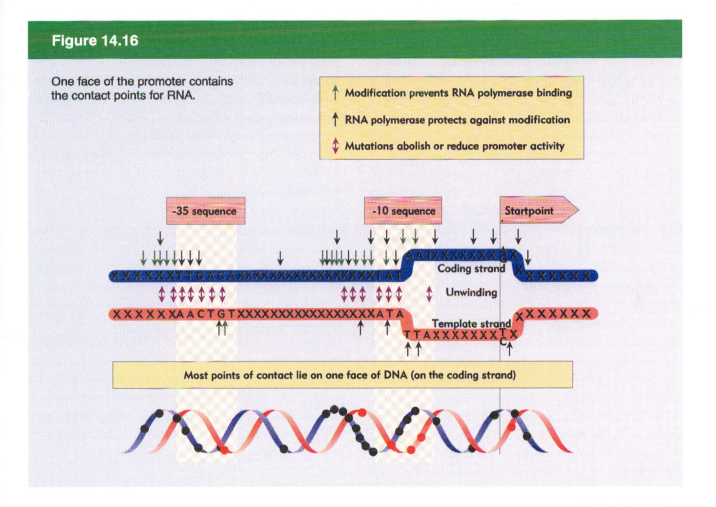

Figure 14.16

One face of the promoter contains the contact points for RNA.

↑ Modification prevents RNA polymerase binding

↑ RNA polymerase protects against modification

↕ Mutations abolish or reduce promoter activity

-35 sequence

-10 sequence

Startpoint

Coding strand

Unwinding

Template strand

Most points of contact lie on one face of DNA (on the coding strand)

It is especially significant that the experiments with prior modification identify *only* sites in the same region that is protected by the enzyme against subsequent modification. These two experiments measure different things. Prior modification identifies all those sites that the enzyme must recognize in order to bind to DNA. Protection experiments recognize all those sites that actually make contact in the binary complex. The protected sites include all the recognition sites and also some additional positions, which suggests that the enzyme first recognizes a set of bases necessary for it to 'touch down', and then extends its points of contact to additional bases.

The region of DNA that is unwound in the binary complex can be identified directly by chemical changes in its availability. When the strands of DNA are separated, the unpaired bases become susceptible to reagents that cannot reach them in the double helix. The susceptibility of sites in the RNA polymerase–DNA binary complex therefore indicates that they lie in an unpaired region. Experiments using methylation of adenine or cytosine have implicated positions between −9 and +3 in the initial melting reaction. The region unwound during initiation therefore includes the right end of the −10 sequence and extends just past the startpoint.

Viewed in three dimensions, the points of contact upstream of the −10 sequence all lie on one face of DNA. This can be seen in the lower drawing in Figure 14.16, in which the contact points lie on one face when the DNA is drawn as a double helix viewed from one side. Most lie on the coding strand (that is, not on the template strand). These bases could be recognized in the initial formation of a closed binary complex. This would make it possible for RNA polymerase to approach DNA from one side and recognize that face of the DNA. As DNA unwinding commences, further sites that originally lay on the other face of DNA might be recognized and bound.

The importance of strand separation in the initiation reaction is emphasized by the effects of supercoiling. Both prokaryotic and eukaryotic RNA polymerases can initiate transcription more efficiently *in vitro* when the template is supercoiled, presumably because the supercoiled structure requires less free energy for the initial melting of DNA in the initiation complex.

The involvement of this effect in controlling promoter activity in bacteria is shown by the effects of interfering with enzymes that influence the degree of supercoiling. Among the relevant enzymes are DNA gyrase, which *introduces* negative supercoils, and topoisomerase I, which *relaxes* (removes) negative supercoils (see Chapter 33).

Inhibitors of these enzymes or mutations in their genes affect transcription. The nature of these effects is not entirely clear; bacteria evidently endeavor to set the degree of supercoiling between certain limits, because mutations in one enzyme that alter the level are compensated by mutations in another to restore the balance. For example, *topA* (topoisomerase I) mutants are only viable if they acquire compensatory mutations that reduce the level of gyrase.

The efficiency of some promoters is influenced by the degree of supercoiling. The most common relationship is for transcription to be aided by negative supercoiling. We understand in principle how this assists the initiation reaction. But why should some promoters be influenced by the extent of supercoiling while others are not? One possibility is that every promoter has a characteristic dependence on supercoiling, determined by its sequence. This would predict that some promoters have sequences that are easier to melt (and are therefore less dependent on supercoiling), while others have more difficult sequences (and have a greater need to be supercoiled). An alternative is that the location of the promoter might be important, if different regions of the bacterial chromosome have different degrees of supercoiling.

A yet more difficult question is why some promoters display the opposite effect: they are transcribed less effectively when supercoiled. Whether this depends solely on an interaction between RNA polymerase and the promoter or is influenced by extraneous features—for example, competition with some protein that recognizes supercoiled DNA—is unknown.

Supercoiling also has a continuing involvement with transcription. As RNA polymerase transcribes DNA, unwinding and rewinding occurs, as illustrated in Figure 14.3. This requires that either the entire transcription complex rotates about the DNA or the DNA itself must rotate about its helical axis. The consequences of the rotation of DNA are illustrated in **Figure 14.17** in the twin supercoiled-domain model for transcription. As RNA polymerase pushes forward along the double helix, it generates positive supercoils (more tightly wound DNA) ahead and leaves negative supercoils (partially unwound DNA) behind. For each helical turn

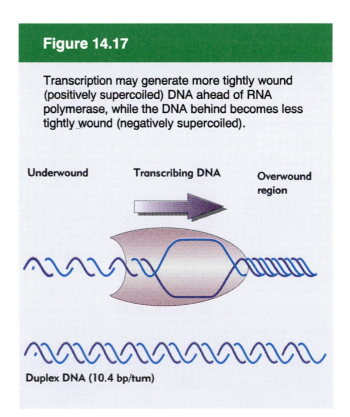

Figure 14.17

Transcription may generate more tightly wound (positively supercoiled) DNA ahead of RNA polymerase, while the DNA behind becomes less tightly wound (negatively supercoiled).

Underwound   Transcribing DNA   Overwound region

Duplex DNA (10.4 bp/turn)

traversed by RNA polymerase, +1 turn is generated ahead and −1 turn behind.

Transcription therefore has a significant effect on the (local) structure of DNA. As a result, the enzymes gyrase and topoisomerase I are required to rectify the situation in front of and behind the polymerase, respectively. If the activities of gyrase and topoisomerase are interfered with, transcription causes major changes in the supercoiling of DNA. For example, in yeast lacking an enzyme that relaxes negative supercoils, the density of negative supercoiling doubles in a transcribed region. A possible implication of these results is that transcription actually is responsible for generating a significant proportion of the supercoiling that occurs in the cell.

A similar situation occurs in replication, when DNA must be unwound at a moving replication fork, so that the individual single strands can be used as templates to synthesize daughter strands. Solutions for the topological constraints associated with such reactions are indicated later in Figure 33.13.

Substitution of sigma factors may control initiation

E. coli RNA polymerase transcribes all the bacterial genes, and must therefore recognize a wide spectrum of promoters, whose associated genes are transcribed at different levels and on different occasions. The ability to initiate at particular promoters is controlled by a seemingly endless series of ancillary factors, assisting or interfering with the enzyme.

In almost none of these instances is any change made in the subunits of the core enzyme itself. Presumably such a mechanism is not favored, because a change that allowed the RNA polymerase specifically to recognize one type of promoter might prevent it from finding another. This would reduce the flexibility of the enzyme.

Yet the division of labors between a core enzyme that undertakes chain elongation and a sigma factor involved in site selection immediately raises the question of whether there is more than one type of sigma, each specific for a different class of promoters.

Changes in sigma factors appear in some cases when there is a wholesale reorganization of transcription, for example, during the change in lifestyle that occurs in a sporulating bacterium (see the next section). *E. coli* does not undergo such dramatic changes, although it does undergo quite complex metabolic changes when it enters a

Table 14.2

E. coli sigma factors recognize promoters with different consensus sequences.

Gene	Mass	Use	-35 Sequence	Separation	-10 Sequence
rpoD	70,000	General	TTGACA	16-18 bp	TATAAT
rpoH	32,000	Heat shock	CCCTTGAA	13-15 bp	CCCGATNT
rpoN	54,000	Nitrogen	CTGGNA	6 bp	TTGCA
fliA	28,000	Flagellar	CTAAA	15 bp	GCCGATAA

stationary phase. For a long time, *E. coli* was thought to have only a single sigma factor, σ^{70}. Now others have been discovered; the best characterized are listed in **Table 14.2**. (They are named either by molecular weight of the product or for the gene.) The alternative sigma factors σ^{32} (also known as σ^{H}) and σ^{54} (or σ^{N}) are activated in response to environmental changes; σ^{28} (or σ^{F}) is used for expression of flagellar genes during normal growth, but its level of expression responds to changes in the environment.

The closest analogy to a sweeping general change in transcription in *E. coli* occurs as a response to 'heat shock', when the bacteria change their transcription pattern as the result of an increase in temperature. A common type of response to heat shock occurs in many organisms, prokaryotic and eukaryotic. Upon an increase in temperature, synthesis of the proteins currently being made is turned off or down, and a new set of proteins is synthesized. The new proteins are the products of the **heat shock genes**. They play a role in protecting the cell against environmental stress, and are synthesized in response to other conditions as well as heat shock. Several of the heat shock proteins are chaperones. In *E. coli*, the expression of 17 heat shock proteins is triggered by changes at transcription. The gene *rpoH* is a regulator needed to switch on the heat shock response. Its product is σ^{32}, which functions as an alternative sigma factor that causes transcription of the heat shock genes.

Another sigma factor is used under conditions of nitrogen starvation. *E. coli* cells contain a small amount of the protein now known as σ^{54}, which is employed when ammonia is absent from the medium. In these conditions, genes are turned on to allow utilization of alternative nitrogen sources. Counterparts to this sigma factor have been found in a wide range of bacteria, so it represents a response mechanism that has been conserved in evolution.

Another case of evolutionary conservation of sigma factors is presented by the factor, σ^{F}, which is present in small amounts and causes RNA polymerase to transcribe genes involved in chemotaxis and flagellar structure. Its counterpart in *B. subtilis* is σ^{D}, which controls flagellar and motility genes;

factors with the same promoter specificity are present in many species of bacteria.

The existence of multiple sigma factors enables us to ask some further questions about the role of sigma vis à vis the core enzyme. What is responsible for controlling the association of sigma factor with core enzyme; is this process regulated to determine which type of holoenzyme is available? And what role does the sigma factor play in controlling the sequence specificity of promoter recognition?

We do not know very much about the control of the relationship between sigma factors in growing bacteria. In the case of response to environmental stress (heat shock or nitrogen starvation), it is the *de novo* production of the new sigma factor that permits its use. We assume that selection of the genes transcribed during heat shock depends on a balance between the general σ^{70} and the heat shock σ^{32}. The two sigma factors can compete for the available core enzyme. The σ^{32} protein is unstable, which should allow its quantity to be increased or decreased rapidly. Its instability is an important feature of how it is regulated.

Each of the sigma factors causes RNA polymerase to initiate at a particular set of promoters. By analyzing the sequences of these promoters, we can show that each set is identified by unique sequence elements. Indeed, the sequence of each type of promoter ensures that it is recognized only by RNA polymerase directed by the appropriate sigma factor. We can deduce the general rules for promoter recognition from the identification of the genes responding to the sigma factors found in *E. coli* and those involved in sporulation in *B. subtilis* (which are discussed in the next section).

A significant feature of the promoters for each enzyme is that *they have the same size and location relative to the startpoint, and they show conserved sequences only around the usual centers of −35 and −10*. The consensus sequences for each set of promoters are different from one another at either or both of the −35 and −10 positions; as a result, an enzyme containing a particular sigma factor can recognize only its own set of promoters, so that transcription of the different groups is mutually exclusive. Thus substitution of one sigma factor by

Figure 14.18

A map of the *E. coli* σ⁷⁰ factor identifies four regions that are conserved in other sigma factors. Sequences within regions 2 & 4 contact the -10 and -35 promoter elements. Another part of region 2 is needed to melt DNA. Residues 361-390 are required for binding to core enzyme. The N-terminal region prevents regions 2 & 4 from binding to DNA in the absence of core enzyme.

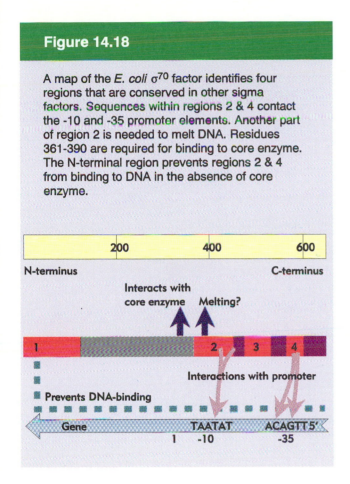

problem of how a small polypeptide can apparently contact sites spanning more than 20 bp of DNA.

An interesting difference in behavior is found with the σ^{54} factor. This causes RNA polymerase to recognize promoters that have a distinct consensus sequence, with a conserved element at -10 and another close by at -20 (given in the '-35' column of Table 14.2). Thus the geometry of the polymerase–promoter complex is different under the direction of this sigma factor. This is associated with a change in the pattern of regulation, in which sites that are rather distant from the promoter influence its activity, unlike the control of other promoters, where the regulator sites are always in close proximity to the promoter itself. In this regard, σ^{54} is more like the eukaryotic regulators we discuss in Chapter 29 than the typical prokaryotic regulators discussed in Chapters 15–17.

Comparisons of the sequences of several bacterial sigma factors identifies regions that have been conserved. We expect that they will be functionally important. There are 4 such regions, and their locations in *E. coli* σ^{70} are summarized in **Figure 14.18**.

Direct evidence that sigma contacts the promoter directly at both the -35 and -10 consensus sequences is provided by mutations in sigma that suppress mutations in the consensus sequences. When a mutation at a particular position in the promoter prevents recognition by RNA polymerase, and a compensating mutation in sigma factor allows the polymerase to use the mutant promoter, we may conclude that the relevant base pair in DNA is contacted by the amino acid that has been substituted. Figure 14.18 shows that two short parts of regions 2 and 4 (named 2.4 and 4.2) are involved in contacting bases in the -10 and -35 elements, respectively. Both of these regions form short stretches of α-helix in the protein.

Other parts of sigma have been less firmly identified with individual functions. Some parts of region 2 have sequences resembling those of proteins that bind single-stranded nucleic acids, and might therefore be involved in the melting reaction. A sequence preceding and overlapping with the start of region 2 seems to be necessary for the interaction with core enzyme.

another turns off transcription of the old set of genes as well as turning on transcription of a new set of genes.

The definition of a series of different consensus sequences recognized at -35 and -10 by holoenzymes containing different sigma factors carries the immediate implication that the sigma subunit must itself contact DNA in these regions. When only a single sigma factor had been identified, it was thought that it might function indirectly, by controlling the conformation of core enzyme. But it is implausible to suppose that sigma could cause core enzyme to take up any one of several conformations, each with specificity for a different promoter sequence. This suggests the general principle that there is a common type of relationship between sigma and core enzyme, in which the sigma factor is positioned in such a way as to make critical contacts with the promoter sequences in the vicinity of -35 and -10. However, we have yet to solve the

When the N-terminal region of σ^{70} is removed, the shortened protein becomes able to bind specifically to promoter sequences, which the intact protein cannot do. This suggests that the N-terminal region occludes the DNA-binding domains when σ^{70} is free, and that association with core enzyme changes the conformation of sigma so that it becomes able to bind to DNA.

The use of α-helical motifs in proteins to recognize duplex DNA sequences is common, as we see in more detail in Chapters 15 and 17. **Figure 14.19** shows diagrammatically how amino acids separated by 3–4 positions in an α-helix might lie on the same face of the helix and therefore be in a position to contact adjacent base pairs. In the case of the major sigma factors of *E. coli* and *B. subtilis* (σ^{70} and σ^{43}, respectively), the amino acids threonine and arginine are found at corresponding positions in the helix, and are responsible for contacting the first two positions of the -10 sequence. It is not clear how the rest of -10 sequence is contacted. Indeed, the extent of the contacts between sigma and DNA, and their relationship to contacts between core enzyme and DNA, remain to be fully established.

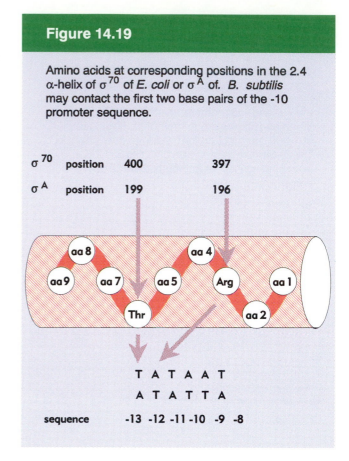

Figure 14.19

Amino acids at corresponding positions in the 2.4 α-helix of σ^{70} of *E. coli* or σ^A of *B. subtilis* may contact the first two base pairs of the -10 promoter sequence.

σ^{70} position   400   397
σ^A position   199   196

T A T A A T
A T A T T A

sequence   -13 -12 -11 -10 -9 -8

Sporulation utilizes a cascade of many sigma factors

Sigma factors are used more extensively to control initiation in the bacterium *B. subtilis*, where ~10 different factors are known. Some are present in vegetative cells; others are produced only in the special circumstances of phage infection or the change from vegetative growth to sporulation.

The major RNA polymerase found in *B. subtilis* cells engaged in normal vegetative growth has the same structure as that of *E. coli*, $\alpha_2\beta\beta'\sigma$. We have mentioned that the sigma factor has a mass of 43,000 daltons and is described as σ^{43} or σ^A. It recognizes promoters with the same consensus sequences used by the *E. coli* enzyme under direction from σ^{70}. Several variants of the enzyme that contain other sigma factors are found in much smaller amounts. The variant enzymes recognize different promoters on the basis of consensus sequences at -35 and -10.

Transitions from expression of one set of genes to expression of another set are a common feature of bacteriophage infection. In all but the very simplest cases, the development of the phage involves shifts in the pattern of transcription during the infective cycle. These shifts may be accomplished by the synthesis of a phage-encoded RNA polymerase or by the efforts of phage-encoded ancillary factors (including new sigma species) that control the bacterial RNA polymerase. A well characterized example of control via the production of new sigma factors occurs during infection of *B. subtilis* by phage SPO1.

The infective cycle of SPO1 passes through three stages of gene expression. Immediately on infection, the **early** genes of the phage are transcribed. After 4–5 minutes, the early genes cease transcription and the **middle** genes are transcribed. Then at 8–12 minutes, middle gene transcription is replaced by transcription of **late** genes.

The early genes are transcribed by the holo-enzyme of the host bacterium. They are essentially indistinguishable from host genes whose promoters have the intrinsic ability to be recognized by the RNA polymerase $\alpha_2\beta\beta' \sigma^{43}$.

Expression of phage genes is required for the transitions to middle and late gene transcription. Three regulatory genes, named *28, 33,* and *34*, control the course of transcription. Their functions are summarized in **Figure 14.20**. The pattern of regulation creates a **cascade**, in which the host enzyme transcribes an early gene whose product is needed

Figure 14.20

Transcription of phage SPO1 genes is controlled by two successive substitutions of the sigma factor that change the initiation specificity.

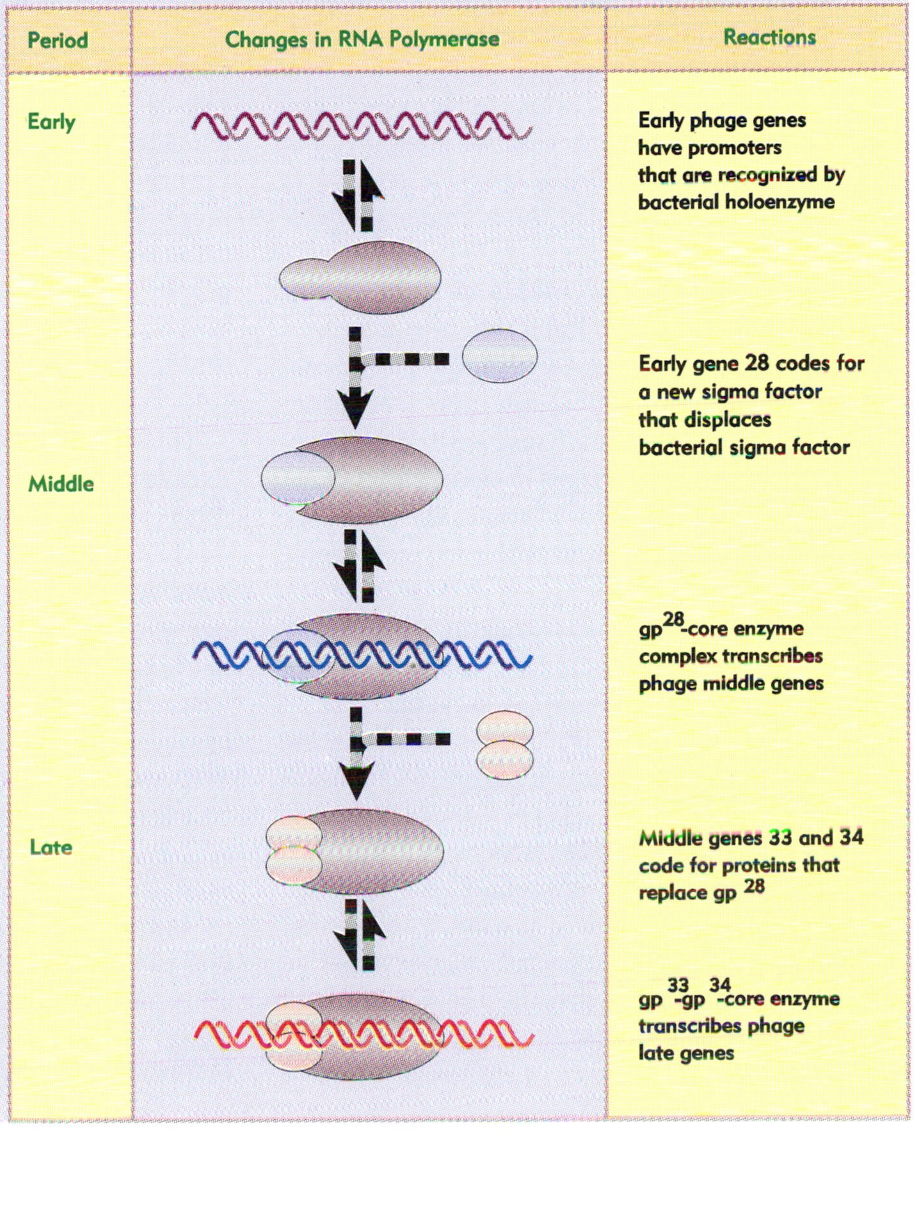

Period	Changes in RNA Polymerase	Reactions
Early		Early phage genes have promoters that are recognized by bacterial holoenzyme
		Early gene 28 codes for a new sigma factor that displaces bacterial sigma factor
Middle		gp^{28}-core enzyme complex transcribes phage middle genes
Late		Middle genes 33 and 34 code for proteins that replace gp 28
		gp^{33}-gp^{34}-core enzyme transcribes phage late genes

to transcribe the middle genes; and then two of the middle genes code for products that are needed to transcribe the late genes.

Mutants in the early gene *28* cannot transcribe the middle genes. The product of gene *28* (called gp28) is a protein of 26,000 daltons that replaces the host sigma factor on the core enzyme. *This substitution is the sole event required to make the transition from early to middle gene expression.* It creates a holoenzyme that can no longer transcribe the host genes, but instead specifically transcribes the middle genes. We do not know how gp28 displaces σ^{43}, or what happens to the host sigma polypeptide.

Two of the middle genes are involved in the next transition. Mutations in either gene *33* or *34* prevent transcription of the late genes. The products of these genes are proteins of 13,000 and 24,000 daltons, respectively, that replace gp28 on the core polymerase. Again, we do not know how gp33 and gp34 exclude gp28 (or any residual host σ^{43}), *but once they have bound to the core enzyme, it is able to initiate transcription only at the promoters for late genes.*

The successive replacements of sigma factor have dual consequences. Each time the subunit is changed, the RNA polymerase becomes able to recognize a new class of genes, *and* no longer recognizes the previous class. These switches therefore constitute global changes in the activity of RNA polymerase. Probably all or virtually all of the core enzyme becomes associated with the sigma factor of the moment; and the change is irreversible.

Perhaps the most extensive example of switches in sigma factors is provided by **sporulation**, an alternative life style available to some bacteria. At the end of the **vegetative phase**, logarithmic growth ceases because nutrients in the medium become depleted. This triggers sporulation, as illustrated in **Figure 14.21**. DNA is replicated, a genome is segregated at one end of the cell, and eventually it is surrounded by the tough spore coat. When the septum forms at stage II, it generates two independent compartments, the mother cell and the forespore. The process takes ~8 hours. It can be viewed as a primitive sort of differentiation, in which a parent cell (the vegetative bacterium) gives rise to two different daughter cells with distinct fates: the mother cell is eventually lysed, while the spore that is released has an entirely different structure from the original bacterium.

Sporulation involves a drastic change in the biosynthetic activities of the bacterium, in which many genes are involved. The basic level of control lies at transcription. Some of the genes that functioned in the vegetative phase are turned off during sporulation, but most continue to be expressed. In addition, the genes specific for sporulation are expressed only during this period. At the end of sporulation, ~40% of the bacterial mRNA is sporulation specific.

New forms of the RNA polymerase become active in sporulating cells; they contain the same core enzyme as vegetative cells, but have different proteins in place of the vegetative σ^{43}. The changes in transcriptional specificity are summarized in **Figure 14.22**. The principle is that in each compartment the existing sigma factor is successively displaced by a new factor that causes transcription of a different set of genes. Communication between the compartments occurs in order to coordinate the timing of the changes in the forespore and mother cell.

The sporulation cascade is initiated when environmental conditions trigger a 'phosphorylation relay', in which a phosphate group is passed along a series of proteins until it reaches SpoOA. (Several gene products are involved in this process, whose complexity may reflect the need to avoid mistakes in triggering sporulation unnecessarily.) SpoOA is a transcriptional regulator whose activity is affected by phosphorylation. In the phosphorylated form, it activates transcription of two operons, each of which is transcribed by a different form of the host RNA polymerase. Under the direction of phosphorylated SpoOA, host enzyme utilizing the general σ^{43} transcribes the gene coding for the factor σ^F; and host enzyme under the direction of a minor factor, σ^H, transcribes the gene coding for the factor pro-σ^E. Both of these new sigma factors are produced before septum formation, but become active later.

Figure 14.21

Sporulation involves the differentiation of a vegetative bacterium into a mother cell that is lysed and a spore that is released. The stages of sporulation are numbered from 0 to VII.

Action	State of bacterium	Stage
Vegetative bacterium		0
DNA replicates		I
Septum forms	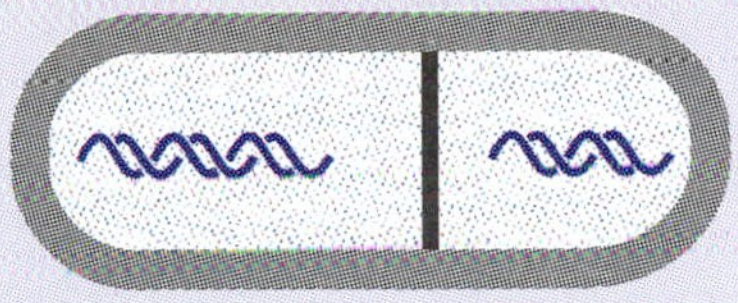	II
Spore is engulfed		III
Spore coat forms		V
Mother cell is lysed		VI
Spore is released		VII

σ^F becomes active only in the forespore compartment, possibly because there is an inhibitor of its action in the mother cell. Sporulation starts when σ^{43} is replaced by σ^F in the forespore. Under the direction of σ^F, RNA polymerase transcribes the first set of sporulation genes instead of the vegetative genes it was previously transcribing. The

replacement reaction probably affects only part of the RNA polymerase population, since σ^F is produced only in small amounts. Some vegetative enzyme remains present during sporulation. The displaced σ^{43} is not destroyed, but can be recovered from extracts of sporulating cells.

Two regulatory events follow from the activity of

Figure 14.22

Sporulation involves successive changes in the sigma factors that control the initiation specificity of RNA polymerase. The cascades in the forespore (left) and the mother cell (right) are related by signals passed across the septum (indicated by the horizontal arrows).

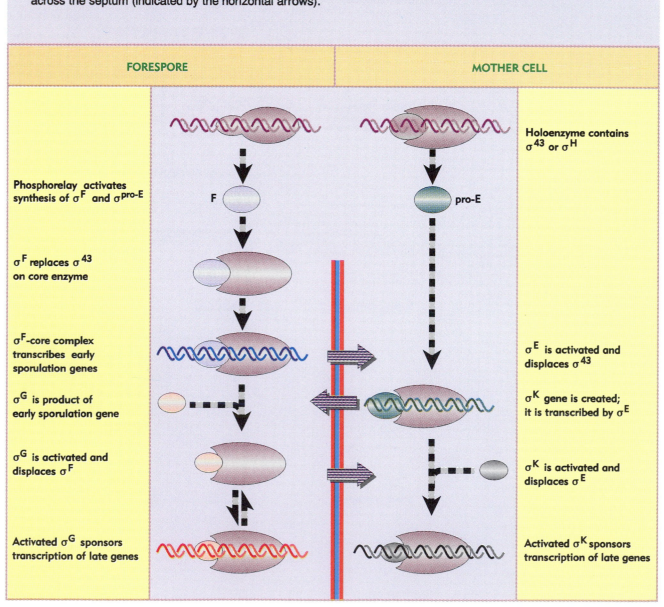

σ^F. In the forespore itself, another sigma factor, σ^G is the product of one of the early sporulation genes. σ^G is the factor that causes RNA polymerase to transcribe the late sporulation genes in the forespore. Another early sporulation gene product is responsible for communicating with the mother cell compartment. A signal is transmitted (through a protein embedded in the septum) to activate the factor σ^E, which was previously synthesized in its precursor form (pro-σ^E); it is activated by cleavage.

The cascade continues when σ^E in turn causes transcription of the gene for σ^K. (Actually the production of σ^K is quite complex, because first its gene must be created by a recombination event!) This factor also is synthesized as an inactive precursor (pro-σ^K) that is activated by a protease. Once σ^K has been activated, it displaces σ^E and causes transcription of the late genes in the mother cell. The timing of these events in the two compartments is coordinated by further signals. The activation of σ^G in the forespore depends on events that occur in the mother cell; and in turn the activity of σ^G is required to generate a signal that is transmitted across the septum to activate σ^K.

Sporulation is thus controlled by two cascades, in which sigma factors in each compartment are successively activated, each directing the synthesis of a particular set of genes. The two cascades are connected by the transmission of signals from one compartment to the other. As new sigma factors become active, old sigma factors are displaced, so that transitions in sigma factors turn genes off as well as on. The incorporation of each factor into RNA polymerase dictates when its set of target genes is expressed; and the amount of factor available influences the level of gene expression. More than one sigma factor may be active at any time, and the specificities of some of the sigma factors overlap. We do not know what is responsible for the ability of each sigma factor to replace its predecessor.

Summary

A transcription unit comprises the DNA between a promoter, where transcription initiates, and a terminator, where it ends. One strand of the DNA in this region serves as a template for synthesis of a complementary strand of RNA. The RNA–DNA hybrid region is very short (2–3 bp) and transient, as the transcription 'bubble' moves along DNA. The RNA polymerase holoenzyme that synthesizes bacterial RNA can be separated into two components. Core enzyme is a multimer of structure $\alpha_2\beta\beta'$ that is responsible for elongating the RNA chain. Sigma factor is a single subunit that is required at the stage of initiation for recognizing the promoter.

Core enzyme has a general affinity for DNA. The addition of sigma factor reduces the affinity of the enzyme for nonspecific binding to DNA, but increases its affinity for promoters. The rate at which RNA polymerase finds its promoters is too great to be accounted for by diffusion and random contacts with DNA; direct exchange of DNA sequences held by the enzyme may be involved.

Bacterial promoters are identified by two short conserved sequences centered at −35 and −10 relative to the startpoint. Most promoters have sequences that are well related to the consensus sequences at these sites. The distance separating the consensus sequences is 16–18 bp. RNA polymerase initially 'touches down' at the −35 sequence and then extends its contacts over the −10 region. The enzyme covers ~77 bp of DNA. The initial 'closed' binary complex is converted to an 'open' binary complex by melting of a sequence of ~12 bp that extends from the −10 region to the startpoint. The A•T-rich base pair composition of the −10 sequence may be important for the melting reaction.

The binary complex is converted to a ternary

410 ■ CHAPTER FOURTEEN

complex by the incorporation of ribonucleotide precursors. There are multiple cycles of abortive initiation, during which RNA polymerase synthesizes and releases RNA chains of 2–9 bases without moving from the promoter. At the end of this stage, sigma factor is released, the core enzyme contracts to cover ~50 bp and loses contacts upstream of −35, and then moves along DNA, synthesizing RNA. A locally unwound region of DNA moves with the enzyme. The enzyme contracts further in size to cover only 30–40 bp when the nascent chain has reached 15–20 nucleotides; then it continues to the end of the transcription unit. RNA polymerase and RNA are released when a terminator is encountered and DNA is then fully restored to duplex condition.

The 'strength' of a promoter describes the frequency at which RNA polymerase initiates transcription; it is related to the closeness with which its −35 and −10 sequences conform to the ideal consensus sequences, but is influenced also by the sequences immediately downstream of the startpoint. Negative supercoiling increases the strength of certain promoters; the opposite effect is seen at other (rarer) promoters. Transcription generates positive supercoils ahead of RNA polymerase and leaves negative supercoils behind the enzyme.

The (single) core enzyme can be directed to recognize promoters with different consensus sequences by alternative sigma factors. In *E. coli*, such sigma factors are activated by adverse conditions, such as heat shock or nitrogen starvation. *B. subtilis* contains a single major sigma factor with the same specificity as the *E. coli* sigma factor, but also contains a variety of minor sigma factors. Another series of factors is activated when sporulation is initiated; sporulation is regulated by two cascades in which sigma factor replacements occur in the forespore and mother cell. A similar mechanism for regulating transcription is also used by phage SPO1 in *B. subtilis*.

The geometry of RNA polymerase–promoter recognition is similar for holoenzymes containing all sigma factors (except σ^{54}). Each sigma factor causes RNA polymerase to initiate transcription at a promoter that conforms to a particular consensus at −35 and −10. Direct contacts between sigma and DNA at these sites have been demonstrated for *E. coli* σ^{70}. It is not clear how contacts made with sigma factor relate to contacts made by subunits of core enzyme. An unanswered question on the use of sigma factors concerns the nature of the mechanism that allows one sigma to replace another.

Further reading

Reviews

A source for historical reviews and original research articles is the volume edited by **Losick & Chamberlin**, *RNA Polymerase* (Cold Spring Harbor Laboratory, New York, 1976). Two chapters that cover the general matters dealt with here were written by **Chamberlin**, giving first a general overview (pp. 17–68) and then an account of the interactions of the bacterial enzyme with its template (pp. 159–192).

The stages of the RNA polymerase–promoter interaction have been reviewed by **Von Hippel *et al.*** (*Ann. Rev. Biochem.* **53**, 389–449, 1984) and **McClure** (*Ann. Rev. Biochem.* **54**, 171–204, 1985). Initiation and its control has been reviewed by **Reznikoff *et al.*** (*Ann. Rev. Genet.* **19**, 355–387, 1985.) The act of transcription has been reviewed by **Yager & von Hippel** (in *E. coli and S. typhimurium,* Ed. Neidhardt, American Society for Microbiology, Washington DC, 1241–1275, 1987).

The structures and functions of sigma factors have been reviewed by **Helmann & Chamberlin** (*Ann. Rev. Biochem.* **57**, 839–872, 1988).

Changes induced by sporulation were reviewed by **Losick *et al.*** (*Ann. Rev. Genet.* **20**, 625–669, 1986). The role of sigma factors in this and other processes was reviewed by **Doi** (*Microbiol. Rev.* **50**, 227–239, 1986) and by **Losick & Stragier** (*Nature* **355**, 601–604, 1992).

Discoveries

Sigma factor was characterized by **Travers & Burgess** (*Nature* **222**, 537–540, 1969). Sigma factor cascades in sporulation were discovered by **Haldenwang & Losick** (*Proc. Nat. Acad. Sci. USA* **77**, 7000–7004, 1980; *Cell* **23**, 615–624, 1981). The first additional sigma factor in *E. coli* was uncovered by **Grossman, Erickson, & Gross** (*Cell* **38**, 383–390, 1984).

The molecular interaction of bacterial RNA polymerase with its promoters was described by **Siebenlist, Simpson & Gilbert** (*Cell* **20**, 269–281, 1980).

Changing views about the initiation and elongation reactions are reflected in the results of **Krummel and Chamberlin** (*Biochemistry* **28**, 7829–7842, 1989) and **Rice, Kane, & Chamberlin** (*Proc. Nat. Acad. Sci.* **88**, 4245–4249, 1991).

The relationship between transcription and supercoiling has been considered by **Liu *et al.*** (*Cell* **53**, 433–°440, 1988).

CHAPTER 15

A panoply of operons: the lactose paradigm and others

The paradigm for control of gene expression derives from the regulation of transcription in *E. coli*. The concept that two different types of functions are involved in genetic expression was proposed by Jacob and Monod in 1961 in their classic formulation of the model for control of bacterial gene expression. Virtually all the subsequent analysis of regulation of transcription in both prokaryotic and eukaryotic cells has followed the distinction between genes that code for products and elements that function exclusively within DNA:

◆ A gene is a sequence of DNA that codes for a diffusible product. This product may be protein (as in the case of the majority of genes) or may be RNA (as in the case of genes that code for tRNA and rRNA). *The crucial feature of the gene in either case is that the product diffuses away from its site of synthesis to act elsewhere.*

◆ An element is a sequence of DNA that is not converted into any other form, but that functions exclusively as a DNA sequence *in situ*. Because it affects only the DNA to which it is physically linked, it is more fully described as a ***cis*-acting** element. (In some cases, a *cis*-acting sequence functions in an RNA rather than DNA molecule.)

Genes may be classified according to the type of function provided by their products. Genes that code for the proteins required by the cell, for enzymatic or structural functions, are called **structural genes**. The overwhelming majority of bacterial genes fall into this category, which therefore represents an enormous variety of protein structures and functions. Genes that code for proteins which in turn regulate the expression of other genes are called **regulator genes**. Because the products of regulator genes are free to diffuse to their appropriate targets, they are described as ***trans*-acting** factors.

The crux of regulation is that *a regulator gene codes for a regulator protein that controls transcription by binding to particular site(s) on DNA.* Recognition of a *cis*-acting sequence by a *trans*-acting factor can regulate a target gene in either a positive manner (the interaction turns the gene on) or in a negative manner (the interaction turns the gene off). The *cis*-acting sites are usually (but not exclusively) located just upstream of the target gene.

Figures 15.1 and 15.2 expand our view of the transcription unit from that of Figure 14.2 to compare schematically the general features of the control of transcription in prokaryotes and eukaryotes.

The sites that mark the beginning and end of the transcription unit, the promoter and terminator, are examples of *cis*-acting sites. *A promoter serves to initiate transcription only of the gene or genes physically connected to it on the same stretch of DNA.* In the same way, a terminator can terminate transcription only by an RNA polymerase that has traversed the gene(s) that physically precede it. In their simplest forms, promoters and terminators are *cis*-acting elements that are recognized by the same *trans*-acting species, that is, by RNA polymerase (although other factors also participate at each site).

Additional *cis*-acting regulatory sites are often juxtaposed to, or interspersed with, the promoter. A bacterial promoter may have one or more such sites located close by, that is, in the immediate vicinity of the startpoint. A eukaryotic promoter is likely to have a greater number of sites, spread out over a longer distance.

The classic mode of control in bacteria is *negative*: a **repressor** protein prevents a gene from being expressed. **Figure 15.1** shows that the default state for such a gene is to be expressed via the recognition of its promoter by RNA polymerase. Close to the promoter is another *cis*-acting site called the **operator**, which is the target for the repressor protein. When the repressor binds to the operator, RNA poly-

merase is prevented from initiating transcription, and *gene expression is therefore turned off.*

The most common mode of control in eukaryotes is *positive*: a **transcription factor** is required to assist RNA polymerase in initiating at the promoter. **Figure 15.2** shows that the typical default state of a eukaryotic gene is inactive: RNA polymerase cannot by itself initiate transcription at the promoter (or perhaps can do so only rather poorly). Several *trans*-acting factors have target sites in the vicinity of the promoter, and binding of some or all of these factors *enables RNA polymerase to initiate transcription.*

No type of regulation is exclusive to any one genome, and examples are found of positive regulation in bacteria and of negative regulation in eukaryotes. The unifying theme is that regulatory proteins are *trans*-acting factors that recognize *cis*-acting elements (usually) upstream of the gene. The consequences of this recognition are to activ-

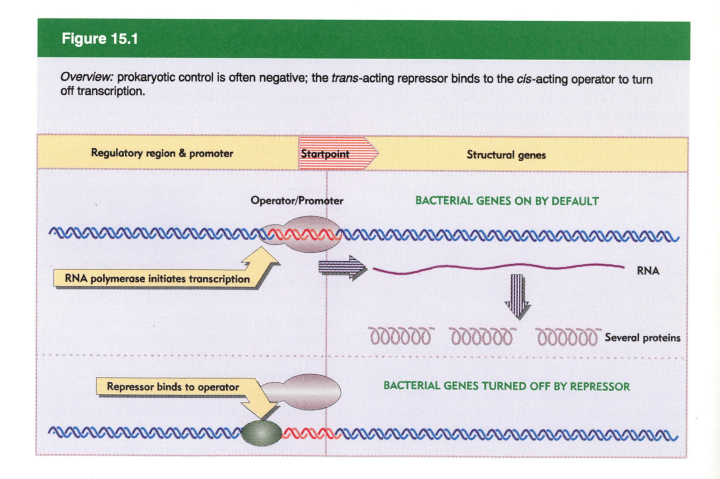

Figure 15.1

Overview: prokaryotic control is often negative; the *trans*-acting repressor binds to the *cis*-acting operator to turn off transcription.

Figure 15.2

Overview: eukaryotic control is often positive; *trans*-acting factors bind to *cis*-acting sites in order for RNA polymerase to initiate transcription at the promoter.

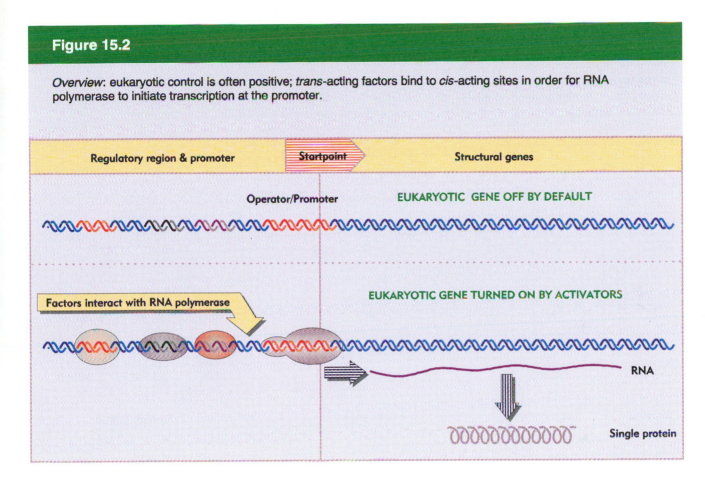

ate or to repress the gene, depending on the individual type of regulatory protein. A typical feature is that the protein functions by recognizing a very short sequence in DNA, usually <10 bp in length, although the protein actually extends over a somewhat greater distance of DNA. The bacterial promoter is an example: although RNA polymerase covers >70 bp of DNA at initiation, the crucial sequences that it recognizes are the hexamers centered at −35 and −10.

A significant difference in gene organization between prokaryotes and eukaryotes is that structural genes in bacteria are organized in clusters, while those in eukaryotes occur individually. Clustering of structural genes allows coordinate control by means of the interactions at a single promoter: as a result of these interactions, the entire set of genes is either transcribed or not transcribed. In this chapter, we discuss this mode of control and its use by bacteria. The means employed to coordinate control of dispersed eukaryotic genes are discussed in Chapter 30.

Structural gene clusters are coordinately controlled

Bacterial structural genes tend to be organized into clusters that include genes coding for proteins whose functions are related. It is common for all the enzymes of a metabolic pathway to be organized in a cluster that is coordinately regulated. In addition to the enzymes actually involved in

Figure 15.3

The *lac* operon occupies ˜6000 bp of DNA. At the left the *lacI* gene has its own promoter and terminator. The end of the *lacI* region is adjacent to the P_{lac} promoter. The O_{lac} operator occupies the first 26 bp of the *lacZ* gene, which is extremely long and is followed by the *lacY* and *lacA* genes and a terminator.

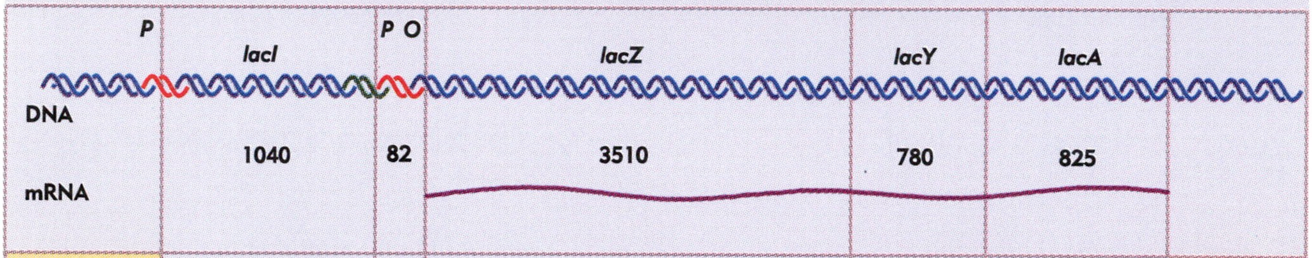

	P	*lacI*	P O	*lacZ*	*lacY*	*lacA*	
DNA							
mRNA		1040	82	3510	780	825	
Polypeptide		360 / 38,000		1021 / 125,000	260 / 30,000	275 / 30,000	Amino acids / Daltons
Protein		Tetramer / 152,000		Tetramer / 500,000	Membrane protein / 30,000	Dimer / 60,000	Structure / Daltons
Function		Repressor		β-galactosidase	Permease	Transacetylase	

the pathway, other related activities may be included in the unit of coordinate control; for example, the protein responsible for transporting the small molecule substrate into the cell.

The cluster of the three *lac* structural genes, *lacZYA*, is typical. The gene products enable cells to take up and metabolize β-galactosides, such as lactose. **Figure 15.3** summarizes the organization of the structural genes, their associated *cis*-acting regulatory elements, and the *trans*-acting regulatory gene. The roles of the three structural genes are:

◆ *lacZ* codes for the enzyme β-galactosidase, whose active form is a tetramer of ~500,000 daltons. The enzyme breaks a β-galactoside into its component sugars. For example, lactose is cleaved into glucose and galactose (which are then further metabolized).

◆ *lacY* codes for the β-galactoside permease, a 30,000 dalton membrane-bound protein constituent of the transport system. This transports β-galactosides into the cell.

◆ *lacA* codes for β-galactoside transacetylase, an enzyme that transfers an acetyl group from acetyl-CoA to β-galactosides.

Mutations in either *lacZ* or *lacY* can create the *lac⁻* genotype, in which cells cannot utilize lactose. The *lacZ⁻* mutations abolish enzyme activity, directly preventing metabolism of lactose. The *lacY⁻* mutants cannot take up lactose from the medium. (No defect is identifiable in *lacA⁻* cells, which is puzzling. It is possible that the acetylation reaction gives an advantage when the bacteria grow in the presence of certain analogs of β-galactosides that cannot be metabolized, because the modification results in detoxification and excretion.)

The entire system, including structural genes and the elements that control their expression, forms a common unit of regulation; this is called an **operon**. The activity of the operon is controlled by regulator gene(s), whose protein products interact with the *cis*-acting control elements.

Transcription of the *lacZYA* genes is controlled by a repressor protein synthesized by the *lacI* gene. It happens that *lacI* is located adjacent to the structural genes, but it comprises an independent transcription unit with its own promoter and terminator. Since *lacI* specifies a diffusible product, in principle it need not be located near the structural genes; it can function equally well if moved elsewhere, or carried on a separate DNA molecule (the classic test for a *trans*-acting regulator).

We can distinguish between structural genes and regulator genes by the effects of mutations. A mutation in a structural gene deprives the cell of the particular protein for which that gene codes. But a mutation in a regulator gene influences the expression of all the structural genes that it controls. The consequences of a regulatory mutation reveal the type of regulation.

The *lac* genes are controlled by **negative regulation**: *they are transcribed unless turned off by the regulator protein.* A mutation that inactivates the regulator causes the structural genes to remain in the expressed condition. The function of the regulator—to prevent the expression of the structural genes—gave rise to its description as a 'repressor' protein.

The lactose repressor is a tetramer of identical subunits of 38,000 daltons each. There are ~10 tetramers in a wild-type cell. The regulator gene is transcribed into a monocistronic mRNA at a rate that appears to be governed simply by the affinity of its promoter for RNA polymerase.

The product of *lacI* is called the **lac repressor,** and it functions by binding to an **operator** (formally denoted O_{lac}) at the start of the *lacZYA* cluster. The operator lies between the promoter (P_{lac}) and the structural genes (*lacZYA*). *When the repressor binds at the operator, its presence prevents RNA polymerase from initiating transcription at the promoter.* **Figure 15.4** expands our view of the

Figure 15.4

Repressor and RNA polymerase bind at sites that overlap around the startpoint of the lac operon.

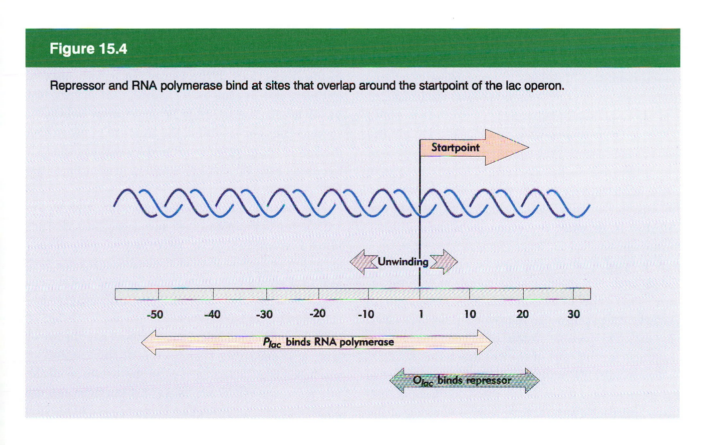

region at the start of the *lac* structural genes. O_{lac} extends from position -5 just upstream of the mRNA startpoint to position $+21$ within the transcription unit. Thus it overlaps the right end of the promoter.

We return later to the issue of the exact effect of repressor upon RNA polymerase, but from the relative locations of the operator and promoter it is evident that repressor binding to DNA is in a position to prevent RNA polymerase from transcribing the structural genes. We note, however, that operators in other operons are in different locations, and that there may therefore be more than one means by which a repressor can block transcription.

The activity of repressor protein is controlled by a small molecule inducer

Bacteria need to respond swiftly to changes in their environment. Capricious fluctuations in the supply of nutrients can occur at any time; survival depends on the ability to switch from metabolizing one substrate to another. Unicellular eukaryotes share this subjection to an incessantly changing world; but more complex, multicellular organisms are restricted to a more constant set of metabolic pathways, and do not have the same need to respond to external circumstances.

Flexibility is therefore at a premium in the bacterial world. Yet economy also is important, since a bacterium that indulges in energetically expensive ways to meet the demands of the environment is likely to be at a disadvantage. Certainly it would be expensive to produce unnecessarily all the enzymes for a metabolic pathway in the absence of the substrates. So the bacterial compromise is to avoid synthesizing the enzymes of a pathway in the absence of their substrate; but to be ready at all times to produce the enzymes if the substrate should appear.

The synthesis of enzymes in response to the appearance of a specific substrate is called **induction**. This type of regulation is widespread in bacteria, and occurs also in lower eukaryotes (such as yeast). The lactose system of *E. coli* provides the paradigm for this sort of control mechanism.

When cells of *E. coli* are grown in the absence of a β-galactoside, there is no need for β-galactosidase, and they contain very few molecules of the enzyme—say, <5. When a suitable substrate is added, the enzyme activity appears very rapidly in the bacteria. Within 2–3 minutes some enzyme is present, and soon there are up to 5000 molecules of enzyme per bacterium. (Under suitable conditions, β-galactosidase can account for 5–10% of the total soluble protein of the bacterium.) If the substrate is removed from the medium, the synthesis of enzyme stops as rapidly as it originally started.

Figure 15.5 summarizes the essential features of induction. Control of transcription of the *lac* genes responds very rapidly to the inducer, as shown in the upper part of the figure. In the absence of inducer, the operon is transcribed at a very low **basal level** (see below). Transcription begins as soon as inducer is added; the amount of *lac* mRNA increases rapidly to an induced level that reflects a balance between synthesis and degradation of the mRNA.

The *lac* mRNA is extremely unstable, and decays with a half-life of only ~3 minutes. This feature allows induction to be reversed rapidly. Transcription ceases as soon as the inducer is removed; and in a very short time all the lactose mRNA has been destroyed, and the cellular content has returned to the basal level.

The production of protein is followed in the lower part of the figure. Translation of the *lac* mRNA produces β-galactosidase (and the products of the other *lac* genes). There is a short lag between the appearance of *lac* mRNA and appearance of the first

Figure 15.5

Addition of inducer results in rapid induction of *lac* mRNA, and is followed after a short lag by synthesis of the enzymes; removal of inducer is followed by rapid cessation of synthesis.

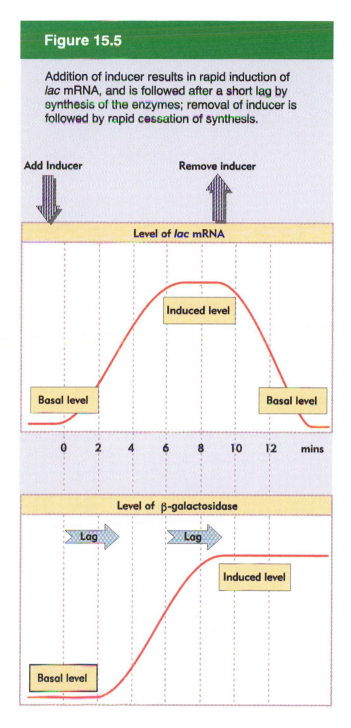

supply not only provides the ability to metabolize new substrates, but also is used to shut off endogenous synthesis of compounds that suddenly appear in the medium. For example, *E. coli* synthesizes the amino acid tryptophan through the action of the enzyme tryptophan synthetase. But if tryptophan is provided in the medium on which the bacteria are growing, the production of the enzyme is immediately halted. This effect is called **repression**. It allows the bacterium to avoid devoting its resources to unnecessary synthetic activities.

Induction and repression represent the same phenomenon. In one case the bacterium adjusts its ability to use a given substrate for growth; in the other it adjusts its ability to synthesize a particular metabolic intermediate. The trigger for either type of adjustment is the small molecule that is the substrate for the enzyme, or the product of the enzyme activity, respectively. Small molecules that cause the production of enzymes able to metabolize them are called **inducers**. Those that prevent the production of enzymes able to synthesize them are called **corepressors**.

The ability to act as inducer or corepressor is highly specific. Only the substrate/product or a closely related molecule can serve. *But the activity of the small molecule does not depend on its interaction with the target enzyme.* Some inducers resemble the natural inducers for β-galactosidase, but cannot be metabolized by the enzyme. The example *par excellence* is isopropylthiogalactoside (IPTG), one of several thiogalactosides with this property. Although it is not recognized by β-galactosidase, IPTG is a very efficient inducer of the *lac* genes.

Molecules that induce enzyme synthesis but are not metabolized are called **gratuitous inducers**. They are extremely useful because they remain in the cell in their original form. (A real inducer would be metabolized, interfering with study of the system.) The existence of gratuitous inducers reveals an important point. *The system must possess some component, distinct from the target enzyme, that recognizes the appropriate substrate; and its ability to recognize related potential substrates is different from that of the enzyme.*

completed enzyme molecules; there is a similar lag in reaching maximal induced levels of mRNA and protein. When inducer is removed, synthesis of enzyme ceases almost immediately (as the mRNA is degraded), but the β-galactosidase in the cell is more stable than the mRNA, so the enzyme activity remains at the induced level for longer.

This type of rapid response to changes in nutrient

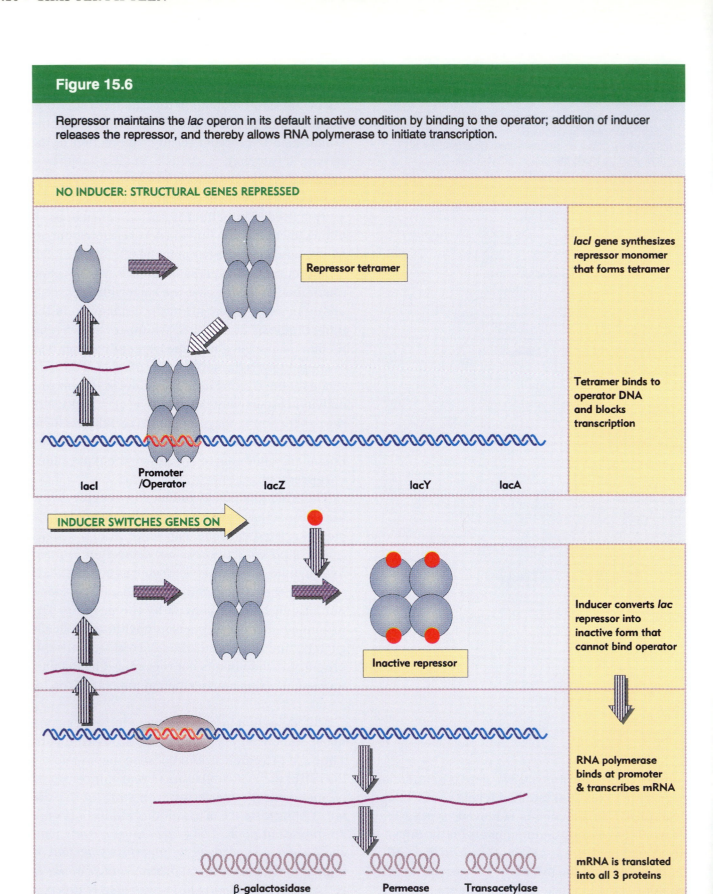

Figure 15.6

Repressor maintains the *lac* operon in its default inactive condition by binding to the operator; addition of inducer releases the repressor, and thereby allows RNA polymerase to initiate transcription.

NO INDUCER: STRUCTURAL GENES REPRESSED

Repressor tetramer

lacI gene synthesizes repressor monomer that forms tetramer

Tetramer binds to operator DNA and blocks transcription

lacI

Promoter /Operator

lacZ

lacY

lacA

INDUCER SWITCHES GENES ON

Inactive repressor

Inducer converts *lac* repressor into inactive form that cannot bind operator

RNA polymerase binds at promoter & transcribes mRNA

β-galactosidase

Permease

Transacetylase

mRNA is translated into all 3 proteins

The component that responds to the inducer is the repressor protein coded by *lacI*. Its role in controlling transcription of the *lacZYA* structural genes in response to the environment is summarized in **Figure 15.6**. The structural genes are transcribed into a single mRNA from a promoter just upstream of *lacZ*. The state of the repressor determines whether this promoter is turned off or on:

◆ In the absence of an inducer, the genes are not transcribed, because the repressor protein is in an active form that is bound to the operator.

◆ When an inducer is added, the repressor is converted into an inactive form which leaves the operator. Then transcription starts at the promoter and proceeds through the genes to a terminator located somewhere beyond *lacA*.

How does the inducer control the activity of the repressor protein? The repressor has a high affinity for the operator; in the absence of inducer, repressor is always bound to the operator to ensure that the adjacent structural genes cannot be transcribed. But when the inducer is present, it binds to the repressor to form a repressor•inducer complex that no longer binds at the operator.

The crucial features of the control circuit reside in the dual properties of the repressor: it can prevent transcription; and it can recognize the small-molecule inducer. The repressor has two binding sites, one for the inducer and one for the operator. When the inducer binds at its site, it changes the conformation of the protein in such a way as to influence the activity of the *other* site. This type of relationship is an example of **allosteric control** (see Figure 1.21).

Induction accomplishes a **coordinate regulation:** *all the genes are expressed (or not expressed) in unison.* The mRNA is translated sequentially from its 5′ end, which explains why induction always causes the appearance of β-galactosidase, β-galactoside permease, and β-galactoside transacetylase, in that order. Translation of a common mRNA explains why the relative amounts of the three enzymes always remain the same under varying conditions of induction.

Induction throws a switch that causes the genes to be expressed. Inducers vary in their effectiveness, and other factors influence the absolute level of transcription or translation, but the relationship between the three genes is predetermined by their organization.

We notice a potential paradox in the constitution of the operon. The lactose operon contains the structural gene (*lacZ*) coding for the β-galactosidase activity needed to metabolize the sugar; it also includes the gene (*lacY*) that codes for the protein needed to transport the substrate into the cell. But if the operon is in a repressed state, how does the inducer enter the cell to start the process of induction?

Two features ensure that there is always a minimal amount of the protein present in the cell, enough to start the process off. There is a basal level of expression of the operon: even when it is not induced, it is expressed at a residual level (0.1% of the induced level). And some inducer enters anyway via another uptake system.

Mutations identify the operator and the regulator gene

A wild-type operon is expressed in a regulated manner. Mutations in the regulatory circuit may either abolish expression or cause it to occur in a manner that does not respond to regulation. Mutants that cannot be expressed at all are called **uninducible.** The ability of a gene to function without responding

Figure 15.7

Operator mutations are constitutive because the operator is unable to bind repressor protein; this allows RNA polymerase to have unrestrained access to the promoter. The O^c mutations are *cis*-acting, because they affect only the contiguous set of structural genes.

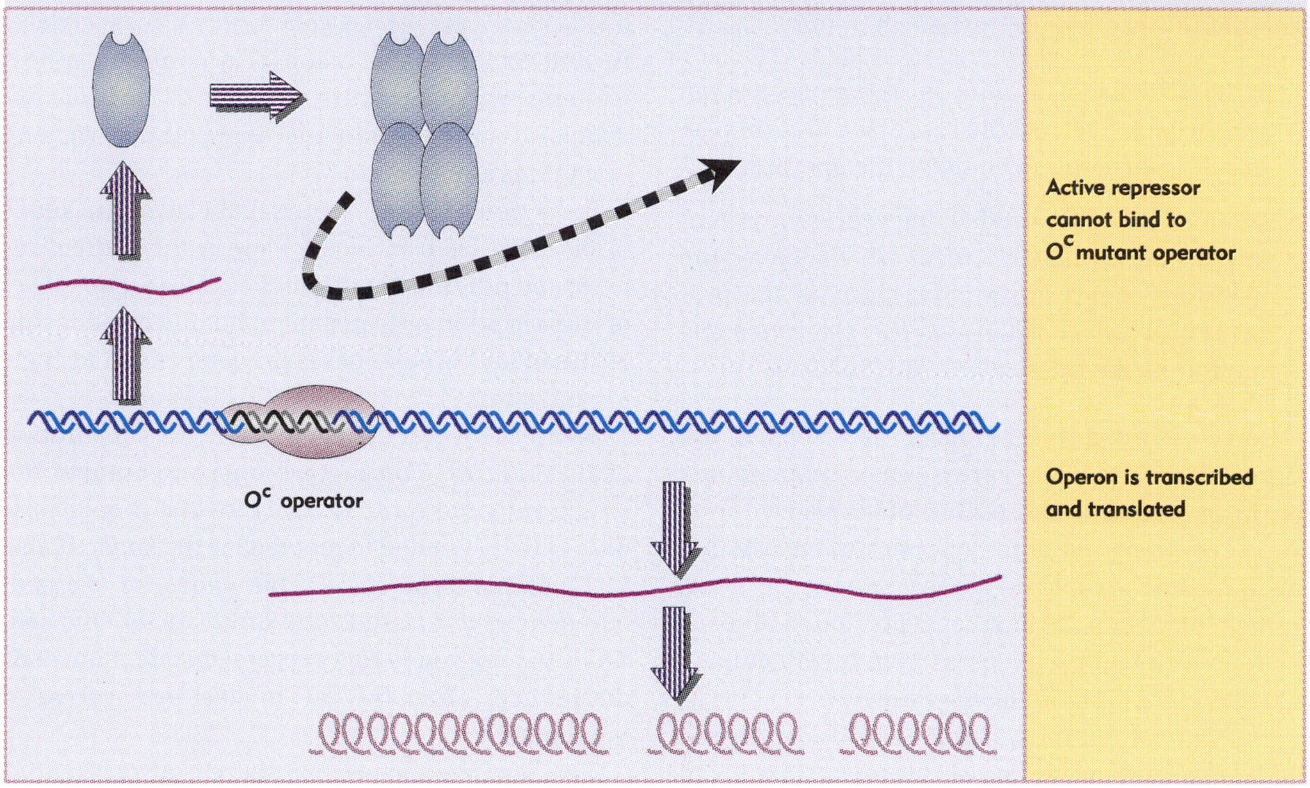

O^c operator

Active repressor cannot bind to O^c mutant operator

Operon is transcribed and translated

to regulation is called **constitutive** gene expression, and mutants with this property are called **constitutive mutants**.

Components of the regulatory circuit of the operon can be identified by mutations that *affect the expression of all the structural genes and map outside them*. They fall into two classes. The promoter and the operator are identified as targets for the regulatory proteins (RNA polymerase and repressor, respectively) by *cis*-acting mutations. And the locus *lacI* is identified as the gene that codes for the repressor protein by *trans*-acting mutations.

The operator was originally identified by constitutive mutations, denoted O^c, whose distinctive properties provided the first evidence for *an ele-*

ment that functions without being represented in a diffusible product.

The structural genes contiguous with an O^c mutation are expressed constitutively because the mutation changes the operator so that the repressor no longer binds to it. Thus the repressor cannot prevent RNA polymerase from initiating transcription. So the operon is transcribed continuously, as illustrated in **Figure 15.7**.

The operator can control only the lac *genes that are adjacent to it.* If a second *lac* operon is introduced into the bacterium on an independent molecule of DNA, it has its own operator. Neither operator is influenced by the other. Thus if one operon has a wild-type operator, it will be repressed under the usual conditions, while a second operon with an O^c mutation will be expressed in its characteristic fashion.

These properties define the operator as a typical *cis*-acting site, whose function is exercised by virtue of recognition of its DNA sequence by some *trans*-acting factor. The operator controls the adjacent genes irrespective of the presence in the cell of other alleles of the site. A mutation in such a site, for example, the O^c mutation, is formally described as **cis-dominant**.

A mutation in a *cis*-acting site cannot be assigned to a complementation group. (The ability to complement is characteristic only of genes expressed as diffusible products.) When two *cis*-acting sites lie close together—for example, a promoter and an operator—we cannot classify the mutations by a complementation test. We are restricted to distinguishing them by their effects on the phenotype.

Cis-dominance is a characteristic of any site that is *physically contiguous with the sequences it controls*. If a control site functions as part of a polycistronic mRNA, mutations in it will display *exactly the same pattern* of *cis*-dominance as they would if functioning in DNA. The critical feature is that the control site cannot be physically separated from the genes that it regulates. From the genetic point of view, it does not matter whether the site and genes are together on DNA or on RNA.

Constitutive transcription is also caused by mutations of the *lacI⁻* type. These can arise from either point mutations or deletions. The latter indicates that it represents *loss* of the usual regulation. **Figure 15.8** shows that the *lacI⁻* mutants express the structural genes all the time,

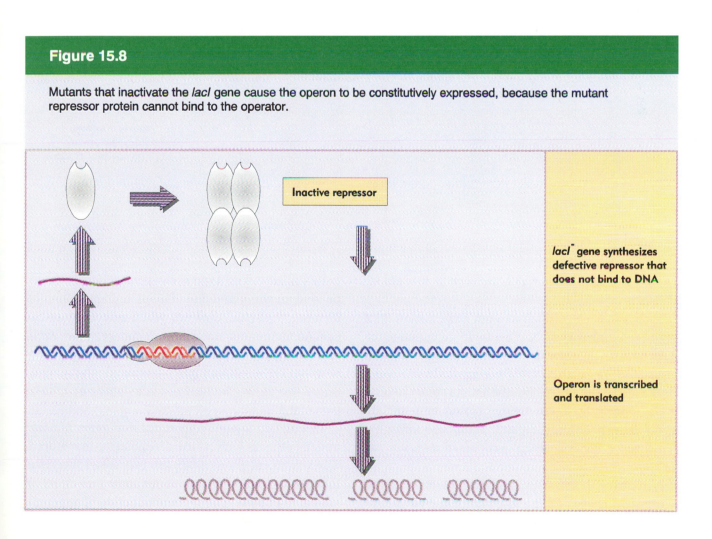

Figure 15.8

Mutants that inactivate the *lacI* gene cause the operon to be constitutively expressed, because the mutant repressor protein cannot bind to the operator.

Inactive repressor

lacI⁻ gene synthesizes defective repressor that does not bind to DNA

Operon is transcribed and translated

Figure 15.9

Constitutive mutants in the *lacI* gene are recessive.

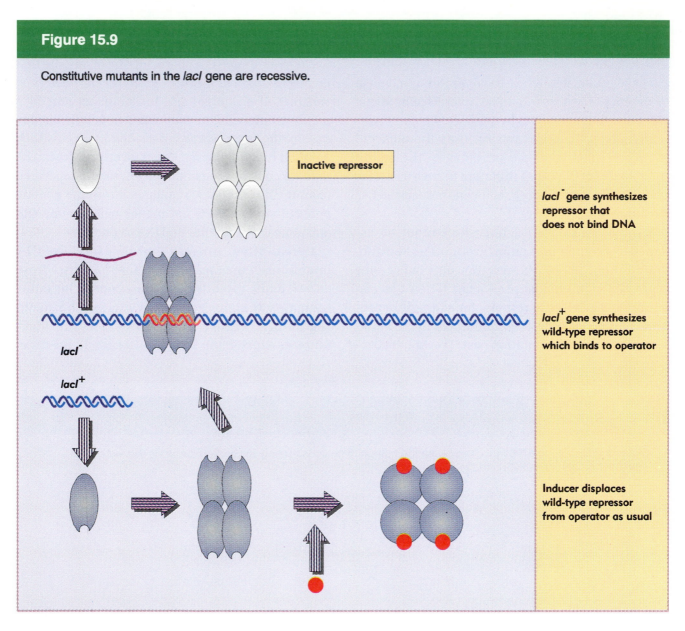

Inactive repressor

lacI⁻ gene synthesizes repressor that does not bind DNA

lacI⁺ gene synthesizes wild-type repressor which binds to operator

Inducer displaces wild-type repressor from operator as usual

lacI⁻

lacI⁺

irrespective of whether the inducer is present or absent.

Such behavior conforms to our expectation for a negative control system. The *lacI⁺* gene codes for a repressor protein that is able to turn off the transcription of the *lacZYA* genes. Constitutive mutations in the repressor abolish its ability to bind to the operator. Then transcription can initiate freely at the promoter. *So lacI⁻ mutations allow the genes to be constitutively expressed because the repressor is inactive.*

We can fortify this conclusion by determining the relationship between the *lacI⁻* constitutive mutant

gene and the wild-type *lacI⁺* gene when both are present in the same cell. This is accomplished by forming a **partial diploid**, when one copy of the operon is present on the bacterial genome itself, and the other is introduced via a **plasmid**, an independent self-replicating DNA molecule that carries only a few genes, including a copy of the operon or part of it.

In cells in which one regulator gene is *lacI⁺* and the other is *lacI⁻*, normal regulation is restored. The structural genes again are repressed when the inducer is removed. We can explain this effect as shown in **Figure 15.9**. The introduction of a second, wild-type regulator gene restores the presence of normal repressor. So once again the operon is turned off in the absence of inducer. In genetic parlance, the wild-type **inducibility** is *dominant* over the mutant constitutivity. This is the hallmark of negative control.

Mutants of the operon that are uninducible cannot be expressed at all. They fall into the same two types of genetic classes as the constitutive mutants. The dominance relationships of each type can be used in the same way to define the nature of the locus:

◆ Promoter mutations are *cis*-acting. If they prevent RNA polymerase from binding at P_{lac}, they render the operon nonfunctional because it cannot be transcribed.

◆ Mutations in *lacI* that abolish the ability of repressor to bind the inducer can cause the same phenotype. Such mutants are described as *lacI*s. **Figure 15.10** shows that they are *trans*-acting and dominant over wild type. The repressor is 'locked in' to the active form that recognizes

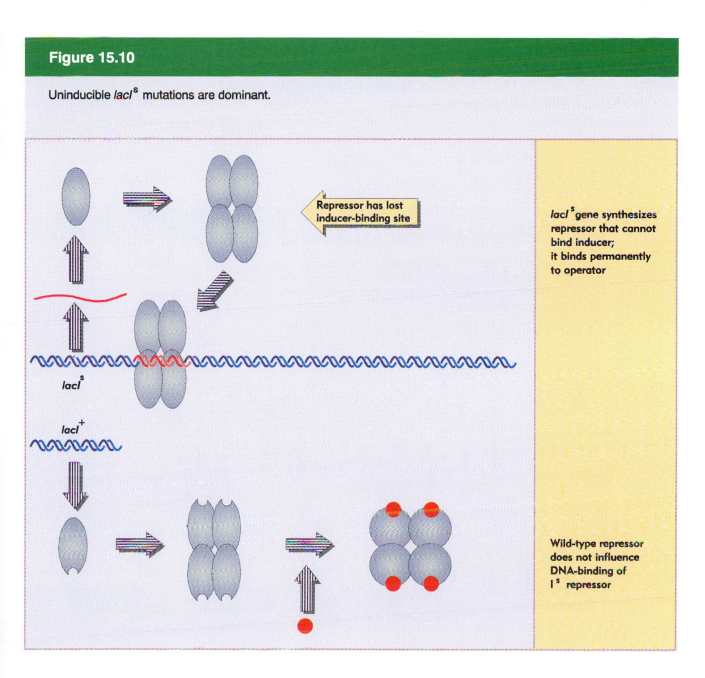

Figure 15.10

Uninducible *lacI* s mutations are dominant.

Repressor has lost inducer-binding site

lacI s

lacI +

lacI s gene synthesizes repressor that cannot bind inducer; it binds permanently to operator

Wild-type repressor does not influence DNA-binding of I s repressor

the operator and prevents transcription. The addition of inducer is to no avail. This happens because the mutant repressor binds to all *lac* operators in the cell to prevent their transcription, and cannot be pried off, irrespective of the properties of any wild-type repressor protein that is present.

The properties of *lacI* mutations can be explained in terms of the structure of the repressor protein. The interaction between the two types of binding site on the repressor controls gene expression in response to the environment. The *DNA-binding site* recognizes the sequence of the operator. The *inducer-binding site* binds the small-molecule inducer; and as a result of this interaction, the DNA-binding site *loses* its ability to hold the operator DNA.

The two binding sites can be identified within the repressor subunit by mutations in *lacI* that inactivate them. Mutations in the DNA-binding site are constitutive (because repressor cannot bind DNA to block RNA polymerase). Mutations in the inducer-binding site are uninducible (because inducer cannot reduce the affinity of repressor for DNA).

An important feature of repressor function is that it is a multimeric protein. Repressor subunits associate at random in the cell to form the active protein tetramer. When two different alleles of the *lacI* gene are present, the subunits made by each can associate to form a heterotetramer, whose properties differ from those of either homotetramer. This type of interaction between subunits is a characteristic feature of multimeric proteins and is described as **interallelic complementation** (see Chapter 1).

Negative complementation occurs between some repressor mutants, as seen in the combination of *lacI*$^{-d}$ with *lacI*$^{+}$ genes. The *lacI*$^{-d}$ mutation

Figure 15.11

Mutations map the regions of the *lacI* gene responsible for different functions. The DNA-binding domain is identified by *lacI*$^{-d}$ mutations at the N-terminal region; *lacI*$^{-}$ mutations unable to form tetramers are located between residues 220-280; other *lacI*$^{-}$ mutations occur throughout the gene; *lac*^{s} mutations occur in regularly spaced clusters between residues 62-300.

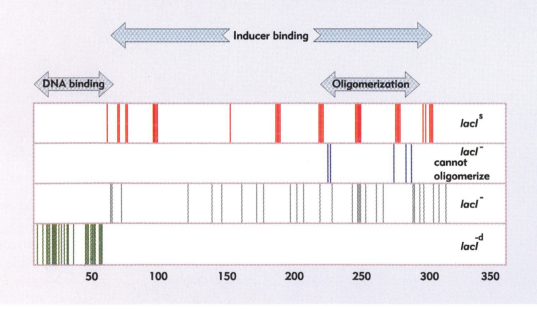

alone results in the production of a repressor that cannot bind the operator, and is therefore constitutive like the *lacI⁻* alleles. Because the *lacI⁻* type of mutation inactivates the repressor, it is recessive to the wild-type. However, the ⁻d notation indicates that this variant of the negative type is dominant when paired with a wild-type allele. Such mutations are said to be ***trans*-dominant**; they are also called **dominant negatives**.

The reason for the dominance is that the *lacI⁻ᵈ* allele produces a 'bad' subunit, which is not only itself unable to bind to operator DNA, but is also able as part of a tetramer to prevent any 'good' subunits from binding. This demonstrates that the repressor tetramer as a whole, rather than the individual monomer, is needed to achieve repression. The poisoning effect also can be produced *in vitro* by mixing appropriate 'good' and 'bad' subunits.

The *lacI⁻ᵈ* mutations lie in the DNA-binding site of the repressor subunit. This explains their ability to prevent mixed tetramers from binding to the operator; a reduction in the number of binding sites reduces the specific affinity for the operator.

The map of the *lacI* gene shown in **Figure 15.11** shows that the *lacI⁻ᵈ* mutations are clustered at the extreme left end of the gene. This identifies the immediate N-terminal region of the protein as the DNA-binding site. Mutations of the recessive *lacI⁻* type also occur elsewhere in the molecule, but could exert their effects on DNA binding indirectly.

The role of the N-terminal region in specifically binding DNA is shown also by its location as the site of occurrence of 'tight binding' mutations. These increase the affinity of the repressor for the operator, sometimes so much that it cannot be released by inducer. They are rare.

Uninducible mutations of the *lacIˢ* type render the repressor unresponsive to the inducer. This could happen either because the protein has lost its inducer-binding site, or because it has become unable to transmit its effect to the DNA-binding site. As can be seen from Figure 15.11, the *lacIˢ* mutations occur in clusters that are rather regularly spaced along the gene. The spacing may represent turns in the polypeptide chain.

Repressor protein binds to the operator and is released by inducer

The repressor was isolated originally by purifying the component able to bind the gratuitous inducer IPTG. (Because the amount of repressor in the cell is so small, in order to obtain enough material it was necessary to use a promoter up mutation to increase *lacI* transcription, and to place this *lacI* locus on a DNA molecule present in many copies per cell. This results in an overall overproduction of 100- to 1000-fold.)

The repressor binds to double-stranded DNA containing the sequence of the wild-type *lac* operator. The repressor does not bind if the DNA has been obtained from an O^c mutant. The addition of IPTG releases the repressor from O_{lac} DNA *in vitro*. The *in vitro* reaction between repressor protein and O_{lac} DNA therefore displays the characteristics of control inferred *in vivo;* so it can be used to establish the basis for repression.

How does the repressor recognize the specific sequence of operator DNA? The interaction between repressor and operator is often taken as a paradigm for sequence-specific DNA-binding reactions.

The operator has a feature common to many

recognition sites for bacterial regulator proteins: it is palindromic. The inverted repeats are high-lighted in **Figure 15.12**.

The same approaches are used to define the points that the repressor contacts in the operator as for analyzing the polymerase–promoter interaction (see Chapter 14). Deletions of material on either side define the end points of the region; constitutive point mutations identify individual base pairs that must be crucial. Experiments in which DNA bound to repressor is compared with unbound DNA for its susceptibility to methylation or UV cross-linking identify bases that are either protected or more susceptible when associated with the protein.

The region of DNA protected from nucleases by bound repressor lies within the region of symmetry,

comprising the 26 bp region from −5 to +21. The area identified by constitutive mutations is even smaller. Within a central region extending over the 13 bp from +5 to +17, there are eight sites at which single base pair substitutions cause constitutivity. This emphasizes the same point made by the promoter mutations summarized earlier in Figure 14.14. *A small number of essential specific contacts within a larger region can be responsible for sequence-specific association of DNA with protein.*

The pattern of enhancement and protection of bases shows some features of symmetry within a general region of close contacts. Figure 15.12 shows that all but two of the thymine residues within the region from +1 to +22 can be crosslinked to repressor. Methylation experiments show that a large pro-

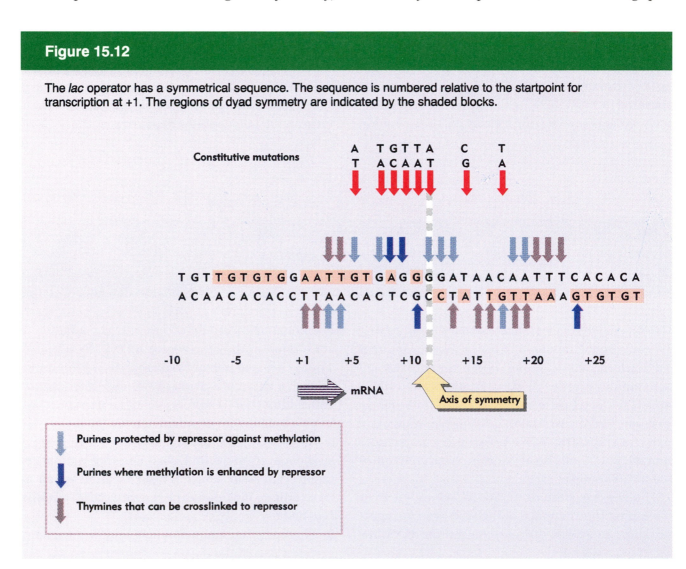

Figure 15.12

The *lac* operator has a symmetrical sequence. The sequence is numbered relative to the startpoint for transcription at +1. The regions of dyad symmetry are indicated by the shaded blocks.

portion of the purines (A and G) between +3 and +19 are protected quite strongly by repressor binding. A few display increased susceptibility, presumably due to the creation of 'hydrophobic pockets' in the repressor–operator complex.

The contacts between +1 and +6 are essentially symmetrical with those made between +21 and +16; contacts closer to the axis of symmetry are not symmetrical. This is reflected in the pattern of constitutive mutations, which are symmetrical at +5/+17 and +8/+14, but not at points closer to the axis of symmetry. Thus the repressor binds to the operator in such a way as to sit symmetrically about the outlying points of contact, while lying asymmetrically in the immediate vicinity of the axis of symmetry.

The inverted repeats of the *lac* operator are not identical. The three differences between them are shown by the breaks in the shaded blocks. The distribution of the sites of mutation suggests that the left side of the operator is more susceptible to damage: it contains six mutations, compared with the two on the right. Also, mutations that occur at equivalent positions on the left side and right side have greater effect on the left. Thus symmetry is clearly relevant to repressor–operator contacts, but the repressor seems to bind to the left side more intimately. Indeed, the affinity of the operator for repressor can be increased ~10-fold by mutations that increase symmetry by making the right inverted repeat a perfect repeat of the left. The 'ideal' operator thus consists of a symmetrical repeat of the left half.

Does the symmetry of the DNA sequence reflect a symmetry in the protein? This is likely to be the case, because the repressor is a tetramer of identical subunits, each of which must therefore have the same DNA-binding site. Each inverted repeat of the operator is probably contacted in the same way by a repressor dimer. Like many other bacterial regulators, the *lac* repressor uses a short α-helical region to contact DNA; the sequence of this α-helix determines the specificity of binding to DNA.

Suppose that a repressor tetramer is bound tightly to the operator. Inducer enters the cell, and reduces the affinity for the operator of any repressor to which it binds. But how does inducer cause the repressor to be released from an operator to which it is already bound?

Various inducers cause characteristic reductions in the affinity of the repressor for the operator *in vitro*. These changes correlate with the effectiveness of the inducers *in vivo*. This suggests that induction results from a reduction in the attraction between operator and repressor. How is this accomplished? The rate at which the repressor dissociates from the operator is rather slow, with a half-life *in vitro* of 10–20 minutes. But when IPTG is added, there is an immediate reduction in the stability of the complex, as seen by a drastic decrease in its half-life.

This result distinguishes between the two models for repressor action illustrated in **Figure 15.13**:

◆ The equilibrium model (left) calls for repressor bound to DNA to be in rapid equilibrium with free repressor; inducer would bind to the free form of repressor, and thus unbalance the equilibrium by preventing reassociation with DNA.

◆ But the rate of dissociation of the repressor from the operator (as measured in the absence of inducer) is much too slow to be compatible with this model. This means that instead the *inducer must bind directly to repressor protein complexed with the operator.* As indicated in the model on the right, inducer binding must produce a change in the repressor that makes it let go of the operator.

Binding of the repressor–IPTG complex to the operator can be studied by using greater concentrations of the protein in the methylation protection/enhancement assay. The large amount compensates for the low affinity of the repressor–IPTG complex for the operator. The complex makes exactly the same pattern of contacts with DNA as the free repressor. An analogous result is obtained with mutant repressors whose affinity for operator DNA is increased; they too make the same pattern of contacts.

Overall, a range of repressor variants whose

Figure 15.13

Does the inducer bind to free repressor to upset an equilibrium (left) or directly to repressor bound at the operator (right)?

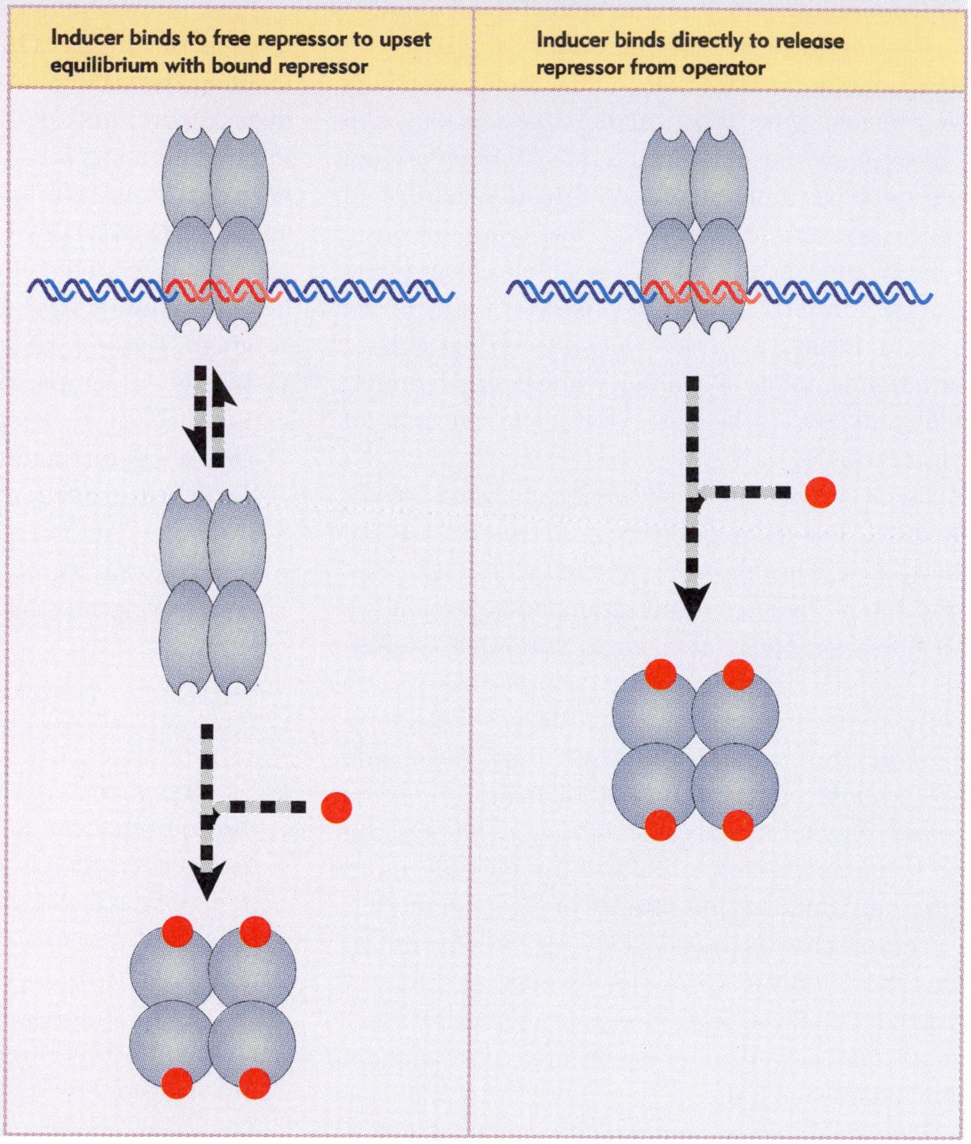

Inducer binds to free repressor to upset equilibrium with bound repressor	Inducer binds directly to release repressor from operator

affinities for the operator span seven orders of magnitude all make the same contacts with DNA. *Changes in the affinity of the repressor for DNA must therefore occur by influencing the general conformation of the protein in binding DNA, not by making or breaking one or a few individual bonds.*

The behavior of the isolated repressor protein *in vitro* shows directly that it possesses a DNA-binding site whose ability to remain attached to operator DNA is influenced by the structure of the rest of the molecule.

When the repressor is treated with trypsin, it is cleaved preferentially at amino acid 59. The C-terminal fragment of the protein, containing residues 60–360, is known as the **trypsin-resistant core**. It retains the ability to aggregate into a tetramer and to bind inducer; but it cannot bind the operator. The amino-terminal fragment, consisting of residues 1–59, is known as the **long headpiece**. It can be cleaved again by trypsin at amino acid 51, generating a fragment of residues 1-51 called the **short headpiece**.

The short and long headpieces retain the ability to bind to DNA; and when presented with operator DNA, they make the same pattern of contacts achieved by intact repressor (although they bind more weakly than the intact protein). The ability of the headpieces to bind to the operator suggests that their structure is independent of the rest of the protein.

This accords with models in which the DNA-binding site lies as an **arm** or **protrusion** of the N-terminal 50 amino acids from the body of the protein. The arm is connected to the core by a **hinge** region, constructed from amino acids 50–80. The remainder of the protein, amino acids 81–360, is responsible for aggregating into the tetrameric structure and for binding inducer.

We know from a variety of techniques that binding of inducer causes an immediate conformational change in the repressor protein. Binding of two molecules of inducer to the repressor tetramer is adequate to release repression. The independence of the headpiece suggests a model for the conformational change that releases repressor from DNA when inducer binds. **Figure 15.14** represents it diagrammatically. The conformation of the DNA-binding site is altered from a state in which it exactly fits the DNA sequence to a state in which the structure of the hinge changes so that the headpiece is held in a different register that cannot contact the DNA tightly enough.

What effect does the binding of repressor at the operator have upon the binding of RNA polymerase at the promoter? The two proteins may be bound to DNA simultaneously, and *the binding of repressor actually enhances the binding of RNA polymerase!* However, the bound enzyme is prevented from initiating transcription.

The equilibrium constant for RNA polymerase binding alone to the *lac* promoter is 1.9×10^7 M^{-1}. The presence of repressor increases this constant by two orders of magnitude to 2.5×10^9 M^{-1}. In terms of the range of values for the equilibrium

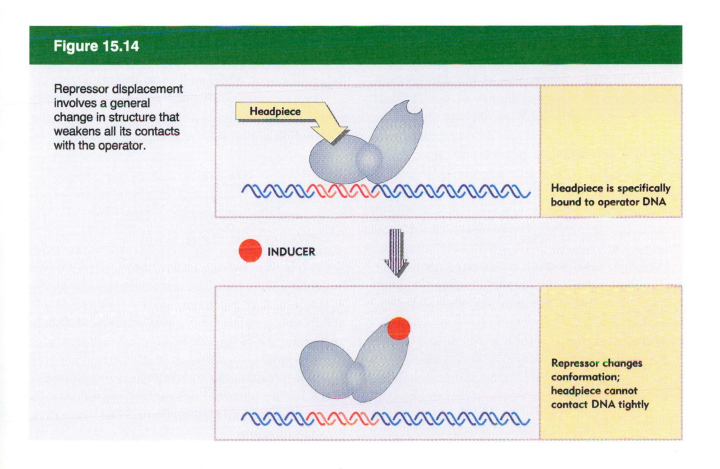

Figure 15.14

Repressor displacement involves a general change in structure that weakens all its contacts with the operator.

Headpiece

Headpiece is specifically bound to operator DNA

INDUCER

Repressor changes conformation; headpiece cannot contact DNA tightly

constant K_B given in Figure 14.8, repressor protein effectively converts the formation of closed complex by RNA polymerase at the *lac* promoter from a weak to a strong interaction.

What does this mean for induction of the operon? The higher value for K_B means that, when occupied by repressor, the promoter is 100 times more likely to be bound by an RNA polymerase (if the concentration of polymerase is limiting). And by allowing RNA polymerase to be bound at the same time as repressor, it becomes possible for transcription to begin immediately upon induction, instead of waiting for an RNA polymerase to be captured.

The repressor in effect causes RNA polymerase to be stored at the promoter. The complex of RNA polymerase•repressor•DNA is blocked at the closed stage. When inducer is added, the repressor is released, and the closed complex is converted to an open complex that initiates transcription. The overall effect of repressor has been to speed up the induction process.

Does this model apply to other systems? The interaction between RNA polymerase, repressor, and the promoter/operator region is distinct in each system, because the operator does not always overlap with the same region of the promoter (see Figure 15.17). For example, in phage lambda, the operator lies in the upstream region of the promoter (see Chapter 17). Thus a bound repressor does not interact with RNA polymerase in the same way in other systems.

The specificity of protein–DNA interactions

Probably all proteins that have a high affinity for a specific sequence also possess a low affinity for any (random) DNA sequence. A large number of low affinity sites will compete just as well for a repressor tetramer as a small number of high-affinity sites. There is only one high-affinity site in the *E. coli* genome: the operator. The remainder of the DNA can be considered to provide low-affinity binding sites. Every base pair in the genome starts a new low affinity site. (Just moving one base pair along the genome, out of phase with the operator itself, creates a low affinity site!) So there are 4.2×10^6 low affinity sites.

The large number of low affinity sites means that, even in the absence of a specific binding site, all or virtually all repressor is bound to DNA; none is free in solution.

We may describe the binding of repressor to DNA by the equilibrium:

$$K_A = \frac{[\text{Repressor–DNA}]}{[\text{Free repressor}]\,[\text{DNA}]}$$

in which [Repressor–DNA] is the concentration of repressor bound to DNA, [Free repressor] is the concentration of free repressor, and [DNA] is the concentration of nonspecific binding sites.

What proportion of total repressor protein is free? By rearranging the equation, we see that the distribution of repressor is given by:

$$\frac{[\text{Free Repressor}]}{[\text{Repressor–DNA}]} = \frac{1}{K_A.\,[\text{DNA}]}$$

Applying the parameters for the *lac* system, we find that:

◆ The nonspecific equilibrium binding constant is $K_A = 2 \times 10^6\ \text{M}^{-1}$.

◆ The concentration of nonspecific binding sites is 4×10^6 in a bacterial volume of 10^{-15} liter, which corresponds to [DNA] = 7×10^{-3} M (a very high concentration).

Substituting these values gives:

Free:Bound repressor = $1 / (2 \times 10^6 \times 7 \times 10^{-3}) = 10^{-4}$

So all but 0.01% of repressor is bound to (random) DNA. Since there are ~10 molecules of repressor per cell, this is tantamount to saying that there is no free repressor protein. This has an important implication for the interaction of repressor with the operator: it means that we are concerned with the *partitioning* of the repressor on DNA, in which the single high affinity site of the operator *competes* with the large number of low affinity sites.

In this competition, the absolute values of the association constants for operator and random DNA are not important; what is important is the ratio of K_{sp} (the constant for binding a specific site) to K_{nsp} (the constant for binding any random DNA sequence), that is, the specificity.

We can define the parameters that influence the ability of a regulator protein to saturate its target site by comparing the equilibrium equations for specific and nonspecific binding. As might be expected intuitively, the important parameters are:

◆ The size of the genome dilutes the ability of a protein to bind specific target sites.

◆ The specificity of the protein counters the effect of the mass of DNA.

◆ The amount of protein that is required increases with the total amount of DNA in the genome and decreases with the specificity.

◆ The amount of protein also must be in reasonable excess of the total number of specific target sites, so we expect regulators with many targets to be found in greater quantities than regulators with individual targets.

Table 15.1 compares the equilibrium constants for *lac* repressor/operator binding with repressor/general DNA binding. From these constants, we can deduce how repressor is partitioned between the operator and the rest of DNA, and what happens to the repressor when inducer causes it to dissociate from the operator.

Repressor binds ~10^7 times better to operator DNA than to any random DNA sequence of the same length. So the operator comprises a single high affinity site that will compete for the repressor 10^7 better than any low affinity (random) site. How does this ensure that the repressor can maintain effective control of the operon?

Using the specificity, we can calculate the distribution between random sites and the operator, and can express this in terms of occupancy of the operator. If there are 10 molecules of *lac* repressor per cell with a specificity for the operator of 10^7, the operator will be bound by repressor 96% of the time. The role of specificity explains two features of the *lac* repressor–operator interaction:

◆ When inducer binds to the repressor, the affinity for the operator is reduced by ~10^3-fold. The affinity for general DNA sequences remains unaltered. So the specificity is now only 10^4, which is insufficient to capture the repressor against competition from the excess of 4.2×10^6 low affinity sites. Only 3% of operators would be bound under these conditions.

◆ Mutations that reduce the affinity of the operator for the repressor by as little as 20–30× have sufficient effect to be constitutive. Within the genome, the mutant operators can be overwhelmed by the preponderance of random sites. The occupancy of the operator is reduced to ~50% if repressor's specificity is reduced just 10×.

Table 15.1

lac repressor binds strongly and specifically to its operator, but is released by inducer.
All equilibrium constants are in M^{-1}.

DNA	Repressor	Repressor + Inducer
Operator	2×10^{13}	2×10^{10}
Other DNA	2×10^6	2×10^6
Specificity	10^7	10^4

The consequence of these affinities is that in an uninduced cell, one tetramer of repressor usually is bound to the operator. All or almost all of the remaining tetramers are bound at random to other regions of DNA, as illustrated in **Figure 15.15**. There are likely to be very few or no repressor tetramers free within the cell.

The addition of inducer abolishes the ability of

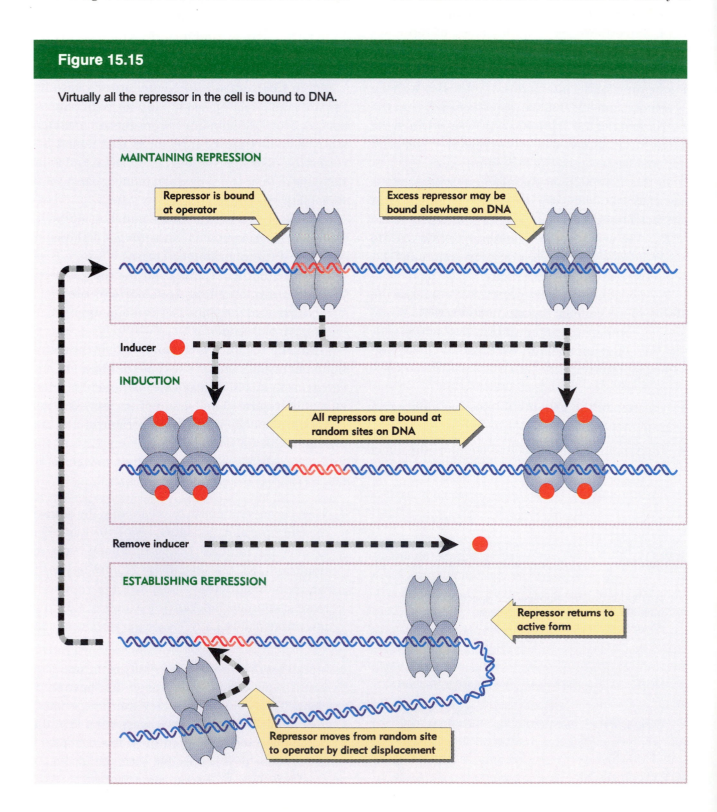

Figure 15.15

Virtually all the repressor in the cell is bound to DNA.

repressor to bind specifically at the operator. Those repressors bound at the operator are released, and bind to random (low affinity) sites. Thus in an induced cell, the repressor tetramers are 'stored' on random DNA sites. In a noninduced cell, a tetramer is bound at the operator, while the remaining repressor molecules are bound to nonspecific sites. *The effect of induction is therefore to change the distribution of repressor on DNA, rather than to generate free repressor.*

When inducer is removed, repressor recovers its ability to bind specifically to the operator, and does so very rapidly. This must involve its movement from a nonspecific 'storage' site on DNA. What mechanism is used for this rapid movement? The ability to bind to the operator very rapidly is not consistent with the time that would be required for multiple cycles of dissociation and reassociation with nonspecific sites on DNA. The discrepancy excludes random-hit mechanisms for finding the operator, suggesting that the repressor can move directly from a random site on DNA to the operator. This is the same issue that we encountered previously with the ability of RNA polymerase to find its promoters (see Figure 14.12). The same solution is likely: movement could be accomplished by direct displacement from site to site (as indicated in Figure 15.15). A displacement reaction might be aided by the presence of more binding sites per tetramer (four) than are actually used to contact DNA at any one time (two).

The parameters involved in finding a high affinity operator in the face of competition from many low affinity sites pose a dilemma for repressor. Under conditions of repression, there must be high specificity for the operator. But under conditions of induction, this specificity must be relieved. Suppose, for example, that there were 1000 molecules of repressor per cell. Then only 0.04% of operators would be free under conditions of repression. But upon induction only 40% of operators would become free. We therefore see an inverse correlation between the ability to achieve complete repression and the ability to relieve repression effectively. We assume that the number of repressors synthesized *in vivo* has been subject to selective forces that balance these demands.

The difference in expression of the lactose operon between its induced and repressed states *in vivo* is actually 10^3-fold. In other words, even when inducer is absent, there is a basal level of expression of ~0.1% of the induced level. This would be reduced if there were more repressor protein present, increased if there were less. Thus it could be impossible to establish tight repression if there were fewer repressors than the 10 found per cell; and it might become difficult to induce the operon if there were too many.

Repression can occur at multiple loci

The *lac* repressor acts only on the operator of the *lacZYA* cluster. Other repressors, however, control dispersed structural genes by binding at more than one operator. An example is the *trp* repressor, which controls three unlinked sets of genes:

◆ An operator at the cluster of structural genes *trpEDBCA* controls coordinate synthesis of the enzymes that synthesize tryptophan from chorismic acid.

◆ An operator at another locus controls the *aroH* gene, which codes for one of the three enzymes that catalyze the initial reaction in the common pathway of aromatic amino acid biosynthesis.

◆ The *trpR* regulator gene is repressed by its own product, the *trp* repressor. Thus the repressor protein acts to reduce its own synthesis. This circuit is an example of **autogenous control**. Such circuits are quite common in regulatory genes, and may be either negative or positive. (We discuss examples later in this chapter and in Chapter 17.)

A related operator sequence, extending over 21 bp, is present at each of the three loci at which the *trp* repressor acts. The conservation of sequence is indicated in **Figure 15.16**. Each operator contains appreciable (but not identical) dyad symmetry. Presumably the features conserved at all three operators include the important points of contact for *trp* repressor. This explains how one repressor protein acts on several loci: *each locus has a copy of a specific DNA-binding sequence recognized by the repressor* (just as each promoter shares consensus sequences with other promoters).

Figure 15.17 summarizes the variety of relationships between operators and promoters. A notable feature of the dispersed operators recognized by TrpR is their presence at different locations within the promoter in each locus. In *trpR* the operator lies between positions −12 and +9, while in the *trp* operon it occupies positions −23 to −3, but in the *aroH* locus it lies farther upstream, between −49 and −29. In other cases, the operator lies downstream from the promoter (as in *lac*), or apparently just upstream of the promoter (as in *gal*, where the nature of the repressive effect is not quite clear).

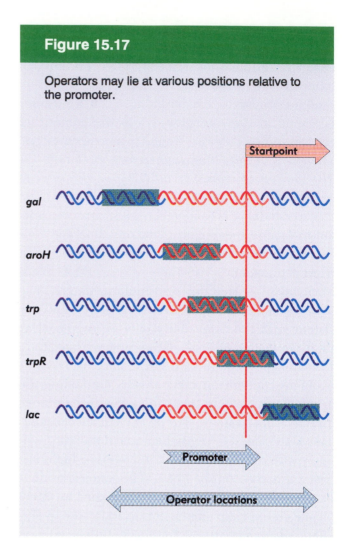

Figure 15.17

Operators may lie at various positions relative to the promoter.

Startpoint

gal

aroH

trp

trpR

lac

Promoter

Operator locations

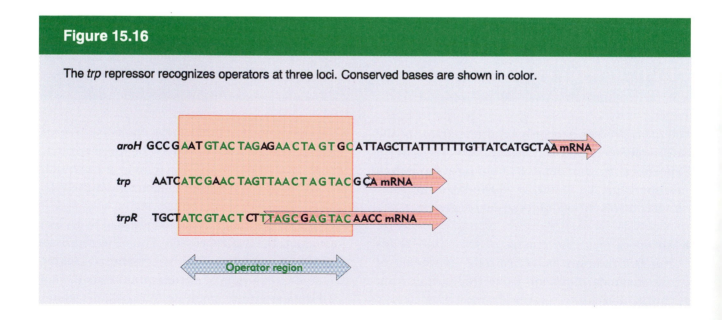

Figure 15.16

The *trp* repressor recognizes operators at three loci. Conserved bases are shown in color.

aroH GCC G AAT GTAC TAGA GAA CTA GT GC ATTAGCTTATTTTTTTGTTATCATGCTAA mRNA

trp AATC ATC GAAC TAGT TAA CT AG TAC G CA mRNA

trpR TGCT ATC GTAC T CT TTAGC GAG TAC AACC mRNA

Operator region

The ability of the repressors to act at operators whose positions are different in each target promoter suggests that there could be differences in the exact mode of repression, the common feature being that RNA polymerase is prevented from initiating transcription at the promoter.

Distinguishing positive and negative control

Positive and negative control systems are defined by the response of the operon when no regulator protein is present. The characteristics of the two types of control system are mirror images.

Genes under negative control are expressed unless they are switched off by a repressor protein. Any action that interferes with gene expression can provide a negative control, but there is a uniformity in these mechanisms: a repressor protein either binds to DNA to prevent RNA polymerase from initiating transcription, or binds to mRNA to prevent a ribosome from initiating translation.

Negative control provides a fail-safe mechanism: if the regulator protein is inactivated, the system functions and so the cell is not deprived of these enzymes. It is easy to see how this might evolve. Originally a system functions constitutively, but then cells able to interfere specifically with its expression acquire a selective advantage by virtue of their increased efficiency.

For genes under positive control, expression is possible only when an active regulator protein is present. The mechanism for controlling an individual operon is an exact counterpart of negative control, but instead of *interfering* with initiation, the regulator protein is *essential* for it. It interacts with DNA and with RNA polymerase to *assist the initiation event.* A positive regulator protein that responds to a small molecule is usually called an **activator**. Other positive controls provide for the global substitution of sigma factors that change the selection of promoters (Chapter 14), or antitermination factors that change the recognition of terminators (Chapter 16).

It is more difficult to see how positive control evolved, since the cell must have had the ability to express the regulated genes even before any control existed. Presumably some component of the control system must have changed its role. Perhaps originally it was used as a regular part of the apparatus for gene expression; then later it became restricted to act only in a particular system or systems.

Operons are defined as **inducible** or **repressible** by the nature of their response to the small molecule that regulates their expression. Just as it is advantageous for a bacterium to induce a set of enzymes only after addition of the inducer substrate that they metabolize, so also it is useful to repress the enzymes that synthesize some compound if it is provided in adequate amounts by the medium. Thus inducible operons function only in the *presence* of the small-molecule inducer. Repressible operons function only in the *absence* of the small-molecule **corepressor** (so called to distinguish it from the repressor protein).

The terminology used for repressible systems describes the active state of the operon as **derepressed**; this has the same meaning as *induced*. The condition in which a (mutant) operon cannot be derepressed is sometimes called **super-repressed**; this is the exact counterpart of *uninducible*.

Either positive or negative control could be used to achieve either induction or repression by utilizing appropriate interactions between the regulator protein and the small-molecule inducer or corepressor. **Figure 15.18** summarizes four simple types of control circuit. Induction is achieved when an inducer inactivates a repressor protein or activates an activator protein. Repression is accomplished when a corepressor activates a repressor protein or inactivates an activator protein.

Figure 15.18

Control circuits are versatile and can be designed to allow positive or negative control of induction or repression.

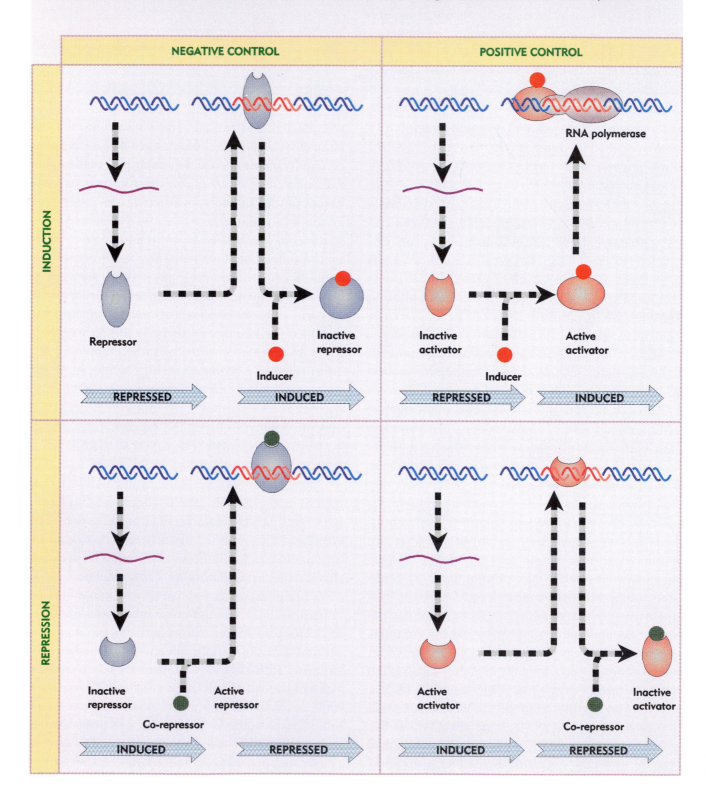

The genetic consequences of inactivating the regulator protein can be used to discriminate between negative and positive control systems. Inactivation of a repressor protein creates the recessive constitutivity (or derepression) typical of negative control systems. On the other hand, a mutation that inactivates an activator causes the recessive uninducibility or super-repression that typifies positive control systems.

The *tryptophan* operon provides an example of a repressible system. Tryptophan is the end product of the reactions catalyzed by a series of biosynthetic enzymes. Both the activity and the synthesis of the tryptophan enzymes are controlled by the level of tryptophan in the cell.

The classic feedback loop of **end-product inhibition** applies to the enzymes: the catalytic activities of the first enzyme of the pathway are inhibited by tryptophan, the ultimate product. This means that when the cell has sufficient tryptophan, it is able to cut off the synthesis of further molecules of the amino acid by inhibiting the beginning of the pathway.

Tryptophan also functions as a corepressor that activates a repressor protein. This is the classic mechanism for repression, one of the examples given in Figure 15.18 (lower left). In conditions when the supply of tryptophan is plentiful, the operon is repressed because the repressor protein•corepressor complex is bound at the operator. When tryptophan is in short supply, the corepressor is inactive, therefore has reduced specificity for the operator, and is stored elsewhere on DNA.

Deprivation of repressor causes ~70-fold increase in the frequency of initiation events at the *trp* promoter. Even under repressing conditions, the structural genes continue to be expressed at a low **basal** or **repressed level**. The efficiency of repression at the operator is much lower than in the *lac* operon (where the basal level is only ~1/1000 of the induced level).

We have treated both induction and repression as phenomena that rely upon allosteric changes induced in regulator proteins by small molecules. Other means also can be used to control the activities of regulator proteins. One amusing example is OxyR, a transcriptional activator of genes induced by hydrogen peroxide. The OxyR protein is directly activated by oxidation, so it provides a sensitive measure of oxidative stress. Another common type of signal is phosphorylation of a regulator protein.

Catabolite repression involves positive regulation at the promoter

So far we have dealt with the promoter as a DNA sequence that is competent to bind RNA polymerase, which then initiates transcription. But there are some promoters at which RNA polymerase cannot initiate transcription without assistance from an ancillary protein. Such proteins are positive regulators, because their presence is necessary to switch on the transcription unit. Several positive regulators are known, some coded by phages and some present within the host cell. One model for their action is that the activator overcomes a deficiency in the promoter, for example, a poor consensus sequence at -35 or -10.

One of the most widely acting activators is a protein that controls the activity of a large set of operons in *E. coli* in response to carbon nutrient conditions. When glucose is available as an energy source, it is used in preference to other sugars. Thus when *E. coli* finds (for example) both glucose and lactose in the medium, it metabolizes the glucose and represses the use of lactose.

This choice is accomplished by preventing expression of several operons, including *lactose,*

Figure 15.19

Glucose causes catabolite repression by reducing the level of cyclic AMP.

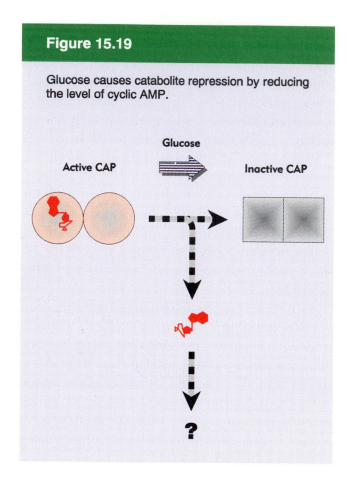

is a positive control factor whose presence is necessary to initiate transcription at dependent promoters. The protein is active *only in the presence of cyclic AMP*, which behaves as the classic small-molecule inducer (see Figure 15.18; upper right). Reducing the level of cyclic AMP renders the protein unable to bind to the control region, which in turn prevents RNA polymerase from initiating transcription. So the effect of glucose in reducing cyclic AMP levels is to deprive the relevant operons of a control factor necessary for their expression.

The CAP factor binds to DNA, and complexes of cyclic AMP•CAP•DNA can be isolated at each promoter at which it functions. The factor is a dimer of two identical subunits of 22,500 daltons, which can be activated by a single molecule of cyclic AMP. A CAP monomer contains a DNA-binding region and a transcription-activating region.

A CAP dimer binds to a site of ~22 bp at a responsive promoter. The binding sites include variations of the consensus sequence given in **Figure 15.21**. Mutations preventing CAP action usually are located within the well-conserved pentamer $\frac{TGTGA}{ACACT}$, which appears to be the essential element in recognition. CAP binds most strongly to sites that contain two (inverted) versions of the pentamer, because both subunits of the dimer bind effectively to the DNA. Many binding sites lack the second pentamer, however, and in these the second subunit must bind a different sequence (if it binds

galactose and *arabinose*. The effect is called **catabolite repression**. It represents a general coordinating system that exercises a preference for glucose by inhibiting the expression of the operons that code for the enzymes of alternative metabolic pathways.

Catabolite repression is set in train by the ability of glucose to reduce (by unknown means) the level of cyclic AMP (cAMP) in the cell. Expression of the catabolite-regulated operons shows an inverse relationship with the level of cyclic AMP. The effect of this relationship is illustrated in **Figure 15.19**.

Cyclic AMP is synthesized by the enzyme **adenylate cyclase**. The reaction uses ATP as substrate and introduces a 3′–5′ link via phosphodiester bonds, generating the structure drawn in **Figure 15.20**. Mutations in the gene coding for adenylate cyclase (*cya⁻*) do not show catabolite repression.

Mutations in the *cap⁻* gene identify the regulator that acts directly on the target operons. The protein is known as CAP (for catabolite activator protein). It

Figure 15.20

Cyclic AMP has a single phosphate group connected to both the 3′ and 5′ positions of the sugar ring.

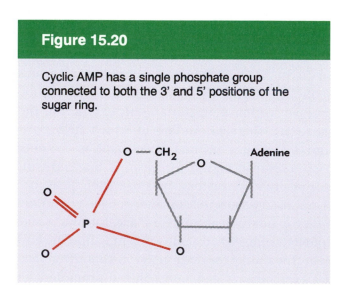

Figure 15.21

The consensus sequence for CAP contains the well conserved pentamer TGTGA and (sometimes) an inversion of this sequence (TCANA).

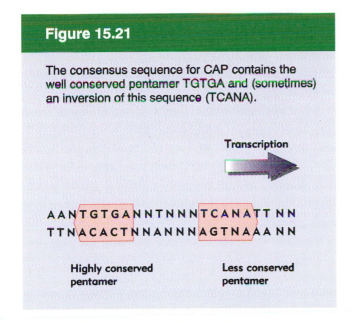

Transcription

AANTGTGANNTNNNTCANATTNN
TTNACACTNNANNNAGTNAAANN

Highly conserved pentamer   Less conserved pentamer

to DNA). The hierarchy of binding affinities for CAP helps to explain why different genes are activated by different levels of cyclic AMP *in vivo*.

The action of CAP has the curious feature that its binding sites lie at different locations relative to the startpoint in the various operons that it regulates. And the TGTGA pentamer may lie in either orientation. The three examples summarized in **Figure 15.22** encompass the range of locations:

◆ Sometimes the CAP-binding site clearly lies within the promoter. An example is the *gal* locus, where the CAP binding site is centered on −41, and extends between −50 and −23. It is likely that only a single CAP dimer is bound, probably in quite intimate contact with RNA polymerase, since the CAP binding site extends well into the region generally protected by the RNA polymerase.

◆ The CAP-binding site is adjacent to the promoter. An example is the *lac* operon, in which the region of DNA protected by CAP is centered on −61 and extends from about −72 to −52. It is possible that two dimers of CAP are bound. The binding pattern is consistent with the presence of CAP largely on one face of DNA, the same face that is bound by RNA polymerase. This location would place the two proteins just about in reach of each other.

◆ In other operons, the CAP-binding site lies well upstream of the promoter. In the example of the *ara* region, the binding site for a single CAP is the farthest from the startpoint, at −107 to −78. Here the CAP cannot be in contact with RNA polymerase, because *another* regulatory protein binds in the region between the CAP and RNA polymerase sites.

Dependence on CAP is related to the intrinsic efficiency of the promoter. No CAP-dependent promoter has a good −35 sequence and some also lack good −10 sequences. Viewed the other way, we might argue that effective control by CAP would be difficult if the promoter had effective −35 and −10 regions that interacted independently with RNA polymerase.

There are in principle two ways in which CAP might activate transcription: it could interact directly with RNA polymerase; or it could act upon DNA to change its structure in some way that assists RNA polymerase to bind. In fact, CAP has effects upon both RNA polymerase and DNA.

Figure 15.22

The CAP protein can bind at different sites relative to RNA polymerase.

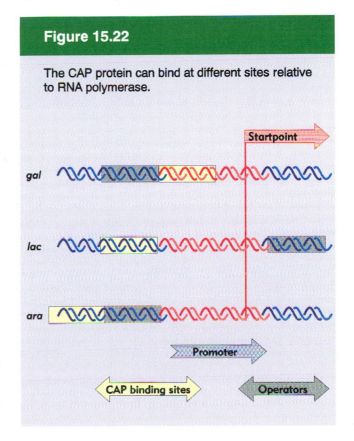

RNA polymerase can be reconstituted from its subunits, and when the α subunit has a deletion in the C-terminal end, transcription appears normal except for the loss of ability to be activated by CAP. This suggests that CAP acts directly upon the α subunit to stimulate RNA polymerase. Experiments using CAP dimers in which only one of the subunits has a functional transcription-activating region shows that, when CAP is bound at the *lac* promoter, only the activating region of the subunit nearer the startpoint is required, presumably because it touches RNA polymerase. This offers an explanation for the lack of dependence on the orientation of the binding site: the dimeric structure of CAP ensures that one of the subunits is available to proffer an activating region to RNA polymerase, no matter which subunit binds to DNA and in which orientation.

As a general principle, formation of a transcription complex can be assisted by protein–protein interactions when two (or more) proteins are bound to DNA. Quite a modest interaction between two proteins would be sufficient to achieve a substantial increase in the ability of RNA polymerase to bind a promoter.

The effect upon RNA polymerase binding de-

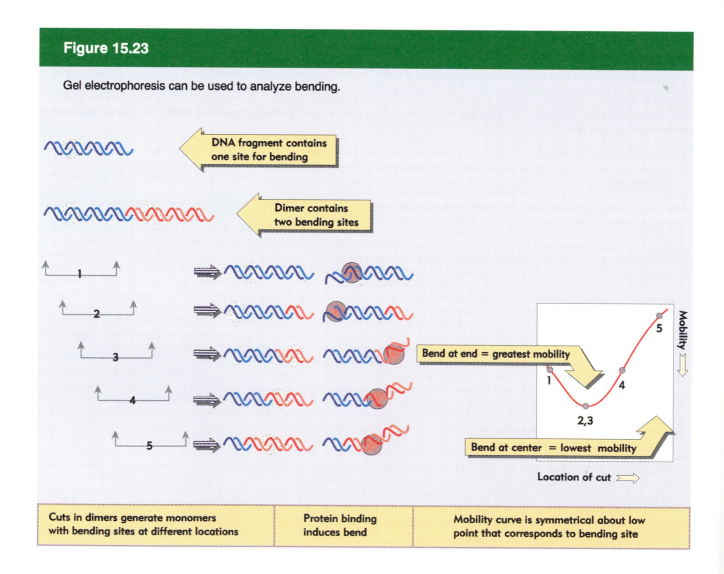

Figure 15.23

Gel electrophoresis can be used to analyze bending.

DNA fragment contains one site for bending

Dimer contains two bending sites

Bend at end = greatest mobility

Bend at center = lowest mobility

Location of cut ⟹

Mobility

| Cuts in dimers generate monomers with bending sites at different locations | Protein binding induces bend | Mobility curve is symmetrical about low point that corresponds to bending site |

pends on the relative locations of the two proteins. When CAP binds within the promoter (as in *gal*), it increases the rate of transition from the closed to open complex. When CAP binds adjacent to the promoter (as in *lac*) its predominant effect is to increase the rate of initial binding to form a closed complex. This suggests that the exact effects of the interaction between CAP and RNA polymerase depend on the geometry of the proteins at the individual promoter. (The interactions at the *ara* promoter may involve different interactions, involving more proteins.)

Binding sites for CAP at most promoters resemble either *gal* (centered at −41 bp) or *lac* (centered at −61). When the distance is changed from either of these two standards, the ability of CAP to activate transcription is reduced. This may reflect the need for CAP and RNA polymerase to be oriented on the same face of DNA.

The structure of the CAP–DNA complex is interesting: *the DNA has a bend.* Proteins may distort the double helical structure of DNA when they bind, and several regulator proteins induce a bend in the axis.

Figure 15.23 illustrates a technique that can be used to measure the extent and location of a bend. A dimer of the target sequence is made, and it is cut with different restriction enzymes to generate a set of circularly permuted fragments each containing a monomeric length of DNA. The protein-binding site therefore lies at a different location in each of these fragments.

The fragments move at different speeds in an electrophoretic gel, depending on the position of the bend. (If there is no bend, all fragments move at the same rate.) The greatest impediment to motion, causing the lowest mobility, happens when the bend is in the center of the DNA fragment. The least impediment to motion, allowing the greatest mobility, happens when the bend is at one end.

The results are analyzed by plotting mobility against the site of restriction cutting. The greatest mobility is given by the low point on the curve,

and this identifies the situation in which the restriction enzyme has cut the sequence immediately adjacent to the site of bending.

For the interaction of CAP with the *lac* promoter, this point lies at the center of dyad symmetry. The bend is quite severe, >90°, as illustrated in the model of **Figure 15.24**. There is therefore a dramatic change in the organization of the DNA double helix when CAP protein binds. It is possible that the bend has some direct effect upon transcription, but it could be the case that it is needed simply to allow CAP to contact RNA polymerase at the promoter.

Whatever the exact means by which CAP activates transcription at various promoters, it accomplishes the same general purpose: to turn off alternative metabolic pathways when they become unnecessary because the cell has an adequate supply of glucose. Again, this makes the point that coordinate control, of either negative or positive type, can extend over dispersed loci by repetition of binding sites for the regulator protein.

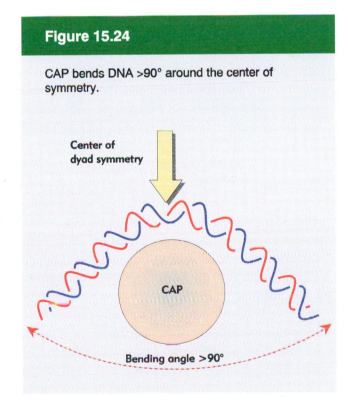

Figure 15.24

CAP bends DNA >90° around the center of symmetry.

Center of dyad symmetry

CAP

Bending angle >90°

Autogenous control may occur at the level of translation

Gene expression can be controlled at any of several stages, which we divide broadly into transcription, processing, and translation:

◆ Transcription often is controlled at the stage of initiation, with the advantage that energy is not wasted producing unnecessary transcripts. Transcription is not usually controlled at elongation, but may be controlled at termination. We see in Chapter 16 that termination may be used to prevent transcription from proceeding past a terminator to the gene(s) beyond.

◆ The RNA primary product may itself be the target of regulation. Availability of the transcript as a whole may be regulated; for example, its stability may determine whether it survives to be translated. Or the ability to process the primary transcript into mature molecules may determine the constitution and functions of the final mRNA. Secondary structure of the RNA may be important in this type of regulation (involving features similar to those involved in regulating termination of transcription). In eukaryotic cells, transport from nucleus to cytoplasm could be a target for regulation, but in bacteria an mRNA is in principle available for translation as soon as it is synthesized.

◆ Translation may be regulated, usually at the stages of initiation and termination (like transcription). Regulation of initiation has a pleasing formal analogy to the regulation of transcription: a regulator molecule determines, directly or indirectly, whether an initiation site for a coding region is available to the ribosomes. Regulators of initiation of translation may be proteins or RNA. Regulation of termination is concerned with the relationship between successive genes or with constructing variants of the same protein that have different C-terminal regions.

Translational control occurs with both monocistronic and polycistronic mRNAs. It is in particular a notable feature of operons coding for components of the protein synthetic apparatus. The operon provides an arrangement for *coordinate* regulation of a group of structural genes. But, superimposed on it, further controls, such as those at the level of translation, may create *differences* in the extent to which individual genes are expressed.

A similar type of mechanism is used to achieve translational control in several systems. *Repressor function is provided by a protein that binds to a target region on mRNA to prevent ribosomes from recognizing the initiation region.* Formally this is equivalent to a repressor protein binding to DNA to

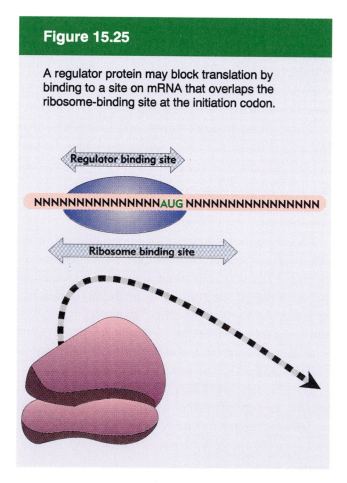

Figure 15.25

A regulator protein may block translation by binding to a site on mRNA that overlaps the ribosome-binding site at the initiation codon.

Regulator binding site

NNNNNNNNNNNNNNNNAUG NNNNNNNNNNNNNNNNN

Ribosome binding site

Table 15.2

Proteins that bind to sequences within the initiation regions of mRNAs may function as translational repressors.

Repressor	Target gene	Site of Action
R17 coat protein	R17 replicase	Hairpin that includes ribosome binding site
T4 RegA	Early T4 mRNAs	Various sequences including initiation codon
T4 DNA polymerase	T4 DNA polymerase	Shine-Dalgarno sequence
T4 p32	Gene 32	Single-stranded 5' leader

---

prevent RNA polymerase from utilizing a promoter. **Figure 15.25** illustrates the most common form of this interaction, in which the regulator protein binds directly to a sequence that includes the AUG initiation codon, thereby preventing the ribosome from binding.

Examples of translational repressors and their targets are summarized in **Table 15.2**. A classic example is the coat protein of the RNA phage R17; it binds to a hairpin that encompasses the ribosome binding site in the phage mRNA. Similarly the T4 RegA protein binds to a consensus sequence that includes the AUG initiation codon in several T4 early mRNAs; and T4 DNA polymerase binds to a sequence in its own mRNA that includes the Shine–Dalgarno element needed for ribosome binding.

About 70 or so proteins constitute the apparatus for bacterial gene expression. The ribosomal proteins are the major component, together with the ancillary proteins involved in protein synthesis. The subunits of RNA polymerase and its accessory factors make up the remainder. The genes coding for ribosomal proteins, protein synthesis factors, and RNA polymerase subunits all are intermingled and organized into a small number of operons. Most of these proteins are represented only by single genes in *E. coli*.

Coordinate controls ensure that these proteins are synthesized in amounts appropriate for the growth conditions: when bacteria grow more rapidly, they devote a greater proportion of their efforts to the production of the apparatus for gene expression. An array of mechanisms is used to control the expression of the genes coding for this apparatus and to ensure that the proteins are synthesized at comparable levels that are related to the levels of the rRNAs.

The organization of six operons is summarized in **Figure 15.26**. About half of the genes for ribosomal proteins (**r-proteins**) map in four operons that lie close together. These are known as *str, spc, S10,* and α (each named simply for the first one of its functions to have been identified). The *rif* and *L11* operons lie together at another location.

Each operon contains a mélange of functions. The *str* operon has genes for small subunit ribosomal proteins as well as for EF-Tu and EF-G. The *spc* and *S10* operons have genes interspersed for both small and large ribosomal subunit proteins. The α operon has genes for proteins of both ribosomal subunits as well as for the α subunit of RNA polymerase. The *rif* locus has genes for large subunit ribosomal proteins and for the β and β' subunits of RNA polymerase.

All except one of the ribosomal proteins are needed in equimolar amounts, which must be coordinated with the level of rRNA. The dispersion of genes whose products must be equimolar, and their intermingling with genes whose products are needed in different amounts, pose some interesting problems for coordinate regulation.

A feature common to all of the operons described in Figure 15.26 is regulation of some of the genes by one of the products. In each case, the gene coding

Figure 15.26

Genes for ribosomal proteins, protein synthesis factors, and RNA polymerase subunits are interspersed in a small number of operons that are autonomously regulated. The proteins subject to regulation are shaded in pink.

Operon	Genes and Proteins	Regulator
str	*rpsL rpsG fusA tufA* S12 S7 EF-G EF-Tu	S7
spc	*rplN rplX rplE rpsN rpsH rplF rplR rpsE rplD rpmO secY-X* L14 L24 L5 S14 S8 L6 L18 S5 L30 L15 Y X	S8
S10	*rpsJ rplC rplB rplD rplW rplS rplV rpsC rpsQ rplP rpmC* S10 L3 L2 L4 L23 S19 L22 S3 S17 L16 L29	L4
α	*rpsM rpsK rpsD rpoA rplQ* S13 S11 S4 α L17	S4
L11	*rplK rplA* L11 L1	L1
rif	*rplJ rplL rpoB rpoC* L10 L7 β β'	L10

for the regulatory product is itself one of the targets for regulation. These are examples of autogenous regulation, which occurs whenever a protein (or RNA) regulates its own production. In the case of the r-protein operons, the regulatory protein inhibits expression of a contiguous set of genes within the operon, so this is an example of negative autogenous regulation.

In each case, *accumulation of the protein inhibits further synthesis of itself and of whatever other gene products are involved.* The effect often is exercised at the level of translation of the polycistronic mRNA, and in several cases can be reproduced *in vitro*. Thus an excess of free ribosomal protein triggers the repression of translation.

Each of the regulators is a ribosomal protein that binds directly to rRNA. *Its effect on translation is a result of its ability also to bind to its own mRNA.* The sites on mRNA at which these proteins bind either overlap the sequence where translation is initiated or lie nearby and probably influence the accessibility of the initiation site by inducing conformational changes. For example, in the S10 operon, protein L4 acts at the very start of the mRNA to inhibit translation of S10 and the subsequent genes. Other regulatory interactions also occur; in the example of this operon, L4 also causes premature termination of transcription in the nontranslated leader upstream of the S10 gene.

The use of r-proteins that bind rRNA to establish autogenous regulation immediately suggests that this provides a mechanism to link r-protein synthesis to rRNA synthesis. A generalized model is depicted in **Figure 15.27**. Suppose that the binding sites for the autogenous regulator r-proteins on rRNA are much stronger than those on the mRNAs. Then so long as any free rRNA is available, the newly synthesized r-proteins will associate with it to start ribosome assembly. There will be no free r-protein available to bind to the mRNA, so its translation will continue. But as soon as the synthesis of rRNA slows or stops, free r-proteins begin to accumulate. Then they are available to bind their mRNAs, repressing further translation. This circuit ensures that each r-protein operon responds in the same way to the level of rRNA: as soon as there is an excess of r-protein relative to rRNA, synthesis of the protein is repressed.

Two objectives are accomplished by this mode of regulation. First, the level of r-proteins corresponds with the growth conditions of the cell. By controlling the level of rRNA, the cell controls production of all ribosomal components. Second, the other proteins coded by these operons are synthesized at their own rates, divorced from the translation of the r-protein genes. Thus in the *rif* operon, the β subunits of RNA polymerase are subject to their own autogenous regulation, but are not controlled by the L10 r-protein. This explains how RNA polymerase is synthesized in smaller amounts than the ribosomal components.

Within the control circuits of these operons, there is provision for allowing disparate rates of synthesis for coordinately regulated proteins. One exceptional ribosomal protein is L7/L12, present in four copies per ribosome; it appears to be translated with increased efficiency. Another exception is EF-Tu, which is present in amounts roughly equimolar with aminoacyl-tRNA—that is, ~10 times greater than the ribosomes. (This is the one case in which there is more than one gene, so the need for extra synthesis is divided between the two genes *tufA* and *tufB*.) The *tuf* genes are regulated by separate circuits that increase synthesis of EF-Tu.

Figure 15.27

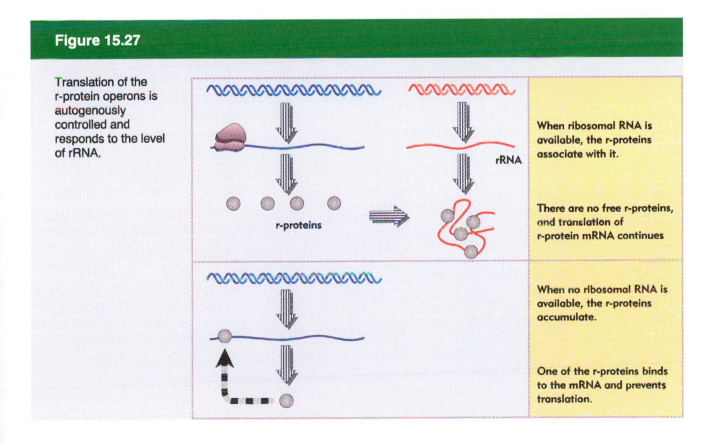

Translation of the r-protein operons is autogenously controlled and responds to the level of rRNA.

When ribosomal RNA is available, the r-proteins associate with it.

rRNA

There are no free r-proteins, and translation of r-protein mRNA continues

When no ribosomal RNA is available, the r-proteins accumulate.

One of the r-proteins binds to the mRNA and prevents translation.

r-proteins

Autogenous regulation has been placed on a quantitative basis for gene *32* of phage T4. The protein (p32) plays a central role in genetic recombination, DNA repair, and replication, in which its function is exercised by virtue of its ability to bind to single-stranded DNA. Nonsense mutations cause the inactive protein to be overproduced. *Thus when the function of the protein is prevented, more of it is made.* This effect occurs at the level of translation; the gene *32* mRNA is stable, and remains so irrespective of the behavior of the protein product.

Figure 15.28 presents a model for the gene *32* control circuit. When single-stranded DNA is present in the phage-infected cell, it sequesters p32. However, in the absence of single-stranded DNA, or at least in conditions in which there is a surplus of p32, the protein prevents translation of its own mRNA. We suspect that the effect is mediated directly by p32 binding to mRNA to prevent initiation of translation. Probably this occurs at an A-T-rich

region that surrounds the ribosome binding site.

Two features of the binding of p32 to the site on mRNA are required to make the control loop work effectively:

◆ The affinity of p32 for the site on gene *32* mRNA must be significantly lower than its affinity for single-stranded DNA. The equilibrium constant for binding RNA is in fact almost two orders of magnitude below that for single-stranded DNA.

◆ But the affinity of p32 for the mRNA must be significantly greater than the affinity for other RNA sequences. It is influenced by base composition and by secondary structure; an important aspect of the binding to gene *32* mRNA is that the regulatory region has an extended sequence lacking secondary structure.

Using the known equilibrium constants, we can plot the binding of p32 to its target sites as a function

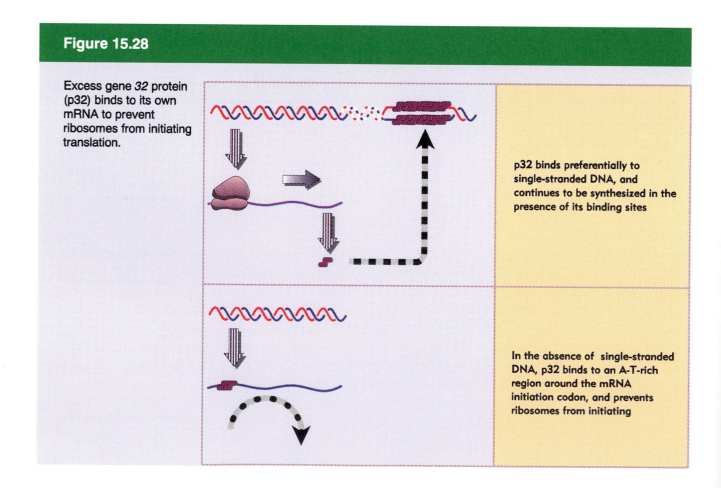

Figure 15.28

Excess gene *32* protein (p32) binds to its own mRNA to prevent ribosomes from initiating translation.

p32 binds preferentially to single-stranded DNA, and continues to be synthesized in the presence of its binding sites

In the absence of single-stranded DNA, p32 binds to an A-T-rich region around the mRNA initiation codon, and prevents ribosomes from initiating

Figure 15.29

Gene 32 protein binding reflects its relative affinities for its various substrates. It binds first to single-stranded DNA, then to its own mRNA, and then to other mRNA sequences. Binding to its own mRNA prevents the level of p32 from rising >10^{-6} M.

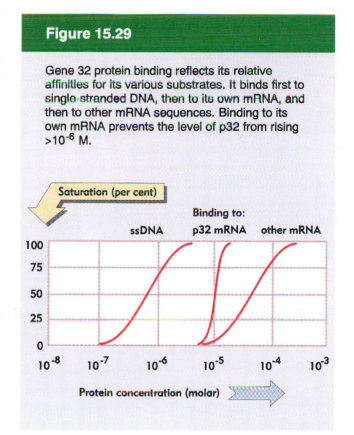

of protein concentration. **Figure 15.29** shows that at concentrations <10^{-6} M, p32 binds to single-stranded DNA. At concentrations >10^{-6} M, it binds to gene *32* mRNA. At yet greater concentrations, it binds to other mRNA sequences, with a range of affinities.

These results imply that the level of p32 should be autoregulated to be <10^{-6} M, which corresponds to ~2000 molecules/bacterium. This fits well with the measured level, 1000–2000 molecules/cell.

A feature of autogenous control is that each regulatory interaction is unique: a protein acts only on the mRNA responsible for its own synthesis. Phage T4 provides an example of a more general translational regulator, coded by the gene *regA,* which represses the expression of several genes that are transcribed during early infection. RegA protein prevents the translation of mRNAs for these genes by competing with 30S subunits for the initiation sites on the mRNA. Its action is a direct counterpart to the function of a repressor protein that binds multiple operators.

Autogenous regulation is a common type of control among proteins that are incorporated into macromolecular assemblies. The assembled particle itself may be unsuitable as a regulator, because it is too large, too numerous, or too restricted in its location. But the need for synthesis of its components may be reflected in the pool of free precursor subunits. If the assembly pathway is blocked for any reason, free subunits accumulate and shut off the unnecessary synthesis of further components.

Eukaryotic cells have a common system in which autogenous regulation of this type occurs. Tubulin is the monomer from which microtubules, a major filamentous system of all eukaryotic cells, are synthesized. The production of tubulin mRNA is controlled by the free tubulin pool. When this pool reaches a certain concentration, the production of further tubulin mRNA is prevented. Again, the principle is the same: tubulin sequestered into its macromolecular assembly plays no part in regulation, but the level of the free precursor pool determines whether further monomers are added to it.

The target site for regulation is a short sequence at the start of the coding region. We do not know yet what role this sequence plays, but we might consider the two models illustrated in **Figure 15.30**. Tubulin may bind directly to the mRNA; or it may bind to the nascent polypeptide representing this region. Whichever model applies, excess tubulin causes tubulin mRNA that is located on polysomes to be degraded, so the consequence of the reaction is to make the tubulin mRNA unstable.

Autogenous control is an *intrinsically* self-limiting system, by contrast with the *extrinsic* control that we discussed previously. Thus a repressor protein's ability to bind an operator may be controlled by the level of an extraneous small molecule, which activates or inhibits its activity. But in the case of autogenous regulation of RNA, the critical parameter is the concentration of the protein itself; the mRNA becomes unavailable to the protein synthetic apparatus, and synthesis of new protein stops, when the free pool of protein reaches a certain molar concentration in the cell.

Figure 15.30

Tubulin is assembled into microtubules when it is synthesized. Accumulation of free tubulin induces instability in the tubulin mRNA by acting at a site at the start of the reading frame in mRNA or at the corresponding position in the nascent protein.

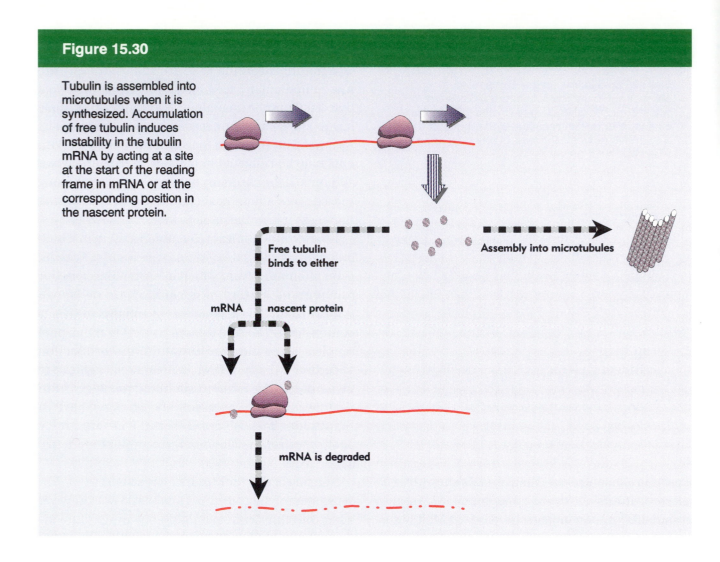

Free tubulin binds to either

mRNA   nascent protein

Assembly into microtubules

mRNA is degraded

Hard times provoke the stringent response

When bacteria find themselves in such poor growth conditions that they lack a sufficient supply of amino acids to sustain protein synthesis, they shut down a wide range of activities. This is called the **stringent response**. We can view it as a mechanism for surviving hard times: the bacterium husbands its resources by engaging in only the minimum of activities until nutrient conditions improve, when it reverses the response and again engages its full range of metabolic activities.

The stringent response causes a massive (10–20×) reduction in the synthesis of rRNA and tRNA. This alone is sufficient to reduce the total amount of RNA synthesis to only 5–10% of its previous level. The synthesis of certain mRNAs is reduced, leading to an overall reduction of ~3× in mRNA synthesis. The rate of protein degradation is increased. Many metabolic adjustments occur, as seen in reduced synthesis of nucleotides, carbohydrates, lipids, etc.

The stringent response causes the accumulation of two unusual nucleotides:

◆ **ppGpp** is guanosine tetraphosphate, with diphosphates attached to both 5′ and 3′ positions)

◆ **pppGpp** is guanosine pentaphosphate, with a 5′ triphosphate group and a 3′ diphosphate).

These nucleotides are typical small-molecule effectors that are expected to function by binding to target proteins to alter their activities. Sometimes they are known collectively as (p)ppGpp.

(p)ppGpp functions to regulate coordinately a large number of cellular activities. Its production is controlled in two ways. A drastic increase in (p)ppGpp is triggered by the stringent response under critical conditions. And there is also a general inverse correlation between (p)ppGpp levels and the bacterial growth rate, which is controlled by some unknown means.

Deprivation of any one amino acid, or mutation to inactivate any aminoacyl-tRNA synthetase, is sufficient to initiate the stringent response. The trigger that sets the entire series of events in train is *the presence of uncharged tRNA in the A site of the ribosome.* Under normal conditions, of course, only aminoacyl-tRNA is placed in the A site by EF-Tu (see Chapter 7). But when there is no aminoacyl-tRNA available to respond to a particular codon, the uncharged tRNA becomes able to gain entry. Of course, this blocks any further progress by the ribosome; and it triggers an **idling reaction**.

The components involved in producing (p)pGpp via the idling reaction have been identified through the existence of **relaxed (*rel*) mutants**. *rel* mutations abolish the stringent response, so that starvation for amino acids does not cause any reduction in stable RNA synthesis or alter any of the other reactions that are usually seen.

The most common site of relaxed mutation lies in the gene *relA*, which codes for a protein called the **stringent factor**. This factor is associated with the ribosomes, although the amount is rather low—say, <1 molecule for every 200 ribosomes. So perhaps only a minority of the ribosomes are able to produce the stringent response.

Ribosomes obtained from stringent bacteria can synthesize ppGpp and pppGpp *in vitro,* provided that the A site is occupied by an uncharged tRNA *specifically responding to the codon.* Ribosomes extracted from relaxed mutants cannot perform this reaction; but they are able to do so if the stringent factor is added.

Figure 15.31 shows the pathways for synthesis of (p)ppGpp. The stringent factor (RelA) is an enzyme that catalyzes the synthetic reaction in which ATP is used to donate a pyrophosphate group to the 3′ position of either GTP or GDP. The formal name for this activity is (p)ppGpp synthetase.

How is ppGpp removed when conditions return to normal? A gene called *spoT* codes for an enzyme that provides the major catalyst for ppGpp degradation. The activity of this enzyme causes ppGpp to be rapidly degraded, with a half-life of ~20 seconds; so the stringent response is reversed rapidly when synthesis of (p)ppGpp ceases. *spoT* mutants have elevated levels of ppGpp, and grow more slowly as a result.

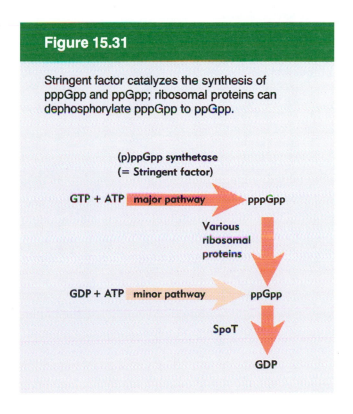

Figure 15.31

Stringent factor catalyzes the synthesis of pppGpp and ppGpp; ribosomal proteins can dephosphorylate pppGpp to ppGpp.

The RelA enzyme uses GTP as substrate more frequently, so that pppGpp is the predominant product. However, pppGpp is converted to ppGpp by several enzymes; among those able to perform this dephosphorylation are the translation factors EF-Tu and EF-G. The production of ppGpp via pppGpp is the most common route, *and ppGpp is the usual effector of the stringent response.*

The response of the ribosome to entry of uncharged tRNA is compared with normal protein synthesis in **Figure 15.32**. When EF-Tu places aminoacyl-tRNA in the A site, peptide bond synthesis is followed by ribosomal movement. But when uncharged tRNA is paired with the codon in the A site, the ribosome remains stationary and engages in the idling reaction.

The purified RelA enzyme is virtually inactive by itself, but becomes active in the presence of ribosomes. Its activity is controlled by the state of the ribosome in protein synthesis. An indication of the nature of this control is revealed by relaxed mutations in another locus, originally called *relC*, which turns out to be the same as *rplK*, which codes for the 50S subunit protein L11.

This protein is located in the vicinity of the A and P sites, in a position to respond to the presence of a properly paired but uncharged tRNA in the A site. A conformational change in this protein or some other component could activate the RelA enzyme, so that the idling reaction occurs instead of polypeptide transfer from the peptidyl-tRNA.

Each round of (p)ppGpp synthesis triggers release of the uncharged tRNA from the A site, so that synthesis of (p)ppGpp is a continuing response to the level of uncharged tRNA. So under limiting conditions, a ribosome stalls when no aminoacyl-tRNA is available to respond to the codon in the A site. Entry of uncharged tRNA triggers the synthesis

Figure 15.32

In normal protein synthesis, the presence of aminoacyl-tRNA in the A site is a signal for peptidyl transferase to transfer the polypeptide chain, followed by movement catalyzed by EF-G; but under stringent conditions, the presence of uncharged tRNA caused RelA protein to synthesize (p)ppGpp and to expel the tRNA.

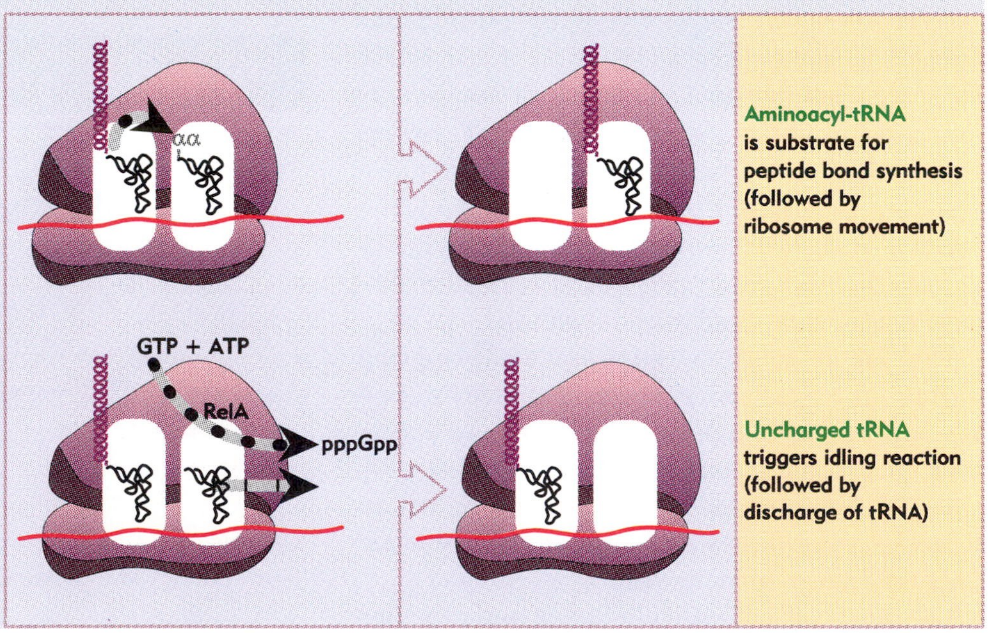

Aminoacyl-tRNA is substrate for peptide bond synthesis (followed by ribosome movement)

Uncharged tRNA triggers idling reaction (followed by discharge of tRNA)

GTP + ATP

RelA

pppGpp

of a (p)ppGpp molecule, and the resulting expulsion of the uncharged tRNA allows the situation to be reassessed. Depending on the availability of aminoacyl-tRNA, the ribosome resumes polypeptide synthesis or undertakes another idling reaction.

What does ppGpp do? It is an effector for controlling several reactions, including the inhibition of transcription. Many effects have been reported, among which two stand out:

◆ *Initiation of transcription is specifically inhibited at the promoters of operons coding for rRNA.* Mutations of stringently regulated promoters can abolish stringent control, which suggests that the effect requires an interaction with specific promoter sequences.

◆ *The elongation phase of transcription of many or most templates is reduced by ppGpp.* The cause is increased pausing by RNA polymerase. This effect is responsible for the general reduction in transcription efficiency when ppGpp is added *in vitro*. We do not yet know the specificity of such inhibition, but it will not be surprising if there is variation in magnitude from operon to operon in such a way that certain operons are more inhibited.

It is interesting that unusual nucleotides are used in at least two control systems with a general coordinating function. Both appear to be specific to bacteria. Guanine nucleotides trigger the stringent response to nutritional deprivation; and cyclic AMP triggers the switch in use of carbon sources caused by the absence of glucose.

Summary

Transcription is regulated by the interaction between *trans*-acting factors and *cis*-acting sites. A *trans*-acting factor is the product of a regulator gene. It is usually protein but can be RNA. Because it diffuses in the cell, it can act on any appropriate target gene. A *cis*-acting site in DNA (or RNA) is a sequence that functions by being recognized *in situ*. It has no coding function and can regulate only those sequences that are physically contiguous with it. Bacterial genes coding for proteins whose functions are related, such as successive enzymes in a pathway, may be organized in a cluster that is transcribed into a polycistronic mRNA from a single promoter. Control of this promoter regulates expression of the entire pathway. The unit of regulation, containing structural genes and *cis*-acting elements, is called the operon.

Initiation of transcription is regulated by interactions that occur in the vicinity of the promoter. The ability of RNA polymerase to initiate at the promoter is prevented or activated by other proteins. Genes that are active unless they are turned off are said to be under negative control. Genes that are active only when specifically turned on are said to be under positive control. The type of control can be determined by the dominance relationships between wild type and mutants that are constitutive/derepressed (permanently on) or uninducible/super-repressed (permanently off).

A repressor protein prevents RNA polymerase either from binding to the promoter or from activating transcription. The repressor binds to a target sequence, the operator, that usually is located around or upstream of the startsite. Operator sequences are short and often are palindromic. The repressor is often a homomultimer whose symmetry reflects that of its target.

The ability of the repressor protein to bind to its operator is regulated by a small molecule. An inducer prevents a repressor from binding; a corepressor activates it. The lactose pathway operates by induction, when an inducer β-galactoside prevents the repressor from binding its operator; transcription and translation of the *lacZ* gene then produces β-galactosidase, the enzyme that metabolizes β-galactosides. The tryptophan pathway operates by repression; the corepressor (tryptophan) activates the repressor protein, so that it binds to the operator and prevents expression of the genes that code for the enzymes that biosynthesize tryptophan. A repressor can control multiple targets that have copies of an operator consensus sequence.

Some promoters cannot be recognized by RNA polymerase (or are recognized only poorly) unless a specific activator protein is present. Activator proteins also may be regulated by small molecules. The CAP activator becomes able to bind to target sequences in the presence of cyclic AMP. All promoters that respond to CAP have at least one copy of the target sequence. Binding of CAP to its target involves bending DNA. Direct contact between one subunit of CAP and RNA polymerase is required to activate transcription.

A protein with a high affinity for a particular target sequence in DNA has a lower affinity for all DNA. The ratio defines the specificity of the protein. Because there are many more nonspecific sites (any DNA sequence) than specific target sites in a genome, a DNA-binding protein such as a repressor or RNA polymerase is 'stored' on DNA; probably none or very little is free. The specificity for the target sequence must be great enough to counterbalance the excess of nonspecific sites over specific sites. The balance for bacterial proteins is adjusted so that the amount of protein and its specificity allow specific recognition of the target in 'on' conditions, but allow almost complete release of the target in 'off' conditions.

Gene expression can be controlled at stages subsequent to transcription. Translation may be controlled by a protein that binds to a region of mRNA overlapping with the ribosome binding site; this prevents ribosomes from initiating translation. Most proteins that repress translation possess this capacity in addition to their other functional roles; in particular, translation is controlled in some cases of autogenous regulation, when a gene product regulates expression of the operon containing its own gene. However, the RegA function of T4 is a general regulator that functions on several target mRNAs at the level of translation. The level of protein synthesis itself provides an important coordinating signal. Deficiency in aminoacyl-tRNA causes an idling reaction on the ribosome, which leads to the synthesis of the unusual nucleotide ppGpp. This is an effector that inhibits initiation of transcription at certain promoters; it also has a general effect in inhibiting elongation on all templates.

Further reading

Reviews

A historical source for reviews of the function of the *lac* operon is **Miller and Reznikoff's** (Eds.) *The Operon* (Cold Spring Harbor Laboratory, New York, 1978). Chapters of special interest include **Beckwith** (pp. 11–30) on the organization of the system, **Zabin and Fowler** (pp. 89–122) on the protein products, **Miller** (pp. 31–88) on the *lacI* gene, **Beyreuther** (pp. 123–154), **Weber and Geisler** (pp. 155–176) on the repressor protein, and **Barkley and Bourgeois** (pp. 177–220) on repressor binding to DNA and inducers. The circuitry of several other operons also was reviewed.

Work on various operons has been brought up to date by chapters in *E. coli and S. typhimurium* (Ed. Neidhardt, American Society for Microbiology, Washington DC, 1241–1275, 1987), including **Schleif** on the *ara* operon (pp. 1473–1481) and **Adhya** on the gal operon (pp. 1503–1512).

Extrapolations from bacterial to eukaryotic regulatory systems are to be found in **Ptashne**, *A Genetic Switch* (Cell Press and Blackwell, 1992).

Regulation of ribosome synthesis and feedback mechanisms in relation to growth control was reviewed by **Nomura** *et al.* (*Ann. Rev. Biochem.* **53**, 75–117, 1984).

The idiosyncracies of translational control mechanisms have been considered by **Gold** (*Ann. Rev. Biochem.* **57**, 199–223, 1988).

Stringent control was reviewed by **Cashel and Rudd** (in *E. coli and S. typhimurium*, Ed. Neidhardt, American Society for Microbiology, Washington DC, 1410–1429, 1987).

Discoveries

The classic formulation of the operon model is still worth reading: **Jacob and Monod** (*J. Mol. Biol.* **3**, 318–356, 1961). **Gilbert and Muller-Hill** isolated the *lac* repressor and showed that it binds to the DNA of the operator (*Proc. Nat. Acad. Sci. USA* **56**, 1891–1898, 1966; **58**, 2415–2421, 1967).

The difficulties involved in extrapolating bacterial regulation to eukaryotes were brought up by **Lin and Riggs** (*Cell* **4**, 107–111, 1975), and parameters obtained with *lac* repressor in a eukaryote were reported by **Hu and Davidson** (*Cell* **48**, 555–566, 1987).

Autogenous control of translation of r-proteins was characterized by **Baughman and Nomura** (*Cell* **34**, 969–988, 1983).

Unusual nucleotides were implicated in the stringent response by **Cashel and Gallant** (*Nature* **221**, 838–841, 1969); the idling reaction was identified by **Haseltine and Block** (*Proc. Nat. Acad. Sci. USA* **70**, 1564–1568, 1973).

CHAPTER 16

Control by RNA structure: termination and antitermination

RNA structure is emerging as a significant opportunity for regulation in both prokaryotes and eukaryotes. Its most common role occurs when an RNA molecule can take up alternative secondary structures by utilizing different schemes for intramolecular base pairing. The properties of the alternative conformations may be different. This type of mechanism can be used to regulate the termination of transcription; an RNA molecule may be driven by external circumstances into structures that do or do not permit termination. Another means of controlling conformation (and thereby function) is provided by the cleavage of an RNA; by removing one segment of an RNA, the conformation of the rest may be altered. Finally, it is possible also for one (small) RNA molecule to control the activity of another by base pairing with it. We might regard the ability of an RNA to shift between different conformations with regulatory consequences as comprising the nucleic acid's alternative to the allosteric changes of conformation that regulate protein function; in analogous ways, these mechanisms allow an interaction at one site in the molecule to affect the structure of another site.

Once RNA polymerase has started transcription, the enzyme moves along the template, synthesizing RNA, until it meets a **terminator (*t*)** sequence. At this point, the enzyme stops adding nucleotides to the growing RNA chain, releases the completed product, and dissociates from the DNA template. (We do not know in which order the last two events occur.) Termination requires that all hydrogen bonds holding the RNA–DNA hybrid together must be broken, after which the DNA duplex reforms.

It is difficult to define the termination point of an RNA molecule that has been synthesized in the living cell. It is always possible that the 3′ end of the molecule has been generated by *cleavage* of the primary transcript, and therefore does not represent the actual site at which RNA polymerase terminated. Unfortunately, the 3′ end looks the same whether generated by termination or cleavage; there is no marker comparable to the triphosphate that is incorporated at the original 5′ end.

The best identification of termination sites is provided by systems in which RNA polymerase terminates *in vitro*. Because the ability of the enzyme to terminate is strongly influenced by parameters such as the ionic strength, its termination at a particular point *in vitro* does not prove that this same point is a natural terminator. But we can identify authentic 3′ ends when the same end is generated *in vitro* and *in vivo*.

Terminators can be identified as sequences that are needed for the termination reaction (*in vitro* or *in vivo*). Terminators in bacteria and their phages have been characterized in detail. They vary widely in both their efficiencies of termination and their dependence on ancillary proteins, at least as seen *in vitro*. Many prokaryotic and some eukaryotic terminators require a hairpin to form in the secondary structure of the RNA being transcribed. *This indicates that termination depends on the RNA product and is not determined simply by scrutiny of the DNA sequence during transcription.*

At some terminators, the termination event can be *prevented* by specific ancillary factors that interact with RNA polymerase. **Antitermination** causes the enzyme to continue transcription past the terminator sequence, an event called **readthrough** (the same term used to describe a ribosome's suppression of termination codons).

Antitermination is used as a control mechanism in both phage regulatory circuits and bacterial operons. The principles of two common mechanisms of antitermination are illustrated in the first two figures:

◆ During phage infection, different ancillary proteins (**antitermination factors**) allow RNA polymerase to bypass specific terminator sequences. **Figure 16.1** shows that such interactions control the ability of the enzyme to read past a terminator into genes lying beyond. In the example shown in the figure, the antitermination factor positively regulates the expression of region 2. Terminators differ in

Figure 16.1

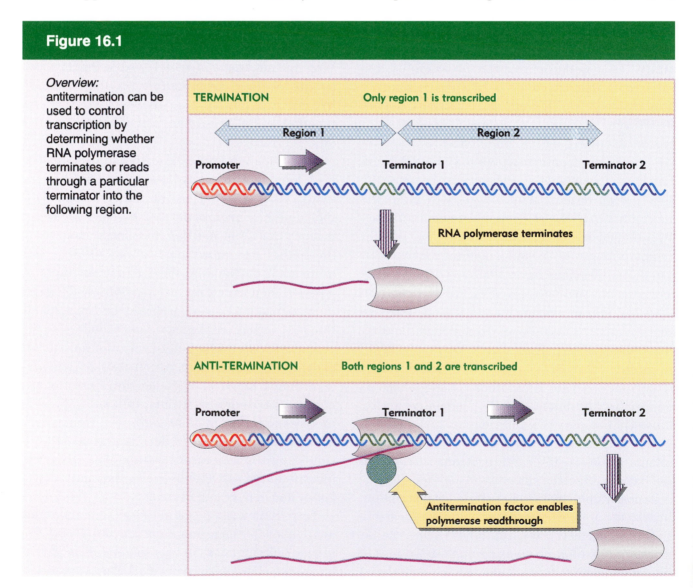

Overview: antitermination can be used to control transcription by determining whether RNA polymerase terminates or reads through a particular terminator into the following region.

TERMINATION Only region 1 is transcribed

Region 1 Region 2

Promoter Terminator 1 Terminator 2

RNA polymerase terminates

ANTI-TERMINATION Both regions 1 and 2 are transcribed

Promoter Terminator 1 Terminator 2

Antitermination factor enables polymerase readthrough

Figure 16.2

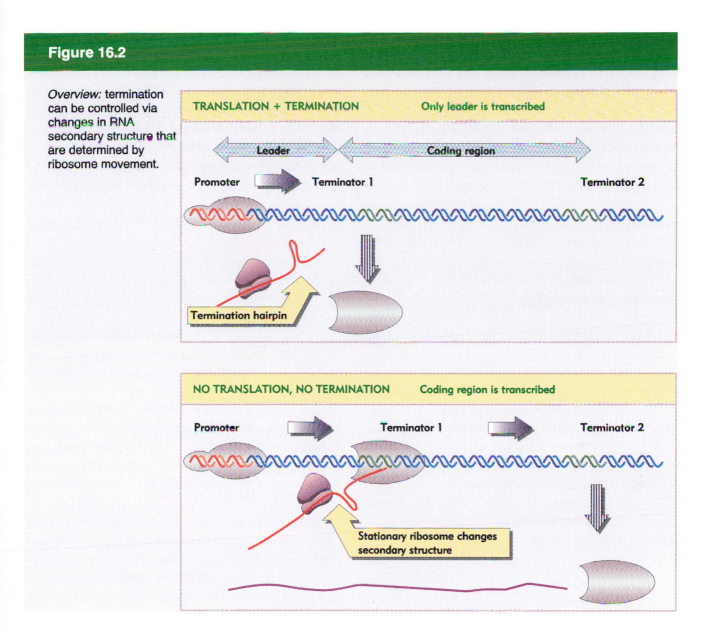

Overview: termination can be controlled via changes in RNA secondary structure that are determined by ribosome movement.

TRANSLATION + TERMINATION   **Only leader is transcribed**

Leader   Coding region

Promoter   Terminator 1   Terminator 2

Termination hairpin

NO TRANSLATION, NO TERMINATION   **Coding region is transcribed**

Promoter   Terminator 1   Terminator 2

Stationary ribosome changes secondary structure

their susceptibility to antitermination; in this example, the antitermination factor can act at terminator 1, but does not prevent RNA polymerase from terminating at terminator 2. *This mechanism of antitermination therefore depends upon the appearance of an anti-termination factor that is specific for a particular terminator.*

◆ Antitermination can provide a link between translation and transcription, in which *termination of transcription is prevented when the ribosome is unable to move along a leader segment of the mRNA.* **Figure 16.2** shows that the basis for this effect rests with changes in RNA secondary structure that are determined by the position of the ribosome on mRNA. When the ribosome can translate the leader region, a termination hairpin forms at terminator 1. But when the ribosome is prevented from translating the leader, a hairpin forms at a different location, with the result that the termination region remains single-stranded, and RNA polymerase transcribes the coding region. *This mechanism of anti–termination therefore depends upon the ability of external circumstances to influence ribosome movement in the leader region.*

In approaching the termination event, we must regard it not simply as a mechanism for generating the 3′ end of the RNA molecule, but as an opportunity to control gene expression. Thus the stages when RNA polymerase associates with DNA (initiation) or dissociates from it (termination) both are subject to specific control. There are interesting parallels between the systems employed in initiation and termination. Both require breaking of hydrogen bonds (initial melting of DNA at initiation, RNA–DNA dissociation at termination); and both require additional proteins to interact with the core enzyme. In fact, they are accomplished by alternative forms of the polymerase. However, whereas initiation relies solely upon the interaction between RNA polymerase and duplex DNA, the termination event involves recognition of signals in the transcript by RNA polymerase or by ancillary factors.

Bacterial RNA polymerase has two modes of termination

The sequences at prokaryotic terminators show no similarities beyond the point at which the last base is added to the RNA. So the responsibility for termination lies with the *sequences already transcribed* by RNA polymerase. Thus termination relies on scrutiny of the template or product that the polymerase is currently transcribing.

Terminators have been distinguished in *E. coli* according to whether RNA polymerase requires any additional factors to terminate *in vitro*:

◆ Core enzymes can terminate *in vitro* at certain sites in the absence of any other factor. These sites are called **intrinsic terminators**, reflecting the fact that (so far) no ancillary factors appear to be necessary for RNA polymerase to terminate at them.

◆ **Rho-dependent** terminators are defined by the need for addition of **rho factor** *in vitro*; and mutations show that the factor is involved in termination *in vivo*.

Intrinsic terminators have the two structural features evident in **Figure 16.3**: a hairpin in the secondary structure; and a run of ~6 U residues at the very end of the unit. Both features are needed for termination.

The hairpin is generated by pairing between inverted repeats (which form the stem) separated by a short distance (which forms the loop). The length of the hairpin is variable: usually it contains a G•C-rich region near the base of the stem.

Point mutations that prevent termination occur within the stem region of the hairpin. What is the effect of a hairpin on transcription? Probably all hairpins that form in the RNA product cause the polymerase to slow or pause in RNA synthesis. The length of the pause varies, but at a typical terminator lasts ~60 seconds.

Termination cannot depend simply on encountering a hairpin. (There is too much secondary structure in RNA.) Pausing creates an opportunity for termination to occur. But if no terminator sequences are present, usually the enzyme moves on again to continue transcription.

The string of U residues probably provides the signal that allows RNA polymerase to dissociate from the template when it pauses at the hairpin. The rU•dA RNA–DNA hybrid has an unusually weak base-paired structure; it requires the least energy of

Figure 16.3

Intrinsic terminators include palindromic regions that form hairpins varying in length from 7-20 bp. The stem-loop structure includes a G-C-rich region and is followed by a run of U residues.

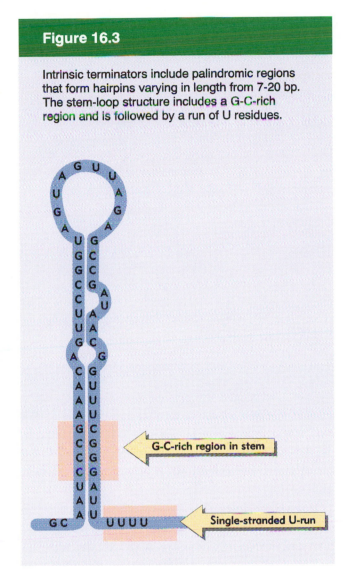

G-C-rich region in stem

Single-stranded U-run

any RNA–DNA hybrid to break the association between the two strands. When the polymerase pauses, the RNA–DNA hybrid unravels from the weakly bonded rU•dA terminal region. Often the actual termination event takes place at any one of several positions toward or at the end of the U-run, as though the enzyme 'stutters' during termination.

The importance of the run of U bases is confirmed by making deletions that shorten this stretch; although the polymerase still pauses at the hairpin, it no longer terminates. The series of U bases corresponds to an A•T-rich region in DNA,

so we see that A•T-rich regions are important in intrinsic termination as well as initiation.

Both the sequence of the hairpin and the length of the U-run influence the efficiency of termination. However, termination efficiency *in vitro* varies from 2–90%, and does not correlate in any simple way with the constitution of the hairpin or the number of U residues in intrinsic terminators. Thus additional parameters, at present undefined, influence the interaction with RNA polymerase.

When certain transcription units are used as templates for RNA synthesis *in vitro*, proper termination does not occur; the polymerase pauses at the terminator, but then resumes RNA synthesis. Rho factor (ρ) was discovered as a protein whose addition to the *in vitro* system allows RNA polymerase to terminate at certain sites, generating RNA molecules with unique 3′ ends. This action is called **rho-dependent termination**. Rho is an essential protein in *E. coli*, although the genome has relatively few rho-dependent terminators; most of the known rho-dependent terminators are found in phage genomes.

The sequences recognized by rho factor have taken a long time to identify. Deletions upstream of rho-dependent termination sites may cause readthrough, which implies that rho recognizes a discrete region preceding the site of termination, 50–90 bases long. Analysis of these sequences shows a common feature: the RNA sequences are rich in C residues and poor in G residues. An example of such a sequence is given in **Figure 16.4**; C is by far the most common base (41%) and G is the least common base (14%). As a general rule the efficiency of a rho-dependent terminator increases with the length of the C-rich/G-poor region.

Less is known about the signals and ancillary factors involved in termination for eukaryotic polymerases. Each class of polymerase uses a different mechanism (see Chapter 29). However, the same features used by bacterial (core) RNA polymerase in intrinsic termination recur in eukaryotes: secondary structure and/or runs of U residues in the transcript are important.

Figure 16.4

A rho-dependent terminator has a sequence rich in C and poor in G preceding the actual site(s) of termination.

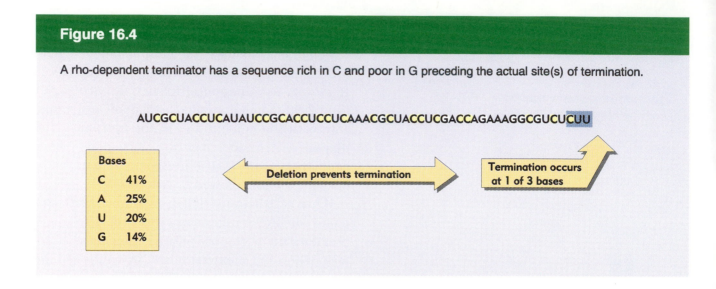

AUCGCUACCUCAUAUCCGCACCUCCUCAAACGCUACCUCGACCAGAAAGGCGUCUCUU

Bases

C	41%
A	25%
U	20%
G	14%

Deletion prevents termination

Termination occurs at 1 of 3 bases

How does rho factor work?

Rho factor functions solely at the stage of termination. It is a protein of ~46,000 daltons, probably active as a hexamer (275,000 daltons). It functions as an ancillary factor for RNA polymerase; typically its maximum activity *in vitro* is displayed when it is present at ~10% of the concentration of the RNA polymerase.

Does rho factor act via recognizing DNA, RNA or RNA polymerase? Rho has an ATPase activity, which is RNA-dependent; it requires the presence of a polyribonucleotide, >50 or so bases long. This suggests that rho binds RNA. Probably an individual rho factor acts processively on a single RNA substrate. The model for rho action shown in **Figure 16.5** supposes that it binds to a nascent RNA chain at some point upstream of the terminator, probably requiring a specific sequence or type of sequence, although attachment could occur just at a free 5′ end. After rho has attached, it proceeds along the RNA.

How does rho catch up with RNA polymerase? One possibility is that rho simply moves along the transcript faster than RNA polymerase moves along the DNA. The enzyme pauses when it reaches a terminator, and termination occurs if rho catches it there.

Does rho release the transcript by acting directly on the DNA–RNA junction or indirectly by causing RNA polymerase to release RNA? In the absence of RNA polymerase, rho can cause an RNA–DNA hybrid to separate; hydrolysis of ATP is used to provide energy for the reaction.

These abilities suggest that rho can directly gain access to the stretch of RNA–DNA hybrid in the transcription bubble and cause it to unwind. Either as a result of this unwinding, or because of some interaction between rho and RNA polymerase, termination is completed by the release of rho and RNA polymerase from the nucleic acids.

The idea that rho moves along RNA leads to an important prediction about the relationship between transcription and translation. Rho must first have access to a binding sequence on RNA, and then must be able to move along the RNA. Either or both of these conditions may be prevented if ribosomes are translating an RNA. Thus the ability of rho factor to reach RNA polymerase at a terminator depends on what is happening in translation.

This model explains a puzzling phenomenon. In some cases, a nonsense mutation in one gene of a transcription unit prevents the expression of subsequent genes in the unit. This effect is called **polarity**. A common cause is the absence of the mRNA corresponding to the subsequent (distal) parts of the unit.

Suppose that there are rho-dependent terminators *within* the transcription unit, that is, before the terminator that *usually* is used. The conse-

Figure 16.5

Rho factor pursues RNA polymerase along the RNA and can cause termination when it catches the enzyme pausing at a rho-dependent terminator.

RNA polymerase transcribes DNA

Rho attaches to recognition site on RNA

Rho moves along RNA, following RNA polymerase

RNA polymerase pauses at terminator and rho catches up

Rho unwinds DNA-RNA hybrid in transcription bubble

Termination: RNA polymerase, rho, and RNA are released

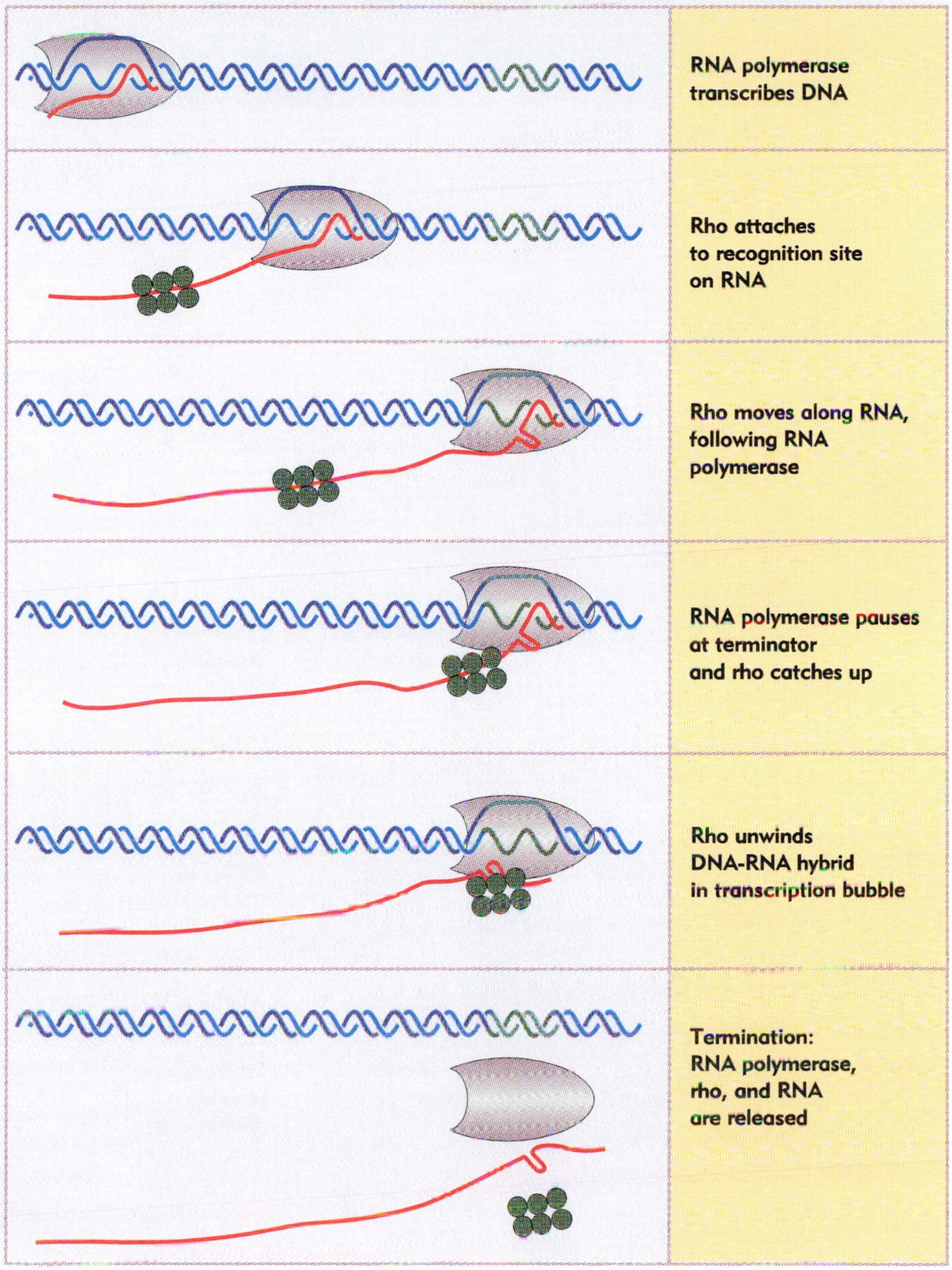

quences are illustrated in **Figure 16.6**. Normally these earlier terminators are not used, because the ribosomes prevent rho from reaching RNA polymerase. But a nonsense mutation releases the ribosomes, so that rho is free to attach to and/or move along the mRNA, enabling it to act on RNA polymerase at the terminator. As a result, the enzyme is released, and the distal regions of the transcription unit are never expressed. (Why should there be internal terminators? Perhaps they are simply sequences that by coincidence mimic the usual rho-dependent terminator. Some stable RNAs that have extensive secondary structure are preserved from polar effects, presumably because the structure impedes rho attachment or movement.)

rho mutations show wide variations in their influence on termination. The basic nature of the effect is a failure to terminate. But the magnitude of the failure, as seen in the percent of readthrough *in vivo*, depends on the particular target locus.

Figure 16.6

The action of rho factor may create a link between transcription and translation when a rho-dependent terminator lies soon after a nonsense mutation.

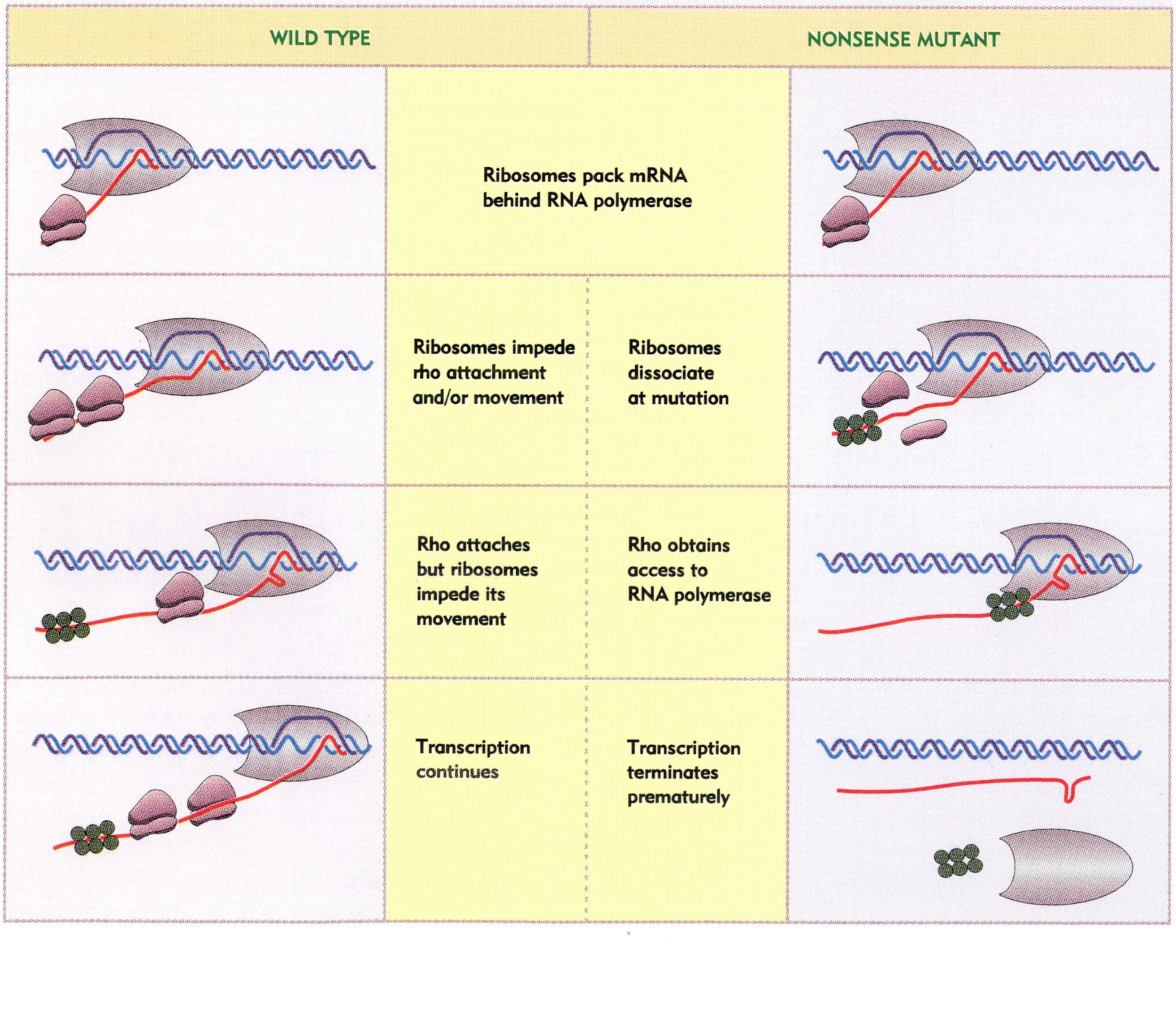

WILD TYPE		NONSENSE MUTANT
	Ribosomes pack mRNA behind RNA polymerase	
	Ribosomes impede rho attachment and/or movement — Ribosomes dissociate at mutation	
	Rho attaches but ribosomes impede its movement — Rho obtains access to RNA polymerase	
	Transcription continues — Transcription terminates prematurely	

Similarly, the need for rho factor *in vitro* is variable. Some (rho-dependent) terminators require relatively high concentrations of rho, while others function just as well at lower levels. This suggests that different terminators require different levels of rho factor for termination, and therefore respond differently to the residual levels of rho factor in the mutants (*rho* mutants are usually leaky).

rho mutations often suppress polarity. This is explained if they reduce the probability that rho will act on the internal terminator that follows the nonsense codon. Then the termination of translation does not cause transcription also to terminate; and the regions of mRNA beyond the mutation can be translated by ribosomes that reattach farther along.

Some *rho* mutations can be suppressed by mutations in other genes. This approach provides an excellent way to identify proteins that interact with rho. The β subunit of RNA polymerase is implicated by two types of mutation. First, mutations in the *rpoB* gene can reduce termination at a rho-dependent site. Second, mutations in *rpoB* can restore the ability to terminate transcription at rho-dependent sites in *rho* mutant bacteria.

Antitermination depends on specific sites

A common feature in the control of bacteriophage infection is that very few of the phage genes (the 'early' genes) can be transcribed by the bacterial host RNA polymerase. Among these genes, however, are regulator(s) whose product(s) allow the next set of phage genes to be expressed. Two common types of action for such a regulator protein are to sponsor initiation at new (phage) promoters or to cause the host polymerase to read through phage terminators. **Figure 16.7** compares the use of new promoters as a control mechanism with the use of antitermination.

One mechanism for recognizing new phage promoters is to replace the sigma factor of the host enzyme with another factor that redirects its specificity in initiation (see Chapter 14). An alternative is to synthesize a new phage RNA polymerase. In either case, the critical feature that distinguishes the new set of genes is their possession of *different promoters from those originally recognized by host RNA polymerase.*

In the case of a switch in promoter recognition, expression of the new set of genes does not depend on expression of the early genes after the critical sigma factor or new polymerase has been synthesized. The two sets of transcripts are independent. Early gene expression can cease when the switch to the next stage is made.

Termination provides an alternative mechanism for phages to control the switch from early genes to the next stage of expression. The use of antitermination depends on a particular arrangement of genes. The early genes lie adjacent to the genes that are to be expressed next, but are separated from them by terminator sites. *If termination is prevented at these sites, the polymerase reads through into the genes on the other side.* Thus in antitermination, the *same promoters* continue to be recognized by RNA polymerase. So the new genes are expressed only by extending the RNA chain to form molecules that contain the early gene sequences at the 5′ end and the new gene sequences at the 3′ end. Since the two types of sequence remain linked, early gene expression inevitably continues.

The best characterized example of antitermination is provided by phage lambda, with which the phenomenon was discovered. The host RNA polymerase initially transcribes two genes, which are called the **immediate early** genes. The transition to the next stage of expression is controlled by

Figure 16.7

Overview: switches in transcriptional specificity can be controlled at initiation or termination.

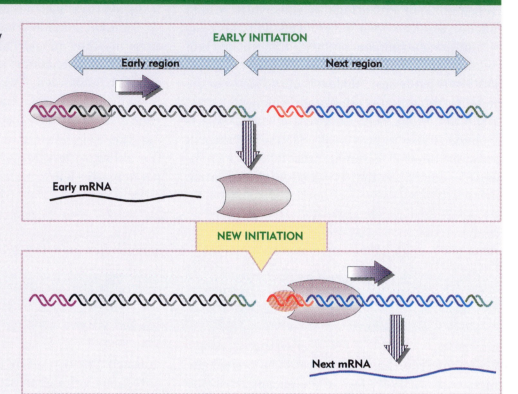

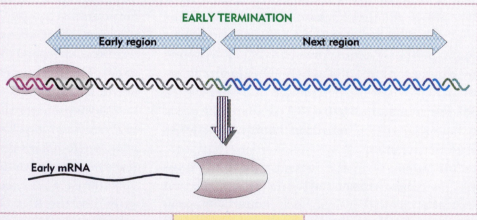

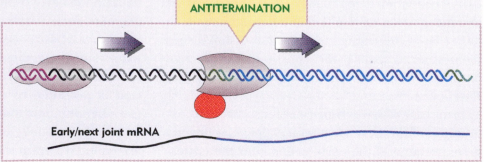

preventing termination at the ends of the immediate early genes, with the result that the **delayed early** genes are expressed. (We discuss the overall regulation of lambda development in more detail in the next chapter.)

The regulator gene that controls the switch from immediate early to delayed early expression is identified by mutations in lambda gene N which can transcribe *only* the immediate early genes; they proceed no further into the infective cycle. The same effect is seen when gene *28* of phage SPO1 is mutated to prevent the production of σ^{gp28}. From the genetic point of view, the mechanisms of new initiation and antitermination are similar. *Both are **positive controls** in which an early gene product must be made by the phage in order to express the next set of genes.*

A map of the early region of phage lambda is drawn in **Figure 16.8**. The immediate early genes, N and *cro*, are transcribed respectively to the left and to the right from the promoters indicated as P_L and P_R. Transcription by *E. coli* RNA polymerase itself stops at the ends of genes N and *cro*, at the terminators t_{L1} and t_{R1}, respectively. Both terminators depend on rho; in fact, these were the terminators with which rho was originally identified.

The situation is changed by expression of the N gene. The product pN is an **antitermination protein** that allows RNA polymerase to read through t_{L1} and t_{R1} into the delayed early genes on either side. Because pN is highly unstable, with a half-life of 5 minutes, continued expression of gene N is needed to maintain transcription of the delayed early genes. This is no problem, since the N gene is part of the delayed early transcription unit, and its transcription must precede that of the delayed early genes.

Like other phages, still another control is needed to express the late genes that code for the components of the phage particle. This switch is regulated by gene Q, itself one of the delayed early genes. Its product, pQ, is another antitermination protein, one that specifically allows RNA polymerase initiating at another site, the late promoter $P_{R'}$, to readthrough a terminator that lies between it and the late genes. Thus by employing antitermination proteins with different specificities, a cascade for gene expression can be constructed.

The different specificities of pN and pQ establish an important general principle: *RNA polymerase interacts with transcription units in such a way that an ancillary factor can sponsor antitermination specifically at some terminators and not others.* Termination can in fact be controlled with the same sort of precision as initiation. What sites are involved in controlling the specificity of termination?

The antitermination activity of pN is highly specific, but *the antitermination event is not determined by the terminators t_{L1} and t_{R1}; the recognition site needed for antitermination lies upstream in the transcription unit, that is, at a different place from the terminator site at which the action eventually is accomplished.* This conclusion establishes a general principle. When we know the site on DNA at which some protein exercises its effect, we cannot assume that this coincides with the DNA sequence that it initially recognizes. They may be separate.

The recognition sites required for pN action are called *nut* (for N *ut*ilization). The sites responsible for determining leftward and rightward antitermination are described as *nutL* and *nutR*, respectively. Mapping of *nut⁻* mutations locates *nutL* between the startpoint of P_L and the beginning of the N coding region; and *nutR* lies between the end of the *cro* gene and t_{L1}. This means that the two *nut* sites lie in different positions relative to the organization of their transcription units. Whereas *nutL* is near the promoter, *nutR* is near to the terminator.

How does antitermination occur? When pN recognizes the *nut* site, it must act on RNA polymerase to ensure that the enzyme can no longer respond to the terminator. The variable locations of the *nut* sites indicate that this event is linked neither to initiation nor to termination, but can occur to RNA polymerase as it elongates the RNA chain past the *nut* site. As illustrated in **Figure 16.9**, the polymerase then becomes a juggernaut that continues past the terminator, heedless of its signal. (This reaction involves antitermination at rho-dependent

Figure 16.8

Phage lambda has two early transcription units; in the "leftward" unit, the "upper" strand is transcribed toward the left; in the "rightward" unit, the "lower" strand is transcribed toward the right. The promoters are indicated by the shaded red or blue arrowheads. The terminators are indicated by the shaded green boxes. Genes *N* and *cro* are the immediate early functions, and are separated from the delayed early genes by the terminators. Synthesis of N protein allows RNA polymerase to pass the terminators t_{L1} to the left and t_{R1} to the right.

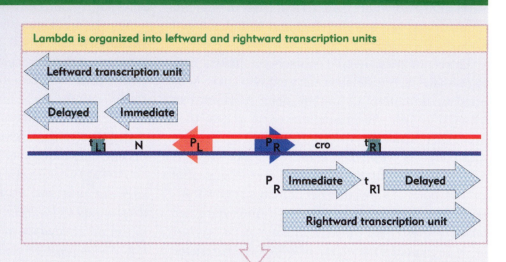

Lambda is organized into leftward and rightward transcription units

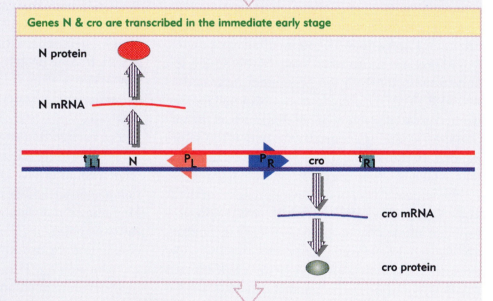

Genes N & cro are transcribed in the immediate early stage

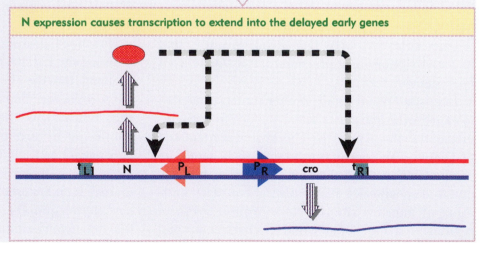

N expression causes transcription to extend into the delayed early genes

terminators, but pN can also act at intrinsic terminators.)

Is the ability of pN to recognize a short sequence within the transcription unit an example of a more widely used mechanism for antitermination? Other phages, related to lambda, have different *N* genes and different antitermination specificities. The region of the phage genome in which the *nut* sites lie has a different sequence in each of these phages, and each phage must therefore have characteristic *nut* sites recognized specifically by its own pN. Each of these pN products must have the same general ability to interact with the transcription apparatus in an antitermination capacity, but has a different specificity for the sequence of DNA that activates the mechanism.

Figure 16.9

Ancillary factors bind to RNA polymerase as it passes certain sites. The *nut* site consists of two sequences. NusB-S10 join core enzyme as it passes *boxA*. Then NusA and pN protein bind as polymerase passes *boxB*. The presence of pN allows the enzyme to readthrough the terminator, producing a joint mRNA that contains immediate early sequences joined to delayed early sequences.

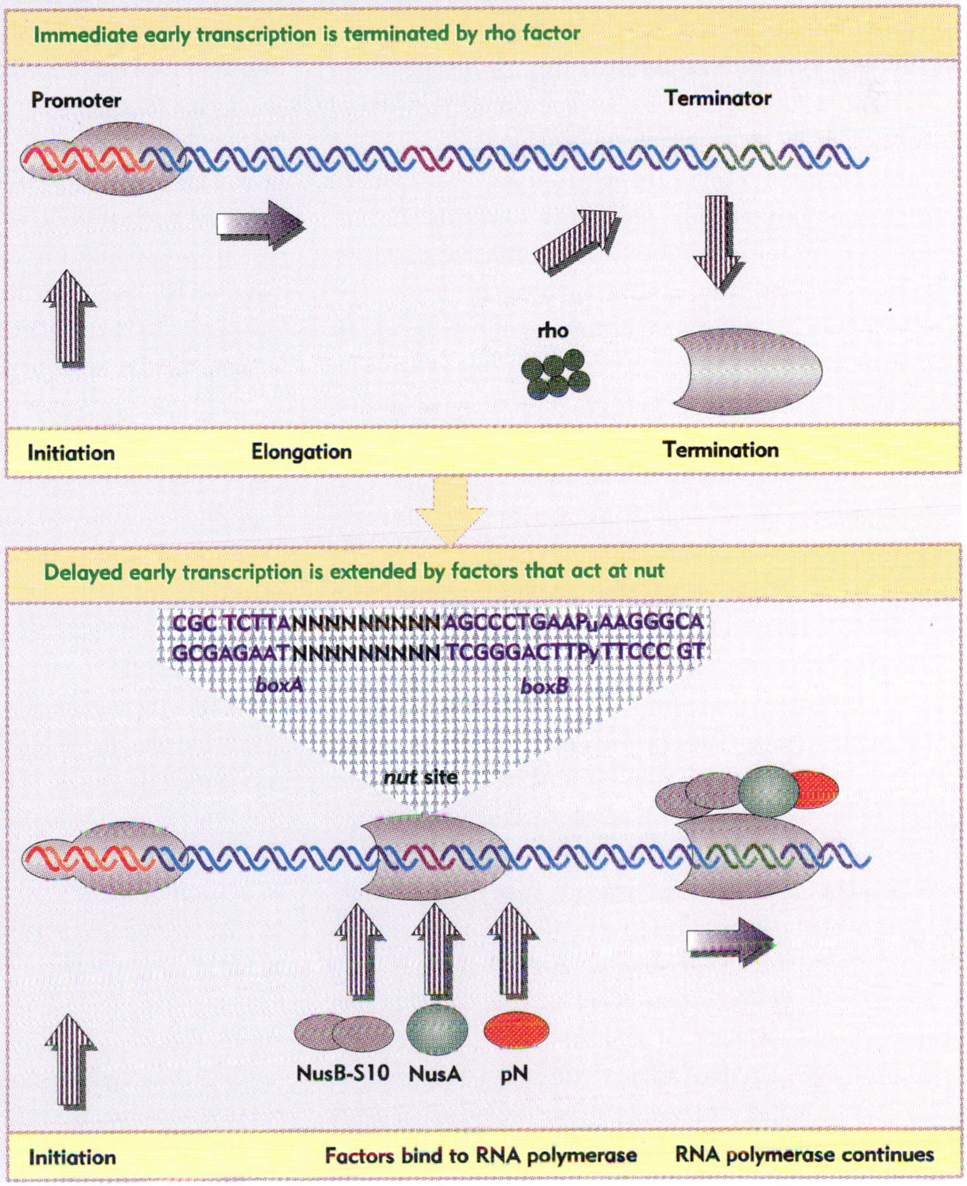

More subunits for RNA polymerase

Termination and antitermination are closely connected, and involve bacterial proteins and phage proteins that interact with RNA polymerase in response to sequences within certain transcription units. A lambda *nut* site consists of two sequence elements, called *boxA* and *boxB*. Sequence elements related to *boxA* are also found in bacterial operons; *boxA* is required for binding bacterial proteins that are necessary for antitermination in both phage and bacterial operons. Mutations in the *boxB* sequence abolish the ability of pN to cause antitermination; *boxB* is specific to the phage genome.

The discovery of antitermination as a phage control mechanism has led to the identification of further components of the transcription apparatus. The bacterial proteins with which pN interacts can be identified by isolating mutants of *E. coli* in which pN is ineffective. The bacterial mutants cannot be infected successfully by lambda, because the phage is limited to expressing only its immediate early genes. Several of these mutations lie in the *rpoB* gene. This argues that pN (like rho factor) interacts with the β subunit of the core enzyme.

Other *E. coli* mutations that prevent pN function identify the *nus* loci: *nusA*, *nusB*, *nusE*, and *nusG*. (The term 'nus' is an acronym for *N* utilization substance.) The *nus* loci code for proteins that form part of the transcription apparatus, but that are not isolated with the RNA polymerase enzyme in its usual form. The *nusA*, *nusB*, and *nusG* functions are concerned solely with the termination of transcription. *nusE* codes for ribosomal protein S10; the relationship between its location in the 30S subunit and its function in termination is not clear.

NusA is a general transcription factor that increases the efficiency of termination, probably by enhancing RNA polymerase's tendency to pause at terminators (and indeed at other regions of secondary structure; see below). NusB and S10 form a dimer that binds specifically to RNA containing a *boxA* sequence. NusG may be concerned with the general assembly of all the Nus factors into a complex with RNA polymerase. A distinction in the requirements for Nus functions is that NusA suffices for pN to prevent termination at intrinsic terminators, whereas all four Nus functions are required for pN action at rho-dependent terminators.

Antitermination occurs in the *rrn* (rRNA) operons of *E. coli*, and involves the same *nus* functions. The leader regions of the *rrn* operons contain *boxA* sequences; NusB–S10 dimers recognize these sequences and bind to RNA polymerase as it elongated past *boxA*. This changes the properties of RNA polymerase in such a way that it can now read through rho-dependent terminators that are present within the transcription unit.

The *boxA* sequence of lambda RNA does not bind NusB–S10, and is probably enabled to do so by the presence of NusA and pN; the *boxB* sequence could be required to stabilize the reaction. So variations in *boxA* sequences may determine which particular set of factors is required for antitermination. The consequences are the same: after RNA polymerase has passed the *nut* site, it has been modified by addition of appropriate factors, and fails to terminate when it subsequently encounters the terminator sites.

NusA binds to the polymerase core enzyme, but does not bind to holoenzyme. When sigma factor is added to the $\alpha_2\beta\beta'$NusA complex, it displaces the NusA protein, thus reconstituting the $\alpha_2\beta\beta'\sigma$ holoenzyme. This suggests that RNA polymerase passes through the cycle illustrated in **Figure 16.10**, in which it exists in the alternative forms of an enzyme ready to initiate ($\alpha_2\beta\beta'\sigma$) and an enzyme ready to terminate ($\alpha_2\beta\beta'$NusA). When the holoenzyme ($\alpha_2\beta\beta'\sigma$) binds to a promoter, it releases its sigma factor, and thus generates the core enzyme ($\alpha_2\beta\beta'$) that synthesizes RNA. Then a NusA protein recognizes the core enzyme and binds to it, generating the $\alpha_2\beta\beta'$NusA complex. While the $\alpha_2\beta\beta'$NusA polymerase is bound to DNA, the Nus components cannot be displaced. But when termination occurs,

Figure 16.10

RNA polymerase may alternate between initiation-competent and termination-competent forms as sigma and Nus factors alternatively replace one other on the core enzyme.

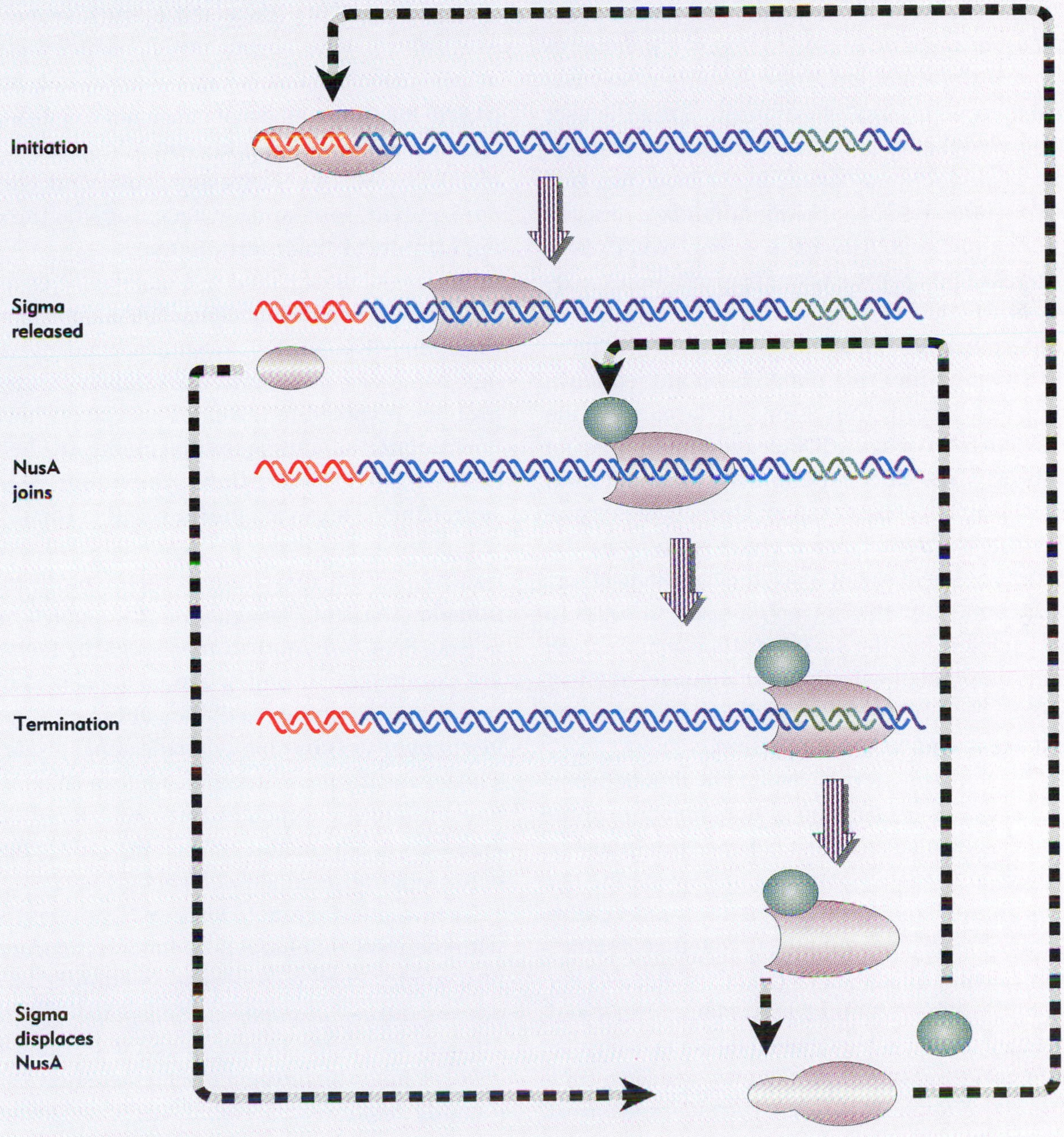

Initiation

Sigma released

NusA joins

Termination

Sigma displaces NusA

the enzyme is released in a state in which the Nus factors either are released or can be displaced by sigma factor.

Core enzyme therefore alternates between associating with sigma for initiation and associating with Nus factors for termination. Sigma and the Nus factors are mutually incompatible associates of the core. There seems no reason to regard either one as any more a component than the other; we may regard them as alternative subunits. Thus the core enzyme represents a minimal form of RNA polymerase, competent to engage in the basic function of RNA synthesis, but lacking subunits necessary for other functions. It is a moot point where RNA polymerase ends and the wider transcription apparatus begins.

Antitermination requires pN to bind to RNA polymerase in a manner that depends on the sequence of the transcription unit. Does pN recognize the *boxB* site in DNA or in the RNA transcript? It does not bind directly to either type of sequence, but does bind to a transcription complex when core enzyme passes the *boxB* site. Probably it recognizes the *boxB* RNA sequence, but also must make protein–protein contacts with RNA polymerase in order to bind. After joining the transcription complex, pN remains associated with the core enzyme, in effect becoming an additional subunit whose presence changes recognition of terminators.

The NusA component provides the link between pN and the core enzyme; pN acts as an antiterminator by preventing NusA from exercising its function. Mutations in *N* can be obtained that overcome the block to antitermination imposed by the *nusA* mutations; and the two proteins bind together *in vitro*. The pN polypeptide is small (13,500 daltons), very basic, and rather asymmetric in shape. NusA and pN probably bind to the transcription complex in rapid succession. In fact, pN cannot act at an RNA polymerase complex passing a *nut* site unless host proteins, including NusA, are present. In the presence of NusA, pN can bind to RNA polymerase.

Further light on the antitermination reaction is cast by the properties of pQ, the phage regulator that prevents termination later in phage infection. The *qut* sequence is required for pQ action, and lies at the start of the late transcription unit. The upstream part of *qut* lies within the promoter, while the downstream part lies at the beginning of the transcribed region. This implies that pQ action involves recognition of DNA.

The basic action of pQ is to interfere with pausing; and once pQ has acted upon RNA polymerase, the enzyme shows much reduced pausing at all sites, including rho-dependent and intrinsic terminators. So pQ does not act directly on termination *per se*, but instead allows the enzyme to hurry past the terminator, thus depriving the core polymerase and/or accessory factor of the opportunity to cause termination.

We do not understand in detail the differences in antitermination at different operons, for example, the ability of NusB–S10 to antiterminate in *rrn* operons, whereas pN is required in lambda. Because the proteins involved in termination and antitermination function as complexes that modify the behavior of RNA polymerase, their individual reactions sometimes appear paradoxical; thus NusA was discovered as a factor needed for antitermination by pN, but may itself be involved in the termination reaction. The general principle, however, is that RNA polymerase may exist in forms that are competent to undertake particular stages of transcription, and its activities at these stages can be changed only by modifying the appropriate form. Thus substitutions of sigma factors may change one initiation-competent form into another; and additions of Nus factors may change the properties of termination-competent forms.

Termination seems to be inextricably connected with the mode of elongation. In its basic transcription mode, core polymerase is subject to many pauses during elongation; and pausing at a terminator site is the prerequisite for termination to occur. Under the influence of factors such as NusA, pausing becomes extended, increasing the efficiency of termination; while under the influence of pN or pQ, pausing is abbreviated, decreasing the efficiency of termination. Because recognition sites for these factors are found only in certain operons, pausing and consequently termination are altered only in those operons.

Alternative secondary structures control attenuation

Several operons are regulated by **attenuation**, a mechanism that controls the ability of RNA polymerase to read through an **attenuator**, which is an intrinsic terminator located at the beginning of a transcription unit. The common feature in attenuation is that some external event controls the formation of the hairpin needed for intrinsic termination. If the hairpin is allowed to form,

termination prevents RNA polymerase from transcribing the structural genes. If the hairpin is prevented from forming, RNA polymerase elongates through the terminator, and the genes are expressed.

Attenuation was discovered in the tryptophan operon, whose five structural genes are arranged in a contiguous series, coding for the three enzymes

Figure 16.11

The *trp* operon consists of five contiguous structural genes preceded by a control region which includes a promoter, operator, leader peptide coding region, and attenuator.

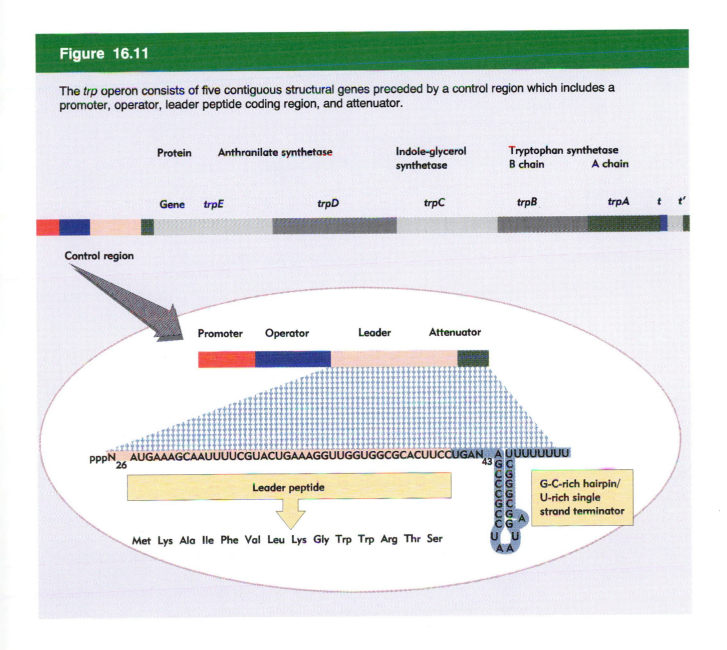

Figure 16.12

An attenuator controls the progression of RNA polymerase into the *trp* genes. RNA polymerase initiates at the promoter and then proceeds to position 90, where it pauses before proceeding to the attenuator at position 140. In the absence of tryptophan, the polymerase continues into the structural genes (*trpE* starts at +163). In the presence of tryptophan there is ˜90% probability of termination to release the 140 base leader RNA.

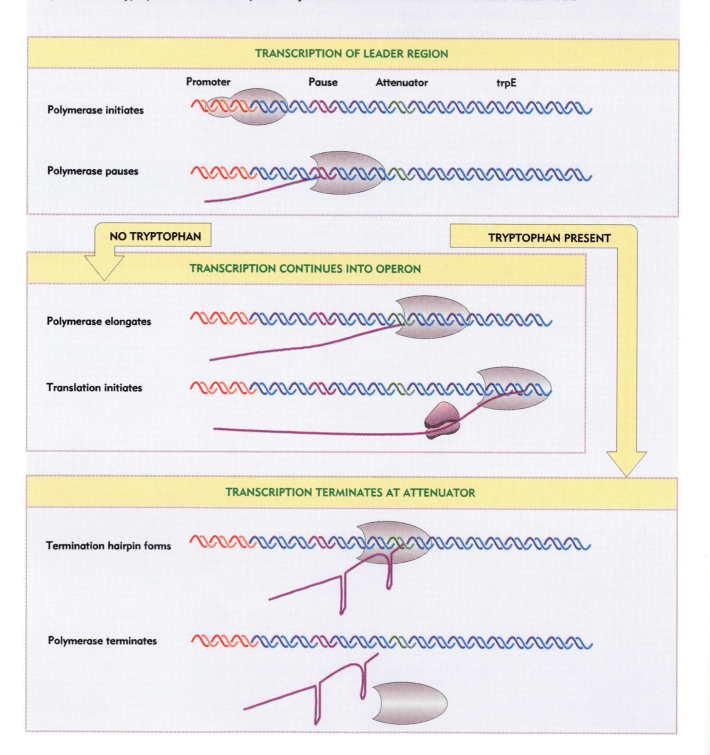

that convert chorismic acid to tryptophan by the pathway given in **Figure 16.11**. Transcription starts at a promoter at the left end of the cluster. Adjacent to it is the operator that binds the repressor protein coded by the unlinked gene *trpR*. A leader sequence lies between the operator and the coding region of the first gene. Transcription of the structural genes is partially terminated at an intrinsic terminator, *trpt*, 36 bp beyond the end of the last coding region. About 250 bp later there is a rho-dependent terminator, *trpt'*, but no requirement of this site for operon function has been demonstrated. Essentially the same operon is present in *E. coli* and in *Salmonella typhimurium*.

In addition to the promoter–operator complex, another site is involved in regulating the *trp* operon. Its existence was first revealed by the observation that deleting a sequence between the operator and the *trpE* coding region can increase the expression of the structural genes. This effect is independent of repression: both the basal and derepressed levels of transcription are increased. So this site influences events that occur *after* RNA polymerase has set out from the promoter (irrespective of the conditions prevailing at initiation).

The leader sequence between the promoter and the *trpE* gene has two interesting features:

◆ A short coding sequence could represent a **leader peptide** of 14 amino acids.

Figure 16.13

The *trp* leader region can exist in alternative base-paired conformations. The center shows the four regions that can base pair. Region 1 is complementary to region 2, which is complementary to region 3, which is complementary to region 4. On the left is the conformation produced when region 1 pairs with region 2, and region 3 pairs with region 4. On the right is the conformation when region 2 pairs with region 3, leaving regions 1 and 4 unpaired.

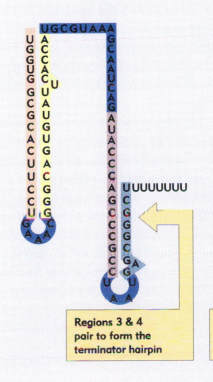

Regions 3 & 4 pair to form the terminator hairpin

ALTERNATIVE STRUCTURES ARE POSSIBLE
Region 2 is complementary to 1 & 3
Region 3 is complementary to 2 and 4

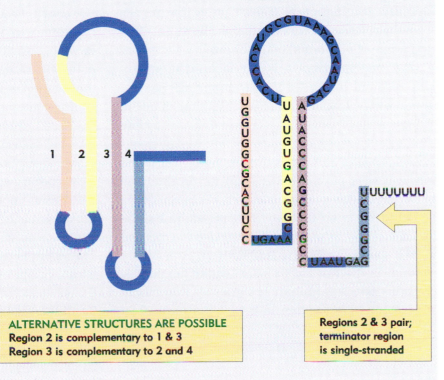

Regions 2 & 3 pair; terminator region is single-stranded

◆ The attenuator (intrinsic terminator) provides a barrier to transcription into the structural genes. RNA polymerase terminates there, either *in vivo* or *in vitro*, to produce a 140-base transcript.

Termination at the attenuator responds to the level of tryptophan, as illustrated in **Figure 16.12**. In the presence of adequate amounts of tryptophan, termination is efficient. But in the absence of tryptophan, RNA polymerase can continue into the structural genes.

Repression and attenuation respond in the same way to the level of tryptophan. When tryptophan is present, the operon is repressed; and most of the RNA polymerases that escape from the promoter then terminate at the attenuator. When tryptophan is removed, RNA polymerase has free access to the promoter, and also is no longer compelled to terminate prematurely.

Attenuation has ~10× effect on transcription. When tryptophan is present, termination is effective, and the attenuator allows only ~10% of the RNA polymerases to proceed. In the absence of tryptophan, attenuation allows virtually all of the polymerases to proceed. Together with the ~70× increase in initiation of transcription that results from the release of repression, this allows an ~700-fold range of regulation of the operon.

How can termination of transcription at the attenuator respond to the level of tryptophan? The sequence of the leader region suggests a mechanism.

Figure 16.11 shows that it contains a ribosome binding site whose AUG codon is followed by a coding region of 13 codons. Is this sequence translated into a leader peptide? Although no product has been detected *in vivo*, probably this is because it is unstable. We know that the ribosome binding site is functional (because when it is fused to a structural gene it sponsors effective translation).

What is the function of the leader peptide? It contains two tryptophan residues in immediate succession. Tryptophan is a rare amino acid in *E. coli* proteins, so this is unlikely to be mere coincidence. When the cell runs out of tryptophan, ribosomes initiate translation of the leader peptide, but stop when they reach the Trp codons. The sequence of the mRNA suggests that this **ribosome stalling** influences termination at the attenuator.

The leader sequence can be written in alternative base-paired structures. The ability of the ribosome to proceed through the leader region controls transitions between these structures. The structure determines whether the mRNA can provide the features needed for termination.

Figure 16.13 draws these structures. In the first, region 1 pairs with region 2; and region 3 pairs with region 4. The pairing of regions 3 and 4 generates the hairpin that precedes the U_8 sequence: this is the essential signal for intrinsic termination. Probably the RNA would take up this structure in lieu of any outside intervention.

A different structure is formed if region 1 is prevented from pairing with region 2. In this case, region 2 is free to pair with region 3. Then region 4 has no available pairing partner; so it is compelled to remain single-stranded. Thus the terminator hairpin cannot be formed.

Figure 16.14 shows that the position of the ribosome can determine which structure is formed, in such a way that termination is attenuated only in the absence of tryptophan. The crucial feature is the position of the Trp codons in the leader peptide coding sequence.

When tryptophan is present, ribosomes are able to synthesize the leader peptide. They continue along the leader section of the mRNA to the UGA codon, which lies between regions 1 and 2. As shown in the lower part of the figure, by progressing to this point, the ribosomes extend over region 2 and prevent it from base pairing. The result is that region 3 is available to base pair with region 4, generating the terminator hairpin. Under these conditions, therefore, RNA polymerase terminates at the attenuator.

Figure 16.14

The alternatives for RNA polymerase at the attenuator depend on the location of the ribosome, which determines whether regions 3 and 4 can pair to form the terminator hairpin.

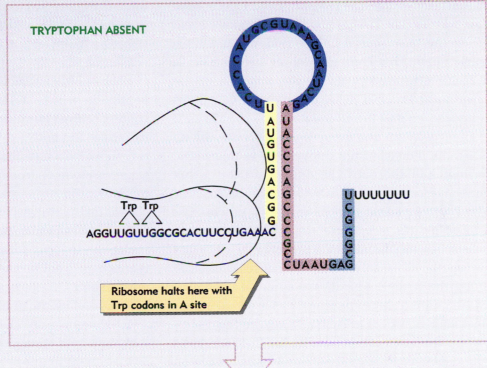

TRYPTOPHAN ABSENT

Trp Trp

AGGUUGUUGGCGCACUUCCUGAAAC

Ribosome halts here with Trp codons in A site

UUUUUUUU

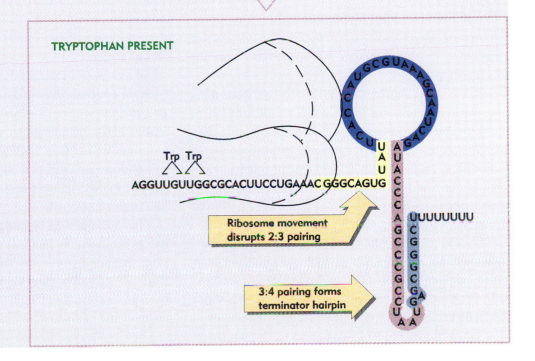

TRYPTOPHAN PRESENT

Trp Trp

AGGUUGUUGGCGCACUUCCUGAAAC GGGCAGUG

Ribosome movement disrupts 2:3 pairing

3:4 pairing forms terminator hairpin

UUUUUUUU

When there is no tryptophan, ribosomes stall at the Trp codons, which are part of region **1**, as shown in the upper part of the figure. Thus region **1** is sequestered within the ribosome and cannot base pair with region **2**. This means that regions **2** and **3** become base paired before region **4** has been transcribed. This compels region **4** to remain in a single-stranded form. In the absence of the terminator hairpin, RNA polymerase continues transcription past the attenuator.

Control by attenuation requires a precise timing of events. For ribosome movement to determine formation of alternative secondary structures that control termination, *translation of the leader must occur at the same time when RNA polymerase approaches the terminator site*. A critical event in controlling the timing is the presence of a site that causes the RNA polymerase to pause at base 90 along the leader. The RNA polymerase remains paused until a ribosome translates the leader peptide. Then the polymerase is released and moves off toward the attenuation site. By the time it arrives there, secondary structure of the attenuation region has been determined.

By providing a mechanism to sense the inadequacy of the supply of Trp-tRNA, attenuation responds directly to the need of the cell for tryptophan in protein synthesis.

How widespread is the use of attenuation as a control mechanism for bacterial operons? It is used in at least six operons that code for enzymes concerned with the biosynthesis of amino acids. So a feedback from the level of the amino acid available for protein synthesis (as represented by the availability of aminoacyl-tRNA) to the production of the enzymes may be common.

The use of the ribosome to control RNA secondary structure in response to the availability of an aminoacyl-tRNA establishes an inverse relationship between the presence of aminoacyl-tRNA and the transcription of the operon, equivalent to a situation in which aminoacyl-tRNA functions as a corepressor of transcription. Since the regulatory mechanism is mediated by changes in the formation of duplex regions, attenuation provides a striking example of the importance of secondary structure in the termination event, and of its use in regulation.

Each operon subject to attenuation has a leader peptide in which stalling of the ribosome at the codons representing the regulator amino acid(s) can cause the mRNA to take up a secondary structure in which a terminator hairpin cannot form. In some cases, an operon is derepressed by starvation for more than one amino acid. In these cases, codons for the various amino acids that regulate the operon are interspersed in the leader sequence in such a way that ribosome stalling at any one of them is able to prevent formation of the terminator hairpin. In some cases (such as the *his* operon), attenuation provides the only regulation of expression.

Attenuation may also be regulated by proteins that bind to RNA, either to stabilize or to destabilize formation of the hairpin required for termination. The activity of such a protein may be intrinsic or may respond to a small molecule in the same manner as a repressor protein responds to corepressor.

An example in which the regulator is required to turn an operon off is provided by the *trp* operon of *B. subtilis*, where the MtrB protein binds to the leader of the transcript to promote formation of the terminator hairpin. MtrB is presumably activated by tryptophan, and prevents expression of the *trp* genes by causing attenuation. It therefore functions as a terminator protein.

The *bgl* operon is controlled by the ability of the protein BglG to bind to a sequence in the 5′ nontranslated leader of the transcript. Binding of BglG sequesters part of the sequence required for the terminator hairpin. BglG therefore acts as an antiterminator.

Small RNA molecules can regulate translation

All the *trans*-acting regulators that we have discussed so far are proteins. Yet the formal circuitry of a regulatory network could equally well be constructed by using an RNA as regulator. In fact, the original model for the operon left open the question of whether the regulator might be RNA or protein.

Several instances now are known in which small RNA molecules regulate gene expression. Like a protein regulator, the RNA is an independently synthesized molecule that diffuses to a target site consisting of a specific nucleotide sequence. The target for a regulator RNA is a single-stranded nucleic acid sequence. The regulator RNA functions by complementarity with its target, at which it can form a double-stranded region.

We can imagine two mechanisms for the action of a regulator RNA:

◆ Formation of a duplex region with the target nucleic acid directly prevents its ability to function, for example, by sequestering a site needed to initiate translation.

◆ Formation of a duplex region in one part of the target molecule changes the conformation of another region, thus indirectly affecting its function.

The feature common to both types of RNA-mediated regulation is that changes in secondary structure of the target control its activity.

A difference between RNA regulators and the proteins that repress operons is that the RNA does not have allosteric properties; it cannot respond to other small molecules by changing its ability to recognize its target. It can be turned on by controlling transcription of its gene or it could be turned off by an enzyme that degrades the RNA regulator product.

RNA–RNA interactions have been implicated in controlling translation (see Chapter 35) and replication (see Chapter 19). The synthesis of a small RNA directly controls translation of the *ompF* gene of *E. coli*. The circuit is shown in **Figure 16.15**.

Expression of the unlinked genes *ompC* and *ompF*, which code for two of the outer membrane proteins of *E. coli*, is controlled by the osmolarity of the medium. Two genes are involved in controlling the response to changes in osmolarity. The locus *envZ* codes for a receptor protein that serves as an osmosensor. When the osmolarity increases, EnvZ in turn activates another protein, the product of *ompR*.

OmpR is a positive regulator that activates transcription of two genes that are transcribed divergently from a central regulatory region: the structural gene *ompC* (not shown in the Figure); and a regulator gene, *micF*.

The product of *micF* is an RNA of 174 bases. The product is called a micRNA, an acronym for mRNA-interfering-complementary RNA. A more general (and common) description is **antisense RNA**. MicF RNA is complementary to a region of OmpF mRNA that includes the ribosome–binding site at which translation is initiated. The MicF RNA could therefore function as a regulator by binding to the OmpF mRNA and preventing its translation. It is also possible that formation of a duplex region destabilizes the OmpF mRNA, for example, by making it susceptible to ribonucleases that act on double-stranded regions.

Increase in osmolarity therefore leads to the synthesis of micRNA, which turns off translation of *ompF* mRNA.

It seems likely that synthesis of a micRNA is sufficient to inactivate a target RNA in either prokaryotic or eukaryotic cells. Artificial genes coding for micRNAs have been introduced into *E. coli*, where they prevent expression of the specific target genes to whose mRNAs they are complementary. Synthesis of RNAs containing anti-sense sequences to mRNAs that code for important regulators is used as an ancillary mechanism to turn off those regulators during the life cycle of phage lambda (see Chapter 17).

Anti-sense genes have been introduced into

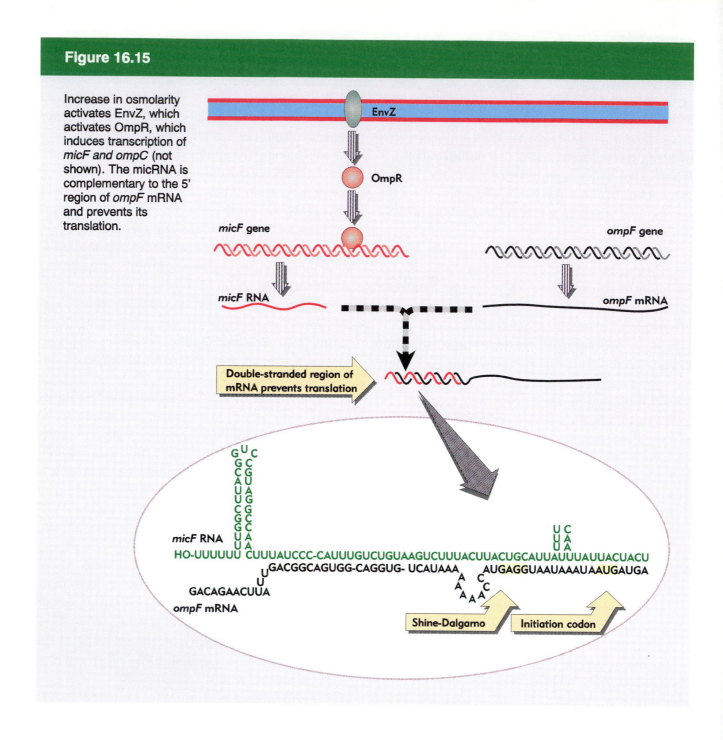

Figure 16.15

Increase in osmolarity activates EnvZ, which activates OmpR, which induces transcription of *micF* and *ompC* (not shown). The micRNA is complementary to the 5' region of *ompF* mRNA and prevents its translation.

eukaryotic cells. Such genes are constructed by reversing the orientation of a gene with regard to its promoter, so that the 'anti-sense' strand is transcribed, as illustrated in **Figure 16.16**.

When introduced into cells, an anti-sense thymidine kinase gene inhibits synthesis of thymidine kinase from the endogenous gene. Quantitation of the effect is not entirely reliable, but it seems that an excess (perhaps a considerable excess) of the anti-sense RNA may be necessary. Only a small part of the target RNA need be recognized; anti-sense sequences representing <100 bases of the

5′ region of a target mRNA inhibit its expression effectively.

At what level does the anti-sense RNA inhibit expression? It could in principle prevent transcription of the authentic gene, processing of its RNA product, or translation of the messenger. Results with different systems show that the inhibition depends on formation of RNA•RNA duplex molecules, but this can occur either in the nucleus or in the cytoplasm. In the case of an anti-sense gene stably carried by a cultured cell, sense–antisense RNA duplexes form in the nucleus, preventing normal processing and/or transport of the sense RNA. In another case, injection of anti-sense RNA into the cytoplasm inhibits translation by forming duplex RNA in the 5′ region of the mRNA.

This technique offers a powerful approach for turning off genes at will; for example, the function of a regulatory gene can be investigated by introducing an anti-sense version. An extension of this technique is to place the anti-sense gene under control of a promoter itself subject to regulation. Then the target gene can be turned off and on by regulating the production of anti-sense RNA. This technique allows investigation of the importance of the timing of expression of the target gene.

Figure 16.16

Anti-sense RNA can be generated by reversing the orientation of a gene with respect to its promoter, and can anneal with the wild-type transcript to form duplex RNA.

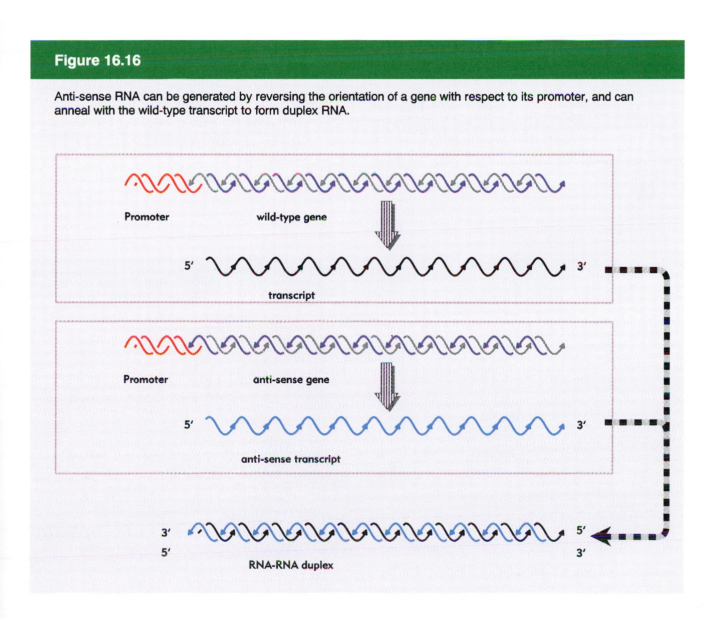

Regulation by cleavage of mRNA

Bacterial mRNAs almost all are translated in the same form in which they are transcribed. Most are polycistronic, some are monocistronic. An exceptional situation, however, is presented by phage T7. Early and late genes are transcribed into polycistronic mRNAs. But then the molecules are cleaved at several sites to release mRNAs for the individual genes.

This form of expression is found in phage T3 as well as in T7. Since the two phage DNA sequences are quite divergent, the retention of a common pathway suggests that it offers some distinct selective advantage. But we do not know why this unique mRNA processing pathway should be favored by these two phages.

The transcription of the early region of phage T7 is illustrated in **Figure 16.17**. Initiation occurs at one of a group of three promoters (*A1, A2, A3*) and continues to a terminator (*t*) ~7000 bp later.

There are six genes within this region, and the enzyme **RNAase III** cleaves the transcript at intercistronic boundaries to release five individual mRNAs (four monocistronic and one bicistronic).

What is the nature of the sites recognized by RNAase III? The targets have been identified by determining the sequences at the 5′ and 3′ ends of the mRNA molecules generated by cleavage. These sequences can be aligned with the known DNA sequence of phage T7 to reveal the complete environment of the cleavage site.

The only known feature common to all RNAase III cleavage sites is their possession of duplex structure. All the T7 sites take the form of a duplex hairpin containing an unpaired bubble. An example is drawn in **Figure 16.18**. All the sites are cleaved by RNAase III, and the existence of the hairpin is the only identified feature required for recognition by the enzyme.

Figure 16.17

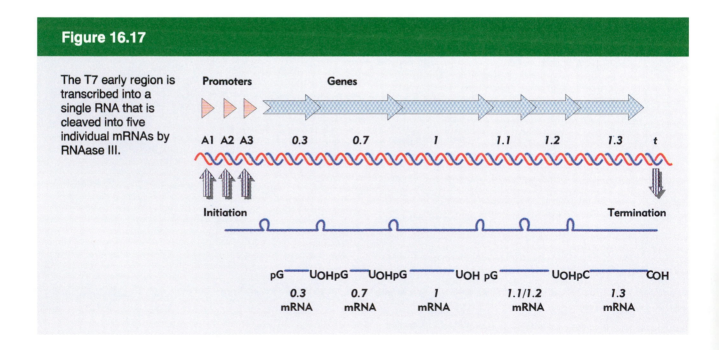

The T7 early region is transcribed into a single RNA that is cleaved into five individual mRNAs by RNAase III.

Figure 16.18

A typical cleavage site for RNAase III in T7 RNA lies within a short hairpin that contains an unpaired bubble.

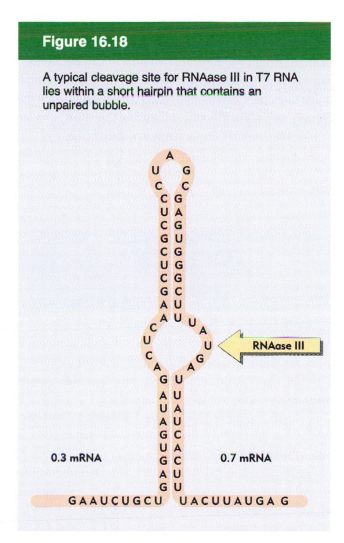

0.3 mRNA   0.7 mRNA

RNAase III

The *sib* mutation prevents RNAase III from recognizing a target site. **Figure 16.19** shows that cleavage of the wild-type mRNA at this site probably releases the 3' end for degradation by other (unknown) ribonucleases. The instability of the mRNA prevents synthesis of the Int protein. Mutations in either *sib* or *rnc* have the same effect; they prevent cleavage, and therefore stabilize the mRNA and allow translation to occur. Because the effect controls expression of the gene upstream of

Figure 16.19

Retroregulation results from cleavage of the *sib* site at the 3' end of *Int* mRNA by RNAase III, which leads to degradation of the mRNA.

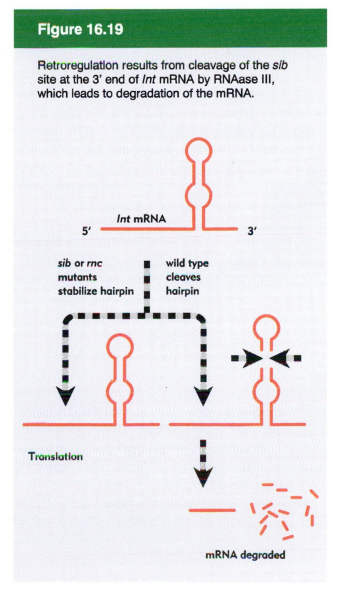

Int mRNA

5'   3'

sib or *rnc* mutants stabilize hairpin   wild type cleaves hairpin

Translation

mRNA degraded

Cleavage at the end of the *1.2* gene is necessary for translation of the bicistronic *1.1/1.2* mRNA. When this cleavage is prevented, neither coding region *1.1* nor *1.2* can be translated. The reason is that the longer mRNA takes up a secondary structure that prevents initiation at the *1.1* coding sequence (and failure to translate the *1.1* coding sequence in turn prevents translation of *1.2*).

Another instance in which RNAase III affects translation is presented by an mRNA of phage lambda that carries the sequence of gene *int*. This mRNA usually is not translated (Int protein is synthesized from another mRNA). Mutations in a site called *sib,* located downstream of the *int* coding sequence, allow Int protein to be translated from the usually inactive mRNA.

the site where the relevant interaction occurs, it is called **retroregulation**.

In both the T7 *1.1/1.2* genes and the lambda *int* system, there is a connection between the ability of the RNA to be translated and its cleavage by RNAase III at a particular site. In both cases, the probable mechanism of the relationship is the effect of cleavage on the secondary structure of the RNA. *By releasing part of the molecule, the cleavage event changes the structure of other regions, including sites that are important in the RNA's functions.* Affected functions can involve initiation of translation, susceptibility to degradation, ability to be spliced (in eukaryotes), and no doubt other events.

Cleavages are needed to release prokaryotic and eukaryotic rRNAs

The maturation of eukaryotic rRNA can be followed in some detail because of the relatively slow rate at which the primary transcript matures through intermediate stages to give the final rRNA molecules. The major processing intermediates can be isolated from a variety of cells, among which the pathway in mammals is well characterized.

The mammalian primary transcript is a 45S RNA containing the sequences of both the 18S and 28S rRNAs, but almost twice their combined length (see Chapter 24). It contains ~110 methyl groups, added during or immediately after its transcription. Almost all of them are attached to ribose moieties, in a variety of oligonucleotide sequences.

The methyl groups are *conserved* during processing of the precursor into mature rRNAs. There are ~39 of the original methyl groups on mature 18S rRNA; another 4 are added later in the cytoplasm. There are ~74 methyl groups (all original) on the 28S rRNA. This suggests that methylation is used to distinguish the regions of the primary transcript that mature into rRNA.

More than one pathway has been found for maturation, as seen from the sizes of the intermediate RNA molecules. But all the pathways can be reconciled by supposing that a small number of cleavage sites can be utilized in varying orders.

Two mammalian pathways, originally characterized in HeLa (human) and L (mouse) cells, are illustrated in **Figure 16.20**. In each case, there are cleavage sites on the 5′ side of the 18S gene, within the spacer between the 18S and 5.8S gene, and in the spacer between the 5.8S and 28S sequences. (The 5.8S sequence becomes a small RNA that associates with the 28S rRNA by base pairing.)

The difference between the two pathways lies solely in the order with which the cleavage sites are utilized. Other variations also are possible. In some cases, more than one pathway is found in a given cell type.

The figure shows the minimum number of likely cleavage sites used to generate all the 5′ and 3′ ends. We do not know whether these cleavages actually generate the mature ends, or whether cleavage releases individual precursors that then are cut or trimmed further. Nor do we have any information about the ribonuclease activities responsible for the process. We do know, however,

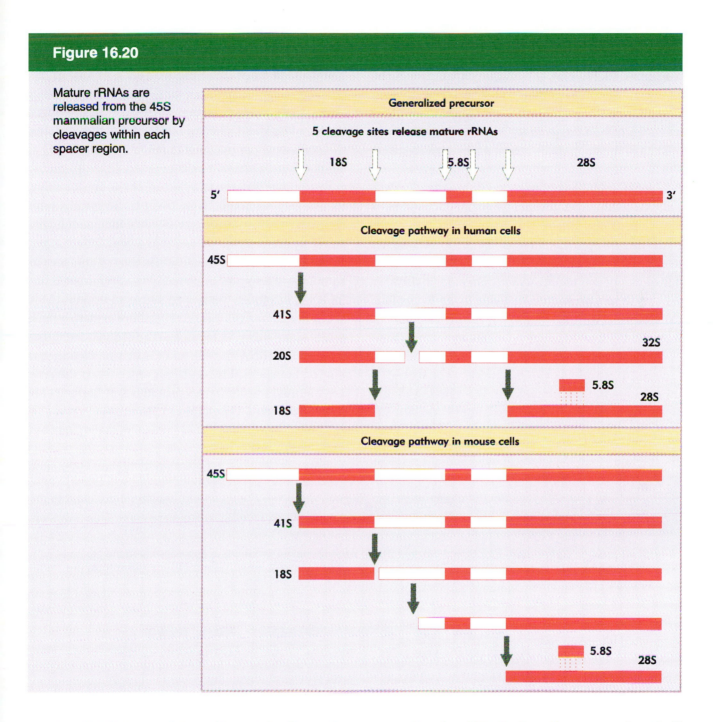

Figure 16.20

Mature rRNAs are released from the 45S mammalian precursor by cleavages within each spacer region.

that the 45S RNA associates with proteins immediately upon its synthesis, so that processing occurs with ribonucleoprotein rather than with free RNA.

Cleavages are also needed to release bacterial rRNAs from their joint precursor. The seven oper-ons coding for rRNA in *E. coli* are named *rrnA–G*. They are not closely linked on the chromosome. Ribosomal RNA sequences are conserved, but we do not know how sequence homogeneity is maintained. The overall organization of all the *rrn* operons is the same, containing all three rRNA molecules in the order 16S–23S–5S, as illustrated in **Figure 16.21**. The two major rRNAs lie in the same order as in the eukaryotic transcription unit; the presence of the 5S RNA in the same transcription unit is unique to the prokaryotes.

Each of these operons is transcribed into a single RNA precursor from which the mature rRNA molecules are released by cleavage. When the cleavage reaction is blocked, the precursor accumulates as an RNA sedimenting at ~30S.

All the *rrn* operons have a dual promoter structure. The first promoter, *P1*, lies ~300 bp upstream of the start of the 16S rRNA sequence. Probably this is the principal promoter. Up to the first 150 bp of the transcription unit are different in the various *rrn* operons. Within this region, ~110 bp from *P1*, is a second promoter, *P2*. The spacer sequences that flank the 16S and 23S rRNA genes are conserved.

Between the sequences for 16S rRNA and 23S rRNA is a transcribed spacer region of 400–500 bp.

Within this region, however, lies a sequence or sequences coding for tRNA. As summarized in the generalized diagram of Figure 16.21, in four of the *rrn* operons the transcribed spacer contains a single tRNA sequence, that of $tRNA_2^{Glu}$. In the other three *rrn* operons, the transcribed spacer contains the sequences of two tRNAs ($tRNA_1^{Ile}$ and $tRNA_{1B}^{Ala}$). Thus on the basis of the 'spacer tRNA' sequences, we can distinguish two types of rRNA operon. The tRNAs are released from the precursor RNA by cleavage; the remaining regions of the spacer presumably are degraded.

Yet another variation is found in the *rrn* operons. In some, the last sequence coded in the precursor is the 5S RNA. In others, there are some additional tRNA sequences. For example, in *rrnC* there are two genes, $tRNA^{Trp}$ and $tRNA_1^{Asp}$, located at the 3' end. (The presence of the $tRNA^{Trp}$ gene makes this particular *rrn* operon indispensable for *E. coli*, since this is the only copy of the gene, and thus provides sole capacity for the utilization of tryptophan in protein synthesis.)

Figure 16.21

The *rrn* operons contain genes for both rRNA and tRNA. Each functional sequence is separated from the next by a transcribed spacer region; the lengths of the leader and trailer depend on which promoters and terminators are used. Each RNA product must be released from the transcript by cuts on either side.

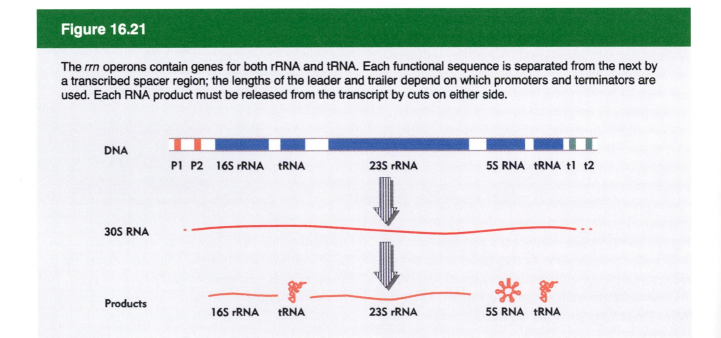

The enzyme responsible for processing the rRNAs is RNAase III. In *rnc⁻* cells, the mature rRNAs do not appear; instead, a 30S precursor containing 16S, 23S, and 5S sequences accumulates. This precursor can be cleaved by RNAase III *in vitro* to generate the molecules that serve as individual precursors to the mature RNA species.

The precursors to the 16S and 23S rRNAs are known as p16 and p23. Each precursor is slightly longer than the rRNA found in the ribosome, because it retains additional sequences at both the 5′ and 3′ termini.

A common mechanism generates the ends of both the p16 and p23 RNAs. In each case, the regions in the primary transcript that contain the ends are complementary. They base pair to generate a duplex structure containing both the 5′– and 3′– terminal regions of the product. A remarkable feature of this reaction is the distance separating the complementary sequences, ~1600 nucleotides for the p16 and ~2900 nucleotides for the p23.

Figure 16.22 shows that RNAase III cleaves on both sides of the double-stranded stem, simultaneously generating the 5′ and 3′ ends of the p16 or p23 molecules. This action of RNAase III differs in two respects from that most commonly seen with T7 early RNA. The target sequence is cut in the stem rather than in a bubble. And two cuts are made rather than one (which also happens in an atypical T7 site).

There is no extensive sequence homology between either the p16 and p23 RNAase III sites or between either of these and any of the T7 sites. The features required by RNAase III for specific cutting of RNA are mysterious, except for the common demand for a double-stranded structure either surrounding or including the actual site of cleavage. The enzyme could depend on a combination of primary sequence and secondary structure.

Since the cleavages accomplished by RNAase III do not release the actual 5′– or 3′– termini of either 16S or 23S rRNA, further processing reactions are

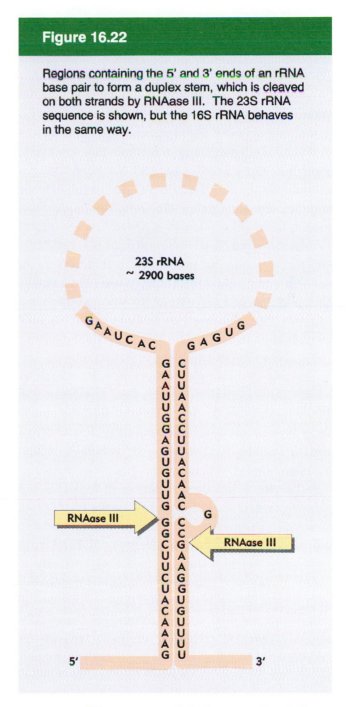

Figure 16.22

Regions containing the 5′ and 3′ ends of an rRNA base pair to form a duplex stem, which is cleaved on both strands by RNAase III. The 23S rRNA sequence is shown, but the 16S rRNA behaves in the same way.

necessary. Not very much is known about them, except that their occurrence is contingent on the prior action of RNAase III. The tRNA sequences included in the *rrn* transcription units are processed by RNAase P and other tRNA-processing enzymes (see Chapter 8).

Summary

Bacterial RNA polymerase terminates transcription at two types of site. Intrinsic terminator sites contain a G•C-rich hairpin followed by a run of U residues. They are recognized *in vitro* by core enzyme alone. Rho-dependent sites require factor both *in vitro* and *in vivo;* the common feature is a stretch of 50–90 nucleotides preceding the site of termination that is rich in C and poor in G residues. Rho factor is an essential protein that acts as an ancillary termination factor, which recognizes RNA and acts at sites where RNA polymerase has paused. The termination activity requires ATP hydrolysis.

The Nus factors increase the efficiency of rho-dependent termination, and provide the means by which antitermination factors act. The NusB–S10 dimer recognizes the *boxA* sequence in the elongating RNA; NusA joins subsequently. Factor NusA and the initiation factor sigma are mutually exclusive associates of core enzyme.

Antitermination occurs rarely in bacteria, but is more widely used by some phages to regulate progression from one stage of gene expression to the next. An example is phage lambda. A rho-dependent terminator lies between two sets of genes. A product of the first set is an antiterminator (pN) that is necessary to allow RNA polymerase to proceed to the next set. N protein recognizes RNA polymerase carrying NusA when the enzyme passes the sequence *boxB*, which lies between the promoter and the terminator site; pN then binds to the complex and prevents termination at the rho-dependent terminator. Various terminators and antiterminators have different (corresponding) specificities.

Attenuation is a mechanism that relies on regulation of termination to control transcription through bacterial operons. It is commonly used in operons that code for enzymes involved in biosynthesis of an amino acid. The polycistronic mRNA of the operon starts with a sequence that can form alternative secondary structures. One of the structures has a hairpin loop that provides an intrinsic terminator upstream of the structural genes; the alternative structure lacks the hairpin. The choice of structure that forms is controlled by the progress of translation through a short leader sequence that includes codons for the amino acid(s) that are the product of the system. In the presence of aminoacyl-tRNA bearing such amino acid(s), ribosomes translate the leader peptide, allowing a secondary structure to form that supports termination. In the absence of this aminoacyl-tRNA, the ribosome stalls, resulting in a new secondary structure in which the hairpin needed for termination cannot form. The supply of aminoacyl-tRNA therefore (inversely) controls amino acid biosynthesis. Attenuation also is controlled in some operons by regulator proteins that bind directly to the leader RNA to assist or inhibit the formation of a terminator hairpin.

An alternative to a regulator protein may be a small RNA that is complementary to a target mRNA; formation of a duplex RNA region may prevent translation by sequestering the initiation site, directly or indirectly.

Further reading

Reviews

The mechanisms and regulation of bacterial termination were reviewed by **Adhya and Gottesman** (*Ann. Rev. Biochem.* **47**, 967–996, 1978), **Platt** (*Ann. Rev. Biochem.* **55**, 339–372, 1986), **Friedman, Imperiale, and Adhya** (*Ann. Rev. Genet.* **21**, 453–488, 1987), and **Das** (*Ann. Rev. Biochem.* **62**, 893–930, 1993).

Attenuation was reviewed by **Bauer** *et al.* (in *Gene Function in Prokaryotes*, Eds. Beckwith, Davies, and Gallant, Cold Spring Harbor, NY, 65–89, 1983). Attenuation of the *trp* operon was reviewed by **Yanofsky** (*Nature* **289**, 751–758, 1981) and **Platt** (*Cell* **24**, 10–23, 1981), and was brought up to date by **Landick and Yanofsky** (in *E. coli and S. typhimurium*, Ed. Neidhardt, American Society for Microbiology, Washington DC, 1276–1301, 1987) and **Yanofsky** (*J. Biol. Chem.* **263**, 609–612, 1988). The functions of the *trp* operon were summarized in terms of its sequence by **Yanofsky** *et al.* (*Nuc. Acids Res.* **9**, 6647–6668, 1981) and in *E. coli and S. typhimurium* (Ed. Neidhardt, American Society for Microbiology, Washington DC, 1453–1472, 1987).

Eukaryotic termination was reviewed by **Birnstiel** (*Cell* **41**, 349–359, 1985).

Use of antisense RNA has been reviewed by **Green, Pines, and Inouye** (*Ann. Rev. Biochem.* **55**, 569–597, 1986).

Few reviews disentangle the enzymes actually involved in tRNA and rRNA processing from the various activities that have been detected. **Gegenheimer and Apirion** attempted to sort out the genetics and biochemistry of bacterial enzymes (*Microbiol. Rev.* **45**, 502–541, 1981). *E. coli* ribonucleases were summarized by **Deutscher** (*Cell* **40**, 731–732, 1985). Types of processing were analyzed by **King, Sirdeskmukh, and Schlessinger** (*Microbiol. Rev.* **50**, 428–451, 1986).

Discoveries

Rho factor and the antagonistic effects of lambda N protein were discovered by **Roberts** (*Nature* **224**, 1168–1174, 1969). Alternative forms of RNA polymerase involved in initiation and termination were identified by **Greenblatt and Li** (*Cell* **24**, 421–428, 1981).

The basis for attenuation was described in the *trp* operon by **Lee and Yanofsky** (*Proc. Nat. Acad. Sci. USA* **74**, 4365–4368, 1977); the link with translation was characterized by **Zurawski** *et al.* (*Proc. Nat. Acad. Sci. USA* **75**, 5988–5991, 1978).

The use of anti-sense RNA in eukaryotes was introduced by **Izant and Weintraub** (*Cell* **36**, 1007–1015, 1984)

A report on RNAase III that also covered earlier ground was from **Bram, Young, and Steitz** (*Cell* **19**, 393–401, 1980).

CHAPTER 17

Phage strategies: lytic cascades and lysogenic repression

Some phages have only a single strategy for survival. On infecting a susceptible host, they subvert its functions to the purpose of producing a large number of progeny phage particles. As the result of this **lytic infection**, the host bacterium dies. In the typical lytic cycle, the phage DNA (or RNA) enters the host bacterium, its genes are transcribed in a set order, the phage genetic material is replicated, and the protein components of the phage particle are produced. Finally, the host bacterium is broken open **(lysed)** to release the assembled progeny particles.

Other phages have a dual existence. They are able to perpetuate themselves via the same sort of lytic cycle in what amounts to an open strategy for producing as many copies of the phage as rapidly as possible. But they also have an alternative, closet existence, in which the phage genome is present in the bacterium in a latent form known as **prophage.** This form of propagation is called **lysogeny.**

In a lysogenic bacterium, the prophage is **integrated** into the bacterial genome, and is inherited in the same way as bacterial genes. By virtue of its possession of a prophage, a lysogenic bacterium has **immunity** against infection by further phage particles of the same type. Immunity is established by a single integrated prophage, so usually a bacterial genome contains only one copy of a prophage of any particular type.

Transitions occur between the lysogenic and lytic modes of existence. **Figure 17.1** shows that when a phage produced by a lytic cycle enters a new bacterial host cell, it either repeats the lytic cycle or enters the lysogenic state. The outcome depends on the conditions of infection and the genotypes of phage and bacterium. A prophage is freed from the restrictions of lysogeny by the process called **induction**, in which it is **excised** from the bacterial genome, to generate a free phage DNA that then proceeds through the lytic pathway.

The form in which such a phage is propagated is determined by the regulation of transcription. Lysogeny is maintained by the interaction of a phage repressor with an operator. The lytic cycle requires a cascade of transcriptional controls. And the transition between the two life styles is

Figure 17.1

Overview: lytic development involves the reproduction of phage particles with destruction of the host bacterium, but lysogenic existence allows the phage genome to be carried as part of the bacterial genetic information.

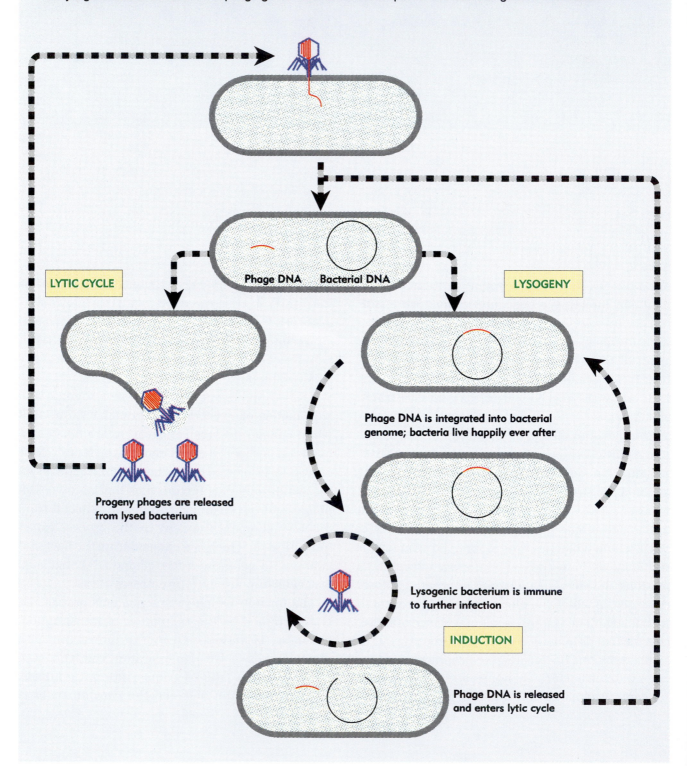

LYTIC CYCLE

Phage DNA   Bacterial DNA

LYSOGENY

Phage DNA is integrated into bacterial genome; bacteria live happily ever after

Progeny phages are released from lysed bacterium

Lysogenic bacterium is immune to further infection

INDUCTION

Phage DNA is released and enters lytic cycle

accomplished by the establishment of repression (lytic cycle to lysogeny) or by the relief of repression (induction of lysogen to lytic phage).

Another type of existence within bacteria is represented by **plasmids.** These are autonomous units that exist in the cell as **extrachromosomal** genomes. Plasmids are self-replicating circular molecules of DNA that are maintained in the cell in a stable and characteristic number of copies; that is, the number remains constant from generation to generation.

Some plasmids also have alternative life-styles. They can exist either in the autonomous extrachromosomal state; or they can be inserted into the bacterial chromosome, and then are carried as part of it like any other sequence. Such units are properly called **episomes** (but the terms 'plasmid' and 'episome' are sometimes used loosely as though interchangeable).

Like lysogenic phages, plasmids and episomes maintain a selfish possession of their bacterium and often make it impossible for another element of the same type to become established. This effect also is called **immunity**, although the basis for plasmid immunity is different from lysogenic immunity. (We discuss the control of plasmid perpetuation in Chapter 18.)

Table 17.1 summarizes the types of genetic units that can be propagated in bacteria as independent genomes. Lytic phages may have genomes of any type of nucleic acid; they transfer between cells by release of infective particles. Lysogenic phages have double-stranded DNA genomes, as do plasmids and episomes. Some plasmids and episomes transfer between cells by a conjugative process (involving direct contact between donor and recipient cells). A feature of the transfer process in both cases is that on occasion some bacterial host genes are transferred with the phage or plasmid DNA, so these events play a role in allowing exchange of genetic information between bacteria.

Table 17.1

Several types of independent genetic units may exist in bacteria.

Type of Unit	Genome Structure	Mode of Propagation	Consequences
Lytic phage	ds- or ss-DNA or RNA linear or circular	Infects susceptible host	Usually kills host
Lysogenic phage	ds-DNA	Linear sequence in host chromosome	Immunity to infection
Plasmid	ds-DNA circle	Replicates at defined copy number May be transmissible	Immunity to plasmids in same group
Episome	ds-DNA circle	Free circle or linear integrated	May transfer host DNA

Lytic development is controlled by a cascade

Phage genomes of necessity are small. Indeed, as with all viruses, a principal restriction is the need to package the nucleic acid within its protein coat. This limitation dictates many of the viral strategies for reproduction. Typically a virus takes over the apparatus of the host cell, which then replicates and expresses phage genes instead of the bacterial genes.

Usually the phage includes genes whose function is to ensure preferential replication of phage DNA. These genes are concerned with the initiation of replication and may even include a new DNA polymerase. Changes are always introduced in the capacity of the host cell to engage in transcription. They involve replacing the RNA polymerase or modifying its capacity for initiation or termination. The result is always the same: phage mRNAs are preferentially transcribed. So far as protein synthesis is concerned, usually the phage is content to use the host apparatus, redirecting its activities principally by replacing bacterial mRNA with phage mRNA.

Lytic development is accomplished by a pathway in which the phage genes are expressed in a particular order. This ensures that the right amount of each component is present at the appropriate time. The cycle can be divided into the two general parts illustrated in **Figure 17.2**:

◆ **Early infection** describes the period from entry of the DNA to the start of its replication.

◆ **Late infection** defines the period from the start of replication to the final step of lysing the bacterial cell to release progeny phage particles.

In the usual order of battle, the early phase is devoted to the production of enzymes involved in the reproduction of DNA. These include the enzymes concerned with DNA synthesis, recombination, and sometimes modification. Their activities cause a **pool** of phage genomes to accumulate. In this pool, genomes are continually replicating and recombining, so that *the events of a single lytic cycle concern a population of phage genomes*.

During the late phase, the protein components of the phage particle are synthesized. Often many different proteins are needed to make up head and tail structures, so the largest part of the phage genome consists of late functions. In addition to the structural proteins, 'assembly proteins' are needed to help construct the particle, although they are not themselves incorporated into it. By the time the structural components are assembling into heads and tails, replication of DNA has reached its maximum rate. The genomes then are inserted into the empty protein heads, tails are added, and the host cell is lysed to allow release of new viral particles.

The organization of the phage genetic map often closely reflects the sequence of lytic development. The concept of the operon is taken to somewhat of an extreme, in which the genes coding for proteins with related functions are clustered to allow their control with the maximum economy. This allows the pathway of lytic development to be controlled with a small number of regulatory switches.

The lytic cycle is under positive control, so that each group of phage genes can be expressed only when an appropriate signal is given.

Usually only a few phage genes can be expressed by the transcription apparatus of the host cell. Their promoters are indistinguishable from those of host genes. The name of this class of genes depends on the phage. In most cases, they are known as the **early genes**. In phage lambda, they are given the evocative description of **immediate early**. Irrespective of the name, they constitute only a preliminary, representing just the initial part of the early period. Sometimes they are exclusively

Figure 17.2

Lytic development takes place by producing phage genomes and protein particles that are assembled into progeny phages. Compare this cycle with the electron microscopy of infected bacteria in Figure 3.2.

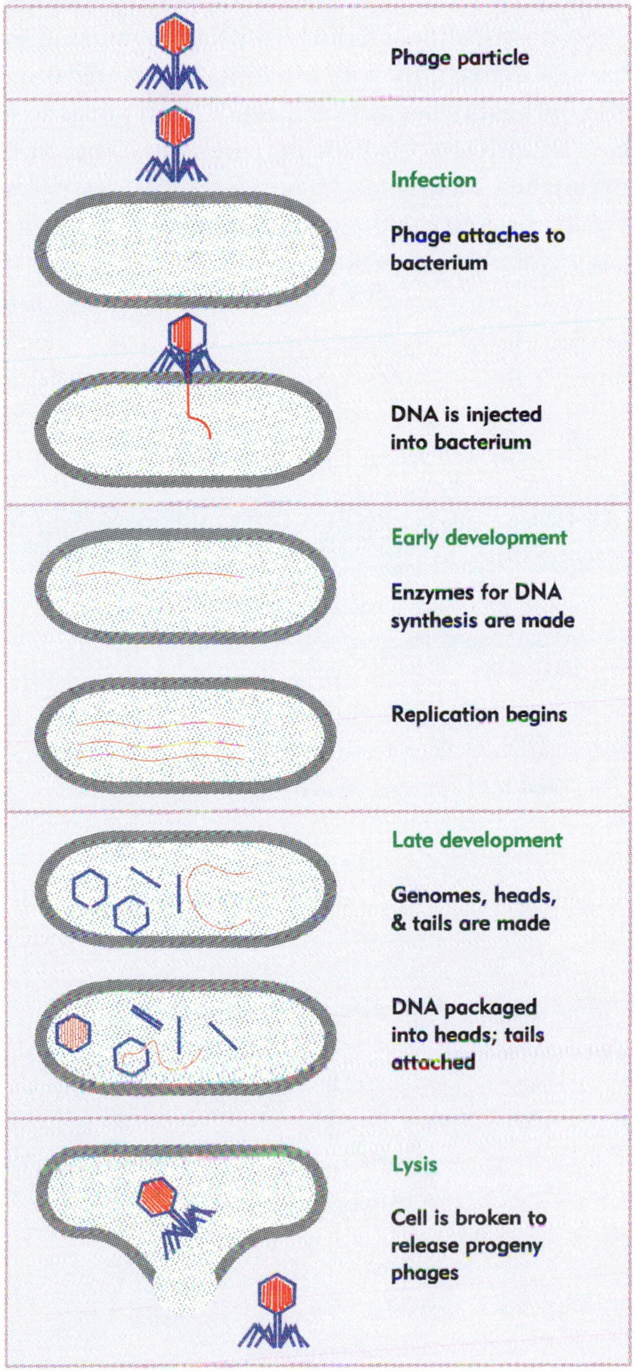

Phage particle

Infection

Phage attaches to bacterium

DNA is injected into bacterium

Early development

Enzymes for DNA synthesis are made

Replication begins

Late development

Genomes, heads, & tails are made

DNA packaged into heads; tails attached

Lysis

Cell is broken to release progeny phages

occupied with the transition to the next period. At all events, *one of these genes always codes for a protein that is necessary for transcription of the next class of genes.*

This second class of genes is known variously as the **delayed early** or **middle** group. Its expression typically starts as soon as the regulator protein coded by the early gene(s) is available. Depending on the nature of the control circuit, the initial set of early genes may or may not continue to be expressed at this stage (see Figure 16.7). Often the expression of host genes is reduced. Together the two sets of early genes account for all necessary phage functions except those needed to assemble the particle coat itself and to lyse the cell.

When the replication of phage DNA begins, it is time for the **late genes** to be expressed. Their transcription at this stage usually is arranged by embedding a further regulator gene within the previous (delayed early or middle) set of genes. This regulator may be another antitermination factor (as in lambda) or it may be another sigma factor (as in SPO1).

The use of these successive controls, in which each set of genes contains a regulator that is necessary for expression of the next set, creates a cascade in which groups of genes are turned on (and sometimes off) at particular times. The means used to construct each phage cascade are different, but the results are similar, as the following sections show.

Functional clustering in phages T7 and T4

The genome of phage T7 has three classes of genes, each constituting a group of adjacent loci. As **Figure 17.3** shows, the class I genes are the immediate early type, expressed by host RNA polymerase as soon as the phage DNA enters the cell. Among the products of these genes are enzymes that interfere with host gene expression and a phage RNA polymerase. The phage enzyme is responsible for expressing the class II genes (concerned principally with DNA synthesis functions) and the class III genes (concerned with assembling the mature phage particle).

Phage T4 has one of the larger genomes (165 kb), organized with extensive functional grouping of genes. **Figure 17.4** presents the genetic map. Genes that are numbered are **essential**: a mutation in any one of these loci prevents successful completion of the lytic cycle. Genes indicated by three-letter abbreviations are **nonessential**, at least under the usual conditions of infection. We do not really understand the inclusion of many nonessential genes, but presumably they confer a selective advantage in some of T4's habitats. (In smaller phage genomes, most or all of the genes are essential.)

There are three phases of gene expression. A summary of the functions of the genes expressed at each stage is given in **Figure 17.5**. The early genes are transcribed by host RNA polymerase. At a slightly later point, some further genes are transcribed; they are called **quasi-late**, because their mode of expression does not fit with the categories described for other phages. The early genes, together with the quasi-late genes, account for virtually all of the phage functions concerned with the synthesis of DNA, modifying cell structure, and transcribing and translating phage genes.

The two essential genes in the 'transcription' category fulfill a regulatory function: their products are necessary for late gene expression. Phage T4 infection depends on a mechanical link between replication and late gene expression. Only actively replicating DNA can be used as template for late gene transcription. The connection is generated by introducing a new sigma factor and also by making other modifications in the host RNA polymerase so that it is active only with a template of replicating DNA (probably because the replication apparatus itself interacts with RNA polymerase). This link establishes a correlation between the synthesis of phage protein components and the number of genomes available for packaging.

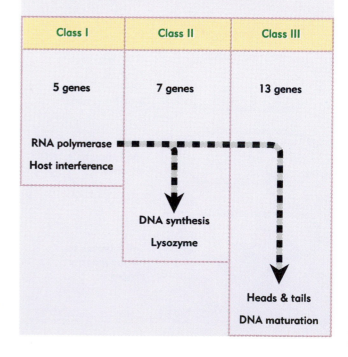

Figure 17.3

Phage T7 contains three classes of genes that are expressed sequentially. The genome is ⁻38,000 bp.

Class I	Class II	Class III
5 genes	7 genes	13 genes
RNA polymerase Host interference	DNA synthesis Lysozyme	Heads & tails DNA maturation

Figure 17.4

The map of T4 is circular. There is extensive clustering of genes coding for components of the phage and processes such as DNA replication, but there is also dispersion of genes coding for a variety of enzymatic and other functions. Essential genes are indicated by numbers; nonessential genes are identified by letters. Only some representative T4 genes are shown on the map.

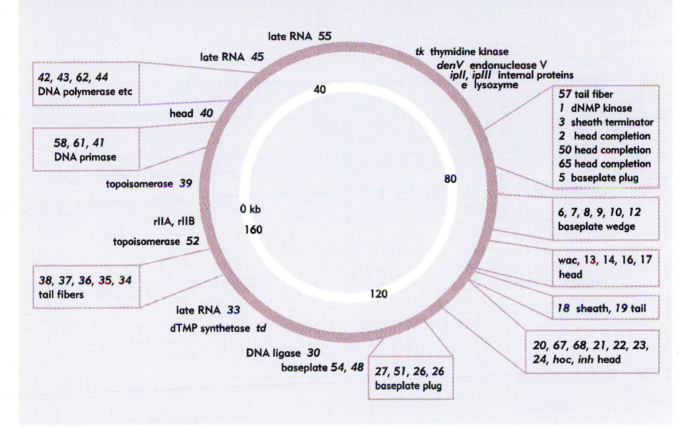

Figure 17.5

The phage T4 lytic cascade falls into two parts: early and quasi-late functions are concerned with DNA synthesis and gene expression; late functions are concerned with particle assembly.

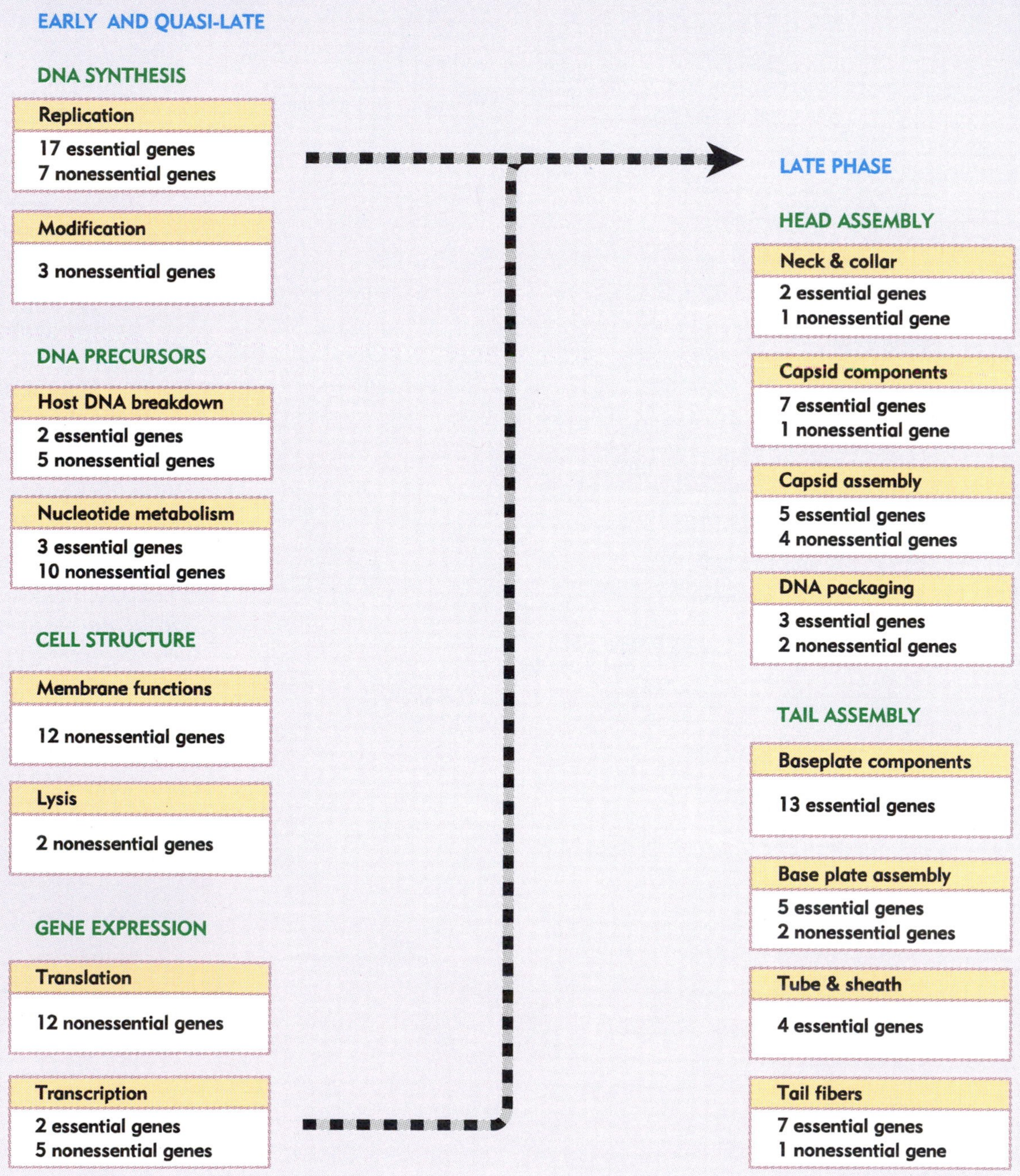

EARLY AND QUASI-LATE

DNA SYNTHESIS

Replication

17 essential genes
7 nonessential genes

Modification

3 nonessential genes

DNA PRECURSORS

Host DNA breakdown

2 essential genes
5 nonessential genes

Nucleotide metabolism

3 essential genes
10 nonessential genes

CELL STRUCTURE

Membrane functions

12 nonessential genes

Lysis

2 nonessential genes

GENE EXPRESSION

Translation

12 nonessential genes

Transcription

2 essential genes
5 nonessential genes

LATE PHASE

HEAD ASSEMBLY

Neck & collar

2 essential genes
1 nonessential gene

Capsid components

7 essential genes
1 nonessential gene

Capsid assembly

5 essential genes
4 nonessential genes

DNA packaging

3 essential genes
2 nonessential genes

TAIL ASSEMBLY

Baseplate components

13 essential genes

Base plate assembly

5 essential genes
2 nonessential genes

Tube & sheath

4 essential genes

Tail fibers

7 essential genes
1 nonessential gene

The lambda lytic cascade relies on antitermination

One of the most intricate cascade circuits is provided by phage lambda. Actually, the cascade for lytic development itself is straightforward, with two regulators controlling the successive stages of development. But the circuit for the lytic cycle is interlocked with the circuit for establishing lysogeny, as summarized in **Figure 17.6**.

When lambda DNA enters a new host cell, the lytic and lysogenic pathways start off the same way. Both require expression of the immediate early and delayed early genes. But then they diverge: lytic development follows if the late genes are expressed; lysogeny ensues if synthesis of the repressor is established.

Lambda has only two immediate early genes, transcribed independently by host RNA polymerase:

◆ *N* codes for an antitermination factor whose action at the *nut* sites allows transcription to proceed into the delayed early genes. (We discussed the mechanisms involved in antitermination in Chapter 16).

◆ *cro* has dual functions: it prevents synthesis of the repressor (a necessary action if the lytic cycle is to proceed); and it turns off expression of the immediate early genes (which are not needed later in the lytic cycle).

The delayed early genes include two replication genes (needed for lytic infection), seven recombination genes (some involved in recombination during lytic infection, two necessary to integrate lambda DNA into the bacterial chromosome for lyso-

geny), and three regulators. The regulators have opposing functions:

◆ The *cII–cIII* pair of regulators is needed to start up the synthesis of repressor.

◆ The *Q* regulator is an antitermination factor that allows host RNA polymerase to proceed into the late genes.

So the delayed early genes serve two masters: some are needed for the phage to enter lysogeny, the others are concerned with controlling the order of the lytic cycle.

To disentangle the two pathways, first consider just the lytic cycle. **Figure 17.7** gives the map of lambda phage DNA. A group of genes concerned with regulation is surrounded by genes needed for recombination and replication. Within the regulatory group are the immediate early genes, *N* and *cro*. They are transcribed from different strands of DNA—*N* toward the left, and *cro* toward the right. In the presence of the pN antitermination factor, transcription continues (through the terminators on either side) to the left of *N* into the recombination genes, and to the right of *cro* into the replication genes (see Figure 16.8).

The map gives the organization of the lambda DNA as it exists in the phage particle. But shortly after infection, the ends of the DNA join to form a circle. **Figure 17.8** shows the true state of lambda DNA during infection. The late genes are welded into a single group, containing the lysis genes *S-R* from the right end of the linear DNA, and the head and tail genes *A-J* from the left end.

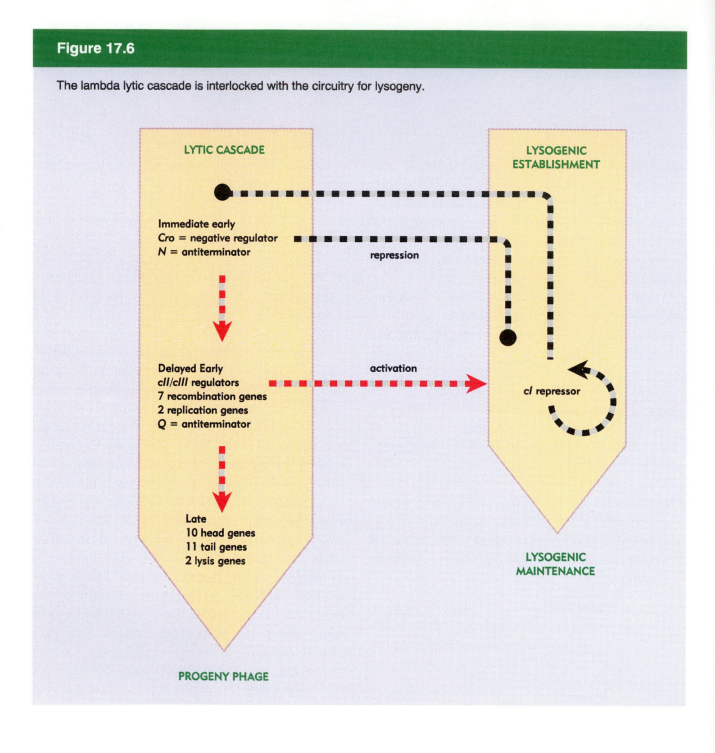

Figure 17.6

The lambda lytic cascade is interlocked with the circuitry for lysogeny.

LYTIC CASCADE

Immediate early
Cro = negative regulator
N = antiterminator

repression

Delayed Early
cII/cIII regulators
7 recombination genes
2 replication genes
Q = antiterminator

activation

Late
10 head genes
11 tail genes
2 lysis genes

PROGENY PHAGE

LYSOGENIC
ESTABLISHMENT

cI repressor

LYSOGENIC
MAINTENANCE

Figure 17.7

The lambda map shows clustering of related functions. The genome is 48,514 bp.

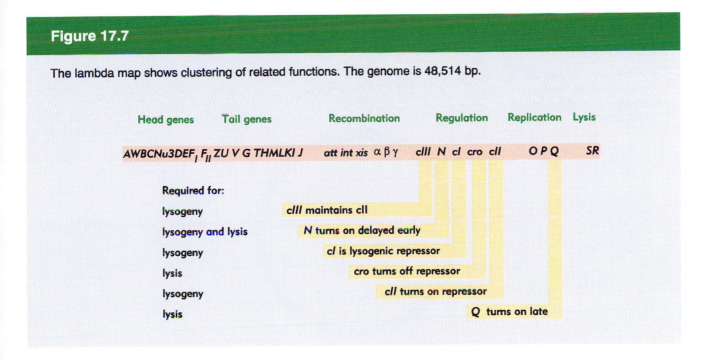

The late genes are expressed as a single transcription unit, starting from a promoter $P_{R'}$ that lies between Q and S. The late promoter is used constitutively. However, in the absence of the product of gene Q (which is the last gene in the rightward delayed early unit), late transcription terminates at a site t_{R3}. The transcript resulting from this termination event is 194 bases long; it is known as 6S RNA. When pQ becomes available, however, it suppresses termination at t_{R3} and the 6S RNA is extended, with the result that the late genes are expressed.

The late antitermination event resembles the action undertaken by pN at the delayed early stage. The site *qut* at which pQ acts is located just downstream of the late promoter, and RNA polymerase pauses there to provide a target for pQ action (see Chapter 16).

Late gene transcription does not seem to terminate at any specific point, but continues through all the late genes into the region beyond. A similar event happens with the leftward delayed early transcription, which continues past the recombination functions. Transcription in each direction is probably terminated before the polymerases could crash into each other.

Figure 17.8

Lambda DNA circularizes during infection, so that the late gene cluster is intact in one transcription unit.

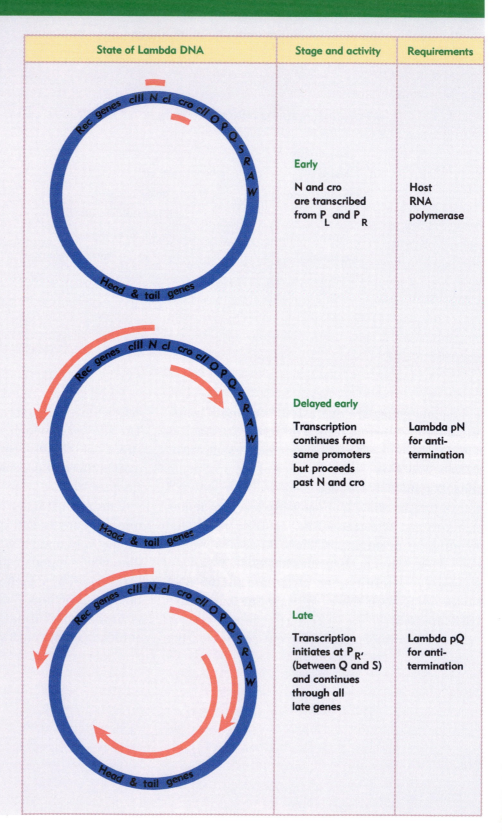

State of Lambda DNA	Stage and activity	Requirements
	Early N and cro are transcribed from P$_L$ and P$_R$	Host RNA polymerase
	Delayed early Transcription continues from same promoters but proceeds past N and cro	Lambda pN for anti-termination
	Late Transcription initiates at P$_{R'}$ (between Q and S) and continues through all late genes	Lambda pQ for anti-termination

Lysogeny is maintained by an autogenous circuit

Looking at the lambda lytic cascade, we see that the entire program is set in train by initiating transcription at the two promoters P_L and P_R for the immediate early genes N and *cro*. Because lambda uses antitermination to proceed to the next stage of (delayed early) expression, the same two promoters continue to be used right through the early period.

The expanded map of the regulatory region drawn in **Figure 17.9** shows that the promoters P_L and P_R lie on either side of the *cI* gene. Associated with each promoter is an operator (O_L, O_R) at which repressor protein binds to prevent RNA polymerase from initiating transcription. The sequence of each operator overlaps with the promoter that it controls; so often these are described as the P_L/O_L and P_R/O_R control regions.

Because of the sequential nature of the lytic cascade, the control regions provide a pressure point at which entry to the entire cycle can be controlled. *By denying RNA polymerase access to these promoters, the repressor prevents the phage genome from entering the lytic cycle.*

The repressor protein is coded by the *cI* gene. Mutants in this gene cannot maintain lysogeny, but are fated always to enter the lytic cycle. The name of the gene reflects the phenotype of the resulting infection.

When a bacterial culture is infected with a phage, the cells are lysed to generate regions that can be seen on a culture plate as small areas of clearing called **plaques**. With wild-type phages, the plaques are turbid or cloudy, because they contain some cells that have established lysogeny instead of being lysed. The effect of a *cI* mutation is to prevent lysogeny, so that the plaques contain only lysed cells. As a result, such an infection generates only **clear plaques**, and three genes (*cI, cII, cIII*) were named for their involvement in this phenotype. **Figure 17.10** compares wild-type and mutant plaques.

The *cI* gene is transcribed from a promoter P_{RM} that lies at its right end. (The subscript 'RM' stands for repressor maintenance). Transcription is terminated at the left end of the gene. The mRNA actually starts with the AUG initiation codon; because of the

Figure 17.9

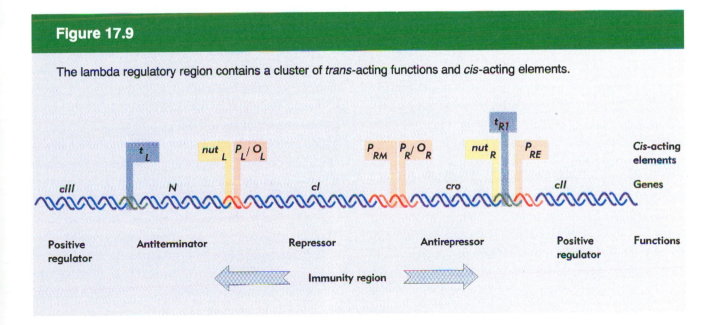

The lambda regulatory region contains a cluster of *trans*-acting functions and *cis*-acting elements.

Figure 17.10

Wild-type lambda cultures form cloudy plaques (left panel); mutants that cannot lysogenize can be detected by their clear plaques (right panel). Photograph kindly provided by Dale Kaiser.

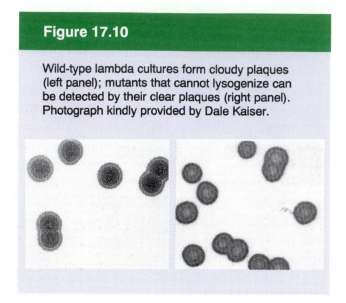

absence of the usual ribosome binding site, the mRNA is translated somewhat inefficiently, producing only a low level of repressor protein.

The repressor binds independently to the two operators. It has a single function at O_L, but has dual functions at O_R. These are illustrated in the upper part of **Figure 17.11**.

At O_L the repressor has the same sort of effect that we have already discussed for several other systems: it prevents RNA polymerase from initiating transcription at P_L. This stops the expression of gene N. Since P_L is used for all leftward early gene transcription, this action prevents expression of the entire leftward early transcription unit. *Thus the lytic cycle is stymied before it can proceed beyond the early stages.*

At O_R, repressor binding prevents the use of P_R. Thus *cro* and the other rightward early genes cannot be expressed. (We see later why it is important to prevent the expression of *cro* when lysogeny is being maintained.)

But the presence of repressor at O_R also has another effect. The promoter for repressor synthesis, P_{RM}, is adjacent to the rightward operator O_R. It turns out that *RNA polymerase can initiate efficiently at* P_{RM} *only when repressor is bound at* O_R. The repressor behaves as a positive regulator protein that is necessary for transcription of the *cI*

gene. *Since the repressor is the product of* cI, *this interaction creates a positive autogenous circuit, in which the presence of repressor is necessary to support its own continued synthesis.*

The nature of this control circuit explains the biological features of lysogenic existence. Lysogeny is stable because the control circuit ensures that, so long as the level of repressor is adequate, there is continued expression of the *cI* gene. The result is that O_L and O_R remain occupied indefinitely. By repressing the entire lytic cascade, this action maintains the prophage in its inert form.

The presence of repressor explains the phenomenon of immunity. If a second lambda phage DNA enters a lysogenic cell, repressor protein synthesized from the resident prophage genome will immediately bind to O_L and O_R in the new genome. This prevents the second phage from entering the lytic cycle.

The operators were originally identified as the targets for repressor action by **virulent** mutations. These mutations prevent the repressor from binding at O_L or O_R, with the result that the phage inevitably proceeds into the lytic pathway when it infects a new host bacterium. And λ*vir* mutants can grow on lysogens because the virulent mutations in O_L and O_R allow the incoming phage to ignore the resident repressor and thus to enter the lytic cycle. Virulent mutations in phages are the equivalent of operator-constitutive mutations in bacterial operons.

Prophage is induced to enter the lytic cycle when the lysogenic circuit is broken. This happens when the repressor is inactivated (see next section). The absence of repressor allows RNA polymerase to bind at P_L and P_R, starting the lytic cycle as shown in the lower part of Figure 17.11.

The autogenous nature of the repressor-maintenance circuit creates a sensitive response. Because the presence of repressor is necessary for its own synthesis, expression of the *cI* gene stops as soon as the existing repressor is destroyed. Thus no repressor is synthesized to replace the molecules that have been damaged. So the lytic cycle can start without interference from the circuit that maintains lysogeny.

Figure 17.11

Lysogeny is maintained by an autogenous circuit (upper). If this circuit is interrupted, the lytic cycle starts (lower).

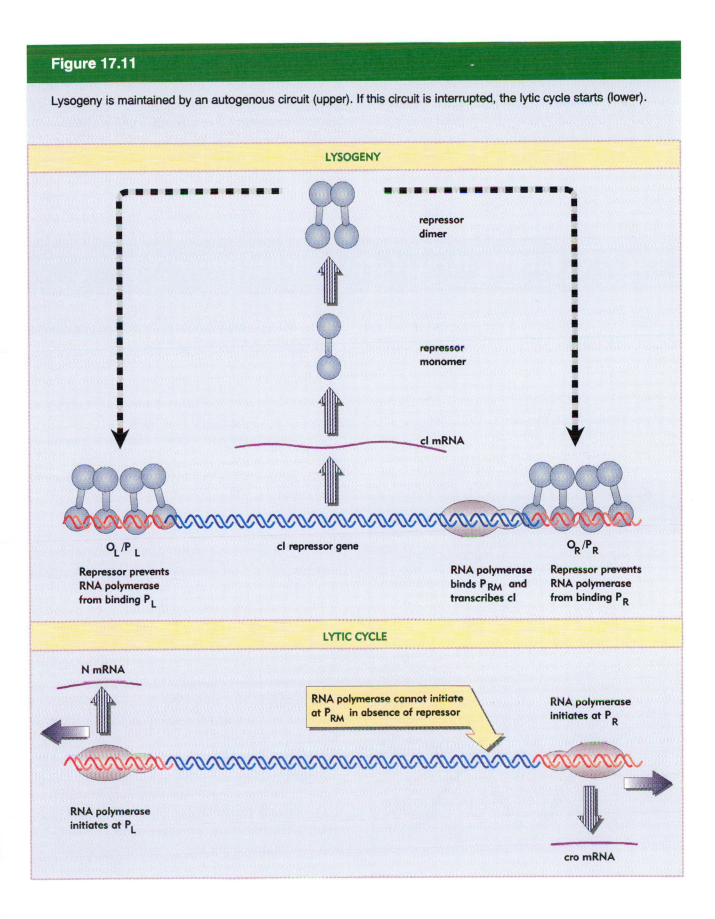

The region including the left and right operators, the *cI* gene, and the *cro* gene determines the immunity of the phage. Any phage that possesses this region has the same type of immunity, because *it specifies both the repressor protein and the sites on which the repressor acts.* Accordingly, this is called the **immunity region** (as marked in Figure 17.9).

Each of the four lambdoid phages ø80, *21*, *434*, and λ has a unique immunity region. When we say that a lysogenic phage confers immunity to any other phage of the same type, we mean more precisely that the immunity is to any other phage that has the same immunity region (irrespective of differences in other regions).

The DNA-binding form of repressor is a dimer

The repressor subunit is a polypeptide of 27,000 daltons with the two distinct domains summarized in **Figure 17.12**.

◆ The N-terminal domain, residues 1–92, provides the operator-binding site.

◆ The C-terminal domain, residues 132–236, is responsible for forming a dimer.

The two domains are joined by a connector of 40 residues. When repressor is digested by a protease, each domain is released as a separate fragment.

Each domain can exercise its function independently of the other. The C-terminal fragment can form oligomers. The N-terminal fragment can bind the operators, although with a lower affinity than the intact repressor. Thus the information for specifically contacting DNA is contained within the N-terminal domain, but the efficiency of the process is enhanced by the attachment of the C-terminal domain.

The dimeric structure of the repressor is crucial in maintaining lysogeny. The induction of a lysogenic prophage to enter the lytic cycle is caused by cleavage of the repressor subunit in the connector region, between residues 111 and 113. (This is a counterpart to the allosteric change in conformation that results when a small-molecule inducer inactivates the repressor of a bacterial operon, a capacity not possessed by the lysogenic repressor.)

In the intact state, dimerization of the C-terminal domains ensures that when the repressor binds to DNA its two N-terminal domains each contact DNA simultaneously. But cleavage releases the C-terminal domains from the N-terminal domains. As

Figure 17.12

The N-terminal and C-terminal regions of repressor form separate domains. The C-terminal domains associate to form dimers; the N-terminal domains bind DNA.

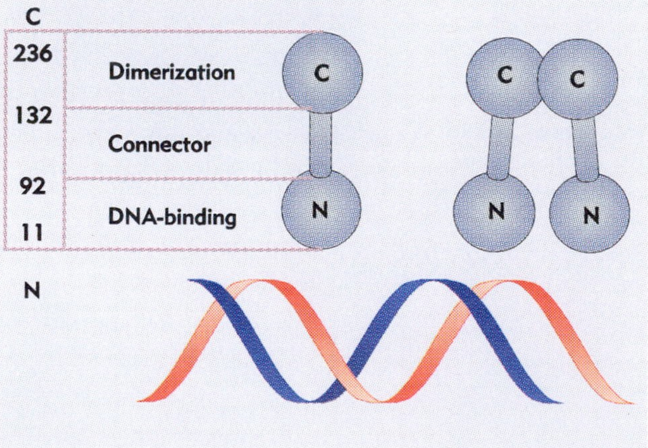

Figure 17.13

Repressor dimers bind to the operator. The affinity of the N-terminal domains for DNA is controlled by the dimerization of the C-terminal domains.

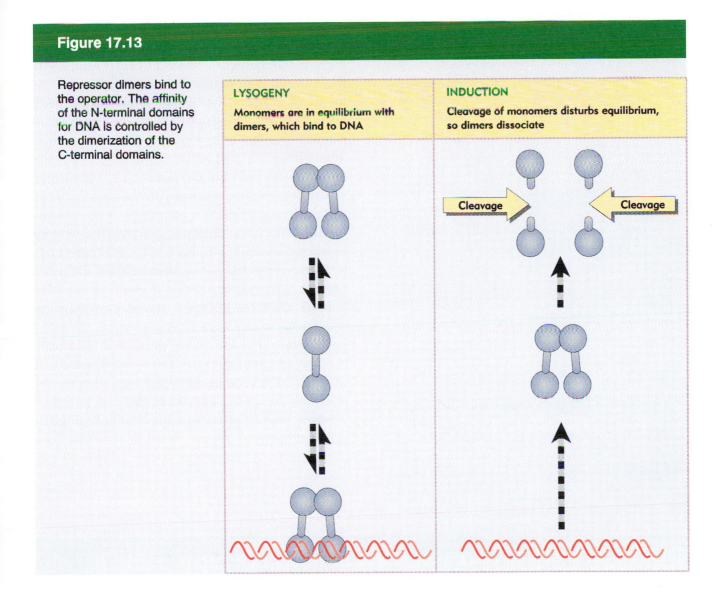

LYSOGENY

Monomers are in equilibrium with dimers, which bind to DNA

INDUCTION

Cleavage of monomers disturbs equilibrium, so dimers dissociate

Cleavage   Cleavage

illustrated in **Figure 17.13**, this means that the N-terminal domains can no longer dimerize; this upsets the equilibrium between monomers and dimers, so that repressor dissociates from DNA, allowing lytic infection to start. (Another relevant parameter is the loss of cooperative effects between adjacent dimers: see later).

The balance between lysogeny and the lytic cycle depends on the concentration of repressor. Intact repressor is present in a lysogenic cell at a concentration sufficient to ensure that the operators are occupied. But if the repressor is cleaved, this concentration is inadequate, because of the lower affinity of the separate N-terminal domain for the operator. Too high a concentration of repressor would make it impossible to induce the lytic cycle in this way; too low a level, of course, would make it impossible to maintain lysogeny.

The dependence of repression on repressor concentration is strongly influenced by the behavior of the repressor. Lambda repressor only binds DNA as a dimer. The equilibrium between monomers and dimers depends on the protein concentration. The consequence of the equilibrium is that repressor

Figure 17.14

Lambda repressor binds to operators by second-order kinetics.

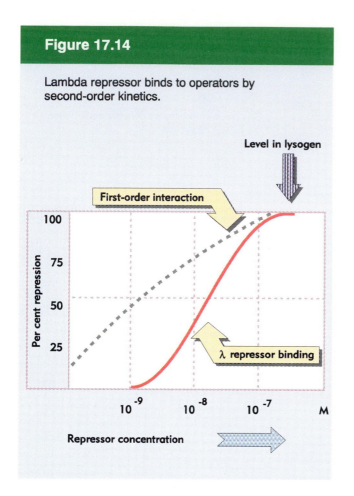

exists as a monomer at low concentrations and as a dimer (the effective DNA-binding form) at higher concentrations. As a result, binding to the operator depends on the square of the monomer concentration, that is, follows a second-order equilibrium, instead of the first order reaction that describes other repressor–operator interactions.

A reduction in lambda repressor concentration shifts the equilibrium toward the monomeric form, which cannot bind DNA effectively, exacerbating the loss of operator-binding capacity. The curve for repressor–operator binding plotted in **Figure 17.14** takes the typical sigmoid form of a second-order reaction. The response to repressor concentration is more sensitive than a first-order reaction (in which repressor structure is constant). The usual concentration of lambda repressor in a lysogen is ~200 molecules/cell, which corresponds to ~4×10^{-7} M, several times the concentration needed to ensure complete repression. But a drop in level of ~$10\times$ will effectively release repression and lead to induction.

Repressor binds cooperatively at each operator using a helix–turn–helix motif

Several DNA-binding proteins that regulate bacterial transcription share a similar mode of holding DNA, in which the active domain contains two short regions of α-helix that contact DNA. (Some transcription factors in eukaryotic cells use a similar motif, as discussed in Chapter 30.)

The N-terminal domain of lambda repressor contains several stretches of α-helix, arranged as illustrated diagrammatically in **Figure 17.15**. The structures of the connector and the C-terminal domain are not known.

Two of the helical regions are responsible for binding DNA, and the **helix-turn-helix model** for

contact is illustrated in **Figure 17.16**. Looking at a single monomer, α-helix-3 consists of 9 amino acids, lying at an angle to the preceding region of 7 amino acids that forms α-helix-2. In the dimer, the two apposed helix-3 regions lie 34Å apart, enabling them to fit into successive major grooves of DNA. The helix-2 regions lie at an angle that would place them across the groove.

Related forms of the α-helical motifs employed in these two regions are found in several DNA-binding proteins, including CAP, the *lac* repressor, and several phage repressors. Each helix plays a distinct role in binding the operators:

Figure 17.15

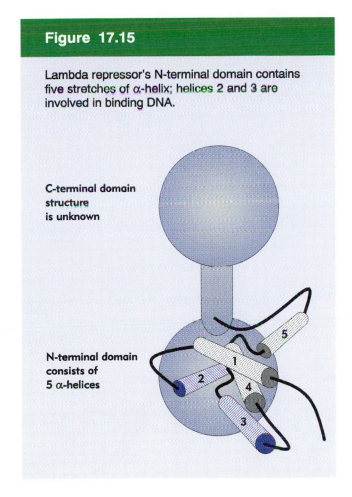

Lambda repressor's N-terminal domain contains five stretches of α-helix; helices 2 and 3 are involved in binding DNA.

C-terminal domain structure is unknown

N-terminal domain consists of 5 α-helices

◆ Contacts between helix-3 and DNA rely on hydrogen bonds between the amino acid side chains and the exposed positions of the base pairs. This helix is responsible for recognizing the specific target DNA sequence, and is therefore also known as the **recognition helix.**

◆ Contacts from helix-2 to the DNA take the form of hydrogen bonds connecting with the phosphate backbone. These interactions are necessary for binding, but do not control the specificity of target recognition. In addition to these contacts, a large part of the overall energy of interaction with DNA is provided by ionic interactions with the phosphate backbone.

What happens if we manipulate the coding sequence to construct a new protein by substituting the recognition helix in one repressor with the cor-responding sequence from a closely related repressor? The specificity of the hybrid protein is that of its new recognition helix. *The amino acid sequence of this short region therefore determines the sequence specificities of the individual proteins, and is able to act in conjunction with the rest of the polypeptide chain.*

Figure 17.17 shows the details of the binding to DNA of two proteins that bind similar DNA sequences. Both lambda repressor and Cro protein have a similar organization of the helix–turn–helix motif, although their individual specificities for DNA are not identical:

◆ Each protein uses similar interactions between hydrophobic amino acids to maintain the relationship between helix-2 and helix-3: repressor has an Ala–Val connection, while Cro has an Ala–Ile association.

◆ Amino acids in helix-3 of the repressor make contacts with specific bases in the operator. Three amino acids in repressor recognize three bases in DNA; the amino acids at these positions and also at additional positions in Cro recognize five (or possibly six) bases in DNA.

Two of the amino acids involved in specific recognitions are identical in repressor and Cro (Gln and Ser at the N-terminal end of the helix), while the other contacts are different (Ala in repressor versus Lys and the additional Asn in Cro). Also, a Thr in helix-2 of Cro directly contacts DNA.

The interactions shown in the figure represent binding to the DNA sequence that each protein recognizes most tightly; the use of overlapping, but not identical contacts between amino acids and bases shows how related recognition helices confer recognition of related DNA sequences. This explains why repressor and Cro both recognize the same set of operators, but have different relative affinities for particular operators. Thus the sequences shown at the bottom of the figure with the contact points in color can be seen to differ at 3 of the 9 base pairs.

The bases contacted by helix-3 of repressor or

Figure 17.16

In the two-helix model for DNA binding, helix-3 of each monomer lies in the wide groove on the same face of DNA, and helix-2 lies across the groove.

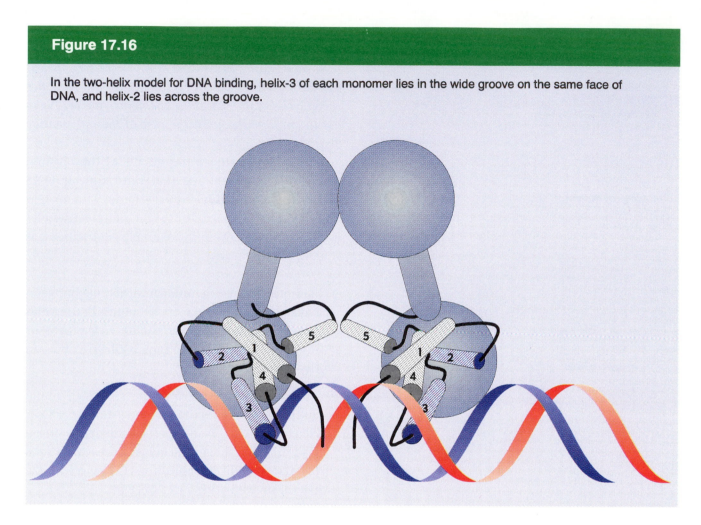

Cro lie on one face of DNA, as can be seen from the positions indicated on the helical diagram in Figure 17.17. However, repressor makes an additional contact with the other face of DNA. The isolated N-terminal domain of repressor makes the same contacts with DNA as the intact protein. Removing the last six N-terminal amino acids (which protrude from helix-1) eliminates some of the contacts. This observation provides the basis for the idea that the bulk of the N-terminal domain contacts one face of DNA, while the last six N-terminal amino acids form an 'arm' extending around the back, as shown by the arrow in Figure 17.17. **Figure 17.18** shows the view from the back. Lysine residues in the arm make contacts with G residues in the major groove, and also with the phosphate backbone. The interaction between the arm and DNA contributes heavily to DNA binding; the affinity of the armless repressor for DNA is reduced by ~1000-fold.

Each operator contains three repressor-binding sites. As can be seen from **Figure 17.19**, each binding site is a sequence of 17 bp displaying partial symmetry about an axis through the central base pair. No two of the six individual repressor-binding sites are identical, but they all conform with a consensus sequence. The binding sites within each operator are separated by spacers of 3–7 bp that are rich in A•T base pairs. The sites at each operator are numbered so that O_R consists of the series of binding sites O_R1-O_R2-O_R3, while O_L consists of the series O_L1-O_L2-O_L3. In each case, site 1 lies closest to the startpoint for transcription in the promoter, and sites 2 and 3 lie farther upstream.

A repressor dimer binds symmetrically at each site, so that each N-terminal domain of the dimer contacts a similar set of bases. Thus each individual

Figure 17.17

Two proteins that use the two-helix arrangement to contact DNA recognize lambda operators with affinities determined by the amino acid sequence of helix-3.

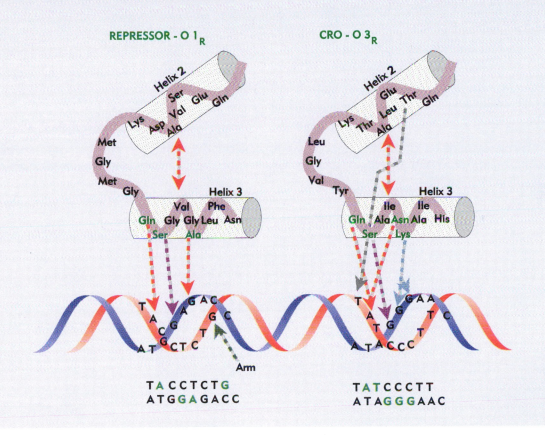

REPRESSOR - O 1_R

CRO - O 3_R

TA CCTCTG
ATGGAGACC

TATCCCTT
ATAGGGAAC

Figure 17.18

A view from the back shows that the bulk of the repressor contacts one face of DNA, but its N-terminal arms reach around to the other face.

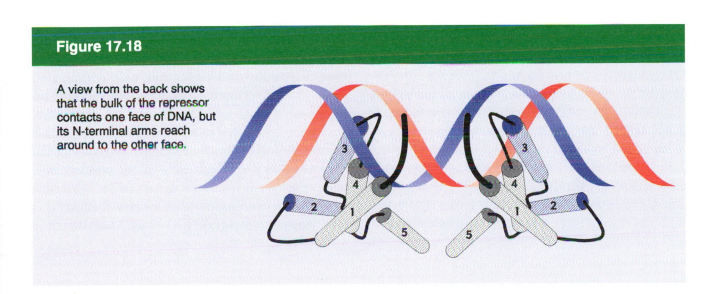

Figure 17.19

Each operator contains three repressor-binding sites, and overlaps with the promoter at which RNA polymerase binds. The orientation of O_L has been reversed from usual to facilitate comparison with O_R.

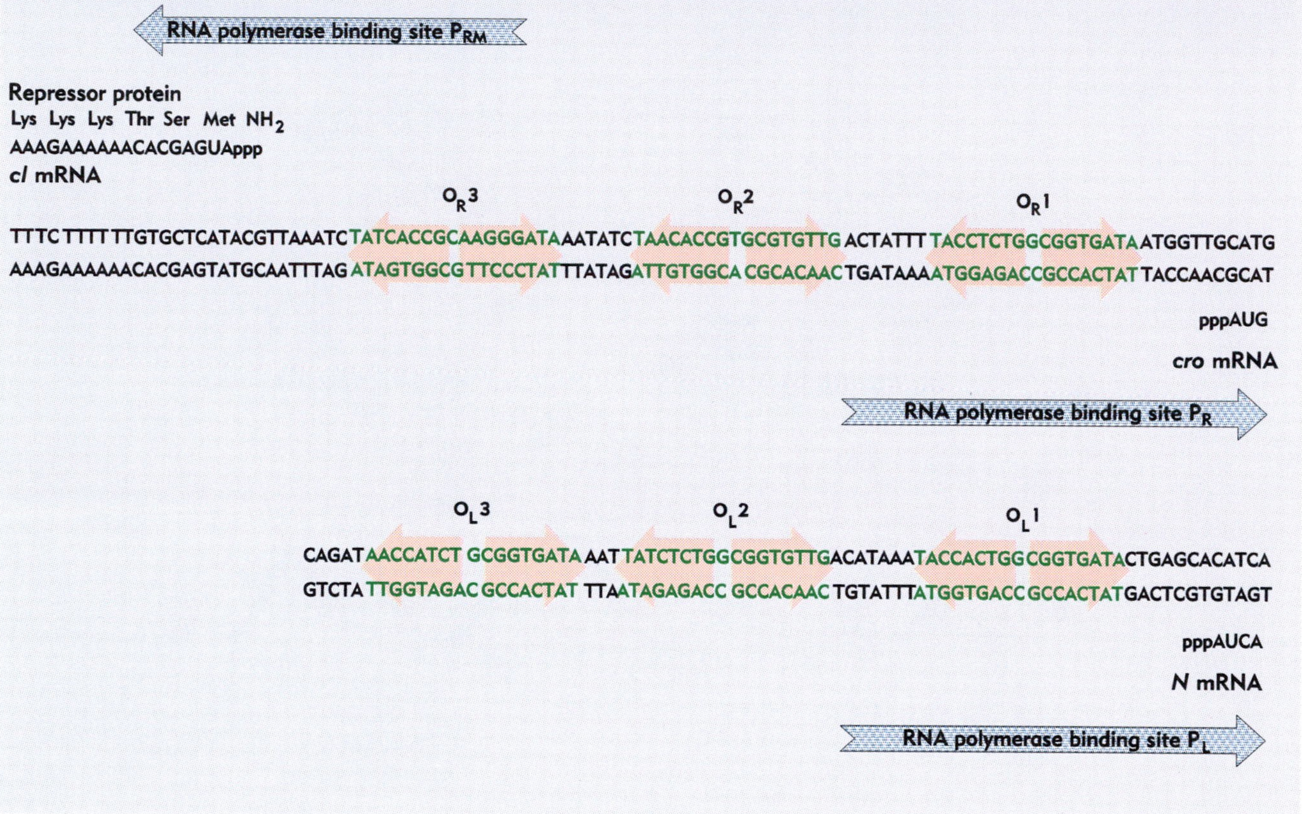

RNA polymerase binding site P_{RM}

Repressor protein
Lys Lys Lys Thr Ser Met NH$_2$
AAAGAAAAAACACGAGUAppp
cI mRNA

O_R3 O_R2 O_R1

TTTC TTTT TTGTGCTCATACGTTAAATC TATCACCGCAAGGGATA AATATC TAACACCGTGCGTGTTG ACTATTT TACCTCTGGCGGTGATA ATGGTTGCATG
AAAGAAAAAACACGAGTATGCAATTTAG ATAGTGGCG TTCCCTAT TTATAG ATTGTGGCA CGCACAAC TGATAAA ATGGAGACCGCCACTAT TACCAACGCAT

pppAUG

cro mRNA

RNA polymerase binding site P_R

O_L3 O_L2 O_L1

CAGAT AACCATCT GCGGTGATA AAT TATCTCTGGCGGTGTTG ACATAAA TACCACTGGCGGTGATA CTGAGCACATCA
GTCTA TTGGTAGAC GCCACTAT TTA ATAGAGACC GCCACAAC TGTATTT ATGGTGACC GCCACTAT GACTCGTGTAGT

pppAUCA

N mRNA

RNA polymerase binding site P_L

N-terminal region contacts a half-binding site.

Bases that are not contacted directly by repressor protein may have an important effect on binding. The phage *434* repressor binds DNA via a helix–turn–helix motif, and the crystal structure shows that helix-3 is positioned at each half site so that it contacts the 5 outermost base pairs but not the inner 2. However, operators with A•T base pairs at the inner positions bind *434* repressor more strongly than operators with G•C base pairs. The reason is that *434* repressor binding slightly twists DNA at the center of the operator, widening the angle between the two half-sites of DNA by ~3°. This is probably needed to allow each monomer of the repressor dimer to make optimal contacts with DNA. A•T base pairs allow this twist more readily than G•C pairs, thus affecting the affinity of the operator for repressor.

Is there a code that relates the sequence of helix-3 in each repressor to the sequence of bases in the operator that it recognizes? This concept has now been abandoned. Cocrystal structures of DNA with λ and *434* repressor or Cro proteins reveal effects that are inconsistent with such a model. More than one amino acid in the protein is involved in generating contacts with a single base. And the structure of the DNA is subject to changes in conformation upon protein binding. The pattern of contacts is generally similar, but by no means identical, so there is no simple pattern that governs contacts between amino acids in particular positions of the protein with bases at particular positions in the DNA.

Faced with the triplication of binding sites at each operator, how does repressor decide where to start binding? At each operator, site 1 has a greater affinity (roughly 10-fold) than the other sites for the repressor. So the repressor always binds first to O_L1 and O_R1.

Lambda repressor binds to subsequent sites within each operator in a cooperative manner. The presence of a dimer at site 1 greatly increases the affinity with which a second dimer can bind to site 2. When both sites 1 and 2 are occupied, this interaction does *not* extend farther, to site 3. At the concentrations of repressor usually found in a lysogen, both sites 1 and 2 are filled at each operator, but site 3 is not occupied.

If site 1 is inactive (because of mutation), then repressor binds cooperatively to sites 2 and 3. That is, binding at site 2 assists another dimer to bind at site 3. This interaction occurs directly between repressor dimers and not via conformational

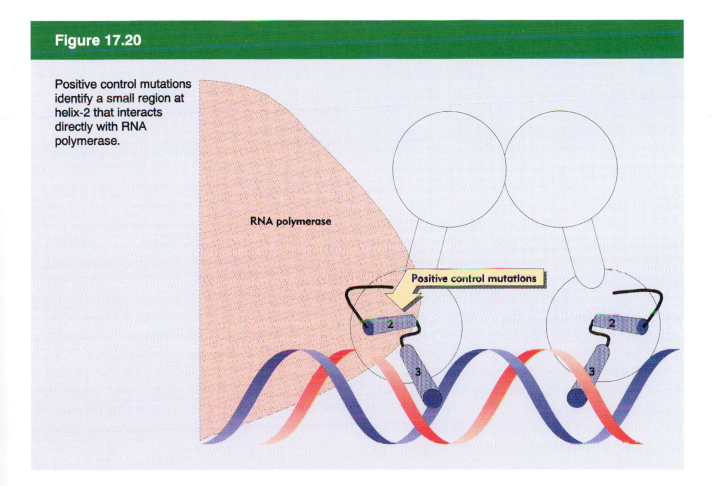

Figure 17.20

Positive control mutations identify a small region at helix-2 that interacts directly with RNA polymerase.

change in DNA. Probably the connector region of the first repressor orients the C-terminal regions of the dimer in such a way that they contact the C-terminal regions of the second dimer.

A result of cooperative binding is to increase the effective affinity of repressor for the operator at physiological concentrations. This enables a lower concentration of repressor to achieve occupancy of the operator. This is an important consideration in a system in which release of repression has irreversible consequences. In an operon coding for metabolic enzymes, after all, failure of repression will merely allow unnecessary synthesis of enzymes. But failure to repress lambda prophage will lead to induction of phage and lysis of the cell.

From the sequences shown in Figure 17.19, we see that O_L1 and O_R1 lie more or less in the center of the RNA polymerase binding sites of P_L and P_R, respectively. Occupancy of O_L1-O_L2 and O_R1-O_R2 thus physically blocks access of RNA polymerase to the corresponding promoter.

A different relationship is shown between O_R and the promoter P_{RM} for transcription of cI. The RNA polymerase binding site is just about adjacent to O_R2. This explains how repressor autogenously regulates its own synthesis. When two dimers are bound at O_R1-O_R2, the dimer at O_R2 interacts with RNA polymerase (see Figure 17.11). Unlike the interaction between repressors, this effect resides in the amino-terminal domain, which can stimulate use of P_{RM} even as an independent fragment when bound to O_R2.

Mutations that abolish positive control map in the cI gene. One interesting class of mutants remain able to bind the operator to repress transcription, but cannot stimulate RNA polymerase to transcribe from P_{RM}. They map within a small group of amino acids, located on the outside of helix-2 or in the turn between helix-2 and helix-3. The mutations reduce the negative charge of the region; conversely, mutations that increase the negative charge enhance the activation of RNA polymerase. This suggests that the group of amino acids constitutes an 'acidic patch' that functions by an electrostatic interaction with a basic region on RNA polymerase.

The location of these 'positive control mutations' in the repressor is indicated on **Figure 17.20**. They lie at a site on repressor that is close to a phosphate group on DNA that is also close to RNA polymerase. Thus the group of amino acids on repressor that is involved in positive control is in a position to contact the polymerase. The interaction between repressor and polymerase is needed for the polymerase to make the transition from a closed complex to an open complex (see Table 17.2 later).

Protein–protein interactions involved in positive control can involve various regions in each protein. Mutations concerned with positive control in phage P22 lie at one end of helix-3, although the DNA-binding structure of the P22 repressor is very similar to that of the lambda repressor. The important common principle is that *protein–protein interactions can release energy that is used to help to initiate transcription.*

What happens if a repressor dimer binds to O_R3? This site overlaps with the RNA polymerase binding site at P_{RM}. Thus if the repressor concentration becomes great enough to cause occupancy of O_R3, the transcription of cI is prevented. This leads in due course to a reduction in repressor concentration; O_R3 then becomes empty, and the autogenous loop can start up again because O_R2 remains occupied.

This mechanism could prevent the concentration of repressor from becoming too great, although it would require repressor concentration in lysogens to reach unusually high levels. In the formal sense, the repressor is an autogenous regulator of its own expression that functions positively at low concentrations and negatively at high concentrations.

Virulent mutations have been found in sites 1 and 2 of both O_L and O_R. The mutations vary in their degree of virulence, according to the extent to which they reduce the affinity of the binding site for repressor, and also depending on the relationship of the afflicted site to the promoter. Consistent with the conclusion that O_R3 and O_L3 usually are not occupied, virulent mutations are not found in these sites.

How is repressor synthesis established?

The control circuit for maintaining lysogeny presents a paradox. *The presence of repressor protein is necessary for its own synthesis.* This explains how the lysogenic condition is perpetuated. But how is the synthesis of repressor established in the first place?

When a lambda DNA enters a new host cell, RNA polymerase cannot transcribe *cI*, because there is no repressor present to aid its binding at P_{RM}. But this same absence of repressor means that P_R and P_L are available. So the first event when lambda DNA infects a bacterium is for genes *N* and *cro* to be transcribed. Then pN allows transcription to be extended farther. This allows *cIII* (and other genes) to be transcribed on the left, while *cII* (and other genes) are transcribed on the right (see Figure 17.9).

The *cII* and *cIII* genes share with *cI* the property

that mutations in them cause clear plaques. But there is a difference. The *cI* mutants can neither establish nor maintain lysogeny. The *cII* or *cIII* mutants have some difficulty in establishing lysogeny, but once established, they are able to maintain it by the *cI* autogenous circuit.

This implicates the *cII* and *cIII* genes as positive regulators whose products are needed for an alternative system for repressor synthesis. The system is needed only to *initiate* the expression of *cI* in order to circumvent the inability of the autogenous circuit to engage in *de novo* synthesis.

The CII protein acts directly on gene expression. Between the *cro* and *cII* genes is another promoter, called P_{RE}. (The subscript 'RE' stands for repressor establishment.) This promoter can be recognized by RNA polymerase only in the presence of CII, whose action is illustrated in **Figure 17.21**.

Figure 17.21

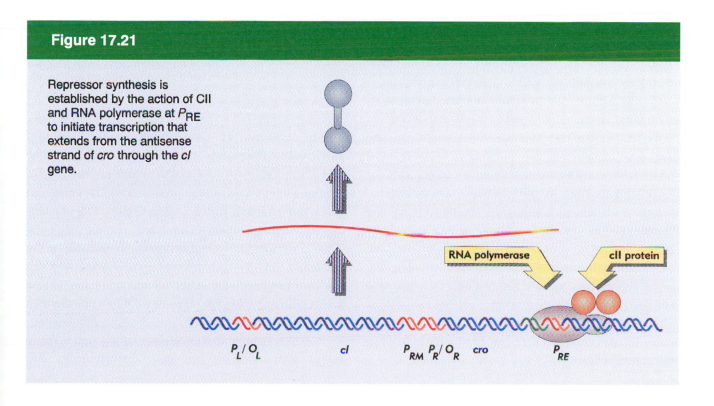

Repressor synthesis is established by the action of CII and RNA polymerase at P_{RE} to initiate transcription that extends from the antisense strand of *cro* through the *cI* gene.

The CII protein is extremely unstable *in vivo*, because it is degraded as the result of the activity of a host protein called HflA. The role of CIII is to protect CII against this degradation.

Transcription from P_{RE} promotes lysogeny in two ways. Its direct effect is that *cI* is translated into repressor protein. An indirect effect is that transcription proceeds through the *cro* gene in the 'wrong' direction. Thus the 5′ part of the RNA corresponds to an antisense transcript of *cro*; in fact, it hybridizes to authentic *cro* mRNA, inhibiting its translation. We see in the next section that this is important because *cro* expression is needed to enter the lytic cycle.

The *cI* coding region on the P_{RE} transcript is very efficiently translated (in contrast with the weak translation of the P_{RM} transcript mentioned earlier). In fact, repressor is synthesized ~7–8 times more effectively via expression from P_{RE} than from P_{RM}. This reflects the fact that the P_{RE} transcript has an efficient ribosome-binding site, whereas the P_{RM} transcript has no ribosome-binding site and actually starts with the AUG initiation codon.

The P_{RE} promoter has a poor fit with the consensus at -10 and lacks a consensus sequence at -35. This deficiency explains its dependence on *cII*. The promoter cannot be transcribed by RNA polymerase alone *in vitro,* but can be transcribed when CII is added. The regulator binds to a region extending from about -25 to -45. When RNA polymerase is added, an additional region is protected, extending from -12 to $+13$. As summarized in **Figure 17.22**, the two proteins bind to overlapping sites.

The importance of the -35 and -10 regions for promoter function, in spite of their lack of resemblance with the consensus, is indicated by the existence of *cy* mutations. These have effects similar to those of *cII* and *cIII* mutations in preventing the establishment of lysogeny; but they are *cis*-acting instead of *trans*-acting. They fall into two groups, *cyL* and *cyR*.

The *cyL* mutations are located around -10, and probably prevent RNA polymerase from recognizing the promoter.

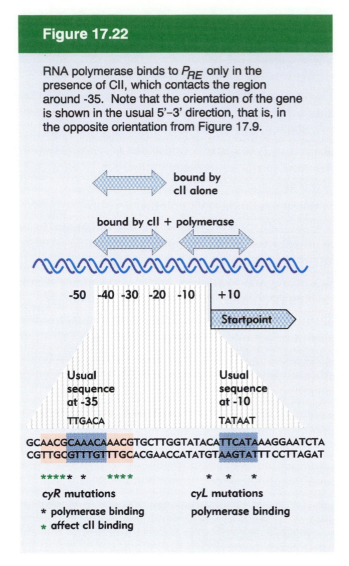

Figure 17.22

RNA polymerase binds to P_{RE} only in the presence of CII, which contacts the region around -35. Note that the orientation of the gene is shown in the usual 5′–3′ direction, that is, in the opposite orientation from Figure 17.9.

bound by
cII alone

bound by cII + polymerase

-50 -40 -30 -20 -10   +10

Startpoint

Usual sequence at -35

TTGACA

Usual sequence at -10

TATAAT

GCAACGCAAACAAACGTGCTTGGTATACATTCATAAAGGAATCTA
CGTTGCGTTTGTTTGCACGAACCATATGTAAGTATTTCCTTAGAT

***** *   ****   * * *

cyR mutations   cyL mutations

* polymerase binding   polymerase binding
* affect cII binding

The *cyR* mutations are located around -35, and fall into two types, affecting either RNA polymerase or CII binding. Mutations in the center of the region do not affect CII binding; presumably they prevent RNA polymerase binding. On either side of this region, mutations in short tetrameric repeats, TTGC, prevent CII from binding. Each base in the tetramer is 10 bp (one helical turn) separated from its homologue in the other tetramer, so that when CII recognizes the two tetramers, it lies on one face of the double helix.

Positive control of a promoter implies that some accessory protein has increased the efficiency with

Table 17.2

Positive regulation can influence RNA polymerase at either stage of initiating transcription.

Locus	Regulator	Polymerase Binding (equilibrium constant, K_B)	Closed-Open Conversion (rate constant, k_2)
λP_{RM}	Repressor	No effect	11X
λP_{RE}	CII	100X	100X

which RNA polymerase initiates transcription. **Table 17.2** summarizes data to show that either or both stages of the interaction between promoter and polymerase—initial binding to form a closed complex (measured by the equilibrium constant $[K_B]$) or its conversion into an open complex (measured by the rate constant $[k_2]$)—can be the target for regulation.

Now we can see how lysogeny is established during a new infection. **Figure 17.23** recapitulates the early stages and shows what happens as the result of expression of *cIII* and *cII*. The presence of CII allows P_{RE} to be used for transcription extending through *cI*. Repressor protein is synthesized in high amounts from this transcript. Immediately it binds to O_L and O_R.

By directly inhibiting any further transcription from P_L and P_R, repressor binding turns off the expression of all phage genes. This halts the synthesis of CII and CIII, which are unstable; they decay rapidly, with the result that P_{RE} can no longer be used. Thus the synthesis of repressor via the establishment circuit is brought to a halt.

But repressor now is present at O_R. It switches on the maintenance circuit for expression from P_{RM}. Repressor continues to be synthesized, although at the lower level typical of P_{RM} function. So the establishment circuit starts off repressor synthesis at a high level; then repressor turns off all other functions, while at the same time turning on the maintenance circuit, which

functions at the low level adequate to sustain lysogeny.

We shall not now deal in detail with the other functions needed to establish lysogeny, but we can just briefly remark that the infecting lambda DNA must be inserted into the bacterial genome (see Chapter 33). The insertion requires the product of gene *int*, which is expressed from its own promoter P_I, at which CII also is necessary. The sequence of P_I shows homology with P_{RE} in the CII binding site (although not in the −10 region). The functions necessary for establishing the lysogenic control circuit are therefore under the same control as the function needed physically to manipulate the DNA. Thus the establishment of lysogeny is under a control that ensures all the necessary events occur with the same timing.

Emphasizing the tricky quality of lambda's intricate cascade, we now know that CII promotes lysogeny in another, indirect manner. It sponsors transcription from a promoter called P_{anti-Q}, which is located within the *Q* gene. This transcript is an anti-sense version of the *Q* region, and it hybridizes with *Q* mRNA to prevent translation of Q protein, whose synthesis is essential for lytic development. So the same mechanisms that directly promote lysogeny by causing transcription of the *cI* repressor gene also indirectly help the lysogenic cause by inhibiting the expression of *cro* (see above) and *Q*, the regulator genes needed for the antagonistic lytic pathway.

Figure 17.23

Summary: a cascade is needed to establish lysogeny, but then this circuit is switched off and replaced by the autogenous repressor-maintenance circuit.

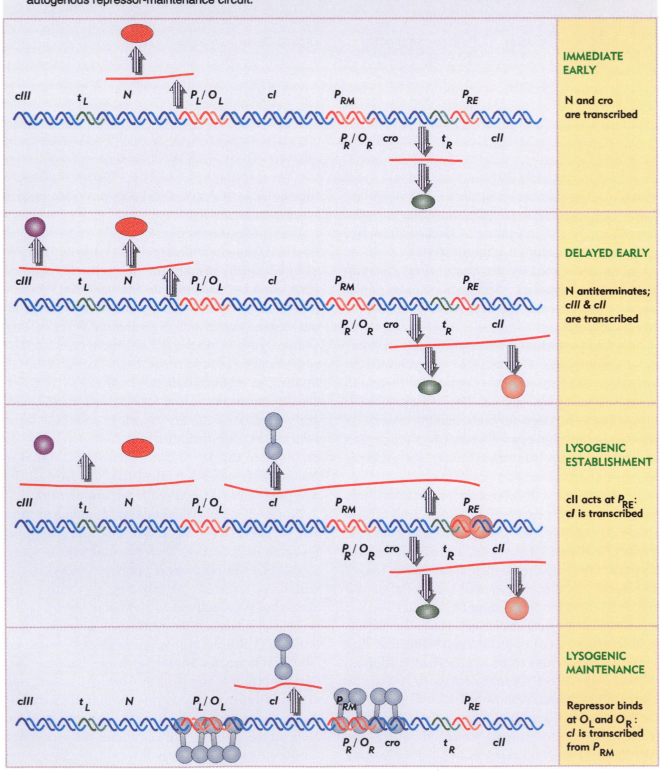

IMMEDIATE EARLY

N and cro are transcribed

DELAYED EARLY

N antiterminates; cIII & cII are transcribed

LYSOGENIC ESTABLISHMENT

cII acts at P_{RE}: cI is transcribed

LYSOGENIC MAINTENANCE

Repressor binds at O_L and O_R: cI is transcribed from P_{RM}

A second repressor is needed for lytic infection

We started this chapter by saying that lambda has the alternatives of entering lysogeny or starting a lytic infection. Lysogeny is initiated by establishing an autogenous maintenance circuit that inhibits the entire lytic cascade through applying pressure at two points. The program for establishing lysogeny actually proceeds through some of the same events that we described earlier in terms of the lytic cascade (expression of delayed early genes via expression of N is needed). We now face a problem. How does the phage enter the lytic cycle?

What we have left out of this account so far is the role of gene *cro,* which codes for another repressor. *Cro is responsible for preventing the synthesis of the repressor protein*; this action shuts off the possibility of establishing lysogeny. Mutants of the *cro⁻* type usually establish lysogeny rather than entering the lytic pathway, because they lack the ability to switch events away from the expression of repressor.

Cro forms a small dimer (the subunit is 9000 daltons) that acts within the immunity region. It has two effects:

◆ It prevents the synthesis of repressor via the maintenance circuit; that is, it prevents transcription via P_{RM}.

◆ It also inhibits the expression of early genes from both P_L and P_R.

This means that, when a phage enters the lytic pathway, Cro has responsibility both for preventing the synthesis of repressor and (subsequently) for turning down the expression of the early genes.

Cro achieves its function by binding to the same operators as (*cI*) repressor protein. Cro includes a region with the same general structure as the repressor; a helix-2 is offset at an angle from recognition helix-3. (The remainder of the structure is different, demonstrating that the helix–turn–helix motif can operate within various contexts.)

Like repressor, Cro binds symmetrically at the operators.

The sequences of Cro and repressor in the helix–turn–helix region are related, explaining their ability to contact the same DNA sequences (see Figure 17.17). Cro makes similar contacts to those made by repressor, but binds to only one face of DNA; it lacks the N-terminal arms by which repressor reaches around to the other side.

How can two proteins have the same sites of action, yet have such opposite effects? The answer lies in the different affinities that each protein has for the individual binding sites within the operators. (Also Cro has no activating region.) Let us just consider O_R, where more is known, and where Cro exerts both its effects. The series of events is illustrated in **Figure 17.24**. (Note that the first two stages are identical to those of the lysogenic circuit shown in Figure 17.23.)

The affinity of Cro for O_R3 is greater than its affinity for O_R2 or O_R1. So it binds first to O_R3. This inhibits RNA polymerase from binding to P_{RM}. Thus Cro's first action is to prevent the maintenance circuit for lysogeny from coming into play.

Then Cro binds to O_R2 or O_R1. Its affinity for these sites is similar, and there is no cooperative effect. Its presence at either site is sufficient to prevent RNA polymerase from using P_R. This in turn stops the production of the early functions (including Cro itself). Because CII is unstable, any use of P_{RE} is brought to a halt. So the two actions of Cro together block *all* production of repressor.

So far as the lytic cycle is concerned, Cro turns down (although it does not completely eliminate) the expression of the early genes. Its incomplete effect is explained by its affinity for O_R1 and O_R2, which is about eight times lower than that of repressor. This effect of Cro does not occur until the early genes have become more or less superfluous, because pQ is present; by this time, the phage has started late gene expression, and is concentrating on the production of progeny phage particles.

Figure 17.24

Summary: the lytic cascade requires Cro protein, which directly prevents repressor maintenance via P_{RE}, as well as turning off delayed early gene expression, indirectly preventing repressor establishment via P_{RM}.

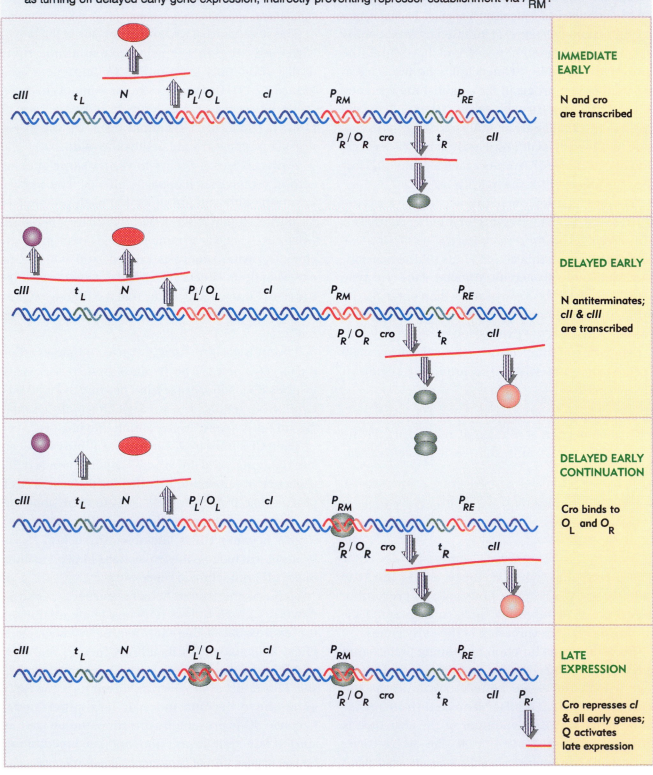

A delicate balance: lysogeny versus lysis

The programs for the lysogenic and lytic pathways are so intimately related that it is impossible to predict the fate of an individual phage genome when it enters a new host bacterium. Will the antagonism between repressor and Cro be resolved by establishing the autogenous maintenance circuit shown in Figure 17.23, or by turning off repressor synthesis and entering the late stage of development shown in Figure 17.24?

The same pathway is followed in both cases right up to the brink of decision. Both involve the expression of the immediate early genes and extension into the delayed early genes. The difference between them comes down to the question of whether repressor or Cro will obtain occupancy of the two operators.

The early phase during which the decision is taken is limited in duration in either case. No matter which pathway the phage follows, expression of all early genes will be prevented as P_L and P_R are repressed; and, as a consequence of the disappearance of CII and CIII, production of repressor via P_{RE} will cease.

The critical question comes down to whether the cessation of transcription from P_{RE} is followed by activation of P_{RM} and the establishment of lysogeny, or whether P_{RM} fails to become active and the pQ regulator commits the phage to lytic development.

The initial event in establishing lysogeny is the binding of repressor at O_L1 and O_R1. Binding at the first sites is rapidly succeeded by cooperative binding of further repressor dimers at O_L2 and O_R2. This shuts off the synthesis of Cro and starts up the synthesis of repressor via P_{RM}.

The initial event in entering the lytic cycle is the binding of Cro at O_R3. This stops the lysogenic-maintenance circuit from starting up at P_{RM}. Then Cro must bind to O_R1 or O_R2, and to O_L1 or O_L2,

to turn down early gene expression. By halting production of CII and CIII, this action leads to the cessation of repressor synthesis via P_{RE}. The shut-off of repressor establishment occurs when the unstable CII and CIII proteins decay.

The critical influence over the switch between lysogeny and lysis is CII. If CII is active, synthesis of repressor via the establishment promoter is effective; and, as a result, repressor gains occupancy of the operators. If CII is not active, repressor establishment fails, and Cro binds to the operators.

The level of CII protein under any particular set of circumstances determines the outcome of an infection. Mutations that increase the stability of CII increase the frequency of lysogenization. Such mutations occur in *cII* itself or in other genes. The cause of CII's instability is its susceptibility to degradation by host proteases. Its level in the cell is influenced by *cIII* as well as by host functions.

The effect of the lambda protein CIII is secondary: it helps to protect CII against degradation. Although the presence of CIII does not guarantee the survival of CII, in the absence of CIII, CII is virtually always inactivated.

Host gene products act on this pathway. Mutations in the host genes *hflA* and *hflB* increase lysogeny—*hfl* stands for *high frequency lysogenization*. The mutations stabilize CII because they inactivate host protease(s) that degrade it.

The influence of the host cell on the level of CII provides a route for the bacterium to interfere with the decision-taking process. For example, host proteases that degrade CII are activated by growth on rich medium, so lambda tends to lyse cells that are growing well, but is more likely to enter lysogeny on cells that are starving (and which lack components necessary for efficient lytic growth).

Summary

Phages have a lytic life cycle, in which infection of a host bacterium is followed by production of a large number of phage particles, lysis of the cell, and release of the viruses. Some phages also can exist in lysogenic form, in which the phage genome is integrated into the bacterial chromosome and is inherited in this inert, latent form like any other bacterial gene.

Lytic infection falls typically into three phases. In the first phase a small number of phage genes are transcribed by the host RNA polymerase. One or more of these genes is a regulator that controls expression of the group of genes expressed in the second phase. The pattern is repeated in the second phase, when one or more genes is a regulator needed for expression of the genes of the third phase. Genes of the first two phases code for enzymes needed to reproduce phage DNA; genes of the final phase code for structural components of the phage particle. It is common for the very early genes to be turned off during the later phases.

In phage lambda, the genes are organized into groups whose expression is controlled by individual regulatory events. The immediate early gene N codes for an antiterminator that allows transcription of the leftward and rightward groups of delayed early genes from the early promoters P_R and P_L. The delayed early gene Q has a similar antitermination function that allows transcription of all late genes from the promoter $P_{R'}$. The lytic cycle is repressed, and the lysogenic state maintained, by expression of the cI gene, whose product is a repressor protein that acts at the operators O_R and O_L to prevent use of the promoters P_R and P_L, respectively. A lysogenic phage genome expresses only the cI gene, from its promoter P_{RM}. Transcription from this promoter involves positive autogenous regulation, in which repressor bound at O_R activates RNA polymerase at P_{RM}.

Each operator consists of three binding sites for repressor. Each site is palindromic, consisting of symmetrical half-sites. Repressor functions as a dimer. Each half binding site is contacted by a repressor monomer. The N-terminal domain of repressor contains a helix–turn–helix motif that contacts DNA. Helix-3 is the recognition helix, responsible for making specific contacts with base pairs in the operator. Helix-2 is involved in positioning Helix-3; it is also involved in contacting RNA polymerase at P_{RM}. The C-terminal domain is required for dimerization. Induction is caused by cleavage between the N- and C-terminal domains, which prevents the DNA-binding regions from functioning in dimeric form, thereby reducing their affinity for DNA and making it impossible to maintain lysogeny. Repressor–operator binding is cooperative, so that once one dimer has bound to the first site, a second dimer binds more readily to the adjacent site.

The helix–turn–helix motif is used by other DNA-binding proteins, including lambda Cro, which binds to the same operators, but has a different affinity for the individual operator sites, determined by the sequence of helix-3. Cro binds individually to operator sites, starting with O_R3, in a noncooperative manner. It is needed for progression through the lytic cycle. Its binding to O_R first prevents synthesis of repressor from P_{RM}; then it prevents continued expression of early genes, an effect also seen in its binding to O_L.

Establishment of repressor synthesis requires use of the promoter P_{RE}, which is activated by the product of the cII gene. The product of $cIII$ is required to stabilize the cII product against degradation. By turning off cII and $cIII$ expression, Cro acts to prevent lysogeny. By turning off all transcription except that of its own gene, repressor acts to prevent the lytic cycle. The choice between lysis and lysogeny depends on whether repressor or Cro gains occupancy of the operators in a particular infection. The stability of CII protein in the infected cell is a primary determinant of the outcome.

Further reading

Reviews

An early view of the rationale underlying lambda repressor function was laid out by **Ptashne** (pp. 325–343) in *The Operon* (Miller and Reznikoff, Eds., Cold Spring Harbor Laboratory, New York, 1978), who has since reviewed the entire body of work from the perspective of regulation by repressor and Cro in *A Genetic Switch* (Cell Press and Blackwell Scientific Publications, San Francisco, 1992).

The lambda life cycle was reviewed by **Friedman and Gottesman** (pp. 43–51) in *Lambda II* (Hendrix, Roberts, Stahl and Weisberg, Eds., Cold Spring Harbor Laboratory, New York, 1983).

The interaction between lambda and its host has been reviewed by **Friedman** *et al.* (*Microbiol. Rev.* 48, 299–325, 1984).

Discoveries

Ptashne isolated lambda repressor and characterized its binding to DNA (*Proc. Nat. Acad. Sci. USA* 57, 306–313, 1967; *Nature* 214, 232–234, 1967).

The helix–turn–helix structure was proposed by **Pabo and Lewis** and by **Sauer** *et al.* (*Nature* 298, 443–447 and 447–451, 1982); the role of helix-3 in specific recognition was investigated by **Wharton, Brown and Ptashne** (*Cell* 38, 361–369, 1984). A crystal structure for Cro-DNA binding was reported by **Brennan et al.** (*Proc. Nat. Acad. Sci.* 87, 8165–8169, 1990).

The dimerization and cooperativity of lambda repressor was reported by **Pirrotta, Chadwick, and Ptashne** (*Nature* 227, 41–44, 1970) and **Johnson, Meyer, and Ptashne** (*Proc. Nat. Acad. Sci. USA* 76, 5061–5065, 1979).

PART 5

Perpetuation of DNA

If it be true that the essence of life is the accumulation of experience through the generations, then one may perhaps suspect that the key problem of biology, from the physicist's point of view, is how living matter manages to record and perpetuate its experiences. Look at a single bacterium in a large volume of fluid of suitable chemical composition. It assimilates substance, grows in length, divides in two. The two daughters do the same, like the broomstick of the Sorcerer's apprentice. Occasionally the replica will be slightly faulty and an individual arises with somewhat different properties, and it perpetuates itself in this modified form. It is quite easy to believe that the game of evolution is on once the trick of reproduction, covariant on mutation, has been discovered, and that the variety of types will be multiplied indefinitely.

Max Delbruck, 1949

CHAPTER 18

The replicon: unit of replication

Whether a cell has only one chromosome (as in prokaryotes) or has many chromosomes (as in eukaryotes), the entire genome must be replicated precisely once for every cell division. How is the act of replication linked to the cell cycle?

Two general principles are used to compare the state of replication with the condition of the cell cycle:

◆ *Initiation of DNA replication commits the cell (prokaryotic or eukaryotic) to a further division.* From this standpoint, the number of descendants that a cell generates is determined by a series of decisions on whether or not to initiate DNA replication.

◆ *If replication proceeds, the consequent division cannot be permitted to occur until the replication event has been completed.* Indeed, the completion of replication may provide a trigger for cell division. Then the duplicate genomes are segregated one to each daughter cell (via mitosis in eukaryotes). The unit of segregation is the chromosome.

Cell cycle regulator genes throw the switches that initiate DNA replication and that trigger division itself. In prokaryotes, the initiation of replication is a single event involving a unique site on the bacterial chromosome, and the process of division is accomplished by the development of a septum. In eukaryotic cells, initiation of replication

is identified by the start of S phase, a protracted period during which DNA synthesis occurs, and which involves many individual initiation events. The act of division is accomplished by the reorganization of the cell at mitosis. In Chapter 13, we discussed the regulatory processes in eukaryotic cells that control entry into S phase and into mitosis, and also the 'checkpoints' that postpone these actions until the appropriate conditions have been fulfilled. In this chapter, we are concerned with the regulation of DNA replication. How is a cycle of replication initiated? What controls its progress and how is its termination signaled?

The unit of DNA in which individual acts of replication occur is called the **replicon**. Each replicon 'fires' no more than once in each cell cycle. The replicon is defined by its possession of the control elements needed for replication. It has an **origin** at which replication is initiated. It may also have a **terminus** at which replication stops.

Any sequence attached to an origin—or, more precisely, not separated from an origin by a terminus—is replicated as part of that replicon. The origin is a *cis*-acting site, able to affect only that molecule of DNA on which it resides.

The original formulation of the replicon (in prokaryotes) viewed it as a unit possessing both the origin *and* the gene coding for the regulator protein. Now, however, 'replicon' is usually applied to eukaryotic chromosomes to describe a unit of replication that contains an origin; *trans*-acting regulator protein(s) may be coded elsewhere.

A genome in a prokaryotic cell constitutes a single replicon; so the units of replication and segregation coincide. The largest such replicon is that of the bacterial chromosome itself. Initiation at a single origin sponsors replication of the entire genome, once for every cell division. Each haploid bacterium has a single chromosome, so this type of replication control is called **single copy**.

In addition to the chromosome, bacteria may contain plasmids. *A plasmid is an autonomous circular DNA genome that constitutes a separate replicon* (see Table 17.1). A plasmid replicon may replicate with the bacterial genome *pari passu* (showing single copy control) or may be under a different type of control. When the number of copies of a plasmid is greater than that of the bacterial chromosome, it is said to replicate under **multicopy control**. Each phage or virus DNA also constitutes a replicon, able to initiate many times during an infectious cycle. Perhaps a better way to view the prokaryotic replicon, therefore, is to reverse the definition: *any DNA molecule that contains an origin can be replicated autonomously in the cell.*

The replicon is a flexible unit. In the case of bacterial chromosomes, it is used to produce copies by duplicating double-stranded DNA (as envisaged in the original Meselson–Stahl experiment discussed in Chapter 4). But it can also be used to generate single-stranded copies of phage or plasmid genomes in monomeric or multimeric forms. The mode of reproduction of the replicon depends on the nature of the interactions that occur during initiation at the origin, and the general principle is that replication is controlled at the stage of initiation. *Once replication has started, it continues until the entire genome has been duplicated.* The frequency of initiation is controlled by the interaction of regulator protein(s) with the origin.

A major difference in the organization of bacterial and eukaryotic genomes is seen in their replication. Each eukaryotic chromosome contains a large number of replicons. So the unit of segregation includes many units of replication. This adds another dimension to the problem of control. All the replicons on a chromosome must be fired during one cell cycle, although they are not active simultaneously, but are activated over a fairly protracted period. *Yet each of these replicons must be activated no more than once in the cell cycle.*

Some signal must distinguish replicated from nonreplicated replicons, so that replicons do not fire a second time. And because many replicons are activated independently, some signal must exist to indicate when the entire process of replicating all replicons has been completed.

We have begun to collect information about the construction of individual replicons, but we still have little information about the relationship between replicons. We do not know whether the pattern of replication is the same in every cell cycle. Are all origins always used or are some origins sometimes silent? Do origins always fire in the same order? If there are different classes of origins, what distinguishes them?

In contrast with the chromosomes, which have a single-copy type of control, the DNA of mitochondria and chloroplasts may be regulated more like plasmids that exist in multiple copies per bacterium. There are multiple copies of each organelle DNA per cell, and the control of organelle DNA replication must be related to the cell cycle.

In all these systems, the key question is to define the sequences that function as origins and to determine how they are recognized by the appropriate proteins of the apparatus for replication. This chapter starts by considering the basic construction of replicons and the various forms that they take; following the consideration of the origin, we turn to the question of how replication of the genome is coordinated with bacterial division and what is responsible for segregating the genomes to daughter bacteria.

Origins can be mapped by autoradiography and electrophoresis

Consider a molecule of DNA engaged in replication. It has two types of regions. **Figure 18.1** demonstrates that the nonreplicated region consists of the parental duplex; this opens into the replicated region where the two daughter duplexes have formed. The point at which replication is occurring is called the **replication fork** (sometimes also known as the **growing point**). *A replication fork moves sequentially along the DNA, from its starting point at the origin.*

Replication may be **unidirectional** or **bidirectional.** The type of event is determined by whether one or two replication forks set out from the origin. In unidirectional replication, one replication fork leaves the origin and proceeds along the DNA. In bidirectional replication, two replication forks are formed; they proceed away from the origin in opposite directions.

When replicating DNA is viewed by electron microscopy, the replicated region appears as an **eye** within the nonreplicated DNA. However, its appearance does not distinguish between unidirectional and bidirectional replication.

As depicted in **Figure 18.2**, the eye can represent either of two structures. If generated by unidirectional replication, the eye represents a fixed origin and a moving replication fork. If generated by bidirectional replication, the eye represents a pair of replication forks. In either case, the progress of replication expands the eye until ultimately it encompasses the whole replicon.

When a replicon is circular, the presence of an eye forms the θ-**structure** drawn in **Figure 18.3**.

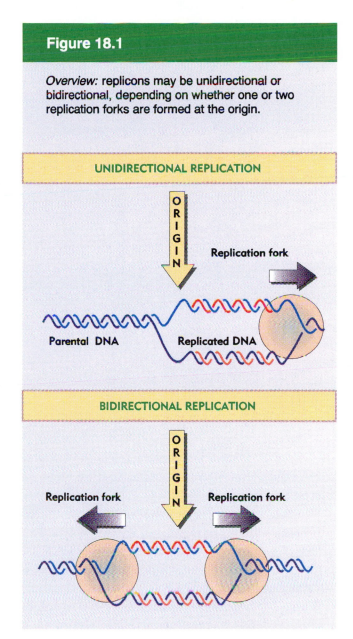

Figure 18.1

Overview: replicons may be unidirectional or bidirectional, depending on whether one or two replication forks are formed at the origin.

UNIDIRECTIONAL REPLICATION

ORIGIN

Replication fork

Parental DNA   Replicated DNA

BIDIRECTIONAL REPLICATION

ORIGIN

Replication fork   Replication fork

Figure 18.2

A replication eye can represent either a unidirectional or bidirectional replicon. The significance of the observed junction points depends upon the type of replicon.

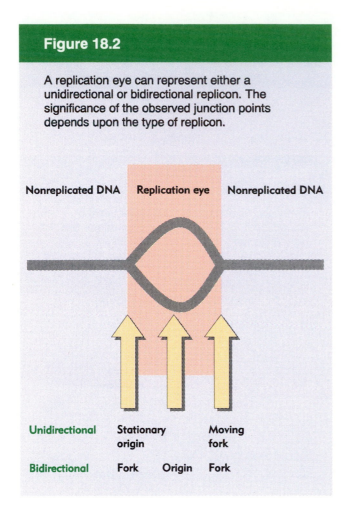

Nonreplicated DNA   Replication eye   Nonreplicated DNA

| Unidirectional | Stationary origin | | Moving fork |
| Bidirectional | Fork | Origin | Fork |

molecules that have eyes of different sizes. If replication is unidirectional, only one of the ends will move; the other is the fixed origin. If replication is bidirectional, both will move; the origin is the point midway between them.

With undefined regions of large genomes, two successive pulses of radioactivity can be used to label the movement of the replication forks. If one pulse has a more intense label than the other, they can be distinguished by the relative intensities of labeling, which are visualized by autoradiography. **Figure 18.5** shows that unidirectional replication causes one type of label to be followed by the other at *one* end of the eye. Bidirectional replication produces a (symmetrical) pattern at *both* ends of the eye. This is the pattern usually observed in replicons of eukaryotic chromosomes.

Figure 18.3

A replication eye forms a theta structure in circular DNA.

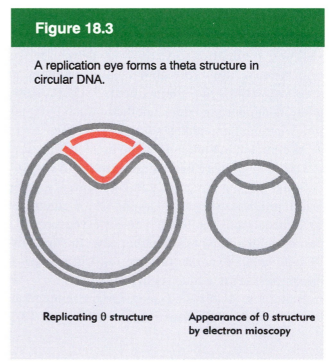

Replicating θ structure   Appearance of θ structure by electron mioscopy

The successive stages of replication of the circular DNA of polyoma virus are visualized by electron microscopy in **Figure 18.4**.

Whether a replicating eye has one or two replication forks can be determined in two ways. The choice of method depends on whether the DNA is a defined molecule or an unidentified region of a cellular genome.

With a defined linear molecule, we can use electron microscopy to measure the distance of each end of the eye from the end of the DNA. Then the positions of the ends of the eyes can be compared in

Figure 18.4

The replication eye becomes larger as the replication forks proceed along the replicon. Note that the "eye" becomes larger than the non-replicated segment. The two sides of the eye can be defined because they are both the same length. Photograph kindly provided by Bernard Hirt.

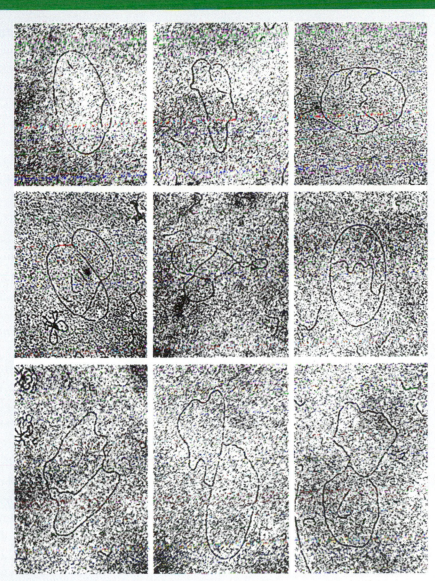

Figure 18.5

Different densities of radioactive labeling can be used to distinguish unidirectional and bidirectional replication.

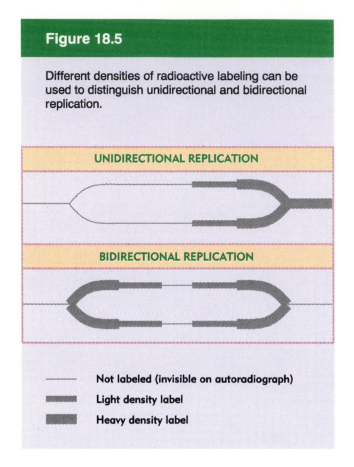

UNIDIRECTIONAL REPLICATION

BIDIRECTIONAL REPLICATION

- - - - - - Not labeled (invisible on autoradiograph)

▬▬▬ Light density label

▬▬▬ Heavy density label

mass, and a second dimension where movement is determined more by shape. Different types of replicating molecules follow characteristic paths, measured by their deviation from the line that would be followed by a linear molecule of DNA that doubled in size.

A simple Y-structure, in which one fork moves along a linear fragment, follows a continuous path, with an inflection point that corresponds to the midsize, when all three branches are the same length, and structure therefore deviates most extensively from linear DNA. Analogous considerations determine the paths of double Y-structures or bubbles. An asymmetric bubble follows a discontinuous path, with a break at the point at which the bubble is converted to a Y-structure as one fork runs off the end.

By analyzing fragments derived from the vicinity of the origin, it is possible to extrapolate from the movement of the replicating forks to identify the location of the origin. This technique has been used to identify origins on plasmids and on chromosomal DNA, and to investigate questions such as whether both replicating forks necessarily move at the same speed in the case of bidirectional replication.

Taken together, the various techniques for characterizing replicating DNA show that origins are most often used to initiate bidirectional replication. From this level of resolution, we must then proceed to the molecular level, to identify the *cis*-acting sequences that comprise the origin, and the *trans*-acting factors that recognize it.

A more recent method for mapping origins with greater resolution takes advantage of the effects that changes in shape have upon electrophoretic migration. **Figure 18.6** illustrates the two-dimensional mapping technique, in which restriction fragments of replicating DNA are electrophoresed in a first dimension that separates by

Figure 18.6

The position of the origin and the number of replicating forks determines the shape of a replicating restriction fragment, which can be assayed from the electrophoretic path followed by a replicating restriction fragment.

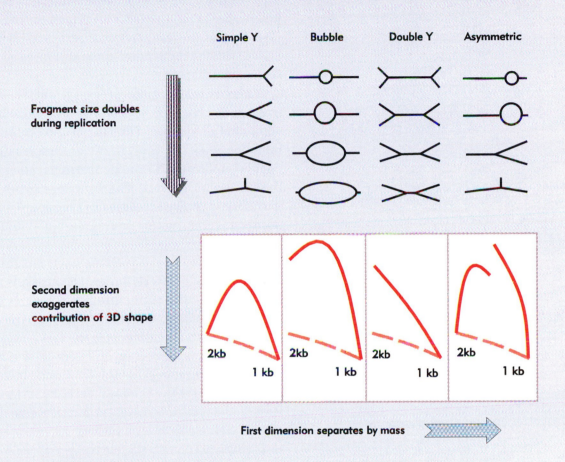

The bacterial genome is a single replicon

To be properly inherited, a bacterial replicon should support several functions:

◆ Initiating a replication cycle.

◆ Controlling the frequency of initiation events.

◆ Segregating replicated chromosomes to daughter cells.

The first two functions must both be exercised by the origin; segregation could be an independent function, but in fact usually rests with sequences in the vicinity of the origin. (Origins in eukaryotes do not function in segregation, but are concerned only with replication.) Isolating mutants that are deficient in any one of these functions should allow us to identify the sequences involved in each activity.

As a general principle, the DNA constituting an origin can be isolated by its ability to support

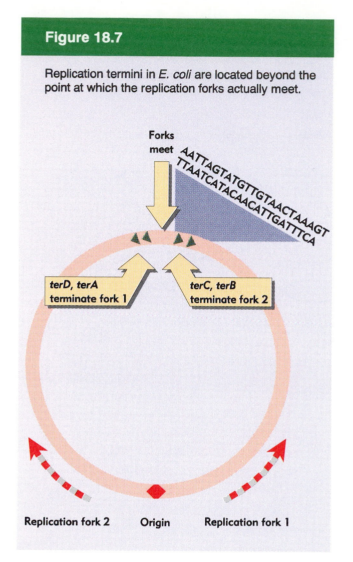

Figure 18.7

Replication termini in *E. coli* are located beyond the point at which the replication forks actually meet.

Forks meet

AATTAGTATGTTGTAACTAAAGT
TTAATCATACAACATTGATTTCA

terD, terA terminate fork 1

terC, terB terminate fork 2

Replication fork 2   Origin   Replication fork 1

replication of any DNA sequence to which it is joined. When DNA from the origin is cloned into a molecule that lacks an origin, the reconstruction will create a plasmid capable of autonomous replication *only if the DNA from the origin contains all the sequences needed to identify itself as an authentic origin for replication.* (A comparable approach has been used to identify centromeric or telomeric DNA in yeast; see Chapter 27.)

Origins now have been identified in bacteria, yeast, chloroplasts, and mitochondria, although not in higher eukaryotes. A general significant feature is that the overall sequence composition is A•T-rich. We assume this is related to the need to melt the DNA duplex to initiate replication.

The genome of *E. coli* is replicated bidirection-

ally from a single origin, identified as the genetic locus *oriC*. Artificial plasmids that are provided with *oriC* can replicate and perpetuate themselves in *E. coli*. By reducing the size of the cloned fragment of *oriC*, the region required to initiate replication has been equated with a fragment of 245 bp. (We discuss the properties of *oriC* and its interaction with the replication apparatus in more detail in the next chapter.)

Plasmids that initiate properly may segregate irregularly, but can be stabilized by introducing additional sequences. Thus the origin required for initiation does not carry sufficient information to enable duplicate DNA molecules to partition when the bacterium divides. The functions involved in partitioning can be identified by characterizing the sequences that confer segregational stability on the plasmid.

So far we have dealt with the bacterial chromosome as though it were linear. Because it is really circular, the two replication forks each move around the genome to a meeting point. What happens at such an encounter? Do the forks crash into each other, or is there a specific terminus at which they stop? Remember that the DNA must be replicated right across the region where the forks meet. How do the enzymes involved in replication disengage from the chromosome?

Termination in *E. coli* has the peculiar features reported in **Figure 18.7**. We know that the replication forks usually meet and halt replication at a point midway round the chromosome from the origin. But two termination regions (*terD,A* and *terC,B*) have been identified, located ~100 kb on either side of this meeting point. Each terminus is specific for one direction of fork movement, and they are arranged in such a way that each fork would have to pass the other in order to reach the terminus to which it is susceptible. This arrangement creates a 'replication fork trap'; if for some reason one fork is delayed, so that the forks fail to meet at the usual central position, the more rapid fork will be trapped at the *ter* region to wait for the arrival of the slow fork.

The *ter* sequences all contain a short (~23 bp) region that causes termination *in vitro*. The termination sequences function in only one orientation. Sequences conforming to a consensus are found also in several plasmids. In all these cases, termination occurs close to the *ter* consensus sequence, but we have yet to define the relationship between the *ter* sequence and the actual site at which replication ceases. Termination requires the product of the *tus* gene, which codes for a protein that recognizes the consensus sequence and prevents the replication fork from proceeding (see Chapter 19).

The effect of replication on proteins that bind to DNA is intriguing. What happens if a replication fork encounters a repressor bound to DNA? We assume that the fork can displace the repressor (after which repressors must bind to both copies of the replicated locus in order to maintain control of gene expression).

A particularly interesting question is what happens when a replication fork encounters an RNA polymerase engaged in transcription. A replication fork moves >10× faster than RNA polymerase. If they are proceeding in the same direction, either the replication fork must displace the polymerase (presumably aborting transcription) or it must slow down as it waits for the RNA polymerase to reach its terminator. We do not know how this situation is resolved in *E. coli*.

An even more serious conflict arises when the replication fork meets an RNA polymerase traveling in the opposite direction, that is, toward it. Can it displace the RNA polymerase? Or do both replication and transcription come to a halt? An indication that these encounters cannot easily be resolved is provided by the organization of the *E. coli* chromosome. Almost all active transcription units are oriented so that they are expressed in the same direction as the replication fork that passes them. The exceptions all comprise small transcription units that are infrequently expressed. The difficulty of generating inversions containing highly expressed genes argues that head-on encounters between a replication fork and a series of transcribing RNA polymerases may be lethal.

So the bacterial chromosome is replicated bidirectionally as a single unit from a unique origin (*oriC*), and termination occurs in a discrete region. Following the termination of DNA replication itself, enzymes that manipulate higher order structure of DNA are required for the two daughter chromosome to be physically separated (see later and Chapter 33).

Each eukaryotic chromosome contains many replicons

In eukaryotic cells, the replication of DNA is confined to part of the cell cycle. S phase occupies part of interphase, and often lasts a few hours in a higher eukaryotic cell. Replication of the large amount of DNA contained in a eukaryotic chromosome is accomplished by dividing it into many individual replicons. Only some of these replicons are engaged in replication at any point in S phase. Presumably each replicon is activated at a specific time during S phase, although the evidence on this issue is not decisive.

The important point is that the start of S phase is signaled by the activation of the first replicons. Over the next few hours, initiation events occur at other replicons. The control of S phase therefore involves two processes: release of the cell from the preceding G1 phase; and initiation of replication at individual replicons in an ordered manner.

Much of our knowledge about the properties of the individual replicons is derived from autoradiographic studies, generally using the types of protocols illustrated in Figures 18.5 and 18.6. Chromosomal replicons usually display bidirectional replication.

How large is the average replicon, and how many are there in the genome? A difficulty in

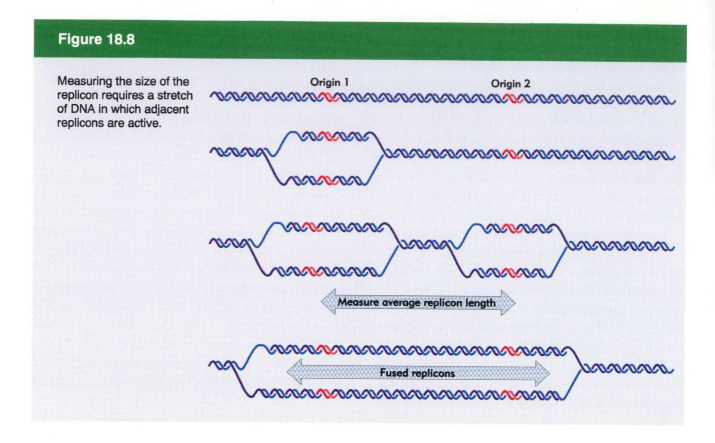

Figure 18.8

Measuring the size of the replicon requires a stretch of DNA in which adjacent replicons are active.

Origin 1   Origin 2

Measure average replicon length

Fused replicons

characterizing the individual unit is that adjacent replicons may fuse to give large replicated eyes, as illustrated in **Figure 18.8**. The approach usually used to distinguish individual replicons from fused eyes is to rely on measurements of stretches of DNA in which several replicons can be seen to be active, presumably captured at a stage when all have initiated around the same time, but before the forks of adjacent units have met.

(There is some evidence that 'regional' controls might produce this sort of activation pattern, in which groups of replicons are initiated more or less coordinately, as opposed to a mechanism in which individual replicons are activated one by

Table 18.1

Eukaryotic replicons are small and replicate more slowly than bacterial DNA.

Organism	No. of Replicons	Average Length	Fork Movement	
Bacterium	1	4200 kb	50,000 bp/min	(E. coli)
Yeast	500	40 kb	3,600 bp/min	(S. cerevisiae)
Fruit fly	3500	40 kb	2,600 bp/min	(D. melanogaster)
Toad	15000	200 kb	500 bp/min	(X. laevis)
Mouse	25000	150 kb	2,200 bp/min	(M. musculus)
Plant	35000	300 kb		(V. faba)

one in dispersed areas of the genome.)

In groups of active replicons, the average size of the unit is measured by the distance between the origins (that is, between the midpoints of adjacent replicons). The rate at which the replication fork moves can be estimated from the maximum distance that the autoradiographic tracks travel during a given time.

Eukaryotic replicons are contrasted with the prokaryotic situation in **Table 18.1**. Individual eukaryotic replicons are relatively small (although they vary >10-fold in length within a species); the rate at which they are replicated is much slower than the rate of bacterial replication fork movement.

We should like to know what constitutes the origin of each replicon. Is it a specific DNA sequence, what is its length, and how are the sequences at different origins related to one another? Is an origin associated with a particular feature of higher order structure in the chromosomal material?

From the speed of replication indicated in Table 18.1, it is evidence that a mammalian genome could be replicated in ~1 hour if all replicons functioned simultaneously. But S phase actually lasts for >6 hours in a typical somatic cell, which implies that no more than 15% of the replicons are likely to be active at any given moment. (There are some exceptional cases, such as the early embryonic divisions of *Drosophila* embryos, where the duration of S phase is compressed by the simultaneous functioning of a large number of replicons.)

How are origins selected for initiation at different times during S phase? Does the sequence of the origin carry information that establishes its time of use? An especially intriguing question is how the replication apparatus distinguishes origins that have already been replicated from those that have yet to be replicated. Is the utilization of each origin once and only once during S phase a property of the DNA (for example, as seen by its state of methylation) or of proteins associated with it?

Available evidence suggests that chromosomal replicons do not have termini at which the replication forks cease movement and (presumably) dissociate from the DNA. It seems more likely that a replication fork continues from its origin until it meets a fork proceeding toward it from the adjacent replicon. We have already mentioned the potential topological problem of joining the newly synthesized DNA at the junction of the replication forks.

Isolating the origins of yeast replicons

Any segment of DNA that has an origin should be able to replicate. So although plasmids are rare in eukaryotes, it may be possible to construct such molecules by suitable manipulation *in vitro*.

Yeast offers the opportunity to isolate discrete origins. Cells of *S. cerevisiae* that are mutant in some function can be 'transformed' by addition of DNA that carries a wild-type copy of the gene. Some yeast DNA fragments (when circularized) are able to transform defective cells very efficiently. These fragments can survive in the cell in the unintegrated (autonomous) state, that is, as self-replicating plasmids.

A high-frequency transforming fragment possesses a sequence that confers the ability to replicate efficiently in yeast. This segment is called an *ARS* (for autonomously replicating sequence). We believe that *ARS* elements are derived from authentic origins of replication; and in some cases initiation has been identified as occurring at the location of an *ARS* element in a chromosome.

Sequences with *ARS* function occur at about the same average frequency as origins of replication. Where *ARS* elements have been systematically mapped over extended chromosomal regions, it seems that only some of them are

Figure 18.9

An ARS extends for ˜50 bp and includes a consensus sequence, some imperfect copies of the consensus sequence, and a transcription factor-binding site. Only mutations in restricted regions adversely affect origin function, and only mutations in the core consensus abolish it entirely.

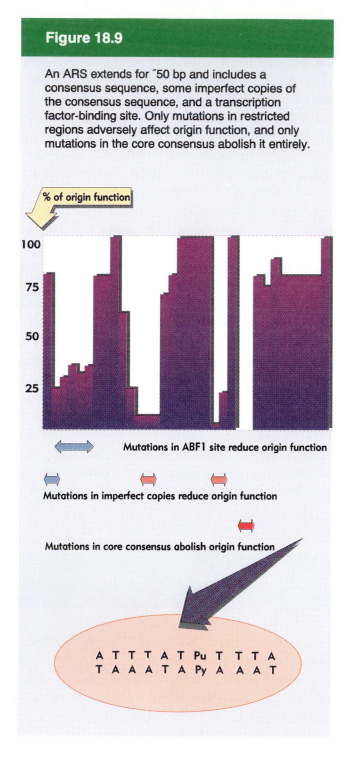

% of origin function

Mutations in ABF1 site reduce origin function

Mutations in imperfect copies reduce origin function

Mutations in core consensus abolish origin function

A T T T A T Pu T T T A
T A A A T A Py A A A T

termini between replicons. In this case, a given region of a chromosome could be replicated from different origins in different cell cycles.

An *ARS* element consists of an A•T-rich region that contains certain discrete sites in which mutations affect origin function. Base composition rather than sequence may be important in the rest of the region. **Figure 18.9** shows a systematic mutational analysis along the length of an origin. Origin function is abolished completely by mutations in a 14 bp 'core' region that contains an 11 bp consensus sequence that consists of A•T base pairs. This consensus sequence is the only homology between known *ARS* elements.

In addition to the consensus sequence in the core, there are additional sequences that have imperfect copies, typically conforming to the consensus in 9/11 positions. Mutations in these sequences reduce, but do not entirely eliminate, origin function. Their functions are presumably partially overlapping, so that no one of them is essential. (Their reduced conformity to the consensus sequence and their duplication together explain why they were not originally defined as components of the origin, but only recently were identified as additional sites that contribute to origin function.)

The origin also contains a binding site for a transcription factor. This factor probably does not function to activate transcription. Its role in initiating replication may be similar to its role in transcription, that is, to interact with other proteins that bind to the site, helping to stabilize a complex that assembles there. A complex of proteins with a mass of ~250,000 daltons binds to the origin; it contains ~6 individual proteins. ATP is required for binding.

Important questions that remain to be answered concern the relationship of the core consensus sequence to the molecular events involved in initiation. Where does the DNA initially melt, and where is synthesis of the new strands actually initiated?

Although some proteins have been found that bind to *ARS* elements, in most cases the function of the element is unaffected by mutations that prevent binding. We have therefore yet to identify the

actually used to initiate replication. The others are silent, or possibly used only occasionally. If it is true that some origins have varying probabilities of being used, it follows that there can be no fixed

proteins that are presumably essential for initiating replication at the origin. An interesting candidate protein binds only to functional *ARS* elements (not to *ARS* elements bearing mutations that inactivate the core consensus); it binds to the T-rich single strand (the upper strand in Figure 18.9) of denatured *ARS* sequences. It may be involved in separating the strands of DNA for replication.

D loops may be maintained at mitochondrial origins

The origins of replicons in both prokaryotic and eukaryotic chromosomes are static structures: they comprise sequences of DNA that are recognized in duplex form and used to initiate replication at the appropriate time. Initiation requires separating the DNA strands and commencing bidirectional DNA synthesis. But a different type of arrangement prevails at origins in other genomes, most notably in mitochondria.

Replication starts at a specific origin in the circular duplex DNA. But initially only one of the two parental strands (the H strand in mammalian mitochondrial DNA) is used as a template for synthesis of a new strand. Synthesis proceeds for only a short distance, displacing the original partner (L) strand, which remains single-stranded, as illustrated in **Figure 18.10**. The condition of this region gives rise to its name as the **displacement** or **D loop**.

A single D loop is found as an opening of 500–600 bases in mammalian mitochondria. The short strand that maintains the D loop is unstable and turns over; it is frequently degraded and resynthesized to maintain the opening of the duplex at this site.

Some mitochondrial DNAs, such as *X. laevis*, possess a single but longer D loop. Others possess several D loops; there may be as many as six in the linear mitochondrial DNA of *Tetrahymena*. The same mechanism is employed in chloroplast DNA, where (in higher plants) there are two D loops.

Replication of mitochondrial DNA starts in the same way as replication of nuclear DNA: with the synthesis of a short RNA that is extended by a DNA polymerase (the mechanism is discussed in Chapter 19). The H strand origin in mammalian mitochondrial DNA is in fact provided by the promoter for transcription of the H strand. RNA synthesis starts at this promoter: if it continues around the template it generates a mitochondrial transcript; if it terminates in the region of the D loop, the RNA is either degraded or used to sponsor DNA replication. We do not know what controls these processes.

To replicate mammalian mitochondrial DNA, the short strand in the D loop is extended. The displaced region of the original L strand becomes longer, expanding the D loop. This expansion continues until it reaches a point about two-thirds of the way around the circle. Replication of this region exposes an origin in the displaced L strand. Synthesis of an H strand initiates at this site, proceeding around the displaced single-stranded L template in the opposite direction from L strand synthesis.

Because of the lag in its start, H strand synthesis has proceeded only a third of the way around the circle when L strand synthesis finishes. This releases one completed duplex circle and one gapped circle, which remains partially single-stranded until synthesis of the H strand is completed. Finally, the new strands are sealed to become covalently intact.

The existence of rolling circles (see p.544) and D loops exposes a general principle. *An origin can be a sequence of DNA that serves to initiate DNA synthesis using one strand as template.* The opening of the duplex does not necessarily lead to the initiation of replication on the other strand. In the case of mitochondrial DNA replication, in fact, the origins for replicating the complementary strands lie at different locations.

Figure 18.10

The D loop maintains an opening in mammalian mitochondrial DNA, which has separate origins for the replication of each strand.

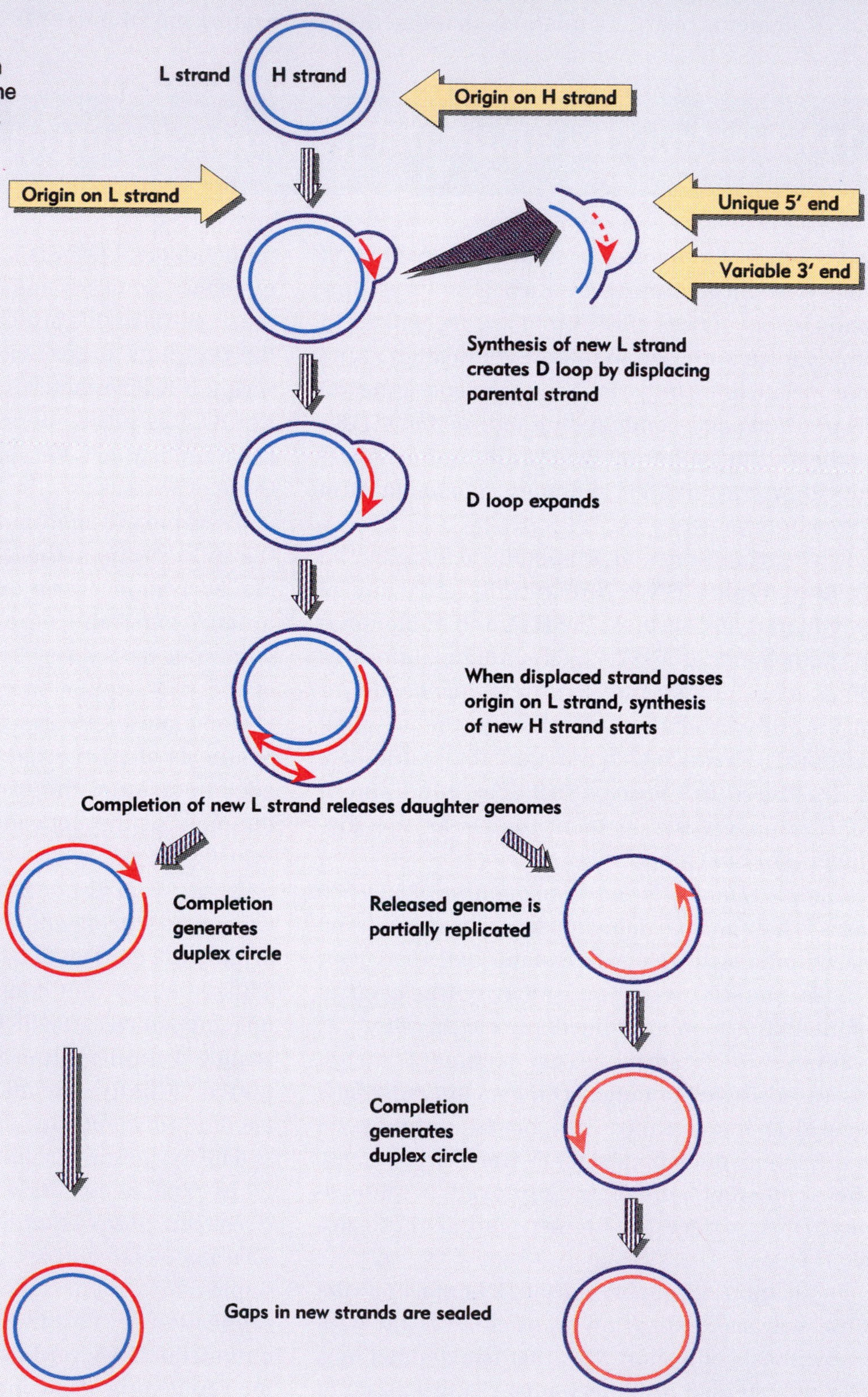

L strand   H strand

Origin on H strand

Origin on L strand

Unique 5' end

Variable 3' end

Synthesis of new L strand creates D loop by displacing parental strand

D loop expands

When displaced strand passes origin on L strand, synthesis of new H strand starts

Completion of new L strand releases daughter genomes

Completion generates duplex circle

Released genome is partially replicated

Completion generates duplex circle

Gaps in new strands are sealed

The problem of linear replicons

None of the replicons that we have considered so far have a linear end: either they are circular (as in the *E. coli* or mitochondrial genomes) or they are part of much longer segregation units (as in eukaryotic chromosomes). But linear replicons occur, in some cases as single extrachromosomal units, and of course at the ends of eukaryotic chromosomes.

The ability of all known nucleic acid polymerases, DNA or RNA, to proceed only in the 5′–3′ direction poses a problem for synthesizing DNA at the end of a linear replicon. Consider the two parental strands depicted in **Figure 18.11**. The lower strand presents no problem: it can act as template to synthesize a daughter strand that runs right up to the end, where presumably the polymerase falls off. But to synthesize a complement at the end of the upper strand, synthesis must start right at the very last base (or else this strand would become shorter in successive cycles of replication).

We do not know whether initiation right at the end of a linear DNA is feasible. We usually think of a polymerase as binding at a site *surrounding* the position at which a base is to be incorporated. So a special mechanism must be employed for replication at the ends of linear replicons. Four types of solution may be imagined to accommodate the need to copy a terminus:

◆ The problem may be circumvented by converting a linear replicon into a circular or multimeric molecule. Phages such as T4 or lambda use such mechanisms (see later).

◆ The DNA may form an unusual structure—for example, by creating a hairpin at the terminus, so that there is in fact no free end. Formation of a cross-link is involved in replication of the linear mitochondrial DNA of *Paramecium*.

Figure 18.11

Replication could run off the 3′ end of a newly synthesized linear strand, but could it initiate at a 5′ end?

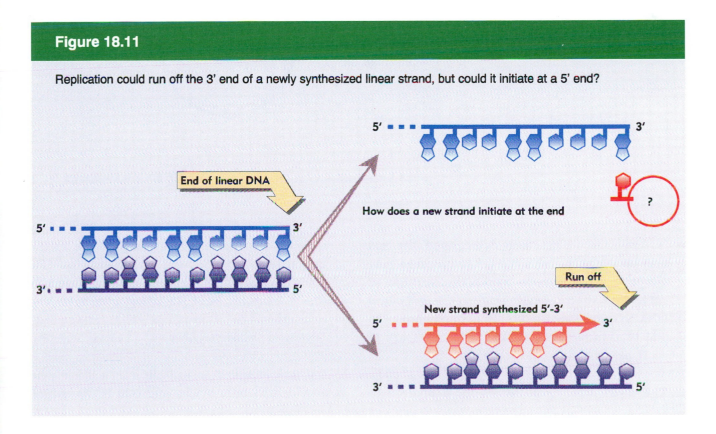

Figure 18.12

Adenovirus DNA replication is initiated separately at the two ends of the molecule and proceeds by strand displacement.

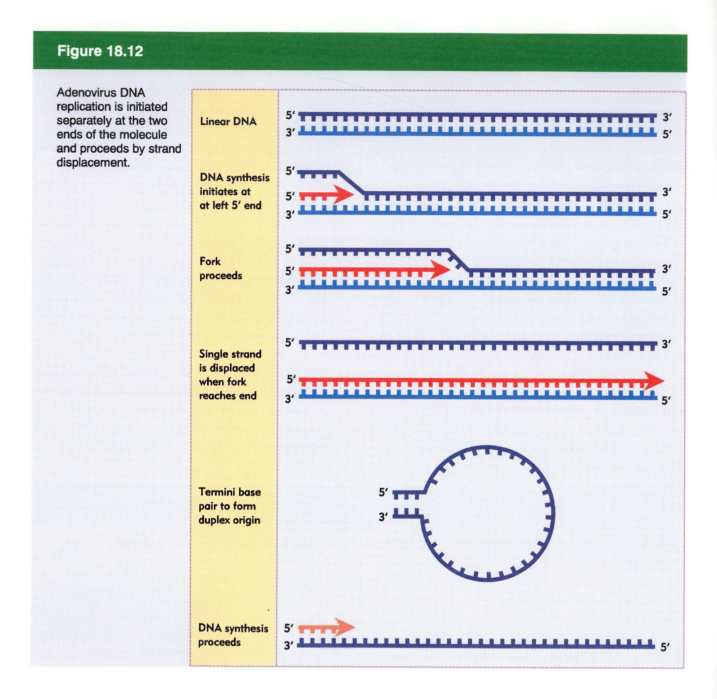

Linear DNA

DNA synthesis initiates at at left 5′ end

Fork proceeds

Single strand is displaced when fork reaches end

Termini base pair to form duplex origin

DNA synthesis proceeds

◆ Instead of being precisely determined, the end may be variable. Eukaryotic chromosomes may adopt this solution, in which the number of copies of a short repeating unit at the end of the DNA changes (see Chapter 27). A mechanism to add or remove units makes it unnecessary to replicate right up to the very end.

◆ The most direct solution is for a protein to

intervene to make initiation possible at the actual terminus. Several linear viral nucleic acids have proteins that are *covalently linked to the 5′ terminal base*. The best characterized examples are adenovirus DNA, phage φ29 DNA, and poliovirus RNA.

A prime example of initiation at a linear end is provided by adenovirus and φ29 DNAs, which

Figure 18.13

The 5' terminal phosphate at each end of adenovirus DNA is covalently linked to serine in the 55,000-dalton Ad-binding protein.

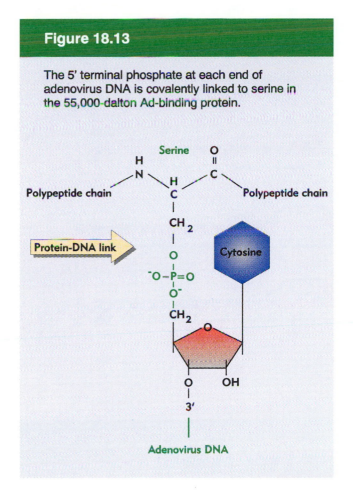

only 22 amino acids is linked via the hydroxyl group of tyrosine to the 5' terminal base. In each case, the attached protein is coded by the virus and is implicated in replication.

The protein present on replicating strands of adenovirus DNA is 80K, larger than the protein found on mature DNA isolated from the virion. The 80K protein is actually the species involved in initiating replication, but, at some point during the maturation of the virus, it is cleaved to the smaller size that remains in the virion.

How does the attachment of the protein overcome the initiation problem? The role of the 80K adenovirus protein is suggested by its ability to become covalently linked to dCTP in extracts of adenovirus-infected cells. The 80K protein is intimately connected with another protein, a 140K DNA polymerase. This suggests the model illustrated in **Figure 18.14**.

The 80K dalton protein forms a complex with the DNA polymerase. The complex of polymerase and terminal protein, bearing dCTP, approaches the end of the adenovirus DNA. The 5' end of the adenovirus is bound to the 55K version of the protein. The 5' end is displaced by the 80K protein; then the free 3'-OH end of the CTP is used to prime the elongation reaction by the DNA polymerase. This generates a new strand whose 5' end is covalently linked to the initiating CTP. A similar model probably applies to φ29 replication.

The entire series of reactions is coordinated. Linkage of the 80K protein to dCTP is undertaken by DNA polymerase in the presence of adenovirus DNA. This suggests that the linkage reaction is catalyzed in a similar way to the synthesis of nucleotide–nucleotide bonds.

The 80K protein binds to the region located between 9 and 18 bp from the end of the DNA. This 'core sequence' is found also in host nuclear DNA. The adjacent region, between positions 17 and 48, is essential for the binding of a host protein, nuclear factor I, which is also required for the initiation reaction. The initiation complex may therefore form between positions 9 and 48, a fixed distance from the actual end of the DNA.

actually replicate from both ends, using the mechanism of **strand displacement** illustrated in **Figure 18.12**. The same events can occur independently at either end. Synthesis of a new strand starts at one end, displacing the homologous strand that was previously paired in the duplex. When the replication fork reaches the other end of the molecule, the displaced strand is released as a free single strand. It is then replicated independently; this requires the formation of a duplex origin by base pairing between some short complementary sequences at the ends of the molecule.

In several viruses that use such mechanisms, a protein is found covalently attached to each 5' end. In the case of adenovirus, a 55Kdalton protein is linked to the mature viral DNA via a phosphodiester bond to serine, as indicated in **Figure 18.13**. The same type of arrangement is found in φ29. In the single-stranded RNA poliovirus, the VPg protein of

Figure 18.14

The 80,000-dalton adenovirus protein displaces the 5' end of DNA (which is bound to the 55,000 dalton protein), and provides CTP to prime synthesis of a new DNA strand.

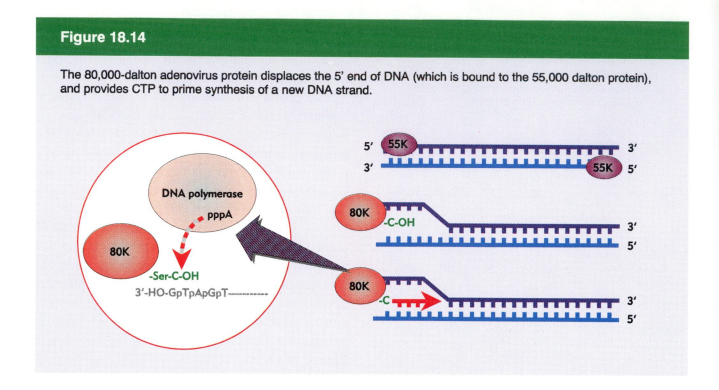

Rolling circles produce multimers of a replicon

The structures generated by replication depend on the relationship between the template and the replication fork. The critical features are whether the template is circular or linear, and whether the replication fork is engaged in synthesizing both strands of DNA or only one.

Replication of only one strand is used to generate copies of some circular molecules. A nick opens one strand, and then the free 3'-OH end generated by the nick is extended by the DNA polymerase. The newly synthesized strand displaces the original parental strand. The ensuing events are depicted in **Figure 18.15**.

This type of structure is called a **rolling circle**, because the growing point can be envisaged as rolling around the circular template strand. It could in principle continue to do so indefinitely. As it moves, the replication fork extends the outer strand and displaces the previous partner.

Because the newly synthesized material is covalently linked to the original material, the displaced strand has the original unit genome at its 5' end. The original unit is followed by any number of unit genomes, synthesized by continuing revolutions of the template. Each revolution displaces the material synthesized in the previous cycle.

An example is shown in the electron micrograph of **Figure 18.16**. The rolling circle is put to several uses *in vivo*. Some pathways that are used to replicate DNA are depicted in **Figure 18.17**.

Cleavage of a unit length tail generates a copy of the original circular replicon in linear form. The linear form may be maintained as a single strand or may be converted into a duplex by synthesis of the complementary strand (which is identical in sequence to the template strand of the original rolling circle).

Some bacteriophages use a mechanism in which cleavage of unit genomes from the displaced tail following each revolution generates a series of monomers, which can be packaged into phage particles or used for further replication cycles. An alternative is to allow continued revolutions of the rolling circle to generate a displaced tail that

Figure 18.15

The rolling circle generates a multimeric single-stranded tail.

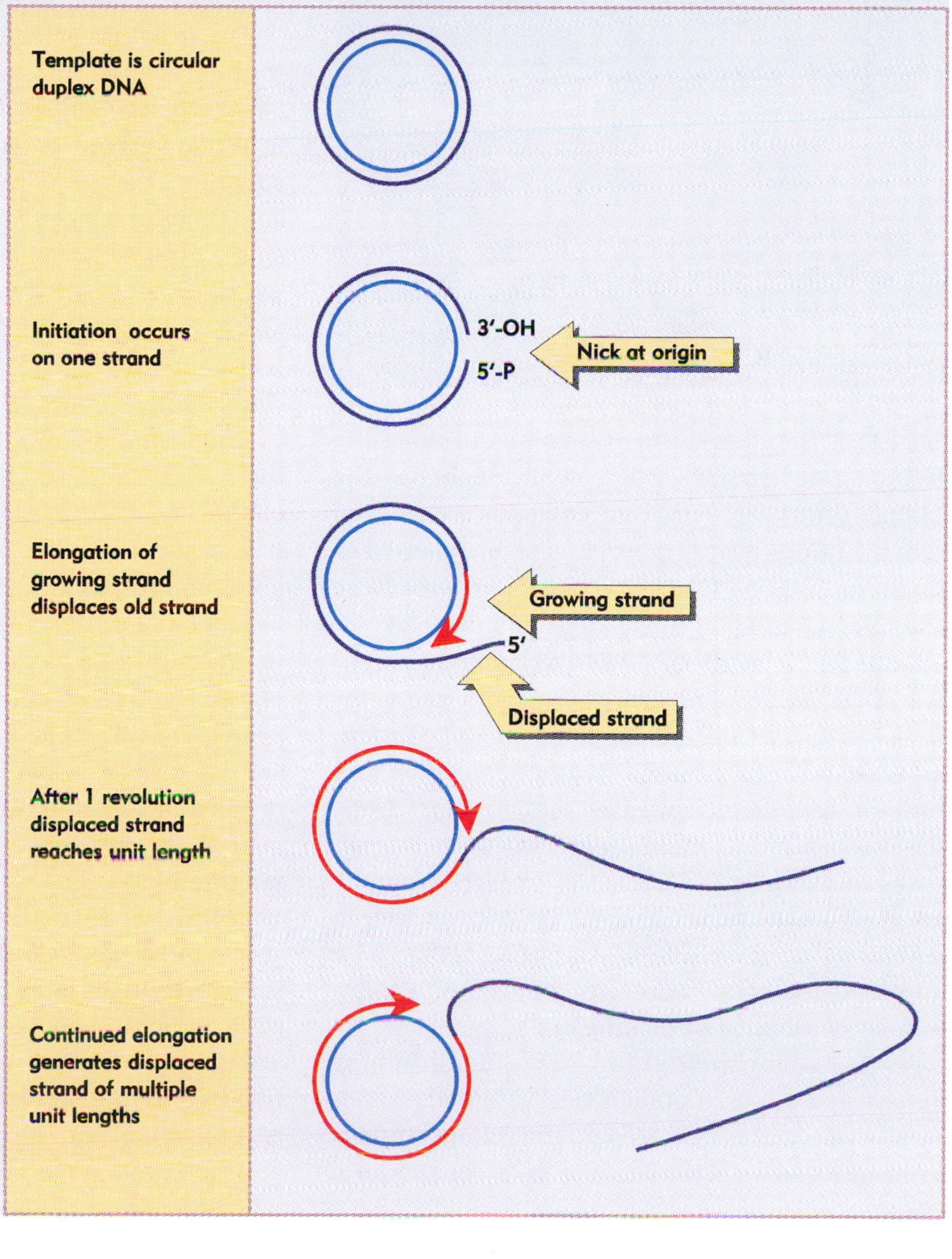

Template is circular duplex DNA

Initiation occurs on one strand

3'-OH
5'-P
Nick at origin

Elongation of growing strand displaces old strand

Growing strand
5'
Displaced strand

After 1 revolution displaced strand reaches unit length

Continued elongation generates displaced strand of multiple unit lengths

Figure 18.16

A rolling circle appears as a circular molecule with a linear tail by electron microscopy. Photograph kindly provided by David Dressler.

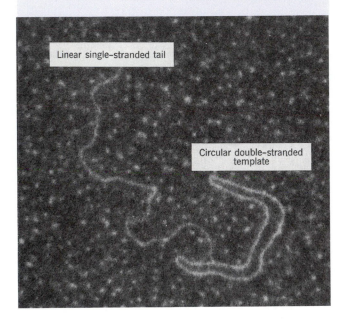

Linear single-stranded tail

Circular double-stranded template

consists of a tandem series of genomes. In effect, the rolling circle provides a means for amplifying the original (unit) replicon. Unit genomes are cleaved from the multimeric duplex tail and are inserted as linear molecules into the phage particles. The example of lambda is illustrated in Figure 27.3.

Rolling circles also are used to generate amplified rDNA in the *Xenopus* oocyte. The genes for rRNA are organized as a large number of contiguous repeats in the genome. A single repeating unit from the genome is converted into a rolling circle. The displaced tail, containing many units, is converted into duplex DNA; later it is cleaved from the circle so that the two ends can be joined together to generate a large circle of amplified rDNA. The amplified material therefore consists of a large number of identical repeating units.

Replication by rolling circles is common among phages, and a more detailed view of a phage replication cycle that is centered on the rolling circle is given in **Figure 18.18**. Phage φX174 consists of a

single-stranded circular DNA, known as the plus (+) strand. A complementary strand, called the minus (−) strand, is synthesized. This action generates the duplex circle shown at the top of the figure, which is then replicated by a rolling circle mechanism.

The duplex circle is converted to a covalently closed form, which becomes supercoiled. A protein coded by the phage genome, the A protein, nicks the (+) strand of the duplex DNA at a specific site that defines the origin for replication. After nicking the origin, the A protein remains connected to the 5′ end that it generates, while the 3′ end is extended by DNA polymerase.

The structure of the DNA plays an important role in this reaction, for the DNA can be nicked *only when it is supercoiled*. The A protein is able to bind to a single-stranded decamer fragment of DNA that surrounds the site of the nick. This suggests that the supercoiling is needed to assist the formation of a single-stranded region that provides the A protein with its binding site. The nick generates a 3′-OH end and a 5′-phosphate end (covalently attached to the A protein), both of which have roles to play in φX174 replication.

Employing the mechanism of the rolling circle, the 3′–OH end of the nick is extended into a new chain. The chain is elongated around the circular (−) strand template, until it reaches the starting point and displaces the origin. Now the A protein functions again. It actually remains connected with the rolling circle as well as to the 5′ end of the displaced tail, and it is therefore in the vicinity as the growing point returns past the origin. Thus the same A protein is available again to recognize the origin and nick it, now attaching to the end generated by the new nick. The cycle can be repeated indefinitely.

Following this nicking event, the displaced single (+) strand is freed as a circle. The A protein is involved in the circularization. In fact, the joining of the 3′ and 5′ ends of the (+) strand product is accomplished by the A protein as part of the reaction by which it is released at the end of one cycle of replication, and starts another cycle.

The A protein has an unusual property that may

Figure 18.17

Rolling circles can be used for varying purposes, depending on the fate of the displaced tail. Cleavage at unit length generates monomers, which can be converted to duplex and circular forms. Cleavage of multimers generates a series of tandemly repeated copies of the original unit. Note that the conversion to double-stranded form could occur earlier, before the tail is cleaved from the rolling circle.

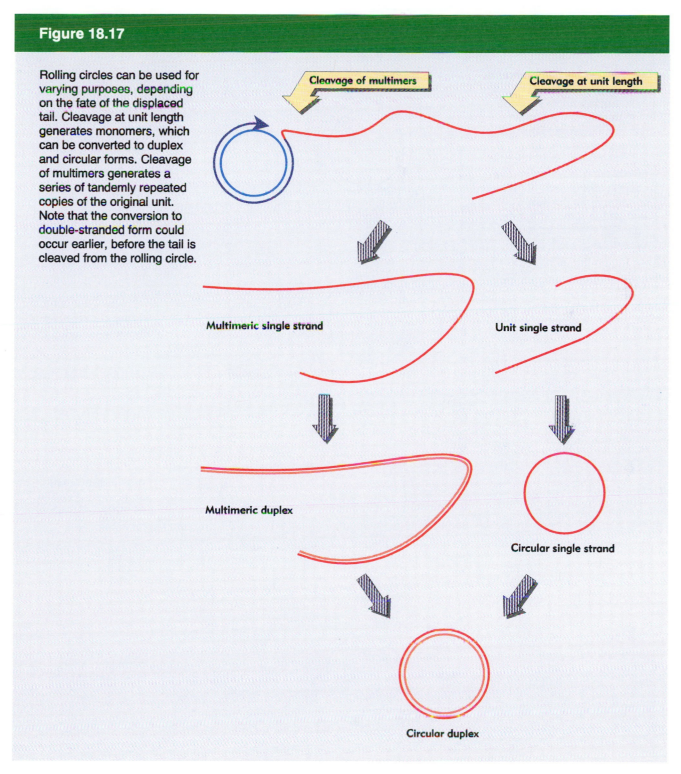

Cleavage of multimers

Cleavage at unit length

Multimeric single strand

Unit single strand

Multimeric duplex

Circular single strand

Circular duplex

be connected with these activities. It is *cis*-acting *in vivo*. (This behavior is not reproduced *in vitro*, as can be seen from its activity on any DNA template in a cell-free system.) *The implication is that* in vivo *the A protein synthesized by a particular genome can attach only to the DNA of that genome.* We do not know how this is accomplished. However, its activity *in vitro* shows how it remains associated

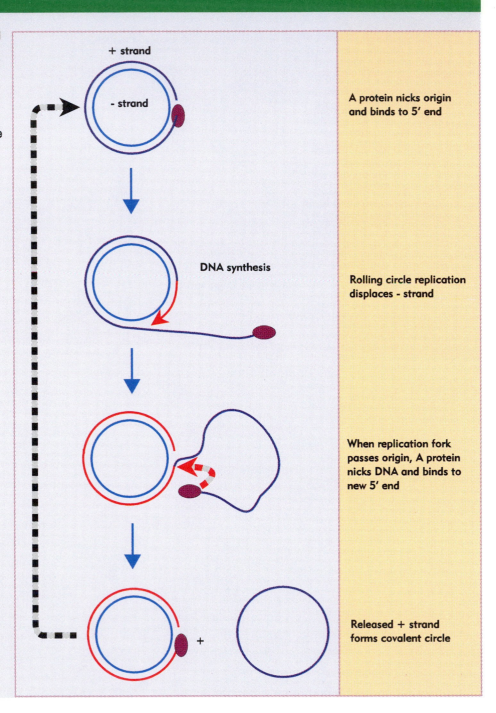

Figure 18.18

φX174 RF DNA can be used as a template for synthesizing single stranded viral circles. The A protein remains attached to the same genome through indefinite revolutions, each time nicking the origin on the viral (+) strand and transferring to the new 5' end. At the same time, the released viral strand is circularized.

+ strand

– strand

DNA synthesis

A protein nicks origin and binds to 5' end

Rolling circle replication displaces – strand

When replication fork passes origin, A protein nicks DNA and binds to new 5' end

+

Released + strand forms covalent circle

with the same parental (s) strand template. The A protein has two active sites; this may allow it to cleave the 'new' origin while still retaining the 'old' origin; then it ligates the displaced strand into a circle.

The displaced (+) strand may follow either of two fates after circularization. During the replication phase of viral infection, it may be used as a template to synthesize the complementary (–) strand. The duplex circle may then be used as a rolling circle to generate more progeny. During phage morphogenesis, the displaced (+) strand is packaged into the phage virion.

Single-stranded genomes are generated for bacterial conjugation

Another example of a connection between replication and the propagation of a genetic unit is provided by bacterial **conjugation**, in which a plasmid genome or host chromosome is transferred from one genome to another.

Conjugation is mediated by the **F plasmid**, which is the classic example of an episome, an element that may exist as a free circular plasmid, or that may become integrated into the bacterial chromosome as a linear sequence (like a lysogenic bacteriophage). The F plasmid is a large circular DNA, ~100 kb in length, and its genetic map is correspondingly described in terms of a system with coordinates from 0–100F. About 60 genes, occupying ~60% of the potential coding sequences, have been identified and mapped.

The F factor can integrate at several sites in the *E. coli* chromosome, often by a recombination event involving certain sequences (called IS sequences; see Chapter 34) that are present on both the host chromosome and F plasmid. In its free (plasmid) form, the F plasmid utilizes its own replication origin (*oriV*) and control system, and is maintained at a level of one copy per bacterial chromosome. When it is integrated into the bacterial chromosome, this system is suppressed, and F DNA is replicated as a part of the chromosome.

The presence of the F plasmid, whether free or integrated, has important consequences for the host bacterium. Bacteria that are F-positive are able to conjugate (or mate) with bacteria that are F-negative. Conjugation involves a contact between donor (F-positive) and recipient (F-negative) bacteria; contact is followed by transfer of the F factor. The F factor may be transferred as a free plasmid, which essentially constitutes an infective process in which the F-negative recipient is converted into an F-positive state. Or in its integrated form, the F plasmid may cause some or all of the bacterial chromosome to be transferred. Many plasmids have conjugation systems that operate in a gen-

erally similar manner, but the F factor was the first to be discovered, and remains the paradigm for this type of genetic transfer.

A large (~33 kb) region of the F plasmid, called the **transfer region**, is required for conjugation. It contains >25 genes that are required for the transmission of DNA; their organization is summarized in **Figure 18.19**. The genes are named as *tra* (and also *trb* loci), and the majority are expressed coordinately as part of a single 32 kb transcription unit (the *traY-I* unit). *traM* and *traJ* are expressed separately. *traJ* is a regulator that turns on both *traM* and *traY-I*. On the opposite strand, *finP* is a regulator that codes for a small antisense RNA that turns off *traJ*. Its activity requires expression of another gene, *finO*. Some of the *tra* genes in the major transcription unit are concerned directly with the transfer of DNA, but most are concerned with the properties of the bacterial cell surface.

F-positive bacteria possess surface appendages called **pili** that are coded by the F factor. The gene *traA* codes for the single subunit protein, **pilin**, which is polymerized into the pilus. At least 12 *tra* genes are required for the modification and assembly of pilin into the pilus. The F-pili are hair-like structures, 2-3 μm long, that protrude from the bacterial surface. A typical F-positive cell has 23 pili. The pilin subunits are polymerized into a hollow cylinder, ~8 nm in diameter, with a 2 nm axial hole.

Mating is initiated when the tip of the F-pilus contacts the surface of the recipient cell. **Figure 18.20** shows an example of *E. coli* cells beginning to mate. A donor cell does not contact other cells carrying the F factor, because the genes *traS* and *traT* code for 'surface exclusion' proteins that make the cell a poor recipient in such contacts. This effectively restricts donor cells to mating with F-negative cells. (Also note that the F-pili have been taken advantage of for other purposes; they provide the sites to which RNA phages and some single-stranded DNA phages attach, so F-positive bacteria are susceptible

Figure 18.19

The *tra* region of the F plasmid contains the genes needed for bacterial conjugation.

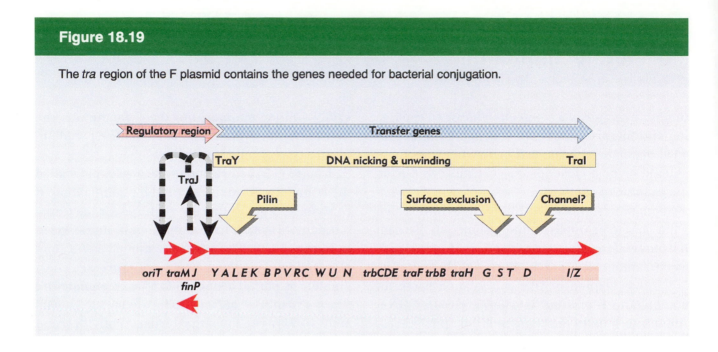

Figure 18.20

Mating bacteria are initially connected when the tops of donor *F pili* contact the recipient strand. Photograph kindly provided by Ron Skurray.

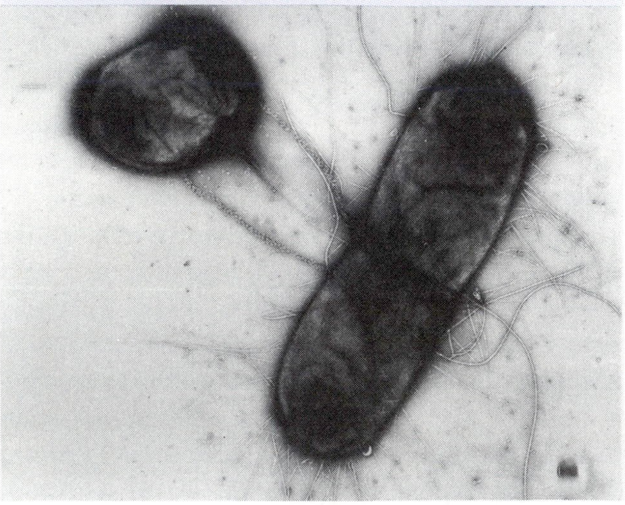

to infection by these phages, whereas F-negative bacteria are resistant.)

The initial contact between donor and recipient cells is easily broken, but other *tra* genes act to stabilize the association, bringing the mating cells closer together. The F pili are essential for initiating pairing, but retract or disassemble as part of the process by which the mating cells are brought into close contact. There must be a channel through which DNA is transferred, but the pilus does not appear to provide it; so far the channel remains unidentified, but TraD may provide it or be part of it.

Transfer of the F factor is initiated at a site *oriT*, the origin of transfer, which is located at one end of the transfer region. *OriT* is nicked on one strand by TraY and/or TraI, after which the proteins bind to the DNA as a large complex that causes ~200 bp of DNA to unwind. The TraY/TraI multimer then migrates along the DNA from the 5′ end, unwinding it at ~1200 bp/second. **Figure 18.21** shows that the freed 5′ end leads the way into the recipient bacterium. A complement for the transferred single strand is synthesized in the recipient bacterium, which as a result is converted to the F-positive state.

A complementary strand must be synthesized in the donor bacterium to replace the strand that has been transferred. If this happens concomitantly with the transfer process, the state of the F plasmid will resemble the rolling circle of Figure 18.15 (and will not generate the extensive single-stranded regions shown in Figure 18.21). A rolling circle

Figure 18.21

Transfer of DNA occurs when the F factor is nicked at *oriT* and a single strand is led by the 5' end into the recipient. Only one unit length is transferred. Complementary strands are synthesized to the single strand remaining in the donor and to the strand transferred into the recipient.

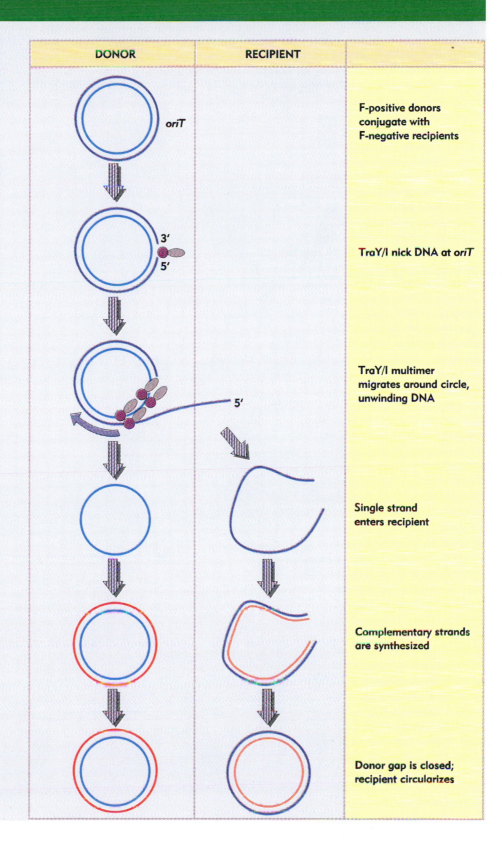

DONOR	RECIPIENT	
		F-positive donors conjugate with F-negative recipients
		TraY/I nick DNA at *oriT*
		TraY/I multimer migrates around circle, unwinding DNA
		Single strand enters recipient
		Complementary strands are synthesized
		Donor gap is closed; recipient circularizes

Figure 18.22

Transfer of chromosomal DNA occurs when an integrated F factor is nicked at *oriT* . Because the direction of transfer is away from the tra region, only a short sequence of F DNA enters the recipient. Transfer of DNA continues until prevented by loss of contact between the bacteria. Following synthesis of a complementary strand, the transferred material may recombine with the bacterial chromosome in the recipient.

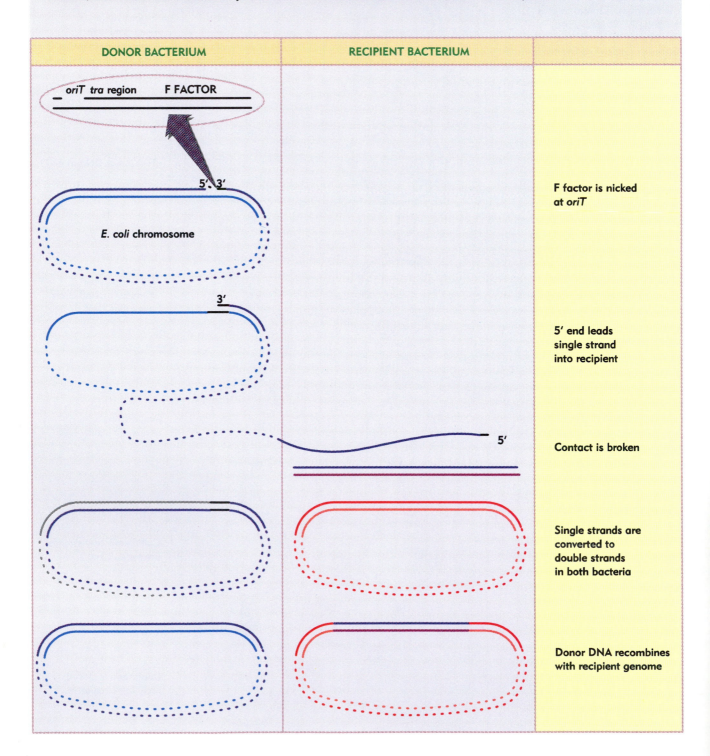

indeed represents the usual state of conjugating DNA. However, although the transferred strand is replaced by the rolling circle mechanism, replication as such is not necessary to provide the driving energy, and single-strand transfer is independent of DNA synthesis. A feature that we do not yet understand is that, even when rolling circle replication is occurring, only a single unit length of the F factor is transferred to the recipient bacterium. This implies that some (unidentified) feature terminates the process after one revolution, after which the covalent integrity of the F plasmid is restored.

When an integrated F plasmid initiates conjugation, the orientation of transfer is directed away from the transfer region, into the bacterial chromosome. **Figure 18.22** shows that, following a short leading sequence of F DNA, bacterial DNA is transferred. The process continues until it is interrupted by the breaking of contacts between the mating bacteria. It takes ~100 minutes to transfer the entire bacterial chromosome, and under standard conditions, contact is often broken before the completion of transfer. The result is that the probability that a region of the bacterial chromosome will be transferred depends upon its distance from *oriT*.

Donor DNA that enters a recipient bacterium is converted to double-stranded form, and may recombine with the recipient chromosome. (Note that two recombination events are required to insert the donor DNA.) Thus conjugation affords a means to exchange genetic material between bacteria (a contrast with their usual asexual growth). A strain of *E. coli* with an integrated F factor supports such recombination at relatively high frequencies (compared to strains that lack integrated F factors); such strains are described as **Hfr** (for high frequency recombination). Each position of integration for the F factor gives rise to a different Hfr strain, with a characteristic pattern of transferring bacterial markers to a recipient chromosome.

Bacterial markers located close to the site of F integration (in the direction of transfer) enter recipient bacteria first, and are therefore found at greater frequencies than those located farther away that enter later. This gives rise to a gradient of transfer frequencies around the chromosome, declining from the position of F integration. Marker positions on the donor chromosome can be assayed in terms of the time at which transfer occurs, and this gave rise to the standard description of the *E. coli* chromosome as a map divided into 100 minutes. The map refers to transfer times from a particular Hfr strain; the gradient of transfer is of course different for each Hfr strain.

Connecting bacterial replication to the cell cycle

Bacteria have two links between replication and cell growth:

◆ The frequency of initiation of cycles of replication is adjusted to fit the rate at which the cell is growing.

◆ The completion of a replication cycle is connected with division of the cell.

The rate of bacterial growth is assessed by the **doubling time**, the period required for the number of cells to double. The lower the doubling time, the faster the growth rate. *E. coli* cells can grow at rates ranging from doubling times as fast as 18 minutes to slower than 180 minutes. Because the bacterial chromosome is a single replicon, the frequency of replication cycles is controlled by the number of initiation events at the single origin. The replication cycle can be defined in terms of two constants:

◆ C is the fixed time of ~40 minutes required to replicate the entire bacterial chromosome. Its duration corresponds to a rate of replication fork movement of ~50,000 bp/minute. (The rate of DNA synthesis is more or less invariant at a

Figure 18.23

The fixed interval of 60 minutes between initiation of replication and cell division produces multiforked chromosomes in rapidly growing cells. Note that only the replication forks moving in one direction are shown; actually the chromosome is replicated symmetrically by two sets of forks moving in opposite directions.

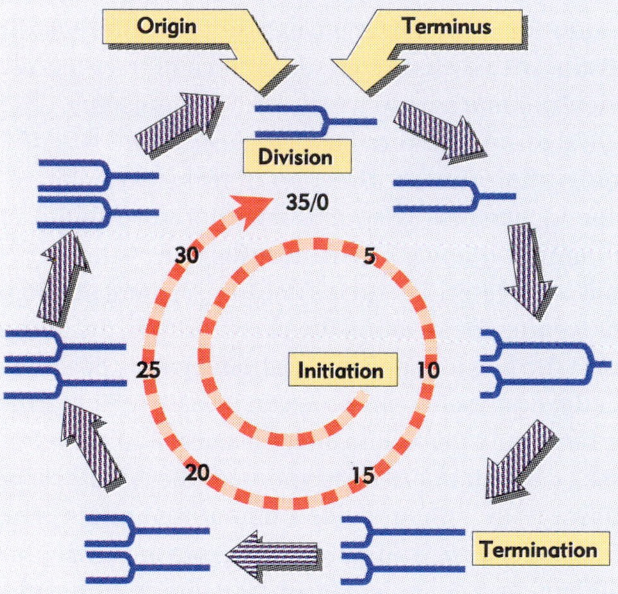

constant temperature; it proceeds at the same speed unless and until the supply of precursors becomes limiting.)

◆ D is the fixed time of ~20 minutes that elapses between the completion of a round of replication and the cell division with which it is connected. This period may represent the time required to assemble the components needed for division.

(The constants C and D can be viewed as representing the maximum speed with which the bacterium is capable of completing these processes; they apply for all growth rates between doubling times of 18 and 60 minutes, but both constant phases become longer when the cell cycle occupies >60 minutes.)

A cycle of chromosome replication must be initiated a fixed time before a cell division, C + D = 60 minutes. For bacteria dividing more frequently than every 60 minutes, a cycle of replication must be initiated *before the end of the preceding division cycle.*

Consider the example of cells dividing every 35 minutes. The cycle of replication connected with a division must have been initiated 25 minutes before the preceding division. This situation is illustrated in **Figure 18.23**, which shows the chromosomal complement of a bacterial cell at 5-minute intervals throughout the cycle.

At division (35/0 minutes), the cell receives a partially replicated chromosome. The replication fork continues to advance. At 10 minutes, when this 'old' replication fork has not yet reached the terminus, initiation occurs at both origins on the partially replicated chromosome. The start of these 'new' replication forks creates a **multiforked chromosome**.

At 15 minutes—that is, at 20 minutes before the next division—the old replication fork reaches the terminus. Its arrival allows the two daughter chromosomes to separate; each of them has already been partially replicated by the new replication forks (which now are the *only* replication forks). These forks continue to advance.

At the point of division, the two partially replicated chromosomes segregate. This recreates the point at which we started. The single replication fork becomes 'old', it terminates at 15 minutes, and 20 minutes later there is a division. We see that the initiation event occurs $1\frac{25}{35}$ cell cycles before the division event with which it is associated.

The general principle of the link between initiation and the cell cycle is that, *as cells grow more rapidly (the cycle is shorter), the initiation event occurs an increasing number of cycles before the related division.* There are correspondingly more chromosomes in the individual bacterium. This relationship can be viewed as the cell's response to its inability to reduce the periods of C and D to keep pace with the shorter cycle.

How does the cell know when to initiate the replication cycle? The initiation event occurs at a constant ratio of cell mass to the number of chromosome origins. Cells growing more rapidly are larger and possess a greater number of origins. In terms of Figure 18.23, it is at the point 10 minutes after division that the cell mass has increased sufficiently to support an initiation event at both available origins.

Two types of model have been proposed for titrating cell mass. An initiator protein could be synthesized continuously throughout the cell cycle; accumulation of a critical amount would trigger initiation. Or an inhibitor protein might be synthesized at a fixed point, and diluted below an effective level by the increase in cell volume. There is evidence to suggest that a titration model actually does regulate initiation, but the data do not distinguish between accumulation of an initiator and dilution of an inhibitor. Current thinking favors an initiator, which is consistent with evidence that protein synthesis is needed for the initiation event.

Whichever type of titration model applies, the growth of the bacterium can be described in terms of the **unit cell**, an entity 1.7 μm long. A bacterium contains one origin per unit cell; a rapidly growing cell with two origins will be 1.7-3.4 μm long. A topological link between the initiation event and the structure of the cell could take the form of a growth site, a physical entity in the cell that provides the only place at which initiation can occur. There should be one growth site per unit cell.

Cell division and chromosome segregation

Chromosome segregation in bacteria is especially interesting because the DNA is likely to be closely involved in the mechanism for partition, which furthermore may involve a relatively small number of components. (This contrasts with eukaryotic cells, in which segregation is achieved by the complex apparatus of mitosis.) The bacterial apparatus is quite accurate, however; anucleate cells form <0.03% of a bacterial population.

The division of a bacterium into two daughter cells is accomplished by the formation of a **septum**, a structure that forms in the center of the cell as an invagination from the surrounding envelope. The septum forms an impenetrable barrier between the two parts of the cell and provides the site at which the two daughter cells eventually separate entirely. Two related questions address the role of the septum in division: what determines the location at which it forms; and what ensures that the daughter chromosomes lie on opposite sides of it?

The formation of the septum is preceded by the organization of the **periseptal annulus**. This is observed as a zone in *E. coli* or *S. typhimurium* in which the structure of the envelope is altered so that the inner membrane is connected more closely to the cell wall and outer membrane layer. As its name suggests, the annulus extends around the cell. **Figure 18.24** illustrates its development.

The annulus is observed in a central position in a new cell. As the cell grows, two events occur. A septum forms at the midcell position defined by the annulus. And new annuli form on either side of the initial annulus. These new annuli are displaced from the center and move along the cell to positions at $\frac{1}{4}$ and $\frac{3}{4}$ of the cell length. These will become the midcell positions after the next division. We do not know how their displacement occurs; it might involve growth of envelope at the center to 'push' the annuli apart; or they might be 'pulled' toward the poles by some unknown mechanism. The displacement of the periseptal annulus to the correct position may be the crucial event that ensures the division of the cell into daughters of equal size.

The behavior of the periseptal annulus suggests that the mechanism for measuring position is associated with the cell envelope. It is plausible to suppose that the envelope could also be used to ensure segregation of the chromosomes. One model suggests that a direct link between DNA and the membrane could account for segregation. If

Figure 18.24

Duplication and displacement of the periseptal annulus gives rise to the formation of a septum that divides the cell.

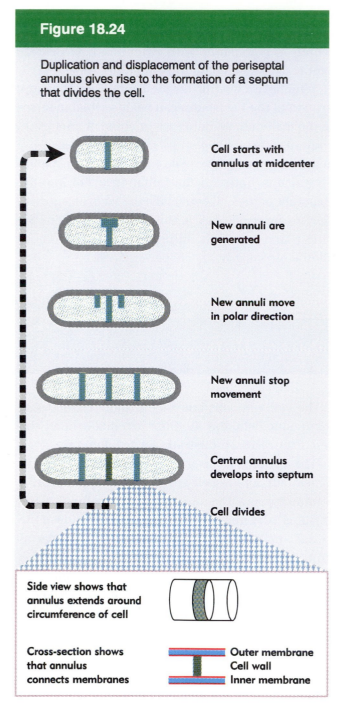

Cell starts with annulus at midcenter

New annuli are generated

New annuli move in polar direction

New annuli stop movement

Central annulus develops into septum

Cell divides

Side view shows that annulus extends around circumference of cell

Cross-section shows that annulus connects membranes

Outer membrane
Cell wall
Inner membrane

The isolation of mutants that are affected in cell division has been sporadic. One difficulty is that mutations in the critical functions may be lethal and/or pleiotropic. For example, if formation of the annulus occurs at a site that is essential for overall growth of the envelope, it would be difficult to distinguish mutations that specifically interfere with annulus formation from those that inhibit envelope growth generally.

Mutations that affect cell division or chromosome segregation cause rather striking phenotypic changes. **Figures 18.26** and **18.27** illustrate the opposite consequences of failure in segregation and failure in the division process:

Figure 18.25

Attachment of bacterial DNA to the membrane could provide a mechanism for segregation.

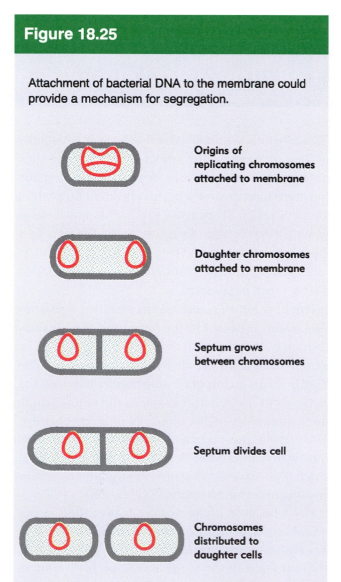

Origins of replicating chromosomes attached to membrane

Daughter chromosomes attached to membrane

Septum grows between chromosomes

Septum divides cell

Chromosomes distributed to daughter cells

daughter chromosomes are attached to the membrane, they could be physically separated as the membrane grows between them. **Figure 18.25** shows that the formation of a septum might then automatically segregate the chromosomes into the different daughter cells. This could be ensured, for example, if the origins were connected to sites that lie on either side of the periseptal annulus.

Figure 18.26

E. coli generate anucleolate cells when chromosome segregation fails. Cells with chromosomes stain blue; daughter cells lacking chromosomes have no blue stain. This field shows cells of the *mukB* mutant; both normal and abnormal divisions can be seen. Photograph kindly provided by Sota Hiraga.

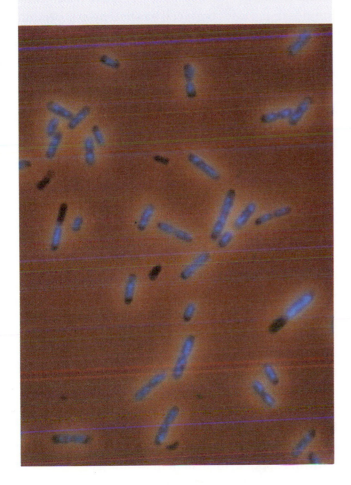

◆ **Minicells** form when septum formation occurs too frequently or in the wrong place, with the result that one of the new daughter cells lacks a chromosome. The minicell has a rather small size, and lacks DNA, but otherwise appears morphologically normal. **Anucleate** cells form when segregation is aberrant; like minicells, they lack a chromosome, but because septum formation is normal, their size is unaltered.

◆ Long **filaments** form when septum formation is inhibited, but chromosome replication is unaffected. The bacteria continue to grow, and even continue to segregate their daughter chromosomes, but septa do not form, so the cell consists of a very long multinucleated filamentous structure.

Mutants that affect cell division have been divided into two general groups (often revealed by the appearance of aberrant cells with one of the above phenotypes):

Figure 18.27

Failure of cell division generates multinucleated filaments. Photograph kindly provided by Sota Hiraga.

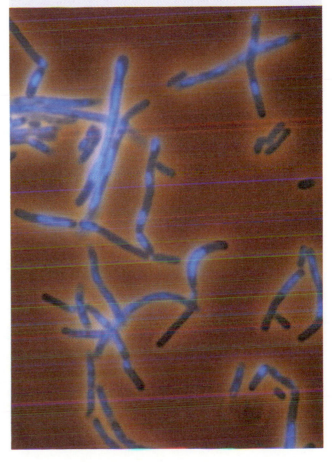

◆ *fts* mutants show *temperature-sensitive filamentation*. Usually the cells form long filaments without septa, but with the nucleoids (bacterial chromosomes) regularly distributed along the length of the cell. This suggests that the defect(s) lie in the division process itself.

◆ *par* (partition) mutants appear to be defective in chromosome segregation, because DNA is aberrantly distributed within the cells; anucleate cells are formed. The defects are likely to lie in the manipulation of DNA.

A candidate for the activator of cell division is the gene *ftsZ*, which is part of a large cluster of genes involved in growth and division. Mutations in *ftsZ* block septum formation and generate filaments; overexpression induces minicells, by causing an increased number of septation events per unit cell mass. The quantity of active FtsZ seems to be directly related to the frequency of cell division, which makes it a prime candidate for the role of division initiator protein.

Some *ftsZ* mutants are defective in septum formation; they act at stages varying from the displacement of the periseptal annuli to septal morphogenesis. However no mutations have yet been identified that affect the first stage, the generation of periseptal annuli. These would be the most interesting, since they would come closer to the question of how annulus formation is regulated, and whether it is connected directly with envelope growth and DNA segregation.

FtsZ has some interesting properties that suggest it could be involved with the early stages of septum formation. Early in the division cycle, FtsZ is localized throughout the cytoplasm. As the cell elongates and begins to constrict in the middle, FtsZ becomes localized in a ring around the circumference, essentially in the position of the midcenter annulus in Figure 18.24. Given the dependence of division on FtsZ concentration, the formation of this ring could be the rate-limiting step in septum formation. FtsZ has GTPase activity, so one possibility is that GTP cleavage is used to support the oligomerization of FtsZ monomers into the ring structure.

Information about the localization of the septum has been provided by **minicell mutants**. The original minicell mutation lies in the locus *minB*; deletion of *minB* generates minicells by allowing septation to occur at the poles as well as (or instead of) at midcell. This suggests that the cell possesses the ability to initiate septum formation either at midcell or at the poles; and the role of the wild-type *minB* locus is to suppress septation at the poles. In terms of the events depicted in Figure 18.24, this implies that a new-born cell has potential septation sites associated with both the annulus at midcenter and with the poles. One pole was formed from the septum of the previous division; the other pole represents the septum from the division before that. Perhaps the poles retain remnants of the annuli from which they were derived, and these remnants can nucleate septation.

The *minB* locus consists of three genes, *minC,D,* and *E*. Their roles are summarized in **Figure 18.28**. The products of *minC* and *minD* form (or are necessary for formation of) a division inhibitor. Expression of *minCD* in the absence of *minE*, or over expression even in the presence of *minE*, causes a generalized inhibition of division. The resulting cells grow as long filaments without septa. Expression of *minE* at levels comparable with *minCD* confines the inhibition to the polar regions, thus restoring normal growth. Thus MinE either protects the midcell sites from inhibition, or specifically directs the MinC/D inhibitor to the poles. Over expression of *minE* induces minicells, because the presence of excess MinE counteracts the inhibition at the poles as well as at midcell, allowing septa to form at both locations.

The important determinant of septation at the proper (midcell) site is therefore the ratio of MinC/D to MinE. The wild-type level prevents polar septation, while permitting midcell septation. The effects of MinC/D and MinE are inversely related; absence of minC/D or too much MinE allows indiscriminate septation, forming minicells; too much MinC/D or absence of MinE allows inhibition of midcell as well as polar sites, causing filamentation.

Partitioning is the process by which the two

◆ The two daughter chromosomes must be released from one another so that they can segregate following termination. This requires disentangling of DNA regions that are coiled around each other in the vicinity of the terminus. Most mutations affecting partitioning map in genes coding for topoisomerases—enzymes with the ability to pass DNA strands through one another. The mutations prevent the daughter chromosomes from segregating, with the result that the DNA is located in a single large mass at midcell. Septum formation then releases an anucleate cell and a cell containing both daughter chromosomes.

◆ Mutations that affect the partition process itself are rare. We expect to find two classes: *cis*-acting mutations in DNA sequences that are the targets for the partition process; and *trans*-acting mutations in genes that code for the protein(s) that cause segregation, which could include proteins that bind to DNA or activities that control the locations on the envelope to which DNA might be attached. (Both types of mutation have been found in the systems responsible for partitioning plasmids—see below—but only *trans*-acting functions have been found in the bacterial chromosome.)

The simplest form of the model for chromosome segregation shown in Figure 18.25 suggests that the envelope grows by insertion of material between the attachment sites of the two chromosomes. This idea is probably too simple, because envelope growth does not seem to be restricted to these sites; and the replicated chromosomes are capable of abrupt moves to their final positions at $\frac{1}{4}$ and $\frac{3}{4}$ cell length. If protein synthesis is inhibited before the termination of replication, the chromosomes fail to segregate and remain close to the midcell position. But when protein synthesis is allowed to resume, the chromosomes move to the quarter positions in the absence of any further envelope elongation. This suggests that there is an active process, requiring protein synthesis, that moves the chromosomes to specific locations.

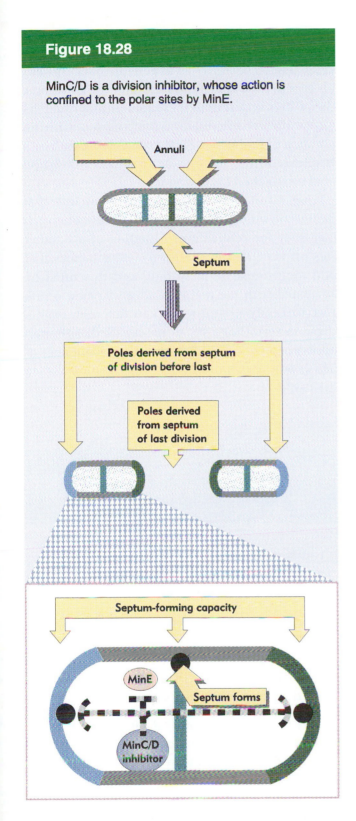

Figure 18.28

MinC/D is a division inhibitor, whose action is confined to the polar sites by MinE.

daughter chromosomes find themselves on either side of the position at which the septum forms. Two types of event are required for proper partitioning:

Segregation is interrupted by mutations of the *muk* class, which give rise to anucleate progeny at a much increased frequency: both daughter chromosomes remain on the same side of the septum instead of segregating. Mutations in the *muk* genes are not lethal, and may identify components of the apparatus that segregates the chromosomes. The gene *mukA* is identical with the gene for a known outer membrane protein (*tolC*), whose product could be involved with attaching the chromosome to the envelope. The gene *mukB* codes for a large (180,000 dalton) globular protein, which has some sequence relationship to the mechanochemical enzyme dynamin, which provides a 'motor' for microtubule-associated objects. This suggests the possibility that MukB is a motor that physically moves the chromosome relative to the envelope.

There have been suspicions for years that a physical link exists between bacterial DNA and the membrane, but the evidence remains indirect. Bacterial DNA can be found in membrane fractions, which tend to be enriched in genetic markers near the origin, the replication fork, and the terminus. The proteins present in these membrane fractions may be affected by mutations that interfere with the initiation of replication. The growth site could be a structure on the membrane to which the origin must be attached for initiation.

What criteria should be satisfied to conclude that membrane binding at the origin (or elsewhere) is necessary for replication? Mutation of *cis*-acting sites on DNA and in the *trans*-acting factors that bind them must be identified, and this must be correlated with the ability of these factors to link DNA (directly or indirectly) to the membrane. These criteria have yet to be satisfied, although more recently the first functional consequences of membrane association have been reported (see Chapter 19).

Multiple systems ensure plasmid survival in bacterial populations

When a segregation unit (a chromosome) consists of many replicons, the responsibility of the individual replicon may be confined to ensuring that its replication responds only once to an initiation signal during each replication cycle. But when a genome consists of a single replicon, responsibility for controlling the number of replication events also determines the number of progeny; and the replicon must also contain whatever system is responsible for segregating the progeny to daughter cells. Here we consider these systems and their connection with replication itself.

We have mentioned that each type of plasmid is maintained in its bacterial host at a characteristic **copy number**:

◆ **Single-copy** control systems resemble that of the bacterial chromosome and result in one replication per cell division. A single-copy plasmid effectively maintains parity with the bacterial chromosome.

◆ **Multicopy** control system allow multiple events per cell cycle, with the result that there are several copies of the plasmid per bacterium. Multicopy plasmids exist in a characteristic number (typically 10–20) per bacterial chromosome.

Copy number is primarily a consequence of the type of replication control mechanism. The system responsible for initiating replication determines how many origins can be present in the bacterium. Since each plasmid consists of a single replicon, the number of origins is the same as the number of plasmid molecules.

Single-copy plasmids have a system for replication control whose consequences are similar to

Figure 18.29

Intermolecular recombination merges monomers into dimers, and intramolecular recombination releases individual units from oligomers.

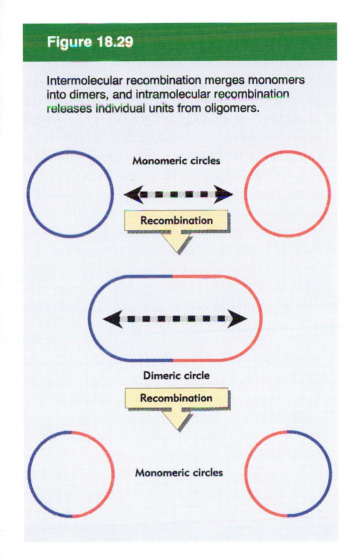

Monomeric circles

Recombination

Dimeric circle

Recombination

Monomeric circles

that governing the bacterial chromosome. A single origin can be replicated once; then the daughter origins are segregated to the different daughter cells.

Multicopy plasmids have a replication system that allows a pool of origins to exist. If the number is great enough (in practice >10 per bacterium), an active segregation system becomes unnecessary, because even a statistical distribution of plasmids to daughter cells will result in the loss of plasmids at frequencies <10^{-6}.

Plasmids are maintained in bacterial populations with very low rates of loss (<10^{-7} per cell division is typical, even for a single copy plasmid). The systems that control plasmid segregation can be identified by mutations that increase the frequency of loss, but that do not act upon replication itself. Several types of mechanism are used to ensure the survival of a plasmid in a bacterial population. It is common for a plasmid to carry several systems, often of different types, all acting independently to ensure its survival. Some of these systems may act indirectly, while others are concerned directly with regulating the partition event. However, in terms of evolution, all serve the same purpose: to help ensure perpetuation of the plasmid to the maximum number of progeny bacteria:

◆ Because the multiple copies of a plasmid in a bacterium consist of the same DNA sequences, they are able to recombine. **Figure 18.29** demonstrates the consequences. A single intermolecular recombination event between two circles generates a dimeric circle; further recombination can generate higher multimeric forms. Such an event reduces the number of physically segregating units. In the extreme case of a single copy plasmid that has just replicated, formation of a dimer by recombination means that the cell only has one unit to segregate, and the plasmid therefore must inevitably be lost from one daughter cell. To counteract this effect, plasmids often have **site-specific recombination systems** that act upon particular sequences to sponsor an *intramolecular recombination* that restores the monomeric condition. Mutations in these systems increase plasmid loss, and therefore have a phenotype that is similar to that of partition mutants.

◆ **Addiction systems**, operating on the basis that 'we hang together or we hang separately', ensure that a bacterium carrying a plasmid can survive *only* so long as it retains the plasmid. There are several ways to ensure that a cell dies if it is 'cured' of a plasmid, all sharing the principle illustrated in **Figure 18.30** that the plasmid produces both a poison and an antidote. The poison is a killer substance that is relatively stable, whereas the antidote consists of a substance that blocks killer action, but is relatively short-lived. When the plasmid is lost, the antidote decays, and then the killer substance causes death of the cell. So bacteria that lose

Figure 18.30

Plasmids may ensure that bacteria cannot live without them by synthesizing a long-lived killer and a short-lived antidote.

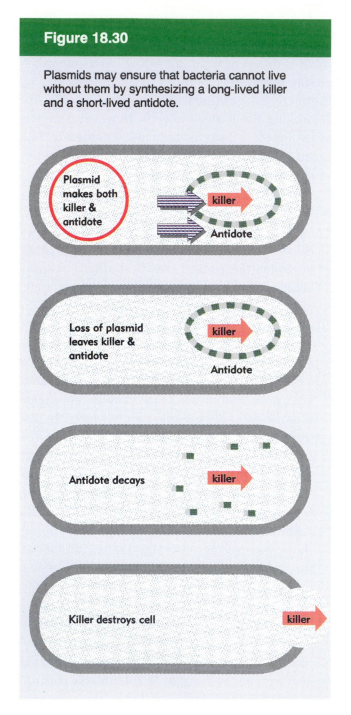

Plasmid makes both killer & antidote

killer

Antidote

Loss of plasmid leaves killer & antidote

killer

Antidote

Antidote decays

killer

Killer destroys cell

killer

the plasmid inevitably die, and the population is condemned to retain the plasmid indefinitely. These systems take various forms. One specified by the F plasmid consists of killer and blocking proteins; the plasmid R1 has a killer that is the mRNA for a toxic protein, while the antidote is a small antisense RNA that prevents expression of the mRNA.

◆ **True partition systems** act upon duplicate DNA molecules to ensure that they reside on either side of the septum at cell division. Probably all low copy number plasmids have such a system. Systems that have been characterized by the plasmids F, P1, and R1 have the generally similar organization depicted in **Figure 18.31**; there are two *trans*-acting proteins and a single *cis*-acting site. In the cases of both P1 and F, the smaller of the two proteins binds to the *cis*-acting site. In spite of their overall similarities, there are no significant sequence homologies between the corresponding genes or *cis*-acting sites.

How does a true partition system segregate replica plasmids to different daughter cells? We may imagine two general types of system:

Extrinsic structures to which the plasmids must bind are limiting. For a single-copy plasmid, such a model closely follows models for the segregation of the bacterial chromosome (see Figure 18.25). The cell must contain only two sites, and each copy of the plasmid binds to one. The sites could be membrane-bound or could simply constitute regions of the bacterial

Figure 18.31

Several plasmids contain partition systems that consist of adjacent genes and a *cis*-acting site.

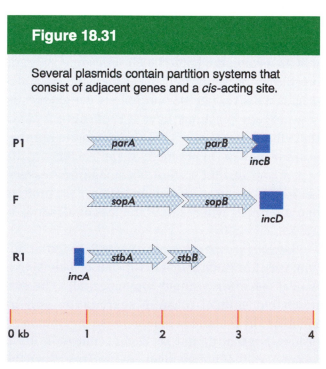

P1   *parA*   *parB*   *incB*

F   *sopA*   *sopB*   *incD*

R1   *incA*   *stbA*   *stbB*

0 kb   1   2   3   4

chromosome itself. A disadvantage of this type of model is that many different types of plasmids are found in bacterial populations, which makes it necessary to suppose that there are separate and exclusive pairs of sites for each type of plasmid.

◆ Segregation is an *intrinsic* process, in which the replica plasmids pass through some stage equivalent to chromosome pairing. Following association of plasmids into pairs, a partition mechanism moves one member of each pair into opposite daughter cells. The difficulty remains that the nature of this mechanism is not easy to envisage, and it is still necessary to explain how the cell can cope with partition of multiple types of plasmids.

Except for the fact that proteins required for partition are coded by the plasmid and bind to *cis*-acting sites, we know little about the mechanism of segregation. The importance to the plasmid of ensuring that all daughter cells gain replica plasmids is emphasized by the existence of multiple, independent systems in individual plasmids, to deal not only with partition *per se*, but also to handle events whose consequences are related to it.

Plasmid incompatibility is connected with copy number

The phenomenon of **plasmid incompatibility** is related to the regulation of plasmid copy number and segregation. A **compatibility group** is defined as a set of plasmids whose members are unable to coexist in the same bacterial cell. The reason for their incompatibility is that they cannot be distinguished from one another at some stage that is essential for plasmid maintenance. DNA replication and segregation are stages at which this may apply.

The negative control model for plasmid incompatibility follows the idea that copy number control is achieved by synthesizing a repressor that measures the concentration of origins. (Formally this is the same as the titration model for regulating replication of the bacterial chromosome.)

The introduction of a new origin in the form of a second plasmid of the same compatibility group mimics the result of replication of the resident plasmid; two origins now are present. Thus any further replication is prevented until after the two plasmids have been segregated to different cells to create the correct prereplication copy number as illustrated in **Figure 18.32**.

A similar effect would be produced if the system for segregating the products to daughter cells could not distinguish between two plasmids. For example, if two plasmids have the same *cis*-acting partition sites, competition between them would ensure that they would be segregated to different cells, and therefore could not survive in the same line.

Viewed in teleological terms, plasmids are selfish. Having obtained possession of a bacterium, the resident plasmid will seek to prevent any other plasmid of the same type from establishing residence. Plasmid incompatibility is a major device used to establish these territorial rights.

The presence of a member of one compatibility group does not directly affect the survival of a plasmid belonging to a different group. Only one replicon of a given compatibility group (of a single-copy plasmid) can be maintained in the bacterium, but it does not interact with replicons of other compatibility groups (although in limiting conditions they compete for *lebensraum*).

The best characterized copy number and incompatibility system is that of the plasmid ColE1, a multicopy plasmid that is maintained at a steady level of ~20 copies per *E. coli* cell. The system for maintaining the copy number depends on the mechanism for initiating replication at the ColE1 origin. The relevant events are illustrated in **Figure 18.33**.

Figure 18.32

Two plasmids are incompatible (they belong to the same compatibility group) if their origins cannot be distinguished at the stage of initiation. The same model could apply to segregation.

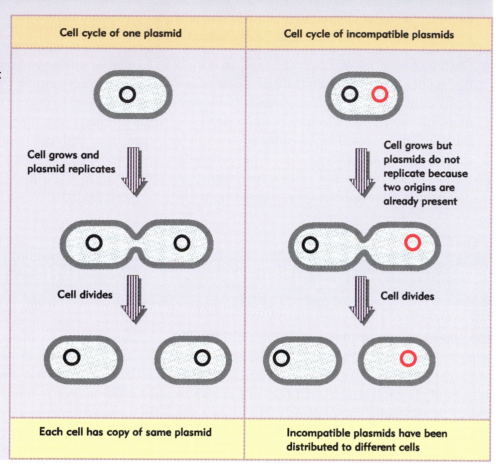

Cell cycle of one plasmid	Cell cycle of incompatible plasmids
Cell grows and plasmid replicates	Cell grows but plasmids do not replicate because two origins are already present
Cell divides	Cell divides
Each cell has copy of same plasmid	Incompatible plasmids have been distributed to different cells

Replication starts with the transcription of an RNA that initiates 555 bp upstream of the origin. Transcription continues through the origin. The enzyme RNAase H (whose name reflects its specificity for a substrate of RNA *hybridized with DNA*) cleaves the transcript at the origin. This generates a 3′-OH end that is used as the 'primer' at which DNA synthesis is initiated (the use of primers is discussed in more detail in Chapter 19).

Two regulatory systems exert their effects on the RNA primer. One involves synthesis of an RNA complementary to the primer; the other involves a protein coded by a nearby locus.

The regulatory species RNA I is a molecule of ~108 bases, coded by the opposite strand from that specifying primer RNA. The relationship between the primer RNA and RNA I is illustrated in **Figure 18.34**. The RNA I molecule is initiated within the primer region and terminates close to the site where the primer RNA initiates. Thus RNA I is complementary to the 5′-terminal region of the primer RNA. *Base pairing between the two RNAs controls the availability of the primer RNA to initiate a cycle of replication.*

An RNA molecule such as RNA I that functions by virtue of its complementarity with another RNA coded in the same region is called a **countertranscript.** This type of mechanism, of course, is another example of the use of antisense RNA (see Chapter 16).

Mutations that reduce or eliminate incompatibility between plasmids can be obtained by selecting plasmids of the same group for their ability to coexist. Incompatibility mutations in ColE1 map in the region of overlap between RNA I and primer RNA. Because this region is represented in two different RNAs, either or both might be involved in the effect.

When RNA I is added to a system for replicating ColE1 DNA *in vitro,* it inhibits the formation of active primer RNA. But the presence of RNA I does not inhibit the initiation or elongation of primer RNA synthesis. This suggests that *RNA I prevents RNAase H from generating the 3' end of the primer RNA.* The basis for this effect lies in base pairing between RNA I and primer RNA.

Both RNA molecules have the same potential secondary structure in this region, with three duplex hairpins terminating in single-stranded loops. Mutations reducing incompatibility are located in these loops, which suggests that the initial step in base pairing between RNA I and primer RNA is contact between the unpaired loops.

How does pairing with RNA I prevent cleavage to form primer RNA? A model is illustrated in **Figure 18.35.** In the absence of RNA I, the primer RNA forms its own secondary structure (involving loops and stems). But when RNA I is present, the two molecules pair, and become completely double-stranded for the entire length of RNA I. The

new secondary structure prevents the cleavage reaction (by indirect means, since the actual site of cleavage lies beyond the paired region).

The model is reminiscent of the mechanism involved in attenuation of transcription, in which the alternative pairings of an RNA sequence permit or prevent formation of the secondary structure needed for termination by RNA polymerase (see Chapter 16). The action of RNA I is exercised by its ability to alter the secondary structure of distant regions of the primer precursor.

Formally, the model is equivalent to postulating a control circuit involving two RNA species. A large RNA primer precursor is a positive regulator, needed to initiate replication. The small RNA I is a negative regulator, able to inhibit the action of the positive regulator.

In its ability to act on any plasmid present in the cell, RNA I provides a repressor that prevents newly introduced DNA from functioning, analogous to the role of the lambda lysogenic repressor (see Chapter 15). Instead of a repressor protein that binds the

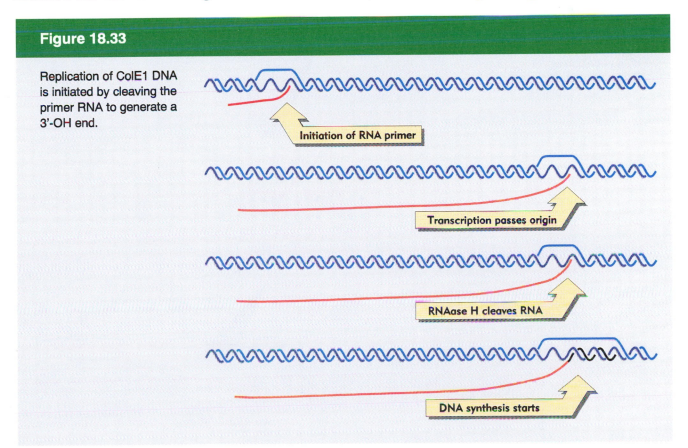

Figure 18.33

Replication of ColE1 DNA is initiated by cleaving the primer RNA to generate a 3'-OH end.

Initiation of RNA primer

Transcription passes origin

RNAase H cleaves RNA

DNA synthesis starts

Figure 18.34

The sequence of RNA I is complementary to the 5' region of primer RNA.

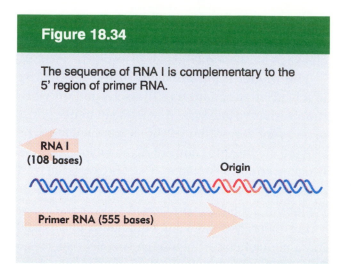

RNA I
(108 bases)

Origin

Primer RNA (555 bases)

new DNA, an RNA binds the newly synthesized precursor to the RNA primer.

Binding between RNA I and primer RNA can be influenced by the Rom protein, coded by a gene located downstream of the origin. The Rom protein enhances binding between RNA I and primer RNA transcripts of >200 bases. The result is to inhibit formation of the primer.

How do mutations in the RNAs affect incompatibility? **Figure 18.36** shows the situation when a cell contains two types of RNA I/primer RNA sequence. The RNA I and primer RNA made from each type of genome can interact, but RNA I from one genome

Figure 18.35

Base pairing with RNA I may change the secondary structure of the primer RNA sequence and thus prevent cleavage from generating a 3'-OH end. (We do not understand the basis for the inhibitory effect. The action of RNAase H has been thought to depend on recognition of RNA-DNA hybrid regions, not on RNA structure.)

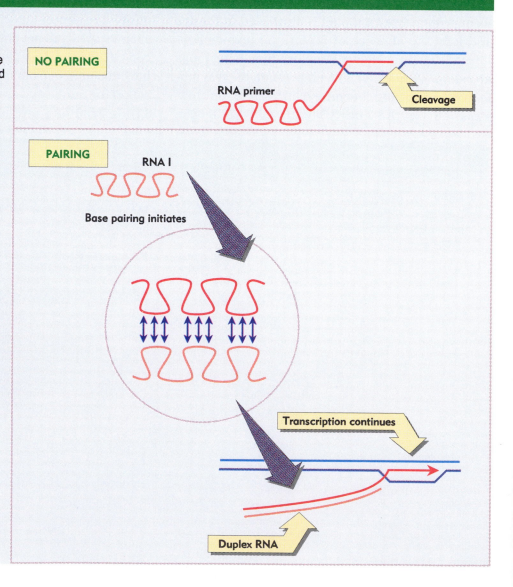

NO PAIRING

RNA primer

Cleavage

PAIRING

RNA I

Base pairing initiates

Transcription continues

Duplex RNA

Figure 18.36

Mutations in the region coding for RNA I and the primer precursor need not affect their ability to pair; but they may prevent pairing with the complementary RNA coded by a different plasmid.

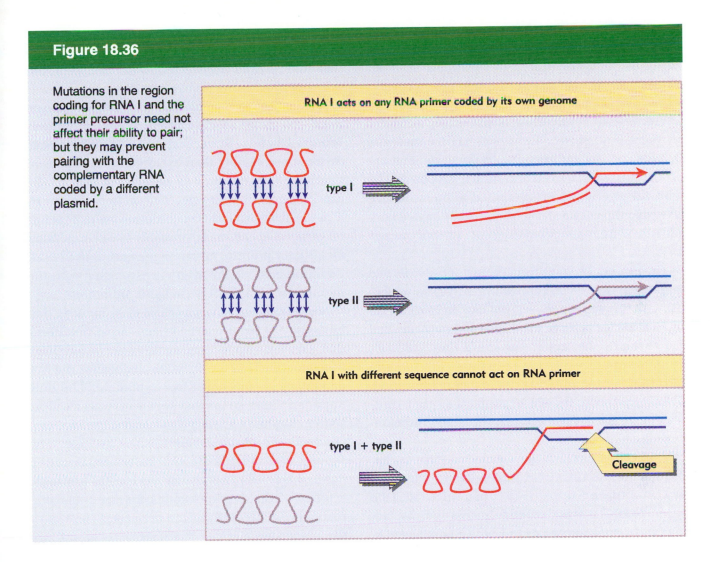

RNA I acts on any RNA primer coded by its own genome

type I

type II

RNA I with different sequence cannot act on RNA primer

type I + type II

Cleavage

does not interact with primer RNA from the other genome. This situation would arise when a mutation in the region that is common to RNA I and primer RNA occurred at a location that is involved in the base pairing between them. Each RNA I would continue to pair with the primer RNA coded by the same plasmid, but might be unable to pair with the primer RNA coded by the other plasmid. This would cause the original and the mutant plasmids to behave as members of different compatibility groups.

Summary

The entire chromosome is replicated once for every cell division cycle. Initiation of replication commits the cell to a cycle of division; completion of replication may provide a trigger for the actual division process. The bacterial chromosome consists of a single replicon, but a eukaryotic chromosome is divided into many replicons that function over the protracted period of S phase. The problem of replicating the ends of a linear replicon is solved in a variety of ways, most often by converting the replicon to a circular form. Some viruses have special proteins that recognize ends. Eukaryotic chromosomes encounter the problem at their terminal replicons, where the telomere provides a special structure.

Eukaryotic replication is (at least) an order of magnitude slower than bacterial replication. Origins sponsor bidirectional replication, and are probably used in a fixed order during S phase. The only eukaryotic origins identified at the sequence level are those of *S. cerevisiae,* which have a consensus sequence consisting of 11 bp, mostly A•T.

The minimal *E. coli* origin consists of ~245 bp and initiates bidirectional replication. Any DNA molecule with this sequence can replicate in *E. coli.* Two replication forks leave the origin and move around the chromosome, apparently until they meet, although sequences that would cause the forks to terminate after meeting have been identified. Transcription units are organized so that transcription usually proceeds in the same direction as replication.

The rolling circle is an alternative form of replication for circular DNA molecules in which an origin is nicked to provide a priming end. One strand of DNA is synthesized from this end, displacing the original partner strand, which is extruded as a tail. Multiple genomes can be produced by continuing revolutions of the circle.

Rolling circles are used to replicate some phages. The A protein that nicks the ϕX174 origin has the unusual property of *cis*-action. It acts only on the DNA from which it was synthesized. It remains attached to the displaced strand until an entire strand has been synthesized, and then nicks the origin again, releasing the displaced strand and starting another cycle of replication.

Rolling circles also are involved in bacterial conjugation, when an F plasmid is transferred from a donor to a recipient cell, following the initiation of contact between the cells by means of the F-pili. A free F plasmid infects new cells by this means; an integrated F factor creates an Hfr strain that may transfer chromosomal DNA. In the case of conjuga-

tion, replication is used to synthesize complements to the single strand remaining in the donor and to the single strand transferred to the recipient, but does not provide the motive power.

A fixed time of 40 minutes is required to replicate the *E. coli* chromosome and a further 20 minutes is required before the cell can divide. When cells divide more rapidly than every 60 minutes, a replication cycle is initiated before the end of the preceding division cycle. This generates multiforked chromosomes. The initiation event depends on titration of cell mass, probably by accumulating an initiator protein. Initiation may occur at the cell membrane, since the origin is associated with the membrane for a short period after initiation.

Segregation involves additional sequences that have not yet been characterized. Most mutations that prevent proper segregation lie in ancillary functions, such as the enzymes needed for the two daughter chromosomes to disentangle. The only mutation found so far that seems directly to affect segregation may identify a motor or a protein that connects the chromosome to the envelope. The septum that divides the cell grows at a location defined by the pre-existing periseptal annulus; a locus of three genes codes for products that regulate whether the midcell periseptal annulus or the polar sites derived from previous annuli are used for septum formation. Absence of septum formation generates multinucleated filaments; excess of septum formation generates anucleate minicells.

Plasmids have a variety of systems that ensure or assist partition, and an individual plasmid may carry systems of several types. The copy number of a plasmid describes whether it is present at the same level as the bacterial chromosome (one per unit cell) or in greater numbers. Plasmid incompatibility can be a consequence of the mechanisms involved in either replication or partition (for single-copy plasmids). Two plasmids that share the same control system for replication are incompatible because the number of replication events ensures that there is only one plasmid for each bacterial genome.

Further reading

Reviews

Mitochondrial DNA replication has been reviewed by **Clayton** (*Cell* **28**, 693–705, 1982; *Ann. Rev. Cell Biol.* **7**, 453–478, 1991).

The control of plasmid regulation has been reviewed by **Scott** (*Microbiol. Rev.* **48**, 1–23, 1984).

The use of origins in yeast has been reviewed by **Umek et al.** (*Biochim. Biophys. Acta* **1007**, 114, 1989) and **Fangman and Brewer** (*Ann. Rev. Cell Biol.* **7**, 375–402, 1991). Other eukaryotic origins have been considered by **De Pamphlis** (*Ann. Rev. Biochem.* **62**, 29–63, 1993). The connection between the directions of replication and transcription was established by **Brewer** (*Cell* **53**, 679–686, 1988).

The act of bacterial cell division has been reviewed by **De Boer, Cook, and Rothfield** (*Ann. Rev. Genet.* **24**, 249–274, 1990). Chromosome segregation was reviewed by **Hiraga** (*Ann. Rev. Biochem.* **61**, 283–306, 1992).

The F factor and bacterial conjugation have been reviewed by **Ippen-Ihler and Minkley** (*Ann. Rev. Genet.* **20**, 493–624, 1986) and **Willetts and Skurray** (in *E. coli* and *S. typhimurium*, Ed. F. C. Neidhart, American Society for Microbiology, Washington DC, pp. 1110–1131, 1987).

Mechanisms for plasmid partition and incompatibility have been unified by **Nordstrom and Austin** (*Ann. Rev. Genet.* **23**, 37–69, 1989).

Discoveries

The model for the bacterial replicon was proposed by **Jacob, Brenner, and Cuzin** (*Cold Spring Harbor Symp. Quant. Biol.* **28**, 329–348, 1963).

The basis for using mutations to analyze cell division was adumbrated by **Hirota, Ryter and Jacob** (*Cold Spring Harbor Symp. Quant. Biol.* **33**, 677–693, 1968).

Minicells were discovered by **Adler et al.** (*Proc. Nat. Acad. Sci.* **57**, 321–326, 1967).

An origin was isolated by **Zyskind and Smith** (*Proc. Nat. Acad. Sci.* **77**, 2460–2464, 1980). A yeast *ARS* was analyzed systematically by **Marahrens and Stillman** (*Science* **255**, 817–823, 1992).

The rolling circle model was proposed by **Gilbert and Dressler** (*Cold Spring Harbor Symp. Quant. Biol.* **33**, 473–484, 1968).

The ColE1 system was elucidated by **Tomizawa and Itoh** (*Proc. Nat. Acad. Sci.* **78**, 6096–6100, 1981).

CHAPTER 19

Primosomes and replisomes: the apparatus for DNA replication

Replication of duplex DNA is a complex endeavor involving a conglomerate of enzyme activities. Different activities are involved in the stages of initiation, elongation, and termination.

◆ Initiation involves recognition of an origin by a complex of proteins. Several events occur at the origin. Before DNA synthesis begins, the parental strands must be separated and (transiently) stabilized in the single-stranded state. Then synthesis of daughter strands can be initiated at the replication fork. The act of initiating synthesis of a DNA strand is accomplished by a protein complex called the **primosome** in E. coli.

◆ Elongation is undertaken by another complex of proteins. The **replisome** is not evident as an independent unit (for example, analogous to the ribosome), but is assembled from its components at the onset of replication. *The replisome exists only as a protein complex associated with the particular structure that DNA takes at the replication fork.* As the replisome moves along DNA, the parental strands unwind and daughter strands are synthesized.

◆ At the end of the replicon, joining and/or termination reactions are necessary. We know least about the enzymatic mechanisms involved in termination. Following termination, the duplicate chromosomes must be separated from one another, which requires manipulation of higher order DNA structure.

All of these individual activities must be undertaken within the constraints imposed by the complex topology of DNA.

Inability to replicate DNA is fatal for a growing cell. Mutants in replication must therefore be obtained as conditional lethals, able to accomplish replication under permissive conditions (provided by the normal temperature of incubation), and displaying their defect only under nonpermissive conditions (provided by the higher temperature of 42°C). A comprehensive series of such temperature-sensitive mutants in E. coli identifies a set of loci called the *dna* genes. The *dna* mutants are divided into two general classes on the basis of their behavior when the temperature is raised. The major class of **quick-stop mutants** cease replication immediately on a temperature rise. Those concerned directly with elongation are defective in the components of the replication apparatus. The smaller class of **slow-stop mutants** complete the current round of replication, but cannot start another. They are defective in the events involved in initiating a cycle of replication at the origin.

Some quick-stop mutants are defective in producing the precursors needed for DNA synthesis.

For example, mutations in *nrdA* were originally called *dnaF*. But the gene codes for a subunit of ribonucleotide reductase, which is essential for providing precursors. This illustrates the importance of the precursor supply chain. The isolation of a mutation with the Dna phenotype therefore does not imply that its gene product is directly involved in replication.

An important assay used to identify the components of the replication apparatus is called *in vitro* **complementation**. An *in vitro* system for replication is fractionated into components that are purified and reconstituted. Then the system is prepared from a *dna* mutant and operated under conditions in which the mutant gene product is inactive. This allows extracts from wild-type cells to be tested for their ability to restore activity; and ultimately the protein coded by the *dna* locus can be purified by this assay.

We have now virtually reached the stage at which we can define a replication apparatus in which each component is available for study *in vitro* as a biochemically pure product, while it is implicated *in vivo* by the effect of mutations in its gene.

With eukaryotes, we have only more recently been able to use completely purified replication systems, but progress has been made in obtaining *in vitro* systems that can replicate defined templates. And some nuclear systems have been obtained that contain all of the components necessary to initiate and continue replication fork movement.

DNA polymerases: the enzymes that make DNA

An enzyme that can synthesize a new DNA strand on a template strand is called a **DNA polymerase**. Both prokaryotic and eukaryotic cells contain multiple DNA polymerase activities. Only some of these enzymes actually undertake replication; sometimes they are called **DNA replicases**. The others are involved in subsidiary roles in replication and/or participate in 'repair' synthesis of DNA to replace damaged sequences (see Chapter 20).

It is convenient to think of DNA-synthesizing activities in terms of the DNA polymerase enzymes; but it is a moot point whether a DNA replicase enzyme exists as a discrete entity. (We consider the same issue for RNA polymerase in Chapter 14.) Bacterial DNA replicase activity is recovered as aggregates containing various 'subunits.' It is not clear which subunits should be considered as components of the replicase *per se* and which constitute accessory factors. The DNA-synthesizing activity is only one of several functions associated in the replisome; and perhaps we should not be unduly influenced to think of it as a separate function because of our ability to assay its particular step in the overall process.

Three DNA polymerase enzymes have been characterized in *E. coli*. DNA polymerase I is involved in the repair of damaged DNA and, in a subsidiary role, in semiconservative replication. DNA polymerase II is also implicated in repair. DNA polymerase III, a multisubunit protein, is the replicase responsible for *de novo* synthesis of new strands of DNA. The enzymatic activities of the three enzymes are summarized in **Table 19.1**.

All prokaryotic and eukaryotic DNA polymerases share the same fundamental type of synthetic activity. Each can extend a DNA chain by adding nucleotides one at a time to a 3'-OH end, as illustrated diagrammatically in **Figure 19.1**. The choice of the nucleotide to add to the chain is dictated by base pairing with the template strand.

The fidelity of replication poses the same sort of

Table 19.1

E. coli has three DNA polymerases.

	DNA Polymerase I	DNA Polymerase II	DNA Polymerase III
Structure			
Daltons	109,000	90,000	ˉ900,000
Constitution	Monomer	Monomer	Heteromultimeric; see Figure 19.11
Number/cell	400	Not known	10-20
Enzymatic activities			
5'-3' elongation	Yes	Yes	Yes
3'-5' exonuclease	Yes	Yes	Yes
5'-3' exonuclease	Yes	No	No
Mutants			
Mutant loci	polA	polB	dnaE (polC), dnaN, dnaX, dnaQ, dnaT
Mutant phenotype	Defective in repair	Repair	Prevents replication
Lethality	Viability reduced only when 5'-3' exonuclease affected	No effect	Conditional lethal

problem we have encountered already in considering (for example) the accuracy of translation. It relies on the specificity of base pairing. Yet when we consider the interactions involved in base pairing, we would expect errors to occur with a frequency of $\sim 10^{-3}$ per base pair replicated. The actual rate in bacteria seems to be $\sim 10^{-8}$-10^{-10}. This corresponds to ~ 1 error per genome per 1000 bacterial replication cycles, or $\sim 10^{-6}$ per gene per generation.

DNA polymerase might improve the specificity of complementary base selection at either (or both) of two stages:

◆ It could scrutinize the incoming base for the proper complementarity with the template base; for example, by specifically recognizing matching chemical features. This would be a **presynthetic** error control.

◆ Or it could scrutinize the base pair *after* the new base has been added to the chain, and, in those cases in which a mistake has been made, remove the most recently added base. This would be a **proofreading** control.

All of the bacterial enzymes possess a 3'–5' exonucleolytic activity that proceeds in the reverse direction from DNA synthesis. This provides the proofreading function illustrated diagrammatically in **Figure 19.2**. In the chain elongation step, a precursor nucleotide enters the position at the end of the growing chain. A bond is formed. The enzyme moves one base pair farther, ready for the next precursor nucleotide to enter. If a mistake has been made, the enzyme moves backward, excising the last base that was added, and creating a site for a replacement precursor nucleotide to enter.

Different DNA polymerases handle the relationship between the polymerizing and proofreading activities in different ways. In some cases, the activities are part of the same protein subunit, but in others they are contained in different subunits. The removal of a mismatched base is actually more complex than drawn in Figure 19.2, since the excision reaction is catalyzed by a different active site from that involved in the addition (see Table 19.2).

The first enzyme to be characterized was DNA

Figure 19.1

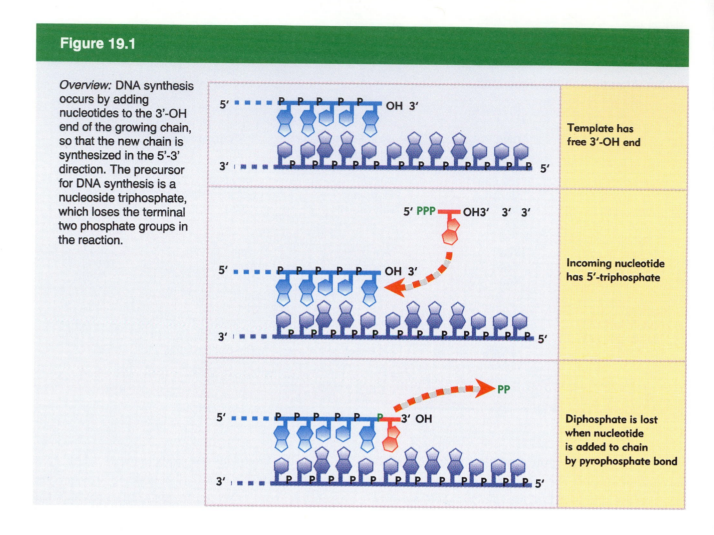

Overview: DNA synthesis occurs by adding nucleotides to the 3'-OH end of the growing chain, so that the new chain is synthesized in the 5'-3' direction. The precursor for DNA synthesis is a nucleoside triphosphate, which loses the terminal two phosphate groups in the reaction.

Template has free 3'-OH end

Incoming nucleotide has 5'-triphosphate

Diphosphate is lost when nucleotide is added to chain by pyrophosphate bond

polymerase I, which is a talented enzyme with several activities. Coded by the locus *polA*, it is a single polypeptide of 103,000 daltons. The chain can be cleaved into two regions by proteolytic treatment.

The larger cleavage product (68,000 daltons) is called the Klenow fragment. It is used in synthetic reactions *in vitro.* It contains the polymerase and the 3'–5' exonuclease activities, which reside in different regions of the fragment. The C-terminal two-thirds of the protein contains the polymerase active site, while the N-terminal third contains the proofreading exonuclease. The active sites appear to be ~30Å apart in the protein, indicating that there is spatial separation between adding a base

and removing one. Mutations that eliminate the proofreading activity increase the error rate ~10×.

The small fragment (35,000 daltons) possesses a 5'–3' exonucleolytic activity, which excises small groups of nucleotides, up to ~10 bases at a time. This activity is coordinated with the synthetic/proofreading activity. It provides DNA polymerase I with a unique ability to start replication *in vitro* at a nick in DNA. At a point where a phosphodiester bond has been broken in a double-stranded DNA, the enzyme extends the 3'-OH end. As the new segment of DNA is synthesized, it *displaces the existing homologous strand in the duplex.*

This process of **nick translation** is illustrated in

Figure 19.2

Bacterial DNA polymerases scrutinize the base pair at the end of the growing chain and excise the nucleotide added in the case of a misfit.

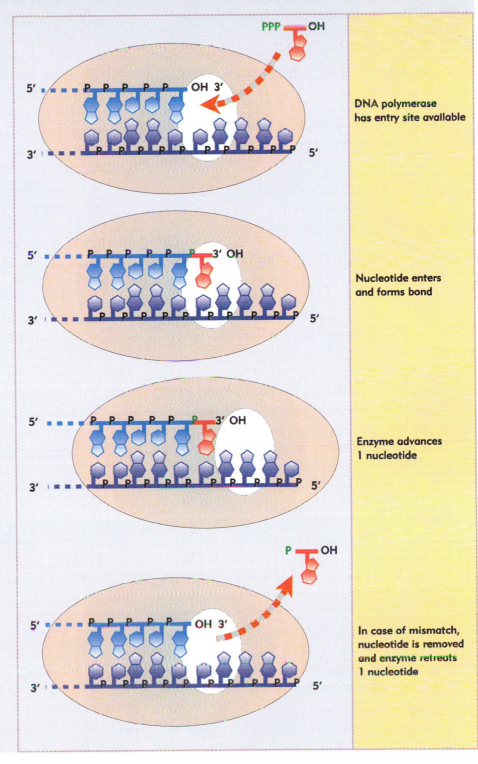

DNA polymerase has entry site available

Nucleotide enters and forms bond

Enzyme advances 1 nucleotide

In case of mismatch, nucleotide is removed and enzyme retreats 1 nucleotide

Figure 19.3

Nick translation replaces part of a pre-existing strand of duplex DNA with newly synthesized material.

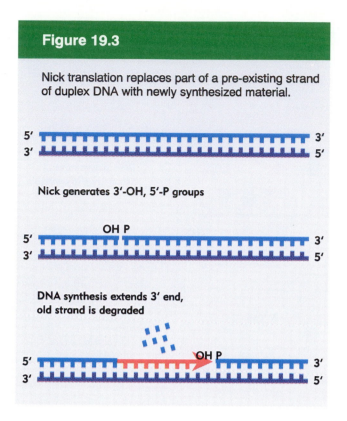

Nick generates 3'-OH, 5'-P groups

DNA synthesis extends 3' end, old strand is degraded

Figure 19.3. The displaced strand is degraded by the 5'–3' exonucleolytic activity of the enzyme. The properties of the DNA are unaltered, except that a segment of one strand has been replaced with newly synthesized material, and the position of the nick has been moved along the duplex. This is of great practical use; nick translation has been a major technique for introducing radioactively labeled nucleotides into DNA *in vitro*.

Although there is a large amount of DNA polymerase I in the bacterium, it is not responsible for replication, which continues in *polA* mutants. The 5'–3' synthetic/3'–5' exonucleolytic action is probably used *in vivo* mostly for filling in short single-stranded regions in double-stranded DNA. These regions arise during replication, and also when bases that have been damaged are removed from DNA (see Chapter 20).

When extracts of *E. coli* are assayed for their ability to synthesize DNA, the predominant enzyme activity is that of DNA polymerase I. In fact, its activity is so great that it makes it impossible to detect the activities of the enzymes actually responsible for DNA replication! To develop *in vitro* systems in which replication can be followed, extracts are therefore prepared from *polA* mutant cells.

The replicase is DNA polymerase III. Purified DNA polymerase III activities have several subunits. We use the plural 'activities' because the enzyme has been obtained in several forms, which we discuss later. The replicase activity was originally discovered by a lethal mutation in the *polC* locus, now renamed *dnaE*, which codes for

Table 19.2

Proofreading reduces the errors made by DNA polymerases.

Enzyme	Synthetic domain	Proofreading domain	Error rate − proof	Error rate + proof
E. coli DNA polymerase I	aa 200-600	N-terminal	10^{-5}	5×10^{-7}
E. coli DNA polymerase III	α subunit	ε subunit	7×10^{-6}	5×10^{-9}
T4 DNA polymerase	C-terminal	N-terminal	5×10^{-5}	10^{-7}
T7 DNA polymerase	?	118-145	10^{-5}	10^{-6}
Reverse transcriptase		None	10^{-5}	

Table 19.3

There are 5 mammalian DNA polymerases (identified by Greek letters; yeast equivalents are given Roman numerals).

DNA Polymerase	α(I)	δ (III)	ε (II)	β	γ
Location	Nuclear	Nuclear	Nuclear	Nuclear	Mitochondrial
Function	Lagging strand Synthesis & priming	Leading strand synthesis	Repair	Repair	Replication
Relative activity	˜80%			10-15%	2-15%
Mass (daltons)	300,000	170-230,000	250,000	40,000	180-300,000
Subunits	Catalytic core 2 primase 1 unknown	Catalytic core 1 unknown requires PCNA	Catalytic core 1 unknown	Catalytic	Catalytic core 2 unknown
3'-5' exonuclease	No	Yes	Yes	No	Yes
Dideoxythymidine	No effect	No effect	Weak	Inhibitory	Inhibitory
Aphidicolin	Inhibitory	Inhibitory	Inhibitory	No effect	No effect

the 130,000 dalton α subunit that possesses the DNA synthetic activity. The 3'–5' exonucleolytic proofreading activity is found in another subunit, ε, coded by *dnaQ*.

The ε subunit usually functions in conjunction with the α subunit, as indicated by the increase in proofreading when the two proteins are combined. The basic role of the ε subunit in controlling the fidelity of replication *in vivo* is demonstrated by the effect of mutations in *dnaQ*: the frequency with which mutations occur in the bacterial strain is increased by >10³-fold.

Proofreading activities have been assessed by various means, including comparing mutation rates under different conditions *in vivo* and examining misincorporation *in vitro*. Each DNA polymerase has a characteristic level of error that is influenced by its 3'–5' proofreading activity, as indicated in **Table 19.2**. Note the high error rate of reverse transcriptase (responsible for synthesizing DNA from retroviral RNA), which has no proofreading activity; its absence may be res-

ponsible for the high rate of change in genomic sequence during a retroviral infection. (We discuss retroviruses in Chapter 35.)

Some phages code for DNA polymerases. They include T4, T5, T7, and SPO1. The enzymes all possess 5'–3' synthetic activities and 3'–5' exonuclease proofreading activities. In each case, a mutation in the gene that codes for a single phage polypeptide prevents phage development. Each phage polymerase polypeptide associates with other proteins, of either phage or host origin.

Several classes of eukaryotic DNA polymerase have been identified. Their properties are summarized in **Table 19.3**. They generally lack any 3'–5' proofreading exonuclease activity, which implies that they must have some other mechanism for controlling the error rate.

Can any of these enzymes be identified as the authentic replicase? DNA polymerase α (called DNA polymerase I in yeast) has the same response to the inhibitors 2'3'-dideoxythymidine and aphidicolin as replication overall *in vitro* and *in vivo*. These results identify DNA polymerase α as a nuclear DNA replicase. However, it synthesizes only one of the daughter strands. DNA polymerase δ (called DNA polymerase III in yeast) is also involved in replication, and probably synthesizes specifically the other daughter strand. The two other nuclear enzymes, DNA polymerases ε and β, are probably involved in repair reactions. And DNA polymerase γ is responsible for replication of mitochondrial DNA.

A common feature of all DNA polymerases is that *they cannot initiate synthesis of a chain of DNA from free nucleotides.* They require a **primer** to provide a free 3′-OH end that can be extended by addition of nucleotides. The primer can take various forms, some of which we have discussed already in Chapter 18 in the context of the replicon. Types of priming reaction are summarized in **Figure 19.4**:

◆ A sequence of RNA is synthesized on the template, so that the free 3′-OH end of the RNA chain is extended by the DNA polymerase. This is commonly used in replication of cellular DNA, and by some viruses. We discussed the example of ColE1 replication previously (see Figure 18.33).

◆ A preformed RNA base pairs with the template, allowing its 3′-OH end to be used to prime DNA synthesis. This mechanism is used by retroviruses to prime reverse transcription of RNA (see Chapter 35).

◆ A primer terminus is generated within duplex DNA. The most common mechanism is the introduction of a nick, as used to initiate rolling circle replication (see Figure 18.15). In this case, the pre-existing strand is displaced by new synthesis. (Note the difference from nick translation shown in Figure 19.3, in which DNA polymerase I simultaneously synthesizes and degrades DNA from a nick.)

◆ A protein primes the reaction directly by presenting a nucleotide to the DNA polymerase. This reaction is used by certain viruses, as discussed previously in Figure 18.14.

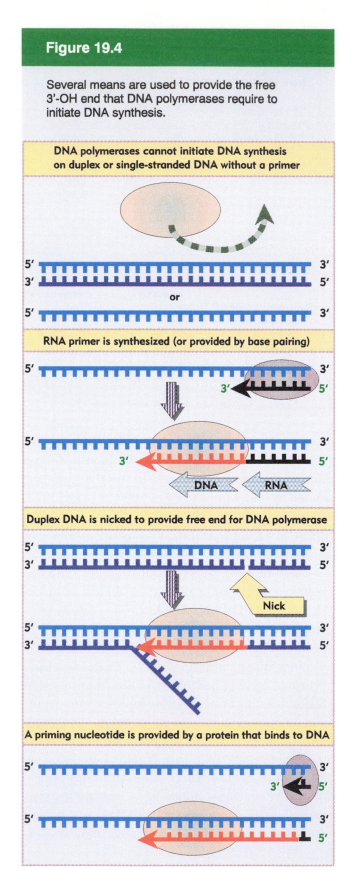

Figure 19.4

Several means are used to provide the free 3′-OH end that DNA polymerases require to initiate DNA synthesis.

DNA synthesis is semidiscontinuous and primed by RNA

The antiparallel structure of the two strands of duplex DNA poses a problem for replication. As the replication fork advances, daughter strands must be synthesized on both of the exposed parental single strands. The fork moves in the direction from 5' to 3' on one strand, and in the direction from 3' to 5' on the other strand. Yet nucleic acids are synthesized only from a 5' end toward a 3' end. *The problem is solved by synthesizing the strand that grows overall from 3' to 5' in a series of short fragments, each actually synthesized in the 'backwards' direction, that is, with the customary 5' to 3' polarity.*

Consider the region immediately behind the replication fork, as illustrated in **Figure 19.5**. We describe events in terms of the different properties of each of the newly synthesized strands:

◆ On the **leading strand** DNA synthesis can proceed continuously in the 5' to 3' direction as the parental duplex is unwound.

◆ On the **lagging strand** a stretch of single-stranded parental DNA must be exposed, and then a segment is synthesized in the reverse direction (relative to fork movement). A series of these fragments are synthesized, each 5' to 3'; then they are joined together to create an intact lagging strand.

Discontinuous replication can be followed by the fate of a very brief label of radioactivity. The label enters newly synthesized DNA in the form of short fragments, sedimenting in the range of 7–11S, corresponding to ~1000–2000 bases in length. These **Okazaki fragments** are found in replicating DNA in both prokaryotes and eukaryotes. After longer periods of incubation, the label enters larger segments of DNA. The transition results from covalent linkages between Okazaki fragments.

The lagging strand *must* be synthesized in the form of Okazaki fragments. For a long time it was

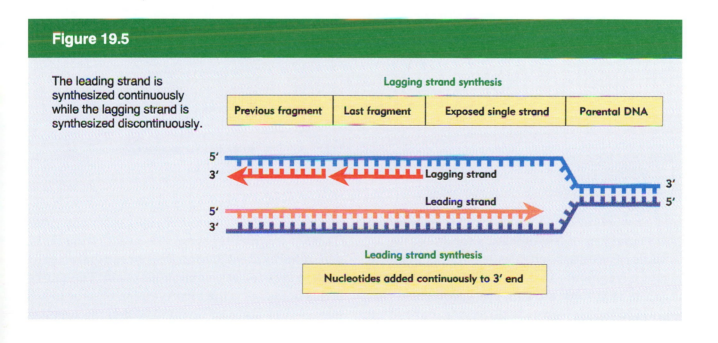

Figure 19.5

The leading strand is synthesized continuously while the lagging strand is synthesized discontinuously.

Lagging strand synthesis

| Previous fragment | Last fragment | Exposed single strand | Parental DNA |

Lagging strand

Leading strand

Leading strand synthesis

Nucleotides added continuously to 3' end

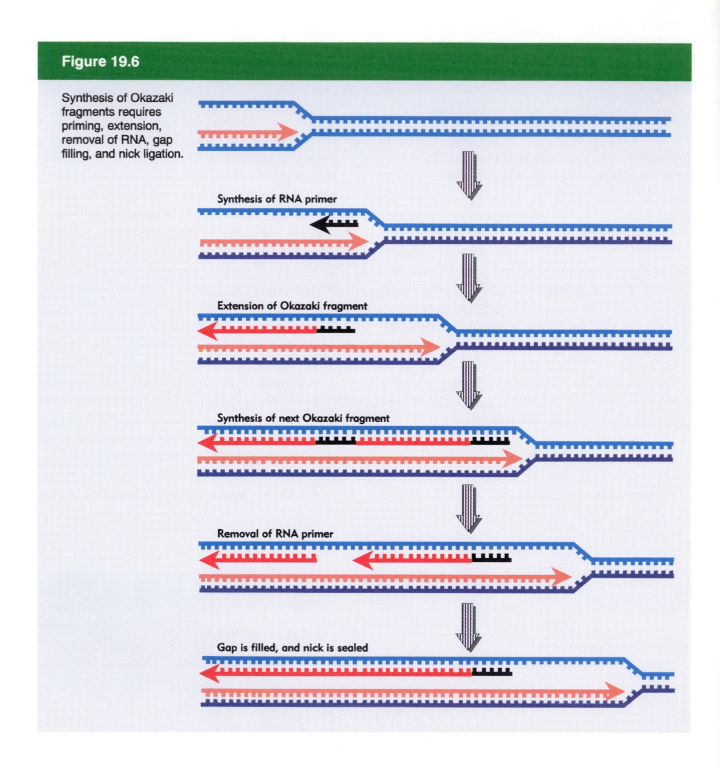

Figure 19.6

Synthesis of Okazaki fragments requires priming, extension, removal of RNA, gap filling, and nick ligation.

Synthesis of RNA primer

Extension of Okazaki fragment

Synthesis of next Okazaki fragment

Removal of RNA primer

Gap is filled, and nick is sealed

unclear whether the leading strand is synthesized in the same way or is synthesized continuously.

All newly synthesized DNA is found in the form of short fragments in *E. coli*. Superficially, this suggests that both strands are synthesized discontinuously. However, it turns out that not all of the fragment population represents *bona fide* Okazaki fragments; some are pseudofragments, generated by breakage in a DNA strand that actually was synthesized as a continuous chain. The source of this breakage is the incorporation of some uracil into DNA in place of thymine. When the uracil is

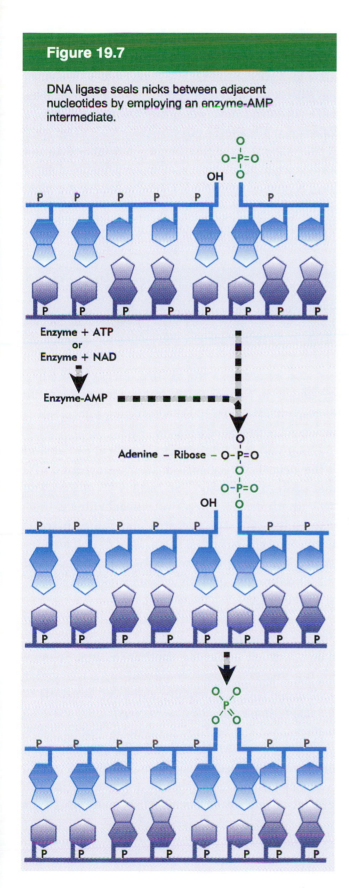

Figure 19.7

DNA ligase seals nicks between adjacent nucleotides by employing an enzyme-AMP intermediate.

removed by a repair system, the leading strand has breaks until a thymine is inserted.

Thus the lagging strand is synthesized discontinuously and the leading strand is synthesized continuously in *E. coli*. The same is probably true in other systems. This mode of synthesis is called **semidiscontinuous replication**.

Some crucial characteristics of replication are dictated by the inability of DNA polymerase to initiate a deoxyribonucleotide chain. An essential feature of the replication apparatus is the provision for a priming activity to provide a 3'-OH end to start off the DNA chain.

The leading strand requires only one such initiation event, which occurs at the origin. But there must be a series of initiation events on the lagging strand, since each Okazaki fragment requires its own start *de novo*. Each Okazaki fragment starts with a primer, a sequence of RNA, ~10 bases long, that provides the 3'-OH end for extension by DNA polymerase.

We can now expand our consideration of the actions involved in joining Okazaki fragments, as illustrated in **Figure 19.6**. The complete order of events is uncertain, but must involve synthesis of RNA primer, its extension with DNA, removal of the RNA primer, its replacement by a stretch of DNA, and the covalent linking of adjacent Okazaki fragments.

The figure suggests that synthesis of an Okazaki fragment terminates just before the start of the RNA primer of the preceding fragment. When the primer is removed, there will be a gap. The gap is filled by DNA polymerase I; *polA* mutants fail to join their Okazaki fragments properly. The 5'–3' exonuclease activity removes the RNA primer while simultaneously replacing it with a DNA sequence extended from the 3'-OH end of the next Okazaki fragment. This is equivalent to nick translation, except that the new DNA replaces a stretch of RNA rather than a segment of DNA. The 5'–3' exonuclease activity is unique to DNA polymerase I, and indeed is essential for removal of the final ribonucleotide from the RNA–DNA junction.

Once the RNA has been removed and replaced,

the adjacent Okazaki fragments must be linked together. The 3′-OH end of one fragment is adjacent to the 5′-phosphate end of the previous fragment. The responsibility for sealing this nick lies with the enzyme **DNA ligase**. Ligases are present in both prokaryotes and eukaryotes. Unconnected fragments persist in *lig⁻* mutants, because they fail to join Okazaki fragments together.

The *E. coli* and T4 ligases share the property of sealing nicks that have 3′-OH and 5′-phosphate termini, as illustrated in **Figure 19.7**. Both enzymes undertake a two-step reaction, involving an enzyme–AMP complex. (The *E. coli* and T4 enzymes use different cofactors. The *E. coli* enzyme uses NAD (nicotinamide adenine dinucleotide) as a cofactor, while the T4 enzyme uses ATP.) The AMP of the enzyme complex becomes attached to the 5′-phosphate of the nick; and then a phosphodiester bond is formed with the 3′-OH terminus of the nick, releasing the enzyme and the AMP.

The primosome initiates synthesis of Okazaki fragments

An Okazaki fragment is synthesized as a complement to the single-stranded region exposed by the movement of the replication fork. So single-stranded DNA phages have been used as templates to provide a model for the events involved in initiating and then elongating the lagging strand. Three phages, M13, G4, and φX174, have illustrated the significant features of priming.

A major point that emerges from the characterization of the phage DNAs is the need for the DNA template to be prepared in the proper structural condition. Two types of function are needed to convert double-stranded DNA to the single-stranded state:

◆ A **helicase** is an enzyme that separates the strands of DNA, usually using the hydrolysis of ATP to provide the necessary energy.

◆ A **single-strand binding protein** binds to the single-stranded DNA, preventing it from reforming the duplex state.

The conversion of φX174 DNA into individual single strands is illustrated in **Figure 19.8**. In peeling a single strand off the circular strand, the reaction resembles the rolling circle described previously in Figure 18.15, but it can occur in the absence of DNA synthesis when the appropriate 3 proteins are provided *in vitro*.

The phage A protein nicks the viral (+) strand at the origin of replication. In the presence of 2 host proteins, Rep and SSB, and ATP, the nicked DNA unwinds. The Rep protein provides a helicase that separates the strands; the SSB traps them in single-stranded form. The *E. coli* **single-strand binding protein** (SSB) is a tetramer of 74,000 daltons that binds cooperatively to single-stranded DNA.

The significance of the cooperative mode of binding is that the binding of one protein molecule makes it much easier for another to bind. So once the binding reaction has started on a particular DNA molecule, it is rapidly extended until *all of the single-stranded DNA is covered with the SSB protein*. Note that this protein is *not a DNA-unwinding protein*; its function is to stabilize DNA that is already in the single-stranded condition.

φX174 DNA is not by itself a substrate for the replication apparatus, because the naked DNA does

Figure 19.8

φX174 DNA can be separated into single-strands by the combined effects of 3 functions: nicking with A protein, unwinding by Rep, and single-strand stabilization by SSB.

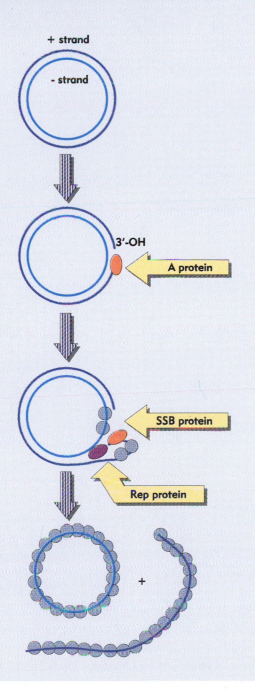

+ strand

- strand

3′-OH

A protein

SSB protein

Rep protein

+

not provide a suitable template. Once it has been converted to single-stranded form and coated with SSB, however, replication can proceed. Under normal circumstances *in vivo,* the unwinding, coating, and replication reactions proceed in tandem. The SSB presumably binds to DNA as the replication fork advances, keeping the two parental strands separate so that they are in the appropriate condition to act as templates. The SSB is needed in stoichiometric amounts at the replication fork. The coordinated action of the helicase and SSB provide a model for the structure of the bacterial replication fork, although a different helicase is used in bacterial DNA replication.

We are now in a position to consider the nature of the priming event. There are two types of priming reaction in *E. coli.* The φX system, named for phage φX174, characterizes one type of primosome. The *oriC* system, named for the bacterial origin, characterizes the other, and is apparently simpler. Sometimes replicons are referred to as being of the φX or *oriC* type.

In the replication of φX174, a **primosome** assembles at a unique site, called the **primosome assembly site (*pas*)**. The *pas* is the equivalent of an origin for synthesis of the complementary strand of φX174. The primosome consists of six proteins: PriA, PriB, PriC, DnaT, DnaB, and DnaC.

Although the primosome forms initially at the *pas* on φX174 DNA, primers are initiated at a variety of sites. This suggests that *the primosome (or some of its components) moves along the single-stranded DNA to additional sites at which priming occurs.* PriA and DnaB are certainly included in the components that end up in the complex at each site where priming occurs; it is not clear whether the other proteins are present. The nature of the sites where the mobile primosome halts is not clear. PriA has an activity that may be important for the ability to move to secondary sites for priming. On partially duplex DNA, it can function as a helicase, moving in the 3′–5′ direction.

Figure 19.9

Priming requires several enzymatic activities, including helicases, single-strand binding proteins, a means of recognizing the primer assembly sequence, and other structural proteins.

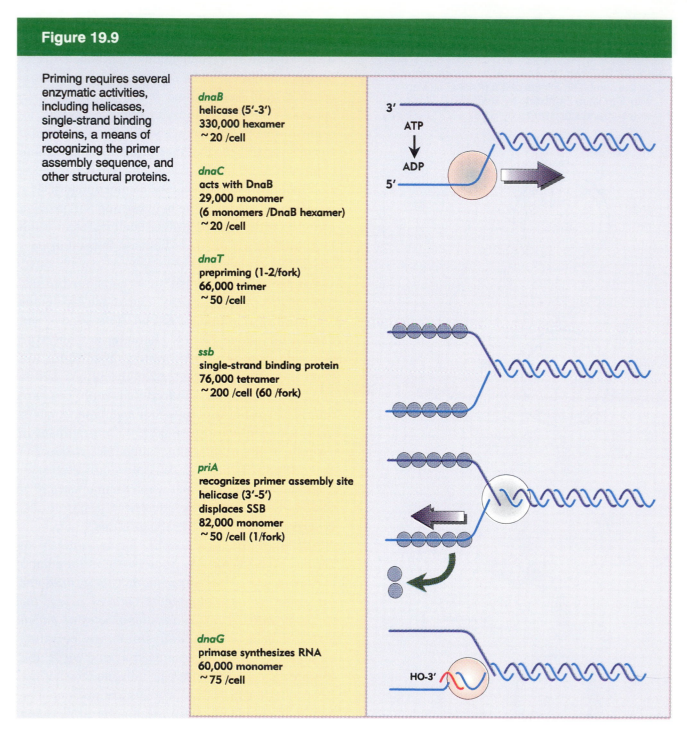

dnaB
helicase (5'-3')
330,000 hexamer
~20 /cell

dnaC
acts with DnaB
29,000 monomer
(6 monomers /DnaB hexamer)
~20 /cell

dnaT
prepriming (1-2/fork)
66,000 trimer
~50 /cell

ssb
single-strand binding protein
76,000 tetramer
~200 /cell (60 /fork)

priA
recognizes primer assembly site
helicase (3'-5')
displaces SSB
82,000 monomer
~50 /cell (1/fork)

dnaG
primase synthesizes RNA
60,000 monomer
~75 /cell

The *pas* is recognized by PriA, which can displace SSB from single-stranded DNA. This may be important for allowing the primosome to assemble. PriA uses cleavage of ATP to provide energy for the reaction. The dual role of PriA as a helicase and in displacing SSB is unique to φX primosomes, since PriA (and also PriB and PriC, whose functions are unknown) is not required for replication of *oriC* primosomes.

The types of activities involved in the φX174 priming reaction are summarized in **Figure 19.9**. Although other replicons in *E. coli* may have alternatives for some of these particular proteins, the same general types of activity are required in every case.

DnaB is the central component in both φX and *oriC* primosomes. It is bound to DNA in the form of a complex with DnaC. DnaT is also required at the prepriming stage. The major difference between the two types of primosome lies in the means that are used to bring DnaB to the priming site (see later).

A **primase** is required to catalyze the actual priming reaction. For both φX and *oriC* replicons, the primase is the product of the *dnaG* gene. The enzyme is a single polypeptide of 60,000 daltons (much smaller than RNA polymerase). In effect, the primase is an RNA polymerase that is used only under specific circumstances, that is, to synthesize

short stretches of RNA that are used as primers for DNA synthesis. DnaG primase associates transiently with the primosome, and is activated by DnaB to initiate synthesis of the primer, which typically is 15–50 bases long. It synthesizes primers starting with the sequence pppAG, opposite the sequence 3'-GTC-5' in the template.

(Some systems use alternatives to the DnaG primase. In the examples of the two phages M13 and G4, which were used for early work on replication, an interesting difference emerged. G4 priming uses DnaG, but M13 priming uses bacterial RNA polymerase. There is another unusual feature in the example of these phages, which is that the site of priming is indicated by a

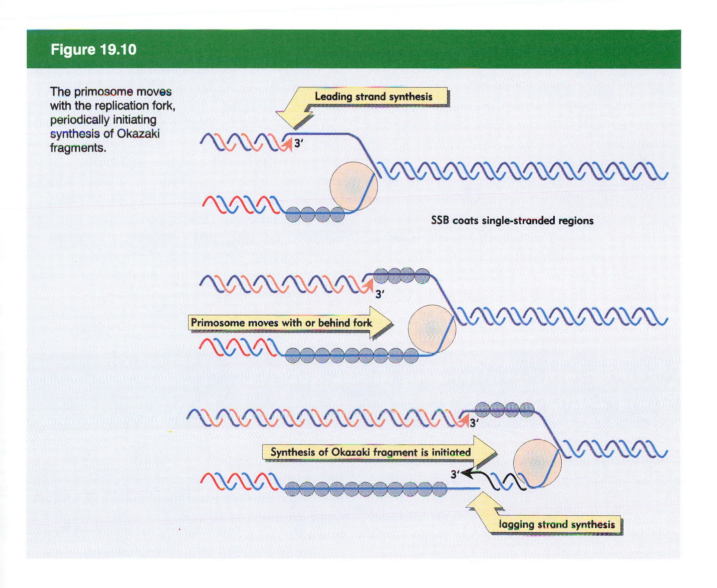

Figure 19.10

The primosome moves with the replication fork, periodically initiating synthesis of Okazaki fragments.

Leading strand synthesis

3'

SSB coats single-stranded regions

3'

Primosome moves with or behind fork

3'

Synthesis of Okazaki fragment is initiated

3'

lagging strand synthesis

region of secondary structure.)

The single-strand binding protein (SSB) is required to maintain DNA in single-stranded form at the replication fork, so that it is a suitable substrate for priming and elongation. It is required for more than one stage of replication; *ssb* mutants have a quick-stop phenotype, and are defective in repair and recombination as well as in replication. (Some phages use different SSB proteins, notably T4; this shows that there may be specific interactions between components of the replication apparatus and the SSB; we discuss this later.)

The primosome moves primarily in the anti-elongation direction—along the parental strand from 5′ toward 3′. **Figure 19.10** shows that this would be suitable for initiating the Okazaki fragments of the lagging strand of a duplex DNA. As the replication fork advances, it creates a single-stranded region ahead of the primosome. After each initiation event, the primosome moves along the single-stranded stretch to the site for starting the next Okazaki fragment. Thus the primosome moves in the same direction as the replication fork, but in the opposite direction from DNA synthesis of the lagging strand.

The reaction at *oriC,* and subsequent priming of individual Okazaki fragments, involve the loading of DnaB by means that do not involve the Pri proteins (see later). We may therefore interpret the figure in terms of the two essential functions that DnaB displays in *oriC* replicons. First, it has a helicase action that generates the actual replication fork in DNA. It

moves in the direction 5′–3′ (along the lower strand as shown in the figure); the unwinding of DNA requires cleavage of ATP. Second, it activates primer synthesis by DnaG primase. So the periodic interaction of DnaB with DnaG may be sufficient to prime Okazaki fragments. The 'primosome' for an *oriC* replicon could consist simply of DnaB, interacting periodically with primase.

Sequences that stop movement of replication forks have been identified in the form of the *ter* elements of the *E. coli* chromosome (see Figure 18.7) or equivalent sequences in the plasmid R6K. The common feature of these elements is a 23 bp consensus sequence that provides the binding site for the product of the *tus* gene, a 36,000 dalton protein that is necessary for termination. Tus binds to the consensus sequence, where it provides a contra-helicase activity and stops DnaB from unwinding DNA. The leading strand continues to be synthesized right up to the *ter* element, while the nearest lagging strand is initiated 50–100 bp before reaching *ter*.

The result of this inhibition is to halt movement of the replication fork and (presumably) to cause disassembly of the replication apparatus. Tus stops the movement of a replication fork in only one direction; presumably a fork proceeding in the opposite direction can displace Tus and thus continue. A difficulty in understanding the function of the system *in vivo* is that it appears to be dispensable, since mutations in the *ter* sites or in *tus* are not lethal.

Coordinating synthesis of the lagging and leading strands

An important point is evident from the semidiscontinuous mode of replication: the two points at which the new DNA strands are growing are moving in opposite directions; they are some distance apart on the DNA double helix (see Figure 19.5). This suggests that there must be two different

enzymatic complexes for synthesizing DNA in the vicinity of the replication fork. One complex is moving with the unwinding point and synthesizing the leading strand. The other complex is moving 'backwards,' relative to the DNA, along the exposed single strand. Each complex has a

Figure 19.11

DNA polymerase III activity can be isolated in several forms.

Catalytic core	Mass = 165,000
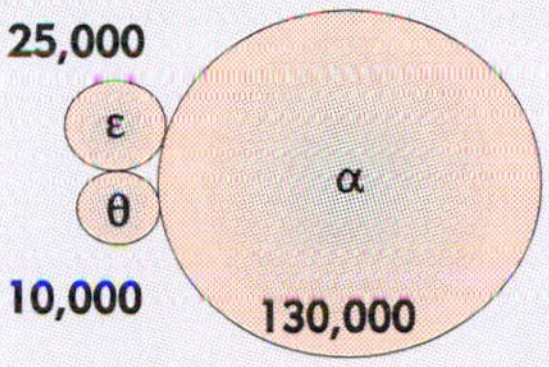 25,000 ε 10,000 θ α 130,000	α (polC) - DNA synthesis ε (dnaQ) - 3'-5' proofreading θ - assembly? Synthesizes DNA
Pol III*	**Mass = 470,000**
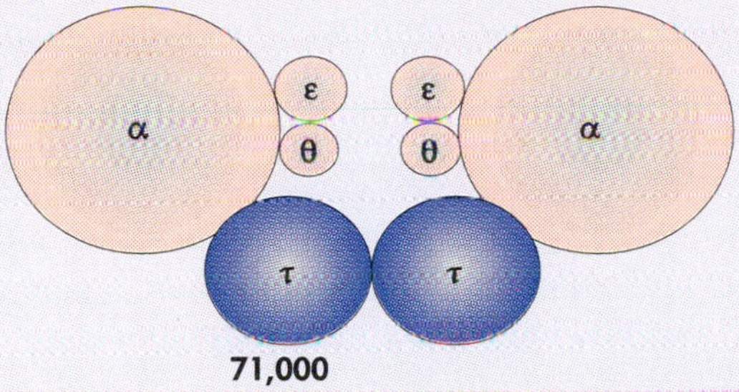 α ε θ ε θ α τ τ 71,000	Consists of 2 cores + 2 τ (dnaX) (τ causes dimer formation) Increased processivity
Pol III'	**Mass = 750,000**
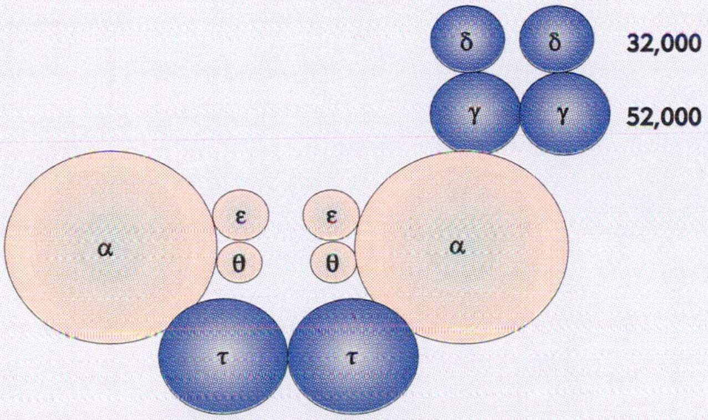 δ δ 32,000 γ γ 52,000 α ε θ ε θ α τ τ	Consists of Pol III* + γδ complex Additional subunits are: γ (dnaX), δ, δ',χ, Ψ γδ complex binds template
Holoenzyme	**Mass = 900,000**
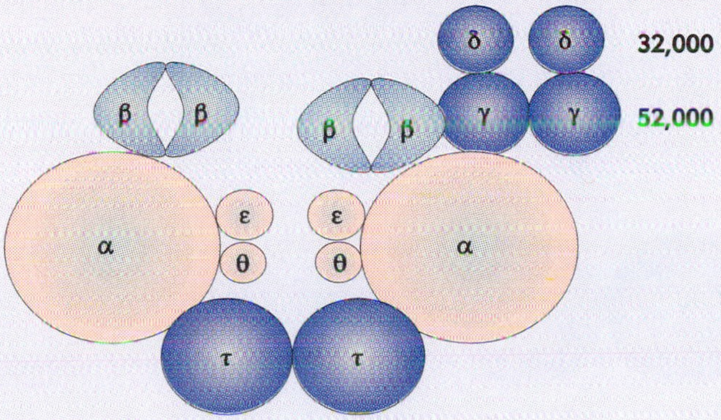 δ δ 32,000 β β β β γ γ 52,000 α ε θ ε θ α τ τ	Consists of Pol III* + β (dnaN) β subunit clamps enzyme to DNA Synthesizes leading & lagging strands

Figure 19.12

DNA polymerase III holoenzyme assembles in stages, generating an enzyme complex that synthesizes the DNA of both new strands.

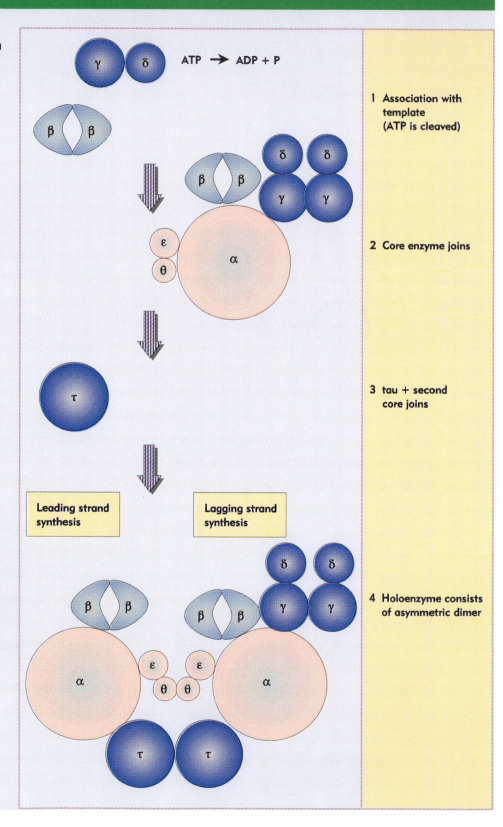

geometry that reflects its particular situation. How is synthesis of the two strands coordinated?

As the replisome moves along DNA, unwinding the parental strands, it elongates the leading strand. Periodically the primosome activity initiates an Okazaki fragment on the lagging strand. We do not know what happens to the DNA replicase when it completes synthesis of an Okazaki fragment; does it dissociate from the template, so that a new complex must be assembled to elongate the next Okazaki fragment; or is the same complex reutilized?

We can now partially relate the subunit structure of DNA polymerase III to the activities required for DNA synthesis. The forms of the enzyme that have been found *in vitro* are summarized in **Figure 19.11**. The 'core enzyme' contains subunits α, ϵ, and θ. The α subunit has a basic ability to synthesize DNA, the ϵ subunit has the 3'–5' proofreading exonuclease, and θ may be required for assembly. However, the processivity of the core (the tendency to remain on a single template rather than to dissociate and reassociate) is low; usually it synthesizes only ~11 bases before it dissociates from the template.

The addition of τ causes the core to dimerize, generating Pol III'. The Pol III* complex results from the addition of the $\gamma\delta$ complex, which consists of several subunits; γ and δ are the best characterized, but there are also some other subunits, less well defined. (τ and γ represent alternative products of the same gene, and have partially overlapping sequences.) The formation of holoenzyme is completed by addition of the β subunit. Processivity increases as each of these complexes is formed. The holoenzyme has about the same mass as a small ribosome subunit.

A speculative model for the assembly of DNA polymerase III is depicted in **Figure 19.12**. The holoenzyme assembles on DNA. In fact, the $\gamma\delta$ complex and the β subunit recognize the primer-template to form a preinitiation complex. In this reaction, the τ subunit cleaves ATP and transfers β subunits to the primed template. The τ

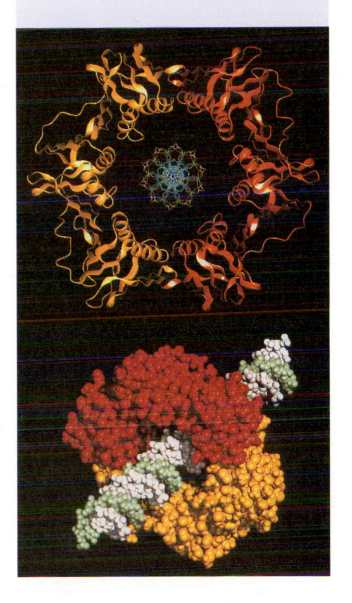

Figure 19.13

The β subunit of DNA polymerase III holoenzyme consists of a head to tail dimer (the two subunits are shown in red and orange) that forms a ring completely surrounding a DNA duplex (shown in the center). The top view shows a cross section through the DNA duplex surrounded by enzyme; the lower photograph shows a space-filling model from the side (with the two strands of DNA shown in white and green). Photograph kindly provided by John Kuriyan.

subunits generate a dimeric complex that forms the functional enzyme. The holoenzyme is asymmetric, because it has only one γδ complex. One model is that each of the cores synthesizes one new strand; because of differences in subunits associated with each core, the two cores could have different abilities to dissociate from DNA, corresponding to the need to synthesize a continuous leading strand and discontinuous lagging strand. The changes in processivity that result from the presence of τ and γ suggest that the leading strand may be synthesized by a core complex associated with the τ dimer, while the lagging strand is synthesized by a core assisted by the γδ complex.

The β subunit makes the holoenzyme highly processive. β is strongly bound to DNA, but can slide along a duplex molecule. The crystal structure of β shows that it forms a ring-shaped dimer. The model in **Figure 19.13** shows the β-ring in relationship to a DNA double helix. The dimer completely surrounds the duplex, providing a 'sliding clamp' that allows the holoenzyme to slide along DNA. The structure explains the high processivity—there is no way for the enzyme to fall off!

The asymmetric dimer model for DNA polymerase III suggests the model for the replication fork that is illustrated in **Figure 19.14**. A catalytic core is associated with each template strand of DNA. The holoenzyme moves continuously along the template for the leading strand; the template for the lagging strand is 'pulled through,' creating a loop in the DNA. DnaB creates the unwinding point, and translocates along the DNA in the 'forward' direction (moving 5′–3′ on the template for the lagging strand).

In φX replicons, PriA translocates in the opposite direction along the same strand, displacing SSB. In *oriC* replicons, where PriA is dispensable, the responsibility for interacting with primase could reside with DnaB itself. In this case, connection between priming and the replication fork is provided by the dual properties of DnaB: it is the helicase that propels the replication fork, and it provides the primosome that interacts with the DnaG primase. We do not fully understand the differences between φX and *oriC* origins, but in

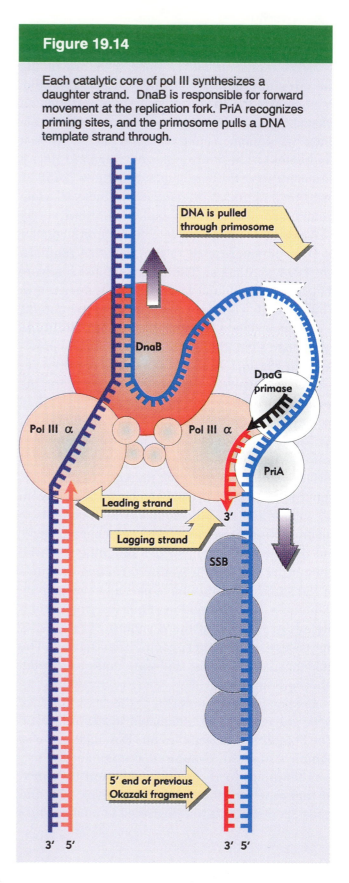

Figure 19.14

Each catalytic core of pol III synthesizes a daughter strand. DnaB is responsible for forward movement at the replication fork. PriA recognizes priming sites, and the primosome pulls a DNA template strand through.

DNA is pulled through primosome

DnaB

DnaG primase

Pol III α

Pol III α

PriA

Leading strand

3′

Lagging strand

SSB

5′ end of previous Okazaki fragment

3′ 5′

3′ 5′

both cases DnaB is a key component that is required for both forward progress and priming of Okazaki fragments.

The primosome in whatever guise recognizes an appropriate site, primase associates with it, primer synthesis occurs, and then the primosome is released. The length of the priming RNA is limited to 8–14 bases; apparently DNA polymerase III is responsible for displacing the primase. The enzyme elongates the growing daughter strand as the template strand is pulled through the holoenzyme, and then it releases the DNA template strand when the end of the previous Okazaki fragment is encountered. This model therefore explains how a single holoenzyme can synthesize both daughter strands, and how components of the primosome can simultaneously translocate in opposite directions. At present, however, it is speculative.

Whatever form of the enzyme is synthesizing the leading strand is assumed to move continuously from one end of the replicon to the other. But synthesis of the lagging strand requires a continuous series of association with the primosome to form a preinitiation complex, followed by conversion to active enzyme. It is possible that this involves some recycling of subunits; for example, dissociation and reassociation between the $\gamma\delta$ complex and the core. The holoenzyme is used with great efficiency; there are ~10 copies in a rapidly growing bacterium, probably all located at replication forks.

The system that synthesizes eukaryotic DNA in animal cells has an interesting difference: different DNA polymerases are used to synthesize the lagging and leading strands. DNA polymerase α exists as a complex consisting of a 180,000 dalton catalytic subunit, associated with other subunits, including two smaller proteins that provide a primase activity. This complex synthesizes the lagging strand. DNA polymerase δ, which is less well characterized, probably synthesizes the leading strand. (It is possible that the primase activity of DNA polymerase α primes the leading strand during initiation at the origin as well as initiating the individual Okazaki fragments.) The counterparts of these enzymes in yeast, DNA polymerases I and III, both are coded by essential genes. This split in responsibilities may therefore be common in the eukaryotes.

We now have purified systems that can replicate eukaryotic DNA *in vitro*, and **Table 19.4** lists the components of a HeLa cell system. With the exception of the topoisomerases, which are needed to maintain the proper topology of replicating templates, the proteins are all components of the replication fork. The functions that they undertake are analogous to those found in *E. coli*.

Among the proteins is one that may coordinate the actions of the two polymerases. PCNA is called proliferating cell nuclear antigen for historical reasons; it acts as a processivity factor for DNA polymerase δ in leading strand synthesis, but also is required for normal synthesis of the lagging strand. It may be the equivalent of subunit β of *E. coli* DNA polymerase III.

Table 19.4

Eukaryotic replication requires two DNA polymerases and other replication factors.

Protein	Subunits	Function
DNA polymerase α/primase	4	Lagging strand synthesis
DNA polymerase δ	2	Leading strand synthesis
PCNA	1	Elongation
Replication factor C	3	Elongation (ATPase)
Replication factor A	3	Single strand binding
Topoisomerases I & II		Maintain DNA topology

The replication apparatus of phage T4

When phage T4 takes over an *E. coli* cell, it provides several functions of its own that either replace or augment the host functions. Because the host DNA is degraded, the phage places little reliance on expression of host functions, except for utilizing enzymes that are present in large amounts in the bacterium. The degradation of host DNA is important in releasing nucleotides that are reused in the synthesis of phage DNA. (The phage DNA differs in base composition from cellular DNA in using hydroxymethylcytosine instead of the customary cytosine.)

The phage-coded functions concerned with DNA synthesis in the infected cell can be identified by mutations that impede the production of mature phages. Essential phage functions are identified by conditional lethal mutations, which fall into three phenotypic classes.

◆ Those in which there is *no DNA synthesis* at all identify genes whose products either are components of the replication apparatus or are involved in the provision of precursors (especially the hydroxymethylcytosine).

◆ Those in which the onset of DNA synthesis is *delayed* are concerned with the initiation of replication.

◆ Those in which DNA synthesis starts but then is *arrested* include regulatory functions, the DNA ligase, and some of the enzymes concerned with host DNA degradation.

◆ There are also nonessential genes concerned with replication; for example, including those involved in glucosylating the hydroxymethylcytosine in the DNA.

Synthesis of T4 DNA is catalyzed by a multienzyme aggregate assembled from the products of seven essential genes. They are described in **Table 19.5**.

The gene *32* protein (gp32) is a highly cooperative single-strand binding protein, needed in stoichiometric amounts. In fact, it was the first example of its type to be characterized. The geometry of the T4 replication fork may specifically require the phage-coded protein, since the *E. coli* SSB cannot substitute. The gp32 forms a complex with the T4 DNA polymerase; this interaction could be important in constructing the replication fork.

The T4 system uses an RNA priming event that is similar to that of its host. With single-stranded T4 DNA as template, the gene *41* and *61* products act together to synthesize short primers. Their behavior is analogous to that of DnaB and DnaG in *E. coli*.

The gene *41* protein is the counterpart to DnaB. It is a helicase, and also has a GTPase activity stimulated by single-stranded DNA. When the protein binds to single-stranded DNA, it can move from its initial binding site, migrating at a rate of ~400 nucleotides per second. It probably functions like DnaB, finding periodic sites at which to initiate primer synthesis. The hydrolysis of GTP is presumed to provide the energy for movement.

The gene *61* protein is needed in much smaller amounts than most of the T4 replication proteins. There are as few as 10 copies of gp61 per cell. (This impeded its characterization. It is required in such small amounts that originally it was missed as a necessary component, because enough was present as a contaminant of the gp32 preparation!) Gene *61* protein has the primase activity, analogous to DnaG of *E. coli*. The primase recognizes the template sequence 3′-TTG-5′ and synthesizes pentaribonucleotide primers that have the general sequence pppApCpNpNpNp. If the complete replication apparatus is present, these primers are extended into DNA chains.

The gene *43* DNA polymerase has the usual 5′–3′ synthetic activity, associated with a 3′–5′

Table 19.5

The T4 replication apparatus consists of 7 defined components.

Gene	Function	Product (daltons)	Relative No.
43	DNA polymerase	110,000	1
32	Single-strand binding protein	35,000	150
44 ⎫	ATPase, increases rate 3-4 fold	4 x 34,000 ⎫	5
62 ⎭		2 x 20,000 ⎭	
41	Helicase required for RNA priming	58,000	20
61	Primase		1
45	Assists DNA polymerase	2 x 27,000	10

exonuclease proofreading activity. The remaining three proteins are referred to as 'polymerase accessory proteins'. The gene 45 product is a dimer. The products of genes 44 and 62 form a tight complex, which has ATPase activity and increases the rate of movement of the DNA polymerase by 3–4-fold, from 250 nucleotides/second to ~800 nucleotides/second, close to the *in vivo* rate of ~1000 nucleotides/second. A similar increase is seen whether the template is single-stranded or double-stranded.

When T4 DNA polymerase uses a single-stranded DNA as template, its rate of progress is uneven. The enzyme moves rapidly through single-stranded regions, but proceeds much more slowly through regions that have a base-paired intrastrand secondary structure. The accessory proteins assist the DNA polymerase in passing these roadblocks, and maintaining its speed.

The presence of the proteins increases the affinity of the DNA polymerase for the DNA, and also its processivity. The combined action of all three proteins is needed for this effect. The product gp45 probably acts as a 'sliding clamp,' equivalent to subunit β of *E. coli* DNA polymerase III, which holds the DNA polymerase subunit more tightly on the template; gp44/62 is the equivalent of the *E. coli* γδ complex. As we have mentioned before,

this type of intimate relationship makes it a moot point what is a component of DNA polymerase and what is an accessory factor.

We have dealt with DNA replication so far solely in terms of the progression of the replication fork. The need for other functions is shown by the DNA-delay and DNA-arrest mutants. The four genes of the DNA-delay mutants include *39*, *52*, and *60*, which code for the three subunits of T4 topoisomerase II, an activity needed for removing supercoils in the template (see Chapter 33). The essential role of this enzyme suggests that T4 DNA does not remain in a linear form, but becomes topologically constrained during some stage of replication. The topoisomerase could be needed to allow rotation of DNA ahead of the replication fork.

Comparison of the T4 apparatus with the *E. coli* apparatus suggests that DNA replication poses a set of problems that are solved in analogous ways in different systems. Common features are that: a helicase is required to unwind DNA; a single-strand binding protein must maintain the single strands; primase initiates synthesis with a short RNA sequence; DNA replicase activity is assisted by proofreading and processivity functions; and ligase seals the breaks between adjacent fragments of newly synthesized DNA on the lagging strand.

Creating the replication forks at an origin

Starting a cycle of replication of duplex DNA requires several successive activities:

◆ The two strands of DNA must suffer their initial separation. This is in effect a melting reaction over a short region.

◆ An unwinding point begins to move along the DNA; this marks the generation of the replication fork, which continues to move during elongation.

◆ The first nucleotides of the new chain must be synthesized into the primer. This action is required once for the leading strand, but is repeated at the start of each Okazaki fragment on the lagging strand.

Some events that are required for initiation therefore occur uniquely at the origin; others recur with the initiation of each Okazaki fragment during the elongation phase.

The slow-stop mutations of *E. coli* identify several loci whose products participate in the events at the origin. We understand the functions of *dnaA, dnaB,* and *dnaC* in the very early stages of initiation, but have less information about the other loci (*dnaT,J,K*). Mutations of either the slow-stop or quick-stop phenotype occur in *dnaB* and *dnaC,* which have related functions in activating the origin and in priming synthesis of Okazaki fragments.

Plasmids carrying the *E. coli oriC* sequence have been used to develop a cell-free system for replication from this origin. Initiation of replication at *oriC in vitro* starts with formation of a complex that requires six proteins: DnaA, DnaB, DnaC, HU, gyrase, and SSB. The reaction converts a circular supercoiled template into a new form in which there is extensive unwinding of the duplex. Of the six proteins involved in prepriming, DnaA draws

our attention as the only one uniquely involved in initiation *vis à vis* elongation. We note that DnaB/DnaC provides the 'engine' of initiation at the origin, as it does in the primosome.

The first stage in complex formation is binding to *oriC* by DnaA protein. The reaction involves action at two different sequences: 9 bp and 13 bp repeats. Together the 9 bp and 13 bp repeats define the limits of the 245 bp minimal origin, as indicated in **Figure 19.15.** An origin is activated by the sequence of events summarized in **Figure 19.16,** in which binding of DnaA is succeeded by association with the other proteins.

The four 9 bp consensus sequences on the right side of *oriC* provide the initial binding sites for DnaA. It binds cooperatively until 20–40 monomers have formed a central core around which *oriC* DNA is wrapped.

Then the DnaA protein acts at three 13 bp tandem repeats located in the left side of *oriC*. The DNA strands are melted at each of these sites to form an open complex. All three 13 bp repeats must be opened for the reaction to proceed to the next stage.

The sequences of the 9 bp and 13 bp repeats are distinct, and we do not know whether they are

Figure 19.15

The minimal origin is defined by the distance between the outside members of the 13-mer and 9-mer repeats.

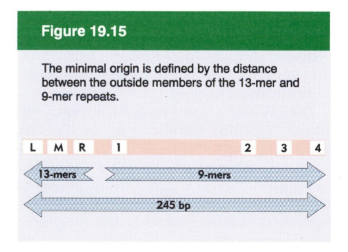

Figure 19.16

Prepriming involves formation of a complex by sequential association of proteins, leading to the separation of DNA strands.

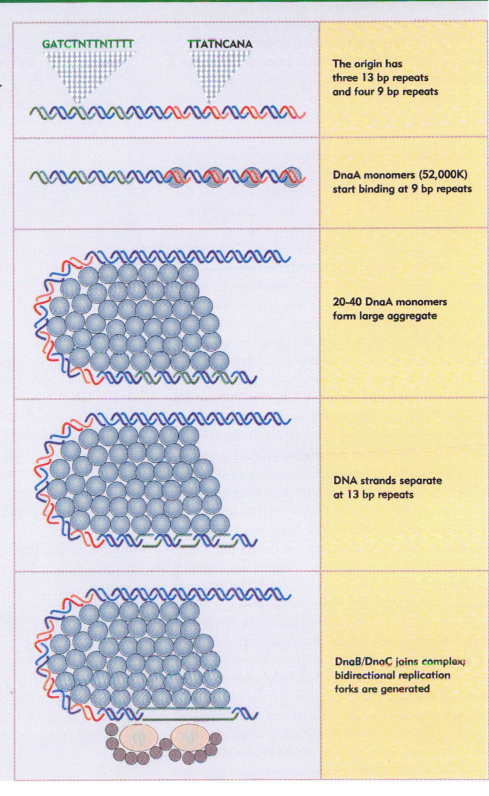

GATCTNTTNTTTT   TTATNCANA

The origin has three 13 bp repeats and four 9 bp repeats

DnaA monomers (52,000K) start binding at 9 bp repeats

20-40 DnaA monomers form large aggregate

DNA strands separate at 13 bp repeats

DnaB/DnaC joins complex; bidirectional replication forks are generated

recognized by the same or different binding sites on the DnaA protein. Both types of consensus sequences are A•T-rich.

The DnaB and DnaC proteins form a complex. As in the primosome, 6 DnaC monomers bind each hexamer of DnaB, giving a protein aggregate of 480,000 daltons, corresponding to a sphere of radius 6 nm. The formation of a complex at *oriC* is detectable in the form of the large protein blob visualized in **Figure 19.17**.

The DnaB•DnaC complex transfers one hexamer of DnaB to form the replication fork. DnaB provides the helicase that actually unwinds the DNA. Probably it recognizes the single-stranded structure of the potential replication fork rather than the actual nucleotide sequence; at any event it displaces DnaA from the 13 bp repeats and commences unwinding. DnaB functions in small amounts (1–2 hexamers) at the origin, where its action is presumed to be catalytic. DnaB also has the same ability here to activate the DnaG primase that it has in the primosome we discussed previously.

Two further proteins are required to support the unwinding reaction. Gyrase provides a swivel that allows one strand to rotate around the other (a reaction discussed in more detail in Chapter 33); without this reaction, unwinding would generate torsional strain in the DNA. The protein SSB stabilizes the single-stranded DNA as it is formed.

The protein HU is a general DNA-binding protein in *E. coli* (see Chapter 27). Its presence is not absolutely required to initiate replication *in vitro*, but it stimulates the reaction. HU has the capacity to bend DNA, and is likely to be involved in some general structural capacity.

In the absence of continuing replication, the prepriming proteins can unwind a considerable length of DNA at *oriC*. The length of duplex DNA that usually is unwound to initiate replication is probably <60 bp.

Input of energy in the form of ATP is required for the prepriming reaction. It is required for unwinding DNA. The helicase action of DnaB depends on ATP hydrolysis; and the swivel action of gyrase requires ATP hydrolysis. ATP is also needed for the action of primase and to activate DNA polymerase III.

Figure 19.17

The complex at *oriC* can be detected by electron microscopy. The upper photograph shows a prepriming complex visualized before the start of replication; the lower photograph shows the complex at a replication bubble 1 min. after the start of replication. Both complexes were visualized with antibodies against DnaB protein. Photographs kindly provided by Barbara Funnell.

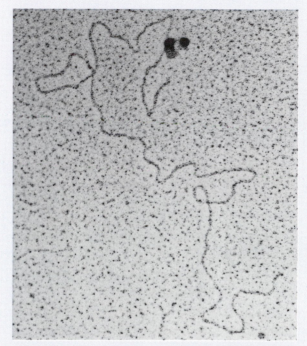

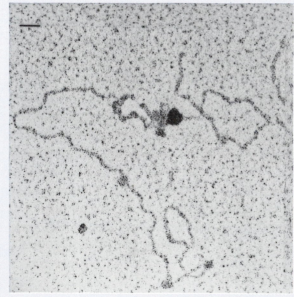

As the only member of the replication apparatus uniquely required at the origin, DnaA has attracted much attention. A complication in analyzing its effects is the absolute requirement for its function at initiation, which makes it difficult to determine whether it plays other roles in addition to the mechanics of activating the origin. Some mutations in *dnaA* render replication asynchronous, suggesting that DnaA could be the 'titrator' or 'clock' that measures the number of origins relative to cell mass. Overproduction of DnaA yields conflicting results, varying from no effect to causing initiation to take place at reduced mass. However, there are no cyclic variations in DnaA concentration or expression, and current opinion has it that those cases in which *dnaA* affects cyclic control are probably due to indirect rather than direct effects.

A protein that could be a new function involved in the control of replication cycles has been identified in the form of a 33,000 dalton species that binds to the 13-mer repeats at *oriC*. Its binding prevents the formation of an open replication complex, that is, it inhibits the melting reaction that is associated with the binding of DnaA. But mutations in the gene *(iciA)* do not affect the control of initiation, so the function of this inhibitor remains to be established.

What would be the properties of a mutation that altered the frequency of initiation? If initiation were caused by accumulation of an activator, a loss-of-function mutant should be in the slow-stop *dna* class, unable to start a new cycle. But if the regulator is an inhibitor, its loss should cause frequent cycles of initiation, perhaps leading to the accumulation of cells of reduced size. Such a mutant would not be found among the *dna* class, and would be more likely to appear among the potential segregation mutants (which we discussed in Chapter 18). However, we have yet to identify the circuit that controls the frequency of initiation events.

Common events in priming replication at the origin

Following generation of a replication fork as indicated in Figure 19.16, the priming reaction occurs to generate a leading strand. We know that synthesis of RNA is used for the priming event, but the details of the reaction are not known. Some mutations in *dnaA* can be suppressed by mutations in RNA polymerase, which suggests that DnaA could be involved in an initiation step requiring RNA synthesis *in vivo*.

RNA polymerase could be required to read into the origin from adjacent transcription units; by terminating at sites in the origin, it could provide the 3'-OH ends that prime DNA polymerase III. Alternatively, the act of transcription could be associated with a structural change that assists initiation. This latter idea is supported by observations that transcription does not have to proceed into the origin; it is effective up to 200 bp away from the origin, and can use either strand of DNA as template *in vitro*. The transcriptional event is inversely related to the requirement for supercoiling *in vitro*, which suggests that it acts by changing the local DNA structure so as to aid melting of DNA.

Another system for investigating interactions at the origin is provided by phage lambda, whose origin sponsors bidirectional replication. A map of the region is shown in **Figure 19.18**. Initiation of replication at the lambda origin requires 'activation' by transcription starting from P_R. As with the events at *oriC*, this does not necessarily imply that the RNA provides a primer for the leading strand. Analogies between the systems suggest that RNA synthesis could be involved in promoting some structural change in the region.

Initiation requires the products of phage genes *O* and *P*, as well as several host functions. The phage O protein binds to the lambda origin; the phage P protein interacts with the O protein and with the bacterial proteins. The origin lies within gene *O*,

Figure 19.18

Transcription initiating at P_R is required to activate the origin of lambda DNA.

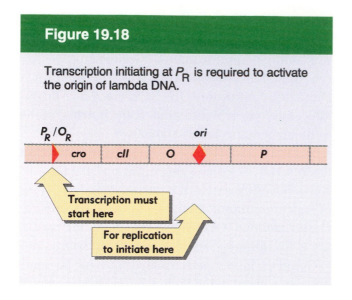

P_R/O_R ori

cro | cII | O | P

Transcription must start here

For replication to initiate here

so the protein acts close to its site of synthesis.

Variants of the phage called λ*dv* consist of shorter genomes that carry all the information needed to replicate, but lack infective functions; λ*dv* DNA survives in the bacterium as a plasmid. The λ*dv* DNA can be replicated *in vitro* by a system consisting of the phage-coded proteins O and

P together with bacterial replication functions.

Lambda proteins O and P form a complex together with DnaB at the lambda origin, *ori*λ. The origin consists of two regions; as illustrated in **Figure 19.19**, a series of four binding sites for the O protein is adjacent to an A•T-rich region.

The first stage in initiation is the binding of O to generate a roughly spherical structure of diameter ~11 nm, sometimes called the O-some. The O-some contains ~100 bp or 60,000 daltons of DNA. There are four 18 bp binding sites for O protein, which is ~34,000 daltons. Each site is palindromic, and probably binds a symmetrical O dimer. The DNA sequences of the O-binding sites appear to be bent, and binding of O protein induces further bending.

If the DNA is supercoiled, binding of O protein causes a structural change in the origin. The A•T-rich region immediately adjacent to the O-binding sites becomes susceptible to S1 nuclease, an enzyme that specifically recognizes unpaired DNA. This suggests that a melting reaction occurs next to the complex of O proteins.

The role of the O protein is analogous to that of

Figure 19.19

The lambda origin for replication comprises two regions. Early events are catalyzed by O protein, which binds to a series of 4 sites; then DNA is melted in the adjacent A-T-rich region. Although the DNA is drawn as a straight duplex, it is actually bent at the origin.

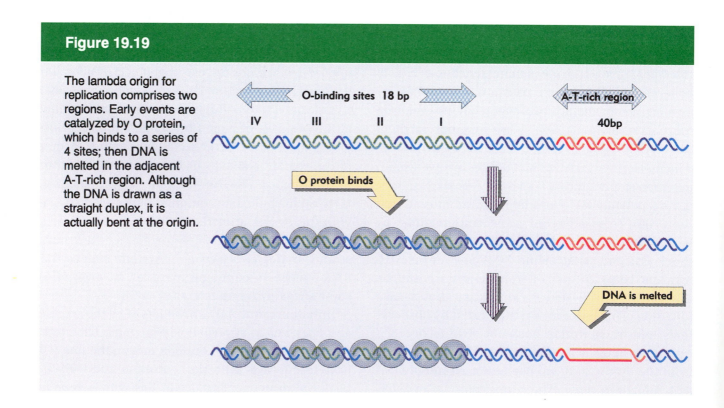

O-binding sites 18 bp   A-T-rich region

IV   III   II   I   40bp

O protein binds

DNA is melted

DnaA at *oriC*: it prepares the origin for binding of DnaB. Lambda provides its own protein, P, which substitutes for DnaC, and brings DnaB to the origin. When lambda P protein and bacterial DnaB proteins are added, the complex becomes larger and asymmetrical. It includes more DNA (a total of ~160 bp) as well as extra proteins. The λ P protein has a special role: it inhibits the helicase action of DnaB. Replication fork movement is triggered when P protein is released from the complex. Priming and DNA synthesis follow.

Some proteins are essential for replication without being directly involved in DNA synthesis as such. Interesting examples are provided by the DnaK and DnaJ proteins. DnaK is a chaperone, related to a common stress protein of eukaryotes. Its ability to interact with other proteins in a conformation-dependent manner plays a role in many cellular activities, including replication. The role of DnaK/DnaJ may be to *disassemble* the prepriming complex; by causing the release of P protein, they allow replication to begin.

The similarities between the initiation reactions at *oriC* and *oriλ* are summarized in **Table 19.6**. The same stages are involved, and rely upon overlapping components. Note that the assignment of proteins to particular stages is tentative. The first step is recognition of the origin by a protein that binds to form complex with the DNA, DnaA for *oriC* and O protein for *oriλ*. A short region of A•T-rich DNA is melted. Then DnaB is loaded; this requires different functions at *oriC* and *oriλ* (and yet other proteins are required for this stage at other origins). When the helicase DnaB joins the complex, a replication fork is created. Finally an RNA primer is synthesized, after which replication begins.

The use of *oriC* and *oriλ* provides a general model for activation of origins. A similar series of events occurs at the origin of the virus SV40 in mammalian cells. Two hexamers of T antigen, a protein coded by the virus, bind to a series of repeated sites in DNA. In the presence of ATP, changes in DNA structure occur, culminating in a melting reaction. In the case of SV40, the melted region is rather short and is not A•T-rich, but it has an unusual composition in which one strand consists almost exclusively of pyrimidines and the other of purines. Near this site is another essential region, consisting of A•T base pairs, at which the DNA is bent; it is underwound by the binding of T antigen. An interesting difference from the prokaryotic systems is that T antigen itself possesses the helicase activity needed to extend unwinding, so that an equivalent for DnaB is not needed.

Table 19.6

Initiation at *oriC* and *oriλ* involves similar reactions.

Stage	oriC	oriλ
Origin recognition and melting	DnaA	O
Helicase association and unwinding	DnaB	DnaB
	DnaC	P
	RNA polymerase	RNA polymerase
Release of complex	DnaJ?	DnaJ
	DnaK?	DnaK
Fork movement	Gyrase	?
	SSB	?
Priming	DnaG	DnaG

Does methylation at the origin regulate initiation?

What feature of a bacterial (or plasmid) origin ensures that it is used to initiate replication only once per cycle? Is initiation associated with some change that marks the origin so that a replicated origin can be distinguished from a nonreplicated origin?

Some sequences that are used for this purpose are included in the origin. *oriC* contains 11 copies of the sequence $^{GATC}_{CTAG}$, which is a target for methylation at the N^6 position of adenine by the *dam* methylase. The reaction is illustrated in **Figure 19.20**.

Before replication, the palindromic target site is methylated on the adenines of each strand. Replication inserts the normal (nonmodified) bases into the daughter strands, generating **hemimethylated** DNA, in which one strand is methylated and one strand is unmethylated. *So the replication event converts* dam *target sites from fully methylated to hemimethylated condition.*

What is the consequence for replication? The ability of a plasmid relying upon *oriC* to replicate in *dam⁻ E. coli* depends on its state of methylation. If the plasmid is methylated, it undergoes a single round of replication, and then the hemimethylated products accumulate, as described in **Figure 19.21**. Thus a hemimethylated origin cannot be used to initiate a replication cycle.

Two explanations suggest themselves. Initiation may require full methylation of the Dam target sites in the origin. Or initiation may be inhibited by hemimethylation of these sites. This latter seems to be the case, because an origin of nonmethylated DNA can function effectively. This explains why *dam⁻* mutants are viable. They show an increased frequency of initiation and therefore have an excess number of origins, but do not lack the essential capacity to initiate.

So hemimethylated origins cannot initiate again until the Dam methylase has converted them into fully methylated origins. The GATC sites at the origin remain hemimethylated for a substantial period after replication; remethylation does not

begin for ~10 minutes. This long period is unusual, because at typical GATC sites elsewhere in the genome, remethylation begins immediately following replication. One other region behaves like *oriC*; the promoter of the *dnaA* gene also shows a delay before remethylation begins.

While it is hemimethylated, the *dnaA* promoter is repressed, which causes a reduction in the level of DnaA protein. Thus the origin itself is inert, and production of the crucial initiator protein is repressed, during this period.

What is responsible for the delay in remethylation at *oriC* and *dnaA*? The most likely explanation is that these regions are sequestered in a form in which they are inaccessible to the *dam* methylase. Probably the sequestration is responsible for the repression of *dnaA*.

Hemimethylation of the GATC sequences in the origin is required for its association with the cell

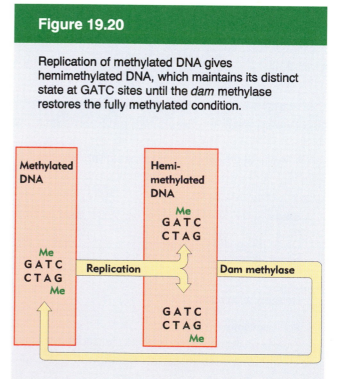

Figure 19.20

Replication of methylated DNA gives hemimethylated DNA, which maintains its distinct state at GATC sites until the *dam* methylase restores the fully methylated condition.

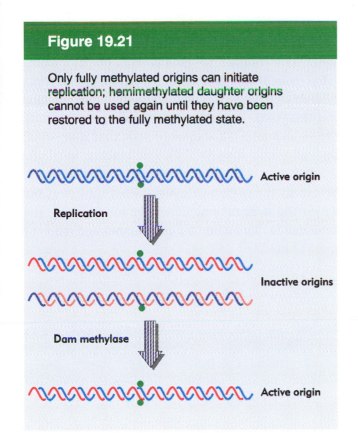

Figure 19.21

Only fully methylated origins can initiate replication; hemimethylated daughter origins cannot be used again until they have been restored to the fully methylated state.

Active origin

Replication

Inactive origins

Dam methylase

Active origin

components that regulate replication. An inhibitor is found in this fraction that competes with DnaA protein. This inhibitor can prevent initiation of replication only if it is added to an *in vitro* system before DnaA protein. This suggests the model of **Figure 19.22**, in which the inhibitor specifically recognizes hemimethylated DNA and prevents DnaA from binding. When the DNA is remethylated, the inhibitor is released, and DnaA now is free to initiate replication. If the inhibitor is associated with the membrane, then association and dissociation of DNA with the membrane may be involved in the control of replication.

The full scope of the system used to control reinitiation is not clear, but several mechanisms may be involved: physical sequestration of the origin; delay in remethylation; inhibition of DnaA binding; repression of *dnaA* transcription. It is not immediately obvious which of these events cause the others, and whether their effects on

membrane *in vitro*. Hemimethylated *oriC* DNA binds to the membranes, but DNA that is fully methylated does not bind. What would be the consequences if this transient binding occurred *in vivo*?

One possibility is that the membrane association could be important for chromosomal segregation. Following the model of Figure 18.25, immediately after a cycle of replication has been initiated, the newly replicated origins would be attached to the membrane, presumably at sites near one another. Perhaps growth of the membrane, movement of these sites, or formation of a septum, occurs during the period of attachment to ensure that the daughter chromosomes are directed to different future daughter cells.

Another possibility is that this association prevents a reinitiation from occurring prematurely, either indirectly because the origins are sequestered or directly because some component at the membrane inhibits the reaction. The properties of the membrane fraction suggest that it includes

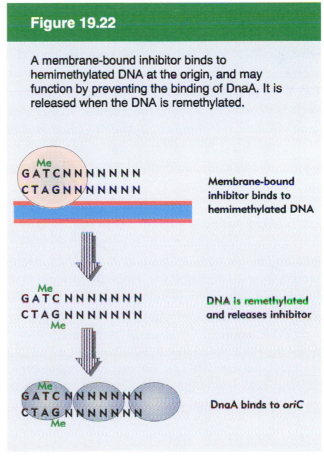

Figure 19.22

A membrane-bound inhibitor binds to hemimethylated DNA at the origin, and may function by preventing the binding of DnaA. It is released when the DNA is remethylated.

Me
GATCNNNNNNN
CTAGNNNNNNN

Membrane-bound inhibitor binds to hemimethylated DNA

Me
GATCNNNNNNN
CTAGNNNNNNN
Me

DNA is remethylated and releases inhibitor

Me
GATCNNNNNNN
CTAGNNNNNNN
Me

DnaA binds to *oriC*

initiation are direct or indirect.

We still have to come to grips with the central issue of which feature has the basic responsibility for *timing*. One possibility is that attachment to the membrane occurs at initiation, and that assembly of some large structure is required to release the DNA. The period of sequestration appears to increase with the length of the cell cycle, which suggests that it directly reflects the clock that controls reinitiation.

It has been extremely difficult to identify the protein component(s) that mediate membrane attachment. A hint that this is a function of DnaA is provided by its response to phospholipids. Phospholipids promote the exchange of ATP with ADP bound to DnaA. We do not know what role this plays in controlling the activity of DnaA (which requires ATP), but the reaction implies that DnaA is likely to interact with the membrane. This would imply that more than one event is involved in associating with the membrane. Perhaps a hemimethylated origin is bound by the membrane-associated inhibitor, but when the origin becomes fully methylated, the inhibitor is displaced by DnaA associated with the membrane.

Summary

DNA synthesis occurs by semidiscontinuous replication, in which the leading strand of DNA growing 5′–3′ is extended continuously, but the lagging strand that grows overall in the opposite 3′–5′ direction is made as short Okazaki fragments, each synthesized 5′–3′. The leading strand and each Okazaki fragment of the lagging strand initiate with an RNA primer that is extended by DNA polymerase. Bacteria and eukaryotes each possess more than one DNA polymerase activity. DNA polymerase III synthesizes both lagging and leading strands in *E. coli*. Different enzymes, DNA polymerase α and DNA polymerase δ, synthesize the strands in eukaryotic nuclei. Many proteins are required for DNA polymerase III action and several constitute part of the replisome within which it functions. Phage T4 codes for a sizeable replication apparatus, consisting of 7 proteins: DNA polymerase, helicase, single strand binding protein, priming activities, and accessory proteins. Similar functions are required in other replication systems, including a HeLa cell system that replicates SV40 DNA.

The common mode of origin activation involves an initial limited melting of the double helix, followed by more general unwinding to create single strands. Several proteins act sequentially at the *E. coli* origin. DnaA binds to a series of 9 bp and 13 bp repeats, forming an aggregate of 20–40 monomers with DNA in which the 13 bp repeats are melted. The helicase activity of DnaB, together with DnaC, unwinds DNA further. DnaT is also needed for DnaB to bind to the prepriming complex. DnaJ and DnaK act indirectly, by releasing the complex to allow replication fork movement. Similar events occur at the lambda origin, where phage proteins O and P are the counterparts of bacterial proteins DnaA and DnaC, respectively. In SV40 replication, several of these activities are combined in the functions of T antigen.

DnaB provides the helicase activity at a replication fork; this depends on ATP cleavage. DnaB may function by itself in *oriC* replicons to provide primosome activity by interacting periodically with DnaG, which provides the primase that synthesizes RNA.

The φX priming event also requires DnaB, DnaC, and DnaT. PriA is the component that defines the primer assembly site (*pas*) for φX replicons; it displaces SSB from DNA in an action that involves cleavage of ATP. PriB and PriC are additional components of the primosome. PriA also has a 3′–5′ helicase activity that may be involved in movement to secondary priming sites.

Several sites that are methylated by the Dam methylase are present in the *E. coli* origin, including those of the 13-mer binding sites for DnaA. The origin remains hemimethylated and is in a sequestered state for ~10 minutes following initiation of a replication cycle. During this period it is associated with the membrane, and reinitiation of replication is repressed.

Further reading

Reviews

An extensive account of the replication apparatus has been given by **Kornberg and Baker** in *DNA Replication* (Freeman and Co., New York, 1992).

Prokaryotic systems for replication have been surveyed by **Nossal** (*Ann. Rev. Biochem.* 53, 581–615, 1983), **Marians** (*Ann. Rev. Biochem.* 61, 673–719, 1992), and **Baker and Wickner** (*Ann. Rev. Genet* 26, 447–477, 1992). DNA polymerase III has been reviewed by **McHenry** (*Ann. Rev. Biochem.* 57, 519–550, 1988). Termination was reviewed by **Hill** (*Ann. Rev. Microbiol.* 46, 603–633, 1992).

Eukaryotic replication has been reviewed by **Campbell** (*Ann. Rev. Biochem.* 55, 733–771, 1986) and by **Stillman** (*Ann. Rev. Cell Biol.* 5, 197–245, 1989).

Proofreading has been reviewed by **Kunkel** (*Cell* 53, 837–840, 1988).

Reviews closely based on research dealt with the T4 system from **Liu** *et al.* (*Cold Spring Harbor Symp. Quant. Biol.* 43, 469–487, 1979), the T7 system from **Richardson** *et al.* (*Cold Spring Harbor Symp. Quant. Biol.* 43, 427–440, 1978), and viral-linked proteins from **Wimmer** (*Cell* 28, 199–201, 1982).

Discoveries

Among the research papers reporting developments in the systems, the *E. coli* system was dealt with by **Arai** *et al.* (*J. Biol. Chem.* 256, 5239–5246, 1981), the primosome by **Arai and Kornberg** (*Proc. Nat. Acad. Sci.* 78, 69–73, 1981), the action of øX174 A protein by **Scott** *et al.* (*Proc. Nat. Acad. Sci. USA* 74, 193–197, 1977), the replication of *oriC* by **Fuller, Kaunagi, and Kornberg** (*Proc. Nat. Acad. Sci.* 78, 7370–7374, 1981), the role of DnaA by **Bramhill and Kornberg** (*Cell* 52, 743–755, 1988), and the connection between transcription and initiation of replication by **Baker and Kornberg** (*Cell* 55, 113–123, 1988). 77, 2460–2464, 1980).

The connection between methylation and membrane association was established by **Ogden, Pratt and Schaechter** (*Cell* 54, 127–135, 1988).

CHAPTER 20

Systems that safeguard DNA

DNA fulfills its hereditary functions via replication and transcription. Each process involves many enzymatic and regulatory activities as well as the appropriate polymerases. In addition to these systems, a variety of enzymes interact with DNA to modify its structure or to repair damage that has occurred to it. These activities are important in understanding how the integrity of the sequence of DNA is maintained; the act of replication itself is insufficient to safeguard its role in evolution.

Although only four types of base are used to synthesize DNA, some of the bases are chemically modified after their incorporation into DNA. Methylation is the most common modification, and provides a signal that marks a segment of DNA. Prokaryotes and eukaryotes both contain enzymes that methylate DNA, although the functions of the methylation are different.

E. coli DNA contains small amounts of 6-methyladenine and 5-methylcytosine. These bases are generated by the action of three methylases on DNA. They identify different types of system:

◆ The *hsd* system confers host specificity by methylating adenine. There are counterparts to this type of system in many bacteria; the specificity is characteristic of each system.

◆ The *dam* system distinguishes the strands of newly replicated DNA, also by methylating adenine. It is involved in control of replication, and in marking DNA strands for repair; it also influences the activity of other elements in the *E. coli* chromosome (see Chapters 19 and 34).

◆ The *dcm* system methylates cytosine; its function is unknown. Bacteria with mutations in all three systems lack methylated bases and are viable, so methylation cannot be an essential event.

Methylation in eukaryotes has a different purpose: distinguishing genes in different functional conditions as described in Chapter 28. There do not appear to be eukaryotic counterparts to any of the bacterial methylation systems.

Host specificity in a bacterial strain is the result of the action of particular enzyme(s) that impose a **modification** pattern on DNA. In reverse perspective, the pattern identifies the source of the DNA. Modification allows the bacterium to distinguish between its own DNA and any 'foreign' DNA, which lacks the characteristic host modification pattern. This difference renders an invading foreign DNA susceptible to attack by **restriction enzymes** that recognize the absence of methyl groups at the appropriate sites.

Such **modification and restriction** systems are widespread in bacteria; in the context of safeguarding DNA, their object is xenophobic—to protect the resident DNA against contamination by sequences of foreign origin (although their presence is not obligatory; some bacterial strains lack any restriction system).

A bacterium contains several systems that protect its DNA against the consequences of damage by external agents or faults in replication. Any event that causes the structure of DNA to deviate irreversibly from its regular double helix is recognized as inadmissible. The change can be a point mutation that converts one base into another, thus creating a pair of bases not related by the Watson–Crick rules. Or it can be a structural change that adds a bulky adduct to DNA or links two bases together.

The sites of damage are recognized by special nucleases that excise the damaged region from DNA; then further enzymes synthesize a replacement sequence. Together, these activities form a **repair system**. As well as the direct repair of damage by excision and replacement, there are systems to cope with the adverse consequences of replicating damaged DNA. These **retrieval systems** are related to those involved in genetic recombination.

The various systems include enzymes with some remarkable abilities to recognize sequences or structures in DNA. Restriction enzymes bind to DNA at specific target sequences, offering another perspective on the nature of protein–nucleic acid interactions. Some of these enzymes bind to DNA at one site, but then cleave at another site far away, offering insights into the ability of proteins to move along duplex DNA. Some repair enzymes recognize damaged sites in DNA apparently by the distortion in the regular structure of the double helix. An enzyme involved in recombination can bind two molecules of DNA to promote base pairing between them.

From these various activities, we learn about the interactions of proteins with DNA in a sequence- or structure-specific manner, an issue at the crux of the molecular biology of the gene. And we begin to obtain a general view of the multifarious nature of the systems needed to preserve and protect DNA against insults from the environment or the errors of other cellular systems.

The consequences of modification and restriction

Modification and restriction systems were discovered by their effects on phage infection. Phage DNA released from a bacterium of one strain can successfully prosecute an infection in another bacterium of the same strain, because it has the same modification pattern as the host DNA. But phage DNA that travels from one bacterial strain to another is attacked by the restriction activity. The phage is 'restricted' to one strain of bacteria.

The restriction is not absolute. Some infecting phages escape restriction, because of mutations in the target sites or negligence by the host system. In the latter case, they acquire the modification type of the new host.

Plasmids contribute to the strain specificity of bacteria. Some plasmids (and phages) possess genes for modification and restriction systems, so their presence in a bacterium determines its specificity. At all events, the existence of these systems serves to counter the mobility of bacterial sequences conferred by the passage of plasmids and phages between bacteria.

The basic feature of a modification and restriction system is that a bacterial strain possesses a DNA methylase activity with the *same sequence specificity* as the restriction activity. The methylase adds methyl groups (to adenine or cytosine residues) in the same target sequence that constitutes the restriction enzyme binding site. The methylation renders the target site resistant to restriction. Thus methylation protects the DNA against cleavage.

Restriction endonucleases all recognize specific, rather short sequences of DNA as binding sites. Binding is followed by cleavage, at the recognition site itself or elsewhere, depending on the enzyme. The cleavage reaction is widely used in mapping and reconstructing DNA *in vitro*. (We discuss these reactions in Chapter 21). Now we must ask how the natural function of these enzymes is exercised.

The resident bacterial DNA is methylated at the appropriate target sites, and so is immune from attack by the restriction activity, as indicated in **Figure 20.1**. Otherwise the restriction enzyme would degrade the DNA of the cell in which it resides! But the protection does not extend to any foreign DNA that gains entry to the cell. Such DNA has unmodified target sites, and therefore is attacked by the restriction enzyme. The combination of modification and restriction allows the cell to distinguish foreign DNA from its own, protected sequences.

In this context, 'foreign' DNA means any DNA derived from a bacterial strain *lacking the same modification and restriction activity*. There is no distinction among *types* of DNA, that is, between bacterial, plasmid, or phage sequences. The same modification pattern is possessed by *all* DNA that is resident in, or has passed through, a particular

Figure 20.1

Overview: methylated sites are perpetuated indefinitely and are safe from restriction, but unmethylated sites are cleaved.

Cycle of Resident DNA	State of DNA	Methylase Activity	Restriction Activity	Result
Replication	Methylated (*)	Inactive	Inactive	None
Dam methylase	Hemimethylated	Active	Inactive	Methylation
	Methylated	Inactive	Inactive	None
Fate of foreign DNA				
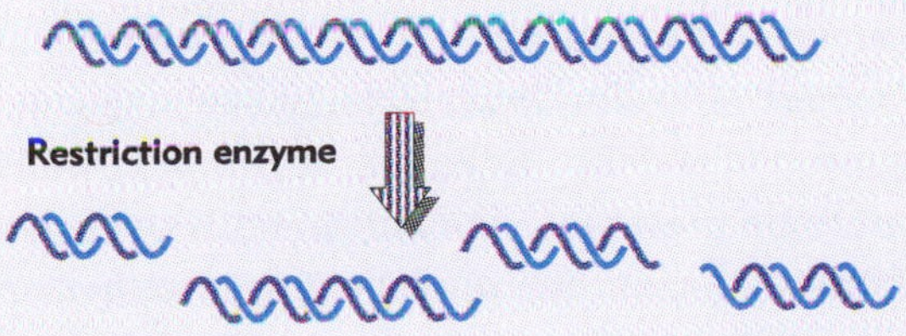 Restriction enzyme	Nonmethylated	Active (?)	Active	Degradation
	Fragmented			

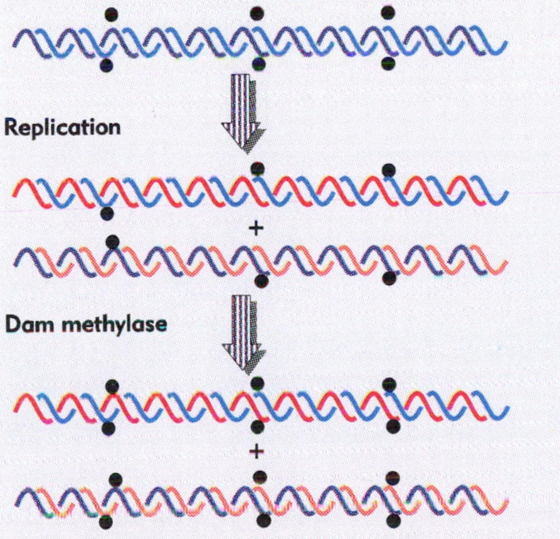

strain of bacteria. The restriction and modification system offers no hindrance to the passage of DNA between bacteria of the same strain.

Restriction enzymes fall into two general classes, which can be further subdivided. Their properties are summarized in **Table 20.1**. The major difference is that type II systems consist of two enzymes, one to undertake modification and one to undertake restriction, whereas in type I and type III systems the same enzyme possesses both activities.

Table 20.1

Restriction and methylation activities may be associated or may be separate.

	Type II Enzyme	Type III Enzyme	Type I Enzyme
Protein structure	Separate endonuclease & methylase	Bifunctional enzyme of 2 subunits	Bifunctional enzyme of 3 subunits
Recognition site	4-6 bp sequence, often palindromic	5-7 bp Asymmetric sequence	Bipartite & asymmetric
Cleavage site	Same as or close to recognition site	24-26 bp Downstream of recognition site	Nonspecific >1000 bp from recognition site
Restriction & methylation	Separate reactions	Simultaneous	Mutually exclusive
ATP needed for restriction?	No	Yes	Yes

Type II restriction enzymes are common

Type II activities include the restriction enzymes whose activities have been discussed previously in Chapter 6 in the context of genetic engineering (see also Chapter 21). They are useful because they recognize a wide variety of target sequences. A type II system is found in about one in three bacterial strains. Each of these enzymes is responsible only for the act of restriction. A separate enzyme is responsible for methylating the same target sequence.

The protein structures tend to be uncomplicated. The best characterized examples are the components of the EcoRI system, where the restriction enzyme is a dimer of identical subunits and the methylase is a monomer. (The recognition of a common target could have arisen either by duplication of the part of the protein that recognizes DNA or by convergent evolution.)

The target sites for type II enzymes are often palindromes of 4–6 bp. Because of the symmetry, the bases to be methylated occur on both strands of DNA. Thus a target site may be *fully methylated* (both strands are modified), *hemimethylated* (only one strand is methylated), or *nonmethylated*. As summarized in Figure 20.1:

◆ A fully methylated site is a target for neither restriction nor modification.

◆ A hemimethylated site is not recognized by the restriction enzyme, but is converted by the methylase into the fully modified condition.

◆ A nonmethylated target site may be a substrate for either restriction or modification *in vitro*. In the cell, unmodified DNA is more likely to be restricted; it is relatively rare for unmodified DNA to survive by gaining the modification pattern of a new host.

In the bacterium, most methylation events are concerned with perpetuating the current state of modification. Replication of fully methylated DNA produces hemimethylated DNA. Recognizing the hemimethylated sites is probably the usual mode of action of the methylase *in vivo*.

Most of the type II restriction enzymes cleave the DNA *at* an unmethylated target site. One bond is cleaved in each strand of DNA; the cleavages occur sequentially. Some enzymes introduce staggered cuts, others generate blunt ends. A subclass consists of enzymes that cleave the DNA a few bases to one side of the recognition site.

The methylase adds only one methyl group at a time. If the substrate is nonmethylated, the enzyme introduces a single methyl group and dissociates from the DNA. A separate binding and methylation event occurs to introduce the second group.

The alternative activities of type I enzymes

The second class of modification and restriction activities consists of the type I and type III enzymes, which comprise multimers that undertake *both* the endonuclease and methylation functions. Their mechanisms of action are somewhat different from each other and from the type II activities. It is uncertain what proportion of bacteria have type I or type III systems, but they are less common than type II.

The **type I** enzymes, represented by the EcoK and EcoB variants of the *hsd* systems of *E. coli* strains K and B, were the first to be discovered. A type I modification and restriction enzyme consists of three types of subunit, as indicated in **Figure** 20.2.

The EcoK enzyme contains two copies of the R subunit, two copies of the M subunit, and one copy of the S subunit. The R subunit is responsible for restriction, the M subunit for methylation, and the S subunit for recognizing the target site on DNA. When a target site has been recognized by the S subunit, the enzyme's binding to DNA maybe succeeded by *either* restriction *or* modification. The activities of the R and M subunits are mutually exclusive.

Each subunit is coded by a single gene. Mut-

ants in *hsdR* are phenotypically r⁻m⁺; they cannot restrict DNA, but can modify it. Mutants in *hsdS* are r⁻m⁻; they can neither modify nor restrict DNA. Mutants in *hsdM* prevent restriction as well as modification. Thus the M subunit is involved in R subunit function. This may be a fail-safe mechanism, ensuring that *hsdM* mutants are r⁻m⁻, rather than the r⁺m⁻ phenotype they would otherwise have, which presumably would be lethal (because the restriction activity would degrade the cell's own unmodified DNA).

The EcoB and EcoK enzymes are allelic. Their target sequences are different, but in diploids, the S subunit of one strain can direct the activities of the R and M subunits of the other bacterial strain. This confirms that the recognition step is independent of the succeeding cleavage or methylation events.

The EcoB enzyme has subunits of similar size, but they are present in different molar ratios. The M and S subunits of strain B can form a 1:1 complex that exercises the methylation function independently of the restriction function. This may be a natural occurrence, because the three genes lie in an operon, in the order: *P1–hsdR–P2–hsdM–hsdS*,

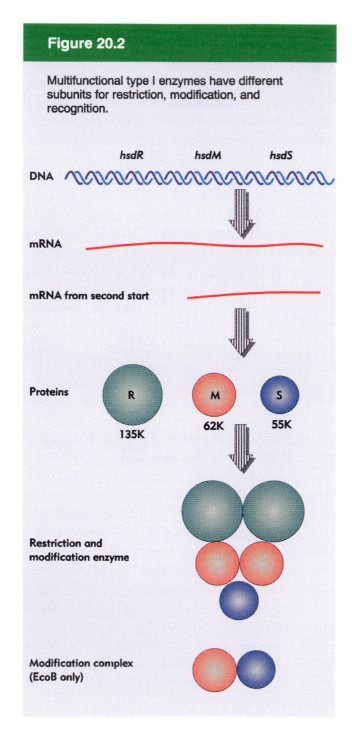

Figure 20.2

Multifunctional type I enzymes have different subunits for restriction, modification, and recognition.

two parts of the recognition site means that both lie on one face of the DNA; the sequence of the intervening region is not important.

Each side of the target site seems to be recognized by a distinct domain of the S subunit. Recombination between *hsdS* genes can generate new recognition subunits containing alternative combinations of domains. The recombinant S subunit recognizes a target site with the left component from one parental strain and the right component from the other.

Neither side of the recognition site is symmetrical, but together they possess adenine residues that are methylated on opposite strands, as indicated by the asterisks in the EcoB recognition sequence.

$$T\ G\ \overset{*}{A}\ N\ N\ N\ N\ N\ N\ N\ N\ T\ G\ C\ T$$
$$A\ C\ T\ N\ N\ N\ N\ N\ N\ N\ N\ \underset{*}{A}\ C\ G\ A$$

Whether a given DNA is to be cleaved or modified is determined by the state of the target site, as indicated in **Figure 20.3**. If the target site is fully methylated, the enzyme binds to it, but is released without any further action. If the target is hemimethylated, the enzyme methylates the unmethylated strand, perpetuating the state of methylation in host DNA. If the target site is unmethylated, its recognition triggers a cleavage reaction.

The groups for methylation are provided by the cofactor S-adenosyl-methionine (SAM), which is converted to S-adenosyl-homocysteine (SAH) in the reaction. SAM binds to the M subunit. In the initial stage of the reaction, SAM acts as an allosteric effector that changes the conformation of S subunit to allow it to bind to DNA.

After binding to DNA, the next step is a reaction with ATP. If the enzyme is bound at a completely methylated site, the arrival of ATP releases it from the DNA. At an unmethylated site, the ATP converts the enzyme into a state ready to sponsor cleavage. This depends on the R subunit. Hydrolysis of ATP is needed to cleave DNA. SAM is released from the enzyme before the restriction step occurs.

The type I enzymes display an extraordinary

where *P1* and *P2* are independent promoters. Thus *hsdM* and *hsdS* can be expressed separately from *hsdR*.

The recognition sites for EcoB and EcoK are bipartite structures, each consisting of a specific sequence of 3 bp separated by a few base pairs from a specific sequence of 4 bp. The separation of the

Figure 20.3

Type I enzymes bind to target sites, after which they are released from fully methylated sites, complete the methylation of hemimethylated sites, or move along DNA from nonmethylated sites to cleave the molecule elsewhere.

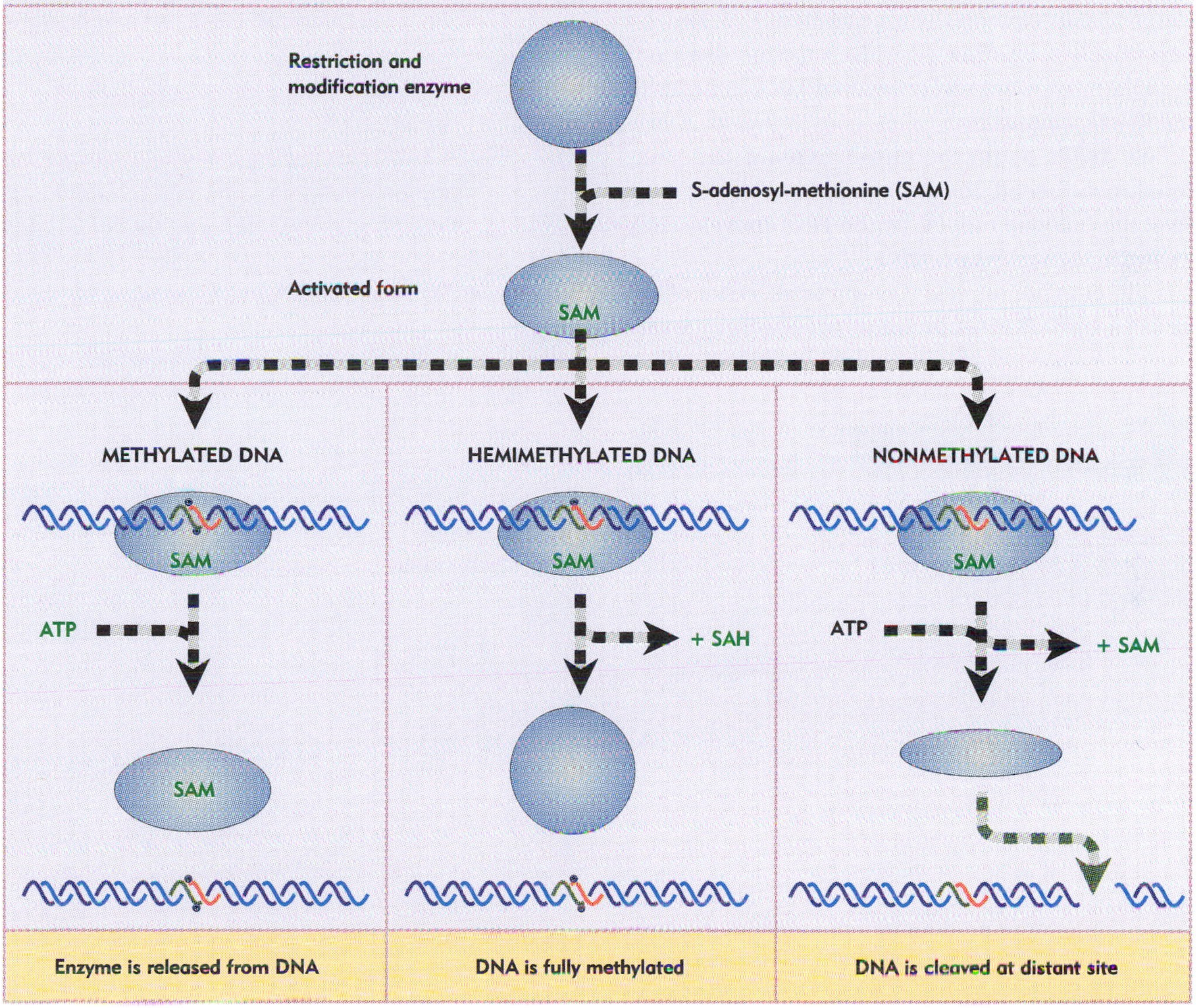

relationship between the sites of recognition and cleavage. *The cleavage site is located >1000 bp away from the recognition site.* Cleavage does not occur at a specific sequence, but the selection of a site for cutting does not seem to be entirely random because some regions of DNA are preferentially cleaved.

(The discrepancy between the sites of recognition and cleavage means that recognition sites cannot be defined by characterizing the breaks in DNA; the target sites are identified by their modification in the methylation event, and by the locations of mutations that abolish recognition.)

The cleavage reaction itself involves two steps. First, one strand of DNA is cut; then the other strand is cut nearby. There is some exonucleolytic degradation in the regions on either side of the site of cleavage. Extensive hydrolysis of ATP occurs; its function is not yet known.

How does the enzyme recognize one site and cleave another so far away? An important feature is that *the protein never lets go of the DNA molecule to which it initially binds.* If the enzyme is incubated with a mixture of modified and unmodified DNA, it preferentially cleaves the unmodified DNA. Having recognized a binding site, the enzyme does not dissociate from the unmethylated DNA in order to find its cleavage site.

Two types of model could explain the relationship between the recognition and cleavage sites: the enzyme moves; or the DNA moves. They are illustrated in **Figure 20.4**.

If the enzyme moves, it could translocate along the DNA until (for some unknown reason) it exer-

Figure 20.4

Does a type I enzyme move along DNA or does it remain at its target site, simultaneously pulling the DNA through the protein?

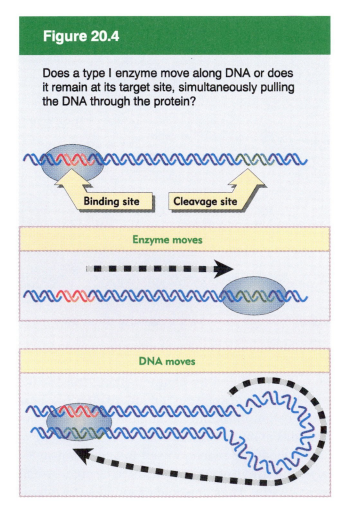

Figure 20.5

A type I restriction enzyme can simultaneously hold two different sites on DNA, creating a loop in the nucleic acid. Figure kindly provided by Robert Yuan.

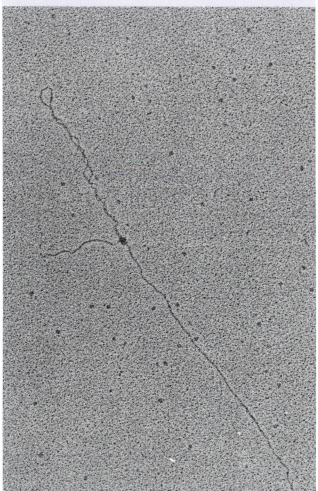

cises the cleavage option. If the DNA moves, the enzyme could remain attached at the recognition site, while it pulls the DNA through a second binding site on the enzyme, winding along until it reaches a cleavage region (again undefined in nature). **Figure 20.5** shows microscopic evidence that the enzyme generates loops in DNA; it seems to remain attached to its recognition site after cleavage, which supports the second model.

The dual activities of type III enzymes

Three type III modification and restriction enzymes have been investigated; EcoP1 and EcoP15 are coded by plasmids P1 and P15 in *E. coli*, and Hinf is found in *H. influenzae* of serotype R_f. Each enzyme consists of two types of subunit. The R subunit is responsible for restriction; the MS subunit is responsible for *both* modification and recognition.

The modification and restriction activities are expressed *simultaneously*. The enzyme first binds to its site on DNA, an action that requires ATP. (Subsequent dependence on ATP varies among the enzymes.) Then the methylation and restriction activities *compete for reaction with the DNA*.

The methylation event takes place at the binding site, consistent with the combination of methylation and recognition in the MS subunit. The restriction cleavage occurs 24–26 bases on one side, probably because the enzyme is large enough for the restriction subunit to contact DNA at this point, as depicted schematically in **Figure 20.6**.

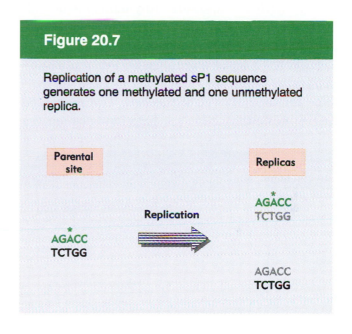

Figure 20.7

Replication of a methylated sP1 sequence generates one methylated and one unmethylated replica.

Restriction involves staggered cuts, 2–4 bases apart.

The enzymes methylate adenine residues, but the target sites of P1 and P15 have an intriguing feature: *they can be methylated only on one strand*. The sequences of the sites are

$$
\text{sP1} \quad \begin{array}{c} \text{AGACC} \\ \text{TCTGG} \end{array} \qquad \text{sP15} \quad \begin{array}{c} \text{CAGCAG} \\ \text{GTCGTC} \end{array}
$$

How is the state of methylation perpetuated? **Figure 20.7** demonstrates that two types of site are generated by replication. One replica has the original methylated strand; in fact it is indistinguishable from the parental site. The other replica is entirely unmethylated. It is therefore in principle a target for either restriction or modification. What determines that it is methylated rather than restricted? We do not know the answer, but one idea is that the modification reaction is linked to the act of replication.

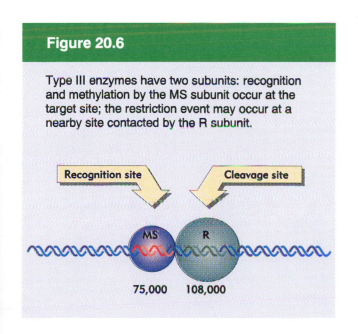

Figure 20.6

Type III enzymes have two subunits: recognition and methylation by the MS subunit occur at the target site; the restriction event may occur at a nearby site contacted by the R subunit.

Dealing with injuries in DNA

Injury to DNA is minimized by damage-containment systems that recognize the occurrence of a change and then rectify it. The repair systems are likely to be as complex as the replication apparatus itself, an indication of their importance for the survival of the cell. The measured rate of mutation reflects a balance between the number of damaging events occurring in DNA and the number that have been corrected (or miscorrected).

'Damage' to DNA consists of any change introducing a deviation from the usual double-helical structure. We can divide such changes into two general classes:

Figure 20.8

Substitutions of individual bases create mismatched pairs that may be corrected by replacing one base; if uncorrected they cause a mutation in one daughter duplex.

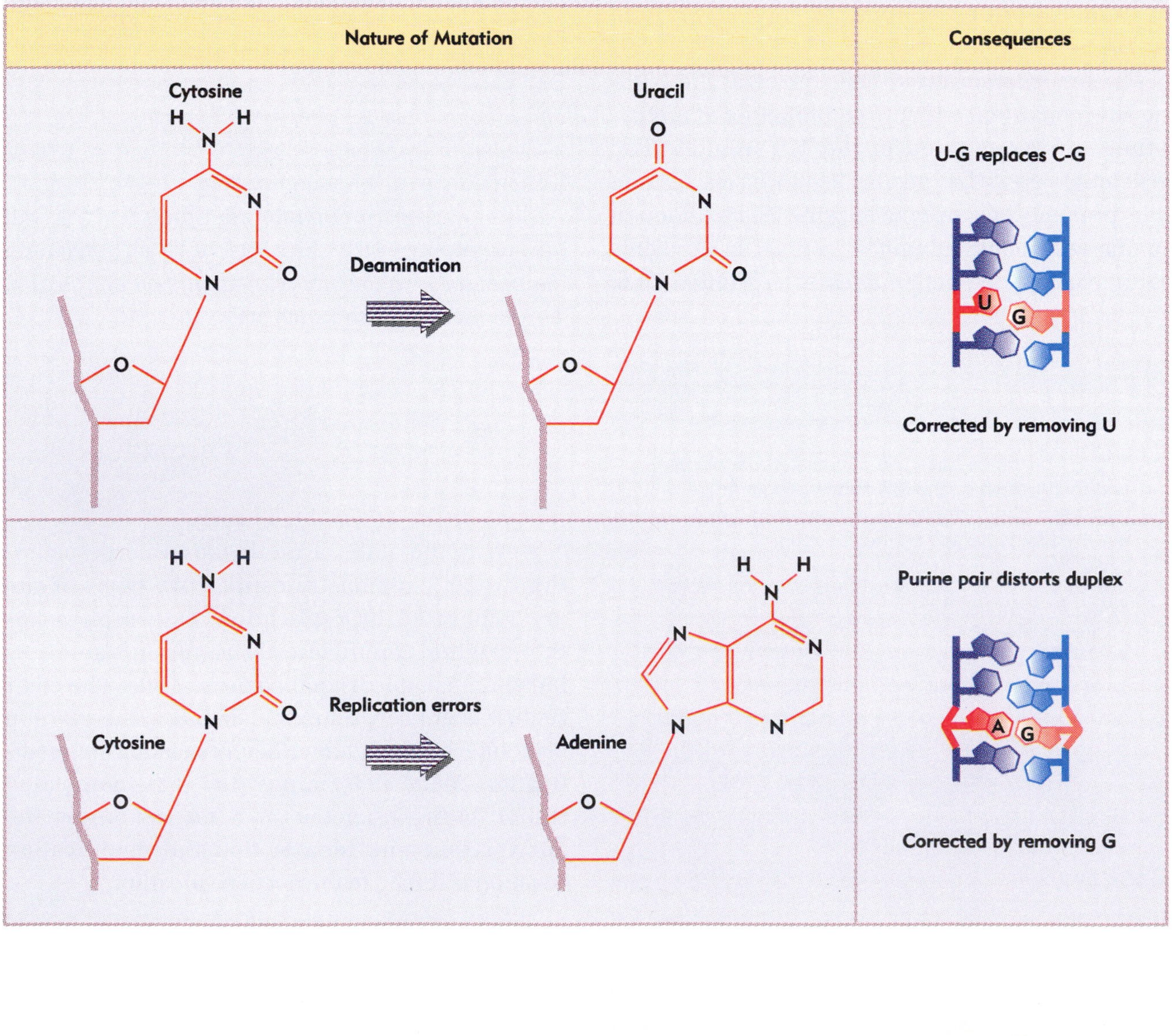

◆ **Single base changes** affect the sequence but not the overall structure of DNA. They do not affect transcription or replication, when the strands of the DNA duplex are separated. Thus these changes exert their damaging effects on future generations through the consequences of the change in DNA sequence (see Chapter 4). Such an effect is caused by the conversion of one base into another that is not properly paired with the partner base. **Figure 20.8** gives two examples: deamination of cytosine (spontaneously or by chemical mutagen) creates a mismatched U•G pair; while a replication error that inserts adenine instead of cytosine creates an A•G pair. Similar consequences could result from covalent addition of a small group to a base that modifies its ability to base pair. These changes may result in very minor structural distortion (as in the case of a U•G pair) or quite significant change (as in the case of an A•G pair), but the common feature is that the mismatch persists only until the next replication.

◆ **Structural distortions** may provide a physical impediment to replication or transcription. **Figure 20.9** shows some examples. Introduction of covalent links between bases on one strand of DNA or between bases on opposite strands inhibits replication and transcription. A well studied example of a structural distortion is caused by ultraviolet irradiation, which introduces covalent bonds between two adjacent thymine bases, giving the intrastrand **pyrimidine dimer** drawn in the figure. Similar consequences could result from addition of a bulky adduct to a base that distorts the structure of the double helix. A single–strand nick or the removal of a base prevents a strand from serving as a proper template for synthesis of RNA or DNA. The common feature in all these changes is that the damaged adduct remains in the DNA, continuing to cause structural problems and/or induce mutations, until it is removed.

Repair systems often can recognize a range of distortions in DNA as signals for action, and a cell may have several systems able to deal with DNA damage. We may divide them into several general types:

◆ **Direct repair** is rare and involves the reversal or simple removal of the damage. **Photo-reactivation** of pyrimidine dimers, in which the offending covalent bonds are simply reversed by a light-dependent enzyme, is the best example. This system is widespread in nature, and appears to be especially important in plants. In *E. coli* it depends on the product of a single gene (*phr*).

◆ **Excision–repair** is initiated by a recognition enzyme that sees an actual damaged base or a change in the spatial path of DNA. Recognition is followed by excision of a sequence that includes the damaged bases; then a new stretch of DNA is synthesized to replace the excised material. Such systems are common; some recognize general damage to DNA, while others act upon specific types of base damage (glycosylases remove specific altered bases; AP endonucleases remove residues from sites at which purine bases have been lost). There are often multiple excision–repair systems in a single cell type, and they probably handle most of the damage that occurs.

◆ **Mismatch repair** is accomplished by scrutinizing DNA for apposed bases that do not pair properly. Mismatches that arise during replication are corrected by distinguishing between the 'new' and 'old' strands and preferentially correcting the sequence of the newly synthesized strand. Other systems deal with mismatches generated by base conversions, such as the result of deamination.

◆ **Tolerance systems** cope with the difficulties that arise when normal replication is blocked at a damaged site. They provide a means for a damaged template sequence to be copied, probably with a relatively high frequency of errors. They are especially important in higher eukaryotic cells.

Figure 20.9

Modifications or removal of bases may cause structural defects that prevent replication or induce mutations in each replication cycle until they are removed.

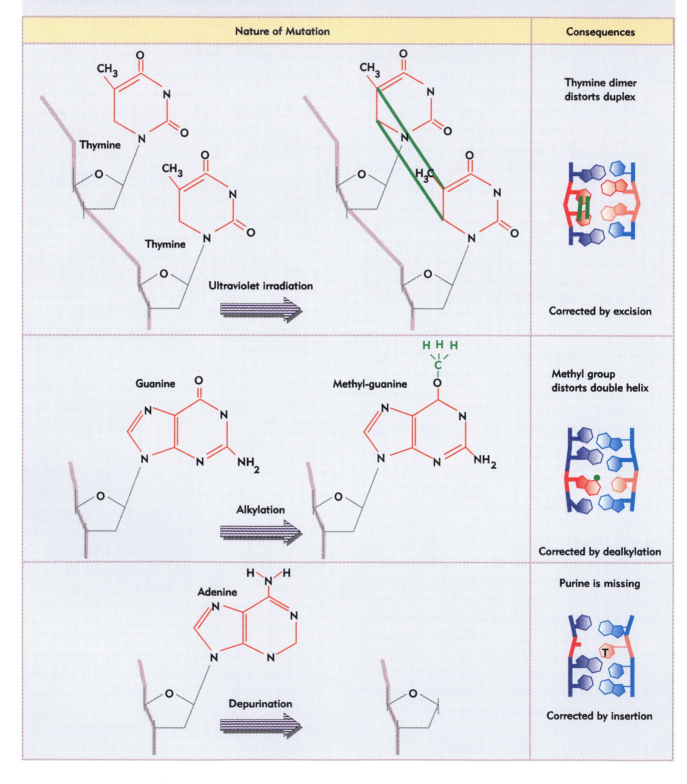

Nature of Mutation	Consequences

Thymine / Thymine — Ultraviolet irradiation → Thymine dimer distorts duplex / Corrected by excision

Guanine — Alkylation → Methyl-guanine / Methyl group distorts double helix / Corrected by dealkylation

Adenine — Depurination → Purine is missing / Corrected by insertion

◆ **Retrieval systems** comprise another type of tolerance system. When damage remains in a daughter molecule, and replication has been forced to bypass the site, a retrieval system uses recombination to obtain another copy of the sequence from an undamaged source. These 'recombination–repair' systems are well characterized in bacteria; it is not clear how important they are elsewhere.

Table 20.2

Many gene products are involved in repairing DNA damage in *E. coli*.

Gene	Effect of Mutation *In vivo*	Gene Product	Function
uvrA	Sensitivity to UV	ATPase subunit of endonuclease	Initiates removal
uvrB	"	Subunit of endonuclease	of thymine dimers
uvrC	"	Subunit of endonuclease	
uvrD	"	DNA helicase II	Unwind DNA
recA	1. deficiency in recombination	RecA protein of 40,000 daltons	1. DNA strand-exchange activity needed for recombination and repair
	2. does not induce repair pathways after damage		2. protease activity initiates SOS induction of repair pathways
recB	Deficiency in recombination	Exonuclease V (ATPase)	Needed for recombination
recC	"	Exonuclease V (DNA binding)	and recombination-repair
recD	"	Exonuclease V (58 K subunit)	
sbcB	Suppresses recBC mutation	Exonuclease I	Not known
recF	Deficiency in recombination-repair; and deficiency in	Not known	Not known
recJ	recombination in recBCsbcB mutants	Single-strand DNA exonuclease	Not known
recNOR	Same as recF	Not known	Part of *recF* pathway
recQ	"	DNA helicase; ATPase activity	"
recE	Deficiency in recombination	Exonuclease VIII	Not known
dam	UV sensitive, increased mutation rate	Methylase acts on GATC	Identifies "old" DNA strand
mutH	"	Endonuclease	Components of repair
mutL	"	Not known	system acting on newly
mutS	"	Recognizes mismatched pairs	synthesized DNA strand
uvrD	"	DNA helicase II (see above)	
mutY	Increases mutation frequency	Adenine glycosylase	Excises A from mismatches
ada	Sensitivity to alkylation	Guanine methyl transferase	Removes CH_3 from guanine
alkA	"	Methyl adenine glycosylase	Removes CH_3 adenine/guanine
umuC	Reduces mutagenic effects of error-prone repair	Not known	Not known
lon	UV-sensitive, septation-deficient	ATP-dependent protease that binds to DNA	May control genes for capsular polysaccharide
lexA	Regulates SOS response	Repressor (22,000 daltons)	Controls many genes

Mutations in many loci of *E. coli* affect the response to agents that damage DNA. Many of these genes are likely to code for enzymes that participate in DNA repair systems, although in the majority of cases the gene product and its function remain to be identified.

Table 20.2 summarizes the properties of some of the mutations that affect the ability of *E. coli* cells to engage in DNA repair. They fall into groups, which correspond to several repair pathways (not necessarily all independent). The major known pathways are the *uvr* excision–repair system, the *dam* replication mismatch–repair system, and the *recB* and *recF* recombination and recombination–repair pathways.

When the repair systems are eliminated, cells become *exceedingly* sensitive to ultraviolet irradiation. The introduction of UV-induced damage has been a major test for repair systems, and so in assessing their activities and relative efficiencies, we should remember that the emphasis might be different if another damaged adduct were studied.

Excision–repair systems in *E. coli*

Excision–repair systems vary in their specificity, but share the same general features. Each system removes mispaired or damaged bases from DNA and then synthesizes a new stretch of DNA to replace them. The main type of pathway for excision–repair is illustrated in **Figure 20.10**.

In the **incision** step, the damaged structure is recognized by an endonuclease that cleaves the DNA strand on both sides of the damage.

In the **excision** step, a 5′–3′ exonuclease removes a stretch of the damaged strand.

In the **synthesis** step, the resulting single-–stranded region serves as a template for a DNA polymerase to synthesize a replacement for the excised sequence. Finally, DNA ligase covalently links the 3′ end of the new material to the old material.

Different excision–repair modes are identified by the heterogeneity of the lengths of the segments of repaired DNA. These pathways are described as **very short patch repair (VSP)**, **short-patch repair** and **long-patch repair**. The VSP system deals with mismatches between specific bases (see later). The latter two excision–repair systems both involve the *uvr* genes.

The *uvr* system of excision–repair includes three genes, *uvrA,B,C*, that code for the components of a repair endonuclease. It functions in the stages indicated in **Figure 20.11**. First, a UvrAB combination recognizes pyrimidine dimers and other bulky lesions. Then UvrA dissociates (this requires ATP), and UvrC joins UvrB. The UvrBC combination makes an incision on each side, one 7 nucleotides from the 5′ side of the damaged site, and the other 3–4 nucleotides away from the 3′ side. This also requires ATP.

UvrD is a helicase that helps to unwind the DNA to allow release of the single strand between the two cuts. Also, several *E. coli* enzymes can excise pyrimidine dimers *in vitro* from DNA in which such incisions have been made. The enzymes include the 5′–3′ exonuclease activity of DNA polymerase I and the single-strand-specific exonuclease VII. Mutation of any one of these putative excision nucleases does not result in a measurable diminution in the ability of *E. coli* cells to remove pyrimidine dimers. DNA polymerase I is likely to perform most of the excision *in vivo*.

The average length of excised DNA is ~12 nucleotides, which gives rise to the description of this mode as the short-patch repair. The enzyme

Figure 20.10

Excision-repair removes and replaces a stretch of DNA that includes the damaged base(s).

Damage

Mutant base is mismatched and/or distorts structure

Incision

Endonuclease cleaves on both sides of damaged base

Excision

Exonuclease removes DNA between nicks

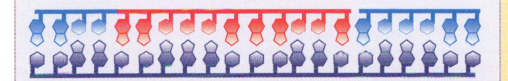

Synthesis

Polymerase synthesizes replacement DNA

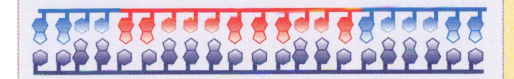

Ligase seals nick

Figure 20.11

The Uvr system operates in stages in which UvrAB recognizes damage, UvrBC nicks the DNA, and UvrD unwinds the marked region.

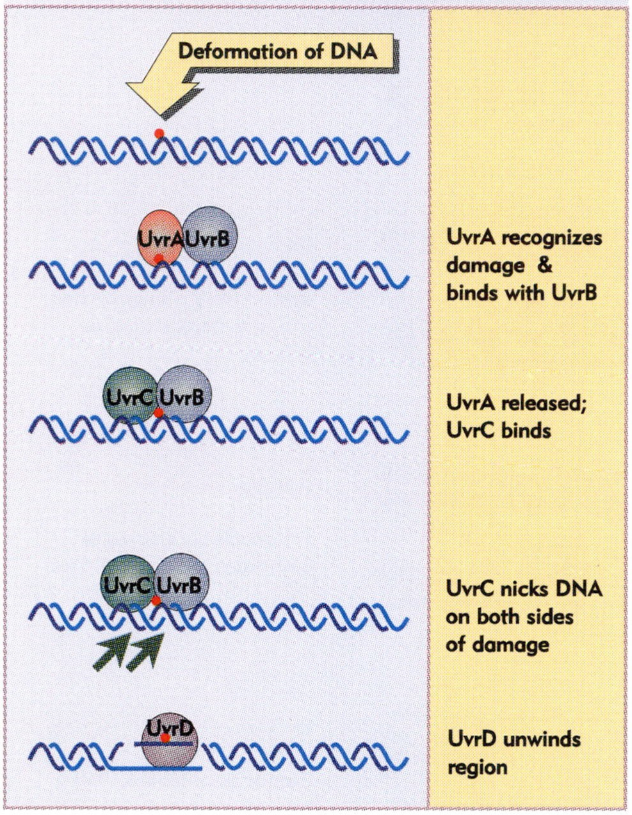

Deformation of DNA

UvrAUvrB — UvrA recognizes damage & binds with UvrB

UvrCUvrB — UvrA released; UvrC binds

UvrCUvrB — UvrC nicks DNA on both sides of damage

UvrD — UvrD unwinds region

involved in the repair synthesis probably also is DNA polymerase I (although polymerases II and III can substitute for it).

Other potential components of the *uvr* system have been identified by mutations that impede short-patch repair, but none has yet been identified with a product or function in the pathway.

For bulky lesions, short-patch repair accounts for 99% of the excision-repair events. The remaining 1% involve the replacement of stretches of DNA mostly ~1500 nucleotides long, but extending up to >9000 nucleotides. This mode also requires the *uvr* genes and involves DNA polymerase I. A difference between the two modes of repair is that short-patch repair is a constitutive function of the bacterial cell, but long-patch repair must be induced by damage

(see later). Long-patch repair probably acts on lesions found in regions near replication forks. We have not yet characterized the differences between these modes in terms of the involvement of different gene products.

The existence of repair systems that engage in DNA synthesis raises the question of whether their quality control is comparable with that of DNA replication. So far as we know, most systems, including *uvr*-controlled excision repair, do not differ significantly from DNA replication in the frequency of mistakes. However, **error-prone** synthesis of DNA occurs in *E. coli* under certain circumstances.

The error-prone feature was first observed when it was found that the repair of damaged λ phage DNA is accompanied by the induction of mutations if the phage is introduced into cells that had previously been irradiated with UV. This suggests that the UV irradiation of the host has activated some protein(s) that do not function in the unirradiated cell, and whose activity generates mutations. The mutagenic response also operates on the bacterial host DNA.

What is the actual error-prone activity? Current thinking focuses on the idea that it is caused by a component of a tolerance pathway that permits or compels replication to proceed past the site of damage. When the replicase passes any site at which it cannot insert complementary base pairs in the daughter strand, it inserts incorrect bases, which represent mutations. The error-prone activity requires DNA polymerase III, the usual replicase, which is consistent with the idea that the relevant function acts in concert with the normal replication apparatus.

Several functions are involved in this error-prone pathway. Mutations in the genes *umuD* and *umuC* abolish UV-induced mutagenesis, but do not interfere with any known enzymatic functions. The genes constitute the *umuDC* operon, whose expression is induced by DNA damage (see below). Some plasmids carry genes called *mucA* and *mucB*, which are homologs of *umuD* and *umuC*, and whose introduction into a bacterium increases resistance to UV killing and susceptibility to mutagenesis.

Genes whose products are involved in controlling the fidelity of DNA synthesis during either replication or repair may be identified by mutations that have a **mutator** phenotype. A mutator mutant has an increased frequency of spontaneous mutation. If identified originally by the mutator phenotype, a gene is described as *mut*; but often a *mut* gene is later found to be equivalent with a known replication or repair activity. Some examples are summarized in **Table 20.3**.

The enzymatic activities coded by *mut* genes fall into groups. The major group consists of enzymes that participate directly or indirectly in the removal of particular types of damage; failure to remove a damaged or mispaired base before replication allows it to induce a mutation. A smaller group, typified by *dnaQ*, is concerned with the accuracy of synthesizing new DNA.

Table 20.3

Mutator genes are concerned with DNA replication or repair.

Mutator locus	Enzyme activity
mutD = dnaQ	Subunit of DNA polymerase III
mutU = uvrD	DNA helicase of uvr repair system
mutH,U	Components of dam repair system
mutL,S	Ditto; also removes T from C-T mismatches
mutY	Removes A from A-G and A-C mismatches

Controlling the direction of mismatch repair

When a structural distortion is removed from DNA, the wild-type sequence is restored. In most cases, the distortion is due to the creation of a base that is not naturally found in DNA, and which is therefore recognized and removed by the repair system.

A problem arises if the target for repair is a mispaired partnership of (normal) bases created when one was mutated. The repair system has no intrinsic means of knowing which is the wild-type base and which is the mutant! All it sees are two improperly paired bases, either of which can provide the target for excision–repair.

If the mutated base is excised, the wild-type sequence is restored. But if it happens to be the original (wild-type) base that is excised, the new (mutant) sequence becomes fixed. Often, however, the direction of excision–repair is not random, but is biased in a way that is likely to lead to restoration of the wild-type sequence.

Some precautions are taken to direct repair in the right direction. For example, for cases such as the deamination of 5-methyl-cytosine to thymine, there is a special system to restore the proper sequence. The deamination generates a G•T pair, and the system that acts on such pairs has a bias to correct them to G•C pairs (rather than to A•T pairs).

The VSP system undertakes this reaction, and it includes the *mutL* and *mutS* genes that also participate in the *dam* repair system. The *mutL,S* system removes T from both G•T and C•T mismatches.

Another repair activity, identified by *mutY*, replaces the A in C•A and G•A mismatches. This system functions by the direct removal of the base from DNA; *mutY* codes for an adenine glycosylase, which creates an apurinic site that is recognized by an endonuclease whose action triggers the involvement of the excision repair system.

When mismatching errors occur during replication in *E. coli*, it may be possible to distinguish the original strand of DNA. Immediately after replication of methylated DNA, only the original parental strand carries the methyl groups (see Figure 20.1). In the period while the newly synthesized strand awaits the introduction of methyl groups, the two strands can be distinguished.

This provides the basis for a system to correct replication errors. The *dam* gene codes for a methylase whose target is the adenine in the sequence $\frac{GATC}{CTAG}$ (see Figure 19.14). The hemimethy-

lated state is used to distinguish replicated origins from nonreplicated origins (see Figure 19.20). The same target sites are used by a replication-related repair system.

Figure 20.12 shows that DNA containing mismatched base partners is repaired preferentially by excising the strand that lacks the methylation. The excision is quite extensive. The result is that the newly synthesized strand is corrected to the sequence of the parental strand.

E. coli dam⁻ mutants show an increased rate of spontaneous mutation. This repair system therefore helps reduce the number of mutations caused by errors in replication. It consists of several proteins, coded by the *mut* genes, and includes one product (MutS) that specifically recognizes mismatched base pairs.

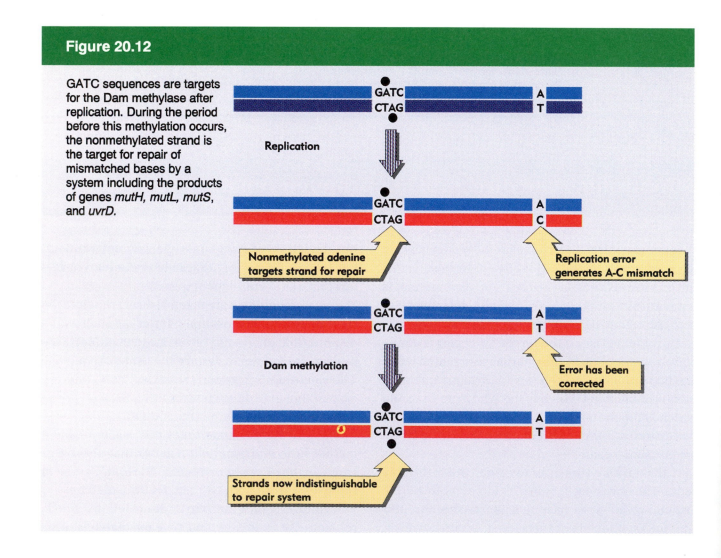

Figure 20.12

GATC sequences are targets for the Dam methylase after replication. During the period before this methylation occurs, the nonmethylated strand is the target for repair of mismatched bases by a system including the products of genes *mutH*, *mutL*, *mutS*, and *uvrD*.

Replication

Nonmethylated adenine targets strand for repair

Replication error generates A-C mismatch

Dam methylation

Error has been corrected

Strands now indistinguishable to repair system

Retrieval systems in *E. coli*

Retrieval systems have variously been termed 'post-replication repair', because they function after replication, or 'recombination-repair', because the activities involved overlap with those involved in genetic recombination. Such systems are effective in dealing with the defects produced in daughter duplexes by replication of a template that contains damaged bases. An example is illustrated in **Figure 20.13**.

Consider a structural distortion, such as a pyrimidine dimer, on one strand of a double helix. When the DNA is replicated, the dimer prevents the damaged site from acting as a template. Replication is forced to skip past it.

DNA polymerase probably proceeds up to or close to the pyrimidine dimer. Then the polymerase ceases synthesis of the corresponding daughter strand. Replication restarts some distance farther along. A substantial gap is left in the newly synthesized strand.

The resulting daughter duplexes are different in nature. One has the parental strand containing the damaged adduct, facing a newly synthesized strand with a lengthy gap. The other duplicate has the undamaged parental strand, which has been copied into a normal complementary strand. The retrieval system takes advantage of the normal daughter.

The gap opposite the damaged site in the first duplex is filled by stealing the homologous single strand of DNA from the normal duplex. Following this **single-strand exchange**, the recipient duplex has a parental (damaged) strand facing a wild-type strand. The donor duplex has a normal parental strand facing a gap; the gap can be filled by repair synthesis in the usual way, generating a normal duplex. Thus the damage is confined to the original distortion (although the same recombination-repair events must be repeated after every replication cycle unless and until the damage is removed by an excision–repair system).

In *E. coli* deficient in excision–repair, mutation in the *recA* gene essentially abolishes all the remaining repair and recovery facilities. Attempts to replicate DNA in *uvr⁻recA⁻* cells produce fragments of DNA whose size corresponds with the expected distance between thymine dimers. This result implies that the dimers provide a lethal obstacle to replication in the absence of RecA function. It explains why the double mutant cannot tolerate >1–2 dimers in its genome (compared with the ability of a wild-type bacterium to shrug off as many as 50).

The *recA* gene, and other *rec* mutations, identify retrieval pathways. The components involved in repair overlap with those involved in recombination, but are not entirely identical with them, since some mutations affect one but not the other activity.

The *recA* mutants are almost completely deficient in genetic recombination as well as in the recovery response. The RecA protein has the function of exchanging strands between DNA molecules, a central activity in recombination (see Chapter 33), and is also involved in the single-strand exchange involved in recombination–repair. RecA appears to be directly involved in the error-prone pathway described above.

The properties of double mutants suggest that *recA* participates in two Rec pathways. To test whether two genes with related functions are part of the same or different pathways, the phenotypes of the single mutants are compared with the phenotype of the double mutant. If the genes are in the same pathway, the phenotype of the double mutant will be no different from that of an individual mutants. If the genes are in different pathways, the double mutant will lack both pathways instead of one, and so has a more severe phenotype than either single mutant. By this criterion, one *rec* pathway involves the *recBC* genes; the other involves *recF*.

The *recBC* genes code for two subunits of exonuclease V, whose action is limited by some other component of the pathway. (In *recA* mutants, the enzyme is uncontrolled, and degrades an excessive amount of DNA. This causes the 'reckless'

Figure 20.13

An *E. coli* retrieval system uses a normal strand of DNA to replace the gap left in a newly synthesized strand opposite a site of unrepaired damage.

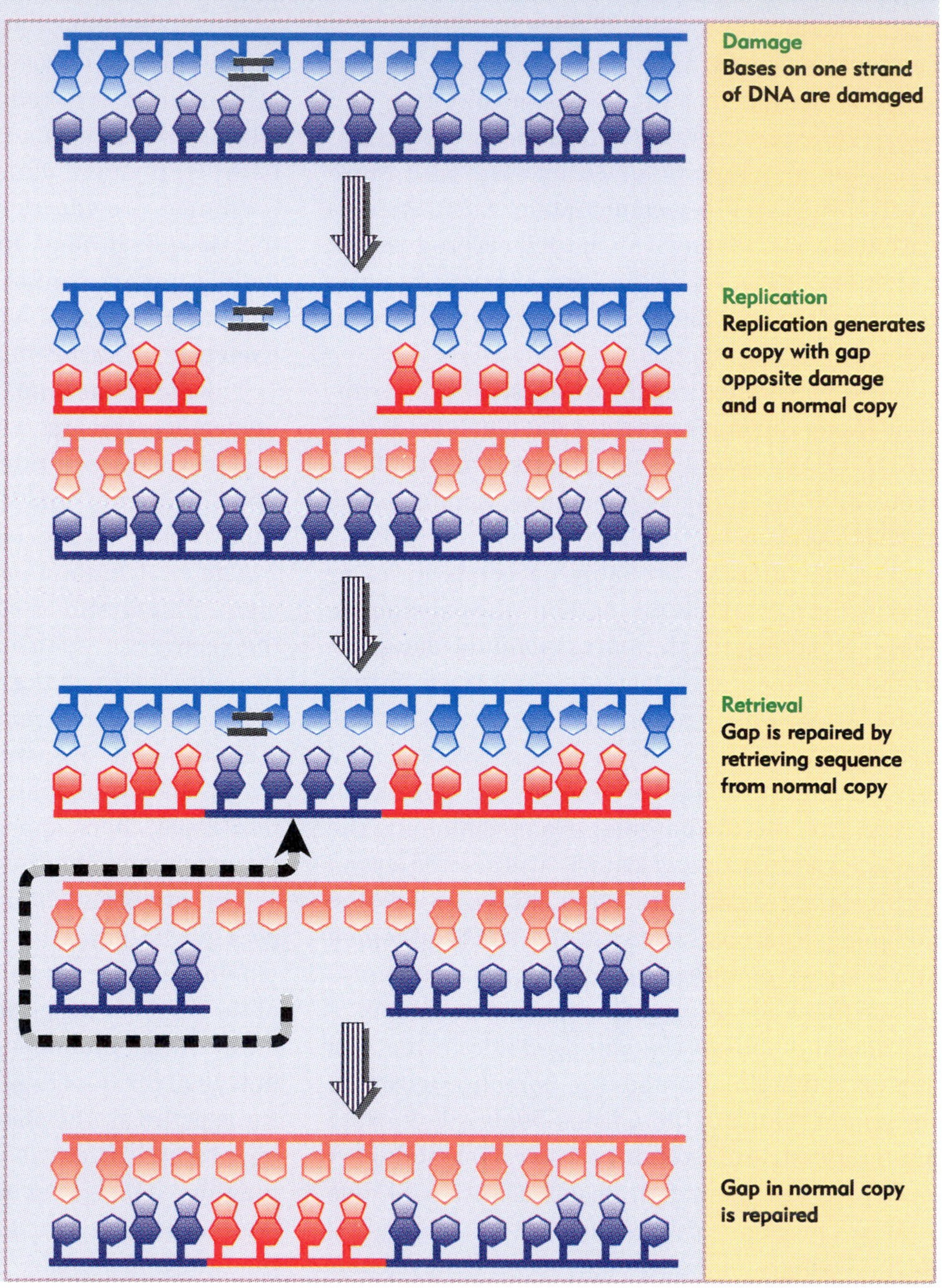

Damage
Bases on one strand of DNA are damaged

Replication
Replication generates a copy with gap opposite damage and a normal copy

Retrieval
Gap is repaired by retrieving sequence from normal copy

Gap in normal copy is repaired

phenotype that gave rise to the name of these loci.) The function of *recF* is unknown. Other *rec* loci have been identified by the effects of mutations on recombination–repair or recombination. Several are identified by their ability to suppress other *rec*

mutants. As with the additional *uvr* loci, they have not been equated with products or functions.

The designations of these genes are based on the phenotypes of the mutants; but sometimes a mutation isolated in one set of conditions and named as a *uvr* locus turns out to have been isolated in another set of conditions as a *rec* locus. This uncertainty makes an important point. We cannot yet define how many functions belong to each pathway or how the pathways interact. The *uvr* and *rec* pathways are not entirely independent, because *uvr* mutants show reduced efficiency in recombination–repair.

We must expect to find a network of nuclease, polymerase, and other activities, constituting repair systems that are partially overlapping (or in which an enzyme usually used to provide some function can be substituted by another from a different pathway).

The direct involvement of RecA protein in recombination–repair is only one of its activities. This extraordinary protein also has another, quite distinct function. When it is activated by ultraviolet irradiation (or by other treatments that block replication), it causes proteolytic cleavage of a series of target proteins. It was originally thought that RecA itself is a protease, but now it seems rather that it triggers latent proteolytic activities that reside in the target proteins. The activation of RecA is responsible for inducing the expression of many genes, whose products include repair functions (including, for example, the long-patch repair mode).

These dual activities of the RecA protein make it difficult to know whether a deficiency in repair in *recA* mutant cells is due to loss of the DNA strand–exchange function of RecA or to some other function whose induction depends on the protease activity.

An SOS system of many genes

Many treatments that damage DNA or inhibit replication in *E. coli* induce a complex series of phenotypic changes called the **SOS response**. This is set in train by the interaction of the RecA protein with the LexA repressor.

The damage can take the form of ultraviolet irradiation (the most studied case) or can be caused by cross-linking or alkylating agents. Inhibition of replication by any of several means, including deprivation of thymine, addition of drugs, or mutations in several of the *dna* genes, has the same effect.

The response takes the form of increased capacity to repair damaged DNA, achieved by inducing synthesis of the components of both the long-patch excision–repair system and the Rec recombination–repair pathways. In addition, cell division is inhibited. Lysogenic prophages may be induced.

The initial event in the response is the activation of RecA by the damaging treatment. We do not know very much about the relationship between the damaging event and the sudden change in RecA activity. Because a variety of damaging events can induce the SOS response, current work focuses on the idea that RecA is activated by some common intermediate in DNA metabolism.

The inducing signal could consist of a small molecule released from DNA; or it might be some structure formed in the DNA itself. *In vitro*, the activation of RecA requires the presence of single-stranded DNA and ATP. Thus the activating signal could be the presence of a single-stranded region at a site of damage. Whatever form the signal takes, its interaction with RecA is rapid: the SOS response occurs within a few minutes of the damaging treatment.

Activation of RecA causes proteolytic cleavage of the product of the *lexA* gene. LexA is a small (22,000 dalton) protein that is relatively stable in untreated cells, where it functions as a repressor at many operons. The cleavage reaction is unusual; LexA has a latent protease activity that is activated by RecA. When RecA is activated, it causes LexA to undertake an autocatalytic cleavage of itself; this inactivates the LexA repressor function, and co-ordinately induces all the operons to which it was bound. The pathway is illustrated in **Figure 20.14**.

The target genes for LexA repression include many repair functions, only some of which are identified at present. A systematic screen for LexA-responsive genes has been constructed by fusing random operons to the *lacZ* gene, and then assaying for increased levels of the product β-galactosidase when cells are treated with an

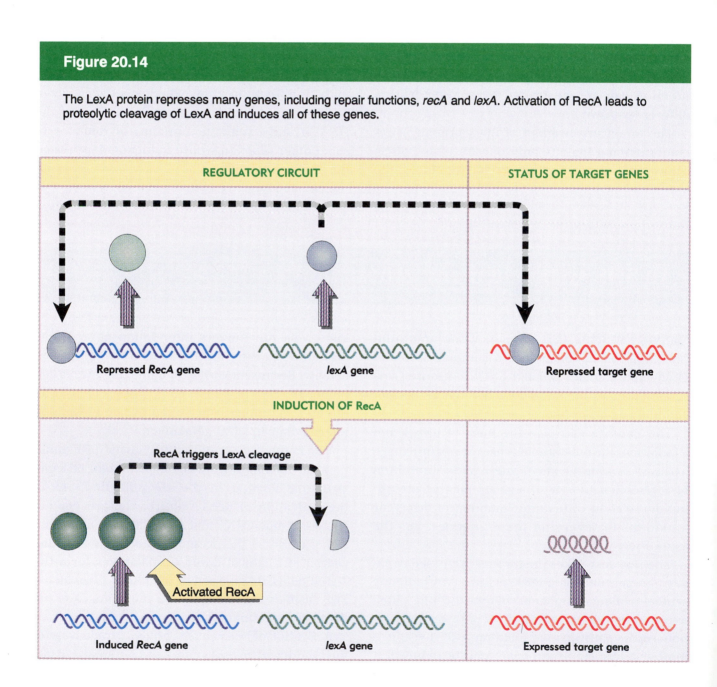

Figure 20.14

The LexA protein represses many genes, including repair functions, *recA* and *lexA*. Activation of RecA leads to proteolytic cleavage of LexA and induces all of these genes.

agent that induces damage. At least five responsive genes have been identified; they are called *din* (damage inducible). Genes previously known to be part of the SOS response by the activation of their products include *recA*, *lexA*, *uvrA*, *uvrB*, *umuC*, and *himA*.

Some of the SOS genes are active only in treated cells; others are active in untreated cells, but the level of expression is increased by cleavage of LexA. In the case of *uvrB*, which is a component of the excision–repair system, the gene has two promoters; one functions independently of LexA, the other is subject to its control. Thus after cleavage of LexA, the gene can be expressed from the second promoter as well as from the first.

LexA represses its target genes by binding to a 20 bp stretch of DNA called an **SOS box**; this sequence displays symmetry, and a copy is present at each target locus. The SOS boxes at different loci are not identical, but conform to a consensus sequence with 8 absolutely conserved positions. Like other operators, the SOS boxes overlap with the respective promoters. At the *lexA* locus, the subject of autogenous repression, there are two adjacent SOS boxes.

RecA and LexA are mutual targets in the SOS circuit: RecA triggers cleavage of LexA, which represses *recA*. The SOS response therefore causes amplification of both the RecA protein and the LexA repressor. The results are not so contradictory as might at first appear.

The increase in expression of RecA protein is necessary (presumably) for its direct role in the recombination–repair pathways. On induction, the level of RecA is increased from its basal level of ~1200 molecules/cell by up to 50-fold. The high level in induced cells means there is sufficient RecA to ensure that all the LexA protein is cleaved. This should prevent LexA from reestablishing repression of the target genes.

But the main importance of this circuit for the cell lies in the ability to return rapidly to normalcy. When the inducing signal is removed, the RecA protein loses the ability to destabilize LexA. At this moment, the *lexA* gene is being expressed at a high level; in the absence of activated RecA, the LexA protein rapidly accumulates in the uncleaved form and turns off the SOS genes. This explains why the SOS response is freely reversible.

RecA also triggers cleavage of other cellular targets, sometimes with more direct consequences. The UmuD protein is cleaved when RecA is activated; the cleavage event activates UmuD and the error-prone repair system.

Activation of RecA also causes cleavage of some other repressor proteins, including those of several prophages. Among these is the lambda repressor (with which the protease activity in fact was discovered). This explains why lambda is induced by ultraviolet irradiation; the lysogenic repressor is cleaved, releasing the phage to enter the lytic cycle.

This reaction is not a cellular SOS response, but instead represents a recognition by the prophage that the cell is in trouble. Survival is then best assured by entering the lytic cycle to generate progeny phages. In this sense, prophage induction is piggybacking onto the cellular system by responding to the same indicator (activation of RecA).

All known proteins that are targets of RecA activation are cleaved at an Ala–Gly dipeptide sequence in the middle of the polypeptide chain. There is only limited amino acid homology on either side of the dipeptide, which suggests that the tertiary structure of the protein is an important feature in target recognition.

The two activities of RecA are relatively independent. The *recA441* mutation allows the SOS response to occur without inducing treatment, probably because RecA remains spontaneously in the activated state. Other mutations abolish the ability to be activated. Neither type of mutation affects the ability of RecA to handle DNA. The reverse type of mutation, inactivating the recombination function but leaving intact the ability to induce the SOS response, would be useful in disentangling the direct and indirect effects of RecA in the repair pathways.

Mammalian repair systems

Biochemical characterization of repair systems in eukaryotic cells is only primitive, for the most part confined to the isolation of the occasional enzyme preparation whose properties *in vitro* suggest that it could be part of a repair system. The existence of excision–repair pathways can be established in cultured cells by following the actual removal of damage from DNA or by detecting the replacement of DNA segments in response to damaging treatments.

Mammalian cells show heterogeneity in the amount of DNA resynthesized at each lesion after damage. However, the longest patches seen in mammalian cells are comparable with those of the short-patch bacterial repair system. This pathway operates on damage caused by ultraviolet irradiation or by treatments that have related effects. Another pathway introduces only 3–4 repair bases at sites of damage generated by X-irradiation or alkylation.

An indication of the existence and importance of the mammalian repair systems is given by certain human hereditary disorders. The best investigated of these is xeroderma pigmentosum (XP), a recessive disease resulting in hyper-sensitivity to sunlight, in particular to ultraviolet. The deficiency results in skin disorders (and sometimes more severe defects).

The disease is explicable in terms of a failing in excision–repair; fibroblasts from XP patients are deficient in the excision of pyrimidine dimers and other bulky adducts. Several (~9) genetic complementation groups have been distinguished, many characterized by a deficiency at the incision step of repair.

Some indirect results suggest that mammalian cells have recombination–repair systems. Again, these systems may be related to genetic recombination itself. An example is the recessive human disorder of Bloom's syndrome; an increased frequency of chromosomal aberrations, including sister chromatid exchanges, could be related to the operation of recombination systems.

Evidence for some general conservation of repair functions in eukaryotes is provided by the ability of cloned segments of mammalian DNA to restore repair activities to yeast strains that have mutations in repair enzymes. This approach may make it possible to define the scope of repair systems in a range of eukaryotes. Some of the deficiencies in XP patients, for example, identify homologs of genes that are involved in repair pathways in yeast.

Summary

Bacteria contain systems that maintain the integrity of their DNA sequences in the face of damage or errors of replication and that distinguish the DNA from sequences of a foreign source. Three types of modification and restriction system share the principle that host DNA is methylated at particular target sites. The state of methylation is perpetuated by a methylase that recognizes hemimethylated DNA following replication. Nonmethylated target sites are recognized by an endonuclease.

Type II systems are the most common and consist of separate methylase and endonuclease enzymes that recognize the same short target sequence, usually a 4–6 bp palindrome. Type I enzymes consist of three subunits, concerned with target sequence recognition, methylation, and cleavage. The cleavage site is not specific and is located >1,000 bp away from the recognition site. Type III enzymes consist of two subunits, one for recognition and modification, another for cleavage.

Repair systems can recognize mispaired, altered, or missing bases in DNA, or other structural distortions of the double helix. Excision–repair systems cleave DNA near a site of damage, remove one strand, and synthesize a new sequence to replace the excised material. Three excision–repair systems in E. coli can be distinguished by the lengths of the regions that are excised. The dam system is involved in correcting mismatches generated by incorporation of incorrect bases during replication, and the uvr genes are involved in both of the other systems for general repair. Recombination–repair systems retrieve information from a DNA duplex and use it to repair a sequence that has been damaged on both strands. The recBC and recF pathways can be distinguished, but have not been characterized in detail. The recA product may be involved in repair pathways in one of its capacities, the ability to synapse molecules of DNA.

The other capacity of recA is the ability to induce the SOS response. RecA is activated by damaged DNA in an unknown manner. It triggers cleavage of the LexA repressor protein, thus releasing repression of many loci, and inducing synthesis of the enzymes of both excision–repair and recombination–repair pathways. Genes under LexA control possess an operator SOS box. RecA also directly activates some repair activities. Cleavage of repressors of lysogenic phages may induce the phages to enter the lytic cycle.

Further reading

Reviews

The separate (type II) modification and restriction activities have been reviewed by **Modrich** (*Quart. Rev. Biophys.* **3**, 315–369, 1979)

A valuable tour of multifunctional (type I and type III) enzymes, emphasizing the enzymatic activities, was provided by **Yuan** (*Ann. Rev. Biochem.* **50**, 285–315, 1981), who developed a model for type I enzyme translocation (*Cell* **20**, 237–244, 1980).

Excision repair systems have been reviewed by **Sancar and Sancar** (*Ann. Rev. Biochem.* **57**, 29–67, 1988). Repair of alkylation damage has been reviewed by **Lindahl** *et al.* (*Ann. Rev. Biochem.* **57**, 133–157, 1988).

The various functions of methylation in bacteria were summarized by **Marinus** (*Ann. Rev. Genet.* **21**, 113–131, 1987).

An integrated view of the systems and mechanisms for DNA repair in prokaryotes and eukaryotes is contained in the book by **Friedberg** (*DNA Repair*, Freeman, New York, 1985).

The SOS control network of E. coli has been reviewed extensively by **Little and Mount** (*Cell* **29**, 11–22, 1982) and **Walker** (*Microbiol. Rev.* **48**, 60–93, 1984).

Organization of the eukaryotic genome

The present day genetics concept visualizes the appearance of an organism as a result of an interaction of the whole set of genes the organism possesses and the environment in which it develops. A change in any of the genes, called a mutation, is liable to upset the balance of that system and show up on the organism as a character, usually as an abnormality, in some respect poorer than the wild type... Studies with deficiencies, *viz.*, material where one or several genes are missing, show that the majority of deficiencies are lethal to the organism when present in a homozygous condition. This suggests that the presence of at least the majority of genes is essential in order that an organism may live. Moreover, the work with *D. melanogaster* deficiencies indicates that many of them are cell-lethal, *viz.*, that even a small group of cells located among normal tissues but containing a homozygous deficiency cannot exist. This suggests that genes are active in every cell and that, probably, the majority of them perform there a function highly important in the vital processes of the cell.

Milislav Demerec, 1935

CHAPTER 21

The extraordinary power of DNA technology

The technology for dealing with DNA has become so powerful that it is now a routine project to obtain the DNA corresponding to any particular gene. At the heart of this technology is the ability to amplify individual DNA sequences. **Cloning** a fragment of DNA allows indefinite amounts to be produced from even a single original molecule. (A **clone** is defined as a large number of cells or molecules all identical with an original ancestral cell or molecule.) Once any particular segment of DNA has been cloned, its properties can be characterized. Its sequence should reveal whether it is likely to code for a protein; and the cloned segment (or parts of it) can be used to test whether it contains sites that are bound by regulatory proteins.

Cloning technology involves the construction of novel DNA molecules by joining sequences from different sources. The product is often described as **recombinant DNA**, and the techniques (more colloquially) as **genetic engineering**. They are applicable equally to prokaryotes and eukaryotes, although the power of this approach is especially evident with eukaryotic genomes.

Cloning of DNA is made possible by the ability of bacterial plasmids and phages to reproduce after additional sequences of DNA have been incorporated into their genomes. An insertion generates a **hybrid** or **chimeric** plasmid or phage, consisting in part of the authentic DNA of the original genome and in part of the additional 'foreign' sequences. These chimeric elements replicate in bacteria just like the original plasmid or phage and so can be obtained in large amounts. Copies of the original foreign fragment can be retrieved from the progeny. Since the properties of the chimeric species usually are unaffected by the particular foreign sequences that are involved, almost any sequence of DNA can be cloned in this way. Because the phage or plasmid is used to 'carry' the foreign DNA as an inert part of the genome, it is often referred to as the **cloning vector**.

Cloning a specific gene requires the ability to identify or characterize particular regions or sequences of the genome. In practical terms, we need a **probe** that will react with the target DNA. If a gene has a known product, in principle it is possible to work back from the protein to the gene, by obtaining the mRNA that codes for the protein and using it (directly or indirectly) as a probe to isolate the gene.

Practical questions are therefore first how to identify the RNA (from the cytoplasm) that represents a particular gene, and then how to obtain the DNA (from the genome) that codes for the RNA. Having identified this nucleic acid, how do we obtain a sufficient amount of material to characterize? Can we direct the isolated coding sequence to synthesize its product *in vitro* or *in vivo*? Can we introduce changes in this sequence that will influence its expression and help reveal the nature of regulatory signals?

One major concern is the need to characterize genes whose products are unknown. When a gene is thought to be expressed in one cell type but not in another, closely related cell, we can attempt to identify those mRNAs found only in one of the cells. In practice, this involves 'subtracting' one mRNA population from the other to find the mRNAs that are unique to one cell.

Another common problem lies with human diseases that are caused by known genetic traits, but where the gene product (and sometimes even the original malfunctioning cell type) is not known. By genetic analysis the chromosome conveying the genetic trait is identified; then the gene is tracked to a region of the chromosome by genetic characterization of individuals with the disease,

and finally we begin to search at the molecular level for a gene within this region that can be associated with the disease. But it is not a trivial problem to identify the correct gene in a large region when the molecular nature of the disease is not well defined.

As well as leading to the isolation of the DNA for genes of interest, cloning technology has significant implications for diagnostic procedures. Once we know that a certain sequence is associated with a particular allele, it is possible to test any individual for the presence of the sequence. In principle, this allows the genotype of an individual to be determined directly for any trait of importance.

Any DNA sequence can be cloned in bacteria or yeast

Hybrid DNA molecules are constructed by using restriction enzymes to cleave DNA at particular, rather short nucleotide sequences (see Chapters 6 and 20). By cleaving both the vector and the target DNA at appropriate sites, we can rejoin them to construct hybrid molecules that can be used to amplify the amount of material or to express a particular sequence.

A critical feature of any cloning vector is that it should possess a site at which foreign DNA can be inserted without disrupting any essential vector function. The simplest approach is to use a restriction enzyme that has only a single target site, at a nonessential location in the vector DNA. The insertion procedure generates only a small proportion of chimeric genomes from the starting material, so it is important to have some means of *selecting* the chimeric genome from the original vector.

Plasmid genomes are circular, so a single cleavage converts the DNA into a linear molecule, as depicted in **Figure 21.1**. Then the two ends can be joined to the ends of a linear foreign DNA, regenerating a circular chimeric plasmid. The

length of foreign DNA that can be inserted is limited only by practical considerations, such as the susceptibility of long DNA molecules to breakage. The chimeric plasmid can be perpetuated indefinitely in bacteria. It can be isolated by virtue of its size or circularity—for example, by gel electrophoresis.

Many plasmids carry genes that specify resistance to antibiotics. This feature is useful in designing cloning systems. A common procedure is to use a plasmid that has genes specifying resistance to two antibiotics. One of the genes is used to identify bacteria that carry the plasmid. The other is used to distinguish chimeric plasmids from parental vectors. If the site used to insert foreign DNA lies within this second gene, the chimeric plasmid *loses* the antibiotic resistance. Thus a parental vector can be identified by its resistance to both antibiotics; and a chimeric plasmid can be selected by its retention of resistance to one antibiotic, but sensitivity to the other.

(As a practical matter, bacteria that are being tested for sensitivities to antibiotics or other agents are maintained by **replica plating**. Replicas made

Figure 21.1

Overview: plasmid vectors can be used to clone any fragment of DNA that is inserted at an appropriate site.

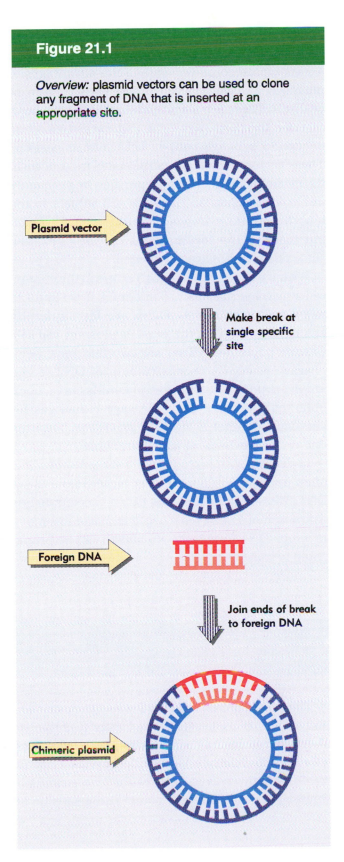

Plasmid vector

Make break at single specific site

Foreign DNA

Join ends of break to foreign DNA

Chimeric plasmid

from a master isolate are tested for sensitivity; if they are killed by the selective agent, the strain can be retrieved from the master, as shown later in Figure 21.11.)

Cloning vectors that have all the desired properties have been developed by making improvements to naturally occurring plasmids. This manipulation may involve the introduction of changes in the replication control system or the addition of genes determining resistance to particular antibiotics. One of the classic cloning vectors is **pBR322**, which was derived by several sequential alterations of earlier cloning vectors. It is a multicopy plasmid carrying genes for resistance to tetracycline and ampicillin; several restriction enzymes have unique cleavage sites at useful locations.

Phages provide another type of vector system.

Figure 21.2

Phage vectors can be used to clone a foreign DNA that is inserted into a nonessential region of the genome.

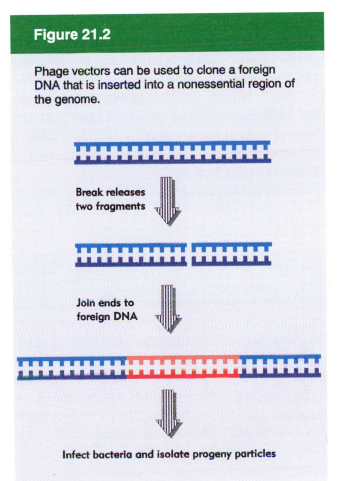

Break releases two fragments

Join ends to foreign DNA

Infect bacteria and isolate progeny particles

Usually the phage is a linear DNA molecule, so that a single restriction break generates two fragments. They are joined together with the foreign DNA to generate a chimeric phage as shown in **Figure 21.2**. Chimeric phage genomes can be conveniently isolated by allowing the phage to proceed through the lytic cycle to produce particles. However, this procedure imposes a limit on the length of foreign DNA that can be cloned, because the capacity of the phage head prevents genomes that are too long from being packaged into progeny particles.

To ameliorate this problem, a fragment of the vector that does not carry any essential phage genes can be *replaced* by the foreign DNA. This approach has been taken to a fine art with phage lambda, where a new vector has been created by manipulating the DNA to produce a shorter genome (lacking nonessential genes) that actually is *too short to be packaged into the phage head*, which has a minimum as well as maximum length requirement. Thus it is *necessary* for a foreign DNA fragment to be inserted into the cleaved parental vector in order to generate a phage that can be perpetuated as progeny particles. This demand creates an automatic selective system for obtaining chimeric phage genomes (with inserted DNA of the right length).

The utility of this type of vector has been in-creased by the development of systems for pack-aging the DNA into the phage particle *in vitro*. An attempt to combine some of the advantages of plasmids and phages led to the construction of **cosmids**. These are plasmids into which have been inserted the particular DNA sequences (*cos* sites) needed to package lambda DNA into its particle. These vectors still can be perpetuated in bacteria in the plasmid form, but can be purified by packaging *in vitro* into phages. They are still subject to the length limitation imposed by the particle head, but more of the foreign DNA can be packaged since phage genes are not needed.

We have dealt with cloning vectors in the context of using bacterial hosts. Sometimes it is useful to use a eukaryotic host. There are few authentic eukaryotic plasmids: the yeast 2μ plasmid and BPV (bovine papilloma virus) are two that have been characterized. By reconstruction of DNA, some 'dual-purpose' or 'shuttle' plasmids have been obtained that have the necessary sequences for surviving in either *E. coli* or *S. cerevisiae*. Thus the one vector can be used with either host.

In dealing with higher eukaryotic genomes, it is often necessary to handle very large fragments of DNA. We see later that this can be accomplished by the use of YACs (yeast artificial chromosomes), in which DNA fragments of ~300 kb can be propa-gated in yeast.

Constructing the chimeric DNA

To join a foreign DNA fragment to a cloning vector requires a reaction between the ends of the frag-ment and the vector. This can be accomplished by generating complementary sequences on the fragment and vector, so that they can recombine into a chimeric DNA when mixed together.

The most common method is to use restriction enzymes that make **staggered cuts** to generate short, complementary single-stranded **sticky ends**. The classic example is provided by the enzyme *Eco*RI, which cleaves each of the two strands of duplex DNA at a different point. These sites lie on

Figure 21.3

Any DNA sequence that lies between EcoRI cleavage sites can be inserted into a plasmid vector that has an EcoRI cleavage site by cutting and annealing the two DNA molecules.

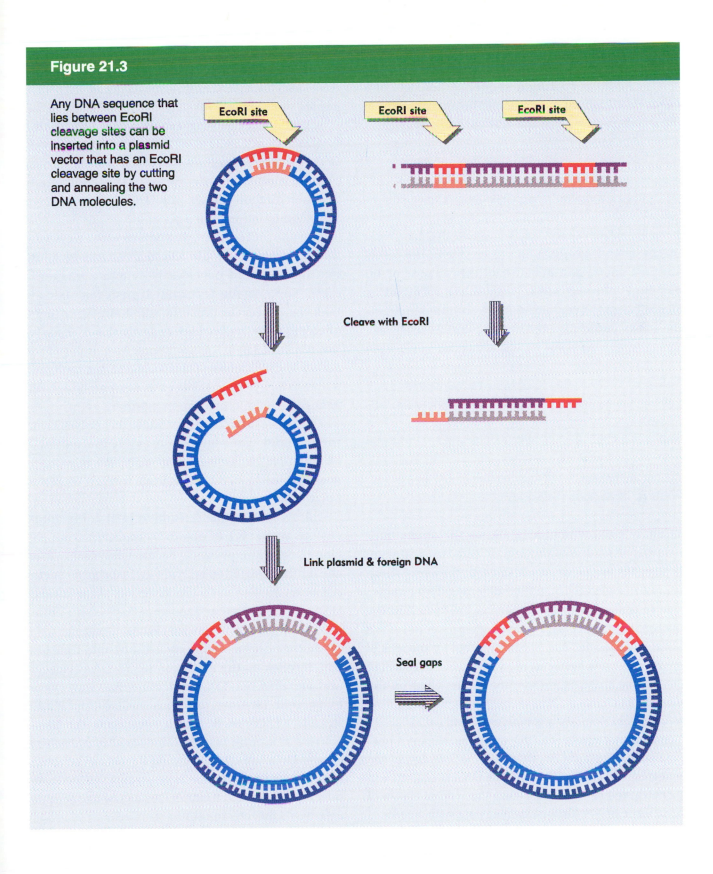

EcoRI site

EcoRI site

EcoRI site

Cleave with EcoRI

Link plasmid & foreign DNA

Seal gaps

either side of a short palindromic sequence that is part of the site recognized by the enzyme.

The *Eco*RI recognition sequence consists of 6 bases (highlighted in color), and the enzyme cuts at the bonds indicated by the vertical arrows:

$$\downarrow$$

5′ ...pNpNpNp GpApApTpTpCpNpNpN ...3′
3′ ... NpNpNp CpTpTpApApGpNpNpNp...5′

$$\uparrow$$

The DNA fragments on either side of the target site fall apart, and because of the stagger of individual cutting sites, they have protruding single-stranded regions that are complementary—the sticky ends indicated by the color:

pApApTpTpCpNpNpN ...3′
GpNpNpNp...5′

and

5′...pNpNpNpG
3′... NpNpNpCp TpTpApAp

Their complementarity allows the sticky ends to anneal by base pairing. When two different molecules are cleaved with EcoRI, the *same* sticky ends are generated on both molecules. This enables one to anneal with the other, as illustrated in the protocol of **Figure 21.3**. The procedure generates a chimeric plasmid that is intact except for the lack of covalent bonds between the vector and the foreign DNA. The missing bonds are made good by the enzyme DNA ligase. This technique for recombining two DNA molecules has both *pros* and *cons*.

An advantage is that the chimeric plasmid possesses regenerated *Eco*RI sites at either end of the inserted DNA. Thus the foreign DNA fragment can be retrieved rather easily from the cloned copies of the chimeric vector, just by cleaving with *Eco*RI.

A disadvantage is that any *Eco*RI sticky end can anneal with any other *Eco*RI sticky end. Thus some vectors reform by direct reaction between their ends, without gaining an insert, while others gain

an insert of several foreign fragments joined end to end. To avoid such complications, it is therefore necessary to select chimeric plasmids that have gained only a single insert.

A problem in relying exclusively on restriction enzymes to generate ends for the joining reaction is that their recognition sites may not lie at convenient points in the foreign DNA sequence. Another method allows any DNA end to be used for recombination with a plasmid.

The plasmid is cleaved as before with an enzyme that recognizes a single site in a suitable location, but this need not necessarily generate staggered ends. The enzyme **terminal transferase** is used with the precursor dATP to add a stretch of polyadenylic acid [poly(dA)] to both the 3′ ends of the plasmid DNA. In the same way, poly(dT) is added to the 3′ ends of the foreign DNA molecule that is to be inserted; these ends also can have been generated in any convenient manner.

As **Figure 21.4** demonstrates, the poly(dA) on the plasmid can anneal *only* with the poly(dT) on the insert fragment. Thus only one reaction is possible: the insertion of a single foreign fragment into the vector.

A drawback of this technique is that it is not so easy to retrieve the inserted fragment from the cloned chimeric plasmid, because the recognition site for the restriction enzyme in the original vector has been abolished by the insertion of the foreign DNA. The inserted sequence is flanked on either side by the poly(dA) : poly(dT) paired region.

Another useful technique is **blunt-end ligation**, which relies on the ability of the T4 DNA ligase to join together two DNA molecules that have blunt ends, that is, they lack any protruding single strands. (This reaction is in addition to the usual activity of joining broken bonds within a duplex). When DNA has been cleaved with restriction enzymes that cut across both strands at the same position, blunt-end ligation can be used to join the fragments directly together.

The great advantage of this technique is that any pair of ends may be joined together, irrespective of sequence. This is especially useful when we want to join two defined sequences without introducing

Figure 21.4

The poly(dA:dT) tailing technique allows any two DNA molecules to be joined by adding poly(dA) to the 3' ends of one and adding poly(dT) to the 3' ends of the other.

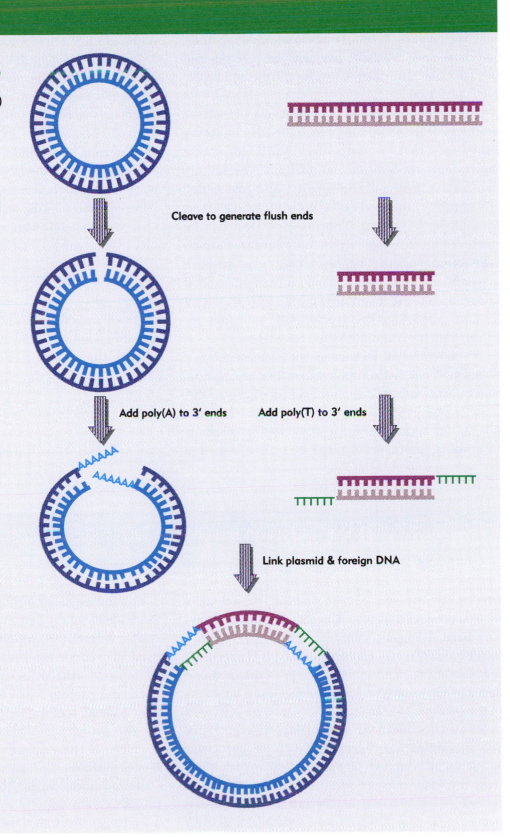

Cleave to generate flush ends

Add poly(A) to 3' ends

Add poly(T) to 3' ends

Link plasmid & foreign DNA

any additional material between them. A problem inherent in this technique is that there is no control over which pairs of blunt ends are joined together, so it is necessary first to perform the reaction and then to isolate the desired products from among the other products.

There are numerous variations of these methods. One technique uses short DNA duplexes ('linkers') that contain the *Eco*RI (or some equivalent) recognition palindrome. The linkers can be synthesized chemically, and are added covalently to the ends of a plasmid or an insert by blunt-end ligation. The inserted DNA can be retrieved by cleavage with *Eco*RI, but there are no restrictions on the original choice of sites to generate the ends. With sufficient manipulation, it is now possible to insert any foreign DNA fragment into any particular vector site, and to arrange for retrieval of the fragment when necessary.

When a foreign DNA fragment is inserted into a plasmid, it can be connected in either orientation, that is, with either of the ends of the foreign DNA joined to either of the ends of the plasmid. This does not matter when the purpose of cloning is simply to amplify the inserted sequence. However, it is important when the experiment is designed to obtain expression of the foreign DNA, which requires insertion in a particular orientation.

In this case, populations of plasmids carrying the plasmid in either orientation are obtained via random insertion, after which they are characterized by restriction mapping to identify the desired class. Or the experiment is designed so as to permit insertion in one orientation only. For example, each of the DNAs, vector and insert, may be cleaved with *two* restriction enzymes that make different sticky ends, to generate the type of pattern where each DNA has the sequence

```
End 1 ———————————End 2
```

Now if only the two end-1 sequences can anneal together, and only the two end-2 sequences can anneal, the insertion can take place only in one orientation, generating the chimeric plasmid

```
End 1———Insert———End 2
  |                    |
End 1———Plasmid———End 2
```

Copying mRNA into cDNA

One of the principal uses of cloning technology is to isolate specific genes directly from the genome. Any particular gene represents only a very small part of a eukaryotic genome. In a typical mammal, the size of the genome is $\sim 10^9$ bp, so that a single gene of (say) 5000 bp represents only 0.0005% of the total nuclear DNA.

To identify such a tiny proportion, we need a very specific probe that reacts *only* with the particular sequence in which we are interested, to pick it out from the vast excess of other sequences. The usual technique is to use a highly labeled radioactive probe of RNA or DNA, whose hybridization with the gene is assayed by autoradiography.

For the purpose of obtaining a DNA sequence that represents a particular protein, the place to start is with mRNA, which, after all, is the template used to produce the protein *in vivo*. But it can be difficult to obtain the mRNA that represents a particular protein when the product is rare. There are several techniques for isolating an mRNA via the properties of its product, but a common problem is the requirement that the RNA must first be purified.

Rather than purify the RNA, a DNA copy of the RNA sequence is made. This has the twin advantages that unlimited amounts of material can be obtained, and the DNA can be radioactively

Figure 21.5

mRNA can be copied into double-stranded DNA.

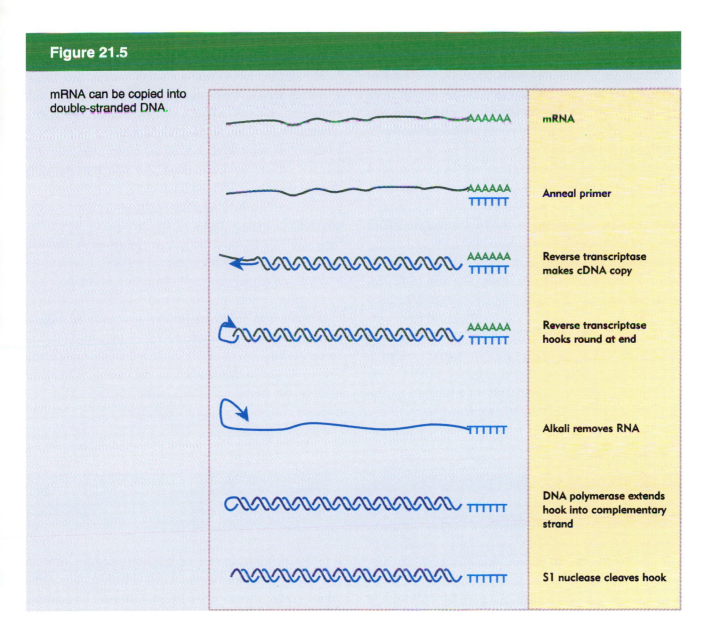

AAAAAA	**mRNA**
AAAAAA TTTTTT	**Anneal primer**
AAAAAA TTTTTT	**Reverse transcriptase makes cDNA copy**
AAAAAA TTTTTT	**Reverse transcriptase hooks round at end**
TTTTTT	**Alkali removes RNA**
TTTTTT	**DNA polymerase extends hook into complementary strand**
TTTTTT	**S1 nuclease cleaves hook**

labeled, providing a much more powerful probe. The existence of reverse transcription makes it possible to synthesize a duplex DNA from any mRNA. This is especially easy for mRNAs that carry a poly(A) tail at the 3′ end, as illustrated in **Figure 21.5**.

First, a **primer** is annealed to the poly(dA). It is a short sequence of oligo(dT), whose purpose is to provide a free 3′ end that can be used for extension by the enzyme reverse transcriptase. The enzyme engages in the usual 5′–3′ elongation, adding deoxynucleotides one at a time, as directed by complementary base pairing with the mRNA template.

The product of the reaction is a hybrid molecule, consisting of a template RNA strand base-paired with the complementary DNA strand. The only practical problem is the propensity *in vitro* of reverse transcriptase to stop before it has reached the 5′ end of the mRNA. In this case, the resulting reverse transcript falls short of representing the entire mRNA, because it lacks some of the sequences complementary to the 5′ end. However, by judicious adjustment of the experimental conditions, usually it is possible to persuade reverse transcriptase to proceed all the way.

A useful reaction tends to occur at the end of the mRNA, where the enzyme causes the reverse transcript to 'loop back' on itself, by using the last few bases of the reverse transcript as a template for synthesis of a complement. That is, the end of the complementary DNA is used to direct synthesis of a short sequence that is identical with the mRNA, and which displaces it. This creates a short hairpin, usually 10–20 bp long.

At this juncture, the original mRNA is degraded by treatment with alkali (a procedure that does not affect DNA). The product is a single-stranded DNA that is complementary to the mRNA; it is called **cDNA**.

The hairpin at the 3′ end of the cDNA provides a natural primer for the next step, the use of *E. coli* DNA polymerase I to convert the single-stranded cDNA into a duplex DNA via synthesis of the complementary strand. In this reaction, the enzyme uses the cDNA as template for synthesis of a sequence identical with the original mRNA. The product is a duplex molecule with a hairpin at one end. The hairpin is cut by the enzyme S1 nuclease (which specifically degrades single-stranded DNA) to generate a conventional DNA duplex.

The duplex DNA can be cloned to generate large amounts of a synthetic gene representing the mRNA sequence. This is called a **cDNA clone**. (From this terminology, a somewhat looser use of the term 'cDNA' has emerged, being taken to describe the duplex insert and not just the original single-stranded reverse transcript.)

The power of sequencing technology has made it possible to bypass most of the problems posed by rare mRNAs by targeting a probe directly for the mRNA sequence. One powerful technique requires knowledge of only a small sequence of the protein. Short oligonucleotides can be synthesized that correspond to this sequence. A variety of oligonucleotides can be made to cover possible alternative codons, especially at third base positions. These oligonucleotides can be used to isolate cDNAs or genomic DNA that include the sequence of the corresponding gene.

Isolating individual genes from the genome

The first step toward identifying the gene corresponding to a particular probe is to break the DNA of the genome into fragments of a manageable size. It is desirable to obtain the gene in as few fragments as possible (ideally only one). Usually the maximum lengths of DNA that can be manipulated directly are in the range of 15–20 kb. Sometimes it is not possible to obtain a gene in the form of a single fragment, and then its structure must be determined by piecing together the information gained from its various fragments. (We have discussed the use of overlapping fragments in Chapter 6.)

The best technique for fragmenting a genome is to make a restriction digest. Then every fragment ends in a site that was recognized by that particular enzyme. However, restriction sites may occur at inconvenient locations—for example, in the middle of a gene that is to be cloned. One way to avoid this is to use more than one restriction enzyme, that is, to repeat the experiment with different enzymes whose recognition sites lie at different locations. But this is time consuming; and when a long sequence is involved, it may be difficult to find an enzyme that does not cleave within it.

When the DNA of an entire genome is digested with a restriction enzyme, the frequency of breakage is controlled by the length of the sequence recognized by the enzyme. The longer the sequence, the less often it occurs by chance. The probability that a particular 4 bp sequence will occur is $0.25^4 = 1/256$, so that an enzyme with such a short recognition sequence will cleave DNA rather frequently. The frequency declines to 1/1000 for a 5 bp sequence and to 1/4000 for a 6 bp sequence.

(This calculation assumes that each base is

equally well represented in DNA, which usually is not the case. The frequency of cutting can be decreased by using an enzyme whose target sequence contains base pairs that are less predominant in the base composition of the DNA, and *vice versa*.)

To make a useful set of fragments from a restriction digest, a trick is employed to reduce the frequency of cutting. An enzyme with a short (4 bp) recognition sequence is used under conditions that generate a *partial* digest. Any particular target site is cleaved only occasionally, so not all target sites are cleaved in any particular DNA molecule. The infrequent cleavage at each site, together with the frequent distribution of sites, means that the fragment distribution approaches a random cleavage of the genome. But each fragment ends in the same sequence, which can be chosen to contain a sticky end and is therefore useful for cloning.

The distribution of sites recognized by an enzyme becomes a matter of chance taken over the genome as a whole. So a restriction digest of eukaryotic DNA generates a continuum of fragments. When electrophoresed on a gel, these fragments form a smear in which no distinct bands are evident (with the exception of some repeated sequences, which are discussed in Chapter 26). However, when a specific probe is available, it is possible to detect the corresponding sequences in the restriction smear.

DNA fragments cannot be handled directly on an agarose gel. The key feature in the protocol for identifying fragments is the ability to transfer the DNA to a medium on which hybridization reactions can occur. So the DNA is denatured to give single-stranded fragments that are transferred from the agarose gel to a nitrocellulose filter on which they become immobilized. The procedure used to transfer the DNA is somewhat akin to blotting, and this is used colloquially as a description of the procedure. When performed with DNA, it is known as **Southern blotting** (named for the inventor of the procedure).

Figure 21.6 illustrates the protocol for transferring the DNA fragments. The agarose gel is placed on a filter paper that has been soaked in a concentrated salt solution. Then the nitrocellulose filter is placed on the gel, and some dry filter paper

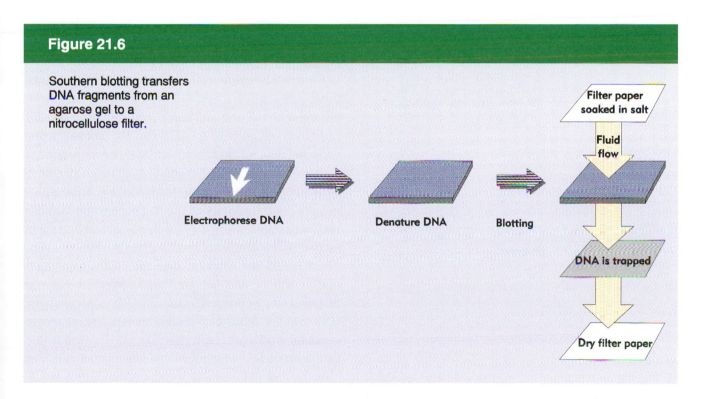

Figure 21.6

Southern blotting transfers DNA fragments from an agarose gel to a nitrocellulose filter.

Electrophorese DNA

Denature DNA

Blotting

Filter paper soaked in salt

Fluid flow

DNA is trapped

Dry filter paper

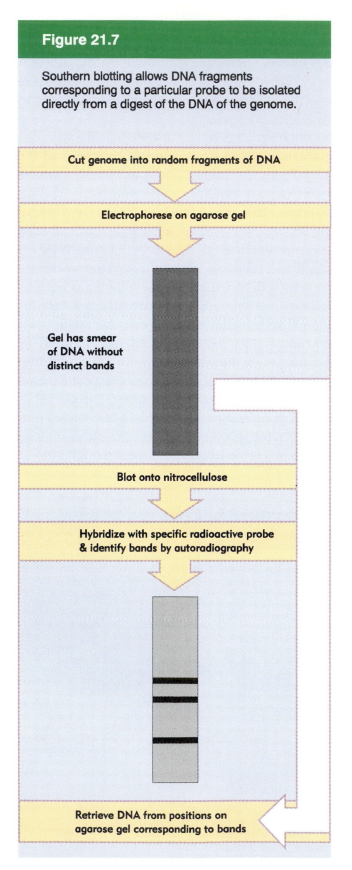

Figure 21.7

Southern blotting allows DNA fragments corresponding to a particular probe to be isolated directly from a digest of the DNA of the genome.

Cut genome into random fragments of DNA

Electrophorese on agarose gel

Gel has smear of DNA without distinct bands

Blot onto nitrocellulose

Hybridize with specific radioactive probe & identify bands by autoradiography

Retrieve DNA from positions on agarose gel corresponding to bands

is placed in contact with the nitrocellulose. The salt solution is attracted to the dry filter paper. To get there, it must pass through the agarose gel and then through the nitrocellulose filter. The DNA is carried along with it, but becomes trapped in the nitrocellulose in the same relative position that it occupied in the gel.

DNA that has been immobilized on the nitrocellulose can be hybridized *in situ* with a radioactive probe, as indicated in **Figure 21.7**. Only those fragments complementary to a particular probe will hybridize with it. Because the probe is radioactive, the hybridization can be visualized by autoradiography. Each complementary sequence gives rise to a labeled band at a position determined by the size of the DNA fragment. Clearly the usefulness of this technique depends on the specificity of the available probes.

The technique can also be performed with RNA. To blot RNA from agarose onto a medium suitable for hybridization, some changes in the technique are necessary. The procedure then is known as **Northern blotting**. (And an analogous procedure that is used for transferring proteins is called **Western blotting**.)

Sometimes we want to identify the sequence in a population that corresponds to a particular probe. A variant of the blotting techniques that is useful for this purpose is called **dot blotting**. As illustrated in **Figure 21.8**, cloned DNAs to be tested (for example, representing fragments of a genome) are spotted adjacent to one another on a filter. The filter is then hybridized with a radioactively labeled probe representing a target sequence. If the sequence is represented in a particular clone, the dot representing that clone will light up by autoradiography. The intensity of the dot corresponds fairly well with the extent to which the probe is represented in the clone.

An interesting technique for isolating a particular mRNA can be used when we have at our disposal two cell types, one of which expresses the RNA and the other of which does not. The protocol used for **subtractive hybridization** is illustrated in **Figure 21.9**. The mRNA of the target cell line

Figure 21.8

Dot blotting allows many cloned sequences to be tested readily for their homology with a labeled probe. The cloned sequences are spotted on a filter paper, which is hybridized with a solution containing the labeled probe. Those spots containing sequences complementary to the probe show up by autoradiography.

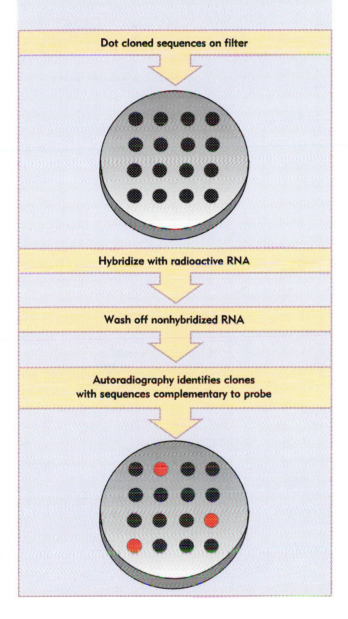

Dot cloned sequences on filter

Hybridize with radioactive RNA

Wash off nonhybridized RNA

Autoradiography identifies clones
with sequences complementary to probe

is used as substrate to prepare a set of cDNA molecules corresponding to all the expressed genes. To remove sequences that are not specific for the target cell, the cDNA preparation is exhaustively hybridized with the mRNA of another, closely related cell. This step removes all the sequences from the cDNA preparation that are common to the two cell types. So the specificity of the technique will depend on the closeness of the relationship between the two cells. After discarding all the cDNA sequences that hybridize with the other mRNA, those that are left (<5% if the technique works well) are hybridized with mRNA from the target cell to confirm that they represent coding sequences. These clones should contain sequences specific to the mRNA population of the target cell, and they can then be characterized.

This technique was used to isolate clones corresponding to the T cell receptor, a protein expressed in T lymphocytes, but not in the closely related B lymphocytes that could therefore be used to provide the subtracting mRNA. Of the original cDNA preparation from T cells, 2.6% was left at the end of the procedure. It led to the isolation of 7 clones, one of which proved to be the specific T cell gene that was sought. The number of clones isolated from this approach tends to be small, but it has proved effective in several cases of this type.

A powerful technique for directly amplifying short segments of the genome is provided by the **polymerase chain reaction (PCR)**. It requires that we know the sequence on either side of the target region, and allows the region between two defined sites to be amplified.

Figure 21.10 summarizes the protocol. A preparation of DNA (usually just an extract of the whole genome) is denatured. The single-stranded preparation is annealed with two short primer sequences (~20 bases each) that are complementary to sites on the opposite strands on either side of the target region. DNA polymerase is used to synthesize a single strand from the 3'-OH end of each primer. The entire cycle can then be repeated by denaturing the preparation and starting again. The number of copies of the target sequence in principle grows exponentially. In practice, it doubles with each cycle until reaching a plateau at which more primer–template accumulates than the

Figure 21.9

Subtractive hybridization can be used to isolate mRNA specific for one cell type when another, closely related cell type is available.

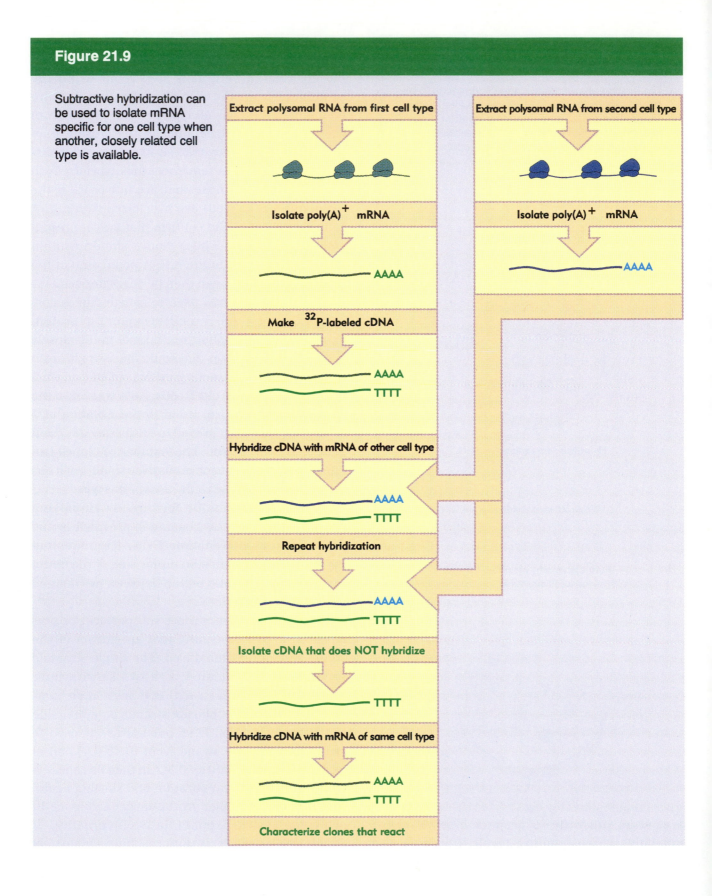

Extract polysomal RNA from first cell type

Extract polysomal RNA from second cell type

Isolate poly(A)$^+$ mRNA

Isolate poly(A)$^+$ mRNA

AAAA

AAAA

Make ^{32}P-labeled cDNA

AAAA
TTTT

Hybridize cDNA with mRNA of other cell type

AAAA
TTTT

Repeat hybridization

AAAA
TTTT

Isolate cDNA that does NOT hybridize

TTTT

Hybridize cDNA with mRNA of same cell type

AAAA
TTTT

Characterize clones that react

enzyme can extend during the cycle; then the increase in target DNA becomes linear.

With recent modifications that include using DNA polymerase from a thermophilic bacterium, so that the same enzyme remains active through the heating steps required for the denaturation–renaturation cycles, a given sequence can be amplified up to 4×10^6 times in 25 cycles. The length of the target sequence is determined by the distance between the two primer sites, and can be up to 40 kb. A restriction on the sensitivity of the technique for examining individual sequences is that the replication event has an error rate of $\sim 2 \times 10^{-4}$, which means that an error occurring in a very early cycle could become prominent through amplification; usually this can be dealt with by taking multiple samples.

The technique offers a powerful approach to distinguishing individual alleles in a genome, and thus to diagnosis of diseases that are defined at the sequence level. If we know that a disease is associated with a particular sequence change, PCR can be used to examine the sequence in the genome of a particular individual, to determine whether the alleles at a given locus are wild-type or mutant. Amplification by PCR is so sensitive that the target sequence in an individual cell can be characterized; this allows the distribution of alleles to be examined directly in (for example) a population of spermatozoa. It also allows DNA to be amplified from very small tissue samples, which is useful for diagnostic and forensic purposes.

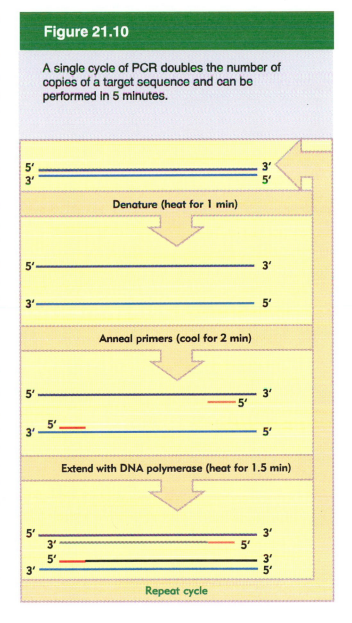

Figure 21.10

A single cycle of PCR doubles the number of copies of a target sequence and can be performed in 5 minutes.

Walking along the chromosome

There are practical difficulties in isolating fragments directly from a genomic digest, and a powerful technique for obtaining a particular gene is to reverse the order of events and to clone the genome first. Then clones containing a particular

sequence are selected. Vectors carrying DNA derived from the genome itself are called **genomic** or **chromosomal DNA clones** (in distinction to the cDNA clones that are representations of the mRNA).

Cloning an entire genome (as opposed to specific

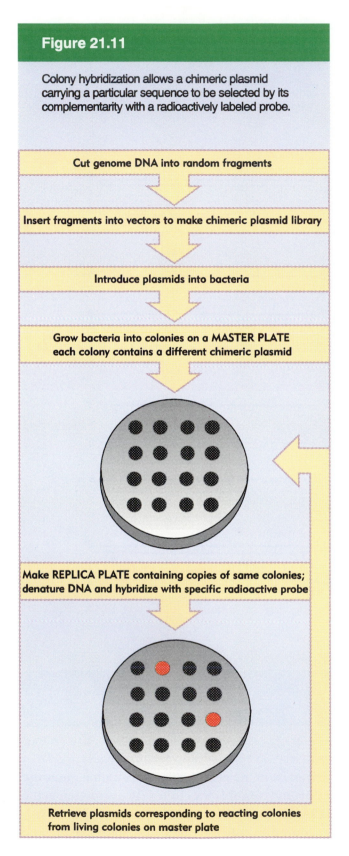

Figure 21.11

Colony hybridization allows a chimeric plasmid carrying a particular sequence to be selected by its complementarity with a radioactively labeled probe.

Cut genome DNA into random fragments

Insert fragments into vectors to make chimeric plasmid library

Introduce plasmids into bacteria

Grow bacteria into colonies on a MASTER PLATE each colony contains a different chimeric plasmid

Make REPLICA PLATE containing copies of same colonies; denature DNA and hybridize with specific radioactive probe

Retrieve plasmids corresponding to reacting colonies from living colonies on master plate

fragments) is often called a **shotgun experiment**. First the genome is broken into fragments of a manageable size. Then the fragments are inserted into a cloning vector to generate a population of chimeric vectors. A set of cloned fragments of this sort is called a **genome library**.

An analogous approach at the level of mRNA is to prepare cDNAs from the entire mRNA population of any cell or tissue; such a set of cDNAs is called a **cDNA library**. In effect it represents the genes that are expressed in the cell or tissue.

Once a genomic or cDNA library has been obtained with either a phage or plasmid vector (more often a phage, because it is easier to store the necessary large numbers of chimeric DNAs), it can be perpetuated indefinitely, and is readily retrieved whenever a new probe is available to seek out some particular fragment.

The number of random fragments that must be cloned to ensure a high probability that *every* sequence of the genome is represented in at least one chimeric plasmid decreases with the fragment size and increases with the genome size and the desired probability. For a probability level of 99%, 1500 cloned fragments are needed with *E. coli*, but the size of the necessary library increases to 4600 with yeast, 48,000 with *D. melanogaster*, and 800,000 with mammals. Libraries of cloned fragments reaching this probability level have been established in all these cases.

How is a particular genomic clone to be selected from the library? The technique of **colony hybridization** is illustrated in **Figure 21.11**. Bacterial colonies carrying chimeric vectors are lysed on nitrocellulose filters. Then their DNA is denatured *in situ* and fixed on the filter. The filter is hybridized with a radioactively labeled probe that represents the desired sequence (usually a cloned cDNA). Any colonies in which it occurs are visualized as dark spots by autoradiography. Then the corresponding chimeric vectors can be recovered from the original reference library.

A genomic clone needs to carry only some of a probe's sequence to react with it (usually the minimum is reckoned to be ~50 bp). Indeed, when a eukaryotic gene is large, it is likely to be fragmented during the formation of a library, and the various fragments will appear in different clones that react with the probe. In a partial digest, the random cleavage generates overlapping fragments, in which different sites have been broken on either side of a given sequence in different genomes. If none of the genomic clones contain the entire sequence of interest, it is necessary to reconstruct the original genomic sequence by taking advantage of the overlaps

Figure 21.12

Chromosome walking is accomplished by successive hybridizations between overlapping genomic clones.

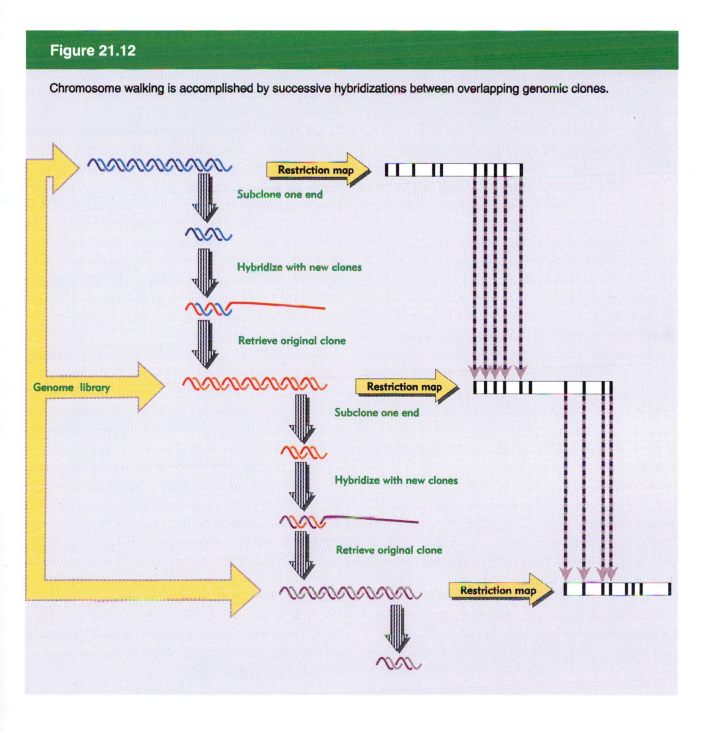

between the individual fragments.

The principle of identifying fragments that overlap is the key not only to reconstructing large genes, but also to characterizing large regions of the chromosome. Some of the most interesting questions about the eukaryotic genome concern the context within which a gene resides and its relationship to adjacent genes. In characterizing a region beyond a gene itself, we venture into unknown territory in the sense that there are no probes derived from gene products. Each successive fragment of the genome is isolated purely by virtue of its relationship with another, partially overlapping fragment.

The technique of **chromosome walking** is illustrated in **Figure 21.12**. We start with a clone that has been isolated because it contains a known gene or because we know (from genetic mapping) that it lies near some region of interest. To make the process easier to manipulate with large fragments, a subfragment from one end of the first clone is used to isolate the next set. This is done by **subcloning** the relevant fragment from the original clone. Then we identify other chimeric vectors in our library that overlap with this clone. These new vectors will extend on one side or the other of the fragment carried in the first clone. We can determine the direction of extension by making a restriction map of each fragment; the map should contain the same fragments as one end of the first clone, extended by new material. The process can be repeated indefinitely.

It is possible in principle to walk for hundreds of kilobases, and regions of the genome exceeding 1 Mb (= 1000 kb) have been mapped in this way. Using conventional cloning techniques to isolate the fragments, it is usually possible to walk on the order of 100 kb a month, making it practical to characterize quite large regions. The distance covered in one step can be increased by extending the range of methods used to isolate the fragments. For example, pulse field gel electrophoresis allows the separation of rather large fragments of DNA (>100 kb), and going from one end of such a fragment to the other end corresponds to more of a 'jump' than a 'walk'.

A critical assumption underlying the entire cloning approach is that a cloned eukaryotic sequence will be perpetuated with complete fidelity in bacteria. This is borne out by the data available. There are some exceptional cases in which deletions have occurred during passage through the bacterial host.

In dealing with chromosomal DNA, the characterization of genomic sequences becomes much more rapid if larger regions can be handled. A cloning technique that makes this possible has been developed with yeast. The crucial elements are illustrated in **Figure 21.13**.

A YAC (yeast artificial chromosome) contains the sequence elements that are needed to propagate a chromosome in yeast: an origin of replication, a centromere to ensure segregation into daughter cells, and telomeres to seal the ends of the chromosome. (The latter elements are discussed in more detail in Chapter 27.) In addition to the elements needed for propagation, a YAC carries three selectable markers. The two markers *TRP1* and *URA3* are shown in the figure; there is also another marker (*SUP4*) that is interrupted by the SmaI cloning site.

Prior to loading with foreign DNA, a YAC is maintained in the circular form shown in the figure. At the ends of the telomere sequences, it has BamHI cleavage sites; cleavage with BamHI generates a linear molecule, whose ends become telomeres which resemble the ends of authentic chromosomes and which confer stability on the linear molecule. Cleavage with SmaI generates flush ends at which foreign DNA can be inserted. Joint cleavage with BamHI and SmaI therefore generates two linear fragments, each of which contains a telomere at one end, and a flush terminus available for ligation at the other end. Each of the 'arms' bears a selectable marker (*TRP1* on the left arm, *URA3* on the right arm), allowing selection for molecules in which both arms have been joined. Loss of the *SUP4* marker at the cloning site distinguishes the religated YACs from the unbroken YACs.

A large amount of DNA can be carried in a YAC—it is not uncommon for the insert to be ~300 kb. This advances the amount that can be carried

Figure 21.13

A YAC contains all the elements needed to propagate a chromosome in yeast, a cloning site, and markers that allow selection for the original DNA and for DNA carrying an insert.

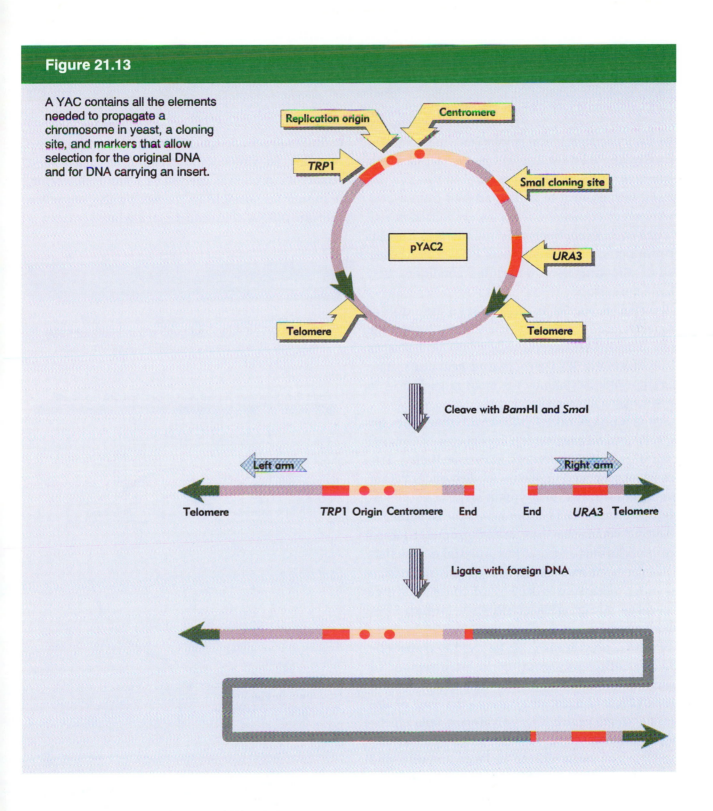

Replication origin Centromere

TRP1

SmaI cloning site

pYAC2

URA3

Telomere Telomere

Cleave with *Bam*HI and *Sma*I

Left arm Right arm

Telomere   *TRP*1 Origin Centromere End   End   URA3 Telomere

Ligate with foreign DNA

compared to a bacterial plasmid by about an order of magnitude. For example, a *D. melanogaster* genomic library can be carried on <2000 YACs. It becomes correspondingly easier to walk or jump along the chromosome. Another important feature is that large eukaryotic genes can be carried on individual YACs (we see in Chapter 23 that some insect and mammalian genes are >100 kb).

Eukaryotic genes can be expressed in prokaryotic systems

The ability to break and rejoin DNA molecules at virtually any desired site allows us to place any sequence (eukaryotic or prokaryotic) under the appropriate control to be transcribed from a new promoter or translated from a new initiation site *in vitro* or *in vivo*. Such manipulations are useful for producing large amounts of RNA or protein as well as allowing us to determine conditions that affect expression.

It is often useful to make large amounts of single-stranded RNA representing a particular gene. The RNA is useful as a probe and also as a substrate for RNA processing reactions. The current system of choice for making such RNA is the SP6 transcription reaction.

SP6 is a phage of *S. typhimurium* that codes for an RNA polymerase which recognizes the phage promoters with extremely high specificity. The transcription reaction occurs *in vitro* under rather simple conditions, and yields large amounts of product. **Figure 21.14** summarizes the structure of a cloning vector that has an SP6 promoter joined to a 'polylinker' region. The polylinker consists of the recognition sequences for many restriction enzymes; cleavage with any of these enzymes generates sticky ends at which a foreign DNA can be inserted.

SP6 RNA polymerase can be used to transcribe RNA very efficiently from the inserted region. To ensure uniform termination, the template is linearized by a single cleavage at the end of the target sequence; the RNA polymerase falls off the end, so the transcript is known as an 'SP6 run-off'. The RNA can be produced in large amounts and can be highly labeled if a radioactive precursor nucleotide is used. The RNA can be translated *in vitro* if a protein product is required.

Because the genetic code is universal, a given coding sequence should always have the same meaning. So when an intact coding sequence for a eukaryotic protein is carried on a chimeric vector, it should be possible to transcribe the sequence into an mRNA that can be translated within a

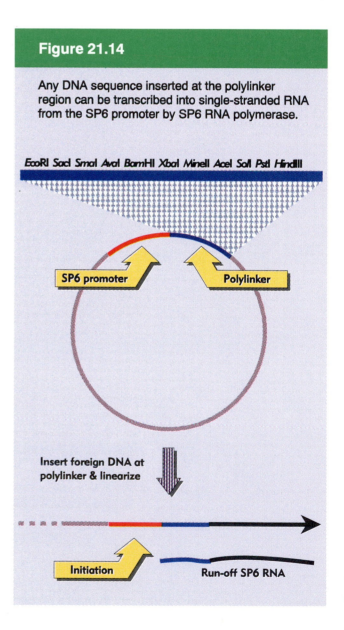

Figure 21.14

Any DNA sequence inserted at the polylinker region can be transcribed into single-stranded RNA from the SP6 promoter by SP6 RNA polymerase.

EcoRI SacI SmaI AvaI BamHI XbaI HincII AccI SaII PstI HindIII

SP6 promoter

Polylinker

Insert foreign DNA at polylinker & linearize

Initiation

Run-off SP6 RNA

Figure 21.15

Any eukaryotic coding sequence can be translated by insertion into a suitable site in an expression cloning vector.

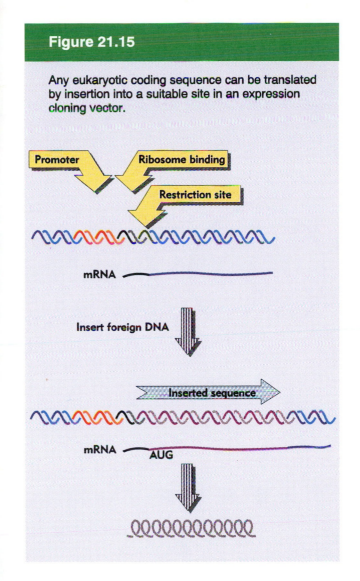

bacterial host or by a prokaryotic *in vitro* system.

The only caveats are that modifications made to the protein in its natural habitat may not occur; and, of course, with an *in vivo* system there is always the risk that the polypeptide chain may be unstable in the bacterium. In an appropriate environment, however, it should be possible to translate any eukaryotic coding sequence into the corresponding polypeptide.

What conditions must be fulfilled to obtain effective expression of eukaryotic genes in bacteria?

Because eukaryotic and prokaryotic promoters differ in sequence, the eukaryotic gene must be placed under the control of a bacterial promoter. To be translated, the mRNA must have a proper ribosome binding site just before the AUG initiation codon.

Standard **expression cloning vectors** have been constructed in which a restriction site for inserting foreign DNA lies just after a region containing a promoter and ribosome binding site. The usual arrangement is for the inserted DNA to contribute the AUG codon. **Figure 21.15** shows that any foreign DNA sequence (prokaryotic or eukaryotic) starting with an AUG codon can be placed into an expression vector so that it is transcribed and translated in bacteria. This produces a protein that corresponds precisely to the inserted coding region.

Another technique is to insert a coding sequence shortly after the start of a *bona fide* protein. Translation produces a hybrid protein; the N-terminal region consists of the bacterial protein whose sequence has been interrupted, and the remainder of the protein corresponds to the foreign coding region. The N-terminal bacterial sequence can be quite short and may be useful. For example, it can provide a signal sequence that ensures secretion of the hybrid protein into the medium; and it may protect the hybrid protein from degradation in the bacterium.

What source should be used for the eukaryotic coding sequence? Because many eukaryotic genes are interrupted, the coding sequence in a genomic DNA clone may not form a continuous region. In such cases, the genomic DNA cannot be translated in bacteria. The coding sequence is therefore best obtained from a cDNA clone prepared from the mRNA template, in which the coding sequence is uninterrupted. It is not necessary for it to include any 5′ and 3′ nontranslated regions; the coding sequence from the initial AUG to the termination codon will suffice.

Sometimes we wish to characterize a promoter,

Figure 21.16

The CAT assay can be used to follow the activity of any eukaryotic promoter. The activity of the coding sequence carried by the plasmid is measured by the ability of a cell extract to convert radioactively labeled chloramphenicol to the acetylated forms. The proportion of acetylated chloramphenicol increases with time and is directly proportional to the number of units of enzyme present.

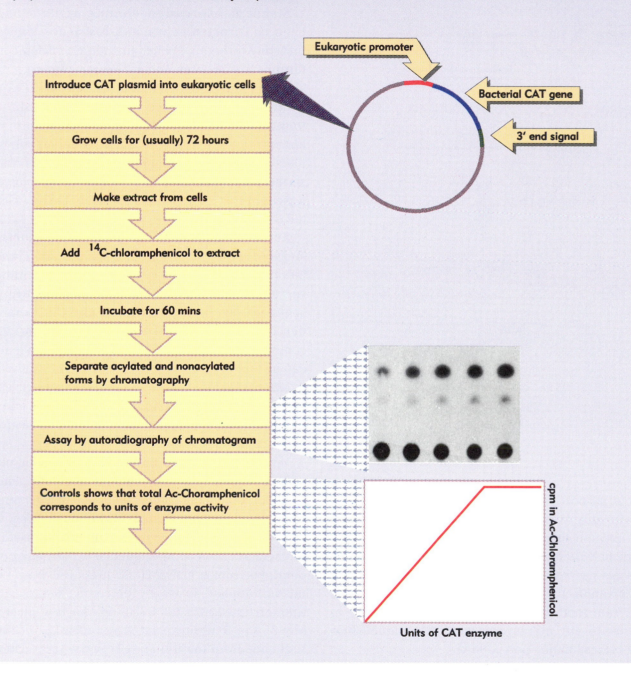

Figure 21.17

Expression of a *lacZ* gene can be followed in the mouse by staining for β-galactosidase (in blue). In this example, *lacZ* was expressed under the control of the promoter of a mouse gene that is usually expressed in the nervous system. Blue staining can be seen in mesodermal tissues throughout the embryo. Photograph kindly provided by Robb Krumlauf.

but the product of its natural gene is difficult to assay. This can be overcome by the reverse type of manipulation, in which a gene whose product can easily be assayed is connected to the promoter. A gene whose product is used in this way to assess other promoters is called a **reporter gene**. A vector that accomplishes this is a variant of phage Mu, called the **Mud phage**; it is constructed to allow insertion of a (bacterial) promoter to drive a standard reporter gene. The same approach can be used with eukaryotic promoters. Two types of reporter gene are in common usage. One that allows accurate quantitation of gene expression is the coding sequence of the bacterial gene *chloramphenicol acetyl transferase (CAT)*. This is particularly useful in eukaryotic systems because there is no counterpart to the *CAT* gene in eukaryotes. This means that the enzyme activity can be directly assayed in a cell extract.

The protocol of this **CAT assay** is illustrated in **Figure 21.16**. A plasmid containing a *CAT* gene under the control of a eukaryotic promoter is introduced into an appropriate cell type. The assay is simple and quantitative. The level of enzyme activity corresponds to the amount of enzyme that was made, which in turn reveals the level of expression from the eukaryotic promoter. Of course, to be sure that the level of expression is determined solely by the promoter (rather than by the stability or translation of the mRNA), we have to measure the level of *CAT* mRNA directly. The CAT assay has been extremely useful in allowing a single procedure to be applied to a variety of situations to determine the factors that influence promoter activity, and *CAT* has been the reporter gene of choice for such studies of eukaryotic gene expression.

Another useful reporter is the *lacZ* gene, which codes for β-galactosidase. Standard assays are available for the enzyme activity, including one in which its cleavage of a suitable substrate generates a compound that gives a blue stain. This has been used in the prokaryotic situation to allow a rapid scan of bacterial colonies to see whether blue color is detectable. It also has been extremely useful in eukaryotes, where the pattern of expression of the enzyme can be visualized by staining tissue sections. When a *lacZ* gene under control of an appropriate promoter is introduced into animals, appropriate tissues or cell types can be examined for the appearance of blue color, thus revealing the specificity of the promoter. A striking example is shown in **Figure 21.17**.

Summary

Any sequence of DNA may be cloned by inserting it into a plasmid or phage vector that can replicate in bacteria. Sequences represented in RNA can be cloned by using reverse transcriptase to make a cDNA copy that is then converted to duplex DNA. Genomic DNA is prepared for cloning by using a restriction enzyme or other means to cleave it into fragments. DNA can be joined to the cloning vector by using restriction enzymes to generate sticky ends, by the poly(dA.dT) tailing technique, by blunt end ligation, or by the use of chemically synthesized linkers. Expression cloning vectors allow any coding sequence to direct synthesis of the corresponding protein in large quantities.

mRNA representing a particular gene can be identified directly by obtaining polysomes synthesizing the protein, by subtractive hybridization to compare the mRNA populations of two closely related cells, or by hybridization with oligonucleotides that represent a potential coding sequence for a protein of known amino acid sequence. The coding sequences are usually characterized by using a cDNA that represents the mRNA.

Southern blotting allows genome fragments that correspond to a particular sequence probe to be identified directly. An unknown gene can be approached by walking along the chromosome from a starting point; overlapping fragments are used to reach the region in which particular mutations identify the target locus. Genomic DNA of known sequence can be amplified by the polymerase chain reaction, allowing the sequences of individual alleles to be determined.

Further reading

Techniques

The notion of constructing libraries was promulgated by **Maniatis *et al*.** (*Cell* 15, 687–701, 1978).

Chromosomal walking was described by **Bender,** **Spierer, and Hogness** (*J. Mol. Biol.* 168, 17–23, 1983).

YACs were developed by **Burke, Carle, and Olson** (*Science* 236, 806–812, 1987).

CHAPTER 22

Genome size and genetic content

The 'eukaryotic genome' should be a somewhat nebulous concept. The essence of being eukaryotic is that the major part of the genome is sequestered in the nucleus, where it is safeguarded by the nuclear membrane from exposure to the cytoplasm. There can be major differences in the types of sequences that constitute the nuclear genome, the total amount of nuclear DNA varies enormously, the number of chromosomes into which it is segregated is extremely variable, and the relatively minor part of the genome contained in organelles also shows wide variations in size.

Yet certain features are unique to eukaryotic genomes compared with prokaryotic genomes. The integrity of individual genes can be interrupted, there may be multiple and (sometimes) identical copies of particular sequences, and there may be large blocks of DNA that do not code for protein. Because of the division between nucleus and cytoplasm, the arrangements for gene expression in eukaryotes must necessarily be different from those in prokaryotes.

Do these features represent any underlying uniformity in the organization of eukaryotic DNA? We should remember that the eukaryotic kingdom is extremely broad, and at present we have detailed information about the genetic organization of only a few types of species. In describing the characteristics of eukaryotic DNA, we are dealing with features that may be represented to widely varying degrees in different

individual genomes. We are therefore concerned with establishing the options available for the organization of eukaryotic DNA, rather than with defining a hypothetical 'typical' pattern.

The major question about eukaryotic DNA concerns the number and types of genes in a genome. We may identify the coding potential of a genome directly, by identifying regions that have open reading frames. Large scale mapping of this nature is complicated by the fact that genes are interrupted in higher eukaryotic genomes, so that many separated open reading frames may be part of a single gene. In principle, however, such an approach allows definition of the coding potential of defined regions of the genome, and by extrapolation, of the whole genome. Since we do not necessarily have information about the functions of the protein products, or indeed proof that they are expressed at all, this approach as such is restricted to defining the potential of the genome (although a strong presumption exists that any conserved open reading frame is likely to be expressed).

Another approach is to define the number of genes directly in terms of their expression in RNA or protein. This gives an assurance that we are dealing with *bona fide* genes that are expressed under known circumstances. It is of course the only approach that allows us to ask how many genes are expressed in a particular tissue or cell type, what variation exists in the relative levels of expression, and how many of the genes expressed in one particular cell are unique to that cell or are also expressed elsewhere.

Concerning the types of genes, we may ask whether a particular gene is essential: what happens to a null mutant that lacks a functional gene? If a null mutation is lethal, or the organism has a visible defect, we may conclude that the gene is essential or at least conveys a selective advantage. But some genes can be deleted without apparent effect on the phenotype. How are we to explain the existence of genes that appear to be dispensable? Why have they not been eliminated in the course of evolution?

The C-value paradox describes variations in genome size

The total amount of DNA in the (haploid) genome is a characteristic of each living species known as its **C-value**. There is enormous variation in the range of C-values, from as little as $<10^6$ bp for a mycoplasma to as much as 10^{11} bp for some plants and amphibians.

Figure 22.1 summarizes the range of C-values found in different evolutionary phyla. There is an increase in the minimum genome size found in each phylum as the complexity increases, but along with the increase in absolute amounts of DNA in the higher eukaryotes go some

Figure 22.1

DNA content of the haploid genome is related to the morphological complexity of lower eukaryotes, but varies extensively among the higher eukaryotes. The range of DNA values within a phylum is indicated by the shaded area.

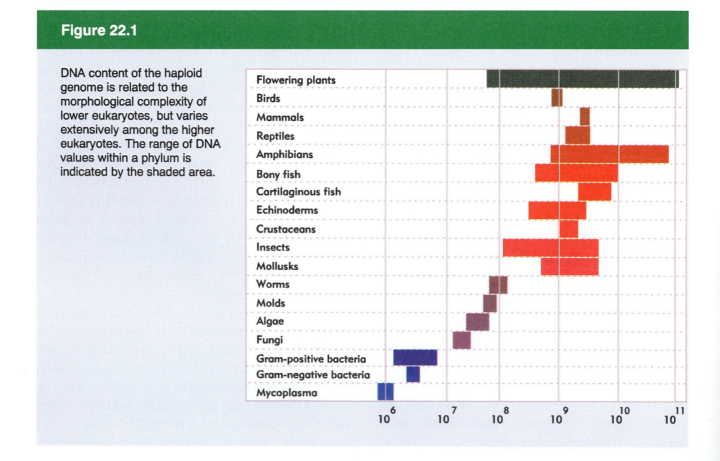

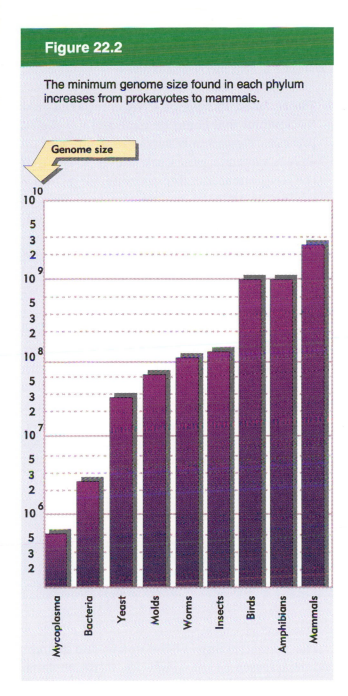

Figure 22.2

The minimum genome size found in each phylum increases from prokaryotes to mammals.

lular alga *Pyrenomas salina*, at 6.6×10^5 bp. (However, these organisms may not be true eukaryotes, but could be an intermediate stage of evolution, representing a primitive presence of a nucleus and chloroplast.) Excluding these algae, mycoplasma are the smallest types of organism known, and also may be constructed from very small genomes. Thus organisms may have genome sizes only ~3× the size of a large bacteriophage (T4 is 1.7×10^5 bp). Bacteria start at ~2×10^6 bp. Unicellular eukaryotes (whose life-styles may resemble the prokaryotic) get by with genomes that are also small, although larger than those of the bacteria. Being eukaryotic *per se* does not imply a vast increase in genome size; a yeast may have a genome size of ~1.3×10^7 bp, only about twice the size of the largest bacterial genomes.

A further twofold increase in genome size is adequate to support the slime mold *D. discoideum,* able to live in either unicellular or multicellular modes. Another increase in complexity is necessary to produce the first fully multicellular organisms; the nematode worm *C. elegans* has a DNA content of 8×10^7 bp.

Climbing further along the evolutionary tree, the relationship between complexity of the wide variations in the genome sizes within some phyla.

Plotting the *minimum* amount of DNA required for a member of each phylum suggests in **Figure 22.2** that an increase in genome size is required to make more complex prokaryotes and lower eukaryotes.

The smallest genome yet identified for a living cell actually belongs to a eukaryotic, the unicel-

Table 22.1

The genome sizes of some common experimental animals.

Phylum	Species	Genome (bp)
Algae	*Pyrenomas salina*	6.6×10^5
Mycoplasma	*M. pneumoniae*	1.0×10^6
Bacterium	*E. coli*	4.2×10^6
Yeast	*S. cerevisiae*	1.3×10^7
Slime mold	*D. discoideum*	5.4×10^7
Nematode	*C. elegans*	8.0×10^7
Insect	*D. melanogaster*	1.4×10^8
Bird	*G. domesticus*	1.2×10^9
Amphibian	*X. laevis*	3.1×10^9
Mammal	*H. sapiens*	3.3×10^9

organism and content of DNA becomes obscure, although it is necessary to increase the genome size in order make insects, birds or amphibians, and mammals.

We can also see the steady increase in genome size with complexity in the listing in **Table 22.1** of some of the most commonly analyzed organisms, although there are some real puzzles. Can the toad *X. laevis* have the same genetic complexity as man, as one might suppose naively from their common genome sizes?

In some phyla, the spread of genome sizes is narrow. Birds, reptiles, and mammals all show little variation within the phylum, with a range of genome sizes in each case about twofold. But in other cases, most notably insects, amphibians, and plants, there is a wide range of values, often more than tenfold. One wonders whether the common housefly (*Musca domestica*) with a genome size of 8.6×10^8 really has a corresponding increase in complexity over the common fruit fly, *D. melanogaster*, with its 6-times smaller genome of 1.4×10^8.

The **C-value paradox** takes its name from our inability to account for the content of the genome in terms of known functions. It expresses the existence of two puzzling features:

◆ There is an excess of DNA compared with the amount that could be expected to code for proteins. We can now account for much of the excess because genes are much larger than the sequences needed to code for proteins (principally because of the intervening sequences that break up a coding region into different segments). We discuss the structure of such genes, and their contribution to overall genome size, in Chapter 23.

◆ There are large variations in C-values between certain species whose apparent complexity does not vary much. In amphibians, the smallest genomes are just below 10^9 bp, while the largest are almost 10^{11} bp. It is hard to believe that this could reflect a 100-fold variation in the number of genes needed to specify different amphibians, so we need to account for the 'excess' DNA in some other way.

Reassociation kinetics depend on sequence complexity

The range of sizes of eukaryotic genomes raises an important question: do larger genomes contain a greater number of different genes or do they instead contain more copies of the same genes that are present in smaller genomes? If the diversity of genes increases with genome size, we should expect the amount of unique DNA sequences in the genome to increase. This will not happen if there are simply more copies of the same genes.

The general nature of the eukaryotic genome can be assessed by the kinetics with which denatured DNA reassociates. Reassociation between complementary sequences of DNA occurs by base pairing, in a reversal of the process of denaturation by which they were separated (see Figure 5.2). The technique can be extended to isolate individual DNA or RNA sequences by their ability to hybridize with a particular probe (see Chapter 21). The kinetics of the reassociation reaction reflect the variety of sequences that are present; so the reaction can be used to quantitate genes and their RNA products.

Renaturation of DNA depends on random collision of the complementary strands, and follows second-order kinetics. The rate of reaction is governed by the equation

$$\frac{dC}{dt} = -kC^2$$

where C is the concentration of DNA that is single-stranded at time t, and k is a reassociation rate constant.

By integrating this equation between the limits of the initial concentration of DNA, C_0 at time $t = 0$, and the concentration C that remains single-stranded after time t, we can describe the progress of the reaction as

$$\frac{C}{C_0} = \frac{1}{1 + k.C_0 t}$$

This equation shows that the parameter controlling the reassociation reaction is the product of DNA concentration (C_0) and time of incubation (t), usually described simply as the **Cot**.

A useful parameter is derived by considering the conditions when the reaction is half complete, at time $t_{\frac{1}{2}}$. Then

$$\frac{C}{C_0} = \frac{1}{2} = \frac{1}{1 + k.C_0 t_{\frac{1}{2}}}$$

so that

$$C_0 t_{\frac{1}{2}} = \frac{1}{k}$$

The value required for half-reassociation ($C_0 t_{\frac{1}{2}}$) is called the **Cot$_{\frac{1}{2}}$**. Since the Cot$_{\frac{1}{2}}$ is the product of the concentration and time required to proceed halfway, a greater Cot implies a slower reaction. The reassociation of any particular DNA can be described by the $C_0 t_{\frac{1}{2}}$ (given in nucleotide-moles × second/liter) or in the form of its reciprocal rate constant k (in units of liter-nucleotide-moles^{-1}-second^{-1}).

The reassociation of DNA usually is followed in the form of a **Cot curve**, which plots the fraction of DNA that has reassociated ($1 - C/C_0$) against the log of the Cot. **Figure 22.3** gives Cot curves for several genomes. The form of each curve is similar, with renaturation occurring over an ~100-fold range of Cot values between the points of 10% reaction and 90% reaction. But the Cot required in each case is different. It is described by the Cot$_{\frac{1}{2}}$.

The Cot$_{\frac{1}{2}}$ is directly related to the amount of DNA in the genome. This reflects a situation in which, as the genome becomes more complex, there are fewer copies of any particular sequence within a given mass of DNA. For example, if the C_0 of DNA is 12 pg, it will contain 3000 copies of each sequence in a bacterial genome whose size is 0.004 pg, but will contain only 4 copies of each sequence present in a eukaryotic genome of size 3 pg. Thus the same *absolute* concentration of DNA measured in moles of nucleotides per liter (the C_0) will provide a concentration of each eukaryotic sequence that is 3,000/4 =750× lower than that of each bacterial sequence.

Since the rate of reassociation depends on the concentration of complementary sequences, for the eukaryotic sequences to be present at the same *relative* concentration as the bacterial sequences, it is necessary to have 750× more DNA (or to incubate the same amount of DNA for 750 times longer). Thus the Cot$_{\frac{1}{2}}$ of the eukaryotic reaction is 750× the Cot$_{\frac{1}{2}}$ of the bacterial reaction.

Figure 22.3

Rate of reassociation is inversely proportional to the length of the reassociating DNA.

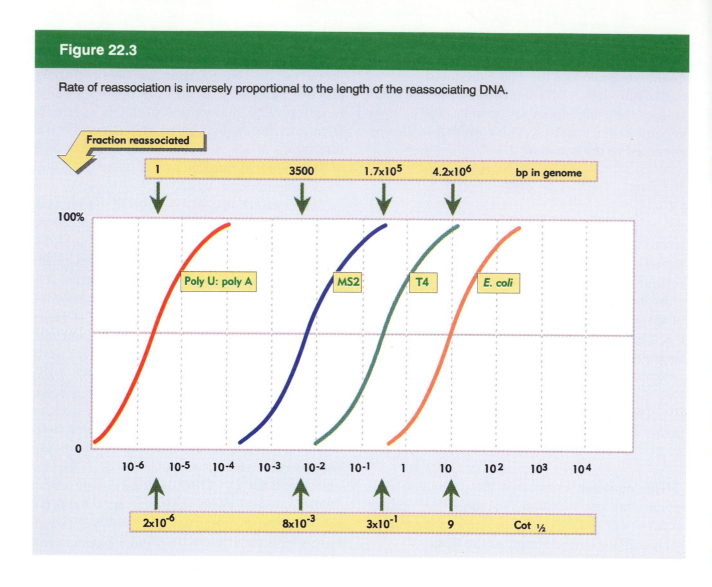

The Cot$_{\frac{1}{2}}$ of a reaction therefore indicates the *total length of different sequences* that are present. This is described as the **complexity**, usually given in base pairs.

The renaturation of the DNA of any genome (or part of a genome) should display a Cot$_{\frac{1}{2}}$ that is proportional to its complexity. Thus the complexity of any DNA can be determined by comparing its Cot$_{\frac{1}{2}}$ with that of a standard DNA of known complexity. Usually *E. coli* DNA is used as a standard. Its complexity is taken to be identical with the length of the genome (implying that every sequence in the *E. coli* genome of 4.2×10^6 bp is unique). So we can write

$$\frac{\text{Cot}_{\frac{1}{2}} \text{ (DNA of any genome)}}{\text{Cot}_{\frac{1}{2}} \text{ (}E.\ coli \text{ DNA)}} = \frac{\text{Complexity of any genome}}{4.2 \times 10^6 \text{ bp}}$$

Eukaryotic genomes contain several sequence components

When the DNA of a eukaryotic genome is characterized by reassociation kinetics, usually the reaction occurs over a range of Cot values spanning up to eight orders of magnitude. This is much broader than the 100-fold range expected from the examples of Figure 22.3. The reason is that each of these curves follows the equation that describes the kinetics of reassociation for a single component. *A genome actually includes several such components, each reassociating with its own characteristic kinetics.*

Figure 22.4 shows the reassociation of a (hypothetical) eukaryotic genome, starting at a Cot of 10^{-4} and terminating at a Cot of 10^4. The reaction falls into three distinct phases, outlined by the shaded boxes. A plateau separates the first two phases, but the second and third overlap slightly. Each of these phases represents a different kinetic component of the genome:

◆ The **fast component** is the first fraction to reassociate. In this case, it represents 25% of the total DNA, renaturing between Cot values of 10^{-4} and ~2×10^{-2}, with a $Cot_{\frac{1}{2}}$ value of 0.0013.

◆ The next fraction is called the **intermediate component**. This represents 30% of the DNA. It renatures between Cot values of ~0.2 and 100, with a $Cot_{\frac{1}{2}}$ value of 1.9.

◆ The **slow component** is the last fraction to renature. This is 45% of the total DNA; it

extends over a Cot range from ~100 to ~10,000, with a $Cot_{\frac{1}{2}}$ of 630.

To calculate the complexities of these fractions, each must be treated as an independent kinetic component whose reassociation is compared with a standard DNA. The slow component represents 45% of the total DNA, so its concentration in the reassociation reaction is 0.45 of the measured C_0 (which refers to the total amount of DNA present). Thus the $Cot_{\frac{1}{2}}$ applying to the slow fraction alone is $0.45 \times 630 = 283$.

So if the slow DNA were isolated as a pure component free of the other fractions, it would renature with a $Cot_{\frac{1}{2}}$ of 283. Suppose that under these conditions, *E. coli* DNA reassociates with a $Cot_{\frac{1}{2}}$ of 4.0. Comparing these two values, we see that the complexity of this fraction is 3.0×10^8 bp. Treating the other components in a similar way shows that the intermediate component has a complexity of 6×10^5 bp, and the fast component has a complexity of only 340 bp. This provides a quantitative basis for our statement that the faster a component reassociates, the lower its complexity.

Reversing the argument, if we took three DNA preparations, each containing a unique sequence of the appropriate length (340 bp, 6×10^5 bp, and 3×10^8 bp, respectively) and mixed them in the proportions 25:30:45, each would renature as though it were a single component, and together the mixture would display the same kinetics as those determined for the whole genome of Figure 22.4.

Figure 22.4

The reassociation kinetics of eukaryotic DNA show three types of component (indicated by the shaded areas). The arrows identify the $Cot_{1/2}$ values for each component.

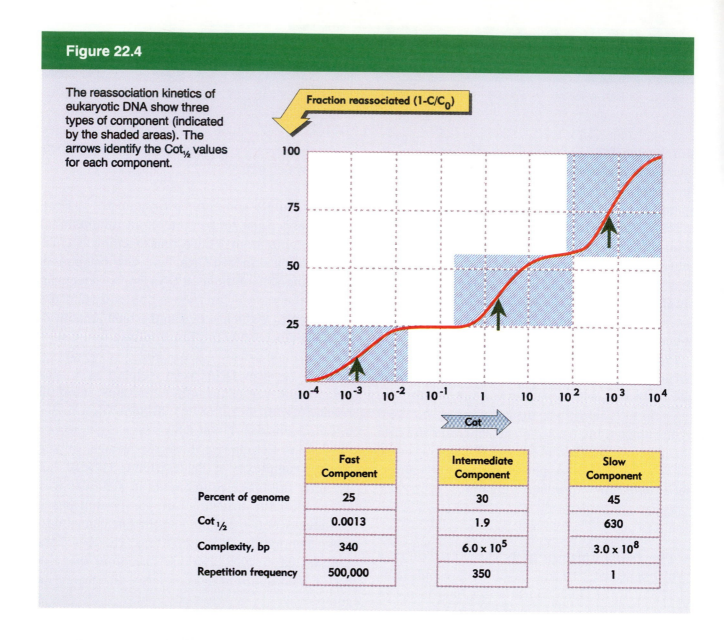

	Fast Component	Intermediate Component	Slow Component
Percent of genome	25	30	45
$Cot_{1/2}$	0.0013	1.9	630
Complexity, bp	340	6.0×10^5	3.0×10^8
Repetition frequency	500,000	350	1

Nonrepetitive DNA complexity can estimate genome size

The complexity of the slow component corresponds with its physical size. Suppose that the genome reassociating in Figure 22.4 has a haploid DNA content of 7.0×10^8 bp, determined by chemical analysis. Then 45% of it is 3.15×10^8 bp, which is only marginally greater than the value of 3.0×10^8 bp measured by the kinetics of reassociation. In fact, given the errors of measurement inherent in both techniques, we can say that the complexity of the slow component is the same whether measured chemically or kinetically. The two values are referred to as the **chemical complexity** and **kinetic complexity.**

The coincidence of these values means that the slow component comprises sequences that are unique in the genome: on denaturation, *each single-stranded sequence is able to renature only with the corresponding complementary sequence.* This part of the genome is the sole component of prokaryotic DNA and is usually a major component in eukaryotes. It is called **nonrepetitive DNA.**

We can use the kinetic complexity of nonrepetitive DNA to estimate the complexity of the genome. This just reverses the calculation that we performed earlier. For the example of Figure 22.4, the complexity of nonrepetitive DNA is 3.0×10^8 bp. If this fraction is unique and represents 45% of the genome, the whole genome should have a size of $3.0 \times 10^8 \div 0.45 = 6.6 \times 10^8$ bp. This provides an independent assessment for genome size that we can compare with the chemical complexity, which in this case is 7.0×10^8 bp.

Figure 22.5 plots the relationship between genome size, as determined by reassociation kinetics of nonrepetitive DNA, and the haploid DNA content, as determined by chemical analysis. The agreement demonstrates that the nonrepetitive DNA component consists of individual sequences present in only one copy per genome. The sole exceptions are some plants that were generated by a recent polyploidization, where as a result there is more than one copy of every sequence.

This is an important point in light of the C-value paradox. The presence of nonrepetitive DNA implies that *larger genomes are not generated simply by increasing the number of copies of the same sequences present in smaller genomes.* If this were the case, larger genomes would behave as though polyploid, so there would be no nonrepetitive DNA.

The reassociation data therefore exclude models to explain the C-value paradox in terms of simple increases in the number of copies of each gene present in the haploid genome. So we must account for differences in genome size on the basis that larger genomes genuinely contain greater diversity of sequences.

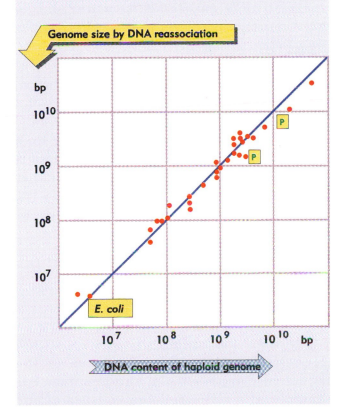

Figure 22.5

There is a good correlation between the kinetic complexity and chemical complexity of eukaryotic genomes, with the exception of polyploid genomes (indicated by P). When determined by the technique of DNA reassociation, DNA amounts are usually given in base pairs or daltons. When measured chemically, the amount of DNA is given in picograms (pg). The equivalence is 1 pg = 0.965×10^9 bp = 6.1×10^{11} daltons.

Eukaryotic genomes contain repetitive sequences that are related but not identical

What is the nature of the components that renature more rapidly than the nonrepetitive (slow) DNA? In the example of Figure 22.4, the intermediate component occupies 30% of the genome. Its chemical complexity is $0.3 \times 7 \times 10^8 = 2.1 \times 10^8$ bp. But its kinetic complexity is only 6×10^5 bp.

The unique length of DNA that corresponds to the Cot for reassociation is much shorter than the total length of the DNA chemically occupied by this component in the genome. In other words, the intermediate component behaves as though consisting of a sequence of 6×10^5 bp that is present in 350 copies in every genome (because $350 \times 6 \times 10^5 = 2.1 \times 10^8$). Following denaturation, *the single strands generated from any one of these copies are able to renature with their complements from any one of the 350 copies.* This effectively raises the concentration of reacting sequences in the reassociation reaction, explaining why the component renatures at a lower $Cot_{\frac{1}{2}}$.

Sequences that are present in more than one copy in each genome are called **repetitive DNA**. The number of copies present per genome is called the **repetition frequency (f)**. Formally, the repetition frequency is defined by the ratio

$$f = \frac{\text{Chemical complexity}}{\text{Kinetic complexity}}$$

Together, the complexity and repetition frequency describe the properties of the sequence components of the genome. For example, if a given genome consists of one sequence that is A bp long, 10 copies of a sequence that is B bp long, and 100 copies of a sequence that is C bp long, the complexity is $A +$ $B + C$. The repetition frequencies of sequences A, B, and C are 1, 10, and 100, respectively. Sequence A is nonrepetitive; sequences B and C are repetitive.

The repetitive DNA fraction includes all components whose repetition frequency is significantly greater than 1. For practical purposes, repetition frequencies between 1 and 2 are taken to indicate sequences that belong in the nonrepetitive component. (It is impossible for a sequence to be represented a nonintegral number of times in the haploid genome!) The usual range of values measured for nonrepetitive DNA in real genomes is ~0.8–1.6.

Repetitive DNA is often classed into two general types, corresponding approximately to the intermediate and fast components of Figure 22.4:

◆ **Moderately repetitive DNA** occupies the intermediate fraction, usually reassociating in a range between a Cot of 10^{-2} and that of nonrepetitive DNA.

◆ **Highly repetitive DNA** occupies the fast fraction, reassociating before a Cot of 10^{-2} is reached.

The behavior of a repetitive DNA component represents only an average that is useful for describing its sequences. The relevant parameters do not necessarily represent the properties of any particular sequence.

The moderately repetitive component of Figure 22.4 includes a total length of 6×10^5 bp of DNA, repeated ~350 times per genome. But this does not correspond to a single, identifiable, continuous length of DNA. *Instead, it is made up of a variety of individual sequences, each much shorter,* whose total length together comes to 6×10^5 bp. These individual sequences are dispersed about the genome. Their average repetition is 350, but some will be present in more copies than this and some in fewer.

When a eukaryotic genome is analyzed by reassociation kinetics, the individual sequence components are rarely so well separated as shown in Figure 22.3. In fact, they often overlap extensively, so that in reality there is probably a continuum of repetitive components, reassociating over a range from >10 times to >20,000 times that of the nonrepetitive component.

The different components of eukaryotic DNA can be isolated in the form of the DNA that enters double-stranded form after renaturation to a particular Cot value. The properties of renatured nonrepetitive and repetitive DNA differ significantly.

Nonrepetitive DNA forms duplex material that behaves very much like the original preparation of DNA before its denaturation. When denatured again, the duplex molecules melt sharply at a T_m only slightly below that of the original native DNA. This shows that strand reassociation has been accurate: each unique sequence has annealed with its exact complement.

Different behavior is shown by renatured repetitive DNA. The reassociated double strands tend to melt gradually over rather a wide temperature range, as shown in **Figure 22.6**. This means that they do not consist of exactly paired molecules. Instead, they must contain appreciable mispairing. The more mispairing in a particular molecule, the fewer hydrogen bonds need be broken to melt it, and thus the lower the T_m.

The breadth of the melting curve shows that renatured repetitive DNA contains a spectrum of sequences, ranging from those that have been formed by reassociation between sequences that are only partially complementary, to those formed by reassociation between sequences that are very nearly or even exactly complementary. How can this happen?

Repetitive DNA components consist of families of sequences that are not exactly the same, but are *related.* The members of each family consist of a set of nucleotide sequences that are sufficiently similar to renature with one another. The differences between the individual members are the result of base substitutions, insertions, and deletions, all creating points within the related sequences at which the complementary strands cannot base pair. The proportion of these changes establishes the relationship between any two sequences. When two closely related members of the family renature, they form a duplex with high T_m; when two more distantly related members associate, they form a duplex

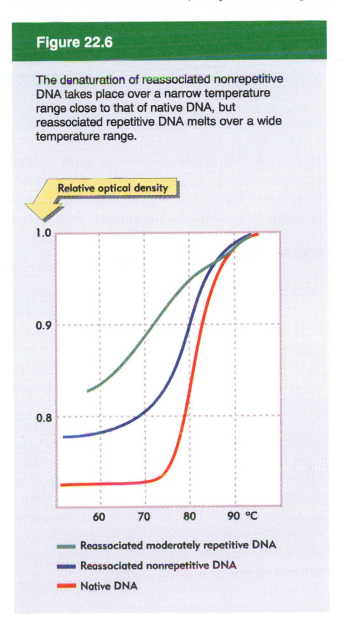

Figure 22.6

The denaturation of reassociated nonrepetitive DNA takes place over a narrow temperature range close to that of native DNA, but reassociated repetitive DNA melts over a wide temperature range.

Relative optical density

— Reassociated moderately repetitive DNA

— Reassociated nonrepetitive DNA

— Native DNA

with a lower T_m. Overall, we see the broad range represented in the figure.

The ability of related but not identical complementary sequences to recognize each other can be controlled by the **stringency** of the conditions imposed for reassociation. A higher stringency is imposed by (for example) an increase in temperature, which requires a greater degree of complementarity to allow base pairing. So by performing the hybridization reaction at high temperatures, reassociation is restricted to rather closely related members of a family; at lower temperatures, more distantly related members may anneal. Note also that, as the stringency of hybridization increases, the proportion of the genome in the nonrepetitive fraction also increases.

There is wide variety in the construction of repetitive families. Some families are well-defined, and consist of quite closely related members. They remain intact even when the stringency of hybridization is increased. Other families consist of a continuum of variously related sequences, so that their apparent membership declines continuously as the stringency is increased. The measured size of such repetitive families is arbitrary, since it is determined by the hybridization conditions.

Most structural genes lie in nonrepetitive DNA

The proportion of the genome occupied by nonrepetitive versus repetitive DNA varies widely. **Figure 22.7** summarizes the genome organization of some representative organisms. Prokaryotes, of course, contain exclusively nonrepetitive DNA. For lower eukaryotes, most of the DNA is nonrepetitive; <20% falls into one or more moderately repetitive components. In animal cells, up to half of the DNA often is occupied by moderately and highly repetitive components. In plants and amphibians, the nonrepetitive DNA may be reduced to a minority of the genome, with the moderately and highly repetitive components accounting for up to 80%.

No single feature characterizes all eukaryotic genomes. One extreme is found in plants that have become polyploid very recently; they have no nonrepetitive DNA, so that the slowest reassociating fraction has a repetition frequency of 2–3 or more. (This is probably a temporary stage in evolution, which usually leads to divergence to regenerate nonrepetitive DNA.) Another pattern is found in crab genomes; there is no moderately repetitive DNA at all—just highly repetitive and nonrepetitive DNA. And in lower eukaryotes, there is no highly repetitive DNA.

The length of the nonrepetitive DNA component tends to increase with overall genome size, as we proceed up to a total genome size $\sim 3 \times 10^9$ (characteristic of mammals). Further increase in genome size, however, generally reflects an increase in the amount and proportion of the repetitive components, so that it is rare for an organism to have a nonrepetitive DNA component $>2 \times 10^9$. The nonrepetitive DNA content of genomes therefore accords better with our sense of the relative complexity of the organism: *E. coli* has 4.2×10^6 bp, *C. elegans* increases an order of magnitude to 6.6×10^7 bp, *D. melanogaster* increases further to $\sim 10^8$ bp, and mammals increase another order of magnitude further to $\sim 2 \times 10^9$ bp.

Mendelian genetics for simple traits imply that there is only one copy of each determining factor in the haploid genome. The factor can be mapped to a particular locus; and the simplest assumption is that each such locus is occupied by a DNA sequence representing a single protein, as exemplified by the definition of the gene in Chapter 3.

This is the classic view of a structural gene:

a unique component of the genome, the only sequence coding for its protein product, and therefore identifiable by mutations that impede the protein function. The sequence of a unique structural gene should form part of the non-repetitive DNA component. If we adopt the view that most genes are present in nonrepetitive DNA, we conclude that genetic complexity increases with the content of nonrepetitive DNA, rather than with genome size overall.

There is a possible bias in the characterization of genes by mutation, because if there are cases in which multiple sequence code for the same or similar proteins, it could be difficult or impossible to identify the genes by point mutations:

inactivation of any one copy might be undetected if the others were able to substitute for it. We must therefore turn to direct analysis of the DNA of the genome to determine the numbers and proportions of unique and repeated genes.

For the purpose of identifying and characterizing structural genes, mRNA provides the ideal intermediate. The protein to which it corresponds can be determined by translating the mRNA. The gene from which it is derived can be obtained by hybridizing the mRNA with the genomic DNA. An individual mRNA provides a handle, as it were, for proceeding back from its protein to the gene.

A population of mRNAs, manifested as the spectrum of sequences found in the polysomes,

Figure 22.7

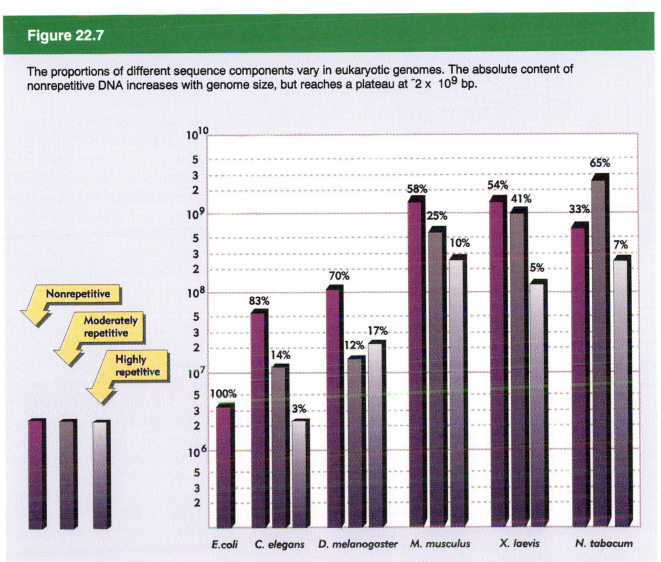

The proportions of different sequence components vary in eukaryotic genomes. The absolute content of nonrepetitive DNA increases with genome size, but reaches a plateau at ˜2 x 10^9 bp.

Figure 22.8

The hybridization of an mRNA tracer preparation in a reassociation curve shows that most mRNA sequences are derived from nonrepetitive DNA, the remainder from moderately repetitive DNA, and none from highly repetitive DNA.

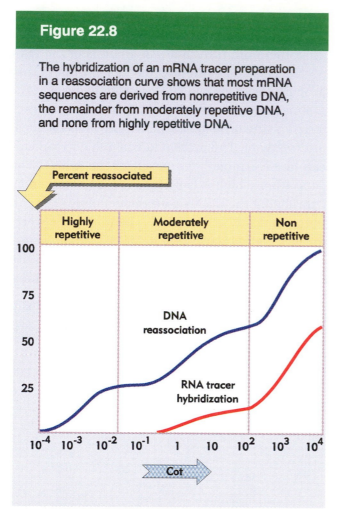

as a **DNA-driven** reaction.

The tracer RNA (or cDNA) participates in the reaction as though it were just another member of the sequence component from which it was transcribed. The component is identified by comparing the radioactive tracer curve with the reassociation curve for whole DNA. Thus the Cot values at which the labeled RNA hybridizes can be taken to identify the repetition frequencies of the corresponding genomic sequences.

When an individual mRNA molecule is used, it hybridizes with a single $Cot_{\frac{1}{2}}$ value determined by the repetition frequency of the gene or genes representing it. For a unique gene, this is the same as nonrepetitive DNA; for a repeated gene, the $Cot_{\frac{1}{2}}$ is less. When a population of mRNA molecules is used, each mRNA hybridizes with a characteristic $Cot_{\frac{1}{2}}$, so that overall the curve is the sum of the individual components. It can be resolved in the same way as the reassociation curve of the genomic DNA itself.

With a population of mRNAs, a typical result resembles that shown in **Figure 22.8**. A small proportion of the RNA, generally 10% or less, hybridizes with a $Cot_{\frac{1}{2}}$ corresponding to moderately repetitive sequences. The major component hybridizes with a $Cot_{\frac{1}{2}}$ identical with or very close to that of nonrepetitive DNA. Usually this represents up to 50% of the total RNA. Most of the material that does not hybridize at all proably represents nonrepetitive sequences.

What is the relationship between the mRNA sequences that hybridize with nonrepetitive DNA and those that hybridize with repetitive DNA? They can be separated into different classes (by retrieving the RNA that hybridizes first). This shows that independent molecules are involved: one class represents genes that lie in nonrepetitive DNA, and the other corresponds to genes that lie in repetitive DNA.

From these results, it is clear that most of the mRNA, perhaps up to 80%, is derived from sequences that reassociate with the nonrepetitive DNA component. Because of the difficulties in detecting very low degrees of repetition (espe-

defines the entire set of genes expressed in a cell or tissue. Thus the constitution of the mRNA reveals both the nature and number of structural genes. By means of nucleic acid hybridization, we can come to grips with some central questions. How many copies of each gene are there? How many genes are expressed in a particular cell type? How much overlap is there between the sets of genes whose expression defines different cell types ?

The genome sequence components represented in mRNA can be determined by using the RNA as a **tracer** in a reassociation experiment. A very small amount of radioactively labeled RNA (or cDNA) is included together with a much larger amount of cellular DNA. The reaction is governed by the reassociation of the complementary cellular DNA strands (as though the tracer RNA were not present). This is described

cially when the repeated copies are related rather than identical), these genes are not necessarily unique, but at least should be present in fewer than three or four copies per genome. Only a small proportion of genes are openly present in repetitive DNA; whether these multiple copies are identical or related is not revealed by the hybridization kinetics.

Note that because these experiments are conducted in terms of the *total mass of mRNA*, and different genes are expressed as different amounts of mRNA, this technique does not prove that 80% of the *genes* lie in the nonrepetitive DNA component. The general conclusion of this analysis, however, is that most genes are part of nonrepetitive DNA, and it is this component of the genome that accounts for the characteristic complexity of the organism.

How many nonrepetitive genes are expressed?

The number of nonrepetitive DNA sequences represented in RNA can be determined directly in terms of the proportion of the DNA able to hybridize with RNA. When a small amount of single-stranded DNA is hybridized with a large amount of RNA, all the sequences in the DNA that are complementary to the RNA should react to form an RNA–DNA hybrid.

In this type of **saturation experiment**, the critical feature is that the excess of RNA should be great enough to ensure that every available complementary DNA sequence actually is hybridized. Because the reaction is **RNA-driven**, the controlling parameter is the product of RNA concentration and time, known as the **Rot**. This is exactly analogous to the use of Cot values to describe DNA-driven reactions.

Saturation hybridization is usually followed by plotting the percent of hybridizing DNA against the Rot value. A curve of this sort is shown in **Figure 22.9**. In this example, the reaction is complete by a Rot value of ~300, but it is necessary to extend it further to be sure that a plateau has been reached. No further DNA is hybridized as the Rot is increased up to 1,200, which demonstrates that all the available nonrepetitive DNA has indeed been hybridized.

From the proportion of hybridized DNA, we can calculate the number of genes represented in the mRNA population used to drive the reaction. At saturation, 1.35% of the available nonrepetitive DNA is hybridized. Because only one strand of DNA is transcribed into RNA, only half of the DNA in principle is potentially able to hybridize with

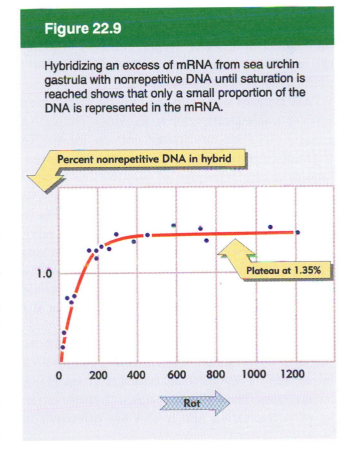

Figure 22.9

Hybridizing an excess of mRNA from sea urchin gastrula with nonrepetitive DNA until saturation is reached shows that only a small proportion of the DNA is represented in the mRNA.

Percent nonrepetitive DNA in hybrid

Plateau at 1.35%

1.0

0   200   400   600   800   1000 1200

Rot

RNA (the other half is identical in sequence with it). Thus 2.70% of the total sequences of nonrepetitive DNA are represented in the mRNA.

The nonrepetitive DNA itself represents 75% of a genome of 8.1×10^8 bp. Thus the complexity of DNA represented in the RNA population is $0.027 \times 0.75 \times 8.1 \times 10^8 = 1.7 \times 10^7$ bp. The mRNA population had an average length of 2000 bases. Thus the total number of different messengers is $1.7 \times 10^7/2000 = 8500$. This corresponds to the total number of genes expressed in the tissue from which the mRNA was taken (the gastrula embryo of the sea urchin).

Similar experiments have been performed for many systems. For a lower eukaryote such as yeast, the total number of expressed genes is relatively low, roughly 4000. For somatic tissues of higher eukaryotes, the number usually is between 10,000 and 15,000. The value is similar for plants and for vertebrates. The total amount of DNA represented in mRNA typically is therefore a very small proportion of the genome, of the order of 1–2%. The only consistent exception to this type of value is presented by mammalian brain, where much larger numbers of genes appear to be expressed, although the exact quantitation is not certain.

This type of experiment can be performed only for nonrepetitive DNA, in which each DNA sequence can be hybridized only by the mRNA originally derived from it. Because of the presence of multiple copies of identical or related sequences in moderately repetitive DNA, an RNA derived from this component might be able to hybridize with other genomic sequences in addition to the particular sequence from which it was originally transcribed. At saturation, therefore, a large number of additional DNA sequences could be contained in the hybrid, overestimating the number of expressed genes.

We can also use the kinetics of an RNA-driven reaction to measure complexity. The mRNA preparation is reverse transcribed to give a cDNA preparation. Then the kinetics of hybridization between the cDNA and mRNA are followed. In effect, we have converted the starting material into a double-stranded form (a single strand of the original RNA and a single strand of complementary DNA), and we can now follow its reassociation reaction just like that of a denatured genomic DNA preparation. The driving parameter is the $Rot_{\frac{1}{2}}$.

Just as with a DNA reassociation curve, a single component hybridizes over about two decades of Rot values, and a reaction extending over a greater range must be resolved by computer curve-fitting into individual components. Again this represents what is really a continuous spectrum of sequences.

An example of an excess mRNA $\times$ cDNA reaction that generates three components is given in **Figure 22.10**. The control for this reaction was the hybridization between purified ovalbumin mRNA (2000 bases long) and its cDNA, which had a $Rot_{\frac{1}{2}}$ of 0.0008. This means that a $Rot_{\frac{1}{2}}$ of 0.0004 is demanded for each 1000 bases of complexity.

◆ The first component has an observed $Rot_{\frac{1}{2}}$ of 0.0015 and represents 50% of the total reaction. Thus the $Rot_{\frac{1}{2}}$ applicable to the purified component should be $0.5 \times 0.0015 = 0.00075$, almost exactly the same as the control Rot for ovalbumin. This suggests that the first component is in fact just ovalbumin mRNA (which indeed occupies about half of the messenger mass in oviduct tissue).

◆ The next component has an observed $Rot_{\frac{1}{2}}$ of 0.04, and provides 15% of the reaction. Thus its purified $Rot_{\frac{1}{2}}$ should be $0.15 \times 0.04 = 0.006$. Its complexity is therefore $0.006 \div 0.0004 \times 1000 = 15,000$ bases. This would correspond to 7–8 mRNA species of average length 2000 bases.

◆ The last component has a $Rot_{\frac{1}{2}}$ of 30 and provides 35% of the reaction; so its purified $Rot_{\frac{1}{2}}$ of $0.35 \times 30 = 10.5$ corresponds to a complexity of $10.5 \div 0.0004 \times 1,000 = 26,000,000$ bases. This corresponds to ~13,000 mRNA species of average length 2,000 bases.

From this analysis, we can see that about half of the mass of mRNA in the cell represents a single mRNA, ~15% of the mass is provided by a mere 7–8 mRNAs, and ~35% of the mass is divided into the large number of 13,000 mRNA species. It is therefore obvious that the mRNAs comprising each component must be present in very different amounts.

Figure 22.10

Hybridization between excess mRNA and cDNA identifies several components in chick oviduct cells, each characterized by the $Rot_{1/2}$ of reaction.

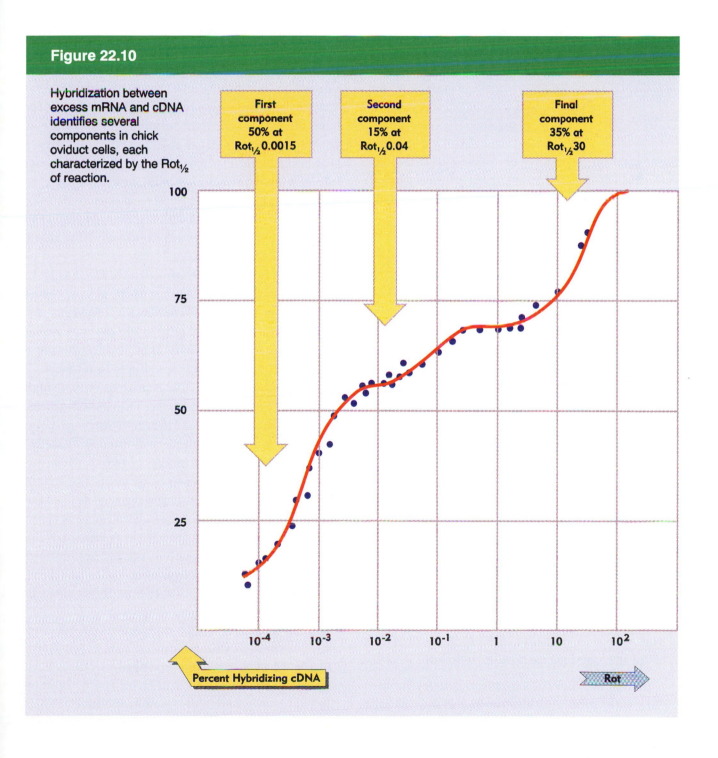

First component 50% at $Rot_{1/2}$ 0.0015

Second component 15% at $Rot_{1/2}$ 0.04

Final component 35% at $Rot_{1/2}$ 30

Percent Hybridizing cDNA

Rot

Genes are expressed at widely varying levels

The average number of molecules of each mRNA per cell is called its **abundance** or **representation**. It can be calculated quite simply if the total mass of RNA in the cell is known. For each component, total mass = abundance × complexity, so that as a general equation,

$$\text{Abundance} = \frac{\text{gm of mRNA in cell} \times \text{fraction in component} \times 6 \times 10^{23}}{\text{Complexity of component in daltons}}$$

The equation is usually expressed in this form, since total mRNA is measured (chemically) in picograms, while complexity is determined (by hybridization) in bases or daltons.

In the example shown in Figure 22.10, there are 0.275 pg mRNA per cell. This corresponds to 100,000 copies of the first component (ovalbumin mRNA), 4000 copies of each of the 7–8 mRNAs in the second component, but only ~5 copies of each of the 13,000 mRNAs that constitute the last component.

We can divide the mRNA population into two general classes, according to their abundance:

◆ The oviduct is an extreme case, with so much of the mRNA represented in only one species, but most cells do contain a small number of RNAs present in many copies each. This **abundant** component typically consists of <100 different mRNAs present in 1000–10,000 copies per cell. It often corresponds to a major part of the mass, approaching 50% of the total mRNA.

◆ About half of the mass of the mRNA consists of a large number of sequences, of the order of 10,000, each represented by only a small number of copies in the mRNA—say, <10. This is the **scarce** or **complex** mRNA class. (It is this class that drives a saturation reaction.)

The total numbers of expressed genes estimated by the saturation and kinetic techniques usually are in fairly good agreement. The kinetic technique provides a lower estimate (because some sequences that are very scarce fail to react). The saturation technique provides a higher estimate (because there are always likely to be some sequences included that are present in more than one copy per genome). Thus for chick oviduct the kinetic technique identifies ~13,000 mRNAs, while the saturation technique corresponds to ~15,000.

Many somatic tissues of higher eukaryotes have an expressed gene number in the range of 10,000 to 20,000. How much overlap is there between the genes expressed in different tissues? For example, the expressed gene number of chick liver is ~11,000–17,000, compared with the value for oviduct of ~13,000–15,000. How many of these two sets of genes are identical? How many are specific for each tissue ?

We see immediately that there are likely to be substantial differences among the genes expressed in the abundant class. Ovalbumin, for example, is synthesized only in the oviduct, not at all in the liver. This means that 50% of the mass of mRNA in the oviduct is specific to that tissue.

But the abundant mRNAs represent only a small proportion of the number of expressed genes. In terms of the total number of genes of the organism, and of the number of changes in transcription that must be made between different cell types, we need therefore to know the extent of overlap between the genes represented in the scarce mRNA classes of different cell phenotypes.

There is no really sensitive technique for measuring overlaps between mRNA populations *en masse,* but some reasonable estimates have been obtained by adapting the methods used to determine the population complexity.

Additive saturation experiments show how many sequences differ between two populations. **Figure 22.11** shows the example of chick liver and oviduct. The mRNA derived from liver by itself saturates 2.05% of the nonrepetitive DNA; a similar experiment with oviduct mRNA saturates 1.80% of the DNA. If the two mRNA populations comprised entirely different sequences, together they should saturate 2.05 + 1.80 = 3.85% of the nonrepetitive DNA. But the actual value is 2.4%, only slightly greater than that of liver alone.

Thus ~75% of the sequences expressed in liver and oviduct are the same (though since this is a saturation experiment, the data do not show whether they are present in the same or different abundances in the two tissues). In other words, ~12,000 genes are expressed in both liver and oviduct, ~5000 additional genes are expressed only in liver, and ~3000 additional genes are expressed only in oviduct.

Another way to estimate the extent of overlap is to hybridize the mRNA of a tissue with nonrepetitive DNA. Then the DNA that reacts is isolated to constitute the **mDNA** preparation, while the DNA that does not react constitutes the **null DNA** preparation. These DNA preparations are then hybridized separately with an excess of mRNA from some other tissue. The proportion of the mDNA preparation that reacts identifies the proportion of

genes expressed in the second as well as in the first tissue. The amount of the null DNA preparation that reacts identifies the number of genes expressed in the second, but not in the first, preparation.

The scarce mRNAs overlap extensively. Between mouse liver and kidney, ~90% of the scarce mRNAs are identical, leaving a difference between the tissues of only 1000–2000 in terms of the number of expressed genes. The general result obtained in several comparisons of this sort is that only ~10% of the mRNA sequences of a cell are unique to it. The majority of sequences are common to many, perhaps even all, cell types.

This suggests that the common set of expressed gene functions, numbering perhaps ~10,000 in a mammal, comprise functions that are needed in all cell types. Sometimes this type of function is referred to as a **housekeeping** or **constitutive** activity. It contrasts with the activities represented by specialized functions (such as ovalbumin or globin) needed only for particular cell phenotypes. These are sometimes called **luxury** genes.

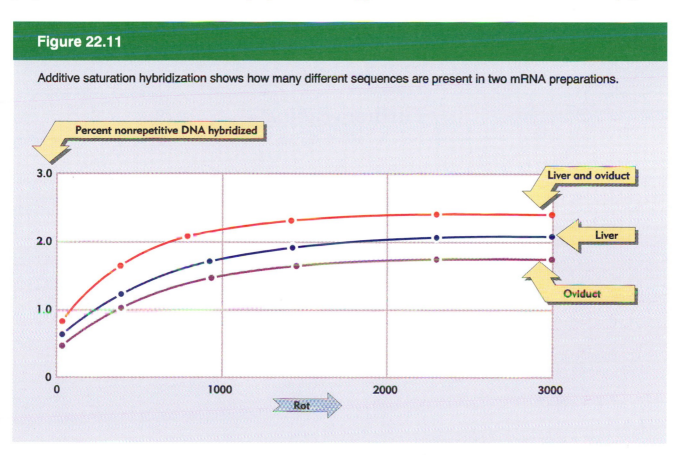

Figure 22.11

Additive saturation hybridization shows how many different sequences are present in two mRNA preparations.

Summary

The sequences comprising a eukaryotic genome can be classified in three groups: nonrepetitive sequences are unique; moderately repetitive sequences are dispersed and repeated a small number of times in the form of related but not identical copies; and highly repetitive sequences are short and usually repeated as a tandem array. The proportions of the types of sequence are characteristic for each genome, although larger genomes tend to have a smaller proportion of nonrepetitive DNA. The complexity of any class describes the length of unique sequences in it; the repetition frequency describes the number of times each sequence is repeated. The C-value paradox refers to discrepancies in the genome sizes of related organisms, and to the apparently excessive overall amount of DNA.

Most structural genes are located in nonrepetitive DNA. The complexity of nonrepetitive DNA is a better reflection of the complexity of the organism than the total genome complexity; nonrepetitive DNA reaches a maximum complexity of ~2×10^9 bp.

Genes are expressed at widely varying levels. There may be 10^5 copies of mRNA for an abundant gene whose protein is the principal product of the cell, 10^3 copies of each mRNA for <10 moderately abundant messages, and <10 copies of each mRNA for >10,000 scarcely expressed genes. Overlaps between the mRNA populations of cells of different phenotypes are extensive; the majority of mRNAs are present in most cells.

The total number of genes expressed in any one cell phenotype is ~10,000, which suggests that the total gene number for a higher eukaryote should be ~100,000. A mammalian genome has 2×10^9 bp of nonrepetitive DNA, which could code for 200,000 genes each 10,000 bp in length.

Further reading

Reviews
The reassociation technique was introduced by **Britten and Kohne** (*Science* **161**, 529–540, 1968) and subsequent results were reviewed by **Britten and Davidson** (*Quart. Rev. Biol.* **48**, 565–613, 1973).

The contemporaneous view of the analysis of gene numbers was summarized by **Lewin** (*Cell* **4**, 77–93, 1975).

CHAPTER 23

The eukaryotic gene: conserved exons and unique introns

A major change in our view of the gene at the molecular level came in 1977 as a result of the discovery that eukaryotic genes may be interrupted. For some years previously, there had been uneasy suspicions that eukaryotic genes might be unusual, that they might differ in some fundamental way from bacterial genes. The roots of this idea lay in the apparently excessive size of eukaryotic genomes described by the C-value paradox.

One escape from this dilemma is to suppose that a large proportion (even a majority) of DNA sequences are not part of the structural genes, but have some other (unknown) function. Another idea is that the structural gene *is much larger than the sequence represented in mRNA*. The additional length must be removed when the RNA is **processed** for transport to the cytoplasm. This idea was supported by the fact that nuclear RNA is much longer than mRNA (see Chapter 31).

The original concept for a large eukaryotic gene supposed that the transcription unit might include extensive sequences on one or the other side of the region represented in mRNA. In addition, if extensive nontranscribed regions were needed for regulation, the unit of gene expression might be rather large relative to the size of the mRNA.

Indeed it turns out that many eukaryotic genes are much longer than their mRNAs. But it was a great surprise that the cause of the discrepancy proved to be the existence of *interruptions that separate different parts of the coding region in DNA*. It is also the case that there may be quite extensive regulatory regions upstream of the startpoint for transcription, but these generally are measured in less than hundreds of base pairs, compared with the thousands of base pairs in the transcription units.

The existence of interrupted genes was revealed by experiments to identify directly the DNA of the gene. The intention of this approach was to proceed back from known mRNAs to their genes, to recover intact transcription units, including flanking sequences such as promoters and other regulatory elements not necessarily represented in the mRNA.

On the basis of experience with bacteria, it was assumed that eukaryotic mRNA would have the same sequence as the DNA from which it is transcribed. As a control to show that an isolated genomic DNA sequence did indeed correspond with the mRNA used to isolate it, their sequences were compared, either by electron microscopy or by restriction mapping. But what this revealed was a discrepancy between the sequences, in the form of *additional regions present in the genomic DNA but absent from mRNA*.

The presence of additional sequences explains the size discrepancy between nuclear RNA and mRNA. The nucleus contains primary transcripts that correspond to the sequence of the gene itself.

Those parts of the sequence not found in mRNA are removed from the primary transcript before the RNA is exported to the cytoplasm to serve as a messenger.

Recall the terminology for describing the relationship between a gene and its RNA product (see Chapter 6). **Figure 23.1** shows that an interrupted gene consists of an alternating series of **exons** and **introns**:

◆ The exons are the sequences represented in the mature RNA. By definition, a gene starts and ends with exons, corresponding to the 5′ and 3′ ends of the RNA.

◆ The introns are the intervening sequences that are removed when the primary transcript is processed to give the mature RNA.

Figure 23.1

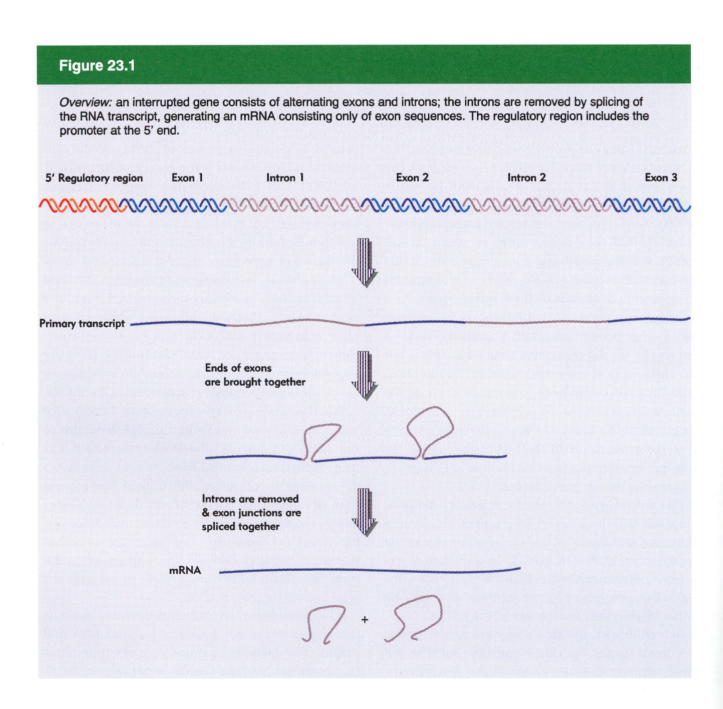

Overview: an interrupted gene consists of alternating exons and introns; the introns are removed by splicing of the RNA transcript, generating an mRNA consisting only of exon sequences. The regulatory region includes the promoter at the 5′ end.

The process by which introns are removed from the primary transcript is called **RNA splicing**. Essentially it involves a precise deletion of an intron from the primary transcript; the ends of the RNA on either side are joined together to form a covalently intact molecule. We discuss the mechanisms and regulation of splicing in Chapter 31.

We define the structural gene as comprising the region in the genome between points corresponding to the 5′ and 3′ terminal bases of mature mRNA. We know that transcription starts at the 5′ end of the mRNA, but probably it extends beyond the 3′ end, which is generated by cleavage of the RNA (see Chapter 29). The definition of the gene can be expanded to include associated regulatory regions: a transcription unit includes a promoter, other regulatory regions in the upstream region, the gene itself, and (sometimes) a terminator.

Proceeding from the structure of the gene itself, the next question is to determine its context. How much material on either side of a gene is involved in its function? How far is it to the next gene, and how much of the genome comprises sequences that lie between transcription units? Do adjacent genes tend to be related and is there any mechanism for regulating regions of the chromosome longer than single genes?

Organization of interrupted genes may be conserved

When a gene is uninterrupted, the restriction map of its DNA corresponds exactly with the map of its mRNA (obtained by characterizing a cDNA reverse transcript).

When a gene possesses an intron, the map at each end of the gene corresponds with the map at each end of the message sequence. But within the gene, the maps diverge, because additional regions are found in the gene, but are not represented in the message. Each such region corresponds to an intron. The example of **Figure 23.2** compares the restriction maps of a β-globin gene and mRNA. Two introns can be recognized. Each intron contains a series of restriction sites that are absent from the cDNA. The exons can be recognized because the pattern of restriction sites coincides with that observed in the cDNA.

It is difficult to work directly with genomic DNA, so genes are characterized by using cloned genomic sequences. Once a cDNA is available, genomic clones can be obtained by the techniques described in Chapter 21. They can be compared with the cDNA by restriction mapping. The only practical limitation of this approach is that it may be impossible to isolate very large genes in the form of single cloned fragments, in which case clones representing parts of the gene must be characterized individually. In such cases, it is essential to use clones whose genomic sequences overlap, to ensure that we do not miss parts of a gene lying between the cloned sequences.

Ultimately a comparison of the nucleotide sequences of the genomic and cDNA clones precisely defines the introns. Resolution at the sequence level is necessary before we can be sure that all the segments of the gene have been identified. Short introns or exons can be missed in restriction maps if they do not happen to contain an appropriate restriction site. (An intron may pass unnoticed if it lies within a long exon, and an exon that is <50 bp long may fail to hybridize with the cDNA probe, and can therefore pass unnoticed within the introns that flank it.) But a sequence comparison is unambiguous. As indicated in **Figure 23.3**, an intron that lies within a coding region usually interrupts the integrity of the reading frame, but an intact reading frame is found in the cDNA sequence.

No particular rhyme or reason yet has been discerned in the extremely varied structures of eukaryotic genes. Some genes are uninterrupted, so that the genomic sequence is colinear with that

Figure 23.2

Comparison of the restriction maps of cDNA and genomic DNA for mouse β-globin shows that the gene has two additional regions not present in the cDNA. The other regions can be aligned exactly between cDNA and gene.

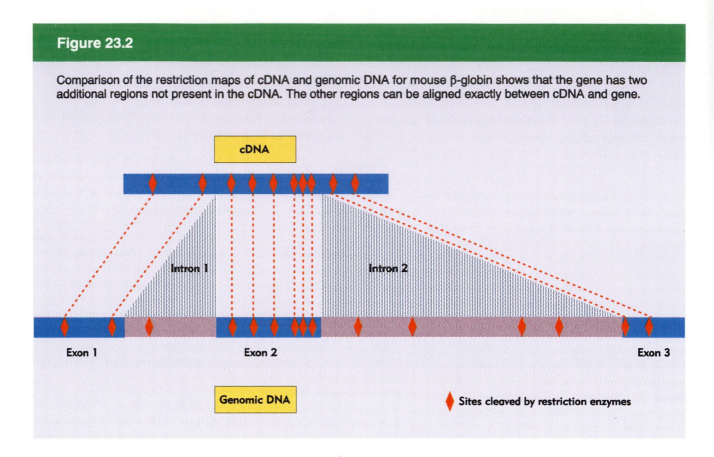

of the mRNA. Most higher eukaryotic genes are interrupted, but the introns vary enormously in both number and size.

Some important features are common to all interrupted genes:

◆ The *order* of the parts of an interrupted gene is the same in the genome as in its mature RNA product. Genes are thus split rather than dispersed.

◆ An interrupted gene retains the *same structure* in all tissues, including the germ line and somatic tissues in which it is or is not expressed. Thus the presence of an intron is an invariant feature.

◆ Introns of nuclear genes generally have termination codons in all reading frames, and have *no coding function.*

All classes of genes may be interrupted: nuclear genes coding for proteins, nucleolar genes coding for rRNA, and genes coding for tRNA. Interruptions also are found in mitochondrial genes in lower eukaryotes, and in chloroplast genes. Interrupted genes do not appear to be excluded from any class of eukaryotes, and have even been found in an archebacterium and a phage of *E. coli.* They appear to be entirely absent only from eubacterial genomes.

In the case of genes coding for mRNA, the terminal exons include the nontranslated leader and trailer, and the internal exons are likely to carry coding sequences. When interrupted genes code for rRNA or tRNA, the exons do not have a protein-coding function.

Some interrupted genes possess only one or a few introns. The globin genes provide an extensively studied example (see Chapter 24). The two general types of globin gene, α and β, share a common type of structure. The consistency of the organization of mammalian globin genes is evident from

Figure 23.3

An intron is a sequence present in the gene but absent from the mRNA (here shown in terms of the cDNA sequence). The reading frame is indicated by the alternating open and shaded blocks; note that all three possible reading frames are blocked by termination codons in the intron.

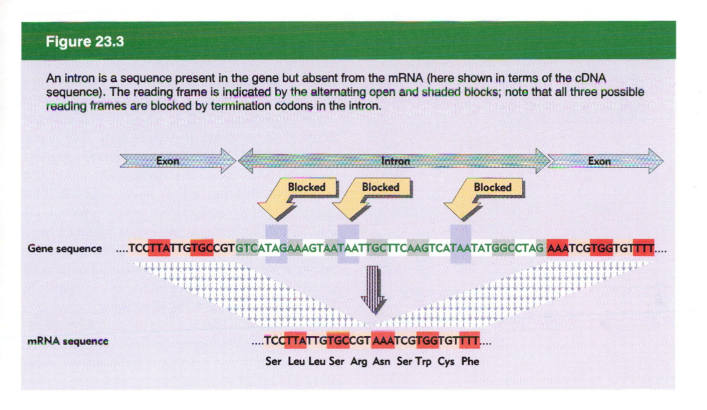

the structure of the 'generic' globin gene summarized in **Figure 23.4**.

Interruptions occur at homologous positions (relative to the coding sequence) in all known active globin genes, including those of mammals, birds, and frogs. The first intron occurs between amino acids 30 and 31 (in a standard mammalian globin sequence), and the second intron is located between amino acids 104 and 105. The first intron is always fairly short, and the second is usually longer, but the actual lengths can vary.

Most of the variation in overall lengths between different globin genes results from the variation in the second intron. In the mouse, the second intron in the α-globin gene is only 150 bp long, so the

Figure 23.4

All functional globin genes have an interrupted structure with three exons. The lengths indicated in the figure apply to the mammalian β-globin genes.

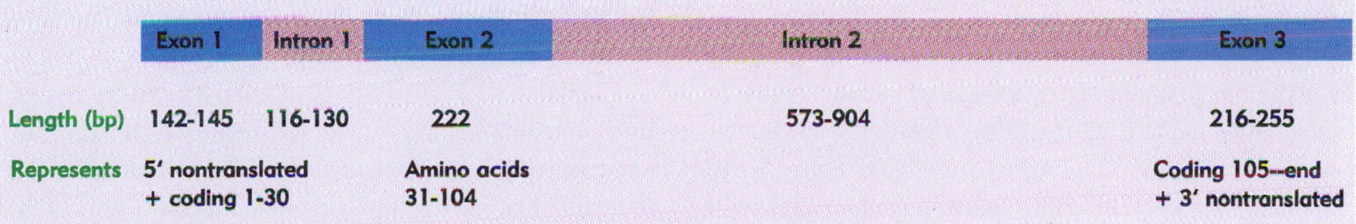

	Exon 1	Intron 1	Exon 2	Intron 2	Exon 3
Length (bp)	142-145	116-130	222	573-904	216-255
Represents	5' nontranslated + coding 1-30		Amino acids 31-104		Coding 105--end + 3' nontranslated

Figure 23.5

Mammalian genes for DHFR have the same relative organization of rather short exons and very long introns, but vary extensively in the lengths of corresponding introns.

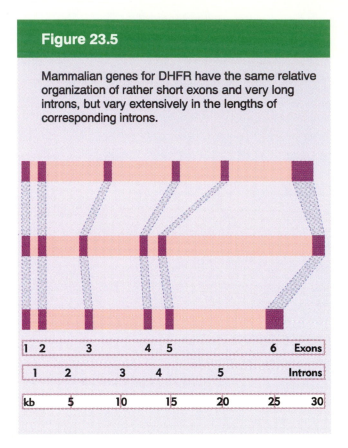

overall length of the gene is 850 bp, compared with the 1382 bp of the major β-globin gene. Thus the variation in length of the genes is much greater than the range of lengths of the mRNAs (α-globin mRNA is 585 bases, β-globin mRNA is 620 bases).

The example of DHFR, which is a somewhat larger gene, is shown in **Figure 23.5**. The mammalian DHFR (dihydrofolate reductase) gene is organized into 6 exons that correspond to the 2000 base mRNA. But they extend over a much greater length of DNA because the introns are exceedingly long. In three mammals the exons remain essentially the same, and the relative positions of the introns are unaltered, but the lengths of individual introns vary extensively, resulting in a variation in the length of the gene from 25 to 31 kb.

The globin and DHFR genes present an example of a general phenomenon: *genes that are related by evolution have related organizations, with conservation of the positions of at least some of the introns. Variations in the lengths of the genes are primarily determined by the lengths of the introns.*

Genes show a wide distribution of sizes

The existence of interrupted genes makes it evident that the gene can be much larger than the unit that codes for protein. With several hundred genes either sequenced or at least compared with their mRNA products in outline, we can explore how far the structure of the gene can account for the size of the eukaryotic genome.

Figure 23.6 shows that the exons coding for stretches of protein tend to be fairly small relative to the size of the gene. The majority code for less than 100 amino acids (often less than 50 in vertebrates), and the general distribution fits well with the idea that genes have evolved by the slow addition of units that code for small, individual domains of proteins (see later). There is no very significant difference in the sizes of exons in different types of organisms, except perhaps for an apparent absence of larger exons in the vertebrates. There are some much larger exons coding for untranslated 5′ and 3′ regions (not included in the figure).

Figure 23.7 shows that introns are longer than exons. Their size distribution extends from approximately the same size as the exons (<200 bp) to lengths measured in tens of kilobases, and in fact extending up to 50–60 kb in extreme cases.

Figure 23.8 shows the overall organization of genes in yeasts, insects and mammals. In *S. cerevisiae*, the great majority of genes in fact are not interrupted, and those that have exons usually remain reasonably compact. There are virtually no *S. cerevisiae* genes with more than four exons.

In insects and mammals, the situation is reversed. Only a small minority of genes consist of uninterrupted coding sequences (6% in mammals). Insect genes tend to have a fairly small number of exons, typically fewer than 10. Mammalian genes are split into more pieces, and some have several tens of exons. The individual introns vary from short interruptions in the coding sequence to very long regions consisting of 10 or more kilobases.

If we now examine the consequences of this type

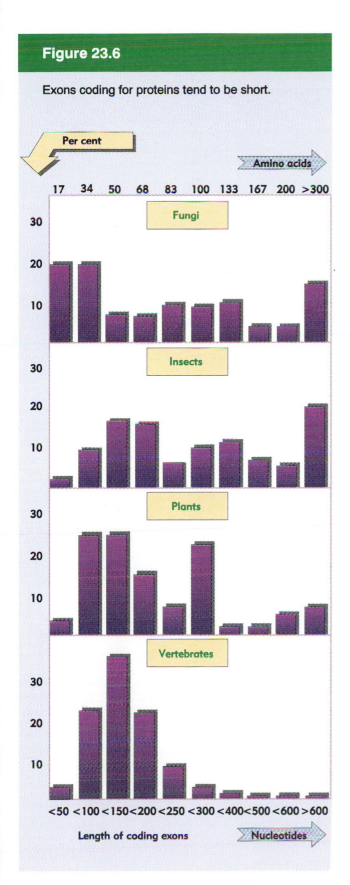

Figure 23.6

Exons coding for proteins tend to be short.

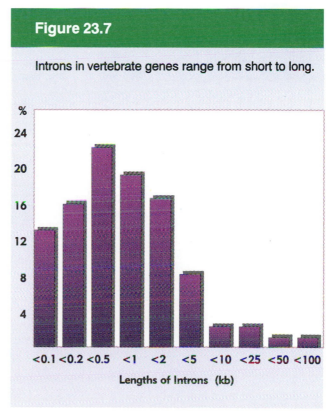

Figure 23.7

Introns in vertebrate genes range from short to long.

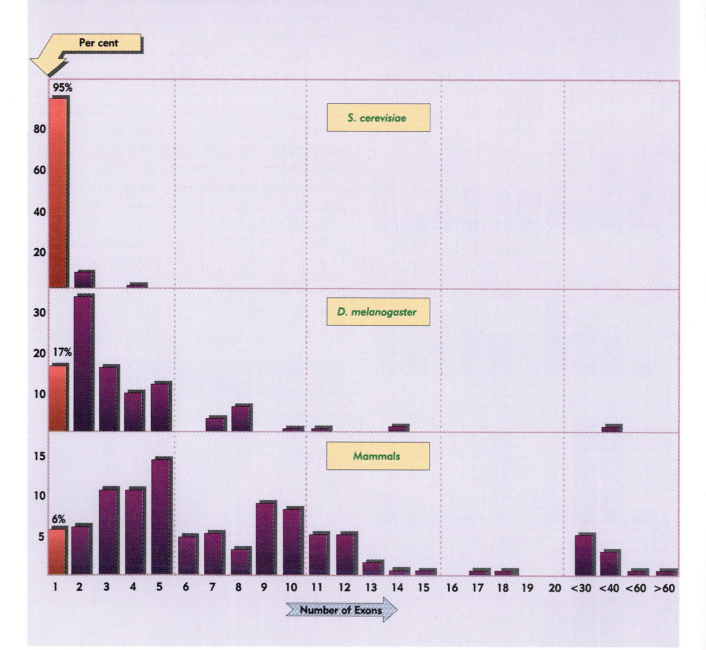

Figure 23.8

Most genes are uninterrupted in yeast, but most genes are interrupted in flies and mammals. (Uninterrupted genes have only 1 exon, and are totalled in the leftmost column.)

Figure 23.9

Yeast genes are fairly small, but genes in flies and mammals have a dispersed distribution extending to very large sizes.

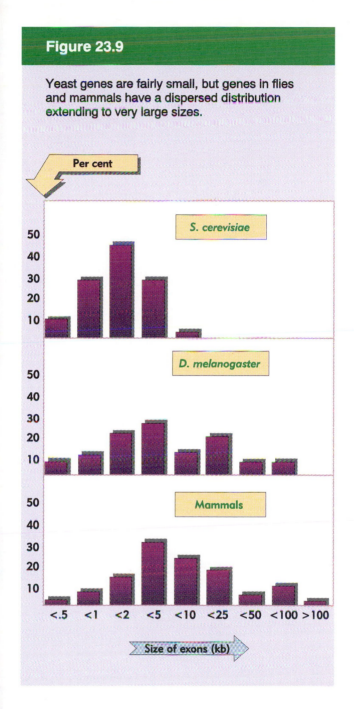

between yeast and the higher eukaryotes. Most yeast genes are shorter than 2 kb, and very few are longer than 5 kb. By contrast, relatively few genes in flies or mammals are shorter than 2 kb, and many have lengths between 5 and 100 kb.

The switch from largely uninterrupted to largely interrupted genes occurs in the lower eukaryotes. In fungi (excepting the yeasts), the majority of genes are interrupted, but they have a relatively small number of exons (<6) and are fairly short (<5 kb). The switch to long genes occurs within the higher eukaryotes, and genes become significantly larger in the insects. Perhaps genes become large at the same point where the relationship between genome complexity and organism complexity is lost (see Figure 22.1).

We still do not know the cause of the variations in DNA size within a phylum; for example, *D. melanogaster* is an example of a fly with a very small genome, and we may ask whether the 6-fold increase in genome size seen in the house fly represents an increase in lengths of individual genes, more space between the genes, or even an increased number of genes.

Comparing the lengths of genes with the lengths of their mRNA products, we see in **Figure 23.10** that in yeast the distribution of messenger sizes is not very different from the distribution of genomic sizes. The only real change in the distribution is an accumulation of short messengers from splicing the minority of interrupted genes. But **Figure 23.11** shows that the situation is very different in mammals, where messengers are essentially all smaller than 10 kb, contrasted with the genes, which extend up to ~100 kb.

Very long genes are the result of very long introns, not the result of coding for longer products. There is no correlation between gene size and mRNA size in higher eukaryotes; nor is

of organization for the overall size of the gene, we see in **Figure 23.9** that there is a striking difference

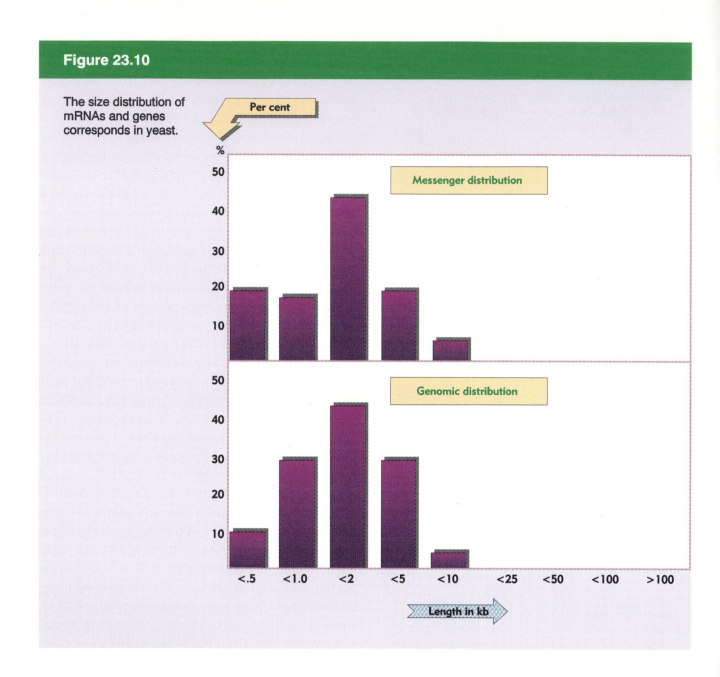

Figure 23.10

The size distribution of mRNAs and genes corresponds in yeast.

there a good correlation between gene size and the number of exons. The size of a gene therefore depends primarily on the lengths of its individual introns.

Table 23.1 summarizes the average gene sizes in a range of organisms. When we compare the average sizes of mRNAs and genes in various species, we see that there is a small increase proceeding from the lower to higher eukaryotes. A large increase in the average number of exons is evident by the start of the higher eukaryotes, and the genes become much larger with the insects. In mammals, insects, and birds, the 'average' gene is approximately 5 times the length of its mRNA.

Figure 23.11

Mammalian genes are much larger than the corresponding mRNAs.

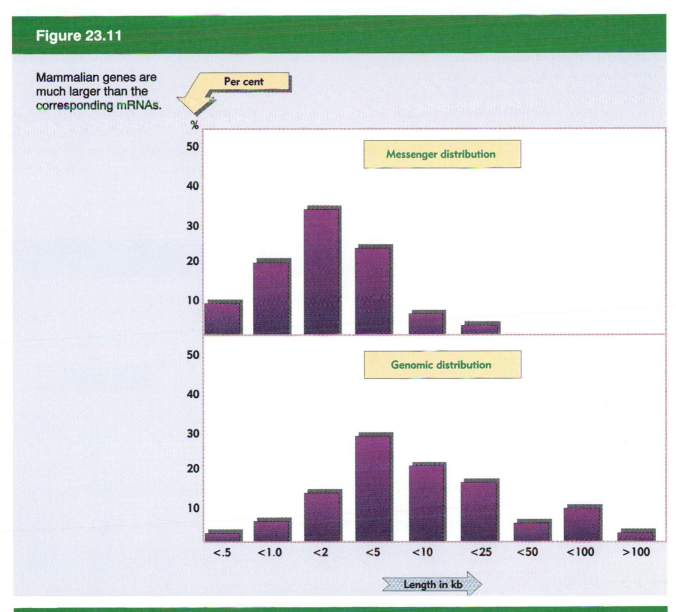

Table 23.1

The average lower eukaryotic gene is similar in size to the average mRNA, but the average size of the higher eukaryotic gene is about 5 times the length of its mRNA product.

Species	Average exon number	Average gene length (kb)	Average mRNA length (kb)
S. cerevisiae	1	1.6	1.6
Fungi	3	1.5	1.5
C. elegans	4	4.0	3.0
D. melanogaster	4	11.3	2.7
Chicken	9	13.9	2.4
Mammals	7	16.6	2.2

One DNA sequence may code for multiple proteins

So far we have dealt with the organization of interrupted genes in terms of an invariant alternating pattern of exons and introns. Although this is scarcely the relationship anticipated from the original definition of the cistron, it is not at odds with it.

A mosaic gene is expressed *only via the splicing together of exons carried by one molecule of RNA,*

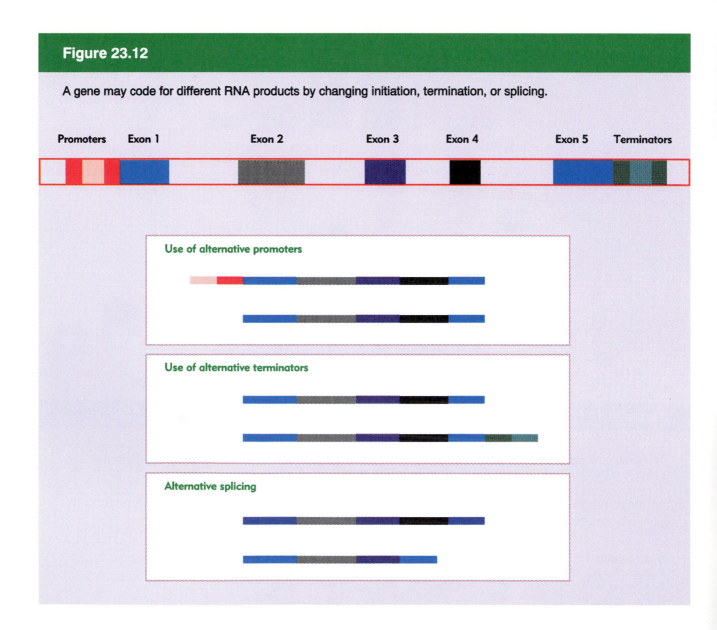

Figure 23.12

A gene may code for different RNA products by changing initiation, termination, or splicing.

Promoters   Exon 1   Exon 2   Exon 3   Exon 4   Exon 5   Terminators

Use of alternative promoters

Use of alternative terminators

Alternative splicing

thus conforming to the original concept of the *cis/trans* test (see Chapter 3). Mutations in the exons of a gene therefore fail to complement one another (see Chapter 6). Mutations in an intron that affect splicing of the RNA behave as part of the same complementation group.

In some cases, however, the fixed relationship between gene and protein does not hold. Here more than one mRNA sequence can be derived from a single stretch of DNA. *The meaning of a particular region of DNA therefore is not invariant, but depends on the pathway selected for its expression.*

Figure 23.12 summarizes the various mechanisms that are used for alternative expression of genes. They include differential use of promoters or of termination sites as well as use of alternative exons. The use of alternative promoters may change the 5′ end, and the use of alternative cleavage sites may change the 3′ end. These changes may be restricted just to the initial and terminal sequences of the mRNA, or they may influence the splicing pattern. It is also possible for alternative splicing to occur entirely within a gene to change the usage of exons.

Alternative splicing may make it impossible to define unambiguously the nature of a particular segment of DNA. In Figure 23.12, exon 4 is present in one mRNA but not in the other. Its presence in the first mRNA defines it as an exon. However, in the splicing of the second mRNA it is treated as part of an intron that extends from exon 3 to exon 5!

In some cases, the alternative means of expression do not affect the sequence of the protein; for example, changes that affect the 5′ nontranslated leader or the 3′ nontranslated trailer may have regulatory consequences, but the same protein is made. In other cases, the substitution of one exon for another, or the addition or deletion of exons, may change the protein sequence as indicated in **Figure 23.13**.

In this example, the proteins produced by the two mRNAs contain sequences that overlap extensively, but that are different within the alternatively

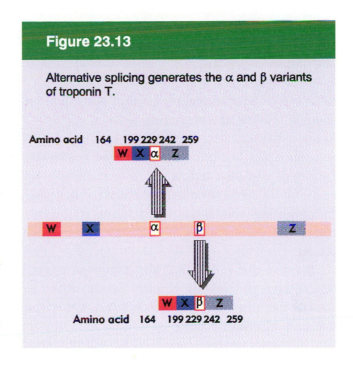

Figure 23.13

Alternative splicing generates the α and β variants of troponin T.

spliced region. The 3′ half of the troponin T gene of rat muscle contains five exons, but only four are used to construct an individual mRNA. Three exons, *WXZ*, are the same in both expression patterns. However, in one pattern the α exon is spliced between *X* and *Z*; in the other pattern, the β exon is used. The α and β forms of troponin T therefore differ in the sequence of the amino acids present between codons 229 and 242, depending on which of the alternative exons, α or β, is used. Either one of the α and β exons can be used to form an individual mRNA, but both cannot be used in the same mRNA.

Thus alternative (or differential) splicing can generate proteins with overlapping sequences from a single stretch of DNA. It is curious that the higher eukaryotic genome is extremely spacious in having large genes that are often quite dispersed, but at the same time it may make multiple products from an individual locus. It is really not possible to say how many genes have alternative modes of expression, but anecdotally, the number would seem to be a few per cent.

Exon sequences are conserved but introns vary

By using restriction fragments that correspond to specific regions of a structural gene, it is possible to test whether these regions are related to other sequences in the genome. A cloned restriction fragment can be used as a probe to detect corresponding sequences in a restriction digest. By obtaining appropriate restriction fragments, regions of either exons or introns can be investigated.

Often it turns out that the exons of a gene are related to those of another gene; when an exonic probe is used, it detects fragments that are part of another gene or genes. This implies that the two genes originated by a duplication of some common ancestral gene, after which differences accumulated between the copies.

Usually the introns are not related to other sequences. And when two genes have related exons, the relationship between their introns is more distant than the relationship between the exons.

Our original description considered nonrepetitive DNA to represent sequences that are unique in the genome. Introns come closer than exons to fitting this definition. In asking whether structural genes are nonrepetitive, we see therefore that the answer is ambiguous: the entire length of the gene is unique as such, but its exons often are related to those of other genes. At least for some genes, the exons constitute slightly repetitive sequences embedded in a unique context of introns.

Figure 23.14

The sequences of the mouse αmaj and αmin globin genes are closely related in coding regions, but differ in the flanking regions and large intron. Data kindly provided by Philip Leder.

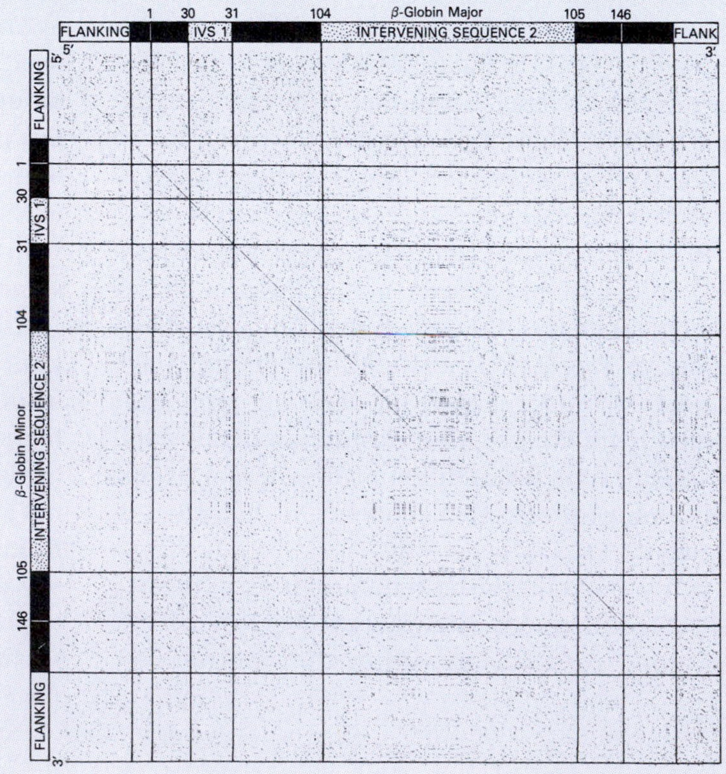

The relationship between two genes can be plotted in the form of the dot matrix comparison of **Figure 23.14**. A dot is placed to indicate each position at which the same sequence is found in each gene. The dots form a line at an angle of 45° if two sequences are identical. The line is broken by regions that lack similarity, and it is displaced laterally or vertically by deletions or insertions in one sequence relative to the other.

When the two β-globin genes of the mouse are compared, such a line extends through the three exons and through the small intron. The line peters out in the flanking regions and in the large intron. This is a typical pattern, in which coding sequences are well related, the relationship can extend beyond the boundaries of the exons, but it is lost in longer introns and the regions on either side of the gene.

The overall degree of divergence between two exons is related to the differences between the proteins. It is caused mostly by base substitutions. In the translated regions, the exons are under the constraint of needing to code for amino acid sequences, so they are limited in their potential to change sequence. Thus many of the changes do not affect codon meanings, because they lie in third-base positions. Changes occur more freely in nontranslated regions (corresponding to the 5' leader and 3' trailer of the mRNA).

In corresponding introns, the pattern of divergence involves both changes in size (due to deletions and insertions) and base substitutions. Introns evolve much more rapidly than exons; in comparisons of the same gene in different species, sometimes the exons are homologous while the introns have diverged so much that corresponding sequences cannot be recognized.

Mutations occur at the same rate in both exons and introns, but are removed more effectively from the exons by adverse selection. However, in the absence of the constraints imposed by a coding function, an intron is able quite freely to accumulate point substitutions and other changes. These changes imply that the intron does not have a sequence-specific function. Whether its presence is at all necessary for gene function is not clear.

Genes can be isolated by the conservation of exons

A useful consequence follows from the contrast between the conservation of exons and the variation of introns. In a region containing a gene whose function has been conserved among a range of species, the sequence actually representing the protein should have two distinctive properties: it must of course have an open reading frame; and it is likely to have a related sequence in other species. These features can be used to isolate genes whose functions we suspect, but of which we otherwise have insufficient knowledge.

Suppose we know by genetic data that a particular genetic trait is located in a given chromosomal region, and we have walked along the chromosome to that region. If we lack knowledge about the nature of the gene product, how are we to identify the gene in a region that may be (for example) >100 kb?

A heroic approach that has proved successful with some genes of medical importance is to screen relatively short fragments from the region for the two properties expected of a conserved gene. First we seek to identify fragments that cross-hybridize with the genomes of other species. Then we examine these fragments for open reading frames.

The first criterion is applied by performing a **zoo blot**. We clone a short fragment from the region and use it as a (radioactive) probe to test for related DNA from a variety of species by Southern blotting. If we find hybridizing fragments in several species related to that of the probe—the probe is usually human—the probe becomes a candidate for an exon of the gene.

Such candidates are sequenced, and if they contain open reading frames, are used to isolate surrounding genomic regions. If these appear to be part of an exon, we may then seek to identify the entire gene, to isolate the corresponding cDNA or mRNA, and ultimately to identify the protein.

This approach is valuable for genes whose existence is implied by genetics, but whose nature is unknown. One example is the gene *zfy* located on the human Y chromosome. **Figure 23.15** shows a zoo blot using a probe from this region. It hybridizes specifically with sex chromosomes of mammals

Figure 23.15

A zoo blot with a probe from the human Y chromosomal gene *zfy* identifies cross-hybridizing fragments on the sex chromosomes of other mammals and birds. There is one reacting fragment on the Y chromosome and another on the X chromosome. Data kindly provided by David Page.

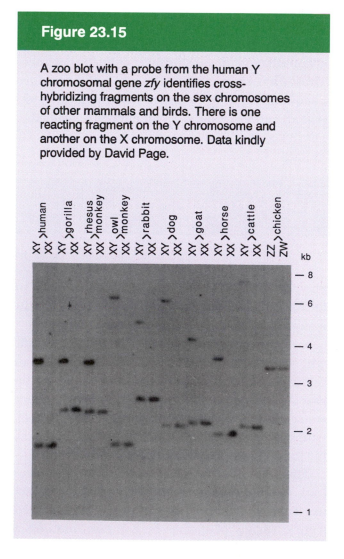

Figure 23.16

The gene involved in Duchenne muscular dystrophy has been tracked down by chromosome mapping and walking to a region in which deletions can be identified with the occurrence of the disease.

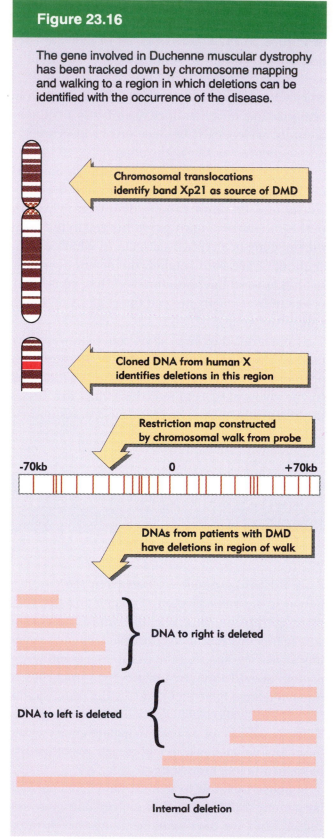

Chromosomal translocations identify band Xp21 as source of DMD

Cloned DNA from human X identifies deletions in this region

Restriction map constructed by chromosomal walk from probe

-70kb   0   +70kb

DNAs from patients with DMD have deletions in region of walk

DNA to right is deleted

DNA to left is deleted

Internal deletion

and also with other species. It contains an open reading frame, which identifies a conserved gene.

This approach is especially important when the target gene is spread out because it has many large introns. This proved to be the case with Duchenne muscular dystrophy (DMD), a degenerative disorder of muscle, which is X-linked and affects 1 in 3,500 of human male births. The steps in identifying the gene are summarized in **Figure 23.16**.

Linkage analysis localized the DMD locus to chromosomal band Xp21. Patients with the disease often have chromosomal rearrangements involving this band. By comparing the ability of X-linked DNA probes to hybridize with DNA from patients and with normal DNA, cloned fragments were obtained that correspond to the region that was rearranged or deleted in patients' DNA.

A chromosomal walk was used to construct a restriction map of the region on either side of the probe, covering a region of >100 kb. Analysis of the DNA from a series of patients identified large deletions in this region, extending in either direction. The most telling deletion is one contained entirely within the region, since this delineates a segment that must be important in gene function and indicates that the gene, or at least part of it, lies in this region.

Having now come into the region of the gene, we need to identify its exons and introns. A zoo blot identified fragments that cross-hybridize with the mouse X chromosome and with other mammalian DNAs. As summarized in **Figure 23.17**, these were scrutinized for open reading frames and the sequences typical of exon–intron junctions. Fragments that met these criteria were used as probes to identify homologous sequences in a cDNA library prepared from muscle mRNA.

The cDNA corresponding to the gene identifies an unusually large mRNA, ~14 kb. Hybridization back to the genome shows that the mRNA is represented in >60 exons, which are spread over ~2,000 kb of DNA. This makes DMD the longest gene identified; in fact, it is 10× longer than any other known gene.

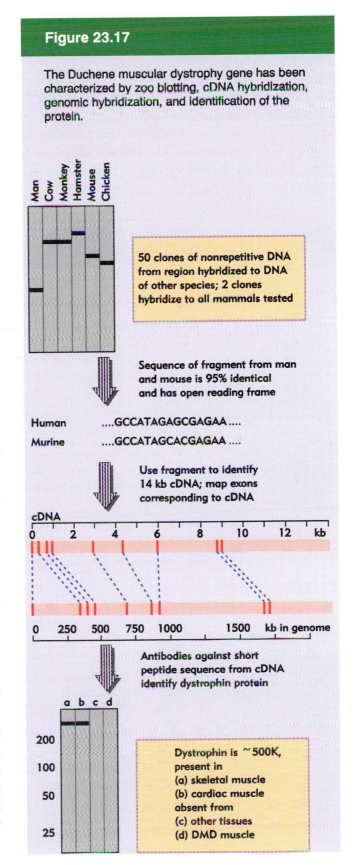

Figure 23.17

The Duchene muscular dystrophy gene has been characterized by zoo blotting, cDNA hybridization, genomic hybridization, and identification of the protein.

50 clones of nonrepetitive DNA from region hybridized to DNA of other species; 2 clones hybridize to all mammals tested

Sequence of fragment from man and mouse is 95% identical and has open reading frame

Human ....GCCATAGAGCGAGAA....
Murine ....GCCATAGCACGAGAA....

Use fragment to identify 14 kb cDNA; map exons corresponding to cDNA

Antibodies against short peptide sequence from cDNA identify dystrophin protein

Dystrophin is ~500K, present in
(a) skeletal muscle
(b) cardiac muscle
absent from
(c) other tissues
(d) DMD muscle

The gene codes for a protein of ~500,000 daltons which has been named dystrophin. It is a component of muscle, present in rather low amounts. All patients with the disease have deletions at this locus, and lack (or have defective) dystrophin.

Another technique that allows genomic fragments to be scanned rapidly for the presence of exons is called **exon trapping. Figure 23.18** shows that it starts with a vector that contains a strong promoter, and has a single intron interrupting the two exons. When this vector is transfected into cells, its transcription generates large amounts of an RNA containing the sequences of the two exons. A restriction cloning site lies within the intron, and is used to insert genomic fragments from a region of interest. If a fragment does not contain an exon, there is no change in the splicing pattern, and the RNA contains only the same sequences as the

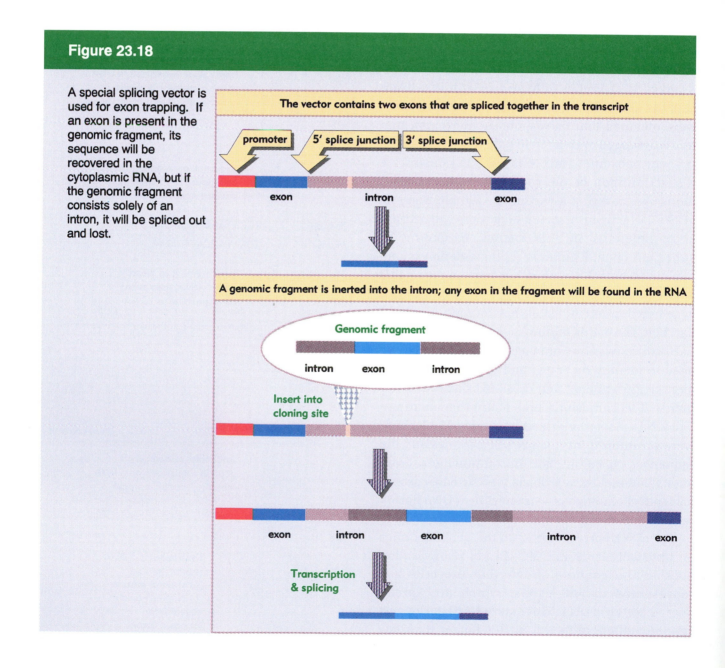

Figure 23.18

A special splicing vector is used for exon trapping. If an exon is present in the genomic fragment, its sequence will be recovered in the cytoplasmic RNA, but if the genomic fragment consists solely of an intron, it will be spliced out and lost.

The vector contains two exons that are spliced together in the transcript

promoter   5' splice junction   3' splice junction

exon   intron   exon

A genomic fragment is inerted into the intron; any exon in the fragment will be found in the RNA

Genomic fragment

intron   exon   intron

Insert into cloning site

exon   intron   exon   intron   exon

Transcription & splicing

parental vector. But if the genomic fragment contains an exon flanked by two partial intron sequences, the splicing sites on either side of this exon are recognized, and the sequence of the exon is inserted into the RNA between the two exons of the vector. This can be detected readily by reverse transcribing the cytoplasmic RNA into cDNA, and using PCR to amplify the sequences between the two exons of the vector. Thus the appearance in the amplified population of sequences from the genomic fragment indicates that an exon has been trapped. Because introns are usually large and exons are small in animal cells, there is a high probability that a large random piece of genomic DNA will contain the required structure of an exon surrounded by partial introns.

How do interrupted genes evolve?

What was the original form of genes that today are interrupted?

◆ Did the ancestral protein-coding units consist of uninterrupted sequences of DNA, into which introns were subsequently inserted?

◆ Or did these genes initially arise as interrupted structures, which since have been maintained in this form?

Another form of this question is to ask whether the difference between eukaryotic and prokaryotic genes is to be accounted for by the acquisition of introns in the eukaryotes or by the loss of introns from the prokaryotes.

Could the mosaic structure be a remnant of an ancient approach to the reconstruction of genes to make novel proteins? Suppose that an early cell had a number of separate protein-coding sequences. One aspect of its evolution is likely to have been the reorganization and juxtaposition of different polypeptide units to build up new proteins.

If the protein-coding unit must be a continuous series of codons, every such reconstruction would require a precise recombination of DNA to place the two protein-coding units in register, end to end in the same reading frame. Furthermore, if this combination is not successful, the cell has been damaged, because it has lost the original protein-coding units.

But if an approximate recombination of DNA could place the two protein-coding units within the same transcription unit, splicing patterns could be tried out at the level of RNA to combine the two proteins into a single polypeptide chain. And if these combinations are not successful, the original protein-coding units remain available for further trials. Such an approach essentially allows the cell to try out controlled deletions in RNA without suffering the damaging instability that could occur from applying this procedure to DNA.

If current proteins evolved by combining ancestral proteins that were originally separate, the accretion of units is likely to have occurred sequentially over some period of time, with one exon added at a time. If this model is realistic, we can ask whether the different functions from which these genes were pieced together are discernible in their present structure. In other words, can we equate particular functions of current proteins with individual exons?

In some cases, there is indeed a clear relationship between the structures of the gene and protein. The example *par excellence* is provided by the immunoglobulin proteins, which are coded by

Figure 23.19

Immunoglobulin light chains and heavy chains are coded by genes whose structures (in their expressed forms) correspond with the distinct domains in the protein. Each protein domain corresponds to an exon; introns are numbered 1-5.

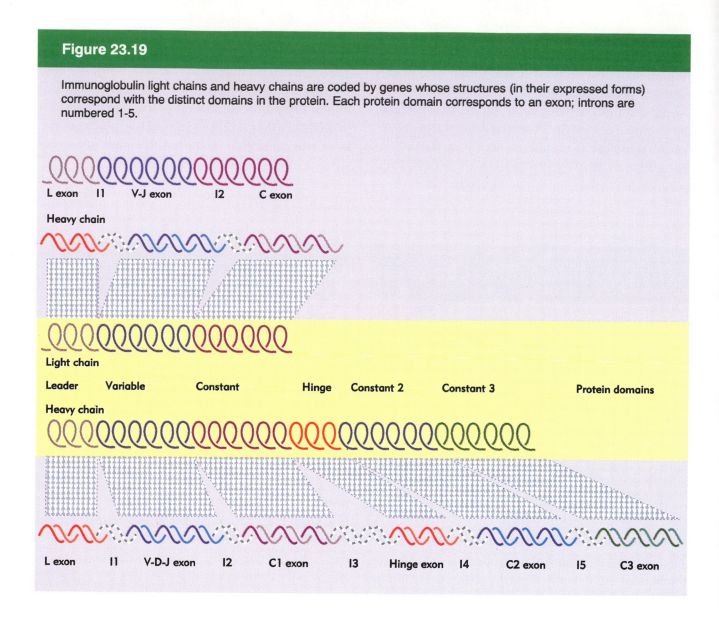

genes in which every exon corresponds exactly with a known functional domain of the protein. **Figure 23.19** compares the structure of an immunoglobulin with its gene.

An immunoglobulin is a tetramer of two light chains and two heavy chains, which aggregate to generate a protein with several distinct domains. Light chains and heavy chains differ in structure, and there are several types of heavy chain. Each type of chain is expressed from a gene that has a series of exons corresponding with the structural domains of the protein. (Chapter 37 explains how the expressed form of the gene is generated from its components.)

In many instances, some of the exons of a gene can be identified with particular functions. In secretory proteins, the first exon, coding for the N-terminal region of the polypeptide, often specifies the signal sequence involved in membrane secretion. An example is insulin.

Sometimes the evolution of a gene involves the duplication of exons, creating an internally repetitious sequence in the protein. In chicken collagen, a 54 bp exon appears to have been multiplied many times, generating a series of exons that are either 54 bp or multiples of 54 bp in length.

Sequences held in common between genes that

are related only in part may represent exons that have migrated or been recruited between genes. **Figure 23.20** summarizes the relationship between the receptor for human LDL (plasma low density lipoprotein) and other proteins.

In the center of the LDL receptor gene is a series of exons related to the exons of the gene for the precursor for EGF (epidermal growth factor). In the N-terminal part of the protein, a series of exons codes for a sequence related to the blood protein complement factor C9. Thus the LDL receptor gene was created by assembling *modules* for its various functions, these modules also being used in other proteins.

The relationship between exons and protein domains is somewhat erratic in known genes. In some cases there is a clear 1:1 relationship; in others no pattern is to be discerned. One possibility is that removal of introns has fused the adjacent exons. A difficulty in this idea is the need to suppose that the intron removal was precise, not changing the integrity of the coding region. An alternative is that some introns arose by insertion into a coherent domain; here the difficulty is that we must suppose that the intron carried with it the ability to be spliced out.

Exons tend to be fairly small (see Figure 23.6), around the size of the smallest polypeptide that can assume a stable folded structure, ~20–40 residues. Perhaps proteins were originally assembled from rather small modules. Each module need not necessarily correspond to a current function; several modules could have combined to generate a function. The number of exons in a gene tends to increase with the length of its protein, which is consistent with the view that proteins acquire multiple functions by successively adding appropriate modules.

This idea might explain another feature of protein structure: it seems that the sites represented at exon-intron boundaries often are located at the surface of a protein. As modules are added to a protein, the connections, at least of the most recently added modules, could tend to lie at the surface.

The equation of at least some exons with pro-

tein domains, and the appearance of related exons in different proteins, supports the idea that 'exon-shuffling' has been a fundamental relationship in the evolution of genes. Certainly the duplication and juxtaposition of exons has played an important role in evolution. We cannot trace the actual events involved in the evolution of every gene; many relationships between exons and protein domains do not conform to a simple equation, but they could be accounted for if events such as exon fusions have modified the ancestral structure during the evolution of nuclear genes. Indeed, one school of thought is that the number of ancestral exons, from which all proteins have been derived by duplication, variation, and recombination, could be relatively small (a few thousands or tens of thousands).

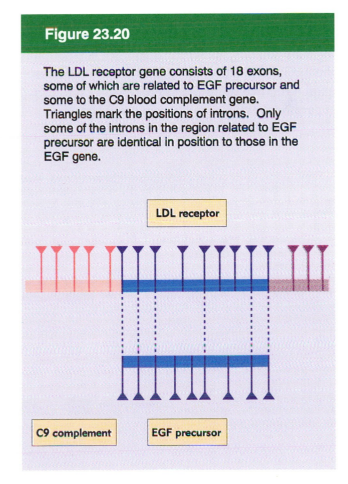

Figure 23.20

The LDL receptor gene consists of 18 exons, some of which are related to EGF precursor and some to the C9 blood complement gene. Triangles mark the positions of introns. Only some of the introns in the region related to EGF precursor are identical in position to those in the EGF gene.

A fascinating case of evolutionary conservation is presented by the globins, all of whose genes have three exons (see Figure 23.2). The two introns are located at constant positions relative to the coding sequence. The central exon appears to represent the heme-binding domain of the globin chain. The active protein is a tetramer containing two globin chains of the α type and two of the β type.

Another perspective on this structure is provided by the existence of two other types of protein that are related to globin. Myoglobin is a monomeric oxygen-binding protein of animals, whose amino acid sequence suggests a common (though ancient) origin with the globin subunits. Leghemoglobins are oxygen-binding proteins present in the legume class of plants; like myoglobin, they are monomeric. They too share a common origin with the other heme-binding proteins. Together, the globins, myoglobin, and leghemoglobin constitute the globin 'super-family', a set of gene families all descended from some (distant) common ancestor.

Myoglobin is represented by a single gene in the human genome, whose structure is essentially the same as that of the globin genes. The three exon structure therefore predates the evolution of separate myoglobin and globin functions.

Leghemoglobin genes contain three introns, the first and last of which occur at points in the coding sequence that are homologous to the locations of the two introns in the globin genes. This remarkable similarity suggests an exceedingly ancient origin for the heme-binding proteins in the form of a split gene, as illustrated in **Figure 23.21**.

The central intron of leghemoglobin separates two exons that together code for the sequence corresponding to the single central exon in globin. Could the central exon of the globin gene have been derived by a fusion of two central exons in the ancestral gene, bringing together the sequences coding for two parts of the protein chain that together form the heme-binding structure?

Cases in which homologous genes differ in structure may provide information about their evolution. An example is insulin. Mammals and birds have only one gene for insulin, except for the rodents, which have two genes. **Figure 23.22** illustrates the structures of these genes.

The principle we use in comparing the organization of related genes in different species is that *a common feature identifies a structure that predated the evolutionary separation of the two species.* In chicken, the single insulin gene has two introns;

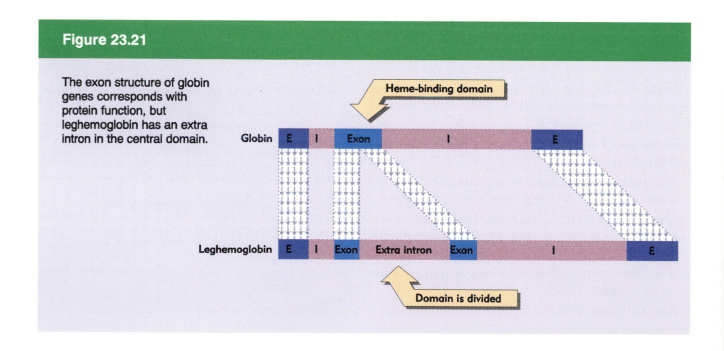

Figure 23.21

The exon structure of globin genes corresponds with protein function, but leghemoglobin has an extra intron in the central domain.

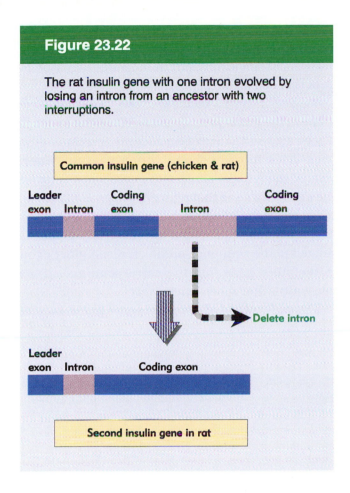

Figure 23.22

The rat insulin gene with one intron evolved by losing an intron from an ancestor with two interruptions.

one of the two rat genes has the same structure. The common structure implies that the ancestral insulin gene had two introns. However, the second rat gene has only one intron. It must have evolved from the first by a process in which a gene duplication in rodents was followed by the precise removal of one intron from one of the copies.

The organization of some genes shows extensive discrepancies between species. In these cases, there must have been extensive removal or insertion of introns during evolution.

The best characterized case is represented by the actin genes. The typical actin gene has a nontranslated leader of <100 bases, a coding region of ~1,200 bases, and a trailer of ~200 bases. Most actin gens are interrupted; the positions of the introns can be aligned with regard to the coding sequence (except for a single intron sometimes found in the leader).

Figure 23.23 shows that almost every actin gene

is different in its pattern of interruptions. Taking all the genes together, introns occur at 12 different sites. However, no individual gene has more than six introns; some genes have only one intron, and one is uninterrupted altogether. How did this situation arise? If we suppose that the primordial actin gene was interrupted, and all current actin genes are related to it by loss of introns, different introns have been lost in each evolutionary branch. Probably some introns have been lost entirely, so the primordial gene could well have had 20 or more. The relationships between the intron locations found in different species may be used ultimately to construct a tree for the evolution of the gene.

Polymorphisms seem common in genes for rRNA and tRNA, where alternative forms can often be found, with and without introns. In the case of the tRNAs, where all the molecules conform to the same general structure, it seems unlikely that evolution brought together the two regions of the gene. After all, the different regions are involved in the base pairing that gives significance to the structure. So here it may be that the introns were inserted into continuous genes.

Organelle genomes provide some striking connections between the prokaryotic and eukaryotic worlds. Because of many general similarities between mitochondria or chloroplasts and bacteria, it seems likely that the organelles originated by an **endosymbiosis** in which an early bacterial prototype was inserted into eukaryotic cytoplasm. Yet in contrast with the resemblances with bacteria—for example, as seen in protein or RNA synthesis—some organelle genes possess introns, and therefore resemble eukaryotic nuclear genes.

Introns are found in several chloroplast genes, including some that have homologies with genes of *E. coli* (see Chapter 25). This suggests that the endosymbiotic event occurred before introns were lost from the prokaryotic line. If a suitable gene can be found, it may therefore be possible to trace gene lineage back to the period when endosymbiosis occurred.

The mitochondrial genome presents a particularly striking case. The genes of yeast and mammalian mitochondria code for virtually identical mitochondrial proteins, in spite of a considerable difference in gene organization. Vertebrate mitochondrial genomes are very small, with an extremely compact organization of continuous genes (see Chapter 25), whereas yeast mitochondrial genomes are larger and have some complex interrupted genes. Which is the ancestral form?

If we accept the endosymbiotic theory, we must conclude that the interrupted genes represent the source. This in turn implies that the ability to splice out introns was lost independently during the evolution of bacteria and vertebrate mitochondria.

Considering the interrupted, sometimes very extensively spread out structure of eukaryotic genes, we can picture the eukaryotic genome as a sea of introns (mostly but not exclusively unique in sequence), in which islands of exons (sometimes very short) are strung out in individual archipelagoes that constitute genes.

Viewed in terms of their exons, few genes are alone in the eukaryotic genome. When the exons of a gene are used as a probe to identify corresponding sequences in the genome, often we find other, related sequences. These sequences may comprise the exons of a gene that represents a protein related to (or occasionally identical with) the product of the first gene. The sequence of each exon of one gene is related to the sequence of the corresponding exon in the other gene, but the corresponding introns generally differ more extensively.

The widespread existence of genes that consist of

Figure 23.23

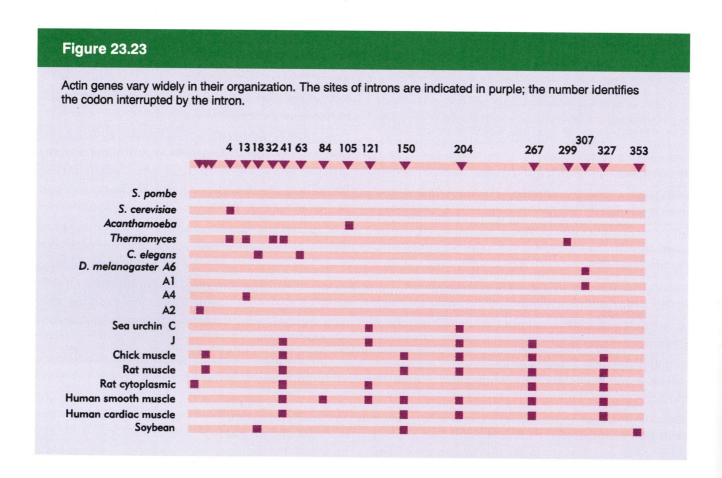

Actin genes vary widely in their organization. The sites of introns are indicated in purple; the number identifies the codon interrupted by the intron.

corresponding exons emphasizes the importance of gene duplication as a mechanism for generating new genes. Duplication of the entire gene allows one copy to evolve via mutation, while the other retains its original function. Such genes are likely to code for proteins that have related functions.

In addition to genes that code for proteins that are related along their entire length, there are also genes that have in common an exon or series of exons coding for related sequences, although the other regions of the genes do not correspond. Such genes could originate by a duplication of part of a gene; the duplicated region could be moved to a new location where its exons combine with another series of exons to generate a new gene. The proteins coded by these genes share features coded by the related exons, but otherwise have different functions.

The history of related genes that have the same organization is likely to encompass a series of events in which first the component exons were brought together to form a coding region; later the entire series of exons and introns constituting the gene may have been duplicated. The duplication will have been followed by lesser divergence in exon sequences and greater divergence in intron sequences. We can attempt to reconstruct the history of their evolution from the sequences of genes related in this manner.

Summary

All types of eukaryotic genome contain interrupted genes. The proportion of genes that is interrupted is low in yeasts and increases in the lower eukaryotes; few genes are uninterrupted in higher eukaryotes. The size of a gene is determined primarily by the lengths of its introns. Introns become larger early in the higher eukaryotes, when gene sizes therefore increase significantly. The range of gene sizes in mammals is generally from 1 to 100 kb, but it is possible to have even larger genes; the longest known case is dystrophin at 2,000 kb.

Introns are found in all classes of genes, including those coding for rRNA and tRNA. The structure of the interrupted gene is the same in all tissues, exons are joined together in RNA in the same order as their organization in DNA, and the introns usually have no coding function. Introns are removed from RNA by splicing. Some genes are expressed by alternative splicing patterns, in which a particular sequence is removed as an intron in some situations, but retained as an exon in others.

Positions of introns are conserved when the organization of homologous genes is compared between species. Intron sequences, however, vary, and may even be unrelated, although exon sequences remain well-related. The conservation of exons can be used to isolate related genes in different species.

Some genes share only some of their exons with other genes, suggesting that they have been assembled by addition of exons representing individual modules of the protein. Such modules may have been incorporated into a variety of different proteins. The idea that genes have been assembled by accretion of exons implies that introns were present in genes of primitive organisms. Some of the relationships between homologous genes can be explained by loss of introns from the primordial genes, with different introns being lost in different lines of descent.

Further reading

Reviews

The first major review of gene structure following the discovery of interrupted genes was provided by **Breathnach and Chambon** (*Ann. Rev. Biochem.* **50**, 349–383, 1981).

The relationship between gene organization and protein structure has been analyzed by **Blake** (*Int. Rev. Cytol.* **95**, 149–185, 1985).

The general characteristics of gene and protein evolution were reviewed by **Wilson *et al***. (*Ann. Rev. Biochem.* **46**, 573–639, 1977). The evolution of multigene families was reviewed by **Maeda and Smithies** (*Ann. Rev. Genet.* **20**, 81–108, 1986).

CHAPTER 24

Gene numbers: repetition and redundancy

From the characteristics of individual genes described in the previous chapter, we can estimate the average size of a gene. By comparing this with the total size of the genome, we obtain a crude estimate for the number of genes. Because some genes give rise to more than one product by alternative means of expression, the number of different proteins is greater, but probably <10% of genes produce more than one type of protein.

Although our information about the sizes and numbers of genes themselves is becoming more extensive, an impediment to determining the number of genes is the lack of sufficient knowledge about their context: we need to define the density of genes in order to determine the total number with any accuracy. Better estimates for gene numbers should become possible when fine structure mapping is achieved for long contiguous stretches of DNA.

If we take the average gene sizes calculated in Chapter 23 and assume further that the separation between adjacent genes will be less than (say half) of the size of the gene itself, we would calculate gene numbers as shown in **Table 24.1**. A bacterium would have ~2,400 genes, the number rises to ~5,000 in a yeast, reaches ~8,000 for the fruit fly, and reaches ~125,000 for a mammal (man).

In two cases where long genomic regions have been sequenced (~100 kb in *E. coli*, >300 kb in *S. cerevisiae*), a feature is that many unsuspected open reading frames are found (some rather short). Also, in these particular stretches of DNA there is very little separation between genes. If these open reading frames represent *bona fide* genes, the total gene number could be increased somewhat, to ~3,500 for *E. coli* and ~7,500 for *S. cerevisiae*.

Table 24.1

Gene numbers can be estimated from the average size of the gene.

Species	Genome	Gene size	Gene number
E. coli	4.2×10^6	1.2kb	2,350
S. cerevisiae	1.3×10^7	1.7kb	5,200
D. melanogaster	1.4×10^8	11.3kb	8,000
H. sapiens	3.0×10^9	16.3kb	125,000

Sequencing of stretches of *C. elegans* DNA totalling 120 kb has identified 32 genes, which would correspond to a total number for the organism of 15,000 (making allowance for the fact that the sequenced region is relatively rich in genes). We lack similar data for other organisms, but an obvious corollary would be that estimates of gene numbers would be increased for fruit fly and man.

Estimates based on the characterization of individual genes (such as in Table 24.1) have the caveat that, without knowing the separation between genes, it is difficult to estimate the overall number. However, identification of individual genes at least gives a minimum estimate for total gene number. Estimates based on large scale sequencing have the deficiency that individual open reading frames may not be utilized (especially if they are very short) and that exons may not be properly connected to identify genes (which would inflate the estimate of gene numbers). Identification of all reading frames gives a maximum estimate for gene number. The range between the two estimates provides a reasonable description for the probable constitution of bacterial and yeast genomes, but at present is too wide to give us a good idea of the likely gene number in higher eukaryotes.

In addition to needing to know the density of genes to estimate the total gene number, we must also ask: is it important in itself? Are there structural constraints that make it necessary for genes to have a certain spacing, and does this contribute to the large size of eukaryotic genomes?

Another insight into gene number is obtained by counting the number of expressed genes. If we rely upon the estimates of the number of different mRNA species that can be counted in a cell, we would conclude that the average vertebrate cell expresses ~10–20,000 genes. The existence of significant overlaps between the messenger populations in different cell types would suggest that the total expressed gene number for the organism should be within a few fold of this, in the range (say) of 50,000–100,000.

When we characterize genes corresponding to individual functions, we often find additional copies representing unsuspected variants of the gene. The extension from individual genes to families of related genes certainly increases the number of genes in the genome. Present evidence is purely anecdotal, but it would not be surprising if on average there were 3–4 genes for every function that is apparently unique. Our total number of genes is therefore likely to include a somewhat smaller number of different types of functions.

A major force in evolution is clearly the duplication of genes, either as intact units or as collections of exons or even individual exons.

When an intact gene is involved, the act of duplication generates two copies of a gene whose activities are indistinguishable, but then the copies diverge as each accumulates different mutations.

A set of genes descended by duplication and variation from some ancestral gene is called a **gene family**. Its members may be clustered together or dispersed on different chromosomes (or a combination of both). The members of a structural gene family usually have related or even identical functions, although they may be expressed at different times or in different cell types. Thus different globin proteins are provided for use in embryonic and adult red blood cells, while different actins are utilized in muscle and nonmuscle cells.

Sometimes a relationship can be discerned in the form of a common overall organization among genes whose roles apparently are not related; this occurs as the result of extensive variation from a common ancestor. The entire set, sometimes including multiple discrete gene families, is called a **super family**. For example, the immunoglobulins constitute a well-defined gene family (comprising the antibodies), and they are part of a super family that includes families of cell surface molecules involved in adhesion between cells.

Some gene families consist of identical members. Clustering is a prerequisite for maintaining identity between genes, although clustered genes are not necessarily identical. **Gene clusters** range from extremes where a duplication has generated two adjacent related genes to cases where hundreds of identical genes lie in a tandem array.

Extensive tandem repetition of a gene may occur when the product is needed in unusually large amounts. Examples are the genes for rRNA or histone proteins.

Situations where related genes are dispersed at different locations must have arisen by **translocation** of one gene at some time after a duplication event. After their separation, the genes usually diverge in sequence.

Sometimes no significance is discernible in a repetition; for example, we know of no difference in the expression or function of the duplicate insulin genes of rodents (and a single insulin gene indeed is adequate in other mammals).

Sometimes all of the family members are functional genes, less or more distantly separated in evolution; sometimes some are nonfunctional **pseudogenes**, relics of evolution, descended from genes that once must have been functional, but that now have become inactive and accumulated many mutational changes.

The organization of some gene clusters appears to have been conserved in evolution, and this raises the question of the extent to which genes are completely independent units. How often does expression of a gene depend on its context? We know that some genes appear capable of full expression so long as they have appropriate regulatory regions, usually in the vicinity of the promoter. But others depend on the organization of a cluster, such as the globin genes. (We discuss their regulation in Chapter 28.)

In terms of characterizing the genome, we should like to know how often related genes are found in clusters under common control. Does the genome have domains that contain regulatory elements controlling several genes? We need to know what proportion of genes are controlled like the globin genes and what proportion function as independent units. And how do such domains relate to the physical structure of the chromosome?

Essential genes and total gene number

How many genes are essential? Another approach to the issue of gene number is to determine the number of essential genes by mutational analysis. If we saturate some specified region of the chromosome with mutations that are lethal, the number of complementation groups into which these mutations map should identify the number of lethal loci in that region. By extrapolating to the genome as a whole, we may calculate the total essential gene number.

The most extensive analyses of essential gene number have been made in *Drosophila* through attempts to correlate visible aspects of chromosome structure with the number of functional genetic units. The notion that this might be possible arose originally from the presence of bands in the polytene chromosomes of *D. melanogaster*.

(These chromosomes are found at certain developmental stages and represent an unusually extended physical form, in which a series of bands (more formally called chromomeres) can be seen. We discuss their properties in Chapter 27.) From the early concept that the bands might represent a linear order of genes, we have come to the attempt to correlate the organization of genes with the organization of bands. There are ~5,000 bands in the *D. melanogaster* haploid set; they vary in size over an order of magnitude, but on average there should be ~20 kb per band.

The basic approach is to saturate a chromosomal region with mutations. Usually the mutations are simply collected as lethals, without analyzing the cause of the lethality. Any mutation that is lethal is taken to identify a locus that is essential for the organism. Sometimes mutations cause visible deleterious effects short of lethality, in which case we also count them as identifying an essential locus. When the mutations are placed into complementation groups, the number can be compared with the number of bands in the region, or individual complementation groups may even be assigned to individual bands. The purpose of these experiments has been to determine whether there is a consistent relationship between bands and genes; for example, does every band contain a single gene?

Figure 24.1 shows an example of one mapped region, in which 12 complementation groups have been mapped into the *rosy* region of 16 bands. In this case, the groups have been mapped against markers of known DNA size, so we can see that the whole region occupies about 300 kb. The distribution of complementation groups generally parallels the distribution of the bands, although it does not necessarily show a 1:1 relationship.

Table 24.2 summarizes analyses along these lines that have been carried out over the past 20 years. There is a reasonable correlation between the number of bands (265) and the number of complementation groups (215). This corresponds to ~5% of the *D. melanogaster genome*. In cases where the groups have been mapped against DNA, there appear to be 10–20 kb per lethal complementation group, which is within a 2-fold range of the size of the average gene. It is an open question whether

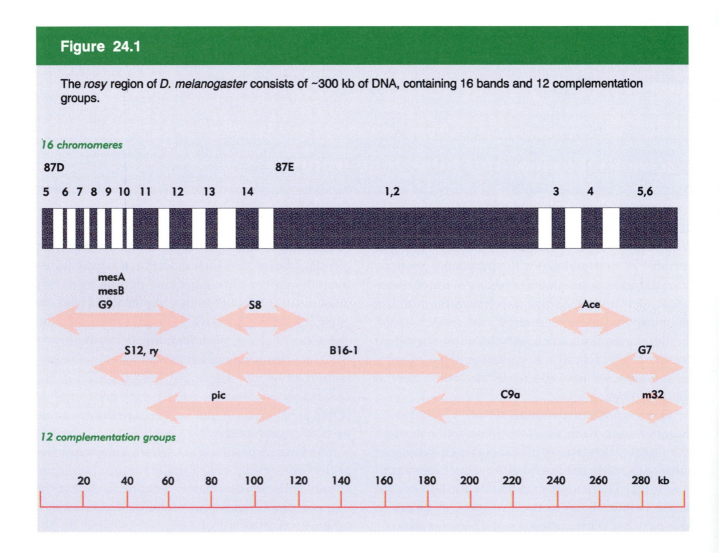

Figure 24.1

The *rosy* region of *D. melanogaster* consists of ~300 kb of DNA, containing 16 bands and 12 complementation groups.

Table 24.2

There is an approximate equivalence between the number of lethal complementation groups and the number of bands in *Drosophila*.

| Chromosome region | Band # | Complementation groups | | | |
		Total	Lethal	Visible	Polymorphic
3A-3C	17	16	16		
34D-35C	34	29	21	8	3
Chromosome IV	50	38	33	5	
37B10-C4	7	12	12		
87D2-87F1	24	21	21		
68C8-69B5	35	37	32	3	2
67A2-D11/13	56	24	24		
84B3-85D5	17	14	14		
85A3-11	9	12	12		
87D5-87E6	16	12	12		
Total	265	215	197	16	5

the equivalence in the number of bands and the number of lethal loci is mere coincidence. Current thinking in the field is that probably this means no more than that the numbers are roughly equivalent over the genome as a whole. But irrespective of the cause, the equivalence gives us a reasonable estimate for the lethal gene number of ~5,000. If we assume that the organization of the *Drosophila* and mammalian genomes is in principle similar, then by comparing the average sizes of their genes and genomes, we would predict >100,000 lethal genes for man. By any measure, the number of lethal loci in *Drosophila* is significantly less than the total number of genes, and presumably the same is true of man.

The difference between lethal and total gene number is paralleled by a discrepancy between genetic and biochemical data on the contents of bands. At the molecular level, we know of many genes that are located in the same vicinity—for example, they can be linked on to the same genomic fragments. It seems likely that such genes are located within the same band. There is therefore a contrast between the apparent equivalence of

lethal loci and bands, and the presence of multiple genes in a band. It may often be the case that when a band contains many genes, several (or all) are not essential. We have no good estimate yet for the number of genes that are defined as nonessential by current criteria, except to note that the average size of the gene would suggest that the number is likely to be of similar magnitude to the number of lethal loci.

An experiment to determine what proportion of genes actually are essential has produced a surprising result in the yeast *S. cerevisiae*. The genome is relatively small, and a large fraction (>50%) is transcribed, compared with higher eukaryotes. When insertions were introduced at random into the genome, only 12% were lethal, and another 14% impeded growth. The majority (70%) of the insertions had no effect. Since the insertions all carried signals that should terminate transcription, we would expect them to prevent expression of any gene in which they reside.

Table 24.3 analyzes the results. Suppose that all of the disruptions that have no effect lie in the nontranscribed regions. This accounts for 50% of

Table 24.3

Disruption analysis of *S. cerevisiae* suggests that more than half the genome is not essential.

Type of Disruption	Total		Proposed distribution		
			Nontranscribed	Transcribed	Genes
Haploid lethal	26	12%		12%	24%
Slow growth	31	14%		14%	28%
Other phenotypes	8	4%		4%	8%
No effect	151	70%	50%	20%	40%
Total disruptions	216	100%	50%	50%	100%

the total disruptions. Thus the remaining 20% of disruptions without effect lie in the transcribed regions. The transcribed regions represent half of the genome, but all of the genes. So a disruption of 20% of these regions represents disruption of 40% of the genes. This implies that 40% of the expressed genes of yeast are nonessential.

How do we explain these results? One possibility is that there is extensive redundancy, that many genes are present in multiple copies. This is certainly true in some cases, in which multiple genes must be knocked out in order to produce an effect. We do not have any means of conducting a systematic survey to determine how many genes actually play unique functions and how many are redundant or simply dispensable.

The idea that some genes are not essential (or at least cannot be shown to have serious effects upon the phenotype) raises some important questions. Does the genome contain genuinely dispensable genes, or do these genes actually have effects upon the phenotype that are significant at least during the long march of evolution? Is each gene function unique (as we would expect if it is essential) or is there **redundancy**: are some genes present in multiple copies?

We do not know whether there are genetic functions that are truly dispensable for higher eukaryotes, but if there are, we may have to rethink some evolutionary questions. We do know that some genes appear to be redundant. There are several cases where mutations of individual loci in yeast or *Drosophila* has no apparent effect, but addition of mutations in other loci then becomes lethal. If there are truly redundant genes, this leads to a difficult question: how is selective pressure applied to these genes?

Key questions are therefore what proportion of the total number of genes is essential, in how many do mutations produce at least detectable effects, and are there genes that are genuinely dispensable? If it is the case that many genes are concerned with specifying proteins that are not essential for the survival of the organism (at least in the sense that mutational damage does not cause any detectable effect), we must ask why these genes continue to be expressed. Subsidiary questions about the genome as a whole are: what is the function (if any) of DNA that does not reside in genes? What effect does a large change in total size have on the operation of the genome, as in the case of the related amphibians?

Globin genes are organized in two clusters

The best characterized example of a gene cluster is presented by the globin genes, which constitute an ancient gene family, concerned with a function that is central to the animal kingdom: the transport of oxygen through the bloodstream. The major constituent of the red blood cell is the globin tetramer, associated with its heme (iron-binding) group in the form of hemoglobin. Functional globin genes in all species have the same general structure, divided into three exons as shown previously in Figures 23.4 and 23.21. We conclude that all globin genes are derived from a single ancestral gene; so by tracing the development of individual globin genes within and between species, we may learn about the mechanisms involved in the evolution of gene families.

In adult cells, the globin tetramer consists of two identical α chains and two identical β chains. The α- and β-globin genes are coded by independent genetic loci whose expression must be coordinated to ensure equivalent production of both polypeptides. This system therefore provides an example of the need for simultaneous control of dispersed genes to generate a particular cell phenotype.

Embryonic blood cells contain hemoglobin tetramers that are different from the adult form. Each tetramer contains two identical α-like chains and two identical β-like chains, each of which is related to the adult polypeptide and is later replaced by it. This is an example of developmental control, in which different genes are successively switched on and off to provide alternative products that fulfill the same function at different times.

The details of the relationship between embryonic and adult hemoglobins vary with the organism. The human pathway consists of three stages: embryonic, fetal, and adult. The distinction between embryonic and adult is common to mammals, but the number of pre-adult stages varies. In man, zeta and alpha are the two α-like chains. Epsilon, gamma, delta, and beta are the β-like chains. The chains expressed at different stages of development are summarized in **Table 24.4**.

ζ is the first α-like chain to be expressed, but is soon replaced by α itself. In the β-pathway, ϵ and γ are expressed first, with δ and β replacing them later. In adults, the $\alpha_2\beta_2$ form provides 97% of the hemoglobin, $\alpha_2\delta_2$ *is* ~2%, and ~1% is provided by persistence of the fetal form $a_2\gamma_2$.

The division of globin chains into α-like and β-like reflects the organization of the genes. Each type of globin is coded by genes organized into a single cluster. The structures of the two clusters in the higher primate genome are illustrated in **Figure 24.2**.

Stretching over 50 kb, the β cluster contains five functional genes (ϵ, two γ, δ and β) and one pseudogene ($\psi\beta$). The two γ genes differ in their coding sequence in only one amino acid; the G variant has glycine at position 136, where the A variant has alanine.

Table 24.4
Human hemoglobins change during development.

Stage of Development	Hemoglobins
Embryonic (<8 weeks)	$\zeta_2\epsilon_2$ $\zeta_2\gamma_2$ $\alpha_2\epsilon_2$
Fetal (3–9 months)	$\alpha_2\gamma_2$
Adult (from birth)	$\alpha_2\delta_2$ $\alpha_2\beta_2$

Figure 24.2

Each of the α-like and β-like globin gene families is organized into a single cluster that includes functional genes and pseudogenes (ψ).
The organization of the clusters in higher primates is conserved; the clusters of man, gorilla, baboon, and orang utan are virtually indistinguishable. In man, the α- cluster lies on chromosome 16, and the β- cluster lies on chromosome 11. All of the active genes are transcribed from left to right.

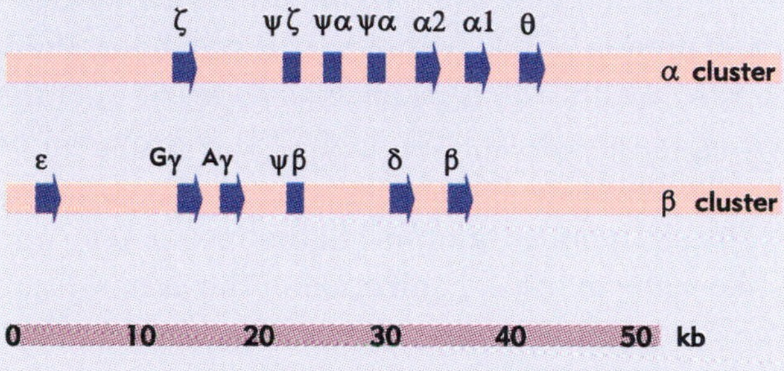

ζ $\psi\zeta$ $\psi\alpha\psi\alpha$ $\alpha2$ $\alpha1$ θ

α cluster

ε Gγ Aγ $\psi\beta$ δ β

β cluster

0   10   20   30   40   50 kb

The more compact α cluster extends over 28 kb and includes one active ζ gene, one ζ pseudogene, two α genes, two α pseudogenes, and the θ gene of unknown function. The two α genes code for the same protein. Two (or more) identical genes present on the same chromosome are described as **nonallelic** copies.

Functional genes are defined by their expression in RNA, and ultimately by the proteins for which they code. **Pseudogenes** are defined as such by their inability to code for proteins; the reasons for inactivity vary, and the deficiencies may be in transcription or translation (or both). Occasionally a gene cannot immediately be placed in either class. After we thought that the structure of the α cluster was defined, the θ gene was discovered at one side. It has no obvious impediment that would classify it as a pseudogene, but neither have we identified a protein product. Its role therefore remains to be ascertained.

Figure 24.3

Clusters of β-globin genes and pseudogenes are found in vertebrates. Seven mouse genes include 2 early embryonic, 1 late embryonic, 2 adult genes, and 2 pseudogenes. Seven β-like genes are also found in the goat. The rabbit has 4 β-like genes: 2 embryonic, 1 pseudo, and 1 adult, lying in order of expression. Chicken also has 4 genes.

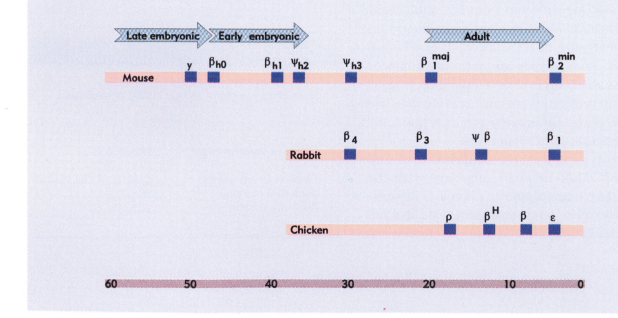

A similar general organization is found in other vertebrate globin gene clusters, but details of the types, numbers, and order of genes all vary. A location that has a pseudogene in one species may have an active gene in another; for example, $\psi\beta$ of the higher primates lies at a position equivalent to an active embryonic gene in goat. Some examples of β-globin clusters are illustrated in **Figure 24.3**.

The characterization of these gene clusters makes an important general point. *There may be more members of a gene family, both functional and nonfunctional, than we would suspect on the basis of protein analysis.* The extra functional genes may represent duplicates that code for identical polypeptides; or they may be related to known proteins, although different from them (and presumably expressed only briefly or in low amounts). Taking these features into account, it is hard to be certain that all the members of a cluster have been identified until flanking regions have been analyzed well beyond the terminal members.

With regard to the question of how much DNA is needed to code for a particular function, we see that coding for the β-like globins requires a range of 20–50 kb in different mammals. This is much greater than we would expect just from scrutinizing the known β-globin proteins or considering the individual genes. The region of the β cluster appears to code only for the β-globins, but until more genes or gene clusters coding for particular proteins have been identified, we shall not be able to tell how often this type of arrangement occurs.

Unequal crossing-over rearranges gene clusters

There are frequent opportunities for rearrangement in a cluster of related or identical genes. We can see the results by comparing the mammalian β clusters included in Figures 24.2 and 24.3. Although the clusters serve the same function, and all have the same general organization, each is different in size, there is variation in the total number and types of β-globin genes, and the numbers and structures of pseudogenes are different. All of these changes must have occurred since the mammalian radiation, ~85 million years ago (the last point in evolution common to all the mammals).

The comparison makes the general point that gene duplication, rearrangement, and variation is as important a factor in evolution as the slow accumulation of point mutations in individual genes. What types of mechanisms are responsible for gene reorganization?

A gene cluster can expand or contract by **unequal crossing-over**, when recombination occurs between nonallelic genes, as illustrated in **Figure 24.4**. Usually, recombination involves corresponding sequences of DNA held in exact aligment between the two homologous chromosomes. However, when there are two copies of a gene on each chromosome, an occasional misalignment allows pairing between them. (This requires some of the adjacent regions to go unpaired.)

When a recombination event occurs between the mispaired gene copies, it generates **nonreciprocal recombinant chromosomes**, one of which has a duplication of the gene and the other a deletion. The first recombinant therefore has an increase in the number of gene copies from 2 to 3, while the second has a decrease from 2 to 1.

In this example, we have treated the noncorresponding gene copies 1 and 2 as though they were entirely homologous. However, unequal crossing-over also can occur when the adjacent genes are well related (although the probability is less than when they are identical).

An obstacle to unequal crossing-over is presented by the interrupted structure of the genes. In a case such as the globins, the corresponding

Figure 24.4

Gene number can be changed by unequal crossing-over. If gene 1 of one chromosome pairs with gene 2 of the other chromosome, the other gene copies are excluded from pairing, as indicated by the extruded loops. Recombination between the mispaired genes produces one chromosome with a single (recombinant) copy of the gene and one chromosome with three copies of the gene (one from each parent and one recombinant).

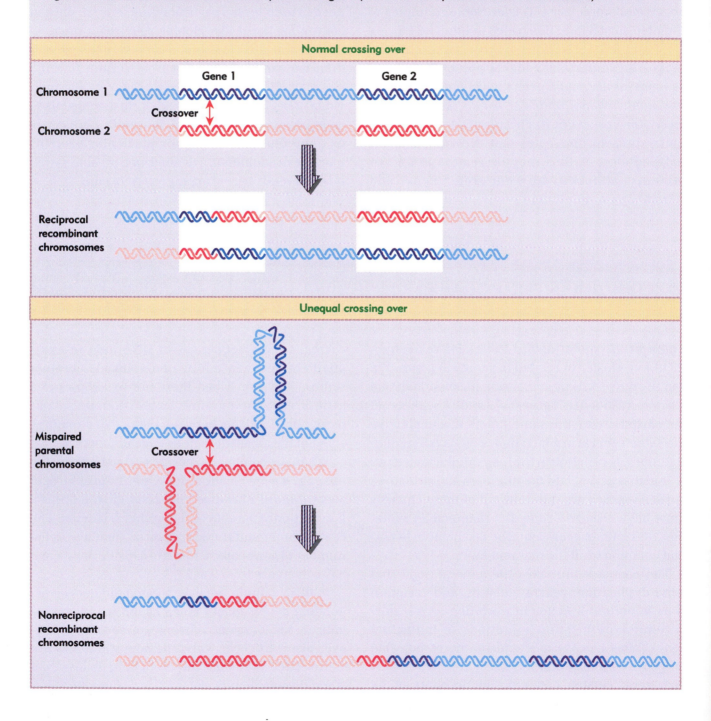

exons of adjacent gene copies are likely to be well enough related to support pairing; but the sequences of the introns have diverged appreciably. The restriction of pairing to the exons considerably reduces the continuous length of DNA that can be involved. This might correspondingly lower the chance of unequal crossing-over. Thus divergence between introns could enhance the stability of gene clusters by hindering the occurrence of unequal crossing-over.

Thalassemias result from mutations that reduce or prevent synthesis of either α- or β-globin. The occurrence of unequal crossing-over in the human globin gene clusters is revealed by the nature of certain thalassemias.

Many of the most severe thalassemias result from deletions of part of a cluster. In at least some cases, the ends of the deletion lie in regions that are homologous, which is exactly what would be expected if it had been generated by unequal crossing-over.

Two general types of deletion are found among the α-thalassemias. The α-**thal-1** deletion eliminates both the α genes. The α-**thal-2** deletion eliminates only one of the two α genes. The name refers to the individual chromosome carrying the deletion. Depending on the diploid combination of thalassemic chromosomes, an affected individual may have any number of α chains from zero to three. There are few differences from the wild type (four α genes) in individuals with three or two α genes. With only one α gene, the excess β chains form the unusual tetramer β_4, which causes **HbH disease**. The complete absence of α genes results in **hydrops fetalis**, which is fatal at or before birth.

Figure 24.5 summarizes the deletions that cause these α-thalassemias. The α-thal-1 deletions are long, varying in the location of the left end, with the positions of the right ends located beyond the known genes. The α-thal-2 deletions are short. The L form removes 4.2 kb of DNA, including the α2 gene. It probably results from unequal crossing-

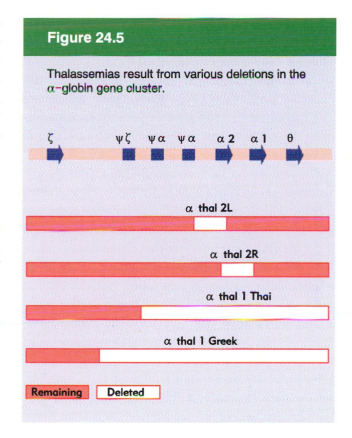

Figure 24.5

Thalassemias result from various deletions in the α–globin gene cluster.

ζ   ψζ ψα ψα   α2   α1   θ

α thal 2L

α thal 2R

α thal 1 Thai

α thal 1 Greek

Remaining   Deleted

over, because the ends of the deletion lie in homologous regions, just to the right of the ψα and α2 genes, respectively. The R form results from the removal of exactly 3.7 kb of DNA, the precise distance between the α1 and α2 genes. It appears to have been generated by unequal crossing-over between the α1 and α2 genes themselves. This is precisely the situation depicted in Figure 24.4.

The same unequal crossing-over that generated the thalassemic chromosome should also have generated a chromosome with three α genes. Individuals with such chromosomes have been identified in several populations. In some populations, the frequency of the triple α locus is about the same as that of the single α locus; in others, the triple α genes are much *less* common than single α genes. This suggests that (unknown) selective factors operate in different populations to adjust the gene levels.

Variations in the number of α genes are found relatively frequently, which argues that unequal crossing-over in the cluster must be fairly common. It occurs more often in the α cluster than in the β cluster, possibly because the introns in α genes are much shorter, and therefore present less impediment to mispairing between nonhomologous genes.

The deletions that cause β-thalassemias are summarized in **Figure 24.6**. In some (rare) cases, only the β gene is affected. These have a deletion of 600 bp, extending from the second intron through the 3′ flanking regions. In the other cases, more than one gene of the cluster is affected. Many of the deletions are very long, extending from the 5′ end indicated on the map for >50 kb toward the right.

The **Hb Lepore** type provided the classic evidence that deletion can result from unequal crossing-over between linked genes. The β and δ genes differ only ~7% in sequence. Unequal recombination deletes the material between the genes, thus fusing them together (see Figure 24.4). The fused gene produces a single β-like chain that consists of the N-terminal sequence of δ joined to the C-terminal sequence of β.

Several types of Hb Lepore now are known, the difference between them lying in the point of transition from δ to β sequences. Thus when the δ and β genes pair for unequal crossing-over, the exact point of recombination determines the position at which the switch from δ to β sequence occurs in the amino acid chain.

The reciprocal of this event has been found in the form of **Hb anti-Lepore**, which is produced by a gene that has the N-terminal part of β and the C-terminal part of δ. The fusion gene lies between normal δ and β genes.

Evidence that unequal crossing-over can occur between more distantly related genes is provided by the identification of **Hb Kenya**, another fused hemoglobin. This contains the N-terminal sequence of the Aγ gene and the C-terminal sequence of the β gene. The fusion must have resulted from unequal crossing-over between Aγ and β, which differ ~20% in sequence.

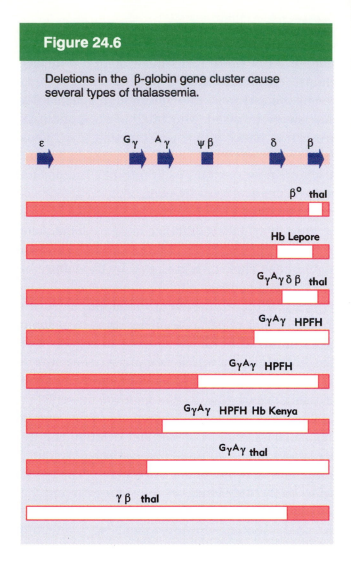

Figure 24.6

Deletions in the β-globin gene cluster cause several types of thalassemia.

A variety of deletions prevent synthesis of both δ and β, which can result in either of two phenotypes. In **HPFH** (hereditary persistence of fetal hemoglobin), there are generally no clinical symptoms; the disease is ameliorated because the synthesis of fetal hemoglobin ($\alpha_2\gamma_2$) continues after the time in development at which it would usually have been turned off. In δβ **thalassemia**, there are anemic symptoms, because, although γ gene expression continues in adult life, it is less effective than in HPFH. The difference between the HPFH and δβ thalassemias probably depends on the deletion of regulatory sequences within the cluster, but outside the δ and β genes themselves.

Gene clusters suffer continual reorganization

From the differences between the globin gene clusters of various mammals, we may infer that duplication followed (sometimes) by variation has been an important feature in the evolution of each cluster. The human thalassemic deletions demonstrate that unequal crossing-over continues to occur in both globin gene clusters. Each such event generates a duplication as well as the deletion, and we must account for the fate of both recombinant loci in the population. Deletions can also occur (in principle) by recombination between homologous sequences lying on the *same* chromosome. This does not generate a corresponding duplication.

It is difficult to estimate the natural frequency of these events, because selective forces rapidly adjust the levels of the variant clusters in the population. There may be a rough correlation between the likelihood of an unequal crossing-over and the relationship of the genes—the more closely related (including both exons and introns), the greater the chance of mispairing. (However, some unequal recombination events do not involve the genes themselves, but rely on repetitive sequences nearby.)

Generally a contraction in gene number is likely to be deleterious and selected against. However, in some populations, there may be a balancing advantage that maintains the deleted form at a low frequency.

What is the result of an expansion? The only examples that have been characterized are the triple α locus and the anti-Lepore. Individuals who possess 5 α-globin genes (one normal locus and one triple) do not display any change in hemoglobin synthesis. However, it is possible that a further increase (to the triple homozygote) could be deleterious, by unbalancing globin synthesis through the production of excess α chains. Individuals with anti-Lepore have the fusion βδ gene as an addition to the normal β and δ genes. It is possible that the additional chain is deleterious because it interferes with the assembly of normal hemoglobin.

These particular changes in gene number are unlikely to have a selective advantage that will cause them to spread through the population. But the structures of the present human clusters show several duplications that attest to the importance of such mechanisms. The *functional* sequences include two α genes coding the same protein, fairly well-related β and δ genes, and two almost identical γ genes. These comparatively recent independent duplications have survived in the population, not to mention the more distant duplications that originally generated the various types of globin genes. Other duplications may have given rise to pseudogenes or have been lost. We expect continual duplication and deletion to be a feature of all gene clusters.

From the organization of globin genes in a variety of species, we should be able eventually to trace the evolution of present globin genes from a single ancestral globin gene. Our present view of the evolutionary descent is pictured in **Figure 24.7**.

Given the propensity of introns to be lost during evolution, the leghemoglobin gene of plants, which is related to the globin genes, may represent the ancestral form. It has an extra intron, dividing the heme-binding domain, as we saw in Figure 23.21.

The furthest back that we can trace a globin gene in modern form is provided by the sequence of the single chain of mammalian myoglobin, which diverged from the globin line of descent ~800 million years ago. The myoglobin gene has the same organization as globin genes, so we may take the three-exon structure to represent their common ancestor, generated at some point after the divergence from plants.

Some 'primitive fish' have only a single type of globin chain, so they must have diverged from the line of evolution before the ancestral globin gene was duplicated to give rise to the α and β variants. This appears to have occurred ~500 million years ago, during the evolution of the bony fish.

The next stage of evolution is represented by the state of the globin genes in the frog *X. laevis*,

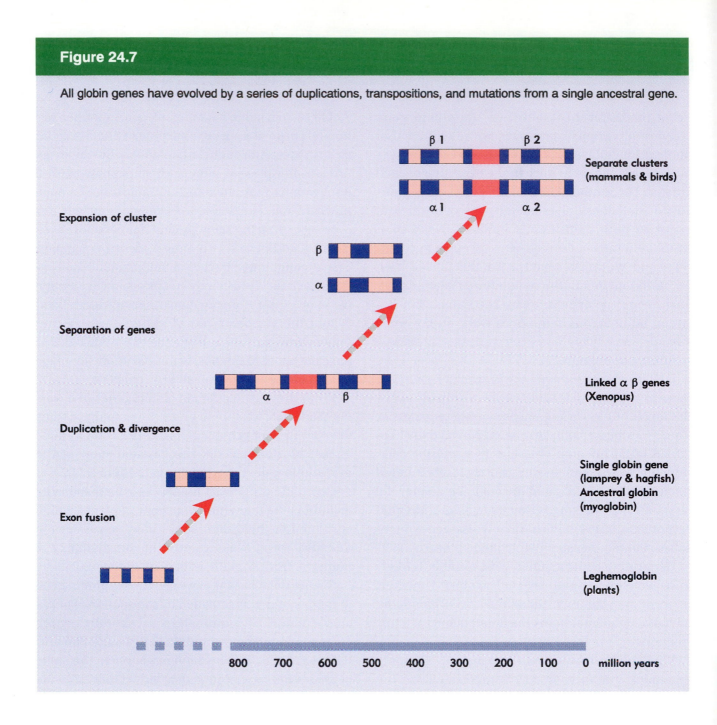

Figure 24.7

All globin genes have evolved by a series of duplications, transpositions, and mutations from a single ancestral gene.

which has two globin clusters. However, each cluster contains *both* α and β genes, of both larval and adult types. The cluster must therefore have evolved by duplication of a linked α–β pair, followed by divergence between the individual copies. Later the entire cluster was duplicated.

The amphibians separated from the mammalian/avian line ~350 million years ago, so the separation of the α- and β-globin genes must have resulted from a transposition in the mammalian/avian forerunner after this time. This probably occurred in the period of early vertebrate evolution. Since there are separate clusters for α- and β-globins in both birds and mammals, the α and β genes must have been physically separated before the mammals and birds diverged from this common ancestor, an event that occurred probably ~270 million years ago.

Changes have occurred within the separate α and β clusters in more recent times, as we see later from the description of the divergence of the individual genes in Figure 24.9.

Sequence divergence distinguishes two types of sites in DNA

Most changes in protein sequences occur by small mutations that accumulate slowly with time. Point mutations and small insertions and deletions occur by chance, probably with more or less equal probability in all regions of the genome, except for hotspots at which mutations occur much more frequently. Most mutations that change the amino acid sequence are *deleterious* and will be eliminated by natural selection.

Few mutations will be advantageous, but those that are may spread through the population, eventually replacing the former sequence. When a new variant replaces the previous version of the gene, it is said to have become **fixed** in the population.

A contentious issue is what proportion of mutational changes in an amino acid sequence are **neutral**, that is, without any effect on the function of the protein, and able therefore to accrue as the result of **random drift and fixation**.

The rate at which mutational changes accumulate is a characteristic of each protein, presumably depending at least in part on its flexibility with regard to change. Within a species, a protein evolves by mutational substitution, followed by elimination or fixation within the single breeding pool. The presence in the population of two (or more) allelic variants is called a **polymorphism**.

A polymorphism can be stable, in which case neither form has any relative advantage. Or the polymorphism may be transient, as must be the case while one variant is replacing another. When we scrutinize the gene pool of a species, we see only the variants that have survived.

When a species separates into two new species, each now constitutes an independent pool for evolution. By comparing the corresponding proteins in two species, we see the differences that have accumulated between them *since the time when their ancestors ceased to interbreed.* Some proteins are highly conserved, showing little or no change from species to species. This indicates that almost any change is deleterious and therefore selected against.

The difference between two proteins is expressed as their **divergence**, the percent of positions at which the amino acids are different. The divergence between proteins can be different from that between the corresponding nucleic acid sequences. The source of this difference is the representation of each amino acid in a three-base codon, in which often the third base has no effect on the meaning.

We may divide the nucleotide sequence of a coding region into potential **replacement sites** and **silent sites**:

◆ At replacement sites, a mutation alters the amino acid that is coded. The effect of the mutation (deleterious, neutral, or advantageous) depends on the result of the amino acid replacement.

◆ At silent sites, mutation only substitutes one synonym codon for another, so there is no change in the protein. Usually the replacement sites account for 75% of a coding sequence and the silent sites provide 25%.

In addition to the coding sequence, a gene contains nontranslated regions. Here again, mutations are potentially neutral, apart from their effects on either secondary structure or (usually rather short) regulatory signals.

Although silent mutations are neutral with regard to the protein, they could affect gene expression via the sequence change in RNA. For example, a change in secondary structure might influence transcription, processing, or translation. Another possibility is that a change in synonym codons calls for a different tRNA to respond, influencing the efficiency of translation.

The mutations in replacement sites should correspond with the amino acid divergence, which essentially is a count of the percent of changes. A nucleic acid divergence of 0.45% at replacement sites corresponds to an amino acid divergence of 1% (assuming that the average number of replacement sites per codon is 2.25). Actually, the measured divergence underestimates the differences that have occurred during evolution, because of the occurrence of multiple events at one codon. Usually a correction is made for this.

To take the example of the human β- and δ-globin chains, there are 10 differences in 146 residues, a divergence of 6.9%. The DNA sequence has 31 changes in 441 residues. However, these changes are distributed very differently in the replacement and silent sites. There are 11 changes in the 330 replacement sites, but 20 changes in only 111 silent sites. This gives (corrected) rates of divergence of 3.7% in the replacement sites and 32% in the silent sites, almost an order of magnitude in difference.

The striking difference in the divergence of replacement and silent sites demonstrates the existence of much greater constraints on nucleotide positions that influence protein constitution relative to those that do not. So probably very few of the amino acid changes are neutral.

If we take the rate of mutation at silent sites to indicate the underlying rate of mutational fixation (this is to assume that there is no selection at all at the silent sites), then over the period since the β and δ genes diverged, there should have been changes at 32% of the 330 replacement sites, a total of 105. All but 11 of them have been eliminated, which means that ~90% of the mutations did not survive.

The evolutionary clock traces the development of globin genes

When a particular protein is examined in a range of species, the divergence between the sequences in each pairwise comparison is (more or less) proportional to the time since they separated. This provides an **evolutionary clock** that measures the accumulation of mutations at an apparently even rate during the evolution of a given protein.

The rate of divergence can be measured as the percent difference per million years, or as its reciprocal, the unit evolutionary period (UEP), the time in millions of years that it takes for 1% divergence to develop. Once the clock has been established by pairwise comparisons between species (remembering the practical difficulties in establishing the actual time of speciation), it can be applied to related genes *within* a species. From their divergence, we can calculate how long it is since the duplication that generated them.

By comparing the sequences of homologous genes in different species, the rate of divergence at both replacement and silent sites can be determined.

In pairwise comparisons, there is an average divergence of 10% in the replacement sites of either the α- or β-globin genes of mammals that have been separated since the mammalian radiation occurred ~85 million years ago. This corresponds to a replacement divergence rate of 0.12% per million years.

The rate is steady when the comparison is extended to genes that diverged in the more distant past. For example, the average replacement divergence between corresponding mammalian and chicken globin genes is 23%. Relative to a separation ~270 million years ago, this gives a rate of 0.09% per million years.

Figure 24.8

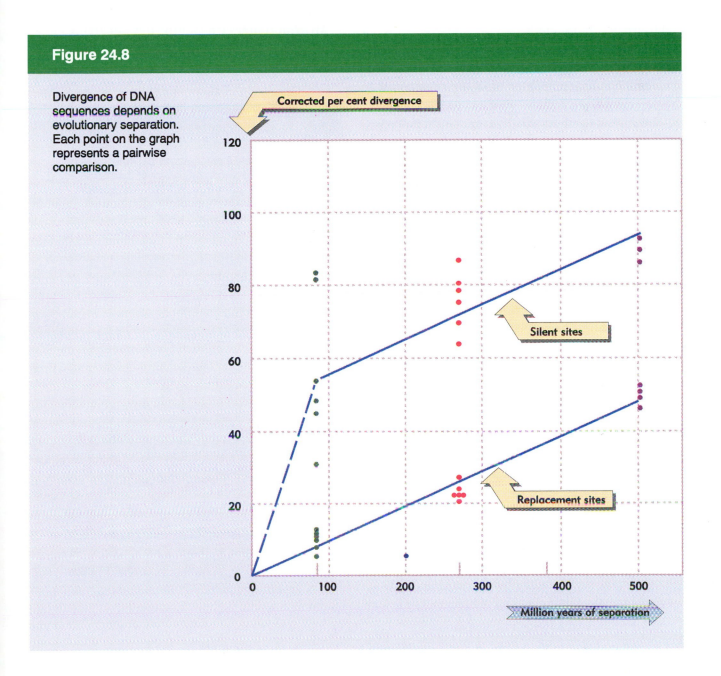

Divergence of DNA sequences depends on evolutionary separation. Each point on the graph represents a pairwise comparison.

Going further back, we can compare the α- with the β-globin genes within a species. They have been diverging since the individual gene types separated ≥500 million years ago (see Figure 24.7). They have an average replacement divergence of ~50%, which gives a rate of 0.1% per million years.

These data are plotted in **Figure 24.8**, which shows that replacement divergence in the globin genes has an average rate of ~0.096% per million years (or a UEP of 10.4). Considering the uncertainties in estimating the times at which the species diverged, the results lend good support to the idea that there is a linear clock.

The data on silent site divergence are much less clear. In every case, it is evident that the silent site divergence is much greater than the replacement site divergence, by a factor that varies from 2 to 10. But the spread of silent site divergences in pairwise comparisons is too great to show whether a clock is applicable (so we must base temporal comparisons on the replacement sites).

From Figure 24.8, it is clear that the rate at silent sites is not linear with regard to time. *If we assume that there must be zero divergence at zero years of separation,* we see that the rate of silent site divergence is much greater for the first ~100 million years of separation. One interpretation is that a fraction of roughly half of the silent sites is rapidly (within 100 million years) saturated by mutations; this fraction behaves as neutral sites. The other fraction accumulates mutations more slowly, at a rate approximately the same as that of the replacement sites; this fraction identifies sites that are silent with regard to the protein, but that come under selective pressure for some other reason.

Now we can reverse the calculation of divergence rates to estimate the times since genes within a species have been apart. The difference between the human β and δ genes is 3.7% for replacement sites. At a UEP of 10.4, these genes must have diverged 10.4 × 3.7 ≈ 40 million years ago—about the time of the separation of the lines leading to

New World monkeys, Old World monkeys, great apes, and man. All of these higher primates have both β and δ genes, which suggests that the gene divergence commenced just before this point in evolution.

Proceeding further back, the divergence between the replacement sites of γ and ε genes is 10%, which corresponds to a time of separation ~100 million years ago. The separation between embryonic and fetal globin genes therefore may have just preceded or accompanied the mammalian radiation.

An evolutionary tree for the human globin genes is constructed in **Figure 24.9**. Features that evolved

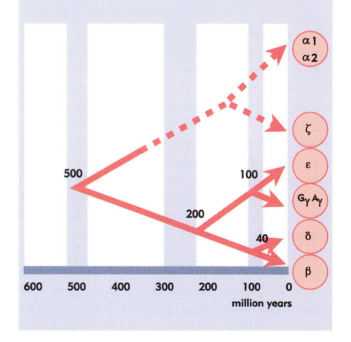

Figure 24.9

Replacement site divergences between pairs of β-globin genes allow the history of the human cluster to be reconstructed. This tree accounts for the separation of classes of globin genes. Duplications of individual genes are of unknown origin. The time of the α–ζ divergence is not known.

before the mammalian radiation—such as the separation of β/δ from γ—should be found in all mammals. Features that evolved afterward—such as the separation of β- and δ-globin genes—should be found in individual lines of mammals.

In each species, there have been comparatively recent changes in the structures of the clusters, since we see differences in gene number (one adult β-globin gene in man, two in mouse) or in type (we are not yet sure whether there are separate embryonic and fetal β-like globins in rabbit and mouse).

When sufficient data have been collected on the sequences of a particular gene, the arguments can be reversed, and comparisons between genes in different species can be used to assess taxonomic relationships.

Pseudogenes are dead ends of evolution

Pseudogenes are defined by their possession of sequences that are related to those of the functional genes, but that cannot be translated into a functional protein. A pseudogene is often denoted by the symbol ψ.

Some pseudogenes have the same general structure as functional genes, with sequences corresponding to exons and introns in the usual locations. They are rendered inactive by mutations that prevent any or all of the stages of gene expression. The changes can take the form of abolishing the signals for initiating transcription, preventing splicing at the exon–intron junctions, or prematurely terminating translation.

Usually a pseudogene has several deleterious mutations, presumably because once it ceased to be active, there was no impediment to the accumulation of further mutations. Pseudogenes that represent inactive versions of currently active genes have been found in many systems, including globin, immunoglobulins, and histocompatibility antigens, where they are located in the vicinity of the gene cluster, often interspersed with the active genes.

A typical example is the rabbit pseudogene, $\psi\beta2$, which has the usual organization of exons and introns, and is related most closely to the functional globin gene $\beta1$. But the deletion of a base pair at codon 20 of $\psi\beta2$ has caused a frameshift that would lead to termination shortly after. Several point mutations have changed later codons representing amino acids that are highly conserved in the β-globins. Neither of the two introns any longer possesses recognizable boundaries with the exons, so probably the introns could not be spliced out even if the gene were transcribed. However, there are no transcripts corresponding to the gene, possibly because there have been changes in the 5' flanking region.

Since this list of defects includes mutations potentially preventing each stage of gene expression, we have no means of telling which event originally inactivated this gene. However, from the divergence between the pseudogene and the functional gene, we can estimate when the pseudogene originated and when its mutations started to accumulate.

If the pseudogene had become inactive as soon as it was generated by duplication from $\beta1$, we should expect both replacement site and silent site divergence rates to be the same. (There is no reason for them to be different if the gene is not translated.) But actually there are fewer replacement site substitutions than silent site substitutions. This suggests that at first (while the gene was expressed) there was selection against replacement site substitution. From the relative extents of substitution in the two types of site, we can calculate that $\psi\beta 2$ diverged from $\beta 1$ ~55 million years ago, remained a functional gene for 22 million years, but has been a pseudogene for the last 33 million years.

Similar calculations can be made for other pseudogenes. Some appear to have been active for some time before becoming pseudogenes, but others appear to have been inactive from the very time of their original generation. The general point made by the structures of these pseudogenes is that each has evolved independently during the development of the globin gene cluster in each species. This reinforces the conclusion that the creation of new genes, followed by their acceptance as functional duplicates, variation to become new functional genes, or inactivation as pseudogenes, is a continuing process in the gene cluster.

The mouse $\psi\alpha 3$ globin gene has an interesting property: it precisely lacks both introns. Its sequence can be aligned (allowing for accumulated mutations) with the α-globin mRNA. The apparent time of inactivation coincides with the original duplication, which suggests that the original inactivating event was associated with the loss of introns. How could the introns have been lost? There is no reason why the systems for reconstructing DNA should recognize the exon–intron boundaries, which argues for the involvement at some level of the mRNA itself. Probably a reverse transcript of the mRNA was inserted into the genome, perhaps carried by a retrovirus (see Chapter 35).

Inactive genomic sequences that resemble the RNA transcript are called **processed pseudogenes**. Supporting the idea that they originated by insertion at some random site of a product derived from the RNA, they may be located anywhere in the genome, not necessarily even on the same chromosome as the active gene.

How common are pseudogenes? Most gene families have members that are pseudogenes.

Usually the pseudogenes represent a small minority of the total gene number. In an exceptional case, however, there is one active gene coding for a mouse ribosomal protein; and it has ~15 processed pseudogene relatives. This type of effect must be taken into account when we try to calculate the number of genes from hybridization data.

If pseudogenes are evolutionary dead ends, simply an unwanted accompaniment to the rearrangement of functional genes, why are they still present in the genome? Do they fulfill any function or are they entirely without purpose, in which case there should be no selective pressure for their retention?

We should remember that we see those genes that have survived in present populations. In past times, any number of other pseudogenes may have been eliminated. This elimination could occur by deletion of the sequence as a sudden event or by the accretion of mutations to the point where the pseudogene can no longer be recognized as a member of its original sequence family (probably the ultimate fate of any pseudogene that is not suddenly eliminated).

Even relics of evolution can be duplicated. In the β-globin genes of the goat, there are two adult species, β^A and β^C. Each of these has a pseudogene a few kilobases upstream of it (called $\psi\beta^Z$ and $\psi\beta^X$, respectively). The two pseudogenes are better related to each other than to the adult β-globin genes; in particular, they share several inactivating mutations. Also, the two adult β-globin genes are better related to each other than to the pseudogenes. This implies that an original $\psi\beta$-β structure was itself duplicated, giving two functional β genes (which diverged further into the β^A and β^C genes) and two nonfunctional genes (which diverged into the current pseudogenes).

The mechanisms responsible for gene duplication, deletion, and rearrangement act on all sequences that are recognized as members of the cluster, whether or not they are functional. It is left to selection to discriminate among the products.

Genes for rRNA comprise a repeated tandem unit

In most of the cases we have discussed so far, there are differences between the individual members of a gene cluster that allow selective pressure to act independently upon each gene. A contrast is provided by two cases of large gene clusters that contain many identical copies of the same gene or genes, a situation that poses some interesting evolutionary questions. Most organisms contain multiple copies of the genes for the histone proteins which are a major component of the chromosomes; and there are very few organisms that do not contain multiple copies of the genes that code for the ribosomal RNAs.

Ribosomal RNA is by far the predominant product of transcription, constituting some 80–90% of the total mass of cellular RNA in both eukaryotes and prokaryotes. There are several tRNA molecules per ribosome, but of course they are much smaller than the rRNAs. Both rRNA and tRNA are represented by multiple genes. **Table 24.5** summarizes the numbers present in a range of genomes.

The number of major rRNA genes varies from 7 in *E. coli,* to between 100 and 200 in lower eukaryotes, to several hundred in higher eukaryotes. In virtually every case, the genes for the large and small rRNA form a tandem pair. (The sole exception is the yeast mitochondrion.)

In bacteria and some lower eukaryotes, the gene for 5S RNA is part of the same unit, so the total number of 5S genes is the same as that of the major rRNAs. In bacteria, the 5S gene is cotranscribed with the major rRNA genes; in eukaryotes it is transcribed independently. In higher eukaryotes, the genes for 5S RNA are separately organized into their own clusters, and their number exceeds that of the major rRNA genes.

The exact number of genes for tRNA is hard to determine, because the extensive secondary structure of the molecule creates technical difficulties in the hybridization reaction. Probably we have underestimated the actual number.

The lack of any detectable variation in the sequences of the rRNA molecules implies that all the copies of each gene must be identical, or at least must have differences below the level of detection in rRNA (~1%). A point of major interest is what mechanism(s) are used to prevent variations from accruing in the individual sequences.

In bacteria, the multiple 16S–23S rRNA gene pairs are dispersed. In most eukaryotic nuclei, the rRNA genes are contained in a tandem cluster or clusters. Sometimes these regions are called **rDNA.** (In some cases, the proportion of rDNA in the total DNA, together with its atypical base composition, is great enough to allow its isolation as a separate fraction directly from sheared genomic DNA.) An important diagnostic feature of a tandem cluster is that it

Table 24.5

All genome contain multiple tRNA and rRNA genes.

Species	18S/28S Genes	5S Genes	tRNA Genes
E. coli	7	7	60
S. cerevisiae	140	140	250
D. discoideum	180	180	?
D. melanogaster		165	850
(X)	250		
(Y)	150		
Man	280	2,000	1,300
X. laevis	450	24,000	1,150

generates a circular restriction map, as shown in **Figure 24.10**.

Suppose that each repeat unit has 3 restriction sites. In the example shown in the figure, fragments A and B are contained entirely within a repeat unit, and fragment C contains the end of one repeat and the beginning of the next. When we map these frag-

ments by conventional means, we find that A is next to B, which is next to C, which is next to A, generating the circular map. If the cluster is large, the internal fragments (A, B, C) will be present in much greater quantities than the terminal fragments (X, Y) which connect the cluster to adjacent DNA. In a cluster of 100 repeats, X and Y would be present

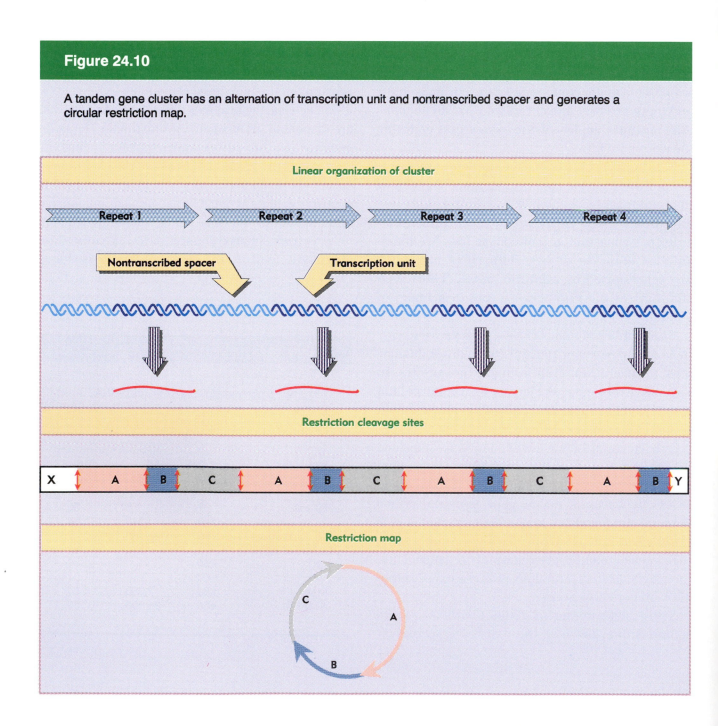

Figure 24.10

A tandem gene cluster has an alternation of transcription unit and nontranscribed spacer and generates a circular restriction map.

at 1% of the level of A, B, C. This can make it difficult to obtain the ends of a gene cluster for mapping purposes.

The region of the nucleus where rRNA synthesis occurs has a characteristic appearance, with a core of fibrillar nature surrounded by a granular cortex. The fibrillar core is where the rRNA is transcribed from the DNA template; and the granular cortex is formed by the ribonucleoprotein particles into which the rRNA is assembled. The whole area is called the **nucleolus**. Its characteristic morphology is evident in **Figure 24.11**.

The particular chromosomal regions associated with a nucleolus are called **nucleolar organizers**. Each nucleolar organizer corresponds to a cluster of tandemly repeated rRNA genes on one chromosome. The concentration of the tandemly repeated rRNA genes, together with their very intensive transcription, is responsible for creating the characteristic morphology of the nucleoli.

The pair of major rRNAs is transcribed as a single precursor in both bacteria and eukaryotic nuclei. Following transcription, the precursor is cleaved to release the individual rRNA molecules. The salient features of the gene organization are described in **Figure 24.12**. The transcript starts with a 5′ leader, followed by the sequence of the small rRNA, then a region called the **transcribed spacer**, and finally the sequence of the large rRNA toward the 3′ end of the molecule.

(The *transcribed spacer* is not to be confused with the *nontranscribed spacer* that separates transcription units. Its name reflects the fact that it is part of the transcription unit, but is not represented in the

Figure 24.11

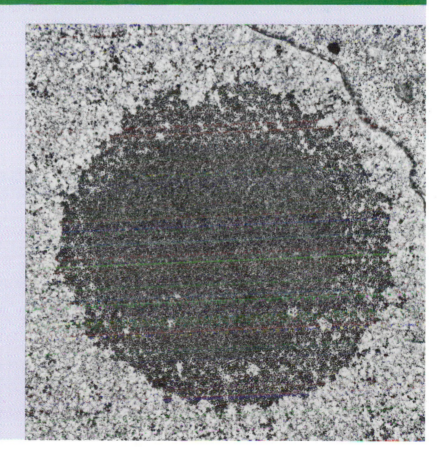

The nucleolar core identifies rDNA under transcription, and the surrounding granular cortex represents assembling ribosomal subunits The photograph shows a thin section through the nucleolus of the newt *Notophthalmus viridescens*; most of the particles exiting around the periphery are precursors to large ribosomal subunits. Photograph kindly provided by Oscar Miller.

Figure 24.12

A single transcription unit contains the sequence for both the small and large rRNA molecules.

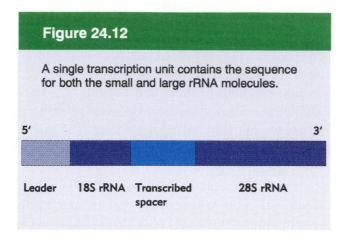

Leader   18S rRNA   Transcribed   28S rRNA
                      spacer

45S RNA, according to its rate of sedimentation). Here the mature rRNA sequences occupy only just over 50% of the length of the complete transcript.

What happens to the nonribosomal parts of the RNA precursor? The leader and the transcribed spacer regions are discarded during the maturation of rRNA. The transcribed spacer contains some short sequences that are released by cleavage. In mammals and amphibians, one short sequence forms the 5.8S RNA, a small molecule that hydrogen bonds with the 28S rRNA in the ribosome. In bacteria, tRNA sequences lie in the transcribed spacer and (sometimes) also at the 3′ end of the primary transcript. The rest of the transcribed spacer, and all of the leader sequences are presumably degraded to nucleotides.

An rDNA cluster contains many transcription units, each separated from the next by a nontranscribed spacer. The alternation of transcription unit and nontranscribed spacer can be seen directly in electron micrographs. The example shown in **Figure 24.13** is taken from the newt *N. viridescens*, in which each transcription unit is intensively expressed, so that many RNA polymerases are simultaneously engaged in transcription on one repeating unit. The polymerases are so closely packed that the increase in the length of their products forms a characteristic matrix moving along the transcription unit.

mature RNA products. Recall the pattern for cleavage of the single rRNA primary transcript into mature rRNAs that we analyzed in Figure 16.20.)

The transcription unit is therefore longer than the combined length of the mature rRNAs. **Table 24.6** summarizes some examples of the relationship between the primary transcript and the rRNAs.

The transcription unit is shortest in bacteria, where the rRNA sequences constitute 80% of its total length of 6 kb bases. No particularly systematic pattern is seen in a range of eukaryotes in which the transcript length varies from 7 to 8 kb, with 70–80% of the sequence representing the rRNAs. The precursor is longest in mammals (where it is known as

Table 24.6

Each rRNA precursor is longer than the two mature rRNAs.

Organism	Species	Length of Transcript (bases)	Length of rRNA Large	Small	rRNA Part of Transcript
Bacterium	E. coli	5,600	2,904	1,542	80%
Yeast	S. cerevisiae	7,200	3,750	2,000	80%
Fruit fly	D. melanogaster	7,750	4,100	2,000	78%
Toad	X. laevis	7,875	4,475	1,925	79%
Plant	N. tabacum	7,900	3,700	1,900	71%
Chicken	G. domesticus	11,250	4,625	1,800	57%
Mouse	M. musculus	13,400	4,712	1,950	52%

Figure 24.13

Transcription of rDNA clusters generates a series of matrices, each corresponding to one transcription unit and separated from the next by the non-transcribed spacer. Photograph kindly provided by Oscar Miller.

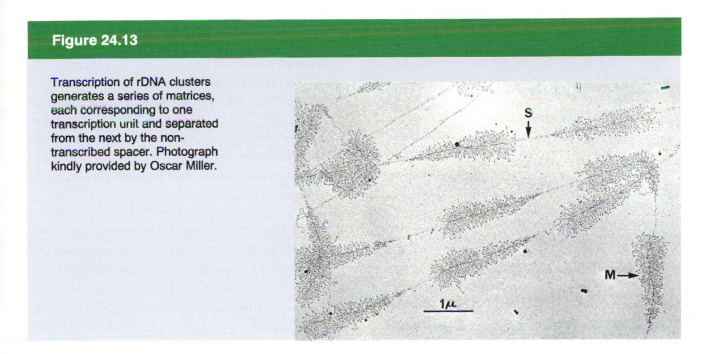

The nontranscribed spacer varies widely in length between and (sometimes) within species. Table 24.7 summarizes the situation.

In yeast there is a short nontranscribed spacer, relatively constant in length. In *D. melanogaster*, there is almost a 2-fold variation in the length of the nontranscribed spacer between different copies of the repeating unit. A similar situation is seen in *X. laevis*. In each of these cases, all of the repeating units are present as a single tandem cluster on one particular chromosome. (In the example of *D. melanogaster*, this happens to be the sex chromosome. The cluster on the X chromosome is larger than that on the Y chromosome, so female flies have more copies of the rRNA genes than male flies—see Table 24.5.)

In mammals the repeating unit is very much larger, comprising the transcription unit of ~13 kb and a nontranscribed spacer of ~30 kb. Usually, the genes lie in several dispersed clusters—in the case of man and mouse residing on five and six chromosomes, respectively. One interesting question is how the corrective mechanisms that presumably function within a single cluster to ensure constancy

Table 24.7

The length of the tandem repeating unit of rDNA clusters is variable.

Species	Repeating Unit Length	Nontranscribed Spacer Length	Transcript Length
S. cerevisiae	8,950 bp	1,750 bp	7,200 bp
D. melanogaster	11,500-14,200 bp	3,750-6,450 bp	7,750 bp
X. laevis	10,500-13,500 bp	2,300-5,300 bp	7,875 bp
M. musculus	44,000 bp	30,000 bp	13,400 bp

of rRNA sequence are able to work when there are several clusters.

The variation in length of the nontranscribed spacer in a single gene cluster contrasts with the conservation of sequence of the transcription unit. In spite of this variation, the sequences of longer nontranscribed spacers remain homologous with those of the shorter nontranscribed spacers. This implies that each nontranscribed spacer is *internally repetitious,* so that the variation in length results from changes in the number of repeats of some subunit.

The general nature of the nontranscribed spacer is illustrated by the example of *X. laevis.* **Figure 24.14** illustrates the situation. Regions that are fixed in length alternate with regions that vary. Each of the three repetitious regions comprises a variable number of repeats of a rather short sequence. One type of repetitious region has repeats of a 97 bp sequence; the other, which occurs in two locations, has a repeating unit found in two forms, 60 and 81 bp long. The variation in the number of repeating units in the repetitious regions accounts for the overall variation in spacer length.

One of the fixed regions (at the start of the unit) is unique in sequence and length. The others are short constant sequences called **Bam islands**. (This description takes its name from their isolation via the use of the BamHI restriction enzyme.) From this type of organization, we see that the cluster has evolved by duplications involving the promoter region.

The nontranscribed spacer has a function in transcription. The 60/81 bp repeats play a role in initiation, probably comparable with the role that enhancers play for genes transcribed by RNA polymerase II (see Chapter 29). The presence of a spacer increases the frequency of initiation, apparently by causing several RNA polymerase I molecules to bind in succession at the promoter. We do not yet understand the exact nature of this relationship, and how it depends on the similarities of sequence between the spacer and the promoter.

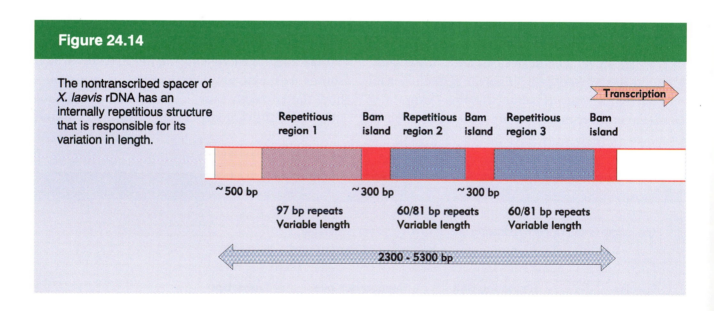

Figure 24.14

The nontranscribed spacer of *X. laevis* rDNA has an internally repetitious structure that is responsible for its variation in length.

Transcription

Repetitious region 1 — Bam island — Repetitious region 2 — Bam island — Repetitious region 3 — Bam island

~500 bp

~300 bp

~300 bp

97 bp repeats
Variable length

60/81 bp repeats
Variable length

60/81 bp repeats
Variable length

2300 - 5300 bp

An evolutionary dilemma: how are multiple active copies maintained?

The same problem is encountered whenever a gene has been duplicated. How can selection be imposed to prevent the accumulation of deleterious mutations?

The duplication of a gene is likely to result in an immediate relaxation of the evolutionary pressure on its sequence. Now that there are two identical copies, a change in the sequence of either one will not deprive the organism of a functional protein, since the original amino acid sequence continues to be coded by the other copy. Thus the selective pressure on the two genes is diffused, until one of them mutates sufficiently away from its original function to refocus all the selective pressure on the other.

Immediately following a gene duplication, changes might accumulate more rapidly in one of the copies, leading eventually to a new function (or to its disuse in the form of a pseudogene). If a new function develops, the gene then evolves at the same, slower rate characteristic of the original function. Probably this is the sort of mechanism responsible for the separation of functions between embryonic and adult globin genes.

Yet there are instances where duplicated genes retain the same function, coding for the identical or nearly identical proteins. Identical proteins are coded by the two human α-globin genes, and there is only a single amino acid difference between the two γ-globin proteins. How is selective pressure exerted to maintain their sequence identity?

Two general types of mechanism have been proposed. They share the principle that nonallelic genes are not independently inherited, but must be continually regenerated from *one* of the copies of a preceding generation. In the simplest case of two identical genes, when a mutation occurs in one copy, either it is by chance eliminated (because the sequence of the other copy takes over), or it is spread to both duplicates (because the mutant copy becomes the dominant version). Spreading exposes a mutation to selection. The result is that the two genes evolve together as though only a single locus existed. This is called **coincidental** or **concerted evolution** (occasionally **coevolution**). It can be applied to a pair of identical genes or (with further assumptions) to a cluster containing many genes.

One mechanism supposes that the sequences of the nonallelic genes are directly compared with one another and homogenized by enzymes that recognize any differences. This can be done by exchanging single strands between them, to form genes one of whose strands derives from one copy and one from the other copy. Any differences show as improperly paired bases, which attract attention from enzymes able to excise and replace a base, so that only A•T and G•C pairs survive. This type of event is called **gene conversion** and is associated with genetic recombination as described in Chapter 33.

We should be able to ascertain the scope of such events by comparing the sequences of duplicate genes. If they are subject to concerted evolution, we should not see the accumulation of silent site substitutions between them (because the homogenization process applies to these as well as to the replacement sites). We know that the extent of the maintenance mechanism need not extend beyond the gene itself, since there are cases of duplicate genes whose flanking sequences are entirely different. Indeed, we may see abrupt boundaries that mark the ends of the sequences that were homogenized.

We must remember that the existence of such mechanisms can invalidate the determination of the history of such genes via their divergence, *because the divergence reflects only the time since the last homogenization/regeneration event, not the original duplication.*

When there are many copies of a gene, the immediate effects of mutation in any one copy must be very slight. For example, in the case of rDNA cluster, there are hundreds of gene copies, all apparently identical. The consequences of an individual mutation are diluted by the large number of copies of the gene that retain the wild-type sequence. This suggests that an appreciable proportion of mutant copies could accumulate before the effect becomes strong enough to be eliminated by evolution.

Lethality becomes quantitative, a conclusion reinforced by the observation that half of the units of the rDNA cluster of *X. laevis* or *D. melanogaster* can be deleted without ill effect. So how are these units prevented from gradually accumulating deleterious mutations? And what chance is there for the rare favorable mutation to display its advantages in the cluster?

The inevitable conclusion is that some mechanism must allow particular variants to be scrutinized by evolution. Two types of mechanism have been proposed.

The **sudden correction** model supposes that every so often the entire gene cluster is replaced by a new set of copies, derived from one or at least from only a very few of the copies present in the previous generation. To impose sufficient selective force, 'every so often' need be only every few generations; but in practical terms, any mechanism must be constructed on a regularly recurring basis—that is, every generation.

The small number of copies that give rise to the new cluster form a 'master' set on which selection acts; the discarded copies are irrelevant to selection. The master copies could constitute a particular set or might be chosen at random. The mechanism would presumably be some form of amplification, or possibly gene conversion.

The difficulty with this model is that it predicts regeneration of the cluster from a few *repeating units*. This means that both the transcription units and spacers should be regenerated, so that the entire repeating unit should be homogeneous in sequence (or at least, should show no more variation than the small number of master copies). But in fact the spacers show a continual range of variation, while only the transcription units are constant.

The **crossover fixation** model supposes that the entire cluster is subject to continual rearrangement by the mechanism of unequal crossing-over. Essentially this applies to the tandem cluster on a grand scale the mechanisms we have already discussed for globin genes on a more restricted and occasional basis. Such events can explain the concerted evolution of multiple genes if unequal crossing-over causes all the copies to be regenerated physically from one copy.

Following the sort of event depicted in Figure 24.4, for example, the chromosome carrying a triple locus could suffer deletion of one of the genes. Of the two remaining genes, $1\frac{1}{2}$ represent the sequence of one of the original copies; only $\frac{1}{2}$ of the sequence of the other original copy has survived. Any mutation in the first region now exists in both genes and is subject to selective pressure.

We explore the consequences of unequal crossing-over in more detail in Chapter 26, but it is immediately evident that tandem clustering provides frequent opportunities for 'mispairing' of genes whose sequences are actually the same, but that lie in different positions in their clusters. By continually expanding and contracting the number of units via unequal crossing-over, it is possible for all the units in one cluster to be derived from rather a small proportion of those in an ancestral cluster. The variable lengths of the spacers are consistent with the idea that unequal crossing-over events take place in spacers that are internally mispaired. This can explain the homogeneity of the genes compared with the variability of the spacers. The genes are exposed to selection when individual repeating units are amplified within the cluster; but the spacers are irrelevant and can accumulate changes.

Summary

The total number of genes is likely to be >2,000 for bacteria, >5,000 for yeast, >8,000 for insects, and >125,000 for mammals. It is likely that not all genes are essential (defining lethal genes by the existence of devastating effects when they are mutated). The numbers of nonessential genes and essential genes could be comparable. In yeast, only 60% of genes appear to be essential; in *D. melanogaster*, there appear to be ~5,000 essential genes. We do not understand how nonessential genes are maintained; they may provide selective advantages that are not evident.

Almost all genes belong to families, defined by the possession of related sequences in the exons of individual members. Families evolve by the duplication of a gene (or genes), followed by divergence between the copies. Some copies suffer inactivating mutations and become pseudogenes that no longer have any function. Pseudogenes also may be generated as DNA copies of the mRNA sequences.

An evolving set of genes may remain together in a cluster or may be dispersed to new locations by chromosomal rearrangement. The organization of existing clusters can sometimes be used to infer the series of events that has occurred. These events act with regard to sequence rather than function, and therefore include pseudogenes as well as active genes.

Mutations accumulate more rapidly in silent sites than in replacement sites (which affect the amino acid sequence); the rate of divergence at replacement sites can be used to establish a clock, calibrated in per cent divergence per million years. The clock can then be used to calculate the time of divergence between any two members of the family.

A tandem cluster consists of many copies of a repeating unit that includes the transcribed sequence(s) and a nontranscribed spacer(s). rRNA gene clusters code only for a single rRNA precursor. Maintenance of active genes in clusters depends on mechanisms such as gene conversion or unequal crossing-over that cause mutations to spread through the cluster, so that they become exposed to evolutionary pressure.

Further reading

Reviews

The globin system was reviewed by **Maniatis** *et al*. (*Ann. Rev. Genet*. 14, 145–178, 1980), and the basis for thalassemic defects was brought up to date by **Weatherall and Clegg** (*Cell* 29, 7–9, 1982).

Discoveries

D. melanogaster was dissected for essential genes by **Judd** *et al*. (*Genetics* 71, 139–156, 1972) and the yeast genome was analyzed by **Petes** (*Cell* 46, 983–992, 1986).

Propaganda about large scale sequencing was published by **Sulston** *et al*. (*Nature* 356, 37–41, 1992) and **Oliver** *et al*. (*Nature* 357, 38–46, 1992).

CHAPTER 25

Genomes sequestered in organelles

The first evidence for the presence of genes outside the nucleus was provided by **nonMendelian inheritance** in plants (observed in the early years of this century, just after the rediscovery of Mendelian inheritance). NonMendelian inheritance is sometimes associated with the phenomenon of **somatic segregation**. They have a similar cause:

◆ NonMendelian inheritance is defined by the failure of the progeny of a mating to display Mendelian segregation for parental characters. It reflects lack of association between the segregating character and the meiotic spindle.

◆ Somatic segregation describes a phenomenon in which parental characters segregate in somatic cells, and therefore display heterogeneity in the organism. This is a notable feature of plant development. It reflects lack of association between the segregating character and the mitotic spindle.

NonMendelian inheritance and somatic segregation are therefore taken to indicate the presence of genes that reside outside the nucleus and do not utilize segregation on the meiotic and mitotic spindles to distribute replicas to gametes or to daughter cells, respectively.

The extreme form of nonMendelian inheritance is uniparental inheritance, when the genotype of only one parent is inherited and that of the other parent is permanently lost. In less extreme examples, the progeny of one parental genotype exceed those of the other genotype. Usually it is the mother whose genotype is preferentially (or solely) inherited. This effect is sometimes described as **maternal inheritance**. The important point is that the genotype contributed by the parent of one particular sex predominates, as seen in abnormal segregation ratios when a cross is made between mutant and wild type. This contrasts with the behavior of Mendelian genetics when reciprocal crosses show the contributions of both parents to be equally inherited (see Chapter 3).

The bias in parental genotypes is established at or soon after the formation of a zygote. There are various possible causes. The contribution of maternal or paternal information to the organelles of the zygote may be unequal; in the most extreme case, only one parent contributes. In other cases, the contributions are equal, but the information provided by one parent does not survive. Combinations of both effects are possible. Whatever the cause, the unequal representation of the information from the two parents contrasts with nuclear genetic information, which derives equally from each parent.

NonMendelian inheritance results from the presence in organelles of DNA that is inherited independently of nuclear genes. Mitochondria and chloroplasts both possess DNA genomes that code for all of the RNA species and for some of the proteins involved in the functions of the organelle.

An organelle genome codes for some, but not all, of the proteins needed to perpetuate the organelle.

The others are coded in the nucleus, expressed via the cytoplasmic protein synthetic apparatus, and imported into the organelle. In effect, the organelle genome comprises a length of DNA that has been physically sequestered in a defined part of the cell, and is accordingly subject to its own form of expression and regulation.

Genes not residing within the nucleus are generally described as **extranuclear**; they are transcribed and translated in the *same* organelle compartment in which they reside. By contrast, *nuclear* genes are expressed by means of *cytoplasmic* protein synthesis. (The term **cytoplasmic inheritance** is sometimes used to describe the behavior of genes in organelles. However, we shall not use this description, since it is important to be able to distinguish between events in the general cytosol and those in specific organelles.)

One type of uniparental inheritance is seen in higher animals. Maternal inheritance can be predicted by supposing that the mitochondria are contributed entirely by the ovum and not at all by the sperm. Thus the mitochondrial genes are derived exclusively from the mother; and in males they are discarded each generation.

Conditions in the organelle are different from those in the nucleus, and organelle DNA therefore evolves at its own distinct rate. If inheritance is uniparental, there can be no recombination between parental genomes; and usually recombination does not occur in those cases where organelle genomes are inherited from both parents. Since organelle DNA is replicated by a different DNA polymerase from that of the nucleus, the error rate during replication may be different; and repair systems and other events that impinge on the fidelity of the sequence are also distinct. Mitochondrial DNA accumulates mutations more rapidly than nuclear DNA in mammals, but in plants the accumulation in the mitochondrion is slower than in the nucleus (the chloroplast is in between).

One consequence of maternal inheritance is that the sequence of mitochondrial DNA is more sensitive than nuclear DNA to reductions in the size of the breeding population. Comparisons of mitochondrial DNA sequences in a range of hu-

man populations allow an evolutionary tree to be constructed. The divergence among human mitochondrial DNAs spans 0.57%. A tree can be constructed in which the mitochondrial variants diverged from a common (African) ancestor. The rate at which mammalian mitochondrial DNA accumulates mutations is 2–4% per million years, >10× faster than the rate for globin. Such a rate would generate the observed divergence over an evolutionary period of 140,000–280,000 years. This implies that the human race is descended from a single female, who lived in Africa ~200,000 years ago.

Another cause of uniparental inheritance is found in the alga *Chlamydomonas reinhardii,* in which gametes of equal size fuse to form the zygote. Both parents contribute chloroplast DNA to a zygote. But the chloroplast DNA provided by one parent is degraded shortly after zygote formation, leaving the DNA of the other parent in exclusive possession of the organelle.

The existence of selective mechanisms makes it impossible to predict the reason for the predominance of one parental genotype in any particular case (for example, in higher plants). Even when the parents contribute different amounts of cytoplasm to the zygote, so that one is likely to provide more copies of the organelle genome than the other, selective mechanisms could operate to alter the balance further.

In plants in which both organelle genotypes survive, the wild and mutant phenotypes segregate during somatic growth. Thus in a single heterozygous plant, some tissues have one parental phenotype, while other tissues display the alternative parental phenotype. This somatic segregation contrasts with the stability of the entire nuclear genotype, as defined by Mendelian genetics.

The association of somatic segregation with nonMendelian inheritance suggests that there are multiple copies of the organelle genes in the zygote, and that they are distributed into different cells during somatic division.

The question of which organelle carries a particular genetic marker is not trivial. Since the only organelle shown to possess DNA in higher animal cells is the mitochondrion, probably it is the sole extranuclear residence of genetic material. But in plants and in some unicellular eukaryotes, both chloroplasts and mitochondria are present. They are the only organelles shown to possess DNA; and it is likely (although not proven) that any marker displaying extranuclear inheritance resides in one or the other compartment. In some cases, the location is clear from the nature of the mutant phenotype; but in other instances, the defect takes a more general nature that does not reveal the compartment in which the relevant gene resides.

Organelle genomes are circular DNA molecules that code for organelle proteins

Most organelle genomes so far characterized take the form of a single circular molecule of DNA of unique sequence (denoted **mtDNA** in the mitochondrion and **ctDNA** in the chloroplast). There are a few exceptions where mitochondrial DNA is a linear molecule, generally in lower eukaryotes.

Usually there are several copies of the genome in the individual organelle. Since there are multiple organelles per cell, there is a large number of organelle genomes per cell. Although the organelle genome itself is unique, it constitutes a repetitive sequence relative to any nonrepetitive nuclear sequence.

Chloroplast genomes are relatively large, usually ~140 kb in higher plants, and <200 kb in lower eukaryotes. This is comparable to the size of a large bacteriophage, for example, T4 at ~165 kb. There are multiple copies of the genome per organelle, typically 20–40 in a higher plant, and multiple copies of the organelle per cell, again 20–40.

Mitochondrial genomes vary in total size by more than an order of magnitude. Animal cells have small mitochondrial genomes, small enough to have been completely sequenced in man, mouse, and cow; all are ~16.5 kb. There are several hundred mitochondria per cell. Each mitochondrion has multiple copies of the DNA. The size of the mitochondrial genome is slightly larger in the fruit fly and frog, where we lack information about the number of genomes and organelles. The total amount of mitochondrial DNA relative to nuclear DNA is small, <1%.

In yeast, the mitochondrial genome is much larger. In *S. cerevisiae,* the exact size varies quite widely among different strains, but is ~84 kb. There are ~22 mitochondria per cell, which corresponds to ~4 genomes per organelle. In growing cells, the proportion of mitochondrial DNA can be as high as 18%. (Some yeasts have much larger mitochondrial genomes.)

Plants show an extremely wide range of variation in mitochondrial DNA size, with a minimum of ~100 kb. The function of the additional DNA has not been decisively defined, but it seems likely that much of it is noncoding. The size of the genome makes it difficult to isolate intact, but restriction mapping in several plants suggests that each has a single sequence, organized as a circle. Within this circle, there are short homologous sequences. Recombination between these elements generates smaller, subgenomic circular molecules that coexist with the complete, 'master' genome, explaining the apparent complexity of plant mitochondrial DNAs.

Both of the organelles concerned with energy conversion run a mixed economy. Most of their constituent proteins are imported from the surrounding cytoplasm, where their synthesis represents the final stage of expression of nuclear genes. But each organelle also engages in its own protein synthesis. As illustrated in **Figure 25.1** for the example of mitochondria, this effort is devoted to the production of a small number of proteins, each of which is a component of an oligomeric

Figure 25.1

Mitochondrial protein aggregates are assembled from the products of expression of nuclear genes and mitochondrial genes.

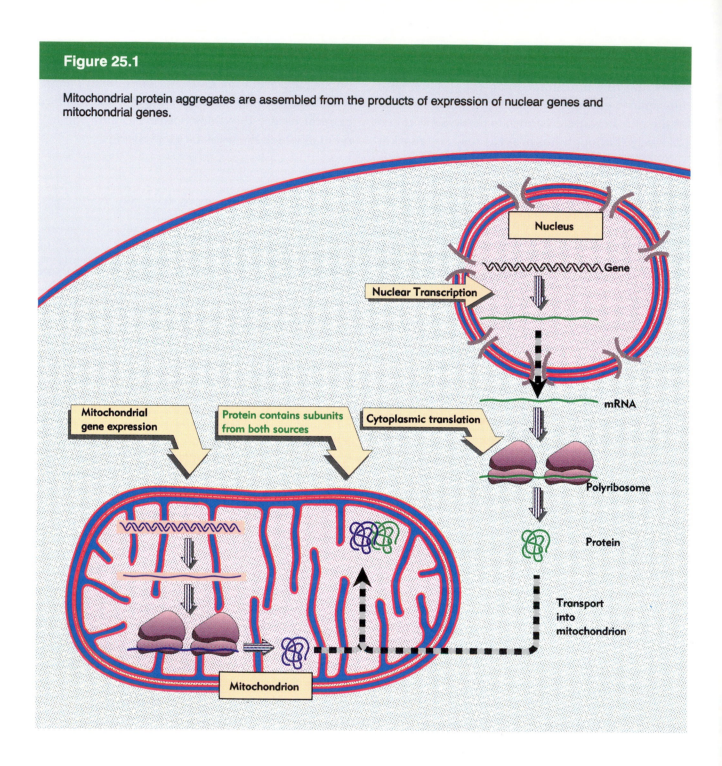

aggregate that includes some protein subunits imported from the cytoplasm.

The apparatus for organelle protein synthesis is itself of mixed origin. Most or all of its protein components are transported into the organelle from the surrounding cytoplasm. But nucleic acids do not pass in either direction through the organelle membrane. Thus among the products of the organelle itself are all of the RNA components of the protein synthetic apparatus.

The origin of the apparatus for gene expression in mitochondria is summarized in **Table 25.1**. The

---

Table 25.1

The mitochondrial apparatus for protein synthesis is assembled from RNA synthesized in the mitochondrion and proteins imported from the cytoplasm.

Component	Synthesized in Mitochondrion	Synthesized in Cytoplasm
Template	~10 mRNAs	RNA polymerase
Ribosome	2 rRNAs	~75 proteins
Adaptor	22 tRNAs	22 aa-tRNA synthetases

RNA components are synthesized in the mitochondrion, and the protein components are coded by the nuclear DNA. The mitochondrial genes are transcribed by an RNA polymerase that (presumably) is coded within the nucleus. Only the mRNAs that have been transcribed from mitochondrial genes are translated in the organelle; and conversely, this is the only compartment in which they can be expressed.

The mitochondrial ribosome consists of the usual larger and smaller subunits, each with a single rRNA coded within the mitochondrion, and with a large number of proteins imported from the cytoplasm. However, the ribosome structures vary extensively between species (see Chapter 9). The set of tRNAs is smaller than needed in the cytoplasm (see Chapter 8). The aminoacyl-tRNA synthetases

responsible for charging the tRNAs are brought in from the cytoplasm. All of the protein components of the synthetic apparatus appear to be unique to the mitochondrion; although coded by nuclear genes, they are different from the proteins of the protein synthetic apparatus in the surrounding cytoplasm.

Organelle protein synthesis has different characteristics from cytoplasmic protein synthesis. Chloroplast and mitochondrial protein synthesis usually is sensitive to antibiotics that inhibit bacterial protein synthesis—for example, erythromycin and chloramphenicol. Cytoplasmic protein synthesis is susceptible to cycloheximide.

The complexes containing the proteins synthesized in the yeast mitochondrion are summarized in **Table 25.2**. The ATPase consists of two units, a membrane factor of 3 subunits coded by the mito-

---

Table 25.2

Mitochondrial (yeast) complexes are assembled from proteins synthesized in different cell compartments.

Protein Complex	Mass (daltons)	Subunits Made in Mitochondrion	Subunits Made in Cytoplasm
Oligomycin-sensitive ATPase	340,000	ATPase 6, 8, 9 (F0 membrane factor)	ATPase 1,2,3,4,7 (F1 ATPase)
Cytochrome c oxidase	137,000	CO 1, 2, 3	CO 4, 5, 6, 7
Cytochrome bc1 complex	160,000	cytochrome b apoprotein	6 subunits

chondrial genome, and the soluble F1 ATPase of 5 subunits synthesized in the cytoplasm. The cytochrome *c* oxidase similarly consists of subunits from both sources. The cytochrome bc_1 complex consists of a single protein from the mitochondrion associated with six subunits from the cytoplasm. The ribosome small subunit contains one protein (Var1) coded in the mitochondrion. Mutations identifying almost all of the mitochondrial genes have been isolated.

Nuclear mutations that abolish each of these complexes also have been found. Some mutations lie in genes coding for components of the complex. Other nuclear mutations prevent complex assembly, but seem to leave the individual subunits unaffected. This suggests that some nuclear-coded products are necessary for assembly, but do not actually comprise part of the final aggregate.

In both organelle and nuclear mutants, the protein complex fails to assemble properly, so that several subunits appear to be absent. This suggests that the assembly of each of these complexes into its usual membrane-associated form is a process in which the absence of one component prevents the assembly of the others. This relationship has the practical consequence of making it difficult to equate nuclear mutations with individual subunits.

If the nuclear mutations reside in unique structural genes, there must be a numerical discrepancy between the representation of organelle-coded and nuclear-coded components, since there are many more copies of the mitochondrial genome. Presumably the nuclear-coded genes are expressed more efficiently.

Is the particular division of labors displayed by the yeast mitochondrion and nucleus typical of other species? The overall constitution of the mixed protein aggregates is similar at least in mammalian and other fungal mitochondria. Genes coding for the same functions are present in yeast and mammalian mitochondrial genomes. Our general expectation is therefore that there will prove to have been a substantial conservation of the coding

functions found in the mitochondrial genomes of different species.

Exceptions to this rule can occur. The constitution of the ATPase is illustrated in **Figure 25.2**. It consists of two components, the F0 membrane factor and the F1 ATPase. The smallest subunit of the F0 membrane factor is coded by the mitochondrial genome in yeast and in *A. nidulans*. But in another fungus, *N. crassa*, the corresponding protein appears to be coded by a nuclear gene. This gene must therefore have been transferred between mitochondrial and nuclear genomes. The corresponding ATPase in chloroplasts is coded in a different way, in which several subunits of the F1 ATPase are imported from the nucleus.

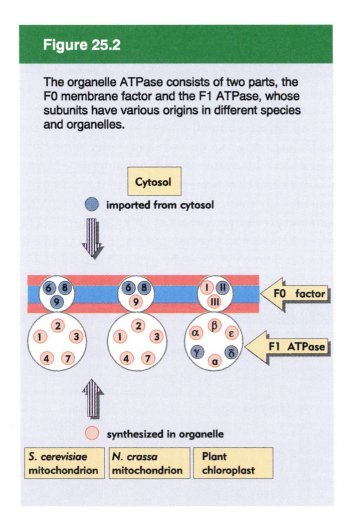

Figure 25.2

The organelle ATPase consists of two parts, the F0 membrane factor and the F1 ATPase, whose subunits have various origins in different species and organelles.

Are mitochondrial and chloroplast genomes unique or are some sequences shared between organelles or by an organelle and the nucleus? Plant mitochondria often have sequences that appear to have been acquired from chloroplast DNA. In some organisms, mitochondrial sequences have homologous regions in the nucleus. Exchanges of DNA between organelles or with the nucleus undoubtedly are rare, but occur in the course of evolution. The mechanism is unknown.

How did a situation evolve in which an organelle contains genetic information for some of its functions, while others are coded in the nucleus? Suppose that these organelles originated in **endosym**biotic events, in which primitive cells captured bacteria that provided the functions that evolved into mitochondria and chloroplasts. At this point, the proto-organelle must have contained all of the genes needed to specify its functions. At some later time, some or most of the organelle genes must have been transferred to the nucleus. Perhaps this occurred at a period when compartments were less rigidly defined. Sequence homologies suggest that both mitochondria and chloroplasts have evolved separately from lineages that are common with eubacteria, with mitochondria sharing an origin with α-purple bacteria, and chloroplasts sharing an origin with cyanobacteria.

The chloroplast genome has similarities to both prokaryotic and eukaryotic DNA

What genes are carried by chloroplasts? **Table 25.3** summarizes a situation generally similar to that of mitochondria, except that more genes are involved. The chloroplast genome codes for all the rRNA and tRNA species needed for protein synthesis. The ribosome includes two small rRNAs in addition to the major species. The tRNA set resembles that of mitochondria in including fewer species than would suffice in the cytoplasm. The chloroplast genome codes for ~50 proteins, including RNA polymerase and some ribosomal proteins. Again the rule is that organelle genes are transcribed and translated by the apparatus of the organelle.

The chloroplast genome of higher plants varies in length, but displays a characteristic landmark. It has a lengthy sequence, 10–24 kb depending on the plant, that is present in two identical copies as an inverted repeat. (Genes that are coded within the inverted repeats are present in two copies per genome, and include the rRNA genes.) The organization of the genome is summarized in **Figure 25.3**.

Recombination between the repeats can generate an inversion of the short single copy sequence (SSC) between them with respect to the long single copy sequence (LSC) that constitutes the rest of the genome.

The complete sequence of chloroplast DNA has been determined for a liverwort (a moss) and for tobacco. In spite of a considerable difference in overall length, between 121 and 155 kb, the gene organization is similar, and the overall number of genes almost identical. Most of their products are components of the thylakoid membranes or concerned with redox reactions, as can be seen from Table 25.3.

The protein synthetic apparatus displays resemblances to the bacterial apparatus. Many of the ribosomal proteins are homologous to those of *E. coli,* and some of the ribosomal protein genes are present in clusters that resemble corresponding clusters in *E. coli.* Genes for RNA polymerase subunits α, β, and β' are homologous to the genes for core enzyme in *E. coli.* Yet the

Table 25.3

The chloroplast genome codes for 4 rRNAs, 30 tRNAs, and ˜50 proteins.

Genes	Types
RNA-coding	
16S rRNA	1
23S rRNA	1
4.5S rRNA	1
5S rRNA	1
tRNA	30
Gene Expression	
r-proteins	19
RNA polymerase	3
Others	2
Thylakoid Membranes	
Photosystem I	2
Photosystem II	7
Cytochrome b/f	3
H⁺-ATPase	6
Others	
NADH dehydrogenase	6
Ferredoxin	3
Ribulose BP Cblase	1
Unidentified	29
Total	110

The role of the chloroplast is to undertake photosynthesis, and many of its genes code for proteins of the complexes of the thylakoid membranes. The constitution of these complexes shows a different balance from that prevailing for mitochondrial

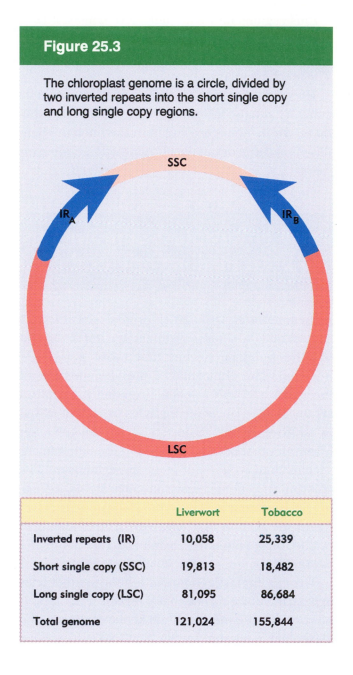

Figure 25.3

The chloroplast genome is a circle, divided by two inverted repeats into the short single copy and long single copy regions.

	Liverwort	Tobacco
Inverted repeats (IR)	10,058	25,339
Short single copy (SSC)	19,813	18,482
Long single copy (LSC)	81,095	86,684
Total genome	121,024	155,844

gene for the β' subunit has an intron.

The presence of interrupted genes suggests that chloroplasts originated before prokaryotes lost their introns. The introns fall into two general classes. Those in tRNA genes are usually (although not inevitably) located in the anticodon loop, like the introns found in yeast nuclear tRNA genes (which we discuss in Chapter 31). Those in protein-coding genes resemble the introns of mitochondrial genes (see Chapter 32).

Table 25.4

Photosynthetic complexes of the thylakoid membranes may be coded by the chloroplast or nuclear genomes or both.

Complex	Subunits Synthesized in Chloroplast	Subunits Synthesized in Cytoplasm
Light-harvesting	None	3 subunits 20-30K
Photosystem I	P700/A1, P700/A2	
Photosystem II	32K, P680, 44K, D2, cyt b559, 10K	
Cytochrome b/f	cyt f, cyt b6, sub 4	
H -ATPase	α, β, ε, I, III, a	γ, δ, II

complexes. **Table 25.4** shows that although some complexes are like mitochondrial complexes in having some subunits coded by the organelle genome and some by the nuclear genome, other chloroplast complexes are coded entirely by one genome.

The identified genes of the chloroplast show the focus of its activities. There are 45 genes coding for RNA, 27 coding for proteins concerned with gene expression, 18 coding for proteins of the thylakoid membrane, and another 10 representing functions concerned with electron transfer. The products of some 30 open reading frames remain to be identified.

The mitochondrial genome is large in yeast but small in mammals

The fivefold discrepancy in size between the *S. cerevisiae* (84 kb) and mammalian (16 kb) mitochondrial genomes alone alerts us to the fact that there must be a great difference in their genetic organization in spite of their common function. The number of endogenously synthesized products concerned with mitochondrial enzymatic functions appears to be similar. Does the additional genetic material in yeast mitochondria represent other proteins, perhaps concerned with regulation, or is it unexpressed?

The map shown in **Figure 25.4** accounts for the major RNA and protein products of the yeast mitochondrion (excluding the ~22 tRNA molecules,

Figure 25.4

The mitochondrial genome of *S. cerevisiae* contains both interrupted and uninterrupted protein-coding genes, rRNA genes, and tRNA genes (positions not indicated). Arrows indicate direction of transcription.

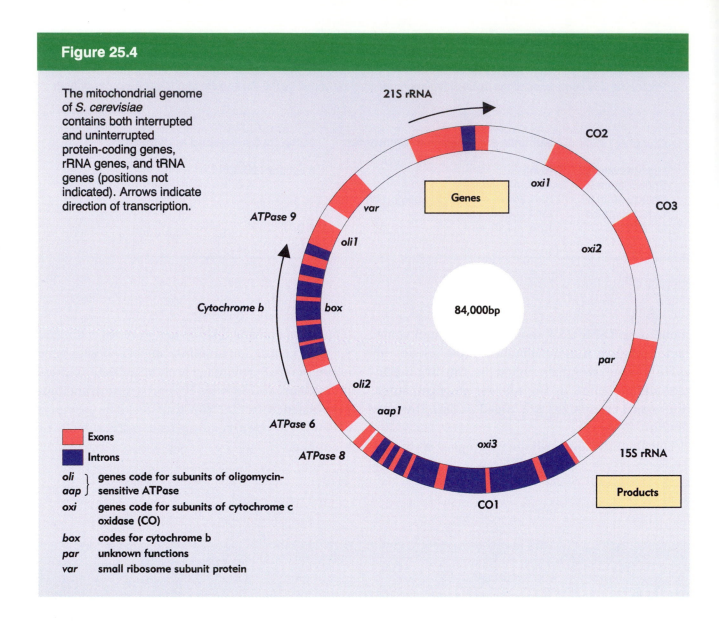

Exons

Introns

oli
aap } genes code for subunits of oligomycin-sensitive ATPase

oxi genes code for subunits of cytochrome c oxidase (CO)

box codes for cytochrome b

par unknown functions

var small ribosome subunit protein

which have not yet been completely mapped). The most notable feature is the dispersion of loci on the map.

This genome has the distinction of providing the extremely rare situation where the genes for rRNA are separated. The gene for 15S rRNA is uninterrupted and lies ~24,000 bp away from the gene for 21S rRNA. In some strains of *S. cerevisiae* the latter

gene has a single intron (as shown on the map); in other strains it is uninterrupted.

The two most prominent loci are the interrupted genes *box* (coding for cytochrome b) and *oxi3* (coding for subunit 1 of cytochrome oxidase). Together these two genes are almost as long as the entire mitochondrial genome in mammals! Many of the long introns in these genes have open reading frames in register with the preceding exon; at least in some cases, there is evidence for translation of the intron. (We discuss this situation in Chapter 32.) This adds several proteins, presumably all synthesized in low amounts, to the complement of the yeast mitochondrion.

The remaining genes appear to be uninterrupted. They correspond to the other two subunits of cytochrome oxidase coded by the mitochondrion, to the subunit(s) of the ATPase, and (in the case of *var1*) to a mitochondrial ribosomal protein.

About 24% of the yeast mitochondrial genome consists of short stretches rich in A-T base pairs, which probably lack any coding function. However, this still leaves a considerable amount of genetic material to be accounted for, and it will be surprising if some further genes are not discovered in the regions that have been left open on the map. Even given this likelihood, though, we can conclude that the total number of yeast mitochondrial genes is unlikely to exceed ~25.

A very different picture is seen in animal mitochondrial DNA, which is rather compact. There are extensive differences in the detailed gene organization found in different animal phyla, but the general principle is maintained of a small genome coding for a restricted number of functions. The most is known about mammalian mitochondrial genomes, which have been sequenced in several species. Here the organization is extremely compact; there are no introns, some genes actually overlap, and almost every single base pair can be assigned to a gene. With the exception of the D loop, a region concerned with the initiation of DNA replication, no more than 87 of the 16,569 bp of the human mitochondrial genome can be regarded as lying in intercistronic regions.

The complete nucleotide sequences of the human and murine mitochondrial genomes show extensive homology in organization. The map of the human genome is summarized in **Figure 25.5**. There are 13 protein-coding regions. In common with yeast, these include cytochrome *b*, the usual 3 subunits of cytochrome oxidase, and one of the subunits of ATPase. In contrast with yeast, the mammalian mitochondrion codes for 7 subunits (or associated proteins) of NADH dehydrogenase.

The organization of the genes has some striking features that reflect the peculiarities of gene expression in the mammalian mitochondrion. Most genes are expressed in the same direction; and tRNA genes lie between the genes coding for rRNA or protein.

In many cases, there is no separation at all between genes. The last base of one gene is adjacent to the first base of the next gene. In some cases, there is an overlap, most commonly of just one base, so that the last base of one gene serves as the first base of the next gene. Five of the reading frames lack a termination codon; these end with U or UA, and so an ochre codon is created by polyadenylation of the transcript (mitochondrial mRNA acquires a short sequence of poly (A) at its 3′ end). In three cases, the termination codon appears to be AGA or AGG, which is usually used as arginine in the genetic code.

The mRNA species corresponding to all the protein-coding regions have been identified; and in each case, an initiation codon lies within six bases of the start. The codon used for initiation apparently can be AUG, AUA, or AUU. Nontranslated 5′ and 3′ regions are virtually absent from human mitochondrial mRNAs.

Almost all of the genes are expressed in the clockwise direction, as indicated by the arrows on the map. In only one case are two clockwise coding regions found in the form of contiguous genes (ATPase 6 and CO3). In every other case, at least one tRNA gene separates adjacent coding regions.

The punctuation of the rRNA- and protein-coding

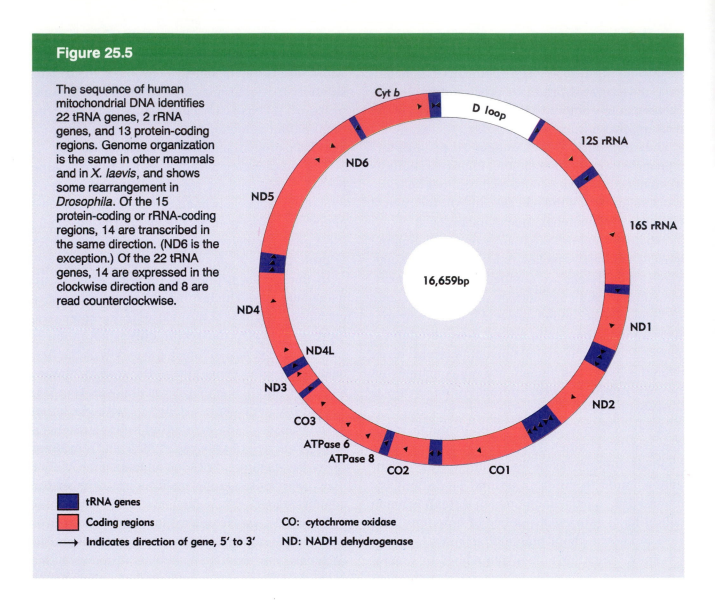

Figure 25.5

The sequence of human mitochondrial DNA identifies 22 tRNA genes, 2 rRNA genes, and 13 protein-coding regions. Genome organization is the same in other mammals and in *X. laevis*, and shows some rearrangement in *Drosophila*. Of the 15 protein-coding or rRNA-coding regions, 14 are transcribed in the same direction. (ND6 is the exception.) Of the 22 tRNA genes, 14 are expressed in the clockwise direction and 8 are read counterclockwise.

16,659bp

■ tRNA genes
■ Coding regions
→ Indicates direction of gene, 5′ to 3′

CO: cytochrome oxidase
ND: NADH dehydrogenase

regions by the tRNA genes does not leave room for promoters comparable to those found in eukaryotic nuclei or in bacteria. A single promoter for clockwise transcription is located in the region of the D loop. This suggests the model illustrated in **Figure 25.6**. Transcription starts just before the tRNA gene in front of the 12S rRNA gene, and continues almost all the way around the circle, to terminate in the D loop. The transcribed strand is

called the **H strand**, and the other strand is called the **L strand** (because mitochondrial DNA can be separated into heavy and light single strands on the basis of their density).

The significance of the alternation of tRNA genes with rRNA- and protein-coding genes is that the tRNAs indicate sites of cleavage. By cleaving the primary transcript on either side of each tRNA gene, all of the genes except ATPase 6 and CO3 give rise

Figure 25.6

Human mitochondrial DNA is transcribed into a single transcript, from which tRNAs are cleaved to release the rRNAs and mRNAs.

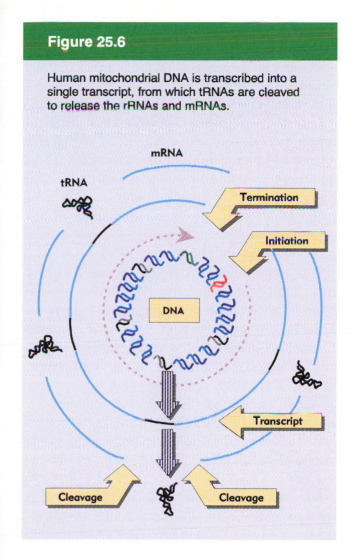

to monocistronic products. The rRNA molecules appear to be synthesized in greater amounts than the mRNAs. This could be caused by premature termination of some proportion of the transcripts at some point after the two rRNA genes have been transcribed.

This mitochondrial DNA therefore presents the closest analogy to a bacterial operon in the eukaryotes. It is transcribed from a single promoter region, but individual tRNAs, rRNAs, and mRNAs are released from the transcript. Processing of the transcript therefore becomes the central event in gene expression. (There are also bacterial operons in which the separation of tRNA from rRNA or protein-coding regions must be accomplished after transcription; see Chapter 15).

How are the tRNA genes coded by the L strand expressed? Probably a similar mechanism applies, with a giant transcript produced for the L strand, from which the tRNA sequences and perhaps some mRNA sequences are conserved, while the rest is degraded. The origin for L strand (counterclockwise) transcription also lies in the region of the D loop. Very large L strand transcripts can be isolated, so their processing probably takes place some time subsequent to transcription. Large H strand transcripts cannot be found, which suggests that they are processed more or less immediately after the transcription of each tRNA junction.

Recombination and rearrangement of organelle DNA

The prerequisite for recombination is the coexistence of genomes of both parents, a condition that is prevented in situations where absolute uniparental inheritance prevails. But the existence of instances where both organelle genomes survive immediately raises the question of whether they can interact. Is complementation possible between mutants; can recombination occur between genomes?

Complementation requires intermingling of the products of gene expression; recombination requires physical exposure of the two genomes to each other. Both demands are somewhat at odds with the static view of the individual organelle implicit in the description of its endogenous gene expression as an event strictly confined to its own compartment. Yet in at least one species, recombination can occur between chloroplast DNAs; and in another, both complementation and recombination can occur with mitochondrial genomes.

In the alga *C. reinhardii,* the chloroplast DNA usually is inherited uniparentally. However, under the unusual circumstance when the genes of the usually predominant parent are irradiated with ultraviolet, the chloroplast genes of the other parent also survive. In the resulting biparental zygotes, recombination occurs between the chloroplast genomes, and this can be used to construct a genetic map.

In the yeast *S. cerevisiae,* recombination between mitochondrial markers is the norm. The nature of the recombination event is unknown, but a cross between two parents whose mitochondrial DNAs have gross differences generates progeny possessing physically recombinant DNAs. This demonstrates directly that the mitochondrial DNA of one parent comes into contact with the mitochondrial DNA provided by the other parent.

How do the parental organelle DNAs reach each other? The most likely explanation is that organelles can undergo fusion, to generate a single compartment containing genomes from both parents. This concept has been taken further in the case of yeast, where it is possible that instead of many individual mitochondria, a large branched structure (or structures) corresponds to many mitochondria. Perhaps there is a continual state of flux, with making and breaking of membranes. This raises the possibility that the presence of a genome is related to some unit smaller than the organelle itself; then the total number of genomes per organelle would be determined by the number of these constituent subunits.

In yeast it takes some hours after the production of a zygote before recombination occurs. But in the interim period, complementation can be detected between different mutations. This provides an opportunity to define the genetic organization of the organelle genome, since mutations can be both mapped and assigned to complementation groups. (A similar approach also has proved successful with the *C. reinhardii* chloroplast.)

The yeast mitochondrial genome has a surprising fluidity of structure. We have already mentioned that different forms of some genes are found in different strains of *S. cerevisiae.* Indeed, the total size of the mitochondrial genome can vary substantially, by as much as 10,000 bp between strains. Extensive deletions of mitochondrial DNA can occur, as found in the existence of **petite** strains of *S. cerevisiae* (and certain other yeasts). They arise spontaneously or are induced by certain treatments, such as addition of ethidium bromide.

All petite mutations lack mitochondrial function. Their existence is possible because the alternative life-styles of yeast allow it to survive aerobically (when respiration is essential) or anaerobically (when it is dispensable). Thus mitochondrial mutations behave as conditional lethals for which aerobic growth provides the nonpermissive condition, while anaerobic growth provides the permissive condition under which the organism survives.

Loss of mitochondrial function forces adoption of the anaerobic life-style. Of course, such a choice is not possible in (for example) animal cells, where loss of mitochondrial function is lethal. (*C. reinhardii* offers an analogous genetic situation with the chloroplast, since photosynthesis is dispensable in the presence of acetate.) Petite mutations of

Table 25.5

Petite mutations remove or rearrange yeast mitochondrial DNA.

Class	Name	Type of mtDNA
Wild-type	grande	~84 kb circle
Nuclear petite	-	Varies
Suppressive petite	rho⁻	Abnormal DNA
Neutral petite	rho⁰	No DNA

S. cerevisiae fall into three types summarized in **Table 25.5**.

◆ **Nuclear petites** have Mendelian (that is, nuclear) mutations that abolish mitochondrial activity.

◆ **Neutral (rho⁰) petites** represent an extreme situation in which all mitochondrial DNA is absent. This is a recessive genotype.

◆ **Suppressive (rho⁻) petites** have grossly abnormal mitochondrial DNA. The rho⁻ mitochondria contain circles of DNA that are much smaller than the usual genome, containing only a small part of the usual complexity, ranging from 0.2% to 36% in different cases. Thus more of the mitochondrial genome is deleted than is retained.

The sequence retained in a *rho⁻* petite often is amplified to generate a large number of copies. Amplification occurs either by increase of the ploidy (the number of DNA molecules) or by formation of multimeric DNA molecules, each containing many copies of the sequence. The multiple copies can be arranged tandemly as direct or inverted repeats. The sequence of the petite DNA need not necessarily represent a formerly contiguous region of the mitochondrial genome, but can have different regions juxtaposed. Some petites are stable; others are unstable, and further rearrangements of the sequence occur.

The DNA retained in a petite strain can recombine with the DNA of another petite or of a wild-type strain (called *grande,* by comparison). This allows the genetic markers present in each petite to be correlated with the sequence of DNA that has been retained. In fact, it is the use of such petites for this deletion mapping that allows construction of the mitochondrial genetic map shown in Figure 25.4.

When a *rho⁻* petite strain of yeast is crossed with a wild-type strain, in a certain proportion of the progeny, only the petite genotype is found. The wild-type mitochondrial DNA has disappeared. This effect is called **suppression**; and the characteristic proportion of progeny in which it occurs gives the degree of **suppressiveness**.

The cause of suppression in highly suppressive petites is preferential replication of the petite mitochondrial DNA. This probably happens because the petite DNA possesses an increased concentration of origin-like sequences, called *rep* regions, relative to wild-type DNA.

Although yeast mitochondrial DNA contains a specific sequence(s) that can be used to initiate replication, this sequence is dispensable, because petite DNA lacking it can survive (although it is less suppressive). In fact, *any sequence of mitochondrial DNA can be retained in a petite.* This implies that all segments of the DNA can be independently replicated. We do not know how this facility is provided and whether it relies on any feature equivalent to an origin for replication (see Chapter 18).

The behavior of petite and wild-type mitochondrial DNA shows that there is extensive flexibility in the organization of the genome, with regard both to its expression and replication. Perhaps the difference from other mitochondrial DNAs lies in the ability of yeast altogether to dispense with mitochondrial functions. It will be interesting to characterize the enzyme systems that act on this DNA.

Summary

NonMendelian inheritance is explained by the presence of DNA in organelles in the cytoplasm. Mitochondria and chloroplasts both represent membrane-bounded systems in which some proteins are synthesized within the organelle, while others are imported. The organelle genome is usually a circular DNA that codes for all of the RNAs and for some of the proteins that are required.

Mitochondrial genomes vary greatly in size from the 16 kb minimalist mammalian genome to the 570 kb genome of higher plants. It is assumed that the larger genomes code for additional functions. Chloroplast genomes range from 120 to 200 kb. Those that have been sequenced have a similar organization and coding functions. In both mitochondria and chloroplasts, many of the major proteins contain some subunits synthesized in the organelle and some subunits imported from the cytosol.

Mammalian mtDNAs are transcribed into a single transcript from the major coding strand, and individual products are generated by RNA processing. Rearrangements can occur in mitochondrial DNA rather frequently in yeast; and recombination between mitochondrial or between chloroplast genomes has been found. There are some tantalizing homologies between mitochondrial and chloroplast genomes.

Further reading

Reviews

A general review of chloroplast and mitochondrial inheritance was provided by **Gillham** in *Organelle Heredity* (Raven Press, 1978).

Expression of the mammalian mitochondrial genome was summarized by **Clayton** (*Ann. Rev. Biochem.* **53**, 573–594, 1984) and by **Attardi** (*Int. Rev. Cytol.* **93**, 93–146, 1985). Evolution of mitochondrial genomes was reviewed by **Gray** (*Ann. Rev. Cell Biol.* **5**, 25–50, 1989).

Advances on chloroplast genomes were reviewed by **Palmer** (*Ann. Rev. Genet.* **19**, 324–354, 1985).

Discoveries

Human evolution was traced from mitochondrial DNA sequences by **Cann, Stoneking, and Wilson** (*Nature* **324**, 31–36, 1987).

The first mitochondrial DNA sequence was reported by **Anderson** *et al.* (*Nature* **290**, 457–465, 1981). Sequences of entire chloroplast genomes were reported by **Ohyama** *et al.* (*Nature* **322**, 572–574, 1986) **and Shinozaki** *et al.* (*EMBO J.*, 5, 2043–2049, 1986)

CHAPTER 26

Organization of simple sequence DNA

Within the eukaryotic genome are many sequences that have no coding function. Very likely constituting a majority of the DNA, these sequences are under different evolutionary constraints from those imposed by the need to represent a series of amino acids. Some of them, of course, are part of transcription units—such as nontranslated flanking regions in mRNA or the introns removed during maturation of mRNA.

Others provide signals that are recognized by proteins, including elements such as promoters for transcription, origins for DNA replication, sites for folding the chromosome, points for attaching the kinetochore, and other cellular functions. Virtually the only elements whose sequences have been delineated systematically are the promoters and enhancers of transcription. (Origins of replication and centromeric sequences have been defined in yeast.) Cis-acting regulatory sequences are usually rather short (too short in fact to be detected as repetitive DNA). It would not be surprising if similar types of elements convey structural information, concerning the organization of chromatin or the condensation of chromosomes, but these remain to be identified.

Repetitive DNA is defined by its renaturation at rates more rapid than the rate of nonrepetitive DNA. By the criteria that we discussed in Chapter 24, a significant proportion of most mammalian genomes consists of repetitive DNA, taking the form of short sequences that are repeated in identical or related copies in the genome. These sequences generally are found outside of coding regions, and are therefore candidates for structural or other roles in genome function.

Renaturation kinetics divide repetitive DNA into two general classes, which have different properties.

Moderately repetitive DNA consists of a variety of sequence families, present in varying degrees of repetition and with differing internal relationships. The members of these families are often interspersed in a more or less regular way with longer stretches of nonrepetitive DNA. A large part of the moderately repetitive component of mammalian genomes turns out to comprise members of a single family, discussed in Chapter 35. No characteristic structural or functional significance can be attributed to these sequences as a class.

Highly repetitive DNA generally consists of very short sequences repeated many times in tandem in large clusters. Because of its short repeating unit, it is sometimes described as **simple sequence DNA**. This type of component is present in almost all higher eukaryotic genomes, but its overall amount is extremely variable. In mammalian genomes it is typically <10%, but in (for example) *Drosophila virilis*, it amounts to ~50%. In addition to the large clusters in which this type of sequence was originally discovered, there are smaller clusters interspersed with nonrepetitive DNA.

Tandemly repeated sequences are especially liable to undergo misalignments during chromosome pairing, and thus the sizes of tandem clusters tend to be highly polymorphic, with wide variations between individuals. In fact, the smaller clusters of such sequences can be used to characterize individual genomes in the technique of 'DNA fingerprinting'.

Comparisons of corresponding regions of simple sequence DNA within and between species are informative about the mechanisms involved in manipulating DNA sequences over evolutionary periods. We may ask whether these sequences have a structural function, although evidence is still difficult to obtain.

Highly repetitive DNA forms satellites

The very short sequences that are tandemly repeated in highly repetitive DNA are identical in some cases, but only related in others. In either case, the tandem repetition of a short sequence often creates a fraction with distinctive physical properties that can be used to isolate it. In some cases, the repetitive sequence has a base composition distinct from the genome average, which allows it to form a separate fraction by virtue of its buoyant density.

The buoyant density of a duplex DNA depends on its G•C content according to the empirical formula

$$\rho = 1.660 + 0.00098 \, (\%G•C) \text{ g/cm}^3$$

Buoyant density usually is determined by centrifuging DNA through a **density gradient** of CsCl. The DNA forms a band at the position corresponding to its own density.

When eukaryotic DNA is centrifuged on a density gradient, two types of material may be distinguished:

◆ Most of the genome forms a continuum of fragments that appear as a rather broad peak centered on the buoyant density corresponding to the average G•C content of the genome. This is called the **main band.**

◆ Sometimes an additional, smaller peak (or peaks) is seen at a different value. This material is called **satellite DNA.**

Fractions of DNA differing in buoyant density by more than ~0.005 g/cm³ can be separated, corresponding to a difference in G•C content of >5%.

A classic example is provided by mouse DNA, shown in **Figure 26.1**. The graph is a quantitative

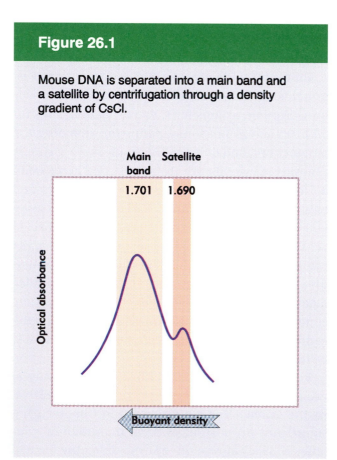

Figure 26.1

Mouse DNA is separated into a main band and a satellite by centrifugation through a density gradient of CsCl.

scan of the bands formed when mouse DNA is centrifuged through a CsCl density gradient. The main band contains 92% of the genome and is centered on a buoyant density of 1.701 g/cm³ (corresponding to its average G•C of 42%, typical for a mammal). The smaller peak represents 8% of the genome and has a distinct buoyant density of 1.690 g/cm³. It contains the mouse satellite DNA, whose G•C content (30%) is much lower than any other part of the genome.

Satellites are present in many eukaryotic genomes. They may be either heavier or lighter than the main band; but it is uncommon for them to represent >5% of the total DNA. The resolution of satellite DNA is often much improved by using centrifugation through gradients of Cs_2O_4 containing silver ions, or certain other reagents, including various dyes. Sometimes a single satellite is more clearly separated from the main band; sometimes multiple satellite sequences can be resolved.

The behavior of satellite DNA on density gradients is anomalous more often than not. When the actual base composition of a satellite is determined, it is often different from what had been predicted from its buoyant density. The reason is that ρ is a function not just of base composition, but of the constitution in terms of nearest neighbor pairs. For simple sequences, these are likely to deviate from the random pairwise relationships needed to obey the equation for buoyant density. Also, satellite DNA may be methylated, which changes its density.

Often most of the highly repetitive DNA of a genome can be isolated in the form of satellites. When a highly repetitive DNA component does not separate as a satellite, on isolation its properties often prove to be similar to those of satellite DNA. That is to say that it consists of multiple tandem repeats with anomalous centrifugation. Material isolated in this manner is sometimes referred to as a **cryptic satellite**. Together the cryptic and apparent satellites usually account for all the tandemly repeated blocks of highly repetitive DNA. When a genome has more than one type of highly repetitive DNA, each exists in its own satellite block (although sometimes different blocks are adjacent).

Satellite DNAs often lie in heterochromatin

Where in the genome are the blocks of highly repetitive DNA located? An extension of nucleic acid hybridization techniques allows the location of satellite sequences to be determined directly in the chromosome complement. In the technique of **in situ** or **cytological hybridization**, the chromosomal DNA is denatured by treating cells that have been squashed on a cover slip. Then a solution containing a radioactively labeled DNA or RNA probe is added. The probe hybridizes with its complements in the denatured genome. The location of the sites of hybridization can be determined by autoradiography. (We describe this technique further in Chapter 38.)

Labeled satellite DNAs often are confined to the heterochromatin present around the centromeres of mitotic chromosomes. **Heterochromatin** is the term used to describe regions of chromosomes that are permanently tightly coiled up and inert, in contrast with the **euchromatin** that represents most of the genome (see Chapter 27).

An example of the localization of satellite DNA for the mouse chromosomal complement is shown in **Figure 26.2**. In this case, one end of each chromosome is labeled, because this is where the centromeres are located in *M. musculus* chromosomes.

The centromeric location of satellite DNA suggests that it has some structural function in the

Figure 26.2

Cytological hybridization shows that mouse satellite DNA is located at the centromeres. Photograph kindly provided by Mary Lou Pardue and Joe Gall.

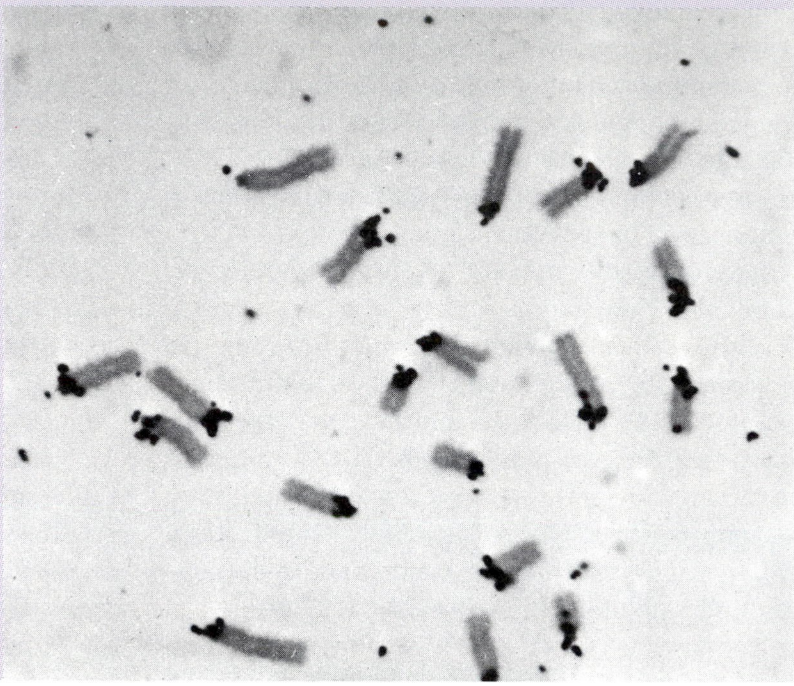

chromosome. Since the centromeres are the regions where the kinetochores are formed at mitosis and meiosis for controlling chromosome movement, this function could be connected with the process of chromosome segregation.

But that is all we know about its role, which apart from this general suggestion remains quite mysterious. Consistent with their simple sequence and condensed structure, satellite DNAs are not transcribed or translated.

Arthropod satellites have very short identical repeats

In the arthropods, as typified by insects and crabs, each satellite DNA appears to be rather homogeneous. Usually, a single very short repeating unit accounts for >90% of the satellite. This makes it relatively straightforward to determine the sequence.

Drosophila virilis has three major satellites and also a cryptic satellite, together representing >40% of the genome. The sequences of the satellites are summarized in **Table 26.1**. The three major satellites have closely related sequences. A single base

Table 26.1

Satellite DNAs of *D. virilis are* related. More than 95% of each satellite consists of a tandem repetition of the predominant sequence.

Satellite	Predominant Sequence	Total length	Part of Genome
I	A C A A A C T T G T T T G A	1.1×10^7	25%
II	A T A A A C T T A T T T G A	3.6×10^6	8%
III	A C A A A T T T G T T T A A	3.6×10^6	8%
Cryptic	A A T A T A G T T A T A T C		

substitution is sufficient to generate either satellite II or III from the sequence of satellite I. Note that satellites II and III have exactly the same base composition (1 G•C pair out of 7 = 14% G•C), but have buoyant densities of 1.688 and 1.671 g/cm³, respectively.

The satellite I sequence is present in other species of *Drosophila* related to *virilis*, and so may have preceded speciation. The sequences of satellites II and III seem to be specific to *D. virilis*, and so may have evolved from satellite I after speciation.

The main feature of these satellites is their very short repeating unit: only 7 bp. Similar satellites are found in other species. *D. melanogaster* has a variety of satellites, several of which have very short repeating units (5, 7, 10, or 12 bp). Comparable satellites are found in the crabs.

The close sequence relationship found among the *D. virilis* satellites is not necessarily a feature of other genomes, where the satellites may have unrelated sequences. *Each satellite has arisen by a lateral amplification of a very short sequence.* This sequence may represent a variant of a previously existing satellite (as in *D. virilis*), or could have some other origin.

Satellites are continually generated and lost from genomes. This makes it difficult to ascertain evolutionary relationships, since a current satellite could have evolved from some previous satellite that has since been lost. The important feature of these satellites is that *they represent very long stretches of DNA of very low sequence complexity, within which constancy of sequence can be maintained.*

One feature of many of these satellites is a pronounced asymmetry in the orientation of base pairs on the two strands. In the example of the *D. virilis* satellites shown in Table 26.1, in each of the major satellites one of the strands is much richer in T and G bases. This increases its buoyant density, so that upon denaturation this **heavy strand** (H) can be separated from the complementary **light strand** (L). This can be useful in sequencing the satellite.

Mammalian satellites consist of hierarchical repeats

In the mammals, as typified by various rodents, the sequences comprising each satellite show appreciable divergence between tandem repeats. Common short sequences can be recognized by their preponderance among the oligonucleotide fragments released by chemical or enzymatic treatment. However, the predominant short sequence usually accounts for only a small minority of the copies. The other short sequences are related to the predominant sequence by a variety of substitutions, deletions, and insertions.

But a series of these variants of the short unit can constitute a longer repeating unit that is itself repeated in tandem with some variation. Thus mammalian satellite DNAs are constructed from a *hierarchy* of repeating units. These longer repeating units constitute the sequences that renature in reassociation analysis. They can also be recognized by digestion with restriction enzymes.

Uncertainty in evaluating the complexity of repetitive DNA is caused by the effect of mismatching on the reassociation reaction. The effect is particularly pronounced with satellite DNA; it means that the length of the reassociating unit can be assessed only within a factor of two or so at best.

When any satellite DNA is digested with an enzyme that has a recognition site in its repeating unit, one fragment will be obtained for every repeating unit in which the site occurs. In fact, when the DNA of a eukaryotic genome is digested with a restriction enzyme, most of it gives a general smear, due to the random distribution of cleavage sites. But satellite DNA generates sharp bands, because a large number of fragments of identical or almost identical size are created by cleavage at restriction sites that lie a regular distance apart.

Determining the sequence of satellite DNA can be difficult. Using the discrete bands generated by restriction cleavage, we can attempt to obtain a sequence directly. However, if there is appreciable divergence between individual repeating units, different nucleotides will be present at the same position in different repeats, so the sequencing gels will be obscure. If the divergence is not too great—say, within ~2%—it may be possible to determine an average repeating sequence.

Individual segments of the satellite can be inserted into plasmids for cloning. A difficulty is that the satellite sequences sometimes tend to be excised from the chimeric plasmid by recombination in the bacterial host. However, when the cloning succeeds, it is possible to determine the sequence of the cloned segment unambiguously. While this gives the actual sequence of a repeating unit or units, we should need to have many individual such sequences to reconstruct the type of divergence typical of the satellite as a whole.

By either sequencing approach, the information we can gain is limited to the distance that can be analyzed on one set of sequence gels. The repetition of divergent tandem copies makes it impossible to reconstruct longer sequences by obtaining overlaps between individual restriction fragments. The satellite DNA of the mouse *M. musculus* is

Figure 26.3

The repeating unit of mouse satellite DNA contains two half-repeats, which are aligned to show the identities (in color).

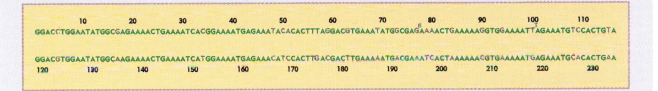

Figure 26.4

The alignment of quarter-repeats identifies homologies between the first and second half of each half-repeat. Positions that are the same in all 4 quarter-repeats are shown in color; identities that extend only through 3 quarter-repeats are indicated by the grey letters in the pink area.

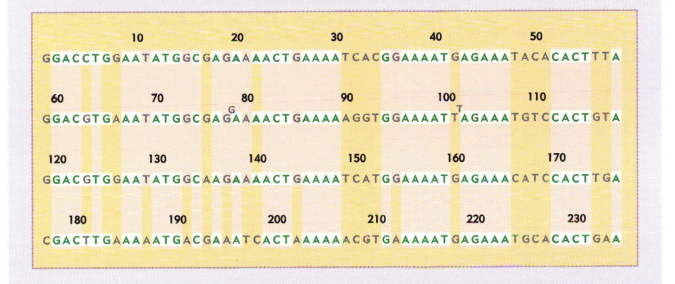

Figure 26.5

The alignment of eighth-repeats shows that each quarter-repeat consists of an α and a β half. The consensus sequence gives the most common base at each position. The "ancestral" sequence shows a sequence very closely related to the consensus sequence, that could have been the predecessor to the α and β units. (The satellite sequence is continuous, so that for the purpose of deducing the consensus sequence, we can treat it as a circular permutation, as indicated by joining the last GAA triplet to the first 6 bp.)

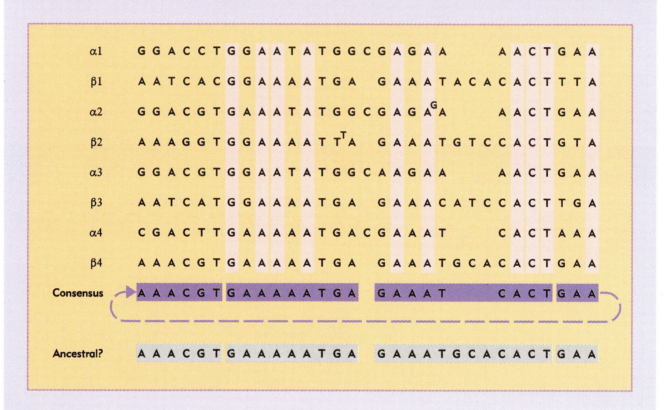

cleaved by the enzyme *Eco*RII into a series of bands, including a predominant monomeric fragment of 234 bp. This sequence must be repeated with few variations throughout the 60–70% of the satellite that is cleaved into the monomeric band. We may analyze this sequence in terms of its successively smaller constituent repeating units.

Figure 26.3 depicts the sequence in terms of two half-repeats. By writing the 234 bp sequence so that the first 117 bp are aligned with the second 117 bp,

we see that the two halves are quite well related. They differ at 22 positions, corresponding to 19% divergence. This means that the current 234 bp repeating unit must have been generated at some time in the past by duplicating a 117 bp repeating unit, after which differences accumulated between the duplicates.

Within the 117 bp unit, we can recognize two further subunits. Each of these is a quarter-repeat relative to the whole satellite. The four quarter-repeats are aligned in **Figure 26.4**. The upper two lines represent the first half-repeat of Figure 26.3; the lower two lines represent the second half-repeat. We see that the divergence between the four quarter-repeats has increased to 23 out of 58 positions, or 40%. Actually, the first three quarter-repeats are somewhat better related, and a large proportion of the divergence is due to changes in the fourth quarter-repeat.

Looking within the quarter-repeats, we find that each consists of two related subunits (one-eighth-repeats), shown as the α and β sequences in **Figure 26.5**. The α sequences all have an insertion of a C, and the β sequences all have an insertion of a trinucleotide, relative to a common consensus sequence. This suggests that the quarter-repeat originated by the duplication of a sequence like the consensus sequence, after which changes occurred to generate the components we now see as α and β. Further changes then took place between tandemly repeated $\alpha\beta$ sequences to generate the individual quarter- and half-repeats that exist today. Among the one-eighth-repeats, the present divergence is $19/31 = 61\%$.

The consensus sequence is analyzed directly in **Figure 26.6**, which demonstrates that the current satellite sequence can be treated as derivatives of a 9 bp sequence. We can recognize three variants of this sequence in the satellite, as indicated at the bottom of Figure 26.5. If in one of the repeats we take the next most frequent base at two positions instead of the most frequent, we obtain three well-related 9 bp sequences.

```
G A A A A A C G T
G A A A A A T G A
G A A A A A A C T
```

The origin of the satellite could well lie in an amplification of one of these three nonamers. The overall consensus sequence of the present satellite is $GAAAAA^{AGT}_{TC}T$, which is effectively an amalgam of the three 9 bp repeats.

Figure 26.6

The existence of an overall consensus sequence is shown by writing the satellite sequence in terms of a 9 bp repeat.

```
            G G A C C T
G G A A T A T G G C
G A G A A A A C T
G A A A A T C A C
G G A A A A T G A
G A A A T C A C T
T T A G G A C G T
G A A A T A T G G C
G A G A^G A A A C T
G A A A A A G G T
G G A A A A T T^T A
G A A A T* C A C T
G T A G G A C G T
G G A A T A T G G C
A A G A A A A C T
G A A A A T C A T
G G A A A A T G A
G A A A C* C A C T
T G A C G A C T T
G A A A A A T G A C
G A A A T C A C T
A A A A A A C G T
G A A A A A T G A
G A A A T* C A C T
G A A
```

G_{20} A_{16} A_{21} A_{20} A_{12} A_{17} T_8 G_{11} A_5

T_7 C_5 A_8 C_9 T_{15}

C_7

* indicates inserted triplet in β sequence

C in position 10 is extra base in α sequence

Evolution of hierarchical variations in the satellite

Having identified the sequence components of each level of the hierarchy of the satellite DNA, we can now reverse the procedure to reconstruct its evolution and explain its properties. A model showing possible steps involved in forming the present satellite is given in **Figure 26.7**.

The general principle of this model is that at various times a group of repeating units is suddenly amplified laterally to generate a large number of identical tandem copies. An event of this sort is called a **saltatory replication**. Then the copies diverge in sequence as mutations accumulate in them. At some subsequent time, a group of these copies is taken for another saltatory replication. The extent of divergence among the copies amplified at each saltation will depend on the period that has passed since the last saltatory replication imposed identity on the satellite. The satellite can evolve by a series of these saltatory replications, alternating with accumulation of mutations.

Suppose that the present satellite originated with the tandem repetition of a sequence such as GAAAAATGT or something closely related to it. (This sequence might have been part of a satellite or could have some quite different origin: its existence is as far back as we can pursue the satellite.) All the original 9 bp units were identical, but with time mutations created differences between them. Then three adjacent units with the ancestral sequence hypothesized in Figure 26.5 were amplified, giving a tandem repeat of 27 bp.

Mutations occurred in this unit, including cases in which one unit gained an additional C while its neighbor gained a triplet insertion. This pair of repeats, now together 58 bp long, was subject to saltatory replication and gave a satellite that we can describe as $(\alpha\beta)_n$.

Once again, the satellite accumulated point mutations, deletions, and insertions to create divergence between its repeating units. Two adjacent $\alpha\beta$ pairs were utilized in the next saltatory replication, giving a 116 bp repeating unit ($\alpha_c\beta_c\alpha_d\beta_d$). After further mutations, two of these adjacent units were amplified to give the present satellite.

The average sequence of the monomeric fragment of the mouse satellite DNA explains its properties. The longest repeating unit of 234 bp is identified by the restriction cleavage. The unit of reassociation between single strands of denatured satellite DNA is probably the 117 bp half-repeat, because the 234 bp fragments can anneal both in register and in half-register (in the latter case, the first half-repeat of one strand renatures with the second half-repeat of the other). In the oligonucleotide digest, the most common fragments, accounting for 4% of the total amount of DNA, are GA_5TGA, GA_4TGA, and GA_4CTGA, all of which can be found in the 234 bp unit and are related to the proposed ancestral units.

So far, we have treated the present satellite as though it consisted of identical copies of the 234 bp repeating unit. Although this unit accounts for the majority of the satellite, variants of it also are present. Some of them are scattered at random throughout the satellite; others are clustered.

The existence of variants is implied by our description of the starting material for the sequence analysis as the 'monomeric' fragment. When the satellite is digested by an enzyme that has one cleavage site in the 234 bp sequence, it also generates dimers, trimers, and tetramers relative to the 234 bp length. They arise when a repeating unit has lost the enzyme cleavage site as the result of mutation.

Figure 26.7

The evolution of mouse satellite DNA can be explained by an alternation of saltatory replications and accumulation of mutations.

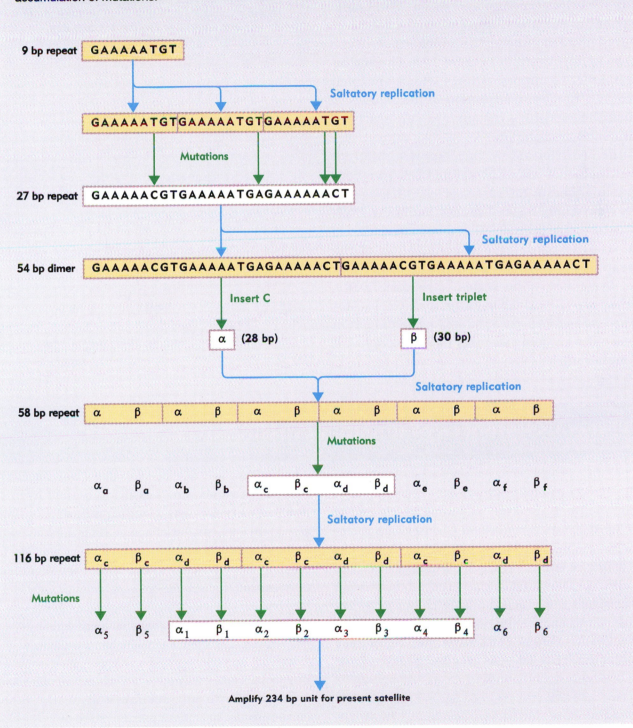

The monomeric 234 bp unit is generated when two adjacent repeats each have the recognition site. A dimer occurs when one unit has lost the site, a trimer is generated when two adjacent units have lost the site, and so on. With some restriction enzymes, most of the satellite is cleaved into a member of this repeating series, as shown in the example of **Figure 26.8**. The declining number of dimers, trimers, etc., shows that there is a random distribution of the repeats in which the enzyme's recognition site has been eliminated by mutation.

Other restriction enzymes show a different type of behavior with the satellite DNA. They continue to generate the same series of bands. But they cleave only a small proportion of the DNA, say 5–10%. This implies that a certain region of the satellite contains a concentration of the repeating units with this particular restriction site. Presumably the series of repeats in this domain are all derived from an ancestral variant that possessed this recognition site (although in the usual way, some members since have lost it by mutation).

Figure 26.8

Digestion of mouse satellite DNA with the restriction enzyme EcoRII identifies a series of repeating units (1, 2, 3) that are multimers of 234 bp and also a minor series (½, 1½, 2½) that includes half-repeats (see text later). The band at the far left is a fraction resistant to digestion.

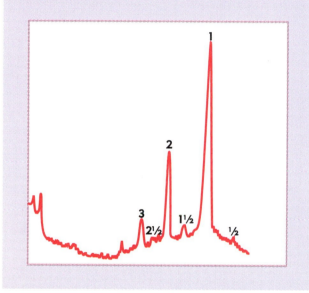

The consequences of unequal crossing-over

In a region of nonrepetitive DNA, recombination occurs between precisely matching points on the two homologous chromosomes, generating reciprocal recombinants. The basis for this precision is the ability of two duplex DNA sequences to align exactly. We know that unequal recombination can occur when there are multiple copies of genes whose exons are related, even though their flanking and intervening sequences may differ (see Chapter 24). This happens because of the mispairing between corresponding exons in *nonallelic* genes.

Imagine how much more frequently misalignment must occur in a tandem cluster of identical or nearly identical repeats. Except at the very ends of the cluster, the close relationship between successive repeats makes it impossible even to define the exactly corresponding repeats! Consider a sequence consisting of a repeating unit 'ab' with ends 'x' and 'y'. If we represent one chromosome in black and the other in color, the exact alignment between 'allelic' sequences would be

xababababababababababababababababy
xababababababababababababababababy

But probably *any* sequence 'ab' in one chromosome could pair with *any* sequence ab in the other chromosome. In a misalignment such as

xababababababababababababababababy
axababababababababababababababababy

the region of pairing is no less stable than in the perfectly aligned pair, although it is shorter. We do not know very much about how pairing is initiated prior to recombination, but very likely it starts between short corresponding regions and then spreads. If it starts within satellite DNA, it is more likely than not to involve repeating units that do not have exactly corresponding locations in their clusters.

Now suppose that a recombination event occurs within the unevenly paired region. The recombinants will have different numbers of repeating units. In one case, the cluster has become longer; in the other, it has become shorter,

xababababababababababababababababy
×
xababababababababababababababababababy
↓
xabababababababababababababababababababababababababy
+
xababababababababababababy

where '×' indicates the site of the crossover.

If this type of event is common, clusters of tandem repeats will undergo continual expansion and contraction. Unfortunately, we do not yet have much data on the extent to which satellite DNA clusters vary in size between different individual genomes in a species.

Unequal recombination has another consequence when there is internal repetition in the repeating unit. In the example above, the two clusters are misaligned with respect to the positions of the repeating units within each cluster, but they are aligned **in register**, as seen by the correspondence between individual 'ab' repeats and ab repeats.

But suppose that the 'a' and 'b' components of the repeating unit are themselves sufficiently well related to pair. Then the two clusters can align in **half-register**, with the 'a' sequence of one aligned with the 'b' sequence of the other. How frequently this occurs will depend on the closeness of the relationship between the two halves of the repeating unit. In mouse satellite DNA, reassociation be-

tween the denatured satellite DNA strands *in vitro* commonly occurs in the half-register.

When a recombination event occurs, it changes the length of the repeating units that are involved in the reaction.

xababababababababababababababy
×
xababababababababababababababy
↓
xabababababababababababababa ababababy
+
xababababababababababababb bababy

In the upper recombinant cluster, an 'ab' unit has been replaced by an 'aab' unit. In the lower cluster, the 'ab' unit has been replaced by a 'b' unit.

This type of event explains a feature of the restriction digest of mouse satellite DNA. In addition to the integral repeating series in Figure 26.8, there is a fainter series of bands at lengths of $\frac{1}{2}$, $1\frac{1}{2}$, $2\frac{1}{2}$, and $3\frac{1}{2}$ repeating units. Suppose that in the preceding example, 'ab' represents the 234 bp repeat of mouse satellite DNA, generated by cleavage at a site in the 'b' segment. The 'a' and 'b' segments correspond to the 117 bp half-repeats.

Then in the upper recombinant cluster, the 'aab' unit generates a fragment of $1\frac{1}{2}$ times the usual repeating length. And in the lower recombinant cluster, the 'b' unit generates a fragment of half of the usual length. (The multiple fragments in the half-repeat series are generated in the same way as longer fragments in the integral series, when some repeating units have lost the restriction site by mutation.)

Turning the argument the other way around, the identification of the half-repeat series on the gel shows us that the 234 bp repeating unit consists of two half-repeats well enough related to pair sometimes for recombination. Also visible in Figure 26.8 are some rather faint bands corresponding to $\frac{1}{4}$- and $\frac{3}{4}$-spacings. These will be generated in the same way as the $\frac{1}{2}$-spacings, when recombination occurs between clusters aligned in a quarter-register. The decreased relationship between quarter-repeats compared with half-repeats explains the reduction in frequency of the $\frac{1}{4}$- and $\frac{3}{4}$-bands compared with the $\frac{1}{2}$-bands.

Crossover fixation could maintain identical repeats

A general presumption about satellite DNA is that its sequence is not under high selective pressure (if indeed under any at all). Unlike a sequence that codes for protein, where mutations inactivate the

---

Figure 26.9

Unequal recombination allows one particular repeating unit to occupy the entire cluster. The numbers indicate the length of the repeating unit at each stage.

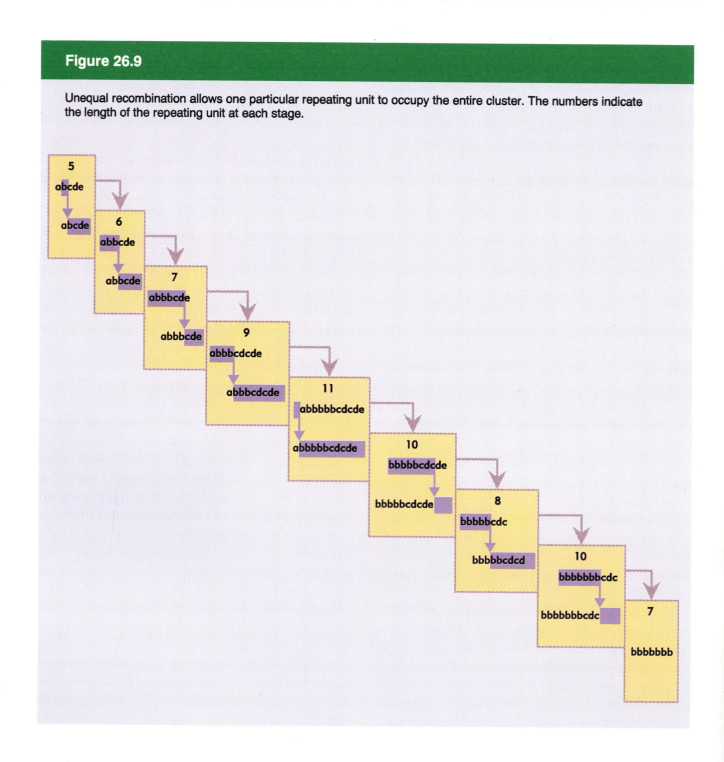

product, satellite DNA seems likely to serve any functions it has by its presence (instead of by virtue of its exact sequence).

This idea fits perfectly well with the structure of satellites whose repeating units are related rather than identical. Mutations have accumulated since the satellite was last rendered uniform in sequence. But how are we to explain the presence in arthropods of satellites most of whose repeating units remain identical? Even if the sequence were important, there still remains the difficulty of how selection could be imposed on so many copies.

Unequal recombination has been proposed as an alternative to saltatory replication to explain the evolution of satellite DNA. The basic idea is that unequal crossing-over occurs frequently at random sites. A series of random events cause one repeating unit to take over the entire satellite. This process is called **crossover fixation**.

The spread of a particular repeating unit through the satellite is illustrated in **Figure 26.9**. Suppose that a satellite consists initially of a sequence *abcde*, where each letter represents a repeating unit. The different repeating units are closely enough related to one another to mispair for recombination. Then by a series of unequal recombination events, the size of the repetitive region increases or decreases, and also one unit spreads to replace all the others.

We may wonder whether the existence of domains that contain particular repeating units (as in mouse satellite DNA) could represent a partial spreading of the unit from its origin. For example, we can see that in the intermediate stages of the spreading illustrated in Figure 26.9, there are clusters with a domain consisting of a series of 'b' variants.

The crossover fixation model actually predicts that *any sequence of DNA that is not under selective pressure will be taken over by a series of identical tandem repeats generated in this way*. The critical assumption is that the process of crossover fixation is fairly rapid relative to mutation, so that new mutations either are eliminated (their repeats are lost) or come to take over the entire cluster.

Minisatellites are useful for genetic mapping

Sequences that resemble satellites in consisting of tandem repeats of a short unit, but that overall are much shorter, consisting of (for example) from 5 to 50 repeats, are common in mammalian genomes. They were discovered by chance as fragments whose size is extremely variable in genomic libraries of human DNA. The variability is seen when a population contains fragments of many different sizes that represent the same genomic region; when individuals are examined, it turns out that there is extensive polymorphism, and that many different alleles can be found.

These sequences are called **minisatellite** or **VNTR** (variable number tandem repeat) regions.

The cause of the variation is that individual alleles have different numbers of the repeating unit. For example, one such minisatellite has a repeat length of 64 bp, and is found in the population with the following distribution:

7%	18 repeats
11%	16 repeats
43%	14 repeats
36%	13 repeats
4%	10 repeats

The cause of this variation is genetic recombination between misaligned repeat units, in the same way that we have discussed already for satellite

Figure 26.10

Alleles may differ in the number of repeats at a minisatellite locus, so that cleavage on either side generates restriction fragments that differ in length. By using a minisatellite with alleles that differ between parents, the pattern of inheritance can be followed.

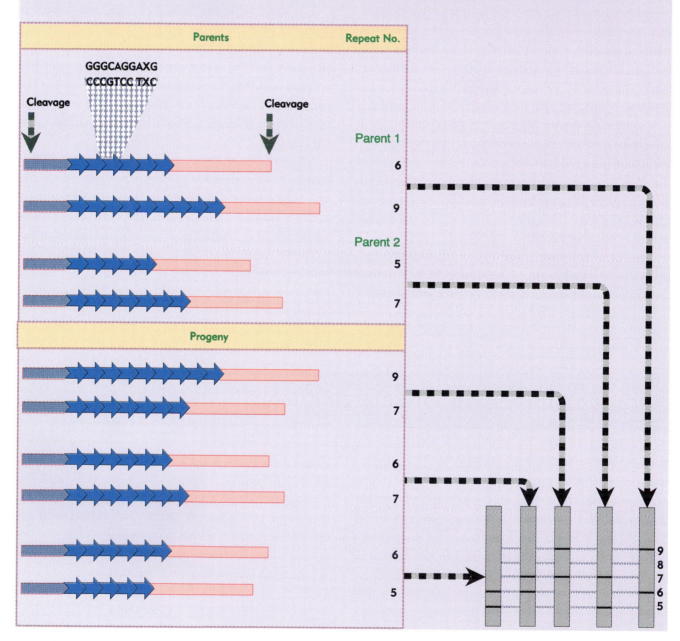

DNA. The rate of genetic exchange at minisatellite sequences is high, ~10⁻⁴ per kb of DNA. (The frequency of exchanges per actual locus is assumed to be proportional to the length of the minisatellite.) This rate is ~10× greater than the rate of homologous recombination at meiosis, that is, in any random DNA sequence. So minisatellites may be hotspots for meiotic recombination.

Sometimes the presence of a minisatellite is correlated with a high rate of exchange in the vicinity, but in some cases the recombination event occurs between sister chromatids (in which case it changes the length of the minisatellite, but has no effect on flanking markers, because these are identical on both recombining molecules of DNA).

The high variability in minisatellites makes them especially useful for genomic mapping, because there is a high probability that individuals will vary in their alleles at such a locus. An example of mapping by minisatellites is illustrated in **Figure 26.10**. This shows an extreme case in which two individuals both are heterozygous at a minisatellite locus, and in fact all four alleles are different. All progeny gain one allele from each parent in the usual way, and it is possible unambiguously to determine the source of every allele in the progeny. In the terminology of human genetics, the meioses described in this figure are highly informative, because of the variation between alleles.

A family of minisatellites has been found in the human genome that share a common 'core' sequence. The core is a GC-rich sequence of 10–15 bp, showing an asymmetry of purine/pyrimidine distribution on the two strands. Each individual minisatellite has a variant of the core sequence, but ~1000 minisatellites can be detected on Southern blot by a probe consisting of the core sequence.

Consider the situation shown in Figure 26.10, but multiplied 1000×. The effect of the variation at individual loci is to create a unique pattern for every individual. This makes it possible to assign heredity unambiguously between parents and progeny, by showing that 50% of the bands in any individual are derived from a particular parent. This is the basis of the technique known as **DNA fingerprinting**.

Summary

Satellite DNA consists of very short sequences repeated many times in tandem. Its distinct centrifugation properties reflect its biased base composition. Satellite DNA is concentrated in constitutive heterochromatin, but its function (if any) is unknown. The individual repeating units of arthropod satellites are identical. Those of mammalian satellites are related, and can be organized into a hierarchy reflecting the evolution of the satellite by the amplification and divergence of randomly chosen sequences. Unequal crossing-over appears to have been a major determinant of satellite DNA organization. Crossover fixation explains the ability of variants to spread through a cluster. Minisatellites have properties similar to satellites, but are much smaller; they are useful for human genomic mapping.

Further reading

Reviews
The implications of unequal crossing-over were developed by **Smith** (*Science* **191**, 525–535, 1976).

Discoveries
The idea of saltatory replication was introduced by **Southern** (*J. Mol. Biol.* **94**, 51–70, 1975).

The variability of minisatellites was established by **Jeffreys, Wilson, and Thein** (*Nature* **314**, 67–73, 1985).

CHAPTER 27

The genome is packaged into chromosomes

A general principle is evident in the organization of all cellular genetic material. It exists as a compact mass, occupying a limited volume; and its various activities, such as replication and transcription, must be accomplished within these confines. The organization of this material must accommodate transitions between inactive and active states.

The condensed state of nucleic acid results from its binding to basic proteins. The positive charges of these proteins neutralize the negative charges of the nucleic acid. The structure of the nucleoprotein complex is determined by the interactions of the proteins with the DNA (or RNA).

A common problem is presented by the packaging of DNA into phages and viruses, into bacterial cells and eukaryotic nuclei. The length of the DNA as an extended molecule would vastly exceed the dimensions of the compartment that contains it. The DNA (or in the case of some viruses, the RNA) must be compressed exceedingly tightly to fit into the space available. *Thus in contrast with the customary picture of DNA as an extended double helix, structural deformation of DNA to bend or fold it into a more compact form is the rule rather than exception.*

The magnitude of the discrepancy between the length of the nucleic acid and the size of its compartment is evident from the examples summarized in **Table 27.1**. For bacteriophages and for eukaryotic viruses, whether rod-like or spherical, the nucleic acid genome, whether DNA or RNA, whether single-stranded or double-stranded, effectively fills the container.

For bacteria or for eukaryotic cell compartments, the discrepancy is hard to calculate exactly, because the DNA is contained in a compact area that occupies only part of the compartment. The genetic material is seen in the form of the **nucleoid** in bacteria and as the mass of **chromatin** in eukaryotic nuclei at interphase (between divisions).

The density of DNA in these compartments is high. In a bacterium it is ~10 mg/ml, in a eukaryotic nucleus it is ~100 mg/ml, and in the phage T4 head it is >500 mg/ml. Such a concentration in solution would be equivalent to a gel of great viscosity. We do not entirely understand the physiological implications, for example, what effect this has upon the ability of proteins to find their binding sites on DNA.

The packaging of chromatin is flexible; it changes during the eukaryotic cell cycle. At the time of division (mitosis or meiosis), the genetic material becomes even more tightly packaged, and individual **chromosomes** become recognizable.

The overall compression of the DNA can be described by the **packing ratio**, the length of the DNA divided by the length of the unit that contains it. For example, the smallest human chromosome contains ~4.6×10^7 bp of DNA (~10 times the genome size of the bacterium *E. coli*). This is equivalent to 14,000 μm (= 1.4 cm) of extended DNA. At the most condensed moment of mitosis, the

Table 27.1

The length of nucleic acid is much greater than the dimensions of the surrounding compartment.

Compartment	Shape	Dimensions	Type of Nucleic Acid	Length		
TMV	Filament	0.008 x 0.3 µm	1 single-stranded RNA	2µm	=	6.4 kb
Phage fd	Filament	0.006 x 0.85 µm	1 single-stranded DNA	2µm	=	6 kb
Adenovirus	Icosahedron	0.07µm diameter	1 double-stranded DNA	11µm	=	35 kb
Phage T4	Icosahedron	0.065 x 0.10 µm	1 double-stranded DNA	55µm	=	170 kb
E. coli	Cylinder	1.7 x 0.65 µm	1 double-stranded DNA	1.3mm	=	4.2×10^3 kb
Mitochondrion (human)	Oblate Spheroid	3.0 x 0.5 µm	~10 identical double-stranded DNAs	50µm	=	16 kb
Nucleus (human)	Spheroid	6µm diameter	46 chromosomes of double-stranded DNA	1.8m	=	6×10^6 kb

chromosome is ~2 μm long. Thus the packing ratio of DNA in the chromosome can be as great as 7000.

Packing ratios cannot be established with such certitude for the more amorphous overall structures of the bacterial nucleoid or eukaryotic chromatin. However, the usual reckoning is that mitotic chromosomes are likely to be 5–10 times more tightly packaged than interphase chromatin, which therefore has a typical packing ratio of 1000–2000.

A major unanswered question concerns the *specificity* of packaging. Is the DNA folded into a *particular* pattern, or is it different in each individual copy of the genome? How does the pattern of packaging change when a segment of DNA is replicated or transcribed?

Condensing viral genomes into their coats

From the perspective of packaging the *individual* sequence, there is an important difference between a cellular genome and a virus. The cellular genome is essentially indefinite in size; the number and location of individual sequences can be changed by duplication, deletion, and rearrangement. Thus it requires a *generalized* method for packaging its DNA, insensitive to the total content or distribution of sequences. By contrast, two restrictions define the needs of a virus. The amount of nucleic acid to be packaged is *predetermined* by the size of the genome. And it must all fit within a coat assembled from a protein or proteins coded by the viral genes.

A virus particle is deceptively simple in its superficial appearance. The nucleic acid genome is contained within a **capsid**, a symmetrical or quasi-symmetrical structure assembled from one or only a few proteins. Attached to the capsid or incorporated into it, are other structures, assembled from distinct proteins, and necessary for infection of the host cell.

The virus particle is tightly constructed. The

internal volume of the capsid is rarely much greater than the volume of the nucleic acid it must hold. The difference is usually less than twofold, and often the internal volume is barely larger than the nucleic acid.

In its most extreme form, the restriction that the capsid must be assembled from proteins coded by the virus means that the entire shell is constructed from a single type of subunit. The rules for assembly of identical subunits into closed structures restrict the capsid to one of two types. The protein subunits stack sequentially in a helical array to form a **filamentous** or rodlike shape. Or they form a pseudospherical shell, a type of structure that conforms to a polyhedron with **icosahedral symmetry**. Some viral capsids are assembled from more than a single type of protein subunit, but although this extends the exact types of structures that can be formed, viral capsids still all conform to the general classes of quasi-crystalline filaments or icosahedrons.

There are two types of solution to the problem of how to construct a capsid that contains nucleic acid:

◆ The protein shell can be assembled around the nucleic acid, condensing the DNA or RNA by protein–nucleic acid interactions during the process of assembly.

◆ Or the capsid can be constructed from its component(s) in the form of an empty shell, into which the nucleic acid must be inserted, being condensed as it enters.

Assembly of the capsid around the genome occurs in the case of single-stranded RNA viruses. The best characterized example is TMV (tobacco mosaic virus). Assembly starts at a duplex hairpin that lies within the RNA sequence. From this **nucleation center**, it proceeds bi-directionally along the RNA, until reaching the ends. The unit of the capsid is a two-layer disk, each layer containing 17 identical protein subunits. The disk is a circular structure, which forms a helix as it interacts with the RNA. The RNA becomes coiled in a helical array on the inside

of the protein shell, as illustrated in **Figure 27.1**.

The structure of the RNA genome is determined by its interaction with the protein shell. An analogous arrangement is presented by spherical capsids that contain single-stranded RNA, for example, TYMV (turnip yellow mosaic virus). The common principle is that *the position of the RNA within the capsid is determined directly by its binding to the proteins of the shell.*

The spherical capsids of DNA viruses are assembled in a different way, as best characterized for the phages lambda and T4. In each case, an empty head shell is assembled from a small set of proteins. *Then the duplex genome is inserted into the head,* a process accompanied by a structural change in the capsid.

Figure 27.2 summarizes the assembly of lambda. It starts with a small head shell that contains a

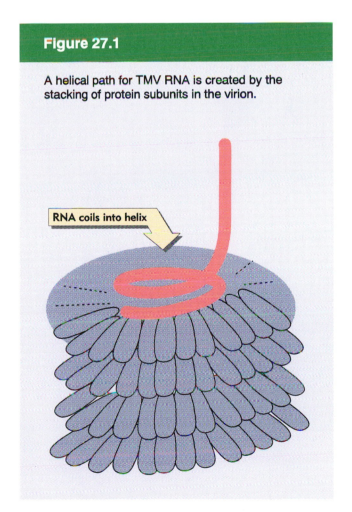

Figure 27.1

A helical path for TMV RNA is created by the stacking of protein subunits in the virion.

RNA coils into helix

Figure 27.2

Maturation of phage lambda passes through several stages. The empty head changes shape and expands when it becomes fiilled with DNA. The electron micrographs show the particles at the start and end of the maturation pathway. Photographs kindly provided by A. F. Howatson.

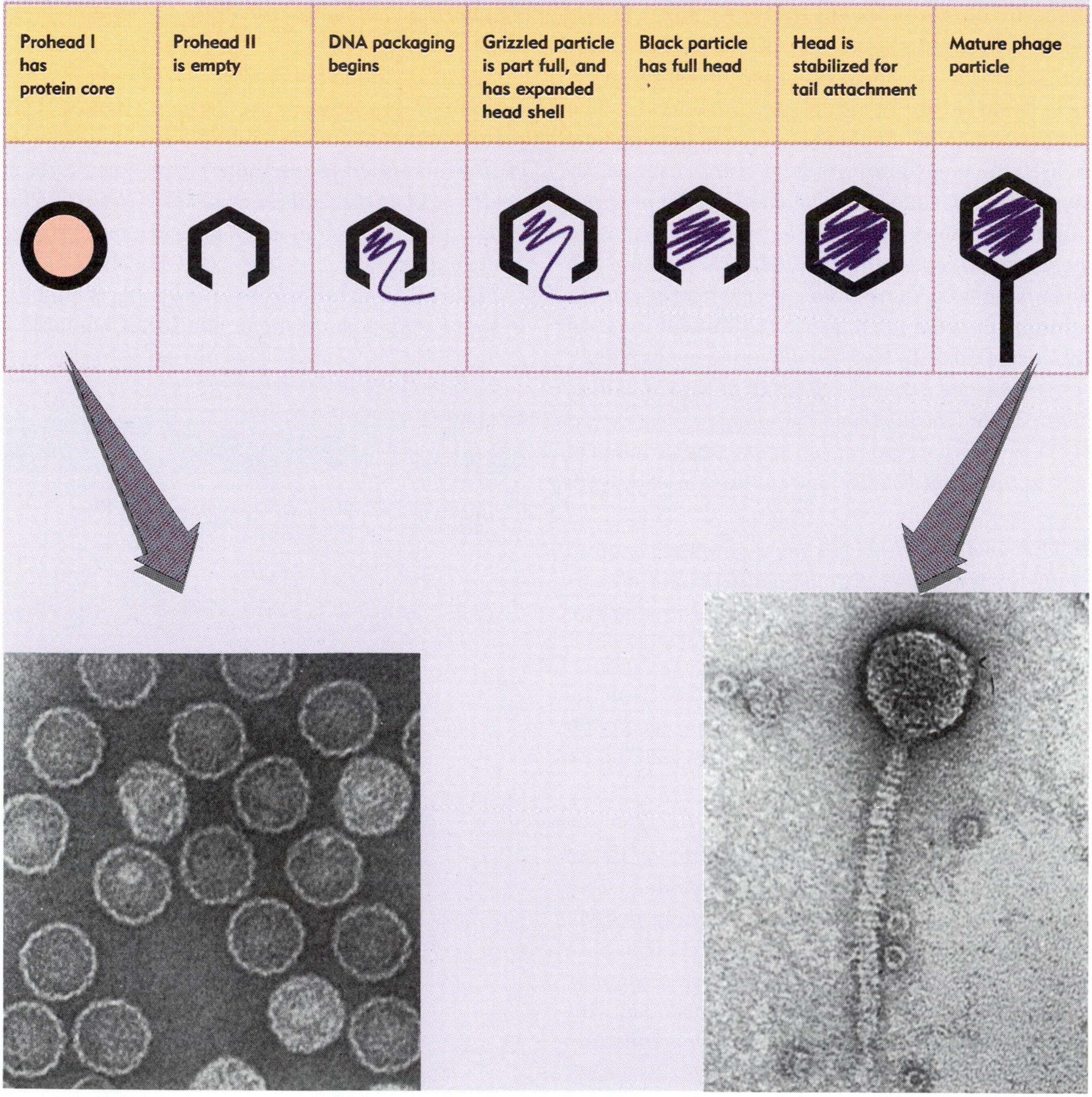

Prohead I has protein core	Prohead II is empty	DNA packaging begins	Grizzled particle is part full, and has expanded head shell	Black particle has full head	Head is stabilized for tail attachment	Mature phage particle

protein 'core'. This is converted to an empty head shell of more distinct shape. Then DNA packaging begins, the head shell expands in size though remaining the same shape, and finally the full head is sealed by addition of the tail.

For both lambda and T4, the DNA that is to be inserted into the empty head takes the form of **concatemeric** molecules. These are multiple genomes joined end to end, as depicted in **Figure 27.3**. Each phage has its own mechanism for recognizing the proper amount of DNA to insert.

The ends of the lambda genome are marked by sequences called *cos* sites. Cleavage occurs at the left *cos* site (as defined on the usual map) to generate a free end that is inserted into the capsid. The insertion of DNA continues until the right *cos* site is encountered, when it is cleaved to generate the other end. The end that goes into the capsid last during assembly comes out first when a new host cell is infected.

Any DNA contained between two *cos* sites can be packaged. (This is the basis of the 'cosmid' cloning technique described in Chapter 21.) Although the sequence of DNA is irrelevant, its length is important: the distance between the *cos* sites can be varied only slightly from the usual length of lambda DNA. Packaging does not occur if the distance is either too great or too small. This demonstrates that there must be *enough* DNA to complete the packaging reaction, as well as showing that there is room in the head only for a very little extra DNA (~15%).

With phage T4, insertion starts at a *random* point in the concatemeric precursor. It continues until a genome's worth of DNA (a 'headful') has been inserted. This implies the existence of some mechanism for measuring the amount of DNA. (Actually, the amount inserted is slightly greater than the length of the unit genome, creating a **terminal redundancy** corresponding to the additional length. In the terms of Figure 27.3, the first virion might contain the DNA from *A* to *A*, the next from *B* to *B*, and so on, so that each genome has a [different] letter repeated at each end.)

Now a double-stranded DNA considered over short distances is a fairly rigid rod. Yet it must be compressed into a compact structure to fit within the capsid. We should like to know whether packaging involves a smooth coiling of the DNA into the head or requires abrupt bends.

Inserting DNA into a phage head involves two types of reaction: translocation and condensation. Both are energetically unfavorable.

Translocation appears to be an active process in

Figure 27.3

Concatemeric DNA consists of a tandem series of phage genomes.

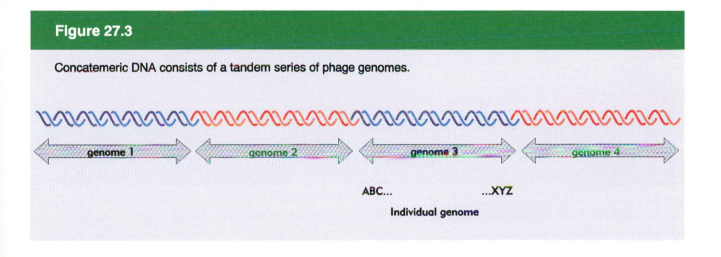

genome 1   genome 2   genome 3   genome 4

ABC...   ...XYZ

Individual genome

which the DNA is driven into the head by an ATP-dependent mechanism. One possibility is that the terminase enzymes that generate the ends of DNA could be involved in pushing it into the head. Another possibility is that the capsid protein(s) could pull the DNA in.

Little is known about the mechanism of condensation, except that the capsid contains 'internal proteins' as well as DNA. One possibility is that they provide some sort of 'scaffolding' onto which the DNA condenses. (This would be a counterpart to the use of the proteins of the shell in the plant RNA viruses.)

How specific is the packaging? It cannot depend on particular sequences, because deletions, insertions, and substitutions all fail to interfere with the assembly process. The relationship between DNA and the head shell has been investigated directly by determining which regions of the DNA can be chemically cross-linked to the proteins of the capsid. The surprising answer is that all regions of the DNA are more or less equally susceptible. This probably means that when DNA is inserted into the head, it follows a general rule for condensing, but the pattern is not determined by particular sequences.

These varying mechanisms of virus assembly all accomplish the same end: packaging a single DNA or RNA molecule into the capsid. However, some viruses have genomes that consist of multiple nucleic acid molecules. Reovirus contains ten double-stranded RNA segments, all of which must be packaged into the capsid. Specific sorting sequences in the segments may be required to ensure that the assembly process selects one copy of each different molecule in order to collect a complete set of genetic information.

Some plant viruses are multipartite: their genomes consist of segments each of which is packaged into a *different* capsid. An example is alfalfa mosaic virus, which has four different single-stranded RNAs, each packaged independently into a coat comprising the same protein subunit. A successful infection depends on the entry of one of each type into the cell.

The four components of the virus exist as particles of different sizes. This means that the same capsid protein can package each RNA into its own characteristic particle. This is a departure from the packaging of a unique length of nucleic acid into a capsid of fixed shape.

The assembly pathway of viruses whose capsids have only one authentic form may be diverted by mutations that cause the formation of aberrant **monster** particles in which the head is longer than usual. These mutations show that a capsid protein(s) has an intrinsic ability to assemble into a particular type of structure, but the exact size and shape may vary. Some of the mutations occur in genes that code for **assembly proteins**, which are needed for head formation, but are not themselves part of the head shell. Such ancillary proteins limit the options of the capsid protein so that it assembles only along the desired pathway. Comparable proteins are employed in the assembly of cellular chromatin (see Chapter 28).

The bacterial genome is a nucleoid with many supercoiled loops

Although bacteria do not display structures with the distinct morphological features of eukaryotic chromosomes, their genomes nonetheless are organized into definite bodies. The genetic material can be seen as a fairly compact clump or series of clumps that occupies about a third of the volume of the cell. **Figure 27.4** displays a thin section through a bacterium in which this **nucleoid** is evident.

In bacteria that have partially replicated their DNA, the nucleoid contains more than one

Figure 27.4

A thin section shows the bacterial nucleoid as a compact mass in the center of the cell. Photograph kindly provided by Jack Griffith.

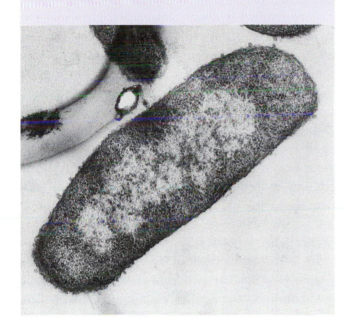

genome's worth of DNA. By the time of cell division, the material has separated into two nucleoids that are partitioned into the daughter cells. A bacterium does not organize a spindle like a eukaryotic cell; the segregation mechanism probably involves attachment of the bacterial genome to the envelope, allowing segregation to occur by active movement or passive separation (see Chapter 18). At all events, when division occurs each nucleoid is in a different compartment.

When *E. coli* cells are lysed, fibers are released in the form of loops attached to the broken envelope of the cell. As can be seen from **Figure 27.5**, the DNA of these loops is not found in the extended form of a free duplex, but is compacted by association with proteins.

Several DNA-binding proteins with a superficial resemblance to eukaryotic chromosomal proteins have been isolated in *E. coli*. What criteria should we apply for deciding whether a DNA-binding protein plays a structural role in the nucleoid? It should be present in sufficient quantities to bind throughout the genome. And mutations in its gene should cause some disruption of structure or of functions associated with genome survival (for example, segregation to daughter cells). None of the candidate proteins yet satisfies the genetic conditions.

Protein HU is a dimer that condenses DNA, possibly wrapping it into a bead-like structure. It stimulates DNA replication (see Chapter 19). It is related to IHF (integration host factor), another dimer, which is involved in some specialized recombination reactions, including the integration and excision of phage lambda (for which it is named), and has a structural role in building a protein complex that holds reacting DNA sequences in apposition (see Chapter 33). Null mutations in either of the genes coding for the subunits of HU (*hupA,B*) have little effect, but loss of both functions causes a cold-sensitive phenotype and some loss of superhelicity in DNA. These results raise the possibility that HU plays a role in nucleoid condensation that is a non-sequence specific counterpart to the specific role played by IHF in lambda recombination.

Protein H1 (also known as H-NS) binds DNA, interacting preferentially with sequences that are bent. Mutations in its gene have turned up in a variety of guises (*osmZ, bglY, pilG*), each identified as an apparent regulator of a different system. These results probably reflect the effect that H1 has on the local topology of DNA, with effects upon gene expression that depend upon the particular promoter.

Protein P has been identified by sequencing a gene whose coding region has an amino acid composition resembling the protamines that bind to DNA in certain sperm. Its sequence suggests it is a DNA-binding protein, but its quantity and functions are unknown.

We might expect that the absence of a protein required for nucleoid structure would have serious effects upon viability. Why then are the effects of deletions in the genes for proteins HU and H1 only relatively restricted? One explanation is that these proteins are *redundant*, that any one can substitute for the others, so that deletions of *all* of them would

Figure 27.5

The nucleoid spills out of a lysed *E. coli* cell in the form of loops of a fiber. Photograph kindly provided by Jack Griffith.

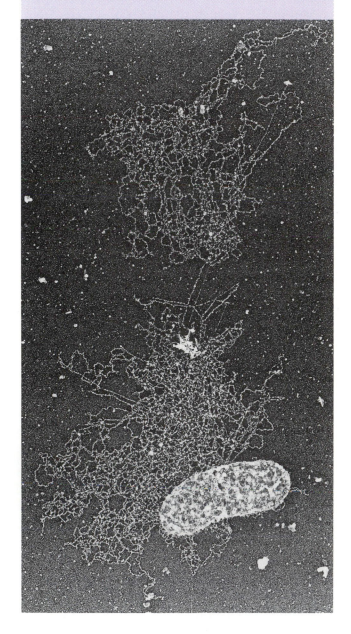

of a very rapidly sedimenting complex, consisting of ~80% DNA by mass. (The analogous complexes in eukaryotes have ~50% DNA by mass; see later.) It can be unfolded by treatment with reagents that act on RNA or protein. The possible role of proteins in stabilizing its structure is evident. The role of RNA has been quite refractory to analysis, and we do not understand it.

The DNA of the compact body isolated *in vitro* behaves as a closed duplex structure, as judged by its response to ethidium bromide. This small molecule intercalates between base pairs to generate *positive* superhelical turns in 'closed' circular DNA molecules, that is, molecules in which both strands have covalent integrity. (In 'open' circular molecules, which contain a nick in one strand, or with linear molecules, the DNA can rotate freely in response to the intercalation, thus relieving the tension.)

In a natural closed DNA that is *negatively* supercoiled, the intercalation of ethidium bromide first removes the negative supercoils and then introduces positive supercoils. The amount of ethidium bromide needed to achieve zero supercoiling is a measure of the original density of negative supercoils.

Some nicks occur in the compact nucleoid during its isolation; they can also be generated by limited treatment with DNAase. But this does not abolish the ability of ethidium bromide to introduce positive supercoils. This capacity of the genome to retain its response to ethidium bromide in the face of nicking means that it must have many independent **domains**; the supercoiling in each domain is not affected by events in the other domains.

This autonomy suggests that the structure of the bacterial chromosome has the general organization depicted diagrammatically in **Figure 27.6**. Each domain consists of a loop of DNA, the ends of which are secured in some (unknown) way that does not allow rotational events to propagate from one domain to another. There are ~100 such domains per genome; each consists of ~40 kb (13 μm) of DNA, organized into some more compact fiber whose structure has yet to be characterized.

The existence of separate domains could permit

be necessary to interfere seriously with nucleoid structure. Another possibility is that we have yet to identify the proteins responsible for the major features of nucleoid integrity.

The nucleoid can be isolated directly in the form

different degrees of supercoiling to be maintained in different regions of the genome. This is a pertinent factor in considering the different susceptibilities of particular bacterial promoters to supercoiling (see Chapter 14).

Supercoiling in the genome can in principle take two forms:

◆ If a supercoiled DNA is free, its path is **unrestrained**, and negative supercoils generate a state of torsional tension that is transmitted freely along the DNA within a domain. It can be relieved by unwinding the double helix, as described in Chapter 5. The DNA is in a dynamic equilibrium between the states of tension and unwinding.

◆ Supercoiling can be **restrained** if proteins are bound to the DNA to hold it in a particular three-dimensional configuration. In this case, the supercoils are represented by the path the DNA follows in its fixed association with the proteins. The energy of interaction between the proteins and the supercoiled DNA stabilizes the nucleic acid, so that no tension is transmitted along the molecule.

Are the supercoils in *E. coli* DNA restrained *in vivo* or is the double helix subject to the torsional tension characteristic of free DNA? Measurements of supercoiling *in vitro* encounter the difficulty that restraining proteins may have been lost during isolation. Various approaches suggest that DNA is under torsional stress *in vivo*, although it is difficult to quantitate the level of supercoiling.

A direct approach is to use the cross-linking reagent psoralen, which binds more readily to DNA when it is under torsional tension. The reaction of psoralen with *E. coli* DNA *in vivo* corresponds to an average density of one negative superhelical turn for every 200 bp ($\sigma = -0.05$).

Another approach is to examine the ability of cells to form alternative DNA structures; for example, to generate cruciforms at palindromic sequences. From the change in linking number that is required to drive such reactions, it is possible to calculate the original supercoiling

density. This approach suggests an average density of $\sigma = -0.025$, or one negative superhelical turn per 100 bp.

These results therefore demonstrate that supercoils *do* create torsional tension *in vivo*. There may be variation about an average level, and although the precise range of densities is difficult to measure, it is clear that the level is sufficient to exert significant effects on DNA structure, for example, in assisting melting in particular regions such as origins or promoters.

Many of the important features of the structure

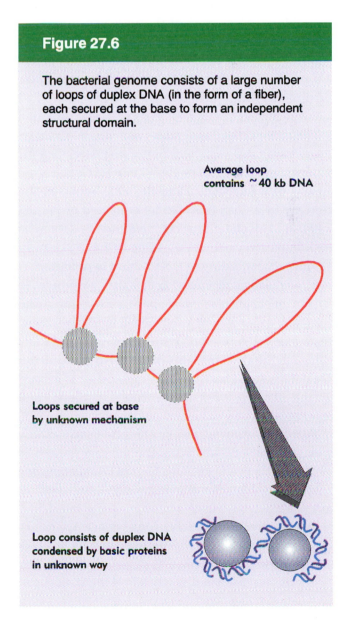

Figure 27.6

The bacterial genome consists of a large number of loops of duplex DNA (in the form of a fiber), each secured at the base to form an independent structural domain.

Average loop contains ~ 40 kb DNA

Loops secured at base by unknown mechanism

Loop consists of duplex DNA condensed by basic proteins in unknown way

of the compact nucleoid remain to be established. What is the specificity with which domains are constructed—do the same sequences always lie at the same relative locations, or can the contents of individual domains shift? How is the integrity of the domain maintained? Biochemical analysis by itself is unable to answer these questions fully, but if it is possible to devise suitable selective techniques, the properties of structural mutants should lead to a molecular analysis of nucleoid construction.

Loops, domains, and scaffolds in eukaryotic DNA

Interphase chromatin appears to be a tangled mass occupying a large part of the nuclear volume, in contrast with the highly organized and reproducible ultrastructure of mitotic chromosomes. What controls the distribution of interphase chromatin within the nucleus?

Some indirect evidence on its nature is provided by the isolation of the genome as a single, compact body. Using the same technique just described for isolating the bacterial nucleoid, nuclei can be lysed on top of a sucrose gradient. This releases the genome in a form that can be collected by centrifugation. As isolated from *D. melanogaster*, it can be visualized as a compactly folded fiber (10 nm in diameter), consisting of DNA bound to proteins.

Supercoiling measured by the response to ethidium bromide corresponds to about one negative supercoil for every 200 bp. These supercoils can be removed by nicking with DNAase, although the DNA remains in the form of the 10 nm fiber. This suggests that the supercoiling is caused by the arrangement of the fiber in space, and represents the existing torsion.

Full relaxation of the supercoils requires one nick for every 85 kb. Thus the average length of 'closed' DNA is ~85 kb. This region could comprise a loop or domain similar in nature to those identified in the bacterial genome. We should like to know whether these loops correspond to specific sequences and whether they have functional significance.

Loops can be seen directly when the majority of proteins are extracted from mitotic chromosomes. The resulting complex consists of the DNA associated with ~8% of the original protein content. As seen in **Figure 27.7**, the protein-depleted chromosomes take the form of a central **scaffold** surrounded by a halo of DNA.

The metaphase scaffold consists of a dense network of fibers. Threads of DNA emanate from the scaffold, apparently as loops of average length 10–30 μm (30–90 kb). The DNA can be digested without affecting the integrity of the scaffold, which consists of a set of specific proteins. This suggests a form of organization in which loops of DNA of ~60 kb are anchored in a central proteinaceous scaffold.

The appearance of the scaffold resembles a mitotic pair of sister chromatids. The sister scaffolds usually are tightly connected, but sometimes are separate, joined only by a few fibers. Could this be the structure responsible for maintaining the shape of the mitotic chromosomes? Could it be generated by bringing together the protein components that usually secure the bases of loops in interphase chromatin?

Interphase cells possess a *nuclear matrix*, a filamentous structure on the interior of the nuclear membrane. Chromatin often appears to be attached to the matrix, and there have been many suggestions that such attachment is necessary for transcription or replication. When nuclei are depleted of proteins, the DNA extrudes as loops from the residual nuclear matrix.

Is DNA attached to the matrix or scaffold via

Figure 27.7

Histone-depleted chromosomes consist of a protein scaffold to which loops of DNA are anchored. Photograph kindly provided by Ulrich K. Laemmli.

specific sequences? DNA sites attached to proteinaceous structures in interphase nuclei are called **MAR** (matrix attachment regions); it is confusing that they are sometimes also called **SAR** (scaffold attachment regions), although they concern the nuclear matrix.

How might we demonstrate that particular DNA regions are genuinely associated with the matrix? *In vivo* and *in vitro* approaches are summarized in **Figure 27.8**. Both start by isolating the matrix as a crude nuclear preparation containing chromatin and nuclear proteins. Different treatments can then be used to characterize DNA

in the matrix or to identify DNA able to attach to it.

To analyze the existing MAR, the chromosomal loops can be decondensed by extracting the proteins. Removal of the DNA loops by treatment with restriction nucleases leaves only the (presumptive) *in vivo* MAR sequences attached to the matrix.

The complementary approach is to remove *all* the DNA from the matrix by treatment with DNAase; then isolated fragments of DNA can be tested for their ability to bind to the matrix *in vitro*.

The same sequences should be associated with the matrix *in vivo* or *in vitro*. Once a potential MAR

Figure 27.8

Matrix-associated regions may be identified by characterizing the DNA retained by the matrix isolated in vivo or by identifying the fragments that can bind to the matrix from which all DNA has been removed *in vivo*.

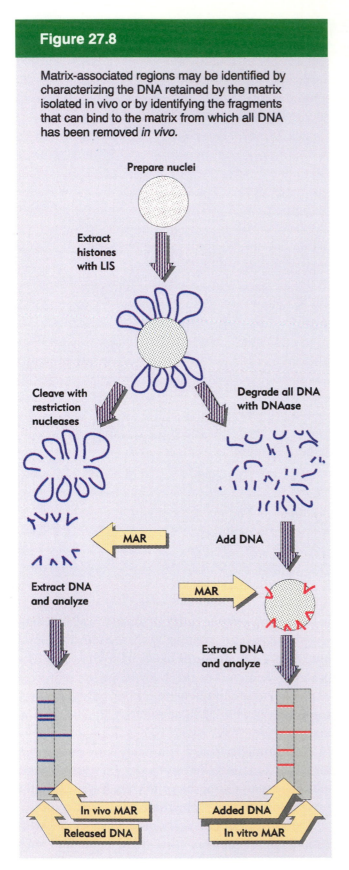

Prepare nuclei

Extract histones with LIS

Cleave with restriction nucleases

Degrade all DNA with DNAase

MAR

Add DNA

MAR

Extract DNA and analyze

Extract DNA and analyze

In vivo MAR

Added DNA

Released DNA

In vitro MAR

has been identified, the size of the minimal region needed for association *in vitro* can be determined by deletions. Point mutations of the MAR should prevent it from associating with the matrix. Specific matrix proteins should bind to the MAR, and it should in principle be possible to identify these proteins via their ability to recognize the specific DNA sequences of the MAR.

Several MAR sequences have been identified by the criteria of matrix binding *in vivo* or *in vitro*. We have not yet reached the stage of systematic mutation of the attachment sequences. Nor have MAR-binding proteins yet been characterized.

What is the relationship between the chromosome scaffold of dividing cells and the nuclear matrix of interphase cells; are the same DNA sequences attached to both structures? In several cases, the same DNA fragments that are found with the nuclear matrix *in vivo* can be retrieved from the metaphase scaffold. And fragments that contain MAR sequences can bind to a metaphase scaffold. It therefore seems likely that DNA contains a single type of attachment site, which in interphase cells is connected to the nuclear matrix, and in mitotic cells is connected to the chromosome scaffold.

The nuclear matrix and chromosome scaffold consist of different proteins, although there are some common components. In particular, topoisomerase II is a prominent component of the chromosome scaffold, and is a constituent of the nuclear matrix. We have yet to quantitate the proportion of the enzyme in the cell that is matrix- or scaffold-associated.

A surprising feature is the lack of conservation of sequence in MAR fragments. They are usually ~70% A•T-rich, but otherwise lack any consensus sequences. However, other interesting sequences often are in the DNA stretch containing the MAR. *Cis*-acting sites that regulate transcription are common. And a recognition site for topoisomerase II is usually present in the MAR. It is therefore possible that an MAR serves more than one function, providing a site for attachment to the matrix, but also containing other sites at which topological changes in DNA are effected.

The contrast between interphase chromatin and mitotic chromosomes

Individual eukaryotic chromosomes come into the limelight for only a brief period, during the act of cell division. Only then can each be seen as a compact unit. **Figure 27.9** is an electron micrograph of a sister chromatid pair, captured at metaphase. (The sister chromatids are daughter chromosomes produced by the previous replication event, still joined together at this stage of mitosis, as described in Chapter 2.) Each consists of a fiber with a diameter of ~30 nm and a nubbly appearance. The DNA is 5–10× more condensed in chromosomes than it is at interphase.

During most of the life cycle of the eukaryotic cell, however, its genetic material occupies an area of the nucleus in which individual chromosomes cannot be distinguished. The structure of the interphase chromatin does not change visibly between divisions. No disruption is evident during the period of replication, when the amount of chromatin doubles. Chromatin is fibrillar, although the overall configuration of the fiber in space is hard to discern in detail. The fiber itself, however, is similar or identical to that of the mitotic chromosomes.

Chromatin can be divided into two types of material, which can be seen in the nuclear section of **Figure 27.10**:

◆ In most regions, the fibers are much less densely packed than in the mitotic chromosome. This material is called **euchromatin**. It has a relatively dispersed appearance in the nucleus, and occupies most of the nuclear region in Figure 27.10.

◆ Some regions of chromatin are very densely packed with fibers, displaying a condition comparable to that of the chromosome at mitosis. This material is called **heterochromatin**. It passes through the cell cycle with relatively little change in its degree of

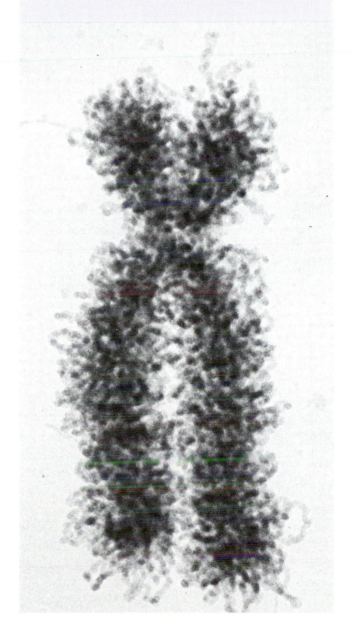

Figure 27.9

The sister chromatids of a mitotic pair each consist of a fiber (˜30 nm in diameter) compactly folded into the chromosome. Photograph kindly provided by E. J. DuPraw.

Figure 27.10

A thin section through a nucleus stained with Feulgen shows heterochromatin as compact regions clustered near the nucleolus and nuclear membrane. Photograph kindly provided by Edmund Puvion.

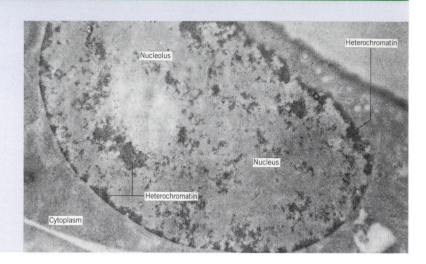

condensation. It forms a series of discrete clumps in Figure 27.10, but often the various heterochromatic regions aggregate into a densely staining **chromocenter**.

The same fibers run continuously between euchromatin and heterochromatin, which implies that these states represent different degrees of condensation of the genetic material. In the same way, euchromatic regions exist in different states of condensation during interphase and during mitosis. Thus the genetic material is organized in a manner that permits alternative states to be maintained side by side in chromatin, and allows cyclical changes to occur in the packaging of euchromatin between interphase and division.

The structural condition of the genetic material is correlated with its transcriptional activity: chromatin is not expressed in the more condensed state. Mitotic chromosomes provide an extreme case; they are transcriptionally inert, as cells virtually cease transcription during the process of division. Interphase cells contain two classes of heterochromatin, each containing a different type of sequence; and in neither type is the DNA transcribed:

◆ **Constitutive heterochromatin** consists of particular regions that are not expressed. They include short repeated sequences of DNA (see Chapter 26), and may play a structural role in the chromosome. Often these sequences are concentrated in characteristic regions (see below).

◆ **Facultative heterochromatin** takes the form of entire chromosomes that are inactive in one cell lineage, although they can be expressed in other lineages. The example *par excellence* is the mammalian X chromosome, one copy of which (selected at random) is entirely inactive in a given female. (This compensates for the presence of two X chromosomes, compared with the one present in males.) The inactive X chromosome is perpetuated in a heterochromatic state, while the active X chromosome is part of the euchromatin. Here it is possible to see a correlation between transcriptional activity and structural organization when the *identical DNA sequences* are involved in both states.

Condensation of the genetic material is thus associated with (perhaps is responsible for) its inactivity. Note, however, that the reverse is not true. Active genes are contained within euchromatin; but only a small minority of the sequences in euchromatin are transcribed at any time. Thus location in euchromatin is *necessary* for gene expression, but is not *sufficient* for it. We may wonder whether the gross changes seen between euchromatin and heterochromatin are mimicked in a lesser manner by changes in the structure of euchromatin, to give transcribed regions a less condensed structure than that of nontranscribed regions.

Because of the diffuse state of chromatin, we cannot directly determine the specificity of its organization. But we can ask whether the structure of the chromosome is ordered. Do particular sequences always lie at particular sites, or is the folding of the fiber into the overall structure a more random event?

At the level of the chromosome, each member of the complement has a different and reproducible ultrastructure. When subjected to certain treatments and then stained with the chemical dye Giemsa, chromosomes generate a series of **G-bands**. An example of the human set is presented in **Figure 27.11**.

Until the development of this technique, chromosomes could be distinguished only by their overall size and the relative location of the centromere (see later). Now each chromosome can

Figure 27.11

G-banding generates a characteristic lateral series of bands in each member of the chromosome set. Photograph kindly provided by Lisa Shaffer.

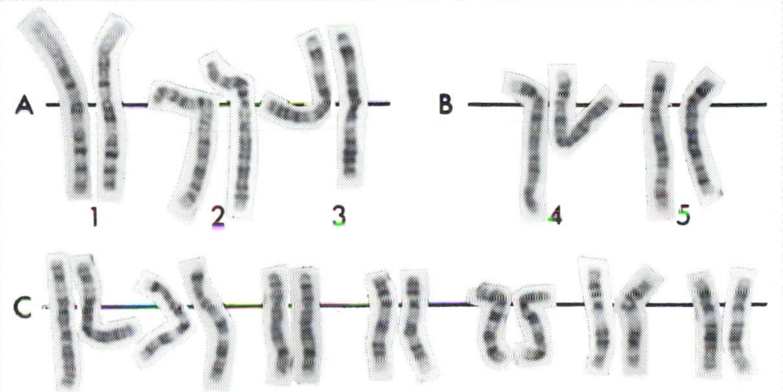

be identified by its characteristic banding pattern. This pattern is reproducible enough to allow translocations from one chromosome to another to be identified by comparison with the original diploid set. **Figure 27.12** shows a diagram of the bands of the human X chromosome.

The banding technique is of enormous practical use, but the mechanism of banding remains a mystery. All that is certain is that the dye stains untreated chromosomes more or less uniformly. So the generation of bands depends on a variety of treatments that change the response of the chromosome (presumably by extracting the component that binds the stain from the nonbanded regions). But the variety of effective treatments is so great that no common cause yet has been discerned. These results imply the existence of a definite long-range structure, but its basis is unknown.

Each chromosome contains a single, very long duplex of DNA. This explains why chromosome replication is semiconservative like the individual DNA molecule (see Figure 4.15); this would not necessarily be the case if a chromosome carried many independent molecules of DNA. The single duplex of DNA is folded into the 30 nm fiber, which runs continuously throughout the chromosome. Thus in accounting for interphase chromatin and mitotic chromosome structure, we have to explain the packaging of a single, exceedingly long molecule of DNA into a form in which it can be transcribed and replicated, and can become cyclically more and less compressed.

Figure 27.12

The human X chromosome can be divided into distinct regions by its banding pattern. The short arm is *p* and the long arm is *q*; each arm is divided into larger regions that are further subdivided. This map shows a low resolution structure; at higher resolution, some bands are further subdivided into smaller bands and interbands, e.g. p21 is divided into p21.1, p21.2, and p21.3. A high resolution map is shown in Figure 6.11.

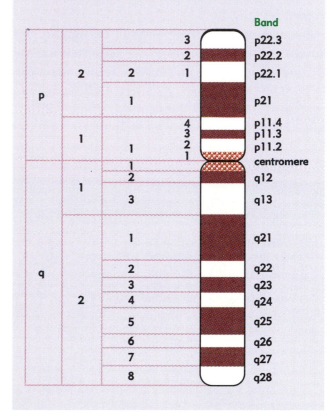

The extended state of lampbrush chromosomes

It would be extremely useful to visualize gene expression in its natural state, to see what structural changes are associated with transcription. But the nature of the material restricts such analysis to some unusual circumstances.

The compression of DNA in chromatin, coupled with the difficulty of identifying particular genes within it, makes it impossible to visualize the

Figure 27.13

A lampbrush chromosome is a meiotic bivalent in which the two pairs of sister chromatids are held together at chiasmata (indicated by arrows). Photograph kindly provided by Joe Gall.

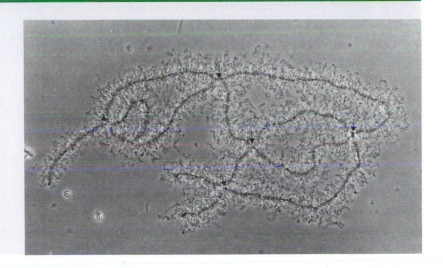

transcription of individual active genes. (However, they do display some distinctive biochemical properties that can be analyzed *in vitro*, as described in Chapter 28.)

Mitotic chromosomes are inert in gene expression, and, in any case, are so compact as to preclude the identification of individual loci. The distinct regions that are rendered discernible by the G-banding technique, as in the example of Figure 27.11, each contain ~10^7 bp of DNA, which could include many hundreds of genes.

Lateral differentiation of structure is evident in many chromosomes when they first appear for meiosis. At this stage, the chromosomes resemble a series of beads on a string. The beads are densely staining granules, properly known as **chromomeres**. However, usually there is little gene expression at meiosis, and it is not practical to use this material to identify the activities of individual genes.

Gene expression can be visualized directly in certain unusual situations, in which the chromosomes are found in a highly extended form that allows individual loci (or groups of loci) to be distinguished. One such situation is presented by **lampbrush chromosomes**, which have been best characterized in certain amphibians.

Lampbrush chromosomes are formed during an unusually extended meiosis, which can last up to several months! During this period, the chromosomes are held in a stretched-out form in which they can be visualized in the light microscope. Later during meiosis, the chromosomes revert to their usual compact size. So the extended state essentially proffers an unfolded version of the normal condition of the chromosome.

The lampbrush chromosomes are meiotic bivalents, each consisting of two pairs of sister chromatids. **Figure 27.13** shows an example in which the sister chromatid pairs have mostly separated so that they are held together only by chiasmata (the sites of crossing-over). Each sister chromatid pair forms a series of ellipsoidal chromomeres, ~1–2 μm in diameter, which are connected by a very fine thread. This thread contains the two sister duplexes of DNA and runs continuously along the chromosome, through the chromomeres.

The lengths of the individual lampbrush chromosomes in the newt *Notophthalmus viridescens* range from 400 to 800 μm, compared with the range of 15–20 μm seen later in meiosis. So the lampbrush chromosomes are ~30 times less tightly packed. The total length of the entire lampbrush chromosome set is 5–6 mm, organized into about 5000 chrommomeres.

The lampbrush chromosomes take their name from the lateral loops that extrude from the chromomeres at certain positions. (These resemble a lampbrush, an extinct object.) The loops extend in pairs, one from each sister chromatid. The loops are continuous with the axial thread, which suggests that they represent chromosomal material extruded from its more compact organization in the chromomere.

The loops are surrounded by a matrix of ribonucleoproteins. These contain nascent RNA chains. Often a transcription unit can be defined by the increase in the length of the RNP moving around the loop. An example is shown in **Figure 27.14**.

So the loop is an extruded segment of DNA that is being actively transcribed. In some cases, loops corresponding to particular genes have been identified. Then the structure of the transcribed gene, and the nature of the product, can be scrutinized *in situ*.

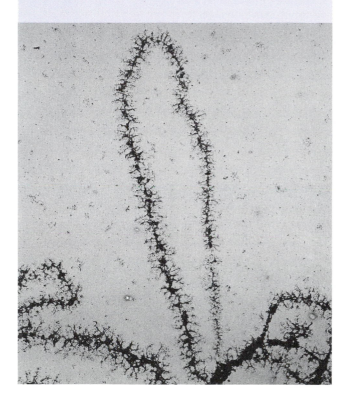

Figure 27.14

A lampbrush chromosome loop is surrounded by a matrix of ribonucleoprotein. Photograph kindly provided by Oscar Miller.

Transcription disrupts the structure of polytene chromosomes

The interphase nuclei of some tissues of the larvae of Dipteran flies contain chromosomes that are greatly enlarged relative to their usual condition. They possess both increased diameter and greater length. **Figure 27.15** shows an example of a chromosome set from the salivary gland of *D. melanogaster*. They are called **polytene chromosomes**.

Each chromosome consists of a visible series of **bands** (more properly, but rarely, described as

Figure 27.15

The polytene chromosomes of *D. melanogaster* form an alternating series of bands and interbands. Photograph kindly provided by Jose Bonner.

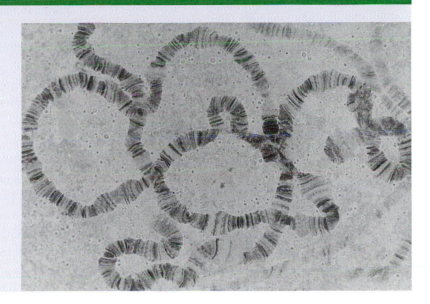

chromomeres). The bands range in size from the largest with a breadth of ~0.5 μm to the smallest of ~0.05 μm. (The smallest can be distinguished only under the electron microscope.) The bands contain most of the mass of DNA and stain intensely with appropriate reagents. The regions between them stain more lightly and are called **interbands**. There are ~5000 bands in the *D. melanogaster* set.

The centromeres of all four chromosomes of *D. melanogaster* aggregate to form a chromocenter that consists largely of heterochromatin (in the male it includes the entire Y chromosome). Allowing for this, ~75% of the haploid DNA set is organized into the band-interband alternation.

The length of the chromosome set is ~2000 μm; 75% of the DNA is 1.3×10^8 bp, which would extend for ~40,000 μm, so the average packing ratio is ~20. This demonstrates vividly the extension of the genetic material relative to the usual states of interphase chromatin or mitotic chromosomes.

What is the structure of these giant chromosomes? Each is produced by the successive replications of a synapsed diploid pair. The replicas do not separate, but remain attached to each other in their extended state. At the start of the process, each synapsed pair has a DNA content of 2C (where C represents the DNA content of the individual chromosome). Then this doubles up to 9 times, at its maximum giving a content of 1024C. The number of doublings is different in the various tissues of the *D. melanogaster* larva.

Each chromosome can be visualized as a large number of parallel fibers running longitudinally, tightly condensed in the bands, less condensed in the interbands. Probably each fiber represents a single (C) haploid chromosome. This gives rise to the name 'polyteny'. The degree of polyteny is the number of haploid chromosomes contained in the giant chromosome.

The banding pattern is characteristic for each strain of *Drosophila*. The constant number and linear arrangement of the bands was first

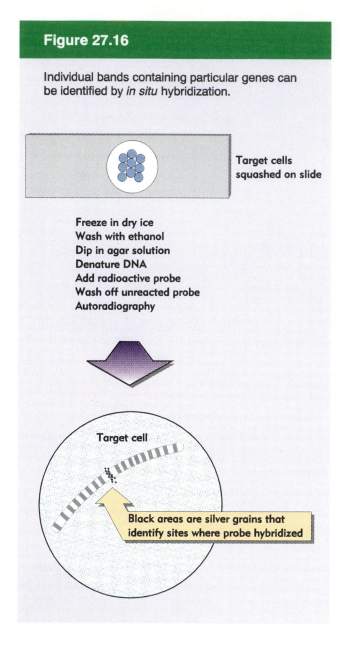

Figure 27.16

Individual bands containing particular genes can be identified by *in situ* hybridization.

Target cells squashed on slide

Freeze in dry ice
Wash with ethanol
Dip in agar solution
Denature DNA
Add radioactive probe
Wash off unreacted probe
Autoradiography

Target cell

Black areas are silver grains that identify sites where probe hybridized

of genes in *D. melanogaster* appears to exceed the number of bands, there are probably multiple genes in most or all bands (see Chapter 24).

The positions of particular genes on the cytological map can be determined directly by the technique of ***in situ*** or **cytological hybridization**. The protocol is summarized in **Figure 27.16**. A radioactive probe representing a gene (most often a labeled cDNA clone derived from the mRNA) is hybridized with the denatured DNA of the polytene chromosomes *in situ*. Autoradiography identifies the position or positions of the corresponding genes by the superimposition of grains at a particular band or bands. An example is shown in **Figure 27.17**. With this type of technique at hand, it is possible to determine directly the band within which a particular sequence lies.

One of the intriguing features of the polytene chromosomes is that active sites can be visualized. Some of the bands pass transiently through an expanded or **puffed** state, in which chromosomal material is extruded from the axis. An example of some very large puffs (called Balbiani rings) is shown in **Figure 27.18**.

What is the nature of the puff? It consists of a region in which the chromosome fibers unwind from their usual state of packing in the band. The fibers remain continuous with those in the chromosome axis. Puffs usually emanate from single bands, although when they are very large, as typified by the Balbiani rings, the swelling may be so extensive as to obscure the underlying array of bands.

The pattern of puffs is related to gene expression. During larval development, puffs appear and regress in a definite, tissue-specific pattern. A characteristic pattern of puffs is found in each tissue at any given time. Puffs are induced by the hormone ecdysone that controls *Drosophila* development. Some puffs are induced directly by the hormone; others are induced indirectly by the products of earlier puffs.

The puffs are *sites where RNA is being synthesized*. The accepted view of puffing has been that expansion of the band is a consequence of the need to relax its structure in order to synthesize

noted in the 1930s, when it was realized that they form a **cytological map** of the chromosomes. Rearrangements—such as deletions, inversions, or duplications—result in alterations of the order of bands.

The linear array of bands can be equated with the linear array of genes. Thus genetic rearrangements, as seen in a linkage map, can be correlated with structural rearrangements of the cytological map. Ultimately, a particular mutation can be located in a particular band. Since the total number

Figure 27.17

A magnified view of bands 87A and 87C shows their hybridization *in situ* with labeled RNA extracted from heat-shocked cells. Photograph kindly provided by Jose Bonner.

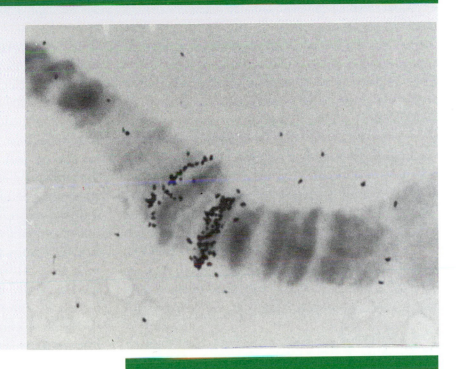

RNA. Puffing has therefore been viewed as a consequence of transcription. A puff can be generated by a single active gene.

The sites of puffing differ from ordinary bands in accumulating additional proteins. Characterization of these proteins at present is only rather primitive. We know that they include RNA polymerase II and other proteins associated with the act of transcription. We should like to analyze the entire set of proteins that accumulate at puffs, in particular to characterize those that are a cause rather than a consequence of the puffing. Then it should be possible to determine the nature of the molecular events responsible for the expansion of material.

The features displayed by lampbrush and polytene chromosomes suggest a general conclusion. In order to be transcribed, the genetic material is dispersed from its usual more tightly packed state. The question to keep in mind is whether this dispersion at the gross level of the chromosome mimics the events that occur at the molecular level within the mass of ordinary interphase euchromatin.

Figure 27.18

Chromosome IV of the insect *C. tentans* has three Balbiani rings in the salivary gland. Photograph kindly provided by Bertil Daneholt.

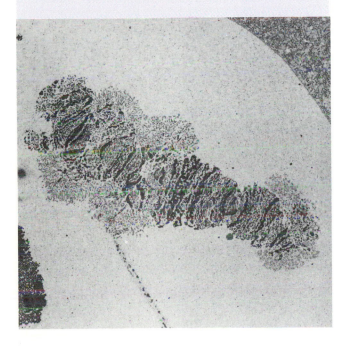

The eukaryotic chromosome as a segregation device

During mitosis, the sister chromatids move to opposite poles of the cell (as illustrated in Figure 2.21). Their movement depends on the attachment of the chromosome to microtubules, which are connected at their other end to the poles. (The microtubules comprise a cellular filamentous system, reorganized at mitosis so that they connect the chromosomes to the poles of the cell.) The sites in the two regions where microtubule ends are organized—in the vicinity of the centrioles at the poles and at the chromosomes—are called **MTOCs** (microtubule organizing centers).

The region of the chromosome that is responsible for its segregation at mitosis and meiosis is called the **centromere**. It is associated with two important features:

◆ It contains the site at which the sister chromatids are held together prior to the separation of the individual chromosomes. This shows as a constricted region connecting all four chromosome arms, as in the photograph of Figure 27.9, which shows the sister chromatids at the metaphase stage of mitosis.

◆ The term 'centromere' historically has been used in both the functional and structural sense to describe the feature of the chromosome responsible for its movement. The centromere is pulled toward the pole during mitosis, and the attached chromosome is dragged along behind, as it were. The chromosome therefore provides a device for attaching a large number of genes to the apparatus for division.

The centromere is essential for segregation, as shown by the behavior of chromosomes that have been broken. A single break generates one piece that retains the centromere, and another, an **acentric fragment**, that lacks it. The acentric fragment does not become attached to the mitotic spindle; and as a result it fails to be included in either of the daughter nuclei.

(When chromosome movement relies on discrete centromeres, there can be *only* one centromere per chromosome. When translocations generate chromosomes with more than one centromere, aberrant structures form at mitosis, since the two centromeres on the *same* sister chromatid can be pulled toward different poles, breaking the chromosome. However, in some species the centromeres are 'diffuse', which creates a different situation. Only discrete centromeres have been analyzed at the molecular level.)

The regions flanking the centromere often are rich in satellite DNA sequences and contain a considerable amount of constitutive heterochromatin. Because the entire chromosome is condensed, centromeric heterochromatin is not immediately evident in mitotic chromosomes. However, it can be visualized by a technique called **C-banding**. In the example of **Figure 27.19**, all the centromeres show as darkly staining regions. Although it is common, constitutive heterochromatin cannot be identified around *every* known centromere, which suggests that it is unlikely to be essential for the division mechanism.

What is the feature of the centromere that is responsible for segregation? Within the centromeric region, a darkly staining fibrous object of diameter or length ~400 nm can be seen. This **kinetochore** appears to be directly attached to the microtubules. Usually it is assumed that a specific sequence of DNA in some way defines the site at which the kinetochore should be established, but so far we have made no progress toward characterizing the molecular location and organization of this structure. The kinetochore provides the MTOC on a chromosome.

If a centromeric sequence of DNA is responsible for segregation, any molecule of DNA possessing this sequence should move properly at cell division, while any DNA lacking it will fail to segregate. This prediction has been used to isolate centromeric DNA in the yeast, *S. cerevisiae*. Yeast

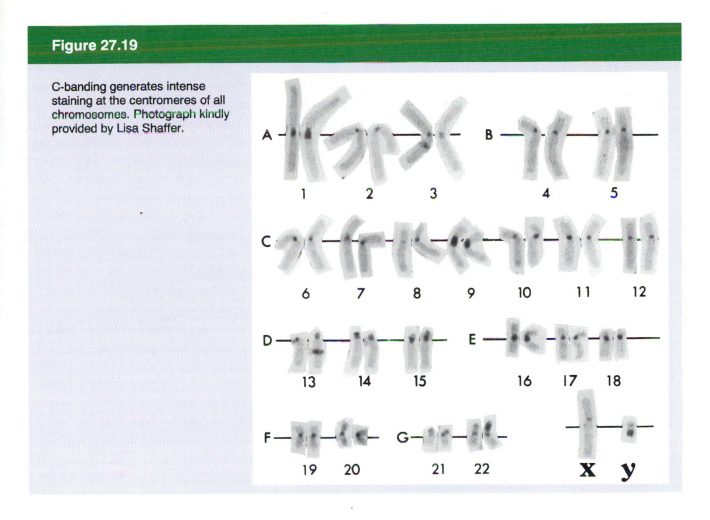

Figure 27.19

C-banding generates intense staining at the centromeres of all chromosomes. Photograph kindly provided by Lisa Shaffer.

chromosomes do not display visible kinetochores comparable to those of higher eukaryotes, but otherwise divide at mitosis and segregate at meiosis by the same mechanisms.

Genetic engineering has produced plasmids of yeast that are replicated like chromosomal sequences (see Chapter 18). However, they are unstable at mitosis and meiosis, disappearing from a majority of the cells because they segregate erratically. Fragments of chromosomal DNA have been isolated by virtue of their ability to confer mitotic stability on these plasmids. Their chromosomal derivations can be identified when they contain genetic markers known to map near a centromere.

A *CEN* fragment is defined by its ability to confer stability upon such a plasmid. By reducing the sizes of the fragments that are incorporated into the plasmid, the minimum region necessary for mitotic centromeric function can be identified. Deletions and other changes can be made to investigate the features involved in centromeric function.

Another way to use the availability of the centromeric sequences is to modify them *in vitro* and then reintroduce them into the yeast cell, where they replace the corresponding centromere on the chromosome. This allows the sequences required for *CEN* function to be defined directly in the context of the chromosome.

A *CEN* fragment derived from one chromosome can replace the centromere of another chromosome with no apparent consequence. This result suggests that centromeres are inter-changeable. *They are used simply to attach the chromosome to the spindle, and play no role in distinguishing one chromosome from another.*

The sequences required for centromeric function fall within a stretch of ~120 bp, in which three types of sequence element may be distinguished, as summarized in **Figure 27.20**:

◆ CDE-I is a sequence of 9 bp that is conserved with minor variations at the left boundary of all centromeres.

◆ CDE-II is a >90% A•T-rich sequence of 80–90 bp found in all centromeres; its function could depend on its length rather than exact sequence. Its constitution is reminiscent of some short tandemly repeated (satellite) DNAs in higher eukaryotes (see Chapter 26). Its base composition may cause some characteristic distortions of the DNA double helical structure.

◆ CDE-III is an 11 bp sequence highly conserved at the right boundary of all centromeres. Sequences on either side of the element are less well conserved, and may also be needed for centromeric function. (CDE-III could be longer than 11 bp if it turns out that the flanking sequences are essential.)

Mutations in CDE-I or CDE-II reduce but do not inactivate centromere function, but point mutations in the central CCG of CDE-III completely inactivate the centromere. Can we identify proteins that are necessary for the function of CEN sequences? Deletion of the gene coding for a protein (called CBF-I) that binds specifically to CDE-I increases the frequency of mitotic chromosome loss from 10^{-5} to 10^{-4}. (An acentric chromosome is lost at a frequency of 10^{-1}.)

A 240,000 dalton complex of three proteins, CBF-IIIA,B,C, binds to CDE-III, but cannot bind a point mutant that lacks centromeric function. Mutations in the components of the genes coding for CBF-III block chromosome movement at mitosis; and the protein complex has a microtubule-based motor activity—it is able to move itself, and presumably attached objects such as chromosomes, along microtubules. Comparable proteins with sequences related to known motor activities have been found at eukaryotic centromeres. Taken together, these observations suggest that a protein complex with motor activity may connect the centromeric region of a chromosome to microtubules, and contribute to movement on the mitotic spindle. The discovery of the CBF-III complex may be the prelude to characterizing the connection between the centromere and the apparatus for chromosome segregation.

Attempts to characterize functional centromeres from the yeast *S. pombe* have been less successful. They cannot be isolated by ability to confer stability on plasmids. However, *S. pombe* has only 3 chromosomes, and the region containing each centromere has been identified by deleting most of the sequences of each chromosome to create a stable minichromosome. This approach locates the centromeres within regions of 40–100 kb that consist largely or entirely of repetitious DNA. It is not clear how much of each of these rather long regions is required for chromosome segregation at mitosis and meiosis.

The significance of the difference between the short centromeric regions in *S. cerevisiae* and the long regions in *S. pombe* is not clear. The common feature is that the DNA consists of noncoding sequences that are repetitive.

The primary motif comprising the constitutive

Figure 27.20

Three conserved regions can be identified by the sequence homologies between yeast *CEN* elements.

TCACATGAT GATATTTGATTTTATTATATTTTTAAAAAAAGTAAAAAATAAAAAGTAGTTTATTTTTAAAAAATAAAATTTAAAATATTTCACAAAATGATTTCCGAA
AGTGTACTA CTATAAACTAAAATAATATAAAAATTTTTTTCATTTTTTATTTTTCATCAAATAAAAATTTTTTATTTTAAATTTTATAAAGTGTTTTACTAAAGGCTT

CDE-I   CDE-II   80-90 bp, >90% A + T   CDE-III

heterochromatin of primate centromeres is the α satellite DNA, which consists of tandem arrays of a 170 bp repeating unit. There is significant variation between individual repeats, although those at any centromere tend to be better related to one another than to members of the family in other locations. It is clear that the sequences required for cen-tromeric function reside within the blocks of a satellite DNA, but it is not clear whether the α satellite sequences themselves provide this func-tion, or whether other sequences are embedded within the α satellite arrays.

Specific chromatin structures are found at cen-tromeres, and we discuss examples in Chapter 28.

Chromosome ends are special

Another essential feature in all chromosomes is the **telomere**. In some way that we do not yet under-stand, this 'seals' the end. We know that the telomere must be a special structure, because chromosome ends generated by breakage are 'sticky' and tend to react with other chromosomes, whereas natural ends are stable.

We can apply two criteria in identifying a telomeric sequence:

◆ It must lie at the end of a chromosome (or, at least, at the end of an authentic linear DNA molecule).

◆ It must confer stability on a linear molecule.

Several telomeric sequences have been obtained from linear DNA molecules present in the genomes of lower eukaryotes. The same type of sequence is found in plants and man, so the construction of the telomere seems to follow a universal principle. Each telomere consists of a long series of short, tandemly repeated sequences. **Table 27.2** lists the repeating units that have been identified at the ends of the linear DNA molecules. All can be written in the general form $C_n(A/T)_m$, where $n>1$ and m is 1–4.

Within the telomeric region is a specific array of discontinuities, taking the form of single-strand breaks whose structure prevents them from being sealed by the ligase enzyme that normally acts

Table 27.2

Telomeres have a common type of short tandem repeat. The repeating unit gives the sequence of one strand, in the direction from the telomere toward the centromere.

Type of Organism	Species	Source of DNA	Repeating unit (5'-3')
Holotrichous ciliates	Tetrahymena, Paramecium	Macronucleus	CCCCAA
Hypotrichous ciliates	Stylonchia, Oxytricha, Euplotes	Macronucleus	CCCCAAAA
Flagellates	Trypanosoma, Leishmania	Minichromosome	CCCTA
Slime molds	Physarum, Dictyostelium	rDNA	CCCTA
Yeast	Saccharomyces	Chromosome	$C_{2\text{-}3}A(CA)_{1\text{-}3}$
Plant	Arabidopsis	Chromosome	C_3TA_3
Man	Homo sapiens	Chromosome	C_3TA_2

upon nicks in one DNA strand. The very terminal bases are blocked in some way—they may be organized in a hairpin—so that they are not recognized by nucleases.

The problem of finding a system that offers an assay for function again has been brought to the molecular level by using yeast. All the plasmids that survive in yeast (by virtue of possessing *ARS* and *CEN* elements) are circular DNA molecules. Linear plasmids are unstable (because they are degraded). Could an authentic telomeric DNA sequence confer stability on a linear plasmid?

Fragments from yeast DNA that prove to be located at chromosome ends can be identified by such an assay. And a region from the end of a known natural linear DNA molecule—the extrachromosomal rDNA of *Tetrahymena*—is able to render a yeast plasmid stable in linear form. The nicks in the telomeric sequence are perpetuated at the same sites in yeast, a remarkable interspecies conservation.

Some indications about how a telomere functions are given by some unusual properties of the ends of linear DNA molecules. In a trypanosome population, the ends are variable in length. When an individual cell clone is followed, the telomere grows longer by 7–10 bp (1–2 repeats) per generation. Even more revealing is the fate of ciliate telomeres introduced into yeast. After replication in yeast, *yeast telomeric repeats are added onto the ends of the Tetrahymena repeats.*

Addition of telomeric repeats to the end of the chromosome in every replication cycle could solve the problem of replicating linear DNA molecules discussed in Chapter 19. The addition of repeats by *de novo* synthesis would counteract the loss of repeats resulting from failure to replicate up to the end of the chromosome. Extension and shortening would be in dynamic equilibrium.

The overall length of the telomere is under genetic control; different strains of yeast have different but characteristic telomeric lengths. Some mechanism must prevent the ends from growing too long, possibly by removing some of the repeats. Mutation of an essential yeast gene causes the telomeres to grow steadily longer; the function of the wild-type gene could be to limit telomere extension.

If telomeres are continually being lengthened (and shortened), their exact sequence may be irrelevant. All that is required is for the present end to be recognized as a suitable substrate for addition. This explains how the ciliate telomere functions in yeast. However, we do not yet understand how the telomeric sequence confers resistance of chromosome ends to damage.

How are the telomeric repeats synthesized? Extracts of *Tetrahymena* contain an enzyme, called **telomerase**, that uses the 3′-OH of the G+T telomeric strand as a primer for synthesis of tandem TTGGGG repeats. Only dGTP and dTTP are needed for the activity. The telomerase is a large ribonucleoprotein. It contains a short RNA component, 159 bases long in *Tetrahymena,* 192 bases long in *Euplotes.* Each RNA includes a sequence of 15–22 bases that is identical to two repeats of the C-rich repeating sequence given in Table 27.2. This RNA provides the template for synthesizing the G-rich repeating sequence, to which it is complementary. Bases are added individually, in the correct sequence, as depicted in **Figure 27.21.** The enzyme progresses discontinuously: the template RNA is positioned on the DNA primer, several nucleotides are added to the primer, and then the enzyme translocates to begin again. The telomerase is a specialized example of a reverse transcriptase, an enzyme that synthesizes a DNA sequence using an RNA template (see Chapter 35). The protein component provides the catalytic activity of reverse transcriptase, and is (presumably) confined to acting upon the RNA template provided by the nucleic acid component.

The structure of the telomere is organized as represented in Figure 27.21, with a single-stranded extension of the G-T-rich strand, usually for 14–16 bases. But isolated telomeric fragments do not

Figure 27.21

Telomerase positions itself by base pairing between the RNA template and the protruding single-stranded DNA primer. It adds G and T bases one at a time to the primer, as directed by the template. The cycle starts again when 1 repeating unit has been added.

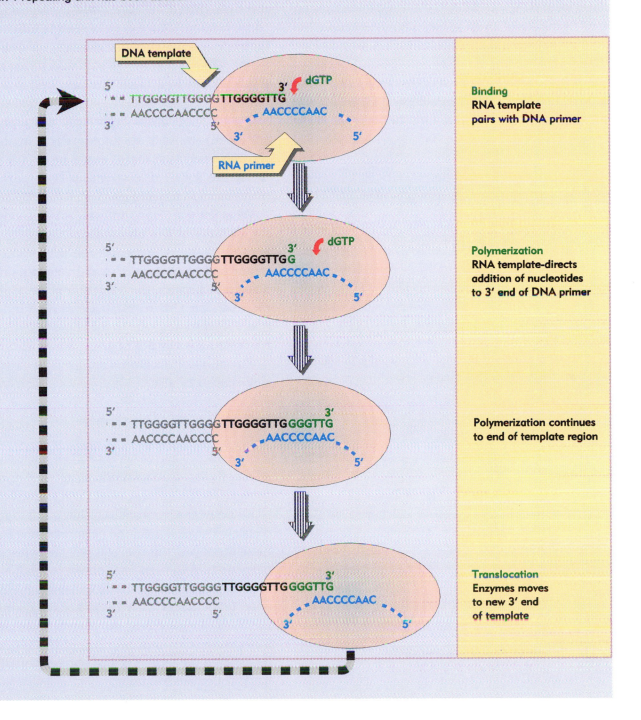

behave as though they contain single-stranded DNA; instead they show aberrant electrophoretic mobility and other properties. Two models for the structure of the end are depicted in **Figure 27.22**.

An early model suggested that a duplex hairpin could form if the extended G-T strand folds back on itself by means of unusual base pair interactions. As few as two repeating units could seal the end of the DNA.

A later model proposed the existence of a 'quartet' of G residues, formed by an association of one G from each repeating unit. In the example in the figure, the second G of each of four successive T_2G_4 units forms a member of the quartet. The rest of the repeating unit is looped out. The association between the G residues requires that two of them change the orientation of the base with regard to the sugar (from the usual *anti* to the unusual *syn* configuration). Since each repeating unit has more than one G, more than one quartet could be formed if other G residues associate, in which case quartets might be stacked upon one another in a helical manner.

We do not know how the complementary (C-A-rich) strand of the telomere is assembled, but we may speculate that it could be synthesized by using the 3'-OH of a terminal G-T hairpin as a primer for DNA synthesis.

What feature of the centromere is responsible for the stability of the chromosome end? Telomere-binding proteins recognize the protruding G-rich strand and protect the DNA at the terminus *in vitro*. They may provide the 'cap' that protects the telomere *in vivo*. Other proteins may bind specifically to the regions adjacent to the repeating units; their role remains to be defined.

The minimum features required for existence as a chromosome are:

◆ Telomeres to ensure survival.

◆ A centromere to support segregation.

◆ An origin to initiate replication (see Chapter 18).

All of these elements have been put together to construct a yeast artificial chromosome (YAC). We have discussed the use of such chromosomes for perpetuating foreign sequences in Chapter 21. It turns out that the synthetic chromosome is stable only if it is longer than 20–50 kb. We do not know the basis for this effect, but the ability to construct a synthetic chromosome offers the potential to investigate the nature of the segregation device in a controlled environment.

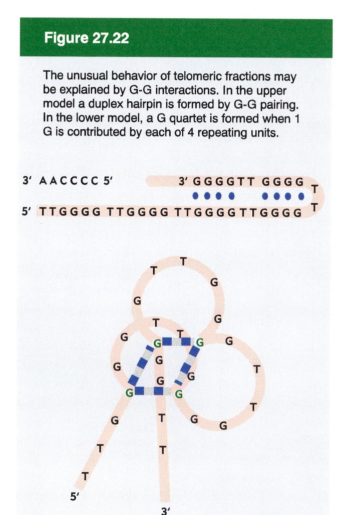

Figure 27.22

The unusual behavior of telomeric fractions may be explained by G-G interactions. In the upper model a duplex hairpin is formed by G-G pairing. In the lower model, a G quartet is formed when 1 G is contributed by each of 4 repeating units.

Summary

The genetic material of all organisms and viruses takes the form of tightly packaged nucleoprotein. Some virus genomes are inserted into preformed virions, while others assemble a protein coat around the nucleic acid. The bacterial genome forms a dense nucleoid, with about 20% protein by mass, but details of the interaction of the proteins with DNA are not known. The DNA is organized into ~100 domains that maintain independent supercoiling, with a density of unrestrained supercoils corresponding to ~1 per 100-200 bp. Interphase chromatin and metaphase chromosomes both appear to be organized into large loops. Each loop may be an independently supercoiled domain. The bases of the loops are connected to a metaphase scaffold or to the nuclear matrix by specific DNA sites.

Transcriptionally active sequences reside within the euchromatin that comprises the majority of interphase chromatin. The regions of heterochromatin are packaged ~5–10× more compactly, and are transcriptionally inert. All chromatin becomes densely packaged during cell division, when the individual chromosomes can be distinguished. The existence of a reproducible ultrastructure in chromosomes is indicated by the production of G-bands by treatment with Giemsa stain. The bands are very large regions, $~10^7$ bp, that can be used to map chromosomal translocations or other large changes in structure.

Lampbrush chromosomes of amphibians and polytene chromosomes of insects have unusually extended structures, with packing ratios <100. Polytene chromosomes of *D. melanogaster* are divided into ~5000 bands, varying in size by an order of magnitude, with an average of ~25 kb. Genetic analysis suggests that each band contains 1 essential locus, but molecular analysis suggests that there are several genes within a band. Transcriptionally active regions can be seen to exist in an even more unfolded ('puffed') structure, in which material is extruded from the axis of the chromosome. This may resemble the changes that occur on a smaller scale when a sequence in euchromatin is transcribed.

The centromeric region contains the kinetochore, which is responsible for attaching a chromosome to the mitotic spindle. The centromere often is surrounded by heterochromatin. Centromeric sequences have been identified only in yeast, where they consist of short conserved elements and a long A•T-rich region. Proteins that bind to these sequences have been identified.

Telomeres make the ends of chromosomes stable. Almost all known telomeres consist of multiple repeats in which one strand has the general sequence $C_n(A/T)_m$, where $n >1$ and $m = 1–4$. The other strand, $G_n(T/A)_m$, has a single protruding end that provides a template for addition of individual bases in defined order. The enzyme telomere transferase is a ribonucleoprotein, whose RNA component provides the template for synthesizing the G-rich strand.

Further reading

Reviews

Packaging of phage DNA has been reviewed by **Black** (*Ann. Rev. Microbiol.* **43**, 267–292, 1989).

The concept of the bacterial nucleoid has been reviewed by **Brock** (*Microbiol. Rev.* **52**, 397–411, 1988).

Proteins potentially involved in bacterial genome structure have been reviewed by **Drlica and Rouviere-Yaniv** (*Microbiol. Rev.* **51**, 301–319, 1987).

Centromeres and telomeres were reviewed by **Blackburn and Szostak** (*Ann. Rev. Biochem.* **53**, 163–194, 1984) and by **Clarke and Carbon** (*Ann. Rev. Genet.* **19**, 29–56, 1985). Centromeres were reviewed by **Schulman and Bloom** (*Ann. Rev. Cell Biol.* **7**, 311–336, 1991). Telomeres have been brought up to date by **Zakian** (*Ann. Rev. Genet.* **23**, 579–604, 1989) and by **Blackburn** (*Nature* **350**, 569–573, 1991). Telomerase was reviewed by **Blackburn** (*Ann. Rev. Biochem.* **61**, 113–129, 1992).

Discoveries

The structure of the centromere was revealed by **Bloom and Carbon** (*Cell* **29**, 305–317, 1982).

The functional elements needed by a chromosome were put together to make a synthetic chromosome by **Murray and Szostak** (*Nature* **305**, 189–193, 1983). Telomerase was discovered by **Greider and Blackburn** (*Cell* **51**, 887–898, 1987) and its catalytic activity characterized by **Shippen-Lentz and Blackburn** (*Science* **247**, 546–552, 1990). The unusual structure of the G-T-rich tail was recognized by **Henderson *et al*.** *Cell* **51**, 899–908, 1987), and a G quartet model was proposed by **Williamson, Raghuraman, and Cech** (*Cell* **59**, 871–880, 1989).

CHAPTER 28

Chromosomes consist of nucleosomes

Chromatin has a compact organization in which most DNA sequences are structurally inaccessible and functionally inactive. Within this mass are the minority of active sequences. What is the general structure of chromatin, and what is the difference between active and inactive sequences?

The fundamental subunit of chromatin has the same type of design in all eukaryotes. It contains ~200 bp of DNA, organized by an octamer of small, basic proteins into a bead-like structure. The protein components form an interior core; the DNA lies on the surface of the particle.

We can define the structure of the particle in terms of the path of the DNA and the contacts between the nucleic acid and the proteins. And we can investigate the protein–protein contacts between the individual basic proteins.

We can ask whether the conformation of the particle can vary, and how such variations might be related to the functions of chromatin. What happens to the particle when chromatin is replicated or transcribed?

When chromatin is replicated, the series of particles must be reproduced on both daughter duplex molecules. As well as asking how the particle itself is assembled, we must inquire what happens to other proteins present in chromatin. Since replication disrupts the structure of chromatin, it both poses a problem for maintaining regions with specific structure and offers an opportunity to change the structure.

The high overall packing ratio of the genetic material immediately suggests that DNA cannot be directly packaged into the final structure of chromatin. There must be *hierarchies* of organization. The first level is the winding of the DNA into the bead-like particles, forming a 10 nm fiber with a packing ratio of ~6. These particles are an invariant component of euchromatin, heterochromatin, and chromosomes.

The second level of organization is the coiling of the series of beads into a helical array to constitute the ~30 nm fiber that is found in both interphase chromatin and mitotic chromosomes (see Figure 27.9). In chromatin this brings the packing ratio of DNA to ~40. The structure of this fiber requires additional proteins.

The final packing ratio is determined by the third level of organization, the packaging of the fiber itself. This gives an overall packing ratio of ≥1000 in euchromatin, cyclically interchangeable with packing into mitotic chromosomes to achieve an overall ratio of ≤10,000. This too is likely to be a function modulated by accessory proteins, as is the difference between euchromatin and heterochromatin.

The mass of chromatin contains up to twice as much protein as DNA. The proteins are divided into two types: histones and nonhistones. The mass of RNA is 10% of the mass of DNA. Much of the RNA consists of nascent chains still associated with the template DNA.

The **histones** are the most basic proteins in chromatin and account for just about the same mass as the DNA. Five classes of histones were originally characterized by their relative proportions of lysine and arginine. The same classes can be recognized in virtually all eukaryotes.

The constancy of histone structures suggests that the histone–DNA interactions, histone–histone interactions, and histone–nonhistone interactions are similar in different species; so we may be able to deduce general rules that govern formation of both the primary particle and the folding of a series of particles into a higher-order structure.

Histones H3 and H4 are among the most conserved proteins known in evolution. They even have identical sequences in species as far distant as the cow and the pea. This suggests that their functions are identical in perhaps all eukaryotes. The types of H2A and H2B can be recognized in all eukaryotes, but show appreciable species-specific variation in sequence. All these histones are small proteins, typically 11,000–15,000 daltons.

Histone H1 comprises a set of several rather closely related proteins, with overlapping amino acid sequences. (A variant in avian red blood cells is called H5.) The H1 histones show appreciable variation between tissues and between species (and this class apparently is absent from yeast). H1 histones are the largest, 23,000 daltons.

As their dysphonious name suggests, the **nonhistones** include all the other proteins of chromatin. They are therefore presumed to be more variable between tissues and species— although good evidence still is lacking on the extent of the variability—and they comprise a relatively smaller proportion of the mass than the histones. They also comprise a much larger number of proteins, so that any individual protein is present in amounts much smaller than any histone.

The nonhistone proteins include functions concerned with gene expression and with higher-order structure. Thus RNA polymerase may be considered to be a prominent nonhistone. The HMG (high-mobility group) proteins comprise a discrete and well-defined subclass of nonhistones (at least some of which are transcription factors). A major problem in working with other nonhistones is that they tend to be contaminated with other nuclear proteins, and so far it has proved difficult to obtain those nonhistone proteins responsible for higher-order structures.

The nucleosome is the subunit of all chromatin

When interphase nuclei are suspended in a solution of low ionic strength, they swell and rupture to release fibers of chromatin. **Figure 28.1** shows a lysed nucleus in which fibers are streaming out. In some regions, the fibers consist of tightly packed material, but in regions that have become stretched, they can be seen to consist of discrete particles, called **nucleosomes**. In especially extended regions, individual nucleosomes are connected by a fine thread, a free duplex of DNA.

A continuous duplex thread of DNA runs through the series of particles.

Individual nucleosomes can be obtained by treating chromatin with the enzyme **micrococcal nuclease**, an endonuclease that cuts the DNA thread at the junction between nucleosomes. First, it releases groups of particles; finally, it releases single nucleosomes. Individual nucleosomes can be seen clearly in **Figure 28.2** as compact particles. They sediment at ~11S.

Figure 28.1

Chromatin spilling out of lysed nuclei consists of a compactly organized series of particles. The bar is 100 nm. Photograph kindly provided by Pierre Chambon.

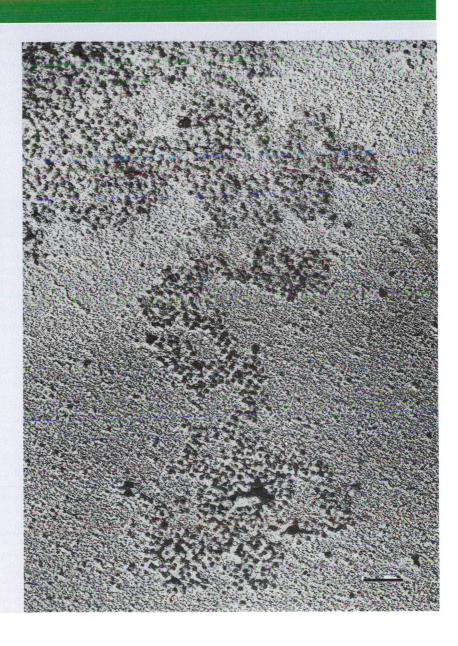

The nucleosome contains ~200 bp of DNA associated with a histone octamer that consists of two copies each of H2A, H2B, H3, and H4. These are known as the **core histones**. Their association is illustrated diagrammatically in **Figure 28.3**. This model explains the stoichiometry of the core histones in chromatin: H2A, H2B, H3, and H4 are present in equimolar amounts, with 1 molecule of each per ~100 bp of DNA.

The role of H1 is different from the core histones. It is present in half the amount of a core histone and can be extracted more readily from chromatin (typically with dilute salt [0.5 M] solution). *All of the H1 can be removed without affecting the structure*

Figure 28.2

Individual nucleosomes are released by digestion of chromatin with micrococcal nuclease. The bar is 100 nm. Photograph kindly provided by Pierre Chambon.

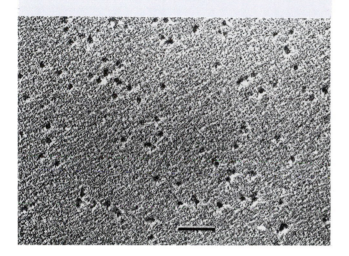

So each step on the ladder represents the DNA derived from a discrete number of nucleosomes. *We therefore take the existence of the 200 bp ladder in any chromatin to indicate that the DNA is organized into nucleosomes.* The micrococcal ladder is generated when only ~2% of the DNA in the nucleus is rendered acid-soluble (degraded to small fragments) by the enzyme. *Thus a small*

of the nucleosome, which suggests that its location is external to the particle.

When chromatin is digested with the enzyme micrococcal nuclease, the DNA is cleaved into integral multiples of a unit length. Fractionation by gel electrophoresis reveals the 'ladder' presented in **Figure 28.4**. Such ladders extend for 10 or more steps, and the unit length, determined by the increments between successive steps, is ~200 bp.

Figure 28.5 shows that the ladder is generated by groups of nucleosomes. When nucleosomes are fractionated on a sucrose gradient, they give a series of discrete peaks that correspond to monomers, dimers, trimers, etc. When the DNA is extracted from the individual fractions and electrophoresed, each fraction yields a band of DNA whose size corresponds with a step on the ladder produced by digestion of whole chromatin. The monomeric nucleosome contains DNA of the unit length, the nucleosome dimer contains DNA of twice the unit length, and so on.

Figure 28.3

The nucleosome consists of approximately equal masses of DNA and histones (including H1). The predicted mass of the nucleosome is 262,000 daltons, with a protein/DNA mass ratio of 1.0. The experimentally measured mass usually lies in the range of 250,000-300,000 daltons, with a protein/DNA ratio of ~1.2. Any additional protein probably represents small amounts of nonhistone proteins associated with the nucleosomes.

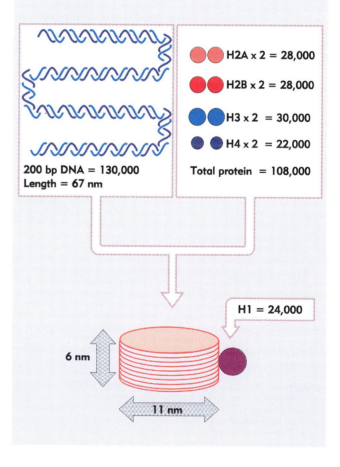

200 bp DNA = 130,000
Length = 67 nm

H2A x 2 = 28,000
H2B x 2 = 28,000
H3 x 2 = 30,000
H4 x 2 = 22,000

Total protein = 108,000

H1 = 24,000

6 nm

11 nm

Figure 28.4

Micrococcal nuclease digests chromatin in nuclei into a multimeric series of DNA bands that can be separated by gel electrophoresis. Photograph kindly provided by Markus Noll.

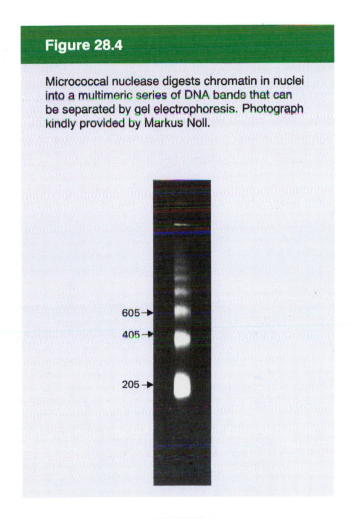

605 →
405 →
205 →

Figure 28.5

Each multimer of nucleosomes contains the appropriate number of unit lengths of DNA. Photograph kindly provided by John Finch.

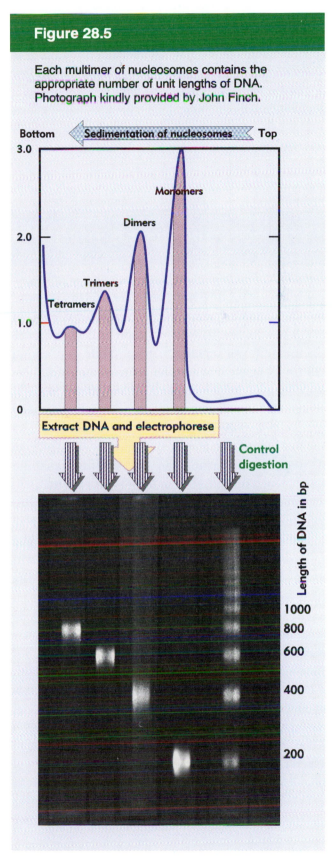

proportion of the DNA is specifically attacked; it must represent especially susceptible regions.

When chromatin is spilled out of nuclei, we often see a series of nucleosomes connected by a thread of free DNA (the beads on a string). However, the need for tight packaging of DNA *in vivo* suggests that probably there is usually little (if any) free DNA.

This view is confirmed by the fact that >90% of the DNA of chromatin can be recovered in the form of the 200 bp ladder. Almost all DNA must therefore be organized in nucleosomes. In their natural state, nucleosomes are likely to be closely packed, with DNA passing directly from one to the next. Free DNA is probably generated by the loss of some histone octamers during isolation.

The length of DNA present in the nucleosome varies somewhat from the 'typical' value of 200 bp.

When entire genomes are characterized from particular cells, each has a fairly well-defined average value (±5 bp). The average most often is between 180 and 200, but there are extremes as low as 154 bp (in a fungus) or as high as 260 bp (in a sea urchin sperm).

The value is not necessarily fixed for a species. It may change during embryonic development. In the case of the sea urchin sperm, it is reduced to a more typical level during the early embryonic cell divisions. The average value may be different in individual tissues of the adult organism. And there can be differences between different parts of the genome in a single cell type; known cases of variation from the genome average include tandemly repeated sequences, such as clusters of 5S RNA genes.

The core particle is highly conserved

A common structure underlies the varying amount of DNA that is contained in nucleosomes of different sources. The association of DNA with the histone octamer forms a **core particle** containing 146 bp of DNA, irrespective of the total length of DNA in the nucleosome. The variation in total length of DNA per nucleosome is superimposed on this basic core structure.

The core particle is defined by the effects of micrococcal nuclease on the nucleosome monomer. The initial reaction of the enzyme is to cut between nucleosomes, but if it is allowed to continue after monomers have been generated, then it proceeds to digest some of the DNA of the individual nucleosome. This occurs by a reaction in which DNA is 'trimmed' from the ends of the nucleosome.

The length of the DNA is reduced in discrete steps, as shown in **Figure 28.6**. With rat liver nuclei, the nucleosome monomers initially have 205 bp of DNA. Then some monomers are found in which the length of DNA has been reduced to 160–170 bp. Finally this is reduced to the length of the DNA of the core particle, 146 bp. (The core is reasonably stable, but if digestion is continued, further cuts generate a **limit digest**, in which the longest fragments are the 146 bp DNA of the core, while the shortest are as small as 20 bp.)

This analysis suggests that the nucleosomal DNA can be divided into two regions:

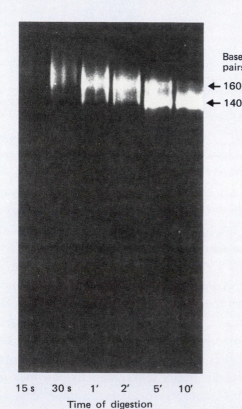

Figure 28.6

Micrococcal nuclease reduces the length of DNA of nucleosome monomers in discrete steps. Photograph kindly provided by Roger Kornberg.

Base pairs
← 160
← 140

15 s   30 s   1'   2'   5'   10'
Time of digestion

◆ **Core DNA** has an invariant length of 146 bp, and is relatively resistant to digestion by nucleases.

◆ **Linker DNA** comprises the rest of the repeating unit. Its length varies from as little as 8 bp to as much as 114 bp per nucleosome.

The discrete nature of the band of DNA generated by the initial cleavage with micrococcal nuclease suggests that the region immediately available to the enzyme is restricted. It represents only part of each linker. (If the entire linker DNA were susceptible, the band would range from 146 bp to >200 bp.) But once a cut has been made in the linker DNA, the rest of this region becomes susceptible, and it can be removed relatively rapidly by further enzyme action. The connection between nucleosomes is represented diagrammatically in **Figure 28.7** (without any implication as to the actual organization of DNA and protein).

Core particles have properties similar to those of the nucleosomes themselves, although they are smaller. Their shape and size are similar to nucleosomes, which suggests that the essential geometry of the particle is established by the interactions between DNA and the protein octamer in the core particle. Because core particles are more readily obtained as a homogeneous population, they are often used for structural studies in preference to nucleosome preparations. (Nucleosomes tend to vary because it is difficult to obtain a preparation in which there has been no end-trimming of the DNA.)

What is the physical nature of the core and the linker regions? *These terms are operational definitions that describe the regions in terms of their relative susceptibility to nuclease treatment.* This description does not make any implication about their actual structure; in particular, it does not imply that the linker DNA has a more extended conformation.

The path of DNA in the nucleosome could be continuous, with no distinction evident between these regions of the monomer. Indeed, this is a

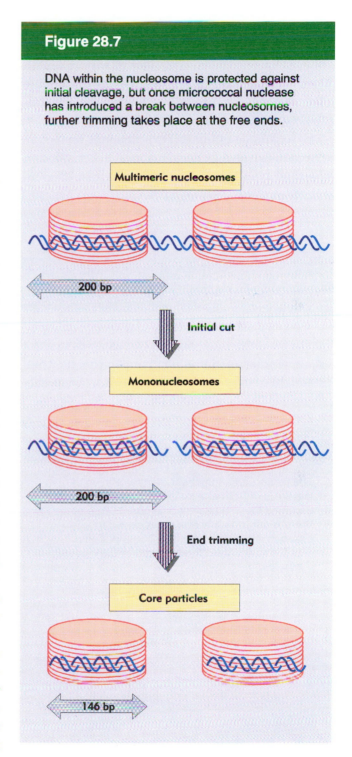

Figure 28.7

DNA within the nucleosome is protected against initial cleavage, but once micrococcal nuclease has introduced a break between nucleosomes, further trimming takes place at the free ends.

convenient working assumption often made in attempts to extrapolate from core particle structure to nucleosome structure. Or it is possible that the path of the linker DNA does differ from the path of the core particle DNA, especially given the pronounced variations that occur in its length. Linker DNA may be constrained to vary in length only by integral numbers of turns of the double helix (~10 bp), presumably reflecting the relationship between adjacent nucleosomes.

The existence of linker DNA depends on factors extraneous to the four core histones. Reconstitution experiments *in vitro* show that histones have an intrinsic ability to organize DNA into core particles, but do not necessarily form nucleosomes with the unit length of DNA characteristic of the *in vivo* state. The degree of supercoiling of the DNA is an important factor. Histone H1 and/or nonhistone proteins influence the length of DNA associated with the histone octamer in a natural series of nucleosomes. And 'assembly proteins' that are not part of the nucleosome structure are involved *in vivo* in constructing nucleosomes from histones and DNA (see later).

DNA is coiled around the histone octamer

The shape of the nucleosome corresponds to a flat disk or cylinder, of diameter 11 nm and height 6 nm. The length of the DNA (200 bp ≈ 67 nm) is roughly twice the ~34 nm circumference of the particle. Immediately these dimensions suggest that DNA could not be squashed within the particle, but must lie on the outside.

Figure 28.8

The nucleosome may be a cylinder with up to two turns of DNA around the surface. This model assumes that the nucleosome has ˜200 bp of DNA. A question that has not received much attention is how the path might be modified to account for the variation in length of DNA on the nucleosome.

DNA "leaves"

DNA "enters"

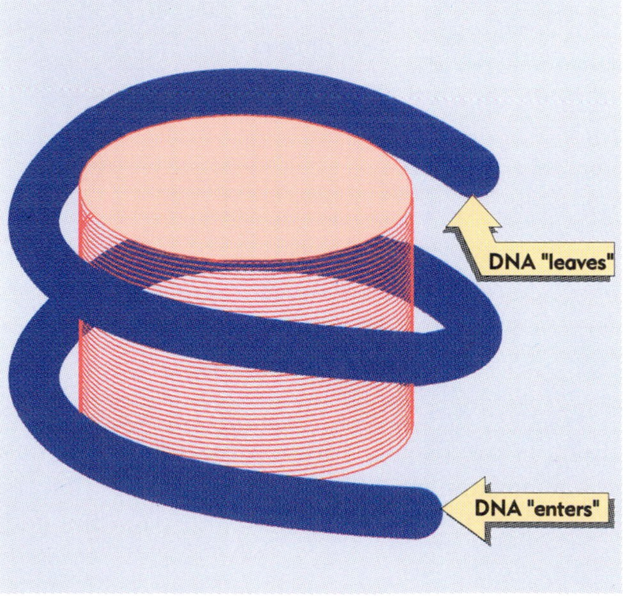

Figure 28.9

The two turns of DNA on the nucleosome lie close together.

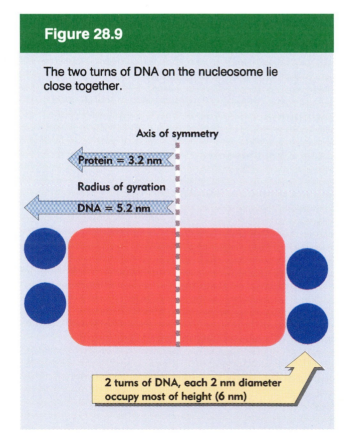

Axis of symmetry

Protein = 3.2 nm

Radius of gyration

DNA = 5.2 nm

2 turns of DNA, each 2 nm diameter occupy most of height (6 nm)

Figure 28.10

A protein could contact sequences on the DNA that lie on different turns around the nucleosome.

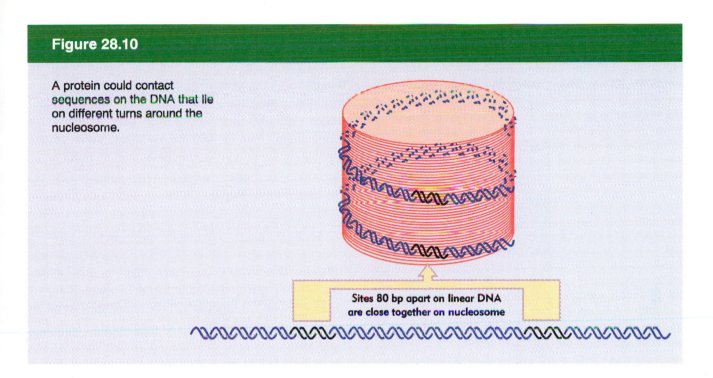

Sites 80 bp apart on linear DNA are close together on nucleosome

Figure 28.11

Nicks in double-stranded DNA are revealed by fragments when the DNA is denatured to give single strands. If the DNA is labeled at (say) 5' ends, only the 5' fragments are visible by autoradiography. The size of the fragment identifies the distance of the nick from the labeled end.

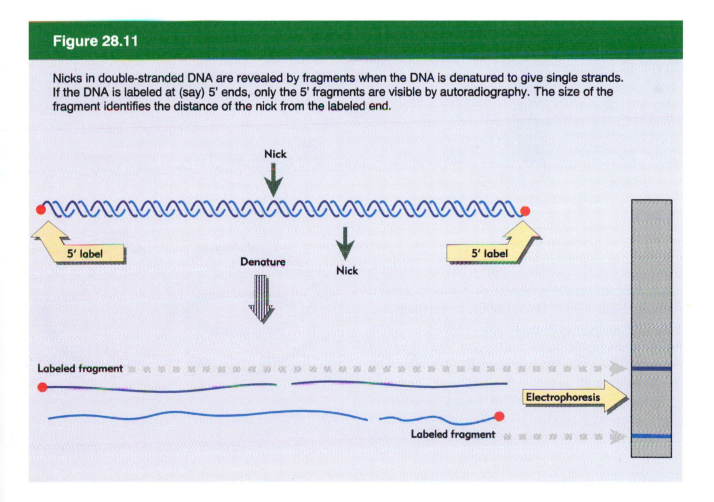

Nick

5' label

Denature

Nick

5' label

Labeled fragment

Electrophoresis

Labeled fragment

Figure 28.12

Sites for nicking lie at regular intervals along core DNA, as seen in a DNAase I digest of nuclei. Photograph kindly provided by Leonard Lutter.

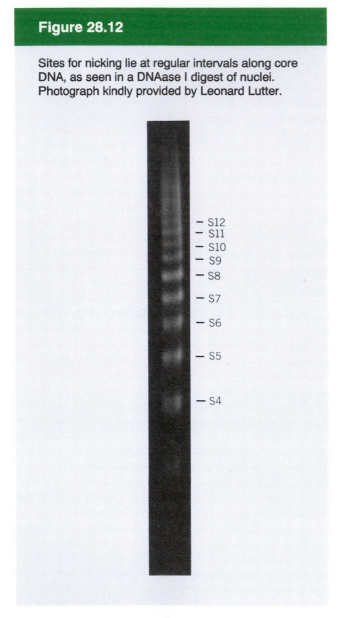

- S12
- S11
- S10
- S9
- S8
- S7
- S6
- S5
- S4

The DNA follows a symmetrical path around the octamer. **Figure 28.8** shows the DNA path diagrammatically as a helical coil that makes two turns around the cylindrical octamer. Note that the DNA 'enters' and 'leaves' the nucleosome at points close to one another.

Considering this model in terms of a cross-section through the nucleosome, in **Figure 28.9** we see that the two circumferences made by the DNA lie close to one another. The height of the cylinder is 6 nm, of which 4 nm is occupied by the two turns of DNA (each of diameter 2 nm).

The pattern of the two turns has a possible functional consequence. Since one turn around the nucleosome takes ~80 bp of DNA, two points separated by 80 bp in the free double helix may actually be rather close on the nucleosome surface, as illustrated in **Figure 28.10**.

The exposure of DNA on the surface of the nucleosome explains why it is accessible to cleavage by certain nucleases. The reaction with nucleases that attack single strands has been especially informative. The enzymes DNAase I and DNAase II make single-strand nicks in DNA; they cleave a bond in one strand, but the other strand remains intact at this point. Thus no effect is visible in the double-stranded DNA. But upon denaturation, short fragments are released instead of full-length single strands. If the DNA has been labeled at its ends, the end fragments can be identified by autoradiography

Figure 28.13

Two numbering schemes divide core particle DNA into 10 bp segments. Sites may be numbered S1 to S13 from one end; or taking S7 to identify coordinate 0 of the dyad symmetry, they may be numbered -7 to +7.

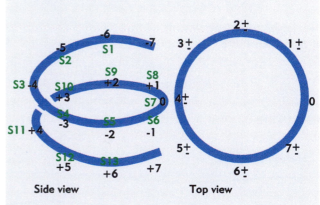

Side view   Top view

Figure 28.14

The most exposed positions on DNA recur with a periodicity that reflects the structure of the double helix. (For clarity, sites are shown for only one strand).

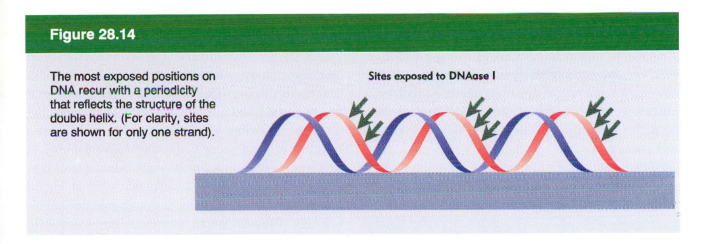

Sites exposed to DNAase I

as summarized in **Figure 28.11**. (This is exactly analogous to the restriction mapping technique shown in Figure 6.5.)

When DNA is free in solution, it is nicked (relatively) at random. The DNA on nucleosomes also can be nicked by the enzymes, *but only at regular intervals*. When the points of cutting are determined by using radioactively end-labeled DNA and then DNA is denatured and electrophoresed, a ladder of the sort displayed in **Figure 28.12** is obtained.

The interval between successive steps on the ladder is approximately 10 bases. The ladder extends for the full distance of core DNA. The cleavage sites are numbered as S1 through S13 (where S1 is ~10 bases from the labeled 5′ end, S2 is ~20 bases from it, and so on). Their positions relative to the DNA superhelix are illustrated in **Figure 28.13**.

Not all sites are cut with equal frequency: some are cut rather effectively, others are cut scarcely at all. The enzymes DNAase I and DNAase II generate the same ladder, although with some differences in the intensities of the bands. This shows that the pattern of cutting represents a unique series of targets in DNA, determined by its organization, with only some slight preference for particular sites imposed by the individual enzyme.

The sensitivity of nucleosomal DNA to nucleases is analogous to a footprinting experiment. Thus we can assign the lack of reaction at particular target sites to the structure of the nucleosome, in which certain positions on DNA are rendered inaccessible.

Since there are two strands of DNA in the core particle, in an end-labeling experiment both 5′ (or 3′) ends are labeled, one on each strand. Thus the cutting pattern includes fragments derived from both strands. This is implied in Figure 28.11, where each labeled fragment is derived from a different strand. The corollary is that, in an experiment, each labeled band in fact represents two fragments, generated by cutting the *same* distance from *either* of the labeled ends.

How then should we interpret discrete preferences at particular sites? One view is that the path of DNA on the particle is symmetrical (about a horizontal axis through the nucleosome drawn in Figure 28.8). Thus if (for example) no 80-base fragment is generated by DNAase I, this must mean that the position at 80 bases from the 5′ end of *either* strand is not susceptible to the enzyme. The second numbering scheme used in Figure 28.13 reflects this view, and identifies S7 = site 0 as the center of symmetry.

The crucial question about the basis for the periodicity of cutting is whether it solely reflects the path of the DNA on the histone octamer or whether it is influenced by other features of nucleosome structure. Because DNAases are large proteins, the periodicity could be influenced by steric hindrance of access to different regions of the nucleosome. However, essentially the same cutting pattern is obtained by using a very small agent—a hydroxyl radical—to produce cleavage, which argues that the pattern genuinely reflects the structure of the DNA itself.

When DNA is immobilized on a flat surface, sites are cut with a regular separation. **Figure 28.14** sug-gests that this reflects the recurrence of the exposed site with the helical periodicity of B-form DNA. The **cutting periodicity** (the spacing between cleavage points) coincides with, indeed, is a reflec-tion of, the **structural periodicity** (the number of base pairs per turn of the double helix). Thus the distance between the sites corresponds to the number of base pairs per turn. Measurements of this type suggest that the average value for double-helical B-type DNA is 10.5 bp/turn.

What is the nature of the target sites on the nucle-osome? **Figure 28.15** shows that each site has 3–4 positions at which cutting occurs; that is, the cutting site is defined ±2 bp. So a cutting site represents a short stretch of bonds on both strands, exposed to nuclease action over 3–4 base pairs. The relative intensities indicate that some sites are preferred to others.

From this pattern, we can calculate the 'average' point that is cut. At the ends of the DNA, pairs of sites from S1 to S4 or from S10 to S13 lie apart a distance of 10.0 bases each. In the center of the particle, the separation from sites S4 to S10 is 10.7 bases. (Because this analysis deals with *average* positions, sites need not lie an integral number of bases apart.)

The variation in cutting periodicity along the core DNA (10.0 at the ends, 10.7 in the middle) means that there is variation in the structural periodicity of core DNA. The DNA has more bp/turn than its solution value in the middle, but has fewer bp/turn at the ends. The average periodicity over the nucleosome is less than the 10.5 base pairs/turn of DNA in solution; it is in the range of 10.2–10.4 base pairs/turn, depending on the method of measurement.

The crystal structure of the core particle sug-gests that DNA is organized as a flat superhelix, with 1.8 turns wound around the histone octamer. The pitch of the superhelix varies, with a discontinuity in the middle. Regions of high curvature are arranged symmetrically, and occur at positions ±1 and ±4. These correspond to S6 and S8 and to S3 and S11, which are the sites least sensitive to DNAase I. The high curvature is probably responsible for these changes, but their precise nature remains to be determined at the molecular level.

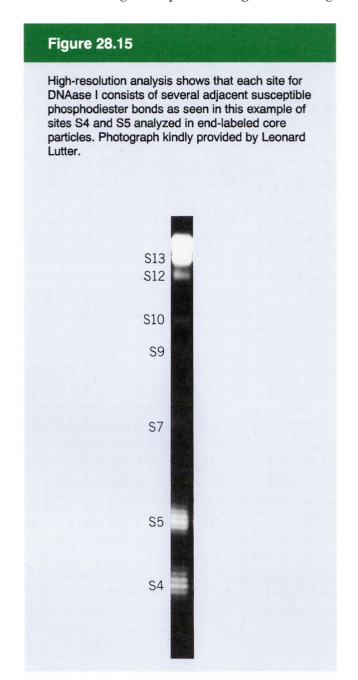

Figure 28.15

High-resolution analysis shows that each site for DNAase I consists of several adjacent susceptible phosphodiester bonds as seen in this example of sites S4 and S5 analyzed in end-labeled core particles. Photograph kindly provided by Leonard Lutter.

Supercoiling and the periodicity of DNA

The path of DNA around the nucleosome in the model of Figure 28.8 shows two turns. If the DNA were not restrained by the histone octamer, this path would generate −1.8 superhelical turns. Can we measure directly the degree of supercoiling in the DNA path?

Much work on the structure of sets of nucleosomes has been carried out with the virus SV40. The

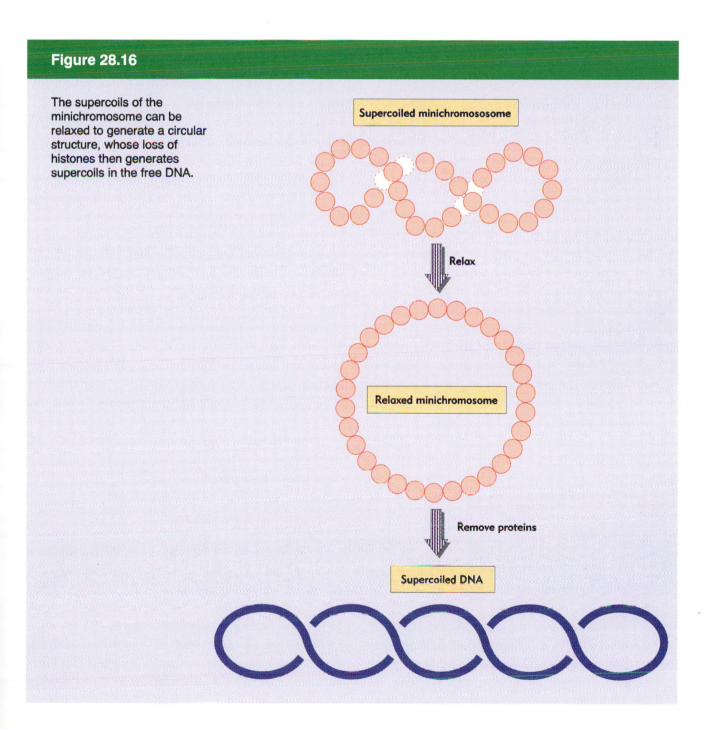

Figure 28.16

The supercoils of the minichromosome can be relaxed to generate a circular structure, whose loss of histones then generates supercoils in the free DNA.

Supercoiled minichromososome

Relax

Relaxed minichromosome

Remove proteins

Supercoiled DNA

DNA of SV40 is a circular molecule of 5200 bp, with a contour length ~1500 nm. In both the virion and infected nucleus, it is packaged into a series of nucleosomes, called a **minichromosome**.

As usually isolated, the contour length of the minichromosome is ~210 nm, corresponding to a packing ratio of ~7 (essentially the same as the nucleosome itself, $67/11 \approx 6$). Changes in the salt concentration can convert it to a flexible string of beads with a much lower overall packing ratio. This emphasizes the point that nucleosome strings can take more than one form *in vitro*, depending on the conditions.

Supercoiling in chromatin can be created at several levels. First, there is supercoiling as a result of the path that DNA follows on the nucleosome. Second, there is supercoiling as a result of the path that the nucleosomes follow in higher-level structures. Some of this supercoiling is restrained by proteins. Direct measurements of the supercoiling density therefore will show the average torsional tension of DNA, resulting from free supercoils, but will not reveal any restrained supercoiling in the DNA path.

The degree of supercoiling on the individual nucleosomes of the minichromosome can be measured as illustrated in **Figure 28.16**. First, the free supercoils of the minichromosome itself are relaxed, so that the nucleosomes form a circular string with a superhelical density of 0. Then the histone octamers are extracted. This releases the DNA to follow a free path. Every supercoil that was present but restrained in the minichromosome will appear in the deproteinized DNA as −1 turn. So now the total number of supercoils in the SV40 DNA is measured.

The value actually observed is close to the number of nucleosomes. The reverse result is seen when nucleosomes are assembled *in vitro* onto a supercoiled SV40 DNA: the formation of each nucleosome removes ~1 negative supercoil.

So the DNA follows a path on the nucleosomal surface that generates ~1 negative supercoiled turn when the restraining protein is removed. The discrepancy between this measurement and the model that shows DNA in a path equivalent to −1.8 superhelical turns is sometimes called the **linking number paradox**.

The discrepancy is at least partly explained by the difference between the 10.2 average bp/turn of nucleosomal DNA and the 10.5 bp/turn of free DNA. When DNA is released from the nucleosome, the structure of the double helix changes, with an average increase of 0.3 bp/turn. This change reduces the apparent degree of supercoiling. If this effect extended over the 19 helical turns of the full 200 bp of DNA in a nucleosome, it could account for $0.3 \times 19 \approx -0.6$ superhelical turns, or most of the discrepancy between the path of −1.8 and the measurement of −1.0 superhelical turns. In effect, some of the torsional strain in nucleosomal DNA goes into increasing the number of bp/turn; only the rest is left to be measured as a supercoil. Given the difficulties of making these measurements, it is possible that the change in periodicity in nucleosomal DNA accounts entirely for the linking number paradox.

The path of nucleosomes in the chromatin fiber

When chromatin is examined in the electron microscope, two types of fiber are seen: the 10 nm fiber and 30 nm fiber. They are described by the approximate diameter of the thread (that of the 30 nm fiber actually varies from ~25–30 nm).

The **10 nm fiber** is essentially a continuous string of nucleosomes. Sometimes, indeed, it runs continuously into a more stretched-out region

Figure 28.17

The 10 nm fiber in partially unwound state can be seen to consist of a string of nucleosomes. Photograph kindly provided by Barbara Hamkalo.

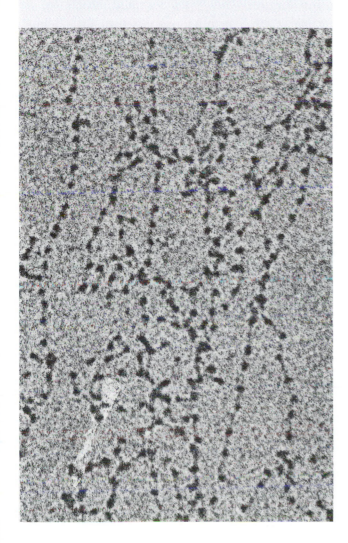

either edge to edge or with their faces touching. These arrangements can be distinguished by biophysical techniques, such as neutron scattering or electric dichroism, that depend on the orientation of the individual subunit relative to the axis of the fiber. The results suggest the type of model illustrated in **Figure 28.18**, in which the cylinders are oriented edge to edge, with their faces parallel (or at least, not much inclined) to the axis of the fiber. The data imply that the angle between the faces and the axis is <20°.

Figure 28.18

The 10 nm fiber consists of a series of nucleosomes organized edge to edge, stacked along the fiber or tilted less than 20° to the axis.

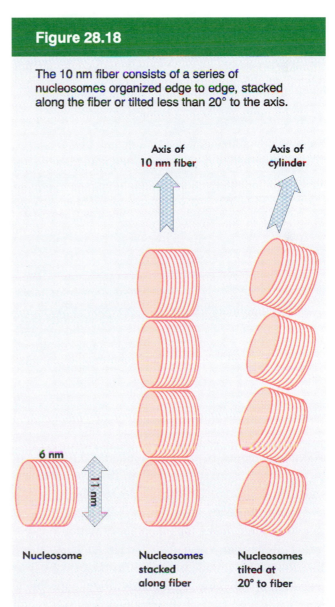

Axis of 10 nm fiber   Axis of cylinder

6 nm   11 nm

Nucleosome   Nucleosomes stacked along fiber   Nucleosomes tilted at 20° to fiber

in which nucleosomes are seen as a string of beads, as indicated in the example of **Figure 28.17**. The 10 nm fibril structure is obtained under conditions of low ionic strength and does not require the presence of histone H1. This means that it is a function strictly of the nucleosomes themselves.

How are the nucleosomes arranged in the 10 nm fiber? Viewing the particle itself as a somewhat flat cylinder, adjacent cylinders might be arranged

Figure 28.19

The 30 nm fiber has a coiled structure (shown at the same magnification as the 10 nm fiber of Figure 28.17.) Photograph kindly provided by Barbara Hamkalo.

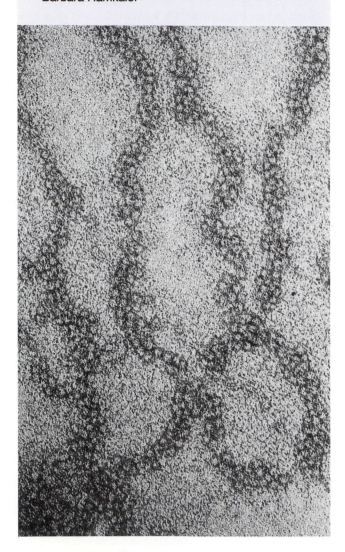

When chromatin is visualized in conditions of greater ionic strength the **30 nm fiber** is obtained. An example is given in **Figure 28.19**. The fiber can be seen to have an underlying coiled structure. It has ~6 nucleosomes for every turn, which corresponds to a packing ratio of 40 (that is, each μm along the axis of the fiber contains 40 μm of DNA). The presence of H1 is required. This fiber is the basic constituent of both interphase chromatin and mitotic chromosomes.

The 30 nm and 10 nm fibers can be reversibly converted by changing the ionic strength. This suggests that the linear array of nucleosomes in the 10 nm fiber is coiled into the 30 nm structure at higher ionic strength and in the presence of H1.

The most likely arrangement for packing nucleosomes into the fiber is a solenoid, illustrated in **Figure 28.20**. The nucleosomes turn in a helical array, with an angle of ~60° between the faces of adjacent nucleosomes. There are six nucleosomes per turn.

It seems likely that the parameters of the 30 nm fiber are not rigidly fixed, but can vary. This would accommodate the variation in the length of DNA per nucleosome, as well as allowing for other changes in the density of packing. It is not certain whether

Figure 28.20

The 30 nm fiber may have a helical coil of 6 nucleosomes per turn, organized radially.

27Å

110Å

57Å

the fiber has the identical structure in interphase chromatin and mitotic chromosomes.

Although the presence of H1 is necessary for the formation of the 30 nm fiber, information about its location is conflicting. Its relative ease of extraction from chromatin seems to argue that it is present on the outside of the superhelical fiber axis; but other data on its accessibility suggest it is harder to find in 30 nm fibers than in 10 nm fibers that retain it, which would argue for an interior location.

How do we get from the 30 nm fiber to the specific structures displayed in mitotic chromosomes? And is there any further specificity in the arrangement of interphase chromatin; do particular regions of 30 nm fibers bear a fixed relationship to one another or is their arrangement random? To such questions we have no answers at present.

Organization of the histone octamer

So far we have considered the construction of the nucleosome from the perspective of how the DNA is organized on the surface. From the perspective of protein, we need to know how the histones interact with each other and with DNA. Do histones react properly only in the presence of DNA, or do they possess an independent ability to form octamers?

We do not know very much about the structures of individual histones in the nucleosome, but we are beginning to deduce their relative locations. Most of the evidence about histone-histone interactions is provided by their abilities to form aggregates, and by cross-linking experiments with the nucleosome.

The core histones form two types of aggregates. H3 and H4 form a tetramer ($H3_2 \bullet H4_2$). Various aggregates are formed by H2A and H2B, in particular a dimer (H2A$\bullet$H2B) that has a tendency to aggregate further. One of the aggregates could be the tetramer ($H2A_2 \bullet H2B_2$).

Intact histone octamers can be obtained either by extraction from chromatin or (with more difficulty) by letting histones associate *in vitro* under conditions of high-salt and high-protein concentration. The octamer can dissociate to generate a hexamer of histones that has lost an H2A$\bullet$H2B dimer. Then the other H2A$\bullet$H2B dimer is lost separately, leaving the $H3_2 \bullet H4_2$ tetramer. This argues for a form of organization in which the nucleosome has a cent-

ral 'kernel' consisting of the $H3_2 \bullet H4_2$ tetramer. The tetramer can organize DNA *in vitro* into particles that display some of the properties of the core particle.

Cross-linking studies extend these relationships to show which pairs of histones lie near each other in the nucleosome. (A difficulty with such data is that usually only a small proportion of the proteins become cross-linked, so it is necessary to be cautious in deciding whether the results typify the major interactions.) From these data, a model has been constructed for the organization of the nucleosome. It is shown in diagrammatic form in **Figure 28.21**.

Structural studies show that the overall shape of the isolated histone octamer is similar to that of the core particle. This suggests that the histone–histone interactions establish the general structure. The positions of the individual histones have been assigned to regions of the octameric structure on the basis of their aggregation behavior and response to cross-linking.

The crystal structure (at 3.1 Å resolution) suggests the detailed model for the histone core octamer shown in **Figure 28.22**. Tracing the paths of the individual polypeptide backbones in the crystal structure suggests that the histones are not organized as individual globular proteins, but that each is interdigitated with its partner, H3 with

Figure 28.21

In a symmetrical model for the nucleosome, the $H3_2H4_2$ tetramer provides a kernel for the shape.

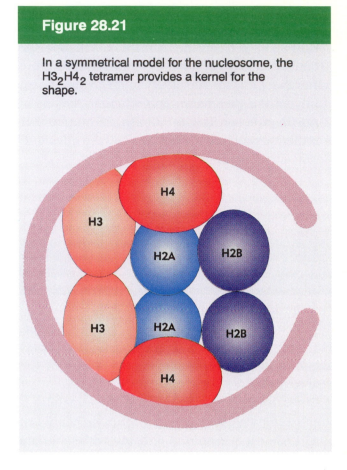

Figure 28.22

The crystal structure of the histone core octamer is represented in a space-filling model with the H3-H4 tetramer shown in white and the H2A-H2B dimers shown in blue. Only one of the H2A-H2B dimers is visible in the top view, because the other is hidden behind. The potential path of the DNA is shown in the top view as a narrow tube (one quarter the diameter of DNA), and in the side view by the parallel lines in a 20Å wide bundle. Photographs kindly provided by Evangelos Moudrianakis.

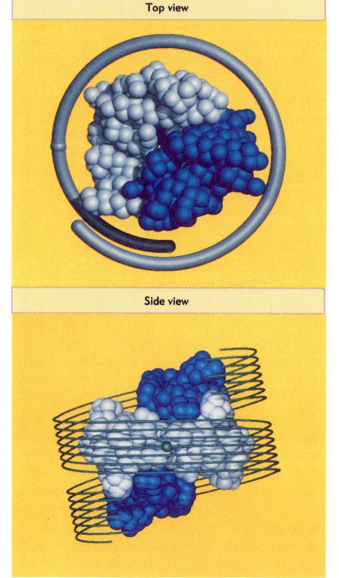

Top view

Side view

H4, and H2A with H2B. Thus the model distinguishes the $H3_2 \bullet H4_2$ tetramer (white) from the H2A•H2B dimers (blue), but does not show individual histones.

The top view represents the same perspective that was illustrated schematically in Figure 28.21. The $H3_2 \bullet H4_2$ tetramer accounts for the diameter of the octamer. The H2A•H2B pairs fit in as two dimers, but they are not distinguished in this view. The side view represents the same perspective that was illustrated in Figure 28.8. Here the responsibilities of the $H3_2 \bullet H4_2$ tetramer and of the separate H2A•H2B dimers can be distinguished. The protein actually forms a sort of spool, with a superhelical path that could correspond to the binding site for DNA, which would be wound in almost two full turns in a nucleosome. The model displays twofold symmetry about an axis that would run perpendicular through the side view.

Where is histone H1 located? The H1 is lost during the degradation of nucleosome monomers. It can be retained on monomers that still have 160–170 bp of DNA; but is always lost with the final reduction to the 146 bp core particle. This suggests that H1 could be located in the region of the linker DNA immediately adjacent to the core DNA.

If H1 is located at the linker, it could 'seal' the DNA in the nucleosome by binding at the point where the nucleic acid enters and leaves. The idea that H1 lies in the region joining adjacent nucleosomes is consistent with old results that H1 is removed the most readily from chromatin, and that H1-depleted chromatin is more readily 'solubilized'; and also with more recent results that show it is easier to obtain a stretched-out fiber of beads on a string when the H1 has been removed.

An important caveat is needed about the material used in these experiments. The *in vitro* data have been obtained with core particles (because of the difficulty of obtaining nucleosome preparations that have a homogeneous size distribution of DNA). Models for nucleosome structure often assume that linker DNA follows the same path as core DNA; but it could be different.

Analysis of the positions occupied by individual histones suggests that the two H2A subunits present near the terminal sites of core DNA (± 7) could block a continuation of the smooth superhelical course. Their presence at the outer edge of the core particle suggests that the linker DNA may have to make a ~50° turn from the path of core DNA. The actual path of linker DNA remains one of the major unsolved questions of nucleosome structure.

Reproduction of chromatin requires assembly of nucleosomes

The description of chromatin as a thread of duplex DNA coiled around a series of nucleosomes is the crucial first step toward visualizing the state of the genetic material in the nucleus. This somewhat static view accounts for the structure of the individual subunit and (to some degree) for its relationship with the adjacent subunit. However, the organization of nucleosomes must be *flexible* enough to satisfy the various structural and functional demands made on chromatin.

Cyclical changes in packing affect the entire mass of euchromatin. During cell division, euchromatin must become more tightly packaged in mitotic chromosomes. The transition is likely to be controlled by changes in proteins that are widely distributed through chromatin.

Replication and transcription are local events that require some dispersion of structure. Replication occurs as a series of individual events in local regions (replicons), generating duplicate double-stranded DNA regions each associated with a set of histone octamers. The events involved in reproducing the nucleosome particle have yet to be defined. We should like to know what happens to the nucleosome during replication, and how new nucleosomes are assembled.

It seems inevitable that the separation of parental DNA strands must disrupt the structure at least of the 30 nm fiber and probably also of the 10 nm fiber. We should like to know the extent of this disruption. Is it confined to the immediate vicinity of the point where DNA is being synthesized, or does it extend farther? Are there discernible structural differences between regions that have replicated and those that have yet to do so? The transience of the replication event is a major difficulty in analyzing the structure of a particular region while it is being replicated.

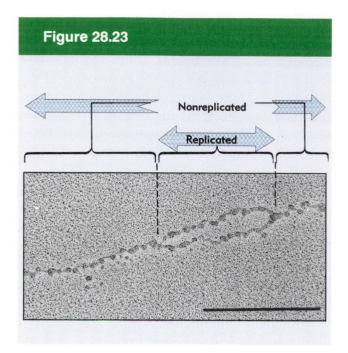

Figure 28.23

Nonreplicated

Replicated

The structure of the replication fork is distinctive. It is more resistant to micrococcal nuclease and is digested into bands that differ in size from nucleosomal DNA. This suggests that a large protein complex is engaged in replicating the DNA, but the nucleosomes reform more or less immediately behind as it moves along.

Reproduction of chromatin does not involve any protracted period during which the DNA is free of nucleosomes. Once DNA has been replicated, nucleosomes are quickly generated on both the duplicates. This point is illustrated by the electron micrograph of **Figure 28.23**, which shows a recently replicated stretch of DNA, already covered with nucleosomes on both daughter duplex segments.

How histones associate with DNA to generate nucleosomes has been a vexed and confusing question. Do the histones *preform* a protein octamer around which the DNA is subsequently wrapped? Or does an $H3_2 \bullet H4_2$ kernel bind DNA, after which $H2A \bullet H2B$ dimers are added?

Self-assembly *in vitro* is a slow process, limited by the tendency of the assembling particles to precipitate. It is difficult to know which conditions mimic the physiological. Both pathways can be used *in vitro* to assemble nucleosomes, as illustrated in **Figure 28.24**.

Accessory proteins are involved in assisting histones to associate with DNA. Candidates for this role have been identified in *Xenopus* eggs, from which extracts can be made that assemble histones and exogenous DNA into nucleosomes. The eggs contain two proteins that bind histones. Nucleoplasmin binds H2A and H2B, and N1 binds H3 and H4. Antibodies directed against nucleoplasmin or N1 can inhibit nucleosome assembly in the *in vitro* extracts.

What is the function of these accessory proteins? They could act as 'molecular chaperones', binding to the histones and releasing either individual histones or aggregates (H3•H4 or H2A•H2B) to the DNA in a controlled manner. This could be necessary because the histones, as basic proteins, have a general high affinity for DNA. Both N1 and nucleoplasmin are acidic proteins, which bind to the target histones to reduce the net positive charge. *Such interactions allow histones to form nucleosomes without becoming trapped in other kinetic intermediates (that is, other complexes resulting from indiscreet binding of histones to DNA).*

Attempts to produce nucleosomes *in vitro* began by considering a process of assembly between free DNA and histones. But nucleosomes form *in vivo* only when DNA is replicated. A system that mimics this requirement has been developed by using extracts of human cells that replicate SV40 DNA and assemble the products into chromatin. The assembly reaction occurs preferentially on replicating DNA. It requires an ancillary factor, CAF-1, that consists of >5 subunits, with a total mass of 238,000 daltons. The nucleosomes have a repeat length of 200 bp, although they do not have any H1 histone, which suggests that proper spacing does not require the presence of H1.

When chromatin is reproduced, a stretch of DNA *already associated with nucleosomes* is replicated, giving rise to two daughter duplexes. What happens to the pre-existing nucleosomes at this point? Are the histone octamers dissociated into free histones for reuse, or do they remain assembled?

The integrity of the octamer can be tested by

cross-linking the histones. **Figure 28.25** summarizes the possible outcomes from an experiment in which cells are grown in the presence of heavy amino acids to identify the histones before replication. Then replication is allowed to occur in the presence of light amino acids. At this point the histone octamers are cross-linked and centrifuged on a density gradient. If the original octamers have been conserved, they will be found at a position of high density, and new octamers will occupy a low

density position; but if the old histones have been released and then reassembled with newly synthesized histones, the octamers will have an intermediate density. Little material actually is found at the high density position, which suggests that histone octamers are not conserved.

The pattern of disassembly and reassembly is far from clear. It could be the case the octamers are entirely dissociated into their constituent histones. However, one possibility is that the octamers lose

Figure 28.24

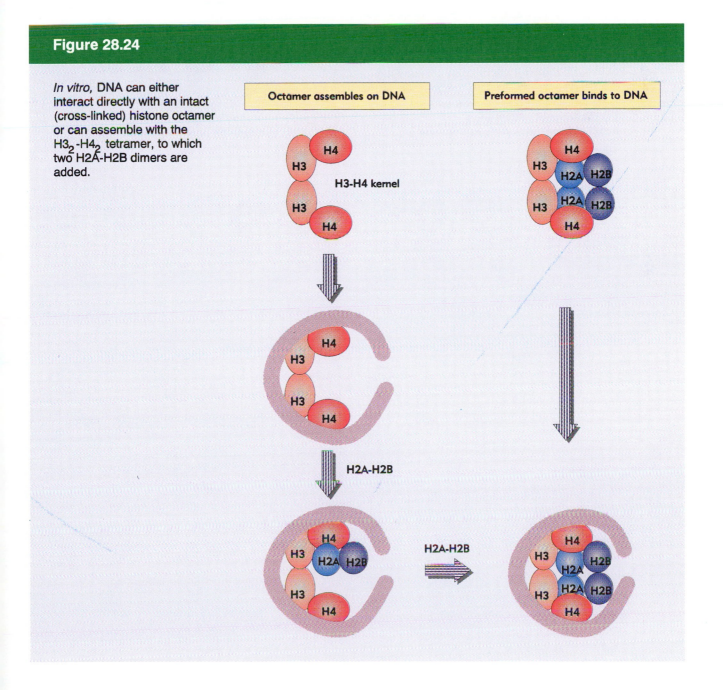

In vitro, DNA can either interact directly with an intact (cross-linked) histone octamer or can assemble with the $H3_2$-$H4_2$ tetramer, to which two H2A-H2B dimers are added.

Figure 28.25

If histone octamers were conserved, old and new octamers would band at different densities when replication of "heavy" octamers occurs in "light" amino acids (left); but actually the octamers band diffusely between heavy and light densities, suggesting dissolution and reassembly (right).

Parental nucleosome

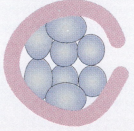

Newly synthesized histones

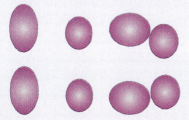

> Conservation of the octamer predicts that nucleosomes contain either exclusively old or exclusively new histones

> Disassembly and reassembly predicts that nucleosomes contain both old and new histones either systematically or randomly reassembled

Conserved old octamer   New octamer

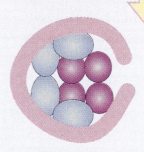

Old H3.H4, new H2A.H2B   Random assortment

> Crosslink histones, extract octamers, and analyze density

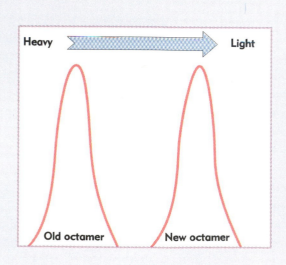

Heavy → Light

Old octamer   New octamer

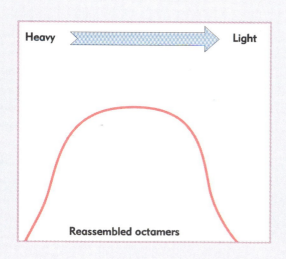

Heavy → Light

Reassembled octamers

one or both H2A•H2B dimers, which are replaced by newly synthesized material. In this case, the $H3_2$•$H4_2$ tetramer might be conserved. It is possible that a similar type of disruption occurs during transcription (see later). The $H3_2$•$H4_2$ tetramer could have an ability to be transiently associated with a single strand of DNA during replication; it may in fact have an increased (3:1) chance of remaining with the leading strand. This would mean that at least some histone octamers do assemble on DNA.

Are nucleosomes arranged in phase?

We know that nucleosomes can be reconstituted *in vitro* without regard to DNA sequence, but this does not exclude the possibility that their formation *in vivo* is controlled in a sequence-dependent manner. Does a particular DNA sequence always lie in a certain position *in vivo* with regard to the topography of the nucleosome? Or are nucleosomes arranged randomly on DNA, so that a particular sequence may occur at any location, for example, in the core region in one copy of the genome and in the linker region in another?

To investigate this question, it is necessary to use a defined sequence of DNA; more precisely, we need to determine the position relative to the nucleosome of a defined point in the DNA. **Figure 28.26** illustrates the principle of a procedure used to achieve this.

Suppose that the DNA sequence is organized into nucleosomes in only one particular configuration, so that each site on the DNA always is located at a particular position on the nucleosome. This type of organization is called **nucleosome phasing**. In a series of phased nucleosomes, the linker regions of DNA comprise unique sites.

Consider the consequences for just a single nucleosome. Cleavage with micrococcal nuclease generates a monomeric fragment that constitutes a *specific sequence*. If the DNA is isolated and cleaved with a restriction enzyme that has only one target site in this fragment, it should be cut at a unique point. This produces two fragments, each of unique size.

The products of the micrococcal/restriction double digest are separated by gel electrophoresis. A probe representing the sequence on one side of the restriction site is used to identify the corresponding fragment in the double digest. This technique is called **indirect end labeling** (not altogether an appropriate name).

Reversing the argument, the identification of a single sharp band demonstrates that the position of the restriction site is uniquely defined with respect to the end of the nucleosomal DNA (as defined by the micrococcal cut). So the nucleosome has a unique sequence of DNA.

What happens if the nucleosomes do *not* lie at a single position? Now the linkers consist of *different* DNA sequences in each copy of the genome. So the restriction site lies at a different position each time; in fact, it lies at all possible locations relative to the ends of the monomeric nucleosomal DNA. **Figure 28.27** shows that the double cleavage then generates a broad smear, ranging from the smallest detectable fragment (~20 bases) to the length of the monomeric DNA.

In discussing these experiments, we have treated micrococcal nuclease as an enzyme that cleaves DNA at the exposed linker regions without any sort of sequence specificity. However, the enzyme actually does have some sequence specificity (biased toward selection of A•T-rich sequences). So we cannot assume that the existence of a specific band in the indirect end-labeling technique represents the distance from a restriction cut to the linker region. It could instead represent the distance from the restriction cut to a preferred micrococcal nuclease site!

Figure 28.26

Nucleosome phasing places restriction sites at unique positions relative to the linker sites cleaved by micrococcal nuclease.

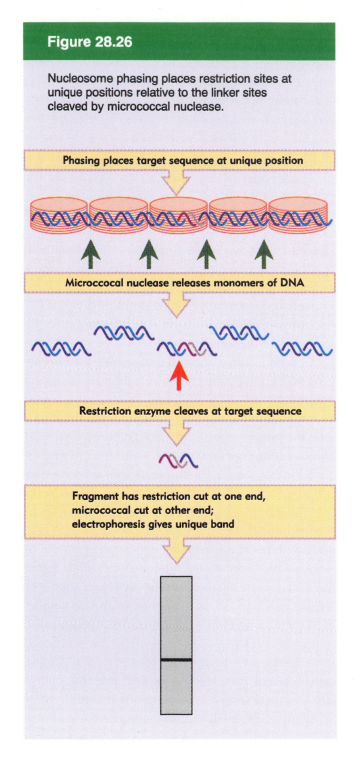

Phasing places target sequence at unique position

Micrococcal nuclease releases monomers of DNA

Restriction enzyme cleaves at target sequence

Fragment has restriction cut at one end, micrococcal cut at other end; electrophoresis gives unique band

This possibility is controlled by treating the naked DNA in exactly the same way as the chromatin. If there are preferred sites for micrococcal nuclease in the particular region, specific bands are found. Then this pattern of bands can be compared with the pattern generated from chromatin.

A *difference* between the control DNA band pattern and the chromatin pattern provides evidence for nucleosome phasing: some of the bands present in the control digest may disappear from the nucleosome digest when preferentially cleaved positions are unavailable; new bands may appear in the nucleosome digest when new sites are rendered preferentially accessible by the nucleosomal organization.

The analysis illustrated in Figures 28.26 and 28.27 applies to a single short sequence. To determine whether nucleosomes are phased over some substantial region, not just over one particular short sequence, early attempts to investigate nucleosome phasing made use of tandemly repeated sequences. This approach effectively amplifies the fragment that is investigated: a single probe will detect whether the corresponding sequence is phased in every one of its genomic repeating units.

In most cases, experimental data have not generated the clear type of result illustrated in Figure 28.26 for a unique disposition of nucleosomes over a protracted region; but sometimes they fall short of the wide bands or smears expected of random location. What we actually see is a confinement of the micrococcal cutting sites to a small number of positions, relative to the defined restriction site. This suggests that nucleosomes are limited so that they lie in only a few (2–4) alternative phases. Furthermore, when shorter regions are investigated, nucleosomes may be organized in a single phase, so that every nucleosome lies at a unique position.

Nucleosome phasing might be accomplished in either of two ways:

◆ It is intrinsic: *every nucleosome is deposited specifically at a particular DNA sequence.* This modifies our view of the nucleosome as a subunit able to form between any sequence of DNA and a histone octamer. In some cases, changes in a region do not alter the pattern of phasing, which argues that every nucleosome is at a particular location.

◆ It is extrinsic: *the first nucleosome in a region is preferentially assembled at a particular site.* A preferential starting point for nucleosome phasing results from the presence of a region from which nucleosomes are excluded. Such regions are created by complexes concerned with generating chromatin higher-order structure or controlling gene expression (see Chapter 30). The excluded region provides a *boundary* that restricts the positions available to the adjacent nucleosome. Then a series of nucleosomes may be assembled sequentially, with a defined repeat length.

It is now clear that the deposition of histone octamers on DNA is not random with regards to sequence. However, the pattern is not influenced directly by sequence-specific interactions between DNA and histones, but represents an effect produced by structural predispositions or variations in DNA.

Certain structural features of DNA will affect placement of histone octamers. DNA has intrinsic tendencies to bend in one direction rather than another; thus A•T-rich runs locate so that the minor groove faces in towards the octamer, whereas G•C-rich runs place so that the minor groove points out. Long runs of dA•dT (>8 bp) avoid positioning in the central superhelical turn of the core. It is not yet possible to sum all of the relevant structural effects and thus entirely to predict the location of a particular DNA sequence with regards to the nucleosome, but it is generally true that the orientation of DNA on the nucleosome is influenced by directional bending preferences of DNA. Of course, sequences that cause DNA to take up more extreme structures may have effects such as the exclusion of nucleosomes, and thus could cause boundary effects.

Phasing of nucleosomes near boundaries is quite common. The boundaries are usually imposed extrinsically, by the binding of some nonhistone protein to DNA. If there is some variability in the construction of nucleosomes—for example, if the length of the linker can vary by, say, 10 bp—the specificity of location would decline proceeding away from the first, defined nucleosome at the

boundary. In this case, we might expect the phasing to be maintained rigorously only relatively near the boundary.

Because DNA lies on the outside of histone

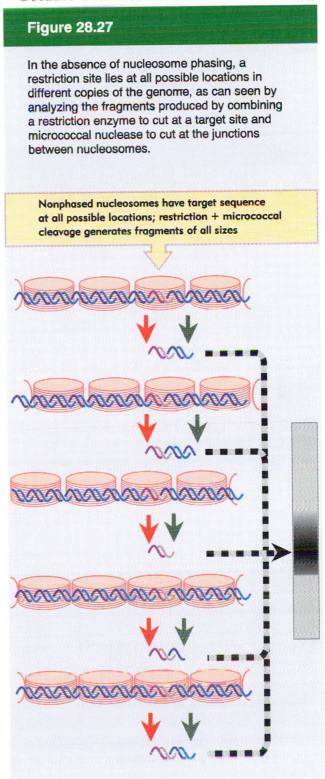

Figure 28.27

In the absence of nucleosome phasing, a restriction site lies at all possible locations in different copies of the genome, as can be seen by analyzing the fragments produced by combining a restriction enzyme to cut at a target site and micrococcal nuclease to cut at the junctions between nucleosomes.

Nonphased nucleosomes have target sequence at all possible locations; restriction + micrococcal cleavage generates fragments of all sizes

octamer, one face of any particular sequence is obscured by the histones, but the other face is accessible. Depending on its positioning with regards to the nucleosome, a site in DNA that must be recognized by a regulator protein could be inaccessible or available. The exact siting of the histone octamer with respect to DNA sequence may therefore be important. In some cases, nucleosomes are positioned nonrandomly at promoters, presumably to

ensure that regulators can bind to DNA, and the histone octamers may be displaced by regulators that bind to the exposed sites, allowing further regulators to bind to sites that previously were obscured. We discuss examples in more detail in Chapter 30. There are other cases where a regulator appears able to bind to DNA irrespective of the position of the nucleosome, but we do not yet understand the basis for this effect.

Are transcribed genes organized in nucleosomes?

There are extensively conflicting data on the nature of transcribed chromatin, and even now it remains impossible to generalize about the relationship of the nucleosomes to transcription. On the one hand, heavily transcribed chromatin can be seen to be rather extended (too extended to be covered in nucleosomes); on the other hand, the presence of histones can be detected on transcribed DNA.

Transcription involves the unwinding of DNA, and may require the fiber to unfold in restricted regions of chromatin. A simple-minded view suggests that some 'elbow-room' must be needed for the process. The features of polytene and lampbrush chromosomes described in Chapter 27 offer hints that a more expansive structural organization is associated with gene expression.

We should like to know what structural changes occur when a gene is being transcribed. Does the overall structure of the region change? Does the transcribed sequence remain organized into nucleosomes; and if so, what happens to them when RNA polymerase transcribes the DNA? What ensures that the promoter is initially accessible to the enzyme?

Some perturbation of structure must occur in a gene when it is being transcribed, if only as a result of the movement of RNA polymerase along the DNA. Thus structural changes could be a consequence of

the act of transcription, rather than a cause of it. So in assessing the properties of transcribed loci, we need to determine which of their particular features occur prior to transcription as a prerequisite for it, and which are induced subsequently as a result of the events involved in the synthesis of RNA.

In thinking about transcription, we must bear in mind the relative sizes of RNA polymerase and the nucleosome. The eukaryotic enzymes are large proteins, typically >500,000 daltons. Compare this with the ~260,000 daltons of the nucleosome. **Figure 28.28** illustrates the approach of RNA polymerase to nucleosomal DNA. Even without detailed knowledge of the interaction, it is evident that it involves the approach of two comparable bodies.

The nucleosome is not an isolated object but is adjacent to others; and consider the two turns that DNA makes around it. Would RNA polymerase have sufficient access to DNA if the nucleic acid were confined to its customary path on the nucleosome? During transcription, as RNA polymerase moves along the template, it binds tightly to a region of ~50 bp, including a locally unwound segment of ~12 bp. The need to unwind DNA makes it seem unlikely that the segment engaged by RNA polymerase could remain on the surface of the nucleosome.

It therefore seems inevitable that transcription must involve a structural change. So the first

Figure 28.28

RNA polymerase is comparable in size to the nucleosome and might encounter difficulties in following the DNA around the nucleosome.

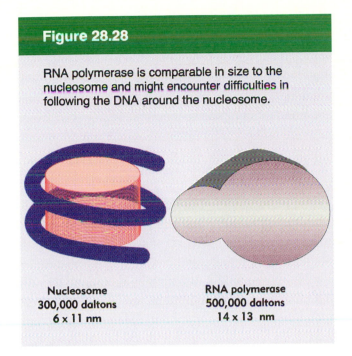

Nucleosome
300,000 daltons
6 x 11 nm

RNA polymerase
500,000 daltons
14 x 13 nm

question to ask about the structure of active genes is whether DNA being transcribed remains organized in nucleosomes. If the histone octamers are displaced, do they remain attached in some way to the transcribed DNA?

Attempts to visualize genes during transcription have produced varying results. In the intensively transcribed genes coding for rRNA, shown in **Figure 28.29**, the extreme packing of RNA polymerases makes it hard to see the DNA. We cannot directly measure the lengths of the rRNA transcripts because the RNA is compacted by proteins, but we know (from the sequence of the rRNA) how long the transcript must be. The length of the transcribed DNA segment, measured by the length of the axis of the 'Christmas tree,' is ~85% of the length of the rRNA. This means that the DNA is almost completely extended, and therefore cannot be organized in nucleosomes.

On the other hand, transcription complexes of SV40 minichromosomes can be extracted from infected cells. They contain the usual complement of histones and display a beaded structure. Chains of RNA can be seen to extend from the minichromosome, as in the example of

Figure 28.30. This argues that transcription can proceed while the SV40 DNA is organized into nucleosomes. Of course, the SV40 minichromosome is transcribed less intensely than the rRNA genes.

Another approach is to digest chromatin with micrococcal nuclease, and then to use a probe to some specific gene or genes to determine whether the corresponding fragments are present in the usual 200 bp ladder at the expected concentration. The conclusions that we can draw from these experiments are limited but important. *Genes that are being transcribed contain nucleosomes at the*

Figure 28.29

The extended axis of an rDNA transcription unit alternates with the only slightly less extended non-transcribed spacer. Photograph kindly provided by Charles Laird.

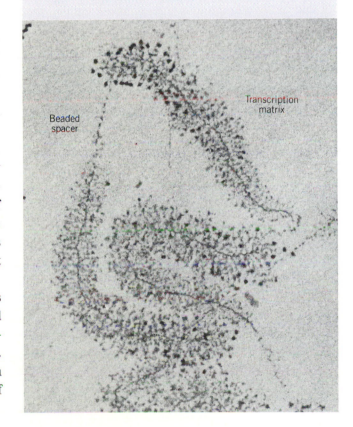

Beaded spacer

Transcription matrix

same frequency as nontranscribed sequences. Thus genes do not necessarily enter an alternative form of organization in order to be transcribed.

But since the proportion of the gene associated with RNA polymerases is rather small, this does not reveal what is happening at the sites actually engaged by the enzyme. Perhaps they retain their nucleosomes; more likely the nucleosomes are temporarily displaced as RNA polymerase passes by, but reform immediately afterward.

These experiments also show that active genes are more rapidly digested into monomeric fragments than are inactive sequences. This suggests that, although organized in nucleosomes, the structure of the active genes is in some way more exposed.

An indication that changes occur in the structure of the chromatin fiber during transcription is offered by the example of some heat-shock genes of *D. melanogaster*. These genes are transcribed rather infrequently prior to a heat shock, and in this condition they display the usual ladder when digested by micrococcal nuclease. When

Figure 28.30

An SV40 minichromosome can be transcribed. Photograph kindly provided by Pierre Chambon.

Figure 28.31

Micrococcal nuclease digestion generates the usual nucleosomal ladder before heat shock, but this is replaced by a continuous smear when intense transcription begins. Photograph kindly provided by Sarah Elgin.

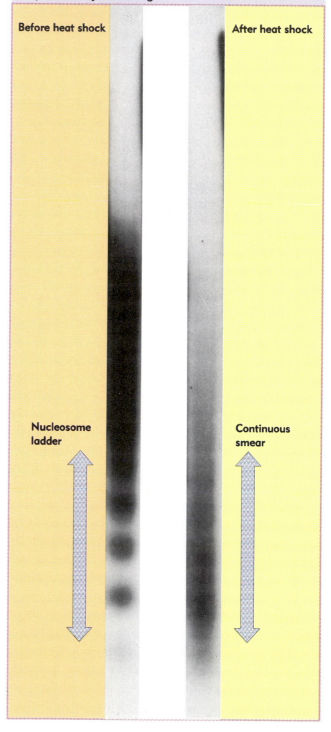

Before heat shock — After heat shock — Nucleosome ladder — Continuous smear

activated by heat shock, they are intensely transcribed. **Figure 28.31** shows that activation results in a considerable smearing of the ladder, implying that the nucleosomal organization has been changed.

We do not know the nature of this change. The ladder may disappear because it is obscured by the presence of RNA polymerase (or other proteins), because there is a disruption of higher-order structure, or because the nucleosomes have been modified or even displaced altogether.

When a heat-shock gene is transcribed, the density of contacts between histones and DNA is substantially reduced. However, histone H4 (and probably also other histones) remain associated with the gene. This raises the possibility that histone octamers remain associated with transcribed genes at the site of transcription in some form modified from the usual nucleosomal organization. One possibility is that the octamer associates with the displaced (nontemplate) single strand of DNA. Another possibility is that the $H3_2 \bullet H4_2$ kernel remains attached to DNA, but the H2A•H2B dimers are displaced.

It is clear that an important structural change occurs when a gene is intensely transcribed. In the case of the rRNA genes, it looks as though the nucleosomes are entirely displaced. But this could be an exceptional case. It might be reconciled with the presence of nucleosomes in less heavily transcribed genes by supposing that RNA polymerase displaces the nucleosome at the point of transcription, but that the histone octamer immediately recaptures its position unless another RNA polymerase is present to prevent it from doing so.

There are some indications that modification of histones is associated with structural changes that occur in chromatin at replication and transcription. All of the histones are modified by covalently linking extra moieties to the free groups of certain amino acids. Acetylation and methylation occur on the free (E) amino group of lysine. As seen in **Figure 28.32**, this removes the positive charge that resides on the NH_3^+ form of the group. Methyl-

ation also occurs on arginine and histidine. Phosphorylation occurs on the hydroxyl group of serine and also on histidine. This introduces a negative charge in the form of the phosphate group.

All of these modifications affect internal residues and are transient. Many of them occur at specific points in the cell cycle and (usually) are reversed at other points. Because they change the charge of the protein molecule, they have been viewed as potentially able to

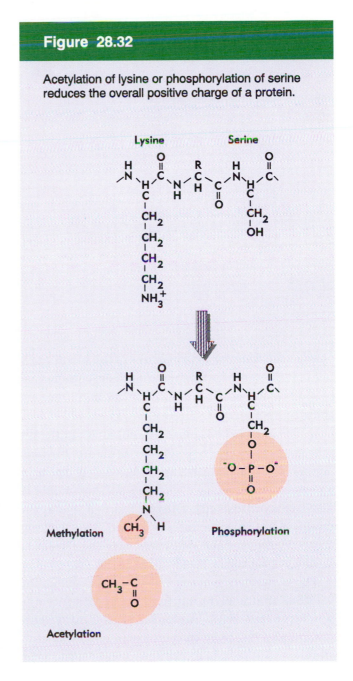

Figure 28.32

Acetylation of lysine or phosphorylation of serine reduces the overall positive charge of a protein.

Figure 28.33

Core histone acetylation occurs during DNA synthesis and then is reversed; the major phosphorylation of HI occurs at the start of mitosis and is reversed at the end of division.

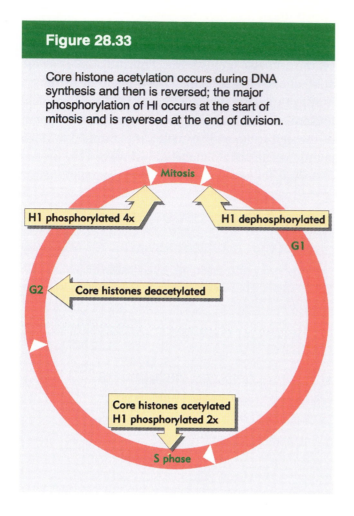

DNA, allowing the reaction to be better controlled. The idea has lost some ground in view of the observation that nucleosomes can be reconstituted, at least *in vitro*, with unmodified histones.

The transience of the acetylation event has been an obstacle to its study. The difficulty can be overcome by adding butyric acid to cells growing in culture. This treatment inhibits the enzyme histone deacetylase, so that acetylated nucleosomes accumulate. All the core histones are acetylated.

Acetylation is associated with changes in chromatin similar to those found on gene activation. The chromatin is more sensitive to DNAase I and (possibly) to micrococcal nuclease. However, it has not been possible to demonstrate any decisive relationship; and we do not have evidence for preferential acetylation of active genes. This result therefore tells us that acetylation can indeed affect the structure of chromatin, but the significance of the change remains to be seen. Acetylated histones can be identified in certain chromatin regions, and it is possible that a general change in acetylation may be associated with the inactivation of one X chromosome in mammalian females.

A cycle of phosphorylation and dephosphorylation occurs with H1, but its timing is different from the modification cycle of the other histones. With cultured mammalian cells, one or two phosphate groups are introduced at S phase. But the major phosphorylation event is the later addition of more groups, to bring the total number up to as many as six. This occurs at mitosis, as indicated in Figure 28.33. All the phosphate groups are removed at the end of the process of division. The phosphorylation of H1 is catalyzed by the M-phase kinase that provides an essential trigger for mitosis (see Chapter 13). In fact, this enzyme is now often assayed in terms of its H1 kinase activity. Not much is known about the phosphatase that removes the groups later.

The timing of the major H1 phosphorylation has prompted speculation that it is involved in mitotic condensation. Certainly this is consistent with the dependence of the 30 nm chromatin fiber on the presence of H1.

change the functional properties of the histones. At present there is no evidence that these changes are related to chromatin functions, although there are some quite provocative correlations.

In synchronized cells in culture, both the pre-existing and newly synthesized core histones appear to be acetylated and methylated during S phase (when DNA is replicated and the histones also are synthesized). During the cell cycle, the modifying groups are later removed. These events are repeated in each cell cycle, and their relative timing is summarized in **Figure 28.33.**

The coincidence of modification and replication suggests that acetylation (and methylation) could be connected with nucleosome assembly. One speculation has been that the reduction of positive charges on histones might lower their affinity for

DNAase hypersensitive sites change chromatin structure

Chromatin consists of DNA wound around a series of nucleosomes, more loosely packaged in regions that contain active genes. In addition to the general changes that characterized active or potentially active regions, structural changes occur at specific sites associated with initiation of transcription or with certain structural features in DNA. These changes were first detected by the effects of digestion with very low concentrations of the enzyme DNAase I.

When chromatin is digested with DNAase I, the first effect is the introduction of breaks in the duplex at specific, **hypersensitive sites**. Since susceptibility to DNAase I reflects the availability of DNA in chromatin, we take these sites to represent chromatin regions in which the DNA is particularly exposed because it is not organized in the usual nucleosomal structure. A typical hypersensitive site is 100× more sensitive to enzyme attack than is bulk chromatin. These sites are also hypersensitive to other nucleases and to chemical agents. They represent stretches of DNA devoid of nucleosomes.

Hypersensitive sites are created by the (tissue-specific) structure of chromatin. Their locations can be determined by the technique of indirect end labeling that we introduced earlier in the context of nucleosome phasing. This application of the technique is recapitulated in **Figure 28.34**.

Many of the hypersensitive sites are related to gene expression. Every active gene has a site of cutting, or sometimes more than one site, in the region immediately upstream of the promoter. *Most hypersensitive sites are found only in chromatin of cells in which the associated gene is being expressed*; they do not occur when the gene is inactive. The DNA sequences contained within the hypersensitive sites are required for gene expression, as seen by mutational analysis.

Although necessary, the presence of a hypersensitive site is not sufficient to ensure transcription. The timing of the appearance of the hypersensitive

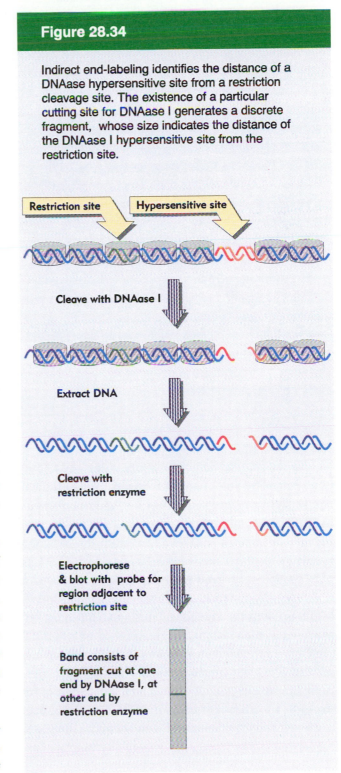

Figure 28.34

Indirect end-labeling identifies the distance of a DNAase hypersensitive site from a restriction cleavage site. The existence of a particular cutting site for DNAase I generates a discrete fragment, whose size indicates the distance of the DNAase I hypersensitive site from the restriction site.

Restriction site   Hypersensitive site

Cleave with DNAase I

Extract DNA

Cleave with restriction enzyme

Electrophorese & blot with probe for region adjacent to restriction site

Band consists of fragment cut at one end by DNAase I, at other end by restriction enzyme

Figure 28.35

The SV40 minichromosome has a nucleosome gap. Photograph kindly provided by Moshe Yaniv.

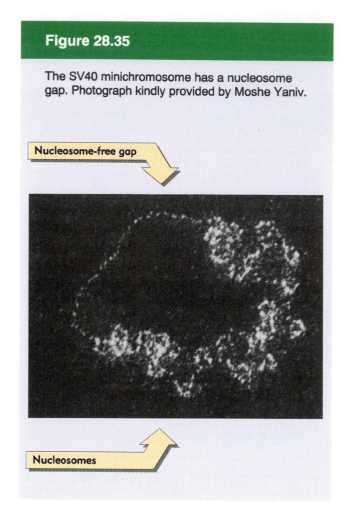

Nucleosome-free gap

Nucleosomes

and other nucleases (including restriction enzymes). The stretch over which the preferential digestion occurs is ~400 bp long, and the segment can be released as a fragment of free DNA.

The state of the SV40 minichromosome can be visualized by electron microscopy. In up to 20% of the samples, a 'gap' is visible in the nucleosomal organization, as evident in **Figure 28.35**. The gap is a region of ~120 nm in length (~350 bp), surrounded on either side by nucleosomes. The visible gap corresponds with the nuclease-sensitive region. This shows directly that increased sensitivity to nucleases is associated with the exclusion of nucleosomes. It does not imply that the DNA is free *in vivo*, because some other proteinaceous structure may have been lost in isolating the minichromosome.

The entire region of the gap is not uniformly sensitive to nucleases. Within a sensitive region of ~260 bp, there are two hypersensitive DNAase I sites and a 'protected' region. The map is given in **Figure 28.36**. The protected region presumably reflects the association of (nonhistone) protein(s) with the DNA.

The region of SV40 or polyoma surrounding

site relative to the onset of transcription is hard to establish in authentic situations, where it is difficult to obtain cells all at the same stage of development. All the indications, however, are that the 5′ hypersensitive site(s) appear before transcription starts, very likely as a prerequisite for initiation.

A hypersensitive site identifies a region of ~200 bp where access is not restricted in the manner typical of nucleosomes. A particularly well-characterized nuclease-sensitive region lies on the SV40 minichromosome. A short segment near the origin of replication, just upstream of the promoter for the late transcription unit, is cleaved preferentially by DNAase I, micrococcal nuclease,

Figure 28.36

The SV40 gap includes hypersensitive sites, sensitive regions, and a protected region of DNA.

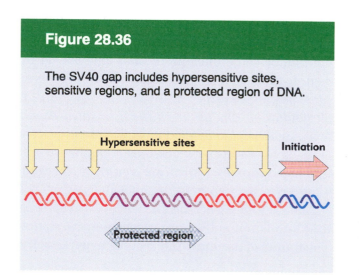

Hypersensitive sites

Initiation

Protected region

Figure 28.37

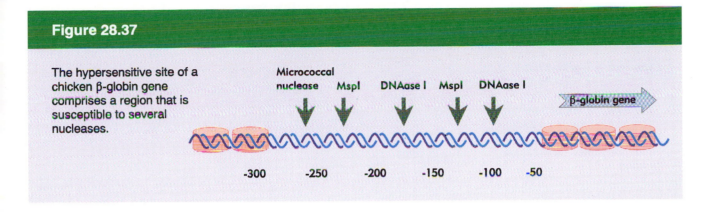

The hypersensitive site of a chicken β-globin gene comprises a region that is susceptible to several nucleases.

the nuclease-sensitive gap has several potential functions, including the initiation of replication and transcription, so the function of the gap cannot simply be equated with a particular activity. The gap is associated with DNA sequence elements which lie in this region and are necessary for promoter function.

An example of another hypersensitive site, at the β-globin promoter, is illustrated in **Figure 28.37**. This site is preferentially digested by several enzymes, including DNAase I, DNAase II, and micrococcal nuclease. The enzymes have preferred cleavage sites that lie at slightly different points in the same general region. Thus a region extending from about −70 to −270 is preferentially accessible to nucleases when the gene is transcribable.

What is the structure of the hypersensitive site? Its preferential accessibility to nucleases indicates that it is not protected by histone octamers, but this does not necessarily imply that it is free of protein. A region of free DNA might be vulnerable to damage; and in any case, why should it exclude nucleosomes? We assume that the hypersensitive site results from the binding of specific regulatory proteins that exclude nucleosomes. Indeed, the binding of such proteins is probably the basis for the existence of the protected region within the hypersensitive site. The acetylation of histones

could be necessary to induce structural changes that permit the formation of hypersensitive sites.

Hypersensitive sites associated with transcription may be generated by transcription factors. When a plasmid containing the hypersensitive region of the adult chick β-globin gene is recombined with histones in the presence of an extract from red blood cell nuclei, the relevant region becomes hypersensitive. The extract cannot confer hypersensitivity if it is added *after* the histones, which suggests that it must recognize DNA directly and in some way change the organization of the region prior to or during the deposition of nucleosomes. In Chapter 30, we discuss similar events at other loci, where transcription factors may have the ability to prevent histones from associating with DNA or (in some cases) to displace them. There is not necessarily a uniform way to generate hypersensitive sites, but the common feature may be the ability of a protein to bind to a specific sequence of DNA and to prevent nucleosomes from forming there.

The proteins that generate hypersensitive sites are likely to be regulatory factors of various types, since hypersensitive sites are found associated with promoters, other elements that regulate transcription, origins of replication, centromeres, and sites with other structural

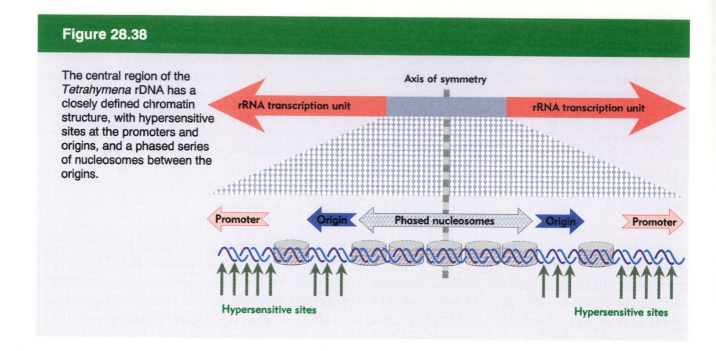

Figure 28.38

The central region of the *Tetrahymena* rDNA has a closely defined chromatin structure, with hypersensitive sites at the promoters and origins, and a phased series of nucleosomes between the origins.

significance. In some cases, they are associated with more extensive organization of chromatin structure. For example, a hypersensitive site may provide a boundary for a series of precisely phased nucleosomes.

An example of such a structure is shown in **Figure 28.38**, which is derived from the extrachromosomal rDNA of the macronucleus of *Tetrahymena pyriformis*. Recall that the rDNA molecule is a linear palindrome, in which the two rRNA transcription units lie some distance apart and are transcribed in opposite directions. Close to the center of the palindrome lie two origins of replication, short sequences at which the replication of DNA is initiated. The structure of the chromatin of this region is shown in the expanded portion of the map.

Several hypersensitive sites are located in the area. Each promoter has one, and so does each origin. Between the two origins are exactly five nucleosomes. The origins lie 1000 bp apart, and the spacing of the nucleosomes is precisely 200 bp, so every nucleosome has a defined posi-

tion. The locations of the histone octamers are probably defined by the boundaries of the hypersensitive sites on either side, rather than by any intrinsic property of the sequence that is phased.

How are hypersensitive (or other) sites propagated through DNA replication? A variety of mechanisms can be imagined. A heritable change might be made in the DNA itself—for example, demethylation. The structure of the DNA might be different from the regular double helix (through the mediation of proteins allowing the structure to reform after each replication). Proteins may form a complex (with duplex DNA) that is able to segregate at replication and then rebuild.

The stability of hypersensitive sites is revealed by the properties of chick fibroblasts transformed with temperature-sensitive tumor viruses. These experiments take advantage of an unusual property: although fibroblasts do not belong to the erythroid lineage, transformation of the cells at the normal temperature leads to activation of the globin genes. The activated genes have hypersensitive sites. If transformation is performed at the higher (nonpermissive) temperature, the globin genes are not activated; and hypersensitive sites do not appear. When the globin genes have been activated by transformation at low temperature, they can be inactivated by raising the temperature. But the hypersensitive sites are retained through at least the next 20 cell doublings.

This result demonstrates that acquisition of a hypersensitive site is only one of the features necessary to initiate transcription; and it implies that the events involved in establishing a hypersensitive site are distinct from those concerned with perpetuating it. Once the site has been established, it is perpetuated through replication in the absence of the circumstances needed for induction. Could some specific intervention be needed to abolish a hypersensitive site?

Regulation of domains

A region of the genome that contains an active gene may have an altered structure that results from the introduction of changes that are necessary for transcription. These changes precede, and are different from, the disruption of nucleosome structure that may be caused by the actual passage of RNA polymerase.

An indication that there is a change in the structure of transcribed chromatin is provided by its increased susceptibility to degradation by DNAase I. DNAase I sensitivity defines a chromosomal **domain**, a region of altered structure including at least one active transcription unit, and perhaps extending farther. (Note that use of the term 'domain' does not imply any necessary connection with the structural domains identified by the loops of chromatin or chromosomes.)

When chromatin is digested with DNAase I, it is eventually degraded into acid-soluble material (very small fragments of DNA). The progress of the overall reaction can be followed in terms of the proportion of DNA that is rendered acid soluble. *When only 10% of the total DNA has become acid soluble, more than 50% of the DNA of an active gene has been lost.* This suggests that active genes are preferentially degraded.

The fate of individual genes can be followed by quantitating the amount of DNA that survives to react with a specific probe. The protocol is outlined in **Figure 28.39**. The principle is that the loss of a particular band indicates that the corresponding region of DNA has been degraded by the enzyme.

Figure 28.40 shows what happens to β-globin genes and an ovalbumin gene in chromatin extracted from chicken red blood cells (in which globin genes are expressed and the ovalbumin gene is inactive). The restriction fragments representing the β-globin genes are rapidly lost, while those representing the ovalbumin gene show little degradation. (The ovalbumin gene in fact is digested at the same rate as the bulk of DNA.)

So the bulk of chromatin is relatively resistant to DNAase I and contains nonexpressed genes (as well as other sequences). *A gene becomes relatively susceptible to the enzyme specifically in the tissue(s) in which it is expressed.*

Is preferential susceptibility a characteristic only of rather actively expressed genes, such as globin, or of all active genes? Experiments using probes representing the entire cellular mRNA population suggest that all active genes, whether coding for abundant or for rare mRNAs, are preferentially susceptible to DNAase I. (However, there are variations in the degree of susceptibility.) Since

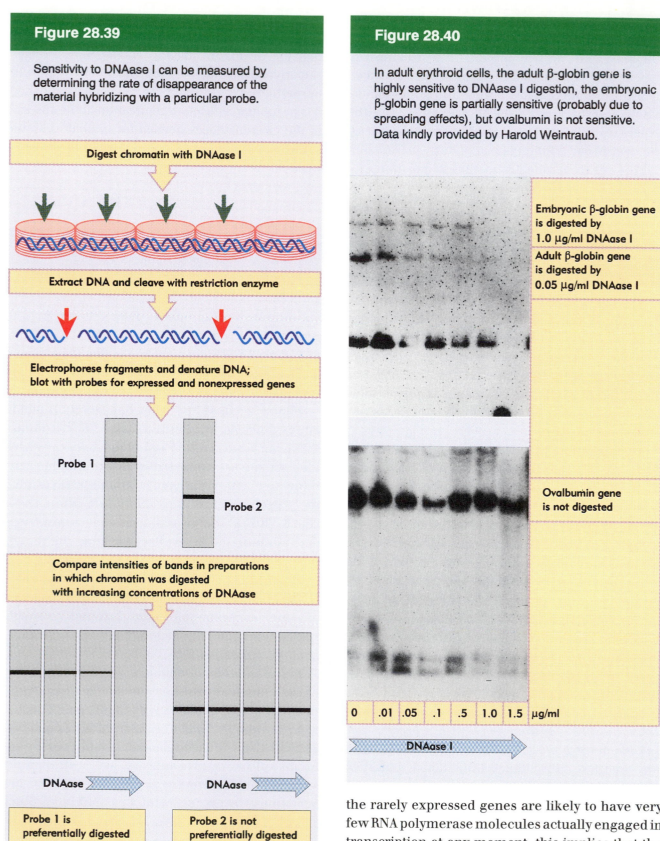

Figure 28.39

Sensitivity to DNAase I can be measured by determining the rate of disappearance of the material hybridizing with a particular probe.

Digest chromatin with DNAase I

Extract DNA and cleave with restriction enzyme

Electrophorese fragments and denature DNA; blot with probes for expressed and nonexpressed genes

Probe 1

Probe 2

Compare intensities of bands in preparations in which chromatin was digested with increasing concentrations of DNAase

DNAase

DNAase

Probe 1 is preferentially digested

Probe 2 is not preferentially digested

Figure 28.40

In adult erythroid cells, the adult β-globin gene is highly sensitive to DNAase I digestion, the embryonic β-globin gene is partially sensitive (probably due to spreading effects), but ovalbumin is not sensitive. Data kindly provided by Harold Weintraub.

Embryonic β-globin gene is digested by 1.0 µg/ml DNAase I

Adult β-globin gene is digested by 0.05 µg/ml DNAase I

Ovalbumin gene is not digested

| 0 | .01 | .05 | .1 | .5 | 1.0 | 1.5 | µg/ml |

DNAase I

the rarely expressed genes are likely to have very few RNA polymerase molecules actually engaged in transcription at any moment, this implies that the sensitivity to DNAase I does not result from the act of transcription, but is a feature of *genes that are able to be transcribed.*

What is the extent of the preferentially sensitive region? This can be determined by using a series of probes representing the flanking regions as well as the transcription unit itself. The sensitive region always extends over the entire transcribed region; an additional region of several kilobases on either side may show an intermediate level of sensitivity (probably as the result of spreading effects).

The critical concept implicit in the description of the domain is that a region of high sensitivity to DNAase I extends over a considerable distance. Often we think of regulation as residing in events that occur at a discrete site in DNA—for example, in the ability to initiate transcription at the promoter. Even if this is true, such regulation must determine, or must be accompanied by, a more wide-ranging change in structure. This is a difference between eukaryotes and prokaryotes.

An insight into the functional significance of such domains is provided by the mammalian β-globin genes. The need for an event that occurs outside of the regions of the structural genes themselves is indicated by the inability of an isolated gene to be properly expressed in an animal.

Recall from Figure 24.3 that the α-globin and β-globin genes in mammals each exist as clusters of related genes, expressed at different times during embryonic and adult development. These genes are provided with a large number of regulatory elements, which have been analyzed in detail. In the case of the adult human β-globin gene, regulatory sequences are located both 5′ and 3′ to the gene and include both positive and negative elements in the promoter region, and additional positive elements within and downstream of the gene.

But a human β-globin gene containing all of these control regions is never expressed in a transgenic mouse within an order of magnitude of wild-type levels. Some further regulatory sequence is required. Regions that provide the additional regulatory function are identified by DNAase I hypersensitive sites that are found upstream and downstream of the cluster. The map of **Figure 28.41** shows that the 20 kb upstream of the ε-gene contains a group of five sites; and there is a single site 30 kb downstream of the β-gene. Transfecting various constructs into mouse erythroleukemia cells shows that sequences between the individual sites in the 5′ region can be removed without much effect, but that removal of any of the sites reduces the overall level of expression. The mouse β-globin

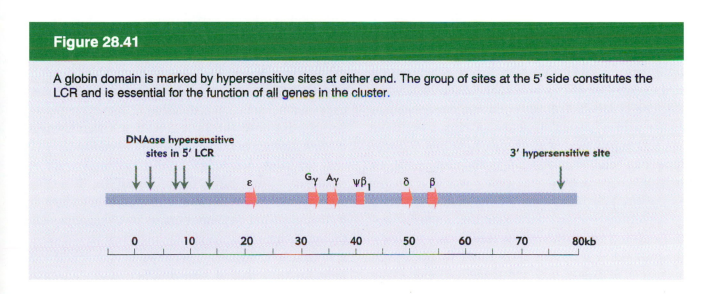

Figure 28.41

A globin domain is marked by hypersensitive sites at either end. The group of sites at the 5′ side constitutes the LCR and is essential for the function of all genes in the cluster.

DNAase hypersensitive sites in 5′ LCR

3′ hypersensitive site

ε G_γ A_γ $\psi\beta_1$ δ   β

0   10   20   30   40   50   60   70   80kb

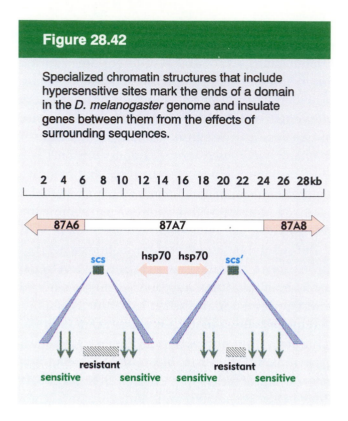

Figure 28.42

Specialized chromatin structures that include hypersensitive sites mark the ends of a domain in the *D. melanogaster* genome and insulate genes between them from the effects of surrounding sequences.

cluster contains a region that seems generally similar to the human upstream region, and which shows good sequence conservation (~70% around the sites examined).

The 5′ regulatory sites are the primary regulators, and the cluster of hypersensitive sites is called the **LCR** (locus control region). The appearance of the 3′ site depends on the 5′ sites: deletion of the 5′ LCR prevents appearance of the 3′ site. The LCR is absolutely required for expression of each of the globin genes in the cluster. Each gene is then further regulated by its own specific controls. Some of these controls are autonomous: expression of the ε- and γ-genes appears intrinsic to those loci in conjunction with the LCR. Other controls appear to rely upon position in the cluster, which provides a suggestion that *gene order* in a cluster is important for regulation.

The entire region containing the globin genes, and extending well beyond them, constitutes a chromosomal domain, as defined by increased sensitivity to digestion by DNAase I. Deletion of the

5′ LCR restores normal resistance to DNAase over the whole region. It is possible that the LCR is required to 'open up' the whole domain and make it available for transcription. The domain extends from an MAR 5′ to the hypersensitive sites to beyond the next genes, which appear to be ~100 kb 3′ from the β gene. The MAR may define the 5′ end of the domain; it is not clear what defines the 3′ end—perhaps an MAR for the adjacent cluster.

Some information about the delineation of domains is provided by the analysis of *D. melanogaster* genes summarized in **Figure 28.42**. Two genes for the 70,000 dalton heat-shock protein lie within an 18 kb region that constitutes band 87A7. The boundaries of the band are marked by sequences, denoted *scs* and *scs′* (specialized chromatin structures). They consist of 350 bp in *scs* or 200 bp in *scs′* of a region highly resistant to degradation by DNAase I, flanked on either side by hypersensitive sites, spaced at about 100 bp. The cleavage pattern at these sites is altered when the genes are turned on by heat shock.

The *scs* elements insulate the hsp70 genes from the effects of surrounding regions. If we take *scs* units and place them on either side of a *white* gene, the gene can function anywhere it is placed in the genome, even in sites where it would normally be repressed by context, for example, in heterochromatic regions.

The *scs* units do not seem to play either positive or negative roles in controlling gene expression, but just restrict effects from passing from one region to the next. If adjacent regions have repressive effects, however, the *scs* elements might be needed to block the spread of such effects, and therefore could be essential for gene expression. In this case, deletion of the *scs* elements could eliminate the function of the band, even though neither of the genes itself is essential.

If we now put together the various types of structures that have been found in different systems, we can think about the possible nature of a chromosomal domain. **Figure 28.43** summarizes the structures that might be involved in defining a domain. A domain (perhaps this might be a band in the case of *Drosophila*) might contain

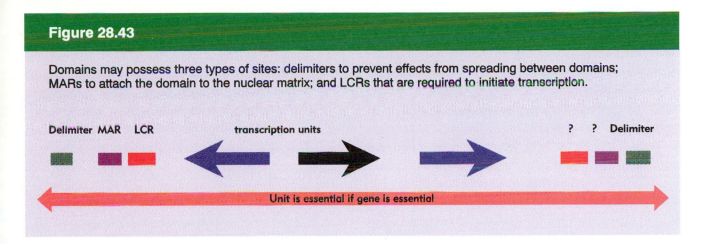

Figure 28.43

Domains may possess three types of sites: delimiters to prevent effects from spreading between domains; MARs to attach the domain to the nuclear matrix; and LCRs that are required to initiate transcription.

Delimiter MAR   LCR   transcription units   ?   ?   Delimiter

Unit is essential if gene is essential

several transcription units. Several types of *cis*-acting structures could be required for function. There may be a unit equivalent to the *scs* of the hsp70 band, which we might call a **delimiter**; something that stops activating or repressing effects spreading beyond the functional unit. This could be the same as, or could be separate from, a matrix attachment site. Either of these could be the same as or separate from an LCR, an essential structure whose function at a distance is essential for the ability of any and all genes in the unit to be expressed. These elements might be found at one end or at both ends of a domain.

Perhaps this type of organization could also help to explain the large size of the genome. A certain amount of space could be required for such a structure to operate, for example, to allow chromatin to become decondensed and to become accessible. Although the exact sequences of much of the unit might be irrelevant, there might be selection for the overall amount of DNA within it, or at least selection might prevent the various transcription units from becoming too closely spaced.

However, the generality of this type of organization is not clear. It does not necessarily apply to a gene cluster, since the α-globin cluster has a different organization, in which *cis*-acting control elements are dispersed over a wide region. Some of the control elements for α-genes are located outside of the α-cluster, and interspersed with adjacent genes that are constitutively expressed. Thus it is not possible in this case to define a discrete domain concerned exclusively with α-globin gene expression.

Gene expression is associated with demethylation

Methylation of DNA occurs at specific sites. In bacteria, it is associated with identifying the particular bacterial strain (see Chapter 20), and also with distinguishing replicated and nonreplicated DNA (see Chapter 19). In eukaryotes, its principal known function is connected with the control of transcription of genes in euchromatin, although methyl groups are also concentrated in the DNA of constitutive heterochromatin. From 2 to 7% of the cytosines of animal cell DNA are methylated (the value varies with the species). Most of the methyl groups are found in CG 'doublets,' and, in fact, the majority of the CG sequences are methylated. Usually the C residues

Figure 28.44

The restriction enzyme MspI cleaves all CCGG sequences whether or not they are methylated at the second C, but HpaII cleaves only nonmethylated CCGG tetramers.

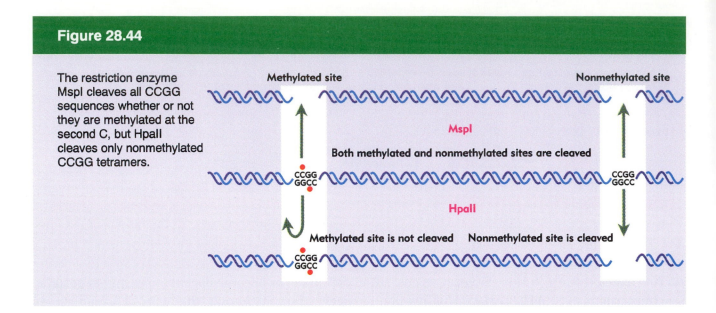

on both strands of this short palindromic sequence are methylated, giving the structure

$$5' \quad ^mCpG \quad 3'$$
$$3' \quad GpC^m \quad 5'$$

A doublet that instead is methylated on only one of the two strands is said to be hemimethylated (see Chapter 19).

The distribution of methyl groups can be examined by taking advantage of restriction enzymes that cleave target sites containing the CG doublet. Two types of restriction activity are compared in **Figure 28.44**. These **isoschizomers** are enzymes that cleave the same target sequence in DNA, but have a different response to its state of methylation.

The enzyme HpaII cleaves the sequence CCGG (writing the sequence of only one strand of DNA). But if the second C is methylated, the enzyme can no longer recognize the site. However, the enzyme MspI cleaves the same target site *irrespective* of the state of methylation at this C. So MspI can be used to identify all the CCGG sequences; and HpaII can be used to determine whether or not they are methylated.

With a substrate of nonmethylated DNA, the two enzymes would generate the same restriction bands. But in methylated DNA, the modified positions are not cleaved by HpaII. For every such position, one larger HpaII fragment replaces two MspI fragments. An example is given in **Figure 28.45**.

Many genes show a pattern in which the state of methylation is constant at most sites, but varies at others. Some of the sites are methylated in all tissues examined; some sites are unmethylated in all tissues. *A minority of sites are methylated in tissues in which the gene is not expressed, but are not methylated in tissues in which the gene is active.* Thus an active gene may be described as *undermethylated.*

As well as examining the state of methylation of resident genes, we can compare the results of introducing methylated or nonmethylated DNA into new host cells. Such experiments show a clear correlation: *the methylated gene is inactive, but the nonmethylated gene is active.*

What is the extent of the undermethylated region? In the chicken α-globin gene cluster in adult erythroid cells, the undermethylation is confined to sites that extend from ~500 bp upstream of the first of the two adult α genes to ~500 bp downstream of the second. Sites of undermethylation are present in the entire region, including the spacer between the genes. The region of undermethylation coincides rather well with the region of maximum sensitivity to DNAase I. This argues that undermethylation is a feature of a domain that contains a transcribed gene or genes.

Methylation at the 5' end of a gene may be directly related to expression. Many genes are not methylated at the 5' end when they are expressed, although they remain methylated at the 3' end. As with other changes in chromatin, it seems likely that the absence of methyl groups is associated with the *ability to be transcribed* rather than with the act of transcription itself.

Our problem in interpreting the general association between undermethylation and gene activation is that only a minority (sometimes a small minority) of the methylated sites are involved. It is likely that the state of methylation is critical at specific sites or in a restricted region; for example, demethylation at the promoter may be necessary to make it available for the initiation of transcription. It is also possible that a reduction in the level of methylation (or even the complete removal of methyl groups from some stretch of DNA) is part of some structural change needed to permit transcription to proceed.

In the γ-globin gene, for example, the presence of methyl groups in the region around the startpoint, between −200 and +90, suppresses transcription. Removal of the three methyl groups located upstream of the startpoint or of the three methyl groups located downstream does not relieve the suppression. But removal of all methyl groups allows the promoter to function. Transcription may therefore require a methyl-free region at the promoter.

There are exceptions to the general relationship we have described. Some genes can be expressed even when they are extensively methylated. Any connection between methylation and expression thus is not universal in an organism, but the general rule is that methylation prevents gene expression

and demethylation is required for expression.

The presence of **CpG-rich islands** in the 5' regions of some genes is connected with the effect of methylation on gene expression. These islands are detected by the presence of an increased density of the dinucleotide sequence, CpG.

The CpG doublet occurs in vertebrate DNA at only ~20% of the frequency that would be expected from the proportion of G•C basepairs. In certain regions, however, the density of CpG doublets reaches the predicted value; in fact, it is increased by ~10× relative to the rest of the genome.

These CpG-rich islands have an average G•C content of ~60%, compared with the 40% average in bulk DNA. They take the form of stretches of DNA typically ~1–2 kb long. There are ~30,000 such islands in a mammalian genome. As diagnosed by restriction analysis, the islands are unmethylated. The nucleosomes at the islands have a reduced content of histone H1 (which probably means that their packaging is relaxed), the other histones are extensively acetylated (a feature that tends to be associated with gene expression), and there are hypersensitive sites (as would be expected of active promoters).

In several cases, CpG-rich islands begin just upstream of a promoter and extend downstream

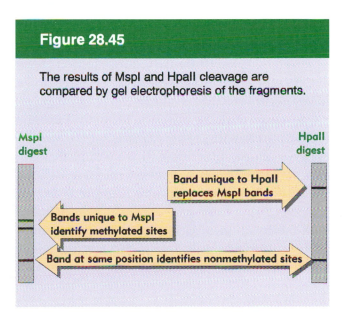

Figure 28.45

The results of MspI and HpaII cleavage are compared by gel electrophoresis of the fragments.

MspI digest

HpaII digest

Band unique to HpaII replaces MspI bands

Bands unique to MspI identify methylated sites

Band at same position identifies nonmethylated sites

into the transcribed region before petering out. **Figure 28.46** compares the density of CpG doublets in a 'general' region of the genome with a CpG island identified from the DNA sequence. Note that the 'general' region represents the 5' half of the γ^G globin gene, a gene under tissue-specific control. The CpG island surrounds the 5' region of the APRT gene, which is constitutively expressed.

Many of the genes with which the islands are associated are 'housekeeping' genes that are constitutively expressed, but CpG islands also occur at the promoters of tissue-regulated genes. In the latter case, the islands are unmethylated irrespective of the state of expression of the gene. The presence of unmethylated CpG-rich islands may be necessary, but therefore is not sufficient, for transcription. Thus the presence of unmethylated CpG islands may be taken as an indication that a gene is potentially active, rather than inevitably transcribed. Many islands that are nonmethylated in the animal become methylated in cell lines in tissue culture, and this could be connected with the inability of these lines to express all of the functions typical of the tissue from which they were derived.

Methylation of a CpG island at a promoter usually prevents expression of the gene. Repression is caused by either of two proteins that bind to methylated CpG sequences. The protein MeCP-1 requires the presence of several methyl groups to bind to DNA, while MeCP-2 can bind to a single methylated CpG base pair. This explains why a methylation-free zone is required for initiation of transcription. Binding of either these proteins prevents transcription *in vitro* by a nuclear extract.

It is clear that the absence of methyl groups is associated with gene expression. However, there are some difficulties in supposing that the state of methylation provides a general means for controlling gene expression. In the case of *D. melanogaster* (and other Dipteran insects), there is no methylation of DNA. The other differences between inactive and active chromatin appear to be the same as in species that display methylation. Thus in *Drosophila*, methylation

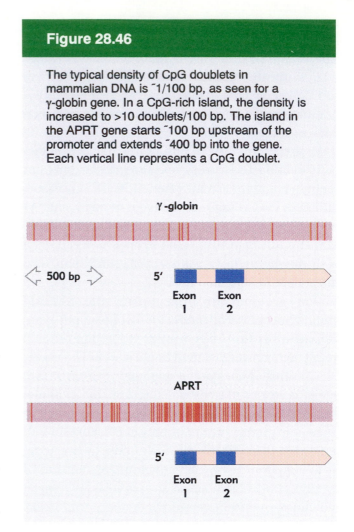

Figure 28.46

The typical density of CpG doublets in mammalian DNA is ˜1/100 bp, as seen for a γ-globin gene. In a CpG-rich island, the density is increased to >10 doublets/100 bp. The island in the APRT gene starts ˜100 bp upstream of the promoter and extends ˜400 bp into the gene. Each vertical line represents a CpG doublet.

γ-globin

500 bp

5′ Exon 1 Exon 2

APRT

5′ Exon 1 Exon 2

either is superfluous or is replaced by some other mechanism.

We have described three changes that occur in active genes:

◆ A hypersensitive site(s) is established near the promoter.

◆ The nucleosomes of a domain including the transcribed region become more sensitive to DNAase I.

◆ The DNA of the same region is undermethylated.

All of these changes are necessary for RNA polymerase to be able to initiate transcription at the promoter as described in Chapter 29.

Methylation is responsible for imprinting

Much attention has been paid to how the state of methylation might be perpetuated or changed. There may be no problem for genes that are constitutively expressed in all tissues: they may exist permanently in the demethylated state, in which the absence of methyl groups could provide a signal for their expression. But genes subject to tissue-specific control show changes in the state of methylation. They often display the inactive state in sperm: they are methylated at both the constant sites (which are methylated in all tissues) and the variable sites (those specifically unmethylated in expressed tissue). Thus the lack of certain methyl groups in the active state represents a *loss* of modifications that were previously present.

We do not know whether cellular genes regain methyl groups if and when they cease to be expressed. A critical question we should like to answer is how sequences are selected as the targets for tissue-specific changes in the state of methylation.

A simple model for the perpetuation of methylated sites is to suppose that the DNA methylase acts on hemimethylated DNA. As can be seen from **Figure 28.47**, replication of a fully methylated CG doublet produces two hemimethylated daughter duplexes. Recognizing each of the hemimethylated sites, the enzyme could convert it to the normal fully methylated state. (Compare with the methylation cycle for bacterial DNA shown in Figure 20.1.)

Such a model accords with the observation that, when methylated DNA is introduced into a cell, it continues to be methylated through an indefinite number of replication cycles, with a fidelity ~95% per site. If nonmethylated DNA is introduced, it is not methylated *de novo*. This implies that the enzyme recognizes *only* the hemimethylated sites. Its action allows the condition of a $\frac{CG}{GC}$ doublet—methylated or nonmethylated—to be perpetuated.

If this model is correct, an entirely different

enzyme activity must be involved in any creation of new sites of methylation (for example, if the methylated condition of a gene is restored when transcription ceases).

The methylated condition might be lost in either of two ways. Methyl groups might be actively removed by a demethylase enzyme. (However, no

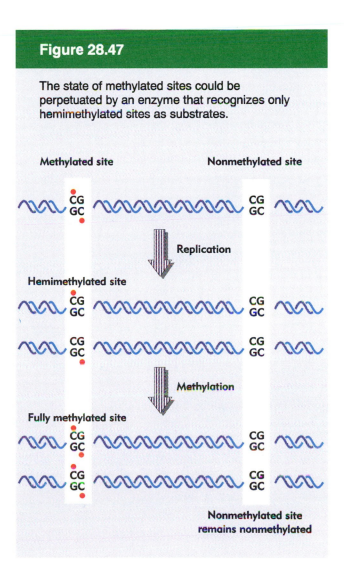

Figure 28.47

The state of methylated sites could be perpetuated by an enzyme that recognizes only hemimethylated sites as substrates.

Methylated site   Nonmethylated site

Hemimethylated site

Methylation

Fully methylated site

Nonmethylated site remains nonmethylated

Figure 28.48

Imprinting distinguishes the paternal and maternal alleles in the mouse embryo, and must be reset during gametogenesis in every generation.

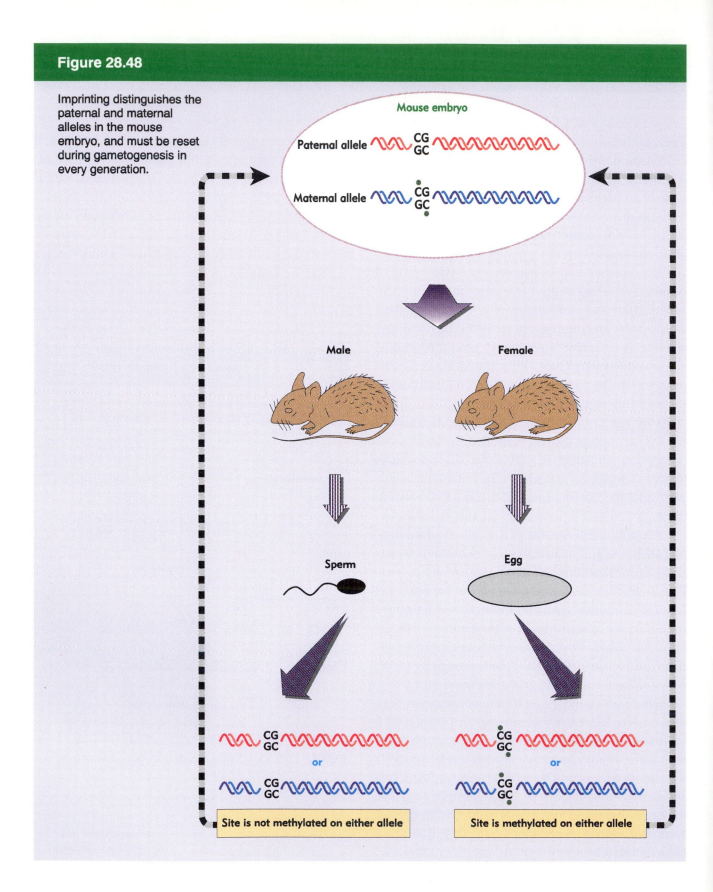

demethylase activity has yet been found.) Or methylation might simply fail to occur at a hemimethylated site generated by replication. One of the DNA duplexes produced by the next replication then would lack the methylated site.

Experiments with the drug 5-azacytidine produce indirect evidence that demethylation can result in gene expression. The drug is incorporated into DNA in place of cytidine, and cannot be methylated, because the 5′ position is blocked. This leads to the appearance of demethylated sites in DNA.

The phenotypic effects of 5-azacytidine include the induction of changes in the state of cellular differentiation; for example, muscle cells are induced to develop from nonmuscle cell precursors. The drug also activates genes on a silent X chromosome, which raises the possibility that the state of methylation could be connected with the condition of chromosomal inactivity.

A specific pattern of methyl groups may be inherited by germ cells, and this may be responsible for the phenomenon of **imprinting**, which describes a difference in behavior between the alleles inherited from each parent. The expression of certain genes in mouse embryos depends upon the sex of the parent from which they were inherited. For example, the allele coding for IGF-II (insulin-like growth factor II) that is inherited from the father is expressed, but the allele that is inherited from the mother is not expressed. The most likely explanation is that the IGF-II gene of oocytes is methylated, but the IGF-II gene of sperm is not methylated, so that the two alleles behave differently in the zygote. This is the most common pattern, but the dependence on sex is reversed for some genes.

This model requires that the pattern of methylation is established specifically during gametogenesis, and must be reversible. The fate of the two IGF-II alleles in a mouse is illustrated in **Figure 28.48**. In the early embryo, the paternal allele is nonmethylated and expressed, and the maternal allele is methylated and silent. What happens when this mouse itself forms gametes? If it is a male, the allele contributed to the sperm must be nonmethylated, irrespective of whether it was originally methylated or not. Thus when the maternal allele finds itself in a sperm, it must be demethylated. If the mouse is a female, the allele contributed to the egg must be methylated; so if it was originally the paternal allele, methyl groups must be added. We therefore require a system that can specifically reset the state of methylation, by either removing or adding methyl groups *de novo*.

The change in methylation pattern involves a series of events in which the state of individual genes is reset during gametogenesis and early development. All allelic differences are lost when primordial germ cells develop in the embryo; irrespective of sex, the previous patterns of methylation are erased, and a typical gene is then unmethylated. Differences in the methylation patterns of the sexes are established at different stages of gametogenesis in each sex.

In males, the pattern develops in two stages. Spermatocytes display the methylation pattern that is characteristic of mature sperm, so no further changes occur during spermatogenesis. But further changes are made in this pattern after fertilization.

In females, the maternal pattern is imposed during oogenesis, when oocytes mature through meiosis after birth.

Only a single methylase that adds methyl groups to CpG doublets has been identified. It is an essential gene, since mouse embryos in which this gene has been disrupted do not survive past early embryogenesis. We may speculate that methylation of the appropriate target genes is essential to prevent their inappropriate expression. We do not have information on the enzyme that is responsible for removing methyl groups. A major question is how the specificity of methylation is determined in the male and female gametes.

Summary

All eukaryotic chromatin consists of nucleosomes. A nucleosome contains a characteristic length of DNA, usually ~200 bp, wrapped around an octamer containing two copies each of histones H2A, H2B, H3, and H4. A single H1 protein is associated with each nucleosome. Virtually all genomic DNA is organized into nucleosomes. Treatment with micrococcal nuclease shows that the DNA packaged into each nucleosome can be divided operationally into two regions. The linker region is digested rapidly by the nuclease; the core region of 146 bp is resistant to digestion. Histones H3 and H4 are the most highly conserved and an H3•H4 tetramer accounts for the diameter of the particle. The H2A and H2B histones are organized as two H2A•H2B dimers. Octamers are assembled by the successive addition of two H2A•H2B dimers to the $H3_2$•$H4_2$ kernel, but it is not known whether this occurs before or after association with DNA.

The path of DNA around the histone octamer creates -1.8 supercoils. The DNA 'enters' and 'leaves' the nucleosome in the same vicinity, and could be 'sealed' by histone H1. Removal of the core histones releases -1.0 supercoils. The difference can be largely explained by a change in the helical pitch of DNA, from an average of 10.2 bp/turn in nucleosomal form to 10.5 bp/turn when free in solution. There is variation in the structure of DNA from a periodicity of 10.0 bp/turn at the nucleosome ends to 10.7 bp/turn in the center. There are kinks in the path of DNA on the nucleosome.

Nucleosomes are organized into a fiber of 30 nm diameter which has 6 nucleosomes per turn and a packing ratio of 40. Removal of H1 allows this fiber to unfold into a 10 nm fiber that consists of a linear string of nucleosomes. The 30 nm fiber probably consists of the 10 nm fiber wound into a solenoid. The 30 nm fiber is the basic constituent of both euchromatin and heterochromatin; nonhistone proteins are responsible for further organization of the fiber into chromatin or chromosome ultrastructure.

Nucleosomes may be excluded from very intensely transcribed genes by the presence of a continuous series of RNA polymerases, but appear to be present in other transcribed regions. What happens to a nucleosome as RNA polymerase passes through the region is not known. Chromatin capable of being transcribed has a generally increased sensitivity to DNAase, possibly reflecting a change in structure.

Hypersensitive sites in DNA are identified by greatly increased sensitivity to DNAase. They occur at sites that regulate transcription, at origins for replication, at centromeres, and other locations. A hypersensitive site consists of a sequence of ~200 bp from which nucleosomes are excluded by the presence of other proteins. A hypersensitive site forms a boundary from which a phased array of nucleosomes emanates. A group of hypersensitive sites upstream of the cluster of β-globin genes forms a locus control region (LCR) that is required for expression of all of the genes in the domain. The ends of domains may be marked by sequences that block the transmission of regulatory effects. Domains may also possess sites for attachment to the nuclear matrix.

CpG islands contain concentrations of CpG doublets and often surround the promoters of constitutively expressed genes, although they are also found at the promoters of regulated genes. The island including a promoter must be unmethylated for that promoter to be able to initiate transcription. A specific protein binds to the methylated CpG doublets and prevents initiation of transcription.

Sperm and eggs contain specific and different patterns of methylation, with the result that paternal and maternal alleles are differently expressed in the embryo. This is responsible for imprinting, which describes the fact that the allele inherited from one parent may be essential, while the other is silent. Patterns of methylation are reset during gamete formation in every generation.

Further reading

Reviews

The development of the nucleosome can be traced from the intuitive leap on which the model was originally propounded by **Kornberg** (*Science* **184**, 868–871, 1974) to the massive weight of evidence now assembled in reviews such as those of **Kornberg** (*Ann. Rev. Biochem.* **46**, 931–954, 1977) and **McGhee and Felsenfeld** (*Ann. Rev. Biochem.* **49**, 1115–1156, 1980).

The paths of DNA and nucleosomes have been succinctly analyzed by **Wang** (*Cell* **29**, 724–726, 1982) and **Felsenfeld and McGhee** (*Cell* **44**, 375–377, 1986). A model for the histone octamer was developed by **Richmond** *et al.* (*Nature* **311**, 532–537, 1984), who also cite earlier studies. The effects of structural variation in DNA on nucleosome positioning have been considered by **Travers and Klug** (*Phil. Trans. Roy. Soc. B* **317**, 537–561, 1987). The structure of active chromatin was reviewed by **Kornberg and Lorch** (*Ann. Rev. Cell Biol.* **8**, 563–587, 1992).

The existence of non-nucleosomal structures in the chromatin fiber has been brought up to date by **Eissenberg** *et al.* (*Ann. Rev. Genet.* **19**, 485–536, 1985). Hypersensitive sites have been reviewed by **Gross and Garrard** (*Ann. Rev. Biochem.* **57**, 159–197, 1988).

Speculations on methylation were the subject of a review from **Razin and Riggs** (*Science* **210**, 604–610, 1980).

Connections between CpG-rich islands, methylation, and gene expression have been explored by **Bird** (*Nature* **321**, 209–213, 1986).

Discoveries

Early models for the nucleosome from **Finch** *et al.* (*Nature* **269**, 29–36, 1977) and **Klug** *et al.* (*Nature* **287**, 509–516, 1980) were refined by **Arents** *et al.* (*Proc. Nat. Acad. Sci. USA* **88**, 10148–10152, 1991).

Hypersensitive sites were discovered in SV40 by **Varshavsky, Sundin, and Bohn** (*Nuc. Acids Res.* **5**, 3469–3479, 1978) and **Scott and Wigmore** (*Cell* **15**, 1511–1518, 1978); they were found in cellular DNA by **Wu** *et al.* (*Cell* **16**, 797–806, 1979).

The propagation of hypersensitive sites was followed by **Groudine and Weintraub** (*Cell* **30**, 131–139, 1982). The effects of the sites controlling the β-globin domain were described by **van Assendelft** *et al.* (*Cell* **56**, 969–977, 1989).

CpG-rich islands were reported by **Bird** *et al.* (*Cell* **40**, 91–99, 1985). The indirect nature of their effect on transcription was described by **Boyes and Bird** (*Cell* **64**, 1123–1134, 1991). Changes in methylation that may be responsible for imprinting were analyzed by **Chaillet** *et al.* (*Cell* **66**, 77–83, 1991).

Eukaryotic transcription and RNA processing

In the genetic programme, therefore, is written the result of all past reproductions, the collection of successes, since all traces of failures have disappeared. The genetic message, the programme of the present-day organism, therefore, resembles a text without an author, that a proof-reader has been correcting for more that two billion years, continually improving, refining and completing it, gradually eliminating all imperfections.

François Jacob, 1973

CHAPTER 29

Building the transcription complex: promoters, factors, and RNA polymerases

Transcription in eukaryotic cells is divided into three categories. Genes are classified by their promoters, and each class is transcribed by a different RNA polymerase. Ribosomal RNA is transcribed by RNA polymerase I, messenger RNA by RNA polymerase II, and tRNAs (and other small RNAs) by RNA polymerase III. As with bacterial RNA polymerase, accessory factors are needed for initiation, but are not required subsequently.

The balance of responsibilities *vis-à-vis* the accessory factors is shifted for eukaryotic RNA polymerases. Bacterial RNA polymerase has a *modus operandi* in which a basic enzyme undertakes transcription, assisted at certain promoters by accessory factors. For eukaryotic RNA polymerases, the *factors*, rather than the enzymes themselves, are principally responsible for recognizing the sequence components of the promoter.

The promoters for RNA polymerases I and II are (mostly) upstream of the startpoint, but some promoters for RNA polymerase III lie downstream of the startpoint. Each promoter contains characteristic sets of short conserved sequences that are recognized by the appropriate class of factors and/or RNA polymerase. RNA polymerases I and III each recognize a relatively restricted set of promoters, and rely upon a small number of accessory factors.

Promoters utilized by RNA polymerase II show more variation in sequence, and are modular in design. Short sequence elements that are recognized by transcription factors lie upstream of the startpoint. These *cis*-acting sites usually are spread out over a region of >100 bp. The sequences between them are not important, although their separation might be. Some of these elements and the factors that recognize them are common: they are found in a variety of promoters and are used constitutively. Others are specific: they identify particular classes of genes and their use is regulated. These elements occur in different combinations in individual promoters.

The number of factors that can act in conjunction with RNA polymerase II is large. We may divide them into three general groups. We consider the first two groups in this chapter, and we consider the control of transcription by the third group in the next chapter:

◆ The **basal factors** are required for the mechanics of initiating RNA synthesis at all promoters. They join with RNA polymerase to form a complex surrounding the startpoint, and they determine the site of initiation.

◆ The **upstream factors** are DNA-binding proteins that recognize specific short consensus elements located upstream of the startpoint. The activity of these factors is not regulated; they are ubiquitous, and act upon any promoter that contains the appropriate binding site on DNA. They increase the efficiency of initiation, and are required for a promoter to function at an adequate level. The precise set of such factors that is required for full expression is characteristic of any particular promoter.

◆ The **inducible factors** function just like the upstream factors, but have a regulatory role. They are synthesized or activated at specific times or in specific tissues, and they are therefore responsible for the control of transcription patterns in time and space. The sequences that they bind are called **response elements**.

A promoter that contains only elements recognized by basal and upstream factors should be transcribed in any cell type. Such promoters may be expected to be responsible for expression of cellular genes that are constitutively expressed (sometimes called housekeeping genes) or for the genes of viruses that can infect a wide range of host cells and that rely on the host transcription apparatus.

No element/factor combination is an essential component of the promoter, which suggests that initiation by RNA polymerase II may be sponsored in many different ways. The common feature is that the upstream or inducible transcription factors bind to sequence elements concentrated upstream of the startpoint. Binding of the factors to DNA is associated with the construction of a complex in which protein–protein interactions are important. The upstream and inducible factors function by interacting with the basal transcription apparatus, that is, by stimulating either basal factors or possibly RNA polymerase itself.

Sequence components of the promoter are defined operationally by the demand that they must be located in the general vicinity of the startpoint and are required for initiation. Another type of site involved in initiation is identified by sequences that enhance initiation, but that are located a considerable distance from the startpoint, either upstream or downstream of it, and in either orientation. Such sequences have been regarded as constituting another type of element, the **enhancer**. The components of an enhancer resemble those of the promoter; they consist of a variety of modular elements. However, the elements are usually contiguous. Enhancer elements are targets for tissue-specific or temporal regulation. They are not unique to RNA polymerase II, but (different) enhancers also act in conjunction with the promoter for RNA polymerase I.

A major question about eukaryotic transcription is how enhancers interact with promoters. How does one sequence affect another when they reside a considerable distance apart? The elements in the enhancer function like those in the promoter, but are less dependent on orientation and distance from the startpoint. Proteins bound at enhancer elements interact with proteins bound at promoter elements. We should therefore assess the construction of the transcription complex in terms of the individual elements and factors that recognize them, rather than be prejudiced by attempts to distinguish between promoters and enhancers. This view is fortified by the fact that some types of elements are found in both promoters and enhancers.

Any protein that is needed for the initiation of transcription, but which is not itself part of RNA polymerase, is defined as a transcription factor. Many transcription factors act by recognizing cis-acting sites that are classified as comprising parts of promoters or enhancers. We must recognize the possibility, however, that binding to DNA is not necessarily the only means of action for a transcription factor. A factor may recognize another factor, or may recognize RNA polymerase, or possibly may be incorporated into an initiation complex only in the presence of several other proteins. The ultimate test for membership of the transcription apparatus is functional: a protein must be needed for transcription to occur at a specific promoter or set of promoters.

A significant difference in the transcription of eukaryotic and prokaryotic mRNAs is that initiation at a eukaryotic promoter involves a large number of factors that bind to a variety of cis-acting elements.

The promoter is defined as the region containing all these binding sites, that is, which can support transcription at the normal efficiency and with the proper control. Thus the major feature defining the promoter for a eukaryotic mRNA is the location of binding sites for transcription factors; RNA polymerase itself binds around the startpoint, but does not directly contact the extended upstream region of the promoter. By contrast, the bacterial promoters we have discussed in Chapter 14 are largely defined in terms of the binding site for RNA polymerase in the immediate vicinity of the startpoint; other sequences nearby regulate (usually by repressing) the promoter, but are generally considered distinct from it.

The common mode of regulation of eukaryotic transcription appears to be positive: a transcription factor is provided under tissue-specific control to activate a promoter or set of promoters that contain a common target sequence. Thus a major class of regulatory proteins are defined as transcription factors (or proteins that regulate transcription factors). Regulation by specific repression of a target promoter is rare.

A eukaryotic transcription unit generally contains a single gene, and termination occurs beyond the end of the coding region. We should like to define the mechanism of termination, but it lacks the regulatory importance that applies in prokaryotic systems. RNA polymerases I and III terminate at discrete sequences in defined reactions, but the mode of termination by RNA polymerase II is not clear. However, the significant event in generating the 3′ end of an mRNA is not the termination event itself, but instead results from a cleavage reaction in the primary transcript.

Eukaryotic RNA polymerases consist of many subunits

The three eukaryotic RNA polymerases have different locations in the nucleus, as befit their responsibilities. Their general properties are defined in **Table 29.1**.

The most prominent activity is the enzyme RNA polymerase I, which resides in the nucleolus and is responsible for transcribing the genes coding for rRNA. It accounts for most cellular RNA synthesis.

The other major enzyme is RNA polymerase II, located in the nucleoplasm (the part of the nucleus excluding the nucleolus). It represents most of the remaining cellular activity and is responsible for synthesizing heterogeneous nuclear RNA (hnRNA), the precursor for mRNA.

Table 29.1

Eukaryotic nuclei have three RNA polymerases.

Enzyme	Location	Product	Relative Activity	α-Amanitin Sensitivity
RNA polymerase I	Nucleolus	Ribosomal RNA	50-70%	Not sensitive
RNA polymerase II	Nucleoplasm	Nuclear RNA	20-40%	Sensitive
RNA polymerase III	Nucleoplasm	tRNA	~10%	Species-specific

Table 29.2

Inhibitors of transcription act preferentially on particular enzymes.

Inhibitor	Target Enzyme	Inhibitory Action
Rifamycin	Bacterial holoenzyme	Binds to β to prevent initiation
Streptolydigin	Bacterial core enzyme	Binds to β to prevent initiation
Actinomycin D	Eukaryotic pol I	Binds to DNA & prevents elongation
α-Amanitin	Eukaryotic pol II	Binds to RNA polymerase II

RNA polymerase III is a minor enzyme activity. This nucleoplasmic enzyme synthesizes tRNAs and other small RNAs.

Inhibitors of transcription have been useful in distinguishing between the enzymes. Different inhibitors act on prokaryotic and eukaryotic enzymes. The properties of some common inhibitors are summarized in **Table 29.2**.

A major distinction between the eukaryotic enzymes is drawn from their response to the bicyclic octapeptide α-**amanitin**. In cells from origins as divergent as animals, plants, and insects, the activity of RNA polymerase II is rapidly inhibited by low concentrations of α-amanitin; RNA polymerase I is not inhibited. The response of RNA polymerase III to α-amanitin has not been so well

Figure 29.1

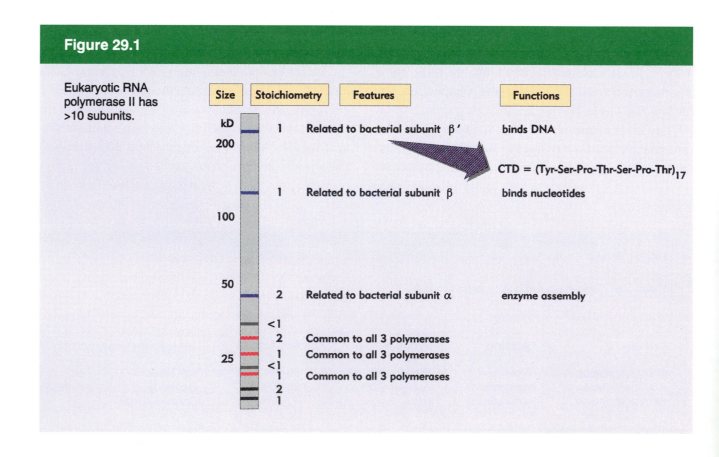

Eukaryotic RNA polymerase II has >10 subunits.

conserved; in animal cells it is inhibited by high levels, but in yeast and insects it is not inhibited.

All eukaryotic RNA polymerases are large proteins, appearing as aggregates of 500,000 daltons or more. They typically have 8–14 subunits. The purified enzyme can undertake template-dependent transcription of RNA, but is not able to initiate selectively at promoters. None of the eukaryotic enzymes has yet been reconstituted from purified subunits, which of course raises the question of whether all of the subunits are essential components. The only case in which all of the subunits have been defined at the level of characterizing their genes is that of *S. cerevisiae*, in which 10–11 subunits are required for transcription.

The general constitution of a eukaryotic RNA polymerase II enzyme as typified in *S. cerevisiae* is illustrated in **Figure 29.1**. The three largest subunits have homology to subunits of bacterial RNA polymerase. They appear to constitute the basic catalytic apparatus, possibly with the particular responsibilities indicated in the figure. The two largest subunits may carry the catalytic site. Three of the remaining subunits are common to all the RNA polymerases, that is, they are also components of RNA polymerases I and III. Most of the subunits are present in single copies, but three are present in two copies each.

The largest subunit in RNA polymerase II has a carboxy-terminal domain (**CTD**), which consists of multiple repeats of the consensus sequence Tyr-Ser-Pro-Thr-Ser-Pro-Ser. The sequence is unique to RNA polymerase II. There are ~26 repeats in yeast and ~50 in mammals. The number of repeats is important, because deletions that remove (typically) more than half of the repeats are lethal. The CTD can be highly phosphorylated on serine or threonine residues; the enzyme with the non-phosphorylated subunit is called II_a, and the phosphorylated form is called II_0. We see later that the CTD is involved in the initiation reaction.

The RNA polymerase activities of mitochondria and chloroplasts are smaller, and resemble bacterial RNA polymerase rather than any of the nuclear enzymes. Of course, the organelle genomes are much smaller, the resident polymerase needs to transcribe relatively few genes, and the control of transcription is likely to be very much simpler (if existing at all). So these enzymes are analogous to the phage enzymes that have a single fixed purpose and do not need the ability to respond to a more complex environment.

Promoter elements are defined by deletions, point mutations, and footprinting

Promoters have been defined in terms of their abilities to initiate transcription in suitable test systems. Having identified sequences that are needed for promoter function, we may then characterize the proteins that bind them. Several types of system have been used:

◆ The **oocyte system** follows the principles established for translation, and relies on injection of a suitable DNA template, this time into the nucleus of the *X. laevis* oocyte. The RNA transcript can be recovered and analyzed. The main limitation of this system is that it is restricted to the conditions that prevail in the oocyte. It allows characterization of DNA sequences, but not of the factors that normally bind them.

◆ **Transfection systems** allow exogenous DNA to be introduced into a cultured cell and expressed. (The procedure is discussed in Chapter 36.) Expression can be followed in a **transient assay**, as soon as the transfected DNA has entered the cell, and while it remains in the form of independent molecules; or **integrants** can be

Figure 29.2

Promoter boundaries can be determined by making deletions that progressively remove more material from one side. When one deletion fails to prevent RNA synthesis but the next stops transcription, the boundary of the promoter must lie between them.

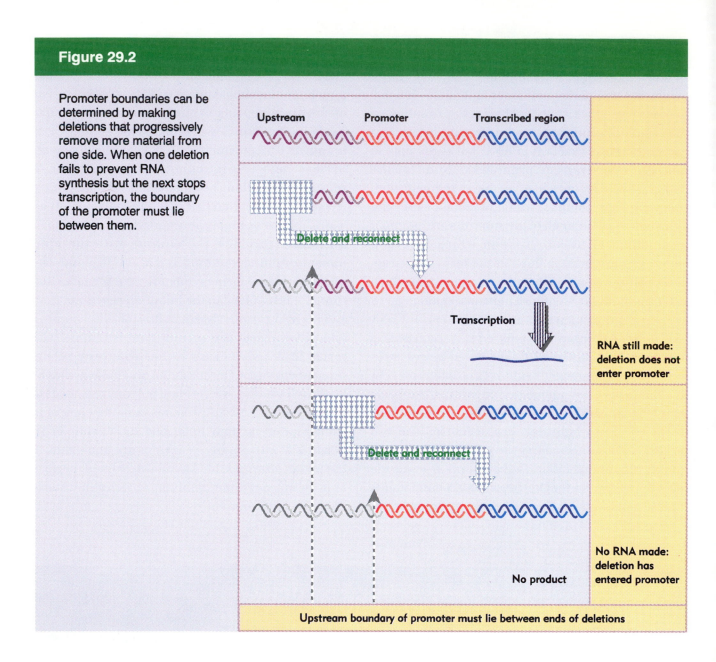

obtained in which the transfected material has become incorporated in the genetic material of the host cell line. The system is genuinely *in vivo* in the sense that transcription is accomplished by the same apparatus responsible for expressing the cell's own genome; but it differs from the natural situation because the template consists of a gene that would not usually be transcribed in the host cell. The usefulness of the system may be extended by using a variety of host cells.

◆ **Transgenic systems** involve the addition of a gene to the germ line of an animal. Expression of the **transgene** can be followed in any or all of the tissues of the animal. Some common limitations apply to transgenic systems and to stable transfection: the additional gene often is present in multiple copies, and is integrated at a different location from the endogenous gene.

◆ The *in vitro* system takes the classic approach of purifying all the components and manipulating conditions until faithful initiation is seen. 'Faithful' initiation is defined as production of an

RNA starting at the site corresponding to the 5′ end of mRNA (or rRNA or tRNA precursors).

The *in vivo* and *in vitro* approaches converge in the demand that a *bona fide* promoter element must be needed for transcription as shown by susceptibility to mutation, while a factor that binds to it must be required to initiate transcription in a suitable test system.

The approach to characterizing the promoter is the same in all these systems. We seek to manipulate the template *in vitro* before it is submitted to the system for transcription. Sometimes this is called 'surrogate genetics'.

We start with a particular fragment of DNA, perhaps quite large, that can initiate transcription in one of these systems. Then the boundaries of the sequence constituting the promoter can be determined by reducing the length of the fragment from either end, until at some point it ceases to be active. The type of protocol is illustrated in **Figure 29.2**. The boundary upstream can be identified by progressively removing material from this end until promoter function is lost. To test the boundary downstream, it is necessary to reconnect the shortened promoter to the sequence to be transcribed (since otherwise there is no product to assay).

Several precautions are required to avoid extraneous effects. To ensure that the promoter is always in the same context, the same long upstream sequence is always placed next to it. Because termination does not occur properly in the *in vitro* systems, the template is cut at some distance from the promoter (usually ~500 bp downstream), to ensure that all polymerases 'run off' at the same point, generating an identifiable transcript.

Once the boundaries of the promoter have been defined, the importance of particular bases within it can be determined by introducing point mutations or other rearrangements in the sequence. As with bacterial RNA polymerase, these can be characterized as *up* or *down* mutations. Some of these rearrangements affect only the *rate* of initiation; others influence the *site* at which initiation occurs, as seen in a change of the startpoint. To be sure that we are dealing with comparable products, in each case it is necessary to characterize the 5′ end of the RNA.

We can apply several criteria in identifying the sequence components of a promoter (or any other site in DNA):

◆ Mutations in the site prevent function *in vitro* or *in vivo*. (Many techniques now exist for introducing point mutations at particular base pairs, and in principle every position in a promoter can be mutated, and the mutant sequence tested *in vitro* or *in vivo*.)

◆ Proteins that act by binding to a site may be footprinted on it. There should be a correlation between the ability of mutations to prevent promoter function and to prevent binding of the factor.

◆ When a site recognized by a particular factor is present at multiple promoters, it should be possible to derive a consensus sequence that is bound by the factor. It should be possible to make a new promoter responsive to this factor by introducing an appropriate copy of the element.

RNA polymerase I has a bipartite promoter

Promoters for RNA polymerase I have the least diversity in the eukaryotic nucleus. RNA polymerase I transcribes only the genes for ribosomal RNA, from a single type of promoter. The transcript includes the sequences of both large and small rRNAs, which are later released by processing (as illustrated previously in Figure 16.20). There are many copies of the transcription unit, alternating with nontranscribed spacers, and organized in a cluster as discussed in Chapter 24. The

Figure 29.3

Promoters for RNA polymerase I consist of a core promoter, separated by ¯70 bp from the upstream control element. UBF1 binds to both regions, after which SL1 can bind. RNA polymerase I then binds to the core promoter. The nature of the interaction between the factors bound at the upstream control element and those at the core promoter is not known.

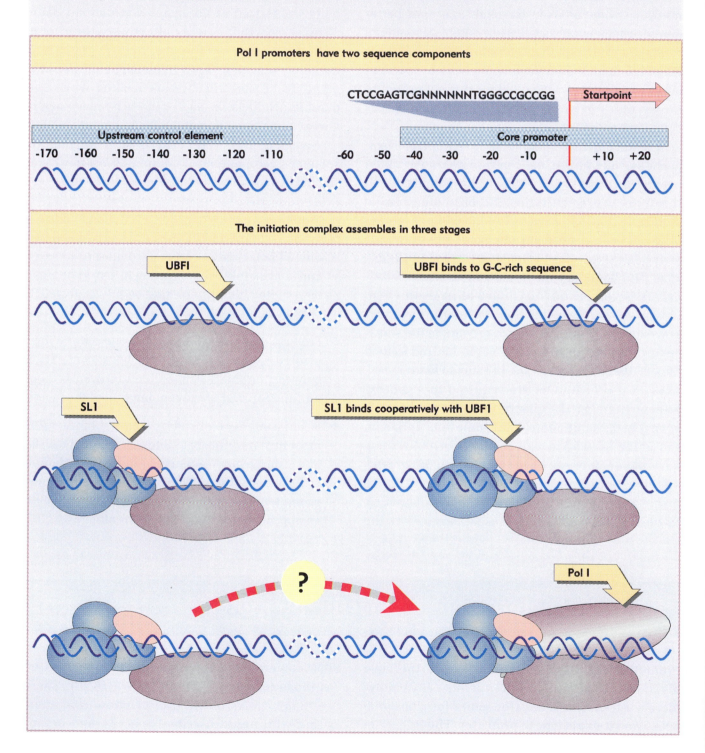

organization of the promoter, and the events involved in initiation, are illustrated in **Figure 29.3**.

The promoter has been best characterized in human cells, in which it consists of a two part sequence in the region preceding the startpoint. The **core promoter** surrounds the startpoint, extending from -45 to $+20$, and is sufficient for transcription to initiate. However, its efficiency is very much increased by the upstream control element (UCE), which extends from -180 to -107. Both regions have an unusual composition for a promoter, being rich in G•C base pairs; they are ~85% identical.

RNA polymerase I requires two ancillary factors. UBF1 binds in a sequence-specific manner to related sequences in the core promoter and UCE. Factor SL1 does not by itself have specificity for the promoter, but once UBF1 has bound, SL1 can bind cooperatively to extend the region of DNA that is covered. Once both factors are bound, RNA polymerase I can bind to the core promoter to initiate transcription. We assume that factors bound at the core promoter interact directly with RNA polymerase I, but we do not know how binding of the same factors at the UCE stimulates initiation in the core region. We see later that action at a distance is a prominent feature of initiation at promoters for RNA polymerase II, in which it has been investigated in more detail.

UBF1 is a single polypeptide that binds specifically to a G•C rich element in the core and UCE. UBF1 and RNA polymerase I can function effectively with heterologous templates; for example, the mouse factor and polymerase can recognize the human genes. Factor SL1 confers species specificity on the initiation reaction; mouse SL1 cannot sponsor initiation on a human template, and vice versa.

SL1 consists of 4 proteins. One of them, called TBP, is a factor that is required also for initiation by RNA polymerases II and III. We discuss its role in initiation by those polymerases shortly. The crucial feature with regards to RNA polymerase I is that TBP is well conserved between species, and is unlikely to be responsible for species specificity. Also, it is unlikely to bind specifically to G•C rich DNA, so probably both species specificity and DNA binding are responsibilities of the other components of SL1. It is likely that TBP interacts with RNA polymerase, possibly with a common subunit or a feature that has been conserved among polymerases.

The behavior of SL1 resembles a bacterial sigma factor. As an isolated protein complex, it does not bind specifically to the promoter, but in conjunction with other components, specific promoter regions are bound. It may have primary responsibility for ensuring that the RNA polymerase is properly localized at the startpoint. We see shortly that a comparable function is provided for RNA polymerases II and III by a factor that consists of TBP associated with other proteins. So a common feature in initiation by all three polymerase is a reliance on a 'positioning' factor that consists of TBP associated with proteins that are specific for each type of promoter.

RNA polymerase III uses both downstream and upstream promoters

Recognition of promoters by RNA polymerase III illustrates strikingly the relative roles of transcription factors and the polymerase enzyme. The promoters fall into two general classes that are recognized in different ways by different groups of factors. The promoters for 5S and tRNA genes are *internal*; they lie downstream of the startpoint. The promoters for snRNA (small nuclear RNA) genes lie

Figure 29.4

Deletion analysis shows that the promoter for 5S RNA genes is internal; initiation occurs a fixed distance (~55 bp) upstream of the promoter.

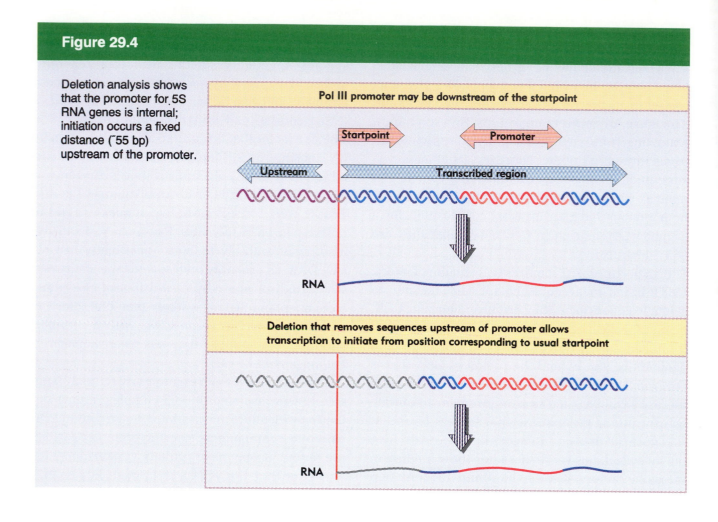

Pol III promoter may be downstream of the startpoint

Startpoint   Promoter

Upstream   Transcribed region

RNA

Deletion that removes sequences upstream of promoter allows transcription to initiate from position corresponding to usual startpoint

RNA

upstream of the startpoint in the more conventional manner of other promoters. In both cases, the individual sequences that are necessary for promoter function consist exclusively of sequences recognized by transcription factors, which in turn direct the binding of RNA polymerase.

Before the promoter of 5S RNA genes was identified in *X. laevis*, all attempts to identify promoter sequences assumed that they would lie upstream of the startpoint. But deletion analysis showed that the 5S RNA product continues to be synthesized when the entire sequence upstream of the gene is removed!

When the deletions continue into the gene, a product very similar in size to the usual 5S RNA continues to be synthesized so long as the deletion ends before about base +55. **Figure 29.4** shows that the first part of the RNA product corresponds to

plasmid DNA; the second part represents the segment remaining of the usual 5S RNA sequence. But when the deletion extends past +55, transcription does not occur. Thus the promoter lies *downstream of position +55*, but causes RNA polymerase III to initiate transcription a more or less fixed distance away. (The wild-type startpoint is unique; in deletions that lack it, transcription initiates at the purine base nearest to the position 55 bp upstream of the promoter.)

When deletions extend into the gene from its distal end, transcription is unaffected so long as the first 80 bp remain intact. Once the deletion cuts into this region, transcription ceases. This places the downstream boundary position of the promoter at about position +80.

So the promoter for 5S RNA transcription lies between positions +55 and +80 within the gene.

A fragment containing this region can sponsor initiation of any DNA in which it is placed, from a startpoint ~55 bp farther upstream.

The structures of three types of promoters for RNA polymerase III are summarized in **Figure 29.5**. Detailed analysis using individual mutations shows that there are two types of internal promoter. Each contains a bipartite structure, in which two short sequence elements are separated by a variable sequence. Comparisons between different promoters show that type 1 consists of a boxA sequence separated from a boxC sequence, and type 2 consists of a boxA sequence separated from a boxB sequence. The distance between boxA and boxB in a type 2 promoter can vary quite extensively, but the boxes usually cannot be brought too close together without abolishing function. We discuss the organization of the upstream type of promoter later.

Figure 29.6 summarizes the stages of reaction at internal promoters. Three accessory factors are involved. TFIIIA has been cloned (and is a member of an interesting class of zinc finger proteins that we discuss later). TFIIIB consists of TBP and two other proteins. TFIIIC is still used as a partially purified preparation; it is a large protein complex (>500,000 daltons), comparable in size to RNA polymerase itself, and containing at least 5 subunits.

At type 2 promoters (on the right side of the figure), TFIIIC recognizes boxB, but binds to a more extensive region including both boxes A and B. At type I promoters (on the left side of the figure), TFIIIA binds to a sequence that includes boxC, and this enables TFIIIC to bind farther downstream. The binding of TFIIIC in turn enables TFIIIB to bind to a sequence surrounding the startpoint in either type of promoter. Two proteins in the TFIIIB complex have been mapped on the DNA, binding together to the region from −40 to +11. TFIIIB binding is associated with additional binding by TFIIIC in type 1 promoters.

A crucial feature in defining the roles of the factors is that, at this point, TFIIIA and TFIIIC can be removed from the promoter (by high salt concentration *in vitro*) without affecting the initiation reaction. *TFIIIB remains bound in the vicinity of the startpoint and its presence is sufficient to allow RNA polymerase III to bind at the startpoint.* Thus TFIIIB is the only true initiation factor required by RNA polymerase III; TFIIIA and TFIIIC are **assembly factors**, whose role is to assist the binding of TFIIIB at the right location. This sequence of events explains how the promoter boxes downstream can cause RNA polymerase

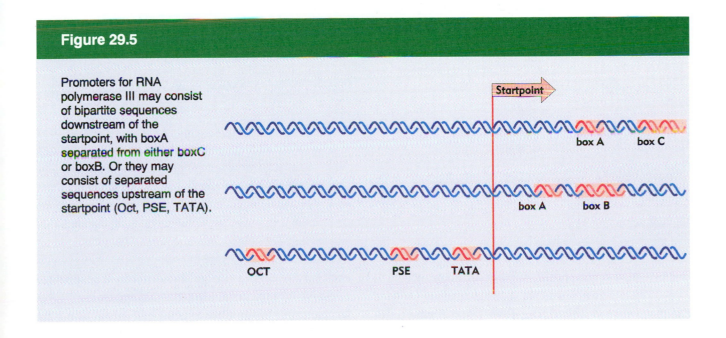

Figure 29.5

Promoters for RNA polymerase III may consist of bipartite sequences downstream of the startpoint, with boxA separated from either boxC or boxB. Or they may consist of separated sequences upstream of the startpoint (Oct, PSE, TATA).

Startpoint

box A   box C

box A   box B

OCT   PSE   TATA

Figure 29.6

Initiation via the internal pol III promoters involves the assembly factors TFIIIA and TFIIIC, the initiation factor TFIIIB, and RNA polymerase III. Note that the binding indicated for each factor is defined by the region of DNA that is protected, and is not defined in terms of the molecular action of the factor; in particular, TFIIIC protects the regions shown, but could act as a single protein or as multiple proteins.

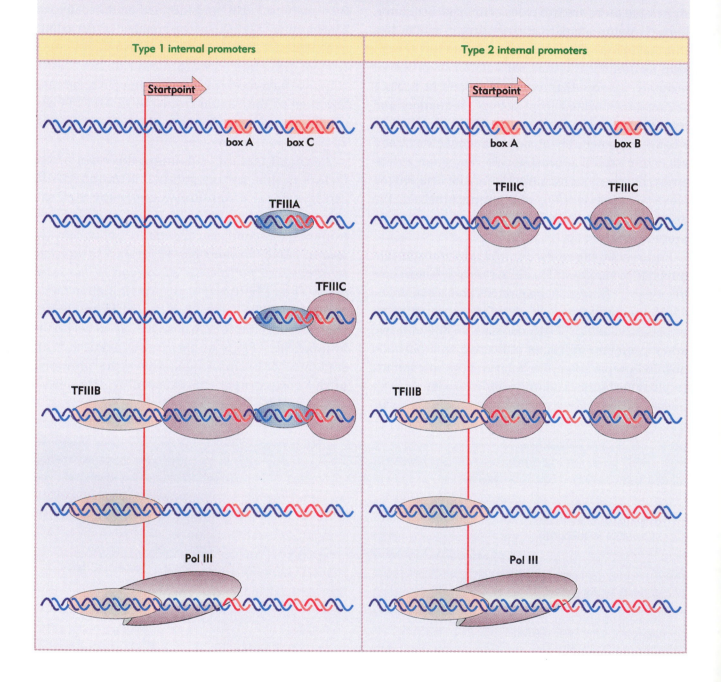

to bind at the startpoint, farther upstream.

So TFIIIB functions as a 'positioning factor,' responsible for localizing RNA polymerase cor-

rectly. Like SL1 at the pol I promoter, it resembles a sigma factor, in lacking the ability to bind DNA by itself, but being able to bind in conjunction with other proteins. Recall that TFIIIB includes the same protein, TBP, that is present in SL1, and this could be the subunit of TFIIIB that interacts directly with RNA polymerase III.

Although the ability to transcribe these genes is conferred by the internal promoter, the startpoint does have some influence. Changes in the region immediately upstream of the startpoint can alter the efficiency of transcription. Thus the primary responsibility for recognition lies with the internal promoter; but some responsibility for establishing the frequency of initiation lies with the region at the startpoint.

The upstream region has a more important role in the third class of polymerase III promoters. In the example shown in Figure 29.5, there are three upstream elements. These elements are also found in promoters for snRNA genes that are transcribed by RNA polymerase II. (Genes for some snRNAs [small nuclear RNAs] are transcribed by RNA polymerase II, while others are transcribed by RNA polymerase III.) The upstream elements function in a similar manner in promoters for both polymerases II and III. The TATA element appears to confer specificity for the type of polymerase.

Initiation at an upstream promoter for RNA polymerase III can occur on a short region that immediately precedes the startpoint and contains only the TATA element. However, efficiency of transcription is much increased by the presence of the PSE and OCT elements. The factors that bind at these elements interact cooperatively. (The PSE element may in fact be essential at promoters used by RNA polymerase II, whereas it is stimulatory in promoters used by RNA polymerase III; its name stands for proximal sequence element.)

The crucial TATA element is recognized by a factor that includes the TBP, the subunit that actually recognizes the sequence in DNA. The TBP is associated with other proteins, some of which are specific for promoters of the pol III type, and this may explain why RNA polymerase III is specifically recruited to these promoters. The function of TBP and its associated proteins is to position RNA polymerase III correctly at the startpoint.

Genetic evidence from yeast, and biochemical analysis of animal transcription extracts, suggest that the factor TBP is required for all initiation by RNA polymerase III. At a promoter that has a TATA element, TBP binds specifically to DNA, but at other promoters it may be incorporated by association with other proteins that bind to DNA. As a component of TFIIIB, it is involved in positioning RNA polymerase III during initiation using internal promoters; as a component of the SL1 factor, it positions RNA polymerase I at the rDNA promoter. We see shortly that it is a crucial member of the set of factors required for initiation by RNA polymerase II (where usually it acts by directly recognizing a TATA element). It is the key component of the 'positioning factor,' incorporated at each type of promoter by a different mechanism. Whatever its means of entry into the initiation complex, it has the common purpose of interaction with the RNA polymerase.

There is an underlying unity in the functions of the factors at RNA polymerase III promoters. *The factors bind at the promoter before RNA polymerase itself can bind.* They form a **preinitiation complex** that directs binding of the RNA polymerase. The alternative modes of promoter recognition by RNA polymerase III emphasize the importance of the factors for establishing the startpoint for initiation. RNA polymerase III does not itself seem to have a great intrinsic affinity for any particular sequence of DNA. It binds adjacent to factors that are themselves bound just upstream of the startpoint. For the type 1 and type 2 internal promoters, the assembly factors ensure that TFIIIB (which includes TBP) is bound just upstream of the startpoint, to provide the positioning information. For the upstream promoters, transcription factors that directly recognize the upstream sites form a complex (including TBP) that is recognized by RNA polymerase III. Thus irrespective of the location of the promoter sequences, factor(s) are bound close to the startpoint in order to direct binding of RNA polymerase III.

The basal transcription apparatus consists of RNA polymerase II and general factors

RNA polymerase II cannot initiate transcription itself, but is absolutely dependent on auxiliary transcription factors. The enzyme together with these factors constitutes the **basal transcription apparatus** that is needed to transcribe any promoter. Our starting point for considering promoter organization is therefore to define a 'generic' promoter, the shortest sequence at which RNA polymerase II can initiate transcription, and to characterize the enzyme subunits and transcription factors that are needed to recognize it.

A generic promoter can in principle be expressed in any cell: it does not depend on sequences whose use is under tissue-specific control. The accessory proteins that are required for polymerase II to initiate at such a promoter define the **general** or **basal transcription factors** involved in the mechanics of binding to DNA and initiating transcription. The basal factors are described as **TFIIX**, where 'X' is a letter that identifies the individual factor. A generic promoter functions at only a low efficiency; usually additional upstream factors would be required for a proper level of function. The upstream and inducible factors are not described systematically, but have casual names reflecting their histories of identification.

We may expect any sequence components involved in the binding of RNA polymerase and general transcription factors to be conserved at most or all promoters. As with bacterial promoters, when promoters for RNA polymerase II are compared, homologies in the regions near the startpoint are restricted to rather short sequences. These elements correspond with the sequences implicated in promoter function by mutation. They are even simpler in sequence than those found in bacterial promoters.

At the startpoint, there is no extensive homology of sequence, but there is a tendency for the first base of mRNA to be A, flanked on either side by pyrimidines. (This description is also valid for the CAT start sequence of bacterial promoters.) This region is called the **initiator** (**Inr**), and may be described in the general form Py_2CAPy_5. The Inr is contained between positions -3 and $+5$. A promoter consisting only of the Inr has the simplest possible form recognizable by RNA polymerase II.

Most promoters have a sequence called the **TATA box**, usually located ~25 bp upstream of the startpoint. It constitutes the only upstream promoter element that has a relatively fixed location with respect to the startpoint. It is found in all eukaryotes. The 8 bp consensus sequence consists entirely of A•T base pairs (at two positions the orientation is variable), and in only a minority of actual cases is a G•C pair present. The TATA box tends to be surrounded by G•C-rich sequences, which could be a factor in its function. It is almost identical with the -10 sequence found in bacterial promoters; in fact, it could pass for one except for the difference in its location at -25 instead of -10.

Single base substitutions in the TATA box act as strong down mutations. Some mutations reverse the orientation of an A•T pair, so base composition alone is not sufficient for its function. Thus the TATA box comprises an element whose behavior is analogous to our concept of the bacterial promoter: a short, well-defined sequence just upstream of the startpoint, which is necessary for transcription. The minority of promoters that do not contain a TATA element are called **TATA-less promoters**.

The major late promoter of adenovirus provides a useful example of a general promoter that contains a TATA box and a startpoint conforming with the Inr sequence. In the presence of the basal transcription factors, RNA polymerase II can initiate transcription at this promoter *in vitro*. This system has been used for much of the work to define the components required for the initiation reaction at a typical promoter. Most of the basal transcription

Figure 29.7

An initiation complex assembles at promoters for RNA polymerase II by an ordered sequence of association with transcription factors.

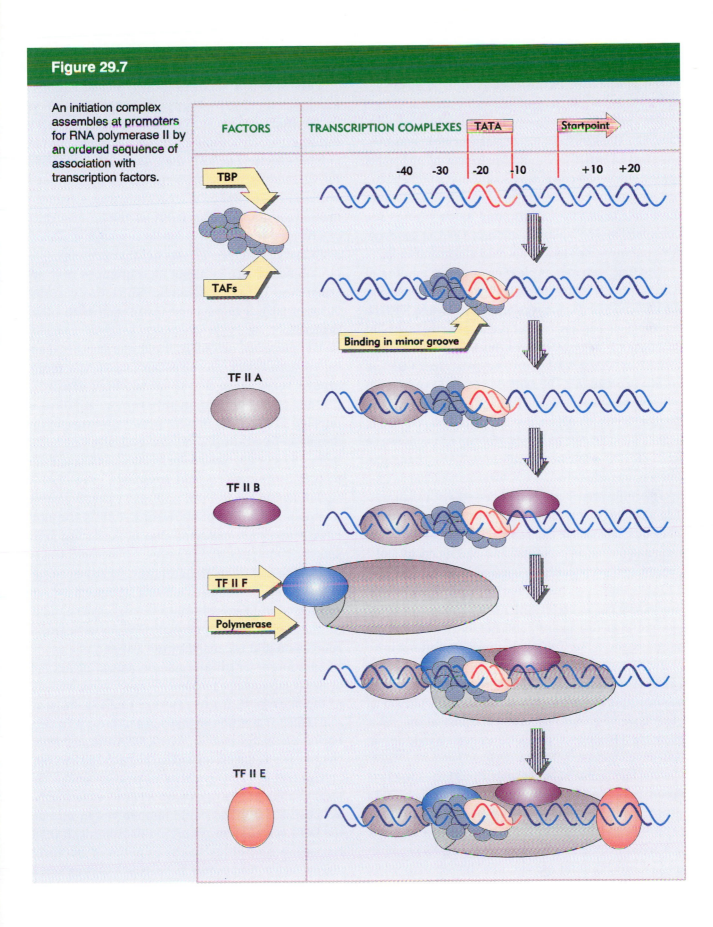

factors have been purified and their genes have been cloned.

Initiation requires the transcription factors to act in a defined order to build a complex that is joined by RNA polymerase. The series of events can be followed by the increasing size of the protein complex associated with DNA. Footprinting of the DNA regions protected by each complex suggests the model summarized in **Figure 29.7**. The TFII factors bind in a prescribed order, and as each joins the complex, an increasing length of DNA is covered. RNA polymerase is incorporated at a late stage.

The first step in complex formation is binding of a factor to a region that extends upstream from the TATA box. The factor was originally called **TFIID**, but now we recognize that it contains two types of component. Recognition of the TATA box is conferred by the **TATA-binding protein (TBP)**, a small protein of ~30,000 daltons. A variety of other subunits are called **TAFs** (for TBP-associated factors). Some TAFs are stoichiometric with TBP; others are present in lesser amounts. It is possible that TFIIDs containing different TAFs recognize different promoters.

The idea that TBP is associated with different TAFs can be extended to the promoters recognized by other polymerases. The factor TFIIIB used by internal pol III promoters, and SL1 used by pol I promoters, may both be viewed as consisting of TBP associated with a particular group of TAFs. Any individual molecule of TBP itself is not necessarily available for all promoters, but may in effect be sequestered by its TAFs to be used continuously by a specific class of promoter. TBP must have the capacity to interact appropriately with the variety of factors and/or polymerases that are employed at each type of promoter, and mutations in the yeast gene for TBP can specifically affect transcription at different promoter classes.

The TBP has attracted much attention as the first factor to contact DNA, and it is sometimes thought of as a 'commitment factor' that in effect consigns a promoter to be transcribed. Because the TATA box is a fixed distance from the startpoint, its recognition is important for 'positioning' RNA polymerase. TBP plays a similar role at promoters for each class

of polymerase, the difference being in how it is incorporated into the transcription complex.

TBP itself is a small protein (~30,000 daltons), and it has the unusual property of binding to DNA in the minor groove. (Virtually all known DNA-binding proteins bind in the wide groove.) The crystal structure of TBP suggests a detailed model for its binding to DNA. **Figure 29.8** shows that it surrounds one face of DNA, forming a 'saddle' around the double helix. In effect, the inner surface of TBP binds to DNA, and the larger outer surface is available to extend contacts to other proteins. The DNA-binding site consists of sequences that are conserved between species, while the variable N-terminal tail is exposed to interact with other proteins.

The presence of TBP in the minor groove, combined with other proteins binding in the major groove, must mean that there is a high density of protein–DNA contacts in this region. Binding of purified TBP to DNA *in vitro* protects ~1 turn of the double helix at the TATA box, typically extending from −37 to −25; but binding of the TFIID complex in the initiation reaction regularly protects the region from −45 to −10, and also extends farther upstream beyond the startpoint. TBP is the only basal transcription factor that makes sequence-specific contacts with DNA.

When TFIIA joins the complex, TFIID becomes able to protect a region extending farther upstream. TFIIA contains several subunits (two in yeast, three in mammals). It may activate TBP by relieving a repression that is caused by the TAFs.

Addition of TFIIB gives some partial protection of the region of the template strand in the vicinity of the startpoint, from −10 to +10. This suggests that TFIIB is bound downstream of the TATA box, perhaps loosely associated with DNA and asymmetrically oriented with regard to the two DNA strands.

The factor TFIIF consists of two subunits. The larger subunit (RAP74) has an ATP-dependent DNA helicase activity that could be involved in melting the DNA at initiation. The smaller subunit (RAP38) has some homology to the regions of bacterial sigma factor that contact the core polymerase; it binds tightly to RNA polymerase II. TFIIF may in

Figure 29.8

A view in cross-section shows that TBP surrounds DNA from the side of the narrow groove. The sequence of TBP consists of two related (40% identical) domains, which are conserved in different species, and are shown in light and dark blue. The N-terminal region varies extensively and is shown in green. The two strands of the DNA double helix are in light and dark grey. Photograph kindly provided by Stephen Burley.

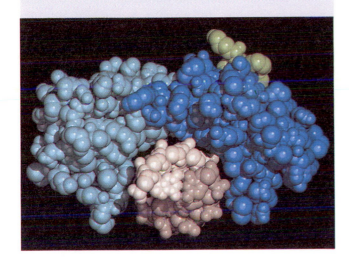

tern of binding to DNA. TFIIH has a kinase activity that can phosphorylate the CTD tail of RNA polymerase II. It is possible that phosphorylation of the tail is needed to release RNA polymerase II from the transcription factors so that it can leave the promoter and start elongation.

The general process of initiation is similar to that catalyzed by bacterial RNA polymerase. Binding of RNA polymerase generates a closed complex, which is converted at a later stage to an open complex in which the DNA strands have been separated. The conversion is associated with increased accessibility of DNA on the downstream side of the complex, perhaps because TFIIE is lost. A separate stage of the initiation reaction requires hydrolysis of ATP, possibly to release some of the transcription factors. The interactions between the basal factors and RNA polymerase II remain to be fully described, but it seems likely that one important reaction occurs between TBP and the CTD tail of the enzyme.

What happens at TATA-less promoters? The same basal transcription factors, including TFIID, are needed. It is not clear how the involvement of TFIID relates to its role in binding the TATA box. The Inr provides the positioning element; perhaps an additional factor binds to the Inr and 'anchors' the complex. At all events, TFIID must enter the complex in some other way than binding the TATA box. The function of TBP at these promoters is more like that at promoters for RNA polymerase I and at internal promoters for RNA polymerase III.

Many of the basal factors consist of multiple subunits, so the total number of polypeptides involved in the basal apparatus is rather large. There are probably ~20 polypeptides with a total mass of ~500,000 daltons. Remember that RNA polymerase II itself has ~10 subunits with a mass of ~500,000 daltons, and we see that initiation involves the assembly of an extremely large complex.

Assembly of the RNA polymerase II initiation complex provides an interesting contrast with prokaryotic transcription. Bacterial RNA polymerase is essentially a coherent aggregate with intrinsic ability to bind DNA; the sigma factor, needed for initiation but not for elongation,

fact bring RNA polymerase II to the assembling transcription complex and provide the means by which it binds. Interaction with TFIIB may be important when TFIIF–polymerase joins the complex.

Polymerase binding extends the sites that are protected downstream to +15 on the template strand and +20 on the nontemplate strand. The enzyme extends the full length of the complex, since additional protection is seen at the upstream boundary.

TFIIE can bind at this point and causes the boundary of the region protected downstream to be extended by another turn of the double helix, to +30.

Two further factors, TFIIH and TFIIJ, join the complex after TFIIE. They do not change the pat-

becomes part of the enzyme before DNA is bound, although it is later released. But RNA polymerase II can bind to the promoter only after separate transcription factors have bound. The factors play a role analogous to that of bacterial sigma factor—to allow the basic polymerase to recognize DNA specifically at promoter sequences—but have evolved more independence. Indeed, the factors are primarily responsible for the specificity of promoter recognition. We assume that these factors are released following initiation. The process of assembling the transcription complex reminds us of ribosome subunit assembly, in which ribosomal proteins must bind to rRNA (or to other proteins in the complex) in a certain order. We note also that only some of the factors participate in protein–DNA contacts (and only TBP makes sequence-specific contacts), and that protein–protein interactions are just as important in the assembly of the complex.

The sequences in the vicinity of the startpoint comprise a 'core' promoter at which the general transcription apparatus is assembled. When a TATA box is present, it determines the location of the startpoint. Its deletion causes the site of initiation to become erratic, although any overall reduction in transcription is relatively small. Indeed, some TATA-less promoters lack unique startpoints; initiation occurs instead at any one of a cluster of startpoints. The TATA box aligns the RNA polymerase (via the interaction with TFIID and other factors) so that it initiates at the proper site. This explains why its location is fixed with respect to the startpoint.

Although assembly can take place just at the core promoter *in vitro*, this reaction is not sufficient for transcription *in vivo*, where interactions with other factors that recognize the more upstream elements are required. These factors interact with the basal apparatus at various stages during its assembly.

The level of binding sponsored directly by the core promoter *in vitro* represents a residual or basal level of activity. Indeed, a general difficulty in characterizing promoters *in vitro* is to know how a given level of expression relates to the efficiency of use *in vivo*.

Promoters for RNA polymerase II contain elements consisting of short sequences

A promoter for RNA polymerase II consists of two types of region. The startpoint itself is identified by the Inr and/or by the TATA box close by. In conjunction with the basal transcription factors, RNA polymerase II forms an initiation complex surrounding the startpoint, as we have just described. The efficiency and specificity with which a promoter is recognized, however, depends upon short sequences, farther upstream, which are recognized by upstream or inducible factors. These sequences

and the factors that recognize them may be quite common, and found in a wide variety of promoters, or they may be rather specific, and particular for transcription in a restricted time or place. Usually these sequences are ~100 bp upstream of the startpoint, but sometimes they are more distant. Binding of factors at these sites may influence the formation of the initiation complex at (probably) any one of several stages.

The short sequences that are needed for

initiation are dispersed over the upstream region. An analysis of a typical promoter is summarized in **Figure 29.9**. Individual base substitutions were introduced at almost every position in the 100 bp upstream of the β-globin startpoint. The striking result is that *most mutations do not affect the ability of the promoter to initiate transcription when it is introduced into a HeLa cell.* Down mutations occur in three locations, corresponding to three short discrete **elements**. The two upstream elements have a greater effect on the level of transcription than the element closest to the startpoint. Up mutations occur in only one of the elements. We conclude that the three short sequences centered at −30, −75, and −90 constitute the promoter. Each of them corresponds

to the consensus sequence for a common type of promoter element.

The TATA box (centered at −30) is the least effective component of the promoter as measured by the reduction in transcription that is caused by mutations. But although initiation is not prevented when a TATA box is mutated, the startpoint varies from its usual precise location. This confirms the role of the TATA box as a crucial positioning component of the core promoter.

The sequence at −75 is the **CAAT box**. Named for its consensus sequence, it was one of the first common elements to be described. It is often located close to −80, but it can function at distances that vary considerably from the startpoint. It functions in either orientation. Susceptibility to

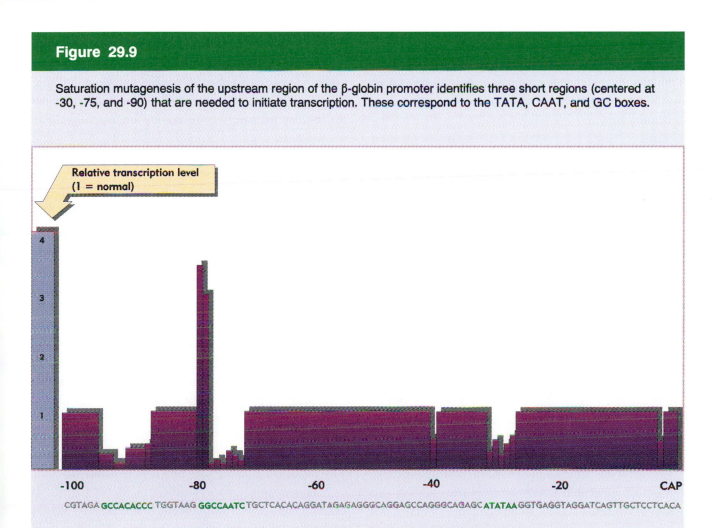

Figure 29.9

Saturation mutagenesis of the upstream region of the β-globin promoter identifies three short regions (centered at -30, -75, and -90) that are needed to initiate transcription. These correspond to the TATA, CAAT, and GC boxes.

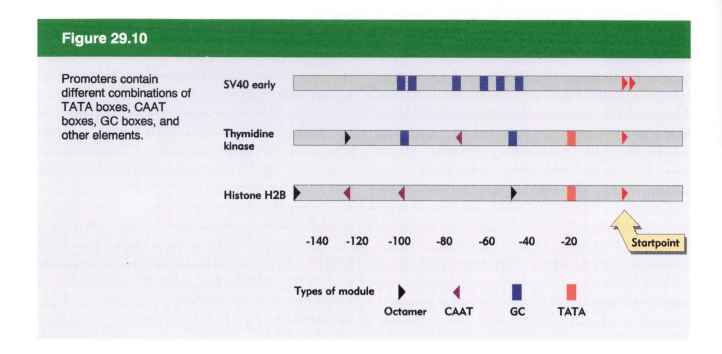

Figure 29.10

Promoters contain different combinations of TATA boxes, CAAT boxes, GC boxes, and other elements.

SV40 early

Thymidine kinase

Histone H2B

−140   −120   −100   −80   −60   −40   −20   Startpoint

Types of module   Octamer   CAAT   GC   TATA

mutations suggests that the CAAT box plays a strong role in determining the efficiency of the promoter. It does not appear to play a direct role in promoter specificity, but its inclusion increases promoter strength.

The **GC box** at −90 contains the sequence GGGCGG. Often multiple copies are present in the promoter, and they occur in either orientation. It too is a relatively common promoter component.

Promoters are organized on a principle of 'mix and match'. A variety of elements can contribute to promoter function, but none is essential for all promoters. Some examples are summarized in **Figure 29.10**. Four types of element are found altogether in these promoters: TATA, GC boxes, CAAT boxes, and the octamer (an 8 bp element). The elements found in any individual promoter differ in number, location, and orientation. No element is common to all of the promoters. One of the puzzles of promoter organization is that the promoter conveys directional information (transcription proceeds only in the downstream direction), but the GC and CAAT boxes seem to be able to function in either orientation (although their sequences are asymmetrical).

Factors that are more or less ubiquitous are assumed to be available to any promoter that has a copy of the element that they recognize. This

common availability distinguishes the upstream factors from the inducible factors that we discuss later. Elements in the upstream category include the CAAT box, GC box, and the octamer. All promoters probably require one or more of these elements in order to function efficiently.

The GC box is recognized by the factor SP1. This interaction illustrates the demands that can be placed on a single factor. The closest GC box usually is 40–70 bp upstream of the startpoint, but the context of the GC boxes is different in every promoter. Thus in the thymidine kinase promoter, GC boxes are adjacent to a CAAT box and a TATA box, but in the SV40 promoter, a tandemly repeated series of GC boxes is upstream of a TATA box. The subunit of SP1 is a monomer of 105,000 daltons, which contacts one strand of DNA over a ~20 bp binding site that includes at least one 6 bp GC box. In the SV40 promoter, the multiple boxes between −70 and −110 all are bound, so that the whole region is protected by SP1. In the thymidine kinase promoter, however, SP1 presumably interacts with a factor bound at the CAAT box on one side, and with TFIID bound at the TATA box on the other side.

The sequences to which the factors bind as characterized by footprinting are typically longer than the consensus sequences identified by comparing promoters. **Table 29.3** summarizes the properties of the most common factors. They usually cover ~20 bp of DNA, whereas the consensus sequences are <10 bp. Given the sizes of the factors, and the length of DNA each covers, we expect that the various proteins will together cover the entire region upstream of the startpoint in which the elements reside.

The diversity of elements from which a functional promoter may be constructed, and the variation in their locations relative to the startpoint, argues that the factors have an ability to interact with one another by protein–protein interactions in multiple ways. There appear to be no constraints on the potential relationships between the elements. The modular nature of the promoter is illustrated by experiments in which equivalent regions of different promoters have been exchanged. Hybrid promoters, for example, between thymidine kinase and β-globin, work well. This suggests that the main purpose of the elements is bring the factors they bind into the vicinity of the initiation complex, where protein–protein interactions determine the efficiency of the initiation reaction.

The basal elements and the more upstream elements have different types of function. We have seen that the basal elements (the TATA box and Inr) primarily determine the location of the startpoint, but can sponsor initiation only at a rather low level.

The sequence elements farther upstream, such as the GC or CAAT boxes, influence the frequency of initiation, most likely by acting directly on the basal transcription factors to enhance their assembly into an initiation complex.

The sequences between the upstream elements are irrelevant, and the distance between them is flexible; the separation between them usually can be changed by 10–30 bp before they become unable to function. How can initiation be influenced by sites spread over a length of DNA that is greater than RNA polymerase could directly contact? Initiation involves a hierarchy of interactions, in which factors bound at upstream elements interact with basal factors, which in turn interact directly with RNA polymerase. This helps to explain the flexibility with which elements may be arranged, and the distance over which they can be dispersed, since it relieves us of the obligation to suppose that factors bound to all these elements must interact directly with RNA polymerase.

The simplicity of the elements recognized by upstream factors disguises a complexity of interactions. We cannot assume *a priori* that a given element will be recognized by a unique transcription factor. The ubiquitous CAAT box provides an example of an element that may be recognized by any one of several factors. Although it is itself ubiquitously used, it may also provide a target for regulation.

Table 29.3

Upstream transcription factors bind to sequence elements that are common to mammalian RNA polymerase II promoters.

Module	Consensus	DNA bound	Factor	Size (daltons)	Abundance (/cell)	Distribution
TATA box	TATAAAA	~10 bp	TBP	27,000	?	General
CAAT box	GGCCAATCT	~22 bp	CTF/NF1	60,000	300,000	General
GC box	GGGCGG	~20 bp	SP1	105,000	60,000	General
Octamer	ATTTGCAT	~20 bp	Oct-1	76,000	?	General
"	"	23 bp	Oct-2	52,000	?	Lymphoid
κB	GGGACTTTCC	~10 bp	NFκB	44,000	?	Lymphoid
"	"	~10 bp	H2-TF1	?	?	General
ATF	GTGACGT	~20 bp	ATF	?	?	General

Figure 29.11

A transcription complex involves recognition of several elements in the sea urchin H2B promoter in testis. Binding of the CAAT displacement factor in embryo prevents the CAAT- binding factor from binding, so an active complex cannot form.

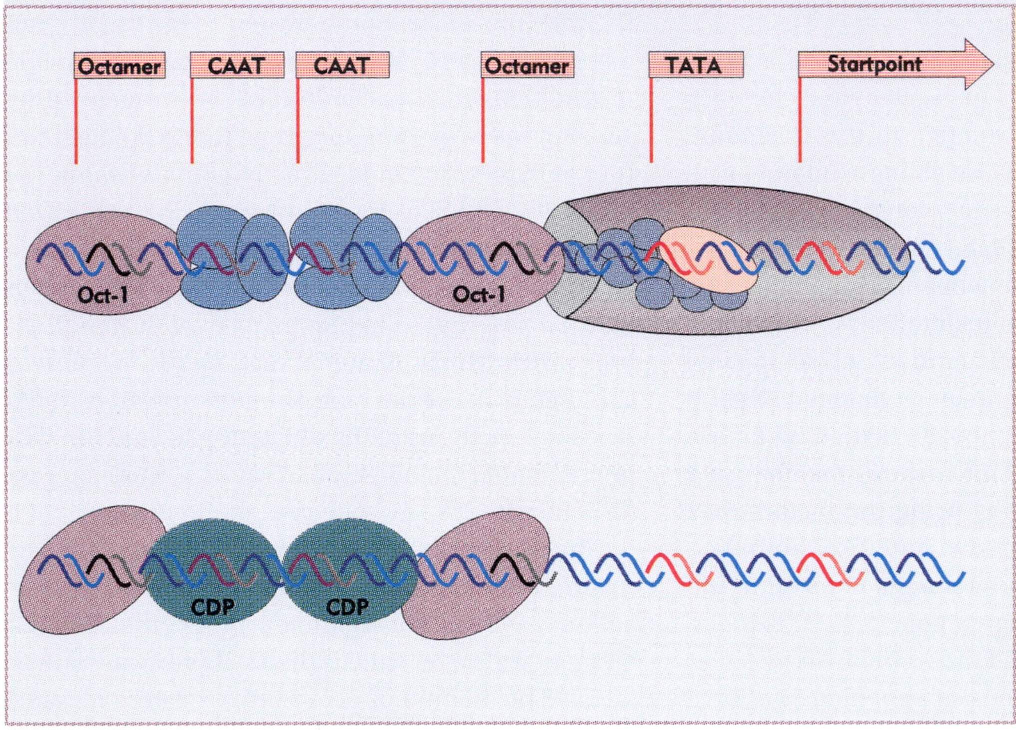

Active

Transcription complex assembles in testis (not all components of basal apparatus are shown)

Inactive

CDP prevents other factors from binding to CAAT box, and basal factors either bind incorrectly or are prevented from binding.

One group of factors that recognizes the CAAT box is the CTF family, which is generated by alternative splicing from a single gene. The CTF members apparently have the same affinities for CAAT boxes. Another family has different properties. CP1 binds with high affinity to the CAAT boxes of α-globin and late adenovirus, CP2 binds with high affinity to a CAAT box in a γ-fibrinogen gene, and CP3 has different specificities. Counterparts to CP1 are found in yeast, and can substitute for it in a mammalian system, suggesting that the use of the CAAT box is an old evolutionary mechanism.

Other proteins also bind CAAT boxes, including two from rat liver; C/EBP prefers the sequence GCAAT, while ACF binds to the more usual sequence CCAAT in the albumin promoter. The exact details of recognition are not so important as the fact that a variety of factors recognize CAAT boxes.

We do not know what other sequences might influence the preferences of these various factors, or whether and how one of these factors rather than another is used to recognize the CAAT box at a particular promoter. Their existence, however, makes the point that *the identification of a conserved element does not inevitably imply that it will always be recognized by the same protein in all promoters.*

The CAAT promoter box may be a target for regulation. Two copies of this element are found in the promoter of a gene for histone H2B (see Figure 29.10) that is expressed only during spermatogenesis in a sea urchin. CAAT-binding factors can be extracted from testis tissue and also from embryonic tissues, but only the former can bind to the CAAT box. In the embryonic tissues, another protein, called the CAAT-displacement protein (CDP), binds to the CAAT boxes, *preventing the transcription factor from recognizing them.*

Figure 29.11 illustrates the consequences for gene expression. In testis, the promoter is bound by transcription factors at the TATA box, CAAT boxes, and octamer sequences. In embryonic tissue, the exclusion of the CAAT-binding factor from the promoter prevents a transcription complex from being assembled. Gene expression is prevented by inhibiting binding of the CAAT transcription factor. The analogy with the effect of a bacterial repressor in preventing RNA polymerase from initiating at the promoter is obvious. These results also make the point that the function of a protein in binding to a known promoter element cannot be assumed: it may be an activator, a repressor, or even irrelevant to gene transcription.

Another example of a element that is recognized by more than one factor is presented by the octamer sequence. A ubiquitous transcription factor, Oct-1, binds to the octamer to activate the H2B (and presumably also other) genes. Oct-1 is the only octamer-binding factor in nonlymphoid cells. But in lymphoid cells, a different factor, Oct-2, binds to the octamer to activate the immunoglobulin κ light gene. Thus Oct-2 is a tissue-specific activator, while Oct-1 is ubiquitous.

The use of the same octamer in the ubiquitously expressed H2B gene and the lymphoid-specific immunoglobulin genes poses a paradox. Why does the ubiquitous Oct-1 fail to activate the immunoglobulin genes in nonlymphoid tissues?

The *context* must be important: Oct-2 rather than Oct-1 may be needed to interact with other proteins that bind at the promoter. These results mean that we cannot predict whether a gene will be activated by a particular factor simply on the basis of the presence of particular elements in its promoter.

It is usually the case that a particular consensus sequence is recognized by a corresponding transcription factor. But in some cases the same sequence (such as the octamer) can be recognized by different proteins (that is, Oct-1 and Oct-2). And there are also cases in which a particular protein can recognize more than one type of sequence. The best characterized example is the protein C/EBP, which binds to the CAAT box, but which also binds to a quite different sequence element. We do not know whether a single domain on the protein recognizes both target sites, or whether a single protein has more than one DNA-binding domain.

A pertinent factor in considering transcription *in vitro* is that the template exists in the form of an accessible DNA molecule. *In vivo* it is organized into nucleosomes, which suggests that its recognition by RNA polymerase is subject to different constraints. This may influence the geometry of the interactions of transcription factors with DNA, with one another, and with RNA polymerase. To investigate the formation of an active transcription complex in natural circumstances, we need really to use a template consisting of DNA assembled into chromatin rather than free DNA. We discuss the effects of chromatin structure on transcription in Chapter 30.

Enhancers contain bidirectional elements that assist initiation

We have considered the promoter so far as an isolated region responsible for binding RNA polymerase. But eukaryotic promoters do not necessarily function alone. In at least some cases, the activity of a promoter is enormously increased by the presence of an **enhancer**, which consists of another group of elements, but located at a variable distance from those regarded as comprising part of the promoter itself.

The concept that the enhancer is distinct from

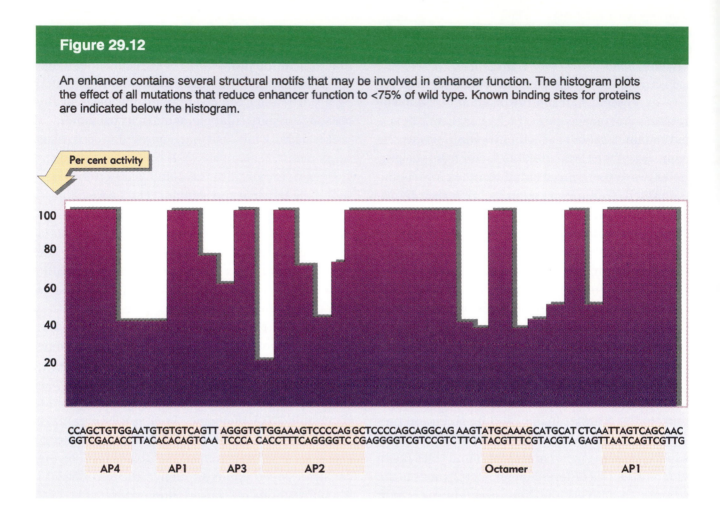

Figure 29.12

An enhancer contains several structural motifs that may be involved in enhancer function. The histogram plots the effect of all mutations that reduce enhancer function to <75% of wild type. Known binding sites for proteins are indicated below the histogram.

Per cent activity

```
CCAGCTGTGGAATGTGTGTCAGTT AGGGTGTGGAAAGTCCCCAG GCTCCCCAGCAGGCAG AAGTATGCAAAGCATGCAT CTCAATTAGTCAGCAAC
GGTCGACACCTTACACACAGTCAA TCCCA CACCTTTCAGGGGTC CGAGGGGTCGTCCGTC TTCATACGTTTCGTACGTA GAGTTAATCAGTCGTTG
```

AP4   AP1   AP3   AP2   Octamer   AP1

the promoter reflects two characteristics. The position of the enhancer relative to the promoter need not be fixed, but can vary substantially. And it can function in either orientation. Manipulations of DNA show that an enhancer can stimulate any promoter placed in its vicinity.

For operational purposes, it is sometimes useful to define the promoter *as a sequence or sequences of DNA that must be in a (relatively) fixed location with regard to the startpoint.* By this definition, the TATA box and other upstream elements are included, but the enhancer is excluded. This is, however, a working definition rather than a description of a rigid classification.

Elements analogous to enhancers, called upstream activator sequences (**UAS**), are found in yeast. They can function in either orientation, at variable distances upstream of the promoter, but cannot function when located downstream. They

have a regulatory role: in several cases the UAS is bound by the regulatory protein(s) that activates the genes downstream.

An enhancer in the virus SV40 was one of the first to be characterized. It is located in a region of the genome that contains two identical sequences of 72 bp each, repeated in tandem ~200 bp upstream of the startpoint of a transcription unit. These **72 bp repeats** lie in a region with an unusual chromatin structure, where the presence of a site hypersensitive to nuclease identifies a region in which DNA is more exposed than usual (see Figures 28.34–28.35). Deletion mapping shows that either one of these repeats is adequate to support normal transcription; but removal of both repeats greatly reduces transcription *in vivo*.

A difference between the enhancer and a typical promoter is presented by the density of regulatory elements. **Figure 29.12** summarizes the susceptibility of the SV40 enhancer to damage by mutation; and we see that a much greater proportion of its sites directly influences its function than is the case with the promoter analyzed in the same way in Figure 29.9. There is a corresponding increase in the density of protein-binding sites. Many of these sites are common elements in promoters; for example, AP1 and the octamer.

Enhancers often show redundancy in function. The SV40 enhancer can be separated into two halves; they function poorly as enhancers by themselves, but constitute an effective enhancer together or even when they are separated by introducing sequences between them. Mutations that inactivate the left element can be compensated by duplicating other regions in the enhancer. Although these regions are different in sequence, they appear to play similar roles, since an active enhancer can be created by the combination of a sufficient *number* of wild-type domains, irrespective of their types. Such redundancy is common in enhancers; the result is that multiple mutations are required, to eliminate more than one element, before an enhancer is inactivated. In the SV40 enhancer, no individual mutation decreases activity by as much as 10×.

Cellular enhancers have similar properties. An enhancer works upon the promoter that is nearest to it, but the enhancer may be either upstream or downstream of the promoter. Responsibility for tissue-specific transcription may lie with either a promoter or enhancer. A promoter may be specifically regulated, and a nearby enhancer used to increase the efficiency of initiation; or a promoter may lack specific regulation, but become active only when a nearby enhancer is specifically activated. An example is provided by immunoglobulin genes, which carry enhancers *within* the transcription unit. The immunoglobulin enhancers appear to be active only in the B lymphocytes in which the immunoglobulin genes are expressed. Such enhancers provide part of the regulatory network by which gene expression is controlled.

Reconstruction experiments in which the enhancer sequence is removed from the DNA and then is inserted elsewhere show that normal transcription can be sustained so long as it is present *anywhere* on the DNA molecule. If a β-globin gene is placed on a DNA molecule that contains an enhancer, its transcription is increased *in vivo* more than 200-fold, even when the enhancer is several kilobase pairs upstream or downstream of the startpoint, in either orientation. We have yet to discover at what distance the enhancer fails to work.

How can an enhancer stimulate initiation at a promoter that can be located at apparently any distance away on either side of it? When enhancers were first discovered, several possibilities were considered for their action as elements distinctly different from promoters:

◆ An enhancer could change the overall structure of the template—for example, by influencing the density of supercoiling.

◆ It could be responsible for locating the template at a particular place within the cell—for example, attaching it to the nuclear matrix.

◆ An enhancer could provide an 'entry site,' a point at which RNA polymerase (or some other essential protein) associates with chromatin.

Now we take the view that enhancer function involves the same sort of interaction with the basal apparatus as the interactions sponsored by upstream promoter elements. Enhancers are modular, like promoters. Some elements are found in both enhancers and promoters. Some individual elements found in promoters share with enhancers the ability to function at variable distance and in either orientation. Thus the distinction between enhancers and promoters is blurred: enhancers might be viewed as containing promoter elements that are grouped closely together, with the ability to function at increased distances from the startpoint.

If the enhancer represents an extreme case of the ability to mix and match promoter elements, it might be considered a part of the promoter in a more distant location. The essential role of the enhancer may be to increase the concentration of transcription factors in the vicinity of the promoter

Figure 29.13

An enhancer may function by bringing proteins into the vicinity of the promoter. An enhancer does not act on a promoter at the opposite end of a long linear DNA, but becomes effective when the DNA is joined into a circle by a protein bridge. An enhancer and promoter on separate circular DNAs do not interact, but can interact when the two molecules are catenated.

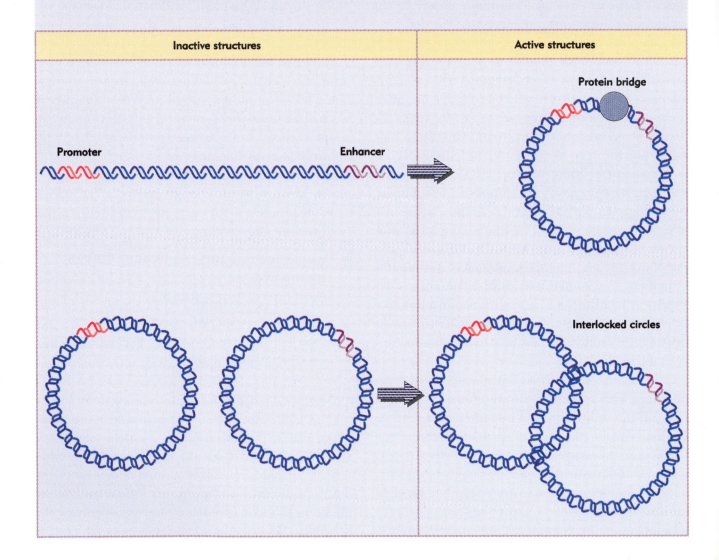

(vicinity in this sense being a relative term). Two types of experiment illustrated in **Figure 29.13** suggest that this may in fact be the case.

A fragment of DNA that contains an enhancer at one end and a promoter at the other is not effectively transcribed, but the enhancer can stimulate transcription from the promoter when they are connected by a protein bridge. Since structural effects, such as changes in supercoiling, could not

be transmitted across such a bridge, this suggests that the critical feature is bringing the enhancer and promoter into close proximity.

A bacterial enhancer provides a binding site for the regulator NtrC, which acts upon RNA polymerase using promoters recognized by σ^{54}. When the enhancer is placed upon a circle of DNA that is catenated (interlocked) with a circle that contains the promoter, initiation is almost as effective as when the enhancer and promoter are on the same circular molecule. But there is no initiation when the enhancer and promoter are on separated circles. Again this suggests that the critical feature is physical restraint of the protein bound at the enhancer, to increase its chance of contacting a protein bound at the promoter.

If proteins bound at an enhancer several kilobases distant from a promoter interact directly with proteins bound in the vicinity of the startpoint, the organization of DNA must be flexible enough to allow the enhancer and promoter to be closely located. This requires the intervening DNA to be extruded as a large 'loop'. Such loops have been directly observed in the case of the bacterial enhancer.

The generality of enhancement is not yet clear. We do not know what proportion of cellular promoters usually rely on an enhancer to achieve their customary level of expression. Nor do we know how often an enhancer provides a target for regulation. Some enhancers are activated only in the tissues in which their genes function, but others could be active in all cells.

3' ends are generated by termination and by cleavage reactions

Information about the termination reaction for eukaryotic RNA polymerases is much less detailed than our knowledge of initiation. RNA polymerases I and III have discrete termination events (like bacterial RNA polymerase), but it is not clear whether RNA polymerase II usually terminates in this way.

For RNA polymerase I, the sole product of transcription is a large precursor that contains the sequences of the major rRNA. The precursor is subjected to extensive processing. Termination occurs at a discrete site >1000 bp downstream of the mature 3' end, which is generated by cleavage. Termination involves recognition of an 18-base terminator sequence by an ancillary factor.

With RNA polymerase III, transcription *in vitro* generates molecules with the same 5' and 3' ends as those synthesized *in vivo*. The termination reaction resembles simple termination by bacterial RNA polymerase. Termination usually occurs at the second U within a run of 4 U bases, but there is heterogeneity, with some molecules ending in 3 or even 4 U bases. The same heterogeneity is seen in molecules synthesized *in vivo*, so it seems to be a *bona fide* feature of the termination reaction.

Just like the prokaryotic terminators, the U run is embedded in a G•C-rich region. Although sequences of dyad symmetry are present, they are not needed for termination, since mutations that abolish the symmetry do not prevent the normal completion of RNA synthesis. Nor are any sequences beyond the U run necessary, since all distal sequences can be replaced without any effect on termination.

The U run itself is not sufficient for termination, because regions of 4 successive U residues exist within transcription units read by RNA polymerase III. (However, there are no internal U_5 runs, which fits with the greater efficiency of termination when the terminator is a U_5 rather than U_4 sequence.) The critical feature in termination must therefore be the recognition of a U_4 sequence in a context that is rich in G•C base pairs.

How does the termination reaction occur? It cannot rely on the weakness of the rU-dA RNA–DNA hybrid region that lies at the end of the transcript, because often only the first two U residues are transcribed. Perhaps the G•C-rich region plays a role in slowing down the enzyme, but there does not seem to be a counterpart to the hairpin involved in prokaryotic termination. We remain puzzled how the enzyme can respond so specifically to such a short signal. And in contrast with the initiation reaction,

Figure 29.14

The sequence AAUAAA is necessary for cleavage to generate a 3' end for polyadenylation.

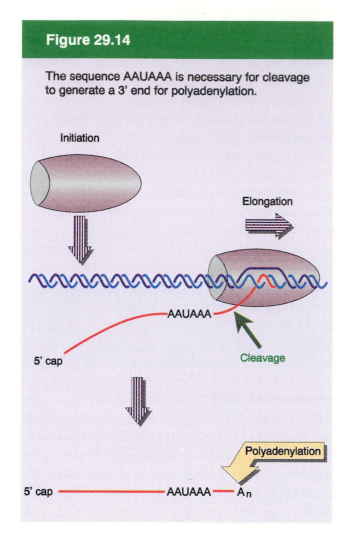

Initiation

Elongation

AAUAAA

5' cap

Cleavage

Polyadenylation

5' cap ———————— AAUAAA —— A$_n$

which RNA polymerase III cannot accomplish alone, termination seems to be a function of the enzyme itself.

Similar termination sites have been found in some RNA polymerase II transcription units, but their significance is not certain. It is possible that termination by RNA polymerase II is only loosely specified. In some transcription units, termination occurs >1000 bp downstream of the site corresponding to the mature 3' end of the mRNA (which is generated by cleavage at a specific sequence). Instead of using specific terminator sequences, the enzyme ceases RNA synthesis within multiple sites located in rather long 'terminator regions.' The nature of the individual termination sites is not known.

The major difficulty in analyzing transcripts of

RNA polymerase II is the lack of certainty about the actual site of termination. Although the RNA molecules identified *in vivo* have defined 3' ends, how are we to know whether they were produced by a termination event or by cleavage of a longer original transcript? The problem is exacerbated by the processing that occurs at the 3' end. Since the 3' terminus for polyadenylation is usually generated by cleavage, the 3' region of the original primary transcript generally remains uncharacterized.

The 3' ends of mRNAs are generated from the transcripts made by RNA polymerase II by cleavage followed by polyadenylation. Addition of poly(A) to nuclear RNA can be prevented by the analog **3'-deoxyadenosine**, also known as **cordycepin**. Although cordycepin does not stop the transcription of nuclear RNA, its addition prevents the appearance of mRNA in the cytoplasm. This shows that polyadenylation is *necessary* for the maturation of mRNA from nuclear RNA.

Generation of the 3' end to which poly(A) is added is illustrated in **Figure 29.14**. RNA polymerase transcribes past the site corresponding to the 3' end, and sequences in the RNA are recognized as targets for an endonucleolytic cut followed by polyadenylation.

A common feature of mRNAs in higher eukaryotes (but not in yeast) is the presence of the sequence AAUAAA in the region from 11 to 30 nucleotides upstream of the site of poly(A) addition. The sequence is highly conserved; only occasionally is even a single base different. Deletion or mutation of the AAUAAA hexamer prevents generation of the usual polyadenylated 3' end. The signal is needed for both cleavage and polyadenylation.

The development of a system in which polyadenylation occurs *in vitro* opens the route to analyzing the reactions. Generation of the proper 3' terminal structure requires an **endonuclease** to cleave the RNA, a **poly(A) polymerase** to synthesize the poly(A) tail, and a **specificity component** that recognizes the AAUAAA sequence and directs the other activities. The endonuclease has not been purified, but the other components have been characterized.

The specificity factor contains 3 subunits, which

Figure 29.15

Generation of the 3' end of histone H3 mRNA depends on two features: a conserved hairpin near the 3' end; and a sequence that base pairs with U7 snRNA.

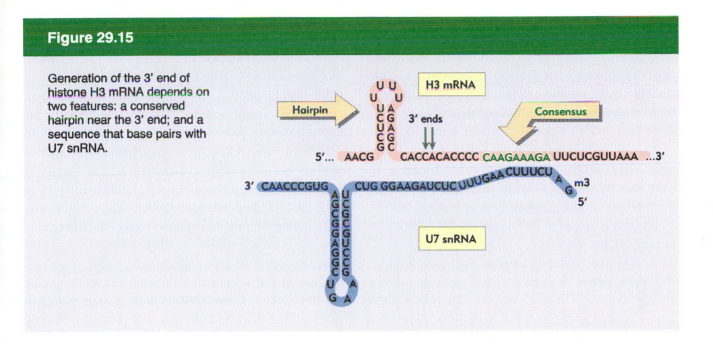

together bind specifically to RNA containing the sequence AAUAAA. The individual subunits are proteins that have common RNA-binding motifs, but which by themselves bind nonspecifically to RNA. Protein–protein interactions between the subunits may be needed to generate the specific AAUAAA-binding site.

The specificity factor is needed for both the cleavage and polyadenylation reactions. It exists in a complex with the endonuclease and poly(A) polymerase, and this complex usually undertakes cleavage followed by polyadenylation in a tightly coupled manner.

The poly(A) polymerase provides a nonspecific catalytic activity. When it is combined with the other components, the synthetic reaction becomes specific for RNA containing the sequence AAUAAA. The polyadenylation reaction passes through two stages. First, a rather short oligo(A) sequence (~10 residues) is added to the 3' end. This reaction is absolutely dependent on the AAUAAA sequence, and poly(A) polymerase performs it under the direction of the specificity factor. In the second phase, the oligo(A) tail is extended to the full ~200 residue length. This reaction does not require the AAUAAA sequence, but instead requires another stimulatory factor that recognizes the

oligo(A) tail and directs poly(A) polymerase to extend it.

Some mRNAs are not polyadenylated. The formation of their 3' ends is therefore different from the coordinated cleavage/polyadenylation reaction. The most prominent members of this mRNA class are the mRNAs coding for histones. Formation of their 3' ends depends upon secondary structure. The RNA terminates in a stem–loop structure, and mutations that prevent formation of the duplex stem prevent formation of the end of the RNA. Secondary mutations that restore duplex structure (though not necessarily the original sequence) behave as revertants. This suggests that *formation of the secondary structure is more important than the exact sequence.*

Either or both of the DNA strands could in principle be involved in forming secondary structure. They can be distinguished by using templates consisting of heteroduplex molecules, in which the two strands of DNA are not identical. It turns out that it is important to be able to write a duplex structure for the *coding strand*, not the strand used as template. This suggests that the secondary structure exerts its effect by forming in the RNA as it is transcribed.

An *in vitro* system has been developed that

generates authentic 3′ ends in *X. laevis* histone mRNAs. When the histone H3 gene is injected into the *Xenopus* oocyte, it is faithfully initiated and transcribed, but termination occurs at variable sites. However, when a nuclear extract from the sea urchin is simultaneously injected with the gene, the transcribed mRNA has the proper 3′ end.

The nuclear extract has three active components: a heat-labile factor (of unknown function); a factor that binds to a hairpin just upstream of the cleavage site; and a 56-base RNA called **U7 snRNA** whose 5′ terminus is complementary to a short sequence just downstream of the 3′ cleavage site.

The reaction between histone H3 mRNA and U7 snRNA is drawn in **Figure 29.15**. The upstream hairpin and the downstream sequence that pairs with U7 snRNA are conserved in histone H3 mRNAs of several species. The U7 snRNA has sequences towards its 5′ end that could pair with the histone mRNA consensus sequences, and has an extensive hairpin at its 3′ end. The sequence and structure of U7 snRNA is well conserved between man and sea urchin.

3′ processing is inhibited by mutations in the downstream histone consensus sequence that reduce ability to pair with U7 snRNA. Compensatory mutations in U7 snRNA that restore complementarity also restore 3′ processing. This suggests that U7 snRNA functions by base pairing with the histone mRNA. We do not yet know the stage at which the snRNA acts: it could be termination or cleavage.

The involvement of the U7 snRNA in 3′ end generation is consistent with the view we develop in Chapter 31 that many—perhaps all—RNA processing events depend on RNA–RNA interactions. The first step of cleavage of pre-rRNA requires another small RNA, called U3 snRNA, and several further RNA molecules of this class are required for splicing. We discuss their properties in more detail in Chapter 31, but note now that the snRNA usually functions in the form of a ribonucleoprotein particle containing several proteins as well as the RNA, and it is common (although not the only means of the action) for the RNA of the particle to base pair with a short sequence in the substrate RNA.

Summary

Of the three eukaryotic RNA polymerases, RNA polymerase I transcribes rDNA and accounts for the majority of activity, RNA polymerase II transcribes structural genes for mRNA and has the greatest diversity of products, and RNA polymerase III transcribes small RNAs. The enzymes have a similar structure, with two large subunits and many smaller subunits; there are some common subunits among the enzymes.

None of the three RNA polymerases recognize their promoters directly. A unifying principle is that transcription factors have primary responsibility for recognizing the characteristic sequence elements of any particular promoter, and they serve in

turn to bind the RNA polymerase and to position it correctly at the startpoint. At each type of promoter, the initiation complex is assembled by a series of reactions in which individual factors join (or leave) the complex. The factor TBP (which recognizes the TATA sequence upstream of the startpoint at polymerase II promoters), is required for initiation by all three RNA polymerases.

Promoters for RNA polymerase II contain a variety of short *cis*-acting elements, each of which is recognized by a *trans*-acting factor. The *cis*-acting elements are located upstream of the TATA box and may be present in either orientation and at a variety of distances with regard to the startpoint.

The TATA box (if there is one) near the startpoint, and the initiator region immediately at the startpoint, are responsible for selection of the exact startpoint. The elements farther upstream determine the efficiency with which the promoter is used.

Promoters may be stimulated by enhancers, sequences that can act at great distances and in either orientation on either side of a gene. Enhancers also consist of sets of elements, although they are more compactly organized. Some elements are found in both promoters and enhancers. Enhancers probably function by assembling a protein complex that interacts with the proteins bound at the promoter, requiring that DNA between is 'looped out.'

The termination capacity of RNA polymerase II has not been characterized, and 3′ ends of its transcripts are generated by cleavage. The sequence AAUAAA, located 10–30 bases upstream of the cleavage site, provides the signal for both cleavage and polyadenylation. An endonuclease and the poly(A) polymerase are associated in a complex with other factors that confer specificity for the AAUAAA signal.

Further reading

Reviews

Eukaryotic RNA polymerases have been reviewed by **Young** (*Ann. Rev. Biochem.* **60**, 689715, 1991). Techniques in characterizing eukaryotic RNA polymerase II promoters were discussed by **Corden et al.** (*Science* **209**, 1406–1414, 1981). The modular nature of enhancers and promoters has been discussed by **Muller, Gerster, and Schaffner** (*Eur. J. Biochem.* **176**, 485–495, 1988). RNA polymerase III and its factors have been reviewed by **Geiduschek and Tocchini-Valentini** (*Ann. Rev. Biochem.* **57**, 873–914, 1988).

Cis-acting sites and *trans*-acting factors for viral and for cellular promoters have been reviewed, respectively, by **McKnight and Tjian** (*Cell* **46**, 795–805, 1986) and by **Maniatis, Goodbourn, and Fischer** (*Science* **236**, 1237–1245, 1987).

Assembly of the basal apparatus has been reviewed by **Reinberg** (*Prog. Nucleic Acids Res. Mol. Biol.* **44**, 67–108, 1993). Motifs in transcription factors have been reviewed by **Mitchell and Tjian** (*Science* **245**, 371–378, 1989).

Processing of 3′ ends was reviewed by **Birnstiel et al.** (*Cell* **41**, 349–359, 1985). Polyadenylation was reviewed by **Wahle and Keller** (*Ann. Rev. Biochem.* **61**, 419–440, 1992).

Discoveries

A thoughtful analysis of the SV40 enhancer was provided by **Banerji, Rusconi, and Schaffner** (*Cell* **27**, 299–308, 1981). A detailed mutational analysis was undertaken by **Zenke et al.** (*EMBO J.* **5**, 387–397, 1986).

The roles of factors for RNA polymerase III were elucidated by **Kassavatis et al.** (*Cell* **60**, 235–245, 1990). Assembly of the general transcription complex was analyzed by **Buratowski et al.** (*Cell* **56**, 549–561, 1989).

Internal promoters for RNA polymerase III were discovered by **Sakonju, Bogenhagen, and Brown** (*Cell* **19**, 13–25, 27–35, 1980).

Antagonistic effects of transcription factors and histones were reported by **Bogenhagen, Wormington, and Brown** (*Cell* **28**, 413–421, 1982) and **Workman and Roeder** (*Cell* **51**, 613–622, 1987).

The involvement of snRNA in generating 3′ ends was discovered by **Galli et al.** (*Cell* **34**, 822–828, 1983).

CHAPTER 30

Regulation of transcription: factors that activate the basal apparatus

The phenotypic differences that distinguish the various kinds of cells in a higher eukaryote are largely due to differences in the expression of genes that code for proteins, that is, those transcribed by RNA polymerase II. In principle, the expression of these genes might be regulated at any one of several stages. The concept of the 'level of control' implies that gene expression is not necessarily an automatic process once it has begun. It could be regulated in a gene-specific way at any one of several sequential steps. We can distinguish (at least) five potential control points, forming the series

Activation of gene structure
↓
Initiation of transcription
↓
Processing the transcript
↓
Transport to cytoplasm
↓
Translation of mRNA

The existence of the first step is implied by the discovery that genes themselves may exist in either of two structural conditions. Relative to the state of most of the genome, genes are found in an 'active' state in the cells in which they are expressed (see Chapter 28). The change of structure is distinct from the act of transcription, and indicates that the gene is 'transcribable'. This suggests that acquisition of the 'active' structure must be the first step in gene expression.

Transcription of a gene in the active state is controlled at the stage of initiation, that is, by the interaction of RNA polymerase with its promoter. This is now becoming susceptible to analysis in the *in vitro* systems (see Chapter 29). For most genes, this is a major control point; probably it is the most common level of regulation.

There is at present no evidence for control at subsequent stages of transcription in eukaryotic cells, for example, via antitermination mechanisms.

The primary transcript always must be modified by capping at the 5′ end, and usually also by polyadenylation at the 3′ end. Introns must be spliced out from the transcripts of interrupted genes. The mature RNA must be exported from the nucleus to the cytoplasm. Regulation of gene expression by selection of sequences at the level of nuclear RNA might involve any or all of these stages, but the one for which we have most evidence concerns changes in splicing; some genes are expressed by means of alternative splicing patterns whose regulation controls the type of protein product. We discuss the examples of sex determination and other systems in Chapter 31.

Finally, the translation of an mRNA in the cytoplasm can be specifically controlled. There is little evidence for the employment of this mechanism in adult somatic cells, but it does occur in some embryonic situations, as described in Chapter 10. The mechanism is presumed to involve the blocking of initiation of translation of some mRNAs by specific protein factors.

But having acknowledged that control of gene expression can occur at multiple stages, and that production of RNA cannot inevitably be equated with production of protein, it is clear that the overwhelming majority of regulatory events occur at the initiation of transcription. Regulation of tissue-specific gene transcription lies at the heart of eukaryotic differentiation; indeed, we see examples in Chapter 38 in which proteins that regulate embryonic development prove to be transcription factors. A regulatory transcription factor serves to provide common control of a large number of target genes, and we seek to answer two questions about this mode of regulation: what identifies the common target genes to the transcription factor; and how is the activity of the transcription factor itself regulated in response to intrinsic or extrinsic signals.

Response elements identify genes under common regulation

The principle that emerges from characterizing groups of genes under common control is that *they share a promoter element that is recognized by a regulatory transcription factor*. An element that causes a gene to respond to such a factor is called a **response element**; examples are the **HSE** (heat shock response element), **GRE** (glucocorticoid response element), **SRE** (serum response element).

The properties of some inducible transcription factors and the elements that they recognize are summarized in **Table 30.1**. Response elements have the same general characteristics as upstream elements of promoters or enhancers. They contain short consensus sequences, and copies of the response elements found in different genes are closely related, but not necessarily identical. The region bound by the factor extends for a short distance on either side of the consensus sequence. The elements are not present at fixed distances from the startpoint, but are usually <200 bp upstream of it. The presence of a single element

Table 30.1

Inducible transcription factors bind to response elements that identify groups of promoters or enhancers subject to coordinate control.

Regulatory Agent	Module	Consensus	DNA bound	Factor	Size (daltons)
Heat shock	HSE	CNNGAANNTCCNNG	27 bp	HSTF	93,000
Glucocorticoid	GRE	TGGTACAAATGTTCT	20 bp	Receptor	94,000
Phorbol ester	TRE	TGACTCA	22 bp	AP1	39,000
Serum	SRE	CCATATTAGG	20 bp	SRF	52,000

Figure 30.1

The regulatory region of a human metallothionein gene contains constitutive elements in its promoter and enhancer. The promoter has elements for metal induction; an enhancer has an element for response to glucocorticoid. Promoter elements are shown above the map, and proteins that bind them are indicated below.

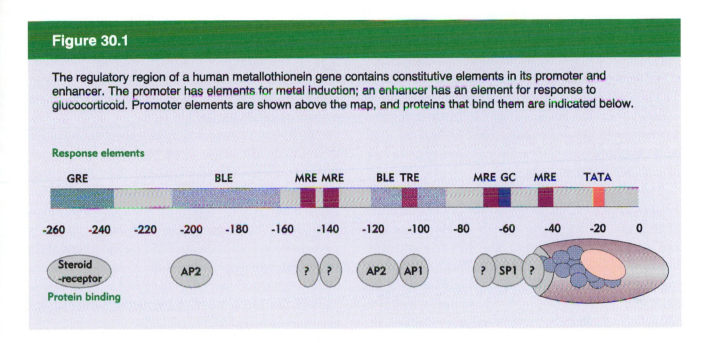

usually is sufficient to confer the regulatory response, but sometimes there are multiple copies.

Response elements may be located in promoters or in enhancers. Some types of elements are typically found in one rather than the other: usually an HSE is found in a promoter, while a GRE is found in an enhancer. We assume that all response elements function by the same general principle. *A gene is regulated by a sequence at the promoter or enhancer that is recognized by a specific protein. The protein functions as a transcription factor needed for RNA polymerase to initiate. Active protein is available only under conditions when the gene is to be expressed; its absence means that the promoter is not activated by this particular circuit.*

An example of a situation in which many genes are controlled by a single factor is provided by the heat shock response. This is common to a wide range of prokaryotes and eukaryotes and involves multiple controls of gene expression: an increase in temperature turns off transcription of some genes, turns on transcription of the **heat shock genes**, and also causes changes in the translation of mRNAs. The control of the heat shock genes illustrates the differences between prokaryotic and eukaryotic modes of control. In bacteria, a new sigma factor is synthesized that directs

RNA polymerase holoenzyme to recognize an alternative −10 sequence common to the promoters of heat shock genes (see Chapter 14). In eukaryotes, the heat shock genes also possess a common consensus sequence (HSE), but it is located at various positions relative to the startpoint, and is recognized by an independent transcription factor, HSTF. The activation of this factor therefore provides a means to initiate transcription at the specific group of ~20 genes that contains the appropriate target sequence at its promoter.

All the heat shock genes of *D. melanogaster* contain multiple copies of the HSE. The HSTF binds cooperatively to adjacent response elements. Both the HSE and HSTF have been conserved in evolution, and it is striking that a heat shock gene from *D. melanogaster* can be activated in species as distant as mammals or sea urchins. The HSTF proteins of fruit fly and yeast appear similar, and show the same footprint pattern on DNA containing HSE sequences. Yeast HSTF becomes phosphorylated when cells are heat-shocked; this modification is responsible for activating the protein.

The metallothionein (MT) gene provides an example of how a single gene may be regulated by many different circuits. The metallothionein protein protects the cell against excess concentrations of heavy metals, by binding the metal and removing it from the cell. The gene is expressed at a basal level, but is induced to greater levels of expression by heavy metal ions (such as cadmium) or by glucocorticoids. The control region combines several different kinds of regulatory element, and suggests the principle that *when a promoter is regulated in more than one way, each regulatory event depends on binding of its own protein to a particular sequence.*

The organization of the promoter for a MT gene is summarized in **Figure 30.1**. A major feature of this map is the high density of elements that can activate transcription. The two 'constitutive' promoter elements are the TATA box and GC box, located at their usual positions fairly close to the startpoint. Also needed for the basal level of constitutive expression are the two basal level elements (BLE), which fit the formal description of enhancers. Although located near the startpoint, they can be moved elsewhere without loss of effect. They contain sequences related to those found in other enhancers, and are bound by proteins that bind the SV40 enhancer.

The TRE is a consensus sequence that is present in several enhancers, including one BLE of metallothionein and the 72 bp repeats of the virus SV40. The TRE has a binding site for factor AP1; this interaction is part of the mechanism for constitutive expression, in line with our previous description of AP1 as an upstream factor. However, AP1 binding also has a second function. The TRE confers a response to phorbol esters such as TPA (an agent that promotes tumors), and this response is mediated by the interaction of AP1 with the TRE. This binding reaction is one (not necessarily the sole) means by which phorbol esters trigger a series of transcriptional changes.

The inductive response to metals is conferred by the multiple MRE sequences. These function as promoter elements. The presence of one MRE confers the ability to respond to heavy metal; a greater level of induction is achieved by the inclusion of multiple elements. The response to steroid hormones is governed by a GRE, located 250 bp upstream of the startpoint, which behaves as an enhancer. Deletion of this region does not affect the basal level of expression or the level induced by metal ions. But it is absolutely needed for the response to steroids.

The regulation of metallothionein illustrates the general principle that *any one of several different elements, located in either an enhancer or promoter, can independently activate the gene.* The absence of a element needed for one mode of activation does not affect activation in other modes. The variety of elements, their independence of action, and the apparently unlimited flexibility of their relative arrangements, suggest that a factor binding to any one element is able independently to increase the efficiency of initiation by the basal transcription apparatus, probably by virtue of protein–protein interactions that stabilize or otherwise assist formation of the initiation complex.

Transcription factors bind DNA and activate transcription through independent domains

Transcription factors and other regulatory proteins require two types of ability:

◆ *They must recognize specific target sequences located in enhancers, promoters, or other regulatory elements that affect a particular target gene.* (For a repressor protein, binding to DNA may itself be a sufficient action to exercise its function: the presence of the protein serves to prevent gene expression.)

◆ For a transcription factor or a positive regulatory protein, more is required; having bound to DNA, *the protein exercises its function by binding to other components of the transcription apparatus.*

Can we characterize domains in the transcription factors that are responsible for these activities? Often there are separate domains that bind DNA and activate transcription, and which function independently. In fact, we might view the function of the DNA-binding domain as having been completed once it has brought the activating domain into the vicinity of the startpoint. This model for transcription factor function is depicted schematically in **Figure 30.2**. In principle, it requires that the connection between the DNA-binding and activation domains is flexible enough to allow the activation domain to find its protein targets irrespective of the exact site held by the DNA-binding domain in the promoter. The protein target is likely to be a basal transcription factor. Experiments with some rather different systems suggest that the principle of independent domains is common in transcriptional activators.

The modular nature of a transcriptional activator has been most thoroughly characterized for the example of the yeast activator GAL4, which controls the transcription of genes whose products are responsible for metabolizing galactose. Regulation is exercised at a UAS$_G$ (an upstream activating sequence that is the yeast equivalent of an enhancer). The GAL4 protein has 4 binding sites in UAS$_G$. Each site has a copy of a palindrome which has a 17 bp consensus. However, a single copy of the consensus sequence is adequate to confer susceptibility to GAL4 regulation.

The GAL4 protein has three functions: it binds to DNA; it activates transcription; and it binds another regulator protein, called GAL80. These functions can be separated, as depicted in **Figure 30.3**.

The DNA-binding domain of GAL4 resides in its 65 amino-terminal residues. An N-terminal fragment can bind the usual consensus sequence, but it does not activate transcription.

The ability to activate transcription can be conferred by two other regions of the proteins; either of the stretches 148–196 or 768–881, if attached to the DNA-binding domain, can activate transcription.

The region 65–94 enables the protein to dimerize. It is quite common for transcription factors to have adjacent domains for DNA-binding and dimerization.

Figure 30.2

DNA-binding and activating domains in transcription factors may function independently. This explains why DNA-binding sites can be located at varying distances from the startpoint.

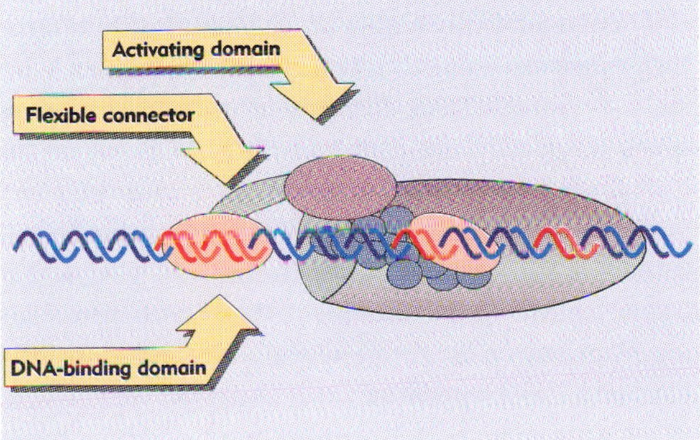

Figure 30.3

The GAL4 protein requires three regions to activate UAS$_G$; residues 1-98 bind DNA, and 148-196 and 768-881 are needed to activate transcription. The C-terminal region of 851-881 is required to bind GAL80.

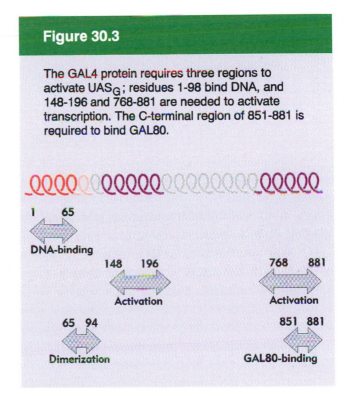

Figure 30.4

The ability of GAL4 to activate transcription is independent of its specificity for binding DNA. When the GAL4 DNA-binding domain is replaced by the LexA DNA-binding domain, the hybrid protein can activate transcription when a LexA operator is placed near a promoter.

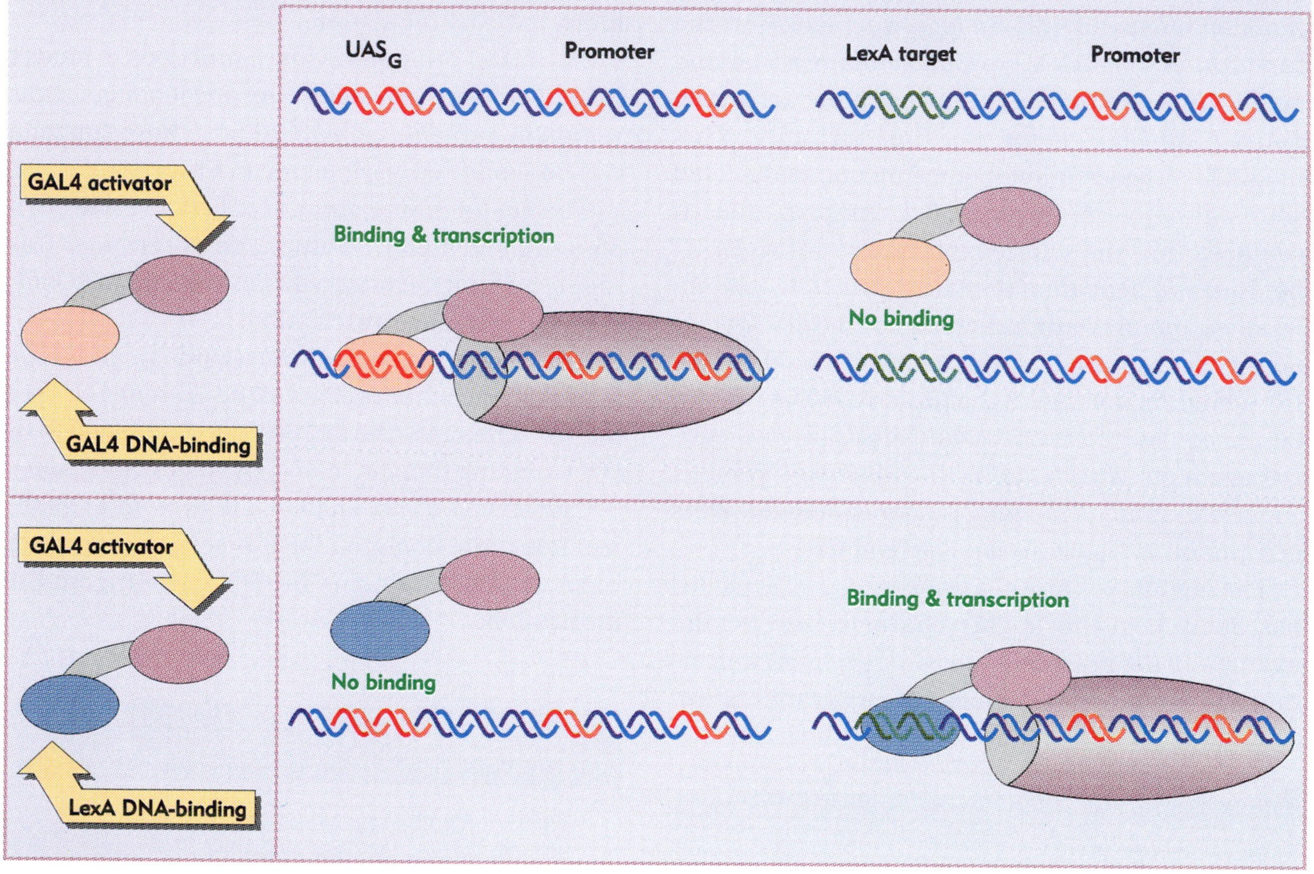

The region 851–881, located within one of the activation domains, binds GAL80. In fact, this is the means by which GAL4 responds to galactose. The *GAL* genes are induced by galactose, which binds to the GAL80 protein. In the absence of galactose, the GAL80 protein binds to GAL4 and prevents it from activating transcription. Galactose induces transcription by releasing GAL80, thus allowing GAL4 to function effectively. This is one variant on a common theme: a protein (or some other ligand) binds to a transcription factor to regulate (directly or indirectly) its transcription-activating domain.

Binding to DNA naturally is prerequisite for activating transcription. But does activation depend on the *particular* DNA-binding domain?

Figure 30.4 illustrates an experiment to answer this question. The bacterial repressor LexA has an N-terminal DNA-binding domain that recognizes a specific operator; binding of LexA at this operator represses the adjacent promoter. In a 'swap' experiment, this sequence can be substituted for the DNA-binding domain of GAL4. The hybrid gene can then be introduced into yeast together with a target gene that contains either the UAS or a LexA operator.

An authentic GAL4 protein can activate a target

gene only if it has a UAS. The LexA repressor by itself of course lacks the ability to activate either sort of target. The LexA–GAL4 hybrid can no longer activate a gene with a UAS, but it can now activate a gene that has a LexA operator!

This result fits the modular view of transcription activators suggested in Figure 30.2. The DNA-binding domain serves to bring the protein into the right location. Precisely how or where it is bound to DNA is irrelevant, but, once it is there, the transcription-activating domains can cause transcription to initiate. According to this view, it does not much matter whether the transcription-activating domains are brought to the vicinity of the promoter by recognition of a UAS via the DNA-binding domain of GAL4 or by recognition of a LexA operator via the LexA specificity module. The ability of the two types of module to function in hybrid proteins suggests that each domain of the protein folds independently into an active structure

Figure 30.5

The activating domain of the tat protein of HIV can stimulate initiation if it is tethered in the vicinity by binding to the RNA product of a previous round of transcription. Activation is independent of the means of tethering, as shown by the maintenance of function when a DNA-binding domain is substituted for the RNA-binding domain.

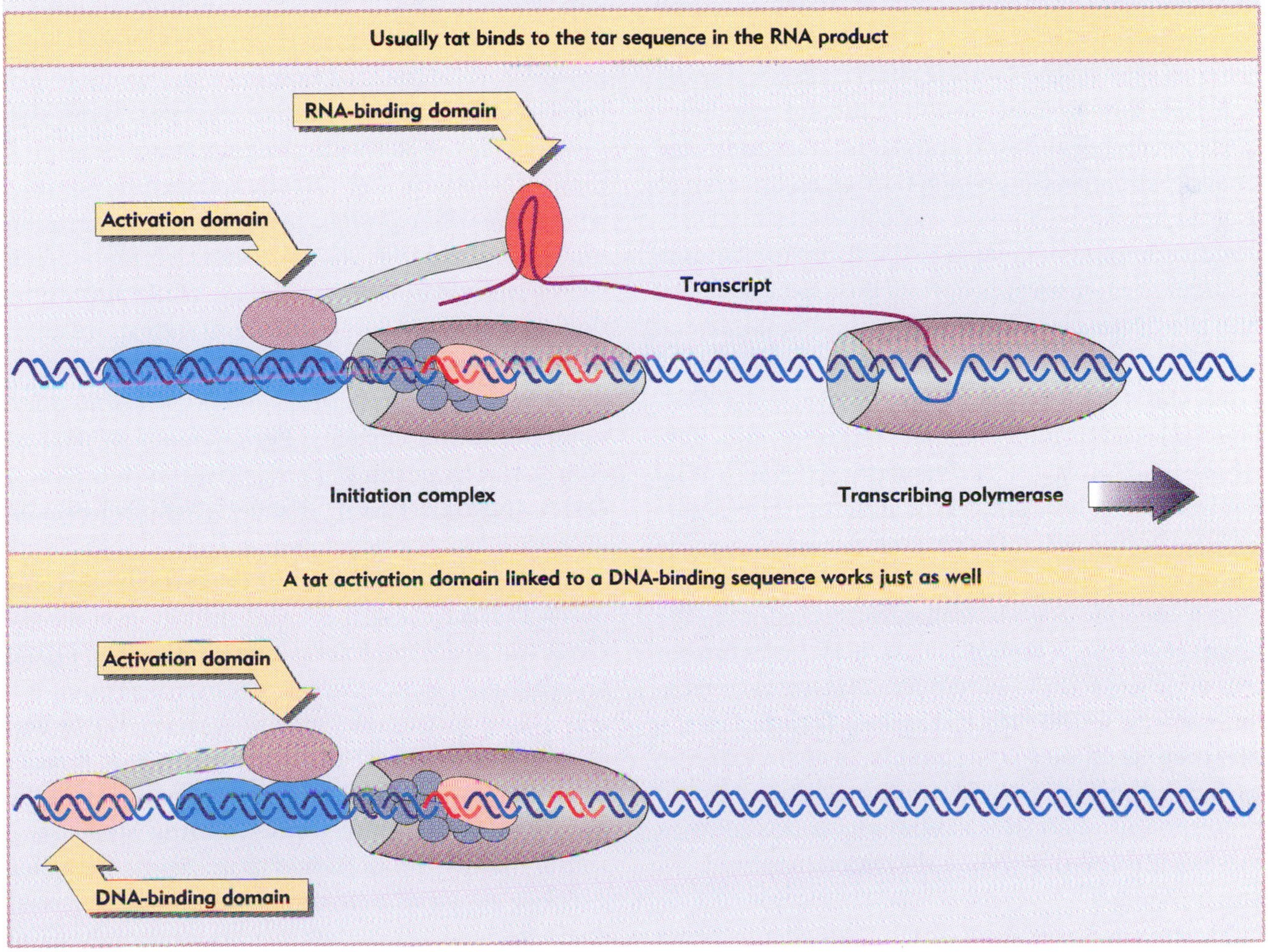

Usually tat binds to the tar sequence in the RNA product

RNA-binding domain

Activation domain

Transcript

Initiation complex

Transcribing polymerase

A tat activation domain linked to a DNA-binding sequence works just as well

Activation domain

DNA-binding domain

and is not influenced by the rest of the protein.

The idea that transcription factors have independent domains that bind DNA and that activate transcription is reinforced by the ability of the tat protein of HIV to stimulate initiation without binding DNA at all. The tat protein binds to a region of secondary structure in the RNA product; the part of the RNA required for tat action is called the *tar* sequence. A model for the role of the tat–*tar* interaction in stimulating transcription is shown in **Figure 30.5**.

The *tar* sequence is located just downstream of the startpoint, so that when tat binds to *tar*, it is brought into the vicinity of the initiation complex. This is sufficient to ensure that its activation domain is in close enough proximity to the initiation complex. The activation domain interacts with one or more of the transcription factors bound at the complex in the same way as an upstream transcription factor. (Note that at least the first transcript must be made in the absence of tat in order to provide the binding site.)

An extreme demonstration of the independence of the localizing and activating domains is indicated by some constructs in which tat was engineered so that the activating domain was connected to a DNA-binding domain instead of to the usual *tar*-binding sequence: when an appropriate target site is placed into the promoter, the tat activating-domain could activate transcription. This suggests that we should think of the DNA-binding (or in this case the RNA-binding) domain as providing a 'tethering' function, whose main purpose is to ensure that the activating domain is in the vicinity of the initiation complex.

The notion of tethering is a more specific example of the general idea that initiation requires a high concentration of transcription factors in the vicinity of the promoter. This may be achieved when factors bind to enhancers that are some distance away on linear DNA, when factors bind to upstream promoter components, or in an extreme case by tethering to the RNA product. The common requirement of all these situations is flexibility in the exact three dimensional arrangement of DNA and proteins.

There are several means by which upstream transcription factors activate transcription, but the common feature is probably a protein–protein interaction with a basal transcription factor. The factors TFIID and TFIIB appear to be the most common targets for activation.

The activating domains of the yeast transcription factors GAL4 and GCN4 have multiple negative charges, giving rise to their description as 'acidic activators'. Another particularly effective activator of this type is carried by the VP16 protein of the Herpes Simplex Virus. (VP16 does not itself have a DNA-binding domain, but interacts with the transcription apparatus via an intermediary protein.) Experiments to characterize acidic activator function have often made use of the VP16 activating region linked to some DNA-binding motif.

It is not clear whether the negative charges are necessary for the function of 'acid' activators. On the one hand, sequences that have activating capacity when linked to the DNA-binding domain of GAL4 can be found in the form of a series of short polypeptide sequences that have in common only their possession of negative charges. This suggested that the activating region is an 'acid blob' that interacts in a relatively nonspecific manner with some other protein involved in transcription. And mutations that remove acidic amino acids of GCN4 reduce activation ability, while mutations that remove basic amino acids generate more effective activators. However, systematic mutation to remove the charged groups in GAL4 has no effect upon activating capacity. Perhaps lack of positive charges is important. The activation domains of GAL4 and GCN4 form β-sheets, which probably provide an interface for contacting other proteins.

Acidic activators function by enhancing the ability of TFIIB to join the basal initiation complex. Experiments *in vitro* show that binding of TFIIB to an initiation complex at an adenovirus promoter is stimulated by the presence of GAL4 or VP16 acid activators; and the VP16 activator can bind directly to TFIIB. Assembly of TFIIB into the complex at this promoter is therefore a rate-limiting step that is stimulated by the presence of an acidic activator.

Other types of activating motifs are less well characterized, but include a glutamine-rich region in Sp1 and a proline-rich region in CTF. Sp1 appears to require the TAF proteins that are associated with TBP for its action, so its target for activation may be the TFIID complex. There are also indications that acidic activators can act on TFIID.

The resilience of an RNA polymerase II promoter to the rearrangement of elements, and even to the particular elements present, suggests that the events by which it is activated are somewhat independent of context and relatively general. In effect, any upstream factor whose activating region is brought within range of the basal initiation complex may be able to stimulate its formation. Some striking illustrations of such versatility have been accomplished by constructing promoters consisting of new combinations of elements. For example, when a yeast UAS_G element is inserted near the promoter of a higher eukaryotic gene, this gene can be activated by GAL4 in a mammalian cultured cell.

Whatever means GAL4 uses to activate the promoter seems therefore to have been conserved between yeast and higher eukaryotes. The GAL4 protein must recognize some feature of the mammalian transcription apparatus that resembles its normal contacts in yeast. One interpretation is that there are in fact common components—either in RNA polymerase or other transcription factors—that can be recognized by any of many transcription factors to activate transcription. The common component could of course be a motif within a protein rather than an entire protein.

There are many types of DNA-binding domains

Comparisons between the sequences of many transcription factors suggest that common types of **motifs** can be found that are responsible for binding to DNA. The motifs are usually quite short and comprise only a small part of the protein structure. Motifs have also been identified that are responsible for activating transcription via interactions between proteins of the transcription apparatus. We have detailed information about several groups of proteins that regulate transcription by using particular motifs to bind DNA:

◆ The **steroid receptors** are defined as a group by a functional relationship: each receptor is activated by binding a particular steroid. The glucocorticoid receptor is the most fully analyzed. Together with other receptors, such as the thyroid hormone receptor or the retinoic acid receptor, the steroid receptors are members of a super-family of transcription factors with the same general *modus operandi*.

◆ The **zinc finger** motif comprises a DNA-binding domain. It was originally recognized in factor TFIIIA, which is required for RNA polymerase III to transcribe 5S rRNA genes. It has since been identified in several other transcription factors (and presumed transcription factors). A distinct form of the motif is found also in the steroid receptors.

◆ The **helix-turn-helix** motif was originally identified as the DNA-binding domain of phage repressors. One α-helix lies in the wide groove of DNA; the other lies at an angle across DNA. A related form of the motif is present in the **homeo domain**, a sequence first characterized in several proteins coded by genes concerned with developmental regulation in *Drosophila*. It has now been identified in genes for mammalian transcription factors.

◆ The amphipathic **helix-loop-helix (HLH)** motif has been identified in some developmental regulators and in genes coding for eukaryotic

Figure 30.6

The activity of regulatory transcription factors may be controlled by synthesis of protein, covalent modification of protein, ligand binding, or binding of inhibitors that sequester the protein or affect its ability to bind to DNA.

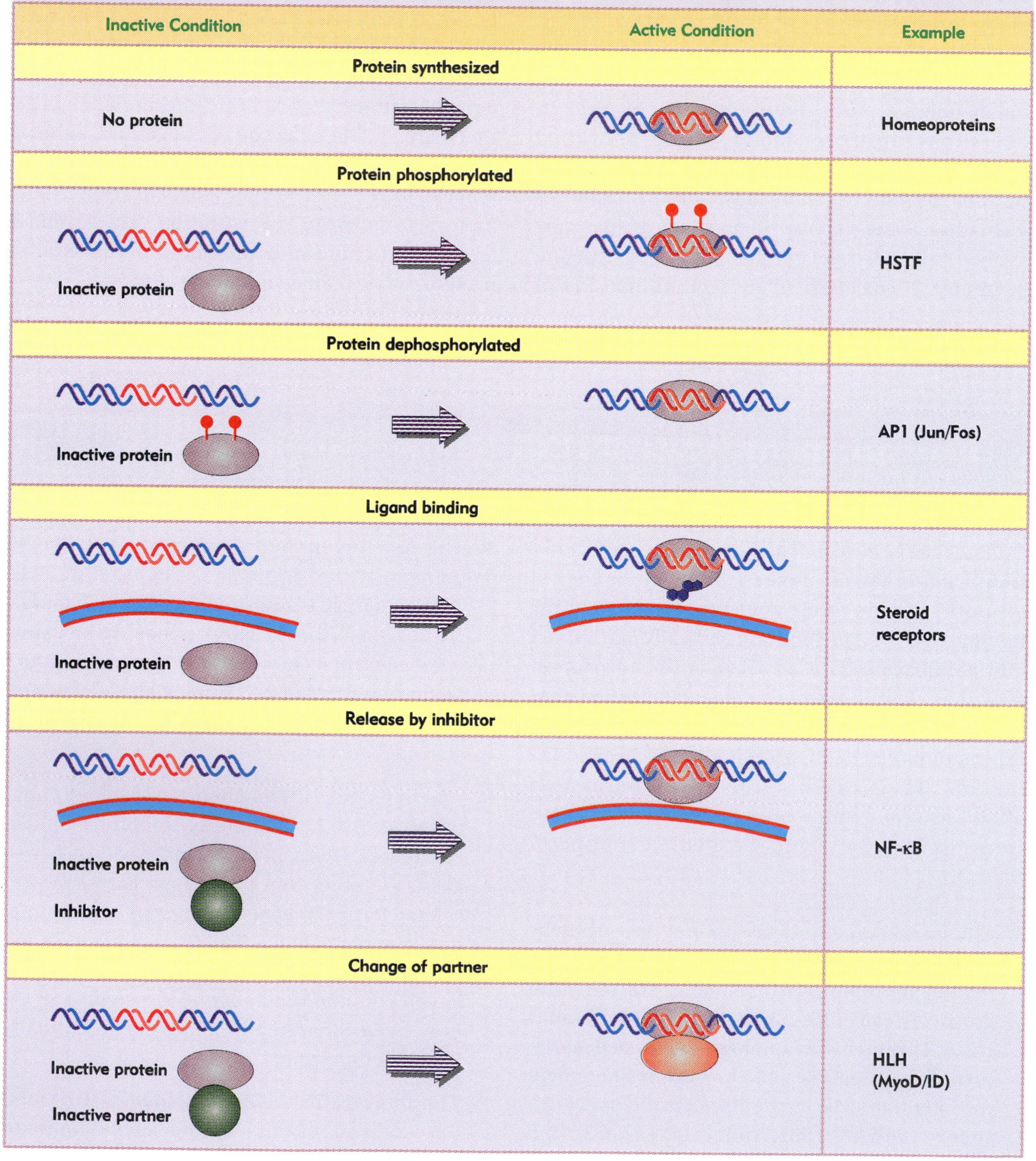

Inactive Condition	Active Condition	Example
Protein synthesized		
No protein		Homeoproteins
Protein phosphorylated		
Inactive protein		HSTF
Protein dephosphorylated		
Inactive protein		AP1 (Jun/Fos)
Ligand binding		
Inactive protein		Steroid receptors
Release by inhibitor		
Inactive protein / Inhibitor		NF-κB
Change of partner		
Inactive protein / Inactive partner		HLH (MyoD/ID)

DNA-binding proteins. Each amphipathic helix presents a face of hydrophobic residues on one side and charged residues on the other side. The length of the connecting loop varies from 12 to 28 amino acids. The motif enables proteins to dimerize, and a basic region near this motif contacts DNA.

◆ **Leucine zippers** consist of a stretch of amino acids with a leucine residue in every seventh position. A leucine zipper in one polypeptide interacts with a zipper in another polypeptide to form a dimer. Adjacent to each zipper is a stretch of positively charged residues that is involved in binding to DNA.

The activity of an inducible transcription factor may be regulated in any one of several ways, as illustrated schematically in **Figure 30.6**:

◆ A factor is tissue-specific because it is synthesized only in a particular type of cell. Several factors of this type are known, including many with a homeobox.

◆ The activity of a factor could be directly controlled by modification. HSTF is converted to the active form by phosphorylation. AP1 (a heterodimer between the subunits Jun and Fos) is converted to the active form by dephosphorylating the Jun subunit.

◆ A factor is activated or inactivated by binding a ligand. The steroid receptors are prime examples. Ligand binding may influence both the localization of the protein (causing transport from cytoplasm to nucleus) and also directly affect its ability to bind to DNA.

◆ Availability of a factor could vary; for example, the factor NFκB (which activates immunoglobulin κ genes in B lymphocytes) is present in many cell types. But it is sequestered in the cytoplasm by the inhibitory protein I-κB. In B lymphocytes, NFκB is released from I-κB and moves to the nucleus, where it activates transcription.

◆ A dimeric factor may have alternative partners. One partner may cause it to be inactive; synthesis of the active partner may displace the inactive partner. We see later that such situations may be amplified into networks in which various alternative partners pair with one another, especially among the HLH proteins.

(We note *en passant* that mutations of the transcription factors in some of these classes give rise to factors that inappropriately activate, or prevent activation, of transcription; their roles in generating tumors are discussed in Chapter 39, and Figure 39.26 can be compared with Figure 30.6.)

We now discuss in more detail the DNA-binding and activation reactions that are sponsored by some of these classes of proteins. In many cases, binding to DNA is undertaken by a short region of α-helix that makes contacts with either the bases or phosphate backbone in the major groove.

A zinc finger motif may provide a DNA-binding domain

Zinc fingers take their name from the structure illustrated in Figure 30.7, in which a small group of conserved amino acids binds a zinc ion, and forms a relatively independent domain in the protein. Two types of DNA-binding proteins have structures of this type: the classic 'zinc finger' proteins; and the steroid receptors.

A 'finger protein' typically has a series of zinc fingers, as depicted in the figure. The consensus sequence of a single finger is:

Cys-X$_{2-4}$-Cys-X$_3$-Phe-X$_5$-Leu-X$_2$-His-X$_3$-His

Figure 30.7

Transcription factor SP1 has a series of three zinc fingers, each with a characteristic pattern of cysteine and histidine residues that constitute the zinc-binding site.

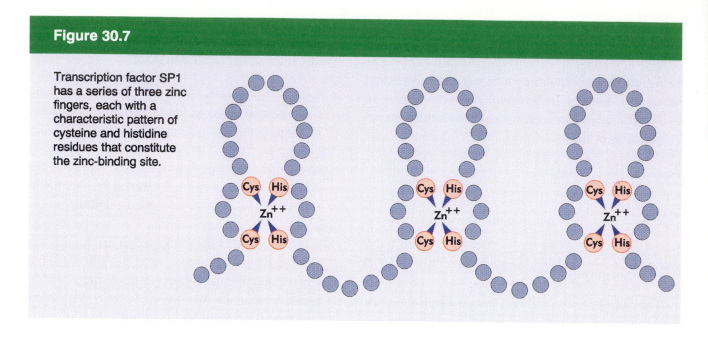

The motif takes its name from the loop of amino acids that protrudes from the zinc-binding site and is described as the Cys_2/His_2 finger. The zinc is held in a tetrahedral structure formed by the conserved Cys and His residues. The finger itself comprises ~23 amino acids, and the linker between fingers is usually 7–8 amino acids.

The transcription factor TFIIIA (required for RNA polymerase III to transcribe 5S RNA genes) has a series of 9 fingers organized as tandem repeats. Three repeats of the finger sequence are present in the transcription factor SP1 that is involved in initiation at many promoters used by RNA polymerase II. The fingers are required for binding to DNA.

It is necessary to be cautious about interpreting the presence of (putative) zinc fingers, especially when the protein contains only a single finger motif. Fingers may be involved in binding RNA rather than DNA or even unconnected with any nucleic acid binding activity.

The prototype finger protein, TFIIIA, binds both to the 5S gene and to the product, 5S rRNA. A translation initiation factor, eIF2β has a zinc finger; and mutations in the finger influence the recognition of initiation codons. Retroviral capsid proteins have a motif related to the finger that may be involved in binding the viral RNA.

Zinc fingers are a common motif in DNA-binding proteins. The fingers usually are organized as a single series of tandem repeats; occasionally there is more than one group of fingers. The stretch of fingers ranges from 9 repeats that occupy almost the entire protein (as in TFIIIA) to providing just one small domain consisting of 2 fingers (as in the *Drosophila* regulator ADR1). The general transcription factor Sp1 has a DNA-binding domain that consists of 3 zinc fingers.

The crystal structure of DNA bound by a protein with three fingers suggests the structure illustrated schematically in **Figure 30.8**. The C-terminal part of each finger forms an α-helix that binds DNA; the N-terminal part forms a β-sheet. (For simplicity, the β-sheet and the location of the zinc ion are not shown in the lower part of the figure.) The three α-helical stretches fit into one turn of the major groove; each α-helix (and thus each finger) makes two sequence-specific contacts with DNA (indicated by the arrows).

If this type of structure is typical of the zinc finger, we should expect that nonconserved amino acids in the C-terminal side of each finger are responsible for recognizing specific target sites. (But note that similar assumptions about the helix-turn-helix motif in prokaryotic repressors have been difficult to sustain, and it is now doubtful whether there will prove to be a code that relates

Figure 30.8

Zinc fingers may form α-helices that insert into the major groove, associated with β-sheets on the other side.

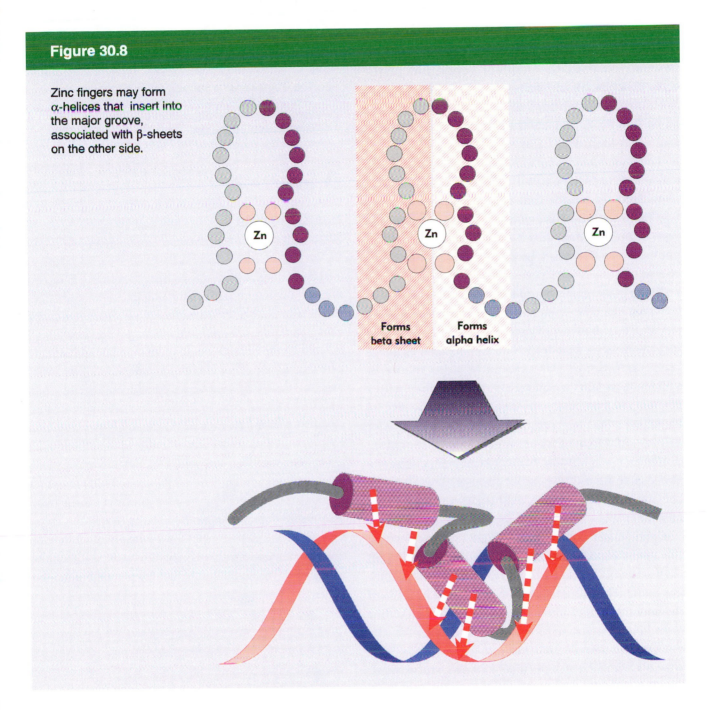

Forms beta sheet

Forms alpha helix

amino acid sequences of the helices to base pair sequence of the DNA target; see Chapter 17.)

Knowing that zinc fingers are found in authentic transcription factors that assist both RNA polymerases II and III, we may view finger proteins from the reverse perspective. When a protein is found to have multiple zinc fingers, there is at least a *prima facie* case for investigating a possible role as a transcription factor. Such an identification has suggested that several loci involved in embryonic development of *D. melanogaster* are regulators of transcription.

Steroid receptors (and some other proteins) have another type of finger. The structure is based on a sequence with the (putative) zinc-binding consensus:

$$\text{Cys-X}_2\text{-Cys-X}_{1-3}\text{-Cys-X}_2\text{-Cys}$$

These are called Cys$_2$/Cys$_2$ fingers. Proteins with fingers of this type often have nonrepetitive fingers, in contrast with the tandem repetition of the Cys$_2$/His$_2$ type. Binding sites in DNA (where known) are short and palindromic. Mutational analyses show that the regions of the regulator proteins that bind DNA include the finger motifs.

The Cys$_2$/Cys$_2$ fingers are distinct from Cys$_2$/His$_2$ fingers. Experiments with the steroid receptors show that mutations that convert the second two Cys residues into His residues, thus changing the type of finger, abolish the ability to activate target genes. Thus the two types of finger are not interchangeable.

The glucocorticoid and estrogen receptors each have two fingers, each with a zinc atom at the center of a tetrahedron of cysteines. The two fingers form α-helices that fold together to form a large globular domain. The aromatic sides of the α-helices form a hydrophobic center together with a β-sheet that connects the two helices. One side of the N-terminal helix makes contacts in the major groove of DNA. Two glucocorticoid receptors dimerize upon binding to DNA, and each engages a successive turn of the major groove. This fits with the palindromic nature of the response element (see later).

Each finger controls one important property of the receptor. **Figure 30.9** identifies the relevant amino acids. Those on the right side of the first finger control binding to DNA; those on the left side of the second finger control the ability to form a dimer (which we discuss in the next section).

Direct evidence that the first finger binds DNA was obtained by a 'specificity swap' experiment. The finger of the estrogen receptor was deleted and replaced by those of the glucocorticoid receptor. The new protein recognized the GRE sequence (the usual target of the glucocorticoid receptor) instead of the ERE (the usual target of the estrogen receptor). This region therefore establishes the specificity with which DNA is recognized.

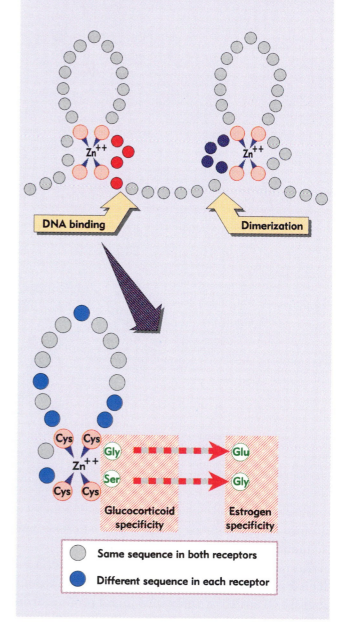

Figure 30.9

The first finger of a steroid receptor controls specificity of DNA-binding (at the positions shown in red); the second finger controls specificity of dimerization (at the positions shown in blue). The expanded view of the first finger shows that discrimination between GRE and ERE target sequences rests on two amino acids at the base of the first finger.

The sequences of the GRE and ERE are rather similar, and further swap experiments show that discrimination between them resides in two amino acids of the first finger. The differences between the sequences of the glucocorticoid receptor and estrogen receptor fingers lie mostly at the base of the finger. The substitution at two positions shown in Figure 30.9 allows the glucocorticoid receptor to bind at an ERE instead of a GRE.

Steroid receptors have domains for DNA binding, hormone binding, and activating transcription

Steroid hormones are synthesized in response to a variety of neuroendocrine activities, and exert major effects on growth, tissue development, and body homeostasis in the animal world. The major groups of steroids and some other compounds with related (molecular) activities are classified in **Figure 30.10**.

The adrenal gland secretes >30 steroids, the two major groups being the glucocorticoids and mineralocorticoids. Steroids provide the reproductive hormones (androgen male sex hormones and estrogen female sex hormones). Vitamin D is required for bone development.

Other hormones, with unrelated structures and physiological purposes, function at the molecular level in a similar way to the steroid hormones. Thyroid hormones, based on iodinated forms of tyrosine, control basal metabolic rate in animals. Steroid and thyroid hormones also may be important in metamorphosis (ecdysteroids in insects, and thyroid hormones in frogs).

Retinoic acid (vitamin A) is a morphogen responsible for development of the anterior–posterior axis in the developing chick limb bud. Its metabolite, 9-cis retinoic acid is found in tissues that are major sites for storage and metabolism of vitamin A.

We may account for these various actions in regulating body development and function in terms of pathways for regulating gene expression. These diverse compounds share a common mode of action: *each is a small molecule that binds to a specific receptor that activates gene transcription.* ('Receptor' may be a misnomer: the protein is a receptor for steroid or thyroid hormone in the same sense that *lac* repressor is a receptor for a β-galactoside: it is not a receptor in the sense of comprising a membrane-bound protein that is exposed to the cell surface.)

We know most about the interaction of glucocorticoids with their receptor, whose action is illustrated in **Figure 30.11**. A steroid hormone can pass through the cell membrane to enter the cell by simple diffusion. Within the cell, a glucocorticoid binds the glucocorticoid receptor. (Work on the glucocorticoid receptor has relied on the synthetic steroid hormone, dexamethasone.) The localization of free receptors is not entirely clear; they may be in equilibrium between the nucleus and cytoplasm. But when hormone binds to the receptor, the protein is converted into an activated form that has a 10× increased affinity for nonspecific DNA; the hormone–receptor complex is always localized in the nucleus.

The activated receptor recognizes a specific consensus sequence that identifies the GRE, the glucocorticoid response element. The GRE is

Figure 30.10

Several types of hydrophobic small molecules activate transcription factors. Corticoids and steroid sex hormones are synthesized from cholesterol, vitamin D is a steroid, thyroid hormones are synthesized from tyrosine, and retinoic acid is synthesized from isoprene (in fish liver).

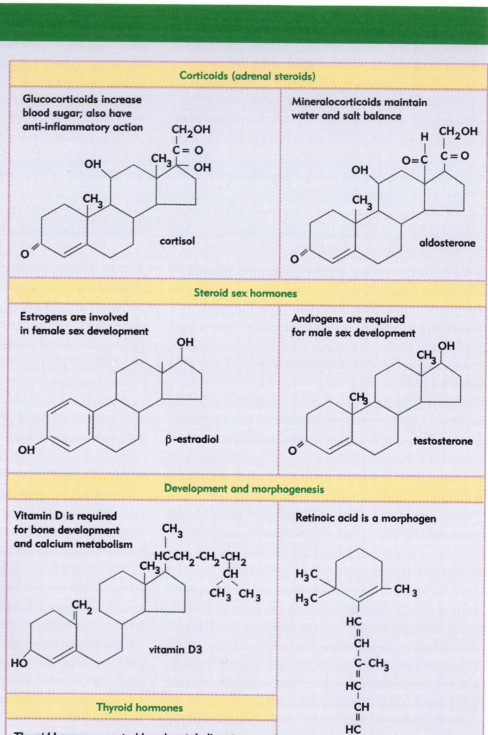

Figure 30.11

Glucocorticoids regulate gene transcription by causing their receptor to bind to an enhancer whose action is needed for promoter function.

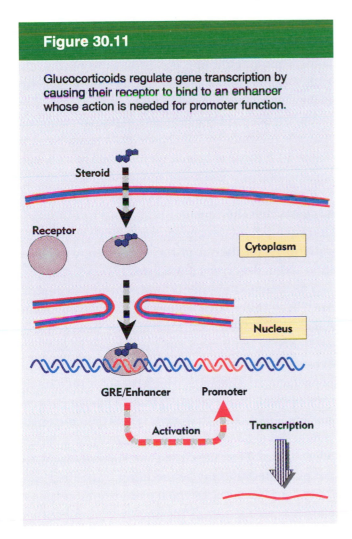

receptor into a cell confers upon that cell the ability to respond to steroid by activating any gene linked to an appropriate response element. This system then can be used to analyze the properties of receptors coded by cDNAs into which mutations have been introduced. Such analysis dissects the receptors into the regions summarized in **Figure 30.12**.

Receptors for the diverse groups of steroid hormones, thyroid hormones, and retinoic acid represent a new 'super-family' of gene regulators, the ligand-responsive transcription factors. A wheel comes full circle with this idea, since the

Figure 30.12

Receptors for many steroid and thyroid hormones have a similar organization, with an individual N-terminal region, conserved DNA-binding region, and a C-terminal hormone-binding region.

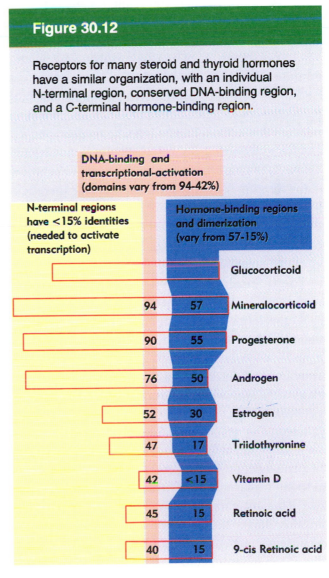

typically located in an enhancer in the general vicinity of a gene that responds to glucocorticoids. The GRE may be several kilobases upstream or downstream of the promoter. When the steroid–receptor complex binds to the enhancer, the nearby promoter is activated, and transcription initiates there. Enhancer activation provides the general mechanism by which steroids regulate a wide set of target genes. This action corresponds formally to the bacterial model for induction by a positive regulator (co-inducer activates inducer protein), illustrated in Figure 15.18.

By cloning steroid receptor cDNAs, the sequences of the corresponding proteins have been deduced. Introducing the cloned cDNA for a

form of these factors of course resembles the classic model for prokaryotic regulators presented by the *lac* operon.

All the receptors have independent domains for DNA binding and hormone binding, in the same relative locations.

The central DNA-binding domains are well related for the various steroid receptors, and remain identifiable in the other receptors. Some ligands have multiple receptors that are closely related, such as the 3 retinoic acid receptors (α, β, γ) and the three receptors for 9-*cis*-retinoic acid (RXRα, β, γ). The conservation of sequence probably reflects the common need to bind to DNA, while the variation is responsible for the selection of different target sequences. The act of binding DNA cannot be disconnected from the ability to activate transcription, because mutations in this domain affect both activities.

The N-terminal regions of the receptors show the least conservation of sequence. They include other regions that are needed to activate transcription.

The C-terminal domains bind the hormones. Those in the steroid receptor family show relationships ranging from 30 to 57%, reflecting specificity for individual hormones. Their relationships with the other receptors are minimal, reflecting specificity for a variety of compounds—thyroid hormones, vitamin D, retinoic acid, etc.

The C-terminal region regulates the activity of the receptor in a way that varies for the individual receptor. If the C-terminal domain of the glucocorticoid receptor is deleted, the remaining N-terminal protein is constitutively active: it no longer requires steroids for activity. This suggests that, in the absence of steroid, the steroid-binding domain prevents the receptor from recognizing the GRE; it functions as an internal negative regulator. The addition of steroid inactivates the inhibition, releasing the receptor's ability to bind the GRE and activate transcription. The basis for the repression could be internal, relying on interactions with another part of the receptor. Or it could result from an interaction with some other protein, which is displaced when steroid binds.

The interaction between the domains is different in the estrogen receptor. If the hormone-binding domain is deleted, the protein is unable to activate transcription, although it continues to bind to the ERE. This region is therefore required to activate rather than to repress activity.

Each receptor recognizes response elements that are related to a consensus. As summarized in Table 30.1, these response elements have a characteristic feature: each consensus consists of two short repeats (or **half sites**). This immediately suggests that the receptor binds as a multimer, so that each half of the consensus is contacted by one subunit (reminiscent of the λ operator–repressor interaction described in Chapter 17).

The response elements for the various receptors may be either palindromes or direct repeats in which the half sites are separated by 0–4 bp whose sequence is irrelevant. Only two types of half site are used by the various receptors. Their orientation and spacing determines which receptor recognizes the response element. This behavior allows response elements that have restricted consensus sequences to be recognized specifically by a variety of receptors. The rules that govern recognition are not absolute, but may be modified by context, and there are also cases in which palindromic response elements are recognized permissively by more than one receptor.

The receptors fall into two groups:

◆ Glucocorticoid (GR), mineralocorticoid (MR), androgen (AR), and progesterone (PR) receptors all form homodimers. They recognize response elements whose half sites have the consensus sequence TGTTCT. The half sites are arranged as palindromes, and the spacing between the sites determines the type of element. The estrogen (ER) receptor functions in the same way, but has the half site sequence TGACCT.

◆ The thyroid (T3R), vitamin D (VDR), retinoic acid (RAR), and 9-*cis*-retinoic acid (RXR) receptors form heterodimers, which recognize half elements with the sequence TGACCT. The half sites are arranged as direct repeats, and recognition is influenced by their separation as follows:

1 bp - RXR
3 bp - VDR
4 bp - T3R
5 bp - RAR

This pattern of recognition in the second group applies for dimeric receptors in which one subunit is the receptor on the list, and the other is RXR (so the first on the list is actually a homodimer). The requirement for heterodimer formation was discovered because retinoic acid and thyroid hormone receptors (RARs and T3Rs) bind much more efficiently to their target sites in the presence of an additional factor, which turns out to be RXR. The heterodimeric receptors respond to the ligand for the first subunit, and apparently do not require RXR to bind its ligand. These receptors can also form homodimers, which recognize palindromic sequences.

Now we are in a position to understand the basis for specificity of recognition. Recall that Figure 30.9 shows how recognition of the sequence of the half site is conferred by the amino acid sequence in the first finger. Specificity for dimer formation is carried by amino acids in the second finger. The formation of dimers in turn probably determines the distance between the subunits that sit in successive turns of the major groove, and thus controls the response to the spacing of half sites.

Homeo domains may bind related targets in DNA

The homeobox is a sequence that codes for a domain of 60 amino acids present in proteins of many or even all eukaryotes. Its name derives from its original identification in *Drosophila* homeotic loci (whose genes determine the identity of body structures). It is present in many of the genes that regulate early development in *Drosophila*, and a related motif is found in genes in a wide range of higher eukaryotes. It is attractive to think that the homeo domain identifies (or at least is common in) genes concerned with developmental regulation (see Chapter 38). Sequences related to the homeo domain are found in several types of animal transcription factors, but with the extension from the original *Drosophila* homeo domains to mammalian transcription factors, the relationship between the conserved regions drops significantly.

In *Drosophila* homeotic genes, the homeo domain often (but not always) occurs close to the C-terminal end. Some examples of genes containing homeoboxes are summarized in **Figure 30.13**. Often the genes have little conservation of sequence except in the homeobox. The conservation of the homeobox sequence varies. A major group of homeobox-containing genes in *Drosophila* has a

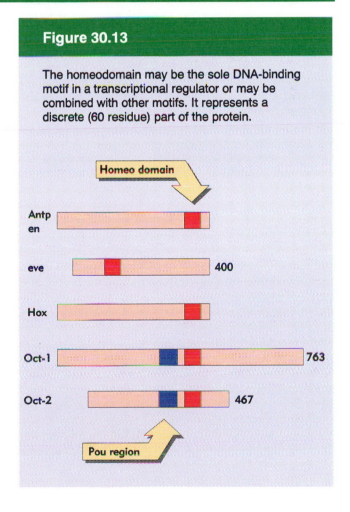

Figure 30.13

The homeodomain may be the sole DNA-binding motif in a transcriptional regulator or may be combined with other motifs. It represents a discrete (60 residue) part of the protein.

well conserved sequence, with 80–90% similarity in pairwise comparisons. Other genes have less well related homeoboxes. The homeo domain is sometimes combined with other motifs in animal transcription factors. One example is presented by the Oct (octamer-binding) proteins, which represent a group in which a conserved stretch of 75 amino acids called the Pou region is located close to a region resembling the homeo domain. The corresponding sequences in homo-boxes of the pou group of proteins are the least well related to the original group, and thus comprise the farthest extension of the family.

The homeo domain is responsible for binding to DNA, and experiments to swap homeo domains between proteins suggest that the specificity of DNA recognition lies within the homeo domain, but (like the situation with phage repressors) no simple code relating protein and DNA sequences can be deduced. The C-terminal region of the homeo domain shows homology with the helix-turn-helix motif of prokaryotic repressors. We recall from Chapter 17 that the lambda repressor has a 'recognition helix' (α-helix-3) that makes contacts in the major groove of DNA, while the other helix (α-helix-2) lies at an angle across the DNA. The homeo domain can be organized into three potential helical regions; the sequences of three examples are compared in **Figure 30.14**. The best conserved part of the sequence lies in the third helix. The difference between these structures and the prokaryotic repressor structures lies in the length of the helix that recognizes DNA, helix 3, which is 17 amino acids long in the homeo domain, compared to 9 residues long in the lambda repressor.

The crystal structure of the homeo domain of the product of the *D.melanogaster engrailed* gene is represented schematically in **Figure 30.15**. Helix 3 binds in the wide groove of DNA and makes the majority of the contacts between protein and nucleic acid. Many of the contacts that orient the helix in the major groove are made with the

Figure 30.14

The homeodomain of the *Antennapedia* gene represents the major group of genes containing homeoboxes in Drosophila; *engrailed (en)* represents another type of homeotic gene; and the mammalian factor Oct-2 represents a distantly related group of transcription factors. The homeodomain is conventionally numbered from 1 to 60. It starts with the N-terminal arm, and the three helical regions occupy residues 10-22, 28-38, and 42-58.

	1	N-terminal arm								10			Helix 1							20		
En	Glu	Lys	Arg	Pro	Arg	Thr	Ala	Phe	Ser	Ser	Glu	Gln	Leu	Ala	Arg	Leu	Lys	Arg	Glu	Phe	Asn	Glu
Antp	Arg	Lys	Arg	Gly	Arg	Gln	Thr	Tyr	Thr	Arg	Tyr	Gln	Thr	Leu	Glu	Leu	Glu	Lys	Glu	Phe	His	Phe
Oct2	Arg	Arg	Lys	Lys	Arg	Thr	Ser	Ile	Glu	Thr	Asn	Val	Arg	Phe	Ala	Leu	Glu	Lys	Ser	Phe	Leu	Ala

							30			Helix 2							40
En	Asn	Arg	Tyr	Leu	Thr	Glu	Arg	Arg	Arg	Glu	Glu	Leu	Ser	Ser	Glu	Leu	Gly	Leu
Antp	Asn	Arg	Tyr	Leu	Thr	Arg	Arg	Arg	Arg	Ile	Glu	Ile	Ala	His	Ala	Leu	Cys	Leu
Oct2	Asn	Glu	Lys	Pro	Thr	Ser	Glu	Glu	Ile	Leu	Leu	Ile	Ala	Glu	Gln	Leu	His	Met

	41									50			Helix 3							60
En	Asn	Glu	Ala	Gln	Ile	Lys	Ile	Trp	Phe	Gln	Asn	Lys	Arg	Ala	Lys	Ile	Lys	Lys	Ser	Asn
Antp	Thr	Glu	Arg	Gln	Ile	Lys	Ile	Trp	Phe	Gln	Asn	Arg	Arg	Met	Lys	Trp	Lys	Lys	Glu	Asn
Oct2	Glu	Lys	Glu	Val	Ile	Arg	Val	Trp	Phe	Cys	Asn	Arg	Arg	Gln	Lys	Glu	Lys	Arg	Ile	Asn

Figure 30.15

Helix 3 of the homeodomain binds in the major groove of DNA, with helices 1 and 2 lying outside the double helix. Helix 3 contacts both the phosphate backbone and specific bases. The N-terminal arm lies in the minor groove, and makes additional contacts.

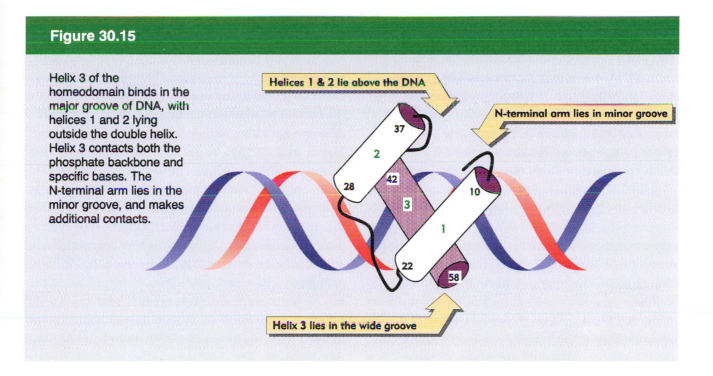

Helices 1 & 2 lie above the DNA

N-terminal arm lies in minor groove

Helix 3 lies in the wide groove

phosphate backbone, so they are not specific for DNA sequence. They lie largely on one face of the double helix, and flank the bases with which specific contacts are made. The remaining contacts are made by the N-terminal arm of the homeo domain, the sequence that just precedes the first helix. It projects into the minor groove. Thus the N- and C-terminal regions of the homeo domain are primarily responsible for contacting DNA.

A striking demonstration of the generality of this model derives from a comparison of the crystal structure of the homeo domain of engrailed with that of the α2 mating protein of yeast. The DNA-binding domain of this protein resembles a homeo domain, and can form three similar helices: its structure in the DNA groove can be superimposed almost exactly on that of the engrailed homeo domain. These similarities suggest that all homeo domains bind to DNA in the same manner. This means that a relatively small number of residues in helix 3 and in the N-terminal arm are responsible for specificity of contacts with DNA.

Helix-loop-helix proteins interact by combinatorial association

Two common features in DNA-binding proteins are the presence of helical regions that bind DNA, and the ability of the protein to dimerize. Both features are represented in the group of **helix-loop-helix** proteins that share a common type of sequence motif: a stretch of 40–50 amino acids contains two amphipathic α-helices separated by a linker region (the loop) of varying length. The proteins in this group form both homodimers and heterodimers by means of interactions between the hydrophobic residues on the corresponding faces of the two helices. The helical regions are 15–16 amino acids long, and each contains several conserved residues. Two examples are compared in **Figure 30.16**. The ability to form dimers resides with these amphipathic helices, and is common to all HLH proteins. The loop is probably important only for allowing the freedom for the two helical regions to interact independently of one another.

Most HLH proteins contain a region adjacent to the HLH motif itself that is highly basic, and which is needed for binding to DNA. There are ~6 conserved residues in a stretch of 15 amino acids. Members of the group with such a region are called **bHLH proteins**. A homodimer or a heterodimer in which both subunits have the basic region can bind to DNA. The HLH domains probably correctly orient the two basic regions contributed by the individual subunits. We do not yet know much about the means by which HLH proteins activate transcription.

The members of the bHLH group include the two proteins E12 and E47 that bind to an element in an immunoglobulin gene enhancer, the myoD, myogenin, and myf-5 transcription factors that are involved in myogenesis (muscle formation), the *myc* genes (which are the cellular counterparts of oncogenes and are involved in growth regulation),

and a group of genes that specify development of the nervous system in *D. melanogaster*.

The bHLH proteins fall into two general groups: class A consists of proteins that are ubiquitously expressed, including mammalian E12/E47 and fly **da**; class B consists of proteins that are expressed in a tissue-specific manner, including mammalian MyoD and fly **AC-S**. The tissue-specific bHLH proteins may be transcriptional regulators that function most efficiently by forming heterodimers with a ubiquitous partner. We do not know whether this is the only type of activity, or whether homodimers formed from ubiquitous bHLH proteins are sometimes used to regulate transcription. The myc proteins form a separate class of bHLH proteins, whose partners and targets are different.

Dimers formed from bHLH proteins differ in their abilities to bind to DNA. For example, E47 homodimers, E12-E47 heterodimers, and MyoD-E47 heterodimers all form efficiently and bind strongly to DNA; E12 homodimerizes well but binds DNA poorly, while MyoD homodimerizes only poorly. Thus both dimer formation and DNA binding may represent important regulatory points. At this juncture, it is possible to define groups of HLH proteins whose members form various pairwise combinations, but not to predict from the sequences the strengths of dimer formation or DNA binding.

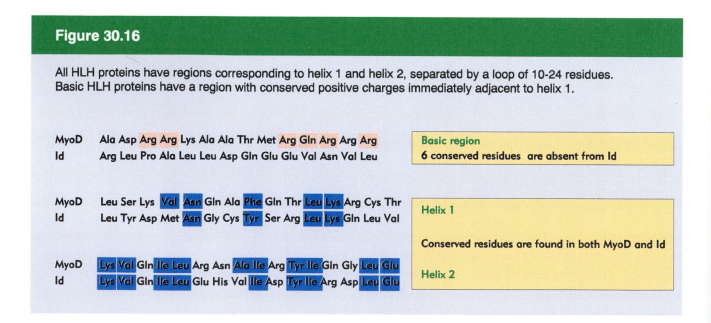

Figure 30.16

All HLH proteins have regions corresponding to helix 1 and helix 2, separated by a loop of 10-24 residues. Basic HLH proteins have a region with conserved positive charges immediately adjacent to helix 1.

MyoD   Ala Asp Arg Arg Lys Ala Ala Thr Met Arg Gln Arg Arg Arg
Id   Arg Leu Pro Ala Leu Leu Asp Gln Glu Glu Val Asn Val Leu

Basic region
6 conserved residues are absent from Id

MyoD   Leu Ser Lys Val Asn Gln Ala Phe Gln Thr Leu Lys Arg Cys Thr
Id   Leu Tyr Asp Met Asn Gly Cys Tyr Ser Arg Leu Lys Gln Leu Val

Helix 1

Conserved residues are found in both MyoD and Id

MyoD   Lys Val Gln Ile Leu Arg Asn Ala Ile Arg Tyr Ile Gln Gly Leu Glu
Id   Lys Val Gln Ile Leu Glu His Val Ile Asp Tyr Ile Arg Asp Leu Glu

Helix 2

All of the dimers in this group that bind DNA recognize the same consensus sequence, but we do not yet know whether different homodimers and heterodimers have preferences for slightly different target sites that are related to their functions.

Differences in DNA binding result from properties of the region in or close to the HLH motif; for example, E12 differs from E47 in possessing an inhibitory region just by the basic region, which inhibits DNA binding by homodimers. Some HLH proteins lack the basic region and/or contain proline residues that appear to disrupt its function. The example of the protein Id is shown in Figure 30.16. Proteins of this type have the same capacity to dimerize as bHLH proteins, but a dimer that contains one subunit of this type can no longer bind to DNA specifically. This is a forceful demonstration of the importance of doubling the DNA-binding motif in DNA-binding proteins.

The importance of the distinction between the nonbasic HLH and bHLH proteins is suggested by the properties of two groups of HLH proteins: the **da-AC-S/emc** group and the MyoD/Id type. A model for their functions in forming a regulatory network is illustrated in **Figure 30.17**.

In *D. melanogaster*, the gene *emc (extramacrochaetae)* is required to establish the normal spatial pattern of adult sensory organs. It functions by *suppressing* the functions of several genes, including *da (daughterless)* and the *achaete–scute complex (AC-S)*, which otherwise cause additional cells to take up this fate. *AC-S* and *da* are genes of the bHLH type. The suppressor *emc* codes for an HLH protein that lacks the basic region. We suppose that, in the absence of *emc* function, the **da** and **AC-S** proteins form dimers that activate transcription of

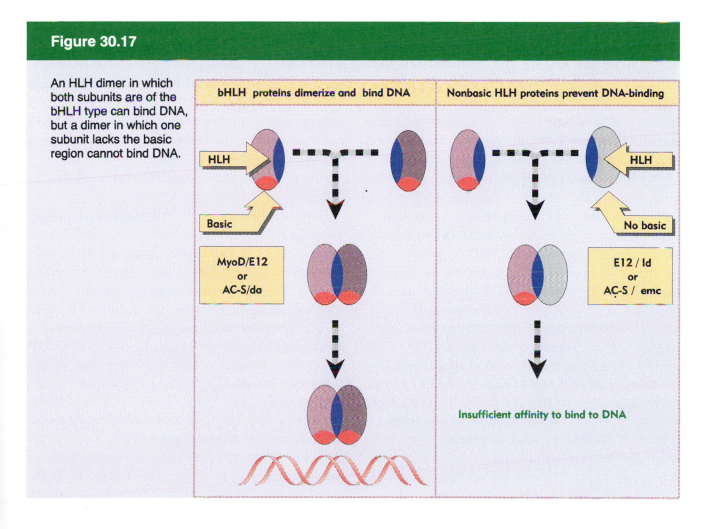

Figure 30.17

An HLH dimer in which both subunits are of the bHLH type can bind DNA, but a dimer in which one subunit lacks the basic region cannot bind DNA.

bHLH proteins dimerize and bind DNA

Nonbasic HLH proteins prevent DNA-binding

HLH

Basic

MyoD/E12 or AC-S/da

HLH

No basic

E12 / Id or AC-S / emc

Insufficient affinity to bind to DNA

appropriate target genes, but the production of **emc** protein causes the formation of heterodimers that cannot bind to DNA. Thus production of **emc** protein in the appropriate cells is necessary to suppress the function of **AC-S/da**.

The formation of muscle cells is triggered by a change in the transcriptional program that requires several bHLH proteins, including MyoD. MyoD is produced specifically in myogenic cells; and, indeed, overexpression of MyoD in certain other cells can induce them to commence a myogenic program. The trigger for muscle differentiation is probably a heterodimer consisting of MyoD-E12 or MyoD-E47, rather than a MyoD homodimer. Before myogenesis begins, a member of the nonbasic HLH type, the Id protein, may bind to MyoD and/or E12 and E47 to form heterodimers that cannot bind to DNA. It binds to E12-E47 better than to MyoD, and

so might function by sequestering the ubiquitous bHLH partner. Overexpression of Id can prevent myogenesis. Thus the removal of Id could be the trigger that releases MyoD to initiate myogenesis.

The behavior of the HLH proteins therefore illustrates two general principles of transcriptional regulation. A small number of proteins form combinatorial associations; it is possible that particular combinations have different functions with regards to DNA binding and transcriptional regulation, but we do not yet entirely understand the relative functions of the tissue-specific and ubiquitous bHLH proteins. And members of the same general class of proteins function as suppressors by participating in the combinatorial associations. Differentiation may depend either on the presence or on the removal of particular suppressor proteins of the nonbasic HLH class.

Leucine zippers may be involved in dimer formation

Interactions between proteins are a common theme in building a transcription complex, and a motif found in several transcription factors (and other proteins) is involved in both homo- and heteromeric interactions. The **leucine zipper** is a stretch of amino acids rich in leucine residues which provide a dimerization motif. Dimer formation itself has emerged as a common principle in the action of proteins that recognize specific DNA sequences, and in the case of the leucine zipper, its relationship to DNA binding is especially clear, because we can see how dimerization juxtaposes the DNA-binding regions of each subunit. The reaction is depicted diagrammatically in **Figure 30.18**.

An amphipathic α-helix has a structure in which the hydrophobic groups (including leucine) face one side, while charged groups face the other side. A leucine zipper forms an amphipathic helix in which the leucines of the zipper on one protein could protrude from the α-helix and interdigitate

with the leucines of the zipper of another protein in parallel to form a coiled coil. The two right-handed helices wind around each other, with 3.5 residues per turn, so the pattern repeats integrally every 7 residues.

How is this structure related to DNA binding? The region adjacent to the leucine repeats is highly basic in each of the zipper proteins, and could comprise a DNA-binding site. The two leucine zippers in effect form a Y-shaped structure, in which the zippers comprise the stem, and the two basic regions bifurcate symmetrically to form the arms that bind to DNA. This is known as the **bZIP** structural motif. It explains why the target sequences for such proteins are inverted repeats with no separation.

Zippers may be used to sponsor formation of homodimers or heterodimers. They are lengthy motifs. Leucine occupies every seventh residue in the potential zipper. There are 4 repeats in the

Figure 30.18

The basic regions of the bZIP motif are held together by the dimerization at the adjacent zipper region when the hydrophobic faces of two leucine zippers interact in parallel orientation.

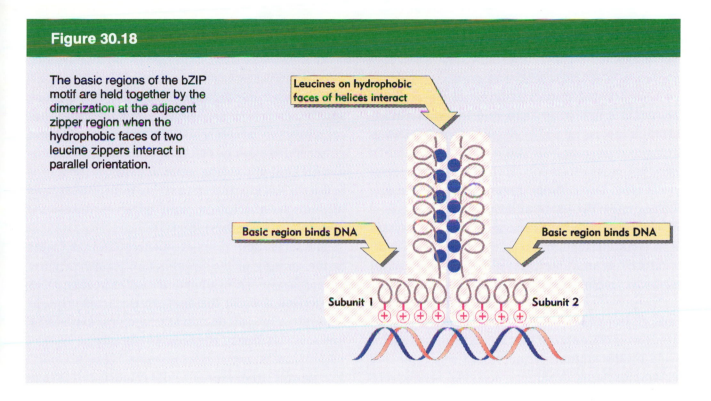

Leucines on hydrophobic faces of helices interact

Basic region binds DNA

Basic region binds DNA

Subunit 1   Subunit 2

protein C/EBP (a factor that binds as a dimer to both the CAAT box and the SV40 core enhancer), and 5 repeats in the factors Jun and Fos (which form the heterodimeric transcription factor, AP1). AP1 was originally identified by its binding to a DNA sequence in the SV40 enhancer (see Figure 29.10). The active preparation of AP1 includes several polypeptides.

A major component is Jun, the product of the gene *c-jun* which was identified by its relationship with the oncogene *v-jun* carried by an avian sarcoma virus (see Chapter 39). The mouse genome contains a family of related genes, *c-jun* (the original isolate) and *junB* and *junD* (identified by sequence homology with jun). There are considerable sequence similarities in the three Jun proteins; they have leucine zippers that can interact to form homodimers or heterodimers.

The other major component of AP1 is the product of another gene with an oncogenic counterpart. The *c-fos* gene is the cellular homologue to the oncogene *v-fos* carried by a murine sarcoma virus. Expression of *c-fos* activates genes whose promoters or enhancers possess an AP1 target site. The *c-fos* product is a nuclear phosphoprotein that is one of a group of proteins. The others are described as Fos-related antigens (FRA); they constitute a family of Fos-like proteins.

Fos also has a leucine zipper. Fos cannot form homodimers, but can form a heterodimer with Jun. A leucine zipper in each protein is required for the reaction. The ability to form dimers is a crucial part of the interaction of these factors with DNA. Fos cannot by itself bind to DNA, possibly because of its failure to form a dimer. But the Jun–Fos heterodimer can bind to DNA with same target specificity as the Jun–Jun dimer; and this heterodimer binds to the AP1 site with an affinity ~10 times that of the Jun homodimer.

One model supposes that, in either the Jun–Jun homodimer or Jun–Fos heterodimer, both subunits contact DNA. Then the increased affinity of the heterodimer for DNA might reflect the ability of the binding sites on the Jun and Fos subunits to contact DNA more strongly than binding sites on two Jun subunits.

Speculations about the nature of gene activation

We have treated transcription so far in terms of interactions involving DNA and individual transcription factors and RNA polymerases. This is an accurate description of the events that occur *in vitro*, but lacks an important feature of transcription *in vivo*. The cellular genome is packaged into nucleosomes; the exact differences in transcribed regions relative to nontranscribed regions remain to be characterized, but it is clear that chromatin structure has an important influence on transcription (see Chapter 28).

This provides a contrast with the regulation of transcription in bacterial cells, which we can view in the terms of Chapters 14–17 as determined exclusively by the actions of individual repressor and activator proteins. Whether a bacterial gene is transcribed can be predicted from the sum of the concentrations of the various factors that either activate or repress the individual gene. Changes in these concentrations at any time will change the state of expression accordingly.

Is this also true in a eukaryotic cell? It may be true of a system in which DNA is transcribed *in vitro* by purified components, but it is not necessarily true for the transcription of chromatin *in vivo*. We may ask two general questions about this relationship that identify the crucial issues:

◆ What changes must occur in chromatin structure at the promoter for initiation to occur?

◆ What effect does replication of DNA have upon the ability to transcribe a gene?

The basic transcriptional state of chromatin is inactive: wrapping DNA into nucleosomes, particularly at the promoter, prevents transcription. In this sense, histones function as generalized repressors of transcription (a rather old idea), although we see below that there are also more specific interactions. The role of transcriptional regulators is therefore usually to activate a target gene. This requires changes in the state of chromatin. Understanding gene activation therefore requires us to understand how nucleosomes impede transcription, and what changes are made to counter this effect.

The formation of hypersensitive DNAase I sites in the vicinity of the promoter of an active gene demonstrates the importance of exposing DNA sequences, free of histones, to the transcription apparatus. We can imagine two general models for making a change in chromatin structure.

◆ A **competitive model** would require that DNA is free of histones in order to allow factors and histones to compete for a given site, but would not allow one to replace the other directly. This would require replication to occur in order to change the state of gene expression.

◆ A **displacement model** would allow factors directly to displace histones, and thus to change the structure of chromatin at any time.

Examples are known that conform to each type of situation.

The example of a competitive model illustrated in **Figure 30.19** suggests that a transcription factor and histones are mutually exclusive residents at a promoter. This provides a molecular basis for the idea that histones are general repressors of gene expression; this repression must be relieved for transcription to initiate. Several examples conform with this general model.

The transcription factor TFIIIA, required for RNA polymerase III to transcribe 5S rRNA genes, cannot activate the target genes *in vitro* if they are com-

Figure 30.19

If nucleosomes form at a promoter, transcription factors (and RNA polymerase) cannot bind, and transcription is precluded. If a transcription factor (and RNA polymerase) bind to the promoter to establish a stable complex for initiation, histones are excluded.

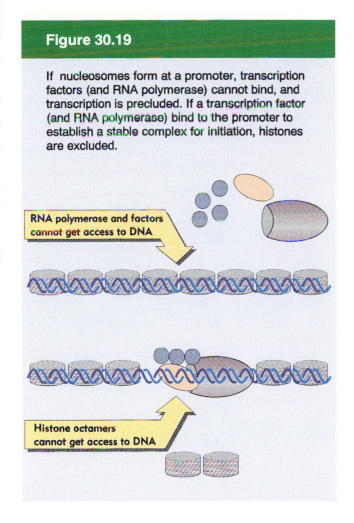

RNA polymerase and factors cannot get access to DNA

Histone octamers cannot get access to DNA

plexed with histones. However, the factor can form the necessary complex with free DNA; then the addition of histones does not prevent the gene from remaining active. Once the factor has bound, it remains at the site, allowing a succession of RNA polymerase molecules to initiate transcription. Whether the factor or histones get to the control site first may be the critical factor.

A similar situation is seen with the TFIID complex at promoters for RNA polymerase II. A plasmid containing an adenovirus promoter can be transcribed *in vitro* by RNA polymerase II in a reaction that requires TFIID and other transcription factors. The template can be assembled into nucleosomes by the addition of histones. If the histones are added

before the TFIID, transcription cannot be initiated. But if the TFIID is added first, the template still can be transcribed in its chromatin form. So TFIID can recognize free DNA, but either cannot recognize or cannot function on nucleosomal DNA. Only the TFIID must be added before the histones; the other transcription factors and RNA polymerase can be added later. This suggests that binding of TFIID to the promoter creates a structure to which the other components of the transcription apparatus can bind.

An important feature of this model is that the continued presence of the factor on DNA may be needed to exclude formation of nucleosomes. If a factor is not available to bind DNA at the time of replication, nucleosomes may form and preclude transcription. In these examples of 'pre-emptive' mechanisms for binding to DNA, the factors do not bind effectively to DNA that has been incorporated into nucleosomes.

In these cases, the basal transcription factor competes with histones for occupancy of DNA. One role of upstream activating factors may be to enhance the ability of basal factors to compete with histones. The ability of the yeast factor GAL4 to stimulate transcription is proportionately greater when the template is packaged into nucleosomes. The acidic activating domain is required for this function. This suggests that the acidic activator acts at a step in transcription initiation that becomes rate limiting when nucleosomes are present; for example, formation of the initiation complex may be mutually exclusive with histone binding to DNA.

Some transcription factors can bind to DNA on a nucleosomal surface. In these cases, there is often a requirement that the nucleosome be positioned precisely, for example, to ensure that the wide groove of the binding site for the factor is exposed. Nucleosomes appear to be precisely positioned at some steroid hormone response elements in such a way that receptors can bind. Receptor binding may alter the interaction of DNA with histones, and even lead to exposure of new binding sites.

Are there circumstances in which histones can

be displaced by activators? The best characterized examples are found in yeast. Induction of the *PHO5* gene (by addition of phosphate) causes the disappearance of 4 nucleosomes that had been precisely positioned at the promoter. Two factors that have binding sites within the region covered by these nucleosomes are essential for the displacement. Because displacement can occur without replication, it seems likely that these (and perhaps other factors) are responsible for directly displacing the histones.

It is important to distinguish between the conditions required to initiate transcription and those needed for its elongation. Although the presence of histones prevents RNA polymerase from initiating transcription at a promoter, RNA polymerases that have initiated transcription seem able to transcribe DNA that is bound to histones. It is controversial whether the passage of polymerase displaces the histones.

In cases in which it is necessary for transcription factors to bind to free DNA, replication of a gene poses both a threat and an opportunity. The threat is that replication inevitably changes (if only transiently) the structure of chromatin; and the opportunity is that this allows changes in the set of transcription factors bound to the gene.

Consider a gene that is activated (or repressed) by the binding to DNA of some specific regulator protein and/or by some change in chromatin structure. How is this particular state to be inherited by the duplicate chromosomes? We can consider two extreme situations:

◆ The condition of associated proteins might be perpetuated by a segregation mechanism. **Figure 30.20** presents a model to explain how one molecule of factor could stimulate many successive rounds of transcription from the promoter. A complex of nonhistone proteins might be established on the DNA, split into half-complexes at replication, and rebuild complete complexes on each daughter duplex. This would perpetuate the structure of chromatin unless and until some event intervened to make a change in it.

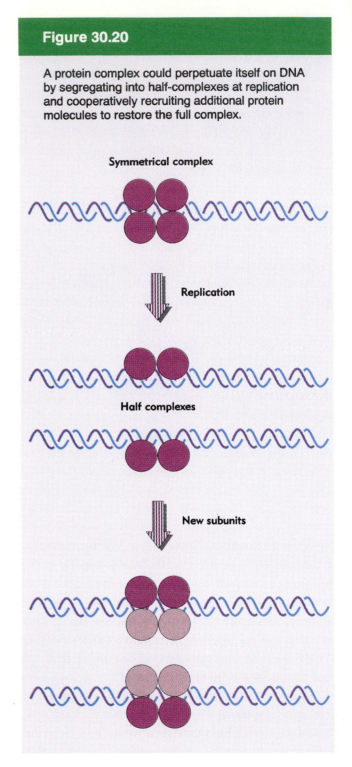

Figure 30.20

A protein complex could perpetuate itself on DNA by segregating into half-complexes at replication and cooperatively recruiting additional protein molecules to restore the full complex.

Symmetrical complex

Replication

Half complexes

New subunits

◆ If all nonhistones dissociate from DNA during replication, any specific state must be re-established in every cell cycle. This would predicate reliance on a model analogous to bacterial repressor–operator interactions, with the proviso that the presence of the relevant factors would be important specifically at the time of replication.

Experiments with the 5S RNA genes of X. laevis show that replication is dominant over transcription. A gene that contains a transcription complex bound at its promoter can be replicated; but the replicated daughter genes lack transcription complexes. The transcription machinery at least in this instance is unable to withstand the passage of replication, and arrangements to transcribe genes must be made anew after replication. To determine whether this is a general situation will require the characterization of systems in which chromatin structure can be activated in vitro.

The existence of a preinitiation complex signals that the gene is in an 'active' state, ready to be transcribed. The complex is stable, and may remain in existence through many cycles of replication. The ability to form a preinitiation complex could be a general regulatory mechanism. By binding to a promoter to make it possible for RNA polymerase in turn to bind, the factor in effect switches the gene on.

Since the reconstitution of the complex is necessary every generation, a corollary is that the gene could be turned off in a dividing cell simply by restricting the amount of factor. When the factor is diluted out by division, the complex becomes unable to reassemble, and the histones can form nucleosomes at the control sites. This could explain how genes are turned off during embryonic development. It poses the question of how an active gene might be turned off in a nondividing cell. Indeed, is it possible to turn off genes in terminally differentiated cells (that is, cells that have reached their final state of phenotypic expression)?

The variety of situations in which hypersensitive sites occur suggests that their existence reflects a general principle. *Sites at which the double helix initiates an activity are kept free of nucleosomes.* In each case, some (unknown) nonhistone proteins, concerned with the particular function of the site, modify the properties of a short region of DNA so that nucleosomes are excluded. The structures formed in each situation need not necessarily be similar (except that each, by definition, creates a site hypersensitive to DNAase I).

We may speculate whether the converse of this principle is true: *genes whose control regions are randomly organized in nucleosomes cannot be expressed.* Suppose that the formation of nucleosomes occurs in a manner independent of sequence to any region of DNA from which histones are not specifically excluded. Then in the absence of specific regulatory proteins, promoters and other regulatory regions will be organized by histone octamers into a state in which *they cannot be activated.* This may explain the need for nucleosomes to be precisely positioned in the vicinity of a promoter, so that essential regulatory sites are appropriately exposed.

Exact mechanisms for chromatin activation (or inactivation) are too speculative to consider now, but the general thread of these views is that active chromatin and inactive chromatin *are not in a dynamic equilibrium.* Sudden, disruptive events are needed to convert one to the other. One implication of the types of model we have discussed is that a greater concentration of a regulatory protein may be needed to initiate the activation event than to maintain the state of active chromatin.

The events involved in activating a gene need not coincide with the initiation of transcription. Formation of the appropriate complex may render the gene *transcribable,* but further action may be needed actually to transcribe it. In the language of developmental biology, complex formation is a *determination* event, preceding by some time the *differentiation* event, when the gene is actually switched on. A possible example is that genes whose expression is hormone-dependent first enter the determined state in the appropriate cell; and then actually start transcription when the hormone arrives. (In these terms, prokaryotic genes are controlled by differentiation-like mechanisms, since the activation event involves the actual start of transcription, rather than creating a state of readiness.)

Summary

Some elements are present in many genes and are recognized by ubiquitous factors; others are present in a few genes and are recognized by tissue-specific factors. Elements that uniquely identify particular groups of genes that are regulated in response to certain transcription factors are called response elements (REs). Binding of factors to specific sequences is followed by protein–protein interactions with other components of the general transcription apparatus. Transcription factors often have a modular construction, in which there are independent domains responsible for binding to DNA and for activating transcription. The main function of the DNA-binding domain may be to tether the activating domain in the vicinity of the initiation complex.

Several groups of transcription factors have been identified by sequence homologies. The homeo domain is a 60-residue sequence found in genes that regulate development in insects and worms and in mammalian transcription factors. It is related to the prokaryotic helix-turn-helix motif and provides the motif by which the factors bind to DNA.

Another motif involved in DNA binding is the zinc finger, which is found in proteins that bind DNA or RNA (or sometimes both). A finger has cysteine residues that bind zinc. One type of finger is found in multiple repeats in some transcription factors; another is found in single or double repeats in others.

Steroid receptors were the first members identi-

fied of a group of transcription factors in which the protein is activated by binding a small hydrophobic hormone. The activated factor becomes localized in the nucleus, and binds to its specific response element, where it activates transcription. The DNA-binding domain has zinc fingers.

The leucine zipper contains a stretch of amino acids rich in leucine that are involved in dimerization of transcription factors. An adjacent basic region is responsible for binding to DNA.

HLH (helix-loop-helix) proteins have amphipathic helices that are responsible for dimerization, adjacent to basic regions that bind to DNA.

Many transcription factors function as dimers, and it is common for there to be multiple members of a family that can engaged in homodimer and heterodimer formation. This creates the potential for complex combinations to govern gene expression. In some cases, a family includes inhibitory members, whose participation in dimer formation prevents the partner from activating transcription.

Transcription factors and histones may be mutually exclusive occupants of the promoter. In some cases, initiation of transcription may require a transcription factor (TFIID for RNA polymerase II or TFIIIA for RNA polymerase III) to bind to DNA before histones form nucleosomes at the promoter. Changing the structure of a site on chromatin could require an act of replication. In other cases, factors can directly displace histones and change the structure of chromatin at the promoter.

Further reading

Reviews

Interactions of regulator elements in yeast promoters and enhancers have been brought together by **Guarente** (*Ann. Rev. Biochem.* **21**, 425–452, 1987). Activation by acidic motifs was reviewed by **Ptashne** (*Nature* **335**, 683–689, 1988).

The functions of genes containing homeoboxes have been summarized by **Hoey and Levine** (*Cell* **55**, 537–540, 1988). Means of DNA binding have been reviewed by **Harrison** (*Nature* **353**, 715–719, 1991) and by **Pabo and Sauer** (*Ann. Rev. Biochem.* **61**, 1053–1095, 1992).

Regulation of enhancers by steroid receptors was reviewed by **Yamamoto** (*Ann. Rev. Genet.* **19**, 209–252, 1985). The interactions of steroid and thyroid hormones with their receptors has been reviewed in a biological context by **Evans** (*Science* **240**, 889–895, 1988; *Neuron* **2**, 1105–1112, 1989).

Regulation by HLH proteins, in particular MyoD, has been reviewed by **Weintraub** *et al.* (*Science* **251**, 760–766, 1991).

A tour d'horizon of the issues involved in establishing and maintaining active gene structures has been given by **Brown** (*Cell* **37**, 359–365, 1984); and connections between structure and function in chromatin were explored by **Weintraub** (*Cell* **42**, 705–711, 1985). The relationship between chromatin structure and transcription has been discussed by **Grunstein** (*Ann. Rev. Cell Biol.* **6**, 643–678, 1990) and by **Felsenfeld** (*Nature* **355**, 219–224, 1992).

Implications of the structures generated by cooperative protein–DNA binding have been discussed by **Ptashne** in *A Genetic Switch* (Cell Press and Blackwell Scientific, 1992).

Discoveries

The ability of ATF to stimulate transcription complex formation was discovered by **Horikoshi** *et al.* (*Cell* **54**, 1033–1042, 1988).

The motif of zinc finger proteins was first noted by **Miller, Rhodes, and Klug** (*EMBO J.* **4**, 1609–1614, 1985). The sequence of SP1 was analyzed by **Kadonaga** *et al.* (*Cell* **51**, 1079–1090, 1987).

Crystal structure have been reported for zinc finger proteins by **Pavelitch and Pabo** (*Science* **252**, 809–817, 1991), and for homeo domains by **Wolberger** *et al.* (*Cell* **67**, 517–528, 1991).

HLH proteins were recognized by **Murre, McCaw and Baltimore** (*Cell* **56**, 777–783, 1989).

The leucine zipper was discovered by **Landschulz** *et al.* (*Science* **240**, 1759–1764, 1988), and its structure further analyzed by **Vinson, Sigler, and McKnight** (*Science* **246**, 911–916, 1989).

Acidic activators were discovered by **Ma and Ptashne** (*Cell* **51**, 113–119, 1987).

CHAPTER 31

The apparatus for nuclear splicing

Interrupted genes are found in all classes of organisms, so the ability to express genes whose coding sequences are not intact is a common requirement. The discrepancy between the interrupted organization of the gene and the uninterrupted organization of its mRNA calls for a change to be made in the RNA at some stage between transcription and translation. Interrupted genes represent a minor proportion of the genes of the very lowest eukaryotes, but the vast majority of genes in higher eukaryotic genomes are interrupted. We have seen in Chapter 23 that there is a rather broad distribution of genes according to the number of interruptions and their overall size, but a typical mammalian gene has 7–8 exons spread out over ~16 kb, because of the presence of relatively large introns. The primary transcript has the same organization as the gene, and is sometimes called the **pre-mRNA**. Removal of the introns from pre-mRNA leaves a typical messenger of ~2.2 kb. The process by which the introns are removed is called **RNA splicing**.

One of the first clues about the nature of the discrepancy in size between nuclear genes and their products in higher eukaryotes was provided by the properties of nuclear RNA. Its average size is much larger than mRNA, it is very unstable, and it has a much greater sequence complexity. Taking its name from its broad size distribution, it was called **heterogeneous nuclear RNA (hnRNA)**. The hnRNA population characterized in these early experiments includes the pre-mRNA population.

The physical form of hnRNA is a ribonucleoprotein particle (**hnRNP**), in which the hnRNA is bound by proteins. As characterized *in vitro,* an hnRNP particle takes the form of beads connected by a fiber. The beads are globular particles of diameter ~200Å, sediment at ~40S, and contain 100–800 bases of RNA associated with a set of proteins that range in size from 34,000 to 120,000 daltons. A group of 6 common proteins (A1, A2, B1, B2, C1, C2) are called the core proteins; there are multiple copies of each protein per bead. There are also some other abundant proteins, as well as some proteins that may be present at lower stoichiometry, making a total of ~20 proteins. The exact structure of the hnRNP, and the functional implications of RNA's packaging in this manner, remain to be determined.

Splicing occurs in the nucleus, together with the other modifications that are made to newly synthesized RNAs. The substrate for these processes is the hnRNP particle. The process of expressing an interrupted gene is reviewed in **Figure 31.1**. The transcript is capped at the 5′ end (as we have seen in Chapter 10), has the introns removed (as we see in this chapter), and is polyadenylated at the 3′ end (which we review in Chapter 29). The RNA is then transported through nuclear pores to the cytoplasm, where it is available to be translated. With regards to the various processing reactions that occur in the nucleus, we should like to know at what point splicing occurs *vis-à-vis* the other modifications of RNA. Does splicing occur at a particular location in the nucleus; and is it connected with

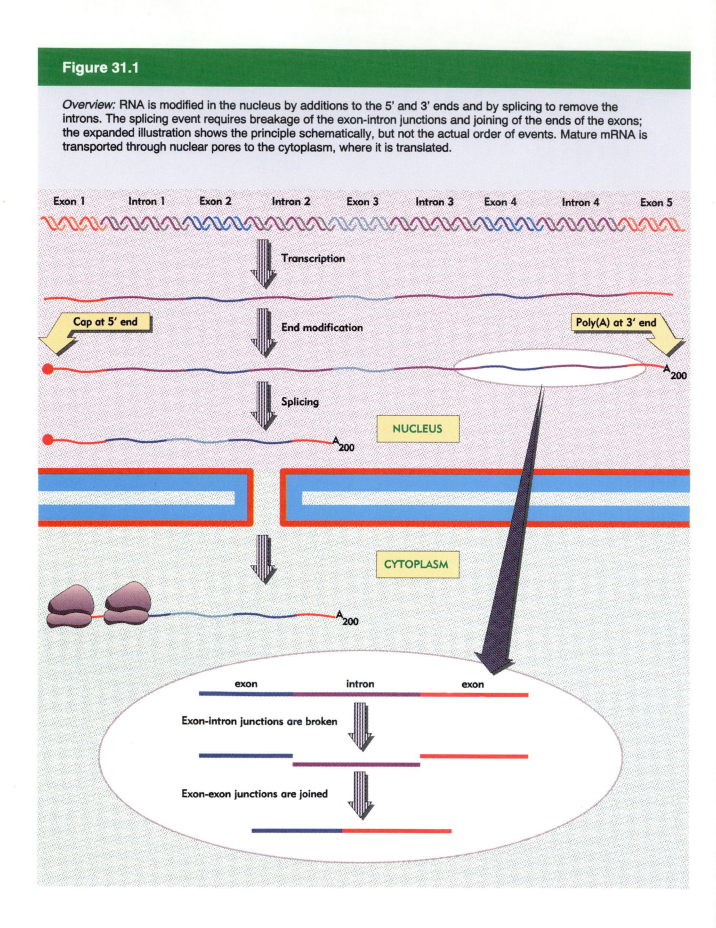

Figure 31.1

Overview: RNA is modified in the nucleus by additions to the 5' and 3' ends and by splicing to remove the introns. The splicing event requires breakage of the exon-intron junctions and joining of the ends of the exons; the expanded illustration shows the principle schematically, but not the actual order of events. Mature mRNA is transported through nuclear pores to the cytoplasm, where it is translated.

other events, for example, nucleocytoplasmic transport? Does the lack of splicing make an important difference in the expression of uninterrupted genes?.

With regards to the splicing reaction itself, one of the main questions is how its specificity is controlled. What ensures that the ends of each intron are recognized in pairs so that the correct sequence is removed from the RNA? Are introns excised from a precursor in a particular order? Is the maturation of RNA used to *regulate* gene expression by discriminating among the available precursors or by changing the pattern of splicing?

There is no unique mechanism for RNA splicing. We can identify several types of splicing systems:

◆ Introns are removed from the nuclear RNAs of higher eukaryotes by a system that recognizes only short consensus sequences conserved at exon–intron boundaries and within the intron. This reaction requires a large splicing apparatus, which takes the form of an array of proteins and ribonucleoproteins that functions as a large particulate complex.

◆ Excision of certain introns is an autonomous property of the RNA itself. Two groups of introns with this capacity are found in diverse locations. Each forms a characteristic type of secondary structure, and has short consensus sequences. The sequences of one group are related to those of nuclear introns. The ability of RNA to perform enzymatic activities is seen also in the self-cleavage of viroid RNAs and the catalytic activity of RNAase P. We discuss these reactions in Chapter 32.

◆ The removal of introns from yeast nuclear tRNA precursors involves enzymatic activities whose dealings with the substrate seem to resemble those of the tRNA processing enzymes, since a critical feature is the conformation of the tRNA precursor.

It is generally agreed that genes originated in the interrupted form, and that uninterrupted genes have lost their introns during evolution. In spite of a probable common origin, divergence since has developed in the mechanism by which the intact coding sequence is reconstituted in RNA transcribed from interrupted genes.

Nuclear splicing junctions are interchangeable but are read in pairs

To home in on the molecular events involved in splicing, we must consider the nature of the **splicing sites**, the two exon–intron boundaries that include the sites of breakage and reunion.

By comparing the nucleotide sequence of mRNA with that of the structural gene, the junctions between exons and introns can be assigned. Two features (or lack thereof) are significant in nuclear genes.

◆ *There is no extensive homology or complementarity between the two ends of an intron.* This excludes the possibility that an extensive secondary structure could form to link them directly together as a preliminary step in splicing.

◆ *The junctions have well-conserved, though rather short, consensus sequences.*

It is possible to assign a specific end to every intron by relying on the homology of exon–intron junctions. They can all be aligned to conform to the consensus sequence given in **Figure 31.2**. (In this as in other cases, we write just the sequence of the DNA strand that is identical with the RNA product.)

The subscripts indicate the percent occurrence of the specified base (or type of base) at each consensus position. The really high conservation is found only *immediately within the intron* at the presumed junctions. This identifies the sequence of a generic intron as:

$$GT......AG$$

Because the intron defined in this way starts with the dinucleotide GT and ends with the dinucleotide AG, the junctions are often described as conforming to the **GT-AG rule**.

Note that the two sites have different sequences and so they define the ends of the intron *directionally*. They are named proceeding from left to right along the intron, that is, as the **left (or 5′)** and **right (or 3′)** splicing sites. Sometimes they are called the **donor** and **acceptor** sites. The consensus sequences are implicated as the sites recognized in splicing by point mutations that prevent splicing *in vivo* and *in vitro*.

The GT-AG rule describes the splicing sites of nuclear genes of many (perhaps all) eukaryotes. This implies that there is a common mechanism for splicing the introns out of RNA. The consensus does not apply to the introns of mitochondria and chloroplasts, nor to the yeast tRNA genes. (We discuss these situations later.)

A typical mammalian mRNA has many introns. What ensures that the correct pairs of junctions are spliced together? We can imagine two types of principle that might be responsible for pairing the appropriate 5′ and 3′ sites:

◆ It could be an *intrinsic property* of the RNA to connect the sites at the ends of a particular intron, for example, because of base pairing involving these regions.

◆ Or all 5′ sites may be functionally equivalent and all 3′ sites may be similarly indistinguishable, but splicing could follow *rules* that ensure a 5′ site is always connected to the 3′ site that comes next in the RNA.

The splicing sites of nuclear RNAs do not themselves have any sequence complementarity, and nor do the surrounding sequences offer the ability to pair. Experiments to construct hybrid RNA precursors show that any 5′ splicing site can in principle be connected to any 3′ splicing site. For example, when the first exon of the early SV40 transcription unit is linked to the third exon of mouse β-globin, the hybrid intron can be spliced out to generate a perfect connection between the SV40 exon and the β-globin exon. (Indeed, this capacity is the basis for the exon trapping technique reviewed previously in Figure 23.18.) Such experiments make two general points:

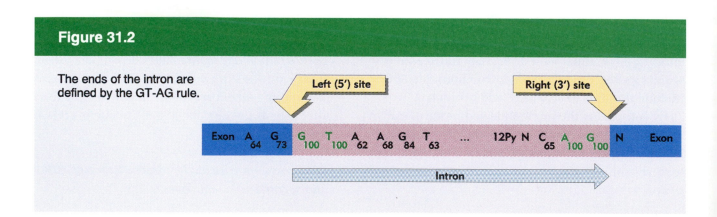

Figure 31.2

The ends of the intron are defined by the GT-AG rule.

Left (5′) site

Right (3′) site

Exon A_{64} G_{73} G_{100} T_{100} A_{62} A_{68} G_{84} T_{63} ... 12Py N C_{65} A_{100} G_{100} N Exon

Intron

◆ Splicing sites are generic: they do not have specificity for individual RNA precursors, and individual precursors do not convey specific information (such as secondary structure) that is needed for splicing.

◆ And the apparatus for splicing is not tissue specific; an RNA can usually be properly spliced by any cell, whether or not it is usually synthesized in that cell. (We discuss some exceptions in which there are tissue-specific alternative splicing patterns later.)

Here is a paradox. Probably all 5′ splicing sites look similar to the splicing apparatus, and all 3′ splicing sites look similar to it. *In principle any 5′ splicing site may be able to react with any 3′ splicing site.* But in the usual circumstances splicing occurs only between the 5′ and 3′ sites of the *same* intron. *What rules ensure that recognition of splicing sites is restricted so that only the 5′ and 3′ sites of the same intron are spliced?*

Are introns removed in a specific *order* from a particular RNA? Using RNA blotting, we can identify nuclear RNAs that represent intermediates from which some introns have been removed. **Figure 31.3** shows a blot of the precursors to ovomucoid mRNA. There is a discrete series of bands, which suggests that splicing occurs via definite pathways. (If the seven introns were removed in an entirely random order, there would be more than >300 precursors with different combinations of introns, and we should not see discrete bands.)

There does not seem to be an *obligatory* pathway, since intermediates can be found in which different combinations of introns have been removed. However, there is evidence for a *preferred* pathway or pathways. When only one intron has been lost, it is virtually always 5 or 6. But either can be lost first. When two introns have been lost, 5 and 6 are again the most frequent, but there are other combinations. Intron 3 is never or very rarely lost at one of the first three splicing steps. From this pattern, we see that there is a preferred pathway in which introns are removed

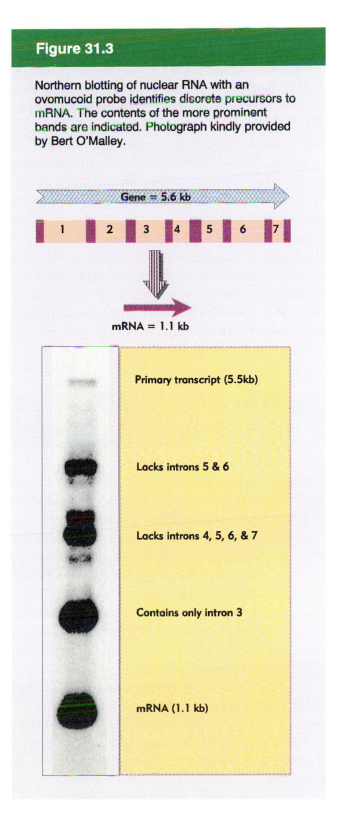

Figure 31.3

Northern blotting of nuclear RNA with an ovomucoid probe identifies discrete precursors to mRNA. The contents of the more prominent bands are indicated. Photograph kindly provided by Bert O'Malley.

Gene = 5.6 kb

1 2 3 4 5 6 7

mRNA = 1.1 kb

Primary transcript (5.5kb)

Lacks introns 5 & 6

Lacks introns 4, 5, 6, & 7

Contains only intron 3

mRNA (1.1 kb)

in the order 5/6, 7/4, 2, 3/1. But clearly there are other pathways, since (for example), there are some molecules in which 4 or 7 is lost last. A caveat in interpreting these results is that we do not have proof that all these intermediates actually lead to mature mRNA, but this is a reasonable assumption.

The general conclusion suggested by this analysis is that the conformation of the RNA influences the accessibility of the splicing sites. As particular introns are removed, the conformation changes, and new pairs of splicing sites become available. But the ability of the precursor to remove its introns in more than one order suggests that alternative conformations are available at each stage. Of course, the longer the molecule, the more structural options become available; and when we consider larger genes, it becomes difficult to see how specific secondary structures could control the reaction. One important conclusion of this analysis is that the reaction

does not proceed sequentially along the precursor.

One model for control of splicing is to suppose that the splicing apparatus acts in a processive manner. Having recognized a 5′ site, the apparatus is compelled to scan the RNA in the appropriate direction until it meets the next 3′ site. This would restrict splicing to adjacent sites.

This model could explain those cases in which the sites are uniquely defined and interact directly (although some introns are very long indeed). It does not explain the existence of alternative splicing patterns, such as those seen in SV40, where a common 5′ site is spliced to more than one 3′ site. We assume that in cases of alternative splicing, some additional event intervenes to make it possible to recognize sites besides those that lie in strict sequence.

Nuclear splicing proceeds through a lariat

A decisive account of the mechanism of splicing requires *in vitro* systems in which introns can be removed from RNA precursors. With such systems, intermediates can be characterized and we can test the dependence of the reaction on particular components. Nuclear extracts can splice purified RNA precursors, which shows that the action of splicing is not linked to the process of transcription. Splicing is also independent of modification of RNA, and can occur to RNAs that are neither capped nor polyadenylated.

Splicing in vitro occurs in two stages, illustrated in the pathway of Figure 31.4. We discuss the reaction in terms of the individual RNA species that can be identified, but we should remember that *in vivo* the species containing exons are not released as free molecules, but remain held together by the splicing apparatus.

In the first stage, a cut is made at the 5′ end of the intron, separating the left exon and the right

intron–exon molecule.

The left exon takes the form of a linear molecule, but the right intron–exon molecule is not linear. The 5′ terminus generated at the left end of the intron becomes linked by a 5′–2′ bond to an A in a sequence that is called the **branch site**, located ~30 bases upstream of the 3′ end of the intron. The linkage holds the intron in a structure that is called a **lariat**.

In the second stage, cutting at the 3′ splicing site releases the free intron in lariat form, while the right exon is ligated (spliced) to the left exon. The reactions are shown separately in the figure for illustrative purposes, but actually occur as one coordinated transfer.

The lariat is then 'debranched' to give a linear excised intron, which is rapidly degraded.

The only three sequences needed for splicing are short consensus sequences at the 5′ and 3′ splicing sites and at the branch site. Together

Figure 31.4

Splicing can be divided into three general stages.

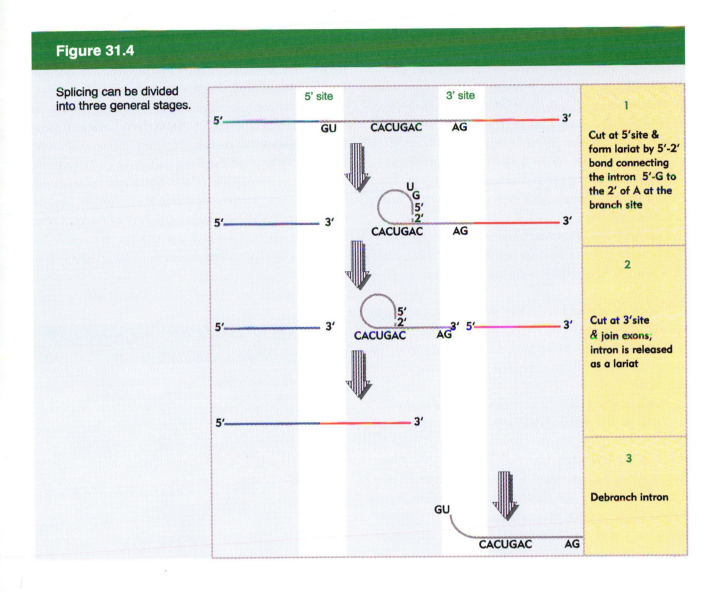

with the knowledge that most of the sequence of an intron can be deleted without impeding splicing, this indicates that there is no demand for specific conformation in the intron (or exon).

The branch site provides the means by which the 3' splicing site is identified. The branch site in yeast is highly conserved, and has the consensus sequence UACUAAC. The target sequence for the branch in higher eukaryotes is not well conserved, but can be recognized as conforming to the constraints in which there is a clear preference for purines or pyrimidines at each position. The consensus is:

$$Py_{80} \, N \, Py_{80} \, Py_{87} \, Pu_{75} \, A \, Py_{95}$$

The branch point lies 18–40 nucleotides upstream of the 3' splicing site. Mutations or deletions of the branch sequence in yeast prevent splicing. In higher eukaryotes, the greater flexibility in its sequence is seen in the ability to use related sequences in the vicinity when the authentic branch is deleted. Proximity to the 3' splicing site appears to be important, since the cryptic site is always close to the authentic site. When a cryptic branch sequence is used in this manner, splicing otherwise appears to be normal; and the exons give the same products as wild type. *The role of the branch site therefore is to identify the nearest 3' splicing site as the target for connection to the 5' splicing site.*

The bond that forms the lariat goes from the 5' position of the invariant G that was at the 5' end of the intron to the 2' position of the invariant A in the branch site, which is the one base that is completely conserved in higher eukaryotes. This corresponds to the third A residue in the yeast UACUAAC box.

Although the reaction proceeds in discrete enzymatic stages, some overall scrutiny of the junction and branch sequences may be made before it is initiated.

Mutations in the UACUAAC box in yeast prevent any reaction from occurring. Thus the target site for branch formation must be recognized in order to make the cut at the 5' intron site that releases the 5' intron end.

Some mutations in the 5' splicing site allow the reaction to proceed only so far as lariat formation. If the initial G of the intron is changed to an A, the site is cut at the left exon, and the lariat is formed; but then the reaction stops. If the 5'–2' branch sequence is incorrect, therefore, cutting does not occur at the 3' site downstream of it. Other mutations simply block the reaction (or divert it so that a substitute splicing site elsewhere is used).

Introns fall into two classes with regards to the role of the 3' site. Mutation of this site in some introns does not affect the preceding stages from occurring at the 5' and branch sites. In other introns, however, an aberrant 3' splice site prevents cleavage at the 5' splice site.

The general message conveyed by the effects of these mutations is that all sequence components involved in splicing are scrutinized before the reaction begins. A serious deficiency at any site may prevent the reaction from initiating. If recognizable forms of each sequence are present in appropriate locations, the reaction is allowed to proceed. However, it is blocked at a subsequent stage if one of the sequence components is aberrant.

The chemical reactions proceed by **transesterification**: a bond is in effect *transferred* from one location to another. **Figure 31.5** shows that the first step is a hydrophilic attack by the A of the UACUAAC sequence on the 5' splicing site. In the second step, the 3' end that was released by the first reaction now attacks the bond at the 3' splicing site. Note that the number of phosphodiester bonds is conserved. There were originally two 5'–3' bonds at the exon–intron splicing sites; one has been replaced by the 5'–3' bond between the exons, and the other has been replaced by the 5'–2' bond that forms the lariat.

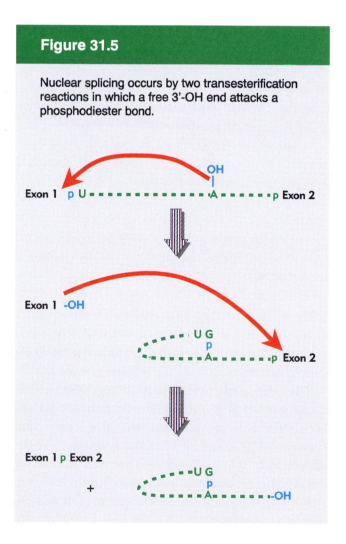

Figure 31.5

Nuclear splicing occurs by two transesterification reactions in which a free 3'-OH end attacks a phosphodiester bond.

Small RNAs are required for splicing and form a spliceosome

How are the splicing sites recognized? We can think of two general types of system. An enzyme could specifically recognize and bring together the consensus sequences. Or an RNA might base pair with them to create a secondary structure that is recognized by an enzyme.

What could be the source of RNA sequences that bind to the splicing sites?

◆ There could be sequences located elsewhere in the same transcript. (This mechanism is used by certain mitochondrial introns, which we discuss later).

◆ Or independent RNA molecule(s) might base pair with the unspliced precursor.

A ready candidate for the role of *trans*-acting RNA is available. Both the nucleus and cytoplasm of eukaryotic cells contain many discrete small RNA species. They range in size from 100 to 300 bases in higher eukaryotes, and extend in length to ~1000 bases in yeast. They vary considerably in abundance. The more abundant species have been detected biochemically in quantities of 10^5–10^6 molecules per cell. Others, present at concentrations too low to be detected directly, have been identified by functional tests for processing reactions in which they are involved.

Those restricted to the nucleus are called **small nuclear RNAs (snRNA)**; those found in the cytoplasm are called **small cytoplasmic RNAs (scRNA)**. In their natural state, they exist as ribonucleoprotein particles (snRNP and scRNP). Colloquially, they are sometimes known as **snurps** and **scyrps**. An snRNP particle usually contains a single RNA and ~10 proteins; some of the proteins are unique to a particular snRNP, while others are common to several types of snRNP. Some of the common proteins are recognized by an autoimmune antiserum called **anti-Sm;** these proteins must be involved in the autoimmune reaction (although their relationship to the phenotype of the autoimmune disease is not clear).

The importance of snRNA molecules can be tested directly in yeast by making mutations in their genes. By this criterion, several of the more abundant snRNAs are dispensable; they appear to play no essential role in the life of yeast. A group of snRNAs involved in splicing, however, is essential. Mutations in their genes interrupt splicing and are lethal. The snRNPs that are abundant in higher eukaryotes and are involved in splicing have counterparts in yeast that are present in lower abundance, but which are essential.

Several of the snRNPs are involved in splicing. Certain snRNAs have sequences that are complementary to the 5′ or 3′ splicing sites or to the branch sequence. Base pairing between snRNA and premRNA, or between snRNAs, plays an important role in splicing.

The splicing apparatus consists of several snRNPs and a large number of additional individual proteins, often called splicing factors. The snRNPs are U1, U2, U5, and U4/U6. They are named according to the snRNAs that are present. The U1, U2, and U5 snRNPs each contain a single snRNA and several proteins; the U4/U6 snRNP contains both the U4 and U6 snRNAs and several proteins. The snRNPs together probably contain ~40 individual proteins, some of which may be involved directly in splicing, others of which may be required in structural roles or just for assembly or interaction of the snRNP particles. The number of splicing factors that participate as individual proteins rather than as members of snRNPs is not known, but could also be ~40 species.

All of the snRNAs can be recognized in conserved

forms in animal, bird, and insect cells. The counterpart RNAs in yeast are often rather larger, but conserved regions include features that are similar to the snRNAs of higher eukaryotes. Do the snRNPs function in splicing by base pairing between the snRNA and a consensus sequence, by some other role of the snRNA, or by some role of one or more of the proteins of the snRNP particle?

The human U1 snRNP contains 8 proteins as well as the RNA. The probable secondary structure of the U1 snRNA is drawn in **Figure 31.6**. It contains several domains. The Sm-binding site is required for interaction with the common snRNP proteins. Domains identified by the individual stem–loop structures provide binding sites for proteins that are unique to U1 snRNP.

The region most directly implicated in splicing consists of the 5'-terminal 11 nucleotides, which are single-stranded and include a stretch complementary to the consensus sequence at the 5' site of the intron. The extent of complementarity between U1 snRNA and actual sites is usually 4–6 bp (because

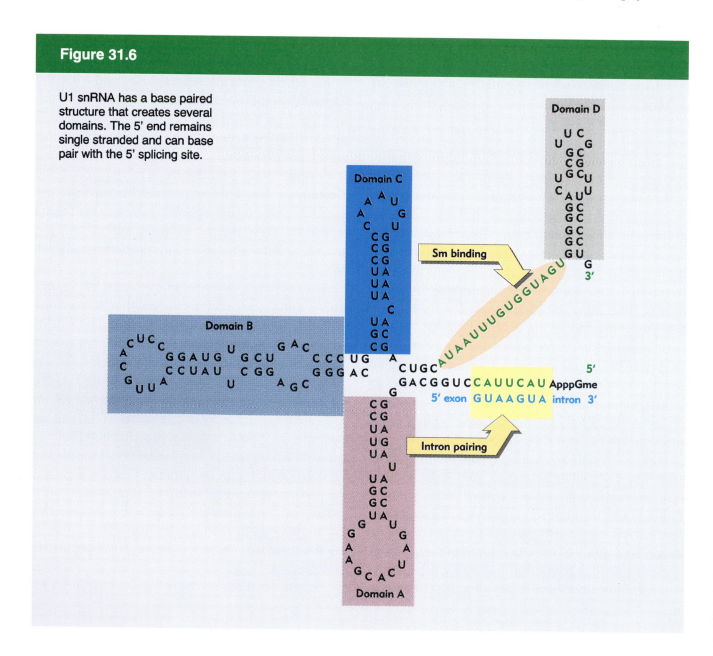

Figure 31.6

U1 snRNA has a base paired structure that creates several domains. The 5' end remains single stranded and can base pair with the 5' splicing site.

actual sites rarely conform perfectly to the consensus). The intact U1 snRNP particle can bind *in vitro* to a 5' splicing site. Binding is a property of the entire particle; purified U1 RNA cannot bind.

We can test directly the necessity for U1 snRNA to pair with the 5' splicing site by mutational analysis. The results of such an experiment are illustrated in **Figure 31.7**. The wild-type sequence of the splice site of the 12S adenovirus pre-mRNA pairs at 5 out of 6 positions with U1 snRNA. A known mutant in the 12S RNA that cannot be spliced has two sequence changes; the GG residues at positions 5–6 in the intron are changed to AU. A particularly informative aspect of this mutant is that this sequence change does not alter the *overall* extent of pairing, because complementarity is lost at one position and gained at the other. This means that the total number of base pairs formed between U1 RNA and the 5' splicing site is not the sole determinant of efficiency, and pairing at some positions may be more important than others.

When a mutation is introduced into U1 snRNA that restores pairing at position 5, normal splicing is regained. Other cases in which corresponding mutations are made in U1 snRNA to see whether they can suppress the mutation in the splicing site suggests the general rule: *complementarity between U1 snRNA and a splicing site is necessary for splicing, but the efficiency of splicing is not determined solely by the number of base pairs that can form.*

The U2 snRNA includes sequences complementary to the branch site. A complex between U2 snRNP and intron sequences including the branch site can be identified by immunoprecipitation with antibodies directed against U2 snRNP. A sequence near the 5' end of the snRNA base pairs with the branch sequence in the intron.

Figure 31.8 relates the stages of splicing to the functions of the snRNPs. Binding of U1 snRNP by base pairing with the 5' splicing site is the first step. However, U1 snRNP also binds to the branch site, because pre-mRNAs with mutations in the branch site cannot bind U1 snRNA to the 5' splicing site. The means by which U1 snRNP recognizes the branch site is not known, and could involve some protein

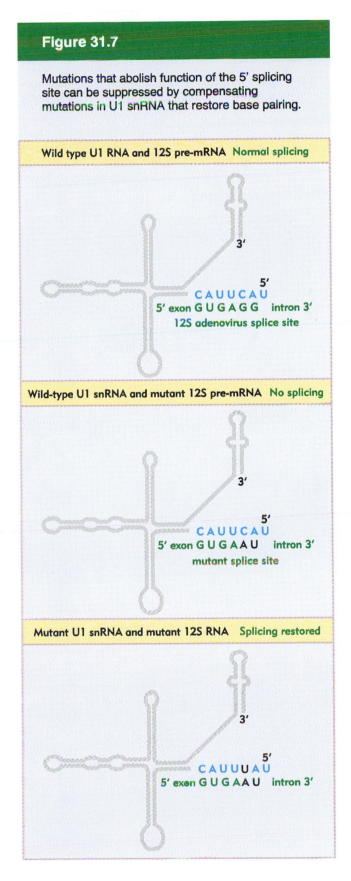

Figure 31.7

Mutations that abolish function of the 5' splicing site can be suppressed by compensating mutations in U1 snRNA that restore base pairing.

Wild type U1 RNA and 12S pre-mRNA   Normal splicing

CAUUCAU
5' exon GUGAGG   intron 3'
12S adenovirus splice site

Wild-type U1 snRNA and mutant 12S pre-mRNA   No splicing

CAUUCAU
5' exon GUGAAU   intron 3'
mutant splice site

Mutant U1 snRNA and mutant 12S RNA   Splicing restored

CAUUUAU
5' exon GUGAAU   intron 3'

Figure 31.8

The splicing reaction proceeds through discrete stages in which spliceosome formation involves the interaction of components that recognize the consensus sequences.

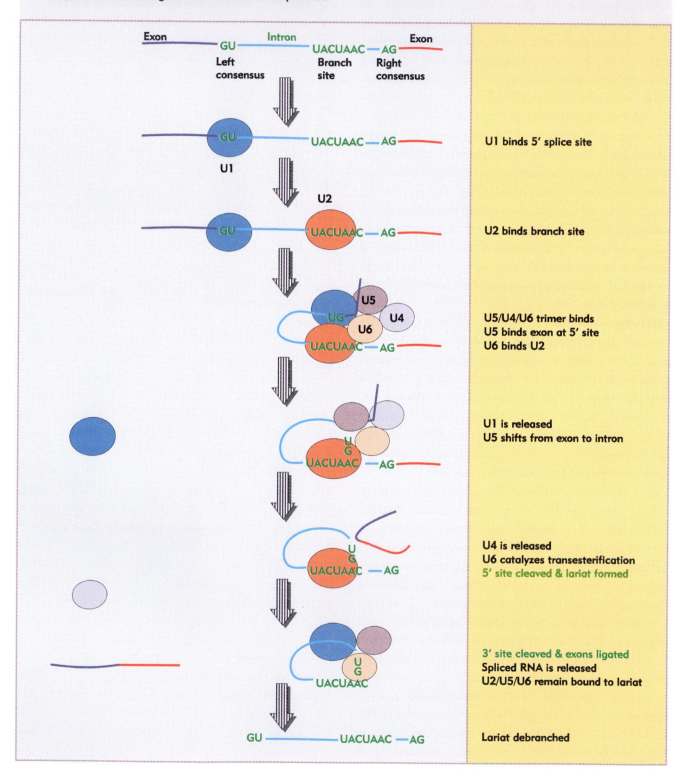

component of the snRNP or even an auxiliary factor.

The presence of U1 is required for the second step, in which U2 snRNP binds to the branch site. The U2 snRNA recognizes the branch site by a base pairing interaction; in yeast this typically involves formation of a 7 bp duplex with the entire UACUACA box (see Figure 31.9 below). The reaction requires the participation of the splicing factor U2AF (U2 auxiliary factor). The binding of U2 snRNP requires ATP hydrolysis, and commits a pre-mRNA to the splicing pathway.

Because the branch site is located close to the 3′ splicing site, the reaction between U1 snRNP (bound at the 5′ site) and U2 snRNP (bound at the 3′ site) brings the two splicing sites close together.

A trimeric complex containing the U5 and U4/U6 snRNPs binds. ATP hydrolysis is required again at this stage. All the components of the splicing apparatus are now present, so the complex has become fully assembled before any splicing reactions occur.

Both U1 and U2 are required for the initial step of splicing, since their inactivation prevents cleavage at the 5′ site and lariat formation. Although base pairing with U1 snRNA is responsible for recognizing the splicing site, it does not control the cleavage event. U1 may dissociate before the reaction begins. At all events, at this point U5 snRNA changes its position; initially it is close to exon sequences at the 5′ splice site, but it shifts to the vicinity of the intron sequences.

The catalytic reaction is triggered by the release of U4; this requires hydrolysis of ATP. The role of U4 snRNA may be to sequester U6 snRNA until it is needed. **Figure 31.9** shows the changes that occur in the base pairing interactions between snRNAs during splicing. In the U6/U4 snRNP, a continuous length of 26 bases of U6 is paired with two separated regions of U4. When U4 dissociates, the region in U6 that is released becomes free to take up another structure. The first part of it pairs with U2; the second part forms an intramolecular hairpin. The interaction between U4 and U6 is mutually incompatible with the interaction between U2 and U6, and thus the release of U4 controls the ability of the spliceosome to proceed.

The components that undertake the transesterifications have not been directly identified, and at this point we do not know what roles are played by the RNA and protein components of the spliceosome, respectively. Base pairing between U2 and U6 creates a structure that resembles the active center of the group II introns that are described in the next section (see Figure 31.13 below). This suggests the possibility that the catalytic component could comprise an RNA structure generated by the U2–U6 interaction.

The formation of the lariat at the branch site is responsible for determining the use of the 3′ splicing site, since the 3′ consensus sequence nearest to the 3′ side of the branch becomes the target for the second transesterification. The second splicing reaction follows rapidly. Binding of U5 snRNP to the 3′ splicing site is needed for this reaction, but we do not know whether the snRNP functions by using its RNA component to recognize the consensus sequence by base pairing. There is no region of complementarity available in single-stranded form. Perhaps the protein components of the snRNP are involved. This idea is supported by the identification of a protein that binds to the site, and which may be part of the snRNP particle, possibly a loosely bound component.

The important conclusion suggested by these results is that *the snRNA components of the splicing apparatus interact both among themselves and with the substrate RNA by means of base pairing interactions, and these interactions allow for changes in structure that may bring reacting groups into apposition and may even create catalytic centers.* Furthermore, the conformational changes in the snRNAs are reversible; for example, U6 snRNA is not used up in a splicing reaction, and at completion must be released from U2, so that it can reform the duplex structure with U4 to undertake another cycle of splicing.

We have described individual reactions in which each snRNP participates, but as might be expected from a complex series of reactions, any particular snRNP may play more than one role in splicing. Thus the ability of U1 snRNP to promote binding of

Figure 31.9

U6-U4 pairing is incompatible with U6-U2 pairing. When U6 joins the spliceosome it is paired with U4. Release of U4 allows a conformational change in U6; one part of the released sequence forms a hairpin (dark grey), and the other part (black) pairs with U2. Because an adjacent region of U2 is already paired with the branch site, this brings U6 into juxtaposition with the branch. Note that the substrate RNA is reversed from the usual orientation and is shown 3'-5'.

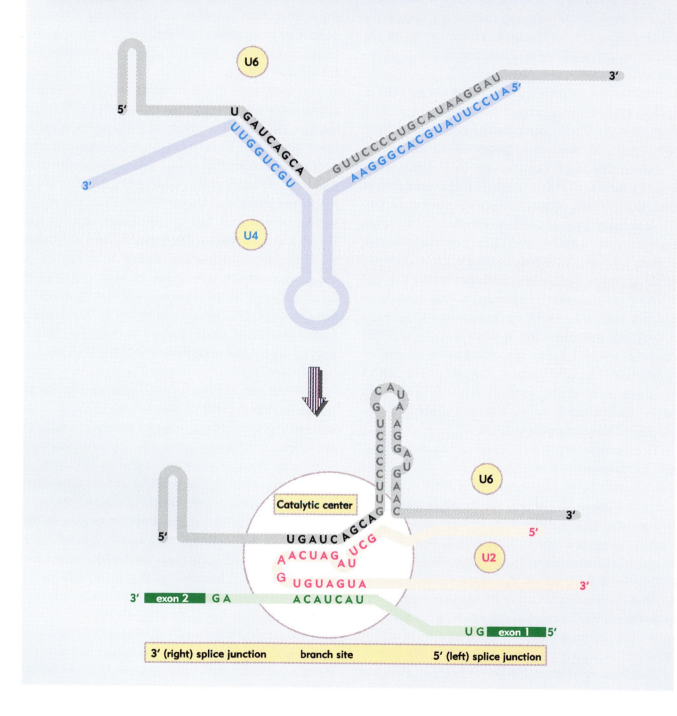

U2 snRNP is independent of its ability to bind to the 5' splicing site. Similarly, different regions of U2 snRNA can be defined that are needed to bind to the branch site and to interact with other splicing components.

The snRNPs involved in splicing, together with protein splicing factors, form a large particulate complex, called the **spliceosome**. Isolated from the *in vitro* splicing systems, it comprises a 50–60S ribonucleoprotein particle. The spliceosome may be formed in stages as the snRNPs join, proceeding through several 'presplicing complexes'. The spliceosome is a large body, equivalent in size to a ribosomal subunit. Formed on the RNA precursor, it includes many proteins as well as the snRNPs.

The spliceosome forms on the intact precursor RNA and passes through an intermediate state in which it contains the individual 5' exon linear molecule and the right lariat intron-exon. Little spliced product is found in the complex, which suggests that cleavage of the 3' site and ligation of the exons is immediately followed by release of the product.

We may think of the snRNP particles as being involved in building the structure of the spliceosome. Like the ribosome, the spliceosome depends on RNA–RNA interactions as well as protein–RNA and protein–protein interactions. Some of the reactions involving the snRNPs require their RNAs to base pair directly with sequences in the RNA being spliced; other reactions require recognition between snRNPs or between their proteins and other components of the spliceosome.

Figure 31.10 shows an electron micrograph of mammalian spliceosomes, which contain U1, U2, U4, U5, and U6 snRNPs, and additional proteins. The spliceosomes appear to be particles ~25 nm × 50 nm, possibly consisting of discrete domains connected by rigid structures. Individual snRNP particles have diameters >8 nm, so that four types of snRNP could between them account for most of the mass and volume of the spliceosome. Could the domains in the particle be individual snRNPs?

An extensive mutational analysis has been

Figure 31.10

Spliceosomes are ellipsoidal particles with several discrete regions. The bar is 50 nm. Photograph kindly provided by Tom Maniatis.

undertaken in yeast to identify both the RNA and protein components of the spliceosome. Mutations in genes needed for splicing are identified by the accumulation of unspliced precursors. A series of loci that identify genes potentially coding for splicing factors were originally called *RNA*, but are now known as *PRP* mutants (for pre-RNA processing). Several of the products of these genes have motifs that identify them as RNA-binding proteins, and some appear to be related to a family of ATP-dependent RNA helicases. We suppose that, in addition to RNA–RNA interactions, protein–RNA interactions are important in creating or releasing structures in the pre-mRNA or snRNA components of the spliceosomes.

Some of the PRP proteins are components of snRNP particles, but others function as independent factors. One interesting example is PRP16, a helicase that hydrolyzes ATP, and associates transiently with the spliceosome to participate in the second catalytic step. Another example is PRP22, another ATP-dependent helicase, which is required to release the mature mRNA from the spliceosome. The conservation of bonds during the splicing reaction means that input of energy is not required to drive bond formation *per se*, which implies that the ATP hydrolysis is required for other purposes. The use of ATP by PRP16 and PRP22 may be examples of a more general phenomenon: the use of ATP hydrolysis to drive conformational changes that are needed to proceed through splicing.

The involvement of snRNPs in splicing is only one example of their involvement in RNA processing reactions. Various snRNPs are involved in poly-adenylation and in the generation of authentic 3' ends of *Xenopus* histone mRNA (see Chapter 29). Other snRNPs are required for the processing of nuclear RNA to mature rRNAs. Especially in view of the demonstration that group I introns are self-splicing, and that the RNA of ribonuclease P has catalytic activity (as discussed in Chapter 32), it is plausible to think that RNA–RNA reactions are important in most or all RNA processing events.

Group II introns autosplice via lariat formation

Introns in protein-coding genes (in fact, in all genes except tRNA-coding genes) can be divided into three general classes. Nuclear introns have in common only the possession of the GT-AG dinucleotides at the 5' and 3' ends. Group I and group II introns are named for the two types of introns found in fungal mitochondrial genes, which are classified according to their internal organization. Each type of intron can be folded into a typical type of secondary structure. **Figure 31.11** shows that all three classes of introns are spliced by a two stage transesterification (shown previously for nuclear introns in Figure 31.5). In the first reaction, the 5' exon–intron junction is attacked by a free hydroxyl group (provided by an internal 2'-OH position in nuclear and group II introns, and by a free guanine nucleotide in group I introns). In the second reaction, the hydroxyl that has been transferred to the end of the 5' exon in turn attacks the 3' intron–exon junction.

Group I introns are more common than group II introns. There is little relationship between the two classes, but they share the striking property that the RNA can perform the splicing reaction by itself, without requiring enzymatic activities provided by proteins. We discuss the catalytic reaction of group I introns in Chapter 32, but now we examine the parallels between group II introns and nuclear splicing.

Group II mitochondrial introns have splicing sites that resemble nuclear introns, and which follow the GT-AT rule. They are spliced by the same mechanism as nuclear pre-mRNAs, via a lariat that is held together by a 5'–2' bond. An example of a lariat produced by splicing a group II intron is shown in **Figure 31.12**. When an isolated group II RNA is incubated *in vitro* in the absence of additional components, it is able to perform the splicing reaction. This means that the two transesterification reactions shown in

Figure 31.11

All classes of splicing reactions proceed by two transesterifications. First, a free OH group attacks the exon 1 - intron junction. Second, the OH created at the end of exon 1 attacks the intron - exon 2 junction.

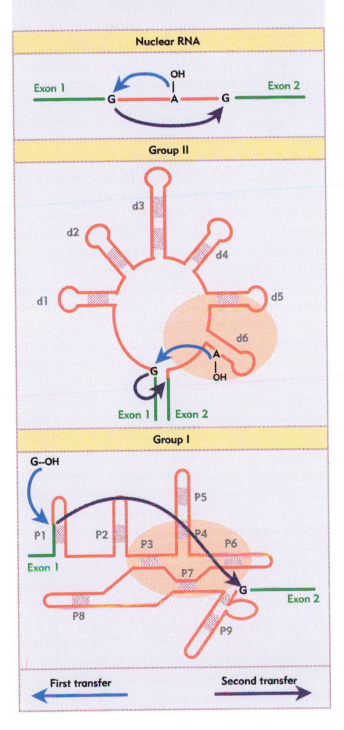

Nuclear RNA

Group II

Group I

First transfer   Second transfer

Figure 31.12

Mitochondrial group II introns are released by splicing in the form of stable lariats. Photograph kindly provided by Leslie Grivell and Annika Arnberg.

Figure 31.11 can be performed by the group II intron RNA sequence itself. Because the number of phosphodiester bonds is conserved in the reaction, an external supply of energy is not required; this could have been an important feature in the evolution of splicing.

The group II intron forms into a secondary structure that involves closely juxtaposing two of its domains. Domain 5 is separated by 2 bases from domain 6, which contains an A residue that donates the 2′-OH group for the first transesterification. This constitutes a catalytic domain in the RNA. **Figure 31.13** compares this secondary structure with the structure formed by the combination of U6 with U2 and of U2 with the branch site. The similarity is the basis for suggesting that U6 may have a catalytic role.

Figure 31.13

Nuclear splicing and group II splicing involve the formation of similar secondary structures. The sequences are more specific in nuclear splicing; group II splicing uses positions that may be occupied by either purine (R) or either pyrimidine (Y).

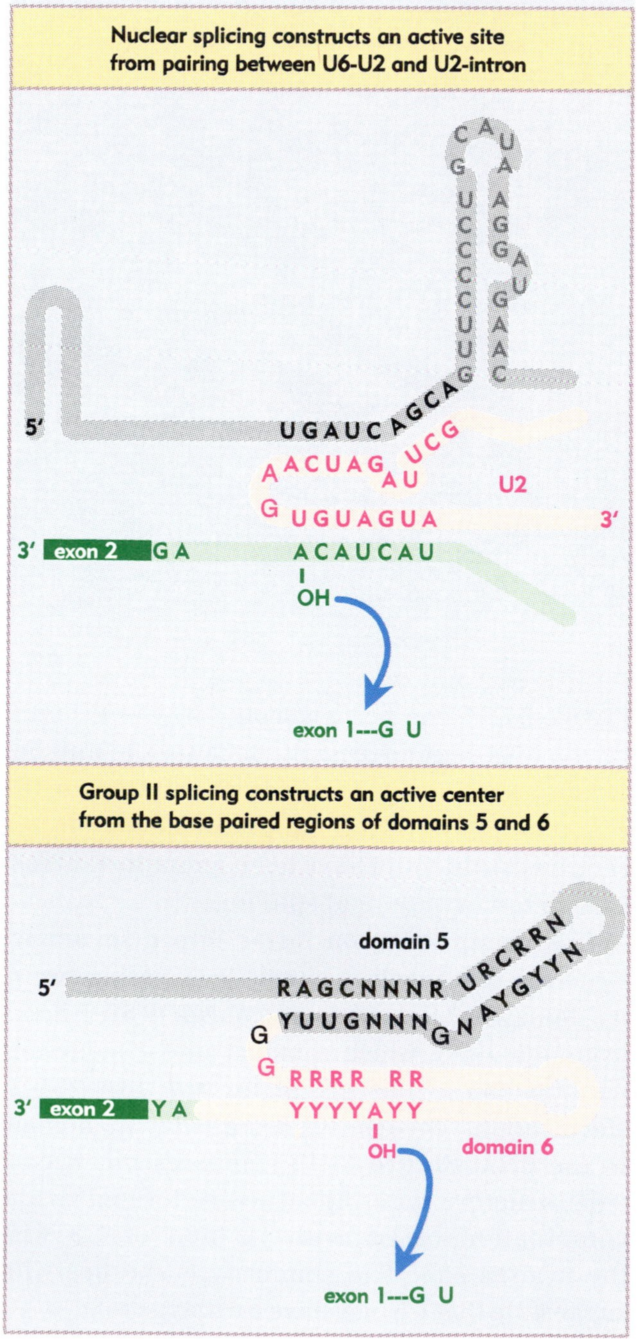

Nuclear splicing constructs an active site from pairing between U6-U2 and U2-intron

Group II splicing constructs an active center from the base paired regions of domains 5 and 6

The features of group II splicing suggest that splicing evolved from an autocatalytic reaction undertaken by an individual RNA molecule, in which it accomplished a controlled deletion of an internal sequence. Probably such a reaction requires the RNA to fold into a specific conformation, or series of conformations, and would occur exclusively in *cis* conformation.

The ability of group II introns to remove themselves by an autocatalytic splicing event stands in great contrast to the requirement of nuclear introns for a complex splicing apparatus. We may regard the snRNAs of the spliceosome as compensating for the lack of sequence information in the intron, and providing the information required to form particular structures in RNA. The functions of the snRNAs may have evolved from the original autocatalytic system. These snRNAs act in *trans* upon the substrate pre-mRNA; we might imagine that the ability of U1 to pair with the 5′ splicing site, or of U2 to pair with the branch sequence, replaced a similar reaction that required the relevant sequence to be carried by the intron. So the snRNAs may undergo reactions with the pre-mRNA substrate and with one another that have substituted for the series of conformational changes that occur in RNAs that splice by group II mechanisms. In effect, these changes have relieved the substrate pre-mRNA of the obligation to carry the sequences needed to sponsor the reaction. Of course, as the splicing apparatus has become more complex (and as the number of potential substrates has increased) some of the reactions that used to be catalyzed by RNA have been taken over by proteins.

Alternative splicing involves differential use of splicing junctions

The majority of interrupted genes are transcribed into an RNA that gives rise to a single type of spliced mRNA: in these cases, there is no variation in assignment of exons and introns. But the RNAs of some genes follow patterns of **alternative spicing**, when a single gene gives rise to more than one mRNA sequence. This is used to particular effect by some viruses. As we saw in Chapter 23, this allows various types of relationships between a gene and its mRNA product. In some cases, the ultimate pattern of expression is dictated by the primary transcript, because the use of different startpoints or termination sequences alters the pattern of splicing. In other cases, a single primary transcript is spliced in more than one way, and internal exons are substituted, added, or deleted. In some cases, the multiple products all are made in the same cell, but in others the process is regulated so that particular splicing patterns occur only under particular conditions.

One of the most pressing questions in splicing is to determine what controls the use of such alternative pathways. Proteins that intervene to bias the use of alternative splice sites have been identified in two ways. In some mammalian systems, it has been possible to characterize alternative splicing *in vitro*, and to identify proteins that are required for the process. In *D. melanogaster*, aberrations in alternative splicing may be caused either by mutations in the genes that are alternatively spliced or in the genes whose products are necessary for the reaction.

Figures 31.14 shows examples in which one splicing site remains constant, but the other varies. The large T / small t antigens of SV40 and the products of the adenovirus E1A region are generated by connecting a varying 5′ site to a constant 3′ site. In the case of the T/t antigens, the 5′ site used for T antigen removes a termination codon that is

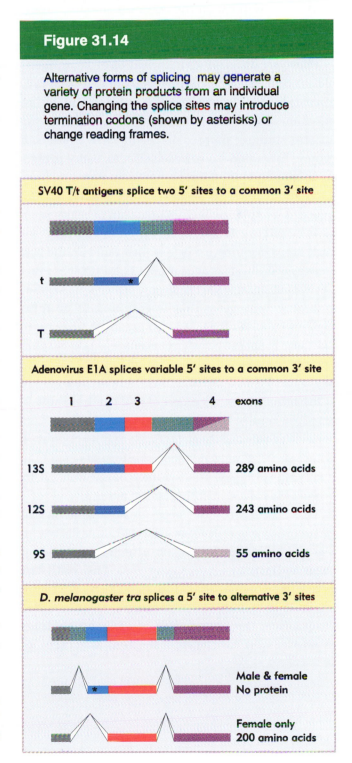

Figure 31.14

Alternative forms of splicing may generate a variety of protein products from an individual gene. Changing the splice sites may introduce termination codons (shown by asterisks) or change reading frames.

SV40 T/t antigens splice two 5′ sites to a common 3′ site

t

T

Adenovirus E1A splices variable 5′ sites to a common 3′ site

1   2   3   4   exons

13S   289 amino acids

12S   243 amino acids

9S   55 amino acids

D. melanogaster tra splices a 5′ site to alternative 3′ sites

Male & female
No protein

Female only
200 amino acids

present in the t antigen mRNA, so that T antigen is larger than t antigen. In the case of the E1A transcripts, one of the 5′ sites connects to the last exon in a different reading frame, again making a significant change in the C-terminal part of the protein. In these examples, all the relevant splicing events take place in every cell in which the gene is expressed, so all the protein products are made.

There are differences in the ratios of T/t antigens in different cell types. Extracts from cells that produce relatively more small t antigen during an infection also produce more RNA with its characteristic splicing pattern in an *in vitro* system. A protein extracted from these cells can cause preferential production of small t RNA in extracts from other cell types. This protein, which was called ASF (alternative splicing factor) turns out to be the same as the splicing factor SF2, which is required for early steps in spliceosome assembly and for the first cleavage–ligation reaction. SF2 is an RNA-binding protein, with an RNA-binding motif rich in Arg-Ser dipeptide sequences that is common in proteins that bind nuclear RNA. When a pre-mRNA has more than one 5′ splicing site preceding a single 3′ splicing site, increased concentrations of SF2 promote use of the 5′ site nearest to the 3′ site at the expense of the other site. This effect of SF2 can be counteracted by another splicing factor, SF5. The exact molecular roles of these factors are not yet known, but we see in general terms that alternative splicing involving different 5′ sites may be influenced by proteins involved in spliceosome assembly.

The pathway of sex determination in *D. melanogaster* involves interactions between a series of genes in which alternative splicing events distinguish ♂s and ♀s. The pathway takes the form illustrated in **Figure 31.15**, in which the ratio of X chromosomes to autosomes determines the expression of *sxl,* and changes in expression are passed sequentially through the other genes to *dsx,* the last in the pathway.

The pathway starts with sex-specific splicing of *sxl,* following the same pattern that is shown below in Figure 31.16 for *dsx.* Exon 3 of the *sxl* gene contains a termination codon that prevents synthesis of functional protein. This exon is included in the mRNA produced in males, but is skipped in females. As a result, only females produce Sxl protein. The protein has a concentration of basic amino acids that resembles other RNA-binding proteins.

The presence of Sxl protein changes the splicing of the *transformer (tra)* gene. Figure 31.14 shows that this involves splicing a constant 5′ site to alternative 3′ sites. One splicing pattern occurs in both ♂ and ♀, and results in an RNA that has an early termination codon. The presence of Sxl protein inhibits usage of the normal 3′ splice site; when this site is skipped, the next 3′ site is used, leaving out a whole exon. This generates a ♀-specific mRNA that codes for a protein.

So *tra* produces a protein only in females; this protein is a splicing regulator which affects the splicing of *tra2,* which in turn produces a protein that affects the splicing of *dsx.* The Tra and Tra2 proteins resemble SF2, and have the Arg-Ser RNA binding motif. They act directly upon the target transcripts. Tra is required for ♀-specific splicing of *tra2,* and both Tra and Tra2 are required for ♀-specific splicing of *dsx.*

Figure 31.16 shows examples of cases in which splicing sites are used to add or to substitute exons or introns, again with the consequence that different protein products are generated. In the *doublesex (dsx)* gene, ♀s splice the 5′ site of intron 3 to the 3′ site of that intron; as a result translation terminates at the end of exon 4. Males splice the 5′ site of intron 3 directly to the 3′ site of intron 4, thus omitting exon 4 from the mRNA, and allowing translation to continue through exon 6. Different proteins are therefore produced in each sex: the ♂

Figure 31.15

Sex determination in *D. melanogaster* involves a pathway in which different splicing events occur in females. Blocks at any stage of the pathway result in male development.

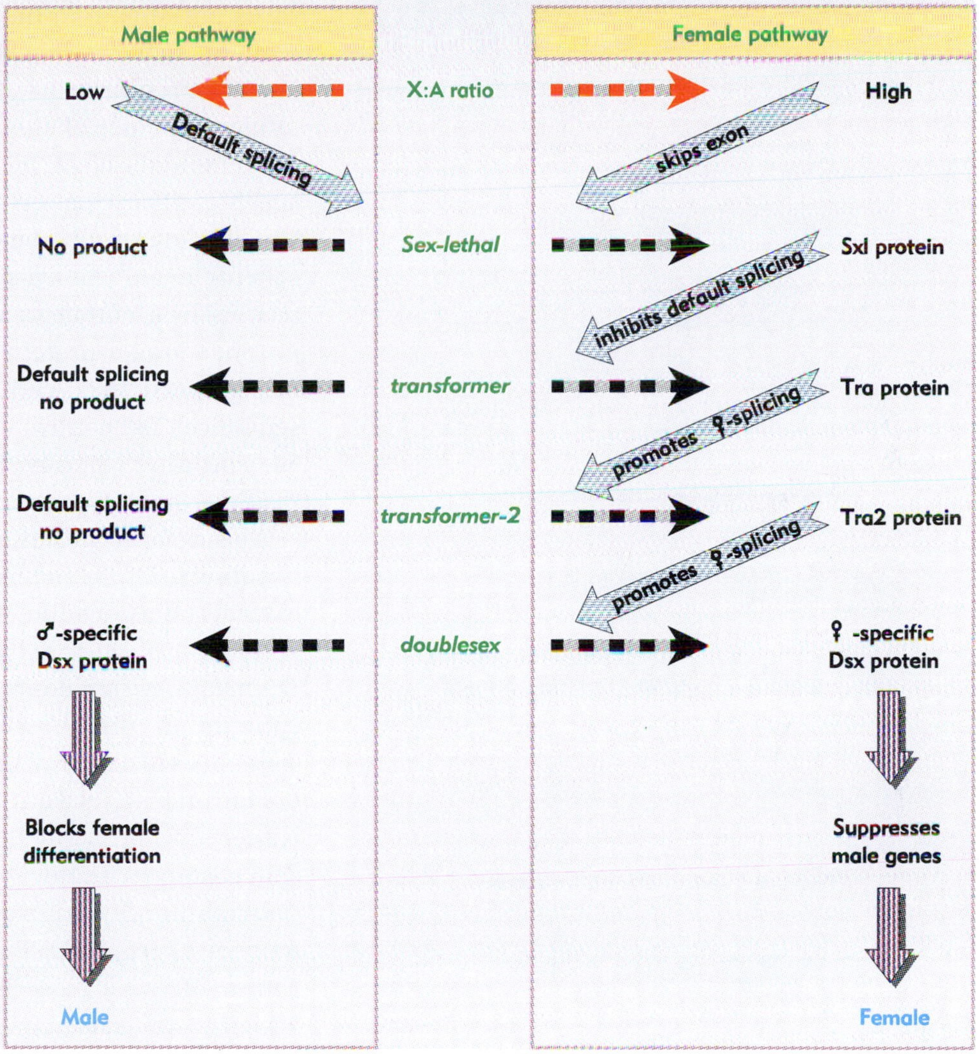

Male pathway		Female pathway
Low	X:A ratio	High
Default splicing		skips exon
No product	Sex-lethal	Sxl protein
	inhibits default splicing	
Default splicing no product	transformer	Tra protein
	promotes ♀-splicing	
Default splicing no product	transformer-2	Tra2 protein
	promotes ♀-splicing	
♂-specific Dsx protein	doublesex	♀-specific Dsx protein
Blocks female differentiation		Suppresses male genes
Male		Female

product blocks ♀ sexual differentiation, while the ♀ product represses expression of ♂-specific genes.

Sex determination therefore has a pleasing symmetry: the pathway both starts and ends by ♀-specific splicing events that cause omission of an exon that has a termination codon. The events have different molecular bases. At the first control point,

Sxl inhibits the default splicing pattern. At the last control point, Tra2 specifically promotes the ♀-specific splice.

Tra2 binds to a short region within the ♀-specific exon of the *dsx* mRNA, and induces ♀-specific splicing *in vitro*, but we do not yet know how it controls the choice of 3′ splicing site. The Tra and Tra2

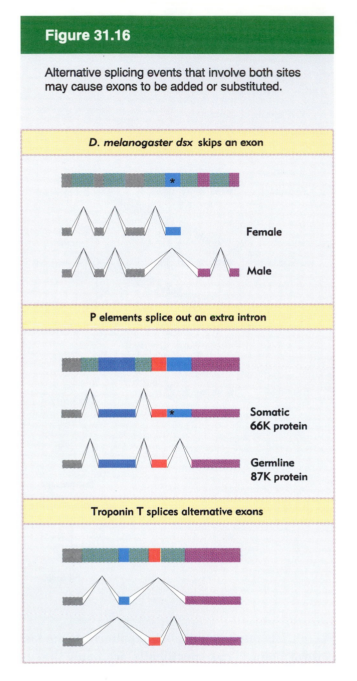

Figure 31.16

Alternative splicing events that involve both sites may cause exons to be added or substituted.

D. melanogaster dsx skips an exon

Female

Male

P elements splice out an extra intron

Somatic
66K protein

Germline
87K protein

Troponin T splices alternative exons

proteins are not needed for normal splicing, because in their absence flies develop as normal males. As specific regulators, they need not necessarily participate in the mechanics of the splicing reaction; in this respect they differ from SF2, which is a factor required for general splicing, but can also influence choice of alternative splice sites.

P elements of *D. melanogaster* show a tissue-specific splicing pattern. In somatic cells, there are two splicing events, but in germline an additional splicing event removes another intron. Because a termination codon lies in the germline-specific intron, a longer protein (with different properties) is produced in germline. We discuss the consequences for control of transposition in Chapter 34, and note for now that the tissue specificity results from differences in the splicing apparatus.

The default splicing pathway when the RNA is subjected to a heterologous (human) splicing extract is the germline pattern, in which intron 3 is spliced out. But extracts of somatic cells of *D. melanogaster* contain a protein that inhibits splicing of this intron. The protein binds to sequences in exon 3; if these sequences are deleted, the intron is spliced out. The function of the protein is therefore probably to repress association of the spliceosome with the 5' site of intron 3.

In mammalian muscle, the alternative splicing pattern of troponin T causes exon substitution. Either of two exons is included, depending on which 3' site is used in the splicing of the leftmost intron shown in the figure (this only shows the part of the gene involved in the alternative splicing event). We do not know how this event is controlled.

Cis-splicing *and trans*-splicing reactions

In both mechanistic and evolutionary terms, splicing has been viewed as an *intramolecular* reaction, amounting essentially to a controlled deletion of the intron sequences at the level of RNA. In genetic terms, splicing occurs only in *cis*. This means that *only sequences on the same molecule of RNA can be spliced together*. The upper part of **Figure 31.17** shows the normal situation. The introns can be

Figure 31.17

Splicing usually occurs only in *cis* between exons carried on the same physical RNA molecule, but *trans* splicing can occur when special constructs are made that support base pairing between introns.

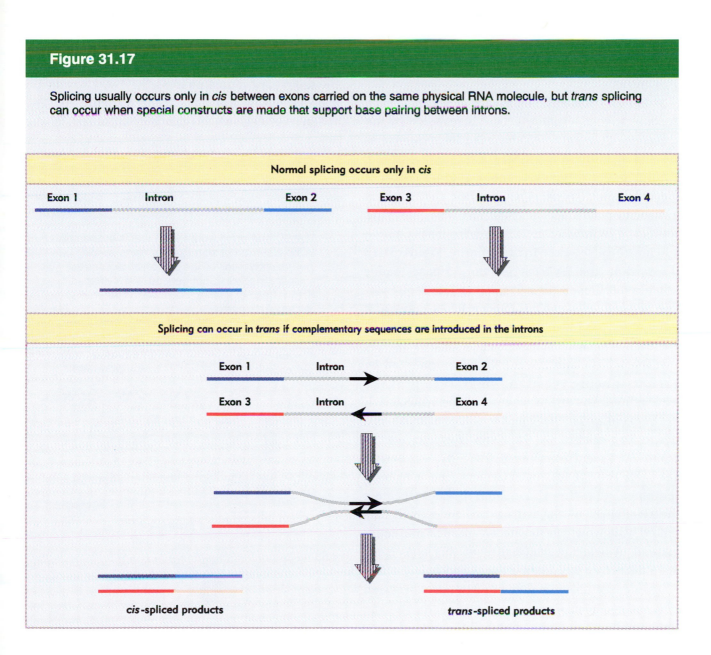

removed from each RNA molecule, allowing the exons of that RNA molecule to be spliced together, but there is no *intermolecular* splicing of exons between different RNA molecules. We cannot say that *trans*-splicing never occurs between pre-mRNA transcripts of the same gene, but we know that it must be exceedingly rare, because if it were prevalent the exons of a gene would be able to complement one another genetically instead of belonging to a single complementation group.

Some manipulations can cause *trans*-splicing to occur. In the example illustrated in the lower part of Figure 31.17, complementary sequences were introduced into the introns of two RNAs. Base pairing between the complements should create an H-shaped molecule. This molecule could be spliced in *cis*, to connect exons that are covalently connected by an intron, or it could be spliced in *trans*, to connect exons of the juxtaposed RNA molecules. Both reactions occur *in vitro*.

This result shows that there is no mechanistic impediment to *trans*-splicing and excludes models for splicing that require processive movement of a spliceosome along the RNA. It must be possible for a spliceosome to recognize the 5′ and 3′ splicing sites of different RNAs because they are in close proximity.

Although *trans*-splicing is rare, it occurs *in vivo* in some special situations. One is revealed by the presence of a common 35 base leader sequence at the end of numerous mRNAs in the trypanosome. But the leader sequence is not coded upstream of the individual transcription units. Instead it is transcribed into an independent RNA, carrying additional sequences at its 3′ end, from a repetitive unit located elsewhere in the genome. **Figure 31.18** shows that this RNA carries the 35 base leader sequence followed by a 5′ splicing site sequence. The sequences coding for the mRNAs carry a 3′ splicing site just preceding the sequence found in the mature mRNA.

If the leader and the mRNA are connected by a *trans*-splicing reaction, the 3′ region of the leader RNA and the 5′ region of the mRNA will in effect comprise the 5′ and 3′ halves of an intron. If splicing occurs by the usual nuclear mechanism, a 5′–2′ link should form by a reaction between the GU of the 5′ intron and a branch sequence near the AG of the 3′ intron. Because the two parts of the intron are not covalently linked, this would generate a Y-shaped molecule instead of a lariat. Molecules of this type have been detected, supporting the model.

A similar situation is presented by the expression of actin genes in *C. elegans*. Three actin mRNAs (and some other RNAs) share the same 22 base leader sequence at the 5′ terminus. The leader sequence is not coded in the actin gene, but is transcribed independently as part of a 100 base RNA coded by a gene elsewhere.

The RNA that donates the 5′ exon for *trans*-splicing is called the **SL RNA** (spliced leader RNA). The SL RNAs found in several species of trypanosomes and also in the nematode (*C. elegans*) have some common features. They fold into a common secondary structure that has three stem–loops and a single-stranded region that resembles the Sm binding site. The SL RNAs therefore exist as snRNPs that count as members of the Sm snRNP class. Trypanosomes possess the U2, U4, and U6 snRNAs, but do not have U1 or U5 snRNAs. The absence of U1 snRNA can be explained by the

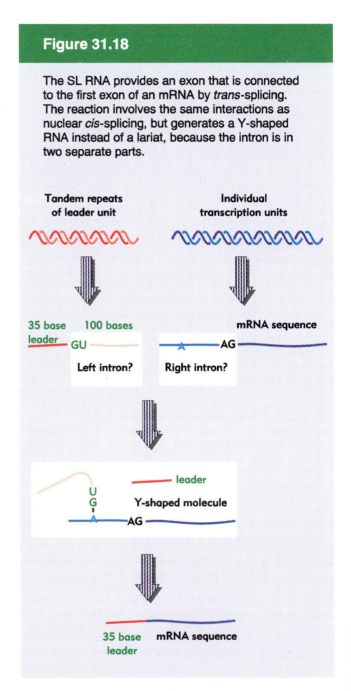

Figure 31.18

The SL RNA provides an exon that is connected to the first exon of an mRNA by *trans*-splicing. The reaction involves the same interactions as nuclear *cis*-splicing, but generates a Y-shaped RNA instead of a lariat, because the intron is in two separate parts.

properties of the SL RNA, which can carry out the functions that U1 snRNA usually performs at the 5′ splicing site; thus SL RNA in effect consist of an snRNA sequence possessing U1 function, linked to the exon–intron site that it recognizes.

The *trans*-splicing reaction of the SL RNA may represent a step towards the evolution of the full nuclear splicing apparatus. The SL RNA provides in *cis* the ability to recognize the 5′ splicing site, and this probably depends upon the specific conformation of the RNA. The remaining functions required for splicing are provided by independent snRNPs. The SL RNA can function without participation of proteins like those in U1 snRNP, which suggests that the recognition of the 5′ splicing site depends directly on RNA.

Some chloroplast genes are *trans*-spliced. One example is shown in **Figure 31.19**. The *psa* gene of the *Chlamydomonas* chloroplast consists of 3 widely separated exons. Exon 1 is located 50 kb away from exon 2, and it is another 90 kb to exon 3. Many other genes lie between these exons, and in any case they cannot be expressed as a common transcript because exon 1 is in the reverse orientation from exons 2 and 3. In fact, the only transcripts that are detected are local RNAs surrounding the individual exons. We assume that the mRNA is pieced together by two *trans*-splicing reactions. Several other genes are required specifically for one or the other of the *trans*-splicing reactions, so the process of piecing the mRNA together is quite complex.

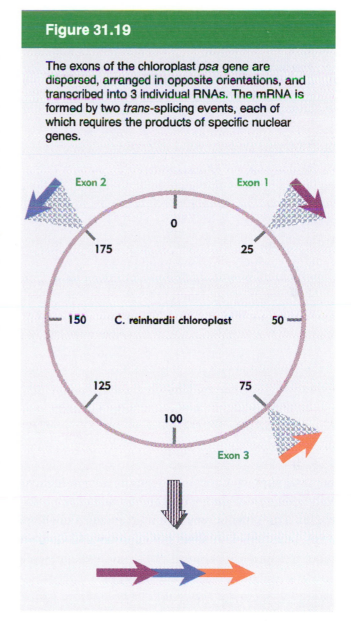

Figure 31.19

The exons of the chloroplast *psa* gene are dispersed, arranged in opposite orientations, and transcribed into 3 individual RNAs. The mRNA is formed by two *trans*-splicing events, each of which requires the products of specific nuclear genes.

Yeast tRNA splicing involves cutting and rejoining

The splicing reactions that we have discussed so far depend upon short consensus sequences and occur by transesterification reactions in which breaking and making of bonds is coordinated. The splicing of tRNA genes is achieved by a different mechanism that relies upon separate cleavage and ligation reactions.

About 40 of the ~400 nuclear tRNA genes in yeast

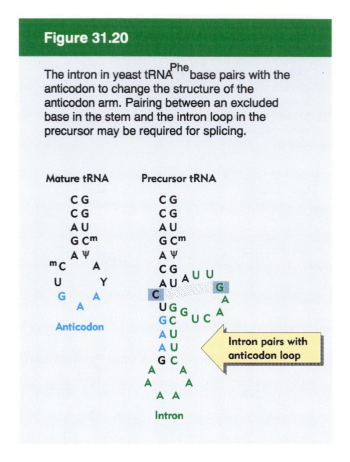

Figure 31.20

The intron in yeast tRNA^{Phe} base pairs with the anticodon to change the structure of the anticodon arm. Pairing between an excluded base in the stem and the intron loop in the precursor may be required for splicing.

Mature tRNA

Precursor tRNA

Anticodon

Intron pairs with anticodon loop

Intron

are interrupted. Each has a single intron, located just one nucleotide beyond the 3′ side of the anticodon. The introns vary in length from 14 to 46 bp. Those in related tRNA genes are related in sequence, but the introns in tRNA genes representing different amino acids are unrelated. *There is no consensus sequence that could be recognized by the splicing enzymes.* The same general rules apply to the nuclear tRNA genes of plants, amphibians, and mammals.

All the introns include a sequence that is complementary to the anticodon of the tRNA. This creates an alternative conformation for the anticodon arm in which the anticodon is base paired to form an extension of the usual arm. An example is drawn in **Figure 31.20**. Only the anticodon arm is affected—the rest of the molecule retains its usual structure.

The exact sequence and size of the intron is not important. Most mutations in the intron do not prevent splicing. *Splicing of tRNA depends principally on recognition of a common secondary structure in tRNA rather than a common sequence of the intron.* Regions in various parts of the molecule are important, including the stretch between the acceptor arm and D arm, in the TψC arm, and especially the anticodon arm. This is reminiscent of the structural demands placed on tRNA for protein synthesis (see Chapter 8).

The intron is not entirely irrelevant, however. Pairing between a base in the intron loop and an unpaired base in the stem is required for splicing. Mutations at other positions that influence this reaction (for example, to generate alternative patterns for pairing) influence splicing. The rules that govern availability of tRNA precursors for splicing resemble the rules that govern recognition by aminoacyl-tRNA synthetases (as discussed in Chapter 8), in the sense that individual mutations in particular tRNAs or groups of tRNAs are important, and it is difficult to devise general rules that apply to all tRNAs.

In a temperature-sensitive mutant of yeast that fails to remove the introns, the interrupted precursors accumulate in the nucleus. The precursors can be used as substrates for a cell-free system extracted from wild-type cells. The splicing of the precursor can be followed by virtue of the resulting size reduction. This is seen by the change in position of the band on gel electrophoresis, as illustrated in **Figure 31.21**. The reduction in size can be accounted for by the appearance of a band representing the intron.

The cell-free extract can be fractionated by assaying the ability to splice the tRNA. The *in vitro* reaction requires ATP. Characterizing the reactions that occur with and without ATP shows that the *two separate stages of the reaction are catalyzed by different enzymes.* The overall series of events is depicted in **Figure 31.22**:

◆ The first step does not require ATP. It involves phosphodiester bond cleavage, taking the form of an atypical nuclease reaction. It is catalyzed by an endonuclease.

Figure 31.21

Splicing of yeast tRNA *in vitro* can be followed by assaying the RNA precursor and products by gel electrophoresis.

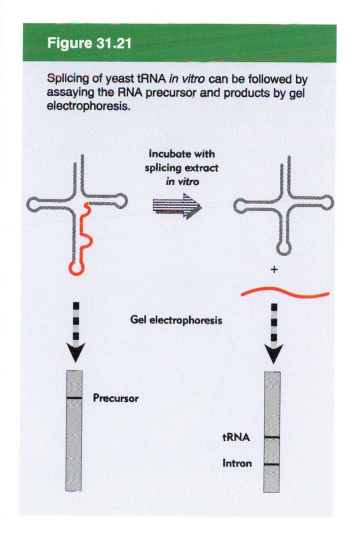

Incubate with splicing extract *in vitro*

Gel electrophoresis

Precursor

tRNA

Intron

◆ The second step requires ATP and involves bond formation; it is a ligation reaction, and the responsible enzyme activity is described as an **RNA ligase**.

The endonuclease cleaves the precursor at both ends of the intron. The products are a linear intron and two half-tRNA molecules. These intermediates have unique ends. Each 5′ terminus ends in a hydroxyl group; each 3′ terminus ends in a 2′,3′-cyclic phosphate group. (All other known RNA processing enzymes cleave on the other side of the phosphate bond.)

The two half-tRNAs base pair to form a tRNA-like structure. When ATP is added, the second reaction occurs. Both of the unusual ends generated by the endonuclease must be altered.

The cyclic phosphate group is opened to generate a 2′-phosphate terminus. This reaction requires a cyclic phosphodiesterase. The product has a 2′-phosphate group and a 3′-OH group.

The 5′-OH group generated by the nuclease must be phosphorylated to give a 5′-phosphate. This generates a site in which the 3′-OH is next to the 5′-phosphate. Then the RNA ligase joins the two halves covalently by making a phosphodiester bond.

The spliced molecule is now uninterrupted, with a 5′–3′ phosphate linkage at the site of splicing, but it also has a 2′-phosphate group marking the event. The surplus group must be removed by a phosphatase. (In the interim it could be useful in marking the site where the ligation occurred.)

Generation of a 2′,3′-cyclic phosphate also occurs during the tRNA-splicing reaction in plants and mammals. The reaction in plants seems to be the same as in yeast, but the detailed chemical reactions are different in mammals.

The yeast tRNA precursors also can be spliced in an extract obtained from the germinal vesicle (nucleus) of *Xenopus* oocytes. This shows that the reaction is not species-specific. *Xenopus* must have enzymes able to recognize the introns in the yeast tRNAs.

The ability to splice the products of tRNA genes is therefore well conserved, but may have a different origin from the other splicing reactions (such as that of nuclear pre-mRNA). The tRNA-splicing reaction relies on external sequences and factors; thus different proteins catalyze the cleavage and synthesis of bonds, and the sequences that are important for the reaction largely lie outside of the intron. Other splicing reactions use transesterification, in which bonds are broken and created by virtue of reactions involving the exon–intron sites and the branch sequence in the intron; and the sequences required for the reaction lie within the intron. This could reflect a different evolutionary origin for the introns in tRNA genes.

Figure 31.22

Splicing of tRNA requires separate nuclease and ligase activities. The exon-intron boundaries are cleaved by the nuclease to generate 2'-3' cyclic phosphate and 5'-OH termini. The cyclic phosphate is opened to generate 3'-OH and 2' phosphate groups. The 5'-OH is phosphorylated. After releasing the intron, the tRNA half molecules fold into a tRNA-like structure that now has a 3'-OH, 5'-P break. This is sealed by a ligase.

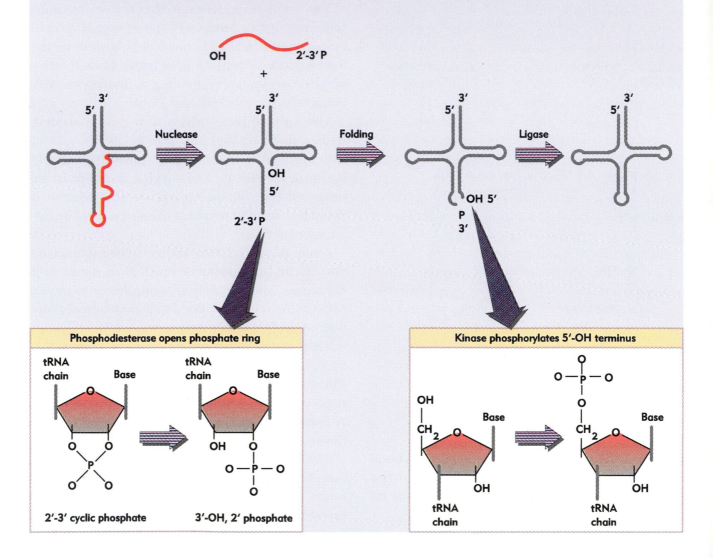

Summary

Splicing accomplishes the removal of introns and the joining of exons into the mature sequence of RNA. There are at least four types of reaction, as distinguished by their requirements *in vitro* and the intermediates that they generate. The systems include eukaryotic nuclear introns, group I and group II mitochondrial introns, and tRNA introns. Each reaction involves a change of organization within an individual RNA molecule, and is therefore a *cis*-acting event.

Nuclear splicing follows preferred but not obligatory pathways. Only very short consensus sequences are necessary; the rest of the intron appears irrelevant. All 5' splicing sites are probably equivalent, as are all 3' splicing sites. We do not know how 5' and 3' sites are linked only in the proper pairs. The required sequences are given by the GT-AG rule, which describes the ends of the intron. The UACUAAC branch sequence of yeast, or a less well conserved consensus in mammalian introns, is also required. The reaction starts with the 5' splicing site, and involves formation of a lariat by joining the GU end of the intron via a 5'–2' linkage to the A at position 6 of the branch sequence. Then the 3'-OH end of the exon attacks the 3' splicing site, so that the exons are ligated and the intron is released as a lariat. Both reactions are transesterifications in which bonds are conserved. Several stages of the reaction require hydrolysis of ATP, probably to drive conformational changes in the RNA and/or protein components. Lariat formation is responsible for choice of the 3' splicing site. Alternative splicing patterns are caused by protein factors that either stimulate use of a new site or that block use of the default site.

Nuclear splicing requires formation of a spliceosome, a large particle that assembles the consensus sequences into a reactive conformation. The spliceosome contains several snRNPs and a large number of additional splicing factors. The U1, U2, and U5 snRNPs each contain a single snRNA and several proteins; the U4/U6 snRNP contains 2 snRNAs and several proteins. Some proteins are common to all snRNP particles. The snRNPs recognize consensus sequences. U1 snRNA base pairs with the 5' splicing site, U2 snRNA base pairs with the branch sequence, U5 snRNP acts at the 5' splicing site. When U4 releases U6, the U6 snRNA base pairs with U2, and this may create the catalytic center for splicing. The existence of these interactions may be a major relic of the original development of nucleic acids as hereditary molecules. It will not be surprising if the snRNA molecules turn out to have catalytic-like roles in splicing and other processing reactions.

Splicing is usually intramolecular, but some cases have been found of *trans-* (intermolecular) splicing. These reactions probably occur by spliceosome formation with the appropriate site sequences on each molecule.

Group II introns share with nuclear introns the use of a lariat as intermediate, but are able to perform the reaction as a self-catalyzed property of the RNA. These introns follow the GT-AG rule, but form a characteristic secondary structure that holds the reacting splice sites in the appropriate apposition.

Yeast tRNA splicing involves separate endonuclease and ligase reactions. The endonuclease recognizes the secondary (or tertiary) structure of the precursor and cleaves both ends of the intron. The two half-tRNAs released by loss of the intron can be ligated in the presence of ATP.

Further reading

Reviews

Nuclear splicing has been reviewed by **Sharp** (*Science* **235**, 766–771, 1987) and (earlier in more length) (*Ann. Rev. Biochem.* **55**, 1119–1150, 1986). The role of splicing components has been reviewed by **Green** (*Ann. Rev. Cell Biol.* **7**, 559–599, 1991).

The splicing apparatus was reviewed by **Guthrie** (*Science* **253**, 157–163, 1991). Spliceosomal snRNAs were reviewed by **Guthrie and Patterson** (*Ann. Rev. Genet.* **22**, 387–419, 1988). Functions of snRNPs in splicing were reviewed by **Maniatis and Reed** (*Nature* **325**, 673–678, 1987).

Spliceosomes and snRNAs were reviewed by **Sharp** (pp. 303–358) and by **Steitz** (pp. 359–382) in *The RNA World* (eds. R. F. Gesteland and J. F. Atkins, Cold Spring Harbor, 1993). Group II splicing was reviewed by **Cech** (*Cell* 207–210, 1986). Unifying developments in splicing were reviewed by **Weiner** (*Cell* **72**, 161–164, 1993).

Discoveries

The first demonstration for base pairing between snRNA and pre-mRNA was accomplished by **Zhuang and Weiner** (*Cell* **46**, 827–835, 1986).

CHAPTER 32

RNA as catalyst: changing the informational content of RNA

The idea that only proteins have enzymatic activity was deeply rooted in biochemistry. (Yet devotées of protein function once thought that only proteins could have the versatility to be the genetic material!) A rationale for the identification of enzymes with proteins lies in the view that only proteins, with their varied three-dimensional structures and variety of side groups, have the flexibility to create the active sites that catalyze biochemical reactions. But the characterization of systems involved in RNA processing has shown this view to be an over-simplification.

Several types of catalytic reactions are now known to reside in RNA. **Ribozyme** has become a general term used to describe an RNA with catalytic activity, and it is possible to characterize the enzymatic activity in the same way as a more conventional enzyme. Some RNA catalytic activities are directed against separate substrates, while others are intramolecular and are described as **auto-cleavage** or **autosplicing**, depending on the type of reaction:

◆ The enzyme ribonuclease P is a ribonucleo-protein that contains a single RNA molecule bound to a protein. The RNA possesses the ability to catalyze cleavage in the tRNA substrate, while the protein component plays an indirect role, probably to maintain the structure of the catalytic RNA.

◆ Small RNAs of the virusoid class have the ability to perform a self-cleavage reaction. Although this reaction is intramolecular, the molecule can be divided into an 'enzymatic' and a 'substrate' part, and engineering of related sequences can create 'enzymes' that function upon independent 'substrates'.

◆ Introns of the group I class possess the ability to splice themselves out of the pre-mRNA that contains them. The reaction can be performed *in vitro* by the RNA alone, but is assisted by ancillary proteins *in vivo*. Engineering of group I introns has generated RNA molecules that have several other catalytic activities related to the original activity.

The common theme of these reactions is that the RNA can perform an intramolecular or intermolecular reaction that involves cleavage or joining of phosphodiester bonds *in vitro*. Although the specificity of the reaction and the basic catalytic activity is provided by RNA, proteins associated with the RNA may be needed for the reaction to occur *in vivo*.

RNA splicing is not the only means by which

changes can be introduced in the informational content of RNA. In the process of **RNA editing**, changes are introduced at individual bases, or bases are added at particular positions within an mRNA. The insertion of bases (usually uridine residues) occurs for several genes in the mitchondria of certain lower eukaryotes; like splicing, it involves the breakage and reunion of bonds between nucleotides, but also requires a template for coding the information of the new sequence.

Group I introns undertake self-splicing by transesterification

Group I introns are found in diverse locations. They occur in the genes coding for rRNA in the nuclei of the lower eukaryotes *Tetrahymena thermophila* (a ciliate) and *Physarum polycephalum* a (slime mold). They are common in the genes of fungal mitochondria. Their presence in three genes of phage T4 extends their evolutionary reach into the prokaryotes. [Introns have been lost entirely from the major class of bacteria (eubacteria), and these T4 genes present the only example where splicing is required in eubacteria.]

Group I introns have two common properties:

◆ The isolated RNA has the ability to splice itself, a reaction called **self-splicing** or **autosplicing**. (This property is not unique, but is found also in the group II introns discussed in Chapter 31.) The reaction *in vitro* occurs by self-splicing, but it may be assisted by proteins *in vivo*.

◆ A group I intron can be organized into a distinct secondary structure, with 9 stem–loops. Some of these stem–loops are generated by reaction between short consensus sequences. The lengths of group I introns vary widely, and the consensus sequences are located a considerable distance from the actual splicing junctions. We discuss the sequences of group I introns in the next section.

Self-splicing was discovered as a property of the transcripts of the rRNA genes in *T. thermophila*. The genes for the two major rRNAs follow the usual organization, in which both are expressed as part of a common transcription unit. The product is a 35S precursor RNA with the sequence of the small rRNA in the 5′ part, and the sequence of the larger (26S) rRNA toward the 3′ end.

In some strains of *T. thermophila*, the sequence coding for 26S rRNA is interrupted by a single, short intron. When the 35S precursor RNA is incubated *in vitro*, splicing occurs as an autonomous reaction. The intron is excised from the precursor and accumulates as a linear fragment of 400 bases, which is subsequently converted to a circular RNA. These events are summarized in **Figure 32.1**.

The reaction requires only a monovalent cation, a divalent cation, and a guanine nucleotide cofactor. No other base can be substituted for G; but a triphosphate is not needed; GTP, GDP, GMP, and guanosine itself all can be used, so there is no net energy requirement. The guanine nucleotide must have a 3′-OH group.

The fate of the guanine nucleotide can be followed by using a radioactive label. The radioactivity initially enters the excised linear intron fragment. The G residue becomes linked to the 5 ′ end of the intron by a normal phosphodiester bond.

Figure 32.2 shows that three transfer reactions

Figure 32.1

Splicing of the Tetrahymena 35S rRNA precursor can be followed by gel electrophoresis. The 35S precursor RNA forms a rather broad band. The removal of the intron is revealed by the appearance of a rapidly moving small band. (No change is seen in the 35S RNA band because of its breadth and the relatively small reduction in size. No free exons are seen.) When the intron becomes circular, it electrophoreses more slowly, as seen by a higher band.

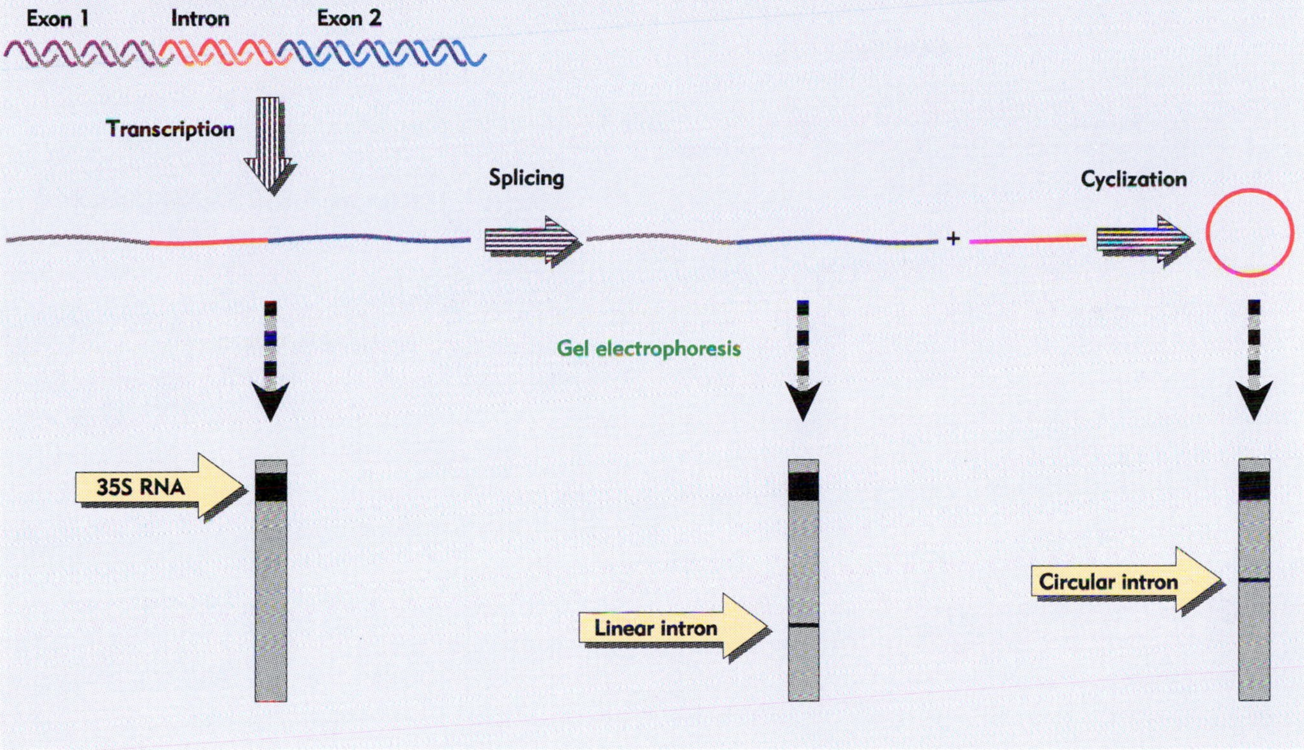

occur. In the first transfer, the guanine nucleotide behaves as a cofactor that provides a free 3′-OH group that attacks the 5′ end of the intron. This reaction creates the G-intron link and generates a 3′-OH group at the end of the exon. The second transfer involves a similar chemical reaction, in which this 3′-OH then attacks the second exon. The two transfers are connected; no free exons have been observed, so their ligation may occur as part of the same reaction that releases the intron. The intron is released as a linear molecule, but the third transfer reaction converts it to a circle.

Each stage of the self-splicing reaction occurs by a transesterification, in which one phosphate ester is converted directly into another, without any intermediary hydrolysis. Because bonds are exchanged directly, energy is conserved; so the reaction does not require input of energy from hydrolysis of ATP or GTP. (There is a parallel for the transfer of bonds without net input of energy in the DNA nicking-closing enzymes discussed in Chapter 33.)

If each of the consecutive transesterification reactions involves no net change of energy, why does the splicing reaction proceed to completion instead of coming to equilibrium between spliced product and nonspliced precursor? The concentration of GTP is high relative to that of RNA, and therefore drives the reaction forward; and a change

Figure 32.2

Self-splicing occurs by transesterification reactions in which bonds are exchanged directly. The bonds that have been generated at each stage are indicated by the shaded boxes.

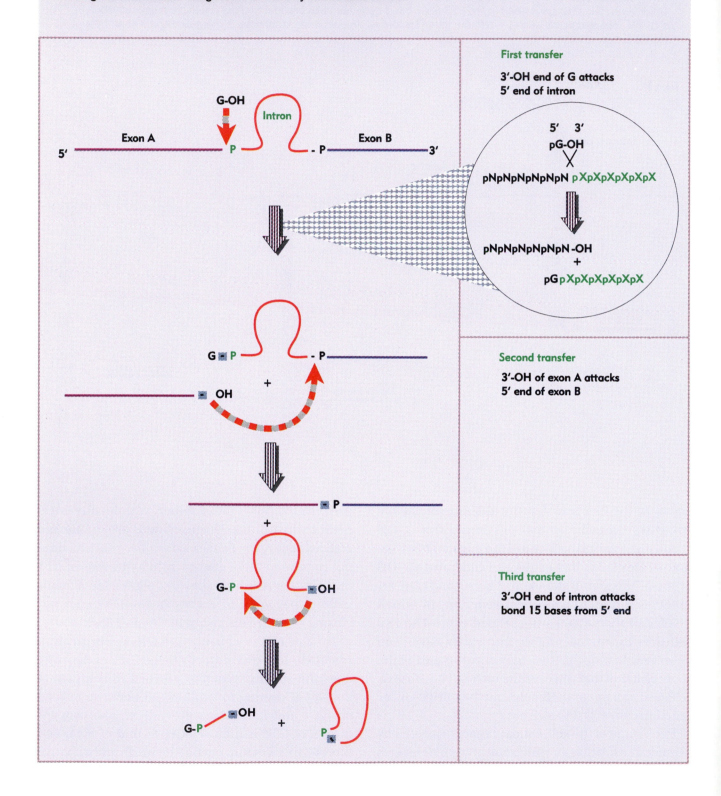

in secondary structure in the RNA prevents the reverse reaction (see later).

The in vitro *system includes no protein so the ability to splice is intrinsic to the RNA.* The RNA forms a specific secondary/tertiary structure in which the relevant groups are brought into juxtaposition so that a guanine nucleotide can be bound to a specific site, and then the bond breakage and reunion reactions shown in Figure 32.2 can occur. Although a property of the RNA itself, the reaction is probably assisted *in vivo* by proteins (which might, for example, stabilize the RNA structure).

The ability to engage in these transfer reactions resides with the sequence of the intron, which continues to be reactive after its excision as a linear molecule. **Figure 32.3** summarizes its activities:

◆ The intron can circularize when the 3′ terminal G attacks either of two positions near the 5′ end. The internal bond is broken and the released 5′ end is transferred to the 3′-OH end of the intron. The *primary cyclization* usually involves reaction between the terminal G^{414} and the A^{16}. This is the most common reaction (shown as the third transfer in Figure 32.2). Less frequently, the G^{414} reacts with U^{20}. Each reaction generates a circular intron and a linear fragment that represents the original 5′ region (15 bases long for attack on A^{16}, 19 bases long for attack on U^{20}). The terminal fragment contains the original added guanine nucleotide.

◆ Either type of circle can regenerate a linear molecule *in vitro* by specifically hydrolyzing the bond (G^{414}–A^{16} or G^{414}–U^{20}) that had closed the circle. This is called a *reverse cyclization*. The linear molecule generated by reversing the primary cyclization at A^{16} remains reactive, and can perform a secondary cyclization by attacking U^{20}.

◆ The final product of the spontaneous reactions following release of the intron is the L-19 RNA, a linear molecule generated by reversing the shorter circular form. This molecule has enzymatic activity, and can catalyze the extension of short oligonucleotides.

◆ The reactivity of the released intron extends beyond merely reversing the cyclization reaction. Addition of the oligonucleotide UUU reopens the primary circle by reacting with the G^{414}–A^{16} bond. The UUU (which resembles the 3′ end of the 15-mer released by the primary cyclization) becomes the 5′ end of the linear molecule that is formed. This is an *intermolecular* reaction, and thus demonstrates the ability to connect together two different RNA molecules.

This series of reactions demonstrates vividly that the autocatalytic activity reflects a generalized ability of the RNA molecule to form an active center that can bind guanine cofactors, recognize oligonucleotides, and bring together the reacting groups in a conformation that allows bonds to be broken and rejoined. Other group I introns have not been investigated in as much detail as the *Tetrahymena* intron, but their properties are generally similar.

The autosplicing reaction is an intrinsic property of RNA *in vitro,* but to what degree are proteins involved *in vivo?* Some indications for the involvement of proteins are provided by mitochondrial systems, where splicing of group I introns requires the *trans*-acting products of other genes. One striking case is presented by the *cyt18* mutant of *N. crassa,* which is defective in splicing several mitochondrial group I introns. The product of this gene turns out to be the mitochondrial tyrosyl-tRNA synthetase! One possible implication is that the intron might take up a tRNA-like tertiary structure that is recognized by the synthetase; binding of the enzyme might be directly involved in splicing or could have an indirect effect such as stabilizing the RNA conformation.

This relationship between the synthetase and splicing is consistent with the idea that splicing originated as an RNA-mediated reaction, subsequently assisted by RNA-binding proteins that originally had other functions. The *in vitro* self-splicing ability may represent the basic biochemical interaction; the RNA structure creates the active site, but is able to function efficiently *in vivo*

Figure 32.3

The excised intron can form circles by using either of two internal sites for reaction with the 5' end, and can reopen the circles by reaction with water or oligonuclotides.

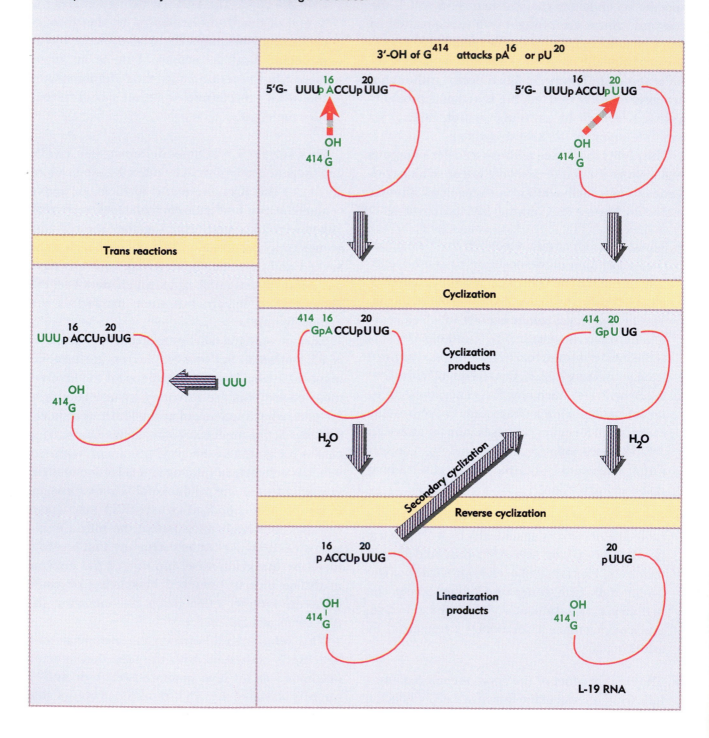

only when assisted by a protein complex.

The conservation of structure and of splicing mechanism among group I introns suggests that the nuclear and mitochondrial members have a common evolutionary origin. A migration must have taken place between nucleus and mitochondrion, although we have no means of knowing in which direction it occurred.

Group I introns form a characteristic secondary structure

All group I introns have a characteristic secondary (and probably also tertiary) structure. **Figure 32.4** shows a model for the secondary structure of the *Tetrahymena* intron that relies upon the formation of 9 short base regions (P1-9). Two of these base paired regions are generated by pairing between

Figure 32.4

Group I introns have a common secondary structure that is formed by 9 base paired regions. The sequences of regions P4 and P7 are conserved, and identify the individual sequence elements P, Q, R, and S. P1 is created by pairing between the end of the left exon and the IGS of the intron; a region between P7 and P9 pairs with the 3' end of the intron.

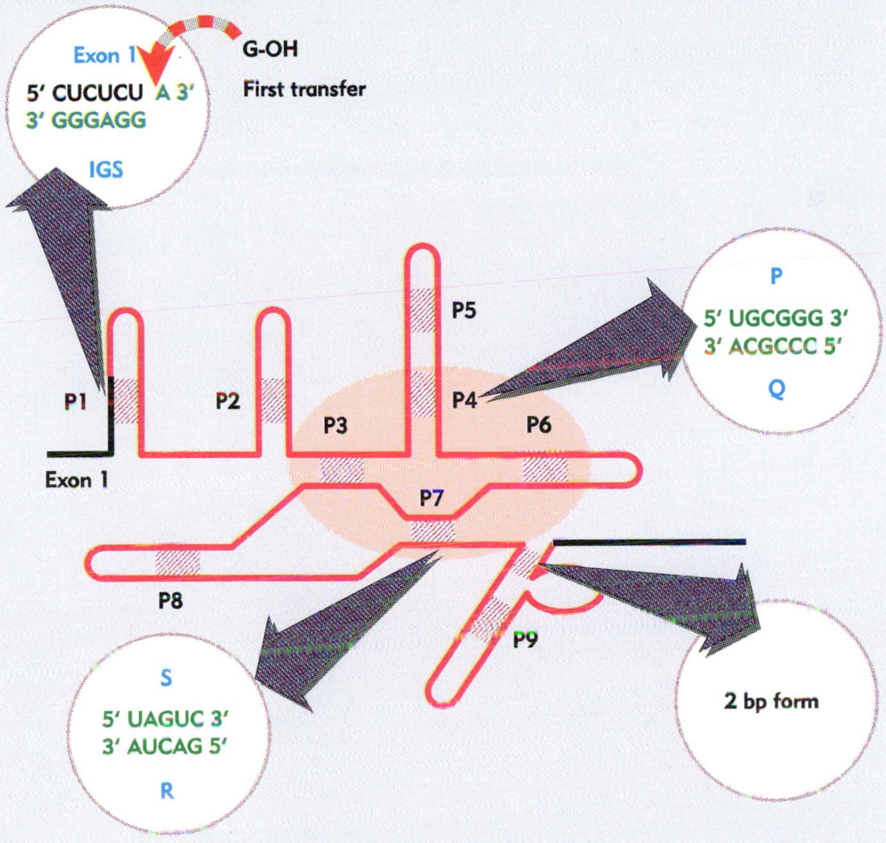

Figure 32.5

Placing the *Tetrahymena* intron within the β-galactosidase coding sequence allows a simple assay to be used to test for its ability to self-splice in *E. coli*. Synthesis of β-galactosidase can be tested by adding a compound that turns blue if the enzyme is present. The entire sequence is carried by a bacteriophage, so the presence of blue plaques indicates successful splicing.

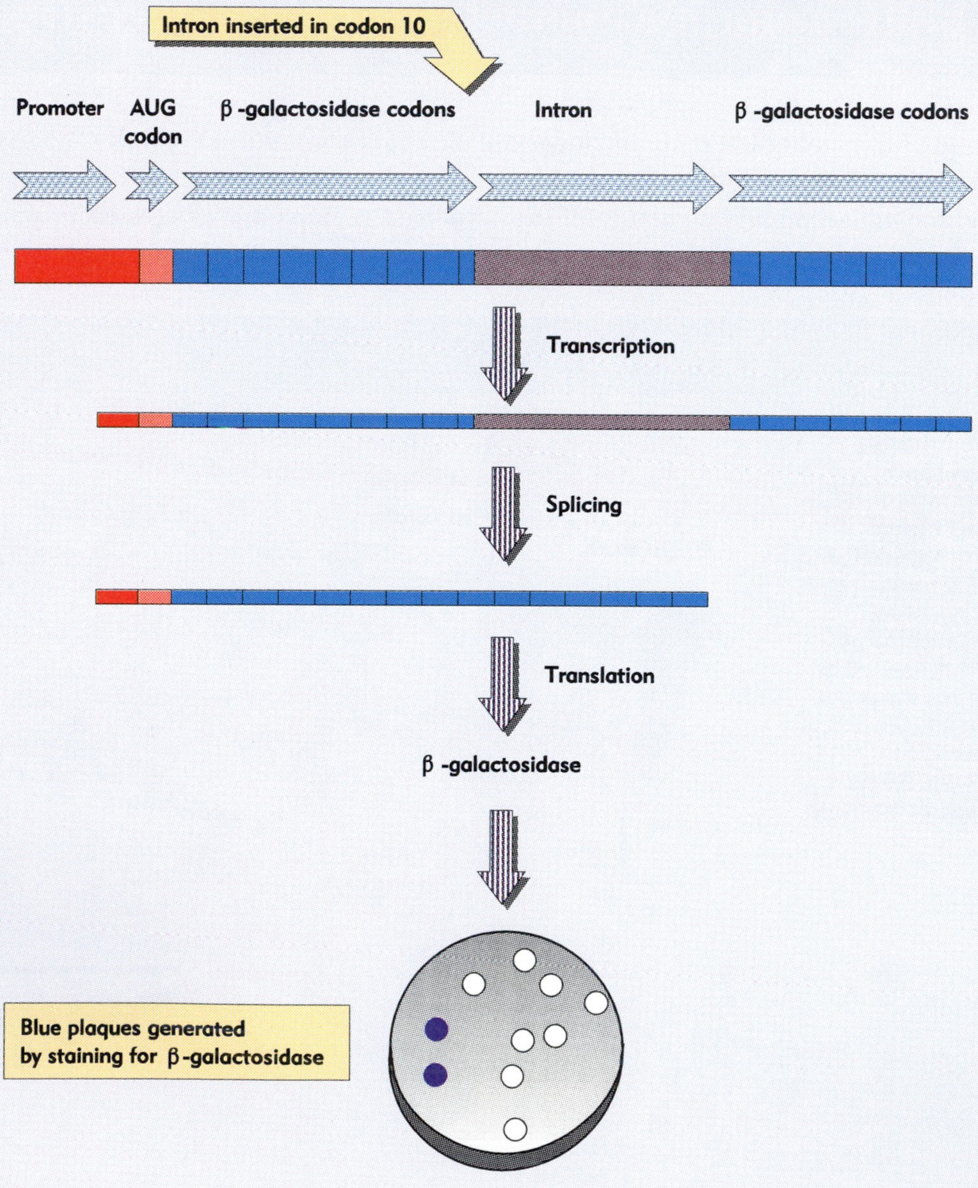

conserved sequence elements that are common to group I introns. P4 is constructed from the sequences *P* and *Q*, which are each 10 bases long; 6–7 bases are involved in pairing. P7 is formed from sequences *R* and *S*, which are 12 bases long, but only 5 bases are involved in pairing. The other base paired regions vary in sequence in individual introns. Mutational analysis identifies an intron 'core', containing P3, P4, P6, and P7, which provides the minimal region that can undertake a catalytic reaction.

Some of the pairing reactions are directly involved in bringing the splicing junctions into a conformation that supports the enzymatic reaction. P1 includes the 3′ end of the left exon. The sequence within the intron that pairs with the exon is called the IGS, or internal guide sequence. (Its name reflects the fact that originally the region immediately 3′ to the IGS sequence shown in the figure was thought to pair with the 3′ splicing junction, thus bringing the two junctions together. This reaction may occur, but does not seem to be essential.) A very short sequence, sometimes as short as 2 bases, between P7 and P9, base pairs with the sequence that immediately precedes the reactive G (position 414 in *Tetrahymena*) at the 3′ end of the intron.

The importance of base pairing in creating the necessary core structure in the RNA is emphasized by the properties of *cis*-acting mutations that prevent splicing of group I introns. Such mutations have been isolated for the mitochondrial introns through mutants that cannot remove an intron *in vivo*, and they have been isolated for the *Tetrahymena* intron by transferring the splicing reaction into a bacterial environment.

The construction shown in **Figure 32.5** allows the splicing reaction to be followed in *E. coli*. The self-splicing intron is placed at a location that interrupts the tenth codon of the β-galactosidase coding sequence. The protein can therefore be successfully translated from an RNA only after the intron has been removed.

The synthesis of β-galactosidase in this system indicates that splicing can occur in conditions quite distant from those prevailing in *Tetrahymena* or even *in vitro*. One interpretation of this result is that self-splicing can occur in the bacterial cell. Another possibility is that there are bacterial enzymes that assist the reaction.

Using this assay, we can introduce mutations into the intron to see whether they prevent the reaction. Mutations in the group I consensus sequences that disrupt their base pairing stop splicing. The mutations can be reverted by making compensating changes that restore base pairing.

Mutations in the corresponding consensus sequences in mitochondrial group I introns have similar effects. A mutation in one consensus sequence may be reverted by a mutation in the complementary consensus sequence to restore pairing; for example, mutations in the R consensus can be compensated by mutations in the S consensus.

Together these results suggest that the group I splicing reaction depends on the formation of secondary structure between pairs of consensus sequences within the intron. The principle established by this work is that *sequences distant from the splicing junctions themselves are required to form the active site that makes self-splicing possible.*

Ribozymes have various catalytic activities

The catalytic activity of group I introns was discovered by virtue of their ability to autosplice, but they are able to undertake other catalytic reactions *in vitro,* giving rise to the name **ribozymes**. All of the reactions that they perform are based on transesterifications. We analyze these reactions in terms of their relationship to the splicing reaction itself.

The catalytic activity of a group I intron is conferred by its ability to generate a particular secondary and tertiary structure that creates active sites, equivalent to the active sites of a conventional (proteinaceous) enzyme. **Figure 32.6** illustrates the splicing reaction in terms of these sites (this is the same series of reactions shown previously in Figure 32.2).

Figure 32.6

Excision of the group I intron in *Tetrahymena* rRNA occurs by successive reactions between the occupants of the guanosine-binding site and substrate-binding site. The left exon is shown in green, and the right exon is purple. The part of the intron that includes the reactive 5′ end is shown in red.

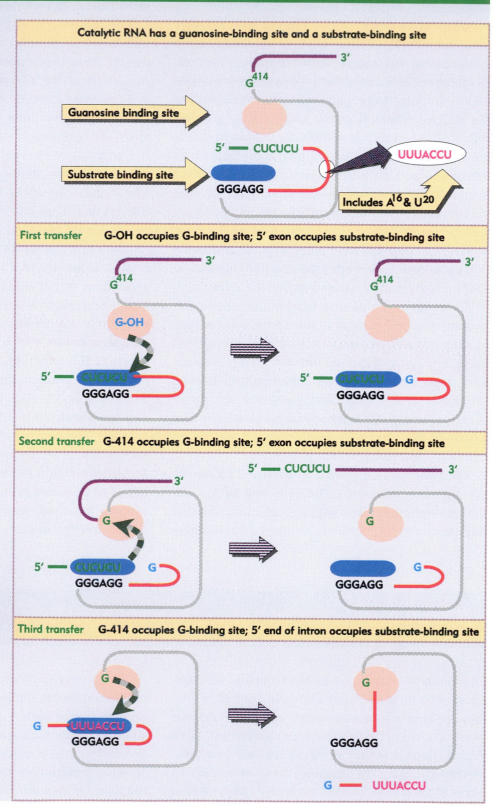

Figure 32.7

The L-19 linear RNA can bind C_5 in the substrate-binding site; the reactive G-OH 3' end is located in the G-binding site, and catalyzes transfer reactions that convert 2 C_5 oligonucleotides into a C4 and a C_6 oligonucleotide.

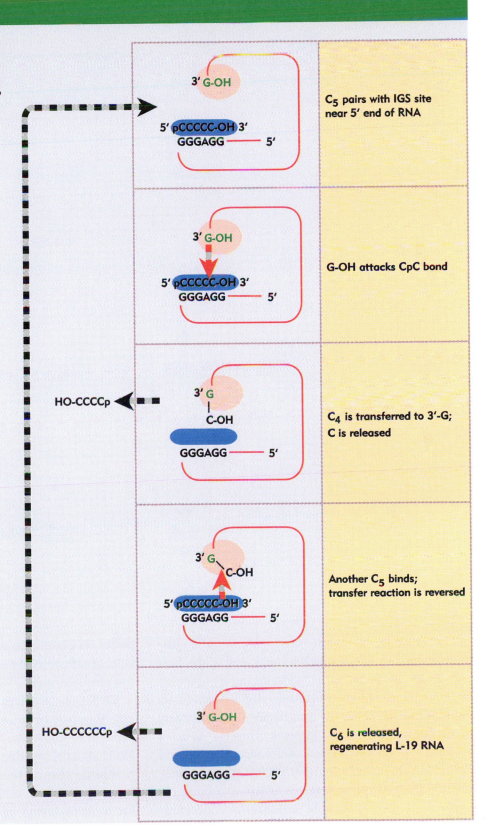

C_5 pairs with IGS site near 5' end of RNA

G-OH attacks CpC bond

C_4 is transferred to 3'-G; C is released

Another C_5 binds; transfer reaction is reversed

C_6 is released, regenerating L-19 RNA

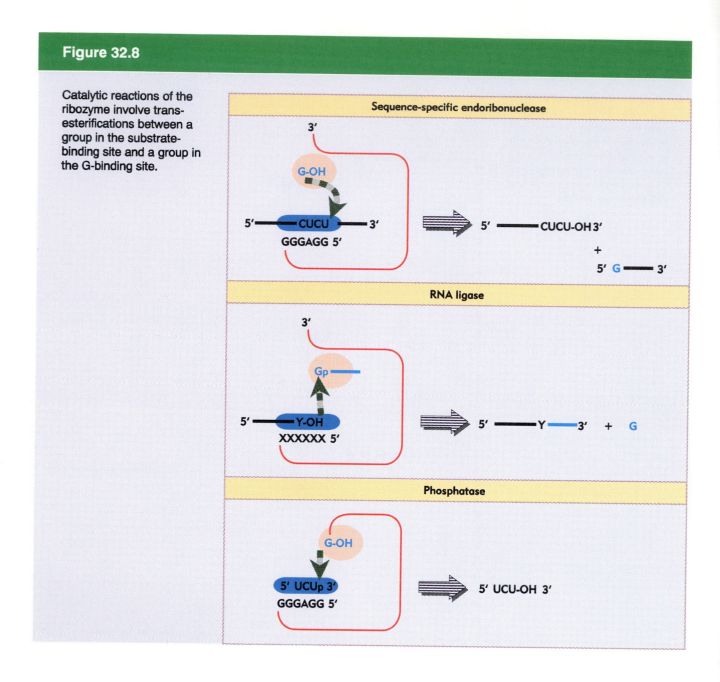

Figure 32.8

Catalytic reactions of the ribozyme involve trans-esterifications between a group in the substrate-binding site and a group in the G-binding site.

The P1 helix represents the formation of a **substrate-binding site**, in which the 3' end of the first intron base pairs with the IGS in an intermolecular reaction. A **guanosine-binding site** is formed by sequences in P7. This site may be occupied either by a free guanosine nucleotide or by the G residue in position 414. In the first transfer reaction, it is used by free guanosine nucleotide; but it is subsequently occupied by G^{414}. Note that this involves a change of conformation (comparable perhaps to a conforma-tional change in an enzyme). The second transfer releases the joined exons. The third transfer creates the circular intron.

The L-19 RNA is generated by opening the circular intron (shown as the last stage of the intramolecular rearrangements given in Figure 32.3). It still retains enzymatic abilities. These resemble the activities involved in the original splicing reaction, and we may consider ribozyme function in terms of the ability to bind an intra-

Table 32.1

Reactions catalyzed by RNA have the same features as those catalyzed by proteins, although the enzyme activity is slower. The K_M gives the concentration of substrate required for half-maximum velocity; this is an inverse measure of the affinity of the enzyme for substrate. The turnover number gives the number of substrate molecules transformed in unit time by a single catalytic site.

Enzyme	Substrate	K_M	Turnover
19 base virusoid	24 base RNA	0.0006 mM	0.5 /min
L-19 Intron	CCCCCC	0.04 mM	1.7 /min
RNAase P RNA	Pre-tRNA	0.0005 mM	1 /min
RNAase P complete	Pre-tRNA	0.0005 mM	2 /min
RNAase T1	GpA	0.05 mM	5,700 /min
β-galactosidase	Lactose	4.0 mM	12,500 /min

molecular sequence complementary to the IGS in the substrate-binding site, while binding either the terminal G^{414} or a free G-nucleotide in the G-binding site.

Figure 32.7 illustrates the mechanism by which it extends the oligonucleotide C_5 to generate a C_6 chain. The C_5 oligonucleotide binds in the substrate-binding site, while G^{414} occupies the G-binding site. By transesterification reactions, a C is transferred from C_5 to the 3′-terminal G, and then back to a new C_5 molecule. Further transfer reactions lead to the accumulation of longer cytosine oligonucleotides. The reaction is a true catalysis, because the L-19 RNA remains unchanged, and is available to catalyze multiple cycles. The ribozyme is behaving as a nucleotidyl transferase.

Some further enzymatic reactions are characterized in **Figure 32.8**. The ribozyme can function as a sequence-specific endoribonuclease by utilizing the ability of the IGS to bind complementary sequences. In this example, it binds an external template containing the sequence CUCU, instead of binding the analogous sequence which is usually contained at the end of the left exon. A guanine-containing nucleotide is present in the G-binding site, and attacks the CUCU sequence in precisely the same way that the exon is usually attacked in the first transfer reaction. This cleaves the target

sequence into a 5′ molecule that resembles the left exon, and a 3′ molecule that bears a terminal G residue. By mutating the IGS element, it is possible to change the specificity of the ribozyme, so that it recognizes sequences complementary to the new sequence at the IGS region.

The cleavage reaction also can be performed with a complementary deoxyribonucleotide as substrate. But the DNA molecule binds more weakly and is cleaved more slowly. Since the only difference between the DNA substrates and RNA substrates is the lack of 2′-OH groups in DNA, this suggests that these groups are important for binding and/or reaction. This makes it seem likely that reactions such as splicing originated in an 'RNA world', in which RNA provided the initial form of genetic material.

Altering the IGS, so that the specificity of the substrate-binding site is changed, and other RNA targets are enabled to enter, can be used to generate a ligase activity. An RNA terminating in a 3′-OH is bound in the substrate site, and an RNA terminating in a 5′-G residue is bound in the G-binding site. An attack by the hydroxyl on the phosphate bond connects the two RNA molecules, with the loss of the G residue.

The phosphatase reaction is not directly related to the splicing transfer reactions. An oligonucleotide sequence that is complementary to the IGS and terminates in a 3′-phosphate can be attacked by the G^{414}. The phosphate is transferred to the G^{414}, and an oligonucleotide with a free 3′-OH end is then released. The phosphate can then be transferred to either an oligonucleotide terminating in 3′-OH (effectively reversing the reaction) or indeed to water (releasing inorganic phosphate and completing an authentic phosphatase reaction).

The reactions catalyzed by RNA can be characterized in the same way as classical enzymatic reactions in terms of Michaelis–Menten kinetics. Table 32.1 summarizes an analysis of reactions catalyzed by RNA. The K_M values for RNA-catalyzed reactions are low, and therefore imply that the RNA can bind its substrate with high specificity. The turnover numbers are low, which reflects a low catalytic rate. In effect, the RNA molecules behave in the same general manner as traditionally defined for enzymes, although they are relatively slow compared to protein catalysts (where a typical range of turnover numbers is 10^3–10^6.)

Introns may code for endonucleases that sponsor mobility

Certain introns of both the group I and group II classes contain open reading frames that may be translated into proteins. We know the most about the function of the protein for certain group I introns. Its role is to help the intron perpetuate itself in crosses in which the alleles for the relevant gene differ with regards to their possession of the intron.

Polymorphisms for the presence or absence of introns are quite common in fungal mitochondria. This is consistent with the view that these introns originated by insertion into the gene. Some light on the process that could be involved is cast by an analysis of recombination in crosses involving the large rRNA gene of the yeast mitochondrion.

This gene has a group I intron that contains a coding sequence. The intron is present in some strains of yeast (called ω$^+$) but absent in others (ω$^-$). Genetic crosses between ω$^+$ and ω$^-$ are *polar:* the progeny are usually ω$^+$.

If we think of the ω$^+$ strain as a donor and the ω$^-$ strain as a recipient, we form the view that in ω$^+$ × ω$^-$ crosses a new copy of the intron is generated in the ω$^-$ genome. As a result, the progeny are all ω$^+$.

Mutations can occur in either parent to abolish the polarity. Mutants show normal segregation, with equal numbers of ω$^+$ and ω$^-$ progeny. The mutations indicate the nature of the process. Mutations in the ω$^-$ strain occur close to the site where the intron would be inserted. Mutations in the ω$^+$ strain lie in the reading frame of the intron and prevent production of the protein. This suggests the model of Figure 32.9, in which the protein coded by the intron in an ω$^+$ strain recognizes the site where the intron should be inserted in an ω$^-$ strain and causes it to be preferentially inherited.

What is the action of the protein? The product of the ω intron is an *endonuclease that recognizes the ω$^-$ gene as a target for a double-strand break.* The endonuclease recognizes an 18 bp target sequence that contains the site where the intron is inserted. The target sequence is cleaved on each strand of DNA 2 bases to the 3′ side of the insertion site. So the cleavage sites are 4 bp apart, and generate overhanging single strands.

This type of cleavage is related to the cleavage characteristic of transposons when they migrate to new sites (see Chapter 34). The double-strand break probably initiates a gene conversion process

Figure 32.9

An intron codes for an endonuclease that makes a double-strand break in DNA. The sequence of the intron is duplicated and then inserted at the break.

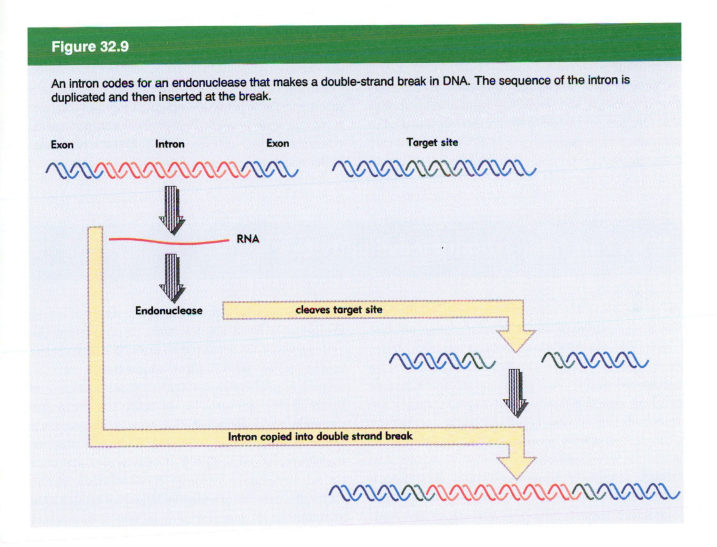

in which the sequence of the ω⁺ gene is copied to replace the sequence of the ω⁻ gene. Note that the insertion of the intron interrupts the sequence recognized by the endonuclease, thus ensuring stability.

Other group I introns that contain open reading frames also are mobile. There are further examples in fungal mitochondrial introns, two of the introns of phage T4 are inherited preferentially like the ω intron, and an intron in *Physarum* nuclear rDNA becomes inserted into all available recipient molecules in appropriate crosses. The general mechanism of intron perpetuation appears to be the same: the intron codes for an endonuclease that cleaves a specific target site where the intron will be inserted. There are differences in the details of insertion; for example, the endonuclease coded by the phage T4 *td* intron cleaves a target site that is 24 bp upstream of the site at which the intron is itself inserted.

In spite of the common mechanism for intron mobility, there is no homology between the sequences of the target sites or the intron coding regions. One assumes that the introns have a common evolutionary origin, but evidently they have diverged greatly. The target sites are among the longest and therefore the most specific known for any endonucleases. The specificity ensures that the intron perpetuates itself only by insertion into a single target site and not elsewhere in the genome.

These results strengthen the view that introns carrying coding sequences originated as independent elements that coded for a function involved in the ability to be spliced out of RNA or to migrate between DNA molecules. Consistent with this idea,

the pattern of codon usage is somewhat different in the intron coding regions from that found in the exons.

Introns of this type therefore differ from nuclear introns in their probable origin. They are likely to have become inserted into pre-existing genes, whereas nuclear introns are likely to be remnants of the original structure of the primeval gene.

RNA can have ribonuclease activities

One of the first demonstrations of the capabilities of RNA was provided by the dissection of ribonuclease P, the *E. coli* tRNA-processing endonuclease (described in Chapter 15). Ribonuclease P can be dissociated into its two components, the 375 base RNA and the 20,000 dalton polypeptide. Under the conditions initially used to characterize the enzyme activity *in vitro*, both components were necessary to cleave the tRNA substrate.

But a change in ionic conditions, an increase in the concentration of Mg^{2+}, renders the protein component superfluous. *The RNA alone can catalyze the reaction!* Analyzing the results as though the RNA were an enzyme, each 'enzyme' catalyzes the cleavage of at least four substrates. In fact, the activity of the RNA is not much less than the activity of crude preparations of ribonuclease P (see Table 32.1).

Because mutations in either the gene for the RNA or the gene for protein can inactivate RNAase P *in vivo*, we know that both components are necessary for natural enzyme activity. Originally it had been assumed that the protein provided the catalytic activity, while the RNA filled some subsidiary role, for example, assisting in the binding of substrate (it has some short sequences complementary to exposed regions of tRNA). But these roles are reversed!

How can RNA provide a catalytic center? Its ability seems reasonable if we think of an active

center as a surface that exposes a series of active groups in a fixed relationship. In a protein, the active groups are provided by the side-chains of the amino acids, which have appreciable variety, including positive and negative ionic groups and hydrophobic groups. In an RNA, the available moieties are more restricted, consisting primarily of the exposed groups of bases. We might suppose that short regions are held in a particular structure by the secondary/tertiary conformation of the molecule, providing a surface of active groups able to maintain an environment in which bonds can be broken and made into another molecule. It seems inevitable that the interaction between the RNA catalyst and the RNA substrate will rely on base pairing to create the environment.

The evolutionary implications of these discoveries are intriguing. The split personality of the genetic apparatus, in which RNA is present in all components, but proteins undertake catalytic reactions, has always been puzzling. It seems unlikely that the very first replicating systems could have contained both nucleic acid and protein.

But suppose that the first systems contained only a self-replicating nucleic acid with primitive catalytic activities, just those needed to make and break phosphodiester bonds. If we suppose that the involvement of 2′ bonds in current splicing reactions is derived from these primitive catalytic activities, we may argue that the original nucleic

The catalytic activity of RNAase P requires RNA to function as a catalyst with an external substrate. Another example of the ability of RNA to function as an endonuclease is provided by some small plant RNAs (~350 bases) that undertake a self-cleavage reaction. As with the case of the group I intron, however, it is possible to engineer constructs that can function on external substrates.

These small plant RNAs fall into two general groups: viroids and virusoids. The **viroids** are infectious RNA molecules that function independently, without enacapsidation by any protein coat (see Chapter 6). The **virusoids** are similar in organization, but are encapsidated by plant viruses, being packaged together with a viral genome. The virusoids cannot replicate independently, but require assistance from the virus. The virusoids are sometimes called **satellite RNAs**.

Viroids and virusoids both appear to replicate via rolling circles (see Chapter 18). The strand of RNA that is packaged into the virus is called the plus strand. The complementary strand, generated during replication of the RNA, is called the minus strand. Multimers of both plus and minus strands are found. Both types of monomer are probably generated by cleaving the tail of a rolling circle; circular plus strand monomers are generated by ligating the ends of the linear monomer.

Both plus and minus strands of viroids and virusoids undergo self-cleavage *in vitro*. The cleavage reaction is promoted by divalent metal cations; it generates 5′-OH and 2′–3′-cyclic phosphodiester termini. Some of the RNAs cleave *in vitro* under physiological conditions. Others do so only after a cycle of heating and cooling; this suggests that the isolated RNA has an inappropriate conformation, but can generate an active conformation when it is denatured and renatured.

Most of the viroids and virusoids for which self-cleavage has been demonstrated can in principle form the 'hammerhead' secondary structure drawn in the upper part of **Figure 32.10**. The sequence of this structure is sufficient for cleavage. When the surrounding sequences are deleted, the need for a heating–cooling cycle is obviated, and the small

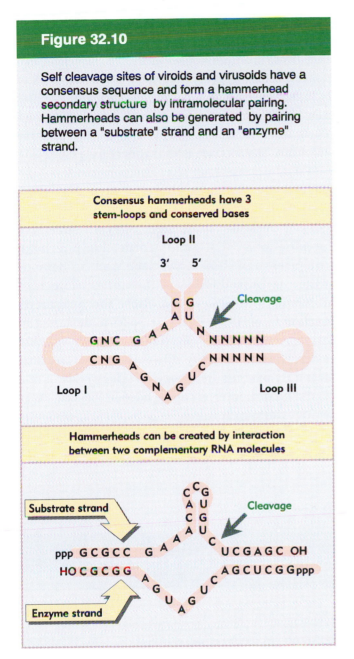

Figure 32.10

Self cleavage sites of viroids and virusoids have a consensus sequence and form a hammerhead secondary structure by intramolecular pairing. Hammerheads can also be generated by pairing between a "substrate" strand and an "enzyme" strand.

acid was RNA, since DNA lacks the 2′-OH group and therefore could not undertake such reactions. Proteins could have been added for their ability to stabilize the RNA structure, which was likely to have been precariously maintained. Then the greater versatility of proteins could have allowed them to take over catalytic reactions, leading eventually to the complex and sophisticated apparatus of modern gene expression.

RNA self-cleaves spontaneously. This suggests that the sequences beyond the hammerhead usually interfere with its formation.

The active site is a sequence of only 58 nucleotides. The hammerhead contains three stem–loop regions whose position and size are constant, and 13 conserved nucleotides, mostly in the regions connecting the center of the structure. The conserved bases and duplex stems generate an RNA with the intrinsic ability to cleave.

It is not necessary for the structure to be formed by intramolecular base pairing. An active hammerhead can be generated by pairing an RNA representing one side of the structure with an RNA representing the other side. The lower part of Figure 32.10 shows an example of a hammerhead generated by hybridizing a 19 base molecule with a 24 base molecule. The hybrid mimics the hammerhead structure, with the omission of loops I and III. When the 19 base RNA is added to the 24 base RNA, cleavage occurs at the appropriate position in the hammerhead.

We may regard the top (24 base) strand of this hybrid as comprising the 'substrate', and the bottom (19 base) strand as comprising the 'enzyme'. When the 19 base RNA is mixed with an excess of the 24 base RNA, multiple copies of the 24 base RNA are cleaved. This suggests that there is a cycle of 19 to 24 base pairing, cleavage, dissociation of the cleaved fragments from the 19 base RNA, and pairing of the 19 base RNA with a new 24 base substrate. The 19 base RNA is therefore a ribozyme with endonuclease activity. The parameters of the reaction are similar to those of other RNA-catalyzed reactions (see Table 32.1).

It is possible to design enzyme–substrate combinations that can form hammerhead structures, and these have been used to demonstrate that introduction of the appropriate RNA molecules into a cell can allow the enzymatic reaction to occur *in vivo*. A ribozyme designed in this way essentially provides a highly specific restriction-like activity directed against an RNA target. By placing the ribozyme under control of a regulated promoter, it can be used in the same way as (for example) antisense constructs to specifically turn off expression of a target gene under defined circumstances.

RNA editing utilizes information from several sources

A prime axiom of molecular biology is that the sequence of an mRNA can only represent what is coded in the DNA. The central dogma envisaged a linear relationship in which a continuous sequence of DNA is transcribed into a sequence of mRNA that is in turn directly translated into protein. The occurrence of interrupted genes and the removal of introns by RNA splicing introduces an additional step into the process of gene expression: the coding sequences (exons) in DNA must be reconnected in RNA. But the process remains one of information transfer, in which the actual coding sequence in DNA remains inviolate.

Changes in the information coded by DNA occur in some exceptional circumstances, most notably in the generation of new sequences coding for immunoglobulins in mammals and birds. These changes occur specifically in the somatic cells (B lymphocytes) in which immunoglobulins are synthesized (see Chapter 37). New information is generated in the DNA of an individual during the process of reconstructing an immunoglobulin gene; and information coded in the DNA is changed by somatic mutation. The information in DNA

Figure 32.11

The sequence of the apo-B gene is the same in intestine and liver, but the sequence of the mRNA is modified by a base change that creates a termination codon in intestine.

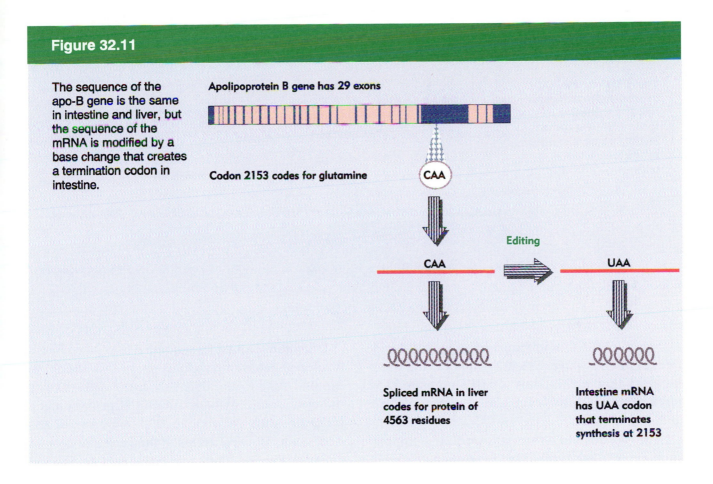

Apolipoprotein B gene has 29 exons

Codon 2153 codes for glutamine

CAA

Editing

CAA ⟹ UAA

Spliced mRNA in liver codes for protein of 4563 residues

Intestine mRNA has UAA codon that terminates synthesis at 2153

continues to be faithfully transcribed into RNA.

RNA editing is a process in which *information changes at the level of mRNA*. It is revealed by situations in which the coding sequence in an RNA differs from the sequence of DNA from which it was transcribed. RNA editing occurs in two different situations, with different causes. In mammalian cells there are cases in which a substitution occurs in an individual base in mRNA, causing a change in the sequence of the protein that is coded. In trypanosome mitochondria, more widespread changes occur in transcripts of several genes, when bases are systematically added or deleted.

Figure 32.11 summarizes the sequences of the apolipoprotein-B (apo-B) gene and mRNA in mammalian intestine and liver. The genome contains a single (interrupted) gene whose sequence is identical in all tissues, with a coding region of 4563 codons. This gene is transcribed into an mRNA that

is translated into a protein of 512,000 daltons representing the full coding sequence in the liver.

A shorter form of the protein, ~250,000 daltons, is synthesized in the intestine. This protein consists of the N-terminal half of the full-length protein. It is translated from an mRNA whose sequence is identical with that of liver except for a change from C to U at codon 2153. This substitution changes the codon CAA for glutamine into the ochre codon UAA for termination.

What is responsible for this substitution? No alternative gene or exon is available in the genome to code for the new sequence, and no change in the pattern of splicing can be discovered. We are forced to conclude that a change has been made directly in the sequence of the transcript. Does a specific enzyme recognize the *apo-B* transcript and change C_{2153} to T? Such an enzyme might recognize a particular region of secondary structure in a manner

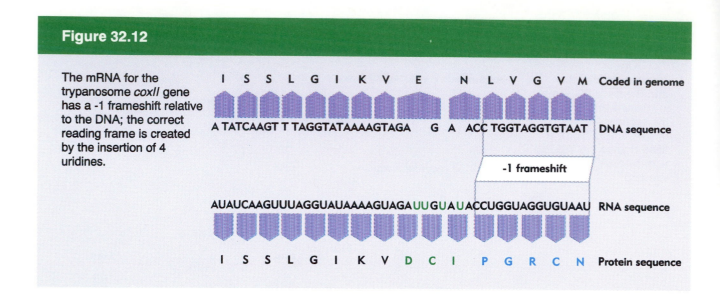

Figure 32.12

The mRNA for the trypanosome *coxII* gene has a -1 frameshift relative to the DNA; the correct reading frame is created by the insertion of 4 uridines.

```
              I  S  S  L  G  I  K  V  E      N  L  V  G  V  M   Coded in genome

       A TATCAAGT T TAGGTATAAAAGTAGA     G  A  ACC TGGTAGGTGTAAT   DNA sequence

                                              -1 frameshift

  AUAUCAAGUUUAGGUAUAAAAGUAGAUUGUAUACCUGGUAGGUGUAAU   RNA sequence

              I  S  S  L  G  I  K  V  D  C  I  P  G  R  C  N   Protein sequence
```

analogous to tRNA-modifying enzymes. The development of an *in vitro* system for this editing event suggests that a relatively small sequence (~50 bases) in the vicinity of the editing site provides a sufficient target.

Editing of this sort is rare, but apo-B is not unique. Another example is provided by glutamate receptors in rat brain. Editing at a single position changes a glutamine codon in DNA into a codon for arginine in RNA; the arginine plays an important role in controlling ion flow through the neurotransmitter, so the editing event is clearly necessary for physiological function.

Dramatic changes in sequence have been found in several genes of trypanosome mitochondria. In the first case to be discovered, the sequence of the cytochrome oxidase subunit II protein has a −1 frameshift relative to the sequence of the *coxII* gene. The sequences of the gene and protein given in **Figure 32.12** are conserved in several trypanosome species. How does this gene function?

The *coxII* mRNA has an insert of an additional four nucleotides (all uridines) around the site of frameshift. The insertion restores the proper reading frame; it inserts an extra amino acid and changes the amino acids on either side. No second

Figure 32.13

Part of the mRNA sequence of *T. brucei* *coxIII* shows many uridines that are not coded in the DNA (shown in green) or that are removed from the RNA (shown as T).

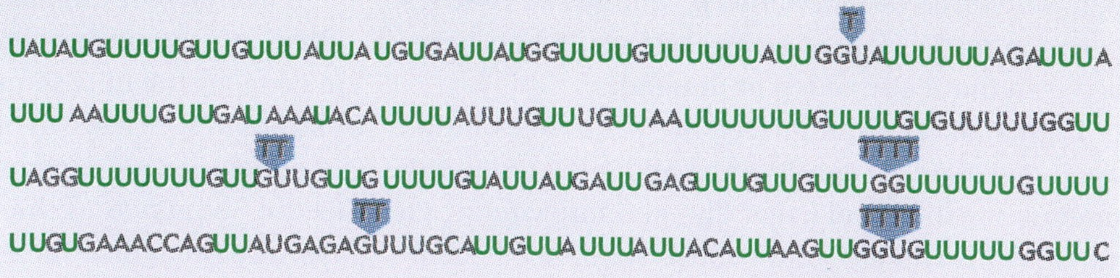

```
UAUAUGUUUUGUUGUUUAUUAUGUGAUUAUGGUUUUGUUUUUUAUUGGUAUUUUUUAGAUUUA

UUU AAUUUGUUGAUAAAUACAUUUUAUUUGUUUGUUAAUUUUUUUGUUUUGUGUUUUUGGUU

UAGGUUUUUUUGUUGUUGUUGUUUUGUAUUAUGAUUGAGUUUGUUGUUUGGUUUUUUGUUUU

UUGUGAAACCAGUUAUGAGAGUUUGCAUUGUUAUUUAUUACAUUAAGUUGGUGUUUUUUGGUUC
```

Figure 32.14

Pre-edited RNA base pairs with a guide RNA on both sides of the region to be edited. The guide RNA provides a template for the insertion of uridines. The mRNA produced by the insertions is complementary to the guide RNA.

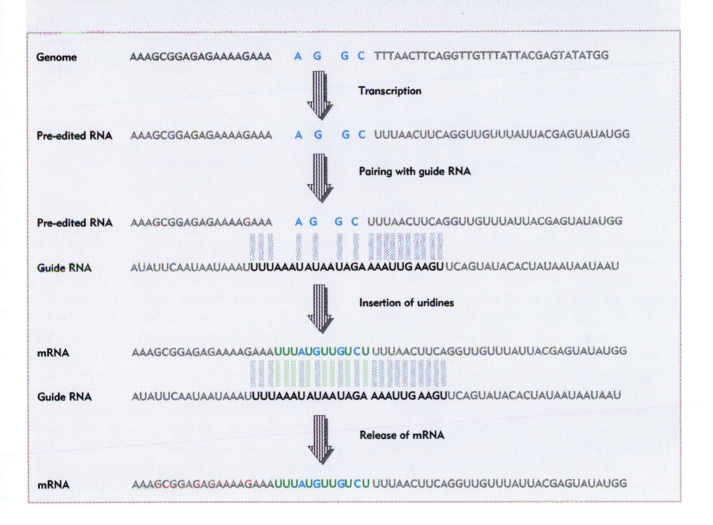

gene with this sequence can be discovered, and we are forced to conclude that the extra bases are inserted during or after transcription. A similar discrepancy between mRNA and genomic sequences is found in genes of the SV5 and measles paramyxoviruses, in these cases involving the addition of G residues in the mRNA.

Similar editing of RNA sequences occurs for other genes, and includes deletions as well as additions of uridine. The extraordinary case of the *coxIII*

gene of *T. brucei* is summarized in **Figure 32.13**. *More than half of the residues in the mRNA consist of uridines that are not coded in the gene.* Comparison between the genomic DNA and the mRNA shows that no stretch longer than 7 nucleotides is represented in the mRNA without alteration; and runs of uridine up to 7 bases long are inserted.

What provides the information for the specific insertion of uridines? A **guide RNA** contains a sequence that is complementary to the correctly

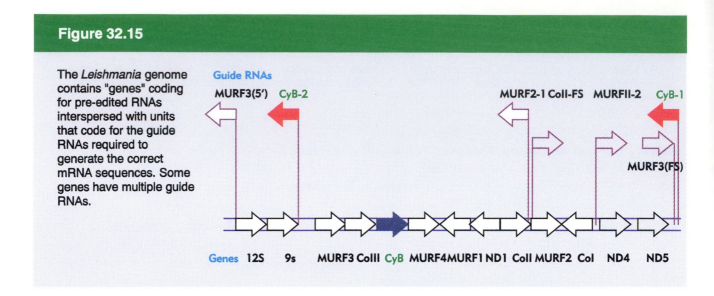

Figure 32.15

The *Leishmania* genome contains "genes" coding for pre-edited RNAs interspersed with units that code for the guide RNAs required to generate the correct mRNA sequences. Some genes have multiple guide RNAs.

Guide RNAs

MURF3(5') CyB-2   MURF2-1 ColI-FS MURFII-2 CyB-1

MURF3(FS)

Genes 12S 9s MURF3 ColII CyB MURF4 MURF1 ND1 ColI MURF2 Col ND4 ND5

edited mRNA. **Figure 32.14** shows a model for its action in the cytochrome *b* gene of *Leishmania*.

The sequence at the top shows the original transcript, or pre-edited RNA. Gaps show where bases will be inserted in the editing process. Eight uridines must be inserted into this region to create the valid mRNA sequence.

The guide RNA is complementary to the mRNA for a significant distance including and surrounding the edited region. Typically the complementarity is more extensive on the 3' side of the edited region and is rather short on the 5' side. Pairing between the guide RNA and the pre-edited RNA leaves gaps where unpaired A residues in the guide RNA do not find complements in the pre-edited RNA. The guide RNA provides a template that allows the missing U residues to be inserted at these positions. When the reaction is completed, the guide RNA separates from the mRNA, which becomes available for translation.

Specification of the final edited sequence can actually be quite complex; in this example, a lengthy stretch of the transcript is edited by the insertion altogether of 39 U residues, and this appears to require two guide RNAs that act at adjacent sites, either simultaneously or successively. We can imagine that there could be even more complicated situations in which, for example, one guide RNA helps to create an edited sequence

that is then a substrate for further editing by another guide RNA.

The guide RNAs are encoded as independent transcription units. **Figure 32.15** shows a map of the relevant region of the *Leishmania* mitochondrial DNA. It includes the 'gene' for cytochrome *b*, which codes for the pre-edited sequence, and two regions that specify guide RNAs. Genes for the major coding regions and for their guide RNAs are interspersed.

In principle, a mutation in either the 'gene' or one of its guide RNAs could change the primary sequence of the mRNA, and thus of the protein. By genetic criteria, each of these units could be considered to comprise part of the 'gene'. Since the units are independently expressed, they should of course complement in *trans*. If mutations were available, we should therefore find that 3 complementation groups were needed to code for the primary sequence of a single protein.

The characterization of intermediates that are partially edited suggests that the reaction proceeds along the pre-edited RNA in the 3'–5' direction. The guide RNA not only dictates specificity of uridine insertions by its pairing with the pre-edited RNA, but also actually provides the U residues that are inserted.

Guide RNAs terminate in stretches of 3–5 U residues, after which further uridines are added to generate 3' oligo-U tails. The tails are heterogeneous in length, varying from 5 to 24 U residues. These uridines are inserted into the edited RNA by a reaction that resembles group I RNA splicing. Pairing between the guide RNA and the pre-edited RNA resembles the pairing between the intron IGS and the 5′ exon. **Figure 32.16** shows that the reaction occurs by two transesterifications.

In the first transfer, the edited region is attacked by the 3′-OH of the U-rich tail of the guide RNA that is paired with it. This breaks the pre-edited RNA into two parts. One part is the 5′ region, terminating in a 3′-OH group. The other part is a hybrid molecule in which the 3′ tail of the guide RNA is covalently linked to the edited RNA. This intermediate is a linear molecule whose structure is held together by base pairing.

In the second transfer, the 3′-OH that was generated in the first transfer now attacks a bond between 2 U residues in the tail region of the guide RNA. This restores the integrity of the pre-edited RNA, which has gained a U residue. And it restores the structure of the guide RNA, which has lost a U residue from its tail. The cycle is then repeated.

The structures of partially edited molecules suggest that the U residues are added one a time, and not in groups. It is possible that the reaction proceeds through successive cycles in which U residues are added, tested for complementarity with the guide RNA, retained if acceptable and removed if not, so that the construction of the correct edited sequence occurs gradually. We do not know whether the same types of reaction are involved in editing reactions that delete U residues or that add C residues.

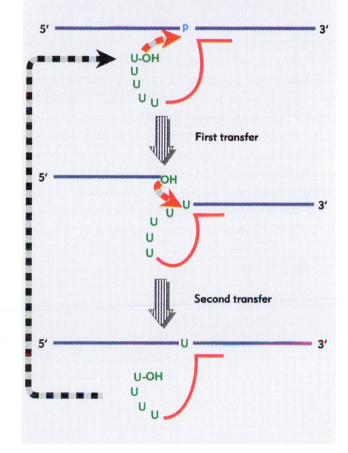

Figure 32.16

Editing involves two transesterification reactions. In the first transfer the 3′-OH of the guide RNA attacks an internal bond in the pre-edited RNA, generating a linear 5′ molecule and a guide-3′ hybrid. In the second transfer, the 3′-OH of the linear 5′ molecule attacks the U-rich region of the guide, releasing the guide RNA and an edited RNA. Cycles of editing occur as dictated by pairing between guide RNA and the edited region.

Summary

Self-splicing is a property of group I introns, found in *Tetrahymena* and *Physarum* nuclei, in fungal mitochondria, and in phage T4. The information necessary for the reaction resides in the intron sequence (although the reaction is actually assisted by proteins *in vivo*). The reaction requires formation of a specific secondary (and presumably tertiary) structure involving short consensus sequences. The RNA creates a structure in which the substrate sequence is held by the IGS region of the intron, and other conserved sequences generate a guanine nucleotide binding site. It occurs by a transesterification involving a guanosine residue as cofactor. No input of energy is required. The guanosine breaks the bond at the 5′ exon–intron junction and becomes linked to the intron; the hydroxyl at the free end of the exon then attacks the 3′ exon–intron junction. The intron cyclizes and loses the guanosine and the terminal 15 bases. A series of related reactions can be catalyzed via attacks by the G-OH residue on internal phosphodiester bonds. By providing appropriate substrates, it has been possible to engineer ribozymes that perform a variety of catalytic reactions, including nucleotidyl transferase activities.

Some group I and some group II mitochondrial introns have open reading frames. The proteins coded by some group I introns are endonucleases that make double-stranded cleavages in target sites in DNA; the cleavage initiates a gene conversion process in which the sequence of the intron itself is copied into the target site. This could mean that these types of introns originated by insertion events.

Catalytic reactions are undertaken by the RNA component of the RNAase P ribonucleoprotein. Virusoid RNAs can undertake self-cleavage at a 'hammerhead' structure. Hammerhead structures can form between a substrate RNA and a ribozyme RNA, allowing cleavage to be directed at highly specific sequences. These reactions support the view that RNA can form specific active sites that have catalytic activity.

RNA editing changes the sequence of an RNA after or during its transcription. The changes are required to create a meaningful coding sequence. Substitutions of individual bases occur in mammalian systems; additions and deletions (most usually of uridine) occur in trypanosome mitochondria and in paramyxoviruses. Extensive editing reactions occur in trypanosomes in which as many as half of the bases in an mRNA are derived from editing. The editing reaction occurs uses a template consisting of a guide RNA that is complementary to the mRNA sequence. The guide RNA pairs with the pre-edited RNA and uses its oligo-U 3′ tail to provide uridine residues that are incorporated into the target RNA by transesterification reactions.

Further reading

Reviews

The behavior of RNA as a catalyst has been treated by **Altman** (*Cell* **36**, 227–229, 1984) and by **Cech** (*Science* **236**, 1532–1539, 1987). Self-cleavage reactions have been reviewed by **Symons** (*Ann. Rev. Biochem.* **61**, 641–671, 1992).

Group I autosplicing was reviewed by **Cech** (*Ann. Rev. Biochem.* **59**, 543–568, 1990). Mobility of introns has been reviewed by **Lambowitz and Belfort** (*Ann. Rev. Biochem.* **62**, 587–622, 1993)

An extensive discussion of the catalytic roles of RNA can be found in *The RNA World* (eds. R. F. Gesteland and J. F. Atkins, Cold Spring Harbor, 1993), in which **Cech** reviewed self-splicing introns (pp. 239–270).

Discoveries

Analyses of splicing and associated processes have perhaps produced more surprises and revealed more fundamental biochemical features than any other topic in recent memory. Many of the original papers are well worth reading.

The *in vitro* autocatalytic system from *Tetrahymena* was discovered by **Cech et al.** (*Cell* **27**, 487–496, 1981).

Catalytic activity of RNA was discovered by **Guerrier-Takada et al.** (*Cell* **35**, 849–857, 1983).

Bacterial splicing was discovered by **Belfort et al.** (*Cell* **41**, 375–382, 1985).

The hammerhead structure for RNA self-cleavage was proposed by **Forster and Symons** (*Cell* **50**, 916, 1987).

Changes in RNA sequence in mammals were discovered by **Powell et al.** (*Cell* **50**, 831–840, 1987). RNA editing in trypanosomes was discovered by **Benne et al.** (*Cell* **46**, 819–826, 1986), and its full extent become apparent with the report of **Feagin, Abraham, and Stuart** (*Cell* **53**, 413–422, 1988). The involvement of guide RNAs was discovered by **Blum, Bakalara, and Simpson** (*Cell* **60**, 189–198, 1990), and the transesterification mechanism was discovered by **Blum et al.** (*Cell* **65**, 543–550, 1991).

The dynamic genome: DNA in flux

Elements carried in the maize chromosomes... serve to control gene action and to induce, at the site of the gene, heritable modifications affecting this action. These elements were initially discovered because they do not remain at one position in the chromosome complement. They can appear at new locations and disappear from previously determined locations. The presence of one such element at or near the locus of a known gene may affect the action of this gene. In so doing, it need not alter the action potential of the genic substances at the locus. Therefore, these elements were called controlling elements... It might be considered that a controlling element represents some kind of extra chromosomal substance that can attach itself or impress its influence in some manner at various positions in the chromosome complement and so affect the action of the genic substances at these positions. The modes of operation of controlling elements do not suggest this, however. Rather they suggest that controlling elements are integral components of the chromosomes themselves, and that they have specific activities and modes of accomplishing them, much as the genes are presumed to have... Transpositions of controlling elements either arise from some yet-unknown mechanism or occur during the chromosome reduplication process itself and are a consequence of it.

Barbara McClintock, 1956

CHAPTER 33

Recombination of DNA

Without genetic recombination, the content of each individual chromosome would be irretrievably fixed in its particular alleles, changeable only by mutation. The length of the target for mutation damage would be increased from the gene to the chromosome. Deleterious mutations would accumulate, eliminating each chromosome (and thereby removing any favorable mutations that have occurred).

By shuffling the genes, recombination allows favorable and unfavorable mutations to be separated and tested as individual units in new assortments. It provides a means of escape and spreading for favorable alleles, and a means to eliminate an unfavorable allele without bringing down all the other genes with which this allele has been associated in the past. From the long perspective of evolution, a chromosome is a bird of passage, a temporary association of particular alleles. It is recombination that makes this possible.

Recombination occurs between precisely corresponding sequences, so that not a single base pair is added to or lost from the recombinant chromosomes. We may consider three types of recombination, which share the feature that the process involves physical exchange of material between duplex DNAs, but which differ in the circumstances.

◆ Recombination involving reaction between homologous sequences of DNA is called **generalized** or **homologous recombination**. In eukaryotes, it occurs at meiosis, usually in both males (during spermatogenesis) and females (during oogenesis). We recall that it happens at the 'four strand' stage of meiosis, and involves only two of the four strands (see Chapter 3).

◆ Another type of event sponsors recombination between *specific* pairs of sequences, and has been best characterized in prokaryotes. **Site-specific recombination** is responsible for the integration of phage genomes into the bacterial chromosome. The recombination event involves specific sequences of the phage DNA and bacterial DNA, which include a short stretch of homology. The enzymes involved in this event act *only* on the particular pair of target sequences. Related reactions are responsible for inverting specific regions of the bacterial chromosome.

◆ A different type of event allows one DNA sequence to be inserted into another without reliance on sequence homology. **Transposition** provides a means by which certain elements move from one chromosomal location to another. The mechanisms involved in transposition depend on breakage and reunion of DNA strands, and thus are related to the pro-

Figure 33.1

Overview: recombination occurs during the first meiotic prophase. The stages of prophase are defined by the appearance of the chromosomes, each of which consists of two replicas (sister chromatids), although the duplicated state becomes visible only at the end. The molecular interactions of any individual crossing over event involve two of the four duplex DNAs, and involve exchanges of single strands.

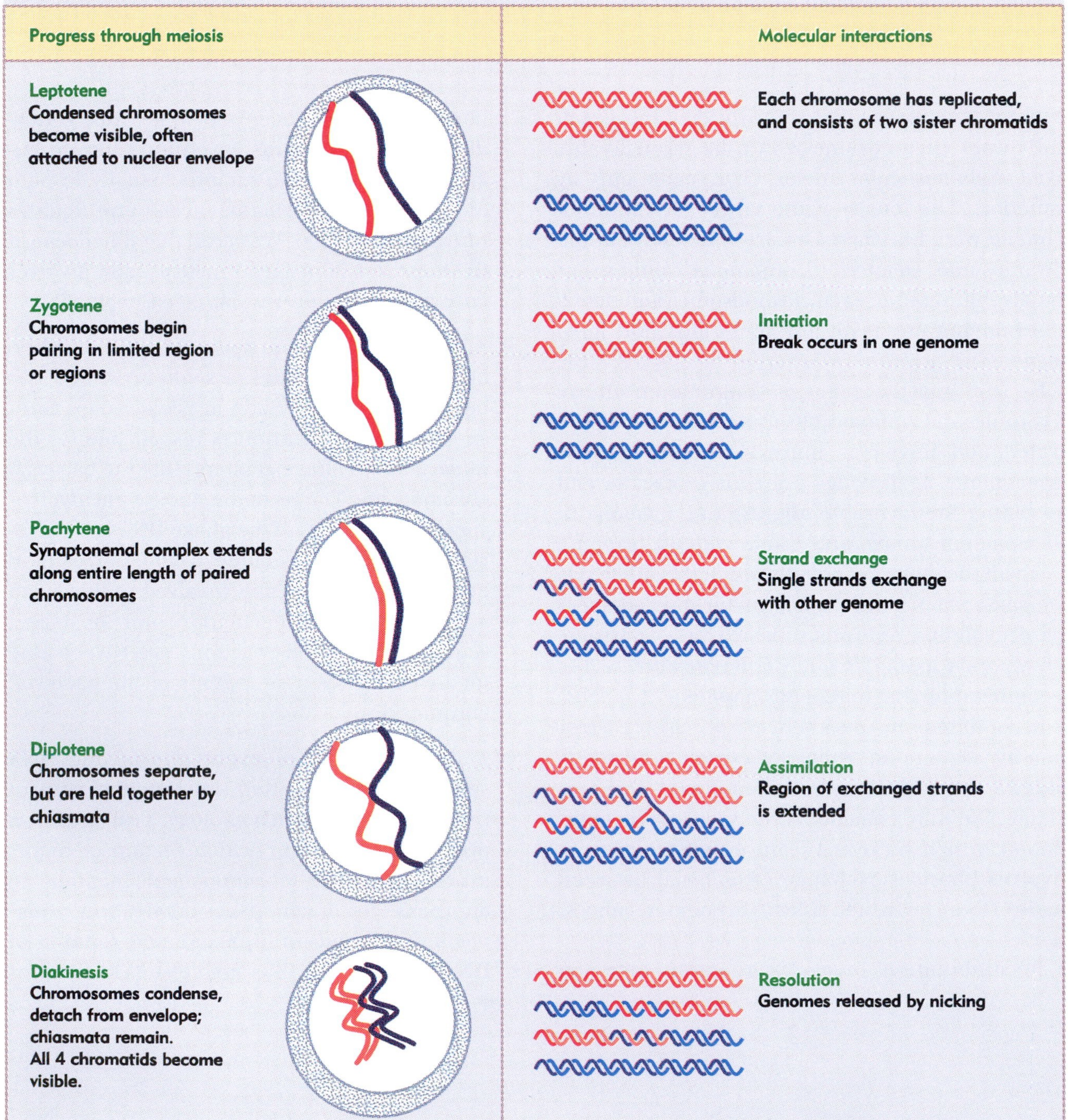

Progress through meiosis		Molecular interactions

Leptotene
Condensed chromosomes become visible, often attached to nuclear envelope

Each chromosome has replicated, and consists of two sister chromatids

Zygotene
Chromosomes begin pairing in limited region or regions

Initiation
Break occurs in one genome

Pachytene
Synaptonemal complex extends along entire length of paired chromosomes

Strand exchange
Single strands exchange with other genome

Diplotene
Chromosomes separate, but are held together by chiasmata

Assimilation
Region of exchanged strands is extended

Diakinesis
Chromosomes condense, detach from envelope; chiasmata remain.
All 4 chromatids become visible.

Resolution
Genomes released by nicking

cesses of recombination. Transposition is the subject of Chapters 34 and 35.

Homologous recombination is a reaction between two duplexes of DNA. Its critical feature is that the enzymes responsible can use *any* pair of homologous sequences as substrates (although some types of sequence may be favored over others). The frequency of recombination is not constant throughout the genome, but is influenced by both global and local effects. The overall frequency may be different in oocytes and in sperm; recombination occurs twice as frequently in female as in male humans. And within the genome its frequency depends on chromosome structure; for example, crossing-over is suppressed in the vicinity of heterochromatin.

Figure 33.1 compares the visible progress of chromosomes through meiosis with the molecular interactions that are involved in exchanging material between duplexes of DNA. We recall that meiosis starts with a protracted prophase (shown previously in Figure 2.22 in the context of the two meiotic divisions). The view of this prophase is expanded in Figure 33.1 into five stages.

The beginning of meiosis is marked by the point at which individual chromosomes become visible. Each of these chromosomes has replicated previously, and consists of two sister chromatids, each of which contains a duplex DNA. The homologous chromosomes approach one another and begin to pair in one or more regions, forming **bivalents**. Pairing extends until the entire length of each chromosome is apposed with its homologue. The process is called **synapsis** or **chromosome pairing**. When the process is completed, the chromosomes are laterally associated in the form of a **synaptonemal complex**, which has a characteristic structure in each species, although there is wide variation in the details between species.

Recombination between chromosomes involves a physical exchange of parts (see Figure 3.7), usually represented as a **breakage and reunion**, in which two nonsister chromatids (each containing a duplex of DNA) have been broken and then linked each with the other. When the chromosomes begin to separate, they can be seen to be held together at discrete sites, the **chiasmata**. The number and distribution of chiasmata parallel the features of genetic crossing-over. Traditional analysis holds that a chiasma represents the crossing-over event (see Figure 3.6), although formal proof of the connection has not been made. The chiasmata remain visible when the chromosomes condense and all four chromatids become evident.

What is the molecular basis for these events? Each sister chromatid contains a single DNA duplex, so each bivalent contains 4 duplex molecules of DNA. Recombination requires a mechanism that allows the duplex DNA of one sister chromatid to interact with the duplex DNA of a sister chromatid from the other chromosome. It must be possible for this reaction to occur between any pair of corresponding sequences in the two molecules in a highly specific manner that allows material to be exchanged with precision at the level of the individual base pair.

We know of only one mechanism for nucleic acids to recognize one another on the basis of sequence: complementarity between single strands. The figure shows a general model for the involvement of single strands in recombination. The first step in providing single strands is to make a break in one DNA duplex. Then one or both of the strands of that duplex can be released; if (at least) one strand displaces the corresponding strand in the other duplex, the two duplex molecules will be specifically connected at corresponding sequences. Furthermore, if this strand exchange is extended, there can be more extensive connection between the duplex; and by exchanging both strands and later cutting them, it is possible to connect the parental duplex molecules by means of a crossover that corresponds to the demands of a breakage and reunion.

We cannot at this juncture relate these molecular events rigorously with the changes that are observed at the level of the chromosomes. There is no detailed information about the molecular events involved in recombination in higher eukaryotic cells (in which meiosis has been most closely observed). However, recently the isolation of mutants in yeast has made it possible to correlate some of the molecular steps with approximate stages of meiosis. Detailed information about the recombination process is available in bacteria, in which molecular activities are known that cause genetic exchange between duplex molecules; but the bacterial reaction involves interaction between restricted regions of the genome, rather than an entire pairing of genomes, and the synapsis of eukaryotic chromosomes remains the most difficult stage to explain at the molecular level.

Breakage and reunion involves heteroduplex DNA

The act of connecting two duplex molecules of DNA is at the heart of the recombination process. Our molecular analysis of recombination therefore starts by expanding the view in Figure 33.1 of the use of base pairing between complementary single strands in recombination. It is useful to imagine the recombination reaction in terms of single strand exchanges (although we shall see that this is not necessarily how it is actually initiated), because the properties of the molecules created in this way are central to understanding the processes involved in recombination.

Figure 33.2 illustrates a process that starts with breakage at the corresponding points of the homologous strands of two paired DNA duplexes. The breakage allows movement of the free ends created by the nicks. Each strand leaves its partner and crosses over to pair with its complement in the other duplex.

The reciprocal exchange creates a connection between the two DNA duplexes. The connected pair of duplexes is called a **joint molecule**. The point at which an individual strand of DNA crosses from one duplex to the other is called the **recombinant joint**.

At the site of recombination, each duplex has a region consisting of one strand from each of the parental DNA molecules. This region is called **hybrid DNA** or **heteroduplex DNA**.

An important feature of a recombinant joint is its ability to move along the duplex. Such mobility is called **branch migration**. **Figure 33.3** illustrates the migration of a single strand in a duplex. The branching point can migrate in either direction as one strand is displaced by the other.

Branch migration is important for both theoretical and practical reasons. As a matter of principle, it confers a dynamic property on recombining structures. As a practical feature, its existence means that the point of branching cannot be established by examining a molecule *in vitro* (because the branch may have migrated since the molecule was isolated).

Branch migration could allow the point of crossover in the recombination intermediate to move in either direction. The rate of branch migration is uncertain, but as seen *in vitro* is probably inadequate to support the formation of extensive regions of heteroduplex DNA in natural conditions. Any extensive branch migration *in vivo* must therefore be catalyzed by a recombination enzyme.

When recombination involves duplex DNA molecules, topological manipulation may be required; either the DNA duplex must be free to rotate, or equivalent relief from topological restraint must be provided (see later). If we imagine that the joint molecule of Figure 33.2 rotates one duplex relative to the other, we can visualize it in one plane as a **Holliday structure** (named for its proposer). This is illustrated in **Figure 33.4**.

Figure 33.2

Recombination between two paired duplex DNAs could be initiated by reciprocal single-strand exchange, extended by branch migration, and resolved by nicking.

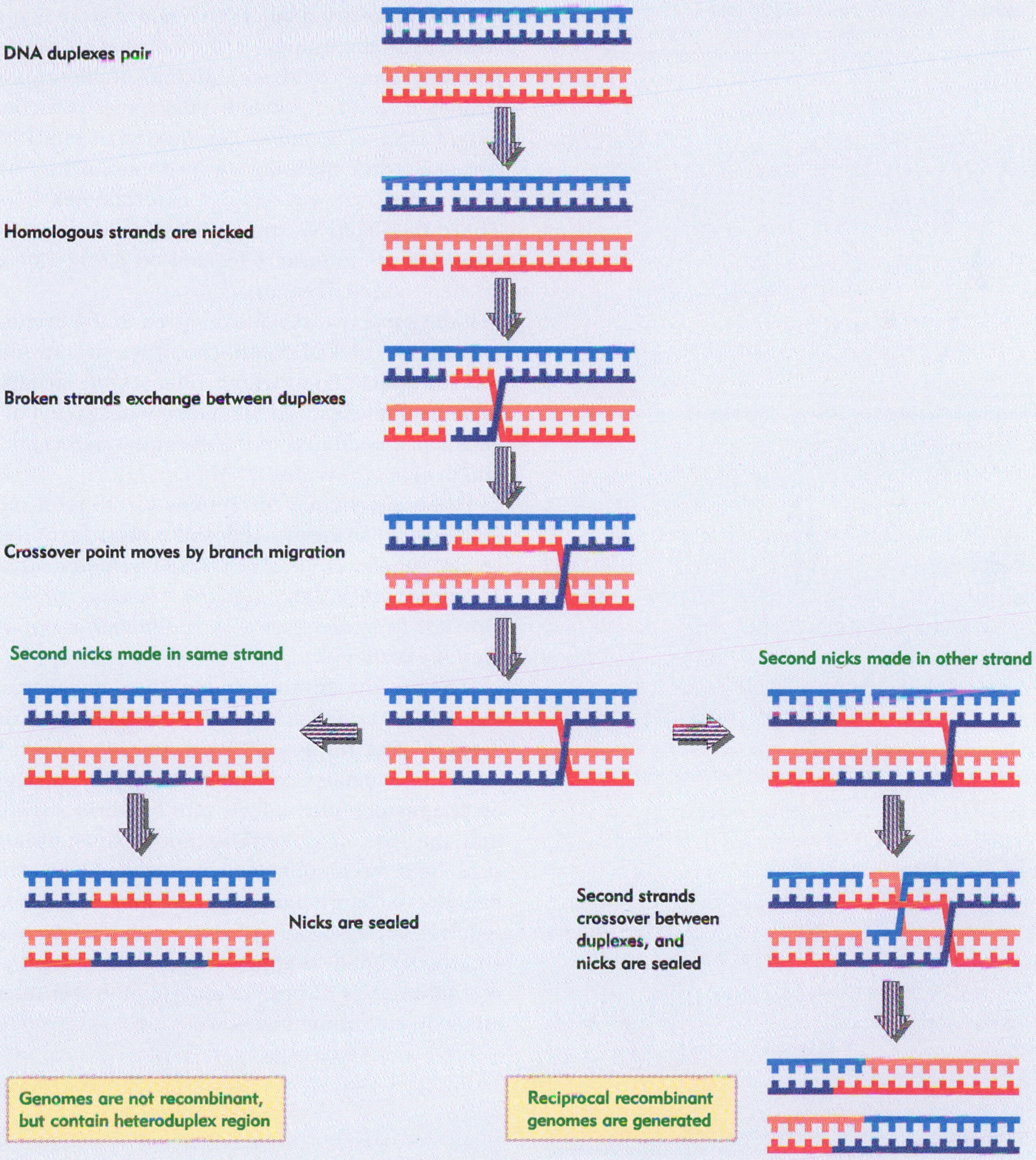

DNA duplexes pair

Homologous strands are nicked

Broken strands exchange between duplexes

Crossover point moves by branch migration

Second nicks made in same strand

Second nicks made in other strand

Nicks are sealed

Second strands crossover between duplexes, and nicks are sealed

Genomes are not recombinant, but contain heteroduplex region

Reciprocal recombinant genomes are generated

Figure 33.3

Branch migration can occur in either direction when an unpaired single strand displaces a paired strand. This structure could be created by a renaturation event in which a single DNA strand anneals at one end with one complementary strand, and anneals at the other end with an independent complementary strand.

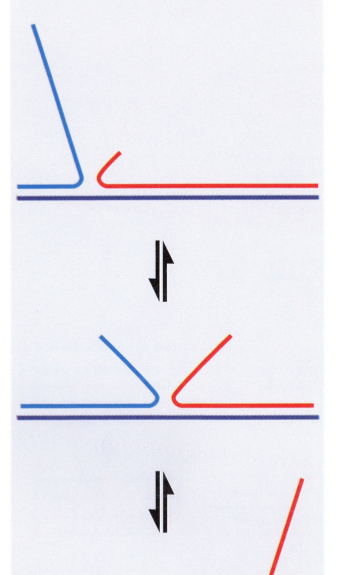

The joint molecule formed by strand exchange must be **resolved** into two separate duplex molecules. Resolution requires a further pair of nicks. The outcome of the reaction depends on which pair of strands is nicked, as can be seen from Figures 33.2 and 33.4.

If the nicks are made in the pair of strands that were *not* originally nicked (the pair that did not initiate the strand exchange), all four of the original strands have been nicked. This releases recombinant DNA molecules. The duplex of one DNA parent is covalently linked to the duplex of the other DNA parent, via a stretch of heteroduplex DNA. There has been a conventional recombination event between markers located on either side of the heteroduplex region.

If the *same* two strands involved in the original nicking are nicked again, the other two strands remain intact. The nicking releases the original parental duplexes, which remain intact except that each has a residuum of the event in the form of a length of heteroduplex DNA.

These alternative resolutions of the joint molecule establish the principle that *a strand exchange between duplex DNAs always leaves behind a region of heteroduplex DNA, but the exchange may or may not be accompanied by recombination of the flanking regions*.

What is the minimum length of the region required to establish the connection between the recombining duplexes? Experiments in which short homologous sequences carried by plasmids or phages are introduced into bacteria suggest that the rate of recombination is substantially reduced if the homologous region is <75 bp. This distance is appreciably longer than the ~10 bp required for association between complementary single-stranded regions, which suggests that recombination imposes demands beyond mere annealing of complements.

Figure 33.4

Resolution of a Holliday junction can generate parental or recombinant duplexes, depending on which strands are nicked. Both types of product have a region of heteroduplex DNA.

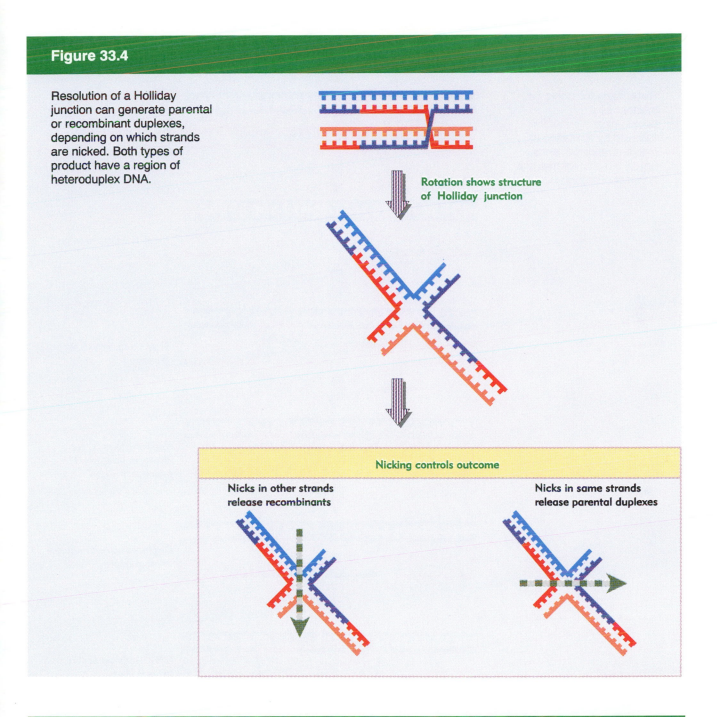

Rotation shows structure of Holliday junction

Nicking controls outcome

Nicks in other strands release recombinants

Nicks in same strands release parental duplexes

Do double-strand breaks initiate recombination?

The general model of Figure 33.1 shows that a break must be made in one duplex in order to generate a point from which single strands can unwind to participate in genetic exchange. Both strands of a duplex must be broken to accomplish a genetic exchange. Figure 33.2 shows a model in which individual breaks in single strands occur successively. However, the current model for recombination

Figure 33.5

Recombination could be initiated by a double-strand break, followed by formation of single-stranded 3' ends, one of which migrates to a homologous duplex.

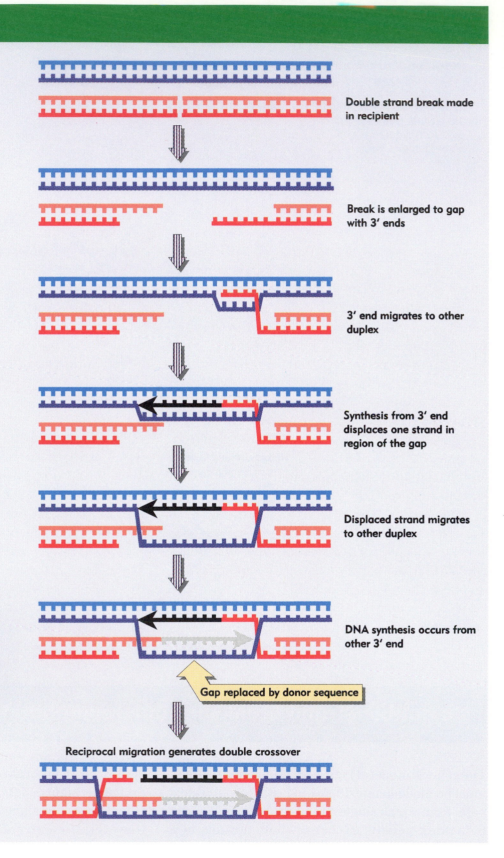

Double strand break made in recipient

Break is enlarged to gap with 3' ends

3' end migrates to other duplex

Synthesis from 3' end displaces one strand in region of the gap

Displaced strand migrates to other duplex

DNA synthesis occurs from other 3' end

Gap replaced by donor sequence

Reciprocal migration generates double crossover

supposes that *genetic exchange is initiated by a double-strand break*. The model is illustrated in **Figure 33.5**.

Recombination is initiated by an endonuclease that cleaves one of the partner DNA duplexes, the 'recipient'. The cut is enlarged to a gap by exonuclease action. The exonuclease(s) nibble away one strand on either side of the break, generating 3′ single-stranded termini. One of the free 3′ ends then invades a homologous region in the other, 'donor' duplex. The formation of heteroduplex DNA generates a D-loop, in which one strand of the donor duplex is displaced. The D-loop is extended by repair synthesis, using the free 3′ end as a primer.

Eventually the D-loop becomes large enough to correspond to the entire length of the gap on the recipient chromatid. When the extruded single strand reaches the far side of the gap, the complementary single-stranded sequences anneal. Now there is heteroduplex DNA on either side of the gap, and the gap itself is represented by the single-stranded D-loop.

The duplex integrity of the gapped region can be restored by repair synthesis using the 3′ end on the left side of the gap as a primer. Overall, the gap has been repaired by two individual rounds of single strand DNA synthesis.

Branch migration converts this structure into a molecule with two recombinant joints. The joints must be resolved by cutting.

If both joints are resolved in the same way, for example, the original noncrossover molecules will be released, each with a region of altered genetic information that is a footprint of the exchange event. If the two joints are resolved in opposite ways a genetic crossover results.

The structure of the two-jointed molecule before it is resolved illustrates a critical difference between the double-strand break model and models that invoke only single-strand exchanges.

◆ Following the double-strand break, heteroduplex DNA has been formed at each end of the region involved in the exchange. *Between the two heteroduplex segments is the region corresponding to the gap, which now has the sequence of the donor DNA in both molecules* (Figure 33.5). So the arrangement of heteroduplex sequences is asymmetric, and part of one molecule has been converted to the sequence of the other (which is why the initiating chromatid is called the recipient).

◆ Following reciprocal single-strand exchange, each DNA duplex has heteroduplex material covering the region from the initial site of exchange to the migrating branch (Figure 33.2). In variants of the single-strand exchange model in which some DNA is degraded and resynthesized, the initiating chromatid is the donor of genetic information.

The double-strand break model does not reduce the importance of the formation of heteroduplex DNA, which remains the only plausible means by which two duplex molecules can interact. However, by shifting the responsibility for initiating recombination from single-strand to double-strand breaks, it influences our perspective about the ability of the cell to manipulate DNA.

The involvement of double-strand breaks seems surprising at first sight. Once a break has been made right across a DNA molecule, there is no going back. Compare the events of Figures 33.2 and 33.5. In the single-strand exchange model, at no point has any information been lost. But in the double-strand break model, the initial cleavage is immediately followed by loss of information. Any error in retrieving the information could be fatal. On the other hand, the very ability to retrieve lost information by resynthesizing it from another duplex provides a major safety net for the cell.

Double-strand breaks may initiate synapsis

A basic paradox in recombination is that the parental chromosomes never seem to be in close enough contact for recombination of DNA to occur. The chromosomes enter meiosis in the form of replicated (sister chromatid) pairs, visible as a mass of chromatin. They pair to form the synaptonemal complex, and it has been assumed for many years that this represents some stage involved with recombination, possibly a necessary preliminary to exchange of DNA. A more recent view is that the synaptonemal complex is a consequence rather than a cause of recombination. However, we do not understand in either case how the structure of the synaptonemal complex relates to molecular contacts between DNA molecules.

Synapsis begins when each chromosome (sister chromatid pair) condenses around a structure called the **axial element**, which is apparently proteinaceous. Then the axial elements of corresponding chromosomes become aligned, and the synaptonemal complex forms as a tripartite structure, in which the axial elements, now called **lateral elements**, are separated from each other by a **central element**. **Figure 33.6** shows an example.

Each chromosome at this stage appears as a mass of chromatin bounded by a lateral element (which

Figure 33.6

The synaptonemal complex brings chromosomes into juxtaposition. This example of *Neotellia* was kindly provided by M. Westergaard and D. Von Wettstein.

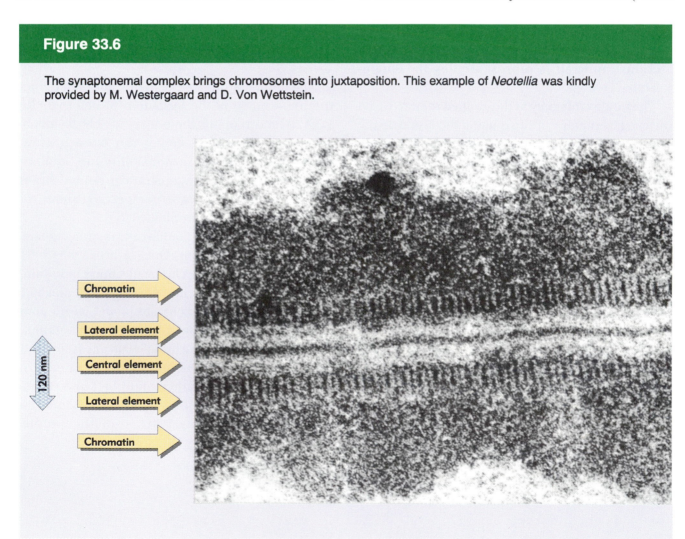

Figure 33.7

Double-strand breaks appear when axial elements form, and disappear during the extension of synaptonemal complexes. DNA recombinants can be detected at the end of pachytene. Time is measured from the end of the replication that precedes meiosis.

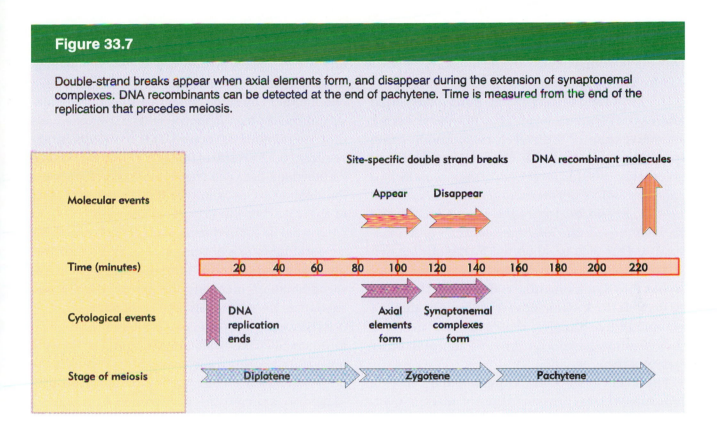

in this case has a striated structure). The two lateral elements are separated from each other by a fine but dense central element. The triplet of parallel dense strands lies in a single plane that curves and twists along its axis. The distance between the homologous chromosomes is considerable in molecular terms, more than 200 nm (the diameter of DNA is 2 nm). Thus a major problem in understanding the role of the complex is that, although it aligns homologous chromosomes, it is far from bringing homologous DNA molecules into contact.

The only visible link between the two sides of the synaptonemal complex is provided by spherical or cylindrical structures observed in fungi and insects. They lie across the complex and are called **nodes** or **recombination nodules**; they occur with the same frequency and distribution as the chiasmata. Their name reflects the hope that they may prove to be the sites of recombination.

The correlation between recombination and synaptonemal complex formation is well established, and recent work has shown that all muta-

tions that abolish chromosome pairing in *Drosophila* or in yeast also prevent recombination. There are few systems in which it is possible to compare molecular and cytological events at recombination, but recently there has been progress in analyzing meiosis in *S. cerevisiae*. The relative timing of events is summarized in **Figure 33.7**.

There is good evidence in yeast that double strands initiate recombination in both homologous and site-specific recombination. Double-strand breaks were initially implicated in the change of mating type, which involves the replacement of one sequence by another (discussed in detail in Chapter 36). Double-strand breaks also occur early in meiosis at specific sites that provide hot spots for recombination; recombination is high in the vicinity of such a site, and the frequency of recombination declines in a gradient on one or both sides of it. The flush ends created by the double-strand break are rapidly converted on both sides into long 3′ single-stranded ends, as shown in the model of Figure 33.5. A yeast mutation (*rad50*) that blocks the conversion of the flush end into the single-stranded protrusion is defective in recombination. This suggests that double-strand breaks are necessary for recombination.

Double-strand breaks appear during the period when axial elements form. They disappear during the conversion of the paired chromosomes into synaptonemal complexes. This relative timing of events suggests that *formation of the synaptonemal complex results from the initiation of recombination via the introduction of double-strand breaks and their conversion into later intermediates of recombination.* This idea is supported by the observation that the *rad50* mutant cannot convert axial elements into synaptonemal complexes. This contrasts with the traditional view of meiosis that the synaptonemal complex represents the need for chromosome pairing to precede the molecular events of recombination.

The generation of the synaptonemal complex coincides with the presumed time of crossing-over. It has been difficult to provide direct evidence that recombination occurs at the stage of synapsis, because recombination is assessed by the appearance of recombinants after the completion of meiosis. However, by assessing the appearance of recombinants in yeast directly in terms of the production of DNA molecules containing diagnostic restriction sites, it has been possible to show that recombinants appear at the end of pachytene. This clearly places the completion of the recombination event after the formation of synaptonemal complexes.

The synaptonemal complex may occupy an intermediate place in the process of recombination. After homologous chromosomes have identified one another (by some unknown means), double-strand breaks may generate single-stranded ends that engage in a search for complementary sequences. The result of this search may be to make connections between the homologues that bring them into close proximity, initiating the formation of the tripartite synaptonemal complex. It is not clear what happens during pachytene, before DNA recombinants are observed. It may be that this period is occupied by the subsequent steps of recombination, involving the extension of strand exchange, DNA synthesis, and resolution.

At the next stage of meiosis (diplotene), the chromosomes shed the synaptonemal complex; then the chiasmata become visible as points at which the chromosomes are connected. This has been presumed to indicate the occurrence of a genetic exchange, but the molecular nature of a chiasma is unknown. It is possible that it represents the residuum of a completed exchange, or that it represents a connection between homologous chromosomes where a genetic exchange has not yet been resolved. Later in meiosis, the chiasmata move toward the ends of the chromosomes. This flexibility suggests that they represent some remnant of the recombination event, rather than providing the actual intermediate.

Bacterial recombination involves single-strand assimilation

To analyze the nature of the events involved in exchange of sequences between DNA molecules, we must turn to bacterial systems. Here the recognition reaction is part and parcel of the recombination mechanism and involves restricted regions of DNA molecules rather than intact chromosomes. But the general order of molecular events is similar: a single strand from a broken molecule interacts with a partner duplex; the region of pairing is extended; and an endonuclease resolves the partner duplexes. Enzymes involved in each stage are known, although they probably represent only some of the components required for recombination.

Bacterial enzymes implicated in recombination have been identified by the occurrence of *rec*⁻ mutations in their genes. The phenotype of Rec⁻

mutants is the inability to undertake generalized recombination. Some 10–20 loci have been identified.

Bacteria do not usually exchange large amounts of duplex DNA, and there may be various routes to initiate recombination in prokaryotes. In some cases, DNA may be available with free single-stranded 3′ ends; DNA may be provided in single-stranded form (as in conjugation, discussed in Chapter 18), single-stranded gaps may be generated by irradiation damage, single-stranded tails may be generated by phage genomes undergoing replication by a rolling circle. However, in circumstances involving two duplex molecules (which is certainly the rule in recombination at meiosis in eukaryotes), single-stranded regions and 3′ ends must be generated.

One mechanism for generating suitable ends has been discovered as a result of the existence of certain hotspots that stimulate recombination. They were discovered in phage lambda in the form of mutants, called *chi*, that have single base-pair changes creating sites that stimulate recombination. These sites lead us to the role of other proteins involved in recombination.

These sites share a constant nonsymmetrical sequence of 8 bp:

$$5'\ \text{GCTGGTGG}\ 3'$$
$$3'\ \text{CGACCACC}\ 5'$$

The *chi* sequence occurs naturally in *E. coli* DNA about once every 5–10 kb. Its absence from wild-type lambda DNA, and also from other genetic elements, shows that it is not essential for recombination.

A *chi* sequence stimulates recombination in its general vicinity, say within a distance of up to 10 kb from the site. A *chi* site can be activated by a double-strand break made several kilobases away *on one particular side* (to the right of the sequence as written above). This dependence on orientation suggests that the recombination apparatus must associate with DNA at a broken end, and then can move along the duplex in only one direction.

Chi sequences identify targets for the enzyme exonuclease V, whose subunits are the products of the *recBCD* genes. This enzyme exercises several activities. It is a potent nuclease that degrades DNA; it can unwind duplex DNA in the presence of SSB; and it has an ATPase activity. Its role in recombination may be to provide a single-stranded region with a free 3′ end.

When exonuclease V binds DNA on the right site of *chi*, it moves along unwinding the DNA. When it reaches the *chi* site, it cleaves one (the top) strand of the DNA at a position between 4 and 6 bases on the right. The nuclease's action is depicted in **Figure 33.8** as rolling along toward the *chi* site and stopping to cleave when it arrives.

This model is based on data obtained from *in vitro* reactions; we do not know exactly how these reactions relate to recombination *in vivo*, but it seems likely that RecBCD-mediated unwinding and cleavage is involved in generating ends that initiate the formation of heteroduplex joints.

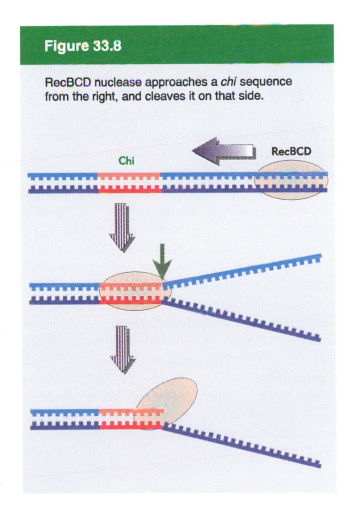

Figure 33.8

RecBCD nuclease approaches a *chi* sequence from the right, and cleaves it on that side.

Figure 33.9

RecA promotes the assimilation of invading single strands into duplex DNA so long as one of the reacting strands has a free end.

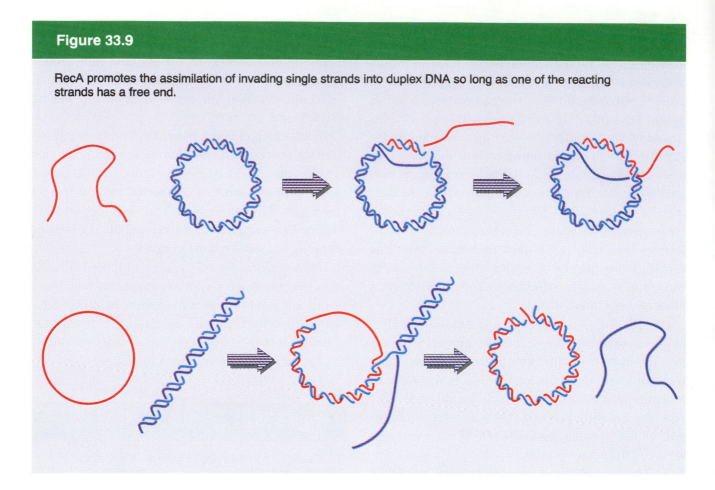

The interaction between DNA molecules is promoted by the enzyme RecA, which has two quite different types of activity: it can stimulate protease activity in the SOS response (see Chapter 20); and can promote base pairing between a single strand of DNA and its complement in a duplex molecule.

RecA requires single-stranded DNA and ATP for ability to stimulate protease activity. The same substrates are required for its ability to manipulate DNA molecules. It is not yet clear exactly how the enzymatic activities of RecA are related to recombination *in vivo*, but they involve several reactions that provide useful paradigms for recombination mechanisms.

The DNA handling activity of RecA enables a single strand to displaces its homolog in a duplex in a reaction that is called **single-strand uptake** or **single-strand assimilation**. The displacement reaction can occur between DNA molecules in several configurations and has three general conditions:

◆ One of the DNA molecules must have a single-stranded region.

◆ One of the molecules must have a free 3′ end.

◆ The single-stranded region and the 3′ end must be located within a region that is complementary between the molecules.

Two forms of the reaction are illustrated in **Figure 33.9**. When a linear single strand invades a circular duplex, it displaces the original partner to its complement, forming a D-loop. When a circular single strand reacts with a linear duplex, the exchange of strands generates an open circular molecule. In both cases, the reaction proceeds 5'–3' along the strand whose partner is being displaced and replaced, that is, the reaction involves an exchange in which (at least) one of the exchanging strands has a 3' end.

Single-strand assimilation is potentially related to the initiation of recombination. All models call for an intermediate in which one or both single strands cross over from one duplex to the other (see Figures 33.2 and 33.5). RecA could catalyze this stage of the reaction.

A mechanism for the activity of RecA in stimulating branch migration is suggested by its ability to aggregate into long filaments with single-stranded or duplex DNA. There are 6 RecA monomers per turn of the filament, which has a helical structure with a deep groove that contains the DNA. The stoichiometry of binding is 3 nucleotides (or base pairs) per RecA monomer. The DNA is held in a form that is extended 1.5 times relative to duplex B DNA, making a turn every 18.6 nucleotides (or base pairs). When duplex DNA is bound, it contacts RecA via its minor groove, leaving the major groove accessible for possible reaction with a second DNA molecule.

The interaction between two DNA molecules occurs within these filaments. When a single strand is assimilated into a duplex, the first step is for RecA to bind the single strand into a filament. Then the duplex is incorporated, probably forming some sort of triple-stranded structure. In this system, synapsis precedes physical exchange of material, because the pairing reaction can take place even in the absence of free ends, when strand exchange is impossible.

A free 3' end is required for strand exchange. The reaction occurs within the filament, and RecA remains bound to the strand that was originally single, so that at the end of the reaction RecA is bound to the duplex molecule. Large amounts of ATP are hydrolyzed during the reaction. The ATP may act through an allosteric effect on RecA conformation.

When bound to ATP, the DNA binding site of RecA has a high affinity for DNA; this is needed to bind DNA and for the pairing reaction. Hydrolysis of ATP converts the binding site to low affinity, which is needed to release the heteroduplex DNA.

We can divide the reaction that RecA catalyzes between single-stranded and duplex DNA into three phases:

◆ a slow presynaptic phase in which RecA polymerizes on single-stranded DNA;

◆ a fast pairing reaction between the single-stranded DNA and its complement in the duplex to produce a heteroduplex joint;

◆ a slow displacement of one strand from the duplex to produce a long region of heteroduplex DNA.

The presence of SSB (single-strand binding protein) stimulates the reaction, by ensuring that the substrate lacks secondary structure. It is not clear yet how SSB and RecA both can act on the same stretch of DNA. Like SSB, RecA is required in stoichiometric amounts, which suggests that its action in strand assimilation involves binding cooperatively to DNA to form a structure related to the filament.

When a single-stranded molecule reacts with a duplex DNA, the duplex molecule becomes unwound in the region of the recombinant joint. The initial region of heteroduplex DNA may not even lie in the conventional double helical form, but could consist of the two strands associated side by side. A region of this type is called a **paranemic joint** (compared with the classical intertwined **plectonemic** relationship of strands in a double helix). A paranemic joint is unstable; further progress of the reaction requires its conversion to the double-helical form. This reaction is equivalent to removing negative supercoils and may require an enzyme that solves the unwinding/rewinding problem by making transient breaks that allow the strands to rotate about each other (see later).

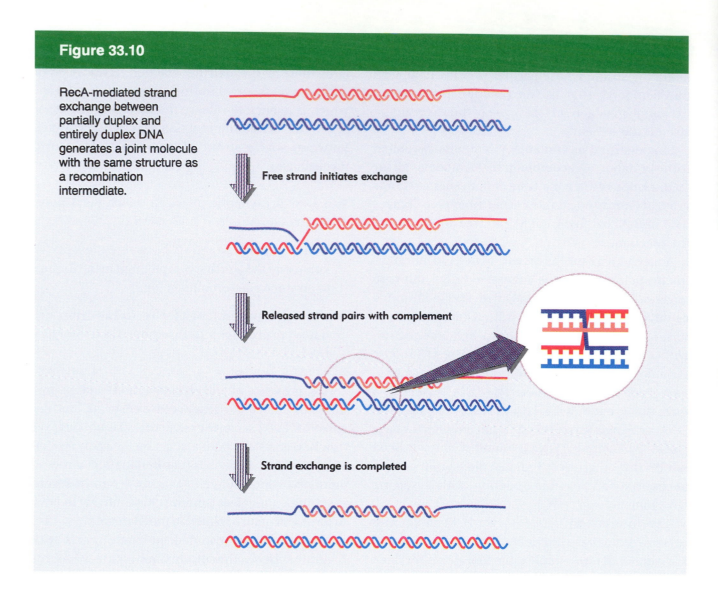

Figure 33.10

RecA-mediated strand exchange between partially duplex and entirely duplex DNA generates a joint molecule with the same structure as a recombination intermediate.

Free strand initiates exchange

Released strand pairs with complement

Strand exchange is completed

All of the reactions we have discussed so far represent only a part of the potential recombination event: the invasion of one duplex by a single strand. Two duplex molecules can interact with each other under the sponsorship of RecA, provided that one of them has a single-stranded region of at least 50 bases. The single-stranded region can take the form of a tail on a linear molecule or of a gap in a circular molecule.

The reaction between a partially duplex molecule and an entirely duplex molecule leads to the exchange of strands. An example is illustrated in **Figure 33.10**. Assimilation starts at one end of the linear molecule, where the invading single strand displaces its homologue in the duplex in the customary way. But when the reaction reaches the region that is duplex in both molecules, the invading strand unpairs from its partner, which then pairs with the other displaced strand.

At this stage, the molecule has a structure indistinguishable from the recombinant joint in Figure 33.4. The reaction sponsored *in vitro* by RecA can generate Holliday junctions, which suggests that the enzyme can mediate reciprocal strand transfer. We know less about the geometry of four-strand intermediates bound by RecA, but presumably two duplex molecules can lie side by side in a way consistent with the requirements of the exchange reaction.

Figure 33.11

Bacterial enzymes can catalyze all stages of recombination in the repair pathway following the production of suitable substrate DNA molecules.

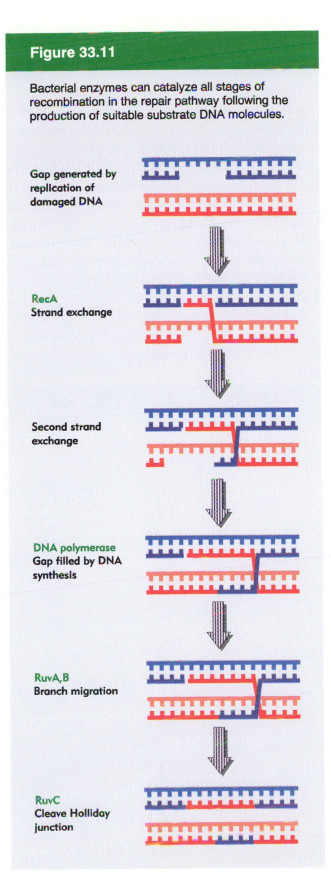

Gap generated by replication of damaged DNA

RecA
Strand exchange

Second strand exchange

DNA polymerase
Gap filled by DNA synthesis

RuvA,B
Branch migration

RuvC
Cleave Holliday junction

A group of three genes in *E. coli* codes for functions involved later in recombination. The products of *ruvA* and *ruvB* increase the formation of heteroduplex structures that RecA catalyzes between one duplex and another with a single-stranded region. RuvA recognizes the structure of the Holliday junction; RuvB is an ATPase that may provide the motor for branch migration. The branch may migrate as fast as 10–20 bp/second.

The third gene, *ruvC*, codes for an endonuclease that specifically recognizes Holliday junctions. It can cleave such junctions *in vitro* to resolve recombination intermediates. Resolution is independent of RecA. It is not clear how the Ruv proteins gain access to DNA within the filament formed by RecA.

We may now account for the stages of recombination in *E. coli* in terms of individual proteins. **Figure 33.11** shows the events that are involved in using recombination to repair a gap in one duplex by retrieving material from the other duplex. The major caveat in applying these conclusions to recombination in eukaryotes is that these events in bacteria largely involve interaction between a fragment of DNA and a whole chromosome, and occur as repair reactions that are stimulated by damage to DNA; they are not entirely equivalent to recombination between genomes at meiosis. Nonetheless, the same molecular activities are involved in manipulating DNA.

Homologues of RecA are ubiquitous among prokaryotes, and related proteins have been found in eukaryotes. Two genes in *S. cerevisiae*, *DMC1* and *RAD51*, code for proteins that are related to RecA. Mutations in these genes cause a similar phenotype; they accumulate double-strand breaks and fail to form normal synaptonemal complexes. This reinforces the idea that exchange of strands between DNA duplexes is involved in formation of the synaptonemal complex, and raises the possibility that chromosome synapsis is related to the bacterial strand assimilation reaction. However, eukaryotic homologs of RecA do not form filaments, so the mechanics of the reaction are likely to be different in eukaryotes.

Gene conversion accounts for interallelic recombination

The involvement of heteroduplex DNA explains the characteristics of recombination between alleles; indeed, allelic recombination provided the impetus for the development of the heteroduplex model. When recombination between alleles was discovered, the natural assumption was that it takes place by the same mechanism of reciprocal recombination that applies to more distant loci. That is to say that an individual breakage and reunion event occurs within the locus to generate a reciprocal pair of recombinant chromosomes. However, in the close quarters of a single gene, the formation of heteroduplex DNA itself is usually responsible for the recombination event.

Individual recombination events can be studied in the Ascomycetes fungi, because the products of a single meiosis are held together in a large cell, the ascus. Even better, the four haploid nuclei produced by meiosis are arranged in a linear order. Actually, a mitosis occurs after the production of these four nuclei, giving a linear series of eight haploid nuclei. **Figure 33.12** shows that each of these nuclei effectively represents the genetic character of one of the eight strands of the four chromosomes produced by the meiosis.

Meiosis in a heterozygote should generate four copies of each allele. This is seen in the majority of spores. But there are some spores with abnormal

Figure 33.12

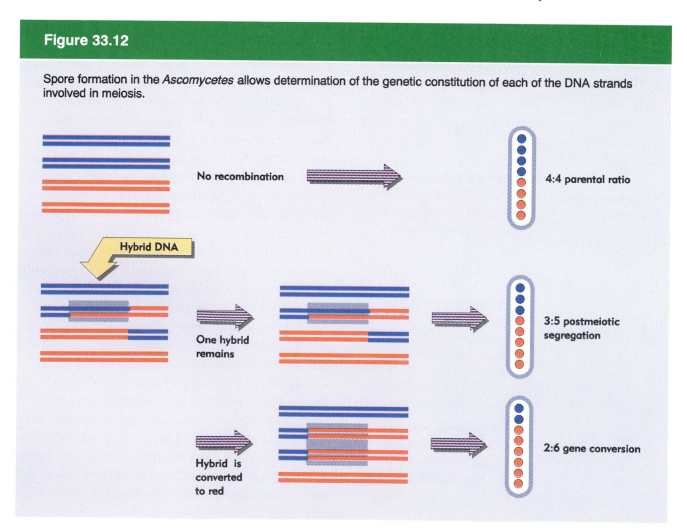

Spore formation in the *Ascomycetes* allows determination of the genetic constitution of each of the DNA strands involved in meiosis.

No recombination — 4:4 parental ratio

Hybrid DNA

One hybrid remains — 3:5 postmeiotic segregation

Hybrid is converted to red — 2:6 gene conversion

ratios. They are explained by the formation and correction of heteroduplex DNA in the region in which the alleles differ. The figure illustrates a recombination event in which a length of hybrid DNA occurs on one of the four meiotic chromosomes, a possible outcome of recombination initiated by a double-strand break.

Suppose that two alleles differ by a single point mutation. When a strand exchange occurs to generate heteroduplex DNA, the two strands of the heteroduplex will be mispaired at the site of mutation. In effect, each strand of DNA carries different genetic information. If no change is made in the sequence, the strands separate at the ensuing replication, each giving rise to a duplex that perpetuates its information. This event is called **postmeiotic segregation**, because it reflects the separation of DNA strands after meiosis. Its importance is that *it demonstrates directly the existence of heteroduplex DNA in recombining alleles*.

Another effect is seen when examining recombination between alleles: *the proportions of the alleles differ from the initial 4:4 ratio*. This effect is called **gene conversion**. It describes a *nonreciprocal transfer of information from one chromatid to another*.

Gene conversion results from exchange of strands between DNA molecules, and the change in sequence may have either of two causes at the molecular level:

◆ As indicated by the double-strand break model in Figure 33.5, one DNA duplex may act as a donor of genetic information that directly replaces the corresponding sequences in the recipient duplex by a process of strand exchange and gap filling.

◆ As part of the exchange process, heteroduplex DNA is generated when a single strand from one duplex pairs with its complement in the other duplex. When a repair system recognizes mispaired bases in heteroduplex DNA, it may excise and replace one of the strands to restore complementarity. Such an event changes the strand of DNA representing one allele into the sequence of the other allele.

Gene conversion does not depend on crossing-over, but is correlated with it. A large proportion of the aberrant asci show genetic recombination between two markers on either side of a site of interallelic gene conversion. This is exactly what would be predicted if the aberrant ratios result from initiation of the recombination process as shown in Figure 33.2 or 33.5, but with an approximately equal probability of resolving the structure with or without recombination (as indicated in Figure 33.4). The implication is that fungal chromosomes initiate crossing-over about twice as often as would be expected from the frequency of recombination between distant genes.

Various biases are seen when recombination is examined at the molecular level. Either direction of gene conversion may be equally likely, or allele-specific effects may create a preference for one direction. Gradients of recombination may fall away from hotspots. We now know that hotspots represent sites at which double-strand breaks are initiated, and the gradient is correlated with the extent to which the gap at the hotspot is enlarged and converted to long single-stranded ends (as discussed previously).

Some information about the extent of gene conversion is provided by the sequences of members of gene clusters. Usually, the products of a recombination event will separate and become unavailable for analysis at the level of DNA sequence. However, if an unequal exchange takes place between two members of a gene cluster, as illustrated previously in Figure 24.5, a heteroduplex may be formed between the two nonallelic genes. Gene conversion effectively converts one of the nonallelic genes to the sequence of the other.

The presence of more than one gene copy on the same chromosome provides a footprint to trace these events. For example, if heteroduplex formation and gene conversion occurred over part of one gene, this part may have a sequence identical with or very closely related to the other gene, while the remaining part shows more divergence. Available sequences suggest that gene conversion events may extend for considerable distances, up to a few thousand bases.

Topological manipulation of DNA

Topological manipulation of DNA is a central aspect of all its functional activities—recombination, replication, and (perhaps) transcription—as well as of the organization of higher order structure. In considering these processes, we might consider the duplex structure of DNA to be an obstacle that must be overcome by any reaction involving strand separation.

All synthetic activities involving double-stranded DNA require the strands to separate. However, the strands do not simply lie side by side; they are intertwined. Their separation therefore requires the strands to rotate about each other in space. Some possibilities for the unwinding reaction are illustrated in **Figure 33.13**.

We might envisage the structure of DNA in terms of a free end that would allow the strands to rotate about the axis of the double helix for unwinding. Given the length of the double helix, however, this would involve the separating strands in a considerable amount of flailing about, which seems unlikely in the confines of the cell.

A similar result is achieved by placing an apparatus to control the rotation at the free end. However, the effect must be transmitted over a considerable distance, again involving the rotation of an unreasonable length of material.

DNA actually behaves as a closed structure lacking free ends, which excludes these models as a matter of principle and brings home the severity of the topological problem. Consider the effects of separating the two strands in a molecule whose ends are not free to rotate. When two intertwined strands are pulled apart from one end, the result is to *increase their winding about each other farther along the molecule.* Thus movement of a replication fork would generate increasing positive supercoiling ahead of it, rapidly generating insuperable resistance to further movement. (Similar consequences ensue during transcription, as described in twin-domain supercoiling model summarized in Figure 14.17.)

The problem can be overcome by introducing a transient nick in one strand. An internal free end allows the nicked strand to rotate about the intact strand, after which the nick can be sealed. Each repetition of the nicking and sealing reaction releases one superhelical turn.

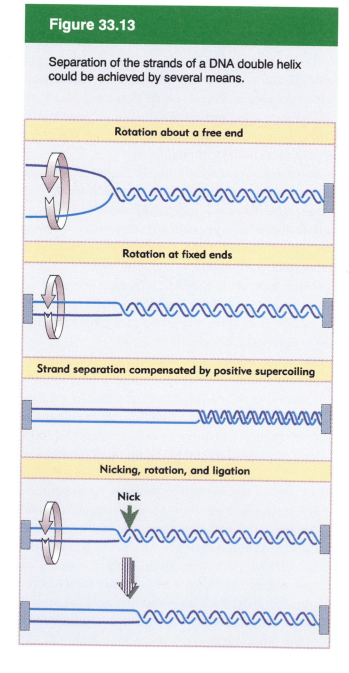

Figure 33.13

Separation of the strands of a DNA double helix could be achieved by several means.

Figure 33.14

Bacterial type I topoisomerases recognize partially unwound segments of DNA and pass one strand through a break made in the other.

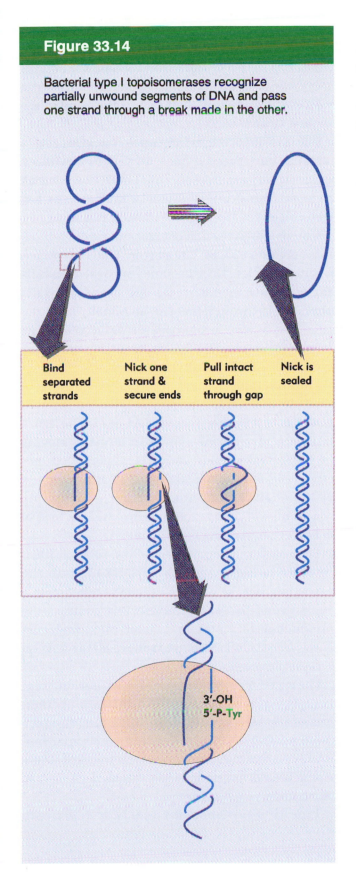

Bind separated strands

Nick one strand & secure ends

Pull intact strand through gap

Nick is sealed

3'-OH
5'-P-Tyr

Recall from Chapter 5 that a closed molecule of DNA can be characterized by its linking number, the number of times one strand crosses over the other in space. Closed DNA molecules of identical sequence may have different linking numbers, reflecting different degrees of supercoiling. Molecules of DNA that are the same except for their linking numbers are called **topological isomers**.

The linking number comprises the sum of the writhing number (W) and the twisting number (T), so that a change in linking number is given by $\Delta L = \Delta W + \Delta T$. In looser terms, a change in linking number is the sum of changes in the coiling of the axis of the duplex in space (ΔW, equivalent to the supercoiling) and the screwing of the double helix itself (ΔT). In a free DNA molecule, W and T are freely adjustable, and a change in linking number is likely to be expressed by a change in W, that is, by a change in supercoiling.

Any change in the linking number requires at least one strand to be broken. Using the free end, one strand can be rotated about the other, after which the break is made good. Such a reaction converts one topological isomer into another. **DNA topoisomerases** catalyze conversions of this type. Some topoisomerases can relax (remove) only negative supercoils from DNA; others can relax both negative and positive supercoils. Some can introduce negative supercoils.

Topoisomerases are divided into two classes, according to the nature of the mechanisms they employ. **Type I** enzymes act by making a transient break in one strand of DNA. **Type II** enzymes act by introducing a transient double-strand break. As well as those enzymes that function as general topoisomerases with DNA irrespective of sequence, enzymes involved in site-specific recombination reactions fit the definition of topoisomerases (see later).

The best characterized type I topoisomerase is the product of the *topA* gene of *E. coli*, which relaxes highly negatively supercoiled DNA. The enzyme does not act on positively supercoiled DNA. Mutations in it cause an increase in the level of

Figure 33.15

Type II topoisomerases can pass a duplex DNA through a double strand break in another duplex.

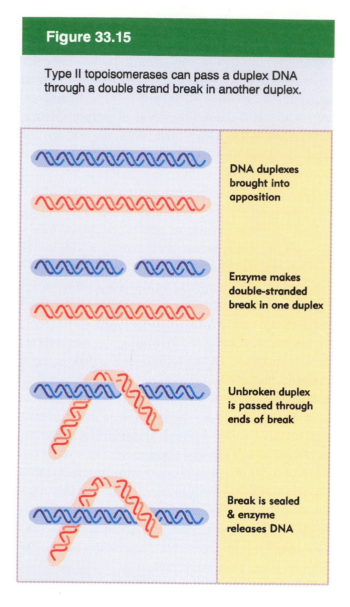

DNA duplexes brought into apposition

Enzyme makes double-stranded break in one duplex

Unbroken duplex is passed through ends of break

Break is sealed & enzyme releases DNA

super-coiling in the nucleoid (and may affect transcription, as described in Chapter 14).

In addition to the relaxation of negative super-coils in duplex DNA, the enzyme interacts with single-stranded DNA. It may like negative super-coils because they tend to stabilize single-stranded regions, which could provide the substrate bound by the enzyme.

When *E. coli* topoisomerase I binds to DNA, it forms a stable complex in which one strand of the DNA has been nicked and its 5'-phosphate end is covalently linked to a tyrosine residue in the enzyme. This suggests a mechanism for the action of the enzyme; it transfers a phosphodiester bond in DNA to the protein, manipulates the structure of the two DNA strands, and then rejoins the bond in the original strand.

The transfer of bonds from nucleic acid to protein explains how the enzyme can function without requiring any input of energy. There has been no irreversible hydrolysis of bonds; their energy has been conserved through the transfer reactions.

A model for the action of topoisomerase I is illustrated in **Figure 33.14**. The enzyme binds to a region in which duplex DNA becomes separated into its single strands; then it breaks one strand, pulls the other strand through the gap, and finally seals the gap. The formal properties of eukaryotic type I topoisomerases are similar, but they can relax positive as well as negative supercoils.

The reaction changes the linking number in steps of 1. Each time one strand is passed through the break in the other, there is a ΔL of +1. The figure illustrates the enzyme activity in terms of moving the individual strands; in a free supercoiled molecule, the interchangeability of W and T should let the change in linking number be taken up by a change of $\Delta W = +1$, that is, by one less turn of negative supercoiling (see Chapter 5).

The reaction is equivalent to the rotation illustrated in bottom part of Figure 33.13, with the restriction that the enzyme limits the reaction to a single strand-passage per event. (By contrast, the introduction of a nick in a supercoiled molecule allows free strand rotation to relieve all the tension by multiple rotations.)

The type I topoisomerase also can pass one segment of a single-stranded DNA through another. This **single-strand passage** reaction can introduce **knots** in DNA and can **catenate** two circular molecules so that they are connected like links on a chain. We do not understand the uses (if any) to which these reactions are put *in vivo*.

Type II topoisomerases generally relax both

negative and positive supercoils. The reaction requires ATP; probably one ATP is hydrolyzed for each catalytic event. As illustrated in **Figure 33.15**, the reaction is mediated by making a double-stranded break in one DNA duplex, and passing another duplex region through it.

A formal consequence of two-strand transfer is that the linking number is always changed in multiples of two. The topoisomerase II activity can be used also to introduce or resolve catenated duplex circles and knotted molecules.

The reaction probably represents a nonspecific recognition of duplex DNA in which the enzyme binds any two double-stranded segments that cross each other. The hydrolysis of ATP may be used to drive the enzyme through conformational changes that provide the force needed to push one DNA duplex through the break made in the other. Because of the topology of supercoiled DNA, the relationship of the crossing segments allows supercoils to be removed from either positively or negatively supercoiled circles.

Gyrase introduces negative supercoils in DNA

Bacterial DNA gyrase is a topoisomerase of type II that is able to *introduce* negative supercoils into a relaxed closed circular molecule. DNA gyrase binds to a circular DNA duplex and supercoils it processively and catalytically: it continues to introduce supercoils into the same DNA molecule. One molecule of DNA gyrase can introduce ~100 supercoils per minute.

The supercoiled form of DNA has a higher free energy than the relaxed form, and the energy needed to accomplish the conversion is supplied by the hydrolysis of ATP. In the absence of ATP, the gyrase can *relax* negative but not positive supercoils, although the rate is more than 10 times slower than the rate of introducing supercoils.

The structure of *E. coli* DNA gyrase is summarized in **Table 33.1**. The enzyme is inhibited by two types of antibiotic, each of which acts on one of the subunits. The drugs inhibit replication, which suggests that DNA gyrase is necessary for DNA synthesis to proceed. Mutations that confer resistance to the antibiotics identify the loci that code for the subunits.

Gyrase binds its DNA substrate around the

Table 33.1

DNA gyrase is a tetramer whose subunits provide targets for different antibiotics.

Subunit	Size	Locus	Antibiotics that Act on Subunit
A	105,000 daltons	gyrA (nalA)	Nalidixic acid & oxilinic acid
B	95,000 daltons	gyrB (cou)	Coumermycin A1 & novobiocin

Figure 33.16

DNA gyrase may introduce negative supercoils in duplex DNA by inverting a positive supercoil.

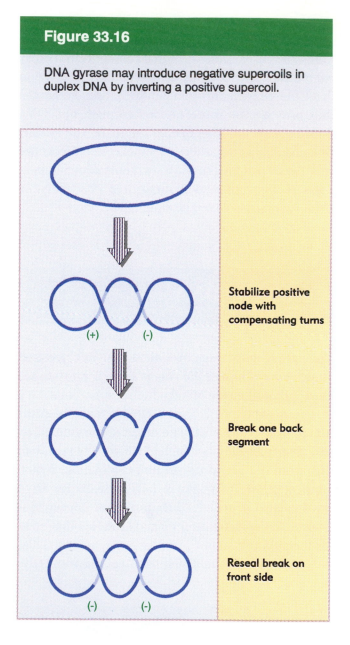

Stabilize positive node with compensating turns

Break one back segment

Reseal break on front side

ing negative supercoil in the unbound DNA. Then the enzyme breaks the double strand at the cross-over of the positive supercoil, passes the other duplex through, and seals the break.

The reaction directly inverts the sign of the supercoil: it has been converted from a +1 turn to a −1 turn. Thus the linking number has changed by $\Delta L = -2$, conforming with the demand that all events involving double-strand passage must change the linking number by a multiple of two.

Gyrase then releases one of the crossing segments of the (now negative) bound supercoil; this allows the negative turns to redistribute along DNA (as change in either T or W or both), and the cycle begins again. The same type of topological manipulation is responsible for catenation and knotting.

On releasing the inverted supercoil, the conformation of gyrase changes. For the enzyme to undertake another cycle of supercoiling, its original conformation must be restored. This process is called **enzyme turnover**. It is thought to be driven by the hydrolysis of ATP, since the replacement of ATP by an analog that cannot be hydrolyzed allows gyrase to introduce only one inversion (−2 super-coils) per substrate. Thus it does not need ATP for the supercoiling reaction, but does need it to undertake a second cycle. Novobiocin interferes with the ATP-dependent reactions of gyrase, by preventing ATP from binding to the B subunit.

The (ATP-independent) relaxation reaction is inhibited by nalidixic acid. This implicates the A subunit in the breakage and reunion reaction. Treating gyrase with nalidixic acid allows DNA to be recovered in the form of fragments generated by a staggered cleavage across the duplex. The termini all possess a free 3'-OH group and a 4-base 5' single-strand extension covalently linked to the A subunit. The covalent linkage retains the energy of the phosphate bond; this can be used to drive the sealing reaction, explaining why gyrase can undertake relaxation without ATP. The sites of cleavage are fairly specific, occurring about once every 100 bp.

outside of the protein tetramer. Gyrase protects ~140 bp of DNA from digestion by micrococcal nuclease (very similar to the protection afforded by the much smaller histone octamer).

The **sign inversion** model for gyrase action is illustrated in **Figure 33.16**. The enzyme binds the DNA in a crossover configuration that is equivalent to a positive supercoil. This induces a compensat-

Specialized recombination involves breakage and reunion at specific sites

The conversion of lambda DNA between its different life forms involves two types of event. The pattern of gene expression is regulated as described in Chapter 17. And the physical condition of the DNA is different in the lysogenic and lytic states:

◆ In the lytic life-style, lambda DNA exists as an independent, circular molecule in the infected bacterium.

◆ In the lysogenic state, the phage DNA is an integral part of the bacterial chromosome (called prophage).

Transition between these states involves site-specific recombination:

◆ To enter the lysogenic condition, free lambda DNA must be **integrated** into the host DNA.

◆ To be released from lysogeny into the lytic cycle, prophage DNA must be **excised** from the chromosome.

Integration and excision occur by recombination at specific loci on the bacterial and phage DNAs called **attachment (att) sites**. The attachment site on the bacterial chromosome is called att^λ in bacterial genetics. The locus is defined by mutations that prevent integration of lambda; it is occupied by prophage λ in lysogenic strains. When the att^λ site is deleted from the *E. coli* chromosome, an infecting lambda phage can establish lysogeny by integrating elsewhere, although the efficiency of the reaction is less than 0.1% of the frequency of integration at att^λ. This inefficient integration occurs at **secondary attachment sites**, which resemble the authentic *att* sequences.

For describing the integration/excision reactions, the bacterial attachment site (att^λ) is called *attB*, consisting of the sequence components *BOB'*. The attachment site on the phage, *attP*, consists of the components *POP'*. **Figure 33.17** outlines the recombination reaction between these sites. The sequence *O* is common to *attB* and *attP*. It is called the **core** sequence; and the recombination event occurs within it. The flanking regions *B, B'* and *P, P'* are referred to as the **arms**; each is distinct in sequence. Because the phage DNA is circular, the recombination event inserts it into the bacterial chromosome as a linear sequence. The prophage is bounded by two new *att* sites, the products of the recombination.

An important consequence of the constitution of the *att* sites is that the integration and excision reactions do not involve the same pair of reacting sequences. Integration requires recognition between *attP* and *attB*; while excision requires recognition between *attL* and *attR*. The directional character of site-specific recombination is thus controlled by the identity of the recombining sites.

Although the recombination event is reversible, different conditions prevail for each direction of the reaction. This is an important feature in the life of the phage, since it offers a means to ensure that an integration event is not immediately reversed by an excision, and *vice versa*.

The difference in the pairs of sites reacting at integration and excision is reflected by a difference in the proteins that mediate the two reactions:

◆ Integration (*attB* × *attP*) requires the product of the phage gene *int* and a bacterial protein called integration host factor (IHF).

Figure 33.17

Circular phage DNA is converted to an integrated prophage by a reciprocal recombination between *attP* and *attB*; the prophage is excised by reciprocal recombination between *attL* and *attR*.

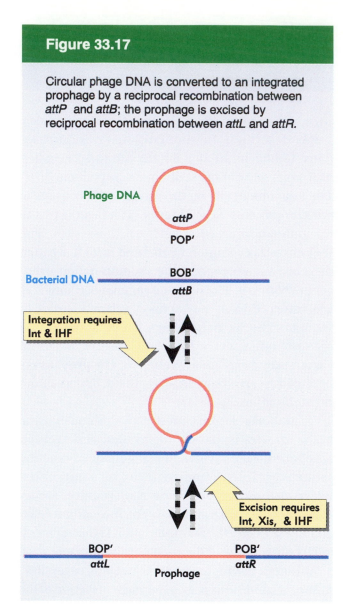

is not an essential protein in *E. coli*, and is not required for homologous bacterial recombination. It is one of several proteins with the ability to wrap DNA on a surface. Mutations in the *him* genes prevent lambda site-specific recombination, and can be suppressed by mutations in λ *int*, which suggests that IHF and Int interact.

Site-specific recombination can be performed *in vitro* by Int and IHF. It involves a precise breakage and reunion in the absence of any synthesis of DNA. The roles of the *att* sites can be investigated by making deletions on either side. It turns out that *attP* is much larger than *attB*. The function of *attP* requires a stretch of 240 bp, but the function of *attB* can be exercised by the 23 bp fragment extending from −11 to +11, in which there are only 4 bp on either side of the core. The disparity in their sizes suggests that *attP* and *attB* play different roles in the

Figure 33.18

Does recombination between *attP* and *attB* proceed by sequential exchange or concerted cutting?

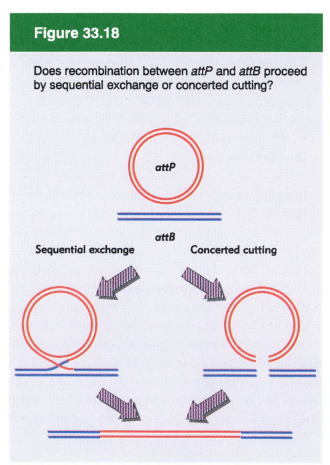

◆ Excision (*attL* × *attR*) requires the product of phage gene *xis*, in addition to Int and IHF.

Thus Int and IHF are required for *both* reactions. Xis plays an important role in controlling the direction; it is required for excision, but inhibits integration.

IHF is a 20,000 dalton protein of two different subunits, coded by the genes *himA* and *himD*. IHF

Figure 33.19

Staggered cleavages in the common core sequence of *attP* and *attB* allow crosswise reunion to generate reciprocal recombinant junctions.

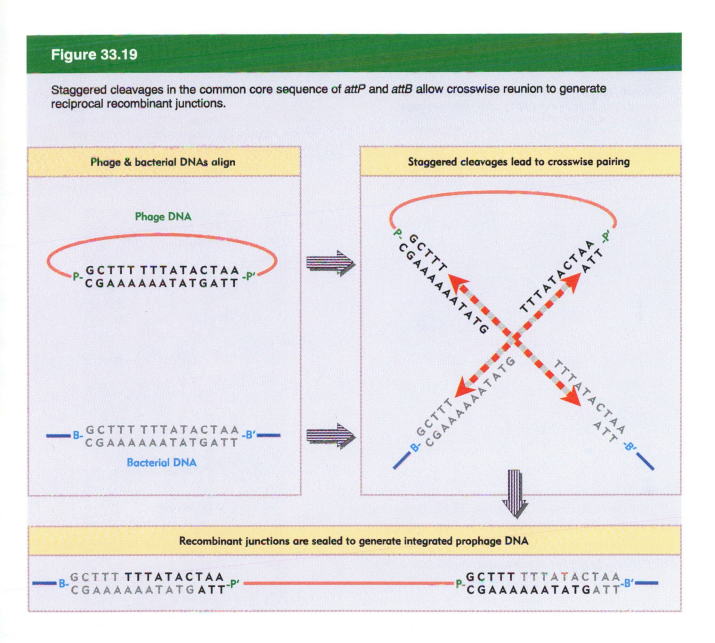

recombination, with *attP* providing additional information necessary to distinguish it from *attB*.

Does the reaction proceed by a concerted mechanism in which the strands in *attP* and *attB* are cut simultaneously and exchanged? Or are the strands exchanged one pair at a time, the first exchange generating a Holliday junction, the second cycle of nicking and ligation occurring to release the structure? The alternatives are depicted in **Figure 33.18**.

The recombination reaction has been halted at intermediate stages by the use of 'suicide substrates', in which the core sequence is nicked. The presence of the nick interferes with the recombination process. This makes it possible to identify molecules in which recombination has commenced but has not been completed. The structures of these intermediates suggest that exchanges of single strands take place sequentially. Int protein can resolve Holliday junctions, and is probably responsible for the cutting and ligation reactions.

Figure 33.20

Int and IHF bind to different sites in *attP*. The Int recognition sequences in the core region include the sites of cutting.

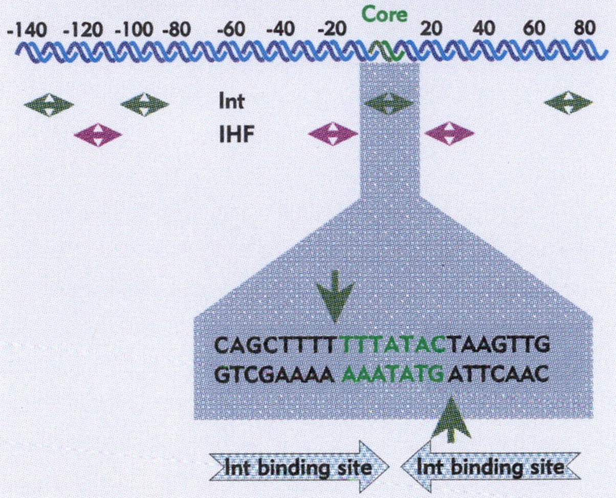

Core

-140 -120 -100 -80   -60 -40 -20   20 40 60 80

Int
IHF

CAGCTTTT TTTATACT AAGTTG
GTCGAAAA AAATATG ATTCAAC

Int binding site   Int binding site

Figure 33.21

The Int binding sites in the core lie on one face of DNA. The large circles indicate positions at which methylation is influenced by Int binding; the large arrows indicate the sites of cutting. Photograph kindly provided by A. Landy.

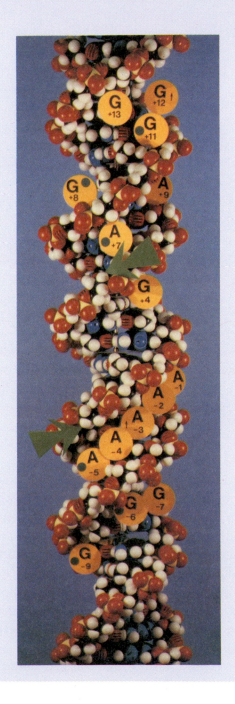

The model illustrated in **Figure 33.19** shows that if *attP* and *attB* sites each suffer the same staggered cleavage, complementary single-stranded ends could be available for crosswise hybridization. The distance between the lambda crossover points is 7 bp, and the reaction generates 3′-phosphate and 5′-OH ends. The reaction is shown for simplicity as generating overlapping single-stranded ends that anneal, but actually occurs by a process akin to the recombination event of Figure 33.2: first the corresponding strands on each duplex are cut at the same position, the free 3′ ends exchange between duplexes, the branch migrates for a distance of 7 bp along the region of homology, and then the structure is resolved by cutting the other pair of corresponding strands.

The *in vitro* reaction requires supercoiling in *attP*, but not in *attB*. When the reaction is performed *in vitro* between two supercoiled

Figure 33.22

Multiple copies of Int protein may organize *attP* into an intasome, which initiates site-specific recombination by recognizing *attB* on free DNA.

Intasome

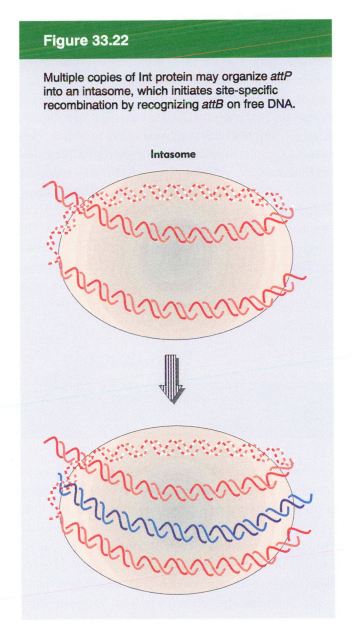

DNA molecules, almost all of the supercoiling is retained by the products. Thus there cannot be any free intermediates in which strand rotation could occur. This is consistent with the idea that the reaction proceeds through a Holliday junction. The breakage and reunion reaction resembles the activity of topoisomerase I, except that nicked strands from different duplexes are sealed together, instead of the ends of a broken strand

from one duplex. (Int indeed has a [rather ineffectual] topoisomerase I ability to relax negatively supercoiled DNA.)

Large amounts of the Int and IHF proteins are needed for recombination *in vitro*. Int and IHF bind cooperatively to *attP*, and their affinity for the site is enhanced by supercoiling. The high stoichiometry suggests that the proteins do not function catalytically, but form some structure that supports only a single recombination event.

The proteins involved in site-specific recombination bind to specific sites in the *att* region. Int has two different modes of binding. It binds to inverted sites at the core sequence, positioning itself to make the cuts on each strand illustrated in **Figure 33.20**. These sites share a consensus sequence. It also binds to sites in the arms of *attP* that have a different consensus sequence. Different domains of Int recognize each type of sequence: an N-terminal domain recognizes the arms of *attP*, while a C-terminal domain recognizes the cores of *attP* and *attB*. The two domains probably bind DNA simultaneously, thus bringing the arms of *attP* close to the core.

IHF binds to sequences of ~20 bp in *attP*; the IHF binding sites are approximately adjacent to sites where Int binds. Xis binds to two sites located close to one another in *attP*, so that the protected region extends over 30–40 bp. Together, Int, Xis, and IHF cover virtually all of *attP*. The binding of Xis changes the organization of the DNA so that it becomes inert as a substrate for the integration reaction.

Figure 33.21 shows that when the core locations bound by Int are mapped on the double helix, virtually all the contacts lie on one face of the DNA. The two sites of cutting are exposed in the major groove. IHF binding sites lie on the same face; and if the spacing between the P1 site for Int and the H1 site for IHF is altered so that it is no longer an integral number of helical turns, then integration is impeded.

When Int and IHF bind to *attP*, they generate a complex in which all the binding sites are pulled together on the surface of a protein oligomer, in a structure analogous to the nucleosome. Supercoiling of *attP* is needed for the formation of this **intasome**.

The only binding sites in *attB* are the two Int sites in the core. But Int does not bind directly to *attB* in the form of free DNA. The intasome is the intermediate that 'captures' *attB*. Probably Int molecules that are part of the intasome bind to the sites in the core of *attB*, as indicated schematically in **Figure 33.22**.

According to this model, the initial recognition between *attP* and *attB* does not depend directly on DNA homology, but instead is determined by the ability of Int proteins to recognize both *att* sequences. The two *att* sites then are brought together in an orientation predetermined by the structure of the intasome. Sequence homology becomes important at this stage, when it is required for the strand exchange reaction.

The asymmetry of the integration and excision reactions is shown by the fact that Int can form a similar complex with *attR* only if Xis is added. This complex can pair with a condensed complex that Int forms at *attL*. IHF is not needed for this reaction.

Much of the complexity of site-specific recombination may be caused by the need to regulate the reaction so that integration occurs preferentially when the virus is entering the lysogenic state, while excision is preferred when the prophage is entering the lytic cycle. By controlling the amounts of Int and Xis, the appropriate reaction will occur.

Summary

Recombination involves the physical exchange of parts between corresponding DNA molecules. This results in a duplex DNA in which two regions of opposite parental origins are connected by a stretch of hybrid (heteroduplex) DNA in which one strand is derived from each parent. Correction events may occur at sites that are mismatched within the hybrid DNA. Hybrid DNA can also be formed without recombination occurring between markers on either side. Gene conversion occurs when an extensive region of hybrid DNA formed during normal recombination (or between nonallelic genes in an aberrant event) is corrected to the sequence of only one parental strand; then one gene takes on the sequence of the other.

Recombination is initiated by a double-strand break in DNA. The break is enlarged to a gap with a single-stranded end; then the free single-stranded end forms a heteroduplex with the allelic sequence. The DNA in which the break occurs actually incorporates the sequence of the chromosome that it invades, so the initiating DNA is called the recipient. Hotspots for recombination are sites where double strand breaks are initiated. A gradient of gene conversion is determined by the likelihood that a sequence near the free end will be converted to a single strand; this decreases with distance from the break.

The recombination event is not well understood at the level of the chromosome, but the properties of yeast mutants, and the relative timing of events during meiosis, suggest that the synaptonemal complex may be the consequence of, rather than a prerequisite for, the initiation of the recombination event.

The only enzymes whose activities have been

characterized in recombination are coded by the *rec* and *ruv* loci of *E. coli*. RecA has the ability to synapse homologous DNA molecules by sponsoring a reaction in which a single strand from one molecule invades a duplex of the other molecule. Heteroduplex DNA is formed by displacing one of the original strands of the duplex. The RecBCD nuclease binds to DNA on one side of a *chi* sequence, and then moves to the *chi* sequence, unwinding DNA as it progresses. A single-strand break is made at the chi sequence. *Chi* sequences provide hotspots for recombination. RuvA and RuvB act on heteroduplexes, and RuvC cleaves Holliday junctions.

Recombination, like replication and (probably) transcription, requires topological manipulation of DNA. Topoisomerases may relax (or introduce) supercoils in DNA, and are required to disentangle DNA molecules that have become catenated by recombination or by replication. The enzymes involved in site-specific recombination have actions related to those of topoisomerases. Phage lambda integration requires the phage Int protein and host IHF protein and involves a precise breakage and reunion in the absence of any synthesis of DNA. The reaction involves wrapping of the *attP* sequence of phage DNA into the nucleoprotein structure of the intasome, which contains several copies of Int and IHF; then the host *attB* sequence is bound, and recombination occurs. Reaction in the reverse direction requires the phage protein Xis.

Further reading

Reviews
Recombination has been a focus for many reviews over the past few years. A splendid view of mechanisms, delving into the role of RecA, was given by **Dressler and Potter** (*Ann. Rev. Biochem.* **51**, 727–761, 1982). More recent work on RecA and other recombination enzymes has been reviewed by **West** (*Ann. Rev. Biochem.* **61**, 603–640, 1992). A review relating mechanisms to earlier observations in fungi, and also considering the role of hotspots, was by **Stahl** (*Ann. Rev. Genet.* **13**, 7–24, 1979). Bacterial recombination has been reviewed by Smith (*Microbiol. Rev.* **52**, 1–28, 1988) and **Cox and Lehman** (*Ann. Rev. Biochem.* **56**, 229–262, 1987).

Topoisomerase activities have been reviewed by **Wang** (*Ann. Rev. Biochem.* **54**, 665–697, 1985), and their biology by **Drlica** (*Microbiol. Rev.* **48**, 273–289, 1984).

The mechanism of lambda site-specific recombination was reviewed by **Weisberg and Landy** (pp. 211–250 in *Lambda II*, ed. Hendrix *et al.*, Cold Spring Harbor Laboratory, New York, 1983). The topic was brought up to date by **Thompson and Landy** (in *Mobile DNA*, ed. Berg and Howe, American Society for Microbiology, Washington DC, 1–22, 1989).

A chapter by **Silverman and Simon** in *Mobile Genetic Elements* (ed. Shapiro, Academic Press, New York, 1983) reviewed phase variation and Mu inversion, which also were previously explored by **Simon *et al.*** (*Science* **209**, 1370–1374, 1980). Bacterial inversion systems have been unified by **Glasgow, Hughes, and Simon** (in *Mobile DNA*, ed. Berg and Howe, American Society for Microbiology, Washington DC, 637–661, 1989).

Discoveries
The influential paper introducing the idea of the double-strand break was written by **Szostak *et al.*** (*Cell* **33**, 25–35, 1983). Further work on the role of double-strand breaks was published by **Sun, Treco, and Szostak** (*Cell* **64**, 1155–1161, 1991). The relative timing of molecular and cytological events was measured by **Padmore, Cao, and Kleckner** (*Cell* **66**, 1239–1256, 1991).

CHAPTER 34

Transposons that mobilize via DNA

Genomes are usually regarded as somewhat static, changing only on the leisurely time scale of evolution. We are accustomed to the idea that the construction of a genetic map identifies the loci at which known genes reside; by implication, other (unidentified) sequences also may be expected to remain at constant positions in the population of genomes. Recombination allows exchange of material between homologous chromosomes, generating new combinations of alleles, but does not otherwise reorganize the genetic material.

The stability of genetic organization is indicated by the retention of linkage relationships even after speciation—for example, between man and the apes. The difference of generation times between prokaryotes and eukaryotes suggests that their evolutionary scales might be different in terms of real time; but even in the prokaryotes, the overall organization of the genome changes only relatively slowly. For example, a similar genetic map describes the different bacterial species *E. coli* and *S. typhimurium.*

In contrast with this picture of overall stability, significant differences between individual genomes are found at the molecular level. We recollect from Chapter 26 that recombination between minisatellites adjusts their lengths so that every individual genome is distinct. Another major cause of variation is provided by **transposable elements** or **transposons**: these are discrete sequences in the genome that are *mobile*—they are able to transport themselves to other locations within the genome. The mark of a transposon is that it does not utilize an independent form of the element (such as phage or plasmid DNA), but moves *directly* from one site in the genome to another.

Sometimes transposons also cause rearrangement of other genomic sequences. Transposons are interesting not just for the mechanisms involved in the manipulation of DNA, but also for the evolutionary consequences of their mobility. They may in fact provide the major source of mutations in the genome, and we should therefore consider them within the context of the other mechanisms that are used to recombine DNA sequences.

Genomes evolve both by acquiring new sequences and by rearranging existing sequences. The sudden introduction of new sequences results from the ability of vectors to carry information between genomes. Extrachromosomal elements move information horizontally by mediating the transfer of (usually rather short) lengths of genetic material. In bacteria, plasmids move by conjugation (see Chapter 18), while phages spread by infection (see Chapter 17). Both plasmids and phages occasionally transfer host genes along with their own replicon. Direct transfer of DNA occurs between some bacteria by means of transformation (see Chapter 4). In eukaryotes, some viruses (notably the retroviruses discussed in Chapter 35) can transfer genetic information during an infective cycle.

Rearrangements may create new sequences and change the functions of existing sequences by placing them in new regulatory situations. *Rearrangements are sponsored by processes internal to the genome.* They result either from recombination undertaken by the cellular systems for homologous recombination and repair or from the transposition sponsored by transposons.

◆ Reciprocal recombination occurs in eukaryotes between corresponding sites on homologous chromosomes; occasionally a nonreciprocal recombination results in duplication or rearrangement of loci. Duplication of sequences within a genome provides a major source of new sequences. One copy of the sequence can retain its original function, while the other may evolve into a new function. Such reorganization within a chromosome is essentially a side effect of the usual mechanisms involved in genetic recombination. Major rearrangements also occur by translocations between nonhomologous chromosomes, but the mechanisms are unknown.

◆ A potent force for change within both prokaryotic and eukaryotic genomes is provided by the mobility of transposons. Unlike most other processes involved in genome restructuring, *transposition does not rely on any relationship between the sequences at the donor and recipient sites.* Transposons are restricted to moving themselves, and sometimes additional sequences of the genome, to new sites elsewhere within the same genome; they are therefore an internal counterpart to the vectors that can transport sequences from one genome to another.

Transposons fall into two general classes. The groups of transposons reviewed in this chapter exist as sequences of DNA coding for proteins that are able directly to manipulate DNA so as to propagate themselves within the genome. The transposons reviewed in the next chapter are related to retroviruses, and the source of their mobility is the ability to make DNA copies of their RNA transcripts; the DNA copies then become integrated at new sites in the genome.

Transposons that mobilize via DNA are found in both prokaryotes and eukaryotes. Each bacterial transposon carries gene(s) that code for the enzyme activities required for its own transposition, although it may also require ancillary functions of the genome in which it resides (such as DNA polymerase or DNA gyrase). Comparable systems exist in eukaryotes, although the enzymatic functions involved in transposition are not so well characterized. A genome may contain both functional and nonfunctional (defective) elements; often the majority of elements in a eukaryotic genome are defective, and have lost the ability to transpose independently, although they may still be recognized as substrates for transposition by the enzymes produced by functional transposons.

Transposable elements can promote rearrangements of the genome, directly or indirectly:

◆ The transposition event itself may cause deletions or inversions or lead to the movement of a host sequence to a new location.

◆ Transposons serve as substrates for cellular recombination systems functioning as 'portable regions of homology'; two copies of a transposon at different locations (even on different chromosomes) may provide sites for reciprocal recombination. Such exchanges result in deletions, insertions, inversions, or translocations.

The intermittent activities of a transposon seem to provide a somewhat nebulous target for natural selection. This concern has prompted suggestions that (at least some) transposable elements confer neither advantage nor disadvantage on the phenotype, but could constitute 'selfish DNA', concerned only with their own propagation. Indeed, in considering transposition as an event that is distinct from other cellular recombination systems, we tacitly accept the view that the transposon is an independent entity that resides in the genome; and although it exists principally *within* a genome, on very rare occasions a transposon may move from one species to another (by unknown means).

Such a relationship of the transposon to the

genome would resemble that of a parasite with its host. Presumably the propagation of an element by transposition is balanced by the harm done if a transposition event inactivates a necessary gene, or if the number of transposons becomes a burden on cellular systems. Yet we must remember that any transposition event conferring a selective advantage—for example, a genetic rearrangement—will lead to preferential survival of the genome carrying the active transposon.

Insertion sequences are simple transposition modules

Transposable elements were first identified in the form of spontaneous insertions in bacterial operons. Such an insertion prevents transcription and/or translation of the gene in which it is inserted. Many different types of transposable elements have now been characterized.

The simplest transposons are called **insertion sequences** (reflecting the way in which they were detected). Each type is given the prefix **IS**, followed by a number that identifies the type. (The original classes were numbered IS1–4; later classes have numbers reflecting the history of their isolation, but not corresponding to the total number of elements so far isolated!)

The IS elements are normal constituents of bacterial chromosomes and plasmids. A standard strain of *E. coli* is likely to contain several (<10) copies of any one of the more common IS elements.

To describe an insertion into a particular site, a double colon is used; thus λ::IS1 describes an IS1 element inserted into phage lambda.

The IS elements are autonomous units, each of which codes only for the proteins needed to sponsor its own transposition. Each IS element is different in sequence, but there are some common features in organization. The parameters of some common IS elements are summarized in **Table 34.1**. They are all short (~1000 bp), end in inverted terminal repeats, and generate short repeats in the target DNA during insertion. The structure of a generic transposon before and after insertion at a target site is illustrated in **Figure 34.1**.

The element ends in short **inverted terminal repeats**; usually the two copies of the repeat are closely related rather than identical. As illustrated in the figure, the presence of the inverted terminal

Table 34.1

Individual elements comprise transposable units or modules of composite transposons.

Element	Length	Inverted Terminal Repeats	Direct Repeats at Target	Target Selection
IS1	768 bp	23 bp	9 bp	Random
IS2	1327 bp	41 bp	5 bp	Hotspots
IS4	1428 bp	18 bp	11 or 12 bp	$AAAN_{20}TTT$
IS5	1195 bp	16 bp	4 bp	Hotspots
IS10R	1329 bp	22 bp	9 bp	NGCTNAGCN
IS50R	1531 bp	9 bp	9 bp	Hotspots
IS903	1057 bp	18 bp	9 bp	Not known

Figure 34.1

Overview: transposons have inverted terminal repeats and generate direct repeats of flanking DNA at the target site. In this example, the target is a 5 bp sequence. The ends of the transposon consist of inverted repeats of 9 bp, where the numbers 1 through 9 indicate a sequence of base pairs.

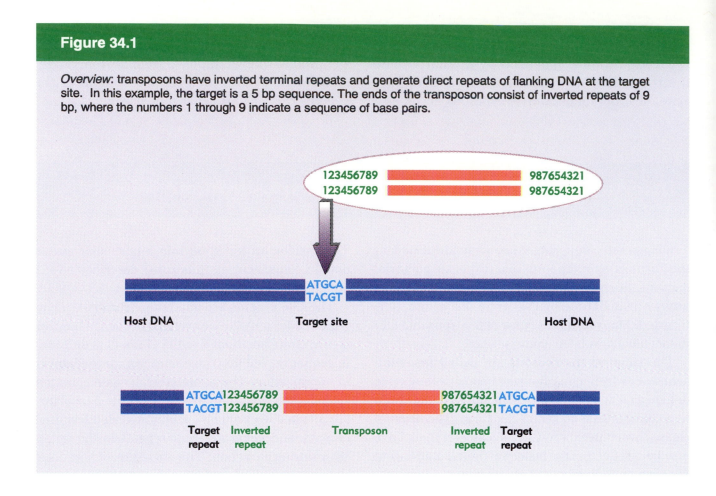

repeats means that the same sequence is encountered proceeding toward the element from the flanking DNA on either side of it.

When an IS element transposes, a sequence of host DNA at the site of insertion is duplicated. The nature of the duplication is revealed by comparing the sequence of the target site before and after an insertion has occurred. Figure 34.1 shows that at the site of insertion, the IS DNA is always flanked by very short **direct repeats**. (In this context, 'direct' indicates that two copies of a sequence are repeated in the same orientation, not that the repeats are adjacent.) But in the original gene (prior to insertion), the target site has the sequence of only *one* of these repeats. In the figure, the target site consists of the sequence $\frac{ATGCA}{TACGT}$. *After transposition, one copy of this sequence is present on either side of the transposon.*

The *sequence* of the direct repeat varies among individual transposition events undertaken by a transposon, but the *length* is constant for any particular IS element (a reflection of the mechanism of transposition). The most common length for the direct repeats is 9 bp.

An IS element therefore displays a characteristic structure in which its ends are identified by the inverted terminal repeats, while the adjacent ends of the flanking host DNA are identified by the short direct repeats. When observed in a sequence of DNA, this type of organization is taken to be diagnostic of a transposon, and makes a *prima facie* case that the sequence originated in a transposition event.

Most IS elements insert at a variety of sites within host DNA. However, some show (varying degrees of) preference for particular hotspots.

The inverted repeats define the ends of a transposon. Recognition of the ends is common to transposition events sponsored by all types of transposon. *Cis*-acting mutations that prevent transposition are located in the ends, which are recognized by a protein(s) responsible for transposition. The protein is called a **transposase**.

All the IS elements except IS1 contain a single long coding region, starting just inside the inverted repeat at one end, and terminating just before or within the inverted repeat at the other end. This codes for the transposase. IS1 has a more complex organization, with two separate reading frames; the transposase is produced by making a frameshift during trans-

lation to allow both reading frames to be used.

The frequency of transposition varies among different elements. The overall rate of transposition is $\sim 10^{-3}-10^{-4}$ per element per generation. Insertions in individual targets occur at a level comparable with the spontaneous mutation rate, usually $\sim 10^{-5}-10^{-7}$ per generation. Reversion (by precise excision of the IS element) is usually infrequent, with a range of rates of 10^{-6} to 10^{-10} per generation, $\sim 10^{3}$ times less frequent than insertion.

Composite transposons have IS modules

Some transposons carry drug resistance (or other) markers in addition to the functions concerned with transposition. These transposons are named **Tn** followed by a number. One class of larger transposons are called **composite elements**, because a central region carrying the drug marker(s) is flanked on either side by 'arms' that consist of IS elements.

The arms may be in either the same or (more commonly) inverted orientation. Thus a composite transposon with arms that are direct repeats has the structure

Arm L   Central Region   Arm R

If the arms are inverted repeats, the structure is

Arm L   Central Region   Arm R

The arrows indicate the orientation of the arms, which are identified as L and R according to an (arbitrary) orientation of the genetic map of the transposon from left to right. The structure of a composite transposon is illustrated in more detail in **Figure 34.2**.

Since arms consist of IS modules, and each module has the usual structure ending in inverted repeats, the composite transposon also ends in the

same short inverted repeats. The properties of some composite transposons are summarized in **Table 34.2**.

In some cases, the modules of a composite transposon are identical, such as Tn9 (direct repeats of IS1) or Tn903 (inverted repeats of IS903). In other cases, the modules are closely related, but not

Figure 34.2

A composite transposon has a central region carrying markers unconnected with transposition (such as drug resistance) flanked by IS modules. The modules have short inverted terminal repeats. If the modules themselves are in inverted orientation (as drawn), the short inverted terminal repeats at the ends of the transposon are identical.

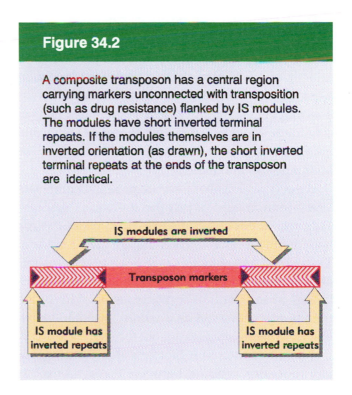

IS modules are inverted

Transposon markers

IS module has inverted repeats

IS module has inverted repeats

Table 34.2

Some transposons are composites bracketed by individual modules.

Element	Length (bp)	Genetic Markers	Terminal Modules	Module Orientation	Module Relationship	Module Functions
Tn903	3100 bp	kan^R	IS903	Inverted	Identical	Both functional
Tn9	2500 bp	cam^R	IS1	Direct	Presumed identical	Presumed functional
Tn10	9300 bp	tet^R	IS10R			Functional
			IS10L	Inverted	2.5% divergence	Nonfunctional
Tn5	5700 bp	kan^R	IS50R			Functional
			IS50L	Inverted	1 bp change	Nonfunctional

identical. Thus we can distinguish the modules in Tn10 or in Tn5.

A functional IS module can transpose either itself or the entire transposon. When the modules of a composite transposon are identical, presumably either module can sponsor movement of the transposon, as in the case of Tn9 or Tn903. When the modules are different, they may differ in functional ability, so transposition can depend entirely or principally on one of the modules, as in the case of Tn10 or Tn5.

We assume that composite transposons evolved when two originally independent modules associated with the central region. Such a situation could arise when an IS element transposes to a recipient site close to the donor site. The two identical modules may remain identical or diverge. The ability of a single module to transpose the entire composite element explains the lack of selective pressure for both modules to remain active.

What is responsible for transposing a composite transposon instead of just the individual module? This question is especially pressing in cases where both the modules are functional. In the example of Tn9, where the modules are IS1 elements, presumably each is active in its own right as well as on behalf of the composite transposon. Why is the transposon preserved as a whole, instead of each insertion sequence looking out for itself?

Two IS elements in fact are able to transpose any sequence residing between them, as well as

themselves. **Figure 34.3** shows that if Tn10 resides on a circular replicon, its two modules can be considered to flank *either* the tet^R gene of the original Tn10 *or* the sequence in the other part of the circle. Thus a transposition event can involve either the original Tn10 transposon or the 'inside-out' transposon with the alternative central region.

Note that both the original and 'inside-out' transposons have inverted modules, but these modules evidently can function in either orientation relative to the central region. The frequency of transposition for composite transposons declines with the distance between the modules. So length dependence is a factor in determining the sizes of the common composite transposons.

A major force supporting the transposition of composite transposons is selection for the marker(s) carried in the central region. An IS10 module is free to move around on its own, and in fact mobilizes an order of magnitude more frequently than Tn10, but Tn10 is held together by selection for tet^R; so that under selective conditions, the relative frequency of intact Tn10 transposition is much increased.

The IS elements code for transposase activities that are responsible both for creating a target site and for recognizing the ends of the transposon. *Only the ends are needed for a transposon to serve as a substrate for transposition.* We know little about the detailed processes and order of events involved in these reactions.

Figure 34.3

Two IS10 modules create a composite transposon that can mobilize any region of DNA that lies between them. When Tn10 is part of a small circular molecule, the IS10 repeats can transpose either side of the circle. Transposition of *tet*R corresponds to the movement of Tn10. Transposition of the markers on the other side creates a new "inside-out" transposon.

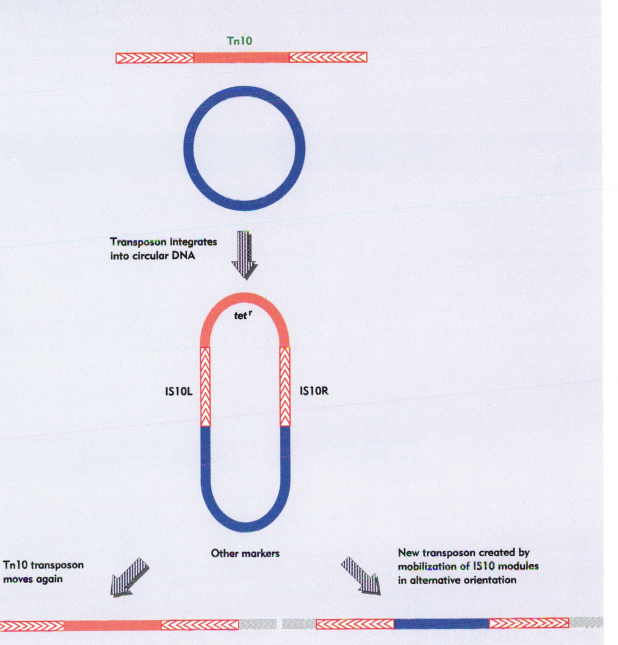

Transposition occurs by both replicative and nonreplicative mechanisms

The general nature of the reaction by which a transposon is inserted into a new site is illustrated in **Figure 34.4**. It consists of making staggered breaks in the target DNA, joining the transposon to the protruding single-stranded ends, and filling in the gaps. The generation and filling of the staggered ends explain the occurrence of the direct repeats of target DNA at the site of insertion. The stagger between the cuts on the two strands determines the length of the direct repeats; thus the target repeat

Figure 34.4

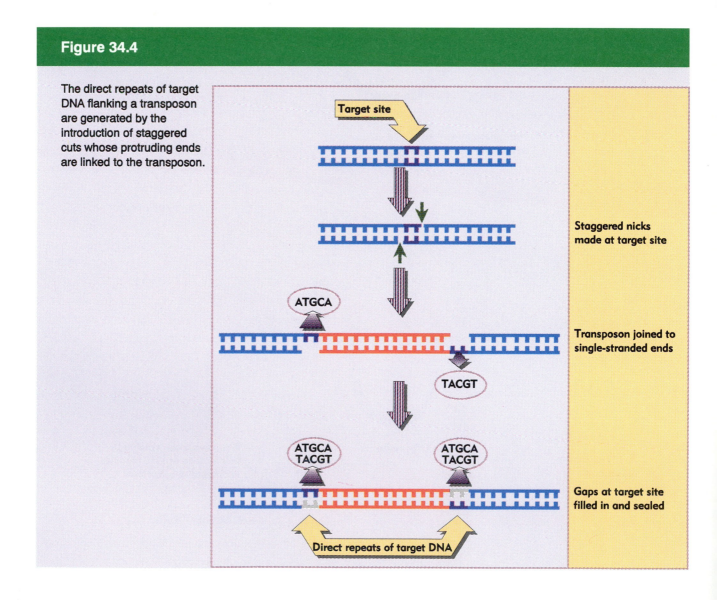

The direct repeats of target DNA flanking a transposon are generated by the introduction of staggered cuts whose protruding ends are linked to the transposon.

Target site

Staggered nicks made at target site

ATGCA

Transposon joined to single-stranded ends

TACGT

ATGCA
TACGT

ATGCA
TACGT

Gaps at target site filled in and sealed

Direct repeats of target DNA

characteristic of each transposon reflects the geometry of the enzyme involved in cutting target DNA.

The use of staggered ends is common to all means of transposition, but we can distinguish three different types of mechanism by which a transposon moves:

◆ In **replicative transposition**, *the element is duplicated during the reaction, so that the transposing entity is a copy of the original element.* **Figure 34.5** summarizes the results of such a transposition. The transposon is copied as part of its movement. One copy remains at the original site, while the other inserts at the new site. Thus transposition is accompanied by an increase in the number of copies of the transposon. Replicative transposition involves two types of enzymatic activity: a **transposase** that acts on the ends of the original transposon; and a **resolvase** that acts on the duplicated copies. A group of transposons related to TnA move only by replicative transposition (see later).

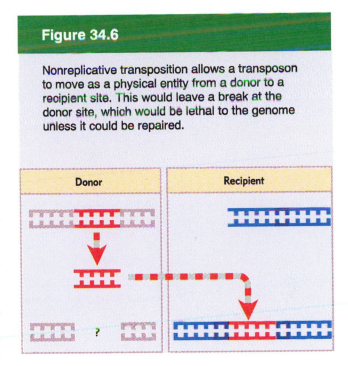

Figure 34.6

Nonreplicative transposition allows a transposon to move as a physical entity from a donor to a recipient site. This would leave a break at the donor site, which would be lethal to the genome unless it could be repaired.

◆ In **nonreplicative transposition**, *the transposing element moves as a physical entity directly from one site to another, and is conserved.* This occurs by either of two types of mechanism. Phage Mu proceeds by connecting donor and target DNA sequences (see next section). The insertion sequences and composite transposons Tn10 and Tn5 use the mechanism shown in **Figure 34.6**, which involves the release of the transposon from the flanking donor DNA during transfer. This type of mechanism requires only a transposase.

Both mechanisms of nonreplicative transposition cause the element to be inserted at the target site and lost from the donor site. What happens to the donor molecule after a nonreplicative transposition? One possibility is that the donor is simply destroyed (which may be tolerated by bacteria that possess more than one copy of the chromosome). Another possibility is that host repair systems recognize the double-strand break and repair it.

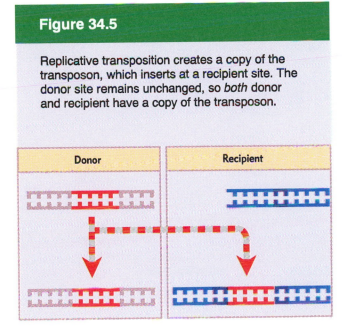

Figure 34.5

Replicative transposition creates a copy of the transposon, which inserts at a recipient site. The donor site remains unchanged, so *both* donor and recipient have a copy of the transposon.

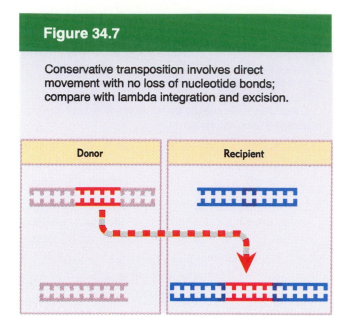

Figure 34.7

Conservative transposition involves direct movement with no loss of nucleotide bonds; compare with lambda integration and excision.

| Donor | Recipient |

A neat experiment that proves Tn10 transposes nonreplicatively made use of an artificially constructed heteroduplex of Tn10 that contained single base mismatches. If transposition involves replication, the transposon at the new site will contain information from only one of the parent Tn10 strands. But if transposition takes place by physical movement of the existing transposon, the mismatches will be conserved at the new site. Conservation of the mismatches was demonstrated by showing that the transposon retains information of both strands after transposition.

◆ **Conservative transposition** describes another sort of nonreplicative event, in which the element is excised from the donor site and inserted into a target site by a series of events in which every nucleotide bond is conserved. **Figure 34.7** summarizes the result of a conservative event. This exactly resembles the mechanism of lambda integration discussed in Chapter 33, and the transposases of such elements are related to

the λ integrase family. The elements that use this mechanism are large, and can mediate transfer not only of the element itself but also of donor DNA from one bacterium to another. Although originally classified as transposons, such elements may more properly be regarded as episomes.

Although some transposons use only one type of pathway for transposition, others may be able to use both pathways. The elements IS1 and IS903 use

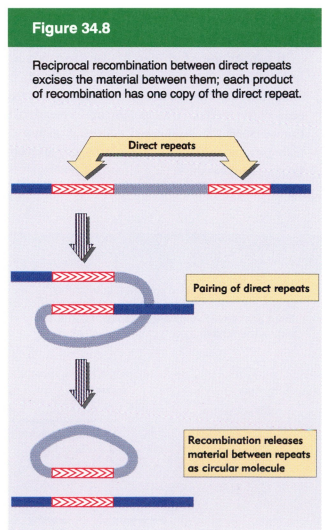

Figure 34.8

Reciprocal recombination between direct repeats excises the material between them; each product of recombination has one copy of the direct repeat.

Direct repeats

Pairing of direct repeats

Recombination releases material between repeats as circular molecule

Figure 34.9

Reciprocal recombination between inverted repeats inverts the region between them.

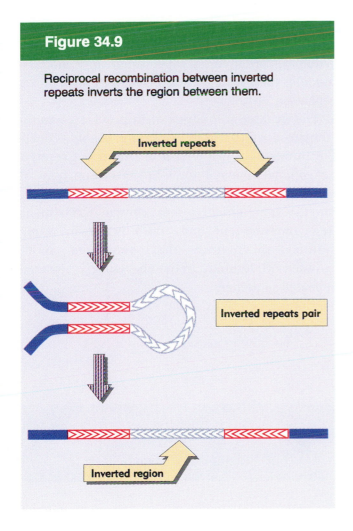

Inverted repeats

Inverted repeats pair

Inverted region

both nonreplicative and replicative pathways, and the ability of phage Mu to turn to either type of pathway from a common intermediate has been well characterized (see later).

In addition to the 'simple' intermolecular transposition that results in insertion at a new site, transposons promote other types of DNA rearrangements. Some of these events are consequences of the relationship between the multiple copies of the transposon. Others represent alternative outcomes of the transposition mechanism, and they leave clues about the nature of the underlying events.

Rearrangements of host DNA may result when a transposon inserts a copy at a second site near its original location. Host systems may undertake reciprocal recombination between the two copies of the transposon; the consequences are determined by whether the repeats are the same or in inverted orientation.

Figure 34.8 illustrates the general rule that recombination between any pair of direct repeats will delete the material between them. The intervening region is excised as a circle of DNA (which is lost from the cell); the chromosome retains a single copy of the direct repeat. Note that a recombination between the directly repeated IS1 modules of the composite transposon Tn9 would therefore replace the transposon with a single IS1 module.

Deletion of sequences adjacent to a transposon could therefore result from a two-stage process; transposition generates a direct repeat of a transposon, and recombination occurs between the repeats. However, the majority of deletions that arise in the vicinity of transposons probably result from a variation in the pathway followed in the transposition event itself.

Figure 34.9 depicts the consequences of a reciprocal recombination between a pair of inverted repeats. The region between the repeats becomes inverted; the repeats themselves remain available to sponsor further inversions. Note that a composite transposon whose modules are inverted is a stable component of the genome, although the direction of the central region with regard to the modules could be inverted by recombination.

Duplicative inversions are promoted by some transposons. They are identified by transposons that lie in inverted orientation on either side of a central region that has been inverted from its usual orientation.

Excision is not supported by transposons

themselves, but may occur when bacterial enzymes recognize homologous regions in the transposons. This is important because the loss of a transposon may restore function at the site of insertion. **Precise excision** requires removal of the transposon *plus one copy of the duplicated sequence*. This is rare; it occurs at a frequency of $\sim 10^{-6}$ for Tn5 and $\sim 10^{-9}$ for Tn10. It probably involves a recombination between the 9 bp duplicated target sites.

Imprecise excision leaves a remnant of the transposon. Although the remnant may be sufficient to prevent reactivation of the target gene, it may be insufficient to cause polar effects in adjacent genes, so that a change of phenotype occurs. Imprecise excision occurs at a frequency of $\sim 10^{-6}$

for Tn10. It involves recombination between sequences of 24 bp in the IS10 modules; these sequences are inverted repeats, but since the IS10 modules themselves are inverted, they form direct repeats in Tn10.

The greater frequency of imprecise excision compared with precise excision probably reflects the increase in the length of the direct repeats (24 bp as opposed to 9 bp). Neither type of excision relies on transposon-coded functions, but the mechanism is not known. Excision is RecA-independent and could occur by some cellular mechanism that generates spontaneous deletions between closely spaced repeated sequences.

Replicative and nonreplicative transposition may pass through common intermediates

The basic route for replicative transposition is illustrated in **Figure 34.10**. The reaction passes through two stages:

◆ First, a replicon containing the transposon becomes fused with a replicon lacking the element. The structure produced by this **replicon fusion** is called a **cointegrate**. The cointegrate has two copies of the transposon, one at each junction between the original replicons, oriented as direct repeats.

◆ Second, a homologous recombination between the two copies of the transposon can regenerate the original donor replicon, releasing a target replicon that has gained a transposon flanked by short direct repeats of the host target sequence. This reaction is called **resolution**; the enzyme activity responsible is called the **resolvase**. It

releases two individual replicons, each of which has a copy of the transposon.

The reactions involved in generating a cointegrate have been defined in detail for the bacteriophage Mu, which uses the process of transposition in two ways. Upon infecting a host cell, Mu integrates into the genome by nonreplicative transposition; during the ensuing lytic cycle, the number of copies is amplified by replicative transposition. Both types of transposition involve the same type of reaction between the transposon and its target, but the manipulation of the structure is different.

The formation of a cointegrate structure is illustrated in **Figure 34.11**. Four single-strand cleavages initiate the process. The donor molecule is cleaved at either end of the transposon by the Mu **transposase**, a site-specific enzyme that recognizes the termini. The target molecule is cleaved at sites staggered by 5 bases.

Figure 34.10

Transposition may fuse a donor and recipient replicon into a cointegrate. Resolution releases two replicons, each containing a copy of the transposon.

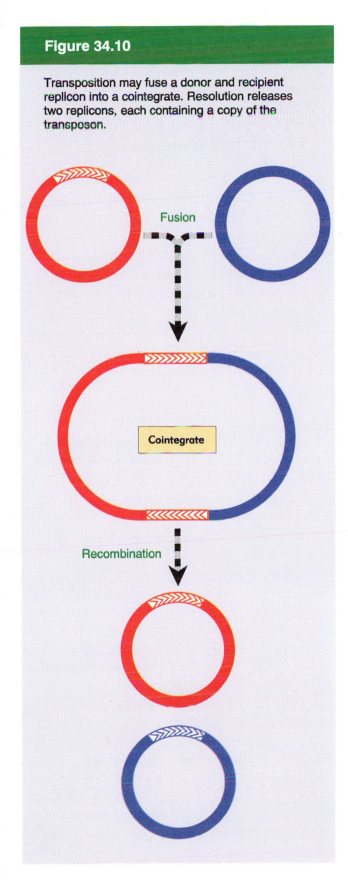

Fusion

Cointegrate

Recombination

The donor and target strands are ligated at the nicks. Each end of the transposon sequence is joined to one of the protruding single strands generated at the target site. The linkage generates a crossover-shaped structure held together at the duplex transposon. The formation of this structure is accomplished by the transposase. The fate of the crossover structure determines the mode of transposition.

The principle of replicative transposition is that replication through the transposon duplicates it, creating copies at both the target and donor sites. The product is a cointegrate.

The crossover structure contains a single-stranded region at each of the staggered ends. These regions are pseudoreplication forks that provide a template for DNA synthesis. (Use of the ends as primers for replication implies that the strand breakage must occur with a polarity that generates a 3′-OH terminus at this point.)

If replication continues from both the pseudo-replication forks, it will proceed through the transposon, separating its strands, and terminating at its ends. Replication is probably accomplished by host-coded functions. At this juncture, the structure has become a cointegrate, possessing direct repeats of the transposon at the junctions between the replicons (as can be seen by tracing the path around the cointegrate).

The crossover structure can also be used in nonreplicative transposition. *The principle of nonreplicative transposition by this mechanism is that a breakage and reunion reaction allows the target to be reconstructed; the donor remains broken. No cointegrate is formed.*

Figure 34.12 shows the cleavage events that generate nonreplicative transposition of phage Mu. Once the unbroken donor strands have been nicked, the target strands on either side of the transposon can be ligated. The single-stranded regions generated by the staggered cuts must be filled in by repair synthesis. The product of this reaction is a target replicon in which the transposon has been inserted between repeats of the sequence created by the original single strand nicks. The donor replicon has a

Figure 34.11

Mu transposition generates a crossover structure, which is converted by replication into a cointegrate.

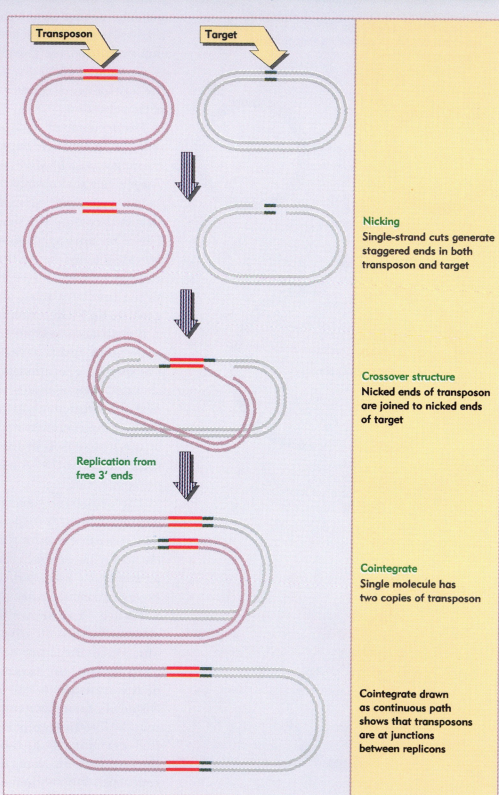

Transposon

Target

Nicking
Single-strand cuts generate staggered ends in both transposon and target

Crossover structure
Nicked ends of transposon are joined to nicked ends of target

Replication from free 3' ends

Cointegrate
Single molecule has two copies of transposon

Cointegrate drawn as continuous path shows that transposons are at junctions between replicons

Figure 34.12

Nonreplicative transposition results when a crossover structure is released by nicking. This inserts the transposon into the target DNA, flanked by the direct repeats of the target, and the donor is left with a double-strand break.

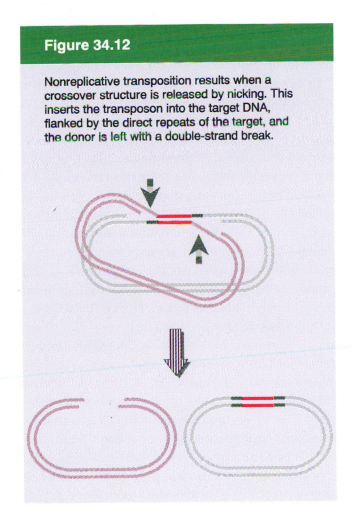

double-strand break across the site where the transposon was originally located.

The crucial feature of Mu movement that allows an outcome of either replicative or nonreplicative transposition is the single-strand cleavage to generate a crossover product where the transposon is connected to both donor and target DNA.

Nonreplicative transposition can also occur by an alternative pathway in which nicks are made at target DNA, but a double-strand break is made on either side of the transposon, releasing it entirely from flanking donor sequences (as envisaged in Figure 34.6). This 'cut and paste' pathway is used by Tn10. In effect, the cleavages on either side of the transposon that occur to generate the cross-over structure and then to release it in Figure 34.12 are made simultaneously. The double-strand break prevents a replicative outcome and forces the reaction to proceed by non-replicative transposition. Although in terms of DNA molecules, the transposon is free, it is not in fact available in the cell as a free form comparable to an episome, but is released in conjunction with the cleavage of target DNA, and presumably is retained within a large proteinaceous structure.

TnA transposition requires transposase and resolvase

Replicative transposition is the only mode of mobility of the TnA family, which consists of large (~5 kb) transposons. They are not composites relying on IS-type transposition modules, but comprise independent units carrying genes for transposition as well as for features such as drug resistance. The TnA family includes several related transposons, of which Tn3 and Tn1000 (formerly called γδ) are the best characterized. They have the usual terminal feature of closely related inverted repeats, generally ~38 bp in length. *Cis*-acting deletions in either repeat prevent transposition of an element. A 5 bp direct repeat is generated

at the target site. They carry resistance markers such as *amp*^{R}.

The two stages of TnA-mediated transposition are accomplished by the transposase and the resolvase, whose genes, *tnpA* and *tnpR*, are identified by recessive mutations. The transposition stage involves the ends of the element, as it does in IS-type elements. Resolution requires a specific internal site, a feature unique to the TnA family.

Mutants in *tnpA* cannot transpose. The gene product is a transposase that binds to a sequence of ~25 bp located within the 38 bp of the inverted terminal repeat. A binding site for the *E. coli* protein

IHF exists adjacent to the transposase binding site; and transposase and IHF bind cooperatively. We believe that the transposase recognizes the ends of the element and also makes the staggered 5 bp breaks in target DNA where the transposon is to be inserted. The role of IHF is not clear, but does not appear to be essential.

The *tnpR* gene product has dual functions, which are revealed by the diverse effects of mutations. The protein both acts as a repressor of gene expression and provides the resolvase function.

Mutations in *tnpR* increase the transposition frequency. The reason is that TnpR represses the transcription of both *tnpA* and its own gene. Thus inactivation of TnpR protein allows increased synthesis of TnpA, which results in an increased frequency of transposition. This implies that the amount of the TnpA transposase must be a limiting factor in transposition.

The *tnpA* and *tnpR* genes are expressed divergently from an A•T-rich intercistronic control region,

indicated in the map of Tn3 given in **Figure 34.13**. Both effects of TnpR are mediated by its binding in this region.

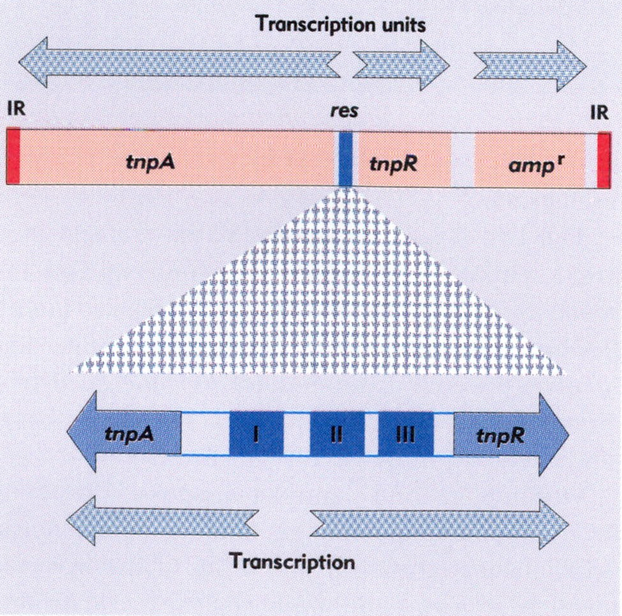

In its capacity as the resolvase, TnpR is involved in recombination between the direct repeats of Tn3 in a cointegrate structure. A cointegrate can in principle be resolved by a homologous recombination between any corresponding pair of points in the two copies of the transposon. But the Tn3 resolution reaction occurs only at a specific site.

The site of resolution is called *res*. It is identified by *cis*-acting deletions that block completion of transposition, causing the accumulation of cointegrates. In the absence of *res*, the resolution reaction can be substituted by RecA-mediated general recombination, but this is much less efficient.

The sites bound by the TnpR resolvase have been determined by footprinting the DNA–protein complex. Their locations are summarized in the lower part of Figure 34.13. Binding occurs independently at each of three sites, each 30–40 bp long. The three binding sites share a sequence homology that defines a consensus sequence with dyad symmetry.

Site I includes the region genetically defined as the *res* site; in its absence, the resolution reaction does not proceed at all. However, resolution also involves binding at sites II and III, since the reaction proceeds only poorly if either of these sites is deleted. Site I overlaps with the startpoint for *tnpA* transcription. Site II overlaps with the startpoint for *tnpR* transcription; an operator mutation maps just at the left end of the site.

Do the sites interact? One possibility is that binding at all three sites is required to hold the DNA in an appropriate topology; the resolution reaction could be triggered when the protein–DNA complex of one transposon interacts with that of the transposon. Binding at a single set of sites may repress *tnpA* and *tnpR* transcription without introducing any change in the DNA.

An *in vitro* resolution assay uses a cointegrate-like DNA molecule as substrate. The substrate must be supercoiled; its resolution produces two catenated circles, each containing one *res* site. The reaction requires large amounts of the TnpR re-

Figure 34.13

Transposons of the TnA family have inverted terminal repeats, an internal *res* site, and three known genes.

solvase; no host factors are needed. Resolution occurs in a sizeable ribonucleoprotein structure. Resolvase binds to each *res* site, and then the bound sites are brought together to form a structure ~10 nm in diameter. Changes in supercoiling occur during the reaction, and DNA is bent at the *res* sites by the binding of transposase.

Resolution is a nonreplicative reaction; bonds are broken and rejoined without demand for input of energy. The products identify an intermediate stage in cointegrate resolution; they consist of resolvase covalently attached to both 5′ ends of double-stranded cuts made at the *res* site. The cleavage occurs symmetrically at a short palindromic region to generate two base extensions. Expanding the view of the crossover region located in site I, we can describe the cutting reaction as:

$$5'\quad T\ T\ A\ T\ A\ A\quad 3'$$
$$3'\quad A\ A\ T\ A\ T\ T\quad 5'$$
$$\downarrow$$

5′ T T A T   +   protein—A A 3′
3′ A A—protein   T A T T 5′

The reaction resembles the action of lambda Int at the *att* sites. Indeed, 20 of the 15 bp of the *res* site are identical to the bases at corresponding positions in *att*. We can align the sequences in terms of the cleavage points as:

$$\downarrow$$
res G A T A A T T T A T A A T A T
att G C T T T T T T A T A C T A A
$$\uparrow$$

The reactions themselves are analogous in terms of manipulation of DNA, although resolution occurs only between intramolecular sites, whereas the recombination between *att* sites is intermolecular and directional (as seen by the differences in *att* sites). However, the mechanism of protein action is different in each case. Resolvase functions in a manner in which four subunits bind to the recombining *res* sites. Each subunit makes a single-strand cleavage. Then a reorganization of the subunits relative to one another physically moves the DNA strands, placing them in a recombined conformation. Then the nicks can be sealed.

Transposition of Tn10 is subject to multiple controls

Control of the frequency of transposition is important for the cell. A transposon must be able to maintain a certain minimum frequency of movement in order to survive; but too great a frequency could be damaging to the host cell. Every transposon appears to have mechanisms that control its own frequency of transposition. A variety of mechanisms has been characterized for Tn10.

Tn10 is a composite transposon in which only a few bases at each end are needed to recognize the element as a substrate for transposition. The inverted terminal repeats are 22 bp long. As indicated by the presence of *cis*-acting mutations, the ends are essential for the transposition reaction.

Tn10 is one of several elements that exhibit a preference for a specific target sequence. The 9 bp direct repeats of flanking DNA generated by transposition display a consensus of a 6 bp sequence symmetrically disposed within the target. Thus the repeats on either side of Tn10 often take the form NGCTNAGCN / NCGANTCGN, where N identifies any base. The stronger the hotspot, the more closely it conforms to the consensus. Probably the same transposase recognizes the target sequence as well as the ends of the transposon.

The element IS10R provides the active module of Tn10. The IS10L module is functionally defective and provides only 1–10% of the transposase activity of IS10R. Although the transposase of IS10L is defective, its ends can be recognized by transposase, as when Tn10 transposes.

The organization of IS10R is summarized in **Figure 34.14**. Two promoters are found close to the outside boundary. The promoter P_{IN} is responsible for transcription of IS10R. The promoter P_{OUT} causes transcription to proceed toward the adjacent flanking DNA. Transcription usually terminates within the transposon, but occasionally continues into the host DNA; sometimes this read-through transcription is responsible for activating adjacent bacterial genes.

The phenomenon of 'multicopy inhibition' reveals that expression of the IS10R transposase gene is regulated. Transposition of a Tn10 element on the bacterial chromosome is reduced when additional copies of IS10R are introduced via a multicopy plasmid. The inhibition requires the P_{OUT} promoter, and is exercised at the level of translation. The basis for the effect lies with the 40 bp overlap in the 5′ terminal regions of the transcripts from P_{IN} and P_{OUT}. *OUT* RNA is a transcript of 69 bases. It is present at >100× the level of *IN* RNA for two reasons: P_{OUT} is a much stronger promoter than P_{IN}; and *OUT* RNA is more stable than *IN* RNA.

RNA_{OUT} functions as an antisense RNA (see Chapter 16). The level of RNA_{OUT} has no effect in a single-copy situation, but has a significant effect when >5 copies are present. There are usually ~5 copies of *OUT* RNA per copy of IS10 (which corresponds to ~150 copies of *OUT* RNA in a typical multicopy situation.) *OUT* RNA base pairs with *IN* RNA; and the excess of *OUT* RNA ensures that *IN* RNA is bound rapidly, before a ribosome can attach. So the paired *IN* RNA cannot be translated.

The quantity of transposase protein is often a critical feature. Tn10, whose transposase is synthesized at the low level of 0.15 molecules per cell per generation, displays several interesting mechanisms. **Figure 34.15** summarizes the various effects that influence transposition frequency.

A continuous reading frame on one strand of IS10R codes for the transposase. The level of the transposase limits the rate of transposition. Mutants in this gene can be complemented in *trans* by another, wild-type IS10 element, but only with some difficulty. This reflects a strong preference of the transposase for *cis*-action; the enzyme functions efficiently only with the DNA template from which

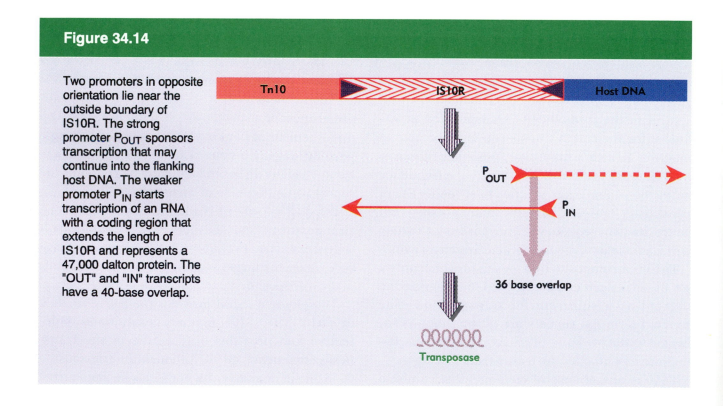

Figure 34.14

Two promoters in opposite orientation lie near the outside boundary of IS10R. The strong promoter P_{OUT} sponsors transcription that may continue into the flanking host DNA. The weaker promoter P_{IN} starts transcription of an RNA with a coding region that extends the length of IS10R and represents a 47,000 dalton protein. The "OUT" and "IN" transcripts have a 40-base overlap.

Tn10 | IS10R | Host DNA

P_{OUT}

P_{IN}

36 base overlap

Transposase

Figure 34.15

Several mechanisms restrain the frequency of Tn10 transposition, by affecting either the synthesis or function of transposase protein. Transposition of an individual transposon is restricted by methylation to occur only after replication. In multicopy situations, *cis*-preference restricts the choice of target, and OUT/IN RNA pairing inhibits synthesis of transposase.

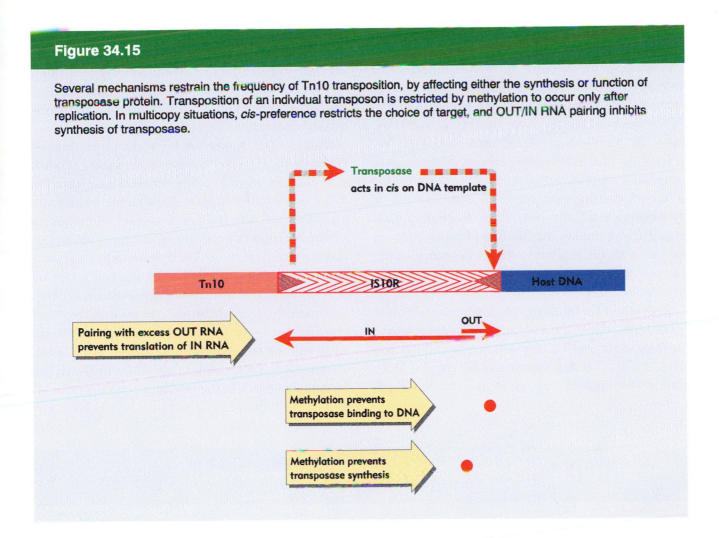

it was transcribed and translated. *Cis*-preference is a common feature of transposases coded by IS elements. (Other proteins that display *cis*-preference include the A protein involved in φX174 replication; see Chapter 18.) The *cis*-preference of the transposase limits the frequency of transposition by ensuring that the presence of more copies of Tn10 in a cell does not increase the effective concentration of transposase.

Does *cis*-preference reflect an ability of the transposase to recognize more efficiently those DNA target sequences that lie nearer to the site where the enzyme is synthesized? One possible explanation is that the transposase binds to DNA so tightly after (or even during) protein synthesis that it has a very low probability of diffusing elsewhere. Another possibility is that the enzyme may be unstable when it is not bound to DNA, so that protein

molecules failing to bind quickly (and therefore nearby) never have a chance to become active.

Together the results of *cis*-preference and multicopy inhibition ensure that an increase in the number of copies of Tn10 in a bacterial genome does not cause an increased frequency of transposition that could damage the genome.

The effects of *dam* methylation provide the most important system of regulation for an individual element. They reduce the frequency of transposition and (more importantly) couple transposition to passage of the replication fork. The ability of IS10 to transpose is related to the replication cycle by the transposon's response to the state of methylation at two sites. One site is within the inverted repeat at the end of IS10R, where transposase is assumed to bind. The other site is in the promoter P_{IN}, from which the transposase gene is transcribed.

Both of these sites are methylated by the *dam* system described in Chapter 20. The Dam methylase modifies the adenine in the sequence GATC on a newly synthesized strand generated by replication. The frequency of Tn10 transposition is increased 1000-fold in *dam⁻* strains in which the two target sites lack methyl groups.

Passage of a replication fork over these sites generates hemi-methylated sequences; this activates the transposon by a combination of transcribing the transposase gene more frequently from P_{IN} and enhancing binding of transposase to the end of IS10R. In a wild-type bacterium, the sites remain hemi-methylated for a short period after replication.

Why should it be desirable for transposition to occur soon after replication? The nonreplicative mechanism of Tn10 transposition places the donor DNA at risk of being destroyed (see Figure 34.6). The cell's chances of survival may be much increased if replication has just occurred to generate a second copy of the donor sequence. The mechanism is effective because only 1 of the 2 newly replicated copies gives rise to a transposition event (because it matters which strand of the transposon is unmethylated at the *dam* sites).

Since a transposon selects its target site at random, there is a reasonable probability that it may land in an active operon. Will transcription from the outside continue through the transposon and thus activate the transposase, whose overproduction may in turn lead to high (perhaps lethal) levels of transposition? Tn10 protects itself against such events by two mechanisms. Transcription across the IS10R terminus decreases its activity, presumably by inhibiting its ability to bind transposase. And the mRNA that extends from upstream of the promoter is poorly translated, because it has a secondary structure in which the initiation codon is inaccessible.

Controlling elements in maize are transposable

Genetic studies of maize initiated in the 1940s identified changes in the genome during somatic cell division. These changes are brought about by **controlling elements**, recognized by their ability to move from one site to another. The name recognized a distinction between the elements and the target genes, on which they impose novel patterns of expression. Originally identified by McClintock using purely genetic means, the controlling elements can now be recognized as transposable elements, directly comparable to those in bacteria.

Insertion of a controlling element may affect the activity of adjacent genes. Deletions, duplications, inversions, and translocations all occur at the sites where controlling elements are present. Chromosome breakage is a common consequence of the presence of some elements. A unique feature of the maize system is that the activities of the controlling elements are regulated during development. The elements transpose and promote genetic rearrangements at characteristic times and frequencies during plant development.

The maize genome contains several families of controlling elements. The numbers, types, and locations of the elements are characteristic for each individual maize strain. The members of each family are divided into two classes:

◆ **Autonomous elements** have the ability to excise and transpose. Because of the continuing activity of an autonomous element, its insertion at any locus creates an unstable or 'mutable' allele. Loss of the autonomous element itself, or of its ability to transpose, converts a mutable allele to a stable allele.

◆ **Nonautonomous elements** are stable; they do not transpose or suffer other changes in condtion spontaneously. They become unstable *only* when an autonomous member of the same family is present elsewhere in the genome. When complemented in *trans* by an autonomous element, a nonautonomous element displays the usual range of activities associated with autonomous elements, including the ability to transpose to new sites. Nonautonomous elements are derived from autonomous elements by loss of *trans*-acting functions needed for transposition.

Families of controlling elements are defined by the interactions between autonomous and non-autonomous elements. A family consists of a single type of autonomous element accompanied by many varieties of nonautonomous elements. A nonautonomous element is placed in a family by its ability to be activated in *trans* by the autonomous elements. The major families of controlling elements in maize are summarized in **Figure 34.16.**

Characterized at the molecular level, the maize transposons share the usual form of organization—inverted repeats at the ends and short direct repeats in the adjacent target DNA—but otherwise vary in size and coding capacity. The best characterized families are those of the *Ac* elements and the *Spm* elements. There are typically several members (~10) of each family in a plant genome.

Two features of maize have helped to follow transposition events. Controlling elements often insert near genes that have visible but nonlethal effects on the phenotype. And because maize displays clonal development, the occurrence and timing of a transposition event can be visualized as depicted diagrammatically in **Figure 34.17.**

The nature of the event does not matter: it may be an insertion, excision, or chromosome break. What is important is that it occurs in a heterozygote to alter the expression of one allele. Then the descendants of a cell that has suffered the event display a new phenotype, while the descendants of cells not affected by the event continue to display the original phenotype.

Mitotic descendants of a given cell remain in the

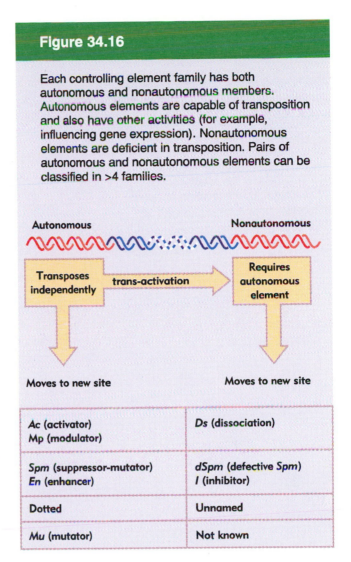

Figure 34.16

Each controlling element family has both autonomous and nonautonomous members. Autonomous elements are capable of transposition and also have other activities (for example, influencing gene expression). Nonautonomous elements are deficient in transposition. Pairs of autonomous and nonautonomous elements can be classified in >4 families.

Autonomous	Nonautonomous
Ac (activator) *Mp* (modulator)	*Ds* (dissociation)
Spm (suppressor-mutator) *En* (enhancer)	*dSpm* (defective *Spm*) *I* (inhibitor)
Dotted	Unnamed
Mu (mutator)	Not known

same location and thus give rise to a **sector** of tissue. A change in phenotype during somatic development is called **variegation**; it is revealed by a sector of the new phenotype residing within the tissue of the original phenotype. The size of the sector depends on the number of divisions in the lineage giving rise to it; so the size of the area of the new phenotype is determined by the timing of the change in genotype. The earlier its occurrence in the cell lineage, the greater the number of descendants and thus the size of patch in the mature tissue. This is seen most vividly in the variation in kernel color, when patches of one color appear within another color.

Figure 34.17

Clonal analysis identifies a group of cells descended from a single ancestor in which a transposition-mediated event altered the phenotype. Timing of the event during development is indicated by the number of cells; tissue specificity of the event may be indicated by the location of the cells.

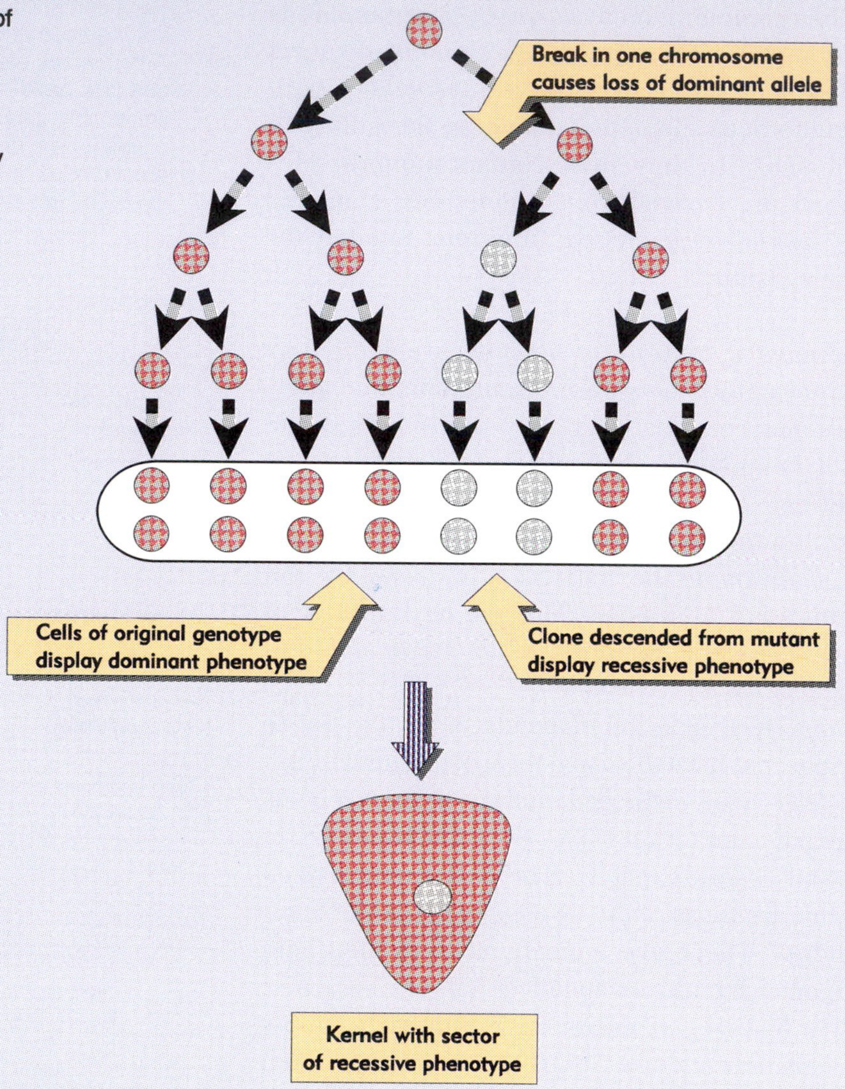

Break in one chromosome causes loss of dominant allele

Cells of original genotype display dominant phenotype

Clone descended from mutant display recessive phenotype

Kernel with sector of recessive phenotype

Autonomous elements give rise to nonautonomous elements

By analyzing autonomous and nonautonomous elements of the *Ac/Ds* family, we have molecular information about many individual examples of these elements. **Figure 34.18** summarizes their structures.

The autonomous *Ac* element is 4563 bp long and is transcribed into a single RNA that is spliced to give a 3500 base mRNA containing a coding sequence of 807 codons. This corresponds to expression of a single gene with 5 exons, whose

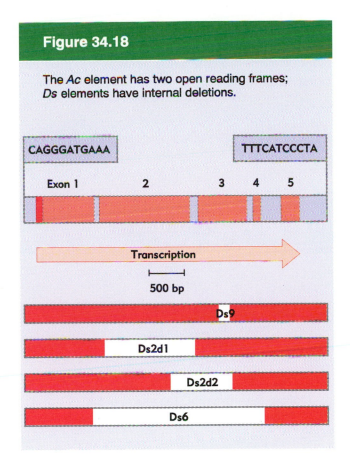

Figure 34.18

The *Ac* element has two open reading frames; *Ds* elements have internal deletions.

CAGGGATGAAA   TTTCATCCCTA

Exon 1   2   3   4   5

Transcription

500 bp

Ds9

Ds2d1

Ds2d2

Ds6

sequence accounts for most of the length of *Ac*. The element itself ends in inverted repeats of 11 bp; and a target sequence of 8 bp is duplicated at the site of insertion.

Ds elements vary in both length and sequence, but are related to *Ac*. They end in the same 11 bp inverted repeats. They are shorter than *Ac*, and the length of deletion varies. At one extreme, the element *Ds9* has a deletion of only 194 bp. In a more extensive deletion, the *Ds6* element retains a length of only 2 kb, representing 1 kb from each end of *Ac*. A complex double *Ds* element has one *Ds6* sequence inserted in reverse orientation into another.

Nonautonomous elements lack internal sequences, but possess the terminal inverted repeats (and possibly other sequence features). *Nonautonomous elements are derived from autonomous elements by deletions (or other changes) that inactivate the trans-acting transposase, but leave the sites intact (including the termini) on which the transposase acts.* Their structures range from minor (but inactivating) mutations of *Ac* to sequences that have major deletions or rearrangements.

At another extreme, the *Ds1* family members comprise short sequences whose only relationship to *Ac* lies in the possession of terminal inverted repeats. Elements of this class need not be directly derived from *Ac*, but could be derived by any event that generates the inverted repeats. Their existence suggests that the transposase recognizes only the terminal inverted repeats, or possibly the terminal repeats in conjunction with some short internal sequence.

Autonomous and nonautonomous elements are subject to a variety of changes in their condition. Some of these changes are genetic, others are epigenetic.

The major change is (of course) the conversion of an autonomous element into a nonautonomous element, but further changes may occur in the nonautonomous element. *Cis*-acting defects may render a nonautonomous element impervious to autonomous elements. Thus a nonautonomous element may become permanently stable because it can no longer be activated to transpose.

Autonomous elements are subject to 'changes of phase', heritable but relatively unstable alterations in their properties. These take the form of a reversible inactivation in which the element cycles between an active and inactive condition during plant development. The phase therefore influences the timing and frequency of transposition.

Phase changes in both the *Ac* and *Mu* types of autonomous element appear to result from methylation of DNA. Comparisons of the susceptibilities of active and inactive elements to restriction enzymes suggest that the inactive form of the element is methylated in the target sequence $\frac{CAG}{GTC}$. There are several target sites in each element, and we do not know which sites control the effect. We should like to know what controls the methylation and demethylation of the elements.

There may be self-regulating controls of transposition, analogous to the immunity effects displayed by bacterial transposons. An increase in the number of *Ac* elements in the genome decreases the frequency of transposition. The *Ac* element may code for a repressor of transposition; the activity could be carried by the same protein that provides transposase function.

Ds may transpose or cause chromosome breakage

Transposition of *Ds* occurs by a nonreplicative mechanism, and is accompanied by its disappearance from the donor location. Clonal analysis suggests that transposition of *Ds* almost always occurs after the donor element has been replicated. These features resemble transposition of the bacterial element Tn10. The recipient site is frequently on the same chromosome as the donor site, and often quite close to it.

Replication generates two copies of a potential *Ac* donor, but usually only one copy actually transposes. What could be responsible for this asymmetry? A transposase may distinguish between the two strands of DNA, as happens with Tn10. Or the *Ac* element could contain an asymmetrical methylation site, such as $\frac{C^*GTG}{G^*CAC}$. After replication, the daughter copies of *Ac* will have different hemimethylated sequences, $\frac{C^*GTG}{G\ CAC}$ and $\frac{C\ GTC}{G^*CAC}$. One sequence may be recognized by the transposase more efficiently than the other, thus initiating the transposition event preferentially on one chromatid.

What happens to the donor site? The rearrangements that are found at sites from which controlling elements have been lost could be explained in terms of the consequences of a chromosome break.

The *Ds* element was originally identified by its ability to provide a site for chromosome breakage upon activation by *Ac*. The consequences are illustrated in **Figure 34.19**. Consider a heterozygote in which *Ds* lies on one homologue between

Figure 34.19

A break at a controlling element causes loss of an acentric fragment; if the fragment carries the dominant markers of a heterozygote, its loss changes the phenotype. The effects of the dominant markers, *C-I*, *Bz*, *Wx*, can be visualized by the color of the cells or by appropriate staining. (*C-I* is a regulator that causes colorless aleurone; *Bz* glycosylates the anthocyanid pigments; *Wx* is starchy and stains blue with I_2-KI.)

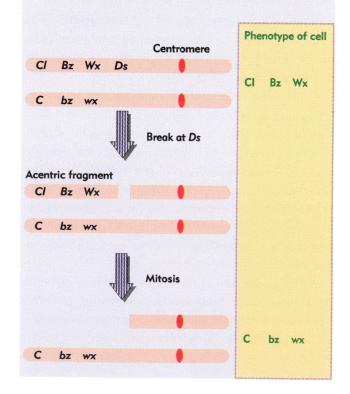

the centromere and a series of dominant markers. The other homologue lacks *Ds* and has recessive markers (*C, bz, wx*). Breakage at *Ds* generates an **acentric fragment** carrying the dominant markers. Because of its lack of a centromere, this fragment is lost at mitosis. Thus the descendant cells have only the recessive markers carried by the intact chromosome. This gives the type of

Figure 34.20

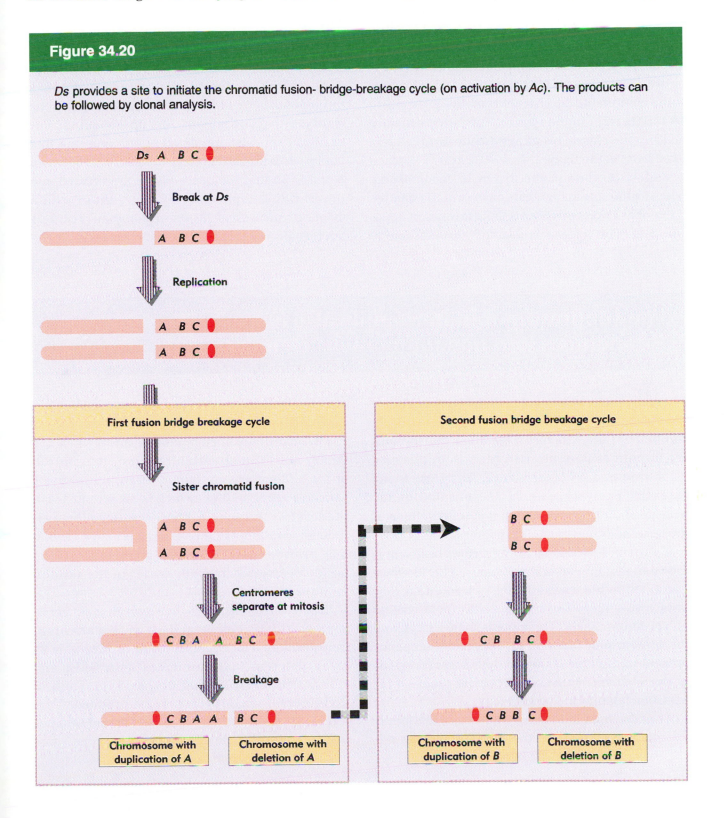

Ds provides a site to initiate the chromatid fusion- bridge-breakage cycle (on activation by *Ac*). The products can be followed by clonal analysis.

situation whose results are depicted in Figure 34.17.

Figure 34.20 shows that breakage at *Ds* leads to the formation of two unusual chromosomes. These are generated by joining the broken ends of the products of replication. One is a U-shaped acentric fragment consisting of the joined sister chromatids for the region distal to *Ds* (on the left as drawn in the figure). The other is a U-shaped **dicentric chromosome** comprising the sister chromatids proximal to *Ds* (on its right in the figure). The latter structure leads to the classic **breakage-fusion-bridge** cycle illustrated in the figure.

Follow the fate of the dicentric chromosome when it attempts to segregate on the mitotic spindle. Each of its two centromeres pulls toward an opposite pole. The tension breaks the chromosome at a random site between the centromeres. In the example of the figure, breakage occurs between loci *A* and *B*, with the result that one daughter chromosome has a duplication of *A*, while the other has a deletion. If *A* is a dominant marker, the cells with the duplication will retain **A** phenotype, but cells with the deletion will display the recessive **a** phenotype.

The breakage-fusion-bridge cycle continues through further cell generations, allowing genetic changes to continue in the descendants. For example, consider the deletion chromosome that has lost *A*. In the next cycle, a break occurs between *B* and *C*, so that the descendants are divided into those with a duplication of *B* and those with a deletion. Successive losses of dominant markers are revealed by subsectors within sectors.

The suppressor–mutator family also consists of autonomous and nonautonomous elements

The *Spm* and *En* autonomous elements are virtually identical; they differ at <10 positions. **Figure 34.21** summarizes the structure. Like other transposons, the termini contain inverted repeats, in this case consisting of a 13 bp sequence. The repeats are essential for transposition, as indicated by the transposition-defective phenotype of deletions at the termini.

A sequence of 8300 bp is transcribed from a promoter in the left end of the element. The 11 exons contained in the transcript are spliced into a 2500 base messenger. The mRNA codes for a protein of 621 amino acids. The gene is called *tnpA*, and the protein binds to a 12 bp consensus sequence present in multiple copies in the terminal regions of the element. Function of *tnpA* is required for excision, but may not be sufficient.

All of the nonautonomous elements of this family (denoted *dSpm* for defective *Spm*) are closely related in structure to the *Spm* element itself. They have deletions that affect the exons of *tnpA*.

Two additional open reading frames (ORF1 and ORF2) are located within the first, long intron of *tnpA*. They are contained in an alternatively spliced 6000 base RNA, which is present at 1% of the level of the *tnpA* mRNA. The (hypothetical) function containing ORFs 1 and 2 is called *tnpB*. It may provide the protein that binds to the 13 bp terminal inverted repeats to cleave the termini for transposition.

In addition to the fully active *Spm* element, there are *Spm-w* derivatives that show weaker activity in transposition. The example given in Figure 34.21 has a deletion that eliminates both ORF1 and ORF2. This suggests that the need for TnpB in transposition can be bypassed or substituted.

Transposons related to *Spm* are found in other plants, and are defined as members of the same

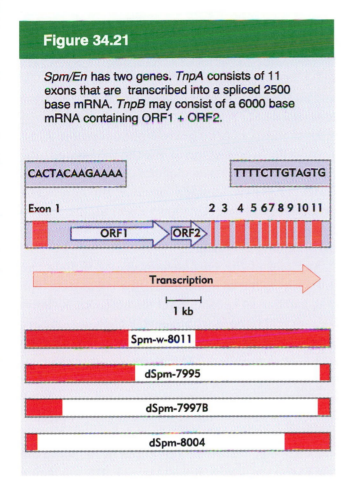

Figure 34.21

Spm/En has two genes. *TnpA* consists of 11 exons that are transcribed into a spliced 2500 base mRNA. *TnpB* may consist of a 6000 base mRNA containing ORF1 + ORF2.

CACTACAAGAAAA

TTTTCTTGTAGTG

Exon 1   2 3 4 5 6 7 8 9 10 11

ORF1   ORF2

Transcription

1 kb

Spm-w-8011

dSpm-7995

dSpm-7997B

dSpm-8004

family by their generally similar organization. They all share nearly identical terminal inverted repeats, and generate 3 bp duplications of target DNA upon transposition. Named for the terminal similarities, they are known as the CACTA group of transposons.

Spm insertions can control the expression of a gene at the site of insertion. A recipient locus may be brought under either negative or positive control. An *Spm-suppressible* locus suffers inhibition of expression. An *Spm-dependent* locus is expressed only with the aid of *Spm*. When the inserted element is a *dSpm*, suppression or dependence responds to the *trans*-acting function supplied by an autonomous *Spm*. What is the basis for these opposite effects?

A *dSpm-suppressible* allele contains an insertion of *dSpm* within an exon of the gene. This structure raises the immediate question of how a gene with a *dSpm* insertion in an exon can ever be expressed! The *dSpm* sequence can be spliced out of the transcript by using sequences at its termini. The splicing event may leave a change in the sequence of the mRNA, thus explaining a change in the properties of the protein for which it codes. A similar ability to be spliced out of a transcript has been found for some *Ds* insertions.

tnpA provides the *suppressor* function for which the *Spm* element was originally named. The presence of a defective element may reduce, but not eliminate, expression of a gene in which it resides. However, the introduction of an autonomous element, possessing a functional *tnpA* gene, may suppress expression of the target gene entirely. Suppression is caused by the ability of TnpA to bind to its target sites in the defective element, which blocks transcription from proceeding.

A *dSpm-dependent* allele contains an insertion near but not within a gene. The insertion appears to provide an enhancer that activates the promoter of the gene at the recipient locus.

Suppression and dependence at *dSpm* elements appear to rely on the same interaction between the *trans*-acting product of the *tnpA* gene of an autonomous *Spm* element and the *cis*-acting sites at the ends of the element. So a single interaction between the protein and the ends of the element either suppresses or activates a target locus depending on whether the element is located upstream of or within the recipient gene.

Spm elements exist in a variety of states ranging from fully active to cryptic. A cryptic element is silent and neither transposes itself nor activates *dSpm* elements. A cryptic element may be reactivated transiently or converted to the active state by interaction with a fully active *Spm* element. Inactivation is caused by methylation of sequences in the vicinity of the transcription start site. The nature of the events that are responsible for inactivating an element by *de novo* methylation or for activating it by demethylation (or preventing methylation) are not yet known.

The role of transposable elements in hybrid dysgenesis

Certain strains of *D. melanogaster* encounter difficulties in interbreeding. When flies from two of these strains are crossed, the progeny display 'dysgenic traits', a series of defects including mutations, chromosomal aberrations, distorted segregation at meiosis, and sterility. The appearance of these correlated defects is called **hybrid dysgenesis**.

Two systems responsible for hybrid dysgenesis have been identified in *D. melanogaster*. In the first, flies are divided into the types I (inducer) and R (reactive). Reduced fertility is seen in crosses of I males with R females, but not in the reverse direction. In the second system, flies are divided into the two types P (paternal contributing) and M (maternal contributing). **Figure 34.22** illustrates the asymmetry of the system; a cross between a P male and an M female causes dysgenesis, but the reverse cross does not.

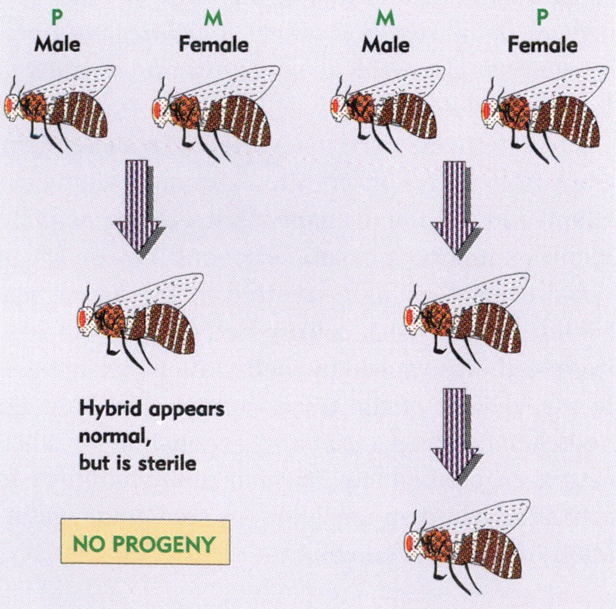

Dysgenesis is principally a phenomenon of the germ cells. In crosses involving the P–M system, the F1 hybrid flies have normal somatic tissues. However, their gonads do not develop. The morphological defect in gamete development dates from the stage at which rapid cell divisions commence in the germ line.

Any one of the chromosomes of a P male can induce dysgenesis in a cross with an M female. The construction of recombinant chromosomes shows that several regions within each P chromosome are able to cause dysgenesis. This suggests that a P male has a large number of **P factors**, sequences occupying many different chromosomal locations. The locations differ between individual P strains. The P factors are absent from chromosomes of M flies.

The events responsible for the induction of mutations in dysgenesis were identified by mapping the DNA of *w* mutants found among the dysgenic hybrids. All the mutations result from the insertion of DNA into the *w* locus. The inserted sequence is called the **P element**. The P insertions form a classic transposable system. Individual elements vary in length but are homologous in sequence. All P elements possess inverted terminal repeats of 31 bp, and generate direct repeats of target DNA of 8 bp upon transposition. The longest P elements are ~2.9 kb long and have four open reading frames. The shorter elements arise, apparently rather frequently, by internal deletions of a full-length P factor. At least some of the shorter P elements have lost the capacity to produce the transposase, but may be activated in *trans* by the enzyme coded by a complete P element.

A P strain carries 30–50 copies of the P factor, about a third of them full length. The factors are absent from M strains. In a P strain, the factors are carried as inert components of the genome. But they become activated to transpose when a P male is crossed with an M female.

Figure 34.22

Hybrid dysgenesis is asymmetrical; it is induced by P male X M female crosses, but not by M male X P female crosses.

P Male   M Female   M Male   P Female

Hybrid appears normal, but is sterile

NO PROGENY

Chromosomes from P–M hybrid dysgenic flies have P factors inserted at many new sites. The chromosome breaks typical of hybrid dysgenesis occur at hotspots that are the sites of residence of P factors. The average rate of transposition of P elements to M chromosomes is ~1 event per generation.

Activation of P elements is tissue-specific: it occurs only in the germ line. But P elements are transcribed in both germ line and somatic tissues. *Tissue-specificity is conferred by a change in the splicing pattern.*

Figure 34.23 depicts the organization of the element and its transcripts. The primary transcript extends for 2.5 or 3.0 kb, the difference probably reflecting merely the leakiness of the termination site. Two protein products can be produced:

◆ In somatic tissues, only the first two introns are spliced, creating an mRNA consisting of ORF0-ORF1-ORF2. Translation of this RNA yields a protein of 66,000 daltons. This protein is a repressor of transposon activity.

◆ In germline tissues, an additional splicing event occurs to remove intron 3. This connects all four open reading frames into an mRNA that is translated to generate a protein of 87,000 daltons. This protein is the transposase.

Two types of experiment have demonstrated that splicing of the third intron is needed for transposition. First, if the splicing junctions are mutated *in vitro* and the P element is reintroduced into flies, its transposition activity is abolished. Second, if the third intron is deleted, so that ORF3 is constitutively included in the mRNA in all tissues, transposition occurs in somatic tissues as well as the germline.

So whenever ORF3 is spliced to the preceding reading frame, the P element becomes active. This is the crucial regulatory event, and usually it occurs only in the germ line. What is responsible for the tissue-specific splicing? We saw in Chapter 31 that somatic cells contain a protein that binds to sequences in exon 3 to prevent splicing of the last intron. The absence of this protein in germline cells

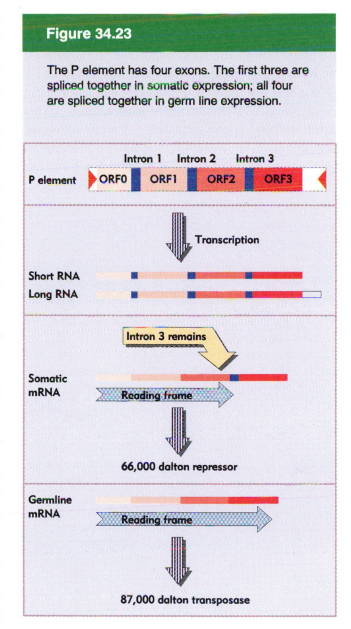

Figure 34.23

The P element has four exons. The first three are spliced together in somatic expression; all four are spliced together in germ line expression.

allows splicing to generate the mRNA that codes for the transposase.

Transposition of a P element requires ~150 bp of terminal DNA. The transposase binds to 10 bp sequences that are adjacent to the 31 bp inverted repeats. Transposition occurs by a nonreplicative 'cut and paste' mechanism resembling that of Tn10. (It contributes to hybrid dysgenesis in two ways. Insertion of the transposed element at a new site may cause mutations. And the break that is left at the donor site [as in the model of Figure 34.6] has a deleterious effect.)

It is interesting that, in a significant proportion of cases, the break in donor DNA is repaired by using the sequence of the homologous chromosome. If the homologue has a P element, the presence of a P element at the donor site may be restored (so the event resembles the result of a replicative transposition). If the homologue lacks a P element, repair may generate a sequence lacking the P element, thus apparently providing a precise excision (an unusual event in other transposable systems).

The dependence of hybrid dysgenesis on the sexual orientation of a cross shows that the cytoplasm is important as well as the P factors themselves. The contribution of the cytoplasm is described as the **cytotype**; a lines of flies containing P elements has P cytotype, while a line of flies lacking P elements has M cytotype. Hybrid dysgenesis occurs only when chromosomes containing P factors find themselves in M cytotype, that is, when the male parent has P elements and the female parent does not.

Figure 34.24

Hybrid dysgenesis is determined by the interactions between P elements in the genome and 66K repressor in the cytotype.

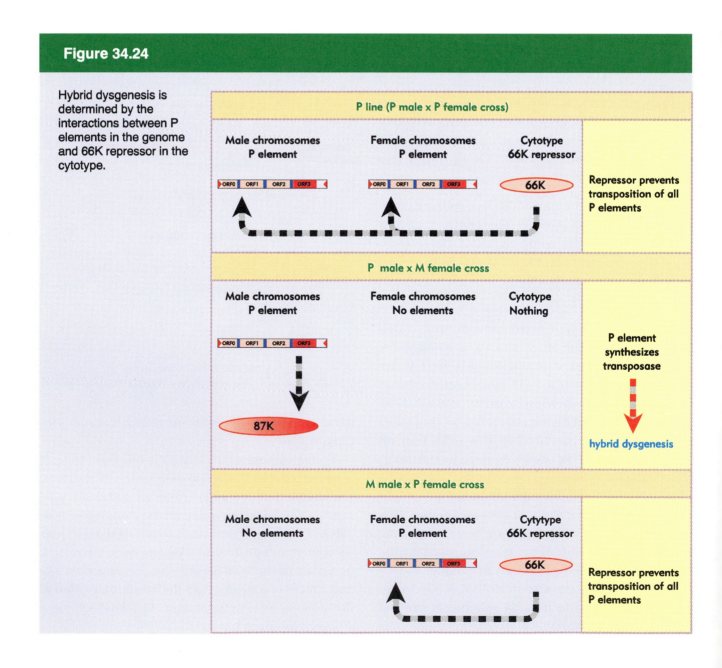

Cytotype shows an inheritable cytoplasmic effect; when a cross occurs through P cytotype (the female parent has P elements), hybrid dysgenesis is suppressed for several generations of crosses with M female parents. Thus something in P cytotype, which can be diluted out over some generations, suppresses hybrid dysgenesis.

The effect of cytotype is explained in molecular terms by the model of **Figure 34.24**. It depends on the ability of the 66K protein to repress transposition. The protein is provided as a maternal factor in the egg. In a P line, there must be sufficient protein to prevent transposition from occurring, even though the P elements are present. In any cross involving a P female, its presence prevents either synthesis or activity of the transposase. But when the female parent is M type, there is no repressor in the egg, and the introduction of a P element from the male parent results in activity of transposase in the germ line. The ability of P cytotype to exert an effect through more than one generation suggests that there must be enough repressor protein in the egg, and it must be stable

enough, to be passed on through the adult to be present in the eggs of the next generation.

Strains of *D. melanogaster* descended from flies caught in the wild more than 30 years ago are always M. Strains descended from flies caught in the past 10 years are almost always P. Does this mean that the P element family has invaded wild populations of *D. melanogaster* in recent years? P elements are indeed highly invasive when introduced into a new population; the source of the invading element would have to be another species.

Because hybrid dysgenesis reduces interbreeding, it is a step on the path to speciation. Suppose that a dysgenic system is created by a transposable element in some geographic location. Another element may create a different system in some other location. Flies in the two areas will be dysgenic for two (or possibly more) systems. If this renders them intersterile and the populations become genetically isolated, further separation may occur. Multiple dysgenic systems therefore lead to inability to mate—and to speciation.

Summary

Prokaryotic and eukaryotic cells contain a variety of transposons that mobilize by moving or copying DNA sequences. The transposon can be identified only as an entity within the genome; its mobility does not involve an independent form. The transposon could be selfish DNA, concerned only with perpetuating itself within the resident genome; if it conveys any selective advantage upon the genome, this must be indirect. All transposons have systems to limit the extent of transposition, since unbridled transposition is presumably damaging, but the molecular mechanisms are different in each case.

The archetypal transposon has inverted repeats at its termini and generates direct repeats of a short sequence at the site of insertion. The simplest types are the bacterial insertion sequences (IS),

which consist essentially of the inverted terminal repeats flanking a coding frame(s) whose product(s) provide transposition activity. Composite transposons have terminal modules that consist of IS elements; one or both of the IS modules provides transposase activity, and the sequences between them (often carrying antibiotic resistance), are treated as passengers.

The generation of target repeats flanking a transposon reflects a common feature of transposition. The target site is cleaved at points that are staggered on each DNA strand by a fixed distance (often 5 or 9 base pairs). The transposon is in effect inserted between protruding single-stranded ends generated by the staggered cuts. Target repeats are generated by filling in the single-stranded regions.

IS elements, composite transposons, and P elements mobilize by nonreplicative transposition, in which the element moves directly from a donor site to a recipient site. A single transposase enzyme undertakes the reaction. It occurs by a 'cut and paste' mechanism in which the transposon is separated from flanking DNA. Loss of the transposon from the donor creates a double-strand break, whose fate is not clear. In the case of Tn10, transposition becomes possible immediately after DNA replication, when sites recognized by the *dam* methylation system are transiently hemimethylated. This imposes a demand for the existence of two copies of the donor site, which may enhance the cell's chances for survival.

The TnA family of transposons mobilize by replicative transposition. After the transposon at the donor site becomes connected to the target site, replication generates a cointegrate molecule that has two copies of the transposon. A resolution reaction, involving recombination between two particular sites, then frees the two copies of the transposon, so that one remains at the donor site and one appears at the target site. Two enzymes coded by the transposon are required: transposase recognizes the ends of the transposon and connects them to the target site; and resolvase provides a site-specific recombination function.

Phage Mu undergoes replicative transposition by the same mechanism as TnA. It also can use its cointegrate intermediate to transpose by a nonreplicative mechanism that differs from the 'cut and paste'.

The best characterized transposons in plants are the controlling elements of maize, which fall into several families. Each family contains a single type of autonomous element, analogous to bacterial transposons in its ability to mobilize. A family also contains many different nonautonomous elements, derived by mutations (usually deletions) of the autonomous element. The nonautonomous elements lack the ability to transpose, but display transposition activity and other abilities of the autonomous element, when an autonomous element is present to provide the necessary *trans*-acting functions.

In addition to the direct consequences of insertion and excision, the maize elements may also control the activities of genes at or near the sites where they are inserted; this control may be subject to developmental regulation. Maize elements inserted into genes may be spliced out of the transcripts, which explains why they do not simply impede gene activity. Control of target gene expression involves a variety of molecular effects, including activation by provision of an enhancer and suppression by interference with post-transcriptional events.

Transposition of maize elements (in particular *Ac*) is nonreplicative, probably requiring only a single transposase enzyme coded by the element. Transposition occurs preferentially after replication of the element, like Tn10, probably because a critical site becomes hemimethylated during DNA synthesis. There are probably mechanisms to limit the frequency of transposition. Advantageous rearrangements of the maize genome may have been connected with the presence of the elements.

P elements in *D. melanogaster* are responsible for hybrid dysgenesis, which could be a forerunner of speciation. A cross between a male carrying P elements and a female lacking them generates hybrids that are sterile. A P element has 4 open reading frames, separated by introns. Splicing of the first 3 ORFs generates a 66K repressor, and occurs in all cells. Splicing of all 4 ORFs to generate the 87K transposase occurs only in the germ line, by a tissue-specific splicing event. P elements mobilize when exposed to cytoplasm lacking the repressor. They transpose by a nonreplicative 'cut and paste' mechanism. The burst of transposition events inactivates the genome by random insertions. Only a complete P element can generate transposase, but defective elements can be mobilized in *trans* by the enzyme.

Further reading

Reviews

The development of early views on transposons can be followed through the reviews of **Kleckner** (*Cell* 11, 11–23, 1977), **Calos and Miller** (*Cell* 20, 579–595, 1980), and **Kleckner** (*Ann. Rev. Genet.* 15, 341–404, 1981).

A collection of reviews on transposons is provided by *Mobile DNA* (ed. Berg and Howe, American Society of Microbiology, 1989). Chapters of particular interest include **Galas and Chandler** on IS elements (pp. 109–162), **Kleckner** on Tn10 (pp. 227–268), **Sherratt** on Tn3 (pp. 163–185), **Berg** on Tn5 (pp. 185–210) and **Pato** on phage Mu (pp. 23–52).

The evolutionary significance of transposons was considered by **Campbell** (*Ann. Rev. Microbiol.* 35, 55–83, 1981).

Mechanisms for transposition have been reviewed by **Grindley and Reed** (*Ann. Rev. Biochem.* 54, 863–896, 1985). Control of transposition frequency has been reviewed by **Kleckner** (*Ann. Rev. Cell Biol.* 6, 297–327, 1990).

Fedoroff brought the topic of maize-controlling elements into the molecular era with reviews in *Mobile Genetic Elements* (ed. Shapiro, Academic Press, New York, 1–65, 1983) and *Mobile DNA* (op. cit., 375–412, 1988). Later progress was summarized by **Gierl, Saedler, and Peterson** (*Ann. Rev. genet.* 23, 71–85, 18989).

Drosophila P elements have been summarized by **Engels** (*Ann. Rev. Genet.* 17, 315–344, 1983) and in *Mobile DNA* (op. cit., pp 437–484).

Discoveries

The nature of the insertion associated with transposition was discovered by **Grindley** (*Cell* 13, 419–426, 1978) and **Johnsrud, Calos, and Miller** (*Cell* 15, 1209–1219, 1978).

Nonreplicative transposition of Tn10 was characterized by **Bender and Kleckner** (*Cell* 45, 801–815, 1986) and by **Haniford, Benjamin, and Kleckner** (*Cell* 64, 171–179, 1881). The effect of methylation on IS10 transposition was discovered by **Roberts** *et al.* (*Cell* 43, 117–130, 1985).

The involvement of cointegrates in replicative transposition of Tn3 was demonstrated by **Grindley** *et al.* (*Cell* 30, 19–27, 1982). The resolution reaction was characterized by **Droge** *et al.* (*Proc. Nat. Acad. Sci. USA* 87, 5336–5340, 1990).

The basis for germ line specificity of P elements was discovered by **Laski, Rio, and Rubin** (*Cell* 44, 7–19, 1986).

CHAPTER 35

Retroviruses and retroposons

Transposable elements propagate in eukaryotes as successfully as in prokaryotes. The genetic background of eukaryotes includes a variety of elements that possess the ability to move to randomly selected new locations within the genome in which they reside.

Eukaryotic transposable elements can be divided into two general groups:

◆ One group of elements is comparable to bacterial transposons. The elements end in short inverted repeats and generate short direct repeats of target DNA at the site of insertion. Like bacterial transposons, these elements have no life outside the genome. Whether valuable components playing a role in cellular survival or selfish parasites concerned only with their own survival, they have no independent existence and do not generate free molecules of DNA. The best characterized are the controlling elements of maize and the P elements of *D. melanogaster* discussed in Chapter 34.

◆ The paradigm for another type of transposition event is provided by the ability of **retroviruses** to insert DNA copies (proviruses) of an RNA viral genome into the chromosomes of a host cell. Some eukaryotic transposons are related to retroviral proviruses in their general organization, and they transpose through RNA intermediates. As a class, these elements are called **retroposons** (or sometimes **retrotransposons**.) They range from the retroviruses themselves, able freely to infect host cells, to sequences that have transposed via RNA, but which do not themselves possess the ability to transpose. They share with the other transposon class the characteristic of generating short direct repeats of target DNA when an insertion occurs.

Even in genomes where active transposons have not been detected, footprints of ancient transposition events are found in the form of direct target repeats flanking dispersed repetitive sequences. The features of these sequences sometimes implicate an RNA sequence as the progenitor of the genomic (DNA) sequence. We think that the RNA must have been converted into a duplex DNA copy that was inserted into the genome by a transposition-like event.

Like any other reproductive cycle, the cycle of a retrovirus or retroposon is continuous; it is arbitrary at which point we interrupt it to consider a 'beginning'. But our perspectives of these elements are biased by the forms in which we usually observe them, indicated in **Figure 35.1**. Retroviruses were first observed as infectious virus particles, capable of transmission between cells, and so the intracellular cycle (involving

Figure 35.1

Overview: the reproductive cycles of retroviruses and retroposons involve alternation of reverse transcription from RNA to DNA with transcription from DNA to RNA. Only retroviruses can generate infectious particles that can be released from a cell to reach other cells. Retroposons are confined to an intracellular cycle.

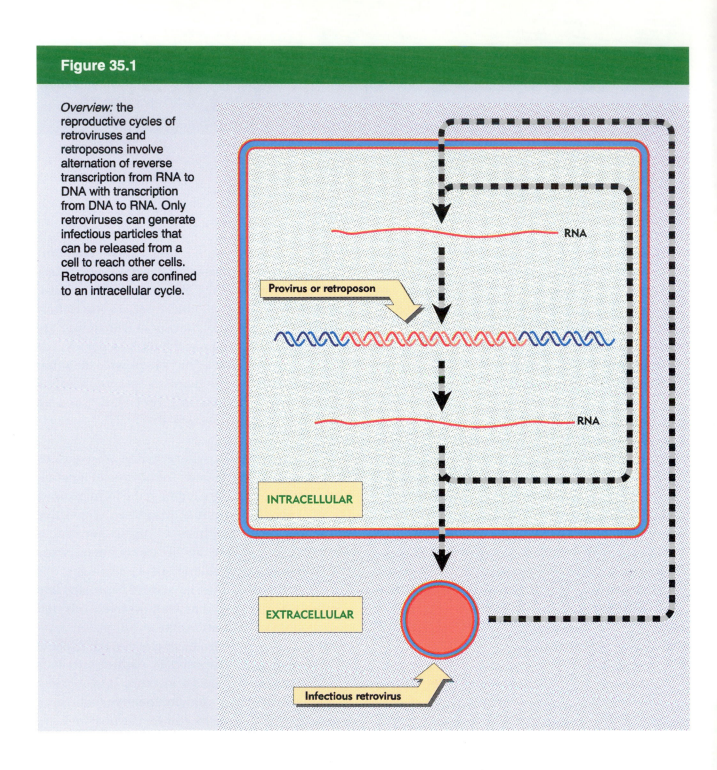

duplex DNA) is thought of as the means of reproducing the RNA virus. Retroposons were discovered as components of the genome; and the RNA forms have been mostly characterized for their functions as mRNAs. So we think of retroposons as genomic (duplex DNA) sequences that occasionally transpose within a genome; they do not migrate between cells.

The retrovirus life cycle involves transposition-like events

Retroviruses have genomes of single-stranded RNA that are replicated through a double-stranded DNA intermediate. The life cycle of the virus involves an obligatory stage in which the double-stranded DNA is inserted into the host genome by a transposition-like event that generates short direct repeats of target DNA.

The significance of this reaction extends beyond the perpetuation of the virus. Some of its consequences are that:

◆ A retroviral sequence that is integrated in the germline remains in the cellular genome as an **endogenous provirus**. Like a lysogenic bacteriophage, a provirus behaves as part of the genetic material of the organism.

◆ Cellular sequences occasionally recombine with the retroviral sequence and then are transposed with it; these sequences may be inserted into the genome as duplex sequences in new locations.

Figure 35.2

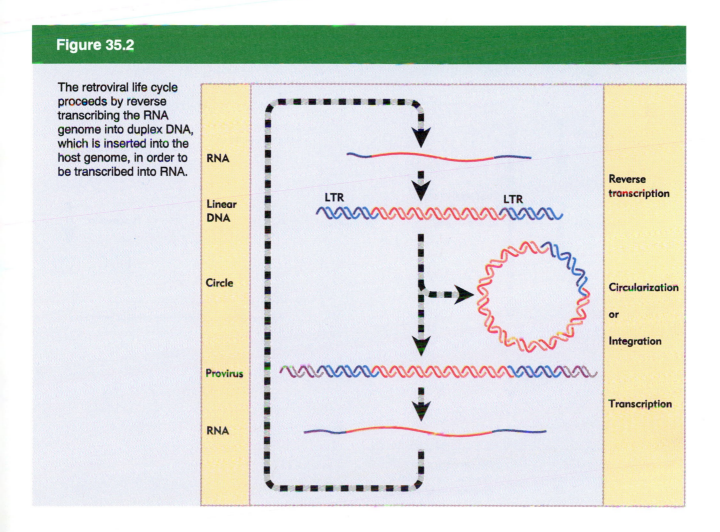

The retroviral life cycle proceeds by reverse transcribing the RNA genome into duplex DNA, which is inserted into the host genome, in order to be transcribed into RNA.

RNA

Linear DNA

Circle

Provirus

RNA

LTR LTR

Reverse transcription

Circularization

or

Integration

Transcription

◆ Cellular sequences that are transposed by a retrovirus may change the properties of a cell that becomes infected with the virus.

The particulars of the retroviral life cycle are expanded in **Figure 35.2**. The crucial steps are that the viral RNA is converted into DNA, the DNA becomes integrated into the host genome, and then the DNA provirus is transcribed into RNA.

The enzyme responsible for generating the initial DNA copy of the RNA is **reverse transcriptase**, which we encountered in Chapter 21 as a powerful tool of recombinant DNA technology. The enzyme converts the RNA into a linear duplex of DNA in the cytoplasm of the infected cell. (The DNA also is converted into circular forms, but these do not appear to be involved in reproduction.)

Figure 35.3

The "genes" of the retrovirus are expressed as polyproteins that are processed into individual products.

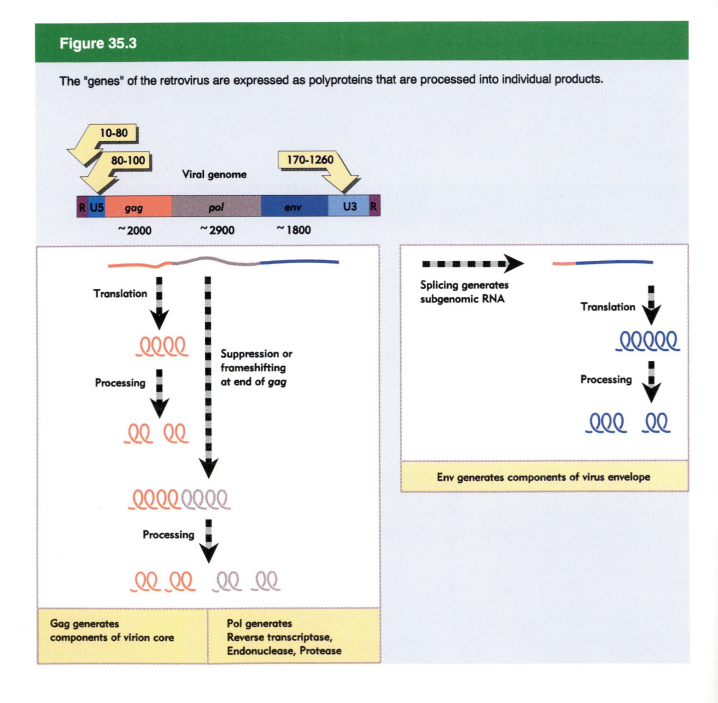

The linear DNA makes its way (by unknown means) to the nucleus. One or more DNA copies become integrated into the host genome. A single enzyme, called **integrase**, is responsible for integration. The integrated **proviral DNA** is transcribed by the host machinery to produce viral RNAs, which serve both as mRNAs and as genomes for packaging into virions. *Integration is a normal part of the life cycle and is necessary for transcription.*

Two copies of the RNA genome are packaged into each virion, making the individual virus particle effectively diploid. When a cell is simultaneously infected by two different but related viruses, it is possible to generate heterozygous virus particles carrying one genome of each type. The diploidy may be important in allowing the virus to acquire cellular sequences. The enzymes reverse transcriptase and integrase are carried with the genome in the viral particle.

A typical retroviral sequence contains three or four 'genes', the term here identifying coding regions each of which actually gives rise to multiple proteins by processing reactions. A typical retrovirus genome with three genes is organized in the sequence *gag-pol-env* as indicated in **Figure 35.3**.

Retroviral mRNA has a conventional structure; it is capped at the 5′ end and polyadenylated at the 3′ end. The mRNA directly representing the genome is translated to give the Gag and Pol polyproteins. The Gag product is translated by reading from the initiation codon to the first termination codon. This termination codon must be bypassed to express Pol.

Different mechanisms are used in different viruses to proceed beyond the *gag* termination codon, depending on the relationship between the *gag* and *pol* reading frames. When *gag* and *pol* follow continuously, suppression by a glutamyl-tRNA that recognizes the termination codon allows a single protein to be generated. When *gag* and *pol* are in different reading frames, a ribosomal frameshift occurs to generate a single protein. Usually the readthrough is about 5% efficient, so Gag protein outnumbers Gag–Pol protein about 20-fold.

The Env polyprotein is expressed by another means: splicing generates a shorter **subgenomic** messenger that is translated into the Env product.

The *gag* gene gives rise to the protein components of the nucleoprotein core of the virion. The *pol* gene codes for functions concerned with nucleic acid synthesis and recombination. The *env* gene codes for components of the envelope of the particle, which also sequesters components from the cellular cytoplasmic membrane.

Both the Gag or Gag–Pol and the Env products are polyproteins that are cleaved by a protease to release the individual proteins that are found in mature virions. The protease activity is coded in various forms: it may be part of *gag* or *pol*, or sometimes takes the form of an additional independent reading frame. **Table 35.1** summarizes the individual proteins released from the polyproteins.

The production of a retroviral particle involves packaging the RNA into a core, surrounding it with capsid proteins, and pinching off a segment of membrane from the host cell. The release of infective particles by such means is shown in **Figure 35.4**. The process is reversed during infection; a virus infects a new host cell by fusing with the plasma membrane and then releasing the contents of the virion.

Retroviruses are called **plus strand viruses**, because the viral RNA itself codes for the protein products. As its name implies, reverse transcriptase is responsible for converting the genome (plus strand RNA) into a complementary DNA strand, which is called the **minus strand DNA**. Reverse transcriptase also catalyzes subsequent stages in the production of duplex DNA. It has a DNA polymerase activity, which enables it to synthesize a duplex DNA from the single-stranded reverse transcript of the RNA. The second DNA strand in this duplex is called **plus strand DNA**. And as a necessary adjunct to this activity, the enzyme has an RNAase H activity, which can degrade the RNA part of the RNA–DNA hybrid. All retroviral reverse transcriptases share considerable similarities of amino acid sequence, and, in fact, homologous sequences can be

Table 35.1

The gag-pol and env polyproteins are cleaved into several small proteins.

Protein	Size	Function
Matrix (MA)	15-20K	Located between nucleocapsid and viral envelope
Capsid (CA)	24-30K	Major structural component of capsid
Nucleocapsid (NC)	10-15K	Packaging the dimer of RNA
Protease (PR)	~15K	Cleaves gag-pol into mature components
Reverse transcriptase (RT)	50-95K	Synthesizes DNA on RNA template
Integrase (IN)	32-46K	Integrating provirus DNA in host genome
Surface protein (SU)	46-120K	Spike on virion surface; interacts with host
Transmembrane (TM)	15-37K	Mediates fusion of viral and host membranes

recognized in some other retroposons (see later).

Like other DNA polymerases, reverse transcriptase requires a primer. The native primer is tRNA. An uncharged host tRNA is present in the virion. A sequence of 18 bases at the 3′ end of the tRNA is base paired to a site 100–200 bases from the 5′ end of one of the viral RNA molecules. The tRNA may also be base paired to another site near the 5′ end of the other viral RNA, thus assisting in dimer formation between the viral RNAs.

The structures of the DNA forms of the virus are compared with the RNA in **Figure 35.5**. The viral RNA has direct repeats at its ends. These **R** segments vary in different strains of virus from 10–80 nucleotides. Following the R segment at the 5′ end of the virus is the **U5** region of 80–100 nucleotides, whose name indicates that it is unique to the 5′ end. Preceding the R segment at the 3′ terminus is the **U3** segment of 170–1350 nucleotides, which is unique to the 3′ end.

Here is a dilemma. Reverse transcriptase starts to synthesize DNA at a site *only 100–200 bases*

Figure 35.4

Retroviruses (HIV) bud from the plasma membrane of an infected cell. Photograph kindly provided by Matthew Gonda.

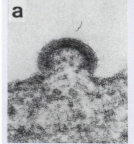

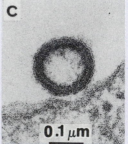
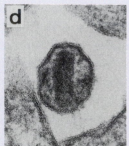
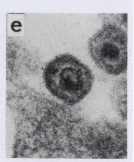

a   b   c   0.1 μm   d   e

downstream from the 5′ end. How can DNA be generated to represent the intact RNA genome? (This is an extreme variant of the general problem in replicating the ends of any linear nucleic acid; see Chapter 18.)

Synthesis *in vitro* proceeds to the end, generating a short DNA sequence called minus strong-stop DNA. This molecule is not found *in vivo* because synthesis continues by the reaction illustrated in **Figure 35.6**. *Reverse transcriptase switches templates,* carrying the nascent DNA with it to the new template.

In this reaction, the R region at the 5′ terminus of the RNA template is degraded by the RNAase H activity of reverse transcriptase. Its removal allows the R region at the 3′ end to base pair with the newly synthesized DNA. Then reverse transcription continues through the U3 region into the body of the RNA.

The result of the switch and extension is to add a U3 segment to the 5′ end. The stretch of sequence U3-R-U5 is called the **long terminal repeat (LTR)** because a similar series of events adds a U5 segment to the 3′ end, giving it the same structure of U5-R-U3.

Because there are two RNA genomes in each retroviral particle, it is possible for recombination to occur during a viral life cycle. Probably two different mechanisms are involved, one during minus strand synthesis and one during plus strand synthesis:

◆ Figure 35.6 shows that reverse transcriptase usually switches from one end of the RNA template to the other end of the same RNA during minus strand synthesis. However, sometimes it switches to the end of a *different* RNA molecule, thus generating a recombinant.

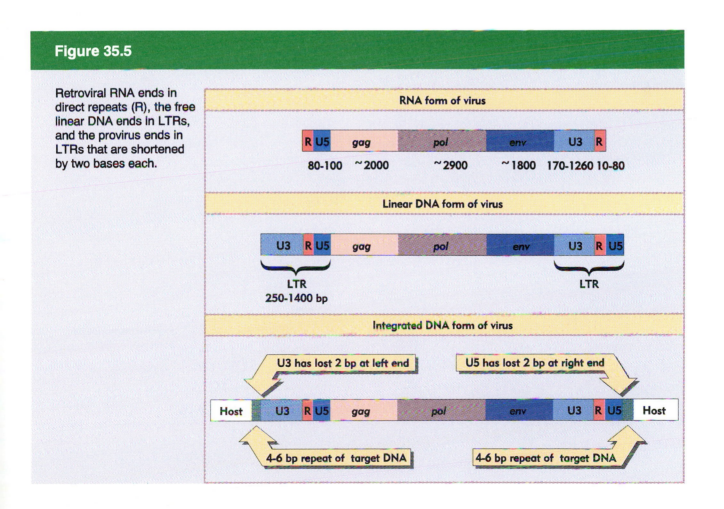

Figure 35.5

Retroviral RNA ends in direct repeats (R), the free linear DNA ends in LTRs, and the provirus ends in LTRs that are shortened by two bases each.

Figure 35.6

The LTR is generated by switching templates during reverse transcription. The new template is usually the other end of the same molecule.

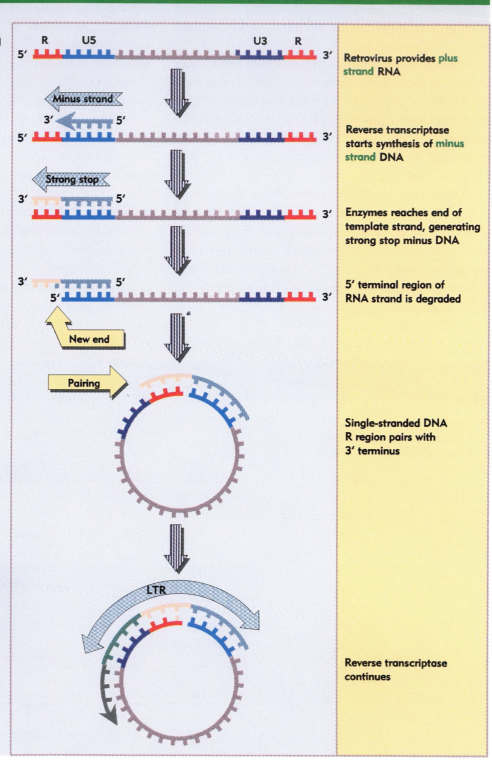

Retrovirus provides plus strand RNA

Reverse transcriptase starts synthesis of minus strand DNA

Enzymes reaches end of template strand, generating strong stop minus DNA

5′ terminal region of RNA strand is degraded

Single-stranded DNA R region pairs with 3′ terminus

Reverse transcriptase continues

♦ Plus strand DNA may be synthesized discontinuously, in a reaction that involves several internal initiations. Strand transfer during this reaction can also occur.

The organization of the integrated provirus resembles that of the linear DNA. Compared with the RNA, the linear provirus DNA therefore has additional sequences at each end, a U3 added to the 5′ end and a U5 added to the 3′ end. The LTRs at each end of the provirus are identical. The 3′ end of U5 consists of a short inverted repeat relative to the 5′ end of U3, so the LTR itself ends in short inverted repeats. The integrated proviral DNA is like a transposon: the proviral sequence ends in inverted repeats and is flanked by short direct repeats of target DNA.

What is the form of the retrovirus involved in the integration reaction? Three candidates are available: the linear DNA; a circular DNA with two adjacent LTR sequences generated by joining the linear ends; and a circular DNA that has only one LTR (presumably generated by a recombination event and actually comprising the majority of circles).

Although for a long time it appeared that the intermediate is a circle (by analogy with the integration of lambda DNA), we now know that the linear form is in fact used for integration. Integration of linear DNA has been characterized *in vitro*, and can be catalyzed by a single viral product, the integrase. Integrase acts on both the retroviral linear DNA and on the target DNA. The reaction is illustrated in **Figure 35.7**.

The ends of the viral DNA are important; as is the case with transposons, mutations in the ends prevent integration. The most conserved feature is the presence of the dinucleotide sequence CA close to the end of each inverted repeat. The integrase brings the ends of the linear DNA together in a ribonucleoprotein complex, and converts the blunt ends into recessed ends by removing the bases

beyond the conserved CA; usually this involves loss of two bases from each 3′ end.

Target sites are chosen at random with respect to sequence. The integrase makes staggered cuts at a target site. In the example of Figure 35.7, the cuts are separated by 4 bp. The length of the target repeat depends on the particular virus; it may be 4, 5, or 6 bp. Presumably it is determined by the geometry of the reaction of integrase with target DNA.

The 5′ ends generated by the cleavage of target DNA are covalently joined to the 3′ recessed ends of the viral DNA. At this point, both termini of the viral DNA are joined by one strand to the target DNA. The single-stranded region is repaired by enzymes of the host cell, and in the course of this reaction the protruding 2 bases at each 5′ end of the viral DNA are removed. The result is that the integrated viral DNA has lost 2 bp at each LTR; this corresponds to the loss of 2 bp from the left end of the 5′ terminal U3 and loss of 2 bp from the right end of the 3′ terminal U5. There is a characteristic short direct repeat of target DNA at each end of the integrated retroviral genome.

The viral DNA integrates into the host genome at randomly selected sites. A successfully infected cell gains 1–10 copies of the provirus.

The U3 region of each LTR carries a promoter. The promoter in the left LTR is responsible for initiating transcription of the provirus. Recall that the generation of proviral DNA is required to place the U3 sequence at the left LTR; so we see that the promoter is in fact generated by the conversion of the RNA into duplex DNA.

Sometimes (probably rather rarely), the promoter in the right LTR sponsors transcription of the host sequences that are adjacent to the site of integration. The LTR also carries an enhancer that can act on cellular as well as viral sequences.

Can integrated proviruses be excised from the genome? Homologous recombination could take place between the LTRs of a provirus; solitary LTRs

Figure 35.7

Integrase is the only viral protein required for the integration reaction, in which each LTR loses 2 bp and is inserted between 4 bp repeats of target DNA

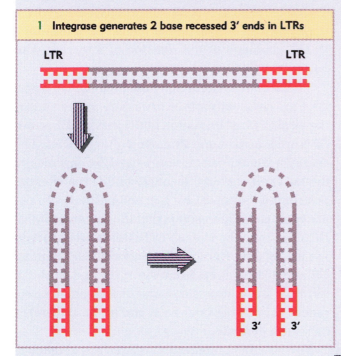

1 Integrase generates 2 base recessed 3′ ends in LTRs

LTR   LTR

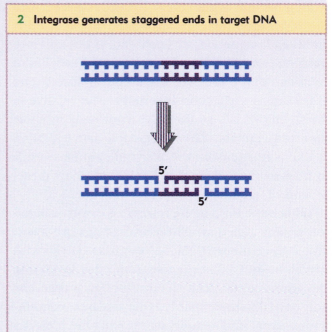

2 Integrase generates staggered ends in target DNA

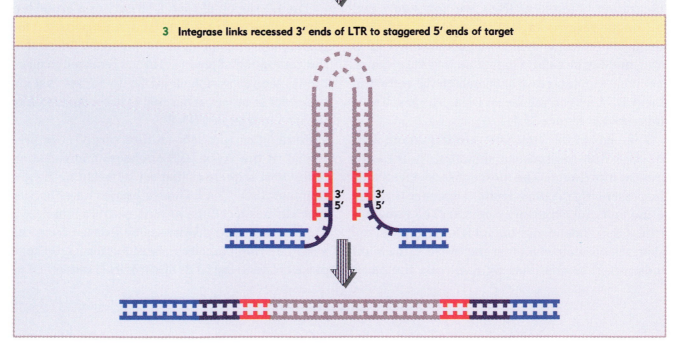

3 Integrase links recessed 3′ ends of LTR to staggered 5′ ends of target

that could be relics of an excision event are present in some cellular genomes.

Integration of a retroviral genome is responsible for some classes of cell transformation—the conversion of the host cell into a tumorigenic state—by activating certain types of cellular genes (see Chapter 39). The ability to switch on flanking host sequences is analogous to the behavior of some bacterial transposons, although often it is due to the provision of enhancer rather than promoter sequences.

We have dealt so far with retroviruses in terms of the infective cycle, in which integration is necessary for the production of further copies of the RNA. However, when a viral DNA integrates in a germ line cell, it becomes an inherited 'endogenous provirus' of the organism. Endogenous viruses usually are not expressed, but sometimes they are activated by external events, such as infection with another virus.

Retroviruses may transduce cellular sequences

An interesting light on the viral life cycle is cast by the occurrence of **transducing viruses**, variants that have acquired cellular sequences in the form illustrated in **Figure 35.8**. Part of the viral sequence has been replaced by the *v-onc* gene. Protein synthesis generates a Gag–v-Onc protein instead of the usual Gag, Pol, and Env proteins. The resulting virus is **replication-defective;** it cannot sustain an infective cycle by itself. However, it can be perpetuated in the company of a **helper virus** that provides the missing viral functions.

Onc is an abbreviation for **oncogenesis**, the ability to **transform** cultured cells so that the usual regulation of growth is released to allow

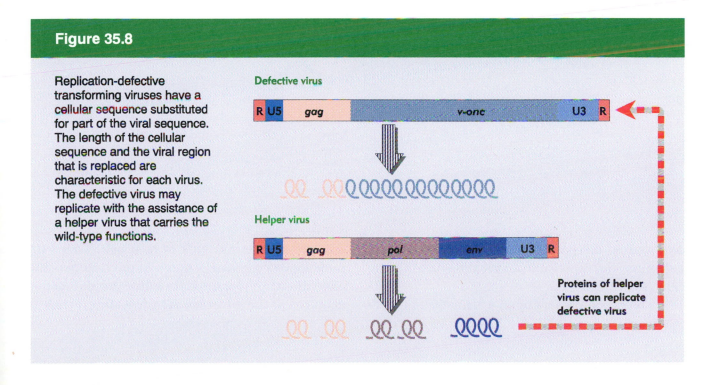

Figure 35.8

Replication-defective transforming viruses have a cellular sequence substituted for part of the viral sequence. The length of the cellular sequence and the viral region that is replaced are characteristic for each virus. The defective virus may replicate with the assistance of a helper virus that carries the wild-type functions.

Defective virus

R U5 gag v-onc U3 R

Helper virus

R U5 gag pol env U3 R

Proteins of helper virus can replicate defective virus

Figure 35.9

Replication-defective viruses may be generated through integration and deletion of a viral genome to generate a fused viral-cellular transcript that is packaged with a normal RNA genome. Nonhomologous recombination is necessary to generate the replication-defective transforming genome.

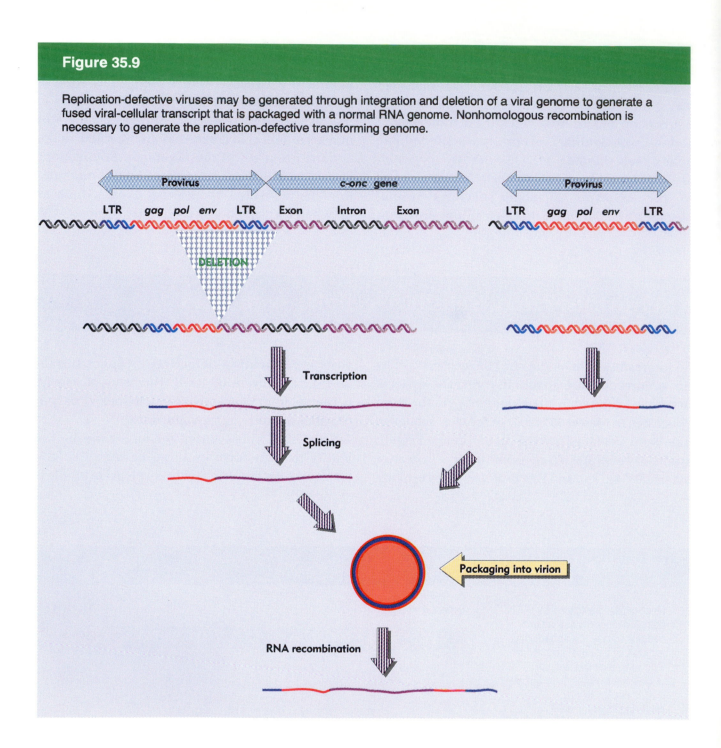

unrestricted division. Both viral and cellular *onc* genes may be responsible for creating tumorigenic cells (see Chapter 39).

A *v-onc* gene confers upon a virus the ability to transform a certain type of host cell. Loci with homologous sequences found in the host genome are called *c-onc* genes. How are the *onc* genes acquired by the retroviruses? A revealing feature is the discrepancy in the structures of c-*onc* and v-*onc* genes. The c-*onc* genes usually are interrupted by introns and are expressed at low levels in the cell. The v-*onc* genes are uninterrupted and are expressed at a high level as part of the viral transcription unit. These structures suggest that *the v-*onc *genes originate from spliced RNA copies of the c-*onc *genes.*

A model for the formation of transforming viruses is illustrated in **Figure 35.9**. A retrovirus has integrated near a c-*onc* gene. A deletion occurs to fuse the provirus to the c-*onc* gene; then transcription generates a joint RNA, containing viral sequences at one end and cellular *onc* sequences at the other end. Splicing removes the introns in both the viral and cellular parts of the RNA. The RNA has the appropriate signals for packaging into the virion; virions will be generated if the cell also contains another, intact copy of the provirus. Then some of the diploid virus particles may contain one fused RNA and one viral RNA.

Recombination between these sequences could generate the transforming genome, in which the viral repeats are present at *both* ends. (Recombination occurs at a high frequency during the retroviral infective cycle, by various means. We do not know anything about its demands for homology in the substrates, but we assume that the nonhomologous reaction between a viral genome and the cellular part of the fused RNA proceeds by the same mechanisms responsible for viral recombination.)

The common features of the entire retroviral class suggest that it may be derived from a single ancestor. Primordial IS elements could have surrounded a host gene for a nucleic acid polymerase; the resulting unit would have the form *LTR-pol-LTR*. It might evolve into an infectious virus by acquiring more sophisticated abilities to manipulate both DNA and RNA substrates, including the incorporation of genes whose products allowed packaging of the RNA. Other functions, such as transforming genes, might be incorporated later. (There is no reason to suppose that the mechanism involved in acquisition of cellular functions is unique for *onc* genes; but viruses carrying these genes may have a selective advantage because of their stimulatory effect on cell growth.)

Yeast Ty elements resemble retroviruses

The Ty elements comprise a family of dispersed repetitive DNA sequences that are found at different sites in different strains of yeast. **Ty** is an abbreviation for 'transposon yeast'. A transposition event creates a characteristic footprint: 5 bp of target DNA are repeated on either side of the inserted Ty element. The frequency of Ty transposition seems to be less than that of bacterial transposons, $\sim 10^{-7} - 10^{-8}$.

There is considerable divergence between individual Ty elements. Most elements fall into one of two major classes, called Ty1 and Ty917. They have the same general organization illustrated in **Figure 35.10**. Each element is 6.3 kb long; the last 330 bp at each end constitute direct repeats, called δ. All elements share the presence of the δ repeats, a long region at the left end, another region in the center, and a short region

Figure 35.10

Ty elements terminate in short direct repeats and are transcribed into two overlapping RNAs. They have two reading frames, with sequences related to the retroviral *gag* and *pol* genes.

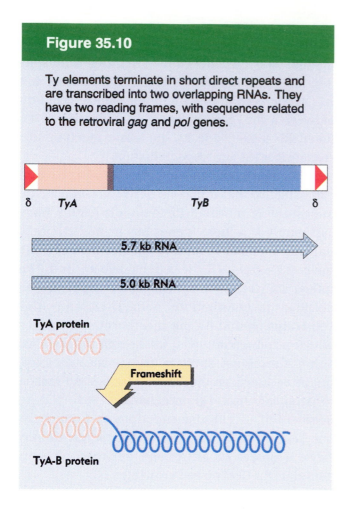

at least very closely related. The delta sequences associated with Ty elements show greater conservation of sequence than the solo δ elements, which suggests that recognition of the repeats is involved in transposition.

The Ty element is transcribed into two poly(A)+ RNA species, which constitute >5% of the total mRNA of a haploid yeast cell. Both initiate ~95 bp from the end of the element. One terminates after 5 kb, within the common region at the other end; the other terminates after 5.7 kb, within 40 bp of the boundary. This longer RNA therefore both starts and stops within the δ sequence and as a result has repeats at its ends.

The sequence of the Ty element has two open reading frames, expressed in the same direction, but read in different phases and overlapping by 13 amino acids. The sequence of *TyA* suggests that it codes for a DNA-binding protein. The sequence of *TyB* contains regions that have homologies with reverse transcriptase, protease, and integrase sequences of retroviruses.

The organization and functions of *TyA* and *TyB* are analogous to the behavior of the retroviral *gag* and *pol* functions. The reading frames *TyA* and *TyB* are expressed in two forms. The TyA protein represents the *TyA* reading frame, and terminates at its end. The *TyB* reading frame, however, is expressed only as part of a joint protein, in which the *TyA* region is fused to the *TyB* region by a specific frameshift event that allows the termination codon to be bypassed (analogous to *gag–pol* translation in retroviruses).

The Ty elements provide regions of portable homology that are targets for recombination events mediated by host systems. The usual tendency of such an event is to damage the chromosome, either by causing a deletion or inversion (by recombination between two Ty elements on a chromosome), or by causing more dramatic changes when recombination

adjacent to the right δ. The two regions of major substitution between the types of element show no homology.

Individual Ty elements of each type have many changes from the prototype of their class, including base pair substitutions, insertions, and deletions. There are ~30 copies of the Ty1 type and ~6 of the Ty917 type in a typical yeast genome. In addition, there are ~100 independent δ elements, called solo δ's.

The δ sequences also show considerable heterogeneity, although the two repeats of an individual Ty element are likely to be identical or

occurs between two Ty elements on different chromosomes.

Recombination between Ty elements seems to occur in bursts; when one event is detected, there is an increased probability of finding others. Gene conversion occurs between Ty elements at different locations, with the result that one element is 'replaced' by the sequence of the other.

Ty elements can excise by homologous recombination between the directly repeated δ sequences. The large number of solo δ elements may be footprints of such events. An excision of this nature may be associated with reversion of a mutation caused by the insertion of Ty; the level of reversion may depend on the exact δ sequences left behind.

A paradox is that both delta elements have the same sequence, yet a promoter is active in the δ at one end and a terminator is active in the δ at the other end. (A similar feature is found in other transposable elements, including the retroviruses.)

Ty elements are classic retroposons, transposing through an RNA intermediate. An ingenious protocol used to detect this event is illustrated in **Figure 35.11**. An intron was inserted into an element to generate a unique Ty sequence. This sequence was placed under the control of a *GAL* promoter on a plasmid and introduced into yeast cells. Transposition results in the appearance of multiple copies

Figure 35.11

A unique Ty element, engineered to contain an intron, transposes to give copies that lack the intron. The copies possess identical terminal repeats, generated from one of the termini of the original Ty element. The frequency of transposition is related to the activity of the promoter controlling its transcription.

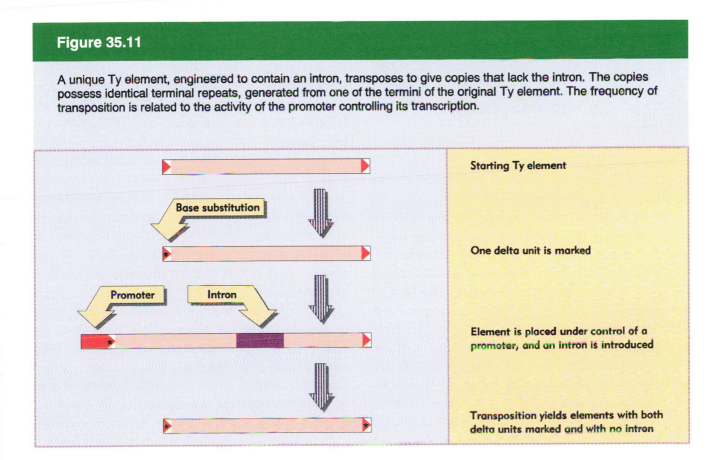

Starting Ty element

Base substitution

One delta unit is marked

Promoter   Intron

Element is placed under control of a promoter, and an intron is introduced

Transposition yields elements with both delta units marked and with no intron

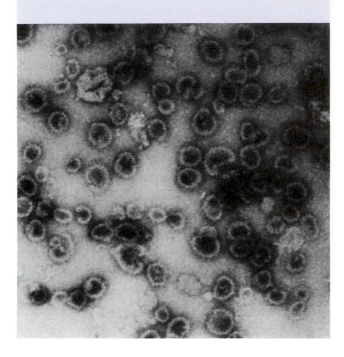

Figure 35.12

Ty elements generate virus-like particles. Photograph kindly provided by Alan Kingsman.

of the transposon in the yeast genome; *but they all lack the intron.*

We know of only one way to remove introns: RNA splicing. This suggests that transposition occurs by the same mechanism as retroviruses. The Ty element is transcribed into an RNA that is recognized by the splicing apparatus. The spliced RNA is recognized by a reverse transcriptase and regenerates a duplex DNA copy.

The analogy with retroviruses extends further. The original Ty element has a difference in sequence between its two δ elements. *But the transposed elements possess identical delta sequences, derived from the 5′ δ of*

the original element. If we consider the δ sequence to be exactly like an LTR, consisting of the regions U3–R–U5, the Ty RNA extends from R region to R region. Just as shown for retroviruses in Figures 35.3–35.6, the complete LTR is regenerated by adding a U5 to the 3′ end and a U3 to the 5′ end.

Transposition is controlled by genes within the Ty element. The *GAL* promoter used to control transcription of the marked Ty element is inducible: it is turned on by the addition of galactose. Induction of the promoter has two effects. It is necessary to activate transposition of the marked element. And its activation also increases the frequency of transposition of Ty elements on the yeast chromosome. This implies that the products of the Ty element can act in *trans* on other elements (actually on their RNAs).

Although the Ty element does not give rise to infectious particles, virus-like particles (VLPs) accumulate within the cells in which transposition has been induced. The particles can be seen in **Figure 35.12**. They contain full-length RNA, double-stranded DNA, reverse transcriptase activity, and a *TyB* product with integrase activity. The *TyA* product is cleaved like a *gag* precursor to produce the mature core proteins of the VLP. This takes the analogy between the Ty transposon and the retrovirus even further. The Ty element behaves in short like a retrovirus that has lost its *env* gene and therefore cannot properly package its genome.

Only some of the Ty elements in any yeast genome in fact are active: most have lost the ability to transpose (and are analogous to inert endogenous proviruses.) Since these 'dead' elements retain the δ repeats, however, they provide targets for transposition in response to the proteins synthesized by an active element.

Many transposable elements reside in *D. melanogaster*

The presence of transposable elements in *D. melanogaster* was first inferred from observations analogous to those that identified the first insertion sequences in *E. coli*. Unstable mutations are found that revert to wild type by deletion, or that generate deletions of the flanking material with an endpoint at the original site of mutation. They are caused by several types of transposable sequence, which are illustrated in **Figure 35.13** and detailed in **Table 35.2**. They include the *copia* retroposon, the FB family of unknown type, and the P elements discussed previously in Chapter 34.

The best characterized family of retroposons is called *copia*. Its name reflects the presence of a large number of closely related sequences that code for abundant mRNAs. The *copia* family is taken as a paradigm for several other types of elements whose sequences are unrelated, but whose structure and general behavior appear to be similar.

The number of copies of the *copia* element depends on the strain of fly; usually it is 20–60. The members of the family are widely dispersed. The locations of *copia* elements show a different (although overlapping) spectrum in each strain of *D. melanogaster*.

These differences have developed over evolutionary periods. Comparisons of strains that have diverged recently (over the past 40 years or so) as the result of their propagation in the laboratory reveal few changes. We cannot estimate the rate of change, but the nature of the underlying events is indicated by the result of growing cells in culture. The number of *copia* elements per genome then increases substantially, up to 2–3 times. The additional elements represent insertions of *copia* sequences at new sites. Adaptation to culture in some unknown way transiently increases the rate

of transposition to a range of 10^{-3}–10^{-4} events per generation.

The *copia* element is 5000 bp long, with identical direct terminal repeats of 276 bp. Each of the direct repeats itself ends in related inverted repeats. A direct repeat of 5 bp of target DNA is generated at the site of insertion. The divergence between individual members of the *copia* family is

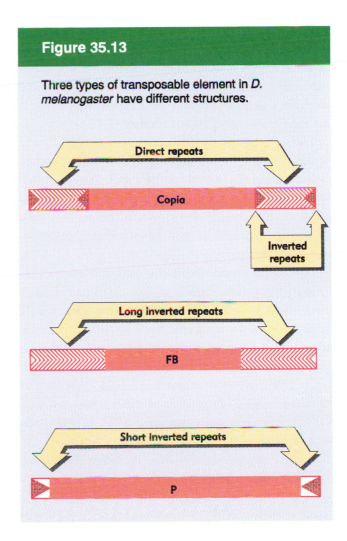

Figure 35.13

Three types of transposable element in *D. melanogaster* have different structures.

Table 35.2

Transposable elements in *D. melanogaster* fall into three classes. Several other types of elements resemble *copia* in organization, including *412, 297, 17.6, mgd1, mgd3, B104, roo,* and *gypsy*, whose copy numbers vary from 10 to 100 per genome.

Element	Copies/Genome	Length	Terminal Direct Repeats	Terminal Inverted Repeats	Target Direct Repeats
copia	20-60	5146 bp	276 bp	13 bp	5 bp
FB	˜30	500-5000 bp	none	250-1250 bp	9 bp
P	˜50 or 0	500-2900 bp	none	31 bp	8 bp

slight, <5%; variants often contain small deletions. All of these features are common to the other *copia*-like families, although their individual members display greater divergence, as seen in the maximum of ~20% for element *297*.

The identity of the two direct repeats of each *copia* element implies either that they interact to permit correction events, or that both are generated from one of the direct repeats of a progenitor element during transposition. As in the similar case of Ty elements, this is suggestive of a relationship with retroviruses.

The *copia* elements in the genome are always intact; individual copies of the terminal repeats have not been detected (although we would expect them to be generated if recombination deleted the intervening material). *copia* elements sometimes are found in the form of free circular DNA: the two major types of molecule in the circular *copia* population are 5000 bp and 4700 bp long. Like retroviral DNA circles, the longer form has two terminal repeats and the shorter form has only one. Particles containing *copia* RNA have been noticed.

The *copia* sequence contains a single long reading frame of 4227 bp. There are homologies between parts of the *copia* open reading frame and the *gag* and *pol* sequences of retroviruses. A notable absence from the homologies is any relationship

with retroviral *env* sequences required for the envelope of the virus, which means that *copia* is unlikely to be able to generate virus-like particles.

Transcripts of *copia* are found in the form of abundant poly(A)$^+$ mRNAs, representing both full-length and part-length transcripts. The mRNAs have a common 5′ terminus, resulting from initiation in the middle of one of the terminal repeats. A variety of proteins are synthesized by translation *in vitro*, ranging from 18,000 to 51,000 daltons. The same proteins are produced from both the short and long mRNAs. The structures of the mRNAs and proteins have not yet been defined, and expression could involve events such as splicing of RNA or cleavage of polyproteins.

Although we lack direct evidence for *copia*'s mode of transposition, there are so many resemblances with retroviral organization that the conclusion seems ineluctable that *copia* must have an origin related to the retroviruses. It is hard to say how many retroviral functions it possesses. We know, of course, that it transposes; but (as is the case with Ty elements) there is no evidence for any infectious capacity.

The members of another family of transposable elements in *D. melanogaster*, called *FB* (an abbreviation for foldback), have inverted terminal repeats of variable length. Some FB elements consist solely of juxtaposed inverted repeats; in others the inverted repeats are separated by a region of nonrepetitive DNA.

In spite of the variation in length, the inverted repeats of all members of the FB family are homologous. This feature is explained by their structure, which consists of tandem copies of a simple-sequence DNA, separated by longer stretches of more diverse sequences. Proceeding from the end into the element, the length of the simple-sequence unit increases; initially it is 10 bp, then expands to 20 bp, and finally expands again to 31 bp.

The two copies of the inverted repeat in a single FB element are not identical. The inverted repeats superficially resemble satellite DNA, and it will be interesting to establish the relationship between the repeats of different members of the family.

The structure of the ends poses a puzzle about FB elements; we have no knowledge of what confers their ability to transpose. Sometimes two (nonidentical) FB elements apparently cooperate to transpose a large intervening segment of DNA, possibly in a manner reminiscent of composite bacterial transposons (although the length of the DNA between FB elements can be much greater, up to 200 kb). It is possible that any sequence of DNA flanked on either end by FB elements could behave as such a unit.

Retroposons fall into two classes

Two classes of retroposons are distinguished in **Table 35.3**.

◆ The retroviruses are the paradigm for retroposons that have the capacity to transpose because they code for reverse transcriptase and/or integrase activities. The retroposons differ from the retroviruses themselves in not passing through an independent infectious form, but otherwise resemble them in the mechanism used for transposition. This group is called the **viral superfamily**.

◆ Another class of retroposons is identified by external and internal features that suggest that they originated in RNA sequences, although in these cases we can only speculate on how a DNA copy was generated. We assume that they were targets for a transposition event by an enzyme system coded elsewhere. They originated in cellular transcripts. They do not code for proteins that have transposition functions. This group is called the **nonviral superfamily**.

A significant part of the moderately repetitive DNA of mammalian genomes consist of retroposons. Two families account for most of this material. They were originally identified as interspersed repeated sequences; each consists of many members dispersed in the genome. The LINES comprise long interspersed sequences, and the SINES comprise short interspersed sequences. A more important distinction may be that LINES are derived from transcripts of RNA polymerase II, while SINES are derived from transcripts of RNA polymerase III.

Mammalian genomes contain 20–50,000 copies of a LINES called L1. The typical member is ~6,500 bp long, terminating in an A-rich tract. Open reading frames may be present. A 6581 bp element that has been sequenced has reading frames of 1137 and 3900 bp that overlap by 14 bp. Transcripts can be found. As implied by its presence in repetitive DNA, the LINES family shows variation among individual members. However, the members of the family within a species are relatively homogeneous

Table 35.3

Retroposons can be divided into the viral or nonviral superfamilies.

	Viral Superfamily	Nonviral Superfamily
Common types	Ty (S. cerevisiae) copia (D. melanogaster) LINES L1 (mammals)	SINES B1/Alu (mammals) Processed pseudogenes of pol III transcripts
Termini	Long terminal repeats	No repeats
Target repeats	4-6 bp	7-21 bp
Reading frames	Reverse transcriptase and/or integrase	None (or none coding for transposon products)
Organization	May contain introns (removed in subgenomic mRNA)	No introns

compared with the variation shown between species.

The LINES elements were originally classed in the nonviral superfamily. However, sequencing identifies homologies between an open reading frame in L1 and the reverse transcriptase. This suggests that L1 could have originated as a mobile gene coding for its own transposition, which would place it as a member of the viral superfamily.

Figure 35.14 compares members of the viral superfamily with the retroviral paradigm. We know that an active Ty element codes for transposition function, we may infer that among the *copia* sequences in a fly genome must be some active elements coding for transposition function, but we do not know whether LINES elements code for functional proteins or merely retain homologies with retroposons as a result of their origin.

Because LINES originate from RNA polymerase II transcripts, the genomic sequences are necessarily inactive: they lack the promoter that was upstream of the original startpoint for transcription. Because they usually possess the features of the mature transcript, they are called **processed pseudogenes**.

The characteristic features of a processed pseudogene are compared with the features of the original gene and the mRNA in **Figure 35.15**. The figure shows *all* the relevant diagnostic features, only some of which are found in any individual example. Any transcript of RNA polymerase II could in principle give rise to such a pseudogene, and there are many examples, including the processed globin pseudogenes that were the first to be discovered (see Chapter 24.)

The pseudogene may start at the point equivalent to the 5′ terminus of the RNA, which would be expected only if the DNA had originated from the RNA. Several pseudogenes consist of precisely joined exon sequences; we know of no mechanism to recognize introns in DNA, so this feature argues for an RNA-mediated stage. The pseudogene may end in a short stretch of A•T base pairs, presumably derived from the poly(A) tail of the RNA. On either side of the pseudogene is a short direct repeat, presumed to have been generated by a transposition-like event.

Supporting the idea that these pseudogenes originated by a mechanism different from those responsible for generating pseudogenes found near the active members of gene clusters, the processed pseudogenes reside at locations unrelated to their presumed sites of origin.

The processed pseudogenes do not carry any information that might be used to sponsor a transposition event (or to carry out the preceding reverse transcription of the RNA). Could the process have been mediated by a retrovirus? Was it accomplished by an aberrant cellular system? Perhaps the ends of the transposed sequence fortuitously resembled sequences at the ends of a transposon.

Are transposition events currently occurring in these genomes or are we seeing only the footprints of ancient systems? Note that for the transpositions to have survived, they must have occurred in the germline; presumably similar events occur in somatic cells, but do not survive beyond one generation.

The most prominent SINES comprises members of a single family. Its short length and high degree of repetition make it comparable to simple sequence DNA, except that the individual

members of the family are dispersed around the genome instead of being confined to tandem clusters. Again there is significant similarity between the members within a species compared with variation between species.

In the human genome, a large part of the moderately repetitive DNA exists as sequences of ~300 bp that are interspersed with nonrepetitive DNA. The duplex DNA corresponding to the moderately repetitive sequence component can be isolated by renaturation at intermediate Cot, followed by degradation of the adjacent regions of nonrepetitive DNA that remain unpaired. At least half of the renatured duplex material is cleaved by the restriction enzyme *Alu*I at a single site, located 170 bp along the sequence.

The cleaved sequences all are members of a single family, known as the **Alu family** after the means of its identification. There are ~300,000 members in the haploid genome (equivalent to one member for every 6 kb of DNA). The individual Alu sequences are widely

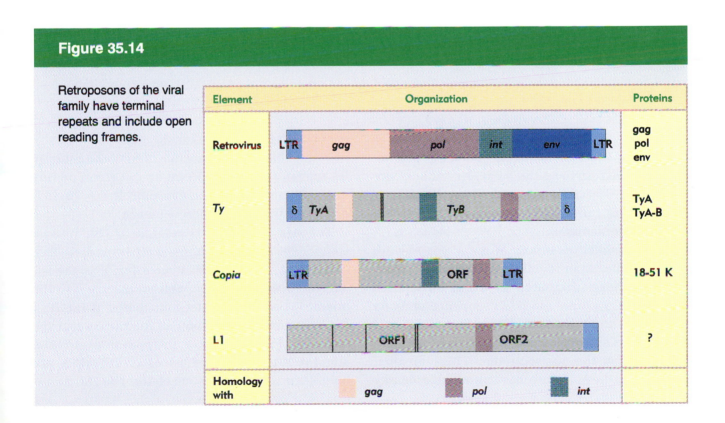

Figure 35.14

Retroposons of the viral family have terminal repeats and include open reading frames.

Figure 35.15

Pseudogenes could arise by reverse transcription of RNA to give duplex DNAs that become integrated into the genome.

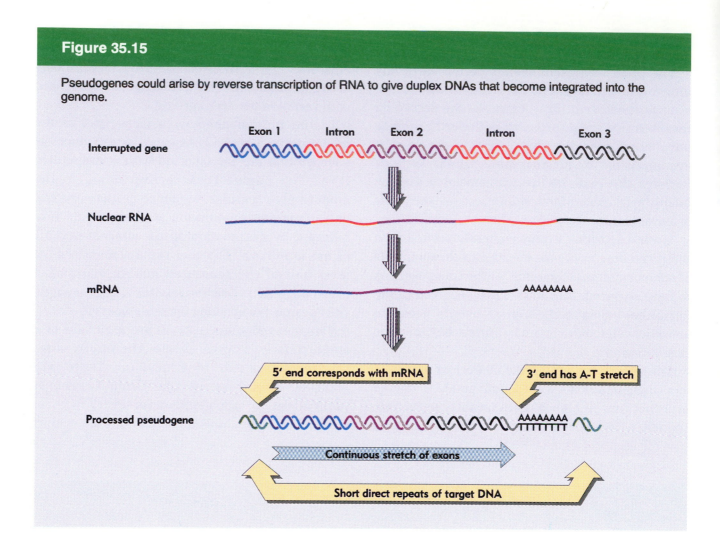

dispersed. A related sequence family is present in the mouse (where the 50,000 members are called the B1 family), in the Chinese hamster (where it is called the Alu-equivalent family), and in other mammals.

The individual members of the Alu family are related rather than identical. The human family seems to have originated by a 130 bp tandem duplication, with an unrelated sequence of 31 bp inserted in the right half of the dimer. The two repeats are sometimes called the 'left half' and 'right half' of the Alu sequence. The individual members of the Alu family have an average identity with the consensus sequence of 87%. The mouse B1 repeating unit is 130 bp long, corresponding to a monomer of the human unit. It has 70–80% homology with the human sequence.

The Alu sequence is related to 7SL RNA, a component of the signal recognition particle (see Chapter 11). The 7SL RNA corresponds to the left half of an Alu sequence with an insertion in the middle. Thus the 90 5' terminal bases of 7SL RNA are homologous to left end of Alu, the central 160 bases of 7SL RNA have no homology to Alu, and the 40 3' terminal bases of 7SL RNA are homologous to the right end of Alu. The 7SL RNA is coded by genes that are actively transcribed by RNA polymerase III. It is possible that these genes (or genes related to them) gave rise to the inactive Alu sequences.

The members of the Alu family resemble transposons in being flanked by short direct repeats. However, they display the curious feature that the lengths of the repeats are different for individual members of the family. Because they derive from RNA polymerase III transcripts, it is possible that individual members carry internal active promoters.

A variety of properties have been found for the Alu family, and its ubiquity has prompted many suggestions on its function, but it is not yet possible to discern its true role.

Part of the Alu sequence is a 14 bp region that is almost identical with a sequence present at the origin of replication in the papova viruses (such as SV40) and in hepatitis B virus. This raises the possibility that the Alu family could be connected with origins of replication for the eukaryotic genome, although the number of members of the family argues against this, since there are about ten times more Alu sequences than the expected number of replication origins.

At least some members of the family can be transcribed *in vitro* into snRNA by the enzyme RNA polymerase III (which is responsible for transcribing small nuclear RNAs, tRNAs, and 5S RNA). In the Chinese hamster, some (although not all) members of the Alu-equivalent family appear to be transcribed *in vivo*. Transcription units of this sort are found in the vicinity of other transcription units.

Members of the Alu family may be included within structural gene transcription units, as seen by their presence in long nuclear RNA. The presence of multiple copies of the Alu sequence in a single nuclear molecule can generate secondary structure. In fact, the presence of Alu family members in the form of inverted repeats is responsible for most of the secondary structure found in mammalian nuclear RNA.

Summary

Reverse transcription is the unifying mechanism for reproduction of retroviruses and perpetuation of retroposons. The cycle of each type of element is in principle similar, although retroviruses are usually regarded from the perspective of the free viral (RNA) form, while retroposons are regarded from the stance of the genomic (duplex DNA) form.

Retroviruses have genomes of single-stranded RNA that are replicated through a double-stranded DNA intermediate. An individual retrovirus contains two copies of its genome. The genome contains the *gag*, *pol*, and *env* genes, which are translated into polyproteins, each of which is cleaved into smaller functional proteins. The Gag and Env components are concerned with packing RNA and generating the virion; the Pol components are concerned with nucleic acid synthesis.

Reverse transcriptase is the major component of *pol*, and is responsible for synthesizing a DNA copy of the viral RNA. The DNA product is longer than the RNA template; by switching template strands, reverse transcriptase copies the 3′ sequence of the RNA to the 5′ end of the DNA, and copies the 5′ sequence of the RNA to the 3′ end of the DNA. This generates the characteristic LTRs (long terminal repeats) of the DNA. Linear duplex DNA is inserted into a host genome by the integrase enzyme. Transcription of the integrated DNA from a promoter in the left LTR generates further copies of the RNA sequence.

During an infective cycle, a retrovirus may exchange part of its usual sequence for a cellular sequence; the resulting virus is usually replication-defective, but can be perpetuated in the course of a joint infection with a helper virus. Many of the defective viruses have gained an RNA version (*v-onc*) of a cellular gene (*c-onc*). The *onc* sequence may be any one of a number of genes whose expression in *v-onc* form causes the cell to be transformed into a tumorigenic phenotype.

The integration event generates direct target repeats (like transposons that mobilize via DNA). An inserted provirus therefore has direct terminal repeats of the LTRs, flanked by short repeats of target DNA. Mammalian and avian genomes have endogenous (inactive) proviruses with such structures. Other elements with this organization have been found in a variety of genomes, most notably in *S. cerevisiae* and *D. melanogaster*. Ty elements of yeast and *copia* elements of flies have coding sequences with homology to reverse transcriptase, and mobilize via an RNA form. They may generate particles resembling viruses, but do not have infectious capability. The LINES sequences of mammalian genomes are further removed from the retroviruses, but retain enough similarities to suggest a common origin.

Another class of retroposons have the hallmarks of transposition via RNA, but have no coding sequences (or at least none resembling retroviral functions). They may have originated as passengers in a retroviral-like transposition event, in which an RNA was a target for a reverse transcriptase. Processed pseudogenes arise by such events. A particularly prominent family apparently originating as such retroposons is the mammalian SINES, including the human Alu family. Some snRNAs, including 7SL snRNA (a component of the SRP) are related to this family.

Further reading

Reviews

Eukaryotic transposable elements have been extensively reviewed in *Mobile Genetic Elements* (ed. Shapiro, Academic Press, New York, 1983) and *Mobile DNA* (ed. Howe and Berg, American Society for Microbiology, 1989). Chapters of particular interest in the first volume: **Roeder and Fink** analyzed Ty elements of yeast (pp. 300–328); and **Rubin** described the *tour de force* of characterizing *D. melanogaster* transposons (pp. 329–362). A chapter of interest in the second volume is **Varmus and Brown** updating work on retroviruses (pp. 53–108). A general summary of eukaryotic transposons has been made by **Finnegan** (*Int. Rev. Cytol.* **93**, 281–326, 1985).

Retroviral integration was reviewed by **Goff** (*Ann. Rev. Genet.* **26**, 527–544, 1992). A *tour d'horizon* of retroposons has been accomplished by **Weiner, Deininger, and Efstratiadis** (*Ann. Rev. Biochem* **55**, 631–661, 1986).

The sequence of *copia* was analyzed by **Mount and Rubin** (*Mol. Cell. Biol.* **5**, 1630–1638, 1985).

The sequence of a LINES was analyzed by **Loeb *et al*.** (*Mol. Cell. Biol.* **6**, 168–182, 1986). LINES have been reviewed by **Hutchison *et al*.** (in *Mobile DNA*, op. cit., pp. 593–617) and SINES have been reviewed by **Deininger** (in *Mobile DNA*, op. cit., pp. 619–636).

Discoveries

Reverse transcription was discovered by **Temin and Mizutani** (*Nature* **226**, 1211–1213, 1970) and **Baltimore** (*Nature* **226**, 1209–1211, 1970). Recombination was characterized by **Hu and Temin** (*Science* **250**, 1227–1233, 1990). Integration *in vitro* was analyzed by **Craigie, Fujiwara, and Bushman** (*Cell* **67**, 829–837, 1990).

The mode of Ty transposition was discovered by **Boeke *et al*.** (*Cell* **40**, 491–500, 1985).

CHAPTER 36

Rearrangement and amplification in the genome

Pressing on the discovery of transposition and other rearrangements of DNA is the knowledge that DNA sequences are surprisingly adjustable. Sequences may be moved within a genome, modified, or even lost, as a natural event; or they may be introduced into cells by experimental means.

The existence of natural mechanisms to adjust the content of the genome seems to contradict the general notion that genome organization is relatively stable, being altered only by genetic recombination in the germ line, and not at all in the soma. Certainly it is true that quantitative or qualitative changes in either the somatic or germ line are the exception rather than the rule, whether measured by frequency of overall occurrence or by the proportion of the genome that is affected. Yet they are interesting not only in themselves, but also for the advantage that we can take of them to introduce particular changes into the genome.

Examples of rearrangement or loss of specific sequences are legion in the lower eukaryotes. Usually these changes involve somatic cells; the germ line remains inviolate. (However, there are organisms whose reproductive cycle involves the loss of whole chromosomes or sets of chromosomes.) Reorganization of particular sequences is rare in animals, although an extensive case is represented by the immune system (see Chapter 37).

Rearrangements of DNA generate diversity in both eukaryotes and prokaryotes. We may distinguish two broad consequences of a rearrangement:

◆ Rearrangement may *create new genes,* needed for expression in particular circumstances, as in the case of the immunoglobulins.

◆ Rearrangement may be responsible for switching expression from one preexisting gene to another. This provides a mechanism for *regulating gene expression.*

Yeast mating-type switching and trypanosome antigen variation share a similar type of plan in which gene expression is controlled by manipulation of DNA sequences. Phenotype is determined by the gene copy present at a particular, active locus. But the genome also contains a store of other, alternative sequences, which are silent. A silent copy can be activated only by a rearrangement of sequences in which it replaces the active gene copy. Such a substitution is equivalent to a unidirectional transposition with a specific target site.

The simplest example of this strategy is found in the yeast, *S. cerevisiae*. Haploid *S. cerevisiae* can have either of two mating types. The type is determined by the sequence present at the active mating type locus. But the genome also contains two other, silent loci, one representing each mating type. Transition between mating types is accomplished by substituting the sequence at the active locus with the sequence from the silent locus carrying the other mating type.

A range of variations is made possible by DNA rearrangement in the African trypanosomes, unicellular parasites that evade the host immune response by varying their surface antigens. The type of surface antigen is determined by the gene sequence at an active locus. This sequence can be changed, however, by substituting a sequence from any one of many silent loci. It seems fitting that the mechanism used to combat the flexibility of the immune apparatus is analogous to that used to generate immune diversity: it relies on physical rearrangements in the genome to change the sequences that are expressed.

Another means of increasing genetic capacity is employed in parasite– or symbiote–host interactions, in which exogenous DNA is introduced from a bacterium into a host cell. The mechanism resembles that of bacterial conjugation. Expression of the bacterial DNA in its new host changes the phenotype of the cell. In the example of the bacterium *Agrobacterium tumefaciens*, the effect is deleterious, since it induces tumor formation by an infected plant cell.

Alterations in the relative proportions of components of the genome during somatic development occur to allow insect larvae to increase the number of copies of certain genes. The occasional **amplification** of genes in cultured mammalian cells is indicated by our ability to select variant cells with an increased copy number of some gene. Initiated within the genome, the amplification event can create additional copies of the gene that survive in either intrachromosomal or extrachromosomal form.

When extraneous DNA is introduced into eukaryotic cells, it may give rise to extrachromosomal forms or may be integrated into the genome. The relationship between the extrachromosomal and genomic forms is irregular, depending on chance and to some degree unpredictable events, rather than resembling the regular interchange between free and integrated forms of bacterial plasmids.

Yet, however accomplished, the process may lead to stable change in the genome; following its injection into animal eggs, DNA may even be incorporated into the genome and inherited thereafter as a normal component, sometimes continuing to function. Injected DNA may enter the germ line as well as the soma, creating a **transgenic** animal. The ability to introduce specific genes that function in an appropriate manner could become a major medical technique for curing genetic diseases.

The converse of the introduction of new genes is the ability to disrupt specific endogenous genes. Additional DNA can be introduced within a gene to prevent its expression and to generate a null allele. Breeding from an animal with a null allele can generate a homozygous 'knockout', which has no active copy of the gene. This is a powerful method to investigate directly the importance and function of a gene.

Considerable manipulation of DNA sequences therefore is achieved both in authentic situations and by experimental fiat. We are only just beginning to work out the mechanisms that permit the cell to respond to selective pressure by changing its bank of sequences or that allow it to accommodate the intrusion of additional sequences.

The mating pathway is triggered by signal transduction

The yeast *S. cerevisiae* can propagate happily in either the haploid or diploid condition. Conversion between these states takes place by mating (fusion of haploid spores to give a diploid) and by sporulation (meiosis of diploids to give haploid spores). The ability to engage in these activities is determined by the **mating type** of the strain.

The properties of the two mating types are summarized in **Table 36.1**. We may view them as resting on the teleological proposition that there is no point in mating unless the haploids are of different genetic types; and sporulation is productive only when the diploid is heterozygous and thus can generate recombinants.

The mating type of a (haploid) cell is determined by the genetic information present at the *MAT* locus. Cells that carry the *MATa* allele at this locus are **a** cells; likewise, cells that carry the *MATα* allele are α cells. Cells of opposite type can mate; cells of the same type cannot.

Recognition of cells of opposite mating type is accomplished by the secretion of **pheromones**. α cells secrete the small polypeptide α-factor; **a** cells secrete **a**-factor. The α-factor is a peptide of 13 amino acids; the **a**-factor is a peptide of 12 amino acids that is modified by addition of a farnesyl (lipid-like) group and carboxymethylation. Each of these peptides is synthesized in the form of a precursor polypeptide that is cleaved to release the mature peptide sequence.

A cell of one mating type carries a surface receptor for the pheromone of the opposite type. When an **a** cell and an α cell encounter one another, their pheromones act on each other to arrest the cells in G1 phase of the cell cycle, and various morphological changes occur, including the induction of agglutinability. In a successful mating, the cell cycle arrest is followed by cell and nuclear fusion to produce an **a**/α diploid cell.

The **a**/α cell carries both the *MATa* and *MATα* alleles and possesses properties very different from those of the haploid cells (see Table 36.1). In particular, the **a**/α cell has the ability to sporulate. **Figure 36.1** demonstrates how this design maintains the

Table 36.1

Mating type controls several activities.

	MATa	MATα	MATa/MATα
Cell type	a	α	a/α
Mating	Yes	Yes	No
Sporulation	No	No	Yes
Pheromone	a factor	α factor	None
Surface receptor	Binds α factor	Binds a factor	None

Figure 36.1

Overview: the yeast life cycle proceeds through mating of *MATa* and *MATα* haploids to give heterozygous diploids that sporulate to generate haploid spores.

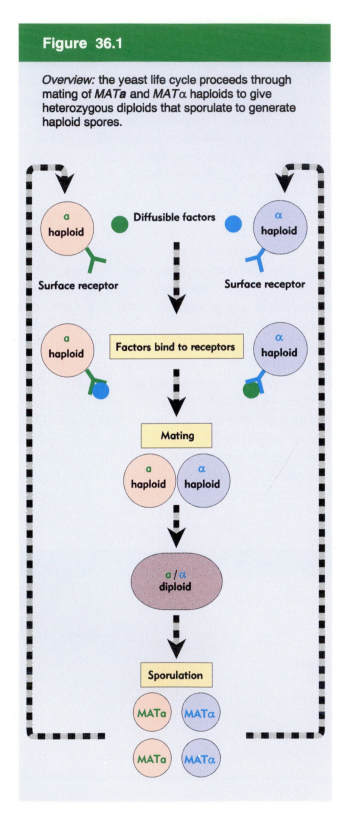

normal haploid/diploid life cycle. Note that *only heterozygous* diploids can sporulate; homozygous diploids (either **a/a** or α/α) cannot sporulate.

Much of the information about the yeast mating-type pathway was deduced from the properties of mutations that eliminate the ability of **a** and/or α cells to mate. The genes identified by such mutations are called *STE* (for sterile). Mutations in the genes *STE2* and *STE3* are specific for individual mating types; but mutations in the other *STE* genes eliminate mating in *both* **a** and α cells. This situation is explained by the idea that *the events that follow the interaction of factor with receptor are identical for both types*.

Mating is a symmetrical process that is set in train by the interaction of pheromone secreted by one cell type with the receptor carried by the other cell type. The only genes that are uniquely required for the response pathway in one mating type are those coding for the receptors: other genes are required for the response in *both* **a** and α cells. Either of the factor–receptor interactions switches on the same response pathway, so mutations that eliminate steps in this common pathway have the same effects in both cell types.

The next step in the mating-type response is summarized in **Figure 36.2**, and involves further proteins located in the cell membrane. They are related to G proteins, which are a common class of membrane receptors, consisting of three sub-units, α, β, and γ (which we discussed in detail in Chapter 12). The α subunit binds a guanine nucleotide. When the nucleotide is GDP, the α subunit is complexed with the βγ subunits. When GTP displaces the GDP, the G protein is activated, and the α subunit is released from the βγ dimer. This separation of subunits allows the G protein to activate the next protein in whatever pathway it is coupled to. Because this interaction effectively transfers the signal from the G protein to a target protein, it is called **signal transduction**. A signal that is received at the surface of the cell can be transmitted by this means to the interior.

The most common form of signal transduction

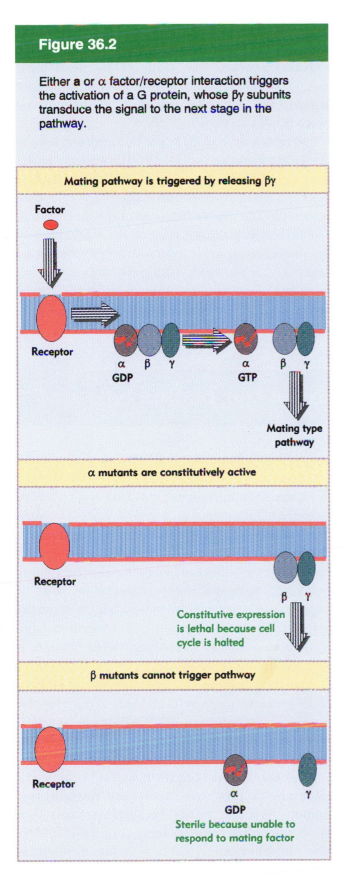

Figure 36.2

Either **a** or α factor/receptor interaction triggers the activation of a G protein, whose βγ subunits transduce the signal to the next stage in the pathway.

Mating pathway is triggered by releasing βγ

Factor

Receptor

α β γ
GDP

α β γ
GTP

Mating type pathway

α mutants are constitutively active

Receptor

β γ

Constitutive expression is lethal because cell cycle is halted

β mutants cannot trigger pathway

Receptor

α γ
GDP

Sterile because unable to respond to mating factor

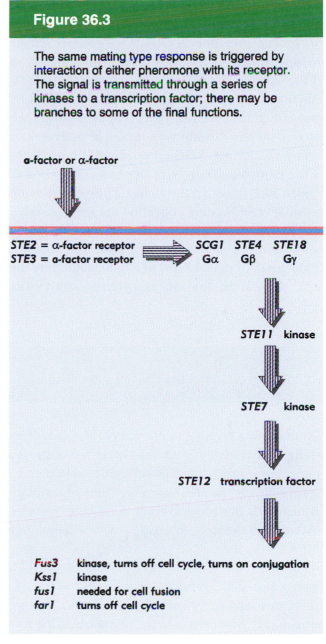

Figure 36.3

The same mating type response is triggered by interaction of either pheromone with its receptor. The signal is transmitted through a series of kinases to a transcription factor; there may be branches to some of the final functions.

a-factor or α-factor

| STE2 = α-factor receptor | SCG1 | STE4 | STE18 |
| STE3 = a-factor receptor | Gα | Gβ | Gγ |

STE11   kinase

STE7   kinase

STE12   transcription factor

Fus3	kinase, turns off cell cycle, turns on conjugation
Kss1	kinase
fus1	needed for cell fusion
far1	turns off cell cycle

is for the activated α subunit to interact with the target protein. However, the situation is different in the mating-type pathway, where the βγ dimer activates the next stage in the pathway. The component proteins of the G-trimer are identified by mutations in three genes, *SCG1*, *STE4*, and *STE18*, that affect the response to binding pheromone. Inactivation of *SCG1*, which codes for the G$_α$ protein, causes constitutive expression of the pheromone response pathway (because G_α is unable to maintain $G_{\beta\gamma}$ in the inactive trimeric form). The mutation is lethal, because its effects include arrest of the cell cycle. Inactivation of *STE4*, which codes for the G_β protein, or of *STE18*, which codes for the G_γ protein, create sterility by abolishing the mating-type response (because the next step in the pathway cannot be activated).

The remaining *STE* genes identify later steps in the pathway. They form the cascade shown in **Figure 36.3**, in which the signal is passed from one to the next, ultimately activating the genes needed for mating. The genes whose products act at the beginning of the cascade code for kinases; kinases such as STE11 and STE7 phosphorylate the next protein in the series, thereby activating

it; eventually STE12, a DNA-binding protein that is a transcription factor, is activated; it in turn activates genes whose products are needed for mating. (Analogous cascades are found in higher organisms, and are compared with the yeast cascade later, in Figure 39.15.)

There are several *STE* genes that have not yet been placed into the cascade, and the order of all the components is not yet certain. Thus genes such as *Fus3* and *Kss1*, which code for kinases, may belong within the cascade or represent branches from it. The principle, however, is clear: the signal created by interaction of pheromone with receptor is passed along a cascade that culminates by repressing functions needed for the normal cell cycle, and by activating functions needed for mating.

Yeast can switch silent and active loci for mating type

A remarkable feature is the ability of some yeast strains to **switch** their mating types. These strains carry a dominant allele *HO* and *change their mating type frequently*, as often as once every generation. Strains with the recessive allele *ho* have a stable mating type, subject to change with a frequency ~10^{-6}.

The presence of *HO* causes the genotype of a yeast population to change. Irrespective of the initial mating type, in a very few generations there are large numbers of cells of both mating types, leading to the formation of *MATa/MATα* diploids that take over the population. The production of stable diploids from a haploid population can be viewed as the *raison d'être* for switching.

The existence of switching suggests that *all cells contain the potential information needed to be either* MATa *or* MATα, *but express only one type*. Where does the information to change mating types come from? Two additional loci are needed for switching. *HMLα* is needed for switching to give a *MATα* type; *HMRa* is needed for switching to give a *MATa* type. These loci lie on the same chromosome that carries

MAT. HML is far to the left, *HMR* far to the right.

The **cassette model** for mating type is illustrated in **Figure 36.4**. It proposes that *MAT* has an **active cassette** of either type α or type a. *HML* and *HMR* have **silent cassettes**. Usually *HML* carries an α cassette, while *HMR* carries an a cassette. All cassettes carry information that codes for mating type, but only the active cassette at *MAT* is expressed. Mating-type switching occurs when the active cassette is replaced by information from a silent cassette. The newly installed cassette is then expressed.

Switching is nonreciprocal; the copy at *HML* or *HMR replaces* the allele at *MAT*. We know this because a mutation at *MAT* is lost permanently when it is replaced by switching—it does not exchange with the copy that replaces it.

The copies present at *HML* or *HMR* also can be mutated. In this case, switching introduces a mutant allele into the *MAT* locus. The mutant copy at *HML* or *HMR* remains there through an indefinite number of switches. Like replicative transposition, the donor element generates a new

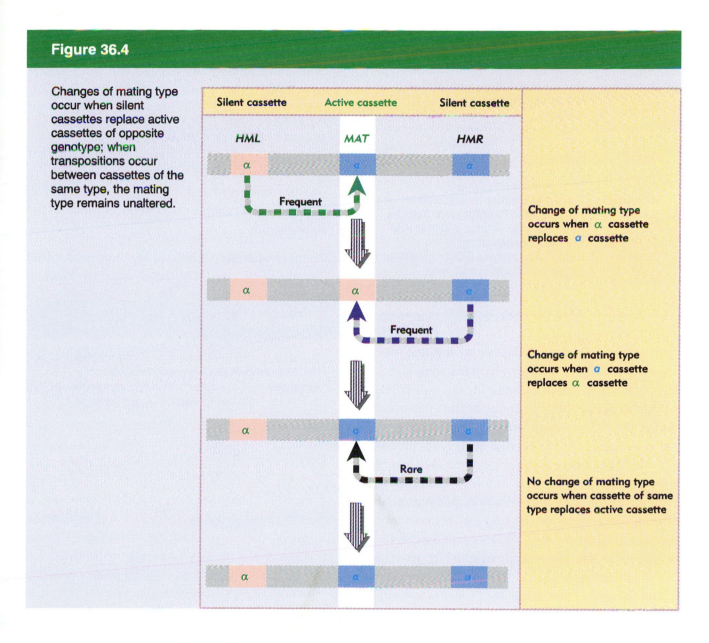

Figure 36.4

Changes of mating type occur when silent cassettes replace active cassettes of opposite genotype; when transpositions occur between cassettes of the same type, the mating type remains unaltered.

copy at the recipient site, while itself remaining inviolate.

Mating-type switching is a directed event, in which there is only one recipient (*MAT*), but two potential donors (*HML* and *HMR*). Switching usually involves replacement of *MATa* by the copy at *HMLα* or replacement of *MATα* by the copy at *HMRa*. In 80–90% of switches, the *MAT* allele is replaced by one of opposite type, an effect apparently determined by the phenotype of the cell. Cells of **a** phenotype preferentially choose *HML* as donor; cells of α phenotype preferentially choose *HMR*.

(It is possible to obtain yeast strains in which the usual orientation of the silent cassettes is reversed. When their genotypes are *HMLa* and *HMRα*, 92% of the replacements are homologous, in which an *a* cassette is replaced by another *a* cassette or an α by another α, because the choice of donor cassettes remains the same regardless of the content of the silent loci.)

Several groups of genes are involved in establishing and switching mating type. They are summarized in **Table 36.2**. As well as the genes that actually determine mating type, they include genes needed to repress the silent cassettes, to switch mating type, or to execute the functions involved in mating.

Table 36.2

Genes that determine mating type are identified by mutations that prevent mating or switching.

Gene	Function
MAT	Determines mating type
HML/HMR	Store silent mating type information
SIR1-4	Repress expression of silent casettes
HO	Endonuclease initiates switch in type
SIN1-5	Repress expression of HO
SWI1-5	Required to express HO for switching
STE1-7, 11,12,18	Sterile mutations: cannot mate

By comparing the sequences of the two silent cassettes (*HMLα* and *HMRa*) with the sequences of the two types of active cassette (*MATa* and *MATα*), we can delineate the sequences that determine mating type. The organization of the mating type loci is summarized in **Figure 36.5**. Each cassette contains common sequences that flank a central region that differs in the *a* and *α* types of cassette (called Ya or Yα). On either side of this region, the flanking sequences are virtually identical, although they are shorter at *HMR*. The silent cassettes are not expressed, but an active cassette is transcribed from a promoter within the Y region.

The basic function of the *MAT* locus is to control expression of pheromone and receptor genes, and other functions involved in mating. Each type of locus codes for regulator proteins. *MATα* codes for two proteins, α1 and α2. *MATa* codes for a single protein, a1. The a and α proteins directly control transcription of various target genes; they function by both positive and negative regulation. They function independently in haploids, and in conjunction in diploids. Their interactions are summarized in the table on the right of **Figure 36.6** in terms of three groups of target genes:

◆ a-specific genes are expressed constitutively in a cells. They are repressed in α cells. The a-specific genes include the a-factor structural

gene, and *STE2*, which codes for the α-factor receptor. Thus the **a** phenotype is associated with readiness to recognize the pheromone produced by the opposite mating type.

◆ α-specific functions are induced in α cells, but are repressed in **a** cells. They include the α-factor structural gene, and the **a**-factor receptor gene, *STE3*. Again, the expression of pheromone of one type is associated with expression of receptor for the pheromone of the opposite type.

◆ Haploid-specific functions include genes that

Figure 36.5

Silent cassettes have the same sequences as the corresponding active cassettes, except for the absence of the extreme flanking sequences in HMRα. Only the Y region changes between **a** and α types.

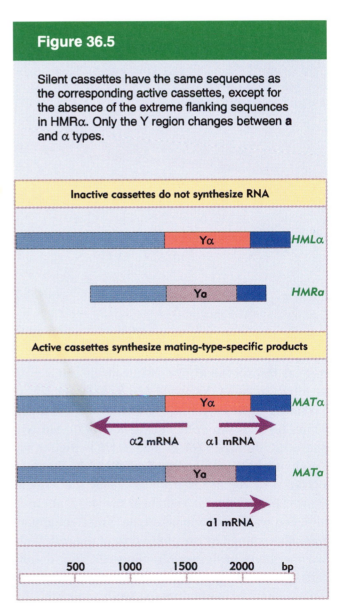

Figure 36.6

In diploids the a1 and α2 functions cooperate to repress haploid-specific functions. In **a** haploids, mating functions are constitutive. In α haploids, the α2 function represses **a** mating functions, while α1 induces **a** mating functions.

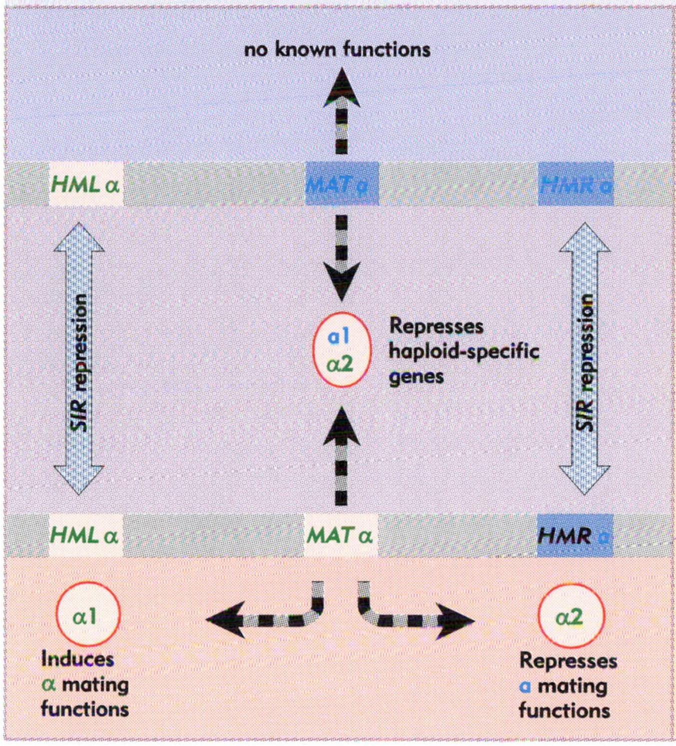

	a functions	α functions	haploid functions
a haploid	constitutive	not expressed	constitutive
diploid	not expressed	not expressed	repressed
α haploid	repressed	induced	constitutive

no known functions

HMLα   MATα   HMRa

SIR repression

a1 / α2 — Represses haploid-specific genes

SIR repression

HMLα   MATα   HMRa

α1 — Induces α mating functions

α2 — Represses a mating functions

are needed for transcription of pheromone and receptor genes, the *HO* gene involved in switching, and *RME*, a repressor of sporulation. They are expressed constitutively in both types of haploid, but are repressed in *a*/α diploids. As a result, the **a**-specific and α-specific functions also remain unexpressed in diploids.

We may now view the functions of the regulators and their targets from the perspective of the *MAT* functions expressed in haploid and diploid yeast cells, as outlined in the diagram on the left of Figure 36.6. The *a* and α mating types are regulated by different mechanisms:

◆ In *a* haploids, *a* mating functions are expressed constitutively. The functions of the products of

MATa in the **a** cell (if any) are unknown. It may be required only to repress haploid functions in diploid cells.

◆ In α haploids, the α1 product turns on α-specific genes whose products are needed for α mating type. The α2 product represses the genes responsible for producing **a** mating type, by binding to an operator sequence located upstream of target genes.

◆ In diploids, the *a1* and α2 products cooperate to repress haploid-specific genes. They combine to recognize an operator sequence different from the target for α2 alone.

The abilities of the α2, **a1**, and α1 proteins to regulate transcription rely upon some interesting

Figure 36.7

Combinations of PRTF, a1, α1 and α2 activate or repress specific groups of genes to correspond with the mating type of the cell.

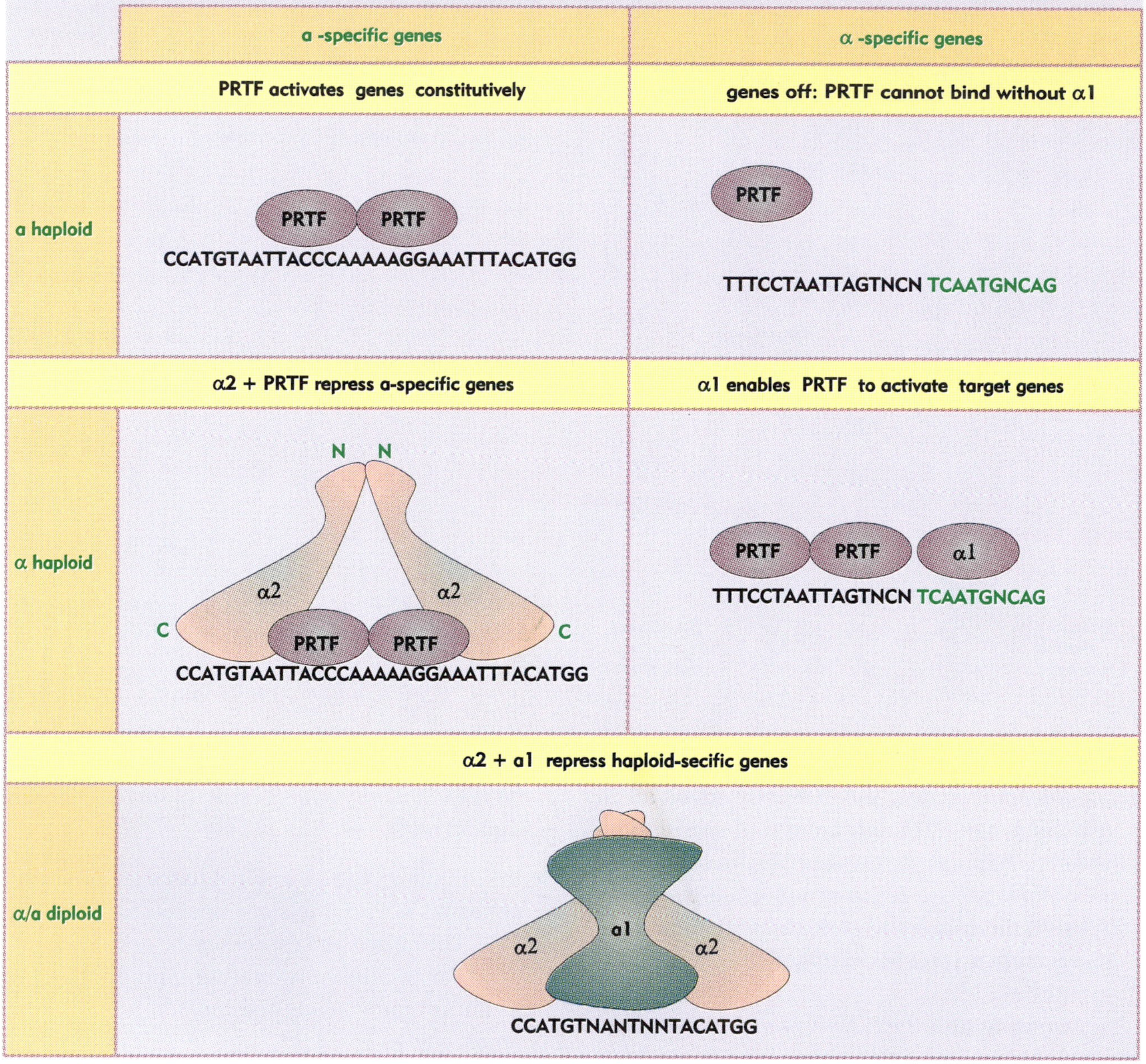

	a -specific genes	α -specific genes
	PRTF activates genes constitutively	**genes off: PRTF cannot bind without α1**
a haploid	PRTF PRTF CCATGTAATTACCCAAAAAGGAAATTTACATGG	PRTF TTTCCTAATTAGTNCN TCAATGNCAG
	α2 + PRTF repress a-specific genes	**α1 enables PRTF to activate target genes**
α haploid	N N α2 α2 C PRTF PRTF C CCATGTAATTACCCAAAAAGGAAATTTACATGG	PRTF PRTF α1 TTTCCTAATTAGTNCN TCAATGNCAG
	α2 + a1 repress haploid-secific genes	
α/a diploid	α2 a1 α2 CCATGTNANTNNTACATGG	

protein–protein interactions between themselves and with other protein(s). The pattern of gene control in **a** cells, α cells, and diploids, is summarized in **Figure 36.7**.

A protein called PRTF (which is not specific for mating type) is involved in many of these interactions. PRTF binds to a short consensus sequence called the P box. The role of PRTF in gene regulation may be quite extensive, because P boxes are found in a variety of locations. In some of these sites, the P box is required for activation of the gene; but at other loci, PRTF is needed for repression. Its effects may therefore depend on the other proteins that bind at sites adjacent to the P box.

Genes that are a-specific may be activated by PRTF alone. This is adequate to ensure their expression in an a haploid.

The a-specific genes are repressed in an α haploid by the combined action of the α2 protein and PRTF. The α2 protein contains two domains. The C-terminal domain binds to short palindromic elements at the ends of an operator consensus sequence of 32 bp. However, binding of this fragment to DNA does not cause repression. The N-terminal domain is needed for repression and is responsible for making contacts with PRTF. The binding site for PRTF is a P box in the center of the operator. In fact, α2 and PRTF bind to the operator cooperatively.

Expression of α-specific genes requires the α1 activator. This is another small protein, 175 amino acids long. Sequences that confer α-specific

transcription are located within *UAS* elements. The consensus sequence is 26 bp long, and can be divided into two parts. The first 16 bp form the P box, where PRTF binds; the adjacent 10 bp sequence forms the binding site for α1. The α1-factor binds *only* when PRTF is present to bind to the P box. In fact, neither protein alone can bind to its target box, but together they can bind to DNA, presumably as a result of protein–protein interactions.

The α-specific genes are controlled by default in a haploids, because in the absence of α1 protein, PRTF is unable to bind to activate them.

The α2 protein can also cooperate with the a1 protein. The combination of these proteins recognizes a different operator. The operator shares the outlying palindromic sequences with the sequence recognized by α2 alone, but is shorter because the sequence between them is different. The α2/a1 combination represses genes with this motif in diploid cells.

Silent cassettes at *HML* and *HMR* are repressed

The transcription map in Figure 36.5 reveals an intriguing feature. Transcription of either *MATa* or *MATα* initiates within the Y region. Only the *MAT* locus is expressed; *yet the same Y region is present in the corresponding nontranscribed cassette (HML or HMR).* This implies that regulation of expression is not accomplished by direct recognition of some site overlapping with the promoter. *A site outside the cassettes must distinguish HML and HMR from MAT.*

Deletion analysis shows that sites ~1 kb upstream of each of *HML* and *HMR* are needed to repress their expression. The target loci are sometimes called E_L (near *HML*) and E_R (near *HMR*). These control sites have two intriguing properties:

◆ They behave like negative enhancers, in the sense that they can function at a distance (up to

2.5 kb away from a promoter) and in either orientation. They are sometimes called **silencers**.

◆ They are associated with *ars* sequences, which probably function as origins of replication.

Can we find the basis for the control of cassette activity by identifying genes that are responsible for keeping the cassettes silent? A convenient assay for mutation in such genes is provided by the fact that, when a mutation allows the usually silent cassettes at *HML* and *HMR* to be expressed, both a and α functions are produced, so the cells behave like *MATa/MATα* diploids.

Four complementation groups have been identified in which mutations lead to expression of *HML* and *HMR*. They are called *SIR* (silent information regulator). The four wild-type *SIR* loci are needed to maintain *HML* and *HMR* in the repressed state; mutation in any one of these loci to give a *sir⁻* allele has two effects. Both *HML* and *HMR* can be transcribed. And both the silent cassettes become targets for replacement by switching. *So the same regulatory event is involved in repressing a silent cassette and in preventing it from being a recipient for replacement by another cassette.*

The existence of temperature-sensitive *sir* mutants has been used to test the involvement of replication in establishing repression of *HML* and *HMR*. Cells carrying a *sir*ᵗˢ mutation are incubated at high temperature; *HML* and *HMR* are expressed. Then the temperature is reduced to restore *SIR* activity. When the cells are blocked in G1 phase, *HML* and *HMR* continue to be expressed. But repression can be established when DNA replication is permitted. This suggests that *replication from the* ars *sequences associated with the E elements is necessary to establish repression of HML and HMR*. This is a novel form of repression, and we do not yet know how it works at the molecular level.

Activation of a cassette results in changes in its chromatin structure. All cassettes possess DNAase I hypersensitive sites, but there are significant differences between the active and silent cassettes. Also, there are changes in the nucleosome ladders and in the degree of supercoiling, which may be associated with gene activation.

The hypersensitive sites are summarized in **Figure 36.8**. The general pattern of hypersensitivity is similar in *MATa* and *MATα*. Hypersensitive sites are found at the promoter and also on either side of the expressed region. Almost all of the sites are absent from *HML* and *HMR*. However, the introduction of a *sir⁻* mutation restores all the missing hypersensitive sites. Thus activation of the cassettes is associated with the presence of hypersensitive sites.

What is the relationship between the four *SIR* loci? We do not know whether they code for subunits of a single regulatory protein or represent individual functions that interact with one another or individually with the target loci. However, although none of these proteins seems to bind DNA,

they appear to exert their effects via changes in chromatin structure.

The dependence on chromatin structure was discovered because deletions of the N-terminus of histone H4 activate the silent cassettes. Mutations of H4 at few specific amino acids in this region have the same effect. The effects of these mutations can be overcome either by introducing new mutations in *SIR3* or by over-expressing *SIR1*, which suggests that there is a specific interaction between histone 4 and the SIR proteins. The nature of the interaction

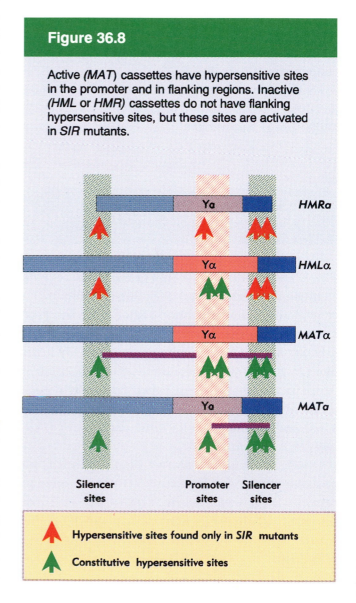

Figure 36.8

Active *(MAT)* cassettes have hypersensitive sites in the promoter and in flanking regions. Inactive *(HML* or *HMR)* cassettes do not have flanking hypersensitive sites, but these sites are activated in *SIR* mutants.

is not known yet, but it could involve influences on acetylation or other modifications of H4.

The general model suggested by these results is that the SIR proteins act on chromatin structure to prevent expression of the genes, by suppressing the formation of essential hypersensitive sites. DNA replication may be needed to permit a structural change to occur in the region, possibly because the SIR proteins only then gain an opportunity to interact with histones and other nucleosomal proteins.

Unidirectional transposition is initiated by the recipient *MAT* locus

A switch in mating type represents a gene conversion in which the recipient site (*MAT*) is converted to the sequence of the donor type (*HML* or *HMR*). Sites needed for transposition have been identified by mutations at *MAT* that prevent switching. The unidirectional nature of the process is indicated by lack of mutations in *HML* or *HMR*.

The mutations identify a site at the right boundary of Y at *MAT* that is crucial for the switching event. Deletions at the right end of Y do not have any effect until they cross the boundary; then they abolish switching. Point mutations that abolish switching occur in the vicinity of the boundary. The nature of the boundary is shown by analyzing the locations of these point mutations relative to the switching event (this is done by examining the results of rare switches that occur in spite of the mutation). Some mutations lie within the region that is replaced (and thus disappear from *MAT* after a switch), while others lie just outside the replaced region (and therefore continue to impede switching). So sequences both within and outside the replaced region are needed for the switching event.

The Y-Z boundary is the site of a change in DNA that marks the initiation of the transposition event. In populations of cells undergoing switching, 1–3% of the DNA of the *MAT* locus has a double-stranded cut at this site. The cut lies close to the boundary and coincides with the DNAase hypersensitive site.

We can now suggest a possible series of events for switching. The hypersensitive site at the Y-Z1 boundary

of *MAT* may be accessible because it lacks a nucleosome. It is recognized by an endonuclease coded by the *HO* locus. In fact, this is equivalent to suggesting that hypersensitivity to DNAase I *in vitro* reflects a natural sensitivity to the *HO* endonuclease *in vivo!*

The *HO* endonuclease makes a staggered double-strand break just to the right of the Y boundary.

Figure 36.9

HO endonuclease cleaves *MAT* just to the right of the Y region, generating sticky ends with a 4 base overhang.

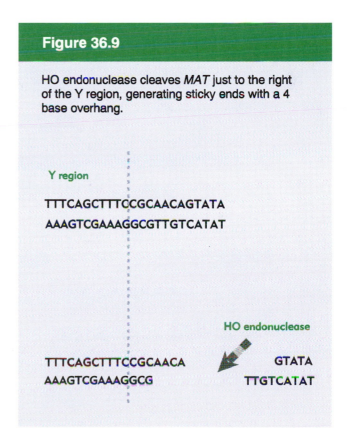

Y region

TTTCAGCTTTCCGCAACAGTATA
AAAGTCGAAAGGCGTTGTCATAT

HO endonuclease

TTTCAGCTTTCCGCAACA   GTATA
AAAGTCGAAAGGCG   TTGTCATAT

Figure 36.10

Cassette substitution is initiated by a double strand break in the recipient (MAT) locus, and may involve pairing on either side of the Y region with the donor (HMR or HML) locus. The sequence that is removed and replaced actually may extend to the left of the Y boundary.

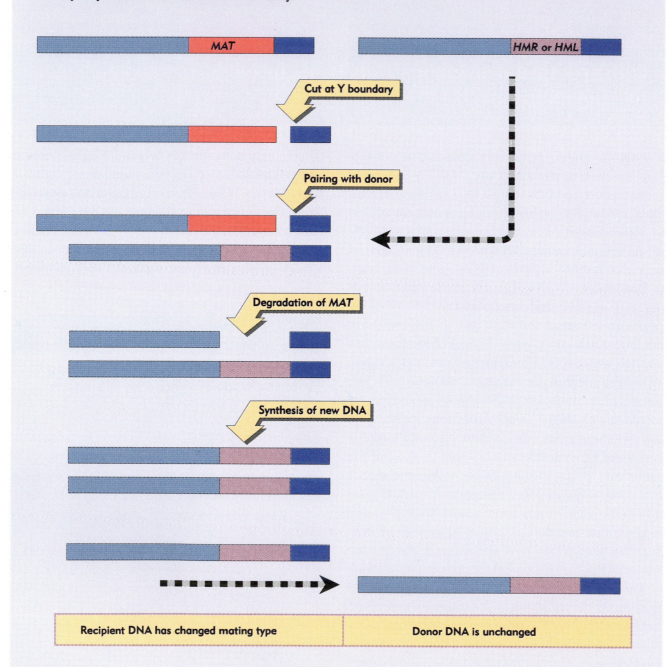

Cleavage generates the single-stranded ends of 4 bases drawn in **Figure 36.9**. The nuclease does not attack mutant *MAT* loci that cannot switch. Deletion analysis shows that most or all of the sequence of 24 bp surrounding the Y junction is required for cleavage *in vitro*. The recognition site is relatively large for a nuclease. Probably the recognition sequence occurs only at the three mating-type cassettes.

It seems plausible that *the same mechanisms that keep the silent cassettes from being transcribed also keep them inaccessible to the* HO *endonuclease. This inaccessibility ensures that switching is unidirectional.*

The reaction triggered by the cleavage is illustrated schematically in **Figure 36.10** in terms of the general reaction between donor and recipient regions. In terms of the interactions of individual strands of DNA, it follows the scheme for recombination via a double-strand break drawn in Figure 33.5; and the stages following the initial cut require the enzymes involved in general recombination. Mutations in some of these genes prevent switching.

Suppose that the free end at *MAT* invades either the *HML* or *HMR* locus and pairs with the Z region. The Y region of *MAT* is degraded until a region with homology on the left side is exposed. At this point, *MAT* is paired with *HML* or *HMR* at both the left side and the right side. The Y region of *HML* or *HMR* is copied to replace the region lost from *MAT* (which might extend beyond the limits of Y itself). The paired loci separate. (The order of events could be different.)

Like the double-strand break model for recombination, the process is initiated by *MAT*, the *locus that is to be replaced*. In this sense, the description of *HML* and *HMR* as donor loci refers to their ultimate role, but not to the mechanism of the process. Like replicative transposition, the donor site is unaffected, but a change in sequence occurs at the recipient; unlike transposition, the recipient locus suffers a substitution rather than addition of material.

Regulation of *HO* expression

Switching is initiated by the *HO* gene, which is itself regulated in an interesting way. Transcription of *HO* responds to several controls:

◆ *HO* is under mating-type control, since it is not synthesized in *MATa/MATα* diploids. In teleological terms, we may think that there is no need for switching when *both MAT* alleles are expressed anyway.

◆ *HO* is transcribed in mother cells but not in daughter cells.

◆ *HO* transcription also responds to the cell cycle. The gene is expressed only at the end of the G1 phase of a mother cell.

The timing of nuclease production explains the relationship between switching and cell lineage. Switching is detected only in the products of a division; *both daughter cells have the same mating type*, switched from that of the parent. **Figure 36.11** demonstrates that the restriction of *HO* expression to G1 phase ensures that the mating type is switched

Figure 36.11

Switching occurs only in mother cells; both daughter cells have the new mating type. A daughter cell must pass through an entire cycle before it becomes a mother cell that is able to switch again.

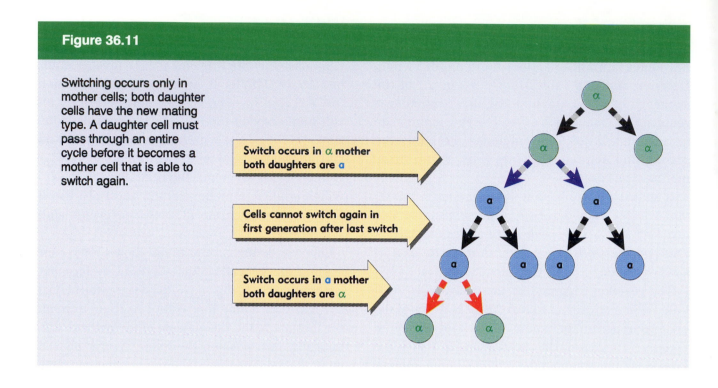

Switch occurs in α mother both daughters are a

Cells cannot switch again in first generation after last switch

Switch occurs in a mother both daughters are α

before the *MAT* locus is replicated, with the result that both progeny have the new mating type.

Cis-acting sites that control *HO* transcription reside in the 1500 bp upstream of the gene. They are summarized in **Figure 36.12**. The general pattern of control is that repression at any one of many sites, responding to several regulatory circuits, may prevent transcription of *HO*.

Mating-type control resembles that of other haploid-specific genes. Transcription is prevented (in diploids) by the **a1**/α2 repressor. There are 10 binding sites (shown in red) for the repressor in the upstream region. These sites vary in their conformity to the consensus sequence; we do not know which and how many of them are required for haploid-specific repression.

Cell-cycle control is conferred by 9 copies of an octanucleotide sequence (shown in blue) that lie in the region between −150 and −900. A copy of the consensus sequence can confer cell-cycle control on a gene to which it is attached. A gene linked to this sequence is repressed except during a transient period toward the end of G1 phase. Its activity depends on the function of the cell-cycle regulator *CDC28*, which executes the START stage of the cycle when the cell becomes committed to dividing.

Some interesting interactions involving the *SWI* and *SIN* genes are involved in cell-cycle and mother–daughter control. The genes *SWI1–5* are required for *HO* transcription. They function by preventing products of the genes *SIN1–6* from repressing *HO*. The *SWI* genes were discovered first, as mutants unable to switch; then the *SIN* genes were discovered for their ability to release the blocks caused by particular *SWI* mutations. The interactions between *SIN* and *SWI* genes are not fully defined, but involve multiple pathways.

SWI5 is required to allow transcription in mother cells. It acts by countering the repression exercised by *SIN3,4*. In mutants that lack these functions, *HO* is transcribed equally well in mother and daughter cells. This system acts on sequences in the far upstream region (−1260 to −1300). *SWI5* is itself subject to cell-cycle control. It is expressed later than *HO*, at some time after the START stage. Daughter cells are born lacking SWI5 function. So if *SWI5* is activated during their first cell cycle, it is not until the second cycle that it gets the chance to activate *HO* and cause switching. This explains the delay in a daughter cell's ability to switch until it has matured to become a mother cell (see Figure 36.11). The product of *SWI4* functions in an analogous manner, by preventing *SIN6* from repressing the *HO* gene.

The *HO* gene is also regulated by *SWI1,2,3*, which code for a group of proteins, possibly functioning as a complex, that are required for expression of a variety of genes. The complex may therefore be a global activator of transcription. Among the genes that prevent *SWI1,2,3* function are *SIN1*, which codes for a nonhistone protein, and *SIN2* which codes for histone H3. This suggests that the *SWI1,2,3* group is needed to make some general change in chromatin structure that is needed for transcription of *HO* and other genes, so their role in mating-type expression may be incidental.

Figure 36.12

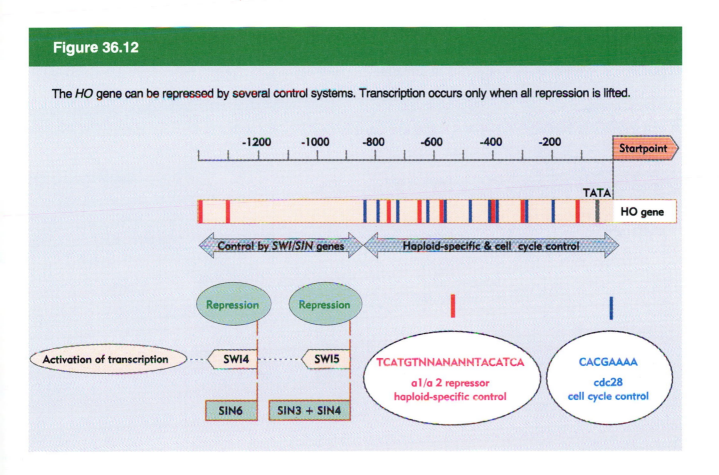

The *HO* gene can be repressed by several control systems. Transcription occurs only when all repression is lifted.

Trypanosomes rearrange DNA to express new surface antigens

Sleeping sickness in man (and a related disease in cows) is caused by infection with African trypanosomes. The unicellular parasite follows the life cycle illustrated in **Figure 36.13**, in which it alternates between tsetse fly and mammal. The trypanosome may be transferred either to or from the fly when it bites a mammal.

During its life cycle, the parasite undergoes several morphological and biochemical changes. The most significant biochemical change is in the **variable surface glycoprotein (VSG)**, the major component of the surface coat. The coat covers the plasma membrane and consists of a monolayer of $5\text{--}10 \times 10^6$ molecules of a single VSG, which is the only antigenic structure exposed on the surface. A trypanosome expresses only one VSG at any time,

and its ability to change the VSG is responsible for its survival through the fly–mammal infective cycle.

Consider the cycle as starting when a fly gains a trypanosome by biting an infected mammal. The trypanosome enters the gut of the fly in the 'procyclic form', and loses its VSG. After about 3 weeks, its progeny differentiate into the 'metacyclic form', which re-acquires a VSG coat. This form is transmitted to the mammalian bloodstream during a bite by the fly. The trypanosome multiplies in the mammalian bloodstream. Its progeny continue to express the metacyclic VSG for about a week. Then a new VSG is synthesized, and further transitions occur every 1–2 weeks.

Each of the successive VSG species is immunologically distinct. As a result, the antigen presented to the

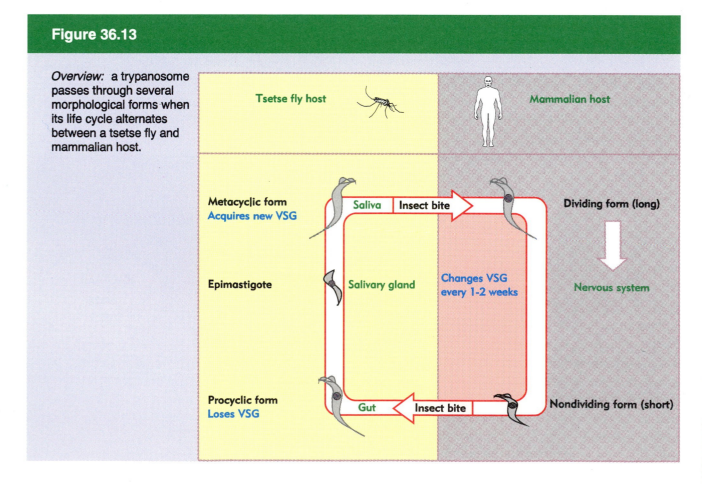

Figure 36.13

Overview: a trypanosome passes through several morphological forms when its life cycle alternates between a tsetse fly and mammalian host.

mammalian immune system is constantly changing. The process of transition is called **antigenic variation**. The immune response always lags behind the change in surface antigen, so that the trypanosome evades immune surveillance, and thereby perpetuates itself indefinitely. Each transition of the VSG is accompanied by a new wave of parasitemia, with symptoms of fever, rash, etc; the parasites eventually invade the central nervous system, after which the mammalian host becomes progressively more lethargic and eventually comatose.

Trypanosomes vary in their host range. The best investigated species is a variety of *Trypanosoma brucei* that grows well on laboratory animals, although not on man. Laboratory strains of *T. brucei* switch VSGs spontaneously at a rate of 10^{-4}–10^{-6} per division. Switching occurs independently of the host immune system. In effect, new variants then are selected by the host, because it mounts a response against the old VSG, but fails to recognize and act against the new VSG.

What is the structure of the VSG? Protein (and cDNA) sequences give the general view of VSG structure depicted in **Figure 36.14**. A nascent VSG is ~500 amino acids long; it has an N-terminal signal sequence, followed by a long *variable region* that provides the unique antigenic determinant, and a C-terminal *homology region* ending in a short hydrophobic tail. The nascent VSG is processed at both ends to give the mature form. The signal sequence is cleaved during secretion. The hydrophobic tail is removed before the VSG reaches the outside surface. The new C-terminus is covalently attached to the trypanosome membrane; three types of homology region are distinguished according to the C-terminal amino acid.

The VSG is attached to the membrane via a phosphoglycolipid. As a result, VSG can be released from the membrane by an enzyme that removes fatty acid. This reaction (which is used in purifying the VSG) may be important *in vivo* in allowing one VSG to be replaced by another on the surface of the trypanosome.

How many varieties of VSG can be expressed by any one trypanosome? It is not clear that any limit is encountered before death of the host. A

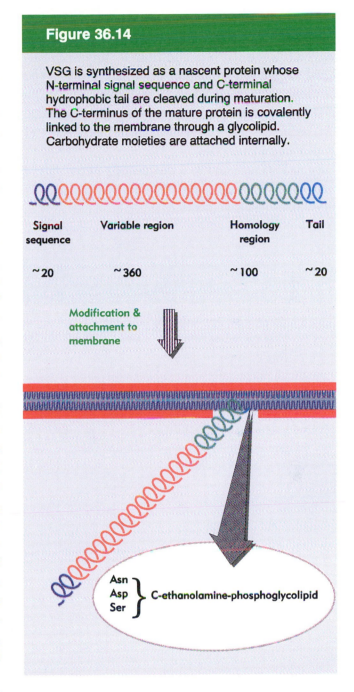

Figure 36.14

VSG is synthesized as a nascent protein whose N-terminal signal sequence and C-terminal hydrophobic tail are cleaved during maturation. The C-terminus of the mature protein is covalently linked to the membrane through a glycolipid. Carbohydrate moieties are attached internally.

Signal sequence	Variable region	Homology region	Tail
~20	~360	~100	~20

Modification & attachment to membrane

Asn
Asp
Ser } C-ethanolamine-phosphoglycolipid

single trypanosome can make at least 100 VSGs sufficiently different in sequence that antibodies against any one do not react against the others.

VSG variation is coded in the trypanosome genome. Every individual trypanosome carries the entire VSG repertoire of its strain. *Diversity therefore depends on changing expression from one preexisting gene to another.*

The trypanosome genome has an unusual organization, consisting of a large number of segregating units. In addition to an unknown number of chromosomes, it contains ~100 'minichromosomes', each containing ~50–150 kb of DNA. Hybridization experiments identify ~1000 VSG genes, scattered among all size classes of chromosomal material.

Each VSG is coded by a **basic-copy gene**. These genes can be divided into two classes according to their chromosomal location:

◆ Telomeric genes lie within 5–15 kb of a telomere. There could be >200 of these genes if every telomere has one.

◆ Internal genes reside within chromosomes (more formally, they lie >50 kb from a telomere).

As might be expected of a large family of genes, individual basic copies show varying degrees of relationship, presumably reflecting their origin by duplication and variation. Genes that are closely

Figure 36.15

VSG genes may be created by duplicative transfer from an internal or telomeric basic copy into an expression site, or by activating a telomeric copy that is already present at a potential expression site.

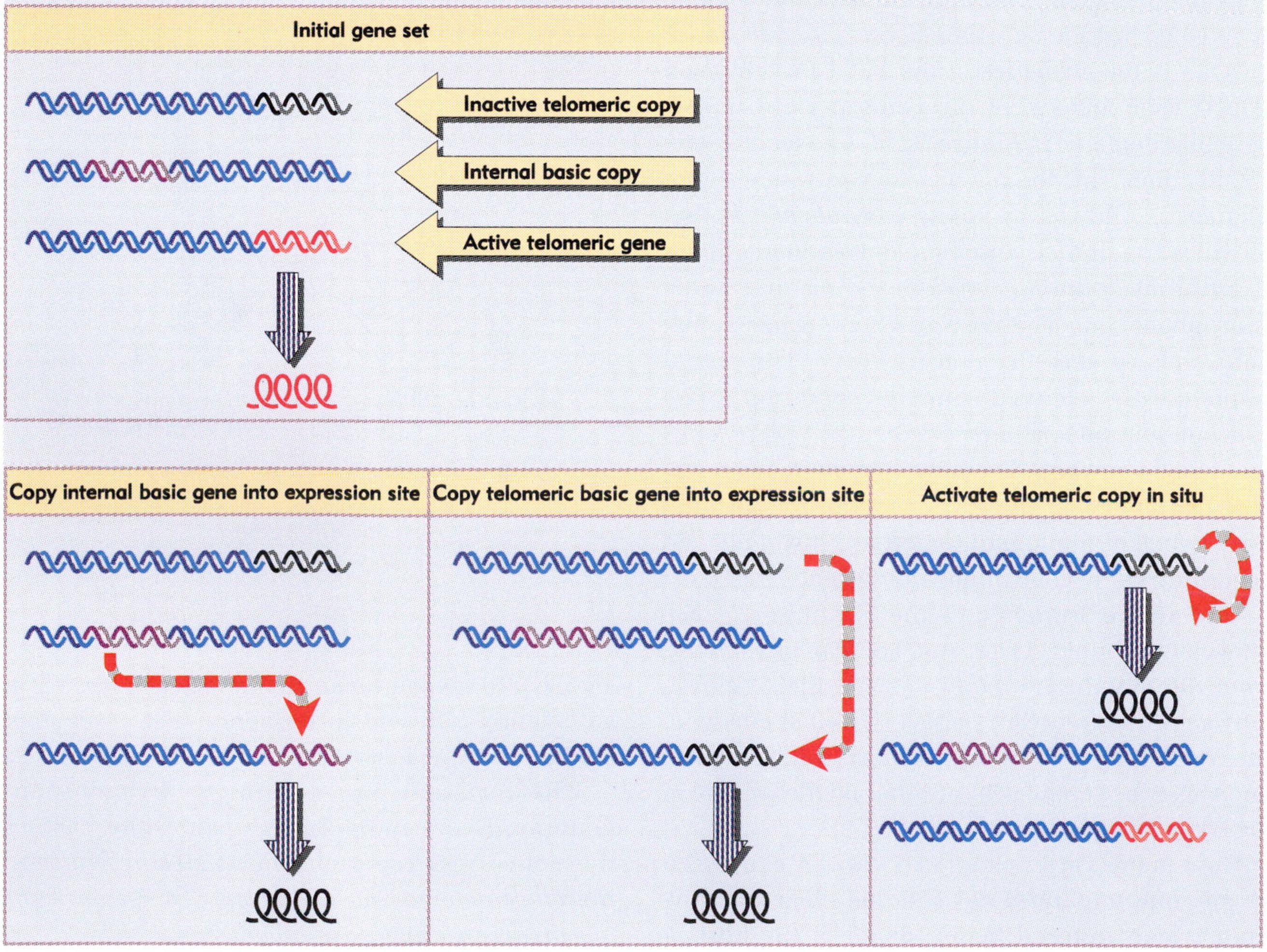

related, and which provoke the same antigenic response, are called isogenes.

How is a single VSG gene selected for expression? Only one VSG gene is transcribed in a trypanosome at a given time. The copy of the gene that is active is called the **expression-linked copy (ELC)**. It is said to be located at an **expression site**. An expression site has a characteristic property: *it is located near a telomere.*

These features immediately suggest that the route followed to select a gene for expression depends on whether the basic copy is itself telomeric or internal. The two types of event that can create an ELC are summarized in **Figure 36.15**.

◆ *The expression site remains the same, but the ELC is changed.* Duplication transfers the sequence of a basic copy to replace the sequence currently occupying the expression site. Either internal or telomeric copies may be activated directly by duplication into the expression site. *The substitution of one cassette for another does not interfere with the activity of the site.*

◆ *The expression site is changed.* Activation *in situ* is available only to a sequence already present at a telomere. *When a telomeric site is activated* in situ, *the previous expression site must cease to be active and the new site now becomes the expression site.*

Internal basic copies probably can be copied into non-expressed telomeric locations as well as into expression sites. Thus an internal gene could be activated by a two-stage process, in which first it is transposed to a non-expressed telomere, and then this site is activated.

We can follow the fate of genes involved in activation by restriction mapping. A probe representing an expressed sequence can be derived from the mRNA. Then we can determine the status of genes corresponding to the probe. We see different results for internal and telomeric basic-copy genes:

◆ *Activation of an internal gene requires generation of new sequences.* **Figure 36.16** shows that when

an internal gene is activated, a new fragment is found. The original basic-copy gene remains unaltered; the new fragment represents an ELC, located close to a telomere. The ELC appears when the gene is expressed and disappears when the gene is switched off. Duplication into the ELC is the *only* pathway by which an internal basic copy can be generated.

◆ *Activation of a telomeric gene can occur in situ.* **Figure 36.17** shows that when a telomeric gene is activated, the gene number need not change. The structure of the gene may be essentially unaffected as detected by restriction mapping. The size of the fragment containing the gene may vary slightly, because the length of the telomere is constantly changing. Telomeric basic copies can also be activated by the same duplication pathway as internal copies; in this case, the basic copy remains at its telomere, while an expression-linked copy appears at another telomere (generating a new fragment as illustrated for internal basic copies in Figure 36.16).

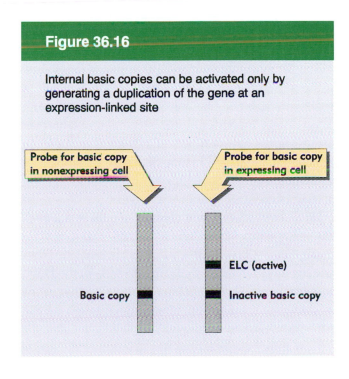

Figure 36.16

Internal basic copies can be activated only by generating a duplication of the gene at an expression-linked site

Probe for basic copy in nonexpressing cell

Probe for basic copy in expressing cell

ELC (active)

Basic copy

Inactive basic copy

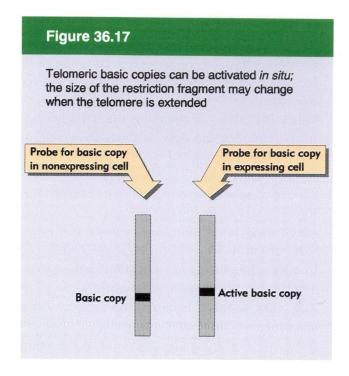

Figure 36.17

Telomeric basic copies can be activated *in situ*; the size of the restriction fragment may change when the telomere is extended

Probe for basic copy in nonexpressing cell

Probe for basic copy in expressing cell

Basic copy

Active basic copy

Formation of the ELC occurs by gene conversion. Like the switch in yeast mating type, it represents the replacement of a 'cassette' at the active (telomeric) locus by a stored cassette. The VSG system is more versatile in the sense that there are many potential donor cassettes (and also more than a single potential recipient site).

Almost all switches in VSG type involve replacement of the ELC by a pre-existing silent copy. Some ·exceptional cases have been found, however, in which the sequence of the ELC does not match any of the repertoire of silent copies in the genome. A new sequence may be created by a series of gene conversions in which short stretches of different silent copies are connected. This resembles the mechanism for generating diversity in chicken λ immunoglobulins (see Chapter 37). Although rare, such occurrences extend VSG diversity.

We assume that some change occurs at a potential expression site when it becomes active. The change could be an alteration in chromatin structure, or might involve a change directly in the DNA, possibly some modification (such as methylation) or some reorganization of sequence (such as the introduction of an enhancer).

How many expression sites are there? We know

that in some cases a series of basic copies may be activated by replacing one another at the same expression site. In other cases (especially when *in situ* activation of telomeric copies is involved), the expression site may change. So far only a few expression sites have been observed, which suggests that only a subset of telomeres can function in this capacity, but it is possible that in fact any telomere can be used (although some may be preferred to others).

The structure of the VSG gene at the ELC is unusual, as illustrated in **Figure 36.18**. The length of DNA transferred into the ELC is 2500–3500 bp,

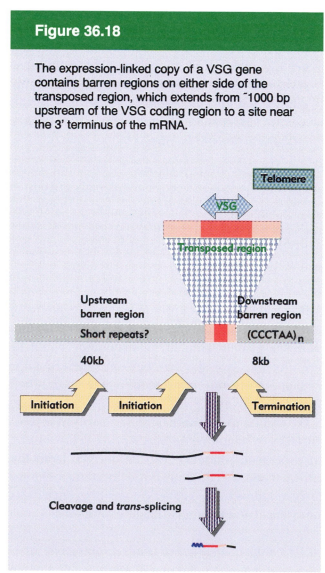

Figure 36.18

The expression-linked copy of a VSG gene contains barren regions on either side of the transposed region, which extends from ˜1000 bp upstream of the VSG coding region to a site near the 3' terminus of the mRNA.

Telomere

VSG

Transposed region

Upstream barren region

Downstream barren region

Short repeats?

(CCCTAA)ₙ

40kb

8kb

Initiation

Initiation

Termination

Cleavage and *trans*-splicing

somewhat longer than the VSG coding region of 1500 bp. Most of the additional length is upstream of the gene. The crossover points at which the duplicated sequence joins the ELC do not appear to be precisely determined.

Analysis of events at the 5′ end of the VSG mRNA is complicated by the fact that the mature RNA starts with a 35 base sequence coded elsewhere, and added in *trans* to the newly synthesized 5′ end (see Chapter 31). However, one model supposes that the ELC is activated because an active promoter lies upstream of the point where the copy is inserted into the expression site.

The signals for initiating and terminating transcription (and sometimes also the end of the coding region itself) are provided by the sequences flanking the transposed region. In fact, transcription may be initiated several kb upstream of the VSG gene itself. Promoters have been mapped at 4 kb and ~60 kb upstream of the VSG sequence. Use of the more distant promoter generates a transcript that contains other genes as well as the active VSG. The VSG sequence (and other gene sequences) must be released by cleavage from the transcript, after which the 35 base spliced leader is added to the 5′ end.

On either side of the transposed region are extensive regions that are not cut by restriction enzymes. These 'barren regions' consist of repetitive DNA; they extend some 8 kb downstream and for up to 40 kb upstream of the ELC. Going downstream, the barren region consists largely of repeats of the sequence CCCTAA, and extends to the telomere. Proceeding upstream, it may also consist of repetitive sequences, but their nature is not yet clear. The existence of the barren regions, however, has been an impediment to characterizing ELC genes by cloning.

The order in which VSG genes are expressed during an infection is erratic, but not completely random. This may be an important feature in survival of the trypanosome. If VSG genes were used in a predetermined order, a host could knock out the infection by mounting a reaction against one of the early elements. The need for unpredictability in the production of VSGs may be responsible for the evolution of a system with many donor sequences and multiple recipients.

Antigenic variation is not a unique phenomenon of trypanosomes. The bacterium *Borrelia hermsii* causes relapsing fever in man and analogous diseases in other mammals. The name of the disease reflects its erratic course: periods of illness are spaced by periods of relief. When the fevers occur, spirochetes are found in the blood; they disappear during periods of relief, as the host responds with specific antibodies.

Like the trypanosomes, *Borrelia* survives by altering a surface protein, called the variable major protein (VMP). Changes in the VMP are associated with rearrangements in the genome. The active VMP is located near the telomere of a linear plasmid. We do not yet know the extent of the coded variants or the mechanisms used to alter their expression. It is intriguing, however, that the eukaryote *Trypanosoma* and the prokaryote *Borrelia* should both rely upon antigenic variation as a means for evading immune surveillance.

Interaction of Ti plasmid DNA with the plant genome

Most events in which DNA is rearranged or amplified occur within a genome, but the interaction between bacteria and certain plants involves the transfer of DNA from the bacterial genome to the plant genome. **Crown gall disease** can be induced in most dicotyledonous plants by the soil bacterium *Agrobacterium tumefaciens*. The bacterium is a parasite that effects a genetic change in the eukaryotic host cell, with consequences for both parasite and host. It improves conditions for

Figure 36.19

An *Agrobacterium* carrying a Ti plasmid of the nopaline type induces a teratoma, in which differentiated structures develop. Photograph kindly provided by Jeff Schell.

survival of the parasite. And it causes the plant cell to grow as a tumor. **Figure 36.19** is a photograph of a crown gall tumor.

Agrobacteria are required to induce tumor formation, but the tumor cells do not require the continued presence of bacteria. Like animal tumors, the plant cells have been transformed into a state in which new mechanisms govern growth and differentiation. Transformation is caused by the expression within the plant cell of genetic information transferred from the bacterium.

The *tumor–inducing principle of Agrobacterium*

resides in the **Ti plasmid**, which is perpetuated as an independent replicon within the bacterium. The plasmid carries genes involved in various bacterial and plant cell activities, including those involved with generating the transformed state, and a set of genes concerned with synthesis or utilization of **opines** (novel derivatives of arginine).

Ti plasmids (and thus the *Agrobacteria* in which they reside) can be divided into four groups, according to the types of opine that are made:

◆ **Nopaline plasmids** carry genes for synthesizing nopaline in tumors and for utilizing it in bacteria. Nopaline tumors can differentiate into shoots with abnormal structures. They have been called **teratomas** by analogy with certain mammalian tumors that retain the ability to differentiate into early embryonic structures.

◆ **Octopine plasmids** are similar to nopaline plasmids, but the relevant opine is different. However, octopine tumors are usually undifferentiated and do not form teratoma shoots.

◆ **Agropine plasmids** carry genes for agropine metabolism; the tumors do not differentiate, develop poorly, and die early.

Table 36.3

Ti plasmids carry genes involved in both plant and bacterial functions.

Locus	Function	Ti Plasmid
Vir	DNA transfer into plant	All
Shi	Shoot induction	All
Roi	Root induction	All
Nos	Nopaline synthesis	Nopaline
Noc	Nopaline catabolism	Nopaline
Ocs	Octopine synthesis	Octopine
Occ	Octopine catabolism	Octopine
Tra	Bacterial transfer genes	All
Inc	Incompatibility genes	All
oriV	Origin for replication	All

Figure 36.20

T-DNA is transferred from *Agrobacterium* carrying a Ti plasmid into a plant cell, where it becomes integrated into the nuclear genome and expresses functions that transform the host cell.

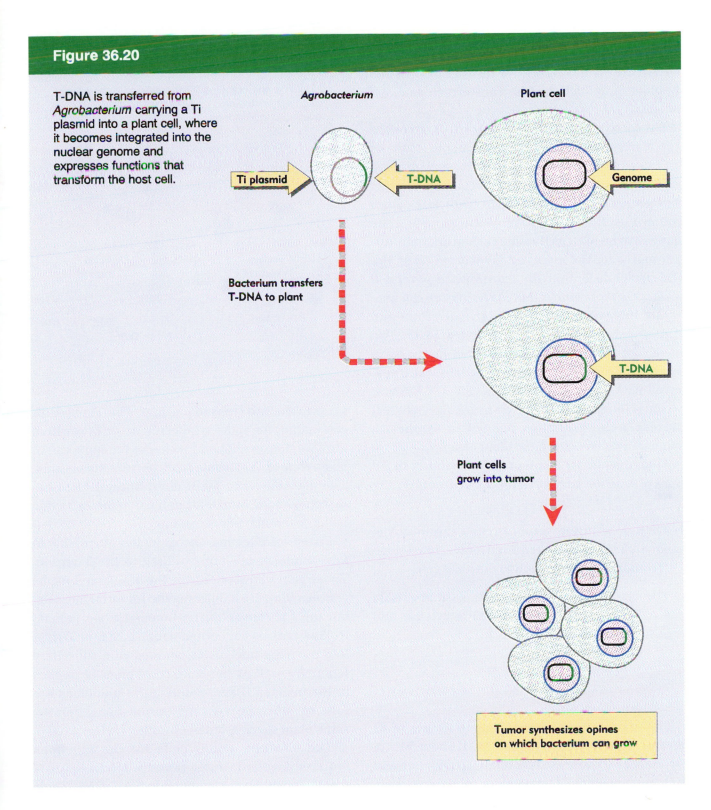

Bacterium transfers T-DNA to plant

Plant cells grow into tumor

Tumor synthesizes opines on which bacterium can grow

◆ **Ri plasmids** can induce hairy root disease on some plants and crown gall on others. They have agropine type genes, and may have segments derived from both nopaline and octopine plasmids.

The types of genes carried by a Ti plasmid are summarized in **Table 36.3**. Genes utilized in bacteria code for plasmid replication and incompatibility, for transfer between bacteria, sensitivity to phages, and for synthesis of other compounds, some of which are toxic to other soil bacteria. Genes used in the plant cell code for transfer of DNA into the plant, for induction of the transformed state, and for shoot and root induction.

The specificity of the opine genes depends on the type of plasmid. Genes needed for opine synthesis are linked to genes whose products catabolize the same opine; thus each strain of *Agrobacterium* causes crown gall tumor cells to synthesize opines that are useful for survival of the parasite. The opines can be used as the sole carbon and/or nitrogen source for the inducing *Agrobacterium* strain. The principle is that *the transformed plant cell synthesizes those opines that the bacterium can use.*

The interaction between *Agrobacterium* and a plant cell is illustrated in **Figure 36.20.** The bacterium does not enter the plant cell, but *transfers part of the Ti plasmid to the plant nucleus.* The transferred part of the Ti genome is called **T-DNA.** It becomes integrated into the plant genome, where it expresses the functions needed to synthesize opines and to transform the plant cell.

Transformation of plant cells requires three types of function carried in the *Agrobacterium*:

♦ Three loci on the *Agrobacterium* chromosome, *chvA, chvB, pscA,* are required for the initial stage of binding the bacterium to the plant cell.

♦ The *vir* region carried by the Ti plasmid outside the T-DNA region is required to release and initiate transfer of the T-DNA.

♦ The T-DNA is required to transform the plant cell.

The organization of the major two types of Ti plasmid is illustrated in **Figure 36.21.** About 30% of the ~200 kb Ti genome is common to nopaline and octopine plasmids. The common regions include genes involved in all stages of the interaction between *Agrobacterium* and a plant host, but considerable rearrangement of the sequences has occurred between the plasmids.

The T-region occupies ~23 kb. Some 9 kb is the

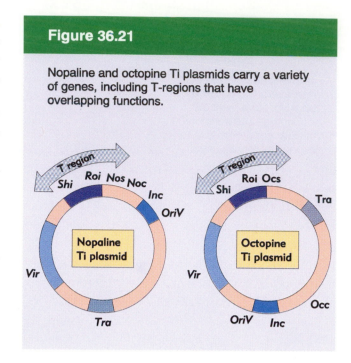

Figure 36.21

Nopaline and octopine Ti plasmids carry a variety of genes, including T-regions that have overlapping functions.

same in the two types of plasmid. The Ti plasmids carry genes for opine synthesis (*Nos* or *Ocs*) within the T-region; corresponding genes for opine catabolism (*Noc* or *Occ*) reside elsewhere on the plasmid. The plasmids code for similar, but not identical, morphogenetic functions, as seen in the induction of the different types of tumors.

Functions affecting oncogenicity—the ability to form tumors—are not confined to the T-region. Those genes located outside the T-region must be concerned with establishing the tumorigenic state, but their products are not needed to perpetuate it. They may be concerned with transfer of T-DNA into the plant nucleus or perhaps with subsidiary functions such as the balance of plant hormones in the infected tissue. Some of the mutations are host-specific, preventing tumor formation by some plant species, but not by others.

The *virulence* genes code for the functions required for the transfer process. Six loci *virA–G* reside in a 40 kb region outside the T-DNA. Their organization is summarized in **Figure 36.22.** Each locus is transcribed as an individual unit; some contain more than one open reading frame.

We may divide the transforming process into (at least) two stages:

◆ *Agrobacterium* contacts a plant cell, and the *vir* genes are induced.

◆ *vir* gene products cause T-DNA to be transferred to the plant cell nucleus, where it is integrated into the genome.

The *vir* genes fall into two groups, corresponding to these stages. Genes *virA* and *virG* are regulators, whose action is required to induce the other genes. Thus mutants in *virA* and *virG* are avirulent and cannot express the remaining *vir* genes. Genes *virB,C,D,E* code for proteins involved in the transfer of DNA. Mutants in *virB* and *virD* are avirulent in all plants, but the effects of mutations in *virC* and *virE* vary with the type of host plant.

Only *virA* and *virG* are expressed constitutively, at a rather low level. The signal that causes the other *vir* genes to be induced is provided by phenolic compounds generated by plants as a response to wounding. **Figure 36.23** presents an example. *N. tabacum* (tobacco) generates the molecules acetosyringone and α–hydroxyacetosyringone. Exposure to these compounds activates *virA*, which in turn causes *virG* to be expressed at a higher level, and induces the expression *de novo* of *virB,C,D,E*. This reaction explains why *Agrobacterium* infection succeeds only on wounded plants.

The product of *virA* is located in the inner membrane. It may respond to the presence of the phenolic compounds in the periplasmic space. It might in principle function in either of two ways.

Figure 36.22

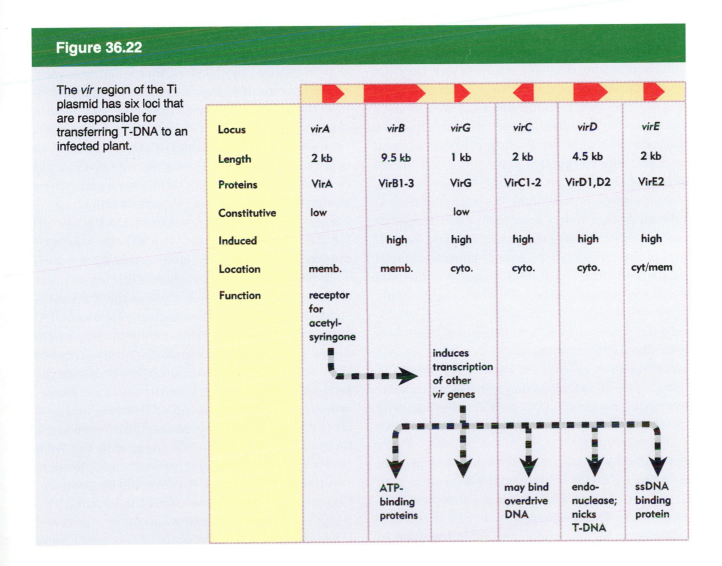

The *vir* region of the Ti plasmid has six loci that are responsible for transferring T-DNA to an infected plant.

Locus	virA	virB	virG	virC	virD	virE
Length	2 kb	9.5 kb	1 kb	2 kb	4.5 kb	2 kb
Proteins	VirA	VirB1-3	VirG	VirC1-2	VirD1,D2	VirE2
Constitutive	low		low			
Induced		high	high	high	high	high
Location	memb.	memb.	cyto.	cyto.	cyto.	cyt/mem
Function	receptor for acetyl-syringone	ATP-binding proteins	induces transcription of other vir genes	may bind overdrive DNA	endo-nuclease; nicks T-DNA	ssDNA binding protein

Figure 36.23

Acetosyringone (4-acetyl-2,6-dimethoxyphenol) is produced by *N. tabacum* upon wounding, and induces transfer of T-DNA from *Agrobacterium*

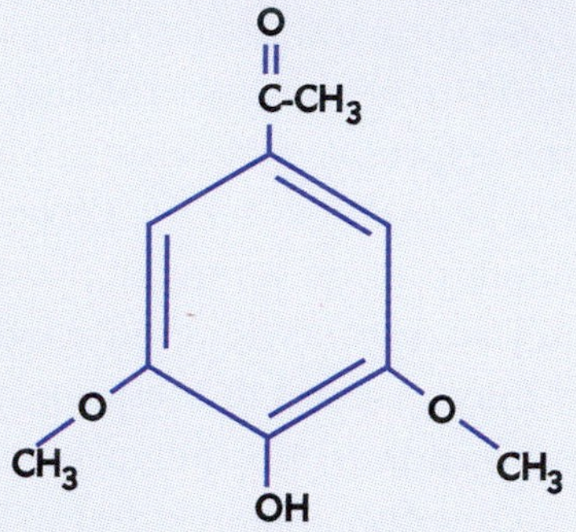

It could be responsible for transporting the compounds into the bacterium, where they act as co-inducers for activating the next protein in the pathway. Or VirA protein may itself be converted into an active state that triggers the next step. The *virA* gene has some similarities of sequence with *envZ* and other regulator genes of *E. coli*, which favors the second model.

When *virG* is activated, its transcription is induced from a new startsite, different from that used for constitutive expression. The sequence of *virG* is closely related to that of *ompR*, a bacterial regulator of transcription. This suggests that VirG is an inducer protein that turns on transcription of the other *vir* genes.

Of the other *vir* loci, *virD* is the best characterized. The *virD* locus has 4 open reading frames. Two of the proteins coded at *virD*, VirD1 and VirD2, provide an endonuclease that initiates the transfer process by nicking T-DNA at a specific site. The proteins form a covalent complex with the DNA substrate.

A typical single-strand DNA-binding protein is coded by *virE2*. This protein is produced in large amounts, and binds single-stranded DNA without regard to sequence. It has a nuclear localization signal, and is probably involved in transporting T-DNA into the plant cell nucleus.

Is the whole Ti plasmid transferred, after which the T-DNA makes its way to the nucleus; or is the transfer process itself responsible for selecting the T-region for entry into the plant? What we know about the DNA structures involved in the processes of integration and transfer is summarized in **Figure 36.24**.

T-DNA is demarcated from the flanking regions in the Ti plasmid by repeats of 25 bp, which differ at only two positions between the left and right ends. When T-DNA is integrated into a plant genome, it has a well-defined right junction, which retains 1–2 bp of the right repeat. The left junction is variable; the boundary of T-DNA in the plant genome may be located at the 25 bp repeat or at one of a series of sites ~100 bp within the T-DNA.

More than one copy of T-DNA may be integrated at a given site in the plant genome. When tandem copies are present, changes may occur in the sequences near the borders; they seem to involve reorganization of Ti plasmid sequences in the vicinity of the ends. This may affect sequences on either side of the 25 bp repeat element in the original Ti plasmid. We do not know whether the tandem copies are generated *in situ* after a single copy has integrated, or whether the T-DNA itself generates a tandem stretch that is integrated as a unit.

Only the right 25 bp repeat of T-DNA is needed for the transfer process. The repeat functions effectively only in its original orientation, which suggests that its recognition initiates a polar transfer process. We might imagine that the DNA is cleaved at this site, and then the sequence to the left is recognized for the transfer or integration reaction.

The transfer process involves single-stranded DNA. Induction of a nopaline plasmid causes the production of a single-stranded molecule corresponding to the T-region. This T-strand is produced at ~1 copy per cell. It is probably transferred in the form of a DNA–protein complex, with the DNA covered by the VirE2 single-strand binding protein.

A model for the transfer process is illustrated in **Figure 36.25**. A nick is made at the right 25 bp repeat. It provides a priming end for synthesis of a DNA single strand. Synthesis of the new strand displaces the old strand, which is used in the transfer process. Transfer is terminated when DNA synthesis reaches a nick at the left repeat. This model explains why the right repeat is essential, and it accounts for the polarity of the process. If the left repeat fails to be nicked, transfer could continue farther along the Ti plasmid.

Outside T-DNA, but immediately adjacent to the right border, is another short sequence, called *overdrive,* which greatly stimulates the transfer process. *Overdrive* functions like an enhancer: it must lie on the same molecule of DNA, but enhances the efficiency of transfer even when located several thousand base pairs away from the border. VirC1, and possibly VirC2, may act at the *overdrive* sequence.

Octopine plasmids have a more complex pattern of integrated T-DNA than nopaline plasmids. The pattern of T-strands is also more complex, and several discrete species can be found, corresponding to elements of T-DNA. This suggests that octopine T-DNA has several sequences that provide targets for nicking and/or termination of DNA synthesis.

This model for transfer of T-DNA closely resembles the events involved in bacterial conjugation, when the *E. coli* chromosome is transferred from one cell to another in single-stranded form (see Chapter 18). A difference is that the transfer of T-DNA is (usually) limited by the boundary of the left repeat, whereas transfer of bacterial DNA is indefinite.

We do not know how the transferred DNA is integrated into the plant genome. At some stage, the newly generated single strand must be converted into duplex DNA. Circles of T-DNA that are found in infected plant cells appear to be generated by recombination between the left and right 25 bp repeats, but we do not know if they are intermediates.

Is T-DNA integrated into the plant genome as

Figure 36.24

T-DNA has almost identical repeats of 25 bp at each end in the Ti plasmid. The right repeat is necessary for transfer and integration to a plant genome. T-DNA that is integrated in a plant genome has a precise junction that retains 1-2 bp of the right repeat, but the left junction varies and may be upto 100 bp short of the left repeat.

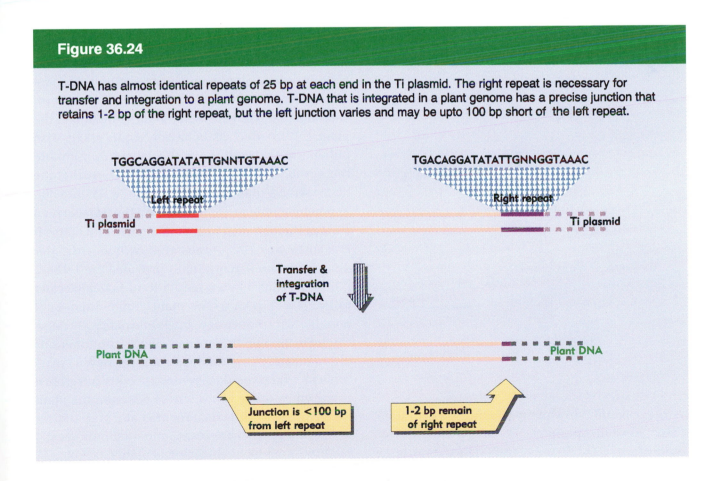

Figure 36.25

T-DNA is generated by displacement when DNA synthesis starts at a nick made at the right repeat. The reaction is terminated by a nick at the left repeat.

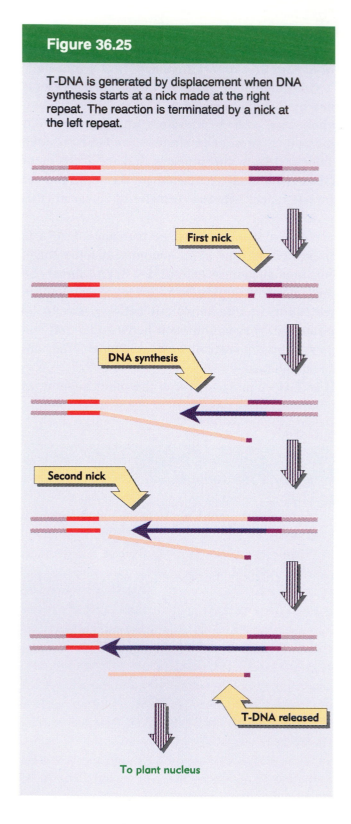

First nick

DNA synthesis

Second nick

T-DNA released

To plant nucleus

an integral unit? How many copies are integrated? What sites in plant DNA are available for integration? Are genes in T-DNA regulated exclusively by functions on the integrated segment? These questions are central to defining the process by which the Ti plasmid transforms a plant cell into a tumor.

What is the structure of the target site? Sequences flanking the integrated T-DNA tend to be rich in A•T base pairs (a feature displayed in target sites for some transposable elements). The sequence rearrangements that occur at the ends of the integrated T-DNA make it difficult to analyze the structure. We do not know whether the integration process generates new sequences in the target DNA comparable with the target repeats created in transposition.

T-DNA is expressed at its site of integration. The region contains several transcription units, each probably containing a gene expressed from an individual promoter. Their functions are concerned with the state of the plant cell, maintaining its tumorigenic properties, controlling shoot and root formation, and suppressing differentiation into other tissues. None of these genes is needed for T-DNA transfer.

The Ti plasmid presents an interesting organization of functions. Outside the T-region, it carries genes needed to initiate oncogenesis; at least some are concerned with the transfer of T-DNA, and we should like to know whether others function in the plant cell to affect its behavior at this stage. Also outside the T-region are the genes that enable the *Agrobacterium* to catabolize the opine that the transformed plant cell will produce. Within the T-region are the genes that control the transformed state of the plant, as well as the genes that cause it to synthesize the opines that will benefit the *Agrobacterium* that originally provided the T-DNA.

As a practical matter, the ability of *Agrobacterium* to transfer T-DNA to the plant genome makes it possible to introduce new genes into plants. Because the transfer/integration and oncogenic functions are separate, it is possible to engineer new Ti plasmids in which the oncogenic functions have been replaced by other genes whose effect on the plant we wish to test. The existence of a natural system for delivering genes to the plant genome should greatly facilitate genetic engineering of plants.

Selection of amplified genomic sequences

The eukaryotic genome has the capacity to accommodate additional sequences of either exogenous or endogenous origin. Exogenous sequences are introduced into cells by **transfection** (as discussed in Figure 4.3 and later in this chapter). Endogenous sequences may be produced by **amplification** of an existing sequence. A common feature in either case is that the additional sequences are likely to take the form of a tandem array, containing many copies of a repeating unit; a gene that is contained within the repeating unit is not necessarily expressed in every copy, but expression tends to increase with the number of copies.

A tandem array of multiple copies produced by either transfection or amplification may exist in either of two forms in a cell. If it takes the form of an extrachromosomal unit, it is inherited in an irregular manner: there is no equivalent in eukaryotic cells to the ability of a bacterial plasmid to be segregated evenly at cell division, so the entire unit is lost at a high frequency. If the array is integrated into the genome, however, it becomes a component of the genotype, and is inherited like any other genomic sequence.

Amplification of endogenous sequences is provoked by selecting cells for resistance to certain agents. The best-characterized example of amplification results from the addition of **methotrexate** (mtx) to certain cultured cell lines. This reagent blocks folate metabolism. Resistance to it is conferred by mutations that change the activity of the enzyme dihydrofolate reductase (DHFR). As an alternative to change in the enzyme itself, the amount of enzyme may be increased. The cause of this increase is an amplification of the number of *dhfr* structural genes. Amplification occurs at a frequency greater than the spontaneous point mutation rate, generally ranging from 10^{-4}–10^{-6}. Similar events have now been observed in >20 other genes.

A common feature in most of these systems is that highly resistant cells are not obtained in a single step, but instead appear when the cells are adapted to gradually increasing doses of the toxic reagent. Thus gene amplification may require several stages. Amplification generally occurs at only one of the two *dhfr* alleles; and increased resistance to methotrexate is accomplished by further increases in the degree of amplification at this locus.

The number of *dhfr* genes in a cell line resistant to methotrexate varies from 40 to 400, depending on the stringency of the selection and the individual cell line. The mtx^r lines fall into two classes, distinguished by their response when cells are grown in the absence of methotrexate, removing the former pressure for high levels of DHFR activity. The basis for the difference is illustrated in **Figure 36.26**.

◆ In **stable** lines, the amplified genes are retained, because they reside on the chromosome, at the site usually occupied by the single *dhfr* gene. Usually the other chromosome retains its normal single copy of *dhfr*.

◆ In **unstable** lines, the amplified genes are at least partially lost when the selective pressure is released, but the amplified genes exist as an extrachromosomal array.

Gene amplification has a visible effect on the chromosomes. In stable lines, the *dhfr* locus can be visualized in the form of a **homogeneously staining region (HSR)**. An example is shown in **Figure 36.27**. The HSR takes its name from the presence of an additional region that lacks any chromosome bands after treatments such as G-banding. This change suggests that some region of the chromosome between bands has undergone an expansion.

In unstable cell lines, no change is seen in the chromosomes carrying *dhfr*. However, large numbers of elements called **double-minute chromosomes** are visible, as can be seen in **Figure 36.28**. In a typical cell line, each double minute carries 2–4 *dhfr* genes. The double minutes appear

Figure 36.26

The dhfr gene can be amplified to give unstable copies that are extrachromosomal (double minutes) or stable (chromosomal). Extrachromosomal copies arise at early times; we do not know whether chromosomal copies arise directly from the chromosomal copy or via the extrachromosomal copies.

1 gene/chromosome

Unstable line

Extrachromosomal

Number of copies changes in daughter cells

Stable line

Amplified region

Genotype is constant in daughter cells

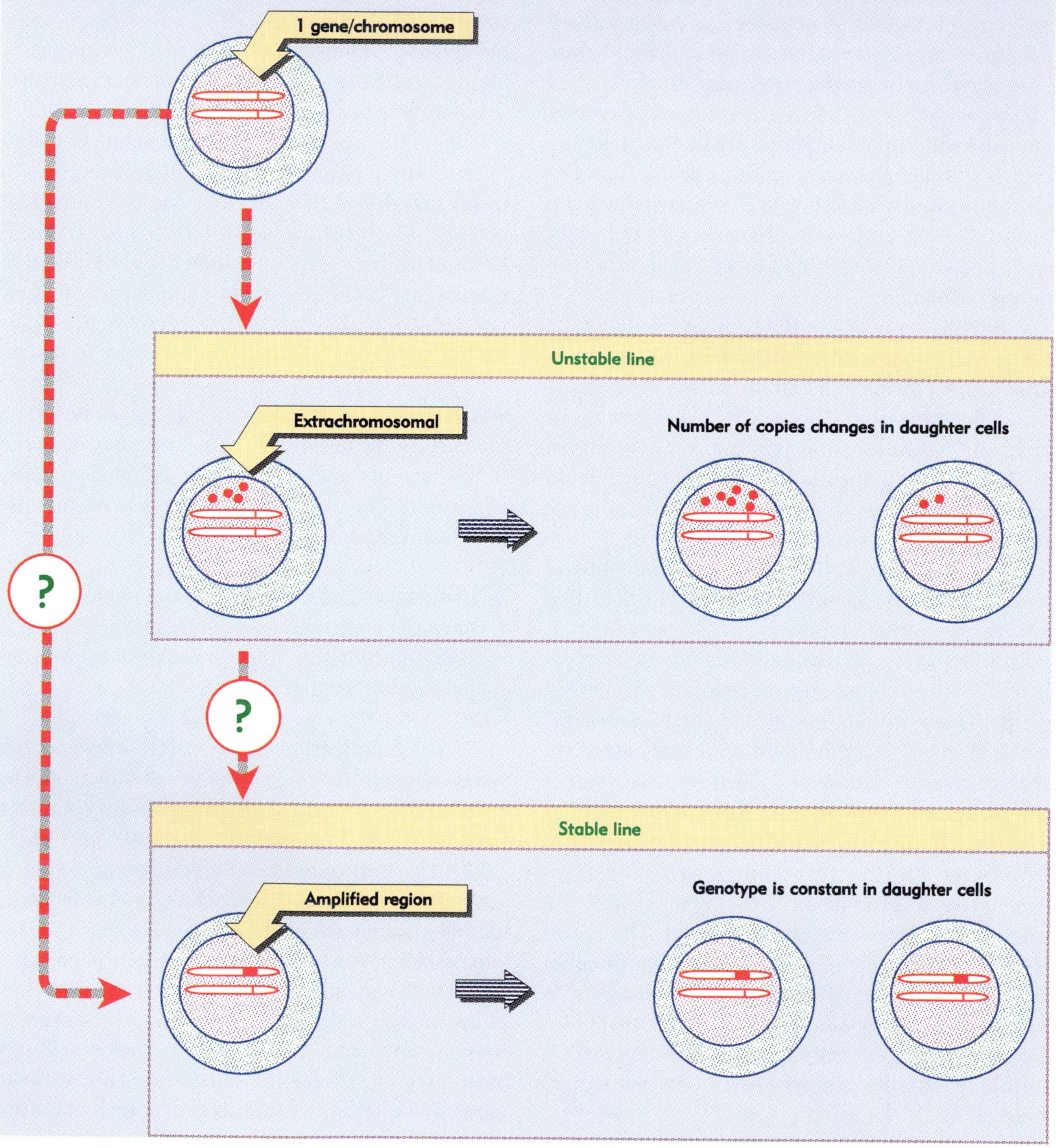

Figure 36.27

Amplified copies of the *dhfr* gene produce a homogeneously staining region (HSR) in the chromosome. Photograph kindly provided by Robert Schimke.

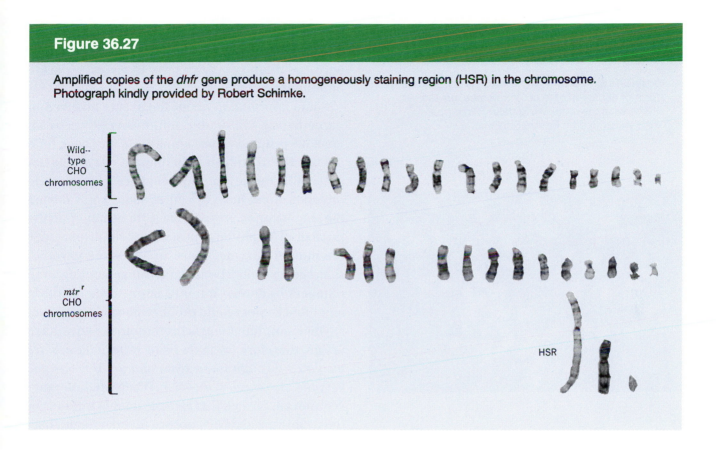

to be self-replicating; but they lack centromeres. As a result, they do not attach to the mitotic spindle and therefore segregate erratically, frequently being lost from the daughter cells. Notwithstanding their name, the actual status of the double minutes is regarded as extrachromosomal.

The irregular inheritance of the double minutes explains the instability of methotrexate resistance in these lines. Double minutes are lost continuously during cell divisions; and in the presence of methotrexate, cells with reduced numbers of *dhfr* genes will die. Only those cells that have retained a sufficient number of double minutes will appear in the surviving population.

The presence of the double minutes reduces the rate at which the cells proliferate. Thus when the selective pressure is removed, cells lacking the amplified genes have an advantage; they generate progeny more rapidly and soon take over the population. This explains why the amplified state is retained in the cell line only so long as cells are grown in the presence of methotrexate.

Because of the erratic segregation of the double minutes, increases in the copy number can occur relatively quickly as cells are selected at each division for progeny that have gained more than their fair share of the *dhfr* genes. Cells with greater numbers of copies are found in response to increased levels of methotrexate. The behavior of the double minutes explains the stepwise evolution of the *mtx^r* condition and the incessant fluctuation in the level of *dhfr* genes in unstable lines.

Both stable and unstable lines are found after long periods of selection for methotrexate resistance. What is the initial step in gene amplification? After short periods of selection, most or all of the resistant cells are unstable. The formation of extrachromosomal copies clearly is a more frequent event than amplification within the chromosome. At very early times in the process, amplified *dhfr*

Figure 36.28

Amplified extrachromosomal *dhfr* genes take the form of double-minute chromosomes, as seen in the form of the small white dots. Photograph kindly provided by Robert Schimke.

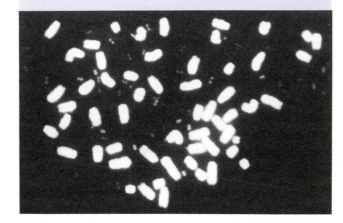

genes can be found as (small) extrachromosomal units before double minutes or any change in chromosomes can be detected. This suggests that the acquisition of resistance is most often due to generation of extrachromosomal repeats.

The amplified region is much longer than the *dhfr* gene itself. The gene has a length of ~31 kb, but the average length of the repeated unit is 500–1000 kb in the chromosomal HSR. The extent of the amplified region is different in each cell line. The amount of DNA contained in a double minute seems to lie in a range of 100–1000 kb.

How do the extrachromosomal copies arise? We know that their generation occurs without loss of the original chromosomal copy. There are two general possibilities. Additional cycles of replication could be initiated in the vicinity of the *dhfr* gene, followed by nonhomologous recombination between the copies. Or the process could be initiated by nonhomologous recombination between alleles. The extra copies could be released from the chromosome, possibly by some recombination-like event. Depending on the nature of this event, it could generate an extrachromosomal DNA molecule

containing one or several copies. If the double minutes contain circular DNA, recombination between them in any case is likely to generate multimeric molecules.

Some information about the events involved in perpetuating the double minutes is given by an unstable cell line whose amplified genes code for a mutant DHFR enzyme. The mutant enzyme is not present in the original (diploid) cell line (so the mutation must have arisen at some point during the amplification process). Despite variations in the number of amplified genes, these cells display *only* the mutant enzyme. Thus the wild-type chromosomal genes cannot be continuously generating large numbers of double minutes anew, because these amplified copies would produce normal enzyme.

Once amplified extrachromosomal genes have arisen, therefore, *changes in the state of the cell are mediated through these genes and not through the original chromosomal copies.* When methotrexate is removed, the cell line loses its double minutes in the usual way. On re-exposure to the reagent, *normal* genes are amplified to give a new population of double minutes. This shows that none of the extrachromosomal copies of the mutant gene had integrated into the chromosome.

Another striking implication of these results is that the double minutes of the mutant line carried *only* mutant genes—so if there is more than one *dhfr* gene per double minute, all must be of the mutant type. This suggests that multicopy double minutes can be generated from individual extrachromosomal genes.

A major question has been whether amplified chromosomal copies arise by integration of the extrachromosomal copies or by an independent mechanism. We do not know whether intrachromosomal amplification simply proceeds less often as a *de novo* step or requires extrachromosomal amplification to occur as an intermediate step. The form taken by the amplified genes is influenced by the cell genotype; some cell lines tend to generate double minutes, while others more readily display the HSR configuration.

The type of amplification event also depends upon the particular locus that is involved. Another case of amplification is provided by resistance to an inhibitor of the enzyme transcarbamylase, which occurs by amplification of the CAD gene. (CAD is a protein which has the first three enzymatic activities of the pathway for UMP synthesis.)

Amplified CAD DNA is always found within the chromosome. In this case, the amplified genes are found in the form of several dispersed amplified regions, often involving more than one chromosome.

Exogenous sequences can be introduced into cells and animals by transfection

The procedure for introducing exogenous donor DNA into recipient cells is called **transfection**. Transfection experiments began with the addition of preparations of metaphase chromosomes to cell suspensions. The chromosomes are taken up rather inefficiently by the cells and give rise to unstable variants at a low frequency. Intact chromosomes rarely survive the procedure; the recipient cell usually gains a fragment of a donor chromosome (which is unstable because it lacks a centromere). Rare cases of stable lines may have resulted from integration of donor material into a resident chromosome.

Similar results are obtained when purified DNA is added to a recipient cell preparation. However, with purified DNA it is possible to add particular sequences instead of relying on random fragmentation of chromosomes. Transfection with DNA yields stable as well as unstable lines, with the former relatively predominant. (These experiments are directly analogous to those performed in bacterial transformation, as indicated in Figure 4.3, but are described as transfection because of the historical use of transformation to describe changes that allow unrestrained growth of eukaryotic cells.)

Unstable transfectants (sometimes called **transient transfectants**) reflect the survival of the transfected DNA in extrachromosomal form; stable lines result from integration into the genome. The transfected DNA can be expressed in both cases. However, the low frequencies of transfection make it necessary to use donor markers whose presence

in the recipient cells can be selected for. Note that the transfected sequence is expressed. Most transfection experiments have used markers representing readily assayed enzymatic functions, but, in principle, any marker that can be selected can be assayed. This allows the isolation of genes responsible for morphological phenomena. Most notably, transfected cells can be selected for acquisition of the transformed (tumorigenic) phenotype. Then we can identify the DNA responsible for conferring the phenotype. This type of protocol has led to the isolation of several cellular *onc* genes (which is discussed in Chapter 39).

Cotransfection with more than one marker has proved informative about the events involved in transfection and has extended the range of questions that we can ask with this technique. A common marker used in such experiments is the *tk* gene, coding for the enzyme thymidine kinase, which catalyzes an essential step in the provision of thymidine triphosphate as a precursor for DNA synthesis.

When *tk⁻* cells are transfected with a DNA preparation containing both a purified *tk⁺* gene and the φX174 genome, *all the* tk⁺ *transformants have both donor sequences*. This is a useful observation, because it allows unselected markers to be introduced routinely by cotransfection with a selected marker.

The arrangement of *tk* and φX174 sequences is different in each transfected line, but remains the same during propagation of that line. Often multiple copies of the donor sequences are present, the number varying with the individual line. Revertants lose the φX174 sequences together with *tk* sequences. Amplification of transfected sequences under selective pressure results in the increase of copy number of all donor sequences *pari passu*. Thus the two types of donor sequence become physically linked during transfection and suffer the same fate thereafter.

To perform a transfection experiment, the mass of DNA added to the recipient cells is increased by including an excess of 'carrier DNA', a preparation of some other DNA (often from salmon sperm). Transfected cells prove to have sequences of the carrier DNA flanking the selected sequences on either side. Transfection therefore appears to be mediated by a large unit, consisting of a linked array of all sequences present in the donor preparation.

Since revertants for the selected marker lose all of this material, it seems likely that the *transfected cell gains only a single large unit*. The unit is formed by a concatemeric linkage of donor sequences in a reaction that is rapid relative to the other events involved in transfection. This transfecting package is ~1000 kb in length.

Because of the size of the donor unit, we cannot tell from blotting experiments whether it is physically linked to recipient chromosomal DNA (the relevant end fragments are present in too small a relative proportion). It seems plausible that the first stage is the establishment of an unstable extrachromosomal unit, followed by the acquisition of stability via integration.

In situ hybridization can be used to show that transfected cells have donor material integrated into the resident chromosomes. Any given cell line has only a single site of integration; but the site is different in each line. Probably the selection of a site for integration is a random event; sometimes it is associated with a gross chromosomal rearrangement.

The sites at which exogenous material becomes integrated usually do not appear to have any sequence relationship to the transfected DNA. The integration event involves a nonhomologous recombination between the mass of added DNA and a random site in the genome. The recombination event may be provoked by the introduction of a double-strand break into the chromosomal DNA, possibly by the action of DNA repair enzymes that are induced by the free ends of the exogenous DNA.

An exciting development of transfection techniques is their application to introduce genes into animals. An animal that gains new genetic information from the addition of foreign DNA is described as **transgenic**. The approach of directly injecting DNA can be used with mouse eggs, as shown in **Figure 36.29**. Plasmids carrying the gene of interest are injected into the germinal vesicle (nucleus) of the oocyte or into the pronucleus of the fertilized egg. The egg is implanted into a pseudopregnant mouse. After birth, the recipient mouse can be examined to see whether it has gained the foreign DNA, and, if so, whether it is expressed.

The first questions we ask about any transgenic animal are how many copies it has of the foreign material, where these copies are located, and whether they are present in the germ line and inherited in a Mendelian manner. The usual result of such experiments is that a reasonable minority (say ~15%) of the injected mice carry the transfected sequence. Usually, multiple copies of the plasmid appear to have been integrated in a tandem array into a single chromosomal site. The number of copies varies from 1 to 150. They are inherited by the progeny of the injected mouse as expected of a Mendelian locus.

An important issue that can be addressed by experiments with transgenic animals concerns the independence of genes and the effects of the region within which they reside. If we take a gene, including the flanking sequences that contain its known regulatory elements, can it be expressed independently of its location in the genome? In other words, do the regulatory elements function independently, or is gene expression (also) controlled by some other effect, for example, location in an appropriate chromosomal domain?

Are transfected genes expressed with the proper developmental specificity? The general rule now

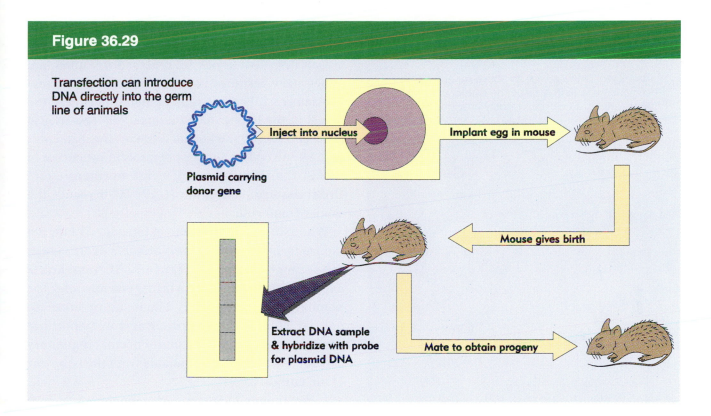

Figure 36.29

Transfection can introduce DNA directly into the germ line of animals

Plasmid carrying donor gene

Inject into nucleus

Implant egg in mouse

Mouse gives birth

Extract DNA sample & hybridize with probe for plasmid DNA

Mate to obtain progeny

appears to be that there is a reasonable facsimile of proper control: the transfected genes are generally expressed in appropriate cells and at the usual time. There are exceptions, however, in which a transfected gene is expressed in an inappropriate tissue.

In the progeny of the injected mice, expression of the donor gene is extremely variable; it may be extinguished entirely, reduced somewhat, or even increased. Even in the original parents, the level of gene expression does not correlate with the number of tandemly integrated genes. Probably only some of the genes are active. In addition to the question of how many of the gene copies are capable of being activated, a parameter influencing regulation could be the relationship between the gene number and the regulatory proteins: a large number of promoters could dilute out any regulator proteins present in limiting amounts.

What is responsible for the variation in gene expression? One possibility that has often been discussed for transfected genes (and which applies also to integrated retroviral genomes), is that the site of integration is important. Perhaps a gene is

expressed if it integrates within an active domain, but not if it integrates in another area of chromatin. Another possibility is the occurrence of epigenetic modification; for example, changes in the pattern of methylation might be responsible for changes in activity. Alternatively, the genes that happened to be active in the parents may have been deleted or amplified in the progeny.

A particularly striking example of the effects of an injected gene is provided by a strain of transgenic mice derived from eggs injected with a fusion consisting of the MT promoter linked to the rat growth hormone structural gene. Growth hormone levels in some of the transgenic mice were several hundred times greater than normal. The mice grew to nearly twice the size of normal mice, as can be seen from **Figure 36.30**.

The introduction of oncogene sequences can lead to tumor formation. Transgenic mice containing the SV40 early coding region and regulatory elements express the viral genes for large T and small t antigen only in some tissues, most often brain, thymus, and kidney. (The T/t antigens are

Figure 36.30

A transgenic mouse with an active rat growth hormone gene (left) is twice the size of a normal mouse (right). Photograph kindly provided by Ralph Brinster.

alternatively spliced proteins coded by the early region of the virus; they have the ability to transform cultured cells to a tumorigenic phenotype; see Chapter 39.) The transgenic mice usually die before reaching 6 months, as the result of developing tumors in the brain; sometimes tumors are found also in thymus and kidney.

Although SV40 is known to cause tumors in hamsters, it had not previously been thought to be oncogenic in mice. In the transgenic strain, however, the SV40 T/t antigens behave as the products of an integrated oncogene. Transgenic mice that develop tumors such as these provide a useful model for investigating the origins of cancer. Different oncogenes may be used to generate mice developing various cancers, thus making possible a range of model systems. For example, introduction of the *myc* gene under control of an active promoter causes the appearance of adenocarcinomas and other tumors.

Can defective genes be replaced by functional genes in the germline using transgenic techniques? One successful case is represented by a cure of the defect in the hypogonadal mouse. The *hpg* mouse has a deletion of >30 kb that removes the distal part of the gene coding for the polyprotein precursor to GnRH (gonadotropin-releasing hormone) and GnRH-associated peptide (GAP). As a result, the mouse is infertile.

When an intact *hpg* gene is introduced into the mouse by transgenic techniques, it is expressed in the appropriate tissues. **Figure 36.31** summarizes experiments to introduce a transgene into *hpg/hpg* homozygous mutant mice. The resulting mice are normal. This provides a striking demonstration that expression of a transgene under normal regulatory control can be indistinguishable from the behavior of the normal allele.

Impediments to using such techniques to cure genetic defects at present are that the transgene must be introduced into the germline of the preceding generation, the ability to express a transgene is not predictable, and an adequate level of expression of a transgene may be obtained in only a small minority of the transgenic animals. Also, the large number of transgenes that may be introduced into the germline, and their erratic expression, could pose problems for the animal in cases in which over-expression of the transgene was harmful.

In the *hpg* murine experiments, for example, only 2 out of 250 mice eggs injected with intact *hpg* genes gave rise to transgenic mice. Each transgenic animal contained >20 copies of the transgene. Only 20 of the 48 offspring of the transgenic mice retained the transgenic trait. When inherited by their offspring, however, the transgene(s) could substitute for the lack of endogenous *hpg* genes. Gene replacement via a transgene is therefore effective only under restricted conditions.

The disadvantage of direct injection of DNA is the introduction of multiple copies, their variable expression, and often difficulty in cloning the

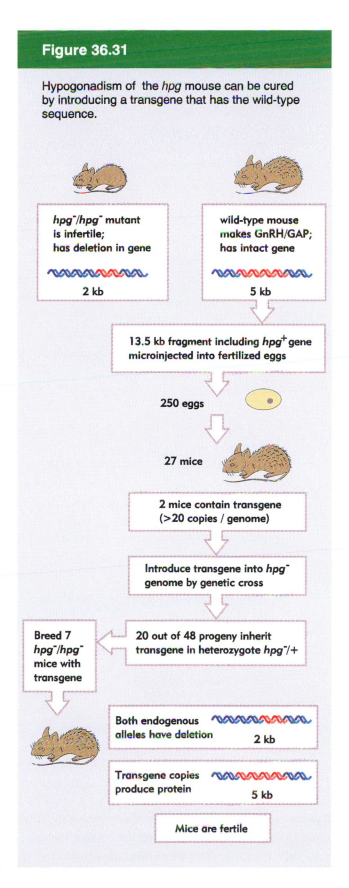

Figure 36.31

Hypogonadism of the *hpg* mouse can be cured by introducing a transgene that has the wild-type sequence.

hpg⁻/hpg⁻ mutant is infertile; has deletion in gene

2 kb

wild-type mouse makes GnRH/GAP; has intact gene

5 kb

13.5 kb fragment including *hpg⁺* gene microinjected into fertilized eggs

250 eggs

27 mice

2 mice contain transgene (>20 copies / genome)

Introduce transgene into *hpg⁻* genome by genetic cross

20 out of 48 progeny inherit transgene in heterozygote *hpg⁻/+*

Breed 7 *hpg⁻/hpg⁻* mice with transgene

Both endogenous alleles have deletion

2 kb

Transgene copies produce protein

5 kb

Mice are fertile

insertion site because sequence rearrangements may have been generated in the host DNA. An alternative procedure is to use a retroviral vector carrying the donor gene. A single proviral copy inserts at a chromosomal site, without inducing any rearrangement of the host DNA. It is possible also to treat cells at different stages of development, and thus to target a particular somatic tissue; however, it is difficult to infect germ cells.

A powerful technique for making transgenic mice takes advantage of embryonic stem (ES) cells, which are derived from the mouse blastocyst (an early stage of development, which precedes implantation of the egg in the uterus). **Figure 36.32** illustrates the principles of this technique.

ES cells are transfected with DNA in the usual way (most often by microinjection or electroporation). By using a donor that carries an additional sequence such as a drug resistance marker or some particular enzyme, it is possible to select ES cells that have obtained an integrated transgene carrying any particular donor trait. An alternative is to use PCR technology to assay the transfected ES cells for successful integration of the donor DNA. By such means, a population of ES cells is obtained in which there is a high proportion carrying the marker.

These ES cells are then injected into a recipient blastocyst. This blastocyst is implanted into a foster mother, and in due course develops into a **chimeric** mouse. Some of the tissues of the chimeric mice will be derived from the cells of the recipient blastocyst; other tissues will be derived from the injected ES cells. The ability of the ES cells to participate in normal development of the blastocyst forms the basis of the technique. The proportion of tissues in the adult mouse that are derived from cells in the recipient blastocyst and from injected ES cells varies widely in individual progeny; if a visible marker (such as coat color gene) is used, areas of tissue representing each type of cell can be seen.

To determine whether the ES cells contributed to the germline, the chimeric mouse is crossed with a mouse that lacks the donor trait. Any progeny that have the trait must be derived from germ

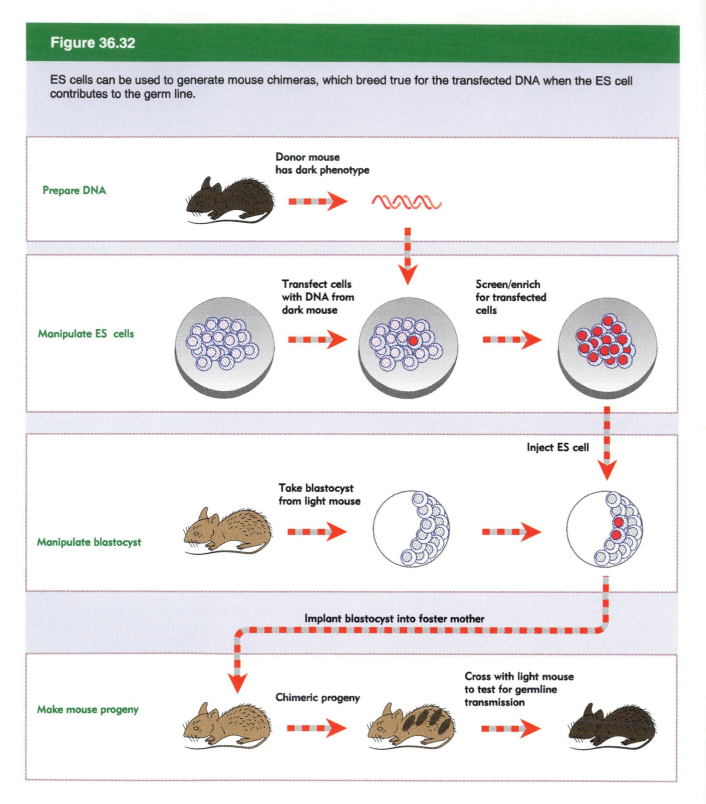

Figure 36.32

ES cells can be used to generate mouse chimeras, which breed true for the transfected DNA when the ES cell contributes to the germ line.

Prepare DNA

Donor mouse has dark phenotype

Manipulate ES cells

Transfect cells with DNA from dark mouse

Screen/enrich for transfected cells

Inject ES cell

Manipulate blastocyst

Take blastocyst from light mouse

Implant blastocyst into foster mother

Make mouse progeny

Chimeric progeny

Cross with light mouse to test for germline transmission

cells that have descended from the injected ES cells. By this means, an entire mouse has been generated from an original ES cell!

A further development of these techniques makes it possible to obtain homologous recombinants. A particular use of homologous recombination is to disrupt endogenous genes, as illustrated in **Figure 36.33**. A wild-type gene is modified by inter-

Figure 36.33

A transgene containing *neo* within an exon and *TK* downstream can be selected by resistance to G418 and loss of TK activity

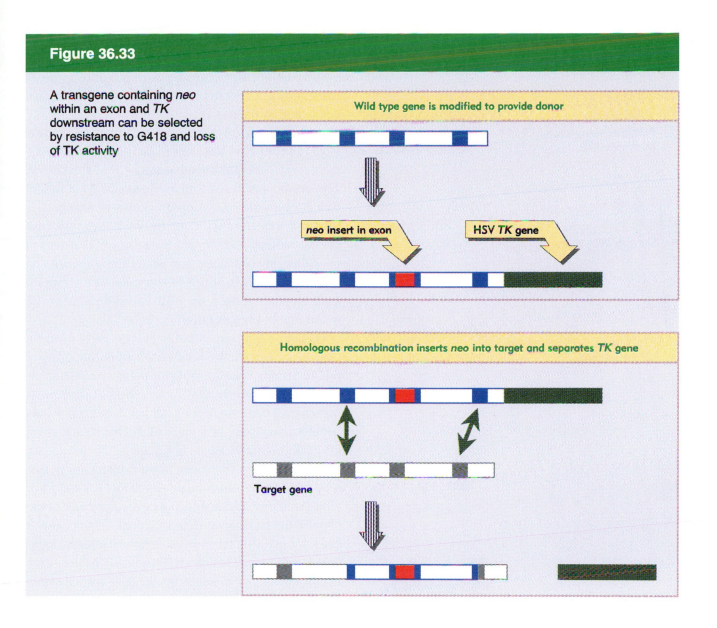

rupting an exon with a marker sequence; most often the *neo* gene that confers resistance to the drug G418 is used. Also, another marker is added on one side of the gene; for example, the *TK* gene of the herpes virus. When this DNA is introduced into an ES cell, it may be inserted into the genome by either nonhomologous or homologous recombination. A nonhomologous recombination inserts the whole unit, including the flanking *TK* sequence. But a homologous recombination requires two exchanges, as a result of which the flanking *TK* sequence is lost. Cells in which a homologous recombination has occurred can therefore

be selected by the gain of *neo* resistance and absence of *TK* activity (which is selected because *TK* causes sensitivity to the drug gancyclovir). If it is not convenient to use a selectable marker such as *TK*, cells can simply be screened by PCR assays for the absence of flanking DNA. The frequency of homologous recombination appears to be $\sim 10^{-7}$, and probably represents <1% of all recombination events.

The presence of the *neo* gene in an exon disrupts transcription, and thereby creates a null allele. A particular target gene can therefore be 'knocked out' by this means; and once a mouse with one null

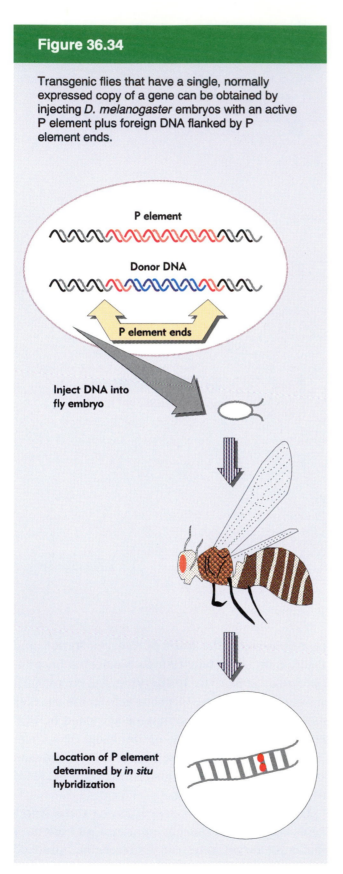

Figure 36.34

Transgenic flies that have a single, normally expressed copy of a gene can be obtained by injecting *D. melanogaster* embryos with an active P element plus foreign DNA flanked by P element ends.

P element

Donor DNA

P element ends

Inject DNA into fly embryo

Location of P element determined by *in situ* hybridization

allele has been obtained, it can be bred to generate the homozygote. This is a powerful technique for investigating whether a particular gene is essential, and what functions in the animal are perturbed by its loss.

A sophisticated method for introducing new DNA sequences has been developed with *D. melanogaster* by taking advantage of the P element. The protocol is illustrated in **Figure 36.34**. A defective P element carrying the gene of interest is injected together with an intact P element into preblastoderm embryos. The intact P element provides a transposase that recognizes not only its own ends but also those of the defective element. As a result, either or both elements may be inserted into the genome.

Only the sequences between the ends of the P DNA are inserted; the sequences on either side are not part of the transposable element. An advantage of this technique is that only a single element is inserted in any one event, so the transgenic flies usually carry only one copy of the foreign gene, a great aid in analyzing its behavior.

Several genes that have been introduced in this way all show the same behavior. They are expressed only in the appropriate tissues and at the proper times during development. These genes have been integrated into various random locations; and, even more strikingly, these features remain true for a gene integrated into heterochromatin. So in *D. melanogaster*, all the information needed to regulate gene expression may be contained within the gene locus itself, and can be relatively impervious to external influence.

With these experiments, we see the possibility of extending from cultured cells to animals the option of examining the regulatory features. The ability to introduce DNA into the genotype allows us to make changes in it, to add new genes that have had particular modifications introduced *in vitro*, or to inactivate existing genes. Thus it may become possible to delineate the features responsible for tissue-specific gene expression. Ultimately it may be possible to replace defective genes in the genotype in a targeted manner.

Summary

Yeast mating type is determined by whether the *MAT* locus carries the *a* or α sequence. Expression in haploid cells of the sequence at *MAT* leads to expression of genes specific for the mating type and to repression of genes specific for the other mating type. Both activation and repression are achieved by control of transcription, and require factors that are not specific for mating type as well as the products of *MAT*. The functions that are activated in either mating type include secretion of the appropriate pheromone and expression on the cell surface of the receptor for the opposite type of pheromone. Interaction between pheromone and receptor on cells of either mating type activates a G protein on the membrane, and sets in train a common pathway that prepares cells for sporulation. Diploid cells do not express mating-type functions.

Additional, silent copies of the mating-type sequences are carried at the loci *HMLa* and *HMRα*. They are repressed by the actions of the *sir* loci. Cells that carry the *HO* endonuclease display a unidirectional transfer process in which the sequence at *HMLa* replaces an α sequence at *MAT* or the sequence at *HMLα* replaces an *a* sequence at MAT. The endonuclease makes a double-strand break at *MAT*, and a free end invades either *HMLa or HMLα*. *MAT* initiates the transfer process, but is the recipient of the new sequence. The *HO* endonuclease is transcribed in mother cells but not daughter cells, and is under cell-cycle control. Thus switching is detected only in the products of a division, and the mating type has been switched in both daughter cells.

Trypanosomes carry >1000 sequences coding for varieties of the surface antigen. Only a single VSG is expressed in one cell, from an active site located near a telomere. The VSG may be changed by substituting a new coding sequence at the active site via a gene conversion process, or by switching the site of expression to another telomere. Switches in expression occur every 10^4–10^6 divisions.

Agrobacteria induce tumor formation in wounded plant cells. The wounded cells secrete phenolic compounds that activate *vir* genes carried by the Ti plasmid of the bacterium. The *vir* gene products cause a single strand of DNA from the T-DNA region of the plasmid to be transferred to the plant cell nucleus. Transfer is initiated at one boundary of T-DNA, but ends at variable sites. The single strand is converted into a double strand and integrated into the plant genome. Genes within the T-DNA transform the plant cell, and cause it to produce particular opines (derivatives of arginine). Genes in the Ti plasmid allow *Agrobacterium* to metabolize the opines produced by the transformed plant cell.

Endogenous sequences may become amplified in cultured cells. Exposure to methotrexate leads to the accumulation of cells that have additional copies of the DHFR gene. The copies may be carried as extrachromosomal arrays in the form of double-minute 'chromosomes', or they may be integrated into the genome at the site of one of the DHFR alleles. Double-minute chromosomes are unstable, and disappear from the cell line rapidly in the absence of selective pressure. The amplified copies may originate by additional cycles of replication that are associated with recombination events.

New sequences of DNA may be introduced into a cultured cell by transfection or into an animal egg by microinjection. The foreign sequences may become integrated into the genome, often as large tandem arrays. The array appears to be inherited as a unit in a cultured cell. The sites of integration appear to be random. A transgenic animal arises when the integration event occurs into a genome that enters the germ-cell lineage. Transgenic mice can be obtained by injecting recipient blastocysts with ES cells that carry transfected DNA. A transgene or transgenic array is inherited in Mendelian manner, but the copy number and activity of the gene(s) may change in the progeny. Often a transgene responds to tissue- and temporal regulation in a manner that resembles the endogenous gene.

Further reading

Reviews
The determination of mating type has been reviewed from a biological perspective by **Nasmyth** (*Ann. Rev. Genet.* **17**, 439–500, 1983). Regulatory events were reviewed by **Nasmyth and Shore** (*Science* **237**, 1162–1170, 1987). The pheromone response was reviewed by **Kurjan** (*Ann. Rev. Biochem.* **61**, 1097–1129, 1992).

The biology of trypanosome infections has been reviewed in terms of antigenic variation by **Boothroyd** (*Ann. Rev. Microbiol.* **39**, 475–502, 1985), and the molecular biology of VSG transitions has been analyzed by **Donelson and Rice-Ficht** (*Microbiol. Rev.* **49**, 107–125, 1985) and **Borst** (*Ann. Rev. Biochem.* **55**, 701–732, 1986).

The infusion of Ti plasmid DNA from *Agrobacterium* into plants was reviewed by **Zambryski** in *Mobile DNA* (Eds. Berg and Howe, American Society for Microbiology, Washington DC, 309–334, 1989) and in *Ann. Rev. Genet.* (**22**, 1–30, 1988).

The events involved in amplification of *dhfr* have been reviewed by **Schimke *et al.*** (*Cold Spring Harbor Symp. Quant. Biol.* **45**, 785–797, 1981) and (*Cell* **37**, 367–379, 1984); other systems and possible mechanisms have been reviewed by **Stark and Wahl** (*Ann. Rev. Biochem.* **53**, 447–491, 1984).

The uses and insights offered by transfection have been reviewed by **Pellicer *et al.*** (*Science* **209**, 1414–1422, 1980). Techniques for obtaining transgenic mice have been reviewed by **Jaenisch** (*Science* **240**, 1468–1474, 1988) and **Capecchi** (*Science* **244**, 1288–1292, 1989).

Discoveries
The cassette model for mating-type switching was proposed by **Hicks, Strathern, and Herskowitz** (in *DNA Insertion Elements*, Eds. Bukhari, Shapiro and Adhya, Cold Spring Harbor Lab., New York, 457–462, 1977). The role of the double-strand break in initiating switch of mating type was uncovered by **Strathern *et al.*** (*Cell* **31**, 183–192, 1982). The identity of the a-factor/receptor and α-factor/ receptor pathways was uncovered by **Bender and Sprague** (*Cell* **47**, 929–937, 1986).

Transfer of T-DNA by conjugation was reported by **Stachel, Timmerman, and Zambryski** (*Nature* **322**, 707–712, 1986).

P–element mediated transfection was developed by **Spradling and Rubin** (*Science* **218**, 341–353, 1982).

Genes in development

In calling the structure of the chromosome fibers a code-script we mean that the all-penetrating mind could tell from their structure whether the egg would develop, under suitable conditions, into a black cock or into a speckled hen, into a fly or a maize plant, a beetle, a mouse or a woman... But the term code-script is, of course, too narrow. The chromosome structures are at the same time instrumental in bringing about the development they foreshadow. They are law-code and executive power—or, to use another simile, they are architect's plan and builder's craft—in one.

Erwin Schrödinger, 1945

CHAPTER 37

Generation of immune diversity by gene reorganization

It is an axiom of genetics that the genetic constitution created in the zygote by the combination of sperm and egg is inherited by all somatic cells of the organism. We look to differential control of gene expression, rather than to changes in DNA content, to explain the different phenotypes of particular somatic cells.

Yet there are exceptional situations in which the reorganization of certain DNA sequences is used to regulate gene expression or to create new genes. The immune system provides a striking and extensive case in which the content of the genome changes, when recombination creates active genes in lymphocytes.

Other cases are represented by the substitution of one sequence for another to change the mating type of yeast or to generate new surface antigens by trypanosomes (see Chapter 36). The gamut of changes in the eukaryotic genome thus runs from unpredictable and rare transpositions to regular tissue-specific reconstructions.

The **immune response** of vertebrates provides a protective system that distinguishes foreign proteins from the proteins of the organism itself. Foreign material (or part of the foreign material) is recognized as comprising an **antigen**. Usually the antigen is a protein (or protein-attached moiety) that has entered the bloodstream of the animal— for example, the coat protein of an infecting virus. Exposure to an antigen initiates production of an immune response that *specifically recognizes the antigen and destroys it.*

Immune reactions are the responsibility of white blood cells—the B and T lymphocytes, and macrophages. The lymphocytes are named after the tissues that produce them. In mammals, **B cells** mature in the bone marrow, while **T cells** mature in the thymus. *Each class of lymphocyte uses the rearrangement of DNA as a mechanism for producing the proteins that enable it to participate in the immune response.*

The immune system has many ways to destroy an antigenic invader, but it is useful to consider them in two general classes. Which type of response the immune system mounts when it encounters a foreign structure depends partly on the nature of the antigen. The response is defined according to whether it is executed principally by B cells or T cells:

◆ The **humoral response** depends on B cells. It is mediated by the secretion of **antibodies**, which are **immunoglobulin** proteins. *Production of an immunoglobulin specific for a foreign molecule is the primary event responsible for recognition of an antigen.* Recognition requires the antibody to bind to a small region or structure on the antigen.

The function of antibodies is represented in **Figure 37.1**. Foreign material circulating in the blood stream, for example, a toxin or pathogenic bacterium, has a surface that presents antigens. The antigen(s) are recognized by the antibodies, which form an antigen–antibody complex. This complex then attracts the

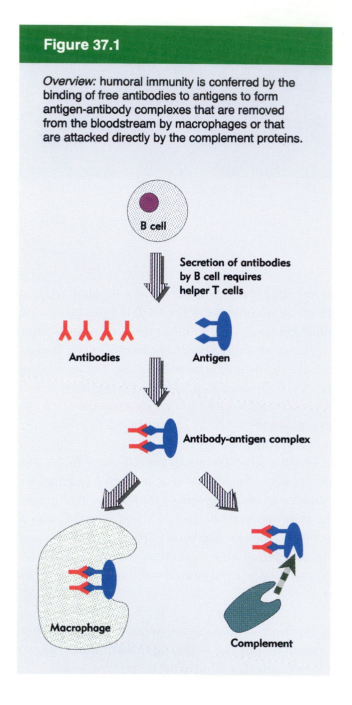

Figure 37.1

Overview: humoral immunity is conferred by the binding of free antibodies to antigens to form antigen-antibody complexes that are removed from the bloodstream by macrophages or that are attacked directly by the complement proteins.

B cell

Secretion of antibodies by B cell requires helper T cells

Antibodies

Antigen

Antibody-antigen complex

Macrophage

Complement

nals provided by T cells to enable them to secrete antibodies. These T cells are called **helper T cells**, because they assist the B cells. Second, antigen–antibody formation is a trigger for the antigen to be destroyed. The major pathway is provided by the action of **complement**, a component whose name reflects its ability to 'complement' the action of the antibody itself. Complement consists of a set of ~20 proteins that function through a cascade of proteolytic actions. If the target antigen is part of a cell, for example, an infecting bacterium, the action of complement complements in lysing the target cell and culminates in lysing the target cell. The action of complement also provides a means of attracting macrophages, which scavenge the target cells or their products. Alternatively, the antigen–antibody complex may be taken up directly by macrophages (scavenger cells) and destroyed.

◆ The **cell-mediated response** is executed by a class of T lymphocytes called **cytotoxic (killer) T cells**. The basic function of the T cell in recognizing a target antigen is indicated in **Figure 37.2**. A cell-mediated response typically is elicited by an intracellular parasite, such as a virus that infects the body's own cells. As a result of the viral infection, foreign (viral) antigens are displayed on the surface of the cell. These antigens are recognized by the **T-cell receptor**, which is the T cells' equivalent of the antibody produced by a B cell.

A crucial feature of this recognition reaction is that *the antigen must be **presented** by a cellular protein that is a member of the MHC (major histocompatibility complex)*. The MHC protein has a groove on its surface that binds a peptide fragment derived from the foreign antigen. The combination of peptide fragment and MHC protein is recognized by the T cell receptor. Every individual has a characteristic

attention of other components of the immune system.

The humoral response depends on these other components in two ways. First, B cells need sig-

Figure 37.2

Overview: in cell-mediated immunity, killer T cells use the T-cell receptor to recognize a fragment of the foreign antigen which is "presented" on the surface of the target cell by the MHC protein.

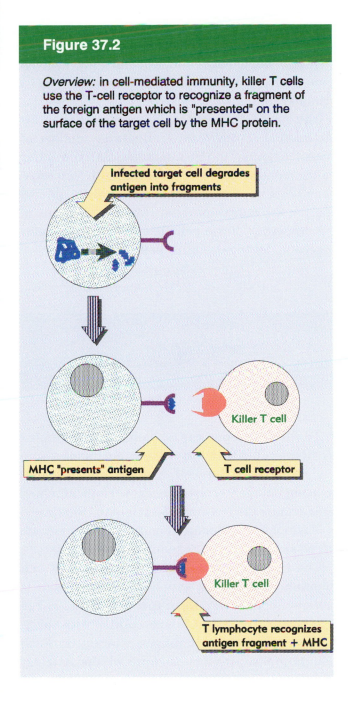

Infected target cell degrades antigen into fragments

Killer T cell

MHC "presents" antigen

T cell receptor

Killer T cell

T lymphocyte recognizes antigen fragment + MHC

an issue of major medical importance. The demand that the T lymphocytes recognize both foreign antigen and MHC protein ensures that the cell-mediated response acts only on host cells that have been infected with a foreign antigen. (We discuss the division of MHC proteins into the general types of Class I and Class II later in this chapter.)

The purpose of each type of immune response is to attack a foreign target. Target recognition is the prerogative of B cell immunoglobulins and T-cell receptors. A crucial aspect of their function lies in the ability to distinguish 'self' from 'nonself'. Proteins and cells of the body itself must *never* be attacked. Foreign targets must be *destroyed entirely*. The property of failing to attack 'self' is called **tolerance**. Loss of this ability results an **autoimmune disease**, in which the immune system attacks its own body, often with disastrous consequences.

What prevents the lymphocyte pool from responding to 'self' proteins? Tolerance probably arises early in lymphocyte cell development when B and T cells that recognize 'self' antigens are destroyed. This is called **clonal deletion**. In addition to this negative selection, there is also positive selection for T cells carrying certain sets of T–cell receptors.

A corollary of tolerance is that it can be difficult to obtain antibodies against proteins that are closely related to those of the organism itself. As a practical matter, therefore, it may be difficult to use (for example) mice or rabbits to obtain antibodies against human proteins that have been highly conserved in mammalian evolution. The tolerance of the mouse or rabbit for its own protein may extend to the human protein in such cases.

set of MHC proteins. They are important in graft reactions; a graft of tissue from one individual to another is rejected because of the difference in MHC proteins between the donor and recipient,

Each of the three groups of proteins required for the immune response—immunoglobulins, T-cell receptors, MHC proteins—is diverse. Examining a large number of individuals, we find many variants of each protein. Each protein is coded by a large family of genes; and in the case of antibodies and the T-cell receptors, the diversity of the population is increased by DNA rearrangements that occur in the relevant lymphocytes.

Immunoglobulins and T-cell receptors are direct counterparts, each produced by its own type of lymphocyte. The proteins are related in structure, and their genes are related in organization. The sources of variability are similar. The MHC proteins also share some common features with the antibodies, as do other lymphocyte-specific proteins. In dealing with the genetic organization of the immune system, we are therefore concerned with a series of related gene families, perhaps a **super-family** that evolved from some common ancestor representing a primitive immune response.

The importance of the immune response for the mammal is indicated by the large number of genes contained in these families; and they offer us the opportunity to characterize the massive gene clusters. Eventually we should be able to analyze the entire immune response in terms of the properties of the gene products, and perhaps to account for the evolution of the system.

Clonal selection amplifies lymphocytes that respond to individual antigens

The name of the immune response describes one of its central features. After an organism has been exposed to an antigen, it becomes *immune* to the effects of a new infection. Before exposure to a particular antigen, the organism lacks adequate capacity to deal with any toxic effects. This ability is acquired during the immune response. After the infection has been defeated, the organism retains the ability to respond rapidly in the event of a re-infection.

These features are accommodated by the **clonal selection theory** illustrated in **Figure 37.3**. The pool of lymphocytes contains B cells and T cells carrying a large variety of immunoglobulins or T-cell receptors. *But any individual B lymphocyte produces one immunoglobulin, which is capable of recognizing only a single antigen; similarly any individual T lymphocyte produces only one particular T-cell receptor.*

In the pool of immature lymphocytes, the unstimulated B cells and T cells are morphologically indistinguishable. But on exposure to antigen, a B cell whose antibody is able to bind the antigen, or a T cell whose receptor can recognize it, is stimulated to divide, probably by some feedback from the surface of the cell, where the antibody/receptor–antigen reaction occurs. The stimulated cells acquire the features characteristic of mature B or T lymphocytes; maturation involves an increase in cell size (especially pronounced for B cells).

The initial expansion of a specific B or T cell population upon first exposure to an antigen is called the **primary immune response**. Large numbers of the B and T lymphocytes with specificity for the offending antigen are produced. Each population represents a clone of the original responding cell. Antibody is secreted from the B cells in large quantities, and it may even come to dominate the antibody population.

After a successful primary immune response has been mounted, the organism retains B cells and T cells carrying the corresponding antibody or receptor. These **memory cells** represent an intermediate state between the immature cell and the mature cell. They have not acquired all of the features of the mature cell, but they are long-lived,

Figure 37.3

The pool of immature lymphocytes contains B cells and T cells making antibodies and receptors with a variety of specificities. Reaction with an antigen leads to clonal expansion of the lymphocyte with the antibody (B cell) or receptor (T cell) that can recognize the antigen.

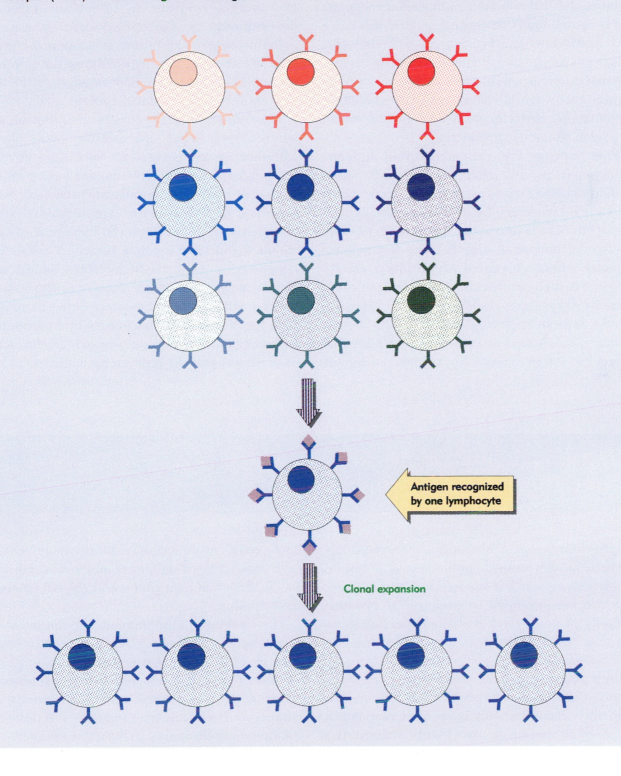

Antigen recognized by one lymphocyte

Clonal expansion

and can rapidly be converted to mature cells. Their presence allows a **secondary immune response** to be mounted rapidly if the animal is exposed to the same antigen again.

The pool of immature lymphocytes in a mammal contains ~10^6–10^8 cells with antibodies or receptors representing many different specificities. The total number of mature lymphocytes is ~10^{12} cells, with some specificities remaining unique (because a corresponding antigen has never been encountered), while others are represented by <10^6 cells (because clonal selection has expanded the pool to respond to an antigen).

What features are recognized in an antigen? Antigens are usually macromolecular. Although small molecules may have antigenic determinants and can be recognized by antibodies, usually they are not effective in provoking an immune response (because of their small size). But they do provoke a response when conjugated with a larger carrier molecule (usually a protein). A small molecule that is used to provoke a response by such means is called a **hapten**.

Only a small part of the surface of a macromolecular antigen is actually recognized by any one antibody. The binding site appears to comprise a region of only 5–6 amino acids. Of course, a particular protein may have more than one such binding site, and therefore provokes antibodies with specificities for different regions. The region provoking a response is called an **antigenic determinant** or **epitope**. When an antigen contains several epitopes, some may be more effective than others in provoking the immune response, in fact, they may be so effective that they entirely dominate the response.

How do lymphocytes find target antigens and where does their maturation take place? Lymphocytes are peripatetic cells. They develop from immature stem cells that are located in the adult bone marrow. They migrate to the peripheral lymphoid tissues (spleen, lymph nodes) either directly via the blood stream (if they are B cells) or via the thymus (where they become T cells). The lymphocytes recirculate between blood and lymph; the process of dispersion ensures that an antigen will be exposed to lymphocytes of all possible specificities. When a lymphocyte encounters an antigen that binds its antibody or receptor, clonal expansion begins the immune response.

Immunoglobulin genes are assembled from their parts in lymphocytes

A mysterious feature of the immune response has been an animal's ability to produce an appropriate antibody whenever it is exposed to a new antigen. How can the organism be prepared to produce antibody proteins each designed specifically to recognize an antigen whose structure cannot be anticipated?

For practical purposes, we usually reckon that a mammal has the ability to produce 10^6–10^8 different antibodies. Each antibody is an immunoglobulin tetramer consisting of two identical **light (L) chains** and two identical **heavy (H) chains**. If any light chain can associate with any heavy chain, to produce 10^6–10^8 potential antibodies requires 10^3–10^4 different light chains and 10^3–10^4 different heavy chains.

The structure of the immunoglobulin tetramer is illustrated in **Figure 37.4**. Each protein chain consists of two principal regions: the N-terminal **variable (V) region**; and the C-terminal **constant (C) region**. They were defined originally by comparing the amino acid sequences of different immunoglobulin chains. As the names suggest, the variable regions show considerable changes in

Figure 37.4

Heavy and light chains combine to generate an immunoglobulin with several discrete domains.

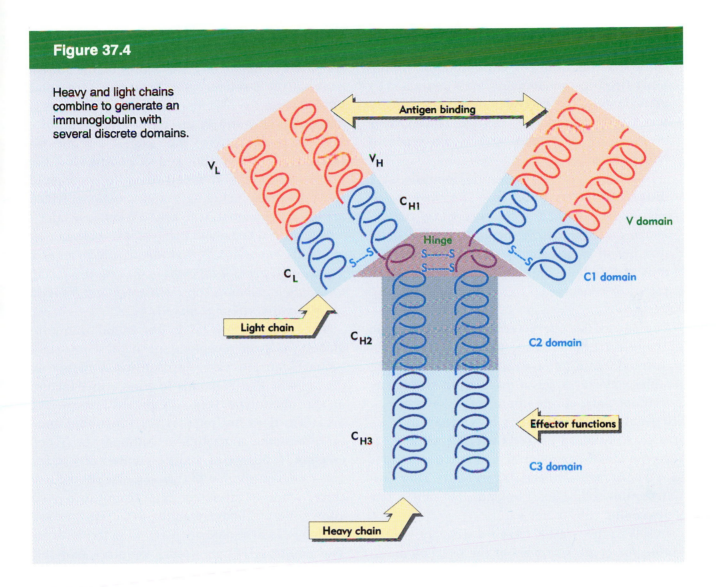

sequence from one protein to the next, while the constant regions show substantial homology.

Corresponding regions of the light and heavy chains associate to generate distinct domains in the immunoglobulin protein:

◆ The variable (V) domain is generated by association between the variable regions of the light chain and heavy chain. *The V domain is responsible for recognizing the antigen.* An immunoglobulin has a Y-shaped structure in which the arms of the Y are identical, and each arm has a copy of the V domain.

Production of V domains of different specificities creates the ability to respond to diverse

antigens. The total number of variable regions for either light or heavy chain proteins is measured in hundreds. *Thus the protein displays the maximum versatility in the region responsible for binding the antigen.*

◆ The association of constant regions generates several individual C domains in the molecule. The first domain results from association of the constant region of the light chain (C_L) with the CH1 part of the heavy chain constant region. The two copies of this domain complete the arms of the Y-shaped molecule.

The hinge between the arms and the foot of the Y depends on the interaction between heavy chains. There are ~10 types of heavy chain. In the example of Figure 37.4, a short hinge region connects the first constant domain to the other two constant domains, each generated by association of the corresponding regions (CH2 and CH3) of the heavy chains. The number of domains is greater with some types of heavy chain.

Different classes of immunoglobulins have different effector functions. *The class is determined by the heavy chain constant region, which exercises the effector function.* The small number of heavy chains means that the regions of the molecule with fixed functions are relatively conserved.

Comparing the characteristics of the variable and constant regions, we see the central dilemma in immunoglobulin gene structure. How does the genome code for a set of proteins in which any individual polypeptide chain must have one of <10 possible constant regions, but can have any one of several hundred possible variable regions? It turns out that the number of coding sequences for each type of region reflects its variability. There are many genes coding for V regions, but only a few genes coding for C regions.

In this context, *'gene' means a sequence of DNA coding for a discrete part of the final immunoglobulin polypeptide* (heavy or light chain). Thus **V genes** code for variable regions and **C genes** code for constant regions, although *neither type of gene is expressed as an independent unit.* To construct a unit that can be expressed in the form of an authentic light or heavy chain, a V gene must be joined physically to a C gene. In this system, two 'genes' code for one polypeptide.

The sequences coding for light chains and heavy chains are assembled in the same way: *any one of many V genes may be joined to any one of a few C genes.* This **somatic recombination** occurs *in the B lymphocyte in which the antibody is expressed.* The large number of available V genes is responsible for a major part of the diversity of immunoglobulins. However, not all diversity is coded in the genome; some is generated by changes that occur during the process of constructing a functional gene.

Essentially the same description applies to the formation of functional genes coding for the protein chains of the T-cell receptor. Two types of receptor are found on T cells, one consisting of two types of chain called α and β, the other consisting of γ and δ chains. Like the genes coding for immunoglobulins, the genes coding for the individual chains in T-cell receptors consist of separate parts, including V and C regions, which are brought together in an active T-cell (see later).

The crucial fact about the synthesis of immunoglobulins, therefore, is that *the arrangement of V genes and C genes is different in the cells producing the immunoglobulins (or T-cell receptors) from all other somatic cells or germ cells.*

The construction of a functional immunoglobulin or T-cell receptor gene might seem to be a Lamarckian process, representing a change in the genome that responds to a particular feature of the phenotype (the antigen). At birth, the organism does not possess the functional gene for producing a particular antibody or T-cell receptor. It possesses a large number of V genes and a smaller number of C genes. The subsequent construction of an active gene from these parts allows the antibody/receptor to be synthesized so that it is available to react with the antigen. The clonal selection theory requires that this rearrangement of DNA occurs *before the exposure to antigen*, which then results in *selection* for those cells carrying a protein able to bind the antigen. The entire process occurs in somatic cells and does not affect the germ line; so the response to an antigen is not inherited by progeny of the organism.

Recombination between V and C genes to give functional loci occurs in a population of immature lymphocytes. A B lymphocyte usually has only one productive rearrangement of light chain genes and one of heavy chain genes; a T lymphocyte productively rearranges an α gene and a β gene, or one δ gene and one γ gene. The antibody or T-cell receptor produced by any one cell is determined by the particular configuration of V genes and C genes that has been joined (and other changes to the coding sequence also may have occurred).

There are two families of immunoglobulin light chains, κ and λ, and one family of heavy chains (H). Each family resides on a different chromosome, and consists of its own set of both V genes and C genes. In the pattern inherited by any animal, the V genes and C genes for each family are a considerable distance apart. This is called the **germ line** pattern, and it is found in the germ-line and in somatic cells of all lineages other than the immune system.

But in a cell expressing an antibody, each of its chains—one light type (either κ or λ) and one heavy type—is coded by a single intact gene. The recombination event that brings a V gene to partner a C gene creates an active gene consisting of exons that correspond precisely with the functional domains of the protein. The introns are removed in the usual way by RNA splicing.

The number of V genes and C genes in a family varies with the species. **Table 37.1** describes the components of each immunoglobulin family in man and mouse. Typically there are a few hundred V genes, but note the difference between man and mouse caused by the loss of λ V genes in the mouse.

The principles by which functional genes are assembled are the same in each family, but there are differences in the details of the organization of the V and C genes, and correspondingly of the recombination reaction between them.

A lambda light chain is assembled from two parts, as illustrated in **Figure 37.5**. The V gene consists of the leader exon (L) separated by a single intron from the variable (V) segment. The C gene consists of the J segment separated by a single intron from the constant (C) exon.

The name of the **J segment** is an abbreviation for joining, since it identifies the region to which the V segment becomes connected. So the joining reaction does not directly involve V and C genes, but occurs via the J segment; when we discuss the joining of 'V and C genes' for light chains, we really mean V–JC joining.

The J segment is short and actually codes for the last few (13) amino acids of the variable region, as defined by amino acid sequences. In the intact gene generated by recombination, the V–J segment therefore constitutes a single exon coding for the entire variable region.

A V_λ gene has a choice of C genes to recombine with; each has the same J–C dipartite structure. In view of this multipartite structure, we shall try to avoid confusion about the nature of the DNA sequences by referring to V, J, or C 'segments' of the genome or 'regions' of the polypeptide.

The consequences of the kappa joining reaction are illustrated in **Figure 37.6**. A kappa light chain also is assembled from two parts, but there is a difference in the organization of the C gene. A group of five J segments is spread over a region of 500–700 bp, separated by an intron of 2–3 kb from the single C_κ exon. In the mouse, the central J segment is nonfunctional (ψJ3). A V_κ segment may be joined to any one of the J segments.

Table 37.1

Each immunoglobulin family consists of a cluster of V genes linked to its C gene(s).

Family	Number of V Genes		Number of C Genes	
	Man	Mouse	Man	Mouse
Lambda	<300	2	>6	4
Kappa	<300	˜1000	1	1
Heavy	˜300	>1000	9	8

Figure 37.5

The lambda C gene is preceded by a J segment, so that V-J recombination generates a functional lambda light-chain gene.

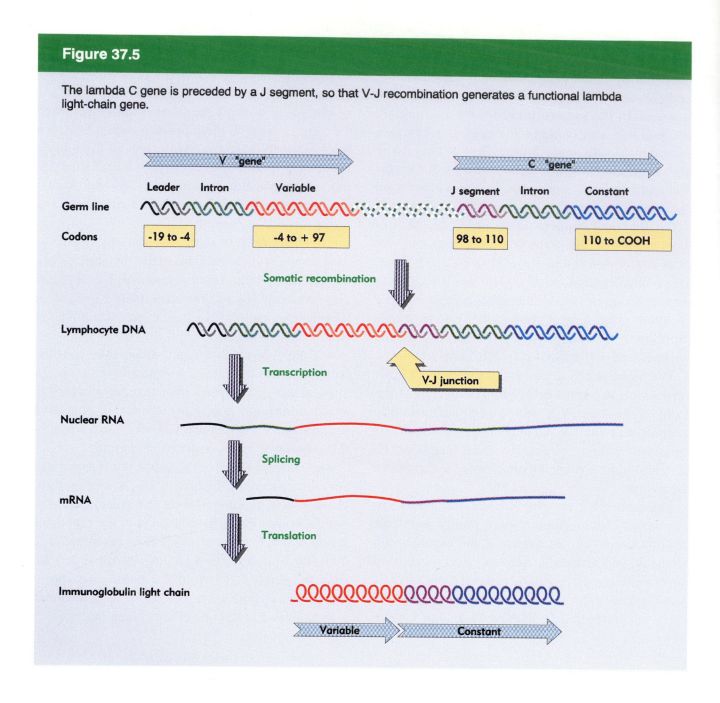

Whichever J segment is used becomes the terminal part of the intact variable exon. Any J segments on the left of the recombining J segment are lost (J1 has been lost in the figure). Any J segment on the right of the recombining J segment is treated as part of the intron between the variable and constant exons (J3 is included in the region that is spliced out in the figure).

All functional J segments possess a signal at the left boundary that makes it possible to recombine with the V segment; and they possess a signal at the right boundary that can be used for splicing to the C exon. Whichever J segment is recognized in DNA joining uses its splicing signal in RNA processing.

Heavy chain construction involves an additional segment. The **D** (for diversity) segment was discovered by the presence in the protein of an extra 2–13 amino acids between the sequences coded by the V segment and the J segment. An array of >10 D segments lies on the chromosome between the V_H segments and the 4 J_H segments (which vary in length between 4 and 6 codons).

V–D–J joining takes place in two stages, as illustrated in **Figure 37.7**. First one of the D segments recombines with a J_H segment; then a V_H segment recombines with the DJ_H combined segment. The reconstruction leads to expression of the adjacent C_H segment (which consists of several exons). (We come later to the use of different C_H genes; it suffices for now to consider the reaction in terms of connection to one of several J segments that precede a C_H gene.)

The D segments are organized in a tandem array. The mouse heavy chain locus contains 12 D

segments of variable length; the human locus has ~30 D segments (not all necessarily active). Some unknown mechanism must ensure that the *same* D segment is involved in the D–J joining and V–D joining reactions. (When we discuss joining of V and C genes for heavy chains, we assume the process has been completed by V–D and D–J joining reactions.)

The V genes of all three immunoglobulin families are similar in organization. The first exon codes for the signal sequence (involved in membrane attachment), and the second exon codes for the major part of the variable region itself (<100 codons long). The remainder of the

---

Figure 37.6

The kappa C gene is preceded by multiple J segments in the germ line; V-J joining may recognize any one of the J segments, which is then spliced to the C region during RNA processing.

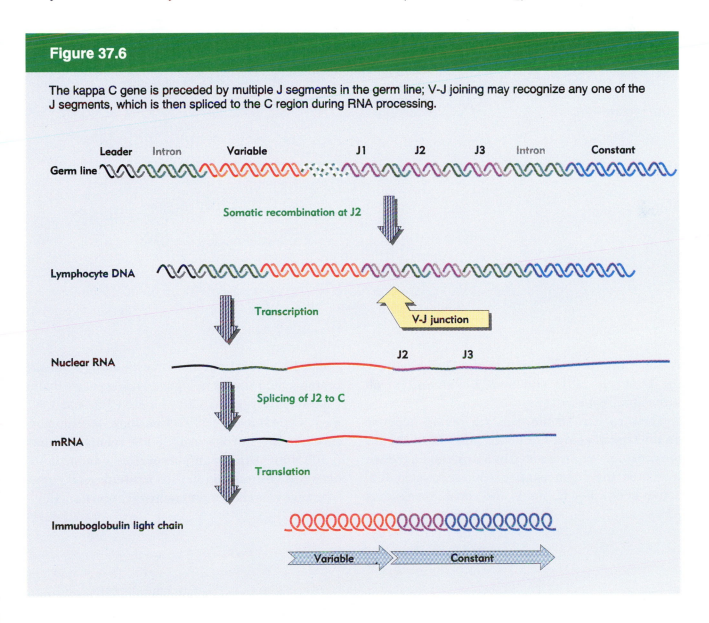

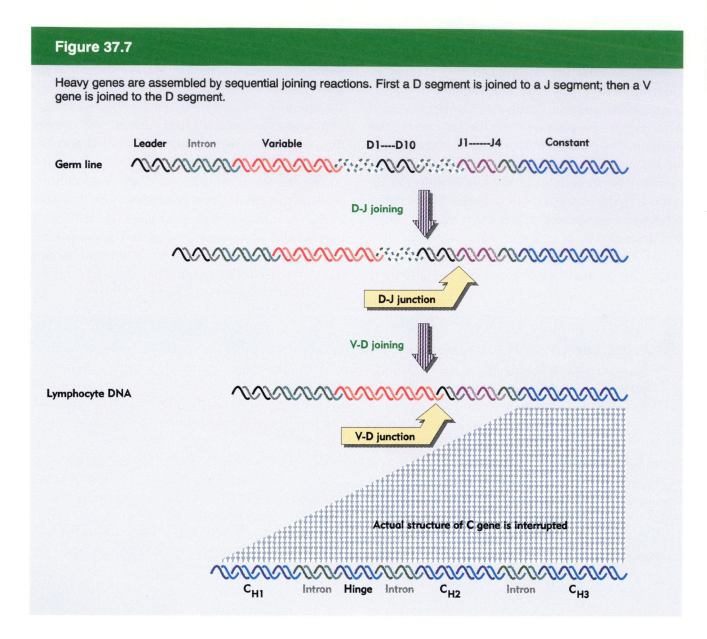

Figure 37.7

Heavy genes are assembled by sequential joining reactions. First a D segment is joined to a J segment; then a V gene is joined to the D segment.

variable region is provided by the D segment (in the H family only) and by a J segment (in all three families).

The structure of the constant region depends on the type of chain. For both κ and λ light chains, the constant region is coded by a single exon (which becomes the third exon of the reconstructed, active gene). For H chains, the constant region is coded by several exons; corresponding with the protein chain shown in Figure 37.4, separate exons code for the regions CH1, hinge, CH2, and CH3. Each CH exon is ~100 codons long; the hinge is shorter. The introns usually are relatively small (~300 bp). (An example of the concordance between immunoglobulin protein and gene structure was shown previously in Figure 23.19.)

The diversity of germ line information

Now we must examine the different types of V and C genes to see how much diversity can be accommodated by the variety of the coding regions carried in the germ line. The structures of the two light Ig gene families of the human genome are summarized in **Figures 37.8.** and **37.9.**

In each case, many V genes are linked to a much smaller number of C genes:

◆ The kappa locus actually has only one C gene, although it is preceded by 5 J segments (one of them inactive).

◆ The lambda locus has ~6 C genes, each preceded by its own J segment.

Because of the varying degrees of divergence between germ-line genes, it is difficult to estimate the number of V_κ and V_λ genes in the germ line. An individual probe may react with several genes; when one of these genes is used as a probe, it reacts with several others; and so on. This is the classic situation of a repetitive gene family, some members of which are closely related and some of which are distantly related. There are probably <300 V germ-line genes in each family in man.

The V_κ genes occupy a large cluster on the chromosome, upstream of the constant region. Its total size is not yet known. The V_κ genes can be subdivided into families, defined by the criterion that members of a family have >80% amino acid identity. The mouse family is unusually large, ~1000 genes, and there are ~18 V_κ families, varying in size from 2 to 100 members. Like other families of related genes, therefore, related V genes form subclusters, generated by duplication and divergence of individual ancestral members.

The lambda locus in mouse is much less diverse than the human locus. The main difference is that there are only two V_λ genes; each is linked to two J–C regions. Of the 4 C_λ genes, one is inactive. At time in the past, the mouse suffered a catastrophic deletion of most of its germ-line V_λ genes.

A given lymphocyte generates *either* a kappa *or* a lambda light chain to associate with the heavy chain. In man, ~60% of the light chains are

Figure 37.8

The human and mouse kappa families consist of <300 V genes linked to 5 J segments connected to a single C gene.

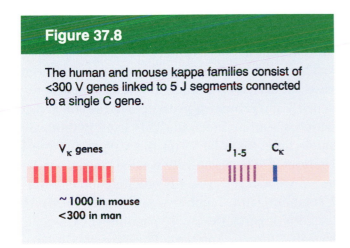

Figure 37.9

The lambda family consists of V genes linked to a small number of J-C genes.

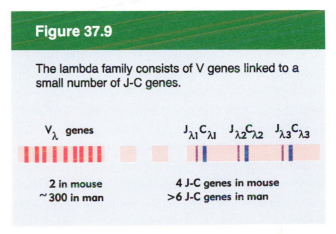

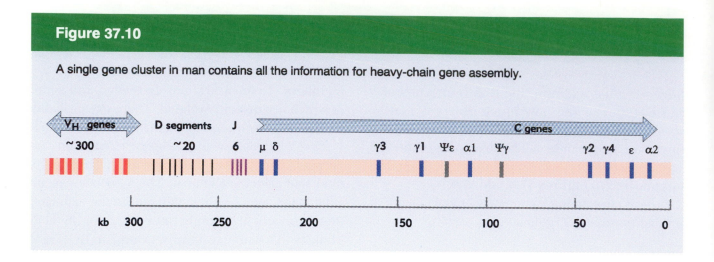

Figure 37.10

A single gene cluster in man contains all the information for heavy-chain gene assembly.

kappa and ~40% are lambda. In mouse, 95% of B cells express the kappa type of light chain, presumably because of the reduced number of lambda genes.

The single locus for heavy chain production consists of several discrete sections, whose structure in the human genome is summarized in **Figure 37.10**. It is similar in the mouse, where there are probably more V_H genes, fewer D and J segments, and a slight difference in the number and organization of C genes. We do not know the total span of the V_H cluster, but the 3' member is separated by only 20 kb from the first D segment. The D segments are spread over ~50 kb, and then comes the cluster of J segments. Over the next 220 kb lie all the C_H genes. There are 9 functional C_H genes and 2 pseudogenes. It seems that a γ gene must have been duplicated to give the sub-cluster of γ–γ–ϵ–α, after which the entire group was then duplicated. There is a tendency for V_H genes at the 3' end of the cluster (nearer the D segments) to be used more frequently early in the immune response. The basis for this effect is unknown, but it is most likely to represent some difference in the way selection operates on the products of recombination in younger and older animals.

To what degree is the diversity of germline information responsible for V region diversity in immunoglobulin proteins? By combining any one

of ~300 V genes with any one of 4–5 J segments, a typical light chain locus has the potential to produce some 1200–1500 chains. There is even greater diversity in the H chain locus; by combining any one of ~300 V_H genes, 20 D segments, and four J segments, the genome potentially can produce 4000 variable regions to accompany any C_H gene. In mammals, this is the starting point for diversity, but additional mechanisms introduce further changes. *When closely related variants of immunoglobulins are examined, there often are more proteins than can be accounted for by the number of corresponding V genes.* The new members are created by somatic changes in individual genes during or after the recombination process. We consider the relevant mechanisms in later sections.

The avian immune system presents a different case. An extreme example of reliance on diversity coded in the genome is presented by the chicken, in which a similar mechanism is used by both the single light chain locus (of the λ type) and the H chain locus. The organization of the λ locus is drawn in **Figure 37.11**. It has only one functional V gene, J segment, and C gene. Upstream of the functional $V_{\lambda 1}$ gene lie 25 V_λ pseudogenes, organized in either orientation. They are classified as pseudogenes because either the coding segment is deleted at one or both ends, or proper signals for recombination are missing (or both). This

Figure 37.11

The chicken lambda light locus has 25 V pseudogenes upstream of the single functional V-J-C region. But sequences derived from the pseudogenes are found in active rearranged V-J-C genes.

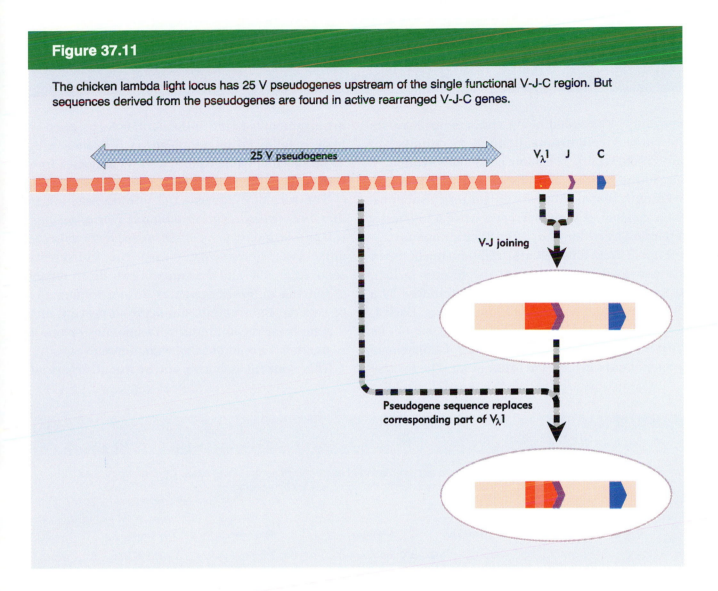

assignment is confirmed by the fact that only the $V_{\lambda1}$ gene recombines with the $J–C_{\lambda}$ gene.

But sequences of active rearranged $V_{\lambda}–J–C_{\lambda}$ genes show considerable diversity! A rearranged gene has one or more positions at which a cluster of changes has occurred in the sequence. A sequence identical to the new sequence can almost always be found in one of the pseudogenes (which themselves remain unchanged). The exceptional sequences that are not found in a pseudogene always represent changes at the junction between the original sequence and the altered sequence.

So a novel mechanism appears to be employed to generate diversity. Sequences from the pseudogenes, between 10 and 120 bp in length, are substituted into the active $V_{\lambda1}$ region, presumably by gene conversion. In fact, the unmodified $V_{\lambda1}$ sequence is not expressed, even at early times during the immune response. A successful conversion event probably occurs every 10–20 cell divisions to every rearranged $V_{\lambda1}$ sequence. At the end of the immune maturation period, a rearranged $V_{\lambda1}$ sequence has 4–6 converted segments spanning its entire length, derived from different donor pseudogenes. If all pseudogenes participate, this allows 2.5×10^8 possible combinations!

Recombination between V and C genes generates deletions and rearrangements

Assembly of light and heavy chain genes involves the same mechanism (although the number of parts is different). The same consensus sequences are found at the boundaries of all germ-line segments that participate in joining reactions. Each consensus sequence consists of a heptamer separated by either 12 or 23 bp from a nonamer.

Figure 37.12 illustrates the relationship between the consensus sequences at the mouse Ig loci. At the kappa locus, each V_κ gene is followed by a consensus sequence with a 12 bp spacing. Each J_κ segment is preceded by a consensus sequence with a 23 bp spacing. The V and J consensus sequences are inverted in orientation. The reverse

arrangement is found at the lambda locus; each V_λ gene is followed by a consensus sequence with 23 bp spacing, while each J_λ gene is preceded by a consensus of the 12 bp spacer type.

The rule that governs the joining reaction is that *a consensus sequence with one type of spacing can be joined only to a consensus sequence with the other type of spacing.* Since the consensus sequences at V and J segments can lie in either order, the different spacings do not impart any directional information, but serve to prevent one V gene from recombining with another, or one J segment from recombining with another.

This concept is borne out by the structure of

Figure 37.12

Consensus sequences are present in inverted orientation at each pair of recombining sites. One member of each pair has a spacing of 12 bp between its components; the other has 23 bp spacing.

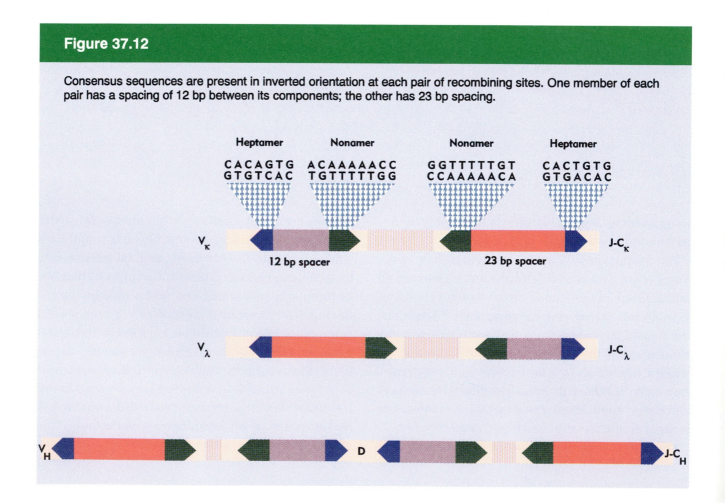

Figure 37.13

Breakage and reunion at consensus sequences generates immunoglobulin genes.

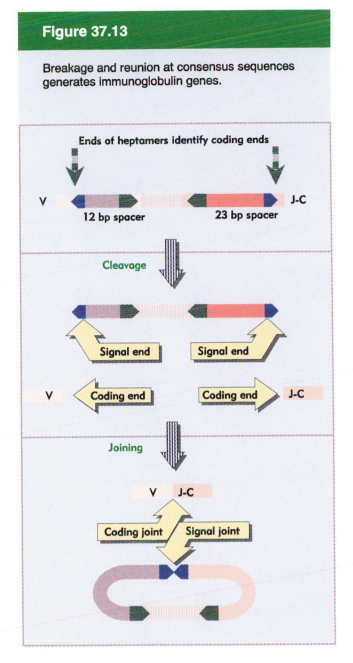

The spacing between the components of the consensus sequences corresponds almost to one or two turns of the double helix. This feature may reflect a geometric relationship in the recombination reaction. For example, the recombination protein(s) may approach the DNA from one side, in the same way that RNA polymerase and repressors approach recognition elements such as promoters and operators.

In using the term 'recombination' to describe the joining of the components of immunoglobulin genes, we do not imply that the reaction involves a reciprocal recombination between homologous sequences. It does represent a physical rearrangement of sequences, involving breakage and reunion, but the mechanism is different from homologous recombination. The general nature of the reaction is illustrated in **Figure 37.13.** for the example of a kappa light chain. (The reaction is similar at a heavy chain locus, except that there are two recombination events: first D–J, then V–DJ.)

Breakage and reunion occur as separate reactions. A double-strand break is made at the heptamers that lie at the ends of the coding units. This releases the entire fragment between the V gene and J–C gene; the cleaved termini of this fragment are called **signal ends**. The cleaved termini of the V and J–C loci are called **coding ends**. The two coding ends are covalently linked to form a coding joint; this is the connection that links the V and J segments. If the two signal ends are also connected, the excised fragment would form a circular molecule.

The joining reaction causes changes in sequence that affect the amino acid coded at the V–J junction in light chains or at the V–D and D–J junctions in heavy chains. These changes are a consequence of the enzymatic mechanisms involved in breaking and rejoining the DNA. Base pairs are lost or inserted at the V_H–D or D–J or both junctions during the recombination process. The sources of additional base pairs are shown in **Figure 37.14**. Deletion also occurs in V_L–J_L joining, but insertion at these joints is unusual.

A major stumbling block in characterizing the mechanism of recombination has been the lack of

the components of the heavy genes. Each V_H gene is followed by a consensus sequence of the 23 bp spacer type. The D segments are flanked on either side by consensus sequences of the 12 bp spacer type. The J_H segments are preceded by consensus sequences of the 23 bp spacer type. Thus the V gene must be joined to a D segment; and the D segment must be joined to a J segment. A V gene cannot be joined directly to a J segment, because both possess the same type of consensus sequence.

Figure 37.14

Breakage and reunion at coding joints allows incorporation of the additional P and N nucleotides.

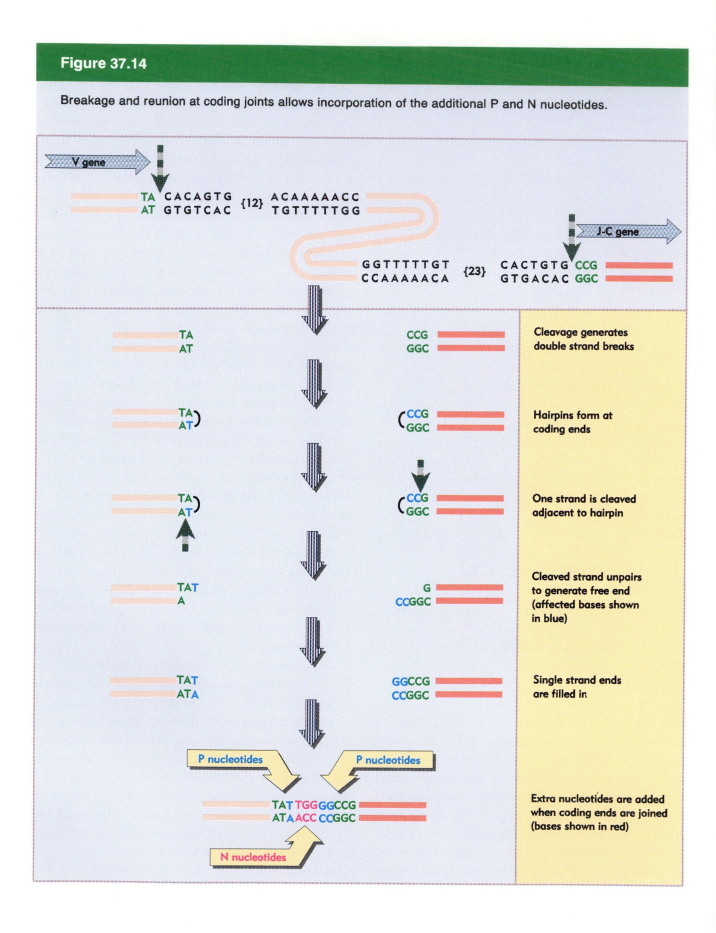

an *in vitro* system. Attempts to isolate extracts from B cells or T cells that can specifically recombine V and C genes so far have failed. A halfway house is represented by introducing particular combinations of germ line sequences into lymphocytes, and characterizing the products of recombination.

The initial stages of the reaction were identified by isolating intermediates from lymphocytes of mice with the *SCID* mutation. The exact cause of the *SCID* mutation is not known, but it results in a much reduced activity in immunoglobulin and TcR recombination. *SCID* mice accumulate broken molecules that terminate in double-strand breaks at the coding ends, and probably the mutation affects some aspect of the joining reaction.

The coding ends have an unusual structure: they have been converted to hairpins, that is, the 3′ end of one strand has been covalently linked to the 5′ end of the other strand. If a single-strand break is then introduced into one strand close to the hairpin, an unpairing reaction at the end generates a single-stranded protrusion. Synthesis of a complement then converts the coding end to an extended duplex. This reaction explains the introduction of **P nucleotides** at coding ends; they consist of a few extra base pairs, related to, but reversed in orientation from, the original coding end.

Some extra bases also may be inserted, apparently with random sequences, between the coding ends. They are called **N nucleotides**. Their insertion occurs via the activity of the enzyme deoxynucleoside transferase (known to be an active component of lymphocytes) at a free 3′ coding end generated during the joining process.

These various mechanisms together ensure that a coding joint may have a sequence that is different from what would be predicted by a direct joining of the coding ends of the V, D, and J regions.

Changes in the sequence at the junction make it possible for a great variety of amino acids to be coded at this site. It is interesting that the amino acid at position 96 is created by the V–J joining reaction. It forms part of the antigen-binding site and also is involved in making contacts between the light and heavy chains. So the maximum diversity is generated at the site that contacts the target antigen.

Changes in the number of base pairs affect the reading frame. The joining process appears to be random with regards to reading frame, so that probably only one-third of the joined sequences retain the proper frame of reading through the junctions. For example, if the V–J region is joined so that the J segment is out of phase, it is translated in the wrong reading frame. The resulting gene is aberrant, since its expression is terminated by a nonsense codon in the incorrect frame. We may think of the formation of aberrant genes as comprising the price the cell must pay for the increased diversity that it gains by being able to adjust the sequence at the joining site.

Similar although even greater diversity is generated in the joining reactions that involve the D segment of the heavy chain. The same result is seen with regards to reading frame; nonproductive genes are generated by joining events that place J and C out of phase with the preceding V gene.

We have shown the V and J–C loci as organized in the same orientation. As a result, the cleavage at each consensus sequence releases the region between them as a linear fragment. If the signal ends are joined, it is converted into a circular molecule, as indicated in Figure 37.13. Deletion to release an excised circle is the predominant mode of recombination at the immunoglobulin and TcR loci.

There are some cases, however, in which the V gene is inverted in orientation on the chromosome relative to the J–C loci. In such a case, breakage and reunion inverts the intervening material instead of deleting it. The outcomes of deletion versus inversion are the same as shown previously for homologous recombination between direct or inverted repeats in Figures 34.8 and 34.9. There is one further proviso, however; recombination with an inverted V gene makes it necessary for the signal ends to be joined, because otherwise there is a break in the locus. Inversion occurs in TcR recombination, and also sometimes in the κ light-chain locus.

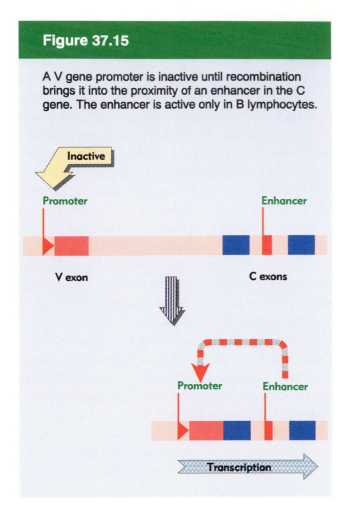

Figure 37.15

A V gene promoter is inactive until recombination brings it into the proximity of an enhancer in the C gene. The enhancer is active only in B lymphocytes.

Inactive

Promoter   Enhancer

V exon   C exons

Promoter   Enhancer

Transcription

Several proteins that may be associated with V(D)J joining have been characterized, including heptamer binding, nonamer binding, and endonuclease activities. The only case in which there is direct evidence for involvement, however, is that of the *RAG* genes. *RAG1* and *RAG2* are two genes, separated by <10 kb on the chromosome, whose transfection into fibroblasts causes a suitable substrate DNA to undergo the V(D)J joining reaction. This suggests that their products either directly form a recombinase or else are able to activate an enzyme that is ubiquitous (but is inactive in their absence). Mice that lack either *RAG1* or *RAG2* are unable to recombine their immunoglobulins or T-cell receptors, and as a result have immature B and T lymphocytes.

What is the connection between joining of V and C genes and their activation? Unrearranged V genes are not actively represented in RNA. But when a V gene is joined productively to a C_κ gene, the resulting unit is transcribed. However, since the sequence upstream of a V gene is not altered by the joining reaction, *the promoter must be the same in unrearranged, nonproductively rearranged, and productively rearranged genes.*

A promoter lies upstream of every V gene, but is inactive. It is activated by its relocation to the C region. The effect must depend on sequences downstream. What role might they play? An enhancer located within or downstream of the C gene activates the promoter at the V gene. The enhancer is tissue-specific; it is active only in B cells. Its existence suggests the model illustrated in **Figure 37.15**, in which the V gene promoter is activated as soon as it is brought into reach of the enhancer.

Allelic exclusion is triggered by productive rearrangement

Each B cell expresses a single type of light chain and a single type of heavy chain, because only a single productive rearrangement of each type occurs in a given lymphocyte, to produce one light and one heavy chain gene. Because each event involves the genes of only *one* of the homologous chromosomes, *the alleles on the other chromosome are not expressed in the same cell.* This phenomenon is called **allelic exclusion**.

The occurrence of allelic exclusion complicates the analysis of somatic recombination. A probe reacting with a region that has rearranged on one homologue will also detect the allelic sequences on the other homologue. We may

Figure 37.16

A successful rearrangement to produce an active light or heavy chain suppresses further rearrangements of the same type, and results in allelic exclusion.

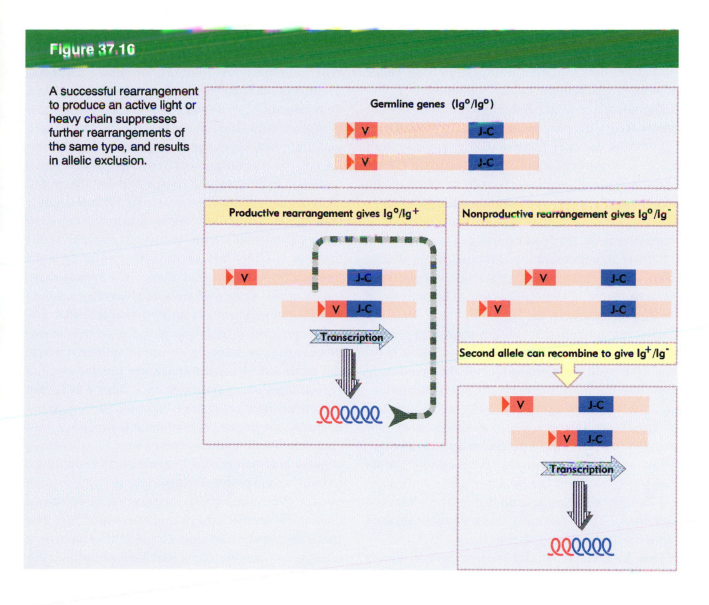

therefore be compelled to analyze the different fates of the two chromosomes together.

The usual pattern displayed by a rearranged active gene can be interpreted in terms of a deletion of the material between the recombining V and C loci.

Two types of gene organization are seen in active cells:

◆ Probes to the active gene may reveal both the rearranged and germ-line patterns of organization. We assume then that joining has occurred on one chromosome, while the other chromosome has remained unaltered.

◆ Two different rearranged patterns may be found,

indicating that the chromosomes have suffered independent rearrangements. In some of these instances, material between the recombining V and C genes is entirely absent from the cell line. This is most easily explained by the occurrence of independent deletions on each chromosome.

When two chromosomes both lack the germ-line pattern, usually only one of them has passed through a **productive rearrangement** to generate a functional gene. The other has suffered a **nonproductive rearrangement**; this may take several forms, but in each case the gene sequence cannot be expressed as an immunoglobulin chain. (It may be incomplete, for example because

D–J joining has occurred but V-D joining has not followed; or it may be aberrant, with the process completed, but not generating a gene that codes for a functional protein.)

The coexistence of productive and nonproductive rearrangements suggests the existence of a feedback loop to control the recombination process. A model is outlined in **Figure 37.16**. Suppose that each cell starts with two loci in the unrearranged germ-line configuration Ig^0. Either of these loci may be rearranged to generate a productive gene Ig^+ or a nonproductive gene Ig^-.

If the rearrangement is productive, the synthesis of an active chain provides a trigger to prevent rearrangement of the other allele. The active cell has the configuration Ig^0/Ig^+.

If the rearrangement is nonproductive, it creates a cell with the configuration Ig^0/Ig^-. There is no impediment to rearrangement of the remaining germ-line allele. If this rearrangement is productive, the expressing cell has the configuration Ig^+/Ig^-. Again, the presence of an active chain suppresses the possibility of further rearrangements.

Two successive nonproductive rearrangements produce the cell Ig^-/Ig^-. In some cases an Ig^-/Ig^- cell can try yet again. Sometimes the observed patterns of DNA can only have been generated by successive rearrangements.

The crux of the model is that the cell keeps trying to recombine V and C genes until a productive rearrangement is achieved. Allelic exclusion is caused by the suppression of further rearrangement as soon as an active chain is produced. The use of this mechanism *in vivo* is demonstrated by the creation of transgenic mice whose germline has a rearranged immunoglobulin gene. Expression of the transgene in B cells suppresses the rearrangement of endogenous genes.

Allelic exclusion is independent for the heavy and light chain loci. Heavy chain genes usually rearrange first. Allelic exclusion for light chains must apply equally to both families (cells may have *either* active kappa or lambda light chains). It is likely that the cell rearranges its kappa genes first, and tries to rearrange lambda only if both kappa attempts are unsuccessful.

There is an interesting paradox in this series of events. The same consensus sequences and the same V(D)J recombinase are involved in the recombination reaction at H, κ, and λ loci. Yet the three loci rearrange in a set order. What ensures that heavy rearrangement precedes light rearrangement, and that κ precedes λ? The loci may become accessible to the enzyme at different times, possibly as the result of transcription. Transcription occurs even before rearrangement, although of course the products have no coding function. The transcriptional event may change the structure of chromatin, making the consensus sequences for recombination available to the enzyme.

DNA recombination causes class switching

The **class** of immunoglobulin is defined by the type of C_H region. **Table 37.2** summarizes the five Ig classes. We do not yet understand the detailed functions of each class. IgM (the first immunoglobulin to be produced by any B cell) and IgG (the most common immunoglobulin by far) possess the central ability to activate complement, which leads to destruction of invading cells. IgA is found in secretions (such as saliva), and IgE is associated with the allergic response and defense against parasites.

All lymphocytes start productive life as immature cells engaged in synthesis of IgM. Cells expressing IgM have the germ-line arrangement of the C_H

Table 37.2

Immunoglobulin type and function is determined by the heavy chain. J is a "joining protein" in IgM; all other Ig types exist as tetramers.

Type	IgM	IgD	IgG	IgA	IgE
Heavy chain	μ	δ	γ	α	ε
Structure	$(\mu_2 L_2)_5 J$	$\delta_2 L_2$	$\gamma_2 L_2$	$\alpha_2 L_2$	$\varepsilon_2 L_2$
Proportion	5%	1%	80%	14%	<1%
Effector function	Activates complement	Development of tolerance (?)	Activates complement	Found in secretions	Allergic response

gene cluster shown in Figure 37.10. The V–D–J joining reaction triggers expression of the C_μ gene. A lymphocyte generally produces only a single class of immunoglobulin at any one time, but the class may change during the cell lineage. A change in expression is called **class switching**. It is accomplished by a substitution in the type of C_H region that is expressed. Switching can be stimulated by environmental affects; for example, the growth factor TGFβ causes switching from C_μ to C_α.

Switching involves only the C_H gene; the same V_H gene continues to be expressed. Thus a given V_H gene may be expressed successively in combination with more than one C_H gene. The same light chain continues to be expressed throughout the lineage of the cell. Class switching therefore allows the type of effector response (mediated by the C_H region) to change, while maintaining a constant facility to recognize antigen (mediated by the V regions).

Changes in the expression of C_H genes are made in two ways. The majority occur via further DNA recombination events, involving a system different from that concerned with V–D–J joining (and able to operate only later during B cell development). Another type of change occurs at the level of RNA processing, but generally this is involved with changing the C-terminal sequence of the C_H region rather than its class (see next section).

Cells expressing later C_H genes have deletions of C_μ and the other genes preceding the expressed C_H gene. Class switching is accomplished by a recombination to bring a new C_H gene into juxtaposition with the expressed V–D–J unit. The sequences of switched V–D–J–C_H units show that the sites of switching lie upstream of the C_H genes themselves. The switching sites are called **S regions**. **Figure 37.17** depicts two successive switches.

In the first switch, expression of C_μ is succeeded by expression of C_{γ_1}. The C_{γ_1} gene is brought into the expressed position by recombination between the sites S_μ and S_{γ_1}, deleting the material between. The S_μ site lies between V–D–J and the C_μ gene. The S_{γ_1} site lies upstream of the C_{γ_1} gene. The region between V–D–J and the C_{γ_1} gene is removed as an intron during processing of the RNA.

The linear deletion model imposes a restriction on the heavy gene locus: *once a class switch has been made, it becomes impossible to express any C_H gene that used to reside between C_μ and the new C_H gene.* In the example of Figure 37.17, cells expressing C_{γ_1} should be unable to give rise to cells expressing C_{γ_3}, which has been deleted.

However, it should in principle be possible to undertake another switch to any C_H gene *downstream* of the expressed gene. The figure

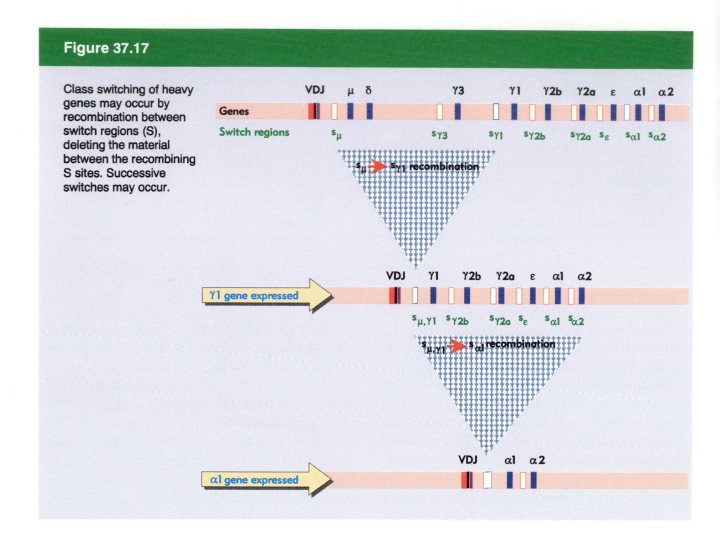

Figure 37.17

Class switching of heavy genes may occur by recombination between switch regions (S), deleting the material between the recombining S sites. Successive switches may occur.

shows a second switch to C_α expression, accomplished by recombination between S_α and the switch region $S_{\mu,\gamma1}$ that was generated by the original switch.

We assume that all of the C_H genes have S regions upstream of the coding sequences. We do not know whether there are any restrictions on the use of S regions. Sequential switches do occur, but we do not know whether they are optional or an obligatory means to proceed to later C_H genes. We should like to know whether IgM can switch directly to *any* other class.

We know that switch sites are not uniquely defined, because different cells expressing the same C_H gene prove to have recombined at different points. When enough of these S sites

have been sequenced, we may be able to define the limits of the regions within which switching can occur. In the meantime, we can scrutinize the sequences in the area for common features that might be involved.

The S regions lie ~2 kb upstream of the C_H genes. S regions vary in length from 1 to 10 kb, and contain groups of short homologous repeats. The recombination reaction is not site-specific, but usually occurs near the conserved repeat elements. Little is known about the molecular basis for the reaction, but it releases the excised material between the switch sites as a circular DNA molecule. Because the S regions lie within the introns that precede the C_H coding regions, switching does not alter the translational reading frame.

Early heavy-chain expression can be changed by RNA processing

The period of IgM synthesis that begins lymphocyte development falls into two parts, during which different versions of the μ constant region are synthesized:

♦ As a stem cell differentiates to a pre-B lymphocyte, an accompanying light chain is synthesized, and the IgM molecule ($L_2\mu_2$) appears at the surface of the cell. This form of IgM contains the μ_m version of the constant region (*m* indicates that IgM is located in the membrane). The membrane location may be related to the need to initiate cell proliferation in response to the initial recognition of an antigen.

♦ When the B lymphocyte differentiates further into a plasma cell, the μ_s version of the constant region is expressed. The IgM actually is secreted as a pentamer IgM_5J, in which J is a joining polypeptide (no connection with the J segment) that forms disulfide linkages with mu chains. Secretion of the protein is followed by the humoral response depicted in Figure 37.1.

The μ_m and μ_s versions of the μ heavy chain differ only at the C-terminal end. The μ_m chain ends in a hydrophobic sequence that probably secures it in the membrane. This sequence is replaced by a shorter hydrophilic sequence in μ_s; the substitution allows the μ heavy chain to pass through the membrane. The change of C-terminus is accomplished by an alternative splicing event, which is controlled by the 3′ end of the nuclear RNA, as illustrated in **Figure 37.18**.

At the membrane-bound stage, the RNA terminates after exon M2, and the constant region is produced by splicing together six exons. The first

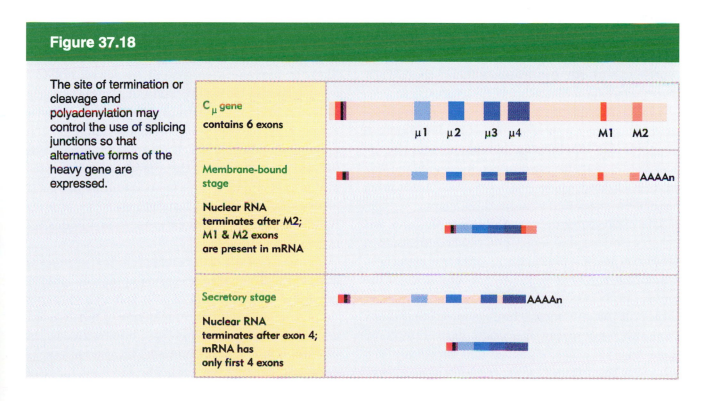

Figure 37.18

The site of termination or cleavage and polyadenylation may control the use of splicing junctions so that alternative forms of the heavy gene are expressed.

C_μ gene contains 6 exons

μ1 μ2 μ3 μ4 M1 M2

Membrane-bound stage

Nuclear RNA terminates after M2; M1 & M2 exons are present in mRNA

AAAAn

Secretory stage

Nuclear RNA terminates after exon 4; mRNA has only first 4 exons

AAAAn

four code for the four domains of the constant region. The last two, M1 and M2, code for the 41-residue hydrophobic C-terminal region and its nontranslated trailer. The 5′ splicing junction within exon 4 is connected to the 3′ splicing junction at the beginning of M1.

At the secreted stage, the nuclear RNA terminates after exon 4. The 5′ splicing junction within this exon that had been linked to M1 in the membrane form is ignored. This allows the exon to extend for an additional 20 codons.

A similar transition from membrane to secreted forms is found with other constant regions. The conservation of exon structures suggests that the mechanism is the same.

An exception to the rule that only one immunoglobulin type is synthesized by any one cell is presented by the simultaneous production of IgM and IgD in mature B lymphocytes. The two immunoglobulins are identical except for the substitution between μ and δ constant regions in the heavy chain. It seems likely that this is the outcome of alternative pathways for RNA processing. The δ constant region is close to μ; if transcription sometimes continues through the region, the VDJ exon could be spliced to the series of delta exons.

Somatic mutation generates additional diversity

Comparisons between the sequences of expressed immunoglobulin genes and the corresponding V genes of the germ line show that new sequences appear in the expressed population. We have seen that some of this additional diversity results from sequence changes at the V–J or V–D–J junctions. However, other changes occur upstream at locations within the variable domain; they represent **somatic mutations** induced specifically in the active lymphocyte.

A probe representing an expressed V gene can be used to identify all the corresponding fragments in the germ line. Their sequences should identify the complete repertoire available to the organism. Any expressed gene whose sequence is different must have been generated by somatic changes.

The main experimental problem in this analysis lies in ensuring that every potential contributor in the germ-line V genes actually has been identified. This problem is overcome by the simplicity of the mouse lambda chain system. A survey of several myelomas producing λ_1 chains showed that many have the sequence of the single germ-line gene. *But others have new sequences that must have been generated by mutation of the germ-line gene.*

To determine the frequency of somatic mutation in other cases, we need to examine a large number of cells in which the same V gene is expressed. A practical procedure for identifying such a group is to characterize the immunoglobulins of a series of cells, all of which express an immune response to a particular antigen.

(Epitopes used for this purpose are small molecules—haptens—whose discrete structure is likely to provoke a consistent response, unlike a large protein, different parts of which provoke different antibodies. A hapten is conjugated with a nonreactive protein to form the antigen. The cells are obtained by immunizing mice with the antigen, obtaining the reactive lymphocytes, and sometimes fusing these lymphocytes with a myeloma [immortal tumor] cell to generate a **hybridoma** that continues to express the desired antibody indefinitely.)

A survey of 19 different cell lines producing antibodies directed against the hapten phosphorylcholine showed that 10 have the same V_H sequence (expressed in conjunction with one of the μ, γ, or α constant regions). The V_H sequence can be identified in the germ line as T15, one of four V_H genes hybridizing with a probe for the expressed V_H sequence. The other 9 expressed genes differ from each other and from all 4 germ-line members of the family. They are more closely related to the T15 germ-line sequence than to any of the others, and their flanking sequences are the same as those around T15. This suggests that they have arisen from the T15 member by somatic mutation.

The sequences of these expressed genes vary from the germ-line sequence in all regions of the variable domain. Sequence changes are found in the downstream flanking regions as well as in the coding regions; they do not extend far upstream. All the variation is due to substitutions of individual nucleotide pairs. The variation is different in each case, and is in the range of 1–4% divergence from the germ line (corresponding to <10 amino acid substitutions in the protein; only some of the mutations affect the amino acid sequence, since others lie in third-base coding positions as well as in nontranslated regions.)

The large proportion of ineffectual mutations suggests that somatic mutation occurs more or less at random in a region including the V gene and extending beyond it. There is a tendency for some mutations to recur on multiple occasions, and these may represent hotspots as a result of some intrinsic preference in the system. The molecular basis for somatic mutation remains unknown.

Somatic mutation is induced by antigenic stimulation, and occurs during clonal proliferation, apparently at a rate $\sim 10^{-3}$ per bp per cell generation. Approximately half of the progeny cells gain a mutation; as a result, cells expressing mutated antibodies become a high fraction of the clone. The consequence is to produce antibodies with increased affinity for the antigen (see below).

There are many cases in which a single family of V genes is used consistently to respond to a particular antigen. The consistency of response indicates the importance of coding a wide range of responses in the V genes. Upon exposure to an antigen, presumably the V region with highest intrinsic affinity provides a starting point. Random mutations have unpredictable effects on protein function; some inactivate the protein, others confer high specificity for a particular antigen. Thus a critical feature of the process is selection

among the lymphocyte population for those cells bearing antibodies in which chance mutation has created a suitable V domain to bind whatever antigen is at hand.

We are now in a position to summarize the relationship between the generation of high affinity antibodies and the differentiation of the B cell. **Figure 37.19** shows that B cells are derived from a self-renewing population of stem cells in the bone marrow. Maturation to give B cells depends upon Ig gene rearrangement, which requires the functions of the SCID and RAG1,2 (and other) genes. If gene rearrangement is blocked, mature B cells are not produced. The antibodies carried by the B cells have specificities determined by the particular combinations of V(D)J regions, and any additional nucleotides incorporated during the joining process.

Exposure to antigen triggers two aspects of the immune response. The primary response occurs by clonal expansion of B cells responding to the antigen. This generates a large number of plasma cells that are specific for the antigen; isotype switching occurs to generate the appropriate type of effector response. The population of cells concerned with the primary response is a dead end; these cells do not live beyond the primary response itself.

Provision for a secondary response is made through the phenomenon of B cell memory. Somatic mutation generates B cells that have increased affinity for the antigen. These cells do not trigger an immune response at this time, although they may undergo isotype switching to select other forms of C_H region. They are stored as memory cells, with appropriate specificity and effector response type, but are inactive. They are activated if there is a new exposure to the same antigen. Because they are pre-selected for the antigen, they enable a secondary response to be mounted very rapidly, simply by clonal expansion; no further somatic mutation or isotype switching occurs during the secondary response.

The pathways summarized in Figure 37.19 show the development of acquired immunity, in response to an antigen. In addition to these cells, there is a

Figure 37.19

Summary: B cell differentiation is responsible for acquired immunity. Pre-B stem cells propagate in the bone marrow, and are converted to B cells by Ig gene rearrangement. Initial exposure to antigen provokes both the primary response and storage of memory cells. Subsequent exposure to antigen provokes the secondary response of the memory cells.

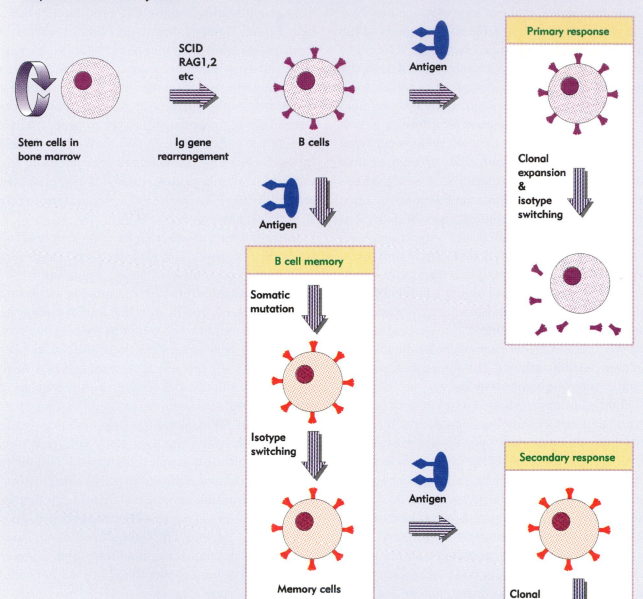

separate set of B cells, named the Ly-1 cells. These cells have gone through the process of V gene rearrangement, apparently selected for expression of a particular repertoire of antibody specificities. They do not undergo somatic mutation or the memory response. They may be involved in natural immunity, that is, an intrinsic ability to respond to certain antigens.

T-cell receptors are related to immunoglobulins

The lymphocyte lineage presents an example of evolutionary opportunism: a similar procedure is used in both B cells and T cells to generate proteins that have a variable region able to provide significant diversity, while constant regions are more limited and account for a small range of effector functions. T cells produce either of two types of T-cell receptor. The different T-cell receptors are synthesized at different times during T cell development, as summarized in **Figure 37.20**.

The $\gamma\delta$ receptor is found on <5% of T lymphocytes. It is synthesized only at an early stage of T cell development. In mice, it is the only receptor detectable before the fifteenth day of gestation, but has virtually been lost by birth at day 20. Its function is not well characterized; it may be involved in an ability of these T cells to lyse target cells in a manner that does *not* depend on the presence of MHC proteins.

TcR $\alpha\beta$ is found on >95% of lymphocytes. It is synthesized later in T cell development than $\gamma\delta$. In mice, it first becomes apparent at 15–17 days after gestation. By birth it is the predominant receptor. It is synthesized by a separate lineage of cells from those involved in TcR$\gamma\delta$ synthesis, and involves independent rearrangement events.

T cells with $\alpha\beta$ receptors are divided into several subtypes that have a variety of functions connected with interactions between cells involved in the immune response. **Cytotoxic (killer) T cells** possess the capacity to lyse an infected target cell. **Helper T cells** assist T cell-mediated target killing or B cell-mediated antibody–antigen interaction.

The immune response requires a T cell to recognize a host cell displaying a fragment of a

Figure 37.20

The $\gamma\delta$ receptor is synthesized early in T-cell development. TcR $\alpha\beta$ is synthesized later and may be responsible for "classical" cell-mediated immunity, in which target antigen and host histocompatibility antigen are recognized together.

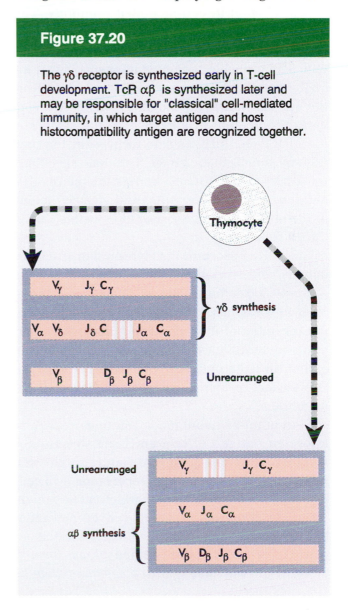

foreign protein on its surface. To do so, the T cell simultaneously recognizes the foreign antigen and an MHC protein carried by the presenting cell, as illustrated previously in Figure 37.2. Both helper T cells and killer T cells work in this way, but they have different requirements for the presentation of antigen; different types of MHC protein are used in each case (see later). Helper T cells require antigen to be presented by an MHC class II protein, while killer T cells require antigen to be presented by an MHC class I protein.

The TcR $\alpha\beta$ receptor is responsible for helper T cell function in humoral immunity and for killer T cell function in cell-mediated immunity. This places upon it the responsibility of recognizing both the foreign antigen and the host MHC protein. The probable sequence of events is that the MHC protein binds a short peptide derived from the foreign antigen, and the TcR then recognizes the peptide in a groove on the surface of the MHC. The MHC is said to **present** the peptide to the TcR. A given TcR has specificity for a particular MHC as well as for the foreign antigen. The basis for this dual capacity is one of the most interesting issues to be defined about the $\alpha\beta$ TcR.

The TcR $\alpha\beta$ receptor is a glycoprotein of ~80,000 daltons, consisting of one α-chain and one β-chain, each ~40,000 daltons. The chains are held together by disulfide bonds. Peptide mapping suggests that the receptors of different T cell lines share part of their sequences in common, but differ in others (like the immunoglobulins). The receptor is a surface protein, probably anchored in the membrane.

Like immunoglobulins, a TcR must recognize a foreign antigen of unpredictable structure. A common view has been that Nature might well solve the problem of antigen recognition by B cells and T cells in the same way, in which case we might expect the organization of the T-cell receptor genes to resemble the immunoglobulin genes in the use of variable and constant regions. *Each locus is organized in the same way as the immunoglobulin genes, with separate segments that are brought together by a recombination reaction specific to the lymphocyte.* The components are the same as those found in the three Ig families.

The organization of the TcR proteins resembles that of the immunoglobulins. The V sequences have the same general internal organization in both Ig and TcR proteins. The TcR C region is related to the constant Ig regions and has a single constant domain followed by transmembrane and cytoplasmic portions. Exon–intron structure is related to protein function.

The resemblance of the organization of TcR genes with the Ig genes is striking. As summarized in **Figure 37.21**, the organization of TcR α resembles that of Ig κ, with V genes separated from a cluster of J segments that precedes a single C gene. The organization of the locus is similar in both man and mouse, with some differences only in the number of V_α genes and J_α segments. (In addition to the V_α segments, this locus also contains δ segments, which we discuss shortly.)

The components of TcR β resemble those of IgH. **Figure 37.22** shows that the organization is different, with V genes separated from two clusters each containing a D segment, several J segments, and a C gene. Again the only differences

Figure 37.21

The human TcRα locus has interspersed α and δ segments. A Vδ segment is located within the Vα cluster. The D-J-C δ segments lie between the V genes and the J-C α segments. The mouse locus is similar, but has more Vδ segments.

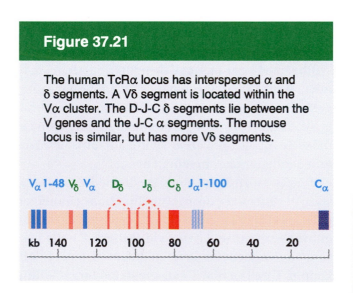

Figure 37.22

The TcRβ locus contains many V genes spread over ~500 kb, and lying ~280 kb upstream of the two D-J-C clusters.

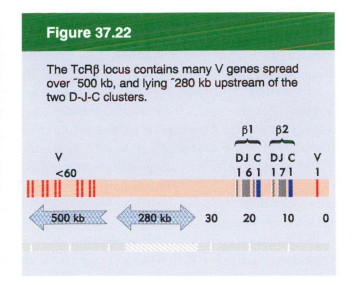

genes. It consists of two subunits, γ and δ.

The organization of the γ locus resembles that of Ig λ, with V genes separated from a series of J–C segments. **Figure 37.23** shows that this locus has relatively little diversity, with ~8 functional V segments. The organization is different in man and mouse. Mouse has 3 functional J–C loci, but some segments are inverted in orientation. Man has multiple J segments for each C gene.

The δ subunit is coded by segments that lie at the TcR α locus, as illustrated previously in Figure 37.21. The segments D_δ–D_δ–J_δ–C_δ lie between the V genes and the J_α–C_α segments. Both of the D segments may be incorporated into the δ chain to give the structure VDDJ. The nature of the V genes used in the δ rearrangement is an interesting question. Very few V sequences are found in active TcR δ chains. In man, only 1 V gene is in general use for rearrangement. In mouse, several V_δ segments are found; some are unique for δ rearrangement, but some are also found in α rearrangements. The basis for specificity in

between man and mouse are in the numbers of the V_β and J_β units.

Diversity is generated by the same mechanisms as in immunoglobulins. Intrinsic diversity results from the combination of a variety of V, D, J, and C segments; some additional diversity results from the introduction of new sequences at the junctions between these components (in the form of P and N nucleotides, as described previously in Figure 37.14). Some TcR β chains incorporate two D segments, generated by D–D joins (directed by an appropriate organization of the nanomer and heptamer sequences). A difference between TcR and Ig is that somatic mutation does not occur at the TcR loci.

The same mechanisms are likely to be involved in the reactions that recombine Ig genes in B cells and TcR genes in T cells. The recombining TcR segments are surrounded by nonamer and heptamer consensus sequences identical to those used by the Ig genes. This argues strongly that the same enzymes are involved. Most rearrangements probably occur by the deletion model (see Figure 37.13). How is the process controlled so that Ig loci are rearranged in B cells, while T-cell receptors are rearranged in T cells?

The second receptor on T cells was identified by virtue of T cell-specific rearrangements of its

Figure 37.23

The TcRγ locus contains a small number of functional V genes (and also some pseudogenes; not shown), lying upstream of the J-C loci. There are still some gaps in the map of each locus, so the total distance could be greater than shown.

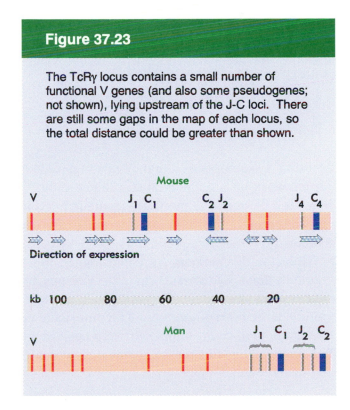

Figure 37.24

The two chains of the T cell receptor associate with the polypeptides of the CD3 complex. The variable regions of the TcR are exposed on the cell surface. The cytoplasmic domains of the ζ chains of CD3 provide the effector function.

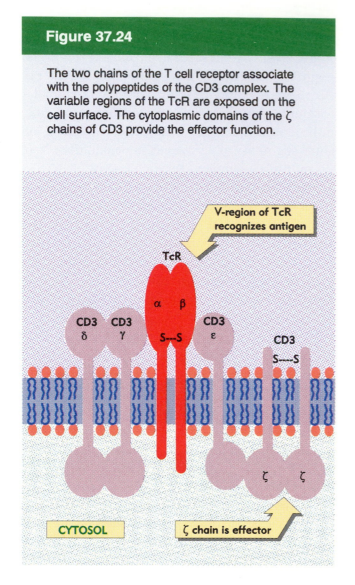

exclusive at any one allele, because the δ locus is lost entirely when the V_α–J_α rearrangement occurs.

Rearrangements at the TcR loci, like those of immunoglobulin genes, may be productive or nonproductive. The β locus shows allelic exclusion in much the same way as immunoglobulin loci; rearrangement is suppressed once a productive allele has been generated. The α locus may be different; several cases of continued rearrangement suggest the possibility that substitution of V_α sequences may continue after a productive allele has been generated.

The T-cell receptor is associated with a complex of proteins called **CD3**, which is involved in transmitting a signal from the surface of the cell to the interior when its associated receptor is activated by binding antigen. Our present picture of the components of the receptor complex on a T cell is illustrated in **Figure 37.24**. The important point is that the interaction of the TcR variable regions with antigen causes the ζ subunits of the CD3 complex to activate the T cell response. The activation of CD3 provides the means by which either αβ or γδ TcR signals that it has recognized an antigen.

A central dilemma about T-cell function remains to be resolved. Cell-mediated immunity requires two recognition processes. Recognition of the foreign antigen requires the ability to respond to novel structures. Recognition of the MHC protein is of course restricted to one of those coded by the genome, but, even so, there are many different MHC proteins. So considerable diversity is required in both recognition reactions. Although helper and killer T cells rely upon different classes of MHC proteins, they use the same pool of α and β gene segments to assemble their receptors. Even allowing for the introduction of additional variation during the TcR recombination process, it is not clear how enough different versions of the T-cell receptor are made available to accommodate all these demands.

choosing V segments in α and δ rearrangement is not known. One possibility is that many of the V_α genes can be joined to the DDJ_δ segment, but that only some (therefore defined as V_δ) can give active proteins.

While for the present we have labeled the V segments that are found in δ chains as V_δ genes, we must reserve judgement on whether they are really unique to δ rearrangement. The interspersed arrangement of genes implies that synthesis of the TcR αβ receptor and the γδ receptor is mutually

The major histocompatibility locus codes for many genes of the immune system

The major histocompatibility locus occupies a small segment of a single chromosome in the mouse (where it is called the *H2* **locus**) and in man (called the *HLA* **locus**). Within this segment are many genes coding for functions concerned with the immune response. At those individual gene loci whose products have been identified, many alleles have been found in the population; the locus is described as highly **polymorphic**, meaning that individual genomes are likely to be different from one another. Genes coding for certain other functions also are located in this region.

Histocompatibility antigens have been classified into three types by their immunological properties. In addition, other proteins found on lymphocytes and macrophages have a related structure and are important in the function of cells of the immune system:

◆ Class I proteins are the **transplantation antigens**. They are present on every cell of the mammal. As their name suggests, these proteins are responsible for the rejection of foreign tissue, which is recognized as such by virtue of its particular array of transplantation antigens. In the immune system, their presence on killer T lymphocytes is required for the cell-mediated response.

The types of class I proteins are defined serologically (by their antigenic properties). The murine class I genes code for the H2-K and H2-D/L proteins. Each mouse strain has one of several possible alleles for each of these functions. The human class I functions include the classical transplantation antigens, HLA-A, B, C.

◆ Class II proteins are found on the surfaces of both B and T lymphocytes as well as macrophages. These proteins are involved in communications between cells that are necessary to execute the immune response; in particular, they are required for helper T cell function. The murine class II functions are defined genetically as I-A and I-E. The human class II region (also called HLA-D) is arranged into four subregions, DR, DQ, DZ/DO, DP.

◆ The **complement proteins** provide the class III MHC. Their genetic locus is also known as the S region; S stands for serum, indicating that the proteins are components of the serum. Their role is to interact with antibody–antigen complexes to cause the lysis of cells in the classical pathway of the humoral response.

◆ The *Qa* and *Tla* loci proteins are found on murine hematopoietic cells. They are known as differentiation antigens, because each is found only on a particular subset of the blood cells, presumably related to their function. They are structurally related to the class I H2 proteins, and like them are polymorphic.

We can now relate the types of proteins to the organization of the genes that code for them.

The murine MHC locus is summarized on the map of **Figure 37.25**. The classical *H2* region occupies 0.3 map units; the adjacent region occupies another 1.0 map units. In molecular terms, this 'small segment' of the chromosome is sizable; the 1.3 map units together potentially represent ~2000 kb of DNA.

The H2^K genes map at the left end, and the H2$^{D/L}$ genes map at the right end. The class II and class III genes map between. Most of the

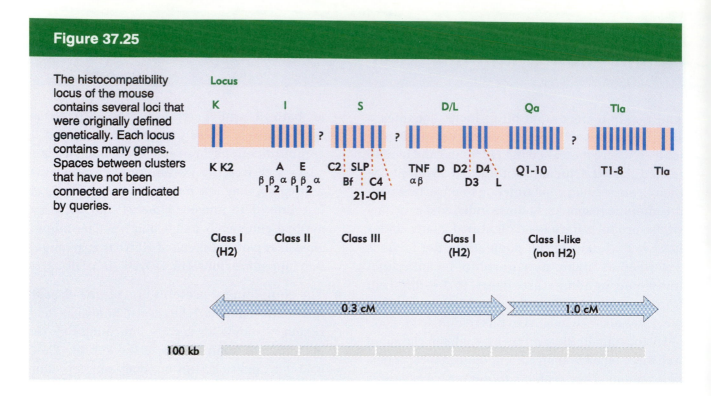

Figure 37.25

The histocompatibility locus of the mouse contains several loci that were originally defined genetically. Each locus contains many genes. Spaces between clusters that have not been connected are indicated by queries.

polymorphism in individual genes occurs in those of the *H2* type. The adjacent region extends for another map unit; within it are genes coding for the differentiation antigens. We may regard them as extending the region of the chromosome devoted to functions concerned with the development of lymphocytes and macrophages. Variation in the number of genes between different mouse strains seems to occur largely in the *Qa* and *Tla* loci.

The class I mouse genes reside in clusters. The genes in each cluster usually are oriented in the same direction; adjacent genes tend to be more closely related, which suggests that they have originated by ancestral tandem duplications. Other genes also lie in the MHC locus. Within the D/L class I region lie the genes for the subunits of

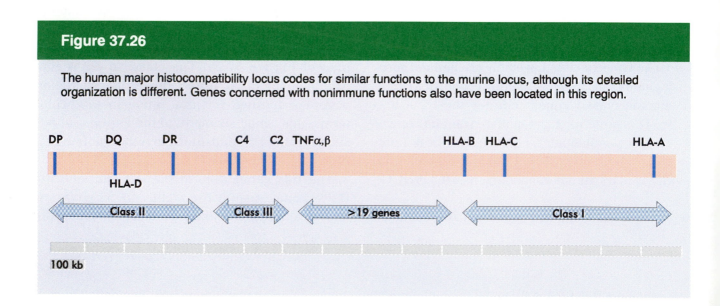

Figure 37.26

The human major histocompatibility locus codes for similar functions to the murine locus, although its detailed organization is different. Genes concerned with nonimmune functions also have been located in this region.

Figure 37.27

Class I and class II histocompatibility antigens have a related structure. Class I antigens consists of a single (α) polypeptide, with three external domains (α1, α2, α3), that interacts with β_2 microglobulin (β_2 m). Class II antigens consist of two (α and β) polypeptides, each with two domains (α1 & α2, β1 & β2) and whose interaction generates a similar overall structure.

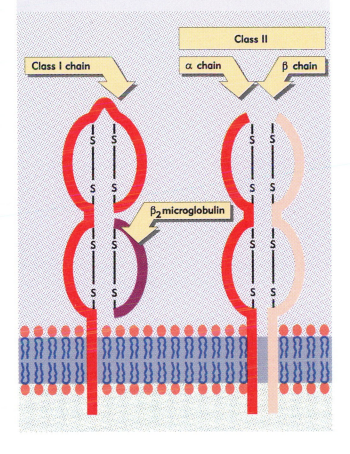

Figure 37.26, it contains similar functions, although not in the identical order. The major difference is that the class I HLA-A, B and C genes all are located in the same region, contrasted with the separation between H2-K and H2-D/L. The relative organization of the class I, II, and III genes is otherwise generally similar.

The genes for HLA-A, B, C are spread over a region of >600 kb. They are separated from the region containing the complement genes by a distance of ~550 kb, which contains other genes, including TNF, factor and steroid 21-hydroxylase. Then come the class II genes for DR, DQ, DP.

All MHC proteins are dimers located in the plasma membrane, with a major part of the protein protruding on the extracellular side. The structures of class I and class II MHC proteins are related, although they have different components, as summarized in **Figure 37.27**.

Class II antigens consist of two chains, α and β, whose combination generates an overall structure in which there are two extracellular domains. The I-A region contains genes for the $A_{\beta 1}$, A_α, and $E_{\beta 1}$ chains; and the I-E region has the gene for E_α. Another class II antigen is coded by genes $A_{\beta 2}$ and $E_{\beta 2}$.

All class I MHC proteins consist of a dimer between the class I chain itself and the $\beta 2$ microglobulin protein. The class I chain is a 45,000 dalton transmembrane component that has three **external domains** (each ~90 amino acids long, one of which interacts with $\beta 2$ microglobulin), a **transmembrane region** of ~40 residues, and a short **cytoplasmic domain** of ~30 residues that resides within the cell.

The $\beta 2$ microglobulin is a secreted protein of 12,000 daltons. It is needed for the class I chain to be transported to the cell surface. Mice that lack the β_2 microglobulin gene have no MHC class I antigen on the cell surface.

The organization of class I genes summarized in **Figure 37.28** coincides with the protein structure. The first exon codes for a signal sequence (cleaved from the protein during membrane passage). The next three exons code for each of the external domains. The fifth exon codes for the transmembrane

tumor necrosis factor (TNF), a protein that is involved in inflammatory diseases. Within the class III S region lie the genes for the subunits of steroid-21-hydroxylase.

Next to the H2-D/L locus are the *Qa* and *Tla* loci. The ~10 *Qa* genes are closely related to the H2 genes. The *Tla* region which contains ~20 genes, but they are less well related to the classical H2 sequences.

The human MHC locus is >3800 kb, about twice the length of the murine locus. As outlined in

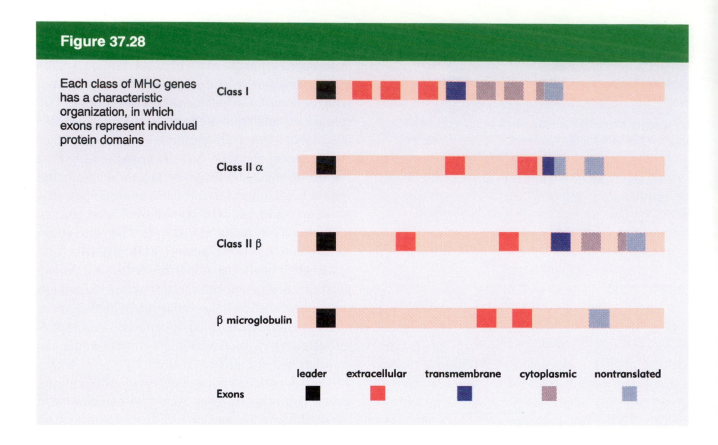

Figure 37.28

Each class of MHC genes has a characteristic organization, in which exons represent individual protein domains

Class I

Class II α

Class II β

β microglobulin

Exons

leader   extracellular   transmembrane   cytoplasmic   nontranslated

domain. And the last three rather small exons together code for the cytoplasmic domain. The only difference in the genes for human transplantation antigens is that their cytoplasmic domain is coded by only two exons.

The exon coding for the third external domain of the class I genes is highly conserved relative to the other exons. The conserved domain probably represents the region that interacts with $\beta2$ microglobulin, which explains the need for constancy of structure. This domain also exhibits homologies with the constant region domains of immunoglobulins.

What is responsible for generating the high degree of polymorphism in these genes? Most of the sequence variation between alleles occurs in the first and second external domains, sometimes taking the form of a cluster of base substitutions in a small region. One mechanism involved in their generation is gene conversion between class I genes.

Pseudogenes are present as well as functional genes; at present we have some way to go before estimating the total number of active genes in the region.

Like the class I genes, the class II and class III genes also are interrupted, with the exons related to protein domains (see Figure 37.28). There are fewer genes in these classes, ~10 each.

The gene for $\beta2$ microglobulin is located on a separate chromosome. It has four exons, the first coding for a signal sequence, the second for the bulk of the protein (from amino acids 3 to 95), the third for the last four amino acids and some of the nontranslated trailer, and the last for the rest of the trailer.

The length of $\beta2$ microglobulin is similar to that of an immunoglobulin V gene; there are certain similarities in amino acid constitution; and there are some (limited) homologies of nucleotide sequence between $\beta2$ microglobulin and Ig constant domains or type I gene third external domains. All the groups of genes that we have discussed in this chapter may have descended from a common ancestor that coded for a primitive domain.

Summary

Immunoglobulins and T-cell receptors are proteins that play analogous functions in the roles of B cells and T cells in the immune system. An Ig or TcR protein is generated by rearrangement of DNA in a single lymphocyte; exposure to an antigen recognized by the Ig or TcR leads to clonal expansion to generate many cells which have the same specificity as the original cell. Many different rearrangements occur early in the development of the immune system, creating a large repertoire of cells of different specificities.

Each immunoglobulin protein is a tetramer containing two identical light chains and two identical heavy chains. A TcR is a dimer containing two different chains. Each polypeptide chain is expressed from a gene created by linking one of many V segments via D and J segments to one of a few C segments. Ig L chains (either κ or λ) have the general structure V–J–C, Ig H chains have the structure V–D–J–C, TcR α and γ have components like Ig L chains, and TcR δ and β are like Ig H chains.

Each type of chain is coded by a large cluster of V genes separated from the cluster of D, J, and C segments. The numbers of each type of segment, and their organization, are different for each type of chain, but the principle and mechanism of recombination appear to be the same. The same nonamer and heptamer consensus sequences are involved in each recombination; the reaction always involves joining of a consensus with 23 bp spacing to a consensus with 12 bp spacing. Although considerable diversity is generated by joining different V, D, J segments to a C segment, additional variations are introduced in the form of changes at the junctions between segments and (in the case of immunoglobulins) by somatic mutation.

Allelic exclusion ensures that a given lymphocyte synthesizes only a single Ig or TcR. A productive rearrangement inhibits the occurrence of further rearrangements. Although the use of the V region is fixed by the first productive rearrangement, B cells switch use of C_H genes from the initial μ chain to one of the H chains coded farther downstream. This process involves a different type of recombination in which the sequences between the VDJ region and the new C_H gene are deleted. More than one switch occurs in C_H gene usage. At an earlier stage of Ig production, switches occur from synthesis of a membrane-bound version of the protein to a secreted version. These switches are accomplished by alternative splicing of the transcript.

Further reading

Reviews

An overall view of the mechanisms involved in generating Ig diversity has been given by **Tonegawa** (*Nature* **302**, 575–581, 1983).

The organization and reconstruction of Ig genes has been summarized by **Yancopoulos and Alt** (*Ann. Rev. Immunol.* **4**, 339–378, 1986), **Alt** *et al.* (*Science* **238**, 1079–1087, 1987), **Blackwell and Alt** (*Ann. Rev. Genetics* **23**, 605–636, 1989), **Schatz, Oettinger, and Schlissel** (*Ann. Rev. Immunol.* **10**, 359–383, 1992), and **Gellert** (*Ann. Rev. Genet.* **22**, 425–426, 1992).

Class switching was reviewed by **Davis, Kim, and Hood** (*Cell* **22**, 1–2, 1980) and **Shimizu and Honjo** (*Cell* **37**, 301–303, 1984).

Somatic mutation was reviewed by **French, Laskov, and Scharff** (*Science* **244**, 1152–1157, 1989).

An excellent review relating B cell differentiation to the underlying molecular events is by **Kocks and Rajewsky** (*Ann. Rev. Immunol.* **7**, 537–559, 1989).

The properties of transgenic mice with new immunoglobulin genes has been summarized by **Storb** (*Ann. Rev. Immunol.* **5**, 151–174, 1987).

Sequences of TcR genes and proteins have been reviewed by **Kronenberg** *et al.* (*Ann. Rev. Immunol.* **4**, 529–591, 1986), **Davis** (*Ann. Rev. Biochem.* **59**, 475–496, 1990), and **Raulet** (*Ann. Rev. Immunol.* **7**, 175–207, 1989). The interaction of T-cell receptor with antigen and MHC has been reviewed by **Marrack and Kappler** (*Science* **238**, 1073–1079, 1987).

The organization of histocompatibility genes was analyzed by **Steinmetz and Hood** (*Science* **222**, 727–732, 1983) and by **Flavell** *et al.* (*Science* **233**, 437–443, 1986).

Discoveries

A deletion model for VJ joining was presented by **Max, Seidman, and Leder** (*Proc. Nat. Acad. Sci. USA* **76**, 3340–3344, 1979); an inversion model was suggested by **Lewis** *et al.* (*Science* **228**, 677–680, 1985).

The use of gene conversion to generate diversity in chicken light chains was analyzed by **Reynaud** *et al.* (*Cell* **48**, 379–388, 1987).

T-cell receptor genes were analyzed by **Hood, Kronenberg, and Hunkapiller** (*Cell* **40**, 225–229, 1985).

Generation of junction sequences was characterized by **Roth** *et al.* (*Cell* **70**, 983–991, 1992).

CHAPTER 38

Gene regulation in development: gradients and cascades

Development begins with a single fertilized egg, but gives rise to cells that have different development fates. The problem of early development is to understand, in terms of molecular events, how this asymmetry is introduced: how a single initial cell gives rise within a few cell divisions to progeny cells that have different properties from one another. The means by which asymmetry is generated varies with the type of organism. The egg itself may be homogeneous, with the acquisition of asymmetry depending on the process of the initial division cycles, as in the case of mammals. Or the egg may have an initial asymmetry in the distribution of its cytoplasmic components, which in turn gives rise to further differences as development proceeds, as in the case of *Drosophila*.

The paradigm for considering the molecular basis for development is to suppose that each cell type may be characterized by its pattern of gene expression, that is, by the particular gene products that it produces. The principal level for controlling gene expression is at transcription. In terms of the mechanism for controlling eukaryotic transcription (as discussed previously in Chapter 30), we suppose that regulation resides with transcription factors. We may generally include within this rubric a variety of regulator proteins, which could act to change the structure of a promoter region, to initiate transcription at a promoter, to regulate

the activity of an enhancer, or indeed sometimes to repress the action of transcription factors. However, the regulators of transcription most often prove to be DNA-binding proteins that activate transcription at particular promoters or enhancers.

The systematic manner in which these regulators are turned on and off to form circuits that determine body parts has been worked out in some detail in *D. melanogaster*. The basic principle is that a series of events resulting from the initial asymmetry of the egg is translated into the control of gene expression so that specific regions of the egg acquire different properties. The details by which asymmetry is translated into control of gene expression differ for each of four systems that have been characterized in the insect egg: it may involve localization within the egg of factors that control transcription or translation or localized control of the activity of such factors. But the end result is the same: spatial and temporal regulation of gene expression.

This initial stage of development is succeeded by a stage at which the identities of parts of the embryo are determined: regions are defined whose descendents will form particular body parts. The genes that regulate this process are identified by the segmentation and homeotic loci of the fly, in which mutations cause one body part to be absent, to be duplicated, or to develop as another. Such loci are therefore prime candidates for genes whose function is to provide regulatory 'switches', analogous (for example) to the switch between the lambda lytic/lysogenic cascades. Described originally in terms of their genetics, these genes now have been isolated, and many of their products prove to be regulators of transcription. These genes act upon one another in a hierarchical manner, but they also act upon other genes whose products are actually responsible for the formation of pattern. These are segmentation genes that code for kinases, cytoskeletal elements, secreted proteins, and transmembrane receptors.

Viewing the process as a whole, we see that the establishment of differences in the patterns of transcription in different regions of the embryo leads to a *cascade* of control, when regulatory events are connected so that a gene turned on (or off) at one stage itself controls expression of other genes at the next stage. Formally, such a cascade resembles those described previously for bacteriophages or for bacterial sporulation (as discussed previously in Chapters 14 and 17), although it is more complex in the case of eukaryotic development. In this paradigm, the common feature of regulatory proteins is that they are transcription factors that regulate the expression of other transcription factors (as well as other target proteins). As in the case of prokaryotic regulation, the basic relationship between the regulator protein and the target gene is that the regulator recognizes a short sequence in the DNA of the promoter (or an enhancer) of a target gene. All of the targets for a particular regulator are identified by their possession of a copy of the appropriate consensus sequence.

The development of an adult organism from a fertilized egg follows a predetermined pathway, in which specific genes are turned on and off at particular times. From the perspective of mechanism, we have most information about the control of transcription. However, subsequent stages of gene expression are also targets for regulation; there may be regulation of processing or transport of RNA, or of stability or translatability.

Of these post-transcriptional stages, we have most information about regulation of translation. But we note that, although we can describe a cascade in which many members are transcription factors, there are stages at which gene expression is controlled at other levels, and the cascade is connected to other types of signalling, including cell–cell interactions that define boundaries between groups of cells.

The mechanics of development in terms of cellular events are different in different types of species, but we assume that the principle established with *Drosophila* will hold in all cases: that a regulatory cascade determines the appropriate pattern of gene expression in cells of the embryo and ultimately of the adult. Indeed, genes in organisms as distant as flies and mammals have a common evolutionary origin, and play related roles in development, in spite of the differences in the process of development.

Genes involved in regulating development are identified by mutations that are lethal early in development or that cause the development of abnormal structures. A mutation that affects the development of a particular body part attracts our attention because a single body part is a complex structure, requiring expression of a particular set of many genes. Single mutations that influence the structure of the entire body part therefore identify potential regulator genes that switch or select between developmental pathways. In *Drosophila*, the body part that is analyzed is the segment, the basic unit that can be seen looking at the adult fly. Mutations fall into (at least) three groups, defined by their effect on the segmental structure:

◆ **Maternal genes** are expressed during oogenesis by the mother. They may act upon or within the maturing oocyte.

◆ **Segmentation genes** are expressed after fertilization. Mutations in these genes alter the number or polarity of segments. Three groups of segmentation genes act sequentially to define increasingly smaller regions of the embryo.

◆ **Homeotic genes** control the identity of a segment, but do not affect the number, polarity, or size of segments. Mutations in these genes cause one body part to develop the phenotype of another part.

The genes in each group act successively to define the properties of increasingly more restricted parts of the embryo. The maternal genes define broad regions in the egg; differences in the distribution of maternal gene products control the expression of segmentation genes; and the homeotic genes act on segment identity at around the time that the last group of segmentation genes are defining the segments.

A gradient must be converted into discrete compartments

The basic question of *Drosophila* development is illustrated in **Figure 38.1** in terms of three stages of development: the egg, the larva, and the adult fly.

At the start of development, **gradients** are established in the egg along two axes, **anterior–posterior** and **dorso-ventral**. The anterior end of the egg will become the head of the adult; the posterior end will become the tail. The dorsal side is on top (looking down on a larva); the ventral side is underneath. The gradients consist of molecules (proteins or RNAs) that are differentially distributed in the cytoplasm. The gradient responsible for anterior–posterior development is established soon after fertilization; the dorso-ventral gradient is established a little later. It is only a modest over-simplification to say that the anterior–posterior systems control positional information along the larva, while the dorso-ventral system regulates tissue differentiation (that is, the specification of distinct embryonic tissues, including mesoderm, neuroectoderm, and dorsal ectoderm).

Insect development involves two quite different types of structure. The first part of development is concerned with elaborating the larva; then the larva metamorphoses into the fly. This means that the structure of the embryo (the larva) is quite different from the structure of the adult (the fly), in contrast with development of (for example) mammals, where the embryo develops the same body parts that are found in the adult. As the larva develops, it forms some body parts that are exclusively larval (they will not give rise to adult tissues; often they are polyploid), while other body parts are the progenitors that will metamorphose into adult structures (they are usually diploid). In spite of this idiosyncratic process, the same general principles appear to govern insect development and vertebrate development, and we discover relationships between *Drosophila* regulators and mammalian regulators.

Discrete regions are determined in the embryo that correspond to parts of the adult body. They are shown in terms of the superficial organization of the larva in the middle panel of Figure 38.1. Bands of **denticles** (small hairs) are found in a particular pattern on the surface (cuticle) of the larva. The cuticular pattern has features that are determined by both the anterior–posterior axis and the dorso-ventral axis:

◆ Along the anterior–posterior axis, the denticles form discrete bands. Each band corresponds to a segment of the adult fly: in fact, the 11 bands of denticles correspond on a 1:1 basis with the 11 segments of the adult.

◆ Along the dorso-ventral axis, the denticles that extend from the ventral surface are coarse; those that extend from the dorsal surface are much finer.

Figure 38.1

Overview: gradients in the egg are translated into segments on the anteroposterior axis and into specialized structures on the dorso-ventral axis of the larva, and then into the segmented structure of the adult fly.

The egg
Polarities are established along anterior-posterior (head-tail) axis and dorsal-ventral (back-abdomen)

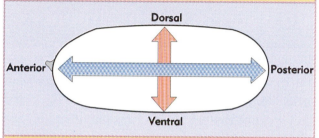

The larva
Mouth parts are at anterior, tail parts at posterior. Bands of denticles extend from the ventral side, and identify segmentation units along the anterior-posterior axis

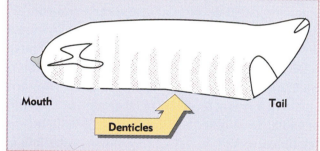

Adult fly
Segmented structure has 3 thoracic segments and 8 abdominal segments

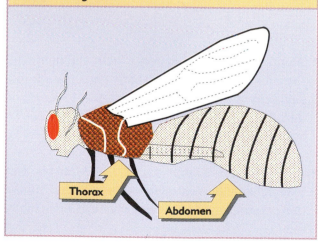

Although the cuticle represents only the surface body layer, its structure is diagnostic of the overall organization of the embryo in both axes. Much of the analysis of phenotypes of mutants in *Drosophila* development has therefore been performed in terms of the distortion of the denticle patterns along one axis or the other.

The difference in form between the gradients of the egg and the segments of the adult poses some prime questions. How are gradients established in the egg? And how is a continuous gradient converted into discrete differences that define individual cell types? How can a large number of separate compartments develop from a single gradient?

The nature of the gradients, and their ability to affect the development of a variety of cell types located throughout the embryo, depend upon some idiosyncratic features of *Drosophila* development. The early stages are summarized in **Figure 38.2**.

At fertilization the egg possesses the two parental nuclei and is distinguished at the posterior end by the presence of a region called polar plasm. For the first 9 divisions, the nuclei divide in the common cytoplasm. Material can diffuse in this cytoplasm (although there are probably constraints imposed by cytoskeletal organization). The region of the polar plasm is set aside; at division 7, some nuclei migrate into the polar plasm, where they become precursors to germ cells. After division 9, nuclei migrate and divide to form a layer at the surface of the egg. Then they divide 4 times, after which membranes surround them to form somatic cells.

Up to the point of cellularization, the nuclei effectively reside in a common cytoplasm. At the stage of the cellular blastoderm, the first discrete compartments become evident, and at this time particular regions of the egg are **determined** to become particular types of adult structures. (Determination is progressive and gradual; over the next few cell divisions, the fate of individual regions of the egg becomes increasingly restricted). At the start of this process, nuclei migrate to the surface to form the

Figure 38.2

The early development of the *Drosophila* egg occurs in a common cytoplasm until the stage of cellular blastoderm.

Fertilized egg
has two nuclei

90 mins

Divisions 1-8
Nuclei divide in common cytoplasm (syncytium). Nuclei in polar plasm (pink) become germ cell precursors.

150 mins

Syncytial Blastoderm
Nuclei migrate to periphery and divide; 4 further divisions occur (in close but not perfect synchrony).

195 mins

Cellular Blastoderm
Membranes surround nuclei to form monolayer of ~6000 somatic cells; there is no ordered pattern to organization of cells.

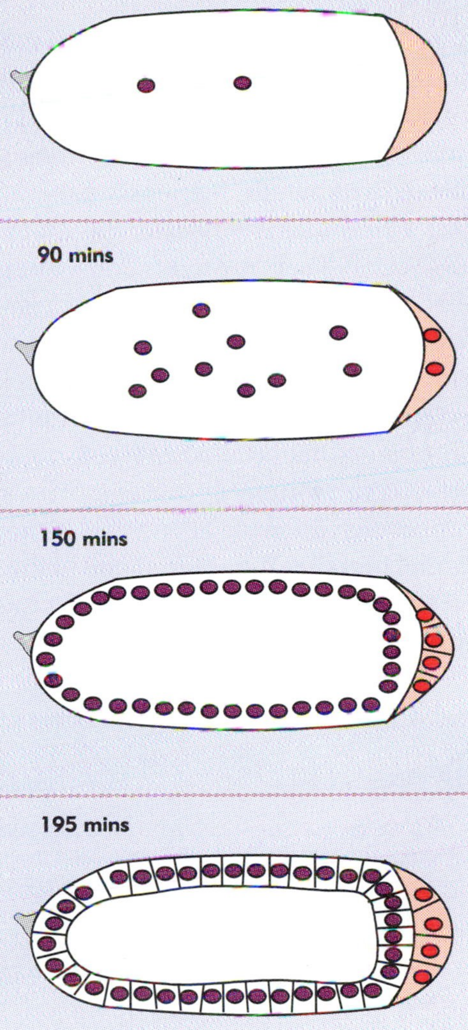

monolayer of the blastoderm, but they do not do so in any predefined manner. It is therefore the location in which the nuclei find themselves at this stage that determines what types of cells their descendants will become. *In effect, a nucleus determines its position in the embryo by reference to the anterior–posterior and dorso–ventral gradients, and behaves accordingly.*

Maternal gene products establish gradients in early embryogenesis

An initial asymmetry is imposed on the *Drosophila* oocyte during oogenesis. **Figure 38.3** illustrates the structure of a follicle in the *Drosophila* ovary. This represents the result of a single meiosis, which is followed by 4 mitotic divisions to generate 16 haploid cells; 1 becomes the oocyte, the other 15 become 'nurse cells'. There are tight connections, sometimes called 'cytoplasmic bridges' or 'ring canals' between the oocyte and the nurse cells that are adjacent to it. Cytoplasmic material, including protein and RNA, passes from the nurse cells to the oocyte; the accumulation of such material accounts for a considerable part of the volume of the egg. The cytoplasmic connections are made at one end of the oocyte, and this end becomes the anterior end of the egg.

Genes that are expressed within the mother fly are important for early development. They are called maternal genes, and are identified by **female sterile** mutations, which is to say that they do not affect the mother itself, but are required in order to have progeny. Females with such mutations lay eggs that fail to develop into adults; the embryos can be recognized by defects in the cuticular pattern, and they die during development.

The common feature in all maternal genes is that they are expressed prior to fertilization (although their products may act either at the time of expression or be stored for later use). The maternal genes are divided into two classes, depending on their site of expression. Genes that are expressed in somatic cells of the mother that affect egg development are called **maternal somatic genes**. For example, they may act in the follicle cells. Genes that are expressed within the germline are called **maternal germline genes**. These genes may act either in the nurse cell or the oocyte (both of which are products of meiosis). Some genes act at both stages.

Four groups of genes concerned with the development of particular regions of the embryo can be identified by mutations in maternal genes. The genes in each group can be organized into a pathway that reflects their order of action, by conventional genetic tests (such as comparing the properties of double mutants with the individual mutants) or by biochemical assays (showing which mutants contain components that can bypass the stages that are blocked in other mutants). Sometimes the genes that act early in such a pathway are said to be 'upstream', while those that act later are said to be 'downstream'.

The components of these pathways are summarized in **Figure 38.4**, which shows that there is a common principle to their operation. *Each pathway is initiated by localized events, which may occur either within or outside the egg; this results in the localization of a signal within*

Figure 38.3

A *Drosophila* follicle contains an outer surface of follicle cells that surround nurse cells that are in close contact with the oocyte. Nurse cells are connected by cytoplasmic bridges to each other and to the anterior end of the oocyte. Follicle cells are somatic; nurse cells and the oocyte are germline in origin.

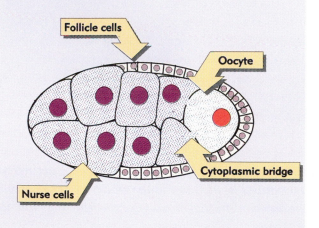

Figure 38.4

Each of the four maternal systems is initiated by genes whose products act outside the egg; the pathway is carried into the egg, where in each pathway the localized product that is the morphogen may either be a receptor or a regulator of gene expression. The final component is a transcription factor, which acts on zygotic targets that are responsible for the next stage of development.

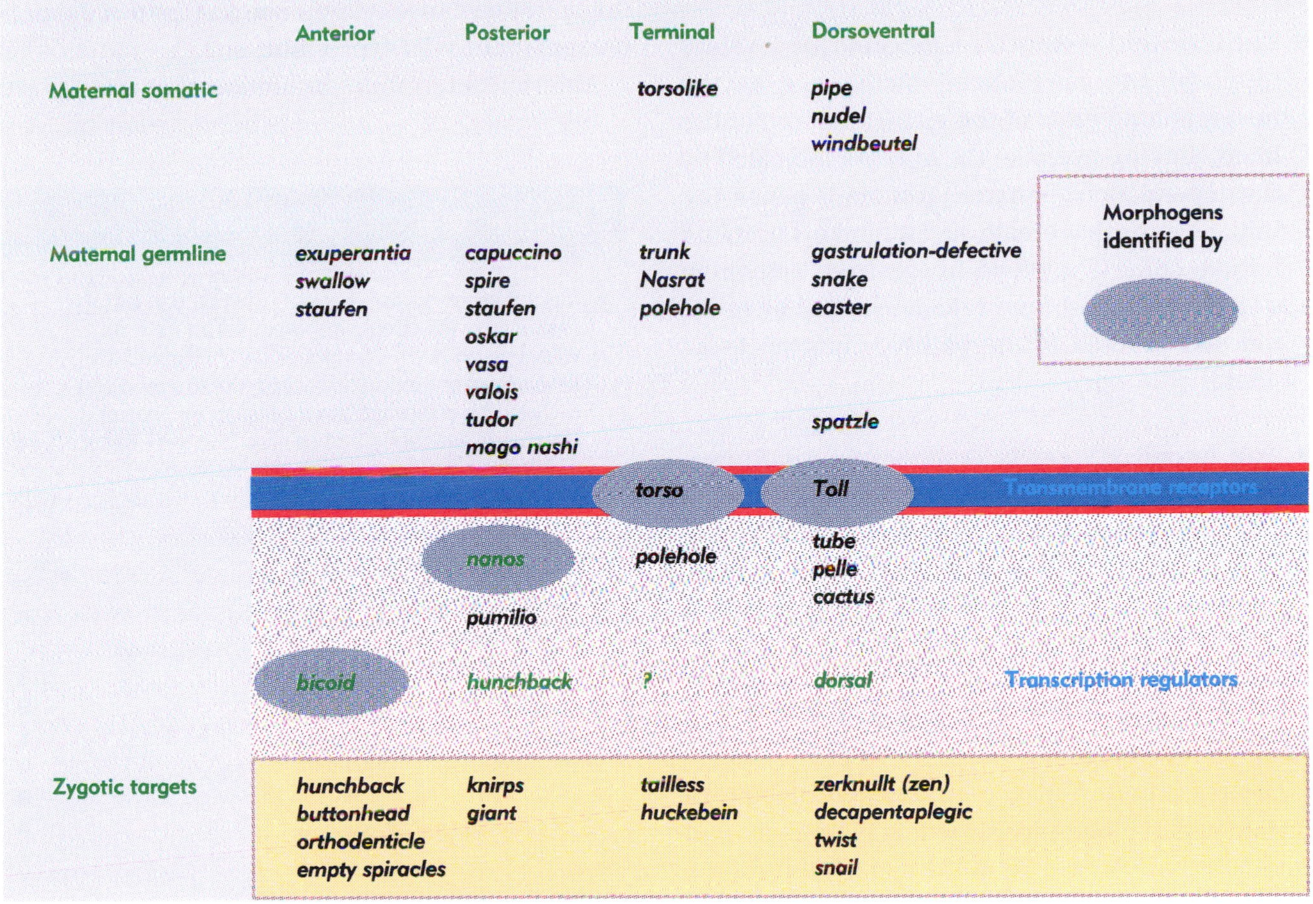

	Anterior	Posterior	Terminal	Dorsoventral	
Maternal somatic			torsolike	pipe	
				nudel	
				windbeutel	
Maternal germline	exuperantia	capuccino	trunk	gastrulation-defective	Morphogens identified by
	swallow	spire	Nasrat	snake	
	staufen	staufen	polehole	easter	
		oskar			
		vasa			
		valois			
		tudor		spatzle	
		mago nashi			
			torso	Toll	Transmembrane receptors
		nanos	polehole	tube	
				pelle	
				cactus	
		pumilio			
	bicoid	hunchback	?	dorsal	Transcription regulators
Zygotic targets	hunchback	knirps	tailless	zerknullt (zen)	
	buttonhead	giant	huckebein	decapentaplegic	
	orthodenticle			twist	
	empty spiracles			snail	

the egg. This signal takes the form of a protein with an asymmetric distribution; this is called a **morphogen**. Formally, we may define a morphogen as a protein whose local concentration (or activity) causes the surrounding region to take up a particular structure or fate. In each of these systems, the morphogen either is a transcriptional regulator or leads to the activation of a transcription factor in the localized region. Three systems are concerned with the anterior–posterior axis, and one with the dorso-ventral axis:

◆ The **anterior system** is responsible for development of the head and thorax. The maternal germline products are required to localize the *bicoid* product at the anterior end of the egg. In fact, *bicoid* RNA is transcribed in nurse cells and transported into the oocyte. Bicoid protein is the morphogen: it functions as a transcriptional regulator, and controls expression of the gene *hunchback* (and probably also other segmentation and homeotic genes).

◆ The **posterior system** is responsible for the

segments of the abdomen. The nature of the initial asymmetrical event is not clear. A large number of products act to cause the localization of the product of *nanos*, which is the morphogen. This leads to localized repression of expression of *hunchback* (via control of translation of the mRNA).

◆ The **terminal system** is responsible for development of the specialized structures at the unsegmented ends of the egg (the acron at the head, and the telson at the tail). As indicated by the dependence on maternal somatic genes, the initial events that create asymmetry occur in the follicle cells. They lead to localized activation of the transmembrane receptor coded by *torso*; the end product of the pathway has yet to be identified.

◆ The fourth system is responsible for **dorso-ventral** development. The pathway is initiated by a signal from a follicle cell on the ventral side of the egg. It is transmitted through the transmembrane receptor coded by *Toll*. This leads to a gradient of activation of the transcription factor produced by *dorsal* (by controlling its localization within the cell).

About 30 maternal genes involved in pattern formation have been identified; it is unlikely that there are many more. All of the components of the four pathways are maternal, so we see that the systems for establishing the initial pattern formation all depend on events that occur prior to fertilization. The two body axes are established independently. Mutations that affect polarity cause posterior regions to develop as anterior structures or ventral regions to develop in dorsal form, but mutations affecting one axis do not affect the other axis. On the anterior–posterior axis, the anterior and posterior systems provide opposing gradients, with sources at the anterior and posterior ends of the larva, respectively, that control development of the segments of the body. Defects in either system affect the body segments. The terminal and dorso-ventral systems operate independently of the other systems.

The existence of localized concentrations of materials needed for development is indicated by the success of the rescue protocol summarized in **Figure 38.5**. Material is removed from a wild-type embryo and injected into the embryo of a mutant that is defective in early development. If the mutant embryo develops normally, we may conclude that the mutation causes a deficiency of material that is present in the wild-type embryo.

Anterior determination has been characterized

Figure 38.5

Mutant embryos that cannot develop can be rescued by injecting cytoplasm taken from a wild-type embryo. The donor can be tested for time of appearance and location of the rescuing activity; the recipient can be tested for time at which it is susceptible to rescue and the effects of injecting material at different locations.

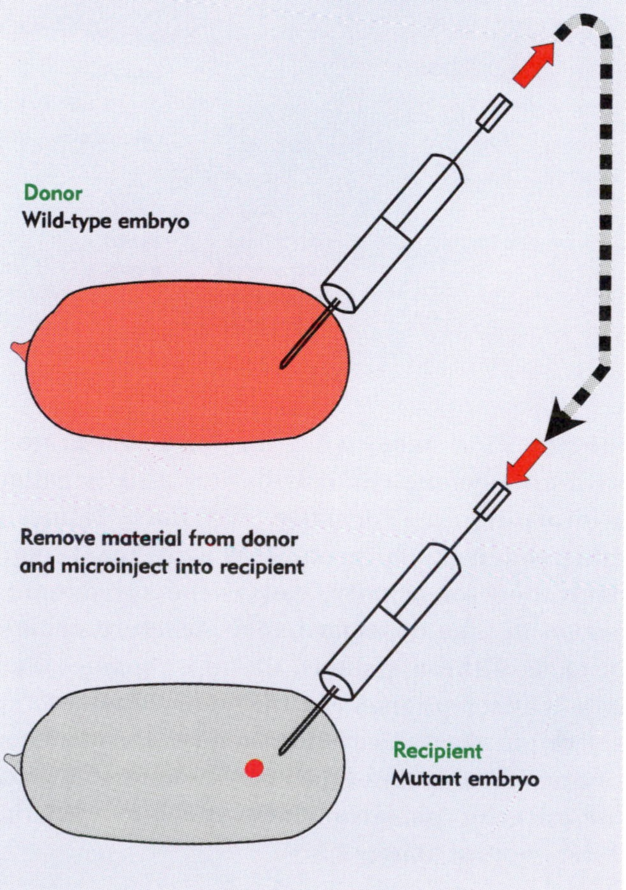

Donor
Wild-type embryo

Remove material from donor and microinject into recipient

Recipient
Mutant embryo

by this technique. *bicoid* mutants do not develop heads; but the defect can be remedied by injecting mutant eggs with cytoplasm taken from the anterior tip of a wild-type embryo. Indeed, anterior structures can be developed elsewhere in the mutant embryo by injection of wild-type anterior cytoplasm. The extent of the rescue depends on the amount of wild-type cytoplasm injected. And the efficacy of the donor cytoplasm depends on the number of wild-type *bicoid* genes carried by the donor. This suggests that the anterior region of a wild-type embryo contains a concentration of the *bicoid* product that depends on the gene dosage. The active component in the preparation can be purified and identified; in the example of *bicoid*, injection of purified *bicoid* mRNA can substitute for the anterior cytoplasm.

The product of *bicoid* establishes a gradient with its source (and therefore the highest concentration) at the anterior end of the embryo. The *bicoid* gene is transcribed in nurse cells; mRNA is transported through the cytoplasmic bridges into the oocyte. The RNA is localized at the anterior tip of the embryo, but it is not translated during oogenesis. Translation begins soon after fertilization. The protein then establishes a gradient along the embryo, as indicated in **Figure 38.6**. The gradient could be produced by diffusion of the protein product from the localized source at the anterior tip. The gradient would take its characteristic exponential form if the protein product has relatively low stability. The gradient is established by division 7, and remains stable until after the blastoderm stage. Gradients in the same direction, but more shallow, are found also for the other genes, *swa* and *exu*, needed for anterior development. The functions of these genes may be concerned with transporting *bicoid* mRNA into the oocyte, or with limiting its diffusion from the anterior end. The localization of the *bicoid* RNA to the anterior end depends upon sequences in the 3′ untranslated region.

What is the consequence of establishing the bicoid gradient? The gradient can be increased or decreased by changing the number of functional gene copies in the mother. The concentration of bicoid protein is correlated with the development

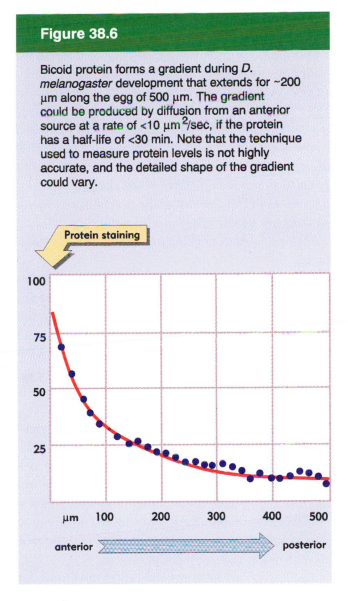

Figure 38.6

Bicoid protein forms a gradient during *D. melanogaster* development that extends for ~200 µm along the egg of 500 µm. The gradient could be produced by diffusion from an anterior source at a rate of <10 µm^2/sec, if the protein has a half-life of <30 min. Note that the technique used to measure protein levels is not highly accurate, and the detailed shape of the gradient could vary.

of anterior structures. Weakening the gradient causes anterior segments to develop more posterior-like characteristics; strengthening the gradient causes anterior-like structures to extend farther along the embryo. Thus the bicoid protein behaves as a morphogen that determines anterior–posterior position in the embryo in a concentration-dependent manner.

The fate of cells in the anterior part of the embryo is determined by the concentration of bicoid protein in which they find themselves. The *bicoid* product is a sequence-specific DNA-binding protein that regulates transcription of genes to whose promoters it binds. The immediate effect of *bicoid* is exercised on other genes that in turn regulate the development of yet further genes. A major target for *bicoid* is the gene *hunchback* (see later); other targets remain to be identified. Transcription of *hunchback* is turned on by *bicoid* in a dose-dependent manner, that is, *hunchback* is activated above a certain threshold of bicoid protein. This allows a gradient to provide a spatial on–off switch that affects gene expression. *In this way, quantitative differences in the amount of the morphogen (bicoid protein) are transformed into qualitatively different states (cell structures) during embryonic development.*

bicoid plays an **instructive** role in anterior development, since it is a positive regulator that is *needed for expression* of genes that in turn determine the synthesis of anterior structures. The effect of bicoid on hunchback is to produce a band of expression that occupies the anterior part of the embryo. The pattern of expression in the anterior system is summarized later, in Figure 38.12.

Posterior development depends on the expression of a large group of genes. Embryos produced by females who are mutant for any one of these genes develop normal head and thoracic segments, but lack the entire abdomen. Some of these genes are concerned with exporting material from the nurse cells to the egg; others are required to transport or to localize the material within the egg.

The posterior system resembles the anterior system in the basic nature of the initial event: a maternal mRNA is localized at the posterior pole. This is the product of *nanos*, and provides the morphogen. There are two important differences between the systems. Localization is more complex than in the case of the anterior system, because posterior determinants that originate in the nurse cells must be transported the full length of the oocyte to the far pole. And nanos protein acts to prevent translation of a transcription factor (hunchback); its role is said to be **permissive**, since it functions to repress genes whose products would interfere with posterior development.

How do we know that *nanos* is the morphogen at the end of the pathway? Rescue experiments

(along the lines shown previously in Figure 38.5) with the mutants in the posterior group showed that in all but one case the cytoplasm of the nurse cell contained the posterior determinant (although it was absent from the posterior end of the oocyte itself). This indicates that these mutants all act in some subsidiary role, most probably concerned with transporting or localizing the morphogen in the egg. The exception was *nanos*, whose mutants did not contain any posterior-rescuing activity. Purified *nanos* RNA can rescue mutants in any of the other posterior genes, indicating that it is the last, or most downstream, component in the pathway. Indeed, injection of *nanos* RNA into other locations in embryos can induce the formation of abdominal structures, showing that it provides the morphogen.

The upper part of **Figure 38.7** shows the localization of *nanos* mRNA at the posterior end of an early embryo. It depends upon sequences in the 3′ untranslated region of the mRNA, which presumably are recognized by the products of other posterior genes. But the localization poses a dilemma: *nanos* activity is required for development of abdominal segments, that is, for structures occupying approximately the posterior half of the embryo. How does *nanos* RNA at the pole control abdominal development? The lower part of Figure 38.7 shows that nanos protein diffuses from the site of translation to form a gradient that extends along the abdominal region.

Both *bicoid* and *nanos* act on the expression of the *hunchback* gene. *hunchback* codes for a repressor of transcription: its presence is needed for formation of anterior structures (in the region of the thorax), and its absence is required for development of posterior structures. It has a complex pattern of expression. It is transcribed during oogenesis to give an mRNA that is uniformly distributed in the egg. After fertilization, the *hunchback* pattern is changed in two ways. The bicoid gradient activates synthesis of *hunchback* RNA in the anterior region. And nanos prevents translation of hunchback mRNA in the posterior region; a result of this inhibition is that the mRNA is degraded. The anterior and posterior systems together therefore

Figure 38.7

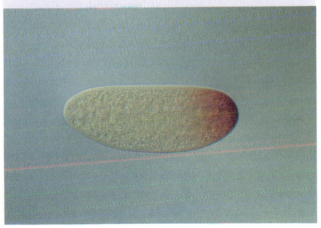

nanos products are localized at the posterior end of a *Drosophila* embryo. The upper photograph shows the tightly localized RNA in the very early embryo (at the time of the 3rd nuclear division). The lower photograph shows the spreading of *nanos* protein at the 8th nuclear division. Photograph kindly provided by Ruth Lehmann.

enhance hunchback levels in the anterior half of the egg, and remove it from the posterior half. We see later that the significance of this distribution lies with the genes that hunchback regulates. It directly represses the genes *knirps* and (probably) *giant*, which are needed to form abdominal structures. Thus *the basic role of* hunchback *is to repress formation of abdominal structures by preventing the expression of* knirps *and* giant *in more anterior regions.*

Two types of pattern-determining event occur at the posterior pole. The polar plasm contains two morphogens: the posterior determinant (nanos) controls abdominal development; and an unidentified signal controls formation of the pole cells, which will give rise to the germline (see Figure 38.2). All of the posterior genes except *nanos* and *pumilio* are required for both processes, that is, they are defective in both abdominal development and pole cell formation. The two defects represent branches of the pathway. *nanos* and *pumilio* identify the abdominal branch. The *pumilio* product may be needed for the diffusion of nanos protein, or to activate or stabilize it, because *pumilio* mutants contain posterior determinant activity, but do not form complete abdomens.

The pole plasm is constructed by a series of events in which one product is responsible for localizing the next. **Figure 38.8** correlates the order of genes in the genetic pathway with the activities of their products in the embryo. The functions *spir* and *capu* are needed for Staufen protein to be localized at the pole. Staufen protein in turn localizes *oskar* RNA; possibly a complex of Staufen protein and *oskar* RNA is assembled. These functions are needed to localize Vasa, which is an RNA-binding protein. Its specificity and targets are not known.

If *oskar* is over-expressed or mislocalized in the embryo, it induces germ cell formation at ectopic (atypically located) sites. It requires only the functions *vasa* and *tudor*. This implies that all of the activities that precede *oskar* in the pathway are needed only to localize *oskar* RNA; the ability to form both pole cells and induce abdominal structures is possessed by *oskar*, in conjunction with *vasa* and *tudor* (and of course any components that are ubiquitously expressed in the egg). One effect of *oskar* function is to localize Vasa protein at the posterior end. The functions of *valois* and *tudor* are not known, but it is possible that *valois* is off the main pathway.

The pathway branches at *tudor*. The localization of *nanos* mRNA, and the effect of *pumilio* on its expression, are sufficient to specify abdominal

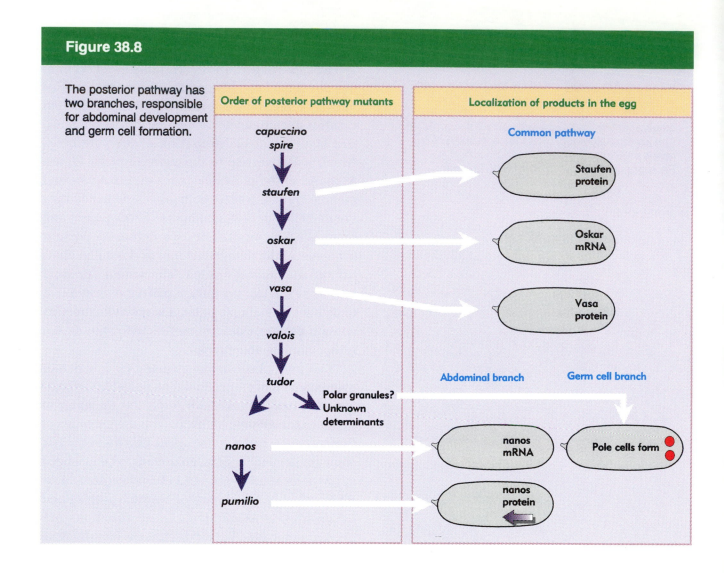

Figure 38.8

The posterior pathway has two branches, responsible for abdominal development and germ cell formation.

Order of posterior pathway mutants

capuccino
spire

staufen

oskar

vasa

valois

tudor

Polar granules?
Unknown
determinants

nanos

pumilio

Localization of products in the egg

Common pathway

Staufen protein

Oskar mRNA

Vasa protein

Abdominal branch   Germ cell branch

nanos mRNA   Pole cells form

nanos protein

segments, as we just described. We do not know whether there are additional functions representing a separate branch for germ cell formation, or whether the pathway up to *tudor* is by itself sufficient. The group of genes that precedes *nanos* and *pumilio* is needed for the formation of structures called polar granules; these structures may play a role in pole cell formation.

Development of the dorsal–ventral pattern of the *Drosophila* embryo requires a group of 11 maternal genes that function to establish the dorsal–ventral axis between the time of fertilization and cellular blastoderm (see Figure 38.4). This system is necessary for the development of ventral structures that comprise particular tissue types including mesoderm and neuroectoderm. Mutants in any of these genes lack ventral structures, and have dorsal structures on the ventral side, as indicated in **Figure 38.9**. But injecting wild-type cytoplasm into mutant embryos rescues the defect and allows ventral structures to develop.

Many of these genes are expressed in the follicle cell and participate in an interaction between the follicle cell and the oocyte. The result of this interaction is to activate a component of the pathway on the ventral side of the oocyte. Thus wild-type cytoplasm rescues the defect because in effect it bypasses the need for the follicle–oocyte interactions.

The pathway is summarized in **Figure 38.10**. The interplay between the oocyte and follicle cells is

Figure 38.9

Wild-type *Drosophila* embryos have distinct dorsal and ventral structures. Mutations in genes of the dorsal group prevent the appearance of ventral structures, and the ventral side of the embryo is dorsalized. Ventral structures can be restored by injecting cytoplasm containing the Toll gene product.

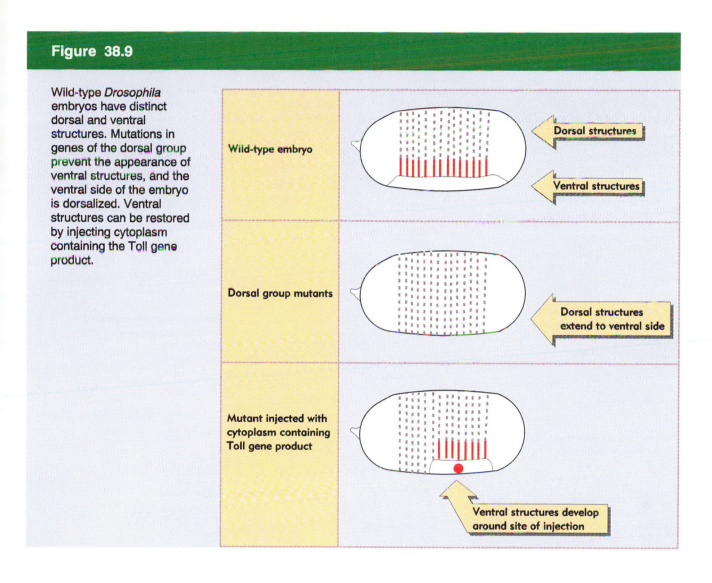

complex. Some genes that function in the oocyte are needed for proper development of the follicle cells; and if follicle cells are defective as a result of mutation in these genes, then dorso-ventral development in turn is aberrant. This happens because three maternal somatic genes (*pipe, nudel, windbeutel*) function in the follicle cell to send a signal to the oocyte. The egg does not receive the signal until after fertilization. The nature of the signal is not known, although we do know some of the events that occur at the oocyte.

The signal leads to a series of proteolytic cleavages that occur at the surface of the egg. The product of *snake* probably cleaves the product of *easter*, which in turn is activated to cleave the *spatzle* product. Cleavage activates spatzle, so that it can provide a ligand for a receptor coded by the *Toll* gene. This is the first component of the pathway that actually functions in the oocyte.

Rescue experiments identify *Toll* as the crucial gene that conveys the signal into the oocyte. *Toll⁻* mutants lack any dorso-ventral gradient, and injection of Toll induces the formation of dorso-ventral structures. The other genes of the dorsal group code for products that either regulate or are required for the action of *Toll*, but they do not establish the primary polarity.

There is a paradox in the distribution of Toll protein. *Toll* gene product activity is found in all parts of a donor embryo when cytoplasm is

Figure 38.10

The dorsoventral pathway involves interactions between follicle cells and the oocyte. The signal that causes spatzle to be available on the ventral side is not known, but when spatzle binds to Toll, it activates the morphogen. The end of the pathway is to transport the transcription factor dorsal into the nucleus.

Before fertilization

Genes in oocyte send (unidentified) signal to torpedo receptor in follicle cell

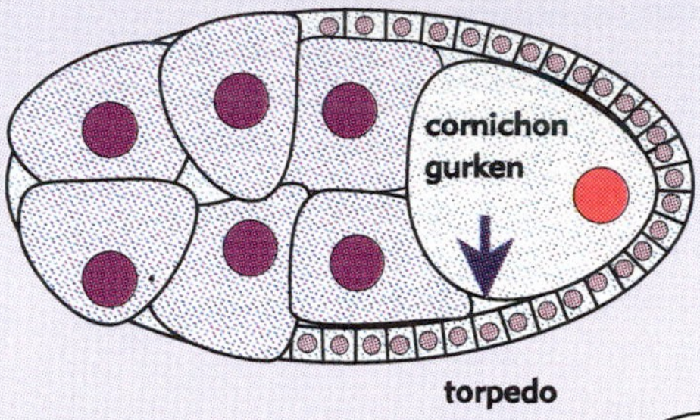

cornichon gurken

torpedo

pipe, nudel, windbeutel in follicle cell send (unidentified) signal on ventral side of oocyte

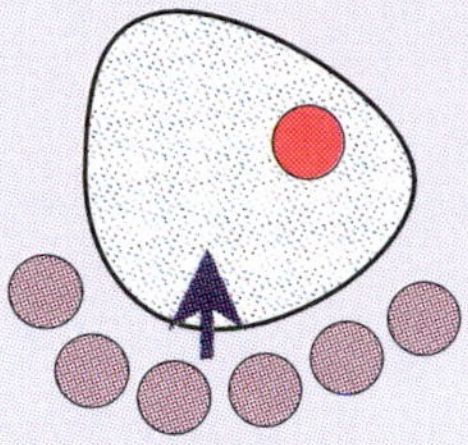

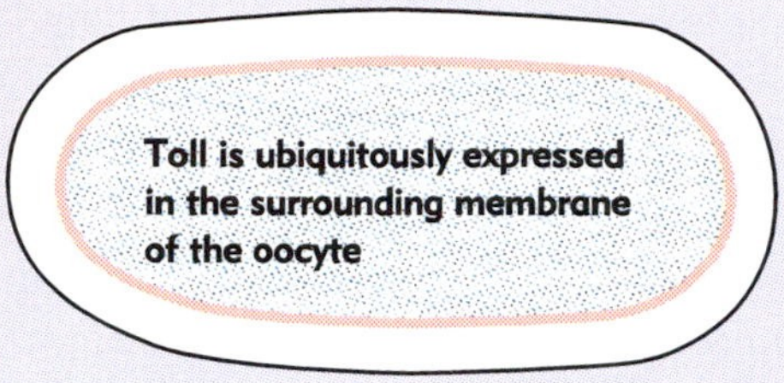

Toll is ubiquitously expressed in the surrounding membrane of the oocyte

After fertilization

gastrulation-defective function is unknown

Oocyte

snake, easter code for proteases that may activate *spatzle*

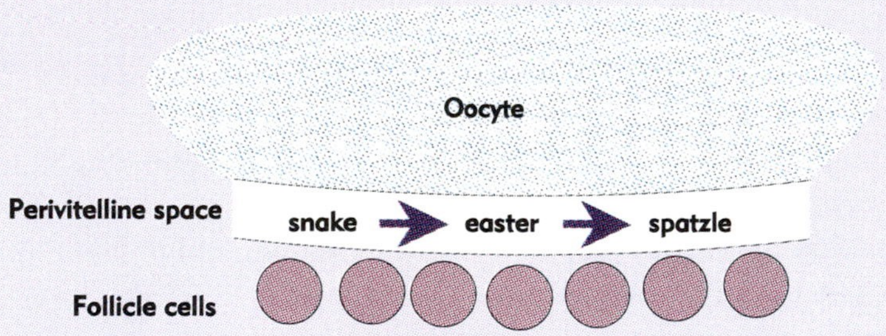

Perivitelline space   snake → easter → spatzle

Follicle cells

Spatzle activates Toll on the ventral side

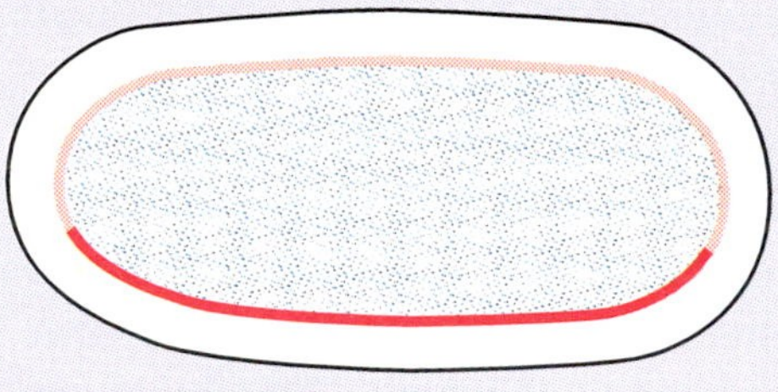

Toll acts through tube (function unknown) to activate pelle, a kinase that phosphorylates cactus

tube → pelle → cactus

Cactus releases dorsal, which enters nucleus

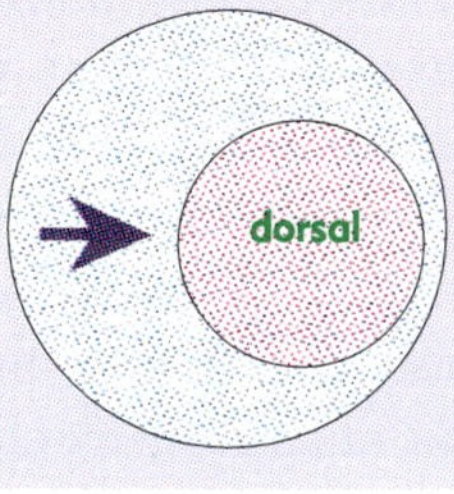

dorsal

extracted and tested by injection. Yet it induces ventral structures only in the appropriate location in normal development. An initial general distribution of *Toll* gene product must therefore in some way be converted into a gradient or axis of active product by local events.

The nature of Toll's activity is not known, but the binding of ligand is all that is needed to activate the ventral-determining pathway. The reaction occurs on the ventral side of the perivitelline space, which is the outermost layer of the oocyte. The spatzle ligand either cannot diffuse far from the site of cleavage, or perhaps it binds to Toll very rapidly, with the result that Toll is activated only on the ventral side of the embryo. Loss of function mutations in Toll are dorsalized, because the receptor cannot be activated. There are also dominant (*Toll*^{D}) mutations, which confer ventral properties on dorsal regions; these are gain-of-function mutations, which are ventralized because the receptor is constitutively active. By whatever means, Toll acts via *tube* and *pelle*. The function of *tube* is unknown; *pelle* codes for a kinase. The target for the kinase is probably the product of *cactus*, which is the final regulator for the transcription factor coded by *dorsal* (but it is possible that the pelle kinase acts as well or instead directly on dorsal protein).

Dorsal and cactus form an interacting pair of proteins that are related to the transcription factor NF-κB and its regulator IκB. We recollect from Chapter 30 that NF-κB is a member of the *rel* gene family; in fact, NF-κB consists of two subunits (related in sequence) which are bound by IκB in the cytoplasm. When IκB is phosphorylated, it releases NF-κB, which then moves into the nucleus, where it functions as a transcription factor of genes whose promoters have the κB sequence motif. (An example of the pathway is illustrated in Figure 30.6.) It seems likely that cactus regulates dorsal in the same way that IκB regulates NF-κB, that is, a cactus–dorsal complex is inert in the cytoplasm, but when cactus is phosphorylated, it releases dorsal, which enters the nucleus.

As a result of the activation of Toll, a gradient of dorsal nuclear activity is established, from ventral to dorsal side. On the ventral side of the embryo, dorsal is released to the nucleus, but on the dorsal side of the embryo it remains in the cytoplasm. The gradient is established at the stage of syncytial blastoderm, and becomes sharper during the transition to cellular blastoderm. The gradient is steep, and the proportion of dorsal protein that is in

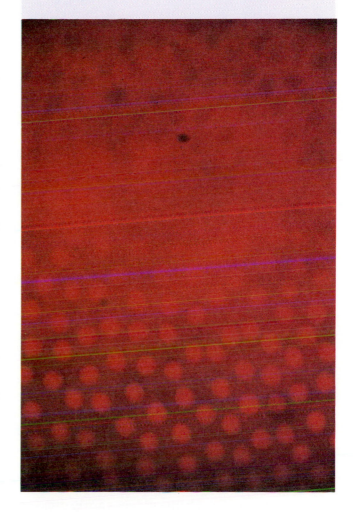

Figure 38.11

Dorsal protein forms a gradient of nuclear localization from ventral to dorsal side of the embryo. On the ventral side (lower) the protein identifies bright nuclei; on the dorsal side (upper) the nuclei lack protein and show as dark holes in the bright cytoplasm. Photograph kindly provided by Michael Levine.

Figure 38.12

In each axis-determining system, localized products in the egg cause other maternal RNAs or proteins to be broadly localized at syncytial blastoderm, and zygotic RNAs are transcribed in bands at cellular blastoderm.

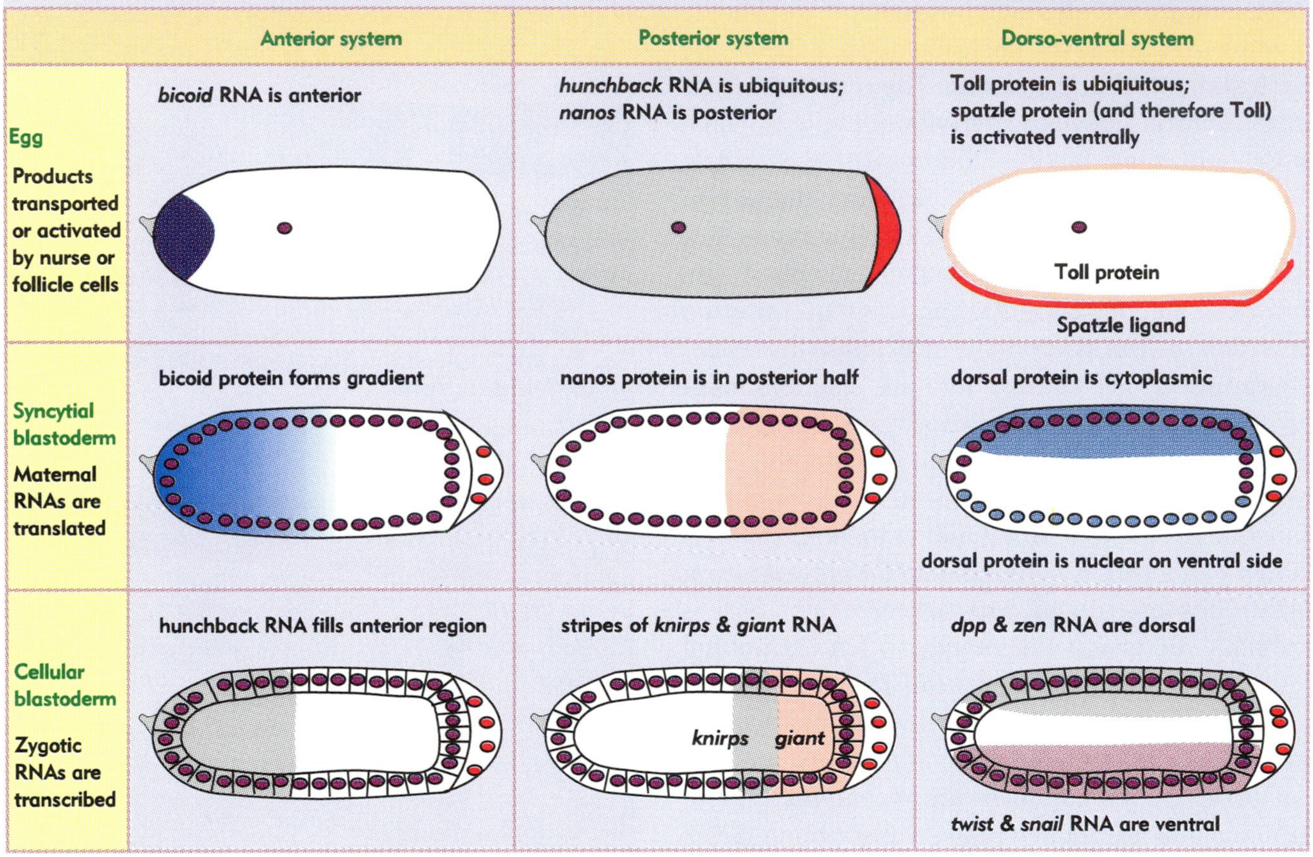

	Anterior system	Posterior system	Dorso-ventral system
Egg Products transported or activated by nurse or follicle cells	*bicoid* RNA is anterior	*hunchback* RNA is ubiquitous; *nanos* RNA is posterior	Toll protein is ubiqiuitous; spatzle protein (and therefore Toll) is activated ventrally Toll protein Spatzle ligand
Syncytial blastoderm Maternal RNAs are translated	bicoid protein forms gradient	nanos protein is in posterior half	dorsal protein is cytoplasmic dorsal protein is nuclear on ventral side
Cellular blastoderm Zygotic RNAs are transcribed	hunchback RNA fills anterior region	stripes of *knirps* & *giant* RNA knirps giant	*dpp* & *zen* RNA are dorsal *twist* & *snail* RNA are ventral

the nucleus correlates with the ventral phenotype that will be displayed by this region. An example of a gradient visualized by staining with antibody against dorsal protein is shown in **Figure 38.11**. The total amount of dorsal protein in the embryo does not change: the gradient is established solely by a redistribution of the protein between nucleus and cytoplasm.

Dorsal both activates and represses gene expression. It activates the genes *twist* and *snail*, which are required for the developmental of ventral structures. And it represses the genes *dpp* and *zen*, which are required for the development of dorsal structures; as a result, these genes are expressed

only in the 40% most dorsal of the embryo, which develops dorsal structures. The product of *dpp* is a member of the TGFβ growth factor family, and is controlled by dorsal so that dpp protein is secreted from cells only across the dorsal side; thus dpp is in effect the morphogen that induces synthesis of dorsal structures.

The terminal system is initiated in a way that is similar to the dorso-ventral system. A transmembrane receptor, coded by the *torso* gene, is produced by translation of a maternal RNA after fertilization. The receptor is localized throughout the embryo. It is activated at the poles by local production of an extracellular ligand. Torso protein has a kinase activity, which initiates a cascade that leads to local expression of the *tailless* and *huckebein* RNAs, which code for factors that regulate transcription.

The pattern of regulators at each stage of development for each of the systems is summarized in **Figure 38.12**. Two types of mechanism are used to create the initial asymmetry. For the anterior–posterior axis, an RNA is localized at one end of the egg (bicoid for the anterior system, nanos for the posterior system); localization depends upon the interaction of sequences in the 3′ end of the RNA with maternal proteins. In the case of the dorsoventral and terminal systems, a receptor protein is specifically activated in a localized manner, as the result of the limited availability of its ligand. All of these interactions depend on RNAs and/or proteins expressed from maternal genes.

The local event leads to the production of a morphogen, which forms a gradient, either quantitatively (bicoid) or by nucleocytoplasmic distribution (dorsal), or is localized in a broad restricted region (nanos); the extent of the region in which the morphogen is active is ~50% across the egg for each of these systems. The morphogens are translated from maternal RNAs, and development is

therefore still dependent on maternal genes up to this stage.

Establishing anterior–posterior and dorsal-ventral gradients is the first step in determining orientation and spatial organization of the embryo. Under the direction of maternal genes, gradients form across the common cytoplasm and influence the behavior of the nuclei located in it. The next step is the development of discrete regions that will give rise to different body parts. This requires the expression of the zygotic genome, and the loci that now become active are called *zygotic genes*. Genes involved at this stage are identified by segmentation mutants.

The products of the segmentation genes form bands that differentiate individual regions on the anterior–posterior axis; when we consider the results of the anterior and posterior systems together, we see that there are several broad regions (the two regions generated by the anterior system are adjacent to the two regions defined by the posterior system; see below). On the dorsoventral axis, there are three rather broad bands that define the regions in which the mesoderm, neuroectoderm, and dorsal ectoderm form (proceeding from the ventral to the dorsal side).

Cell fate is determined by compartments that form by the blastoderm stage

A 'fate map' of the embryo can be drawn at the blastoderm stage to identify regions in terms of the adult segments that will develop from the descendants of the embryonic cells. This is possible because, by this stage, cells have begun to acquire information about the pathways they will follow and the structures they will therefore form. This information derives from the maternal regulators described above, and then it is further refined by the actions of zygotic genes. The concept that is intrinsic in the fate map is that a region identified at blastoderm consists of a 'compartment' of cells

that will give rise specifically to a particular adult structure.

We can consider the development of *D. melanogaster* in terms of the two types of unit depicted in **Figure 38.13**: the segment and parasegment.

◆ The **segment** is a visible morphological structure. The adult fly consists of a series of clearly demarcated segments, and the larva has a series of corresponding segments separated by grooves. We are concerned primarily with the

Figure 38.13

Drosophila development proceeds through formation of compartments that form parasegments and segments.

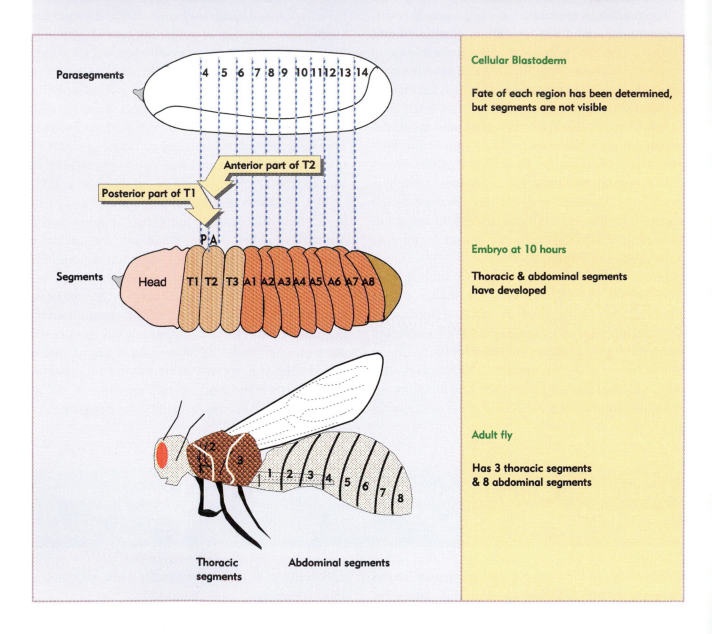

Parasegments

4 5 6 7 8 9 10 11 12 13 14

Anterior part of T2

Posterior part of T1

P A

Segments Head T1 T2 T3 A1 A2 A3 A4 A5 A6 A7 A8

Thoracic segments   Abdominal segments

Cellular Blastoderm

Fate of each region has been determined, but segments are not visible

Embryo at 10 hours

Thoracic & abdominal segments have developed

Adult fly

Has 3 thoracic segments & 8 abdominal segments

three thoracic (T) and eight abdominal (A) segments, about whose development most is known. The pattern of segmental units is determined by blastoderm, when the main mass of the embryo is divided into a series of alternating anterior (A) and posterior (P) compartments. Thus a segment consists of an A compartment succeeded by a P compartment; segment A3, for example, consists of compartments A3A and A3P.

◆ Another type of classification originates earlier, when divisions can first be seen at gastrulation. The embryo can be divided into **parasegments**, each consisting of a P compartment succeeded by an A compartment. Parasegment 8, for example, consists of compartments A2P and A3A. In the 5–6 hour embryo, shallow grooves on the surface separate the adjacent parasegments. When segments form at around 9 hours, the grooves deepen and move, so that each segmental boundary represents the center of a parasegment. Thus the anterior part of the segment is derived from one parasegment and the posterior part of the segment is derived from the next parasegment. In effect, the segmental units are initially evident as P–A pairs in parasegments, and then are recognized as A–P pairs in segments.

How are these compartments defined during embryogenesis? The general nature of segmentation mutants suggests that the functions of segmentation genes are to establish 'rules' by which segments form; *a mutation changes a rule in such a way as to cause many or all segments to form improperly.* The drastic consequences of segment malformation make these mutants embryonic lethals—they die at various stages before metamorphosis into adults.

Probably ~30 loci are involved in segment formation. **Figure 38.14** shows that they can be classified according to the size of the unit that they affect:

◆ **Gap** mutants have a group of several adjacent segments deleted from the final pattern. Four

Figure 38.14

Segmentation genes affect the number of segments and fall into three groups.

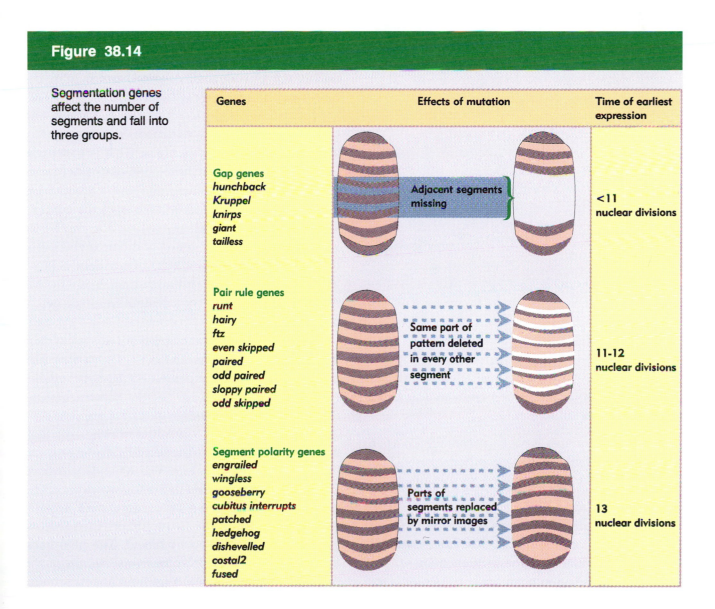

Genes	Effects of mutation	Time of earliest expression
Gap genes *hunchback* *Kruppel* *knirps* *giant* *tailless*	Adjacent segments missing	<11 nuclear divisions
Pair rule genes *runt* *hairy* *ftz* *even skipped* *paired* *odd paired* *sloppy paired* *odd skipped*	Same part of pattern deleted in every other segment	11-12 nuclear divisions
Segment polarity genes *engrailed* *wingless* *gooseberry* *cubitus interrupts* *patched* *hedgehog* *dishevelled* *costal2* *fused*	Parts of segments replaced by mirror images	13 nuclear divisions

Figure 38.15

Maternal and segmentation genes act progressively on smaller regions of the embryo.

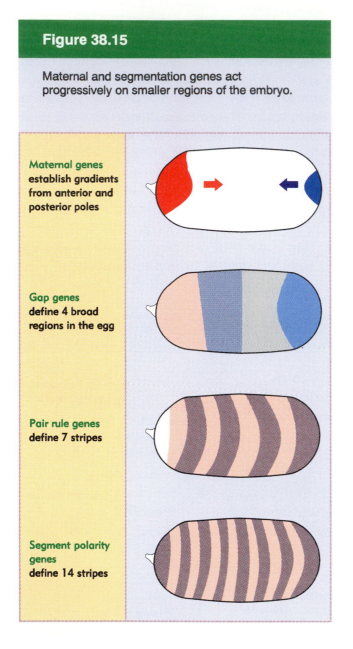

Maternal genes establish gradients from anterior and posterior poles

Gap genes define 4 broad regions in the egg

Pair rule genes define 7 stripes

Segment polarity genes define 14 stripes

gap genes are involved in formation of the major body segments, and others are concerned with the head and tail structures.

◆ **Pair-rule** mutants have corresponding parts of the pattern deleted in every other segment. The afflicted segments may be even-numbered or odd-numbered. There are 8 pair-rule genes.

◆ **Segment polarity** mutants most often lose part of the P compartment of each segment, and it is replaced by a mirror image duplication of the A compartment. Some mutants cause loss of A

compartments or middle segments. There are 16 known segment polarity genes, and probably more remain to be discovered.

These groups of genes are expressed at successive periods during development; and they define increasingly restricted regions of the egg, as can be seen from **Figure 38.15**. The maternal genes establish gradients from the anterior and posterior ends. The maternal gradients either activate or repress the gap genes, which are amongst the earliest to be transcribed following fertilization (following the 11th nuclear division); they divide the embryo into 4 broad regions. The gap genes regulate the pair-rule genes, which are transcribed slightly later; their target regions are restricted to *pairs* of segments. The pair-rule genes in turn regulate the segment polarity genes, which are expressed during the 13th nuclear division, and by now the target size is the *individual* segment.

Many of the maternal genes, the gap genes, and the pair-rule genes are regulators of transcription. Their effects may be either to activate or to repress transcription; in some cases, a given protein may activate some target genes and repress other target genes, depending on either its level or the context. The genes in any one class regulate one another as well as regulating the genes of the next class. It is only when we reach the level of segment polarity genes that products are found that act in other ways; for example, secretion of a protein from one cell to influence its neighbor.

The principle that emerges from this analysis is that at each stage *a small number of maternal, gap, and pair-rule regulator proteins is used in combinatorial associations to specify the pattern of gene expression in a particular region of the embryo.*

The gap genes are controlled in two ways: they may respond directly to the bicoid morphogen; and they regulate one another. The 4 bands shown in Figure 38.15 are created in the following way. The most anterior band consists of hunchback protein, transcribed from the hunchback RNA whose synthesis was activated by bicoid. The next band consists of Kruppel protein; transcription of the *Kruppel* gene is activated by hunchback protein.

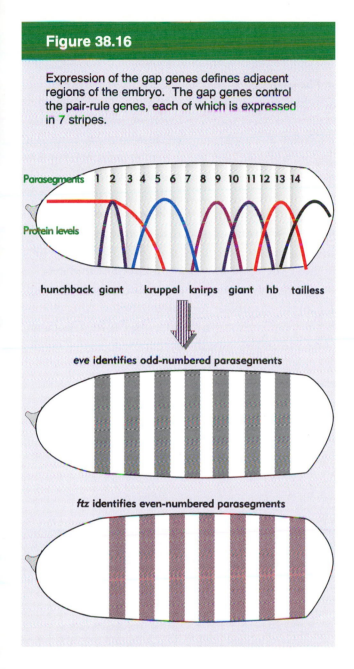

Figure 38.16

Expression of the gap genes defines adjacent regions of the embryo. The gap genes control the pair-rule genes, each of which is expressed in 7 stripes.

Parasegments 1 2 3 4 5 6 7 8 9 10 11 12 13 14

Protein levels

hunchback giant   kruppel   knirps   giant   hb   tailless

eve identifies odd-numbered parasegments

ftz identifies even-numbered parasegments

especially important role. It is expressed in a broad anterior region, with a gradient of decline in the middle of the embryo. High levels of hunchback repress *Kruppel*; this determines the anterior boundary of Kruppel expression, which rises just as hunchback falls off, in parasegment 3. But some level of hunchback is needed for *Kruppel* expression, so when the level of hunchback decreases further, *Kruppel* is turned off, around parasegment 5. In the same way, expression of *giant* responds to successive changes in the level of hunchback; and *knirps* expression requires the absence of hunchback.

The control is refined further by interactions among the proteins. *The general principle is that one interaction may be required to express a protein in a particular region, and other interactions may be required to repress its expression at the boundaries.* The effects are worked out by examining pairwise interactions. For example, over-expression of *giant* causes the Kruppel band to become much narrower, suggesting that giant contributes to repressing the boundaries of Kruppel. The posterior margins of knirps and giant are determined by the operation of the terminal system. Altogether, these interactions mean that, as we proceed along the egg from anterior to posterior, any particular position can be defined by the levels of the various gap proteins.

All the gap proteins are regulators of transcription, and in addition to regulating one another, they regulate the expression of the pair-rule genes that function at the next stage. Each pair-rule protein is found in a pattern of 7 'stripes' along the embryo, and Figure 38.16 shows the approximate positions that these stripes will take as the result of expression of the gap genes. (Of course, the parasegments have not developed yet, and are shown just to relate their positions to the protein distribution.) The 7 stripes of a pair-rule gene identify either all the odd-numbered parasegments (like *eve*) or all the even-numbered parasegments (like *ftz*). Two of the pair-rule genes, *hairy* and *eve*, are called primary pair-rule genes, because they are expressed first, and their pattern of expression influences the expression of the other pair-rule genes.

Recall that mutations in pair-rule genes delete

The next two bands consists of knirps and giant proteins; transcription of these genes is repressed by hunchback, and they are expressed in the posterior part of the embryo because nanos has prevented the expression of hunchback there.

Figure 38.16 examines the transition from the 4-band to the 7-striped stage in more detail. The detailed interactions among the gap proteins are determined by examining the pattern of the distribution of other gap proteins in a mutant lacking one particular gap protein. Hunchback plays an

Figure 38.17

ftz mutants have half the number of segments present in wild-type. Photographs kindly provided by Walter Gehring.

Wild type

ftz mutant

must be formed. In other words, expression of *ftz* is required for survival of the cells in the regions in which it is expressed.

The expression of *ftz* is an example of the general rule that the stripes in which a pair-rule gene is expressed correspond to the regions that are missing from the embryo when the gene is mutated. *Compartments are therefore determined by the pattern of expression of segmentation genes.* The width of the stripe in which a gene is expressed corresponds to the size of the segmental unit that it affects. Different mechanisms are used to specify the expression patterns of different pair-rule genes; we have the most information about *ftz* and *eve*.

In the early embryo, *ftz* is uniformly expressed. If protein synthesis is blocked before the stripes develop, the embryo retains the initial pattern. Thus the development of stripes depends on the specific degradation of *ftz* RNA in the regions between the bands and at the anterior and posterior ends of the embryo. Once the stripes have developed, transcription of *ftz* ceases in the interbands and at the ends of the embryo. The specificity of transcription depends on regions upstream of the

half the segments. **Figure 38.17** compares the segmentation patterns of wild-type and *fushi tarazu* (*ftz*) larvae; the mutant has only half the number of segments, because every other segment is missing.

The *ftz* mRNA is present from early blastoderm to gastrula stages of development. **Figure 38.18** shows the locations of the transcripts, visualized *in situ* at blastoderm in wild type. *The gene is expressed in 7 stripes, each 3–4 cells wide, running across the embryo.* As shown previously in Figure 38.16, the stripes correspond to even-numbered parasegments (4 = T1P/T2A, 6 = T3P/A1A, 8 = A2P/A3A, etc.).

This pattern suggests a function for the *ftz* gene: *it must be expressed at blastoderm for the structures that will be descended from the even-numbered parasegments to develop.* Mutants in which *ftz* is defective lack these parasegments because the gene product is absent during the period when they

Figure 38.18

Transcripts of the *ftz*⁺ gene are localized in stripes corresponding to even-numbered parasegments. The expressed regions in wild type correspond to the regions that are missing in the *ftz* mutant of the previous figure. Photograph kindly provided by Walter Gehring.

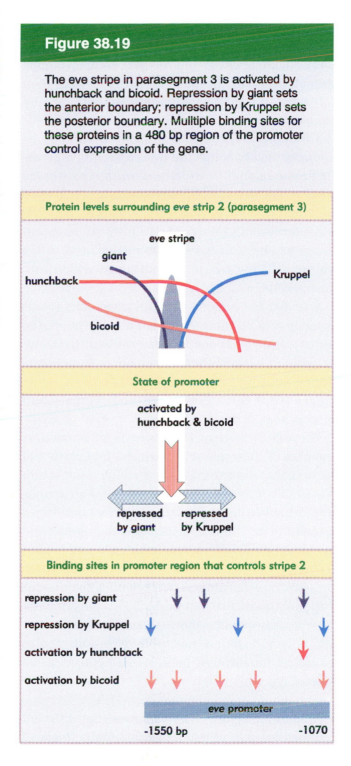

Figure 38.19

The eve stripe in parasegment 3 is activated by hunchback and bicoid. Repression by giant sets the anterior boundary; repression by Kruppel sets the posterior boundary. Multiple binding sites for these proteins in a 480 bp region of the promoter control expression of the gene.

Protein levels surrounding eve strip 2 (parasegment 3)

eve stripe

giant

hunchback

Kruppel

bicoid

State of promoter

activated by
hunchback & bicoid

repressed   repressed
by giant   by Kruppel

Binding sites in promoter region that controls stripe 2

repression by giant

repression by Kruppel

activation by hunchback

activation by bicoid

eve promoter

-1550 bp   -1070

separately in each stripe. A detailed reconstruction using subregions of the *eve* promoter shows that the information for localization in each stripe is coded in a separate part of the promoter; the promoter can be divided into regions that correspond to the local levels of gap gene products in particular parasegments. For example, the promoter region that is responsible for *eve* expression in parasegment 3 has binding sites for the gap proteins bicoid, hunchback, giant and Kruppel. **Figure 38.19** shows that this part of the promoter extends for 480 bp, and works in the following way: *eve* transcription is activated by hunchback and bicoid, and the two boundaries are determined because it is repressed by giant on the anterior side and by Kruppel on the posterior side (see also Figure 38.16). Other parts of the promoter respond to the protein levels in other parts of the embryo. Thus the different stripes of the primary pair-rule gene products are regulated by separate pathways, each of which is susceptible to activation by a particular combination of gap gene products and other regulators.

This illustrates in miniature the principle that combinations of proteins control gene expression in local areas. The general principle is that generally distributed proteins (such as bicoid or hunchback) are needed for activation, whereas the borders are formed by selective repression (by giant and Kruppel in this particular example). We should emphasize that the hierarchy of gene control is not exclusively restricted to interactions between successive stages of control; for example, the involvement of bicoid protein in regulating *eve* transcription in parasegment 3 shows that a maternal gene may have a direct effect on a pair-rule gene.

The stripes of eve and ftz are fuzzy to begin with, and become sharper as development proceeds, corresponding to more finely defined units. **Figure 38.20** shows an example of an embryo simultaneously stained for expression of *ftz* and *eve*. Initially there is a series of alternating fuzzy stripes, but the stripes narrow from the posterior margin and sharpen on the anterior side as they intensify during development. This may depend on an autoregulatory loop, in which the expression

ftz promoter, and also on the function of several other segmentation genes. The transcription of *ftz* responds to other pair-rule genes (and perhaps gap genes) through elements that act on all stripes.

The expression pattern of *eve* is complementary to *ftz*, but has a different basis: it is controlled

Figure 38.20

Simultaneous staining for *ftz* (brown) and *eve* (grey) shows that they are first expressed as broad alternating stripes at the time of blastoderm (upper), but narrow during the next 1 hour of development (lower). Photograph kindly provided by Peter Lawrence.

of the gene is regulated by its own product.

The pair-rule genes control the expression of the segment polarity genes, which are expressed in 14 stripes. Each stripe identifies a segment. The compartmental pattern in which segment polarity genes are expressed is exceedingly precise. Perhaps the ultimate demonstration of precision is provided by the pattern gene *engrailed*. The function coded by *engrailed* is needed in all segments and is concerned with the distinction between the A and P compartments. *engrailed* is expressed in every P compartment, but not in A compartments. Mutants in this gene do not distinguish between anterior and posterior compartments of the segments.

Antibodies against the protein coded by *engrailed* react against the nuclei of cells expressing it. The regions in which *engrailed* is expressed form a pattern of stripes. When the stripes of engrailed protein first become apparent, *they are only one cell wide*. **Figure 38.21** shows the pattern at a stage when each segment has a stripe just 1 cell in width, with the stripe beginning to widen into several cells.

Actually, the pattern of stripes becomes established over a 30 minute period, moving along the embryo from anterior to posterior. Initially one stripe is apparent; then every other segment is striped; and finally the complete pattern has a stripe 3–4 cells wide corresponding to the P compartment of every segment.

The expression of *engrailed* is of particular importance, because it defines the boundaries of the actual compartments from which adult structures will be derived. The initial 1-cell-wide stripes of engrailed protein form at the anterior boundaries of both the ftz and eve stripes, and delineate what will become the anterior boundary of every P compartment. Why is *engrailed* initially transcribed exclusively in this anterior edge, within the broader stripes of ftz and eve expression? This question is a specific example of a more general question: how can a broad stripe be subdivided into more restricted, narrower stripes? We can consider two general types of model:

◆ A combinatorial model supposes that different genes are expressed in overlapping patterns of stripes. A pattern of stripes develops for each of the pair-rule genes. The different pair-rule gene stripes overlap, because they are out of phase with one another. As a result of these patterns, different cells in the cellular blastoderm express different combinations of pair-rule genes. Each

Figure 38.21

Engrailed protein is localized in nuclei and forms stripes precisely delineated as 1 cell in width. Photograph kindly provided by Patrick O'Farrell.

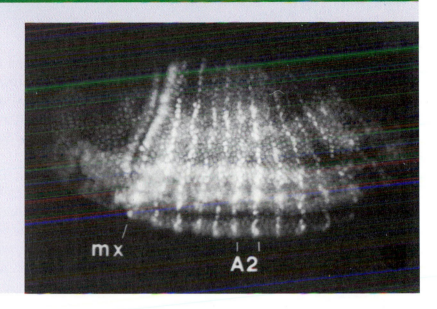

compartment is defined by the particular combination of the genes that are expressed, and these combinations determine the responses of the cells at next stage of development. In other words, the segmentation genes are controlled by the pair-rule genes in the same general manner that the pair-rule genes are controlled by the gap genes.

◆ A boundary model supposes that a compartment is defined by the striped pattern of expression, but that interactions involving cell–cell communication at the boundaries cause subdivisions to arise within the compartment. In the case of *engrailed,* we would suppose that some unique event is triggered by the juxtaposition of cells possessing ftz (or eve) with cells that do not, and this is necessary to trigger *engrailed* expression.

Each of the 14 segments is subdivided further by functions of the segment polarity genes. The type of subdivision imposed by a segment polarity gene is repeated in every segment. For example, *engrailed*

distinguishes the A and P compartments. *engrailed* is a transcription factor, but other segment polarity genes have different types of functions. *wingless* codes for a protein that is secreted and taken up by nearby cells, so this allows cell–cell interactions to play a role. *wingless* is initially expressed in a row of cells immediately adjacent to the anterior side of the cells expressing *engrailed*; thus *wingless* comes to identify the posterior boundary of the preceding parasegment. The initial expression of *engrailed* in response to ftz and eve is shortly replaced by an autoregulatory loop in which secretion of wingless protein from the adjacent cells is needed for expression of *engrailed*; and expression of *engrailed* is needed for expression of *wingless*. This keeps the boundary sharp. We do not yet have a good understanding of how the other segment polarity genes function to delineate the segments or parts of segments. However, their products include secreted proteins, transmembrane proteins, kinases, cytoskeletal proteins, as well as transcription factors, suggesting that at this stage cell–cell communications become important.

Complex loci are extremely large and involved in regulation

Segment polarity genes control the pattern within each segment, including the polarity. **Homeotic genes** impose the program that determines the unique differentiation of each segment. Homeotic mutations cause cells of one compartment to develop the phenotype of a different cell compartment. Some homeotic mutants 'transform' part of a segment or an entire segment into another type of segment. Most homeotic genes are expressed in a spatially restricted manner that corresponds to parasegments.

Homeotic genes interact in complicated interlocking patterns that are not yet fully defined. Many homeotic genes code for transcription factors that act upon other homeotic genes as well as upon other target loci. As a result, a mutation in one homeotic gene influences the expression of other homeotic genes. The consequence is that *the final appearance of a mutant depends not only on the loss of one homeotic gene function, but also on how other homeotic genes change their spatial patterns in response to the loss.*

Homeotic genes act during embryogenesis. Their expression depends on the prior expression of the segmentation genes; we might regard the homeotic genes as integrating the pattern of signals established by the segmentation genes. Homeotic mutations cause one segment of the abdomen to develop as another, legs to develop in place of antennae, or wings to develop in place of eyes. Note that homeotic genes do not *create* patterns *de novo*; they modify cell fates that are determined by genes such as the segment polarity genes, by switching the set of genes that functions in a particular place. Indeed, the segment polarity genes are active at about the same time as the peak of expression of the homeotic genes.

The genetic properties of some homeotic genes are unusual and gave rise to their description as **complex loci**. A conventional gene—even an interrupted one—is identified at the level of the genetic map by a cluster of noncomplementing mutations. In the case of a large gene, the mutations might map into individual clusters corresponding to the exons. A hallmark of complex loci is that, in addition to rather well-spaced groups of mutations, extending over a relatively large map distance, there are complex patterns of complementation, in which some pairwise combinations complement but others do not. The individual mutations may have different and complex morphological effects on the phenotype. We now know that these relationships can be caused by two effects: the genes are extremely large; and there is an array of regulatory elements extending far beyond the transcription unit itself. Many of the bizarre results that are obtained in complementation assays turn out to result from mutations in promoters or enhancers that affect expression in one cell type but not another. We now recognize that complex loci do not have any novel features of genetic organization, apart from the fact that they have many regulatory elements that control expression in different parts of the embryo.

Two of the complex loci are involved in regulating development of the adult insect body. The *ANT-C* and *BX-C* complex loci together provide a continuum of functions that specify the identities of all of the segmented units of the fly. Each of these complexes contains several homeotic genes. The two separate complexes may have evolved from a split in a single ancestral complex, as suggested by observations of the evolution of the corresponding genes in other species. In the beetle *Tribolium*, the *ANT-C* and *BX-C* complexes are found together at a single chromosomal location. The individual genes may have been derived from duplications and mutations of an original ancestral gene. And in mammals, there are arrays of related genes whose individual members are related sequentially to the

Figure 38.22

The homeotic genes of the ANT-C complex confer identity on the most anterior segments of the fly. The genes vary in size, and are interspersed with other genes. The *antp* gene is very large and has alternative forms of expression.

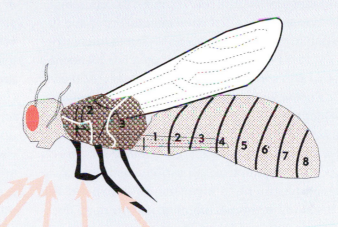

Loss-of-function phenotypes

lab   affects head
pb   labial structures replaced by antennal
Dfd   deletes mandibular segments
Scr   transforms T1 to T2
Antp   transforms T2-T3 to T1

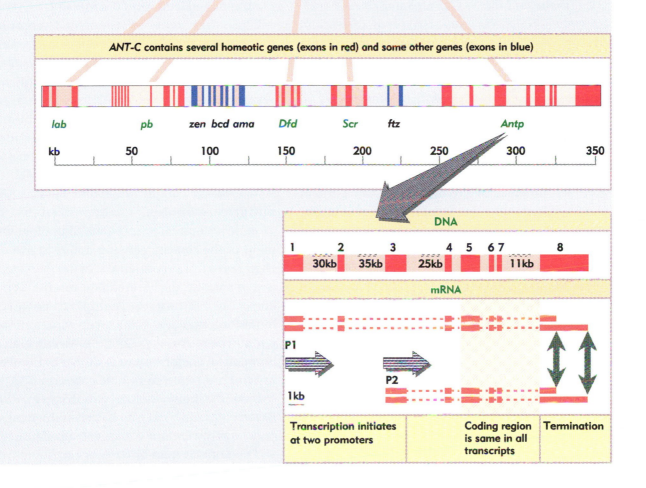

ANT-C contains several homeotic genes (exons in red) and some other genes (exons in blue)

lab   *pb*   *zen bcd ama*   *Dfd*   *Scr*   *ftz*   *Antp*

kb   50   100   150   200   250   300   350

DNA

1   2   3   4   5   6 7   8
     30kb   35kb   25kb   11kb

mRNA

P1

P2

1kb

| Transcription initiates at two promoters | Coding region is same in all transcripts | Termination |

genes of the *ANT-C* and *BX-C* complexes (as we describe in the next section).

The individual genes in the *ANT-C* and *BX-C* complexes code for a set of transcription factors that have related DNA-binding domains. The homeotic genes clustered at the *ANT-C* and *BX-C* complexes show a relationship between genetic order and the position in which they are expressed in the body of the fly. *Proceeding from left to right, each homeotic gene in the complex acts upon a more posterior region of the fly. The basic principle is that formation of a compartment requires the gene product(s) expressed in the previous compartment, plus a new function coded by the next gene along the cluster.* Thus loss-of-function mutations usually cause one compartment to have the phenotype of the corresponding compartment on its anterior side.

The identities of the most anterior parts of the fly (parasegments 1–4) are specified by the complex locus, *ANT-C*. *ANT-C* contains several homeotic genes, including *labial (lab), proboscipedia (pb), Deformed (Dfd), Sex combs reduced (Scr),* and *Antennapedia (Antp)*. The homeotic genes lie in a cluster over a region of ~350 kb, but several other genes are interspersed; most of these genes are regulators that function at different stages of development.

Figure 38.22 correlates the organization of *ANT-C* with its effects upon body parts. Proceeding from left to right along the cluster, the genes are expressed in successively more posterior parts of the embryo, ranging from *labial* (the most anterior acting, which affects the head) to *Antp* (the most posterior acting, which affects segments T2–T3).

The *Antp* gene gave its name to the complex, and among the mutations in it are alleles that change antennae into second legs or second and third legs into first legs. *Antp* usually functions in the thorax; it is needed both to promote formation of segments T2–T3 and to suppress formation of head structures. Loss of function therefore causes T2–T3 to resemble the more anterior structure of T1; gain of function, for example, by overexpression in the head, causes the anterior region to develop structures of the thorax.

The figure summarizes the organization of the gene. The locus is large at the molecular level, spanning ~103 kb. There are 8 exons, with very large introns. The single open reading frame begins only in exon 5, and apparently gives rise to a protein of 43,000 daltons. Alternative spliced forms may generate variants of the protein, but these are not yet characterized. The discrepancy between the length of the locus and the size of the protein means that only 1% of its DNA codes for protein.

Transcription starts at either of two promoters, located ~70 kb apart! One promoter is located upstream of exon 1, the other upstream of exon 3. Use of the first promoter is associated with omission of exon 3. The transcripts generated from either promoter end either within or after exon 8. All the transcripts appear to code for the same protein. No difference has yet been identified in the use of the two promoters, which suggests that their significance could lie in the different structures of the nontranslated leaders of the mRNAs.

The other genes of the *ANT-C* complex are expressed in the head and first thoracic segment. In the most anterior compartments, *lab, pb, Dfd,* have unique patterns of expression, so that deletions of segmental regions can result from loss-of-function mutations. An exception to the left–right/anterior–posterior order of action is that loss of *Scr* allows the overlapping *Antp* to function, that is, the direction of transformation is opposite from usual.

The classic complex homeotic locus is *BX-C*, the **bithorax complex**, characterized by several groups of homeotic mutations that affect development of the thorax, causing major morphological changes in the abdomen. When the whole complex is deleted, the insect dies late in embryonic development. Within the complex, however, are mutations that are viable, but which change the phenotype of certain segments. An extreme case of homeotic transformation is shown in **Figure 38.23**, in which a triple mutation converts T3A (which carries the halteres [truncated wings]) into the tissue type of the T2 (which carries the wings). This creates a fly with four wings instead of the usual two.

The genetic map of *BX-C* is correlated with the

Figure 38.23

A four-winged fly is produced by a triple mutation in *abx*, *bx*, and *pbx* at the *BX-C* complex. Photograph kindly provided by Ed Lewis.

body structures that it controls in the fly in **Figure 38.24**. The body structures extend from T2 to A8. The *BX-C* complex is therefore concerned with the development of the major part of the body of the fly. Like *ANT-C*, a crucial feature of this complex is also that mutations affecting particular segments lie in the same order on the genetic map as the corresponding segments in the body of the fly. Proceeding from left to right along the genetic map, mutations affect segments in the fly that become successively more posterior.

A difference between *ANT-C* and *BX-C* is that *ANT-C* functions largely or exclusively via its protein-coding loci, but *BX-C* displays a complex pattern of *cis*-acting interactions in addition to the effects of mutations in protein-coding regions. Thus the individual mutations fall into two classes:

◆ Three transcription units (*Ubx*, *abdA*, *AbdB*) produce mRNAs that code for proteins. (The *bxd* unit produces an RNA apparently without coding function.)

◆ There are *cis*-acting mutations at intervals throughout the entire cluster. They control expression of the transcription units. *Cis*-acting mutations of any particular type may occur in a large region.

As a historical note, the complex was originally defined in terms of two 'domains'. Mutations in the *Ultrabithorax* domain were characterized first; they have the thoracic segments T2P–T3P and the abdominal compartment A1A as their targets (this corresponds to parasegments 5–6). These mutations lie either in the *Ubx* transcription unit or in the *cis*-acting sites that control it. The mutations within the ultrabithorax domain are named for their phenotypes. The *bx* and *bxd* types are identified by a series of mutations, in each case spreading over ~10 kb. The *abx* and *pbx* mutations are caused by deletions, which vary from 1 to 10 kb.

Mutations in the *Infraabdominal* domain were found later; they have the abdominal segments A1P–A8P (parasegments 7–14) as targets.) These mutations lie either in the *AbdA,B* transcription units, or in the *cis*-acting sites that control them. Within the infra–abdominal domain, *cis*-acting mutations are named systematically as *iab2–8*. These mutations affect individual compartments, or sometimes adjacent sets of compartments, as shown at the top of the figure.

Proceeding from left to right along the cluster, transcripts are found in increasingly posterior parts of the embryo, as shown at the bottom of the figure. The patterns overlap. *Ubx* has an anterior boundary of expression in compartment T2P (parasegment 5), *abdA* is expressed from compartment A1P (parasegment 7), and *AbdB* is expressed in compartments posterior to compartment A4P (parasegment 10).

Transcription of *Ubx* has been studied in the most detail. The *Ubx* transcription unit is ~75 kb, and has alternative splicing patterns that give rise to several short RNAs. A transient 4.7 kb RNA appears first, and then is replaced by RNAs of 3.2 and 4.3 kb. A feature common to both the latter two

Figure 38.24

The *bithorax (BX-C)* locus has 3 coding units. A series of regulatory mutations affects successive segments of the fly. The sites of the regulatory mutations show the regions within which deletions, insertions, and translocations confer a given phenotype.

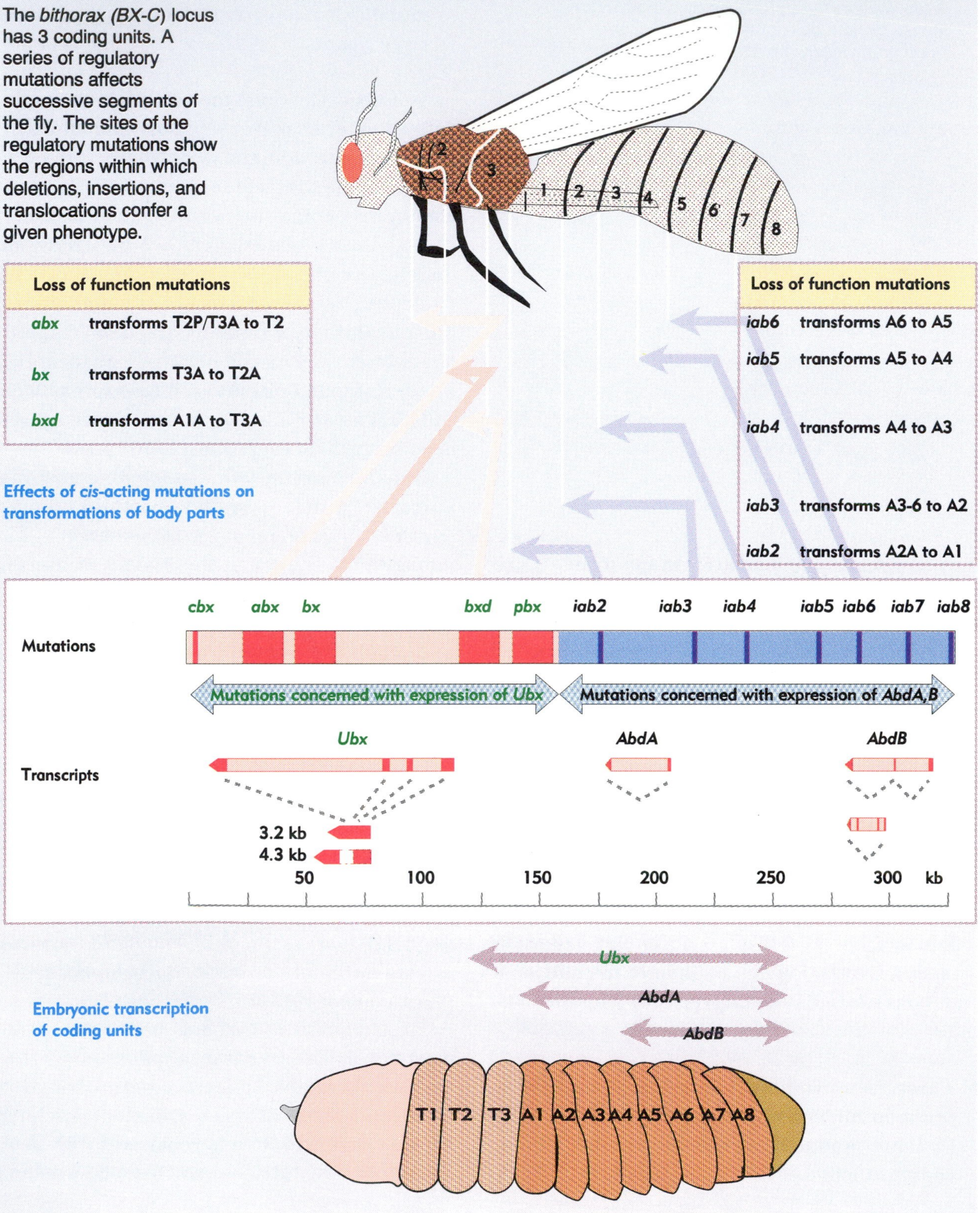

Loss of function mutations

abx   transforms T2P/T3A to T2

bx   transforms T3A to T2A

bxd   transforms A1A to T3A

Effects of *cis*-acting mutations on transformations of body parts

Loss of function mutations

iab6   transforms A6 to A5

iab5   transforms A5 to A4

iab4   transforms A4 to A3

iab3   transforms A3-6 to A2

iab2   transforms A2A to A1

Mutations: cbx   abx   bx   bxd   pbx   iab2   iab3   iab4   iab5   iab6   iab7   iab8

Mutations concerned with expression of *Ubx*   Mutations concerned with expression of *AbdA,B*

Transcripts: *Ubx*   *AbdA*   *AbdB*

3.2 kb
4.3 kb

50   100   150   200   250   300   kb

Embryonic transcription of coding units

Ubx

AbdA

AbdB

T1 T2 T3 A1 A2 A3 A4 A5 A6 A7 A8

RNAs is their inclusion of sequences from both ends of the primary transcript. Of course, there may be other RNAs that have not yet been identified.

We do not yet have a good idea of whether there are significant differences in the coding functions of these RNAs. The first and last exons are quite lengthy, but the interior exons are rather small. Small exons from within the long transcription unit may enter mRNA products by means of alternative splicing patterns. So far, however, we do not know of any functional differences in the Ubx proteins produced by the various modes of expression.

A working model for the expression of the Ubx transcript supposes that alternative splicing gives rise to RNAs (including but not necessarily restricted to the 3.2 and 4.3 kb species) which share the beginning and end of the coding region, but differ by including different exons internally. On this premise, the entire set of proteins coded by this unit have been localized by using antibodies directed against the common first exon. The Ubx protein(s) are the only products coded by the *Ubx* region.

Ubx proteins are found in the compartments that correspond to the sites of transcription, that is in T2P-A1 and at lower levels in A2-A8. No staining is found in larvae with a small deletion that introduces a termination codon near the 5' end of the coding sequence. From these results, we conclude that the Ubx unit codes for a set of related proteins that are concentrated in the compartments affected by mutations in the *Ultrabithorax* domain. Ubx proteins are located in the nucleus, and they fall into the general type of transcriptional regulators whose DNA-binding region consists of a homeodomain (see later).

We can understand the general function of the *BX-C* complex by considering the effects of loss-of-function mutations. If the entire complex is deleted, the larva cannot develop the individual types of segments. In terms of parasegments (which are probably the affected units), all the parasegments differentiate in the same way as parasegment 4; the embryo has 10 repetitions of the repeating structure T1P/T2A all along its length, in place of the usual

compartments between parasegments 5 and 14. In effect, the absence of *BX-C* functions allows *Antp* to be expressed throughout the abdomen, so that all the segments take on the characteristic of a segment determined by *Antp*; *BX-C* functions are needed to add more posterior-type information.

Each of the transcription units affects successive segments, according to its pattern of expression. Thus if *Ubx* alone is present, the larva has parasegment 4 (T1P/T2A), parasegment 5 (T2P/T3A), and then 8 copies of parasegment 6 (T3P/A1A). This suggests that the expression of *Ubx* is needed for the compartments anterior to A1A; *Ubx* is also expressed in the more posterior segments, but in the wild type, *abdA* and *AbdB* are also present; if they are removed, the expression of *Ubx* alone in all the posterior segments has the same effect that it usually has in parasegment 6 (T3P/A1A).

The addition of *abdA* to *Ubx* adds the wild-type pattern to parasegments 7, 8, and 9. In other words, *Ubx* plus *abdA* can specify up to compartments A3P/A4A, and in the absence of *AbdB*, this continues to be the default pattern for all the more posterior compartments. The addition of *AbdB* is needed to specify parasegments 10–14.

The general model for the function of the *ANT-C* and *BX-C* complexes is to suppose that additional functions are added to define successive segments proceeding in the posterior direction. It functions by reliance on a *combinatorial pattern* in which the addition of successive gene products confers new specificities. This explains the rule that a loss-of-function mutation in one of the genes of the *ANT-C/BX-C* complexes generally allows the gene on the more anterior side of the mutated gene to determine phenotype, that is, loss-of-function results in homeotic transformation of posterior regions into more anterior phenotypes.

Expression of *Ubx* in a more anterior segment than usual should have the opposite effect to a loss of function; the segment develops a more anterior phenotype. When this is tested by arranging for *Ubx* to be expressed in regions that are more anterior, that is, in the head, its expression converts these compartments to the phenotype of parasegment 6. Thus lack of expression of *Ubx* causes a homeotic transformation in which posterior segments acquire more anterior phenotypes; and over-expression of *Ubx* causes a homeotic transformation in which anterior segments acquire more posterior phenotypes.

This type of relationship is true generally for the cluster as a whole, and explains the properties of *cis*-acting mutations as well as those in the transcription units. These regulatory mutations cause loss of the protein in part of its domain of expression or cause additional expression in new domains. Thus they may have either loss-of-function or gain-of-function phenotypes (or sometimes both). The most common is loss-of-function in an individual compartment. For example, *bx* specifically controls expression of *Ubx* in compartment T3A; a *bx* mutation loses expression of *Ubx* in that compartment, which is therefore transformed to the more anterior type of T2A. This example is typical of the general rule for individual *cis*-acting mutations in the complex; each converts a target compartment *so that it develops as though it were located at the corresponding position in the previous segment.*

The order of the *cis*-acting sites of mutation on the chromosomes reflects the order of the compartments in which they function. Thus the expression of *Ubx* in parasegments 4, 5 and 6 is controlled sequentially by *abx* (affects parasegment 5), *bx* (affects T3A), etc. Most mutants in the *Ubx* domain have cytologically visible rearrangements of the chromosome, which at the molecular level turn out to represent deletions eliminating varying lengths of the left half of the cluster. A typical *Ubx* mutation consists of a large deletion that removes the transcription unit and many of the regulatory sites; it behaves as though possessing multiple mutations in the individual functions identified by recessive mutations in the *Ultrabithorax* domain.

The presence of only 3 genes within the *BX-C* complex poses two major questions. First, how do the combinations of 3 proteins specify the identity of 10 parasegments? One possibility is that there are quantitative differences in the various regions, allowing for the same sort of varying responses in target genes that we have described previously for the combinatorial functioning of the segmentation genes. Second, how do the proteins function in different tissue types? The pattern of expression described above refers generally to the epidermis; the development of other tissues is controlled in a way that is parallel, but not identical. For example, although *Ubx* is expressed in all posterior segments up to A8 in the epidermis, in mesoderm, it is repressed posterior of segment A7; the posterior boundary reflects repression by *abdA*, since in *abdA* mutants, *Ubx* expression extends posterior in the mesoderm.

Why are loci involved in regulating development of the adult insect from the embryonic larva different from genes coding for the everyday proteins of the organism? Is their enormous length necessary to generate the alternative products? Could it be connected with some timing mechanism, determined by how long it takes to transcribe the unit? At a typical rate of transcription, it would take ~100 minutes to transcribe *Antp*, which is a significant proportion of the 22 hour duration of *D. melanogaster* embryogenesis.

Proceeding from anterior to posterior along the embryo, we encounter the changing patterns of expression of the genes of the *ANT-C* and *BC-C* loci. What controls their transcription? As in the case of the segmentation loci, the homeotic loci are controlled *partially by the genes that were expressed at the previous stage of development, and partially by interactions among themselves.* For example, the expression of *Ubx* is changed by mutations in *bicoid, hunchback,* or *Kruppel.* The anterior boundary of expression respects the parasegment border defined by *ftz* and *eve.* The general principle is that all of these regulatory genes function by controlling transcription, either by activating it or by repressing it, and that the gene products may exert specific effects by both qualitative and quantitative combinations.

The homeobox is a common coding motif in homeotic genes

The three groups of genes that control *D. melanogaster* development—maternal genes, segmentation genes, and homeotic genes—regulate one another and (presumably) target genes that code for structural proteins. Interactions between the regulator genes have been defined by analyses that show defects in expression of one gene in mutants of another. However, we have identified rather few of the structural targets on which these groups of genes act to cause differentiation of individual body parts.

Consistent with the idea that the segmentation genes code for proteins that regulate transcription, the genes of 3 gap loci (*hb, Kr, kni*) contain zinc finger motifs. As first identified in the transcription factors TFIIIA and Sp1 (see Figure 30.7), these motifs are responsible for making contacts with DNA. The products of other loci in the gap class also have DNA-binding motifs; *giant* encodes a protein with a basic zipper motif, and *tailless* encodes a protein that resembles the steroid receptors. This suggests that the general function of gap genes is to function as transcriptional regulators.

Many of the homeotic and segmentation genes include motifs consisting of rather simple repeating sequences. The *opa* box, which codes for a stretch of polyglutamic acid, is one such example. We do not know what function such stretches have. If we accept the idea that the regulator genes function by regulating expression of other genes, then we may expect their proteins to bind to DNA. One possible function for common structural motifs in the proteins is to provide regions that bind DNA. Another could be to bind other proteins involved in gene expression, for example, RNA polymerase or perhaps some structural component of chromatin.

The most common of the conserved motifs is the **homeobox**, a 180 bp region located near the 3′ end of the transcription unit of each of several segmentation and homeotic loci. This sequence is part of an open reading frame, in which many of the base changes are located at third-base positions in the codons. The homeobox is constant in size and fairly basic in sequence. There are ~40 genes that contain a homeobox, and almost all are known to be involved in developmental regulation. (The homeobox was first identified by its predominance in the homeotic genes, from which it took its name.) The protein sequence coded by the homeobox is called the homeodomain; we have described in Chapter 30 how it has been characterized in detail as a DNA-binding motif in transcription factors (see Figure 30.15).

The fly homeodomains fall into several groups. A major group in *Drosophila* consists of the homeotic genes in the *BX-C/ANT-C* complexes; they are called the *Antennapedia* group. Their homeodomains are 70–80% conserved, and usually occur at the C-terminal end of the protein (see Figure 30.13 earlier). A distinct homeodomain sequence is found in the related genes *engrailed* and *invected*; it has only 45% sequence conservation with the *Antennapedia* group. (Their sequences were compared earlier in Figure 30.14.) Other types of homeodomain sequences are represented in 2–4 genes each.

Many of the *Drosophila* genes that contain homeoboxes are organized into clusters. Three of the homeotic genes in the *BX-C* cluster have homeoboxes, the *ANT-C* complex contains a group of 5 homeotic genes with homeoboxes, and 4 other genes at *ANT-C* also contain homeoboxes. The homeotic loci at *BX-C* and *ANT-C* are sometimes described under the general heading of *HOM-C* genes.

What is the basic function of the *HOM-C* genes in determining identity on the anterior–posterior axis? We assume that homeodomains of different amino

Figure 38.25

Mouse and human genomes each contain 4 clusters of genes that have homeoboxes. The order of genes reflects the regions in which they are expressed on the anterior-posterior axis. The *Hox* genes are aligned with the fly genes according to homology, which is strong for groups 1, 2, 4, and 9. The genes are named according to the group and the cluster, e.g., *HoxA1* is the most anterior gene in the *HoxA* group. All *Hox* genes are present in both man and mice except for some mouse genes missing from cluster C (indicated by half boxes).

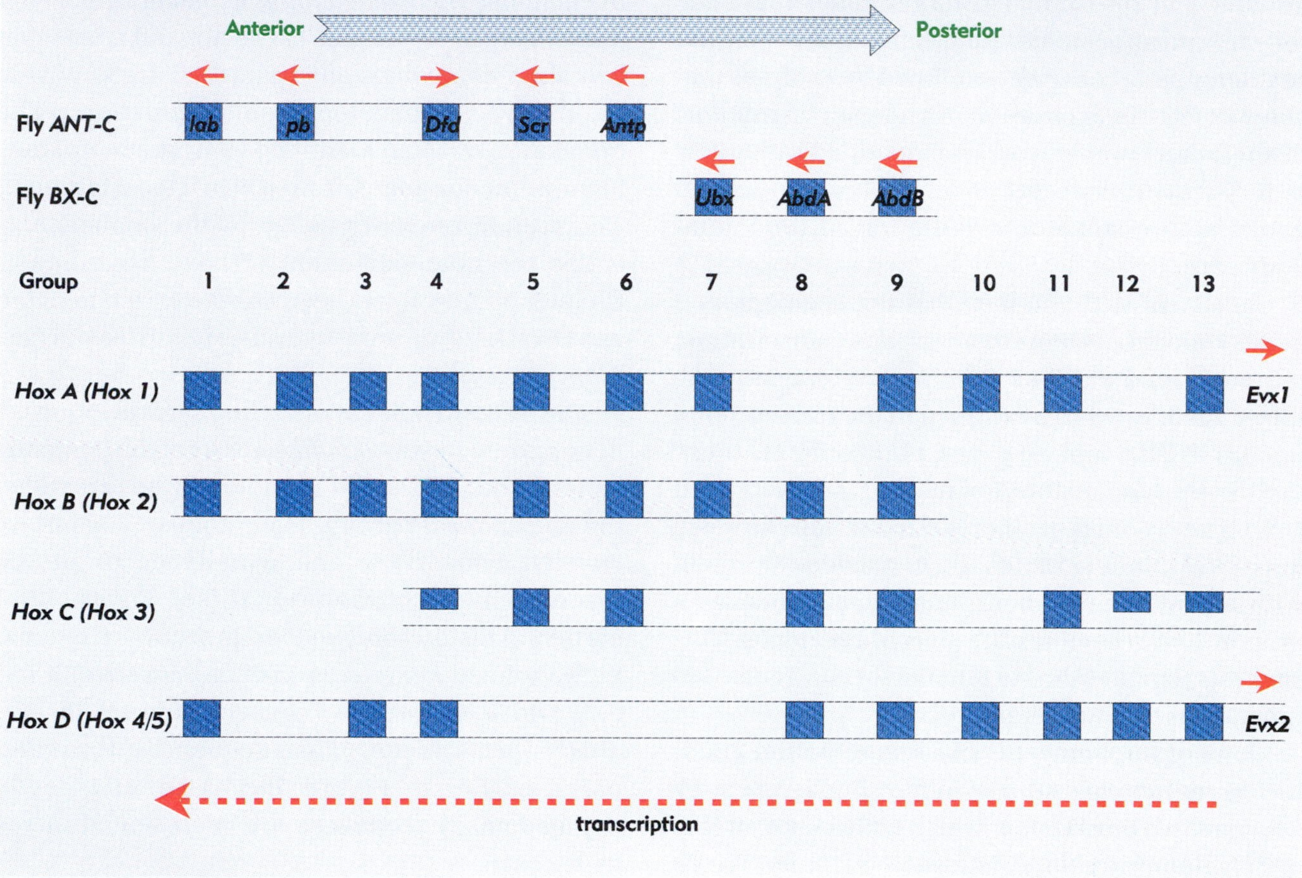

acid sequence recognize different target sequences in DNA. Experiments in which regions have been swapped between different proteins suggest that a major part of the specificity of these proteins rests with the homeodomain. However, the ability to bind to a particular DNA target site may not account entirely for their properties. For example, some of these proteins alternatively activate or repress transcription in response to the context, that is, their actions depend on the set of other proteins that are bound, not just on recognition of the DNA-binding site.

The similarities between the homeodomains of the more closely related members of the group suggest that they could recognize overlapping patterns of target sites. This would open the way for combinatorial effects that could be based on quantitative as well as qualitative differences, that is, there could be competition between proteins with related homeodomains for the same sites. Similarities in target sites recognized by different homeoproteins pose a puzzle with regards to defining their specificity of action; we assume that there are subtle differences in DNA binding yet to be discovered, or there are other interactions, such as protein–protein interactions, that play a role.

The homeobox motif is extensively represented in evolution. A striking extension of the significance of homeoboxes is provided by the discovery that a DNA probe representing the homeobox hybridizes with the genomes of many eukaryotes. Genes containing homeoboxes have been characterized in particular in frog, mouse, and human DNA. The frog and mammalian genes are expressed in early embryogenesis, which strengthens the parallel with the fly genes, and suggests the possibility that genes containing homeoboxes are involved in regulation of embryogenesis in a variety of species.

Genes in mammals (and possibly all animals) that are related to the *HOM-C* group have a striking property: like those of the *BX-C/ANT-C* complexes, they are organized in clusters. The individual mammalian genes are called *Hox* genes. A cluster of *Hox* genes may extend 20–100 kb and contain up to 10 genes. Four *Hox* clusters of genes containing homeoboxes have been characterized in the mouse and human genomes. Their organization is compared with the two large fly clusters in **Figure 38.25**.

By comparing the sequences of the homeoboxes (and sometimes other short regions), the mammalian genes can be placed into groups that correspond with the fly genes. This is shown by vertical alignment in the figure. For example, *HoxA4* and *HoxB4* are best related to *Dfd*. When these relationships are defined for the cluster as a whole, it appears that within each cluster we can recognize a series of genes that are related to the genes in the *ANT-C* and *BX-C* clusters. Groups 1–9 in the mammalian loci are defined as corresponding to the genes of the *ANT-C* and *BX-C* loci organized end to end in anterior–posterior orientation. (However, exact relationships have not been established within groups 5–8.) Groups 10–13 are defined simply as lying sequentially farther along the cluster. The corresponding loci in each cluster are sometimes called **paralogs** (for example, *HoxA4* and *HoxB4* are paralogous).

This situation could have arisen if the fly and mammalian loci diverged at a point when there was only a single complex, containing all of the genes

that define anterior–posterior polarity. Subsequent to their divergence, the *Drosophila* genes broke into two separate clusters, while the entire group of mammalian genes became duplicated, some

Figure 38.26

A comparison of *ANT-C/BX-C* and *HoxB* expression patterns shows that the individual gene products share a progressive localization of expression towards the more posterior of the animal proceeding along the gene cluster from left to right. Expression patterns show the regions of transcription in the fly epidermis at 10 hours, and in the central nervous system of the mouse embryo at 12 days.

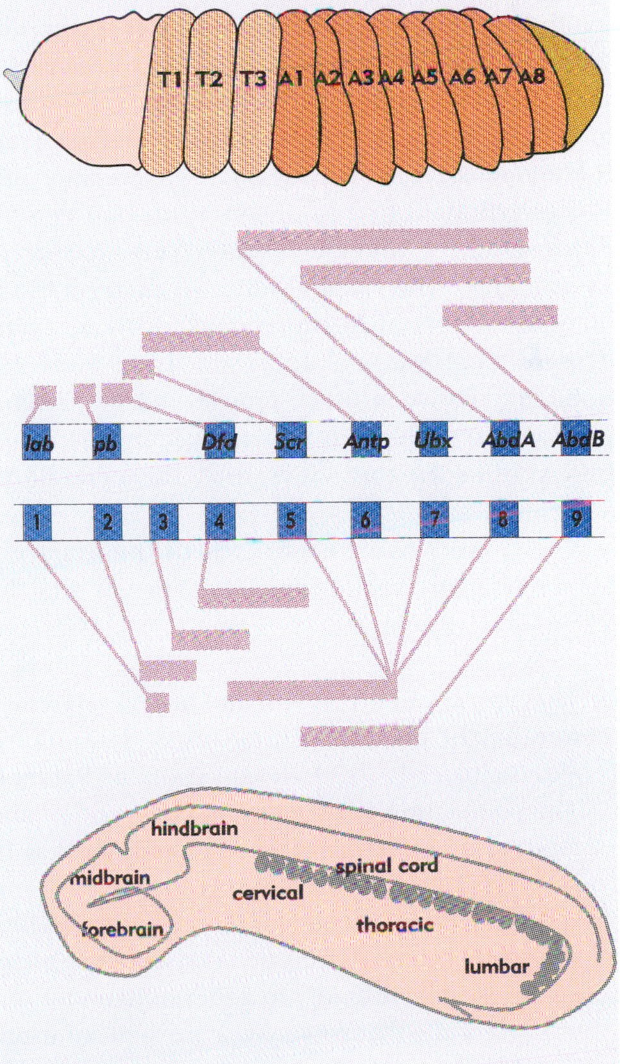

individual members being lost from each complex after the duplication.

The parallel between the mouse and fly genes extends to their pattern of spatial expression. The genes within a *Hox* cluster are expressed in the embryo in a manner that matches their organization in the genome. Progressing from the left toward the right end of the cluster drawn in Figure 38.25, genes are expressed in the embryo in locations progressively more restricted to the posterior end. The patterns of expression for fly and mouse are compared schematically in **Figure 38.26**. The domain of expression extends strongly to the posterior boundary shown in the figure, and then tails off into more posterior segments.

These results raise the extraordinary possibility that the clusters of genes not only share a common evolution, but also have maintained a common general function in which genome organization is related to spatial expression in fly and mouse, and there is some correspondence between the homologous genes. The idea of such a relationship is strengthened by the observation that ectopic expression of mouse *HoxD4* or *HoxB6* in *Drosophila* causes homeotic transformations virtually identical to those caused by homeotic expression of *Dfd* or *Antp*, respectively! Since the homology between these mouse and fly proteins rests almost exclusively with their homeodomains, this reinforces the view that these domains determine specificity.

There are some differences in the apparent behavior of the vertebrate *Hox* clusters and the *ANT-C/BX-C* fly clusters:

◆ The *Hox* genes are small, and there is a greater number of protein-coding units. The mouse *HoxB* cluster is ~120 kb and contains 9 genes. The connection between genomic position and embryonic expression is analogous to that in *Drosophila*, but describes only the genes themselves; we have no information about *cis*-acting sites. Of course, this may be a consequence of the much greater difficulty in generating mutations in vertebrates. However, our present information identifies the control of *Hox* genes only by promoters and enhancers in the region upstream of the startpoint. It remains to be seen whether there is any counterpart to the very extensive and complex regulatory regions of the *Drosophila* homeotic genes.

◆ In *Drosophila*, each gene is unique; but in vertebrates, the duplication of the clusters enables multiple genes (paralogs) to have the same or very similar patterns of expression. If the paralogs have redundant or partially redundant functions, so that the absence of one product may be at least partially substituted by the corresponding protein of another cluster, the effects of mutations will be minimized.

Hox genes can be disrupted in mice (by the techniques described in Chapter 36; see Figures 36.32–33). The disruptions often generate recessive lethals. In the examples of *HoxA1* and *HoxA3* various structures of the head and thorax are absent. Not all of the structures that usually express the mutant genes are missing, suggesting that there is indeed some functional redundancy, that is, other *Hox* genes of group 1 or group 3 can substitute in some but not all other tissues for the absence of the *HoxA* gene.

Homeotic transformations are less common with mutants in mice than in *Drosophila*, but sometimes occur. Loss of *HoxC8*, for example, causes some skeletal segments to show more anterior phenotypes. It remains to be seen whether this is a general rule.

Ectopic expression of *Hox* genes has been used successfully to demonstrate that gain-of-function can transform the identity of a segment towards the identity usually conferred by the gene. The most common type of effect in *Drosophila* is to transform a segment into a phenotype that is usually more posterior; in effect, the expression of the homeotic gene has added additional information that confers a more posterior identity on the segment. Similar effects are observed in some cases in the mouse. However, the pattern is not completely consistent.

Taken together, these results make it clear that the *Hox* genes resemble their counterparts in

Drosophila in determining segment identity. It may be the case that there is a combinatorial code of *Hox* gene expression, or there may be differences in degree of functional redundancy between paralogs, but we cannot yet provide a systematic model for their role in determining identity.

The most striking feature of organization of the *Hox* loci still defies explanation: why has the organization of the cluster, in which genomic position correlates with embryonic expression, been maintained in evolution? The obvious explanation is that there is some overall control of gene expression to ensure that it proceeds through the cluster, with the result that a gene could be properly expressed *only* when it is within the cluster. But this does not appear to be true, at least for those individual cases in which

genes have been removed from the cluster. Analysis of promoter regions suggest that a *Hox* gene may be controlled by a series of promoter or enhancer elements that together ensure its overall pattern of expression. Usually these elements are in the region upstream of the startpoint. For example, *HoxB4* expression can be reconstructed as the sum of the properties of a series of such elements, tested by introducing appropriate constructs to make transgenic mice. But then why should there have been evolutionary pressure to retain genes in an ordered cluster? One possibility is that an enhancer for one gene might be embedded within another gene, in such a way that, even if an individual gene could function when translocated elsewhere, its removal would impede the expression of other gene(s).

Summary

The development of segments in *Drosophila* occurs by the action of segmentation genes that delineate successively smaller regions of the embryo. Asymmetry in the distribution of maternal gene products is established by interactions between the oocyte and surrounding cells. This leads to the expression of the gap genes, in 4 broad regions of the embryo. The gap genes in turn control the pair-rule genes, each of which is distributed in 7 stripes; and the pair-rule genes define the pattern of expression of the segment polarity genes, which delineate individual compartments. At each stage of expression, the relevant genes are controlled both by the products of genes that were expressed at the previous stage, and by interactions among themselves. The segmentation genes act upon the homeotic genes, which determine the identities of the individual compartments.

Each of the 4 maternal systems consists of a cascade which generates a locally distributed or locally active morphogen. The morphogen either is

a transcription factor or causes the activation of a transcription factor. The transcription factor is the last component in each pathway.

The major anterior–posterior axis is determined by two systems: the anterior system establishes a gradient of bicoid from the anterior pole; and the posterior system produces nanos protein in the posterior half of the egg. These systems function to define a gradient of hunchback protein from the anterior end, with broad bands of knirps and giant in the posterior half. The terminal system acts to produce localized events at both termini. The dorso–ventral system produces a gradient of nuclear localization of dorsal protein on the ventral side, which represses expression of *dpp* and *zen*; this leads to the ventral activation of *twist* and *snail*, and the dorsal-side activation of *dpp* and *zen*.

The early embryo consists of a syncytium, in which nuclei are exposed to common cytoplasm. It is this feature that allows all 4 maternal systems to control the function of a nucleus according to the coordinates of its position on the anterior–posterior and dorso–ventral axes. At cellular blastoderm, zygotic RNAs are transcribed, and the developing embryo becomes dependent upon its own genes. Cells form at the blastoderm stage, after which successive interactions involve a cascade of transcriptional regulators.

Three gap genes are zinc-finger proteins, and one is a basic zipper protein. Their concentrations control expression of the pair-rule genes, which are also transcription factors. In particular, the expression of *eve* and *ftz* controls the boundaries of compartments, functioning in every other segment. The segment polarity genes represent the first step in the developmental cascade that involves functions other than transcription factors. Interactions between the segmentation gene products define unique combinations of gene expression for each segment.

Homeotic genes impose the program that determines the unique differentiation of each segment. The complex loci *ANT-C* and *BX-C* each contain a cluster of functions, whose spatial expression on the anterior–posterior axis reflects its genetic position in the cluster. Each cluster contains one exceedingly large transcription unit as well as other, shorter units. Many of the transcription units (including the largest genes, *Ubx* and *Antp*) have patterns of alternative splicing, but no significance has been attributed to this yet. Proceeding from left to right in each cluster, genes are expressed in more posterior tissues. The genes are expressed in overlapping patterns in such a way that addition of a function confers new features of posterior identity; thus loss of a function results in a homeotic transformation from posterior to more anterior phenotype. The genes are controlled in a complex manner by a series of regulatory sites that extend over large regions; mutations in these sites are *cis*-acting, and may cause either loss-of-function or gain-of-function. The *cis*-acting mutations tend to act on successive segments of the fly, by controlling expression of the homeotic proteins.

The genes of the *ANT-C* and *BX-C* loci, and many segmentation genes (including the maternal gene *bicoid* and most of the pair-rule genes) contain a conserved motif, the homeobox. Homeoboxes are also found in genes of other eukaryotes, including worms, frogs, and mammals. In each case, these genes are expressed during early embryogenesis. In mammals, the *Hox* genes (which specify homeodomains in the *Antennapedia* class) are organized in clusters. There are 4 *Hox* clusters in both man and mouse. These clusters can be aligned with the *ANT-C/BX-C* clusters in such a way as to recognize homologies between genes at corresponding positions. Proceeding towards the right in a *Hox* cluster, a gene is expressed more towards the posterior of the embryo. The *Hox* genes have roles in conferring identity on segments of the brain and skeleton (and other tissues). The analogous clusters represent regulators of embryogenesis in mammals and flies. *Hox* clusters may be a characteristic of all animals.

Drosophila genes containing homeoboxes form an intricate regulatory network, in which one gene may activate or repress another. The definition of these interactions and of the sites at which the proteins act has only just begun. The relationship between the sequence of the homeodomain, the DNA target it recognizes, and the regulatory consequences, remains to be fully elucidated. Specificity in target choice appears to reside largely in the homeodomain; we have yet to explain the abilities of a particular homeoprotein to activate or to repress gene transcription at its various targets. The general principle is that segmentation and homeotic genes act in a transcriptional cascade, in which a series of hierarchical interactions between the regulatory proteins is succeeded by the activation of structural genes coding for body parts.

Further reading

Historical

The mutant screen that initiated the field was performed by **Nusslein-Vollhard and Wieschaus** (*Nature* **287**, 795–801, 1980).

Reviews

A general review of *Drosophila* development has been provided by **Lawrence** (*The Making of a Fly*, Blackwell Scientific, Oxford, 1992). The background of genes involved in *Drosophila* development has been reviewed by **Mahowald and Hardy** (*Ann. Rev. Genet.* **19**, 149–177, 1985) and by **Scott and Carroll** (*Cell* **51**, 689–698, 1987). The molecular activities of these genes were reviewed by **Ingham** (*Nature* **335**, 25–34, 1988). The underlying rationale for the development of pattern in the *Drosophila* embryo was developed by **St Johnston and Nusslein-Vollhard** (*Cell* **68**, 201–219, 1992).

Complex loci have been analyzed by **Scott** (*Ann. Rev. Biochem.* **56**, 195–227, 1987). An extensive review of homeobox sequences and organization has been made by **Scott *et al.*** (*Biochim. Biophys. Acta* **989**, 25–48, 1989).

The connection between homeobox genes and axes in flies and vertebrates was reviewed by **McGinnis and Krumlauf** (*Cell* **68**, 283–302, 1992). Hox genes were reviewed further by **Krumlauf** (*Ann. Rev. Cell Biol.* **8**, 227–256, 1992).

Discoveries

The *BX-C* complex was cloned by **Bender *et al.*** (*Science* **221**, 23–29, 1983) and characterized by **Beachy *et al.*** (*Nature* **313**, 545–551, 1985) and **Karch *et al.*** (*Cell* **43**, 81–96, 1985). *ANT-C* was characterized by **Scott *et al.*** (*Cell* **35**, 763–766, 1983).

The relationship between gradients and morphogenesis in the anterior–posterior systems was explored by **Anderson *et al.*** (*Cell* **42**, 791–798, 1985) and by **Driever and Nusslein-Volhard** (*Cell* **54**, 83–93 and 95–104, 1988). The dorso–ventral gradient was characterized by **Roth, Stein, and Nusslein-Vollhard** (*Cell* **59**, 1189–1202, 1989). The role of the hunchback gradient at the next stage was reported by **Struhl, Johnston, and Lawrence** (*Cell* **69**, 237–250, 1992). The idea that follicle cells imprint pattern on the embryo emerged from the work of **Schupback** (*Cell* **49**, 699–707, 1987).

Expression of a mouse *Hox* gene cluster was systematically examined by **Graham, Papalopulu, and Krumlauf** (*Cell* **57**, 367–378, 1989). The relationship between fly and mammalian genes was explored by **Malicki, Schughart, and McGinnis** (*Cell* **63**, 961–967, 1990)

CHAPTER 39

Oncogenes: gene expression and cancer

A major feature of all higher eukaryotes is the defined life span of the organism, a property that extends to the individual somatic cells, whose growth and division is highly regulated. A notable exception is provided by cancer cells, which arise as variants that have lost their usual growth control. Their ability to grow in inappropriate locations or to propagate indefinitely may be lethal for the individual organism in which they occur.

Three types of changes that occur when a cell becomes tumorigenic are summarized in **Figure 39.1**:

◆ **Immortalization** describes the property of indefinite growth (without any other changes in the phenotype necessarily occurring). It involves changes in growth control that are not well defined in molecular terms.

◆ **Transformation** describes the failure to observe the normal constraints of growth; for example, transformed cells become independent of factors usually needed for cell growth. Features that are affected by transformation can be described quite well in molecular and cellular terms.

◆ **Metastasis** describes another feature—one of the most damaging to the organism—in which the cancer cell gains the ability to invade normal tissue, so that it can move away from the tissue of origin and establish a new colony elsewhere in the body. Molecular events involved in metastasis (such as the secretion of proteases) are just beginning to be described.

To characterize the aberrant events that enable cells to bypass normal control and generate tumors, we need to compare the growth characteristics of normal and transformed cells *in vitro*. Transformed cells can be grown readily, but it is much more difficult to grow their normal counterparts.

When cells are taken from a vertebrate organism and placed in culture, they grow for several (~50) divisions, but then enter **crisis**, in which most of the cells die. The survivors that emerge are capable of dividing indefinitely, but by definition their properties have changed in the act of emerging from crisis. The nature of crisis is poorly defined, and we have little understanding of the molecular changes that adapt a cell to growth in culture, but in principle this comprises the process of immortalization.

The limitation of the life span of most cells by crisis restricts us to two options in studying nontransformed cells, neither entirely satisfactory:

◆ **Primary cells** are the immediate descendants of cells taken directly from the organism. They faithfully mimic the *in vivo* phenotype, but in most cases survive for only a relatively short period, because the culture dies out at crisis.

◆ Cells that have passed through crisis become **established** to form a (nontumorigenic) cell line. They can be perpetuated indefinitely, but their properties have changed in passing through crisis, and may indeed continue to change during adaptation to culture. These changes may partly resemble those involved in tumor formation, which reduces the usefulness of the cells.

Figure 39.1

Overview: three types of properties distinguish a cancer cell from a normal cell. Sequential changes in cultured cells can be correlated with changes in tumorigenicity.

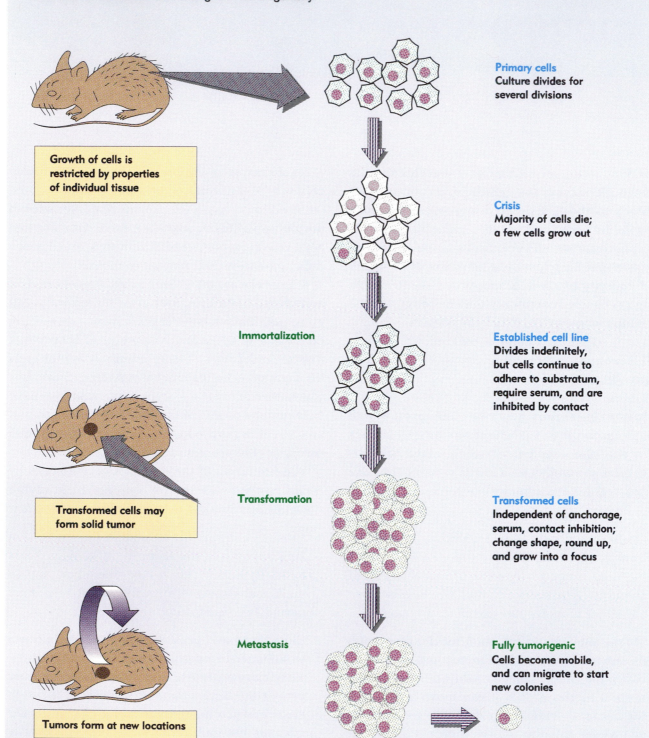

Growth of cells is restricted by properties of individual tissue

Primary cells
Culture divides for several divisions

Crisis
Majority of cells die; a few cells grow out

Immortalization

Established cell line
Divides indefinitely, but cells continue to adhere to substratum, require serum, and are inhibited by contact

Transformation

Transformed cells may form solid tumor

Transformed cells
Independent of anchorage, serum, contact inhibition; change shape, round up, and grow into a focus

Metastasis

Fully tumorigenic
Cells become mobile, and can migrate to start new colonies

Tumors form at new locations

An established cell line by definition has become immortalized, but is usually not tumorigenic. Nontumorigenic established cell lines remain subject to growth control, and display characteristic features similar to those of primary cultures, often including:

◆ **Anchorage dependence**—a solid or firm surface is needed for the cells to attach to.

◆ **Serum (or growth factor) dependence**—serum is needed to provide essential growth factors.

◆ **Density-dependent inhibition**—cells grow only to a limited density, because growth is inhibited, perhaps by processes involving cell–cell contacts.

◆ **Cytoskeletal organization**—cells are flat and extended on the surface on which they are growing, and have a typical elongated network of stress fibers (consisting of actin filaments).

The consequence of these properties is that the cells grow only as a thin layer or **monolayer** on a substratum.

These properties provide parameters by which the normality of the cell may be judged. Of course, any established cell line provides only an approximation of *in vivo* control. The need for caution in analyzing the genetic basis for growth control in such lines is emphasized by the fact that almost always they suffer changes in the chromosome complement and are not true diploids. A cell whose chromosomal constitution has changed from the true diploid is said to be **aneuploid**.

Cells cultured from tumors instead of from normal tissues show changes in some or all of these properties. They have passed through the second stage of conversion and are said to be transformed. A transformed cell grows in a much less restricted manner: usually it does not need a solid surface to which to attach (so that individual cells 'round-up' instead of spreading out), has reduced serum dependence, piles up into a thick mass of cells (called a **focus**) instead of being restricted to a surface monolayer, and may induce tumors when injected into appropriate test animals. **Figure 39.2** compares a 'normal' fibroblast growing in culture with a 'transformed' variant.

It would be naive to suppose that there is a uniform basis for cancer cell formation; many types of changes in the cellular constitution confer the ability to form a tumor. However, the joint changes of immortalization and transformation of cells in culture provide a paradigm for the formation of animal tumors. By comparing transformed cell lines with normal cells, we hope to identify the genetic basis for tumor formation and also to understand the phenotypic processes that are involved in the conversion.

Certain events convert normal cells into transformed cells, and are therefore models for the processes involved in tumor formation. These events may be triggered by environmental or genetic changes. Usually multiple changes are necessary to create a cancer; and sometimes tumors gain increased virulence as the result of a progressive series of changes. The incidence of human cancers with age suggests that typically 6–7 events are required over a span of 20–40 years to induce a cancer.

A variety of agents increase the frequency with which cells (or animals) are converted to the transformed condition; they are said to be **carcinogenic**. Sometimes these **carcinogens** are divided into those that 'initiate' and those that 'promote' tumor formation, implying the existence of different stages in cancer development. Do the carcinogens cause epigenetic changes or act, directly or indirectly, to change the genotype of the cell?

In certain (rare) cases, propensity to cancer is inherited as a Mendelian trait, implying that a single genetic change is sufficient. In some tumors, multiple genetic changes have been identified; some could trigger or be essential for tumor growth, but others merely assist growth.

Until recently, we had no insight into the molecular basis for cancer. However, we now know of the existence of two classes of genes in which mutations cause transformation:

Figure 39.2

Normal fibroblasts grow as a layer of flat, spread-out cells, whereas transformed fibroblasts are rounded up and grow in cell masses. The cultures on the left contain normal cells, those on the right contain transformed cells. The top views are by conventional microscopy, the bottom by scanning electron microscopy. Photographs kindly provided by Hidesaburo Hanafusa and J. Michael Bishop.

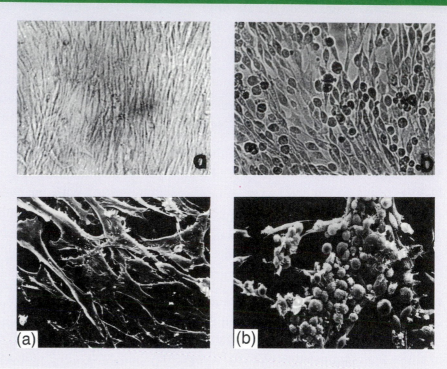

◆ **Oncogenes** were initially identified as genes carried by viruses that cause transformation of their target cells. A major class of the viral oncogenes have cellular counterparts that are involved in normal cell functions. The cellular genes are called **proto-oncogenes**, and in certain cases their mutation or aberrant activation in the cell is associated with tumor formation. About 100 oncogenes have been identified. The oncogenes fall into several groups, representing different types of activities ranging from trans-membrane proteins to transcription factors, and the definition of these functions may therefore lead to an understanding of the types of changes that are involved in tumor formation. The generation of an oncogene represents a gain-of-function in which a cellular proto-oncogene is either acquired by a virus or activated in some other way. The change can involve constitutive activation of a function that usually is regulated, expression of the gene in a cell type where usually it is not expressed, or over-expression in the usual tissue. Often it involves mutational changes in the protein product.

◆ **Tumor suppressors** are detected in the form of deletions (or other inactivating mutations) that are tumorigenic. The most compelling evidence for their nature is provided by certain hereditary cancers, in which patients with the disease develop tumors that have lost both alleles, and therefore lack an active gene. There is also now evidence that changes in these genes may be associated with the progression of a wide range of cancers. About 10 tumor suppressors are known at present. They represent loss-of-function in genes that usually impose some constraint on the cell cycle or cell growth; the release of the constraint is tumorigenic.

In the first part of this chapter, we consider how these two classes of genes are identified, and we ask how oncogenes are activated, and tumor suppressors are inactivated. Then we consider the molecular basis for these events, and how oncogenes are connected into pathways that extend from signal transduction at the cell surface to activation of transcription factors in the nucleus.

Transforming viruses may carry oncogenes

Transformation may occur spontaneously, may be caused by certain chemical agents, and, most notably, may result from infection with **tumor viruses**. There are many classes of tumor viruses, including both DNA and RNA viruses, and they occur widely in the avian and animal kingdoms.

What allows a tumor virus to transform its target cell? The virus does not carry an extensive set of genes directly coding for products that introduce the many necessary changes in the cell phenotype. Rather does the virus initiate a series of events that is executed by cellular proteins. We assume that in some sense the virus throws a regulatory switch that changes the growth properties of its target cell.

The transforming activity of a tumor virus resides in a particular gene or genes carried in the viral genome. Oncogenes were given their name by virtue of their ability to convert cells to a tumorigenic (or oncogenic) state. **Table 39.1** summarizes the general properties of the major classes of transforming viruses.

Polyomaviruses and adenoviruses have been isolated from a variety of mammals. Although perpetuated in the wild on a single species, a virus may be able to grow in culture on a variety of cells from different species. The response of a cell to infection depends on its species and phenotype, and falls into one of two classes, as illustrated in **Figure 39.3**:

◆ **Permissive cells** are productively infected. The virus proceeds through a lytic cycle that is divided into the usual early and late stages. The cycle ends with release of progeny viruses and (ultimately) cell death.

Table 39.1

Transforming viruses may carry oncogenes.

Viral Class	Type of Virus	Genome Size	Oncogenes	Origin of Oncogene
Polyoma	DNA duplex	5-6 kb	T antigens	Early viral gene
HPV	DNA duplex	˜8 kb	E6 & E7	Early viral gene
Adeno	DNA duplex	˜37 kb	E1A & E1B	Early viral gene
Epstein-Barr	DNA duplex	˜160 kb	BNLF-1	Latent viral gene
Retrovirus (acute)	RNA single strand	6-9 kb	Individual	Cellular

Figure 39.3

Permissive cells are productively infected by a DNA tumor virus that enters the lytic cycle, while nonpermissive cells are transformed to change their phenotype.

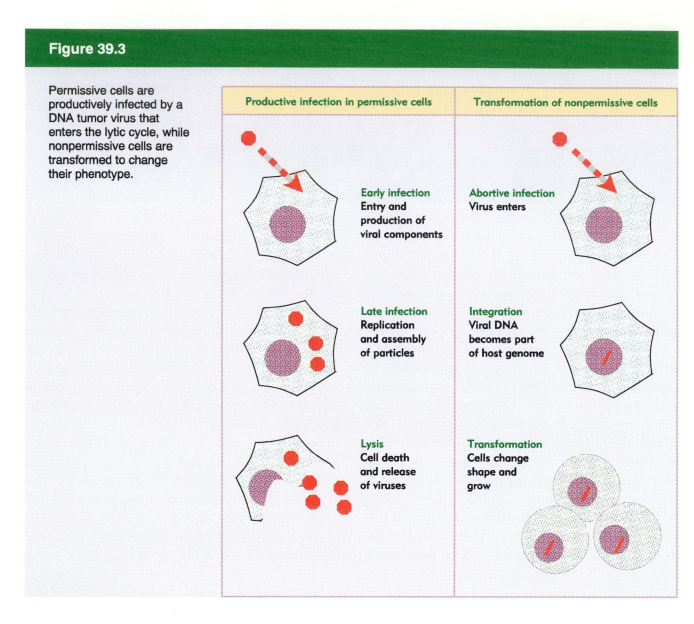

Productive infection in permissive cells	Transformation of nonpermissive cells
Early infection Entry and production of viral components	**Abortive infection** Virus enters
Late infection Replication and assembly of particles	**Integration** Viral DNA becomes part of host genome
Lysis Cell death and release of viruses	**Transformation** Cells change shape and grow

◆ **Nonpermissive cells** cannot be productively infected, and viral replication is abortive. Some of the infected cells are transformed; in this case, the phenotype of the individual cell changes and the culture is perpetuated in an unrestrained manner.

Polyomaviruses viruses are small. Polyoma virus itself is common in mice, the analogous virus SV40 (simian virus 40) was isolated from rhesus monkey cells, and more recently the human viruses BK and JC have been characterized. All of the polyomaviruses can cause tumors when injected into newborn rodents.

During a productive infection, the early region of each virus uses alternative splicing to synthesize overlapping proteins called T antigens. (The name reflects their isolation originally as the proteins found in *t*umor cells.) The various T antigens have a variety of functions in the lytic cycle. They are required for expression of the late region and for DNA replication of the virus.

Cells transformed by polyomaviruses contain integrated copies of part or all of the viral genome. The integrated sequences always include the early region. The responsibility of T antigens for transformation has been confirmed by showing that mutants in the corresponding genes do not transform, and by introducing the wild-type genes to demonstrate directly that they can transform nonpermissive cells. The transforming activity of T antigens rests upon their ability to interact with cellular proteins to stimulate quiescent cells to divide; this is separate from their ability to interact directly with the viral genome.

Papillomaviruses are small DNA viruses that cause epithelial tumors; there are ~60 human papillomaviruses (HPVs); most are associated with benign growths (such as warts), but some are associated with cancers, in particular cervical cancers. Two viral-associated products are expressed in cervical cancers; these are the E6 and E7 proteins, which can immortalize target cells.

Adenoviruses were originally isolated from human adenoids; similar viruses have since been isolated from other mammals. They comprise a large group of related viruses, with >80 individual members. Human adenoviruses remain the best characterized, and are associated with respiratory diseases. They can infect a range of cells from different species.

Human cells are permissive and are therefore productively infected by adenoviruses, which replicate within the infected cell. But cells of some rodents are nonpermissive. All adenoviruses can transform nonpermissive cultured cells, but the oncogenic potential of the viruses varies; the most effective can cause tumors when they are injected into newborn rodents. The genomes of cells transformed by adenoviruses have gained a part of the early viral region that contains the E1A and E1B genes.

The E1A and E1B regions code for several nuclear proteins that in each case are related by alternative splicing. The E1A and E1B proteins act upon other viral proteins to further the lytic cycle, and upon a variety of cellular targets in transformation. Whether a cell is permissive or nonpermissive presumably depends upon its response to expression of these early adenovirus functions.

A common mechanism underlies the ability of DNA tumor viruses to transform target cells. *Oncogenic potential resides in a single function or group of related functions that are active early in the viral lytic cycle. When transformation occurs, the relevant gene(s) are integrated into the genomes of transformed cells and expressed constitutively.* This suggests the general model for transformation by these viruses illustrated in **Figure 39.4**, in which the constitutive expression of the oncogene generates transforming protein(s) (sometimes called onco-proteins). The polyomaviruses and adenoviruses tend to have broad host ranges, so we assume that the cellular targets for their oncogenes are relatively non tissue-specific.

Epstein–Barr is a human herpes virus associated with a variety of diseases, including infectious mononucleosis, nasopharyngeal carcinoma, African Burkitt lymphoma, and other lymphoproliferative disorders. EBV has a limited host range for both species and cell phenotype. Human B lymphocytes that are infected *in vitro* become immortalized, and some rodent cell lines can be transformed. Viral DNA is found in transformed cells, although it has been controversial whether it is integrated. It has been difficult to establish which, if any, viral function is required for transformation, but now it seems that a single virus gene is associated with transformation; BNLF-1 codes for a membrane protein.

Retroviruses present a different situation from the DNA tumor viruses. They exist in two forms:

◆ Replication-competent viruses perpetuate themselves, but do not damage the host cells that carry them. They are therefore described as transformation-incompetent or nontransforming.

◆ Transformation-competent retroviruses have the ability to change the phenotype of an infected cell so that it becomes tumorigenic. They may or may not be able to replicate independently (see below).

Retroviruses can transfer genetic information both horizontally and vertically, as illustrated in **Figure 39.5**. Horizontal transfer is accomplished by the normal process of viral infection, in which

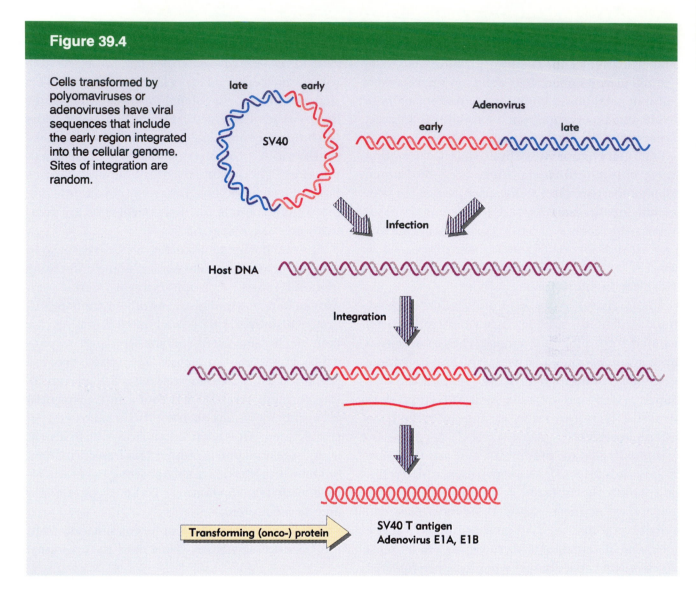

Figure 39.4

Cells transformed by polyomaviruses or adenoviruses have viral sequences that include the early region integrated into the cellular genome. Sites of integration are random.

late   early

SV40

Adenovirus

early   late

Infection

Host DNA

Integration

Transforming (onco-) protein

SV40 T antigen
Adenovirus E1A, E1B

increasing numbers of cells become infected in the same host. Vertical transfer results whenever a virus becomes integrated in the germline of an organism as an endogenous provirus; like a lysogenic bacteriophage, it is inherited as a Mendelian locus by the progeny (see Chapter 35).

The retroviral life cycle propagates genetic information through both RNA and DNA templates. A retroviral infection proceeds through the stages illustrated previously in Figure 35.2, in which the RNA is reverse-transcribed into single-stranded DNA, then converted into double-stranded DNA, and finally integrated into the genome, where it may be transcribed again into infectious RNA.

Integration into the genome leads to vertical transmission of the provirus; expression of the provirus may generate retroviral particles that are horizontally transmitted. Integration is a normal part of the life cycle of every retrovirus, whether it is nontransforming or transforming.

The tumor retroviruses fall into two general groups with regards to the origin of their tumorigenicity:

◆ **Nondefective viruses** have the same general structure as replication-competent retroviruses. Their tumorigenic ability is conferred by individual mutations and other genetic changes relative to the nontransforming counterparts. As indicated by their name, the nondefective viruses follow the usual retroviral life cycle. They provide infectious agents that have a long latent period, and often are associated with the induction of leukemias. Two classic models are FeLV (feline leukemia virus) and MMTV (mouse mammary tumor virus). *Tumorigenicity does not rely upon an individual viral oncogene, but upon the ability of the virus to activate a cellular proto-oncogene(s).*

◆ **Acute transforming viruses** have gained new genetic information in the form of an oncogene.

Figure 39.5

Retroviruses transfer genetic information horizontally by infecting new hosts; information is inherited vertically if a virus integrates in the genome of the germ line.

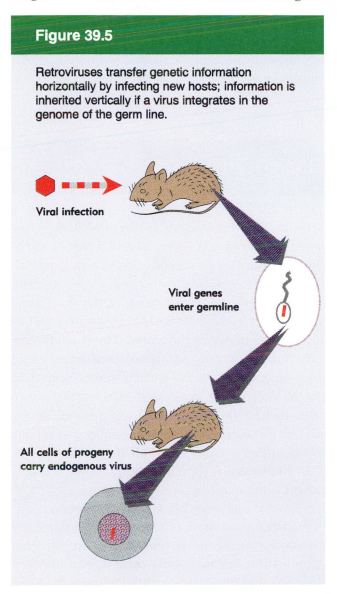

Viral infection

Viral genes enter germline

All cells of progeny carry endogenous virus

Figure 39.6

A transforming retrovirus carries a copy of a cellular sequence in place of some its own gene(s).

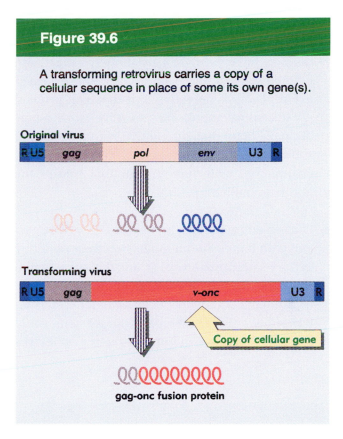

Original virus

Transforming virus

Copy of cellular gene

gag-onc fusion protein

This gene is not present in the ancestral (nontransforming virus); it originated as a cellular gene that was captured by the virus by means of a transduction event during an infective cycle. These viruses usually induce tumor formation *in vivo* rather rapidly, and they can transform cultured cells *in vitro*.

Reflecting the fact that each acute transforming virus has specificity toward a particular type of target cell, these viruses are divided into classes according to the type of tumor that is caused in the animal: leukemia, sarcoma, carcinoma, etc. Specificity may be a property of the particular oncogene carried by a transforming virus.

How does a retrovirus capture a cellular gene? During an infectious cycle, a retrovirus may exchange part of its own sequence for a cellular sequence, as illustrated previously in Figure 35.8 and summarized again in **Figure 39.6**. This type of event is rare, but generates a transducing virus that has two important properties:

◆ Usually it cannot replicate by itself, because (*trans*-acting) viral functions needed for reproduction have been lost by the exchange with cellular sequences. So almost all of these viruses are replication-defective. But they can propagate in a simultaneous infection with a wild-type 'helper' virus that provides the functions that were lost in the recombination event. (RSV is an exceptional transducing virus that retains the ability to replicate).

◆ During an infection, the transducing virus carries with it the cellular genes that were obtained in the recombination event, and their expression may alter the phenotype of the infected cell. Any transducing virus whose cellular genetic information assists the growth of its target cells could have an advantage in future infective cycles. If a virus gains a gene whose product stimulates cell growth, the acquisition may enable the virus to spread by stimulating the growth of the particular cells that it infects. After a virus has collected a cellular gene, the gene may gain mutations that enhance its ability to influence cell phenotype.

Of course, transformation is not the only mechanism by which retroviruses affect their hosts. A notable example is the HIV-1 retrovirus, which belongs to the retroviral group of lentiviruses. The virus infects and kills T lymphocytes carrying the CD4 receptor, devastating the immune system of the host, and inducing the disease of AIDS. The virus carries the usual *gag-pol-env* regions, and also has an additional series of reading frames, which overlap with one another, to which its lethal actions are attributed.

Retroviral oncogenes have cellular counterparts

Oncogenes of some retroviruses are summarized in **Table 39.2**. It is striking that usually the oncogenic activity resides in a single gene. AEV is one of a very few exceptions in which a retrovirus carries more than one oncogene.

The new sequences present in an acute transforming retrovirus can be delineated by comparing the sequence of the virus with that of the parental (nontumorigenic) virus. Invariably there is a new region that is closely related to a sequence in the cellular genome. The normal cellular sequence itself is not oncogenic—if it were, the organism could scarcely have survived—but it defines a proto-oncogene, a cellular gene whose capture by the retrovirus and subsequent modification created the oncogene.

The viral oncogenes and their cellular counterparts are described by using prefixes *v* for viral and *c* for cellular. Thus the oncogene carried by Rous sarcoma virus is called *v-src*, and the proto-oncogene related to it in cellular genomes is called *c-src*.

As described previously in Figure 35.9, the *v-onc* gene is captured in the form of RNA during a retroviral infection. When we compare a *v-onc* sequence with a corresponding *c-onc* sequence, we find inevitably that the organization of the viral gene corresponds to the mRNA of the *c-onc* gene rather than to its genomic organization. Thus *v-onc* genes consist of uninterrupted coding sequences, while the *c-onc* genes have the usual genomic organization of alternating exons and introns. When the *v-onc* coding sequence is closely related to the sequence of a gene or part of a gene in the cellular host genome, we may be able to define the boundaries where the recombination event occurred to insert the cellular sequence in the virus.

Some 25 *c-onc* genes have been identified so far by their representation in retroviruses. Sometimes the same *c-onc* gene is represented in different transforming viruses; for example, the monkey virus SSV and the PI strain of the feline virus FeSV both carry a *v-onc* derived from *c-sis*. Some viruses

Table 39.2

Each transforming retrovirus carries an oncogene derived from a cellular gene. Viruses have names and abbreviations reflecting the history of their isolation and the types of tumor they cause. This list shows some representative examples of the retroviral oncogenes.

Virus	Name	Species	Tumor	Oncogene
Rous sarcoma	RSV	Chicken	Sarcoma	*src*
Harvey murine sarcoma	Ha-MuSV	Rat	Sarcoma & erythroleukemia	*H-ras*
Kirsten murine sarcoma	Ki-MuSV	Rat	Sarcoma & erythroleukemia	*K-ras*
Moloney murine sarcoma	Mo-MuSV	Mouse	Sarcoma	*mos*
FBJ murine osteosarcoma	FBJ-MuSV	Mouse	Chondrosarcoma	*fos*
Simian sarcoma	SSV	Monkey	Sarcoma	*sis*
Feline sarcoma	PI-FeSV	Cat	Sarcoma	*sis*
Feline sarcoma	SM-FeSV	Cat	Fibrosarcoma	*fms*
Feline sarcoma	ST-FeSV	Cat	Fibrosarcoma	*fes*
Avian sarcoma	ASV-17	Chicken	Fibrosarcoma	*jun*
Fujinami sarcoma	FuSV	Chicken	Sarcoma	*fps*
Avian myelocytomatosis	MC29	Chicken	Carcinoma, sarcoma, & myelocytoma	*myc*
Abelson leukemia	MuLV	Mouse	B cell lymphoma	*abl*
Reticuloendotheliosis	REV-T	Turkey	Lymphatic leukemia	*rel*
Avian erythroblastosis	AEV	Chicken	Erythroleukemia & fibrosarcoma	*erbB, erbA*
Avian myeloblastosis	AMV	Chicken	Myeloblastic leukemia	*myb*

carry related *v-onc* genes, such as in the Harvey and Kirsten strains of MuSV, which carry *v-ras* genes derived from two different members of the cellular *c-ras* gene family. In other cases the *v-onc* genes of related viruses represent unrelated cellular progenitors; for example, three different isolates of FeSV may have been derived from the same original (nontransforming) virus, but have transduced the *sis*, *fms*, and *fes* oncogenes. The events involved in formation of a transducing virus can be complex; some viruses include sequences derived from more than one cellular gene.

Given the rarity of the transducing event, it is significant that multiple independent isolates occur representing the same *c-onc* gene. For example, several viruses carry *v-myc* genes, differing in their exact ends, and related by a rather small number of point mutations. The existence of such isolates probably means that we have identified most of the genes of the *c-onc* type that can be activated by viral transduction.

Two general types of theory might explain the difference in properties between *v-onc* genes and *c-onc* genes:

◆ A quantitative model proposes that viral genes are functionally indistinguishable from the cellular genes, but are oncogenic because they are expressed in much greater amounts or in inappropriate cell types, or because their expression cannot be switched off.

◆ A qualitative model supposes that the *c-onc* genes intrinsically lack oncogenic properties, but may be converted by mutation into oncogenes whose devastating effects reflect the acquisition of new properties (or loss of old properties).

Of course, these models are not mutually exclusive; and each may be correct for some oncogenes. A wide variety of mechanisms activate proto-oncogenes. Some proto-oncogenes are activated by increases in the level of protein product.

The level of expression of a *v-onc* gene is typically much greater than that of the *c-onc* gene, given the level of expression of the retroviral genome during an infection. In the case of a cellular gene, a regulatory change in the genome may increase expression so that a proto-oncogene is activated. Other proto-oncogenes lack effect when over-expressed, and become oncogenic only as the result of changes in the oncoprotein sequence itself.

How closely related are *v-onc* genes to the corresponding *c-onc* genes? **Table 39.3** summarizes the relationships between some typical *v-onc* and *c-onc* genes.

The *mos, ras, sis*, and *myc* genes identify cases in which the entire *c-onc* gene has been gained by the virus, where it has suffered relatively few changes that affect its product. This suggests that the *v-onc* product is likely to fulfill the same enzymatic or other functions as the *c-onc* product, but with some change in its regulation. In the example of *ras*, which has been studied in detail, changes in the regulation of activity can be directly attributed to the individual point mutations that have occurred in the *v-onc* gene. In the cases of *mos, sis*, and *myc*, the mutations in the sequence are not required for transforming activity, and over-expression is responsible for oncogenicity.

In some cases, the *v-onc* gene has been truncated by the loss of sequences from the N-terminus or C-terminus (or both) of the *c-onc* gene. Loss of these regions may remove some regulatory constraint that normally limits the activity of the *c-onc* product. Thus the viral and cellular *src* genes are coextensive, but *v-src* has replaced the C-terminal 19 amino acids of *c-src* with a different sequence of 12 amino acids. This has an important regulatory consequence, in activating the Src protein constitutively. In cases where *v-onc* genes are truncations of *c-onc* genes, point mutations may also contribute to the oncogenicity of the *v-onc* product.

One notable difference between c-onc proteins and v-onc proteins is that the sequence carried by the virus may be expressed by readthrough from a preceding viral sequence. As a result, the v-onc protein sequence may be part of a larger protein also carrying other viral functions (often part of the *gag* sequence which binds the viral RNA in the core of the retroviral particle). And because the *v-onc* sequence may correspond to only part of the *c-onc*

Table 39.3

The sequences of *v-onc* coding regions are derived from *c-onc* genes by substitutions and (sometimes) loss of terminal regions.

Gene	Changes in corresponding regions	Identity	Region Missing in *v-onc*
mos	11 / 369 substitutions	97%	None
H-ras	3 / 189	98%	None
K-ras	7 / 189	96%	None
sis	18 / 220	92%	None
myc	2 / 417	99%	None
src	16 / 514 substitutions	97%	C-terminal 19 amino acids
fms	20 / 930	98%	C-terminal 50 amino acids
erbB	99 / 600	83%	N-terminal half & C-terminus (610 total)
erbA	22 / 396	95%	N-terminal 12 amino acids
myb	11 / 372	97%	N- & C-termini (268 residues)

sequence, the change in size (representing loss of *c-onc* domains and/or gain of viral domains) may influence the activity of the *v-onc* function. In principle, such changes could alter accessibility or localization of the protein.

What is the evidence that the *v-onc* sequence is in fact responsible for the oncogenic potential of a transforming virus? Two approaches have been taken to identify the features that confer oncogenicity:

◆ Mutating the *v-onc* sequence to abolish transforming activity.

◆ Finding conditions under which the *c-onc* gene becomes oncogenic, for example, by changing the *c-onc* sequence (often by exchanging parts with *v-onc*).

Direct evidence that expression of the *v-onc* sequence accomplishes transformation was first obtained with RSV. Temperature-sensitive mutations in *v-src* allow the transformed phenotype to be reverted by increase in temperature, and regained by decrease in temperature. This shows clearly that the *v-src* gene is needed both to initiate and maintain the transformed state.

Src is an exceptional case, in which the viral protein is synthesized independently, and thus closely resembles the cellular protein. Specific sequence changes are required for oncogenicity. Thus *v-src* is oncogenic at low levels of protein, but *c-src* is not oncogenic at high protein levels (>10× normal).

Some proto-oncogenes are weakly oncogenic when the *c-onc* is expressed at increased levels, but sequence changes found in *v-onc* or occurring spontaneously in *c-onc* activate the gene more effectively, as in the case of *ras* and possibly *myc*.

In other cases, most notably *c-myc,* changes in protein sequence do not appear to be essential for oncogenicity, and over-expression or altered regulation is responsible for the oncogenic phenotype.

Ras proto-oncogenes can be activated by mutation

Are oncogenes responsible for transformation events that do not involve viral infection? We can assay directly for transforming genes by transfecting 'normal' recipient cells with DNA obtained from animal tumors. (Actually the established mouse 3T3 fibroblast line usually is used as recipient.) The procedure is illustrated diagrammatically in **Figure 39.7**. If a specific gene contributes to the transformed state, its introduction into new cells transforms them. Another assay that can be used is to inject cells into 'nude' mice (which lack the ability to reject such transplants immunologically). The ability to form tumors can then be measured directly in the animal.

When a cell is transformed in a 3T3 culture (or some other 'normal' culture), its descendants pile up into a **focus**. The appearance of foci is used as a measure of the transforming ability of a DNA preparation. Starting with a preparation of DNA isolated from tumor cells, the efficiency of focus formation is low. However, once the transforming gene has been isolated and cloned, greater efficiencies can be obtained. In fact, the transforming 'strength' of a gene can be characterized by the efficiency of focus formation by the cloned sequence. A highly transforming gene might have an efficiency of >100 foci/ng DNA/10^6 cells, while a weakly transforming gene's efficiency might be <10 foci/ng DNA/10^6 cells.

DNA with transforming activity can be isolated only from tumorigenic cells; it is not present in normal DNA. The transforming genes isolated by this assay have two revealing properties:

Figure 39.7

The transfection assay allows (some) oncogenes to be isolated directly by assaying DNA of tumor cells for its ability to transform normal cells into tumorigenic cells.

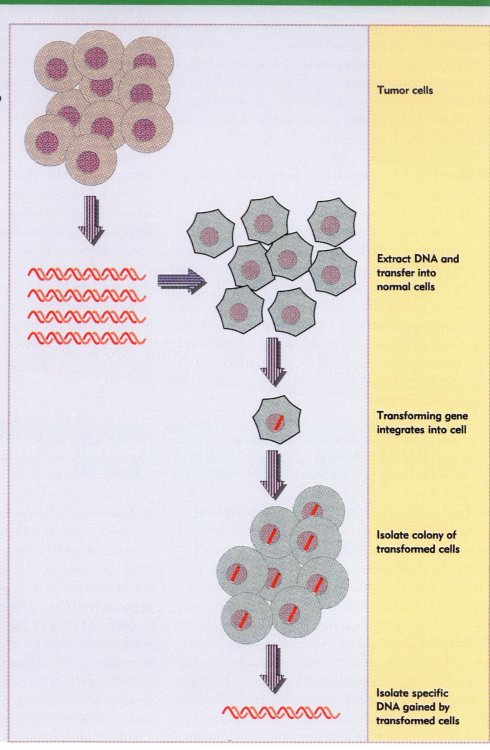

Tumor cells

Extract DNA and transfer into normal cells

Transforming gene integrates into cell

Isolate colony of transformed cells

Isolate specific DNA gained by transformed cells

◆ *They have closely related sequences in the DNA of normal cells.* This argues that transformation was caused by mutation of a normal cellular gene (a proto-oncogene) to generate an oncogene. The change may take the form of a point mutation or more extensive reorganization of DNA around the *c-onc* gene.

◆ *They may have counterparts in the oncogenes carried by known transforming viruses.* This suggests that the repertoire of proto-oncogenes is limited, and probably the same genes are targets for mutations to generate oncogenes in the cellular genome or to become viral oncogenes.

Oncogenes derived from the *c-ras* family are often detected in the transfection assay. The family consists of three active genes in both man and rat, dispersed in the genome. (There are also some pseudogenes.) The individual genes, *N-ras, H-ras,* and *K-ras* are closely related, and code for protein products ~21,000 daltons and known collectively as p21^{ras}. We discuss their functions in signal transduction pathways later.

The *H-ras* and *K-ras* genes have *v-ras* counterparts, carried by the Harvey and Kirsten strains of murine sarcoma virus, respectively (see Table 39.2). Each *v-ras* gene is closely related to the corresponding *c-ras* gene, with only 3 or 7 amino acid changes, respectively (see Table 39.3). The Harvey and Kirsten virus strains must have originated in independent recombination events in which a progenitor virus gained one of the *c-ras* sequences.

Oncogenic variants of the *c-ras* genes are found in transforming DNA preparations obtained from various primary tumors and tumor cell lines. Each of the *c-ras* proto-oncogenes can give rise to a transforming oncogene by a single base mutation. *The mutations in several independent human tumors cause substitution of a single amino acid, most commonly at position 12 or 61, in one of the ras proteins.*

Position 12 is one of the residues that is mutated in the *v-H-ras* and *v-K-ras* genes. *Thus mutations at the same positions altered in the retrovirus are found in mutant ras genes in multiple rat and human tumors.* This suggests that the normal Ras protein has a high potential to be converted into a tumorigenic form by a mutation in one of a few codons in rat or man (and perhaps any mammal). We discuss the molecular effects of these mutations on protein function later.

The 'tumorigenic' form of *c-ras* is defined as such by its ability to carry this property when transfected into a new cell. The fact that several different cell lines derived from human or rodent tumors have mutant *c-ras* genes provides a powerful correlation between the presence of an active oncogene and the existence of a tumor. But it does not prove that the mutated oncogene was responsible for creating the original tumor; its function may be one of several that must be activated for tumor formation or propagation.

The general principle established by this work is that *amino acid substitution can convert a cellular proto-oncogene into an oncogene.* Such an oncogene can be associated with the appearance of a spontaneous tumor in the organism. It may also be carried by a retrovirus, in which case a tumor is induced by viral infection.

The *ras* genes appear to be finely balanced at the edge of oncogenesis. Almost any mutation at either position 12 or 61 can convert a *c-ras* proto-oncogene into an active oncogene. All three *c-ras* genes have glycine at position 12. If it is replaced *in vitro* by any other of the 19 amino acids except proline, the mutated *c-ras* gene can transform cultured cells. The particular substitution influences the strength of the transforming ability.

Position 61 is occupied by glutamine in wild-type *c-ras* genes. Its change to another amino acid usually creates a gene with transforming potential. Some substitutions are less effective than others; proline and glutamic acid are the only substituents that have no effect.

When the expression of a normal *c-ras* gene is increased, either by placing it under control of a more active promoter or by introducing multiple copies into transfected cells, recipient cells are transformed. Some mutant *c-ras* genes that have changes in the protein sequence also have a mutation in an intron that increases the level of expression (by increasing processing of mRNA

~10×). Also, some tumor lines have amplified *ras* genes. A 20-fold increase in the level of a nontransforming Ras protein is sufficient to allow the transformation of some cells. The effect has not been fully quantitated, but it suggests the general conclusion that oncogenesis depends on overactivity of Ras protein, and is caused either by increasing the amount of protein or (probably more efficiently) by a variety of mutations that increase the activity of the protein.

Transfection by DNA can be used to transform only certain cell types. Although transforming oncogenes have been isolated from both rodent and human cells, most targets for transformation by transfection with oncogenes have been rodent fibroblasts in culture. (In fact, the difference in the source of the oncogene [human] and the recipient cell [rodent] is an important factor in allowing the donor gene to be distinguished unequivocally from recipient DNA.) Limitations of the assay explain why relatively few oncogenes have been detected by transfection. This system has been most effective with *ras* genes, where there is extensive correlation between mutations that activate *c-ras* genes in transfection and the occurrence of tumors.

Insertion, translocation, or amplification may activate proto-oncogenes

A variety of genomic changes can activate proto-oncogenes, sometimes involving a change in the target gene itself, sometimes activating it without changing the protein product. In cases of insertion and translocation, there is evidence that the genomic change is the causative event; in cases of amplification there is a correlation with tumorigenesis, but no direct proof for a causative role.

Many tumor cell lines have visible regions of chromosomal amplification, as shown by homogeneously staining regions (see Figure 36.27) or double minute chromosomes (see Figure 36.28). In some cases, the amplified region contains a known oncogene or a gene related to one. In other cases, where amplification is not visible, the use of batteries of probes representing oncogenes shows that a particular oncogene is amplified. Examples of oncogenes that are amplified in various tumors include *c-myc, c-abl, c-myb, c-erbB, and c-K-ras.*

Established cell lines are prone to amplify genes (it is one of several karyotypic changes to which they are susceptible). All the same, the presence of known oncogenes in the amplified regions, and the consistent amplification of particular oncogenes in many independent tumors of the same type, again strengthens the correlation between increased expression and tumor growth. Of course, it is possible that the gene amplification gives an advantage to growth of the established tumor; it is not necessarily an event involved in its initiation.

Some proto-oncogenes are activated by events that change their expression, but which leave their coding sequence unaltered. The best characterized is *c-myc*, whose expression is elevated by several mechanisms. One common mechanism is the insertion of a nondefective retrovirus in the vicinity of the gene.

The ability of a retrovirus to transform without expressing a *v-onc* sequence was first noted during analysis of the bursal lymphomas caused by the transformation of B lymphocytes with avian virus. Similar events occur in the induction of T cell lymphomas by murine leukemia virus. In each case, the transforming potential of the retrovirus seems to lie with its LTR rather than with a coding sequence.

In many independent tumors, the virus has integrated into the cellular genome within or close to the *c-myc* gene. The gene consists of three exons; the first represents a long nontranslated leader, and the second two code for the c-Myc protein. **Figure 39.8** summarizes the types of insertion at this locus. The retrovirus may be inserted at a variety of locations relative to the *c-myc* gene.

The simplest insertions to explain are those that occur within the first intron. The LTR provides a promoter, and transcription reads through the two coding exons. Transcription of *c-myc* under viral control differs in two ways from its usual expression: the level of expression is increased (because the LTR provides an efficient promoter); and the transcript lacks its usual nontranslated leader (which may usually limit expression).

Activation of *c-myc* in the other two classes of insertions reflects different mechanisms. The retroviral genome may be inserted within or upstream of the first intron, but in reverse orientation, so that its promoter points in the wrong direction. Probably the LTR provides an enhancer that acts on an upstream sequence that fortuitously resembles a promoter. The retroviral genome also may be inserted downstream of the *c-myc* gene, in which case transcription initiates at the usual *c-myc* promoter(s), but is increased by the enhancer in the retroviral LTR.

In all of these cases, the coding sequence of c-myc *is unchanged, so oncogenicity is attributed to the loss of normal control, and increased expression, of the gene.*

Other oncogenes that are activated in tumors by the insertion of a retroviral genome include *c-erbB*, *c-myb*, *c-mos*, *c-H-ras*, and *c-raf*. Up to 10 other cellular genes (not previously identified as oncogenes by their presence in transforming viruses) are implicated as potential oncogenes by this criterion. The best characterized among this latter class are *wnt1* and *int2*. The *wnt1* gene codes for a protein involved in early embryogenesis that is related to the *wingless* gene of *Drosophila*; *int2* codes for a growth factor related to FGF.

Translocation to a new chromosomal location is another of the mechanisms by which oncogenes are activated. Certain chromosomal translocations are consistently associated with activation of oncogenes that lie near the breakpoints. This situation was originally discovered via a connection between the loci coding immunoglobulins and the

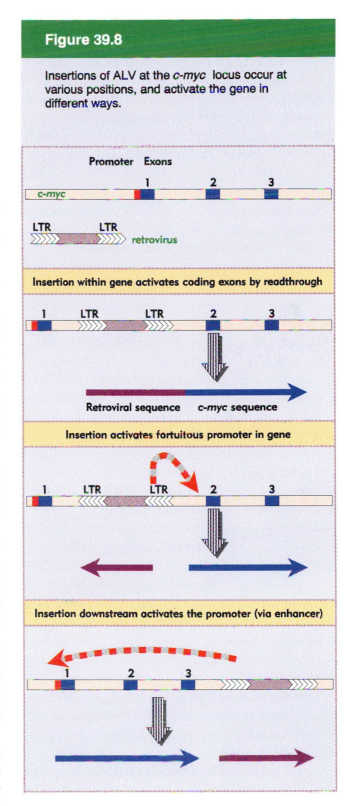

Figure 39.8

Insertions of ALV at the *c-myc* locus occur at various positions, and activate the gene in different ways.

occurrence of certain tumors. Specific chromosomal translocations are often associated with plasmacytomas in the mouse and with Burkitt lymphomas in man. These tumors arise from undifferentiated B lymphocytes. The common feature in both species is that an oncogene on one chromosome is brought into the proximity of an Ig locus on another chromosome. Similar events occur in T lymphocytes to bring oncogenes into the proximity of a TcR locus.

The basic cause of the translocation event is a malfunction of the system responsible for recombining V and C genes or switching between IgH C genes. Instead of acting on two sites within the Ig or TcR locus, the system recombines the immune locus with an unrelated region on a different chromosome. This results in a **reciprocal translocation**, which is illustrated in **Figure 39.9**.

We do not know the basis for the involvement of the nonimmune partner, but in both man and mouse it is often the *c-myc* locus. In man, the translocations in B cell tumors usually involve chromosome 8, which carries *c-myc,* and chromosome 14, which carries the IgH locus; ~10% involve chromosome 8 and either chromosome 2 (κ locus) or chromosome 22 (lambda locus). The translocations in T cell tumors often involve chromosome 8, and either chromosome 14 (which has the TcRα locus at the other end from the Ig locus) or chromosome 7 (which carries TcR β locus). Analogous translocations occur in the mouse.

Translocations at the IgH locus in B cells fall into two classes. One type is similar to those observed at other Ig loci and at TcR loci, involving the consensus sequences used for V-D-J somatic recombination of active Ig genes. In the other type, the translocation occurs at a switching site, so these cases may be associated with function of the system that switches expression from one C_H gene to another.

When *c-myc* is translocated to the Ig locus, its level of expression is usually increased. The increase varies considerably among individual tumors, generally being in the range from 2 to 10. Why does translocation activate the *c-myc* gene? The translocation event does not involve fixed sites,

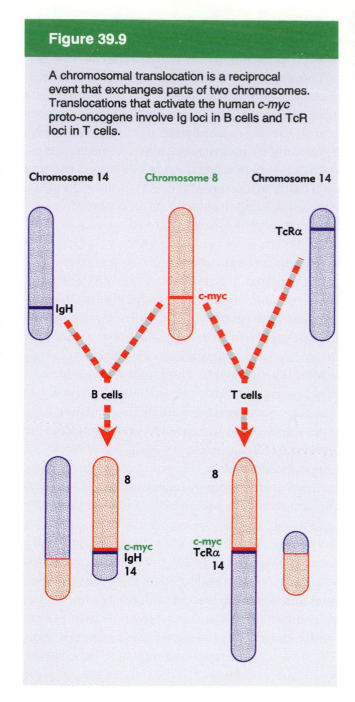

Figure 39.9

A chromosomal translocation is a reciprocal event that exchanges parts of two chromosomes. Translocations that activate the human *c-myc* proto-oncogene involve Ig loci in B cells and TcR loci in T cells.

but occurs at a variety of locations within a general region on each recombining chromosome. The event has two consequences: *c-myc* is brought into a new region, one in which an Ig or TcR gene was actively expressed; and the structure of the *c-myc* gene may itself be changed (but usually not involving the coding regions). It seems likely that several different mechanisms can activate the *c-myc* gene in its new location (just as retroviral insertions activate *c-myc* in a variety of ways).

The correlation between the tumorigenic phenotype and the activation of *c-myc* by either insertion or translocation suggests that continued high expression of the c-Myc protein is oncogenic. Expression of *c-myc* must be switched off to enable immature lymphocytes to differentiate into mature B and T cells; failure to turn off *c-myc* maintains the cells in the undifferentiated (dividing) state. The oncogenic potential of *c-myc* has been demonstrated directly by the creation of transgenic mice carrying a normal *c-myc* gene linked to an enhancer. Transgenic mice carrying a *c-myc* gene linked to a B lymphocyte-specific enhancer (the IgH enhancer) develop lymphomas. The tumors represent both immature and mature B lymphocytes, suggesting that over-expression of *c-myc* is tumorigenic throughout the B cell lineage. Transgenic mice carrying a *c-myc* gene under the control of the LTR from a mouse mammary tumor virus, however, develop a variety of cancers, including mammary carcinomas. This suggests that increased or continued expression of *c-myc* transforms the type of cell in which it occurs into a corresponding tumor. Specificity of the tumor type may therefore depend on the mechanism used to activate *c-myc*; it is not an intrinsic property of the gene.

c-myc exhibits three means of oncogene activation: retroviral insertion, chromosomal translocation, and gene amplification. The common thread among them is increased expression of the oncogene rather than a qualitative change in its coding function, although in at least some cases the transcript has lost the usual (and possibly regulatory) nontranslated leader. *c-myc* provides the paradigm for oncogenes that may be effectively activated by increased (or possibly altered) expression.

Every translocation generates reciprocal products; sometimes a known oncogene is activated in one of the products, but in other cases it is not evident which of the reciprocal products has responsibility for oncogenicity. Also, it is not axiomatic that the gene(s) at the breakpoint have responsibility; for example, the translocation could provide an enhancer that activates another gene nearby.

A variety of translocations found in B and T cells have identified new oncogenes. In some cases, the translocation generates a hybrid gene, in which an active transcription unit is broken by the translocation. This has the result that the exons of one gene may be connected to another. In such cases, there are two potential causes of oncogenicity: the proto-oncogene part of the protein may be activated in some way that is independent of the other part, for example, because it is over-expressed under its new management (a situation directly comparable with the example of *c-myc*); or the other partner in the hybrid gene may have some positive effect that generates a gain-of-function in the part of the protein coded by the proto-oncogene.

One of the best characterized cases in which a translocation creates a hybrid oncogene is provided by the *Philadelphia (PH[1])* chromosome present in patients with chronic myelogenous leukemia (CML). This reciprocal translocation is too small to be visible in the karyotype, but links a 5,000 kb region from the end of chromosome 9 carrying *c-abl* to the *bcr* region of chromosome 22. The *bcr* (*breakpoint cluster region*) was originally named to describe a region of ~5.8 kb within which breakpoints occur on chromosome 22. Different cases of CML have breakpoints at different locations within this region.

The consequences of this translocation are summarized in **Figure 39.10**. The *bcr* region lies within a large (>90 kb) gene, which is now known as the *bcr* gene. The breakpoints in CML usually occur within one of two introns in the middle of the gene. The same gene is also involved in translocations that generate another disease, ALL (acute lymphoblastic leukemia); in this case, the breakpoint in the *bcr* gene occurs in the first intron.

The *c-abl* gene is expressed by alternative splicing that uses either of the first two exons. The breakpoints in both CML and ALL occur in the intron that precedes the first common exon. Although the exact breakpoints on both chromosomes 9 and 22 vary in individual cases, the

Figure 39.10

Translocations between chromosome 22 and chromosome 9 generate Philadelphia chromosomes that synthesize *bcr-abl* fusion transcripts that are responsible for two types of leukemia.

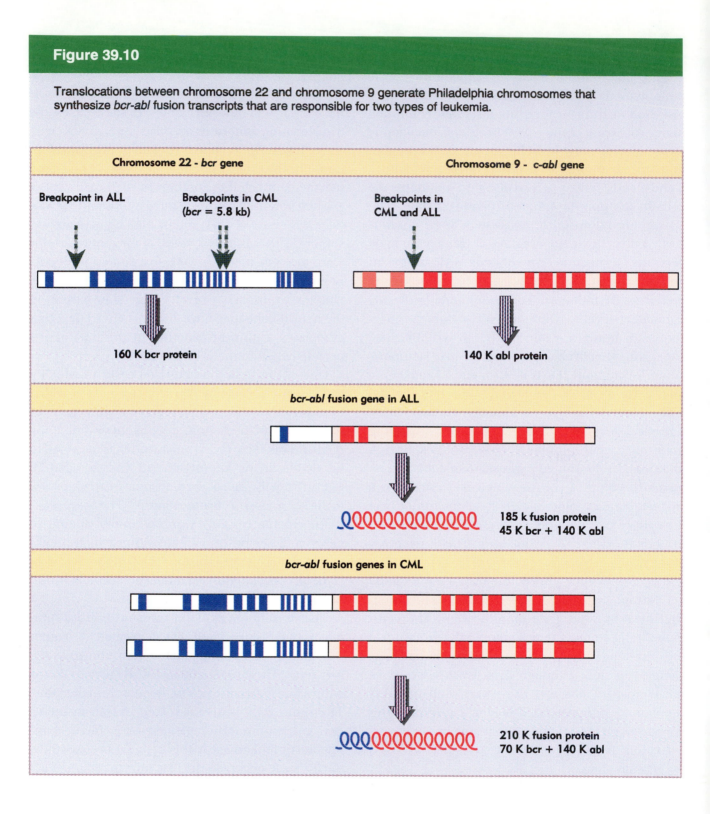

common outcome is the production of a transcript coding for a Bcr–Abl fusion protein, in which N-terminal sequences derived from *bcr* are linked to *c-abl* sequences. In ALL, the 185,000 dalton fusion protein has ~45,000 daltons of the Bcr protein linked to c-Abl; in CML the fusion protein of 210,000 daltons has ~70,000 daltons of the Bcr protein. In each case, the fusion protein contains ~140,000 daltons of the usual ~145,000 c-Abl protein, that is, it has lost just a few N-terminal amino acids of the c-Abl sequence.

Why is the fusion protein oncogenic? It relies on an interaction between the N-terminal region provided by *bcr* with the c-Abl protein. The *bcr* gene has a variety of sequence motifs related to proteins involved in signalling pathways, but the pertinent one is a serine/threonine kinase activity that is coded within the first exon. This autophosphorylates residues in this part of the protein, and the phosphorylation confers the ability to interact with a region of the c-Abl protein called the SH2 domain (we discuss the nature of SH2 domains later). This raises the possibility that the Bcr part of the protein interacts with the c-Abl sequences, perhaps changing their conformation and activating a latent oncogenic potential.

Changes at the N-terminus are involved in the activation of oncogenic activity of *v-abl*, a transforming version of the gene carried in a retrovirus. The *c-abl* gene codes for a tyrosine kinase activity; this activity is essential for transforming potential in oncogenic variants. Deletion (or replacement) of the N-terminal region activates the kinase activity and transforming capacity. So the N-terminus provides a domain that usually regulates kinase activity; its loss may cause inappropriate activation.

Loss of tumor suppressors causes tumor formation

The common theme in the role of oncogenes in tumorigenesis is that increased or altered activity of the gene product is oncogenic. Whether the oncogene is introduced by a virus or results from a mutation in the genome, it is dominant over its allelic proto-oncogene(s). A mutation that activates a single allele is tumorigenic. Tumorigenesis then results from gain of a function.

Certain tumors are caused by a different mechanism: loss of both alleles at a locus is tumorigenic. Propensity to form such tumors may be inherited through the germline; it also occurs as the result of somatic change in the individual. Tumorigenesis then results from loss of function. Such cases identify tumor suppressors: genes whose products are needed for normal cell function, and whose loss of function causes tumors. The two best characterized genes of this class code for the proteins RB and p53.

Retinoblastoma is a human childhood disease, involving a tumor on the retina. It occurs both as a heritable trait and sporadically (by somatic mutation). It is often associated with deletions of band q14 of human chromosome 13. The *RB* gene has been localized to this region by molecular cloning. **Figure 59.11** illustrates the situation.

Retinoblastoma arises when both copies of the *RB* gene are inactivated. In the inherited form of the disease, one parental chromosome carries an alteration in this region, usually a deletion. A somatic event in retinal cells that causes loss of the other copy of the *RB* gene causes a tumor. In the sporadic form of the disease, the parental chromosomes are normal, and both *RB* alleles are lost by (individual) somatic events.

Almost half the cases of retinoblastoma show deletions at the *RB* locus. In other cases, transcripts of the locus are either absent or altered in length. The protein product is absent from retinoblastoma cells. The cause of the tumor is therefore loss of protein function, usually resulting from mutations that prevent gene expression (as opposed to point mutations that affect function of the protein product). Loss of *RB* is involved also in other forms of cancer, including osteosarcomas and small cell lung cancers.

What is the molecular function of RB? It interacts with a variety of other proteins, including several tumor antigens: SV40 T antigen, adenovirus E1A, human papilloma virus E7. One possibility is that part of the oncogenicity of these proteins is due to

Figure 39.11

Retinoblastoma is caused by loss of both copies of the RB gene in chromosome band 13q14. In the inherited form, one chromosome has a deletion in this region, and the second copy is lost by somatic mutation in the individual. In the sporadic form, both copies are lost by individual somatic events.

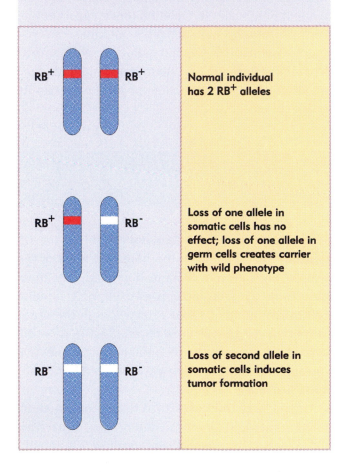

RB⁺ RB⁺	Normal individual has 2 RB⁺ alleles
RB⁺ RB⁻	Loss of one allele in somatic cells has no effect; loss of one allele in germ cells creates carrier with wild phenotype
RB⁻ RB⁻	Loss of second allele in somatic cells induces tumor formation

their ability to interact with, and presumably inhibit, some function of RB that constrains cell proliferation.

RB is a nuclear phosphoprotein that influences the cell cycle. In resting (G0/G1) cells, RB is not phosphorylated. RB is phosphorylated during the cell cycle by cyclin–CDK complexes, most particularly at the end of G1; it is dephosphorylated during mitosis. SV40 T antigen binds only to the nonphosphorylated form. This suggests the model shown in **Figure 39.12**. Nonphosphorylated RB prevents cell

proliferation; this activity must be transiently suppressed in order to pass through the cell cycle, which is accomplished by the cyclic phosphorylation. And it may also be suppressed when a tumor antigen sequesters the nonphosphorylated RB, presumably thereby preventing its function.

The unphosphorylated form of RB specifically binds several proteins, and these interactions therefore occur only during part of the cell cycle

Figure 39.12

A block to the cell cycle is released when RB is phosphorylated (in the normal cycle) or when it is sequestered by a tumor antigen (in a transformed cell).

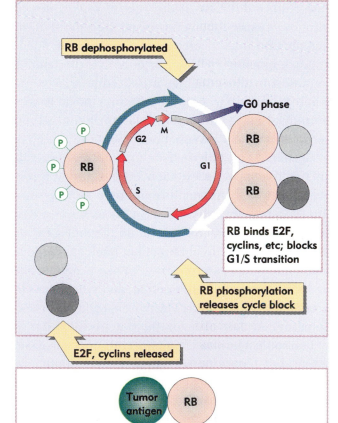

(prior to S phase). The target proteins include E2F (a transcription factor) and cyclins of the D and E types. Binding to RB inhibits the ability of E2F to activate transcription, which suggests that RB may repress the expression of genes dependent on E2F. We do not have direct evidence on the functional consequences of the interactions with cyclins; in particular, we should like to show whether RB is controlling the activities of its partners or whether they are controlling its activities. These complexes are disrupted when tumor antigens bind to RB.

Over-expression of RB impedes cell growth. An indication of the importance of RB for cell proliferation is given by the properties of an osteosarcoma cell line that lacks RB; when RB is introduced into this cell line, its growth is impeded. However, the inhibition can be overcome by expression of certain cyclins. The detailed manner in which RB influences the cell cycle is not known, but it seems plausible that its basic function may lie in some interaction that controls progression through the cycle. It is not the only protein of its type: a protein with related sequences, called p107, has similar properties.

The p53 gene is a tumor suppressor, but is inherited as an autosomal dominant in Li–Fraumeni syndrome, which is a rare form of inherited cancer. Affected individuals display cancers in a variety of tissues, and they have missense mutations in the gene. These mutations confer a new property on p53, and explains a long history of its association with oncogenic properties.

p53 is a nuclear phosphoprotein, which was originally discovered in SV40-transformed cells, where it is associated with T antigen. A large increase in the amount of p53 protein is found in many transformed cells or lines derived from tumors. In early experiments, the introduction of cloned p53 was found to immortalize cells. These experiments caused p53 to be classified as an oncogene, with the usual trait of dominant gain-of-function.

But all the transforming forms of p53 turned out to be mutant forms of the protein! They fall into the category of dominant negative mutants, which function by overwhelming the wild-type protein and preventing it from functioning. The most common form of a dominant negative mutant is one that forms a heteromeric protein containing both mutant and wild-type subunits, in which the wild-type subunits are unable to function. p53 probably exists as a tetramer. When mutant and wild-type subunits of p53 associate, the tetramer takes up the mutant conformation.

Figure 39.13 shows that the same phenotype is produced either by the deletion of both alleles or by a missense point mutation in one allele that produces a dominant negative subunit. Both situations are found in human cancers. Mutations in p53 accumulate in many types of human cancer, probably because loss of p53 provides a growth advantage to cells; that is, wild-type p53 is required to restrain growth in some way. The diversity of these cancers suggests that p53 is not involved in a tissue-specific event, but in some general and rather common control of cell proliferation; and the loss of this control may be a secondary event that occurs to assist the growth of many tumors.

This interpretation implies that a normal cell has a capacity to grow in an unrestrained manner that usually is inhibited by p53. It is certainly clear that loss of p53 activity (because of deletions in both parental alleles) causes unrestrained growth. If the dominant (missense) mutations work by directly preventing the wild-type subunits in heteromeric complexes from functioning, they could abolish all p53 function if there is over-expression of the mutant allele; or it may be that a reduction in the quantity of wild-type p53 is deleterious. Of course, it is difficult entirely to exclude the possibility that some of the missense mutations confer a gain-of-function that acts directly to cause unrestrained cell growth. p53 is defined as a tumor suppressor also by the fact that wild-type p53 can suppress or inhibit the transformation of cells in culture by various oncogenes.

What is the function of p53 at the molecular level? Two theories have been proposed. p53 was discovered as a protein that binds to the large T tumor antigen of SV40 virus. The ability to bind is a property of the wild-type protein; all mutant forms of p53 fail to bind T antigen. Binding of wild-type p53 to T antigen interferes with its ability to

Figure 39.13

Wild-type p53 is required to restrain cell growth. Its activity may be lost by deletion of both wild-type alleles or by a dominant mutation in one allele.

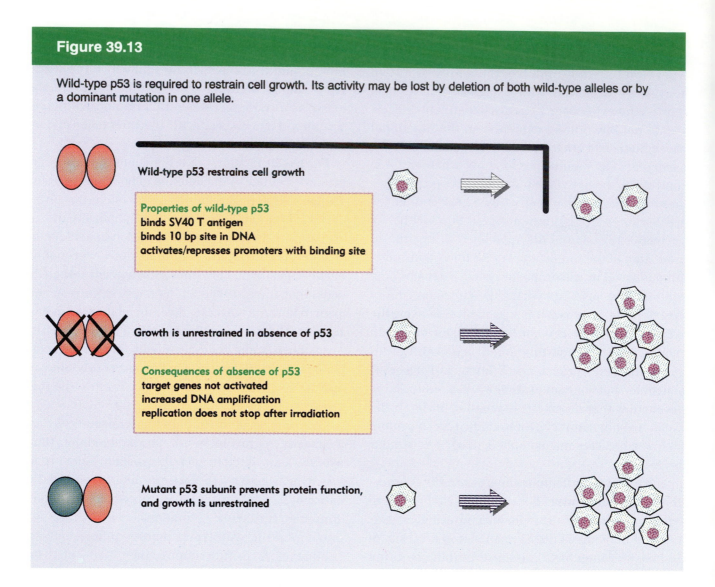

Wild-type p53 restrains cell growth

Properties of wild-type p53
binds SV40 T antigen
binds 10 bp site in DNA
activates/represses promoters with binding site

Growth is unrestrained in absence of p53

Consequences of absence of p53
target genes not activated
increased DNA amplification
replication does not stop after irradiation

Mutant p53 subunit prevents protein function, and growth is unrestrained

promote replication of SV40 DNA. One possibility is that wild-type p53 binds to some cellular analog of T antigen, inhibiting its activity; this function has been lost in the mutants. p53 is also a DNA-binding protein, recognizing a symmetrical 10 bp motif, and it activates transcription at promoters that contain multiple copies of this motif. At some loci it may repress its target. Perhaps it regulates the expression of genes that control the cell cycle; loss of this regulation could lead to unrestrained growth.

Most of the missense mutations in p53 occur at positions where the amino acid residue is well-conserved in many organisms, suggesting that the mutations may interfere with conserved function(s). The mutations that have been characterized have various effects upon the properties of p53, including increasing its half-life from 20 minutes to several hours, causing a change in conformation that can be detected with an antibody, changing its location from the nucleus to the cytoplasm, preventing binding to SV40 T antigen, and preventing DNA binding.

Cells with defective p53 function have a variety of phenotypes; this pleiotropy makes it difficult to determine which (if any) of these effects is directly connected to the tumor suppressor function. We cannot at the moment prove whether the basic effects in loss of tumor suppressor function stem from affecting p53's ability to interact with other proteins or to bind to DNA.

Mutant p53 cells have an increased propensity to amplify DNA, as detected with a CAD test gene. There has been speculation that this could be a cause of the characteristic instability of the genome that is found in cancer cells.

The general definition of their properties shows that both RB and p53 are tumor suppressors that in some way usually control cell proliferation; their absence removes some control, and contributes to tumor formation. Both of these proteins have functions that are connected with cell cycle progression, but it remains to be proven formally that this is the activity responsible for tumor suppression, and to characterize how its absence permits unrestrained growth.

Immortalization and transformation

Most tumors arise as the result of multiple events. It is likely that some of these events involve the activation of oncogenes, while others take the form of inactivation of tumor suppressors. The requirement for multiple events reflects the fact that normal cells have multiple mechanisms to regulate their growth and differentiation, and several separate changes may be required to bypass these controls. Indeed, the existence of many genes in which single mutations were tumorigenic would no doubt be deleterious to the organism, and has been selected against. Nonetheless, several genes now are known in which mutations create a strong predisposition to tumors, that is, they represent one of the necessary events. The identification of oncogenes with transforming potential as measured in available assays may not be sufficient to account entirely for the occurrence of cancers, but we look to the oncogenes to explain the nature of (at least) some of these events.

The need for multiple functions fits with the pattern established by some DNA tumor viruses, in which (at least) two functions are needed to transform the usual target cells. In the same way, expression of two or more oncogenes in the cellular transfection assay is usually needed to convert a primary cell (one taken directly from the organism) into a tumor cell. The need for multiple functions of different types is sometimes described as the requirement for **cooperativity**. The division of functions is summarized in **Table 39.4**, which assigns oncoproteins to immortalization or transformation functions (or both).

Adenovirus carries the *E1A* region, which allows primary cells to grow indefinitely in culture, and the *E1B* region, which causes the morphological changes characteristic of the transformed state.

Polyoma produces three T antigens; large T elicits indefinite growth, middle T is responsible for morphological transformation, and small T is without known function. Large T and middle T together can transform primary

Table 39.4

Some oncoproteins can be characterized as possessing immortalizing or transforming activity (or both).

System	Immortalizing Function	Transforming Function
SV40	Large T	Large T
Polyoma	Large T	Middle T
HPV	E7 & E6	
Adenovirus	E1A	E1B
Retroviral/cellular	*v-src*	*v-src*
"		ras
"	*v-myc, v-fos, v-jun*	

cells. Polyoma middle T may act through binding to c-Src.

Consistent with the classification of oncogenic functions, adenovirus E1A together with polyoma middle T can transform primary cells. This suggests that one function of each type is needed.

Yet the separation of functions is not inevitable; SV40 large T can transform alone. (All of the T proteins, however, are multifunctional, and cannot really be compared directly with cellular oncogenes. For example, SV40 T antigen binds to the origin of replication of SV40 DNA, but mutations that eliminate the DNA-binding activity do not prevent T antigen from immortalizing and transforming cells. The E1A product(s) of adenovirus activate certain viral and cellular promoters. Again, however, when this activity is mutated, the ability to immortalize is not lost. Thus these oncoproteins may have different domains responsible for their known molecular actions and for their immortalizing/transforming effects.)

Several cellular oncogenes have been identified by transforming ability in the 3T3 transfection assay; 10–20% of spontaneous human tumors have DNA with detectable transforming activity in this assay. Of course, 3T3 cells have been adapted to indefinite growth in culture over many years, and have passed through some of the changes characteristic of tumor cells; the exact nature of these changes is not clear, but generally they can be classified as involving functions concerned with immortalization. Oncogenic activity in the transfection assay therefore depends on the ability to induce morphological and other phenotypic changes in an established cell line. The principal products of 3T3 transfection assays are mutated *c-ras* genes. They do not have the ability to transform primary cells *in vitro*, and this supports the implication that their functions are concerned with the act of transforming cells that have previously been immortalized. *ras* oncogenes clearly provide one major pathway for transforming immortalized cells; we do not know how many other pathways there may be that are independent of *ras*.

Whatever functions are required for immortalization can be provided (or circumvented) by other oncogenes. Although *ras* oncogenes alone cannot transform primary fibroblasts, dual transfection with *ras* and another oncogene can do so. The ability to transform primary cells in conjunction with *ras* provides a general assay for oncogenes that have an immortalization-like function. This group includes several retroviral oncogenes, *v-myc*, *v-jun*, and *v-fos*. It also includes adenovirus E1A and polyoma large T. Mutant p53 genes have the same effect, suggesting that loss of p53 constitutes one route to immortalization. However, in many cases the distinctions between immortalizing and transforming proteins is blurred. For example, although E1A is classified as an immortalizing function, it has (some) of the functions usually attributed to transforming proteins, and loss of p53 confers some properties that are usually considered transforming.

One way to investigate the oncogenic potential of individual oncogenes free from the constraints that usually are involved in their expression is to create transgenic animals in which the oncogene is placed under control of an inducible promoter. A general pattern is that increased proliferation often occurs in the tissue in which the oncogene is expressed. Oncogenes whose expression have this effect with a variety of tissues include SV40 T antigen, *v-ras*, and *c-myc*. The pattern is not universal; there may be certain tissues in which oncogene expression is ineffectual.

Increased proliferation (hyperplasia) is often damaging and sometimes fatal to the animal (usually because the proportion of one cell type is increased at the expense of another). However, in relatively few cases does the expression of a single oncogene cause malignant transformation (neoplasia), with the production of tumors that kill the animal. The minority of such cases is probably due to the occurrence of a second event.

The need for two types of event is indicated by the difference between transgenic mice that carry either the *v-ras* or activated *c-myc* oncogene, and mice that carry both oncogenes. Mice carrying either oncogene develop malignancies at rates of 10% for *c-myc* and 40% for *v-ras*; mice carrying both oncogenes develop 100% malignancies over the same period. These results with transgenic mice are even more striking than the comparable results on cooperation between oncogenes in cultured cells.

An interesting convergence is seen in the properties of the tumor antigens of the DNA tumor viruses: the antigens can bind to the cellular tumor suppressor products Rb and p53. The two cellular proteins are recognized independently. Either different tumor antigens of the virus bind separately to RB and to p53, or different domains of the same antigen do so. Thus adenovirus E1A binds RB, while E1B binds p53; HPV E7 binds RB, while E6 binds p53. SV40 T antigen can bind both RB and p53. The consequences for function of the tumor suppressor are completely clear only in the case of HPV E6, which targets p53 for degradation. In effect, HPV converts a target cell into a p53⁻ state. It seems likely that the loss of p53 (and/or RB) is a major step in the transforming action of DNA tumor viruses, and explains some significant part of the action of the tumor antigens. Loss of the tumor suppressors may be one major route in the immortalization pathway.

We do not really understand what is involved in 'immortalization' in terms of changes in cellular properties. In some systems, it may be connected with an inability of the cells to differentiate. Growth and differentiation are often mutually exclusive, because a cell must stop dividing in order to differentiate. An oncoprotein that blocks differentiation may allow a cell to continue proliferating (in a sense resembling the immortalization of cultured cells);

continued proliferation in turn may provide an opportunity for other oncogenic mutations to occur. This may explain the occurrence among the oncoproteins of products that usually regulate differentiation.

A connection between differentiation and tumorigenesis is shown by avian erythroblastosis virus (AEV). The AEV-H strain carries only *v-erbB*, but the AEV-E54 strain carries two oncogenes, *v-erbB* and *v-erbA*. The major transforming activity of AEV is associated with *v-erbB*, a truncated form of the EGF receptor, which is equivalent to the single oncogene carried by other tumor retroviruses: it can transform erythroblasts and fibroblasts. The other gene, *v-erbA*, cannot transform target cells alone, but it increases the transforming efficacy of *v-erbB*. Expression of *v-erbA* itself has two phenotypic effects upon target cells: it prevents the spontaneous differentiation (into erythrocytes) of erythroblasts that have been transformed by *v-erbB*; and it expands the range of conditions under which transformed erythroblasts can propagate. *v-erbA* may therefore contribute to tumorigenicity by a combination of inhibiting differentiation and stimulating proliferation. In fact, *v-erbA* has a similar effect in extending the efficacy of transformation by other oncogenes that induce sarcomas, notably *v-src*, *v-fps*, and *v-ras*.

Correlations between the activation of oncogenes and the successful growth of tumors are strong in some cases, but by and large the nature of the initiating event remains open. It seems clear that oncogene activity assists tumor growth, but activation could occur (and be selected for) after the initiation event and during early growth of the tumor. We hope that the functions of *c-onc* genes will provide insights into the regulation of cell growth in normal as well as aberrant cells, so that it will become possible to define the events needed to initiate and establish tumors.

Oncogenes code for components of signal transduction cascades

Whether activated by quantitative or qualitative changes, oncogenes may be presumed to influence (directly or indirectly) functions connected with cell growth. Transformed cells lack restrictions imposed on normal cells, such as dependence on serum or inhibition by cell–cell contact. They may acquire new properties, such as the ability to metastasize. Many phenotypic properties are changed when we compare a normal cell with a tumorigenic counterpart, and it is therefore striking indeed that individual genes can be identified that are associated with this transformation.

We assume that oncogenes, individually or in concert, set in train a series of phenotypic changes that involve the products of many genes. In this description, we see at once a similarity with genes that regulate developmental pathways: they do not themselves necessarily code for the products that characterize the differentiated cells, but they may direct a cell and its progeny to enter a particular pathway. The same analogy suggests itself for oncogenes and developmental regulators: do they provide switches responsible for causing transitions between one discrete phenotypic state and another?

Taking this argument further, we may ask what activities the products of proto-oncogenes play in the normal cell, and how are they changed in the transformed cell? Could some proto-oncogenes be regulators of normal development whose malfunction results in aberrations of growth that are manifested as tumors? We have stumbled across some examples of such relationships, but do not yet have any systematic understanding of the connection.

Oncoproteins are organized according to their types of function in **Figure 39.14**. The left part of the figure groups the oncogenes according to the locations of their products. The boxes on the right give details of the corresponding proto-oncogenes. The functions of many oncogenes remain unknown, and further groups will no doubt be identified:

◆ Growth factors are proteins secreted by one cell that act on another. In none of these cases do we have any information about how the function of the oncoprotein is related to the normal function of the proto-oncoprotein.

◆ The growth factor receptors are transmembrane proteins that are activated by binding an extracellular ligand (usually a polypeptide). In all these examples except mas, the receptor is a protein tyrosine kinase, that is, when it is activated it phosphorylates target proteins on tyrosine residues. Oncogenicity may result from constitutive activation of the kinase activity.

◆ Steps in signal transduction pathways are identified by several groups. The best characterized protein is c-Ras, an example of the class of small GTP-binding proteins, which plays a central role in transmitting the signal from receptor tyrosine kinases (see later). Oncogenic mutations change the regulation of Ras activity. Other stages in signal transduction are identified by Gsp and Gip, which are mutant forms of the α subunits of the G_s and G_i trimeric G proteins. Crk and Vav are proteins associated with later stages of signalling.

◆ One group of intracellular protein kinases phosphorylate tyrosine residues in target proteins. c-Src, which may associate with the cytoskeleton as well as with the membrane, is the prototype of a family of kinases with similar catalytic activities (including Yes, Fgr, Lck, Fps, Fes, Fyn). We understand the effects of oncogenic mutations on the Src kinase activity in some detail, although we have yet to explain why the altered kinase activity is oncogenic. Other protein tyrosine kinases in the intracellular group, including c-Fps and c-Abl, are cytosolic.

Figure 39.14

Oncogenes may code for secreted proteins, transmembrane proteins, cytoplasmic proteins, or nuclear proteins.

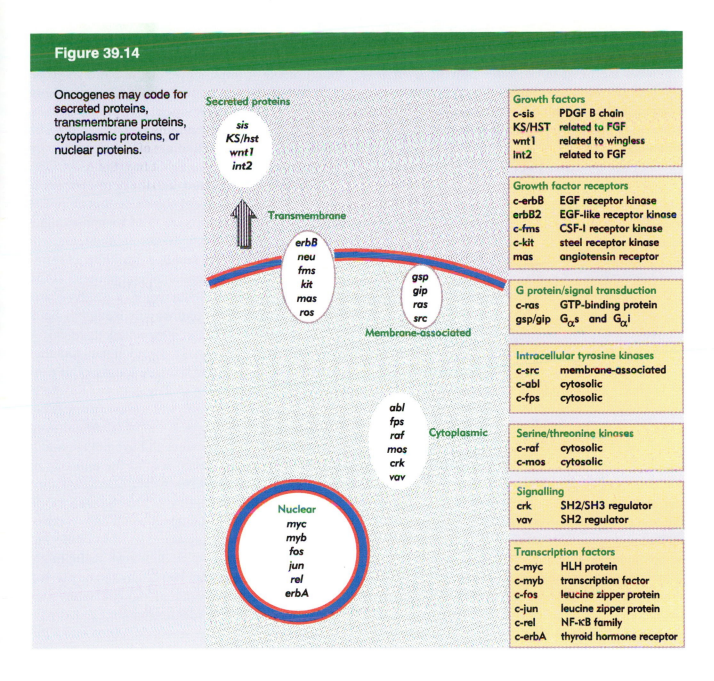

◆ A group of cytosolic enzymes are protein serine/threonine kinases, that is, they phosphorylate target proteins on serine or threonine. Little is known about the effects of oncogenic mutations beyond the fact they probably increase or constitutively activate the kinase activities.

◆ Nuclear proteins include transcription factors of several types. The functions of the proto-oncoproteins are rather well described (see Chapter 30). Generally we understand what effects the oncogenic mutations have on the factors, but we cannot yet relate these changes to the activation or repression of particular target genes that contribute to the oncogenic state.

The common feature is that each type of protein is in a position to trigger general changes in cell phenotypes, either by initiating or responding to changes associated with cell growth, or by changing gene expression directly. Before we consider in detail the potential of each group for initiating a series of events that has an oncogenic outcome, we need to consider how many independent pathways are identified by these factors.

Suppose that in principle a pathway for activating a series of changes in cell phenotype consists of the following stages:

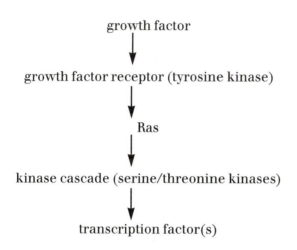

growth factor

↓

growth factor receptor (tyrosine kinase)

↓

Ras

↓

kinase cascade (serine/threonine kinases)

↓

transcription factor(s)

When a growth factor interacts with its receptor, it activates a tyrosine kinase activity. The signal passes along a cascade of kinases, in which one kinase phosphorylates a target protein, which is itself a kinase that is activated by the phosphorylation. At some point during this cascade, there is a switch from tyrosine phosphorylation to serine/threonine phosphorylation. In the cascade drawn above, Ras is involved in making this switch. The targets at the end of the pathway may be controlled directly or indirectly by phosphorylation, and include transcription factors, which are in a position to make widespread changes in the pattern of gene expression.

If a pathway functions in a linear manner, in which the signal passes directly from one component to the next, the same results should be achieved by constitutive activation of any component. The relationship between components of a pathway can be tested by investigating the effects of

one component upon the action of another. For example, a mutation that inactivates a component should make it impossible for the pathway to be activated by any components that act earlier. Using such tests allows components to be ordered in a pathway; those that act earlier are said to be 'upstream', and those that act later are said to be 'downstream'.

A signal transduction pathway, of course, is likely to branch at several stages, so that an initial stimulus may trigger a variety of responses. The activation of more downstream components will therefore activate a smaller number of end-functions than the activation of components at the start of the pathway. But we can analyze any individual part of the pathway by tracing it back to the beginning as though it were strictly linear.

Several of the stages of a common pathway that is activated by receptor tyrosine kinases can now be filled in by such means. The pathway has been characterized in several situations: in terms of biochemical components in mammalian cultured cells, as the pathway involved in eye development in the fly *D. melanogaster,* and as the pathway of vulval development in the worm *C. elegans* (see below). The striking feature is that the pathway is activated by different means in each case (appropriate to the individual system), and it has different end effects in each system, but many of the intermediate components can be recognized as playing analogous roles. It is much as though Nature has developed a signal transduction cascade that can be employed wholesale by means of connecting the beginning to an appropriate stimulus and the end to an appropriate effector. Several of the components of this pathway (in mammals) are proto-oncogenes, which suggests that the aberrant activation of this pathway at any stage has a powerful potential to cause tumors.

Although there are still some gaps in the pathway to fill in, and branches that have not yet been identified, the broad outline is clear, as illustrated in **Figure 39.15**. In mammalian cells, the cascade is initiated by activation of a tyrosine kinase receptor; the EGF and PDGF receptors have been the best studied. The receptor phosphorylates various

Figure 39.15

A common signal transduction cascade includes several components that are represented in oncoproteins (named in green). Missing components are indicated by successive arrows.

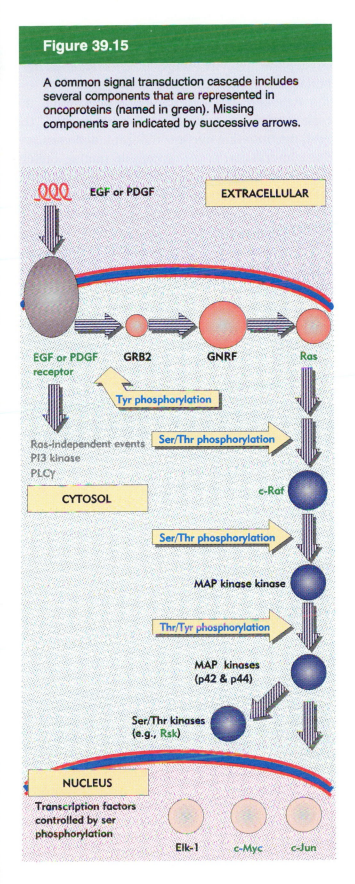

EGF or PDGF

EXTRACELLULAR

EGF or PDGF receptor

GRB2

GNRF

Ras

Tyr phosphorylation

Ras-independent events
PI3 kinase
PLCγ

Ser/Thr phosphorylation

CYTOSOL

c-Raf

Ser/Thr phosphorylation

MAP kinase kinase

Thr/Tyr phosphorylation

MAP kinases
(p42 & p44)

Ser/Thr kinases
(e.g., Rsk)

NUCLEUS

Transcription factors controlled by ser phosphorylation

Elk-1   c-Myc   c-Jun

target proteins, and activates various pathways. One of these is the Ras-dependent pathway. We discuss the properties of Ras in the next section, and then return to the interactions involving the receptor itself, but note for now that a small protein called GRB2 binds to the activated receptor. The interaction with GRB2 does not require its phosphorylation; in fact, tyrosine kinase receptors activate some substrates without phosphorylating them (another example is PI3 kinase in the Ras-independent pathway), while other substrates are activated directly by phosphorylating tyrosine (such as PLCγ in the Ras-independent pathway).

GRB2 in turn acts on a protein called GNRF, which then activates Ras. Ras is required to activate the Raf serine/threonine kinase, although we do not know how this is accomplished. The experiments that identify the roles of Ras and Raf are a good example of the genre. We know that Ras and Raf are on the same pathway, because both of them are required for the phosphorylation of the proteins later in the pathway (such as MAP kinase). Ras must be upstream of Raf because it is required for the activation of Raf in response to extracellular ligands. Similarly, Raf must be downstream of Ras because transformation by Ras can be suppressed by expression of a dominant-negative (kinase deficient) mutant of Raf.

Raf activity leads to the activation of MAP kinase kinase, probably by a direct interaction. MAP kinase kinase takes its name because it is the enzyme that phosphorylates MAP kinase. It is activated by phosphorylation on serine or threonine. It is an unusual enzyme with dual specificity, which can phosphorylate both threonine and tyrosine.

Both types of phosphorylation are necessary to convert MAP kinase into the active state. There are at least 2 MAP kinases, and they appear to be important switching points in the pathway. The name of the MAP kinases reflects their identification as mitogen-activated kinases; they have also been called ERKs, for extracellular signal-regulated. They are activated in response to a wide variety of stimuli, including stimulation of cell growth, differentiation, etc., and appear to play central roles in controlling changes in cell phenotype. The MAP kinases are serine/threonine kinases. They have been seen as the point in the pathway at which the cascade switches to serine/threonine phosphorylation.

The MAP kinases have several targets, including other kinases, such as Rsk, which extend the cascade along various branches. MAP kinases are cytosolic, but can translocate into the nucleus after activation; this extends the range of substrates. The direct end of one branch of the cascade is provided by the phosphorylation of transcription factors, including, c-Myc, c-Jun, and Elk-1 (which cooperates with SRF [serum response factor]). This enables the cascade to regulate the activity of a wide variety of genes.

The counterparts for this cascade in other organisms are summarized in **Figure 39.16**. In mammals,

fly, and worm, it starts by the activation of a tyrosine kinase receptor; in mammals the ligand is a polypeptide growth factor, in *D. melanogaster* retina it is a surface transmembrane protein on an adjacent cell (a 'counter-receptor'), and in *C. elegans* vulval induction it is not known. In yeast, the initiating event is different, and consists of the interaction of a polypeptide mating factor with a trimeric G protein; the yeast pathway was discussed previously in more detail in Figure 36.3.

The pathway continues through GRB2 in mammals, and through close homologs in the worm and fly. At the next stage, a homolog of GNRF functions in the fly in the same way as in mammals. The pathway continues through Ras-like proteins (that is, monomeric guanine nucleotide-binding proteins) in all three higher eukaryotes. Mammalian Ras activity can be controlled by a protein called

---

Figure 39.16

Homologous proteins are found in signal transduction cascades in a wide variety of organisms.

	Mammal	Fly		Worm	Yeast	
	Polypeptide growth factor	Membrane counter-receptor		?	Polypeptide pheromone	
	Tyrosine kinase receptor	Sevenless receptor		let-23 receptor	Mating factor receptor	
	GRB2	Drk	?	sem-5	G-protein	
		activates	*inactivates*			
	GNRF	Sos (GNRF)	GAP1			
	Ras	Ras1		let-60		
					STE5	
					STE11	
	c-Raf	D-raf				
	MAPKK				STE7	
	MAP kinase				FUS3 KSS1	
	Transcription factors	Sina nuclear protein			STE12 transcription factor	FAR1 halts cell cycle

GAP as well as by GNRF (which we discuss in the next section). Mutations in a homolog of GAP also may influence the pathway in *D. melanogaster*, suggesting that there are alternative regulatory circuits, at least in flies. An interesting feature is that, although the Ras-dependent pathway is utilized in a variety of cells, the mutations in the GNRF and GAP functions in *Drosophila* are specific for eye development; this implies that a common pathway may be regulated by components that are tissue-specific.

The mammalian MAP kinases have direct counterparts in yeast. *STE7* is homologous to MAPKK, and *FUS3* and *KSS1* code for kinases that share with MAP kinase the requirement for activation by phosphorylation on both threonine and tyrosine. Their targets in turn directly execute the consequences of the cascade.

Oncogenic variants of ras proteins are constitutively active

Ras proteins are small monomeric proteins that bind GTP. Ras proteins have a GTPase activity that may be important for their function. The model for Ras function illustrated in **Figure 39.17** is based on an analogy with (trimeric) G proteins. Ras protein bound to GDP is inactive; Ras carrying GTP is active and acts upon its target molecule. Following this interaction, the GTPase activity hydrolyzes the GTP to GDP, returning Ras to the inactive condition.

The left side of the figure illustrates the functions of two proteins that regulate the state of guanine nucleotide bound to Ras; the right side of the figure illustrates ways in which a pathway could activate Ras. GNRF (guanine nucleotide release factor) stimulates the exchange of GDP with GTP, and therefore activates Ras. GAP (GTPase activating protein) stimulates the hydrolysis of GTP, and therefore inactivates Ras. Several different GAPs have been found, with specificity for different GTP-binding proteins; the GAP that acts on Ras in mammalian cultured cells is called Ras-GAP. A tissue-specific GAP protein that functions in the *Drosophila* retina system is called GAP1.

Constitutive activation of Ras could be caused by mutations that allow the GDP-bound form of Ras to be active or that prevent hydrolysis of GTP. What are the effects of the mutations that create oncogenic *ras* genes? Many mutations that confer transforming activity inhibit the GTPase activity. Although interaction with GAP increases the GTPase activity of proto-Ras proteins, it has no effect on Ras proteins that have already been activated by oncogenic mutations. In other words, Ras has become refractory to the usual ability of GAP to turn off its activity. Inability to hydrolyze GTP could cause Ras to remain in a permanently activated form; its continued action upon its target protein could be responsible for its oncogenic activity.

The general structure of mammalian Ras proteins is illustrated in **Figure 39.18**. Three groups of regions are responsible for the characteristic activities of Ras:

◆ The regions between residues 5–22 and 109–120 are implicated in guanine nucleotide binding by their homology with other G-binding proteins. Some mutations that activate the oncogenic potential of Ras (notably at position 12) lie within these regions.

◆ Ras is attached to the cytoplasmic face of the membrane by farnesylation close to the C-terminus. Mutations that prevent the modification abolish oncogenicity, showing that membrane location is important for Ras function.

Figure 39.17

Ras is active when bound to GTP and inactive when bound to GDP. The relative amounts of each form are controlled by two proteins. GNRF activates Ras by stimulating replacement of GDP by GTP. GAP inactivates Ras by stimulating hydrolysis of GTP. Pathways that rely on Ras could function by controlling either GNRF or GAP. Oncogenic Ras mutants are refractory to control, because Ras remains in the active form.

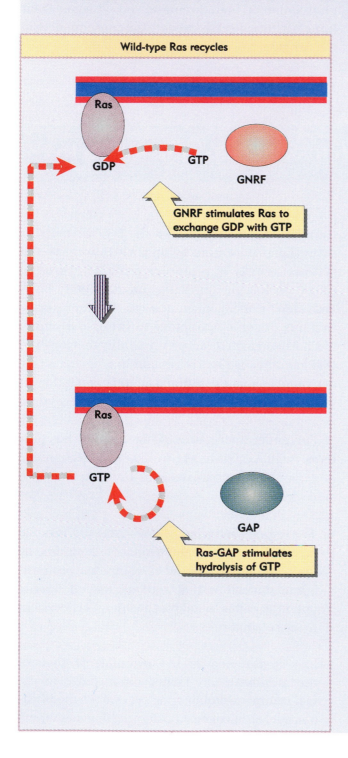

Wild-type Ras recycles

GNRF stimulates Ras to exchange GDP with GTP

Ras-GAP stimulates hydrolysis of GTP

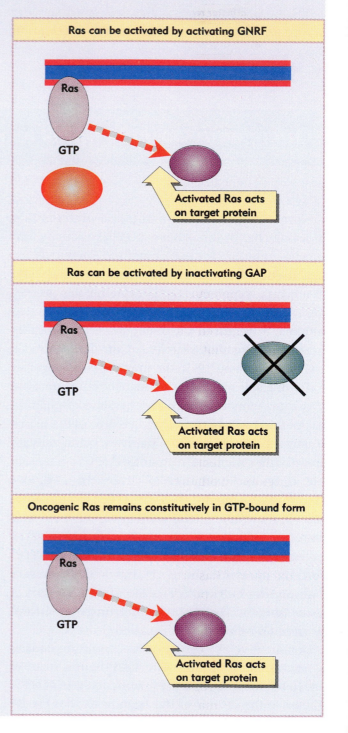

Ras can be activated by activating GNRF

Activated Ras acts on target protein

Ras can be activated by inactivating GAP

Activated Ras acts on target protein

Oncogenic Ras remains constitutively in GTP-bound form

Activated Ras acts on target protein

Figure 39.18

Discrete domains of Ras proteins are responsible for guanine nucleotide binding, effector function, and membrane attachment.

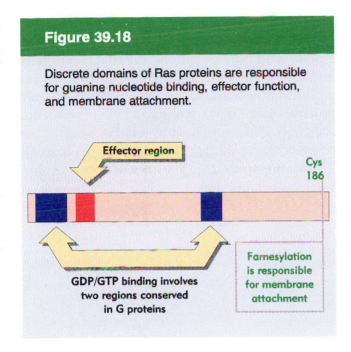

Effector region

Cys 186

GDP/GTP binding involves two regions conserved in G proteins

Farnesylation is responsible for membrane attachment

◆ The effector domain is the region that reacts with the target molecule when Ras has been activated. Activity of the region between residues 30–40 is required for the oncogenic activity of Ras proteins that have been activated by mutation at position 12. The same region is required for the interaction with Ras-GAP.

The crystal structure of Ras protein is illustrated schematically in **Figure 39.19**. The regions close to the guanine nucleotide include the domains that are conserved in other GTP-binding proteins. The potential effector loop is located near the phosphates; it consists of hydrophilic residues, and is potentially exposed in the cytoplasm.

When GTP is hydrolyzed, there is a change in the conformation of Ras protein. The change involves L4, which includes position 61, at which some oncogenic mutations occur. The oncogenic mutations at position 12 occur in loop 1, and directly affect binding to GTP. The conformational changes between the wild-type and oncogenic forms are restricted to these regions, and impede the ability of the mutant Ras to make the conformational change.

The primary basis for the oncogenic property, therefore, lies in the reduced ability to hydrolyze GTP.

The yeast *S. cerevisiae* contains two *ras* genes, related to *H-ras*, but somewhat larger. Their products stimulate the enzyme adenylate cyclase, which is responsible for catalyzing the formation of cyclic AMP, a well-characterized regulatory small molecule. Mammalian and yeast Ras proteins are at least

Figure 39.19

The crystal structure of Ras protein has 6 β strands, 4 α helices, and 9 connecting loops. The GTP is bound by a pocket generated by loops L9, L7, L2 and L1; the amino acids in these loops are close to, but do not exactly coincide with, the guanine nucleotide-binding regions previously identified. The effector region from 30-40 is relatively well exposed, but so are other regions.

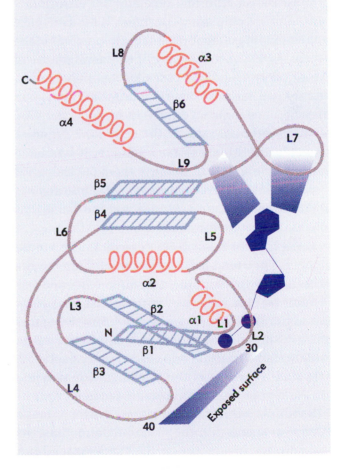

to some degree interchangeable functionally. The functions of yeast ras genes can be substituted by *v-ras* or *c-ras*. And mutant yeast Ras protein is oncogenic in the transfection assay.

We do not know how the functions of the yeast Ras proteins are related to oncogenicity in higher eukaryotes, but their properties strengthen the analogy with G proteins. Two of the mammalian G proteins are the stimulatory and inhibitory G proteins, which differ in their effects upon adenylate cyclase. The G protein with a $G_s\alpha$ subunit stimulates adenylate cyclase, while the protein with the $G_i\alpha$ subunit inhibits it. These G proteins are targets for stimulatory and inhibitory receptors, respectively, and thus transmit the signal from the receptor to the adenylate cyclase. In having adenylate cyclase as a target, the yeast Ras proteins and mammalian G proteins share the features of location and biochemical function.

Growth factor receptor kinases and cytoplasmic tyrosine kinases

The protein tyrosine kinases constitute a major class of oncoproteins, and fall into two general groups: transmembrane receptors for growth factors, and cytoplasmic enzymes. We have more understanding about the biological functions of the receptors, because we know the general nature of the signal transduction cascades that they initiate, and we can see how their inappropriate activation may be oncogenic. We have little understanding about the normal roles in the cell of the cytoplasmic tyrosine kinases—for example, it has been extremely difficult to identify their physiological targets—but we do have a great deal of information about the enzymatic activities and the molecular effects of oncogenic mutations.

Receptors for many growth factors have kinase activity. They tend to be large integral membrane proteins, with domains assembled in modular fashion from a variety of sources. We discussed the general nature of transmembrane receptors and the means by which they are activated to initiate signal transduction cascades in Chapter 12. The EGF receptor is the paradigm for tyrosine kinase receptors. It is a group I receptor with its N-terminus on the extracellular surface, has a single transmembrane region, and has its C-terminus inside the cell. The extracellular N-terminal region binds the ligand that activates the receptor. The intracellular C-terminal region includes a domain that is responsible for the tyrosine kinase activity. Most of the receptors that are coded by cellular proto-oncogenes have a similar form of organization.

A working model for receptor function is that binding of ligand to the extracellular domain activates the tyrosine kinase activity of the intracellular domain. Various forms of this reaction were summarized previously in Figure 12.12. Binding of ligand triggers the formation of dimers, which brings the cytoplasmic domains of the monomers into contact. This triggers an autophosphorylation reaction, probably because each monomer phosphorylates the other, as shown previously in Figure 12.13.

The relationship between the EGF receptor and an oncogenic variant is illustrated in **Figure 39.20**. The EGF receptor is regulated by ligand binding. In the free receptor, the N-terminal extracellular domain represses interaction between monomers. Binding EGF releases this inhibition, allowing the receptor to form dimers. This in turn activates the receptor, and triggers signal transduction.

The oncogene *v-erb* is a truncated version of *c-erbB*, the gene coding for the EGF receptor. The oncogene retains the tyrosine kinase and transmembrane domains, but lacks the N-terminal half

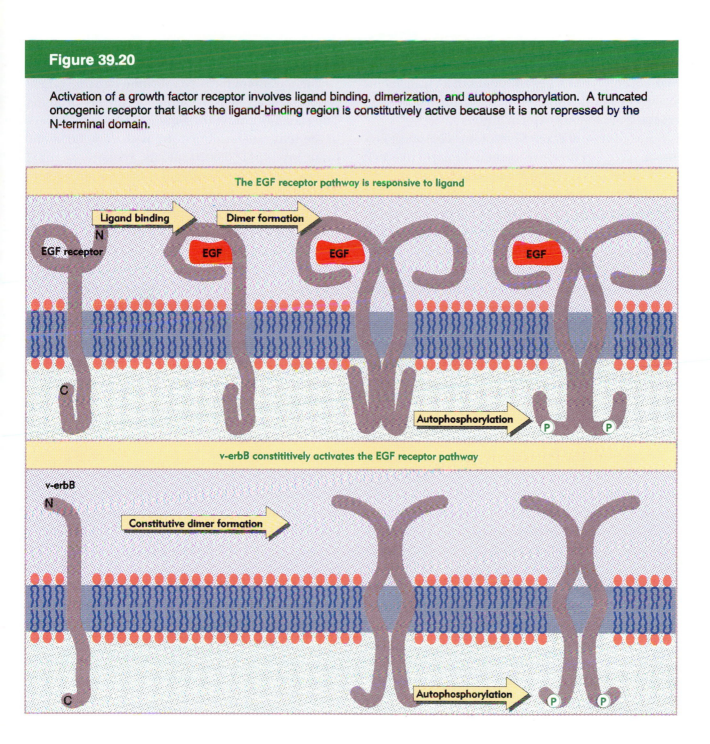

Figure 39.20

Activation of a growth factor receptor involves ligand binding, dimerization, and autophosphorylation. A truncated oncogenic receptor that lacks the ligand-binding region is constitutively active because it is not repressed by the N-terminal domain.

The EGF receptor pathway is responsive to ligand

Ligand binding → Dimer formation

EGF receptor

N

EGF

EGF

EGF

C

Autophosphorylation → P P

v-erbB constititively activates the EGF receptor pathway

v-erbB

N

Constitutive dimer formation →

C

Autophosphorylation → P P

of the protein that binds EGF, and does not have the C-terminus. The deletions at both ends may be needed for oncogenicity. The change in the extracellular N-terminal domain allows the protein to dimerize spontaneously; and the C-terminal deletion may be necessary to remove a cytosolic domain (which contains autophosphorylation sites) that inhibits transforming activity. Thus the basis for oncogenicity is the combination of mutations that activate the receptor constitutively, and possibly also change its interaction with intracellular components.

The general principle that constitutive or altered activity may be responsible for oncogenicity applies to the group of growth factor receptors summarized previously in Figure 39.14. The exact nature of the oncogenic mutation may vary; for example, *erbB2*, which codes for a receptor closely related to the EGF receptor, has a key mutation in its transmembrane region; this increases the propensity of the receptor monomers to form dimers.

Some proto-oncogenes code for receptors or factors involved in the development of particular cell types. Such a receptor (or a growth factor) may be mutated in such a way as to promote unrestricted growth of cells of that type. For example, the receptor coded by *c-fms* is the CSF-I receptor, which mediates the action of colony stimulating factor I, a macrophage growth factor that stimulates the growth and maturation of myeloid precursor cells. *c-fms* can be rendered oncogenic by a mutation in the extracellular domain; perhaps this makes the protein constitutively active in the absence of CSF-I. Oncogenicity is enhanced by C-terminal mutations, which could be inactivating an inhibitory intracellular domain.

The cellular action and basis for oncogenicity of the cytoplasmic group of protein tyrosine kinases is more obscure. The cytoplasmic group is characterized by the viral oncogenes *src, yes, fgr, fps/fes, abl, ros*. (c-Src is actually associated with membranes.) A major stretch of the sequences of all these genes is related, corresponding to residues 250–516 of *c-src*. This includes the 'catalytic domain' responsible for kinase activity. Presumably the regions outside this domain control the activities of the individual members of the family. In few cases,

however, do we know the cellular function of a *c-onc* member of this group.

The paradigm for a cytoplasmic tyrosine kinase in search of a role is presented by the Src proteins. The transforming *v-src* sequence is closely related to the nontransforming *c-src*. Both code for membrane proteins of 60,000 daltons. Actually, there are several *v-src* sequences. Since its isolation by Rous in 1911, RSV has been perpetuated under a variety of conditions, and there are now several 'strains', carrying variants of *v-src*. The common feature is that the C-terminal sequence of *c-src* has been replaced, by a different sequence in each strain. The various strains contain different point mutations within the *src* sequence.

Src proteins have several interesting features. **Figure 39.21** summarizes their activities in terms of protein domains.

Both v-Src and c-Src have an unusual modification at the N-terminus. The N-terminal amino acid is cleaved, and myristic acid (a rare fatty acid of 14 carbon residues) is covalently added to the N-terminal glycine. Myristylation enables Src proteins to attach to the cytosolic face of membranes in the cytoplasm, although they lack stretches of hydrophobic amino acids. Most of the protein is associated with the cytoplasmic face of the endosomes, and it is enriched in regions of cell to cell contact and adhesion plaques.

Amino acids 2–14 are required for myristylation. The modification is essential for oncogenic activity

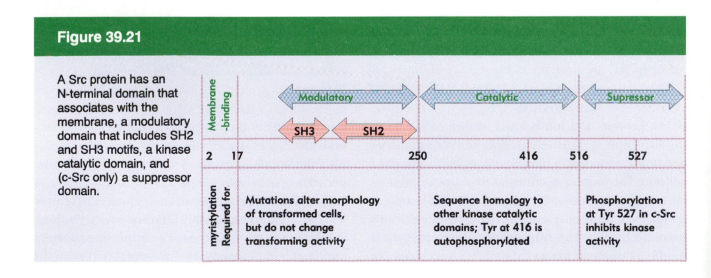

Figure 39.21

A Src protein has an N-terminal domain that associates with the membrane, a modulatory domain that includes SH2 and SH3 motifs, a kinase catalytic domain, and (c-Src only) a suppressor domain.

of v-Src, since N-terminal mutants that cannot be myristylated have reduced tumorigenicity. Two possible reasons have been proposed for the dependence of transformation on the membrane location of v-Src: important substrates for Src may be located in the membrane; or other proteins that associate with Src may use it to position themselves at the membrane, and in turn recognize targets in the vicinity.

Proteins in the Src family were the first oncoproteins of the kinase type to be characterized. Src was also the first example of a kinase whose target is a tyrosine residue in protein. The level of phosphotyrosine is increased about 10× in cells that have been transformed by RSV. In addition to acting on other proteins, Src is able to phosphorylate itself. Catalytic activity resides in the C-terminal half of the protein.

The major difference between v-Src and c-Src lies in their kinase activities. The activity of v-Src is ~20× greater than that of c-Src. The transforming activity of *src* mutants is correlated with the level of kinase activity, and we believe that oncogenicity results from phosphorylation of target protein(s). We do not know whether the increased activity is itself responsible for oncogenicity or whether there is also a change in the specificity with which target proteins are recognized.

Kinase activity plays two roles in Src function. First, attempts to identify a function for the phosphorylation in cell transformation have concentrated on identifying cellular substrates that may be targets for v-Src (especially those that may not be recognized by c-Src). A variety of substrates has been identified, but none has yet been equated with the cause of transformation. Second, the autophosphorylation reaction may be important for Src's transforming activity. Differences between c-Src and v-Src are summarized in **Figure 39.22**.

The c-Src protein is phosphorylated *in vivo* at tyrosine residue 527. This amino acid is located in the C-terminal region, and is part of the sequence of 19 amino acids that is missing from v-Src (where it has been replaced by an unrelated sequence of 12 amino acids).

The v-Src protein is phosphorylated *in vivo* at tyrosine 416. At steady state, 10–30% of v-Src molecules are phosphorylated at this site; the phosphate turns over rapidly, but is not transferred to other proteins. This amino acid residue is present in c-Src, but is not phosphorylated *in vivo*, although it can be phosphorylated *in vitro*.

The importance of these autophosphorylations can be tested by mutating the tyrosine residues at 416 and 527 to prevent addition of phosphate groups. The mutations have opposite effects:

◆ Mutation of tyrosine 527 to the related amino acid phenylalanine activates the transforming potential of c-Src. The protein c-Src$^{Phe-527}$ becomes phosphorylated on tyrosine 416, has its kinase activity increased ~10×, and it transforms target cells, although not as effectively as v-Src. *Phosphorylation of tyrosine 527 therefore represses the oncogenicity of c-src. Removal of this residue when the C-terminal region was lost in generating v-src contributes significantly to the oncogenic activity of the transforming protein.* Thus the Src proteins may be an example in which loss of a function contributed to oncogenicity.

◆ Mutation of tyrosine 416 in c-Src eliminates its residual ability to transform. This mutation also greatly reduces the activity of the c-Src$^{Phe-527}$ mutant. It also reduces the transforming potential of v-Src, but less effectively. *Phosphorylation at tyrosine 416 therefore activates the oncogenicity of Src proteins.*

Point mutations at other residues in c-Src that increase oncogenicity show the same correlation: phosphorylation is decreased at tyrosine 527 and increased at tyrosine 416. The state of these tyrosines may therefore be a general indicator of the oncogenic potential of c-Src. The reduced phosphorylation at tyrosine 527 may be responsible for the increased phosphorylation at tyrosine 416, which may be the crucial event. However, v-Src protein is less dependent on the state of tyrosine 416, and mutants retain transforming activity; presumably *v-src* has accumulated other mutations that increase transforming potential.

Figure 39.22

Two tyrosine residues are targets for autophosphorylation in Src proteins. Phosphorylation at Tyr 527 of c-Src suppresses phosphorylation at Tyr 416, which is associated with transforming activity. Only Tyr 416 is present in v-Src. Transforming potential of c-Src may be activated by removing Tyr 527 or repressed by removing Tyr 416.

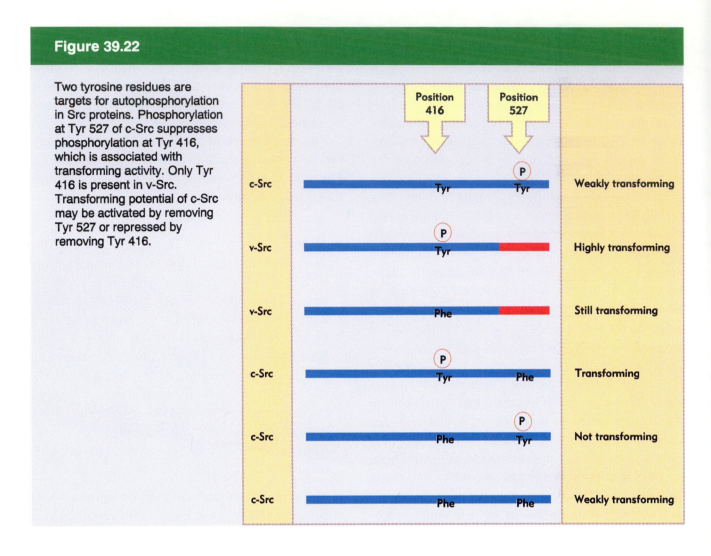

The ability of phosphorylation events both to repress and to activate oncogenic potential makes the important point that *distribution of phosphate groups may be more critical than the overall level of phosphorylation*. This conclusion could apply to the substrates of v-Src that are involved in transformation, as well as to the transforming protein itself.

What is the function of c-Src; and how is it related to the oncogenicity of v-Src? The c-Src and v-Src proteins are very similar: they share N-terminal modification, cellular location, and protein tyrosine kinase activity. c-Src is expressed at high levels in several types of terminally differentiated cells, which suggests that it is not involved in regulating cell proliferation. But we have so far been unable to determine the normal function of c-Src.

The modulatory region of c-Src contains two motifs that are found in a variety of other cytoplasmic proteins that are involved in signal transduction: these may connect a protein to the components that are upstream and downstream of it in a signalling pathway. The domains are named **SH2** and **SH3**, for *Src* homology.

Their presence in various proteins is summarized in **Figure 39.23**. The cytoplasmic tyrosine kinases comprise one group of proteins that have these domains; other prominent members are phospholipase Cγ and the regulatory subunit (p85) of PI3 kinase (both of which are found in the Ras-independent pathways triggered by receptor tyrosine kinases). Ras-GAP has two SH2 domains. The extreme example of a protein with these domains is GRB2/sem-5, which consists *solely* of an SH2 domain flanked by two SH3 domains.

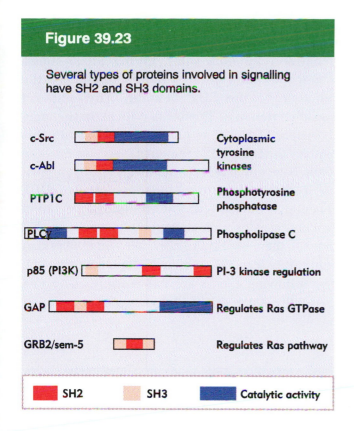

Figure 39.23

Several types of proteins involved in signalling have SH2 and SH3 domains.

c-Src	Cytoplasmic tyrosine kinases
c-Abl	
PTP1C	Phosphotyrosine phosphatase
PLCγ	Phospholipase C
p85 (PI3K)	PI-3 kinase regulation
GAP	Regulates Ras GTPase
GRB2/sem-5	Regulates Ras pathway

SH2   SH3   Catalytic activity

sine kinase, such as c-Src or c-Abl may be activated by the interaction; other types of proteins that are activated by such interactions include phosphatases and phospholipases. Ras-GAP binds via its SH2 domains to the activated EGF receptor, which may provide one link from receptor tyrosine kinases to Ras activity. Another route, as summarized previously in Figure 39.15, is for the activated receptor to bind the SH2 domain of GRB2; this in turn interacts with GNRF, which in turn controls Ras activity. GRB2 is one example of a protein containing an SH2 domain that does not have a catalytic activity. Another is provided by p85, the regulatory subunit of PI3 kinase; binding the p85 SH2 domains to a phosphoprotein in turn activates the associated catalytic subunit.

The SH2 domain is a region of ~100 amino acids that binds to phosphotyrosine in other proteins. The target site is called an **SH2-binding site**. **Figure 39.24** shows an example of a reaction in which SH2 domains are involved. Activation of a tyrosine kinase receptor typically causes it to autophosphorylate a site in the cytosolic tail. The nonphosphorylated sequence is inert; but phosphorylation converts it into an SH2-binding site. Each SH2 domain specifically binds to a particular SH2-binding site; the typical SH2-binding site is only 4 amino acids long. The specificity of each SH2 domain is different, except for a group of kinases related to Src, which seem to share the same specificity. Some proteins contain multiple SH2 domains, which increases their affinity for binding to phosphoproteins or confers the ability to bind to different phosphoproteins.

A protein that contains an SH2 domain is activated when it binds to an SH2-binding site. The consequences of binding depend on the type of protein that contains the SH2 domain. We expect that a tyro-

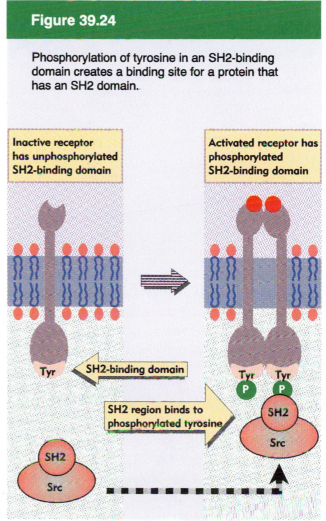

Figure 39.24

Phosphorylation of tyrosine in an SH2-binding domain creates a binding site for a protein that has an SH2 domain.

Inactive receptor has unphosphorylated SH2-binding domain

Activated receptor has phosphorylated SH2-binding domain

Tyr   SH2-binding domain

Tyr   Tyr

SH2 region binds to phosphorylated tyrosine

SH2   Src

SH2   Src

The SH3 domain may provide an effector function by which some of these proteins bind to a downstream component. The case of GRB2/sem-5 strengthens this idea, since SH3 domain is the only region available to provide effector function (and is responsible for binding to GNRF). In fact, it suggests the general model that SH3 domains may provide a connection to small GTP-binding proteins (of which Ras is the paradigm). Another role that has been proposed for SH3 domains (and in particular for the SH3 domain of c-Src) is the ability to interact with proteins of the cytoskeleton, thus triggering changes in cell structure.

How is c-Src usually activated? Most mutations in the SH2 region reduce transforming activity (suggesting that the SH2 function is required to activate c-Src), and most mutations in SH3 increase transforming activity (suggesting that SH3 has a negative regulatory role). **Figure 39.25** shows a more detailed autoregulatory model for the function of the SH2 domain. The state of phosphorylation at Tyr-527 is critical. In the inactive state, Tyr-527 is phosphorylated, and this enables the C-terminal region of c-Src itself to bind to the N-terminal SH2 domain. When an appropriate receptor tyrosine kinase (such as PDGF receptor) is activated, the autophosphorylation reaction creates a phosphopeptide sequence that binds to the c-Src SH2 region, releasing the region containing Tyr-527, which is dephosphorylated. The following events are not entirely clear; one possibility is that this leads to the phosphorylation

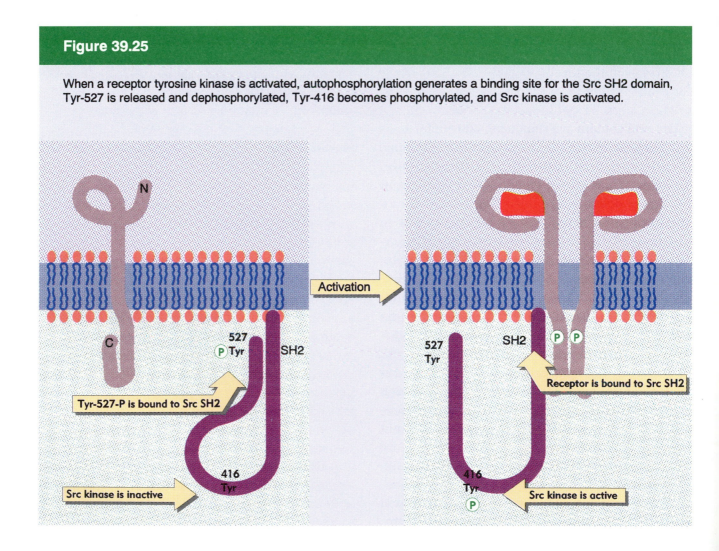

Figure 39.25

When a receptor tyrosine kinase is activated, autophosphorylation generates a binding site for the Src SH2 domain, Tyr-527 is released and dephosphorylated, Tyr-416 becomes phosphorylated, and Src kinase is activated.

of Tyr-416, and activation of kinase activity. In this way, c-Src kinase activity responds to the activation of the receptor kinase.

Alternative ways for activating c-Src may be involved in oncogenic reactions. For example, the polyoma middle T antigen activates c-Src by binding to Tyr-527 and preventing its phosphorylation. The oncogenic v-Src protein, of course, lacks Tyr-527 and is constitutively active. Some mutations in the SH2 domain of c-Src can activate the kinase activity (with oncogenic consequences), presumably because they prevent it from sequestering Tyr-527. Mutations in the SH2 and SH3 domains of c-Src can influence its specificity with regards to transforming different types of target cells, which suggests that these regions provide the connections to other (cell specific) proteins in the pathway.

Oncoproteins may regulate gene expression

It goes almost without saying that it is necessary to make changes in gene expression in order to convert a cell to the transformed phenotype. Many oncogenes act at early stages in pathways that ultimately lead to changes in gene expression. Some act directly at the level of transcription. Retroviral oncogenes include examples derived from the major classes of cellular transcription factors. Several prominent gene families coding for transcription factors are identified by v-onc genes: rel, jun, fos, erbA, myc, and myb. In the cases of Rel, Jun, and ErbA proteins, there are differences in transcriptional activity between the c-Onc and v-Onc proteins that may be related to transforming capacity.

DNA tumor viruses code for oncoproteins that affect transcription, but these oncoproteins have pleiotropic effects; and in none of these cases has the ability to influence transcription been directly correlated with transforming activity. The large T antigens of polyomaviruses are nuclear proteins that bind to the viral genomes, where they are needed to stimulate DNA replication and late transcription. However, these activities can be divorced from transforming activity. The E1A proteins of adenovirus affect transcription indirectly, but it remains to be seen whether this is necessary for transforming ability.

As oncogenes carried by retroviruses, the v-onc genes are defined as dominant oncogenes. Their actions may in principle be quantitative or qualitative; and those that affect transcription might either increase or decrease expression of particular genes. By virtue of increased expression or activity they could turn up transcription of genes whose products can be tolerated only in small amounts. Failure to respond to normal regulation of activity by other cellular factors also might lead to increased gene expression. A less likely possibility is the acquisition of specificity for new target promoters. Alternatively, if the oncoproteins are defective in the ability to activate transcription, they might function as dominant negative suppressors of the cellular transcription factors. The first steps towards distinguishing these possibilities lie with determining which functions are altered in v-Onc compared with c-Onc proteins: is DNA-binding altered either quantitatively or quantitatively; is the ability to activate transcription altered? **Figure 39.26** summarizes the properties of some of these oncoproteins.

The oncogene v-rel was identified as the transforming function of the avian (turkey) reticuloendotheliosis virus. The retrovirus is highly oncogenic in chickens, where it causes B cell lymphomas. v-rel is a truncated version of c-rel, lacking the ~100 C-terminal amino acids, and has a small number of point mutations in the remaining sequence.

The rel gene defines a family with several

Figure 39.26

Oncogenes that code for transcription factors have mutations that inactivate transcription (*v-erbA* and possibly *v-rel*) or that activate transcription (*v-jun* and *v-fos*).

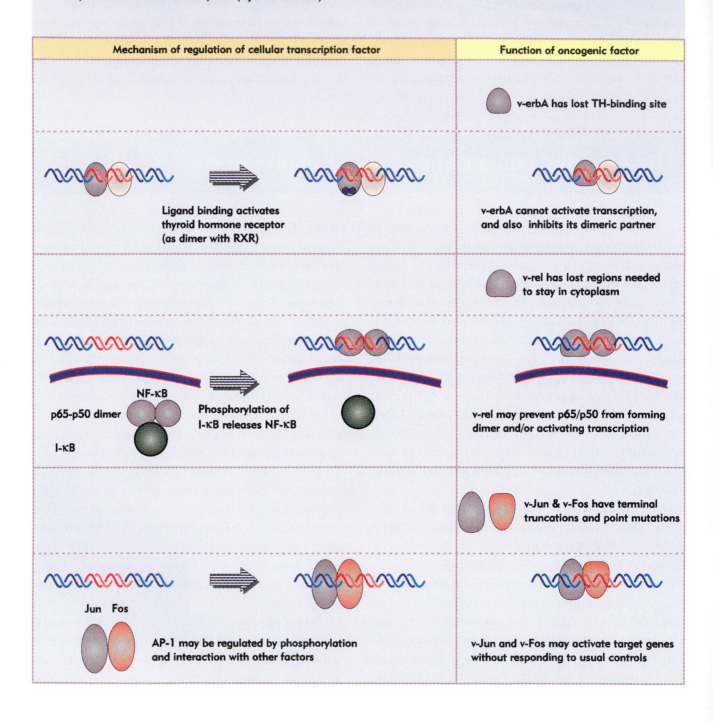

Mechanism of regulation of cellular transcription factor	Function of oncogenic factor
	v-erbA has lost TH-binding site
Ligand binding activates thyroid hormone receptor (as dimer with RXR)	v-erbA cannot activate transcription, and also inhibits its dimeric partner
	v-rel has lost regions needed to stay in cytoplasm
NF-κB p65-p50 dimer I-κB Phosphorylation of I-κB releases NF-κB	v-rel may prevent p65/p50 from forming dimer and/or activating transcription
	v-Jun & v-Fos have terminal truncations and point mutations
Jun Fos AP-1 may be regulated by phosphorylation and interaction with other factors	v-Jun and v-Fos may activate target genes without responding to usual controls

members, which form various pairwise combinations that regulate transcription. For example, at least 4 proteins related to v-Rel immunologically are activated when a T lymphocyte is stimulated. The best characterized family member is the transcription factor NF-κB. This is a dimer of two subunits, p65 and p50, which is held in the cytoplasm by a regulator, I-κB. When I-κB is phosphorylated, it releases NF-κB, which enters the nucleus and activates transcription of target genes whose promoters or enhancers have the κB motif (see Figure 30.6). This regulatory story is essentially recapitulated in *Drosophila* development, where *dorsal* codes for an NF-κB homolog that is held in the cytoplasm by the *cactus* product, an Iκ-B homolog (see Chapter 38). There is sequence relationship between the two subunits of NF-κB, and *c-rel* has 60% similarity with p50.

NF-κB is one of the most pleiotropic transcription factors; indeed, it has been suggested that it may constitute a general second messenger. Many types of stimulus to the cell result in activation of NF-κB; and a broad range of genes are activated via the presence of κB binding sites. We have not yet identified the activities of all the family members. It remains to be seen what changes in *v-rel* make it oncogenic, and how they affect the transcriptional regulation exercised by the Rel-related proteins. One possibility is that the *v-rel* product forms dimers with cellular family members, and that these dimers are inactive, so that v-Rel prevents NF-κB or related factors from functioning.

The transcription factor AP-1 recognizes a short sequence (TGACTCA) originally identified in the enhancer of SV40, but since found in promoters and enhancers of cellular genes. It is the nuclear factor required to mediate transcription induced by phorbol ester tumor promoters (such as TPA), and an AP-1 binding site confers TPA inducibility upon a target gene. Preparations of the factor include a variety of polypeptides, amongst which are the products of the genes *c-jun* and *c-fos*. Jun-Fos dimers form a transcription factor that activates genes whose promoters or enhancers have an AP-1 binding site.

Jun and Fos are transcription factors of the leucine zipper class. Each protein is a member of a family, and a series of pairwise interactions between Jun family members and Fos family members may generate a series of transcription factors. Mutations of *v-jun* or *v-fos* that abolish the ability to bind DNA also render the product nontransforming, providing a direct proof that ability to bind to DNA is required for transforming activity. Beyond this, however, it is not clear what sort of changes in transcription are required for transformation: they could involve quantitative changes (over-expression or under-expression of particular target genes) or qualitative changes (alteration in the pattern of genes that responds to the factor).

The cellular gene *c-erbA* codes for a thyroid hormone receptor, a member of the general class of steroid hormone receptors (see Figures 30.10–30.12). Upon binding its ligand, a steroid receptor activates expression of particular target genes by binding to its specific response element in a promoter or enhancer. The mode of action for thyroid hormone is distinct: it is located permanently in the nucleus, and, indeed, may bind its response element whether or not ligand is present. The effect of hormone binding may therefore be to activate transcription by previously bound receptor.

Ability to bind DNA is required for transforming capacity. *v-erbA* is truncated at both ends and has a small number of substitutions relative to *c-erbA*. Hormone binding may be altered; the *c-erbA* product binds triiodothyronine (T_3) with high affinity, but the *v-erbA* product has little or no affinity for the ligand in mammalian cells. This suggests that loss of the ligand-binding capacity (perhaps together with other changes) may create a protein whose function has become independent of the hormone.

The consequence of losing the response to ligand is that the factor can no longer be stimulated to activate transcription.

These results place *v-erbA* as a dominant negative oncogene, one that functions by overcoming the action of its normal cellular counterpart. The implication is that genes usually activated by c-ErbA act to *suppress* transformation. In this particular case, it seems likely that these genes usually promote differentiation; blocking this action leaves the cells free to proliferate.

c-jun, c-fos, c-rel, and also *c-myc* are 'immediate early' genes, members of a class of genes that are rapidly induced when resting cells are treated with mitogens, which suggests that they may be involved in a cascade that initiates cycling. Thus their targets are likely to be concerned with initiating or promoting growth. We should therefore expect an increase in their activities to be associated with oncogenesis, an expectation that may be fulfilled for *v-fos* and *v-myc*, but does not explain the behavior of *v-rel*.

The adenovirus oncogene E1A provides an example of a protein that regulates gene expression indirectly, that is, without itself binding to DNA. The E1A region is expressed in the form of three transcripts, derived by alternative splicing, as indicated in **Figure 39.27**. The 13S and 12S mRNAs code for closely related proteins and are produced early in infection. They possess the ability to immortalize cells, and can cooperate with other oncoproteins (notably Ras) to transform primary cells (see Table 39.4). No other viral function is needed for this activity.

The E1A proteins exercise a variety of effects on gene expression. They activate the transcription of some genes, but repress others. Loci that are activated include genes transcribed by RNA polymerase III as well as RNA polymerase II. Some of these effects reside in particular protein domains.

Mutation of the E1A proteins suggests that transcriptional activation requires only the short region of domain 3, found only in the 289 amino acid protein coded by 13S mRNA. This conclusion has been confirmed by showing that an isolated 49-mer

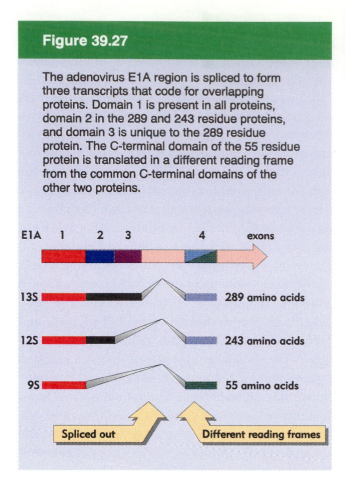

Figure 39.27

The adenovirus E1A region is spliced to form three transcripts that code for overlapping proteins. Domain 1 is present in all proteins, domain 2 in the 289 and 243 residue proteins, and domain 3 is unique to the 289 residue protein. The C-terminal domain of the 55 residue protein is translated in a different reading frame from the common C-terminal domains of the other two proteins.

peptide corresponding to domain 3 can activate transcription of target genes.

Repression of transcription, induction of DNA synthesis, and morphological transformation all require domains 1 and 2, common to both the 289 and 243 amino acid proteins. This suggests the possibility that repression of target genes is required to cause transformation. The E1A products are phosphoproteins located in the nucleus. They do not bind to DNA. Since other viral proteins are not needed for their actions on cellular genes, it seems that E1A proteins must act by binding to other proteins that in turn repress or activate transcription of appropriate target genes. Common features have not been identified in genes known to be activated (or repressed) by E1A, and it therefore seems likely that E1A interacts with several different cellular proteins. Most of these target proteins are unidentified at present.

Summary

A tumor cell is distinguished from a normal cell by its immortality, transformation, and (sometimes) ability to metastasize. Oncogenes are identified by genetic changes that represent gain-of-functions associated with the acquisition of these properties. An oncogene may be derived from a proto-oncogene by mutations that affect its function or level of expression. Oncogenic viruses either carry oncogenes or activate cellular proto-oncogenes.

DNA tumor viruses carry oncogenes without cellular counterparts. Their oncogenes may function by inhibiting the activities of cellular tumor suppressors. RNA tumor viruses carry v-onc genes that are derived from the mRNA transcripts of cellular (c-onc) genes. Some v-onc oncogenes represent the full length of the c-onc proto-oncogene, but others are truncated at one or both ends. All v-onc genes carry point mutations relative to the c-onc coding sequence. Most are expressed as fusion proteins with a retroviral product. Src is an exception in which the retrovirus (RSV) is replication-competent, and the protein is expressed as an independent entity. Some v-onc genes are qualitatively different from their c-onc counterparts, since the v-onc gene is oncogenic at low levels of protein, while the c-onc gene is not active even at high levels. Deletions and point mutations of c-onc genes may both be important in activating such v-onc genes.

Some c-onc proto-oncogenes can be activated in situ by point mutations, translocations, retroviral insertions, or amplification. Some proto-oncogenes are activated efficiently only by changes in the protein coding sequence. Others can be activated by large (>10×) increases in the level of expression; c-myc is an example that can be activated quantitatively by a variety of means, including translocations with the Ig or TcR loci when their recombination systems malfunction, or insertion of retroviruses.

Many of the c-onc genes have counterpart v-onc genes in retroviruses, but some have been identified only by their association with cellular tumors. The transfection assay detects some activated c-onc sequences by their ability to transform rodent fibroblasts. Ras genes are the predominant type identified by this assay. The creation of transgenic mice directly demonstrates the transforming potential of certain oncogenes.

Tumor suppressors are identified by loss-of-function mutations that allow increased cell proliferation. Retinoblastoma (RB) arises when both copies of the RB gene are deleted or inactivated. The RB product is a nuclear phosphoprotein whose nonphosphorylated form blocks progression through S phase. The block is released by phosphorylation by cyclin–CDK complexes. Nonphosphorylated RB binds to the E1A or T antigen oncoproteins of DNA tumor viruses; this interaction could prevent RB from regulating the cell cycle.

p53 was originally classified as an oncogene because missense mutations in it are oncogenic. It is now classified as a tumor suppressor because the missense mutants in fact function by inhibiting the activity of wild-type p53. The same phenotype is produced by loss of both wild-type alleles. p53 is bound by viral oncogenes such as SV40 T antigen and by cellular proteins; it binds a ~10 bp sequence in DNA and activates (or sometimes represses) genes whose promoters have this sequence. Mutant p53 lacks these activities. Loss of p53 may be associated with increased amplification of DNA sequences.

Cellular oncoproteins may be derived from several types of genes. The only common feature is that each type of gene product is likely to be involved in regulating the activities or synthesis of other proteins involved in growth regulation.

Growth factor receptors located in the plasma membrane usually are represented by truncated versions in v-onc genes. The cellular receptors often have protein tyrosine kinase activity. The oncogenic versions have constitutive activity or altered regulation. In the same way, mutation of genes for polypeptide growth factors gives rise to oncogenes, because a receptor becomes inappropriately activated. The kinase activity of the receptors is important in two ways: autophosphorylation of the cytoplasmic domain creates a phosphopeptide sequence (SH2-binding site) that associates with protein that have SH2 domains, causing them to be activated; and phosphorylation of target proteins may activate them.

Some oncoproteins are cytoplasmic tyrosine kinases; their targets are largely unknown. They may be activated in response to the autophosphorylation of tyrosine kinase receptors. A common feature is the presence of an SH2 domain. The SH2 domain of Src recognizes the phosphopeptide sequence created by autophosphorylation of PDGF receptor; the PDGF receptor competes with the C-terminal region of Src for the SH2 domain, thus causing a conformational change that activates the kinase activity of Src.

Ras proteins can bind GTP and are related to the α subunits of G proteins involved in signal transduction across the cell membrane. Oncogenic variants have altered GTPase activity. Activation of Ras is an obligatory step in a signal transduction cascade that is initiated by activation of a tyrosine kinase receptor such as the EGF receptor; the cascade passes to the MAP kinase, which is a serine/threonine kinase, and terminates with the nuclear phosphorylation of transcription factors including Jun and Fos.

Nuclear oncoproteins may be involved directly in regulating gene expression, and include Jun and Fos, which are part of the AP1 transcription factor. v-ErbA is derived from another transcription factor, the thyroid hormone receptor, and is a dominant negative mutant that prevents the cellular factor from functioning. v-Rel is related to the common factor IF-κB, but its mode of oncogenic action is not known.

Further reading

Reviews

Cellular and viral oncogenes were reviewed by **Bishop** (*Ann. Rev. Biochem.* **52**, 301–354, 1983; *Cell* **42**, 23–38, 1983). Activation of oncogenes was analyzed by **Varmus** (*Ann. Rev. Genet.* **18**, 553–612, 1984). Oncogenes in transgenic animals were reviewed by **Cory and Adams** (*Ann. Rev. Immunol.* **6**, 25–48, 1988), **Hanahan** (*Ann. Rev. Genet.* **22**, 479–519, 1988), and **Adams and Cory** (*Science* **254**, 1161–1167, 1991). The nature of the various classes of oncogenes, and cooperation between oncogenes was reviewed by **Hunter** (*Cell* **64**, 249–270, 1991).

Ras oncogenes were reviewed by **Barbacid** (*Ann. Rev. Biochem.* **56**, 779–827, 1987), and by **Lowy and Willumsen** (*Ann. Rev. Biochem.* **62**, 851–891, 1993).

The roles of E1A proteins were reviewed by **Berk** (*Ann. Rev. Genet* **20**, 45–79, 1986) and **Nevins** (*Microbiol. Rev.* **51**, 419–430, 1987).

Tumor suppressors were reviewed by **Marshall** (*Cell* **64**, 313–326, 1991). Changing perspectives on p53 were reviewed by **Levine, Momand, and Finlay** (*Nature* **351**, 453–456, 1991).

The connection between translocations involving the Ig loci or TcR loci and tumor formation in the immune system has been reviewed by **Showe and Croce** (*Ann. Rev. Immunol.* **5**, 253–277, 1987) and **Haluska, Tsujimoto, and Croce** (*Ann. Rev. Genet.* **21**, 321–345, 1987).

Relationships between oncogene products and growth factors were summarized by **Heldin and Westermark** (*Cell* **37**, 9–20, 1984). Src proteins and the roles of kinase activity were reviewed by **Jove and Hanafusa** (*Ann. Rev. Cell Biol.* **3**, 31–56, 1987).

Tyrosine kinases have been reviewed by **Hunter and Cooper** (*Ann. Rev. Biochem.* **54**, 897–930, 1985).

Oncogenes and signal transduction pathways were reviewed by **Cantley** *et al.* (*Cell* **64**, 281–303, 1991). The functions of SH2 domains were reviewed by **Koch** *et al.* (*Science* **252**, 668–674, 1991). Relationships between oncogenesis and development were reviewed by **Cross and Dexter** (*Cell* **64**, 271–302, 1991)

Discoveries

The induction of tumors by oncogenes in transgenic mice was first observed by **Stewart, Pattengale, and Leder** (*Cell* **38**, 627–637, 1984). The synergistic effects of combining oncogenes was reported by **Sinn** *et al.* (*Cell* **49**, 465–475, 1987).

Dependence of retinoblastoma on loss of the *RB* gene was reported by **Cavanee** *et al.* (*Nature* **305**, 779–784, 1983). The occurrence of deletions in p53 in hereditary cancers was reported by **Malkin** *et al.* (*Science* **250**, 1233–1238, 1990). p53 was characterized as a tumor suppressor by **Finlay, Hinds, and Levine** (*Cell* **57**, 1083–1093, 1989).

Src and its kinase activity were discovered by **Brugge and Erikson** (*Nature* **269**, 346–348, 1977) and **Collet and Erikson** (*Proc. Nat. Acad. Sci. USA* **75** 2021–2024, 1978).

The link between oncogenes and receptors/growth factors was made when **Waterfield** *et al.* (*Nature* **304**, 35–39, 1983) demonstrated that *c-sis* codes for PDGF.

Interactions in the *bcr–c-abl* fusion oncogene were analyzed by **Maru and Witte** (*Cell* **67**, 459–468, 1991). The first example of a proto-oncogene that codes for a transcription factor was reported by **Bohmann** *et al.* (*Science* **238**, 1386–1392, 1987).

EPILOGUE

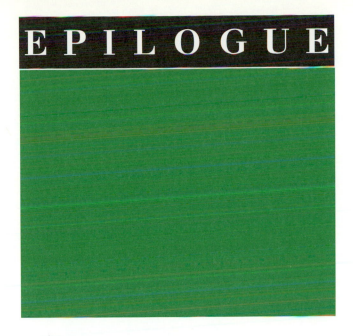

Reflecting on the prospects for further advance, one may be tempted to take an attitude of romantic pessimism: all that remains is either applications or epistemological disquisition. Such was the mood in physics around 1900 — after Maxwell and Boltzmann, and just before Curie, Planck, Rutherford, Einstein and Bohr entered the picture. And so there is hope for the young biologists who dream of discovery.

Salvador Luria, 1986

EPILOGUE

Landmark shifts in perspectives

Looking back from the vantage point of the early 1990s, we can see the shifts in perspective since the initial genetic and biochemical discoveries of the nature of the gene in the 1940s. Some of the landmark years for discoveries of major intellectual or practical import were:

1941 Beadle and Tatum show that a gene codes for a single protein.

1944 Avery proves that DNA is the genetic material.

1953 Watson and Crick propose the double helical structure for DNA.

1958 Meselson and Stahl demonstrate that DNA replicates semiconservatively.

1961 The triplet nature of the genetic code is discovered.
Messenger RNA is uncovered.
Jacob and Monod propose the operon model for gene regulation.

1970 Temin and Baltimore report the discovery of reverse transcriptase in retroviruses.

1974 Eukaryotic genes are cloned in bacterial plasmids.

1976 Retroviral oncogenes are identified as the causative agents of transformation.

1977 DNA sequencing becomes possible.
Interrupted genes are discovered and splicing of their transcripts is inferred.

1979 Cellular oncogenes are discovered by transfection.

1981 Catalytic activity of RNA is discovered.
Transgenic mice and flies are obtained by introducing new DNA into the germ line.

The pace of discovery remains active. What appear to be major discoveries may now deal with more restricted issues than the major discoveries of the earlier period—we are concerned now with showing that DNA can be rearranged or transfected rather than with elucidating its fundamental structure or mode of replication. Yet there is at present no end in sight to the series of discoveries, both intellectual and practical, about DNA and its expression.

We have come a long way from the concept that the gene is an isolated and fixed locus coding for a single protein. Genes may reside in huge clusters containing related sequences. They may be expressed in terms of alternative versions of a protein. They may be reconstructed by physical rearrangement of sequences during somatic development. New genes may even be added to the germ line by introducing DNA into eggs. Seemingly at odds with our perspective at the time, each discovery has been assimilated by an extension in our view of the capacity of the gene.

The most pressing questions of the moment concern gene expression, in particular the nature of the network of regulatory events that must be involved in the development of an adult eukaryote from its embryo. Are genes controlled in a temporal cascade? How are the geographical distinctions between parts of the organism originally established and then perpetuated and extended? How far, indeed, can networks involved in gene regulation explain the process of development?

Our ideas on possible solutions to these problems are certainly more diffuse than were our thoughts about the problems of the time twenty or thirty years ago—we lack the focus of supposing there is a single developmental code to decipher, analogous to the genetic code. Yet fundamental discoveries may remain to be made and we must wonder which shibboleths of genetics will tumble next.

GLOSSARY

Abundance of an mRNA is the average number of molecules per cell.

Abundant mRNAs consist of a small number of individual species, each present in a large number of copies per cell.

Acceptor splicing site—*see* right splicing junction.

Acentric fragment of a chromosome (generated by breakage) lacks a centromere and is lost at cell division.

Acrocentric chromosome has the centromere located nearer one end than the other.

Active site is the restricted part of a protein to which a substrate binds.

Allele is one of several alternative forms of a gene occupying a given locus on a chromosome.

Allelic exclusion describes the expression in any particular lymphocyte of only one allele coding for the expressed immunoglobulin.

Allosteric control refers to the ability of an interaction at one site of a protein to influence the activity of another site.

Alu family is a set of dispersed, related sequences, each ~300 bp long, in the human genome. The individual members have Alu cleavage sites at each end (hence the name).

Alu-equivalent family is the set of sequences in a mammalian genome that is related to the human Alu family.

α-Amanitin is a bicyclic octapeptide derived from the poisonous mushroom *Amanita phalloides*; it inhibits transcription by certain eukaryotic RNA polymerases, especially RNA polymerase II.

Amber codon is the nucleotide triplet UAG, one of three codons that cause termination of protein synthesis.

Amber mutation describes any change in DNA that creates an amber codon at a site previously occupied by a codon representing an amino acid in a protein.

Amber suppressors are mutant genes that code for tRNAs whose anticodons have been altered so that they can respond to UAG codons as well as or instead of to their previous codons.

Aminoacyl-tRNA is transfer RNA carrying an amino acid; the covalent linkage is between the NH_2 group of the amino acid and either the 3'- or 2'-OH group of the terminal base of the tRNA.

Aminoacyl-tRNA synthetases are enzymes responsible for covalently linking amino acids to the 2' or 3'-OH position of tRNA.

Amphipathic structures have two surfaces, one hydrophilic and one hydrophobic. Lipids are amphipathic; and some protein regions may form amphipathic helices, with one charged face and one neutral face.

Amplification refers to the production of additional copies of a chromosomal sequence, found as intra-chromosomal or extrachromosomal DNA.

Anchorage dependence describes the need of normal eukaryotic cells for a surface to attach to in order to grow in culture.

Aneuploid chromosome constitution differs from the usual diploid constitution by loss or duplication of chromosomes or chromosomal segments.

Annealing is the pairing of complementary single strands of DNA to form a double helix.

Antibody is a protein (immunoglobulin) produced by B lymphocyte cells that recognizes a particular foreign 'antigen,' and thus triggers the immune response.

Anticoding strand of duplex DNA is used as a template to direct the synthesis of RNA that is complementary to it.

Antigen is any molecule whose entry into an organism provokes synthesis of an antibody (immunoglobulin).

Antiparallel strands of the double helix are organized in opposite orientation, so that the 5′ end of one strand is aligned with the 3′ end of the other strand.

Antitermination proteins allow RNA polymerase to transcribe through certain terminator sites.

AP endonucleases make incisions in DNA on the 5′ side of either apurinic or apyrimidinic sites.

Apoinducer is a protein that binds to DNA to switch on transcription by RNA polymerase.

Archebacteria comprise a minor line of prokaryotes, and may have introns in the genome.

Ascus of a fungus contains a tetrad or octad of the (haploid) spores, representing the products of a single meiosis.

att sites are the loci on a phage and the bacterial chromosome at which recombination integrates the phage into, or excises it from, the bacterial chromosome.

Attenuation describes the regulation of termination of transcription that is involved in controlling the expression of some bacterial operons.

Attenuator is the terminator sequence at which attenuation occurs.

Autogenous control describes the action of a gene product that either inhibits (negative autogenous control) or activates (positive autogenous control) expression of the gene coding for it.

Autonomous controlling element in maize is an active transposon with the ability to transpose (*cf* nonautonomous controlling element).

Autoradiography detects radioactively labeled molecules by their effect in creating an image on photographic film.

Autosomes are all the chromosomes except the sex chromosomes; a diploid cell has two copies of each autosome.

B lymphocytes (or **B cells**) are the cells responsible for synthesizing antibodies.

Backcross is another (earlier) term for a **testcross**.

Back mutation reverses the effect of a mutation that had inactivated a gene; thus it restores wild type.

Bacteriophages are viruses that infect bacteria; often abbreviated as **phages**.

Balbiani ring is an extremely large puff at a band of a polytene chromosome.

Bands of polytene chromosomes are visible as dense regions that contain the majority of DNA; **bands of normal chromosomes** are relatively much larger and are generated in the form of regions that retain a stain on certain chemical treatments.

Base pair (bp) is a partnership of A with T or of C with G in a DNA double helix; other pairs can be formed in RNA under certain circumstances.

Bidirectional replication is accomplished when two replication forks move away from the same origin in different directions.

Bivalent is the structure containing all four chromatids (two representing each homologue) at the start of meiosis.

Blastoderm is a stage of insect embryogenesis in which a layer of nuclei or cells around the embryo surround an internal mass of yolk.

Blocked reading frame cannot be translated into protein because it is interrupted by termination codons.

Blunt-end ligation is a reaction that joins two DNA duplex molecules directly at their ends.

bp is an abbreviation for base pairs; distance along DNA is measured in bp.

Branch migration describes the ability of a DNA

strand partially paired with its complement in a duplex to extend its pairing by displacing the resident strand with which it is homologous.

Breakage and reunion describes the mode of genetic recombination, in which two DNA duplex molecules are broken at corresponding points and then rejoined crosswise (involving formation of a length of heteroduplex DNA around the site of joining).

Buoyant density measures the ability of a substance to float in some standard fluid, for example, CsCl.

C banding is a technique for generating stained regions around centromeres.

C genes code for the constant regions of immunoglobulin protein chains.

C value is the total amount of DNA in a haploid genome.

CAAT box is part of a conserved sequence located upstream of the startpoints of eukaryotic transcription units; it is recognized by a large group of transcription factors.

Cap is the structure at the 5′ end of eukaryotic mRNA, introduced after transcription by linking the terminal phosphate of 5′ GTP to the terminal base of the mRNA. The added G (and sometimes some other bases) are methylated, giving a structure of the form $^{7Me}G5'ppp5'Np\ldots$

CAP (CRP) is a positive regulator protein activated by cyclic AMP. It is needed for RNA polymerase to initiate transcription of certain (catabolite-sensitive) operons of *E. coli.*

Capsid is the external protein coat of a virus particle.

Catabolite repression describes the decreased expression of many bacterial operons that results from addition of glucose. It is caused by a decrease in the level of cyclic AMP, which in turn inactivates the CAP regulator.

cDNA is a single-stranded DNA complementary to an RNA, synthesized from it by reverse transcription *in vitro.*

cDNA clone is a duplex DNA sequence representing an RNA, carried in a cloning vector.

Cell cycle is the period from one division to the next.

Cell hybrid is a somatic cell containing chromosomes derived from parental cells of different species (e.g. a man–mouse somatic cell hybrid), generating by fusing the cells to form a heterokaryon in which the nuclei subsequently fused.

Centrioles are small hollow cylinders consisting of microtubules that become located near the poles during mitosis. They reside within the centrosomes.

Centromere is a constricted region of a chromosome that includes the site of attachment to the mitotic or meiotic spindle (*see also* kinetochore).

Centrosomes are the regions from which microtubules are organized at the poles of a mitotic cell. In animal cells, each centrosome contains a pair of centrioles surrounded by a dense amorphous region to which the microtubules attach. See also MTOC.

Molecular **chaperone** is a protein that is needed for the assembly or proper folding of some other protein, but which is not itself a component of the target complex.

Chemical complexity is the amount of a DNA component measured by chemical assay.

Chi sequence is an octamer that provides a hotspot for RecA-mediated genetic recombination in *E. coli.*

Chi structure is a joint between two duplex molecules of DNA revealed by cleaving an intermediate of two joined circles to generate linear ends in each circle. It resembles a Greek chi in outline, hence the name.

Chiasma (*pl.* chiasmata) is a site at which two homologous chromosomes appear to have exchanged material during meiosis.

Chromatids are the copies of a chromosome produced by replication. The name is usually used to describe them in the period before they separate at the subsequent cell division.

Chromatin is the complex of DNA and protein in the nucleus of the interphase cell. Individual chromosomes cannot be distinguished in it. It was originally recognized by its reaction with stains specific for DNA.

Chromocenter is an aggregate of heterochromatin from different chromosomes.

Chromomeres are densely staining granules visible in chromosomes under certain conditions, especially early in meiosis, when a chromosome may appear to consist of a series of chromomeres.

Chromosome is a discrete unit of the genome carrying many genes. Each chromosome consists of a very long molecule of duplex DNA and an approximately equal mass of proteins. It is visible as a morphological entity only during cell division.

Chromosome walking describes the sequential isolation of clones carrying overlapping sequences of DNA, allowing large regions of the chromosome to be spanned. Walking is often performed in order to reach a particular locus of interest.

cis-**acting locus** affects the activity only of DNA sequences on its own molecule of DNA; this property usually implies that the locus does not code for protein.

cis-**acting protein** has the exceptional property of acting only on the molecule of DNA from which it was expressed.

cis **configuration** describes two sites on the same molecule of DNA.

cis/trans **test** assays the effect of relative configuration on expression of two mutations. In a double heterozygote, two mutations in the same gene show mutant phenotype in *trans* configuration, wild-type in *cis* configuration.

Cistron is the genetic unit defined by the *cis/trans* test; equivalent to **gene** in comprising a unit of DNA representing a protein.

Class switching is a change in the expression of the C region of an immunoglobulin heavy chain during lymphocyte differentiation.

Clone describes a large number of cells or molecules identical with a single ancestral cell or molecule.

Cloning vector is a plasmid or phage that is used to 'carry' inserted foreign DNA for the purposes of producing more material or a protein product.

Closed reading frame contains termination codons that prevent its translation into protein.

Coated vesicles are vesicles whose membrane has on its surface a layer of the protein clathrin.

Coconversion is the simultaneous correction of two sites during gene conversion.

Coding strand of DNA has the same sequence as mRNA.

Codominant alleles both contribute to the phenotype; neither is dominant over the other.

Codon is a triplet of nucleotides that represents an amino acid or a termination signal.

Coevolution—*see* concerted evolution.

Cognate tRNAs are those recognized by a particular aminoacyl-tRNA synthetase.

Coincidental evolution—*see* concerted evolution.

Cointegrate structure is produced by fusion of two replicons, one originally possessing a transposon, the other lacking it; the cointegrate has copies of the transposon present at both junctions of the replicons, oriented as direct repeats.

Cold-sensitive mutant is defective at low temperature but functional at normal temperature.

Colony hybridization is a technique for using *in situ* hybridization to identify bacteria carrying chimeric vectors whose inserted DNA is homologous with some particular sequence.

Compatibility group of plasmids contains members unable to coexist in the same bacterial cell.

Complementation refers to the ability of independent (nonallelic) genes to provide diffusible products that produce wild phenotype when two mutants are tested in *trans* configuration in a heterozygote.

In vitro **complementation assay** consists of identifying a component of a wild-type cell that can confer activity on an extract prepared from a mutant cell. The assay identifies the component rendered inactive by the mutation.

Complementation group is a series of mutations unable to complement when tested in pairwise combinations in *trans*; defines a genetic unit (the cistron) that might better be called a *noncomplementation* group.

Complex locus (of *D. melanogaster*) has genetic properties inconsistent with the function of a gene representing a single protein. Complex loci are usually very large (>100 kb) at the molecular level.

Complexity is the total length of different sequences of DNA present in a given preparation.

Composite transposons have a central region

flanked on each side by insertion sequences, either or both of which may enable the entire element to transpose.

Concatemer of DNA consists of a series of unit genomes repeated in tandem.

Concatenated circles of DNA are interlocked like rings on a chain.

Concerted evolution describes the ability of two related genes to evolve together as though constituting a single locus.

Condensation reaction is one in which a covalent bond is formed with loss of a water molecule, as in the addition of an amino acid to a polypeptide chain.

Conditional lethal mutations kill a cell or virus under certain (nonpermissive) conditions, but allow it to survive under other (permissive) conditions.

Conjugation describes 'mating' between two bacterial cells, when (part of) the chromosome is transferred from one to the other.

Consensus sequence is an idealized sequence in which each position represents the base most often found when many actual sequences are compared.

Conservative recombination involves breakage and reunion of preexisting strands of DNA without any synthesis of new stretches of DNA.

Conservative transposition refers to the movement of large elements, originally classified as transposons, but now considered to be episomes. The mechanism of movement resembles that of phage lambda.

Constant regions of immunoglobulins are coded by C genes and are the parts of the chain that vary least. Those of heavy chains identify the type of immunoglobulin.

Constitutive genes are expressed as a function of the interaction of RNA polymerase with the promoter, without additional regulation; sometimes also called household genes in the context of describing functions expressed in all cells at a low level.

Constitutive heterochromatin describes the inert state of permanently nonexpressed sequences, usually satellite DNA.

Constitutive mutations cause genes that usually are regulated to be expressed without regulation.

Contractile ring is a ring of actin filaments that forms around the equator at the end of mitosis and is responsible for pinching the daughter cells apart.

Controlling elements of maize are transposable units originally identified solely by their genetic properties. They may be autonomous (able to transpose independently) or nonautonomous (able to transpose only in the presence of an autonomous element).

Coordinate regulation refers to the common control of a group of genes.

Cordycepin is $3'$ deoxyadenosine, an inhibitor of polyadenylation of RNA.

Core DNA is the 146 bp of DNA contained on a core particle.

Core particle is a digestion product of the nucleosome that retains the histone octamer and has 146 bp of DNA; its structure appears similar to that of the nucleosome itself.

Corepressor is a small molecule that triggers repression of transcription by binding to a regulator protein.

Cosmids are plasmids into which phage lambda *cos* sites have been inserted; as a result, the plasmid DNA can be packaged *in vitro* in the phage coat.

Cot is the product of DNA concentration and time of incubation in a reassociation reaction.

Cot$_{\frac{1}{2}}$ is the Cot required to proceed to half completion of the reaction; it is directly proportional to the unique length of reassociating DNA.

Cotransfection is the simultaneous transfection of two markers.

Crossing-over describes the reciprocal exchange of material between chromosomes that occurs during meiosis and is responsible for genetic recombination.

Crossover fixation refers to a possible consequence of unequal crossing-over that allows a mutation in one member of a tandem cluster to spread through the whole cluster (or to be eliminated).

Cruciform is the structure produced at inverted repeats of DNA if the repeated sequence pairs with its complement on the same strand (instead of with its regular partner in the other strand of the duplex).

Cryptic satellite is a satellite DNA sequence not

identified as such by a separate peak on a density gradient; that is, it remains present in main-band DNA.

ctDNA is chloroplast DNA.

Cyclic AMP (cAMP) is a molecule of AMP in which the phosphate group is joined to both the 3' and 5' positions of the ribose; its binding activates the CAP, a positive regulator of prokaryotic transcription.

Cyclins are proteins that accumulate continuously throughout the cell cycle and are then destroyed by proteolysis during mitosis. (*See also* MPF).

Cytokinesis is the final process involved in separation and movement apart of daughter cells at the end of mitosis.

Cytological hybridization—*see in situ* hybridization.

Cytoplasm describes the material between the plasma membrane and the nucleus.

Cytoplasmic inheritance is a property of genes located in mitochondria or chloroplasts (or possibly other extranuclear organelles).

Cytoplasmic protein synthesis is the translation of mRNAs representing nuclear genes; it occurs via ribosomes attached to the cytoskeleton.

Cytoskeleton consists of networks of fibers in the cytoplasm of the eukaryotic cell.

Cytosol describes the general volume of cytoplasm in which organelles (such as the mitochondria) are located.

D loop is a region within mitochondrial DNA in which a short stretch of RNA is paired with one strand of DNA, displacing the original partner DNA strand in this region. The same term is used also to describe the displacement of a region of one strand of duplex DNA by a single-stranded invader in the reaction catalyzed by RecA protein.

Degeneracy in the genetic code refers to the lack of an effect of many changes in the third base of the codon on the amino acid that is represented.

Deletions are generated by removal of a sequence of DNA, the regions on either side being joined together.

Denaturation of DNA or RNA describes its conversion from the double-stranded to the single-stranded state; separation of the strands is most often accomplished by heating.

Denaturation of protein describes its conversion from the physiological conformation to some other (inactive) conformation.

Derepressed state describes a gene that is turned on. It is synonymous with *induced* when describing the normal state of a gene; it has the same meaning as *constitutive* in describing the effect of mutation.

Dicentric chromosome is the product of fusing two chromosome fragments, each of which has a centromere. It is unstable and may be broken when the two centromeres are pulled to opposite poles in mitosis.

Diploid set of chromosomes contains two copies of each autosome and two sex chromosomes.

Direct repeats are identical (or related) sequences present in two or more copies in the same orientation in the same molecule of DNA; they are not necessarily adjacent.

Discontinuous replication refers to the synthesis of DNA in short (Okazaki) fragments that are later joined into a continuous strand.

Disjunction describes the movement of members of a chromosome pair to opposite poles during cell division. At mitosis and the second meiotic division, disjunction applies to sister chromatids; at first meiotic division it applies to sister chromatid pairs.

Divergence is the percent difference in nucleotide sequence between two related DNA sequences or in amino acid sequences between two proteins.

Divergent transcription refers to the initiation of transcription at two promoters facing in the opposite direction, so that transcription proceeds away in both directions from a central region.

dna mutants of bacteria are temperature-sensitive; they cannot synthesize DNA at 42°C, but can do so at 37°C.

DNAase is an enzyme that attacks bonds in DNA.

DNA-driven hybridization involves the reaction of an excess of DNA with RNA.

DNA polymerase is an enzyme that synthesizes a daughter strand(s) of DNA (under direction from a DNA template). May be involved in repair or replication.

DNA replicase is a DNA-synthesizing enzyme required specifically for replication.

Domain of a chromosome may refer *either* to a discrete structural entity defined as a region within which supercoiling is independent of other domains; *or* to an extensive region including an expressed gene that has heightened sensitivity to degradation by the enzyme DNAase I.

Domain of a protein is a discrete continuous part of the amino acid sequence that can be equated with a particular function.

Dominant allele determines the phenotype displayed in a heterozygote with another (recessive) allele.

Donor splicing site—*see* left splicing junction.

Down promoter mutations decrease the frequency of initiation of transcription.

Downstream identifies sequences proceeding farther in the direction of expression, for example, the coding region is downstream of the initiation codon.

Early development refers to the period of a phage infection before the start of DNA replication.

Extopic expression describes the expression of a gene in a tissue in which it is not usually expressed; for example, in a transgenic animal.

Elongation factors (EF in prokaryotes, eEF in eukaryotes) are proteins that associate with ribosomes cyclically, during addition of each amino acid to the polypeptide chain.

End labeling describes the addition of a radioactively labeled group to one end (5′ or 3′) of a DNA strand.

End-product inhibition describes the ability of a product of a metabolic pathway to inhibit the activity of an enzyme that catalyzes an early step in the pathway.

Endocytosis is a process by which proteins at the surface of the cell are internalized, being transported into the cell within membranous vesicles.

Endocytic vesicles are membranous particles that transport proteins through endocytosis; also known as clathrin-coated vesicles.

Endonucleases cleave bonds within a nucleic acid chain; they may be specific for RNA or for single-stranded or double-stranded DNA.

Endoplasmic reticulum is a highly convoluted sheet of membranes, extending from the outer layer of the nuclear envelope into the cytoplasm.

Enhancer element is a *cis*-acting sequence that increases the utilization of (some) eukaryotic promoters, and can function in either orientation and in any location (upstream or downstream) relative to the promoter.

Envelopes surround some organelles (for example, nucleus or mitochondrion) and consist of concentric membranes, each membrane consisting of the usual lipid bilayer.

Epigenetic changes influence the phenotype without altering the genotype. They consist of changes in the properties of a cell that are inherited but that do not represent a change in genetic information.

Episome is a plasmid able to integrate into bacterial DNA.

Epistasis describes a situation in which expression of one gene wipes out the phenotypic effects of another gene.

Essential gene is one whose deletion is lethal to the organism (*see also* lethal locus).

Established cell lines consist of eukaryotic cells that have been adapted to indefinite growth in culture (they are said to be immortalized).

Eubacteria comprise the major line of prokaryotes.

Euchromatin comprises all of the genome in the interphase nucleus except for the heterochromatin.

Evolutionary clock is defined by the rate at which mutations accumulate in a given gene.

Excision of phage or episome or other sequence describes its release from the host chromosome as an autonomous DNA molecule.

Excision-repair systems remove a single-stranded sequence of DNA containing damaged or mispaired bases and replace it in the duplex by synthesizing a sequence complementary to the remaining strand.

Exocytosis is the process of secreting proteins from a cell into the medium, by transport in membranous vesicles from the endoplasmic reticulum, through the Golgi, to storage vesicles, and finally (upon a regulatory signal) through the plasma membrane.

Exocytic vesicles (also secretory vesicles) are membranous particles that transport and store proteins during exocytosis.

Exon is any segment of an interrupted gene that is

represented in the mature RNA product.

Exonucleases cleave nucleotides one at a time from the end of a polynucleotide chain; they may be specific for either the 5′ or 3′ end of DNA or RNA.

Expression vector is a cloning vector designed so that a coding sequence inserted at a particular site will be transcribed and translated into protein.

Extranuclear genes reside outside the nucleus in organelles such as mitochondria and chloroplasts.

F factor is a bacterial sex or fertility plasmid.

F1 generation is the first generation produced by crossing two parental (homozygous) lines.

Facultative heterochromatin describes the inert state of sequences that also exist in active copies— for example, one mammalian X chromosome in females.

Fast component of a reassociation reaction is the first to renature and contains highly repetitive DNA.

Fate map is a map of an embryo showing the adult tissues that will develop from the descendants of cells that occupy particular regions of the embryo.

Figure eight describes two circles of DNA linked together by a recombination event that has not yet been completed.

Filter hybridization is performed by incubating a denatured DNA preparation immobilized on a nitrocellulose filter with a solution of radioactively labeled RNA or DNA.

Fingerprint of DNA is a pattern of polymorphic restriction fragments that differ between individual genomes.

Fingerprint of a protein is the pattern of fragments (usually resolved on a two dimensional electrophoretic gel) generated by cleavage with an enzyme such as trypsin.

Fluidity is a property of membranes; it indicates the ability of lipids to move laterally within their particular monolayer.

Focus formation describes the ability of transformed eukaryotic cells to grow in dense clusters, piled up on one another.

Focus forming unit (ffu) is a quantitative measure of focus formation.

Foldback DNA consists of inverted repeats that have renatured by intrastrand reassociation of denatured DNA.

Footprinting is a technique for identifying the site on DNA bound by some protein by virtue of the protection of bonds in this region against attack by nucleases.

Forward mutations inactivate a wild-type gene.

Founder effect refers to the presence in a population of many individuals all with the same chromosome (or region of a chromosome) derived from a single ancestor.

Frameshift mutations arise by deletions or insertions that are not a multiple of 3 bp; they change the frame in which triplets are translated into protein.

G banding is a technique that generates a striated pattern in metaphase chromosomes that distinguishes the members of a haploid set.

G1 is the period of the eukaryotic cell cycle between the last mitosis and the start of DNA replication.

G2 is the period of the eukaryotic cell cycle between the end of DNA replication and the start of the next mitosis.

Gamete is either type of reproductive (germ) cell— sperm or egg—with haploid chromosome content.

Gap in DNA is the absence of one or more nucleotides in one strand of the duplex.

Gene (cistron) is the segment of DNA involved in producing a polypeptide chain; it includes regions preceding and following the coding region (leader and trailer) as well as intervening sequences (introns) between individual coding segments (exons).

Gene conversion is the alteration of one strand of a heteroduplex DNA to make it complementary with the other strand at any position(s) where there were mispaired bases.

Gene dosage gives the number of copies of a particular gene in the genome.

Gene family consists of a set of genes whose exons are related; the members were derived by duplication and variation from some ancestral gene.

Gene cluster is a group of adjacent genes that are identical or related.

Genetic code is the correspondence between triplets in DNA (or RNA) and amino acids in protein.

Genetic marker—*see* marker.
Genomic (chromosomal) DNA clones are sequences of the genome carried by a cloning vector.
Genotype is the genetic constitution of an organism.
Golgi apparatus consists of individual stacks of membranes near the endoplasmic reticulum; involved in glycosylating proteins and sorting them for transport to different cellular locations.
G proteins are guanine nucleotide-binding trimeric proteins that reside in the plasma membrane. When bound by GDP the trimer remains intact and is inert. When the GDP bound to the α subunit is replaced by GTP, the α subunit is released from the $\beta\gamma$ dimer. One of the separated units (either the α monomer or the $\beta\gamma$ dimer) then activates or represses a target protein.
Gratuitous inducers resemble authentic inducers of transcription but are not substrates for the induced enzymes.
GT-AG rule describes the presence of these constant dinucleotides at the first two and last two positions of introns of nuclear genes.
Gyrase is a type II topoisomerase of *E. coli* with the ability to introduce negative supercoils into DNA.

Hairpin describes a double-helical region formed by base pairing between adjacent (inverted) complementary sequences in a single strand of RNA or DNA.
Haploid set of chromosomes contains one copy of each autosome and one sex chromosome; the haploid number *n* is characteristic of gametes of diploid organisms.
Haplotype is the particular combination of alleles in a defined region of some chromosome, in effect the genotype in miniature. Originally used to describe combinations of MHC alleles, it now may be used to describe particular combinations of RFLPs.
Hapten is a small molecule that acts as an antigen when conjugated to a protein.
Helper virus provides functions absent from a defective virus, enabling the latter to complete the infective cycle during a mixed infection.

Hemizygote is a diploid individual that has lost its copy of a particular gene (for example, because a chromosome has been lost) and which therefore has only a single copy.
Heterochromatin describes regions of the genome that are permanently in a highly condensed condition and are not genetically expressed. May be constitutive or facultative.
Heteroduplex (hybrid) DNA is generated by base pairing between complementary single strands derived from the different parental duplex molecules; it occurs during genetic recombination.
Heterogametic sex has the diploid chromosome constitution 2A + XY.
Heterogeneous nuclear (hn) RNA comprises transcripts of nuclear genes made by RNA polymerase II; it has a wide size distribution and low stability.
Heteromultimeric proteins consist of nonidentical subunits (coded by different genes).
Heterokaryon is a cell containing two (or more) nuclei in a common cytoplasm, generated by fusing somatic cells.
Heterozygote is an individual with different alleles at some particular locus.
Highly repetitive DNA is the first component to reassociate and is equated with satellite DNA.
Histones are conserved DNA-binding proteins of eukaryotes that form the nucleosome, the basic subunit of chromatin.
Homeobox describes the conserved sequence that is part of the coding region of *D. melanogaster* homeotic genes; it is also found in amphibian and mammalian genes expressed in early embryonic development.
Homeotic genes are defined by mutations that convert one body part into another; for example, an insect leg may replace an antenna.
Homogametic sex has the diploid chromosome constitution 2A + XX.
Homologues are chromosomes carrying the same genetic loci; a diploid cell has two copies of each homologue, one derived from each parent.
Homomultimeric protein consists of identical subunits.
Homozygote is an individual with the same

allele at corresponding loci on the homologous chromosomes.

Hotspot is a site at which the frequency of mutation (or recombination) is very much increased.

Hox genes are clusters of mammalian genes containing homeoboxes; the individual members are related to the genes of the complex loci *ANT-C* and *BX-C* in *D. melanogaster*.

Housekeeping (constitutive) genes are those (theoretically) expressed in all cells because they provide basic functions needed for sustenance of all cell types.

Hybrid-arrested translation is a technique that identifies the cDNA corresponding to an mRNA by relying on the ability to base pair with the RNA *in vitro* to inhibit translation.

Hybrid dysgenesis describes the inability of certain strains of *D. melanogaster* to interbreed, because the hybrids are sterile (although otherwise they may be phenotypically normal).

Hybrid DNA—*see* heteroduplex DNA.

Hybridization is the pairing of complementary RNA and DNA strands to give an RNA–DNA hybrid.

Hybridoma is a cell line produced by fusing a myeloma with a lymphocyte; it continues indefinitely to express the immunoglobulins of both parents.

Hydrolytic reaction is one in which a covalent bond is broken with the incorporation of a water molecule.

Hydropathy plot is a measure of the hydrophobicity of a protein region and therefore of the likelihood that it will reside in a membrane.

Hydrophilic groups interact with water, so that hydrophilic regions of protein or the faces of a lipid bilayer reside in an aqueous environment.

Hydrophobic groups repel water, so that they interact with one another to generate a nonaqueous environment.

Hyperchromicity is the increase in optical density that occurs when DNA is denatured.

DNAase I hypersensitive site is a short region of chromatin detected by its extreme sensitivity to cleavage by DNAase I and other nucleases; probably comprises an area from which nucleosomes are excluded.

Hypervariable regions of an immunoglobulin are the parts of the variable region that show maximum alteration when different antibodies are compared.

Ideogram is a diagrammatic representation of the G-banding pattern of a chromosome.

Idling reaction is the production of pppGpp and ppGpp by ribosomes when an uncharged tRNA is present in the A site; triggers the stringent response.

Immortalization describes the acquisition by a eukaryotic cell line of the ability to grow through an indefinite number of divisions in culture.

Immunity in phages refers to the ability of a prophage to prevent another phage of the same type from infecting a cell. It results from the synthesis of phage repressor by the prophage genome.

Immunity in plasmids describes the ability of a plasmid to prevent another of the same type from becoming established in a cell. It results usually from interference with the ability to replicate.

Immunity in transposons refers to the ability of certain transposons to prevent others of the same type from transposing to the same DNA molecule. It results from a variety of mechanisms.

Imprinting describes a change in a gene that occurs during passage through the sperm or egg with the result that the paternal and maternal alleles have different properties in the very early embryo. May be caused by methylation of DNA.

***In situ* hybridization** is performed by denaturing the DNA of cells squashed on a microscope slide so that reaction is possible with an added single-stranded RNA or DNA; the added preparation is radioactively labeled and its hybridization is followed by autoradiography.

Incompatibility is the inability of certain bacterial plasmids to coexist in the same cell. It is a cause of plasmid immunity.

Indirect end-labeling is a technique for examining the organization of DNA by making a cut at a specific site and isolating all fragments containing the sequence adjacent to one side of the cut; it reveals the distance from the cut to the next break(s) in DNA.

Induced mutations result from the addition of a mutagen.

Inducer is a small molecule that triggers gene transcription by binding to a regulator protein.

Induction refers to the ability of bacteria (or yeast) to synthesize certain enzymes only when their substrates are present; applied to gene expression, refers to switching on transcription as a result of interaction of the inducer with the regulator protein.

Induction of prophage describes its excision from the host genome and entry into the lytic (infective) cycle as a result of destruction of the lysogenic repressor.

Initiation factors (IF in prokaryotes, eIF in eukaryotes) are proteins that associate with the small subunit of the ribosome specifically at the stage of initiation of protein synthesis.

Insertion sequence (IS) is a small bacterial transposon that carries only the genes needed for its own transposition.

Insertions are identified by the presence of an additional stretch of base pairs in DNA.

Integral membrane protein is a protein (noncovalently) inserted into a membrane; it retains its membranous association by means of a stretch of ~25 amino acids that are uncharged and/or hydrophobic.

Integration of viral or another DNA sequence is its insertion into a host genome as a region covalently linked on either side to the host sequences.

Interallelic complementation describes the change in the properties of a heteromultimeric protein brought about by the interaction of subunits coded by two different mutant alleles; the mixed protein may be more or less active than the protein consisting of subunits only of one or the other type.

Interbands are the relatively dispersed regions of polytene chromosomes that lie between the bands.

Intercistronic region is the distance between the termination codon of one gene and the initiation codon of the next gene.

Intermediate component(s) of a reassociation reaction are those reacting between the fast (satellite DNA) and slow (nonrepetitive DNA) components; contain moderately repetitive DNA.

Interphase is the period between mitotic cell divisions; divided into G1, S, and G2.

Intervening sequence is an intron.

Intron is a segment of DNA that is transcribed, but removed from within the transcript by splicing together the sequences (exons) on either side of it.

Inversion is a chromosomal change in which a segment has been rotated by 180° relative to the regions on either side and reinserted.

Inverted repeats comprise two copies of the same sequence of DNA repeated in opposite orientation on the same molecule. Adjacent inverted repeats constitute a palindrome.

Inverted terminal repeats are the short related or identical sequences present in reverse orientation at the ends of some transposons.

IS is an abbreviation for **insertion sequence**, a small bacterial transposon carrying only the genetic functions involved in transposition.

Isoaccepting tRNAs represent the same amino acid.

Isotype is a group of closely related immunoglobulin chains.

Karyotype is the entire chromosomal complement of a cell or species (as visualized during mitosis).

kb is an abbreviation for 1000 base pairs of DNA or 1000 bases of RNA.

Kinase is an enzyme that phosphorylates (adds a phosphate group) to a substrate; the substrates for **protein kinases** are amino acids in other proteins, and they are divided into those specific for tyrosine and those specific for threonine/serine.

Kinetic complexity is the complexity of a DNA component measured by the kinetics of DNA reassociation.

Kinetochore is the structural feature of the chromosome to which microtubules of the mitotic spindle attach (*see also* centromere).

Lagging strand of DNA must grow overall in the 3'-5' direction and is synthesized discontinuously in the form of short fragments (5'-3') that are later connected covalently.

Lampbrush chromosomes are the large meiotic chromosomes found in amphibian oocytes.

Lariat is an intermediate in RNA splicing in which a circular structure with a tail is created by a 5'-2' bond.

Late period of phage development is the part of infection following the start of DNA replication.

Leader is the nontranslated sequence at the 5′ end of mRNA that precedes the initiation codon.

Leader sequence of a protein is a short N-terminal sequence responsible for passage into or through a membrane.

Leading strand of DNA is synthesized continuously in the 5′-3′ direction.

Leaky mutations allow some residual level of gene expression.

Left splicing junction is the boundary between the right end of an exon and the left end of an intron.

Lethal locus is any gene in which a lethal mutation can be obtained (usually by deletion of the gene).

Library is a set of cloned fragments together representing the entire genome.

Ligation is the formation of a phosphodiester bond to link two adjacent bases separated by a nick in one strand of a double helix of DNA. (The term can also be applied to blunt-end ligation and to joining of RNA.)

LINES are long period interspersed sequences in mammalian genomes that are retroposons generated from RNA polymerase II transcripts.

Linkage describes the tendency of genes to be inherited together as a result of their location on the same chromosome; measured by percent recombination between loci.

Linkage group includes all loci that can be connected (directly or indirectly) by linkage relationships; equivalent to a chromosome.

Linkage disequilibrium describes a situation in which some combinations of genetic markers occur more or less frequently in the population than would be expected from their distance apart. It implies that a group of markers has been inherited coordinately. It can result from reduced recombination in the region or from a founder effect, in which there has been insufficient time to reach equilibrium since one of the markers was introduced into the population.

Linker DNA is all DNA contained on a nucleosome in excess of the 146 bp core DNA.

Linker fragment is short synthetic duplex oligonucleotide containing the target site for some restriction enzyme; may be added to ends of a DNA fragment prepared by cleavage with some other enzyme during reconstructions of recombinant DNA.

Linker scanner mutations are introduced by recombining two DNA molecules *in vitro* at a restriction fragment added to the end of each; the result is to insert the linker sequence at the site of recombination.

Linking number is the number of times the two strands of a closed DNA duplex cross over each other.

Linking number paradox describes the discrepancy between the existence of −2 supercoils in the path of DNA on the nucleosome compared with the measurement of −1 supercoil released when histones are removed.

Lipids have polar heads, containing phosphate (phospholipid), sterol (such as cholesterol), or saccharide (glycolipid) connected to a hydrophobic tail consisting of fatty acid(s).

Lipid bilayer is the form taken by concentration of lipids in which the hydrophobic fatty acids occupy the interior and the polar heads face the exterior.

Liquid (solution) hybridization is a reaction between complementary nucleic acid strands performed in solution.

Locus is the position on a chromosome at which the gene for a particular trait resides; locus may be occupied by any one of the alleles for the gene.

LOD score is a measure of genetic linkage, defined as the $\log_{10}$ ratio of the probability that the data would have arisen if the loci are linked to the probability that the data could have arisen fom unlinked loci. The conventional threshold for declaring linkage is a LOD score of 3.0, that is, a 1000:1 ratio (which must be compared with the 50:1 probability that any random pair of loci will be unlinked).

Long-period interspersion is a pattern in the genome in which long stretches of moderately repetitive and nonrepetitive DNA alternate.

Loop is a single-stranded region at the end of a hairpin in RNA (or single-stranded DNA); corresponds to the sequence between inverted repeats in duplex DNA.

LTR is an abbreviation for **long-terminal repeat**, a

sequence directly repeated at both ends of a retroviral DNA.

Lumen described the interior of a compartment bounded by membranes, usually the endoplasmic reticulum or the mitochondrion.

Luxury genes are those coding for specialized functions synthesized (usually) in large amounts in particular cell types.

Lysis describes the death of bacteria at the end of a phage infective cycle when they burst open to release the progeny of an infecting phage. Also applies to eukaryotic cells, for example, infected cells that are attacked by the immune system.

Lysogen is a bacterium that possesses a repressed prophage as part of its genome.

Lysogenic immunity is the ability of a prophage to prevent another phage genome of the same type from becoming established in the bacterium.

Lysogenic repressor is the protein responsible for preventing a prophage from reentering the lytic cycle.

Lysogeny describes the ability of a phage to survive in a bacterium as a stable prophage component of the bacterial genome.

Lysosomes are small bodies, enclosed by membranes, that contain hydrolytic enzymes.

Lytic infection of bacteria by a phage ends in destruction of bacteria and release of progeny phage.

Main band of genomic DNA consists of a broad peak on a density gradient, excluding any visible satellite DNAs that form separate bands.

Major histocompatibility locus is a large chromosomal region containing a giant cluster of genes that code for transplantation antigens and other proteins found on the surfaces of lymphocytes.

Map distance is measured as cM (centiMorgans) = percent recombination (sometimes subject to adjustments).

MAR (matrix attachment site; also known as SAR for scaffold attachment site) is a region of DNA that attaches to the nuclear matrix.

Marker (DNA) is a fragment of known size used to calibrate an electrophoretic gel.

Marker (genetic) is any allele of interest in an experiment.

Maternal inheritance describes the preferential survival in the progeny of genetic markers provided by one parent.

Meiosis occurs by two successive divisions (meiosis I and II) that reduce the starting number of 4n chromosomes to 1n in each of four product cells. Products may mature to germ cells (sperm or eggs).

Melting of DNA means its denaturation.

Melting temperature (T_m) is the midpoint of the temperature range over which DNA is denatured.

Membranes consist of an asymmetrical lipid bilayer that has lateral fluidity and contains proteins.

Membrane proteins have hydrophobic regions that allow part or all of the protein structure to reside within the membrane; the bonds involved in this association are usually noncovalent.

Metastasis describes the ability of tumor cells to leave their site of origin and migrate to other locations in the body, where a new colony is established.

Micrococcal nuclease is an endonuclease that cleaves DNA; in chromatin, DNA is cleaved preferentially between nucleosomes.

Microsomes are fragmented pieces of endoplasmic reticulum associated with ribosomes.

Microtubules are filaments consisting of dimers of tubulin; interphase microtubules are reorganized into spindle fibers at mitosis, when they are responsible for chromosome movement.

Microtubule associated proteins (MAPs) are proteins associated with microtubules and responsible for influencing their stability and organization.

Microtubule organizing center (MTOC) is a structure from which microtubules may be extended.

Minicell is an anucleate bacterial (*E. coli*) cell produced by a division that generates a cytoplasm without a nucleus.

Minichromosome of SV40 or polyoma is the nucleosomal form of the viral circular DNA.

Mitosis is the division of a eukaryotic somatic cell.

Modification of DNA or RNA includes all changes made to the nucleotides after their initial incorporation into the polynucleotide chain.

Modified bases are all those except the usual four

from which DNA (T, C, A, G) or RNA (U, C, A, G) are synthesized; they result from postsynthetic changes in the nucleic acid.

Monocistronic mRNA codes for one protein.

Monolayer describes the growth of eukaryotic cells in culture as a layer only one cell deep.

Morphogen is a factor that induces development of particular cell types in a manner that depends on its concentration.

MPF (maturation- or M phase-promoting factor) is a dimeric kinase, containing the p34 catalytic subunit and a cyclin regulatory subunit, whose activation triggers the onset of mitosis.

mtDNA is mitochondrial DNA.

MTOC (microtubule organizing center) is a region from which microtubules emanate. The major MTOCs in a mitotic cell are the centrosomes.

Multicopy plasmids are present in bacteria at amounts greater than one per chromosome.

Multiforked chromosome (in bacterium) has more than one replication fork, because a second initiation has occurred before the first cycle of replication has been completed.

Multimeric proteins consist of more than one subunit.

Mutagens increase the rate of mutation by inducing changes in DNA.

Mutation describes any change in the sequence of genomic DNA.

Mutation frequency is the frequency at which a particular mutant is found in the population.

Mutation rate is the rate at which a particular mutation occurs, usually given as the number of events per gene per generation.

Myeloma is a tumor cell line derived from a lymphocyte; usually produces a single type of immunoglobulin.

Negative complementation occurs when inter-allelic complementation allows a mutant subunit to suppress the activity of a wild-type subunit in a multimeric protein.

Negative regulators function by switching off transcription or translation.

Negative supercoiling comprises the twisting of a duplex of DNA in space in the opposite sense to the turns of the strands in the double helix.

Neutral substitutions in a protein are those changes of amino acids that do not affect activity.

Nick in duplex DNA is the absence of a phospho-diester bond between two adjacent nucleotides on one strand.

Nick translation describes the ability of *E. coli* DNA polymerase I to use a nick as a starting point from which one strand of a duplex DNA can be degraded and replaced by resynthesis of new material; is used to introduce radioactively labeled nucleotides into DNA *in vitro*.

Nonautonomous controlling elements are defective transposons that can transpose only when assisted by an autonomous controlling element of the same type.

Nondisjunction describes failure of chromatids (duplicate chromosomes) to move to opposite poles during mitosis or meiosis.

Nonpermissive conditions do not allow conditional lethal mutants to survive.

Nonrepetitive DNA shows reassociation kinetics expected of unique sequences.

Nonreplicative transposition describes the movement of a transposon that leaves a donor site (usually generating a double strand break) and moves to a new site.

Nonsense codon is any one of three triplets (UAG, UAA, UGA) that cause termination of protein synthesis. (UAG is known as amber; UAA as ochre.)

Nonsense mutation is any change in DNA that causes a (termination) codon to replace a codon representing an amino acid.

Nonsense suppressor is a gene coding for a mutant tRNA able to respond to one or more of the termination codons.

Nontranscribed spacer is the region between transcription units in a tandem gene cluster.

Northern blotting is a technique for transferring RNA from an agarose gel to a nitrocellulose filter on which it can be hybridized to a complementary DNA.

Nuclear envelope is a layer of two membranes surrounding the nucleus. It is penetrated by nuclear pores and bounded on the interior by the nuclear laminin.

Nuclear lamina consists of a proteinaceous layer on the inside of the nuclear envelope. It consists of (upto) three lamin proteins.

Nuclear matrix is a network of fibers surrounding and penetrating the nucleus.

Nuclear pores represent holes in the nuclear envelope and are presumed to be used for transport of macromolecules.

Nucleolar organizer is the region of a chromosome carrying genes coding for rRNA.

Nucleoid is the compact body that contains the genome in a bacterium.

Nucleolus is a discrete region of the nucleus created by the transcription of rRNA genes.

Nucleosome is the basic structural subunit of chromatin, consisting of ~200 bp of DNA and an octamer of histone proteins.

Nucleolytic reactions involve the hydrolysis of a phosphodiester bond in a nucleic acid.

Null mutation completely eliminates the function of a gene, usually because it has been physically deleted.

Ochre codon is the triplet UAA, one of three codons that cause termination of protein synthesis.

Ochre mutation is any change in DNA that creates a UAA codon at a site previously occupied by another codon.

Ochre suppressor is a gene coding for a mutant tRNA able to respond to the UAA codon to allow continuation of protein synthesis; ochre suppressors also suppress amber codons.

Okazaki fragments are the short stretches of 1000–2000 bases produced during discontinuous replication; they are later joined into a covalently intact strand.

Oncogenes are genes whose products have the ability to transform eukaryotic cells so that they grow in a manner analogous to tumor cells. Oncogenes carried by retroviruses have names of the form *v-onc. See also* proto-oncogene.

Open reading frame (ORF) contains a series of triplets coding for amino acids without any termination codons; sequence is (potentially) translatable into protein.

Operator is the site on DNA at which a repressor protein binds to prevent transcription from initiating at the adjacent promoter.

Operon is a unit of bacterial gene expression and regulation, including structural genes and control elements in DNA recognized by regulator gene product(s).

Organelles are compartments located in the cytoplasm and surrounded by a membrane.

Origin (*ori*) is a sequence of DNA at which replication is initiated.

Orphons are isolated individual genes found in isolated locations, but related to members of a gene cluster.

Overwinding of DNA is caused by positive super-coiling (which applies further tension in the direction of winding of the two strands about each other in the duplex).

Packing ratio is the ratio of the length of DNA to the unit length of the fiber containing it.

Pairing of chromosomes—*see* synapsis.

Palindrome is a sequence of DNA that is the same when one strand is read left to right or the other is read right to left; consists of adjacent inverted repeats.

Papovaviruses are a class of animal viruses with small genomes, including SV40 and polyoma.

Paranemic joint describes a region in which two complementary sequences of DNA are associated side by side instead of being intertwined in a double helical structure.

pBR322 is one of the standard plasmid cloning vectors.

PCR (polymerase chain reaction) describes a technique in which cycles of denaturation, annealing with primer, and extension with DNA polymerase, are used to amplify the number of copies of a target DNA sequence by $>10^6$ times.

Perinuclear space lies between the inner and outer membranes of the nuclear envelope.

Periodicity of DNA is the number of base pairs per turn of the double helix.

Permissive conditions allow conditional lethal mutants to survive.

Petite strains of yeast lack mitochondrial function.

Phage (bacteriophage) is a bacterial virus.

Phase variation describes an alternation in the type of flagella produced by a bacterium.

Phenotype is the appearance or other characteristics of an organism, resulting from the interaction of its genetic constitution with the environment.

Phosphatase is an enzyme that removes phosphate groups from substrates.

Plasma membrane is the continuous membrane defining the boundary of every cell.

Plasmid is an autonomous self-replicating extrachromosomal circular DNA.

Playback experiment describes the retrieval of DNA that has hybridized with RNA to check that it is nonrepetitive by a further reassociation reaction.

Plectonemic winding describes the intertwining of the two strands in the classical double helix of DNA.

Pleiotropic gene affects more than one (apparently unrelated) characteristic of the phenotype.

Ploidy refers to the number of copies of the chromosome set present in a cell; a haploid has one copy, a diploid has two copies, etc.

Point mutations are changes involving single base pairs.

Polarity refers to the effect of a mutation in one gene in influencing the expression (at transcription or translation) of subsequent genes in the same transcription unit.

Polyadenylation is the addition of a sequence of polyadenylic acid to the 3′ end of a eukaryotic RNA after its transcription.

Polycistronic mRNA includes coding regions representing more than one gene.

Polymorphism refers to the simultaneous occurrence in the population of genomes showing allelic variations (as seen either in alleles producing different phenotypes or—for example—in changes in DNA affecting the restriction pattern).

Polyploid cell has more than two sets of the haploid genome.

Polyprotein is a gene product that is cleaved into several independent proteins.

Polysome (polyribosome) is an mRNA associated with a series of ribosomes engaged in translation.

Polytene chromosomes are generated by successive replications of a chromosome set without separation of the replicas.

Position effect refers to a change in the expression of a gene brought about by its translocation to a new site in the genome; for example, a previously active gene may become inactive if placed near heterochromatin.

Positive regulator proteins are required for the activation of a transcription unit.

Positive supercoiling describes the coiling of the double helix in space in the same direction as the winding of the two strands of the double helix itself.

Postmeiotic segregation describes the segregation of two strands of a duplex DNA that bear different information (created by heteroduplex formation during meiosis) when a subsequent replication allows the strands to separate.

Primary cells are eukaryotic cells taken into culture directly from the animal.

Primary transcript is the original unmodified RNA product corresponding to a transcription unit.

Primer is a short sequence (often of RNA) that is paired with one strand of DNA and provides a free 3′-OH end at which a DNA polymerase starts synthesis of a deoxyribonucleotide chain.

Primosome describes the complex of proteins involved in the priming action that initiates synthesis of each Okazaki fragment during discontinuous DNA replication; the primosome may move along DNA to engage in successive priming events.

Procentriole is an immature centriole, formed in the vicinity of a mature centriole.

Processed pseudogene is an inactive gene copy that lacks introns, contrasted with the interrupted structure of the active gene. Such genes presumably originate by reverse transcription of mRNA and insertion of a duplex copy into the genome.

Processive enzymes continue to act on a particular substrate, that is, do not dissociate between repetitions of the catalytic event.

Prokaryotic organisms (bacteria) lack nuclei.

Promoter is a region of DNA involved in binding of RNA polymerase to initiate transcription.

-10 sequence is the consensus sequence TATAATG centered about 10 bp before the startpoint of a bacterial gene. It is involved in the initial melting of DNA by RNA polymerase.

-35 sequence is the consensus sequence centered about 35 bp before the startpoint of a bacterial gene. It is involved in initial recognition by RNA polymerase.

Proofreading refers to any mechanism for correcting errors in protein or nucleic acid synthesis that involves scrutiny of individual units *after* they have been added to the chain.

Prophage is a phage genome covalently integrated as a linear part of the bacterial chromosome.

Proteolytic reactions comprise the hydrolysis of peptide bonds in protein.

Proto-oncogenes are the normal counterparts in the eukaryotic genome to the oncogenes carried by some retroviruses. They are given names of the form *c-onc*.

Provirus is a duplex DNA sequence in the eukaryotic chromosome corresponding to the genome of an RNA retrovirus.

Pseudogenes are inactive but stable components of the genome derived by mutation of an ancestral active gene.

Puff is an expansion of a band of a polytene chromosome associated with the synthesis of RNA at some locus in the band.

Pulse-chase experiments are performed by incubating cells very briefly with a radioactively labeled precursor (of some pathway or macromolecule); then the fate of the label is followed during a subsequent incubation with a nonlabeled precursor.

Quaternary structure of a protein refers to its multimeric constitution.

Quick-stop *dna* mutants of *E. coli* cease replication immediately when the temperature is increased to 42°C.

R loop is the structure formed when an RNA strand hybridizes with its complementary strand in a DNA duplex, thereby displacing the original strand of DNA in the form of a loop extending over the region of hybridization.

Rapid lysis (*r*) mutants display a change in the pattern of lysis of *E. coli* at the end of an infection by a T-even phage.

Reading frame is one of three possible ways of reading a nucleotide sequence as a series of triplets.

Reassociation of DNA describes the pairing of complementary single strands to form a double helix.

RecA is the product of the *recA* locus of *E. coli*; a protein with dual activities, activating proteases and also able to exchange single strands of DNA molecules. The protease-activating activity controls the SOS response; the nucleic acid handling facility is involved in recombination-repair pathways.

Receptor is a transmembrane protein, located in the plasma membrane, that binds a ligand in a domain on the extracellular side, and as a result has a change in activity of the cytoplasmic domain. (The same term is sometimes used also for the steroid receptors, which are transcription factors that are cativated by binding ligands that are steroids or other small molecules.)

Recessive allele is obscured in the phenotype of a heterozygote by the dominant allele, often due to inactivity or absence of the product of the recessive allele.

Recessive lethal is an allele that is lethal when the cell is homozygous for it.

Reciprocal recombination is the production of new genotypes with the reverse arrangements of alleles according to maternal and paternal origin.

Reciprocal translocation exchanges part of one chromosome with part of another chromosome.

Recombinant progeny have a different genotype from that of either parent.

Recombinant joint is the point at which two recombining molecules of duplex DNA are connected (the edge of the heteroduplex region).

Recombination nodules (nodes) are dense objects present on the synaptonemal complex; could be involved in crossing-over.

Recombination-repair is a mode of filling a gap in one strand of duplex DNA by retrieving a homologous single strand from another duplex.

Regulatory gene codes for an RNA or protein product whose function is to control the expression of other genes.

Relaxed mutants of *E. coli* do not display the stringent response to starvation for amino acids (or other nutritional deprivation).

Relaxed replication control refers to the ability of some plasmids to continue replicating after bacteria cease dividing.

Release (termination) factors respond to termination codons to cause release of the completed polypeptide chain and the ribosome from mRNA.

Renaturation is the reassociation of denatured complementary single strands of a DNA double helix.

Repeating unit in a tandem cluster is the length of the sequence that is repeated; appears circular on a restriction map.

Repetition frequency is the (integral) number of copies of a given sequence present in the haploid genome; equals 1 for nonrepetitive DNA, >2 for repetitive DNA.

Repetitive DNA behaves in a reassociation reaction as though many (related or identical) sequences are present in a component, allowing any pair of complementary sequences to reassociate.

Replacement sites in a gene are those at which mutations alter the amino acid that is coded.

Replication-defective virus has lost one or more genes essential for completing the infective cycle.

Replication eye is a region in which DNA has been replicated within a longer, unreplicated region.

Replication fork is the point at which strands of parental duplex DNA are separated so that replication can proceed.

Replicative transposition describes the movement of a transposon by a mechanism in which first it is replicated, and then one copy is transferred to a new site.

Replicon is a unit of the genome in which DNA is replicated; contains an origin for initiation of replication.

Replisome is the multiprotein structure that assembles at the bacterial replicating fork to undertake synthesis of DNA. Contains DNA polymerase and other enzymes.

Reporter gene is a coding unit whose product is easily assayed (such as chloramphenicol transacetylase); it may be connected to any pro-

moter of interest so that expression of the gene can be used to assay promoter function.

Repression is the ability of bacteria to prevent synthesis of certain enzymes when their products are present; more generally, refers to inhibition of transcription (or translation) by binding of repressor protein to a specific site on DNA (or mRNA).

Repressor protein binds to operator on DNA or RNA to prevent transcription or translation, respectively.

Resolvase is enzyme activity involved in site-specific recombination between two transposons present as direct repeats in a cointegrate structure.

Restriction enzymes recognize specific short sequences of (usually) unmethylated DNA and cleave the duplex (sometimes at target site, sometimes elsewhere, depending on type).

Restriction fragment length polymorphism (RFLP) refers to inherited differences in sites for restriction enzymes (for example, caused by base changes in the target site) that result in differences in the lengths of the fragments produced by cleavage with the relevant restriction enzyme. RFLPs are used for genetic mapping to link the genome directly to a conventional genetic marker.

Restriction map is a linear array of sites on DNA cleaved by various restriction enzymes.

Retroposon is a transposon that mobilizes via an RNA form; the DNA element is transcribed into RNA, and then reverse-transcribed into DNA, which is inserted at a new site in the genome.

Retroregulation describes the ability of a sequence downstream to regulate translation of an mRNA.

Retrovirus is an RNA virus that propagates via conversion into duplex DNA.

Reverse transcription is synthesis of DNA on a template of RNA; accomplished by reverse transcriptase enzyme.

Reverse translation is a technique for isolating genes (or mRNAs) by their ability to hybridize with a short oligonucleotide sequence prepared by predicting the nucleic acid sequence from the known protein sequence.

Reversion of mutation is a change in DNA that either reverses the original alteration (true reversion) or compensates for it (second site reversion in the same gene).

Revertants are derived by reversion of a mutant cell or organism.

Rho factor is a protein involved in assisting *E. coli* RNA polymerase to terminate transcription at certain (rho-dependent) sites.

Rho-independent terminators are sequences of DNA that cause *E. coli* RNA polymerase to terminate *in vitro* in the absence of rho factor.

Rifamycins (including rifampicin) inhibit transcription in bacteria.

Right splicing junction is the boundary between the right end of an intron and the left end of the adjacent exon.

RNAase is an enzyme whose substrate is RNA.

RNA-driven hybridization reactions use an excess of RNA to react with all complementary sequences in a single-stranded preparation of DNA.

RNA polymerase is an enzyme that synthesizes RNA using a DNA template (formally described as DNA-dependent RNA polymerase).

RNA replicase is an enzyme that synthesizes RNA using an RNA template (used for replication by RNA viruses).

Rolling circle is a mode of replication in which a replication fork proceeds around a circular template for an indefinite number of revolutions; the DNA strand newly synthesized in each revolution displaces the strand synthesized in the previous revolution, giving a tail containing a linear series of sequences complementary to the circular template strand.

Rot is the product of RNA concentration and time of incubation in an RNA-driven hybridization reaction.

Rough ER consists of endoplasmic reticulum associated with ribosomes.

S phase is the restricted part of the eukaryotic cell cycle during which synthesis of DNA occurs.

S1 nuclease is an enzyme that specifically degrades unpaired (single-stranded) sequences of DNA.

Saltatory replication is a sudden lateral amplification to produce a large number of copies of some sequence.

Satellite DNA consists of many tandem repeats (identical or related) of a short basic repeating unit.

Saturation density is the density to which cultured eukaryotic cells grow *in vitro* before division is inhibited by cell–cell contacts.

Saturation hybridization experiment has a large excess of one component, causing all complementary sequences in the other component to enter a duplex form.

Scaffold of a chromosome is a proteinaceous structure in the shape of a sister chromatid pair, generated when chromosomes are depleted of histones.

Scarce (complex) mRNA consists of a large number of individual mRNA species, each present in very few copies per cell.

scRNA is any one of several small cytoplasmic RNAs, molecules present in the cytoplasm and (sometimes) nucleus. **scRNPs** are small cytoplasmic ribonucleoproteins (scRNAs associated with proteins).

Segmentation genes are concerned with controlling the number or polarity of body segments in insects.

Selection describes the use of particular conditions to allow survival only of cells with a particular phenotype.

Semiconservative replication is accomplished by separation of the strands of a parental duplex, each then acting as a template for synthesis of a complementary strand.

Semidiscontinuous replication is mode in which one new strand is synthesized continuously while the other is synthesized discontinuously.

Septum constitutes the material that forms in the center of a bacterium to divide it into two daughter cells at the end of a division cycle.

Serum dependence describes the need of eukaryotic cells for factors contained in serum in order to grow in culture.

Sex chromosomes are those whose contents are different in the two sexes; usually labeled X and Y (or W and Z), one sex has XX (or WW), the other sex has XY (or WZ).

Sex linkage is pattern of inheritance shown by genes carried on a sex chromosome (usually the X).

Sex plasmid is actually an episome; it is able to initiate the process of conjugation, by which chromosomal material is transferred from one bacterium to another.

Shine-Dalgarno sequence is part or all of the polypurine sequence AGGAGG located on bacterial mRNA just prior to an AUG initiation codon; is complementary to the sequence at the 3′ end of 16S rRNA; involved in binding of ribosome to mRNA.

Short-period interspersion is a pattern in a genome in which moderately repetitive DNA sequences of ~300 bp alternate with nonrepetitive sequences of ~1000 bp.

Shotgun experiment is cloning of an entire genome in the form of randomly generated fragments.

Shuttle vector is a plasmid constructed to have origins for replication for two hosts (for example, *E. coli* and *S. cerevisiae*) so that it can be used to carry a foreign sequence in either prokaryotes or eukaryotes.

Sigma factor is the subunit of bacterial RNA polymerase needed for initiation; is the major influence on selection of binding sites (promoters).

Signal hypothesis describes the role of the N-terminal sequence of a secreted protein in attaching nascent polypeptide to membrane; that is, mRNA and ribosome are attached to membrane via the N-terminal end of the protein under synthesis.

Signal sequence is the region of a protein (usually N-terminal) responsible for co-translational insertion into membranes of the endoplasmic reticulum.

Signal transduction describes the process by which a receptor interacts with a ligand at the surface of the cell and then transmits a signal to trigger a pathway within the cell.

Silent mutations do not change the product of a gene.

Silent sites in a gene describe those positions at which mutations do not alter the product.

Simple-sequence DNA equals satellite DNA.

SINES are a class of retroposons found as short interspersed repeats in mammalian genomes; derived from transcripts of RNA polymerase III.

Single-copy plasmids are maintained in bacteria at a ratio of one plasmid for every host chromosome.

Single-strand assimilation describes the ability of RecA protein to cause a single strand of DNA to displace its homologous strand in a duplex; that is, the single strand is assimilated into the duplex.

Single-strand exchange is a reaction in which one of the strands of a duplex of DNA leaves its former partner and instead pairs with the complementary strand in another molecule, displacing its homologue in the second duplex.

Sister chromatids are the copies of a chromosome produced by its replication.

Site-specific recombination occurs between two specific (not necessarily homologous) sequences, as in phage integration/excision or resolution of cointegrate structures during transposition.

Slow component of a reassociation reaction is the last to reassociate; usually consists of nonrepetitive DNA.

Slow-stop *dna* mutants of *E. coli* complete the current round of bacterial replication but cannot initiate another at 42°C.

Smooth ER consists of a regions of endoplasmic reticulum devoid of ribosomes.

snRNA (small nuclear RNA) is any one of many small RNA species confined to the nucleus; several of the snRNAs are involved in splicing or other RNA processing reactions.

snRNPs are small nuclear ribonucleoproteins (snRNAs associated with proteins).

Solution hybridization is the same as liquid hybridization.

Somatic cells are all the cells of an organism except those of the germ line.

Somatic mutation is a mutation occurring in a somatic cell, and therefore affecting only its descendants; it is not inherited.

SOS box is the DNA sequence (operator) of ~20 bp recognized by LexA repressor protein.

SOS response in *E. coli* describes the coordinate induction of many enzymes, including repair activities, in response to irradiation or other damage to DNA; results from activation of protease activity by RecA to cleave LexA repressor.

Southern blotting describes the procedure for transferring denatured DNA from an agarose gel to a nitrocellulose filter where it can be hybridized with a complementary nucleic acid.

Spheroplast is a bacterial or yeast cell whose wall has been largely or entirely removed.

Spindle describes the reorganized structure of a eukaryotic cell passing through division; the nucleus has been dissolved and chromosomes are attached to the spindle by microtubules.

Splicing describes the removal of introns and joining of exons in RNA; thus introns are spliced out, while exons are spliced together.

Splicing junctions are the sequences immediately surrounding the exon–intron boundaries.

Spontaneous mutations are those that occur in the absence of any added reagent to increase the mutation rate.

Sporulation is the generation of a spore by a bacterium (by morphological conversion) or by a yeast (as the product of meiosis).

SSB is the single-strand protein of *E. coli,* a protein that binds to single-stranded DNA.

Staggered cuts in duplex DNA are made when two strands are cleaved at different points near each other.

Startpoint (startsite) refers to the position on DNA corresponding to the first base incorporated into RNA.

Stem is the base-paired segment of a hairpin.

Sticky ends are complementary single strands of DNA that protrude from opposite ends of a duplex or from ends of different duplex molecules; can be generated by staggered cuts in duplex DNA.

Stop codons are the three triplets (UAA, UAG, UGA) which terminate protein synthesis.

Strand displacement is a mode of replication of some viruses in which a new DNA strand grows by displacing the previous (homologous) strand of the duplex.

Streptolydigins inhibit the elongation of transcription by bacterial RNA polymerase.

Stringent replication describes the limitation of single-copy plasmids to replication *pari passu* with the bacterial chromosome.

Stringent response refers to the ability of a bacterium to shut down synthesis of tRNA and ribosomes in a poor-growth medium.

Structural gene codes for any RNA or protein product other than a regulator.

Supercoiling describes the coiling of a closed duplex DNA in space so that it crosses over its own axis.

Superrepressed means the same as uninducible.

Suppression describes the occurrence of changes that eliminate the effects of a mutation without reversing the original change in DNA.

Suppressor (extragenic) is usually a gene coding a mutant tRNA that reads the mutated codon either in the sense of the original codon or to give an acceptable substitute for the original meaning.

Suppressor (intragenic) is a compensating mutation that restores the original reading frame after a frameshift.

Synapsis describes the association of the two pairs of sister chromatids representing homologous chromosomes that occurs at the start of meiosis; resulting structure is called a bivalent.

Synaptonemal complex describes the morphological structure of synapsed chromosomes.

Syntenic genetic loci lie on the same chromosome.

T cells are lymphocytes of the T (thymic) lineage; may be subdivided into several functional types. They carry TcR (T cell receptor) and are involved in the cell-mediated immune response.

T_m is the abbreviation for melting temperature.

Tandem repeats are multiple copies of the same sequence lying in series.

TATA box is a conserved A•T-rich septamer found about 25 bp before the startpoint of each eukaryotic RNA polymerase II transcription unit; may be involved in positioning the enzyme for correct initiation.

Telomerase is the ribonucleoprotein enzyme that creates repeating units of one strand at the telomere, by adding individual bases.

Telomere is the natural end of a chromosome; the DNA sequence consists of a simple repeating unit with a protruding single-stranded end that may fold into a hairpin.

Temperature-sensitive mutation creates a gene product that is functional at low temperature but inactive at higher temperature (the reverse relationship is usually called cold-sensitive).

Terminal redundancy describes the repetition of the same sequence at both ends of (for example) a phage genome.

Termination codon is one of three triplet sequences, UAG (amber), UAA (ochre), or UGA that cause termination of protein synthesis; they are also called 'nonsense' codons.

Terminator is a sequence of DNA, represented at the end of the transcript, that causes RNA polymerase to terminate transcription.

Tertiary structure of a protein describes the organization in space of its polypeptide chain.

Testcross involves crossing an unknown genotype to a recessive homozygote so that the phenotypes of the progeny correspond directly to the chromosomes carried by the parent of unknown genotype.

Thalassemia is disease of red blood cells resulting from lack of either α or β globin.

Thymine dimer comprises a chemically cross-linked pair of adjacent thymine residues in DNA, a result of damage induced by ultraviolet irradiation.

Topoisomerase is an enzyme that can change the linking number of DNA (in steps of 1 by type I; in steps of 2 by type II).

Topological isomers are molecules of DNA that are identical except for a difference in linking number.

Tracer is a radioactively labeled nucleic acid component included in a reassociation reaction in amounts too small to influence the progress of reaction.

Trailer is a nontranslated sequence at the 3' end of an mRNA following the termination codon.

Trans configuration of two sites refers to their presence on two different molecules of DNA (chromosomes).

Transcribed spacer is the part of an rRNA transcription unit that is transcribed but discarded during maturation; that is, it does not give rise to part of rRNA.

Transcription is synthesis of RNA on a DNA template.

Transcription unit is the distance between sites of initiation and termination by RNA polymerase; may include more than one gene.

Transduction refers to the transfer of a bacterial gene from one bacterium to another by a phage; a phage carrying host as well as its own genes is called transducing phage. Also describes the acquisition and transfer of eukaryotic cellular sequences by retroviruses.

Transfection of eukaryotic cells is the acquisition of new genetic markers by incorporation of added DNA.

Transformation of bacteria describes the acquisition of new genetic markers by incorporation of added DNA.

Transformation of eukaryotic cells refers to their conversion to a state of unrestrained growth in culture, resembling or identical with the tumorigenic condition.

Transgenic animals are created by introducing new DNA sequences into the germ line via addition to the egg.

Transit peptide is the short leader sequence cleaved from proteins that are imported into cellular organelles by post-translational passage of the membrane.

Transition is a mutation in which one pyrimidine is substituted by the other or in which one purine is substituted for the other.

Translation is synthesis of protein on the mRNA template.

Translocation of a chromosome describes a rearrangement in which part of a chromosome is detached by breakage and then becomes attached to some other chromosome.

Translocation of a gene refers to the appearance of a new copy at location in the genome elsewhere from the original copy.

Translocation of a protein refers to its movement across a membrane.

Translocation of the ribosome is its movement one codon along mRNA after the addition of each amino acid to the polypeptide chain.

Transmembrane protein is a component of a membrane; a hydrophobic region or regions of the protein resides in the membrane, and hydrophilic regions are exposed on one or both sides of the membrane.

Transplantation antigen is protein coded by a major histocompatibility locus, present on all mammalian cells, involved in interactions between lymphocytes.

Transposase is the enzyme activity involved in insertion of transposon at a new site.

Transposition immunity refers to the ability of certain transposons to prevent others of the same type from transposing to the same DNA molecule.

Transposon is a DNA sequence able to insert itself at a new location in the genome (without any sequence relationship with the target locus).

Transposition refers to the movement of a transposon to a new site in the genome. See also nonreplicative transposition, replicative transposition, and conservative transposition.

Transvection describes the ability of a locus to influence activity of an allele on the other homologue only when two chromosomes are synapsed.

Transversion is a mutation in which a purine is replaced by a pyrimidine or vice versa.

True-breeding organisms are homozygous for the trait under consideration.

Twisting number of a DNA is the number of base pairs divided by the number of base pairs per turn of the double helix.

Underwinding of DNA is produced by negative supercoiling (because the double helix is itself coiled in the opposite sense from the intertwining of the strands).

Unequal crossing-over describes a recombination event in which the two recombining sites lie at nonidentical locations in the two parental DNA molecules.

Unidirectional replication refers to the movement of a single replication fork from a given origin.

Uninducible mutants cannot be induced.

Unscheduled DNA synthesis is any DNA synthesis occurring outside the S phase of the eukaryotic cell.

Up promoter mutations increase the frequency of initiation of transcription.

Upstream identifies sequences proceeding in the opposite direction from expression; for example, the bacterial promoter is upstream from the transcription unit, the initiation codon is upstream of the coding region.

URF is an open (unidentified) reading frame, presumed to code for protein, but for which no product has been found.

V gene is sequence coding for the major part of the variable (N-terminal) region of an immunoglobulin chain.

Variable region of an immunoglobulin chain is coded by the V gene and varies extensively when different chains are compared, as the result of multiple (different) genomic copies and changes introduced during construction of an active immunoglobulin.

Variegation of phenotype is produced by a change in genotype during somatic development.

Vector—*see* cloning vector.

Vesicles are small bodies bounded by membrane, derived by budding from one membrane, often able to fuse with another membrane.

Virion is the physical virus particle (irrespective of its ability to infect cells and reproduce).

Virulent phage mutants are unable to establish lysogeny.

Wobble hypothesis accounts for the ability of a tRNA to recognize more than one codon by unusual (non-G•C, A•T) pairing with the third base of a codon.

Writhing number is the number of times a duplex axis crosses over itself in space.

Zero time-binding DNA enters the duplex form at the start of a reassociation reaction; results from intramolecular reassociation of inverted repeats.

Zinc finger protein has a repeated motif of amino acids with characteristic spacing of cysteines that may be involved in binding zinc; is characteristic of some proteins that bind DNA and/or RNA.

Zoo blot describes the use of Southern blotting to test the ability of a DNA probe from one species to hybridize with the DNA from the genomes of a variety of other species.

Zygote is produced by fusion of two gametes—that is, it is a fertilized egg.

INDEX